W0018298

ADVANCES IN ADDITIVE MANUFACTURING TECHNOLOGIES

ADVANCES IN ADDITIVE MANUFACTURING TECHNOLOGIES

PROCEEDINGS OF 5TH INTERNATIONAL CONFERENCE ON ADVANCES IN ADDITIVE MANUFACTURING TECHNOLOGIES (ICAAMT 2023), NOVEMBER, 27-29TH, 2023, CHENNAI, INDIA

Edtitor
Gurusamy Pathinettampadian
Professor & Head, Department of Mechanical Engineering,
Chennai Institute of Technology, Chennai, Tamilnadu, India

CRC Press
Taylor & Francis Group
Boca Raton London New York

CRC Press is an imprint of the
Taylor & Francis Group, an **informa** business

First edition published 2024
by CRC Press
4 Park Square, Milton Park, Abingdon, Oxon, OX14 4RN

and by CRC Press
2385 NW Executive Center Drive, Suite 320, Boca Raton FL 33431

© 2024 selection and editorial matter, Gurusamy Pathinettampadian; individual chapters, the contributors

CRC Press is an imprint of Informa UK Limited

The right Gurusamy Pathinettampadian to be identified as the authors of the editorial material, and of the authors for their individual chapters, has been asserted in accordance with sections 77 and 78 of the Copyright, Designs and Patents Act 1988.

All rights reserved. No part of this book may be reprinted or reproduced or utilised in any form or by any electronic, mechanical, or other means, now known or hereafter invented, including photocopying and recording, or in any information storage or retrieval system, without permission in writing from the publishers.

For permission to photocopy or use material electronically from this work, access www.copyright.com or contact the Copyright Clearance Center, Inc. (CCC), 222 Rosewood Drive, Danvers, MA 01923, 978-750-8400. For works that are not available on CCC please contact mpkbookspermissions@tandf.co.uk

Trademark notice: Product or corporate names may be trademarks or registered trademarks, and are used only for identification and explanation without intent to infringe.

British Library Cataloguing-in-Publication Data
A catalogue record for this book is available from the British Library

ISBN: 978-1-032-90013-1 (pbk)
ISBN: 978-1-003-54577-4 (ebk)

DOI: 10.1201/9781003545774

Typeset in Times LT Std
by Aditiinfosystems

Advances in Additive Manufacturing Technologies – Gurusamy Pathinettampadian (eds)
© 2024 Taylor & Francis Group, London, ISBN 978-1-032-90013-1

Contents

Advances in Additive Manufacturing Technologies – Gurusamy Pathinettampadian (eds)
© 2024 Taylor & Francis Group, London, ISBN 978-1-032-90013-1

List of Figures

Advances in Additive Manufacturing Technologies – Gurusamy Pathinettampadian (eds)
© 2024 Taylor & Francis Group, London, ISBN 978-1-032-90013-1

List of Tables

Advances in Additive Manufacturing Technologies – Gurusamy Pathinettampadian (eds)
© 2024 Taylor & Francis Group, London, ISBN 978-1-032-90013-1

Preface

We are delighted to present the proceedings of the 5th International Conference on Advances in Additive Manufacturing Technologies (ICAAMT 2023). This conference serves as a premier forum for researchers, practitioners, and industry experts to share their latest findings, innovations, and insights in the field of additive manufacturing. The rapid advancements and the increasing adoption of these technologies across various sectors underscore the importance of this gathering. The conference was held from November 27-29, 2023, in Chennai, India and organized by the Department of Mechanical Engineering, Chennai Institute of Technology, Chennai, India.

The ever-increasing demands for new products accelerate the growth of manufacturing technologies. Additive manufacturing is the most popular due to its ability to print complex customized geometries. Technological advancements and research take additive manufacturing to the next level of customized functional end-use products. The applications are diversified into various fields, including aerospace, medical, automobiles, etc.

Some of the critical features of additive manufacturing are lower start-up costs, ease of learning, reduced raw material wastage, customization to the individual, digital design integration, speed of prototype, speed from prototype to production, lower energy and environmental costs, low volume production runs, distributed manufacturing etc.,

We deeply acknowledge that the success of the conference is a collective achievement, a testament to the collaborative spirit of our community. We are grateful to our esteemed Patron, Chairs, and Principal for their invaluable guidance and support in organizing this event. The program committee's meticulous and timely review of the papers played a pivotal role in the success of this international conference.

We extend our heartfelt gratitude to the authors for their significant contributions, the reviewers for their insightful feedback, and the organizing committee for their relentless efforts in making this conference a resounding success. We are confident that the knowledge shared through these proceedings will not only inspire but also drive further research and innovation in the dynamic and rapidly evolving field of additive manufacturing.

We are optimistic that these proceedings will prove to be a valuable resource, enriching your work and contributing to the advancement of additive manufacturing. We hope you find them informative and insightful.

Sincerely,

Dr. P. Gurusamy
Organizing Secretary (ICAAMT 2023)
Professor & HOD, Mechanical Engineering,
Chennai Institute of Technology, Chennai-600 069,
Tamilnadu, India

Advances in Additive Manufacturing Technologies – Gurusamy Pathinettampadian (eds)
© 2024 Taylor & Francis Group, London, ISBN 978-1-032-90013-1

Technical Program Committee

PROGRAM COMMITTEE

CHIEF PATRONS

Shri. P. Sriram, Chairman, Chennai Institute of Technology, Chennai

Smt. S. Sridevi, Secretary, Chennai Institute of Technology, Chennai

Shri. S. Gokulakrishnan, Director-Innovations, Chennai Institute of Technology, Chennai

PARTON

Dr. A. Ramesh, Principal, Chennai Institute of Technology, Chennai

CO PATRONS

Dr. S. Neethi, Dean – Curriculum Development & Enrichment, Chennai Institute of Technology, Chennai

Dr. P. Partheepan, Dean – Planning & Development, Chennai Institute of Technology, Chennai

CONVENER

Dr. P. Gurusamy, Professor & Head, Mechanical Engineering, Chennai Institute of Technology, Chennai

CO-CONVENER

Dr. M.D. Vijayakumar, Professor, Mechanical Engineering, Chennai Institute of Technology, Chennai

ORGANISING COMMITTEE

Dr. S. Ravi, Professor in Mechanical Engineering, Chennai Institute of Technology, Chennai

Dr. R. Ganesamoorthy, Professor in Mechanical Engineering, Chennai Institute of Technology, Chennai

Dr. T. Arun Kumar, Professor in Mechanical Engineering, Chennai Institute of Technology, Chennai

Dr. R. Sathiyamoorthi, Professor in Mechanical Engineering, Chennai Institute of Technology, Chennai

Dr. S. Rajarasalnath, Associate Professor in Mechanical Engineering, Chennai Institute of Technology, Chennai

Dr. D. Jayabalakrishnan, Associate Professor in Mechanical Engineering, Chennai Institute of Technology, Chennai

Dr. J. Rajaparthiban, Asst. Prof in Mechanical Engineering, Chennai Institute of Technology, Chennai

Dr. B. Yokesh Kumar, Asst. Prof in Mechanical Engineering, Chennai Institute of Technology, Chennai

Dr. R. Vijayan, Asst. Prof in Mechanical Engineering, Chennai Institute of Technology, Chennai

Dr. S. Mohan Kumar, Asst. Prof in Center for Additive Manufacturing, Chennai Institute of Technology, Chennai

Dr. S. Ravi, Asst. Prof in Mechanical Engineering, Chennai Institute of Technology, Chennai

Dr. Biplab Bhattacharjee, Asst. Prof in Center for Additive Manufacturing, Chennai Institute of Technology, Chennai

Dr. Ramasamy, Asst. Prof in Center for Additive Manufacturing, Chennai Institute of Technology, Chennai

Dr. N. Shivakumar, Asst. Prof in Mechanical Engineering, Chennai Institute of Technology, Chennai

Dr. S.D. Dhanesh Babu, Asst. Prof in Mechanical Engineering, Chennai Institute of Technology, Chennai

Dr. S.M. Muthu, Asst. Prof in Mechanical Engineering, Chennai Institute of Technology, Chennai

Dr. S. Balamurugan, Asst. Prof in Mechanical Engineering, Chennai Institute of Technology, Chennai

Dr. M.J. Hepsi Beaula, Asst. Prof in Mechanical Engineering, Chennai Institute of Technology, Chennai

Advisory Committee

Dr. Michel Fillon, Mantamise, *Nouvelle-Aquitaine, France*

Dr. Joao Pedro Oliveira, *Nova University Lisban, Portugal*

Dr. Uday Venkatadri, *Dalhousie University, Halifa, Canada*

Dr. Muthukumaran Pakirisamy, *CIADI, Concordia*

Dr. Kannan Govindan, *University of Southern Denmark, Denmark*

Dr. Joel Anderson, *University of West, Sweden*

Dr. Srikanth Joshi, *University of West, Sweden*

Dr. Senthil Kumar, *National University of Singapore, Singapore*

Dr. Rajkumar Sundaram, *Government Technical Institute, Oman*

Dr. Kesavan Ulaganathan, *Colombo Plan Staff College, Manila, The Philipines*

Dr. Haarindra Prasad, *AIMST, Malaysia*

Dr. S. Balasivanandha Prabu, *Anna University, Chennai*

Dr. R.Velmurugan, *Department of Aerospace Engineering, IIT Madras*

Dr. N. Sivashanmugan, *Department of Manufacturing, NIT, Trichy*

Dr A. Arockiarajan, *Dept. of Applied Mechanics, Solid Mechanics Division, IIT Madras*

Dr. K. Senthilkumaran, *Mechanical Engineering Department, IITDM-Kancheepuram*

Dr. M. Kamaraj, *SRM Institute of Science and Technology, Kattankulathur, Chennai*

Dr.J. Jancirani, *Department of Production Tech., MIT Campus, Anna University, Chennai*

Dr.Sivakumar Narayanasamy, *CIADI, Concordia*

Dr.Arunachalam, *Department of Manufacturing, IIT Madras*

Dr.S. N. Jaisankar, *Center Leather Research Institute, Chennai*

Dr.T.P.D. Rajan, *CSIR- NIIST, Trivandrum*

Dr.E. Vijayaragavan, *SRM Institute of Science and Technology, Kattankulathur, Chennai*

Advances in Additive Manufacturing Technologies – Gurusamy Pathinettampadian (eds)
© 2024 Taylor & Francis Group, London, ISBN 978-1-032-90013-1

About the Editor

Dr. P. Gurusamy is a distinguished academician and a reputed researcher in Mechanical Engineering with a remarkable career spanning over two decades. He commenced his academic journey with a Bachelor of Engineering (B.E.) in Mechanical Engineering from the Government College of Engineering, Tirunelveli, in 1996. His educational pursuits continued with a Master of Engineering (M.E.) in Production Engineering from Thiagarajar College of Engineering, Madurai, which he completed in 1998.

Dr Gurusamy's quest for knowledge led him to Anna University, CEG Campus, Chennai, where he earned his PhD in 2014. His doctoral research, titled "Influence of Squeeze Casting Parameters on the Mechanical Properties and the Solidification Behaviour of the Al/SiCp Composites," received a 'Highly Recommended' rating from the examiner, highlighting the significant contribution of his work to the field.

With two years of Industrial Experience and a 25-year teaching career, Dr. Gurusamy has become a pivotal figure in Mechanical Engineering education and research. He is a recognized supervisor for guiding PhD and M.S. research scholars at Anna University, Faculty of Mechanical Engineering. Currently, eleven scholars are pursuing research under his guidance.

Dr. Gurusamy's research interests are diverse and impactful, encompassing Composite Materials, Foundry Technology, Nanomaterials, Surface Technology & Advanced Materials Processing, Carbon Nanotubes, and Polymer Nanoclay Composites. His prolific contributions to the academic community include 79 publications in International Journals and numerous participations in International Conferences across India and abroad.

In addition to his academic and research endeavours, Dr Gurusamy has also demonstrated his innovative capabilities by publishing many patents in Material Science. His professional affiliations include memberships in esteemed organizations such as the American Society of Mechanical Engineers (ASME), the Society of Automotive Engineers (SAE), and the Indian Society for Technical Education (ISTE).

Dr. P. Gurusamy's enduring dedication to education, research, and innovation continues to inspire his students and peers, making him a respected and influential figure in the field of Mechanical Engineering.

Advances in Additive Manufacturing Technologies – Gurusamy Pathinettampadian (eds)
© 2024 Taylor & Francis Group, London, *ISBN 978-1-032-90013-1*

1 A Process of Additive Component with Machine Learning

Dhandapani.S[1], Muthiah.C.T[2], Shanthi.R[3]
Mechanical Dept, Rajalakshmi Engineering College,
Thandalam, Tamil Nadu, India
Eswaran.A[4]
Robotics & Automtion Dept, Rajalakshmi Engineering College,
Thandalam, Tamil Nadu, India

ABSTRACT: Industry 5.0 has resulted in an explosion of machine learning (ML) frameworks, especially auto-ML techniques that are easy for non-experts to use. To do away with the trial-and-error process of determining the right input parameters in 3D printers. Specimens were prepared for the process using 3D printing and the Fusion Deposition Method (FDM). The result of the software was used to estimate which of the several machine learning algorithms was the best. Surface testing on the vehicle component specimen was used to forecast several output metrics, such as drilling accuracy and roughness, with the use of machine learning in 3D printing. The program has an accuracy of 93.56% and an error of 6.43%, according to the final results.

KEYWORDS: Machine learning, 3D printing, Algorithm, Accuracy with variance

1. INTRODUCTION

The most recent machine learning techniques used for clean data processing go beyond simple learning. Time series forecasting, supply chain management, energy, healthcare, sales, environmental research, manufacturing, agriculture, and many more fields have a wide range of industrial and scientific applications. Similar to this, a Python library is used for a uniform interface in machine learning, which makes it simple to carry out model building, inspection, and application. Many industries, including self-driving cars, space research, and medicine, use machine learning techniques. This study focuses on the area of additive manufacturing specifically related to 3D printers.

1.1 Artificial Intelligence

Artificial intelligence (AI) many subdivisions, including machine learning. AI is the fusion of machine intelligence and other components. Prominent AI researchers investigate "intelligent agents," which include a feature to accurately assess the environment and optimize execution to successfully accomplish objectives. AI is used to explain machines that replicate cognitive processes like "learning" and "problem solving," which are associated with the human mind. The basic ideas of AI research's computer algorithms, which automatically get better with experience, are becoming more and more capable of handling machines

[1]dhandapani.s@rajalakshmi.edu.in, [2]muthiah.ct@ rajalakshmi.edu.in, [3]shanthi.r@ rajalakshmi.edu.in, [4]eswaran@ rajalakshmi.edu.in

DOI: 10.1201/9781003545774-1

1.2 Machine Learning

Machine learning that automatically advances through practice with data and an understanding of computer techniques. As a component of artificial intelligence, it is impractical to create traditional algorithms for the necessary computational statistics-based activities[1]. ktrain is a low-code Python module that uses machine learning to make machine learning (ML) more approachable and simpler to use. Created a unified interface that required only three or four "commands" or lines of code to swiftly solve a variety of jobs in his research, also noted that there has been a lot of attention focused on unified machine learning. [2] Because of the variations in the approaches, the many approaches that have been released are not well documented and unclear. The goal of this work is to construct a design that fills in the gaps and completes the missing links in the areas of auto machine learning, hyper parameter tweaking, and metal earning process. Darts can solve a Python machine learning framework for time series problems like forecasting. This can take the form of a series, meta-learning across several series, training on huge datasets, combining outside data, putting together models, and offering robust support for probabilistic forecasting. [3] With the help of strong parameter adjustment and Auto ML approaches, the ML frameworks can be used by systems that are not experts. The directed cyclic graphs utilized in mlr3pipelines are a foundation for defining both simple and complicated non-linear machine learning operations. [4]. The framework uses bench marking, tweaking, and convenient resemblance components as part of the mlr3 ecosystem. A few engines, including the Knowledge Base, GUI, Visualization, and Human engines, were introduced by Penn AI. Even while these are intriguing ideas for creating a user-friendly auto ML system, they don't offer a significant contribution to the cutting edge of auto ML research. [5].

1.3 Deep Learning

Artificial neural networks, on which machine learning techniques are based, provide the basis for deep learning. The process of learning may be partly, fully, or not seen. Deep learning algorithms generate substantial credit assignment path (CAP) depth. The number of layers [6] is used to more precisely convert this data. Deep learning models in the medical field have grown exceptionally well as a result of the rapid advancements in machine learning, graphics processing technology, and medical imaging data. [7] The difficulties in applying 3D CNNs to the realm of medical imaging, deep learning models, and potential future developments. Machine learning uses a variety of techniques when there isn't a completely functional algorithm.

1.4 Researcher's Outlook

The biometric utility is necessary to identify facial analysis and run the systems, and it can also be helpful in removing low-quality data. This body of literature offers survey analysis using visible wavelength facial photographs within the face biometrics paradigm. [6] For instance, the picture selection and facial image quality evaluation that are covered in his research indicate the infrared quality assessment as a tendency towards deep learning. To provide high-quality and effective services, customer service is changed by converting from traditional to automated service using various computational data. [8] Big data is needed since it requires knowledge of data analysis. In small and medium-sized businesses, a lack of resources and scale leads to a lack of information and experience. In order to address this problem, open and automated customer service platforms like block chains, automated machine learning (AutoML), and the Internet of things (IoT) are suggested. This creates a third party dependency, which increases costs. These platforms take into account data collection, data security, and economic factors.AutoML has advanced neural network architecture significantly by using complex, expertly-designed layers as building blocks [9]. By proposing a unique framework, we may eliminate human bias through a generic search space of two-layer neural networks trained by back propagation.

2. Materials and Methods

Additive manufacturing, also known as 3D printing, is the process of creating thin layers. Above the resources by lying down, we can eventually obtain a 3D digital replica of the physical object. Layer by layer of materials, it transforms the computer object of the CAD model into a real form. The reagent, powder, filament, and pellets are the materials delivered in various states and employed for this purpose. There are numerous production methods available for the AM process. In this undertaking, the merged. The specimen is made using the fused deposition modeling (FDM) technique. This procedure is conducted using temperature-controlled head sequentially extrudes thermoplastic material, allowing the path using 3D CAD information. Starting with the model's STL file, pre-processing software is used. Mathematically divided horizontal layers with thicknesses ranging from +/- 0.137 to 0.264 mm.When necessary, support structure is required. A 3D model is produced using the tool path and data.

The specimen used is printed using the following setup parameters. 1. Layer Height: 0.01mm, 2. Fan Speed: 25%, 3. Nozzle temperature:250 deg 4.Bed temperature 50 deg, 5. Print speed: 120 (mm/s), 6. Materials: Polylactic Acid (PLA)

Figure 1.1 shows the specimen created by the AM method as well as the experimental procedure. The Fused Deposition Modelling (FDM) procedure is used to 3D print the object.

Fig. 1.1 Specimen created by AM method

3. Results and Discussion

The 3D printing technology' Fusion deposition method was used to prepare the specimen. Next, a surface roughness tester was used to measure the specimen's roughness value. Roughness testing was also done at a few drilling locations on the sample. Using the appropriate machine learning program, the algorithm was created. The process's appropriate algorithmic approach was predicted with accuracy and error value. This is measured in real time and compared. Additionally, ML code was designed and implemented successfully utilizing a Python coder.

The sequence of the machine learning algorithm (MLA) is as follows:

(a) Bringing in the necessary packages.

(b) Importing the dataset and determining which values are NULL

(c) Maintaining the variables, both independent and dependent partitioning the dataset (d).

(e) Maintaining a variable containing machine learning algorithms (MLA)

(f) Making a box plot to assess how accurate they are

(g) Evaluating every algorithm for machine learning.

The above mentioned procedures are taken when creating a program and designing an algorithm. There was testing and comparing done The above steps are followed in the part program formation and algorithm was designed. The testing and comparison was carried out.

3.1 Calculation and Algorithm

Define Cost Function

$$.J(\theta) = \frac{1}{2n}\sum_{i=1}^{n}(y' - y)^2 \tag{1}$$

$.J(\theta)$ - Cost function

n - Number of samples

y - True value

y' - Predicted value

Θ - weights (trainable parameters)

$$y' - \theta_0 + \theta_1 x_1 + ... + b_k x_k \tag{2}$$

b_0 - Y-intercept

b_i - Slope of the i th variable

x_i - value corresponding to feature i

$$\theta^{n+1} = \theta^n - \alpha\frac{\partial J(\theta)}{\partial \theta} \tag{3}$$

θ^{n+1} - updated weights (weights at next step)

θ^n - previous weights

α - learning rate ($0 < \alpha < 1$)

$\frac{\partial J(\theta)}{\partial \theta}$ - change of cost function $J(\theta)$ with respect to α

$$0 \leq J(\theta^{n+1}) - J(\theta^n) \leq t \tag{4}$$

t - Threshold value ($0 < t < 1$)

$J(\theta^{n+1})$ - Total cost of updated weights

$\mathrm{J}(\theta^n)$ - Total cost of previous weights

Algorithm

- Establish the cost function $J(\theta)$, which quantifies the discrepancy between the expected and realized results.
- Set the model's random weights to zero.
- Using the gradient descent technique, update the weights θ iteratively: $\theta := \theta - \alpha * \nabla J(\theta)$.
- Using a stopping condition, repeat the weight update until convergence is reached.such as the cost function difference threshold or the maximum number of repetitions

4. Conclusion

According to the project's goal, the FDM method for creating 3D printed specimens was tested by gathering surface measurement data, which is displayed in Fig. 1.2 at 0.84 nm and 0.0.87. Which information is contrasted in the MLA result. The last window is created using the 3D printer data set following training. To train the data set, multivariate linear regression is employed. Referring to Fig. 1.3, the model predicts the valueon the training data set with a variance of 2.0840360280300282e-26% and an accuracy of 100%. The program yields 0.095, however the average roughness measured with the "Surface Roughness Tester" is 0.0855. The program has an accuracy of 93.56% and an error of 6.43%, it may be inferred.The accuracy

Fig. 1.2 Roughness check

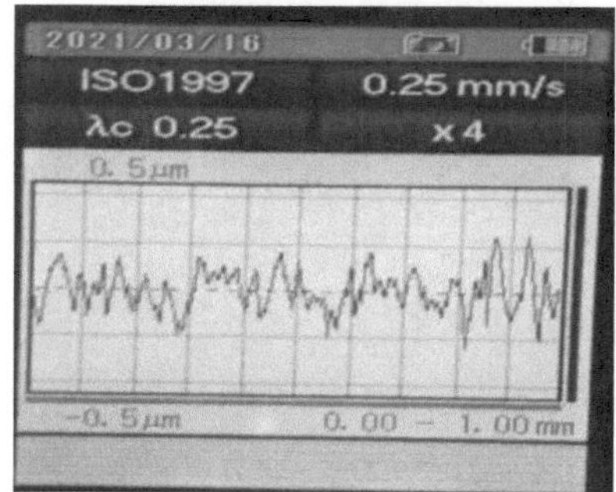

Fig. 1.3 Roughness graph

computation was reported along with the percentage of error.

Figure 1.2–1.3 : Roughness Measurement (a) Position one on Auto Specimen (b) Positon two (Drill hole area) on Specimen. (c) Roughness graph for position one.

Accuracy calculation:

- Average roughness value = 0.087 + 0.0842
- Average roughness value = 0.0855 µm
- Accuracy = 100 – 0.091 – 0.0855
- Accuracy = 100 – 6.43 Accuracy = 93.57%.

The bar chart of variance s in Fig. 1.4 for several machine learning algorithms, including K-nearest neighbour, decision tree classifier, Gaussian Naive Bayes, Random Forest classifier, and adaptive boosting classifier, is displayed in the graph that compares the various algorithms. We may deduce from the above chart that the decision tree classifier is most appropriate for the prediction phase because it has the lowest variance.

Comparably, Fig. 1.4 displays an accuracy bar chart for several machine learning algorithms, including Gaussian Naive Bayes, Random Forest, Adaptive Boosting, K-nearest neighbour, and decision tree classifiers. The decision tree classifier is the most accurate and, thus, most appropriate for the prediction phase, according to the above figure. Figure 1.4 displays the accuracy vs. variance scatter plot for a variety of machine learning algorithms, including Random Forest, Adaptive Boosting, K-Nearest Neighbour, Decision Tree, and Gaussian Naive Bayes.

The decision tree classifier is the most accurate and least variable, making it the most appropriate for the prediction phase, according to the above figure. The Decision Tree Classifier has a low variance with good accuracy, according to the data analysis of the algorithm selection. After that, a surface roughness tester is used to test the second spot on the specimen, and the results are compared with software that uses machine learning algorithms. The outcome as discovered in the drill bit data set. The program displays the appropriate algorithm and accuracy level.

The drill bit data set is used for the last window following training. Five distinct models are used to train the data set, and accuracy and variance are displayed. Additionally, the models' output is pooled using the maximum vote technique, increasing the average accuracy from 89.73% to 100%. The K-Fold cross validation approach is used to identify the optimal model.

Without any prior computer programming experience, a wide range of engineering areas can benefit from the machine learning project. using less than 10% variance, 90% accuracy may be achieved using the aforementioned

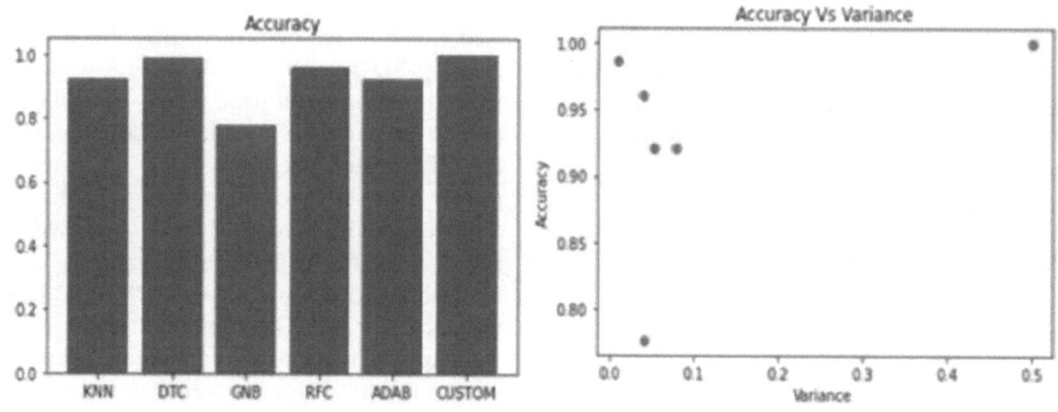

Fig. 1.4 Bar chart and scatter plot of a accuracy for different machine learning

program, and a 3D printed object is analyzed. While the program yields 0.091, the average roughness measured with the "Surface Roughness Tester" is 0.0855. The program has an accuracy of 93.56% and an error of 6.43%, it may be inferred.

It can measure a 3D printed object's roughness simply and without the need for trial and error. By employing this strategy, time is saved. Everyone can simply access it. to enable machine learning in fields other than technology. to use zero code to create machine learning models. to use machine learning with a 3D printer to get precise parameter settings.

5. Acknowledgement

The authors gratefully acknowledge the students, staff, and authority of department for their cooperation in the research.

References

1. Arun S. Maiya, ktrain, (2022) *A Low-Code Library for Augmented Machine Learning*. Journal of Machine Learning Research 23 1–6.
2. Thiloshon Nagarajah and Guhanathan Poravi, (2019), *A Review on Automated Machine Learning (AutoML)* Systems. 5th International Conference for Convergence in Technology(I2CT),. DoI:10.1109/I2CT45611.2019.9033810.
3. Julien. Herzen, Francesco L¨assig, Samuele Giuliano Piazzetta Thomas Neuer L´eo Tafti Guillaume Raille, Tomas Van Pottelbergh, Marek Pasieka†, Andrzej Skrodzki, Nicolas Huguenin† Maxime Dumonal, Jan Ko´scisz, Dennis Bader†, Fr´ed´erick Gusset, Mounir Benheddi Camila WilliamsonMichal Kosinski, Matej Petrik, Ga¨el Grosch. Darts,(2021), *User-Friendly Modern Machine Learning forTime Series Unit 8 SA,* Switzerland Journal of Machine Learning Research 23, 1–6.
4. R. Martin Binder, Florian Pfisterer, Michel Lang, Lennart Schneider, Lars Kotthoff (2022), mlr3pipelines – *Flexible Machine Learning Pipelines.*
5. Ronan Perry, Gavin Mischler, Richard Guo, Theodore Lee, Alexander Chang, Arman Koul, Cameron Franz, Hugo Richard, Iain Carmichael, Pierre Ablin, Alexandre Gramfort, Joshua T. Vogelsteinmvlearn, (2019), *Multiview Machine Learning in Python.*
6. Torsten Schlett, Christian Rathgeb, Olaf Henniger, Igd Javier Galbally, Julian Fierrez, Christoph Busch, Face Image uality Assessment: (2022). A Literature Survey.
7. S. P. Singh, Lipo Wang, Sukrit Gupta, Haveesh Goli Parasuraman Padmanabhan, Balázs Gulyás, Lee Kong Chian; (2020) Published: *date3D Deep Learning on Medical Images: A Review.*
8. ZhLi; HanyangGuo; Wai Ming Wang; Yijiang Guan; Ali Vatankhah Barenji; George Q. Huang"(2019), ABlockchain and AutoML *Approach for Open and Automated Customer* Service.IEEE Transactions on Industrial Informatics.
9. S.Esteban Real, Chen Liang, David R AutoML-Zero: (2020). Evolving Machine Learning Algorithms FromSo.
10. Yougesh R and - Yashwanth M (2021) *Analysis of 3D printed object using machine learning* - A project report -Rajalalakshmi Engineering College, chennai, india.

Note: Every figures in this chapter was created using the experimental work; none of them were taken from any publications or online.

Advances in Additive Manufacturing Technologies – Gurusamy Pathinettampadian (eds)
© 2024 Taylor & Francis Group, London, ISBN 978-1-032-90013-1

2 Risk Assessment and Management Strategies for Turbine Blade Failures in Power Generation Systems

Shankha Shubhra Goswami*,
Bikash Banerjee, Subhadip Das, Surajit Mondal, Rohit Halder
Department of Mechanical Engineering,
Abacus Institute of Engineering and Management,
Hooghly, India

ABSTRACT: Turbine blade failures in power generation systems pose significant risks to operational safety, reliability, and economic viability. This research article investigates the complex factors contributing to turbine blade failures and proposes comprehensive risk assessment and management strategies to mitigate these risks. Through a combination of empirical analysis, computational modeling, and industry case studies, this study aims to enhance the understanding of turbine blade failure mechanisms and develop effective strategies for proactive risk management in power generation systems.

KEYWORDS: Turbine blade failures, Power generation, Risk assessment, Management strategies

1. INTRODUCTION

Turbine blades play a critical role in power generation systems, particularly in gas turbines and steam turbines. However, turbine blade failures can lead to costly downtime, maintenance expenses, and potential safety hazards. Understanding the factors contributing to turbine blade failures and implementing robust risk assessment and management strategies are essential for ensuring the reliability and longevity of power generation systems [1,2]. This article explores the various aspects of turbine blade failures and presents proactive approaches to mitigate associated risks.

2. FACTORS CONTRIBUTING TO TURBINE BLADE FAILURES

Turbine blade failures are often the result of a combination of factors that affect the structural integrity and performance of the blades over time. Understanding these contributing factors is crucial for developing effective risk mitigation strategies [3]. In this section, we delve into the various elements that play a significant role in turbine blade failures.

2.1 Material Degradation

Turbine blades are subjected to extreme conditions, including high temperatures, mechanical stresses, and aggressive environments. These conditions can cause material degradation processes such as fatigue, creep, and corrosion. Fatigue occurs when cyclic loading leads to progressive weakening of the material, ultimately resulting in crack initiation and propagation. Creep is the gradual deformation of the material under sustained high temperatures and mechanical loads, leading to dimensional changes and microstructural alterations [4]. Corrosion, whether from chemical reactions with the surrounding environment or

*Corresponding author: ssg.mech.official@gmail.com

DOI: 10.1201/9781003545774-2

contaminants in the working fluid, can weaken the material and promote crack initiation.

2.2 Operational Conditions

The operating environment of turbine blades can vary significantly, influencing their degradation mechanisms and failure modes. Factors such as temperature fluctuations, transient loads during start-up and shutdown sequences, and the frequency of operational cycles can all impact the fatigue life and integrity of turbine blades. Thermal gradients across the blade section can induce thermal stresses, leading to thermal fatigue and thermal mechanical fatigue [5]. Additionally, aerodynamic forces acting on the blades can cause vibration, flutter, and dynamic loading, further exacerbating fatigue and stress-related failure mechanisms.

2.3 Manufacturing Defects

Despite advancements in manufacturing processes, turbine blades may still contain inherent defects or imperfections that can compromise their structural integrity. These defects may arise from casting processes, machining operations, or material inconsistencies. Common defects include porosity, inclusions, surface irregularities, and grain boundary anomalies. Even minor defects can serve as stress concentration points, accelerating crack initiation and propagation under operational loading conditions [6]. Quality control measures and inspection protocols are essential for detecting and mitigating manufacturing defects before they lead to catastrophic failures.

2.4 Maintenance Practices

The effectiveness of maintenance practices significantly influences the reliability and longevity of turbine blades in power generation systems. Inadequate maintenance procedures, such as infrequent inspections, improper repair techniques, and deferred component replacements, can increase the likelihood of turbine blade failures. Routine inspections, non-destructive testing (NDT), and condition monitoring programs are essential for identifying early signs of degradation, assessing the health status of turbine blades, and scheduling preventive maintenance interventions [7]. Proactive maintenance strategies, including component refurbishment, blade coating applications, and operational adjustments, can mitigate the effects of degradation and extend the service life of turbine blades.

In summary, turbine blade failures result from a combination of material degradation, operational conditions, manufacturing defects, and maintenance practices. Addressing these contributing factors requires a holistic approach that integrates materials science, engineering design, operational optimization, and maintenance management [8]. By understanding the complex interplay of these factors, power generation operators can develop proactive strategies to mitigate the risks associated with turbine blade failures and ensure the reliability and efficiency of their systems.

3. Risk Assessment Methodologies

Effective risk assessment methodologies are essential for identifying, evaluating, and mitigating the risks associated with turbine blade failures in power generation systems. By systematically analyzing potential failure scenarios, quantifying their likelihood and consequences, and prioritizing mitigation efforts, risk assessment methodologies provide valuable insights for decision-making and resource allocation [9]. In this section, we explore several key methodologies commonly used in assessing turbine blade failure risks.

3.1 Probabilistic Risk Assessment (PRA)

Probabilistic Risk Assessment (PRA) is a quantitative method used to evaluate the likelihood and consequences of turbine blade failures by considering various failure modes and their associated probabilities. PRA relies on statistical techniques to integrate data from historical failure records, empirical models, and expert judgment to estimate the probability of failure occurrence under different operating conditions. By quantifying the uncertainties inherent in turbine blade failure mechanisms, PRA enables stakeholders to prioritize risk mitigation measures and allocate resources effectively. Additionally, PRA facilitates sensitivity analyses to identify the most influential factors contributing to turbine blade failures and inform decision-making processes.

3.2 Failure Mode and Effects Analysis (FMEA)

Failure Mode and Effects Analysis (FMEA) is a systematic method for identifying potential failure modes of turbine blades, analyzing their causes and effects, and assessing the severity of their impact on power generation systems. FMEA involves multidisciplinary teams comprising engineers, maintenance personnel, and subject matter experts who collaboratively examine the failure modes, their associated risks, and the effectiveness of existing mitigation measures. By assigning risk priority numbers (RPNs) based on the likelihood, severity, and detectability of each failure mode, FMEA enables stakeholders to prioritize mitigation actions and implement targeted interventions to reduce the overall risk profile of turbine blade failures [10]. FMEA is particularly valuable during the design phase of power generation systems to identify design weaknesses and enhance the reliability of turbine blades.

3.3 Reliability-Centered Maintenance (RCM)

Reliability-Centered Maintenance (RCM) is a proactive approach to maintenance management that aims to optimize maintenance activities, enhance equipment reliability, and minimize the risk of turbine blade failures. RCM integrates risk-based decision-making principles to identify critical components, failure modes, and maintenance tasks based on their impact on system performance and safety. By assessing the consequences of turbine blade failures in terms of operational downtime, production losses, and safety hazards, RCM enables operators to develop customized maintenance strategies tailored to the specific needs and operating conditions of power generation systems. RCM emphasizes preventive and predictive maintenance techniques, such as condition monitoring, predictive analytics, and reliability-centered inspections, to detect early signs of degradation and address potential failure risks before they escalate.

In summary, risk assessment methodologies such as Probabilistic Risk Assessment (PRA), Failure Mode and Effects Analysis (FMEA), and Reliability-Centered Maintenance (RCM) provide valuable tools for evaluating and managing the risks associated with turbine blade failures in power generation systems [11,12]. By adopting a systematic and proactive approach to risk assessment, operators can enhance the reliability, safety, and efficiency of their turbine assets while minimizing the potential for costly downtime and operational disruptions.

4. Management Strategies for Turbine Blade Failures

Effectively managing turbine blade failures requires a proactive approach that integrates advanced technologies, robust maintenance practices, and continuous improvement processes. By implementing comprehensive management strategies, power generation operators can minimize the likelihood and consequences of turbine blade failures, enhance operational reliability, and optimize asset performance. In this section, we elaborate on various management strategies for mitigating turbine blade failures.

4.1 Advanced Materials and Coatings

Developing and implementing advanced materials and coatings is a key strategy for mitigating turbine blade failures. High-performance materials with enhanced resistance to corrosion, erosion, and thermal fatigue can significantly improve the durability and reliability of turbine blades. Additionally, protective coatings applied to blade surfaces can mitigate the effects of environmental degradation and extend the service life of turbine components [13]. Research and development efforts focus on exploring novel material compositions, nanostructured coatings, and surface treatments tailored to the specific operating conditions and performance requirements of power generation systems.

4.2 Condition Monitoring and Predictive Maintenance

Condition monitoring and predictive maintenance techniques enable early detection of turbine blade degradation and impending failures, allowing for timely intervention and mitigation. Real-time monitoring systems, sensor technologies, and predictive analytics collect data on blade health parameters such as vibration levels, temperature profiles, and stress distributions [14]. By analyzing this data using machine learning algorithms and statistical models, operators can identify abnormal trends, anticipate potential failure modes, and schedule preventive maintenance activities proactively. Condition-based maintenance strategies optimize inspection intervals, prioritize maintenance tasks, and minimize downtime while maximizing the reliability and availability of turbine assets.

4.3 Design Optimization and Retrofitting

Design optimization and retrofitting initiatives aim to enhance the structural integrity, aerodynamic performance, and operational flexibility of turbine blades. Engineers employ computational modeling, parametric optimization techniques, and experimental validation to refine blade geometries, improve material utilization, and reduce stress concentrations. Aerodynamic enhancements such as airfoil modifications, vortex generators, and boundary layer controls optimize flow patterns around the blades, reducing aerodynamic loading and mitigating erosion and fatigue [15]. Retrofitting existing turbine blades with upgraded components, such as advanced cooling systems or damping mechanisms can extend their service life and enhance performance without requiring complete replacement.

4.4 Training and Education

Training and education programs are essential for equipping operators, maintenance personnel, and engineers with the knowledge and skills necessary to prevent and mitigate turbine blade failures effectively. Training initiatives cover a wide range of topics, including turbine operation principles, maintenance procedures, failure analysis techniques, and risk management practices. Hands-on training exercises, workshops, and simulation-based scenarios enhance participants' understanding of turbine blade failure mechanisms and empower them to implement best practices for risk mitigation [5]. Continuous professional development and knowledge-sharing forums foster a culture of innovation, collaboration, and continuous improvement within the organization, ensuring the long-term success and sustainability of turbine asset management initiatives.

In summary, management strategies for turbine blade failures encompass a holistic approach that addresses material selection, maintenance practices, design optimization, and human factors [16]. By integrating advanced technologies, proactive maintenance strategies, and ongoing training efforts, power generation operators can mitigate the risks associated with turbine blade failures, optimize asset performance, and ensure the reliability and longevity of their turbine assets.

5. Case Studies and Industry Applications

Case studies and industry applications provide valuable insights into real-world experiences with turbine blade failures, root cause analysis, failure investigations, and the effectiveness of risk mitigation strategies. By examining specific incidents, lessons learned, and best practices from diverse sectors of the power generation industry, operators can gain practical knowledge and guidance for enhancing turbine blade reliability and performance [17]. In this section, we elaborate on several case studies and industry applications relevant to turbine blade failures.

Case Study 1: Root Cause Analysis of High-Temperature Corrosion in Gas Turbines

- *Background:* A gas turbine power generation facility experienced a series of turbine blade failures attributed to high-temperature corrosion. The facility operated in a coastal environment with high humidity and salt content in the air, exacerbating corrosion rates in turbine components.
- *Root cause analysis:* Engineers conducted a thorough investigation to identify the root cause of the corrosion-related failures. Metallurgical analysis revealed that the turbine blade material, primarily composed of nickel-based superalloys, was susceptible to oxidation and sulfidation at elevated temperatures [18]. The corrosive environment within the combustion chamber, characterized by high-temperature flue gases and combustion byproducts, accelerated the degradation of turbine components.
- *Mitigation strategies:* Based on the findings of the root cause analysis, the operator implemented several mitigation strategies to address high-temperature corrosion.
    1. Upgrading turbine components with corrosion-resistant materials, such as advanced nickel-based alloys or ceramic coatings, to improve resistance to oxidation and sulfidation.
    2. Implementing improved cooling systems and thermal barrier coatings to maintain lower surface temperatures on turbine blades, reducing the rate of corrosion.
    3. Optimizing fuel composition and combustion parameters to minimize the formation of corrosive byproducts and mitigate the impact of high-temperature corrosion on turbine performance.
- *Outcome:* The implementation of these mitigation strategies resulted in a significant reduction in turbine blade failures attributed to high-temperature corrosion. Operational reliability improved, leading to fewer unplanned outages and reduced maintenance costs for the power generation facility.

Case Study 2: Failure Investigation of Fatigue Cracking in Steam Turbine Blades

- *Background:* A steam turbine plant experienced fatigue cracking in turbine blades during startup and shutdown cycles, leading to unplanned downtime and production losses. The turbine blades were subjected to cyclic loading conditions, thermal stresses, and aerodynamic forces during operation.
- *Failure investigation:* Engineers conducted a detailed failure investigation to determine the root cause of fatigue cracking in steam turbine blades. Finite element analysis (FEA) and computational fluid dynamics (CFD) simulations were employed to model the mechanical behavior and thermal response of turbine blades under operational conditions [19]. The investigation identified stress concentrations at blade root attachments and trailing edges as critical areas prone to fatigue initiation.
- *Mitigation strategies:* Based on the findings of the failure investigation, the operator implemented several mitigation strategies to address fatigue cracking in steam turbine blades.
    1. Adjusting startup and shutdown procedures to minimize thermal gradients and mechanical stresses, reducing the cyclic loading on turbine components.
    2. Retrofitting turbine blades with improved cooling channels, thermal barrier coatings, and damping mechanisms to enhance fatigue resistance and thermal management.
    3. Enhancing blade design features, such as fillet radii and smooth transitions, to reduce stress concentrations and improve structural integrity.

- *Outcome:* The implementation of these mitigation strategies resulted in a significant reduction in fatigue-related failures and improved operational reliability for the steam turbine plant. Downtime associated with turbine blade failures decreased, leading to increased productivity and efficiency in power generation operations.

Case Study 3: Proactive Maintenance Program for Wind Turbine Blades

- *Background:* A wind farm operator sought to optimize the performance and reliability of wind turbines by implementing a proactive maintenance program. Wind turbine blades were susceptible to leading edge erosion, lightning strike damage, and adhesive bond failures, which could compromise their aerodynamic efficiency and structural integrity.
- *Proactive maintenance program:* The operator deployed a proactive maintenance program focused on continuous condition monitoring and predictive maintenance techniques.
    1. Regular inspections using unmanned aerial vehicles (UAVs) equipped with high-resolution cameras and sensors to detect early signs of blade degradation, including erosion, damage, and delamination.
    2. Structural health monitoring systems installed on wind turbine towers to track changes in blade vibration frequencies, modal characteristics, and fatigue response.
    3. Deployment of predictive analytics software to analyze data from condition monitoring systems, forecast the remaining useful life of wind turbine blades, and schedule maintenance activities proactively.
- *Outcome:* The proactive maintenance program resulted in improved reliability, reduced downtime, and optimized energy production from the wind farm. Early detection of blade defects allowed for timely intervention and repair, minimizing the impact of blade failures on operational performance and maximizing the lifespan of wind turbine assets. The operator achieved greater operational efficiency and cost savings through proactive maintenance practices.

In summary, these case studies highlight the importance of root cause analysis, failure investigation techniques, and proactive maintenance programs in mitigating turbine blade failures and optimizing the performance of power generation systems [20]. By implementing appropriate mitigation strategies tailored to specific failure modes and operational conditions, operators can enhance the reliability, safety, and efficiency of their turbine assets while minimizing the impact of blade failures on operational performance.

6. CONCLUSION

Effective risk assessment and management are essential for minimizing the likelihood and consequences of turbine blade failures in power generation systems. By understanding the complex interplay of factors contributing to blade failures and implementing proactive mitigation strategies, operators can enhance the reliability, safety, and economic efficiency of their power generation assets. Continued research and innovation in materials science, computational modeling, and maintenance practices will further advance the field of turbine blade reliability and contribute to sustainable energy production.

Research on turbine blade failures in power generation systems holds several practical implications for operators, maintenance personnel, and engineers. Understanding and mitigating the factors contributing to blade failures can enhance reliability, reduce downtime, and optimize performance, leading to cost savings and improved competitiveness in the energy market. By implementing proactive maintenance strategies based on predictive analytics and advanced materials, operators can ensure operational safety, meet regulatory requirements, and drive technological advancements in turbine design and maintenance practices. Knowledge transfer through training programs facilitates the dissemination of best practices, empowering personnel to effectively manage turbine assets and minimize the impact of failures on operational performance.

While research on turbine blade failures in power generation systems offers valuable insights and practical implications, it also has several limitations and future scope for exploration. One limitation is the inherent complexity of turbine blade failure mechanisms, which may involve multifaceted interactions among material degradation, operational conditions, and maintenance practices. Future research could focus on developing more comprehensive models and simulation techniques to capture the dynamic behavior of turbine blades under diverse operating scenarios. Additionally, there is a need for further investigation into emerging technologies, such as advanced materials, sensor networks, and artificial intelligence, to enhance predictive maintenance capabilities and optimize turbine performance. Moreover, interdisciplinary collaboration between researchers, industry stakeholders, and regulatory bodies is essential to address broader challenges related to turbine reliability, safety, and sustainability in a rapidly evolving energy landscape.

ACKNOWLEDGEMENT

We would like to express our sincere gratitude to all individuals and organizations who contributed to the completion of this research paper. Special thanks to our research team members for their dedication and hard work throughout the project.We are also grateful to Abacus Institute of Engineering and Management for providing the necessary resources and support.

REFERENCES

1. Molinari, J. F. and Ortiz, M. (2002). A study of solid-particle erosion of metallic targets. Int. J. Impact Eng. 27:347-358. https://doi.org/10.1016/S0734-743X(01)00055-0
2. Mittal, U. and Panchal, D. (2023). AI-based evaluation system for supply chain vulnerabilities and resilience amidst external shocks: An empirical approach. Rep. Mech. Eng. 4(1):276-289.https://doi.org/10.31181/rme040122112023m
3. Kablov, E. N. and Muboyadzhyan, S. A. (2017). Erosion-resistant coatings for gas turbine engine compressor blades. Russ. Metall. 494-504. https://doi.org/10.1134/S0036029517060118
4. Sahoo, S. K., Goswami, S. S.and Halder, R. (2024). Supplier selection in the age of industry 4.0: A review on MCDM applications and trends. Decis. Mak. Adv. 2(1):32-47.https://doi.org/10.31181/dma21202420
5. Hamed, A. A., Tabakoff, W., Rivir, R. B., Das, K. and Arora, P. (2005). Turbine blade surface deterioration by erosion. J. Turbomach. 127:445–452. https://doi.org/10.1115/1.1860376
6. Mittal, U., Yang, H., Bukkapatnam, S. T. and Barajas, L. G. (2008). Dynamics and performance modeling of multi-stage manufacturing systems using nonlinear stochastic differential equations. IEEE International Conference on Automation Science and Engineering, IEEE, pp. 498–503. https://doi.org/10.1109/COASE.2008.4626530
7. Branco, J. R. T., Gansert, R., Sampath, S., Berndt, C. C. and Herman, C. (2004). Solid particle erosion of plasma sprayed ceramic coatings. Mater. Res. 7:147–153. https://doi.org/10.1590/S1516-14392004000100020
8. Sahoo, S. K., Goswami, S. S., Sarkar, S. and Mitra, S. (2023). A review of digital transformation and industry 4.0 in supply chain management for small and medium-sized enterprises. Spectr. Eng. Manag. Sci. 1(1):58–72. https://doi.org/10.31181/sems1120237j
9. Mittal, U. (2023). Detecting hate speech utilizing deep convolutional network and transformer models. International Conference on Electrical, Electronics, Communication and Computers, IEEE, pp. 1–4. https://doi.org/10.1109/ELEXCOM58812.2023.10370502
10. Shin, D. and Hamed, A. (2018). Influence of micro–structure on erosion resistance of plasma sprayed 7YSZ thermal barrier coating under gas turbine operating conditions. Wear 396:34–47. https://doi.org/10.1016/j.wear.2017.11.005
11. Sethy, N. K., Yenugula, M., Goswami, S. S., Bhola, A. and Behera, D. K. (2023). Selection of ideal IoT based overhead conductor for optimizing the performance of a small hydropower project. J. Nano. Electron. Phys. 15(4). https://doi.org/10.21272/jnep.15(4).04006
12. Rajabinezhad, M., Bahrami, A., Mousavinia, M., Seyedi, S. J. and Taheri, P. (2020). Corrosion-fatigue failure of gas-turbine blades in an oil and gas production plant. Mater. 13:900. https://doi.org/10.3390/ma13040900
13. Sahoo, S. K.and Goswami, S. S. (2024). Green supplier selection using MCDM: A comprehensive review of recent studies. Spectr. Eng. Manag. Sci. 2(1):1–16. https://doi.org/10.31181/sems1120241a
14. Taherkhani, B., Anaraki, A. P., Kadkhodapour, J., Farahani, N. K. and Tu, H. (2019). Erosion due to solid particle impact on the turbine blade: experiment and simulation. J. Fail. Anal. Prev. 19:1739–1744. https://doi.org/10.1007/s11668-019-00775-y
15. Goswami, S. S., Sarkar, S., Gupta, K. K. and Mondal, S. (2023). The role of cyber security in advancing sustainable digitalization: Opportunities and challenges. J. Decis. Anal. Intell. Comput. 3(1):270–285. https://doi.org/10.31181/jdaic10018122023g
16. Chowdhury, T. S., Mohsin, F. T., Tonni, M. M., Mita, M. N. H. and Ehsan, M. M. (2023). A critical review on gas turbine cooling performance and failure analysis of turbine blades. Int. J. Thermofluids. 100329. https://doi.org/10.1016/j.ijft.2023.100329
17. Goswami, S. S. and Behera, D. K. (2023). Developing fuzzy-AHP-integrated hybrid MCDM system of COPRAS-ARAS for solving an industrial robot selection problem. Int. J. Decis. Support Syst. Technol. 15(1):1–38. http://doi.org/10.4018/IJDSST.324599
18. Rodriguez, E., Flores, M., Pérez, A., Mercado-Solis, R. D., González, R., Rodriguez, J. and Valtierra, S. (2009). Erosive wear by silica sand on AISI H13 and 4140 steels. Wear. 267:2109–2115. https://doi.org/10.1016/j.wear.2009.08.009
19. Sahoo, S. K., Das, A. K., Samanta, S. and Goswami, S. S. (2023). Assessing the role of sustainable development in mitigating the issue of global warming. J. Process Manag. New Technol. 11(1-2):1–21. https://doi.org/10.5937/jpmnt11-44122
20. Alshahrani, R., Yenugula, M., Algethami, H., Alharbi, F., Goswami, S. S., Naveed, Q. N., Lasisi, A., Islam, S., Khan, N. A. and Zahmatkesh, S. (2023). Establishing the fuzzy integrated hybrid MCDM framework to identify the key barriers to implementing artificial intelligence-enabled sustainable cloud system in an IT industry. Exp. Syst. Appl. 121732. https://doi.org/10.1016/j.eswa.2023.121732

Advances in Additive Manufacturing Technologies – Gurusamy Pathinettampadian (eds)
© 2024 Taylor & Francis Group, London, ISBN 978-1-032-90013-1

3 Experimental Investigation of Mango Wood Powder and Aghil Tree Resin Reinforcement with Combined Epoxy Hybrid Matrix Bio-Composite

A. Latha*
Department of Civil Engineering,
Panimalar Engineering College, Chennai-123, Tamilnadu, India

K. Yesuraj, A. Shyaam Siddharth,
Department of Mechanical Engineering,
Panimalar Engineering College, Chennai-123, Tamilnadu, India

R. Ganesan
Department of Civil Engineering, Velammal college of Engineering and Technology,
Madurai-09, Tamil Nadu, India

R. Sathiyamoorthi
Department of Mechanical Engineering,
Chennai Institute of Technology, Chennai, Tamilnadu, India

ABSTRACT: This paper seeks to investigate the utilisation of natural filler resources to reduce the reliance on polymer matrix materials, while simultaneously improving tensile strength and heat resistance properties. This research aims to examine the utilisation of Mango wood powder as a particulate reinforcement in Aghil tree resin mixed epoxy bio-composites, while altering the volume fractions. The tensile characteristics were examined following the manual fabrication process using a layup technique. The incorporation of organic filler into the hybrid composite material partially improved its tensile properties. The epoxy resin composites have a Modulus of Elasticity of 34 MPa and a flexural strength of 1700 MPa. The incorporation of Aghil Epoxy Hybrid resin resulted in a 15.37% improvement in flexural strength and a 22.70% rise in flexural modulus within the composite containing 30% volume of mango wood powder filler with Aghil Epoxy Hybrid Matrix. The addition of mango wood powder shell powder filler to the Aghil Epoxy coupled epoxy with epoxy hybrid Matrix composite resulted in a 30% increase in flexural qualities, which is equivalent to improved tensile capabilities.

KEYWORDS: Aghil tree resin, Epoxy resin, Mechanical properties, Mango wood powder,Viscoelastic properties, Bio-composite

1. INTRODUCTION

Polymers are used as natural fibres and fillers. Investigators all over the globe today view composites as a viable substitute for traditional polymer composites [1]. Because of the detrimental ecological effects of using artificial fabrics and polymers, numerous natural fibres and additives are being explored for their environmental conduciveness,

*Corresponding author: lathaganesan.a@gmail.com

DOI: 10.1201/9781003545774-3

renewability, affordability, and reliability. The use of carbon black as an adhesive in an epoxy composite. The substance is comprised of 5% carbon black derived from empty oil palm fruit bunches, bamboo stalks, and coconut husks [2,3]. The mechanical hardness and elasticity of the empty fruit bunches and bamboo stem-generated carbon black integrated hybrid components were enhanced in the evaluations. The aggregate X-ray Diffraction (XRD) characteristics indicated a heterogeneous crystal structure as well as a polymorphic architecture [4,5].

The impact of extrinsic additives such as kaolin, talc, and zinc borate on the tensile as well as the water-absorbing capabilities of wood powder-included adhesive [5, 6] The findings demonstrated that the use of additive materials increased water retention capabilities. In terms of mechanical strength and moisture resistance, the kaolin filler beat the other additives employed in the research due to its fixed plate design and tiny particle size [7–10].

The tensile characteristics of polymer matrix composites using particle fillers such as talc, silicon carbide, aluminium powder, and steel strands [10, 11] According to the research, Noryl has superior properties to the other types of fillers. The characteristics of polyvinyl chloride nano clay composites were examined [12, 13]. Nano clay diffused efficiently, leading to enhanced mechanical qualities as well as a small decrease in heat resilience. The impact of additives on adhesive epoxy combinations [14]. Nano fillings like alumina and silicon carbide are used as granular injectors in DGEBA, often described as elastomeric glue. The healing activity, heat resistance, and dynamical tensile characteristics of Al_2O_3 hybrids were tested, and the findings indicated that the disintegration temperature rose from 630 °C to 853 °C and from 858 °C to 853 °C for SiC hybrids. The glass transition temperature of the adhesive polymer was elevated by 10 °C above that of the raw adhesive polymer. The effect of a bilayered calcium phosphate binder on sustainable PLA. The additive substance was raised to 25% v/v [15]. At greater capacity percentages, SEM examination demonstrated that the supplement material was densified. The inclusion of additive substances was said to have raised the optical transition frequency of the composite [16, 17, 18].

An organic filler substance for particle strengthening in recovered polypropylene [19, 20] Wooden sawdust was utilised as an additive in different loading circumstances as well as in different particulate dimensions. The tensile as well as fluid-absorbing characteristics were studied, and the findings indicated that a particulate density of 30 percent as well as an additive dimension of 100 mm gave improved tensile characteristics [21, 22–23]. The properties of wooden polymer hybrids, including sawdust particles for an additive as well as polypropylene as a framework [24, 25].

The consumption of water grew in lockstep with the rise of sawdust. The coconut shell particles in adhesive resins are a sustainable binder [26]. The influence of filler quantity on composite tensile attributes was examined. The experiments revealed that increasing the filler concentration enhanced the mechanical capabilities of the hybrid owing to increased interface connections among the additive as well as the framework. Similarly, other studies investigated the application of fillers [27, 28–29]. Several investigations [30, 31, 32] found that additives may be employed to enhance the tensile characteristics of composite polymers to varied levels. There have only been a few investigations on the mango wood powder reinforced with Aghil hybrid resin bio-composite material [33–35].

2. Materials and Methods

Mango wood powder, being a natural filler, was combined with Aghil Epoxy Hybrid resin. After grinding the mango wood powder to a finer particle of 5–10 μ inside a ball mill for five hours, the filler material was mixed into the Aghil Epoxy matrix at various volume percents (5, 10, 15, 20, and 25 percent). To achieve uniform distribution, a mechanical mixer continuously stirred both the filler and the matrix for 1 hour. The hand lay-up method was used to make the Prunus Dulcis shell powder filler-reinforced Aghil Epoxy Hybrid composite laminates. A silicon rubber mould with a length of 300mm, a breadth of 300mm, and a thickness of 3mm was used. The mould's exteriors were covered with rigid plates. Following the addition of a predetermined amount of resin filler combination, the hard sample was positioned on top, and a weight of 500–600 N was located to compress the laminate. Next, they were cut to ASTM specifications for various procedures.

2.1 Aghil Tree Resin

Aquilaria agallocha, commonly known as Aghil, Agaru, or Agarwood Eaglewood, is a resinous heartwood of the Aquilaria tree. The aghil tree belongs to the Thymelaeaceae family and is found predominantly in South Asian regions. The resinous agarwood is produced as a self-defence mechanism against the fungus growing in the injured part of the tree. The heartwood of the affected tree turns darker and heavier with the passage of time, as opposed to being lighter and brighter at the time of infection. The cut heartwood of the Aghil tree is desiccated and ground to a fine powder by means of the ball milling process. The powder is sieved to maintain uniformity of size, and the size of the powder is analysed by a particle size analyzer. Aquilaria agallocha heartwood and black Aquilaria agallocha resin are depicted in Figs. 3.1(a) and (b), respectively.Aquilaria agallocha, or Aghil resin for this study, is obtained from Mrs. Evergreen Fibres and Chemicals, Chennai, India.

(a)

(b)

Fig. 3.1 (a) Aquilaria agallocha heartwood (b) Black Aquilaria agallocha resin

3. Results and Discussion

3.1 Mechanical Properties

The tensile evaluation was carried out following the ASTM D638 criteria at a rate of 2 mm/min. Flexural as well as impact assessments were carried out following D790 and ASTM D256 regulations, respectively. The structural characteristics of the Aghil Epoxy Hybrid Resin Bio Composite with varied volumetric percentages of mango wood powder are shown in Figs. 3.2–3.4. It should be noted that the inclusion of filler elements enhances the structural characteristics of combination elements to a degree. The greatest tensile strength of the composite specimen with 30% v/v mango wood powder, including Aghil Epoxy Hybrid Resin Bio-composite, was 30.2 MPa. Similarly, the bending and impacting characteristics of the composite with 30% v/v mango wood powder incorporation were greate

The impact analysis is performed to measure the toughness of a material, which is described as the amount of power received throughout the rupture. Natural fillers are frequently used to increase the stiffness (modulus) and endurance of polymer composites. The bonding strength of the filler and the hybrid matrix bio-composite material determines the amount of energy required when the framework and innate filler separate as a result of loading. The addition of more than 35% mango wood powder to the Aghil Epoxy Hybrid Matrix bio composites resulted in a decrease in mechanical characteristics. The reason is the clustering of natural filler in themango wood powder with *Aghil Epoxy Hybrid* Matrix bio composite, as shown by the SEM images in Figs. 3.4 and 3.5. The reduced strain demonstrates how the composite material becomes more brittle as a result of the clustering of the filling components. Furthermore, the filler could stop swilling and lead to relinquishing in acicular areas, as if an anti-adhesive had been used on the particles and messed up their interaction

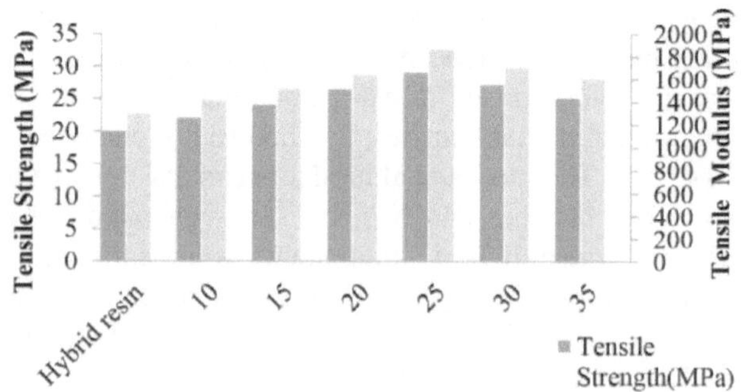

Fig. 3.2 Tensile properties of mango wood powder with aghil epoxy hybrid resin matrix bio-composite

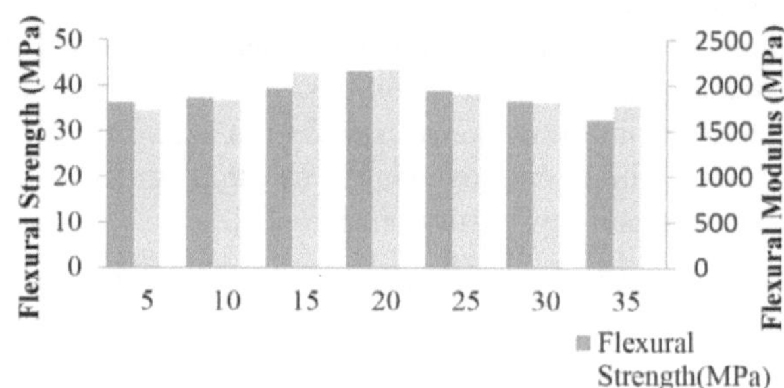

Fig. 3.3 Flexural properties of mango wood powder with aghil epoxy hybrid matrix bio-composite

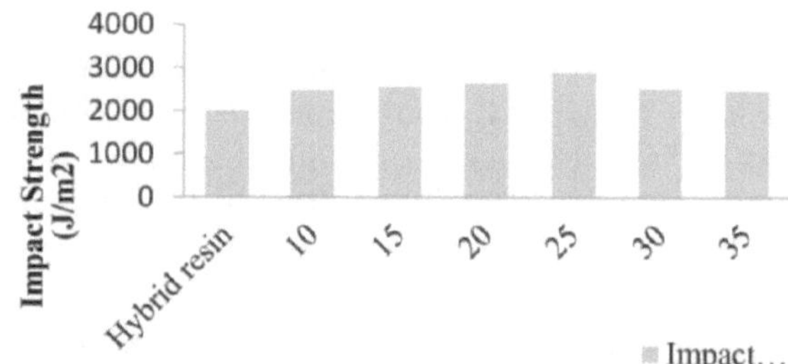

Fig. 3.4 Impact strength of mango wood powder with aghil epoxy hybrid matrix bio-composite

3.2 SEM Analysis

The addition of more than 35% mango wood powder with Aghil Epoxy Hybrid Matrix bio composite resulted in a decrease in mechanical characteristics. This is owing to natural filler aggregation in the epoxy matrix, as seen by the SEM images in Fig. 3.5. Because of the clustering of the filling components the composite material becomes more brittle, as demonstrated by the lower strain to failure. Furthermore, the filling might inhibit swilling as well as induce surrendering in clustered areas.

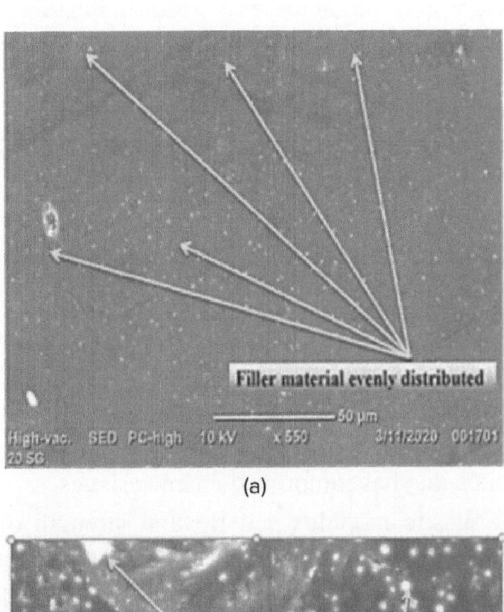

(a)

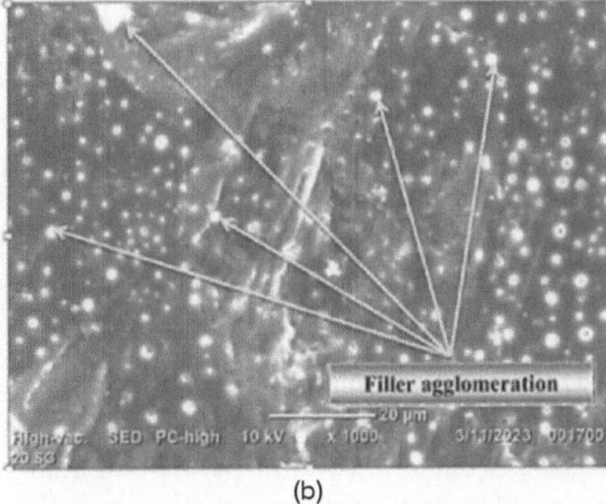

(b)

Fig. 3.5 SEM representation of 35vol% Mango wood powder with aghil epoxy hybrid matrix biocomposite (a) Tensile specimen (b) Flexural specimen

3.3 Dynamic Mechanical Analysis (DMA)

During a dynamic mechanical analysis test, the user usually configures the testing parameters, such as temperature range, frequency, and amplitude, using the control software. The apparatus subsequently applies a varying load on the sample while simultaneously measuring its mechanical response, so yielding significant insights into its viscoelastic characteristics. The test temperature ranged from 40° C to 180° C, with a frequency of 10 Hz. The elastic, as well as viscous characteristics, are determined using dynamic mechanical analysis. This study evaluated the influence of temperature and filler level on the storage modulus of mango wood particles with Aghil Epoxy Hybrid Matrix bio Composite, as depicted in Fig. 3.6.

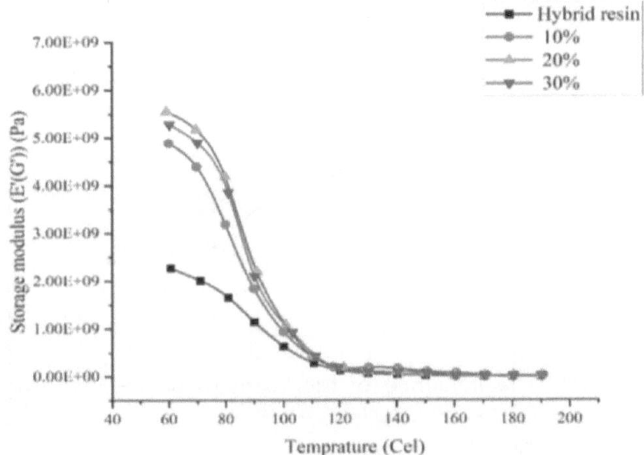

Fig. 3.6 Storage modulus of mango wood powder with aghil epoxy hybrid matrix bio composite at 10 Hz frequency

As shown in Fig. 3.6, the retention value is greater for the sample with 20% v/v filler than for the exhibit with no filler.The power received by the epoxy and composite is approximately equal as the testing temperature increases. This indicates that the mango wood filler is effectively transferring the load until 90°C. Following that, the filler's link with the matrix weakens, causing the collapse of the desired functionality. Using groundnut filler in the combined adhesive composite substrate increases the storage modulus, or stiffness factor, of the composite. The material's molecular reordering results in a minimization of localized stress. The particles fail to resonate because they are so stationary at low temperatures. Because the particles are so stationary at small temperatures, they are incapable of resonance with the cyclical stresses and hence stay inflexible. Approximately 20 atoms in cementitious polymeric materials are interconnected. As this relationship is maintained at different temperatures, they are resilient.

The viscous and elasticity characteristics of the component are determined using DMA. The influence of temperature and filler volume on the retention value of mango wood powder with Aghil Epoxy Hybrid Matrix bioComposite was analyzed in this research, as shown in Fig. 3.7. The temperature ranged from 40° C to 180° C, with 10 Hz. The storage modulus is higher for the exhibit having 30% v/v filler compared to the exhibit minus filler, as depicted in Fig. 3.7. This implies that the mango wood filler is efficiently distributing the weight until the temperature reach-

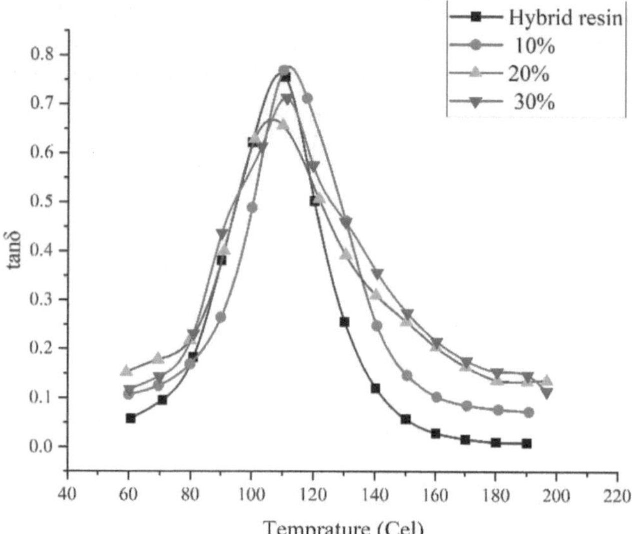

Fig. 3.7 The influence of filler content on the damping factor Tan at 10 Hz

es 90°C. After that, the filler's link with the matrix weakens, causing the breakdown of the desired application. The use of mango wood filler in the integrated adhesive hybrid matrix improves the storage modulus or stiffness factor of the composite. The molecular reorganization of the material reduces localized stress. The molecules are unable to resonate since they are so stationary at lower temperatures. Because the particles are so stationary at small temperatures, they are incapable of resonance with the cyclical stresses and hence stay inflexible. Approximately 20 atoms in cementitious polymeric materials are interconnected. As this relationship is maintained at different temperatures, they are resilient.

3.4 Damping Factor (Tan δ)

This is expressed as a percentage of the material's damage and storage modulus. This demonstrates the level of molecule movement in the polymer link as well as the substance's energy waste under pressure. Figure 3.7 depicts a plot of the damping factor of mango wood powder. The results show that incorporating strengthening reduces the dampening rate. As a consequence, the composite surpasses the matrix in terms of energy dispersion. The substance exists in the glass phase or elastic energy phase at reduced temperatures. Whenever the temperature increases, the substance phase transforms to a polymer or entropy elastic level; the glass crossover is the shift from the glass phase to the latex flexible condition. The temperature of the maximum loss modulus (E") is frequently assumed to be the Glass Transition temperature (Tg). The Tg is calculated in this work using the peak of the tan value. While contrasted with epoxy resin, the peak of tan decreases regardless of the magnitude of applied force, while an understandable alteration in the Tg is observed along. This demonstrates that mango wood powder filler has a synergistic effect.

The attenuation point is connected with the glass transitional zone when the substance transitions from an inflexible to a pliable condition. Initially, the molecules are stable. But, the displacement of tiny clusters as well as sequences of molecules inside the polymeric framework initiates as the temperature increases. As a consequence, the higher the tan peaks, the better the molecular movement. The decrease in peak height and the increase in width indicate a decrease in molecule mobility and an increase in homogeneity phases. Peak height decrease implies an enhancement in the matrix-reinforcement interfacial adherence. Property degradation occurs as the material changes phases due to polymer chain breakage.

4. CONCLUSIONS

This experimental inquiry enhances comprehension of sustainable and environmentally-friendly materials, yielding valuable discernment for prospective implementation in industries such as building, automobile, and packaging. This discovery could also facilitate the creation of new biocomposites that have improved characteristics.

- The elastic modulus and flexural strength of epoxy resin composites are 34 and 1700 MPa, respectively.
- The flexural properties of mango wood powder shell powder filler with Aghil Epoxy combined epoxy with epoxy hybrid Matrix composite increased by up to 30% volume percent, equivalent to tensile capabilities.
- SEM image of an Aghil Epoxy hybrid composite with a volume of 35%. A hybrid polymer composite with 30% natural resin.
- The DMA demonstrated that the composite material's storage modulus and glass transition temperature are increased when 30 volume percent of mango wood filler is added, indicating improved composite material toughness.

REFERENCES

1. Okokpujie, I. P., Tartibu, L. K., Babaremu, K., Akinfaye, C., Ogundipe, A. T., & Akinlabi, E. T. (2023). Study of the corrosion, electrical, and mechanical properties of aluminium metal composite reinforced with coconut rice and eggshell for wind turbine blade development. *Cleaner Engineering and Technology*, *13*, 100627.
2. Sulthana, R., Taqui, S. N., Syed, U. T., Soudagar, M. E. M., Mujtaba, M. A., Mir, R. A., & Hossain, N. (2022). Biosorption of crystal violet by nutraceutical industrial fennel seed spent equilibrium, kinetics, and thermodynamic

studies. *Biocatalysis and Agricultural Biotechnology, 43*, 102402.

3. Senthilkumar, R., Natarajan, M. P., Ponnuvel, S., & Sathyamurthy, R. (2022). Mechanical and visco elastic analysis of Sal tree gum incorporated epoxy bio composite. *Journal of Materials Research and Technology, 17*, 819–827.
4. Jagadeesh, P., Puttegowda, M., ThyavihalliGirijappa, Y. G., Rangappa, S. M., &Siengchin, S. (2022). Effect of natural filler materials on fiber reinforced hybrid polymer composites: An Overview. *Journal of Natural Fibers, 19*(11), 4132–4147.
5. Kumar, B. R., Alphin, M. S., Santhanam, V., & Palanikumar, V. (2022). Mechanical, Vibration and Visco-elastic Behavior of Abelmoschus Esculentus Fiber Reinforced Epoxy Composite. *Materiale Plastice, 59*(4), 70–81.
6. Zhou, S., Li, J., Kang, S., & Zhang, D. (2022). Effect of carbonized ramosissima nanoparticles on mechanical properties of bamboo fiber/epoxy composites. *Journal of Natural Fibers, 19*(4), 1239–1248.
7. Nanni, A., Parisi, M., Colonna, M., &Messori, M. (2021). Thermo-mechanical and morphological properties of polymer composites reinforced by natural fibers derived from wet blue leather wastes: a comparative study. *Polymers, 13*(11), 1837.
8. Pashaei, S., Hosseinzadeh, A., Mohammadi-Aghdam, S., Saadat, Y., & Hosseinzadeh, S. (2021). Fabrication and characterization of carboxylated and aminolated multiwalled carbon nanotube/polyethersulfone (PES) membranes for the removal of heavy metals from wastewater. *Polymer-Plastics Technology and Materials, 60*(9), 994–1004.
9. Bhaskar, K. B., Devaraju, A., & Paramasivam, A. (2021). Experimental investigation of glass powder reinforced polymer composite. *Materials Today: Proceedings, 39*, 484–487.
10. Santhanam, V., Dhanaraj, R., Chandrasekaran, M., Venkateshwaran, N., & Baskar, S. (2021). Experimental investigation on the mechanical properties of woven hybrid fiber reinforced epoxy composite. *Materials Today: Proceedings, 37*, 1850–1853.
11. Yesuraj, K., Pazhanivel, K., Srinivasan, S. P., Santhanam, V., & Muruganantham, K. (2021). Static investigation of almond shell particulate reinforced aquilariaagallocharoxb blended epoxy hybrid matrix composite. *Digest Journal of Nanomaterials and Biostructures, 16*(2), 359–365.
12. Ashok, B., Naresh, S., Reddy, K. O., Madhukar, K., Cai, J., Zhang, L., &Rajulu, A. V. (2014). Tensile and thermal properties of poly (lactic acid)/eggshell powder composite films. *International journal of polymer analysis and characterization, 19*(3), 245–255.
13. Venkateshwaran, N., Santhanam, V., &Alavudeen, A. (2019). Feasibility study of fly ash as filler in banana fiber-reinforced hybrid composites. *Processing of green composites*, 31–47.
14. Nwanonenyi, S. C., Obidiegwu, M. U., &Onuegbu, G. C. (2013). Effects of particle sizes, filler contents and compatibilization on the properties of linear low density polyethylene filled periwinkle shell powder. *The International Journal of Engineering and Science, 2*(2), 1–8.
15. Jagath Narayana Kamineni , G. Anbuchezhiyan J. Anichai, Ravish Sharma,R. Ganesan A. Latha, B. Nagaraj Gou, (2023) Investigation of the microstructure and mechanical properties of borosilicate reinforced magnesium nano composites. *Materials Today: Proceedings.*https://doi.org/10.1016/j.matpr.2023.05.193
16. Sudheer, M., Prabhu, R., Raju, K., & Bhat, T. (2014). Effect of filler content on the performance of epoxy/PTW composites. *Advances in materials science and engineering, 2014.*
17. Abdul Khalil, H. P. S., Jawaid, M., Firoozian, P., Amjad, M., Zainudin, E., &Paridah, M. T. (2013). Tensile, electrical conductivity, and morphological properties of carbon black–filled epoxy composites. *International Journal of Polymer Analysis and Characterization, 18*(5), 329–338.
18. Arunkumar D, Latha A, Suresh Kumar S,Jasgurpreet SinghChohan,VelmuruganG,Nagaraj M.(2023), Experimental Investigations of Flammability, Mechanical and Moisture Absorption Properties of Natural Flax/NanoSiO2 Based Hybrid Polypropylene Composites, *Silicon,* https://doi.org/10.1007/s12633-023-02611-3
19. Anjan Kumar Reddy G.Rajesh,G. Anbuchezhiyan, A. Ponshanmugakumar,R. Ganesan , A. Latha ,M. Satyanarayana Gupta, .(2023) Investigating the mechanical properties of titanium dioxide reinforced magnesium composites, *Materials Today:Proceedings.*https://doi.org/10.1016/j.matpr.2023.05.628

Note: Every figures in this chapter was created using the experimental work; none of them were taken from any publications or online.

Advances in Additive Manufacturing Technologies – Gurusamy Pathinettampadian (eds)
© 2024 Taylor & Francis Group, London, ISBN 978-1-032-90013-1

4 Mechanical Properties of Novel Polylactic Acid-3D Printed Core/Aloevera/Glass Fiber Reinforced Hybrid Composites

R. Karthick*, V. Anbumalar
Department of Mechanical Engineering,
Velammal College of Engineering and Technology,
Madurai, Tamilnadu, India

R. Jeyakumar
Department of Mechanical Engineering,
Sri Krishna College of Engineering and Technology,
Coimbatore, Tamilnadu, India

M. Kubendiran
Department of Mechanical Engineering,
Jeppiaar Institute of Technology,
Kanchipuram, Tamilnadu, India

S. Kevin Jasper, V. Sankara Pandian, D. Manish Kumar
Department of Mechanical Engineering,
Velammal College of Engineering and Technology,
Madurai, Tamilnadu, India

ABSTRACT: In this work, the composites were created by consistently stacking glass fiber, aloevera fiber, and PLA-3D (Polylactic Acid) printed core into the polyester matrix and then the entire structure was subjected to compression moulding, ensuring the creation of a uniformly structured composite. PLA-3D printed core was created by using 3D printer. Then, evaluated various mechanical characteristics such as hardness, impact, tensile and compressive strength. From the results, it is found that, PLA-3D printed core (S1) shows the lowest hardness, impact, tensile and compressive strength of 42.57, 0.85 KJ/m2, 15.24 MPa and 42.57 MPa respectively. Further, S4 composite which has glass fiber to the PLA-3D printed core shows the highest hardness, impact, tensile and compressive strength of 67.81, 6.87 KJ/m2, 45.89 MPa and 71.58 MPa respectively. SEM analysis shows fiber breakages, fiber pullout, good adhesion between fiber and matrix and broken in PLA-3D printed core. Overall, S3 and S4 composites were performed well and recommended for various engineering applications.

KEYWORDS: Polylactic acid, Glass fiber, Aloevera fiber, Mechanical properties

*Corresponding author: rkarthick33@gmail.com

DOI: 10.1201/9781003545774-4

1. INTRODUCTION

The process of making three-dimensional items from a computer model by layering on material is called additive manufacturing, or 3D printing. This technology has gained widespread use in various industries and has become accessible to individuals and small businesses[1]. The use of fibers (natural and synthetic) in 3D printing is an emerging trend that combines the benefits of additive manufacturing with the sustainability and eco-friendliness of natural materials. Natural fibers are derived from various sources such as minerals, plants, and animals and can be used in 3D printing to create more environmentally friendly and biodegradable products. These fibers, often referred to as fiber reinforcement, are added to the 3D printing materials to improve the mechanical properties and performance of the printed objects[2]. In order to facilitate 3D printing, Zhuang et al. [2]created various prepreg continuous fibre filament types and looked into the effects of fibre types and content on the tensile characteristics of filaments. The samples that were printed at 280°C exhibit improved shear and tensile strengths of 29 MPa and 189 MPa, respectively.The additive manufacturing of continuous carbon composites with fibre reinforcement was introduced by Zhang et al. [3] to enable the design and development of auxetic structures with 4 Poisson's ratios. Experimental investigations are conducted to analyze the compressive properties of the 3D-printed composites. The findings show that the modified arrangement considerably affects the compressive characteristics and deformation behaviours at different Poisson's ratios.In the study by Zhang et al. [4], an orthopedic plate was initially fabricated using (PLA) through 3D printing, and then it was coated with nano-fibers composed of polycaprolactone (PCL) and Akermanite (AKT). The thermomechanical characteristics of coated and uncoated specimens were compared and examined in this study, covering pressure, 3-point flexural, and thermal conductivity. Incorporation of nAKT into the PLA + PCL sample results in a significant 21.39% increase in the maximum bending flexural force.Zarna et al. [5] conducted a study focusing on characterizing and comparing bending characteristics of different cellular configurations for wood fiber &PLAbiocomposite panels. Material extrusion 3-D printing was employed for five distinct cell designs were put through mechanical testing. Notably, the research discovered that the elastic&plastic bending behaviour in 3D-printed sandwich panels may be well described by a linear elastic, bimodular, and completely plastic material model.

Senthamaraikannan et al. [6] conducted a study involving the design and 3D printing of an I-shaped beam using Fused Filament Fabrication. They made use of PLA and ABS reinforced with carbon fibre. The outcomes showed that, when compared with PLA with the same carbon fibre reinforcement, ABS reinforced with carbon fibre had better damping properties. The objective of Reddy et al.'s study [7] was to evaluate the influence of temperature, the speed, and the thickness of layers on the breaking strength of parts made from 3D printing. Their research aimed to comprehend the effects of layer thickness, printing speed, and infill density on the tensile characteristics of PLA samples. Paz-González et al. [8] introduced an innovative structural composite created through additive manufacturing of PLA layers and lamination of carbon fiber, with possible applications in prosthetic implants. The findings suggest that the CFRC/PLA composite holds promise as a suitable material for manufacturing hip femoral stem prostheses. A novel 3D printing technique for continuous carbon fibre composites with high heat conductivity was presented by Olcun et al. [9]. With this technique, PLA coated pitch carbon fibres may be printed on demand using a twin extruder. The study experimentally characterized the thermal conductivity of unidirectional 3D-printed samples with varying fiber properties and volume fractions. The measured thermal conductivities fell short of predictions due to post-printing fiber breakage, indicating room for improvement in fiber handling during printing.High-performance three-dimensional printed CFFRCs with a significant fiberwt. fraction can be made using the method developed by Long et al. [10]. The results indicated that CFFRCs created with the fiber prepreg method achieved a 44.1% fiber volume fraction and a low void content of approximately 1.9%. Consequently, these CFFRCs demonstrated the highest tensile modulus and strength compared to previouslyreported 3D printed composites.

Lau et al. [11] examined how various kenaf volume ratios affected the 2003D PLA filament's tensile characteristics for 3D printing. The results revealed that a 15% kenaf fiber delivered the best tensile behaviour among various filler loadings, confirming the enhanced tensile properties resulting from the addingkenaf fiber. In a study led by Kajbič et al. [12] biodegradable PLA composites with continuous flax fibers were 3D printed. They conducted static and mechanical fatigue tests on plain specimens, notched matrix specimens, and two composite configurations across a range of load cycles. Depending on the fibre content, notched composites with fibres at the notch tip had fatigue strength that was between 1.9 and 2.9 times greater than that of the notched matrix. Estakhrianhaghighi et al. [13] investigates the incorporation of cellulose-derived substances into a biobased polymer to produce sustainable materials with enhanced characteristics that are high-performing. We look into the elastic flexural properties of 3D printed composites made of a PLA biopolymer reinforced by different weight percentages (2.5%–15%) of waste wood fibre. In order to advance additive manufacturing, this study introduces WF-PLA tailored cellular composites as a sustainable means of adjusting structural qualities. Awad et al. [14] conducted a study in which they produced Polylactic Acid (PLA) composites by melting PLA with various locally sourced natural fibers. The tensile strength, strain at break and young's modulus of the PLA-natural fibre composites were improved by these natural fibres in comparison to plain PLA. The results affirm the suitability of these 3D printed biocomposites for biomedical applications due to their impressive stiffness, tensile properties, and dimensional stability.Avci et al. [15]carried out a study to look into the impacts of three different boron compounds—ulexite, zinc borate, and borax boric acid combine—on the microstructural and mechanical features of flax fiber/PLA biocomposites at different water exposure times. Notably, the study revealed a slight improvement in the tensile value at break of these composites following instant water absorption.

Ye et al. [16] made 3D printing filament from PLA/TPU and used it to craft a constant carbon fiber reinforced honeycomb structure using a 3D printer. They explored the structure's ability to detect strain during cyclic compression, opening doors for the development of smart constructions made of 3D-printed carbon fibre composite for monitoring structural health.Suresha et al. [17] investigated how annealing affected carbon fibre (CF) reinforced Mah-g-PLA (CF/Mah-g-PLA) with maleic anhydride (Mah) grafted PLA and fused filament deposition (Mah-g-PLA). For dynamic mechanical analysis, they employed a dynamic mechanical analyzer. According to their findings, adding a Mahcompatibilizer to grafted PLA strengthens the composite materials' dynamic mechanical qualities and increases the interfacial bonding of CF/Mah-g-PLA. In their research by Rajapandian et al. [18]calculated the mechanical characteristics of Carbon/PLA and Glass/PLA composites. They manufactured specimens using a 3-D printer and performed tensile and flexural tests. When compared with glass fiber-reinforced PLA partial biocomposites, the results indicated that the improved carbon fiber-reinforced composites had a 53% stronger tensile strength and a 59% greater flexural strength.Maqsood et al. [19]sought to use FDM 3D printing to combine continuous carbon thermoplastic materials with CCF to make continuous carbon fiber-reinforced polyester composite (CCFRTC). Following fabrication, their specimens were tested for flexural bending, which resulted in delamination.Chen et al. [20]presented a novel approach to producing CGF/PLA composites using 3D printing. The printed goods were remarkable in terms of their flexural modulus (21.5 GPa) and strength (312 MPa).

Based on the existing literature and research pathways, it is evident that no prior study has explored the impact of hybridization involving a PLA core with both glass fiber and aloe vera fiber in polyester composites. Indeed, the development of composites with these compositions holds the potential to yield materials with desirable properties, making them suitable for various engineering applications. This unexplored combination may open up new possibilities for innovative and high-performing materials. As a result, the primary objective of this research is to manufacture lightweight 3D printed PLA cores combined with glass fiber and aloevera fiber polyester hybrid composites. Interestingly, the mechanical characteristics of these hybrid composites can be assessed using ASTM standards, providing a standardized and reliable method for evaluating their performance.

2. Materials and Methods

2.1 Materials

PLA, or Polylactic Acid, is a popular thermoplastic material commonly used in 3D printing. It's favored for its ease of use, biodegradability, and versatility. PLA filament is fed into a 3D printer, which then melts and extrudes the material layer by layer to create 3D printed core. PLA is known for producing high-quality, detailed prints and is often used for prototyping, hobbyist projects, and educational purposes.

Aloevera fiber and glass fiber were purchased as a mat from GVR Enterprises, Madurai, Tamilnadu, India. In addition to this, polyester resin and hardener was obtained from the local chemical stores.

2.2 Fabrication of Composites

Before creating the composite, the aloevera fiber underwent a silane treatment process, following the procedure outlined in previous journal publications[21-23]. Furthermore, mild steel moulds with cavity dimensions of 3 mm in thickness and 300 mm in length were employed to produce the specimens needed for characterization. To facilitate the easy removal of the composite material, a thin layer of wax was applied to the surface of the mould. Before manufacturing the composite, the mould's surface was thoroughly cleaned to eliminate any residual polymeric material [24]. Prior to the fabrication of composites, PLA-3D printed core was manufactured using 3D printing machine. The composite was created by consistently stacking glass fiber, aloevera fiber, and PLA-3D printed core into the polyester matrix, as shown in the Fig. 4.1. Also it shows the process of printing PLA-3D core and S2 specimen. Subsequently, the entire structure was subjected to compression moulding,

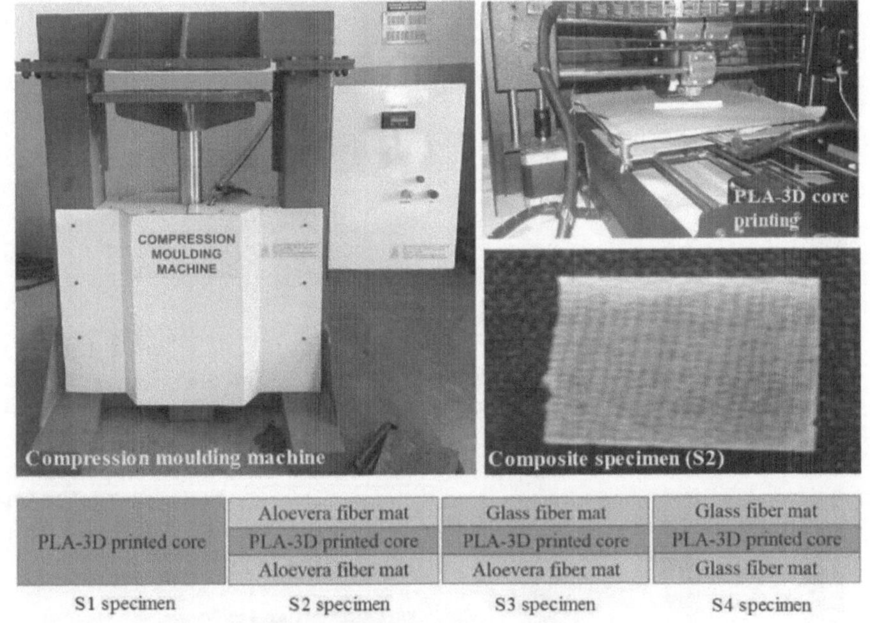

Fig. 4.1 Fabrication of composites

ensuring the creation of a uniformly structured composite. The composites were subjected to a compression process for 1 hour at 50°C under an applied pressure of 75 KPa. Subsequently, they underwent a 24-hour curing period at room temperature, followed by an additional 48 hours of post-curing at 160°C.

2.3 Mechanical Characterization

All the fabricated composites were subjected to various mechanical characterizations, including assessments of hardness, tensile strength, compressive and impact strength, in accordance with ASTM standards. This ensures that the testing procedures followed established industry standards for accuracy and consistency in the results.

3. Results and Discussions

3.1 Hardness Number

Hardness value of PLA-3D printed composites is illustrated in Fig. 4.2. From the findings, it is found that, PLA-3D printed composite which contains PLA content only revealed lowest hardness number of 42.57 than other composites. Among fiber filled composites (S2, S3 and S4), S2 composite which have aloevera fiber content only showed lowest hardness number of 51.54. This is due to less density of aloevera fiber and absence of glass fiber. The composite (S4) which contains glass fiber onlyshowed highest hardness value of 67.81 than other composites. This fact is owing to high strength of glass fiber [22]. Further, hybrid composite (S3) which have aloeverafiber and glass fiber showed reasonably good hardness number of 59.25.

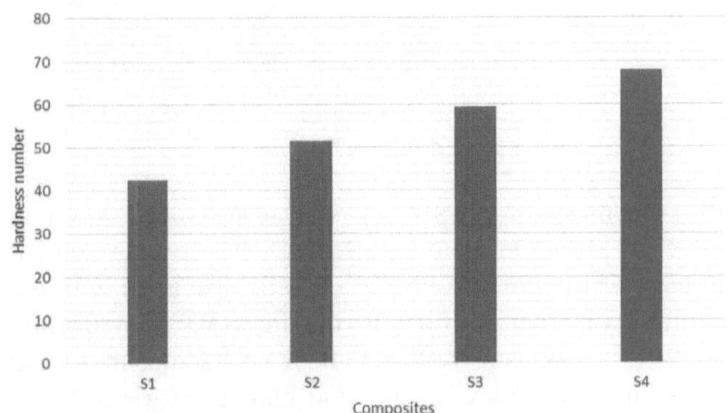

Fig. 4.2 Hardness number

3.2 Impact Strength

Impact strength of PLA-3D printed composites is presented in Fig. 4.3. From the findings, it is came to know that, PLA-3D printed core shown the lowest impact strength of 0.85 KJ/m2 which is 87.63% less than S4 composite. This is due to the absence of natural and synthetic fibers. Meanwhile in fiber filled composites, S2 specimen showed the lowest impact strength of 2.48 KJ/m2. This is 65.73% higher than S1 composite. S4 composite showed the highest impact strength of 6.87 KJ/m2 than remaining composites. This is 33.63% higher than S3 composite. So, from the findings, it is the evidence that, glass fiber led to increase the impact strength of the composites.

3.3 Tensile Strength

Figure 4.4 presents the tensile strength of the PLA-3D printed composites. From the results, it is measured that,

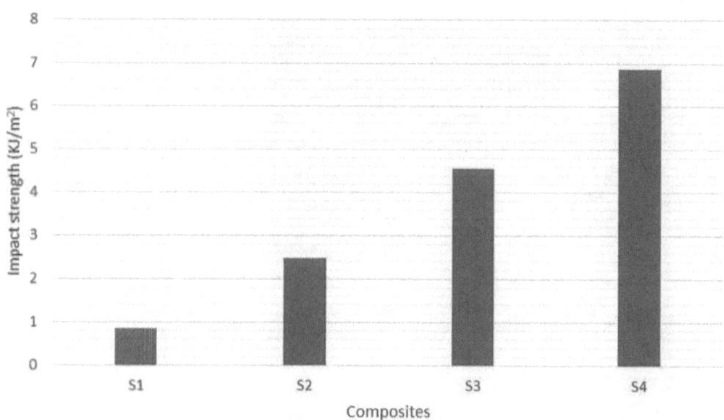

Fig. 4.3 Impact strength

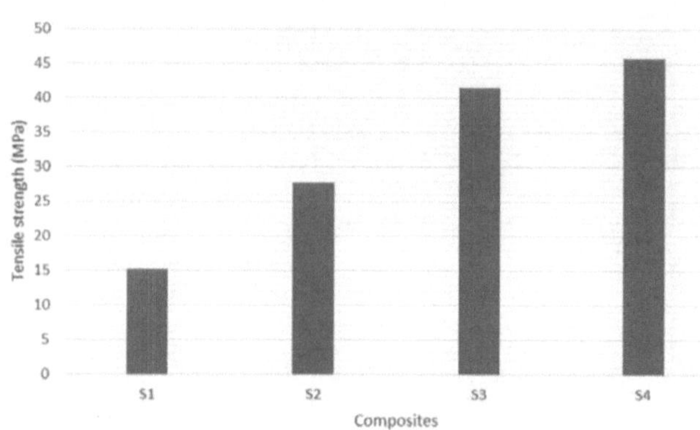

Fig. 4.4 Tensile strength

PLA-3D printed core (S1) shows the lowest tensile strength of 15.24 MPa. On the other hand, 45.2% of improvement in tensile strength was observed on S2 specimen with the addition of aloevera fiber to the PLA-3D printed core. Further, it was increased to 33.08% on S3 specimen with the addition of aloevera fiber and glass fiber to the PLA-3D printed core. Furthermore improvement of 9.44% in tensile strength was observed on S4 composite with the addition of glass fiber. Among fiber filled composites, S4 composite proved the maximum tensile strength of 45.89 MPa than other composites and this is 39.39% higher than S2 composite and 9.44% higher than S3 composite. Also it is observed that, there is no much difference between S3 and S4 composites.

3.4 Compressive Strength

Compressive strength of composites is a fundamental property that is necessary to ensure the safety, reliability, and performance of composite materials in various engineering and manufacturing applications. Compressive strength of PLA-3D printed composites is illustrated in Fig. 4.5. From the findings, it is seen that, PLA-3D printed core (S1) showed the lowest compressive strength of 42.57 MPa and S4 specimen which have glass fiber with PLA-3D printed core proved the superior compressive strength

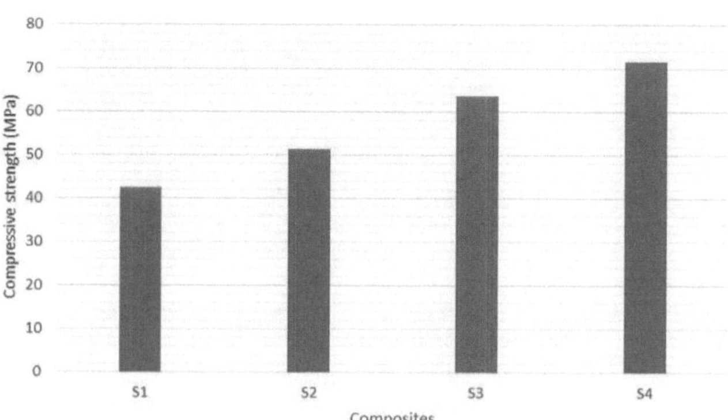

Fig. 4.5 Compressive strength

of 71.58 MPa. This is 40.53% higher than S1 specimen. In between, S2 and S3 composites showed the reasonable compressive strength of 51.25 MPa and 63.54 MPa respectively. Tensile, impact and hardness tested specimen is presented in Fig. 4.6.

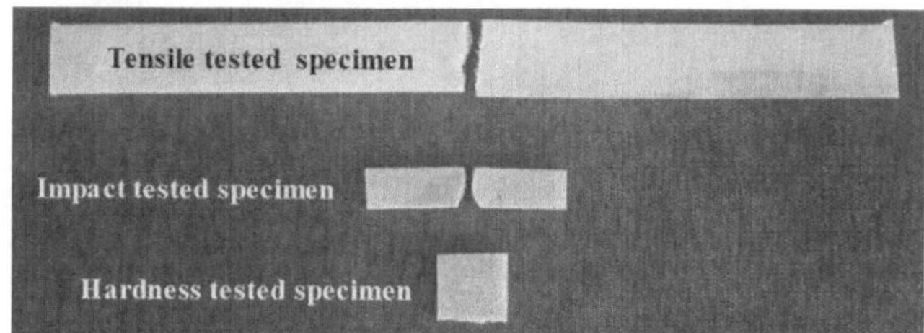

Fig. 4.6 Tensile, impact and hardness tested specimen (S2)

3.5 SEM Analysis

SEM images of tensile tested specimen is illustrated in Fig. 4.7. S2, S3 and S4 specimens were taken for SEM analysis. From the SEM analysis, it is observed that, more

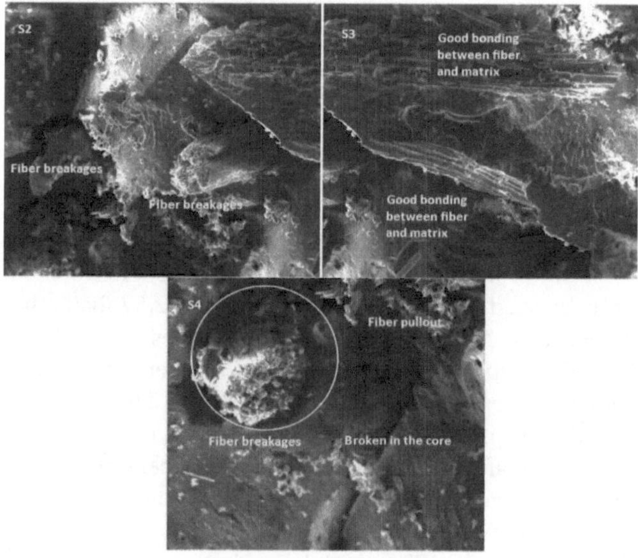

Fig. 4.7 SEM image

fiber breakages has been found in S2 specimen. Generally natural fibers are having less tensile strength than synthetic fiber. This is the fact for poor mechanical characteristics of S2 specimen. Whereas, in S3 specimen, notably good adhesion between fiber and matrix has been seen and this was the reason for good mechanical characteristics. This is due to hybridization of glass fibers to aloevera fiber. In addition to this, pullout in the glass fiber and broken in the PLA-3D printed core was noted in S4 specimen. Further, fiber pullout was not noted in S4 specimen due to good tensile strength of glass fiber. This is the reason for good mechanical characteristics of S4 specimen.

4. Conclusions

This research investigated the creation of hybrid composites by combining PLA-3D printed core with glass fiber and aloevera fiber polyester matrix. The following specific findings have emerged from this research.

PLA-3D printed core (S1) shows the lowest hardness, impact, tensile and compressive strength of 42.57, 0.85 KJ/m2, 15.24 MPa and 42.57 MPa respectively.S4 composite which has glass fiber to the PLA-3D printed core shows the highest hardness, impact, tensile and compressive strength of 67.81, 6.87 KJ/m2, 45.89 MPa and 71.58 MPa respectively. SEM analysis shows fiber breakages, fiber pullout, good adhesion between fiber and matrix and broken in PLA-3D printed core.Considering the strong mechanical properties exhibited by S3 and S4 composites, they are well-suited for a wide range of engineering applications.

References

1. Avci, Ali, AysegulAkdoganEker, Mehmet SafaBodur, and ZekiCandan. (2023). Water Absorption Characterization of Boron Compounds-Reinforced PLA/Flax Fiber Sustainable Composite. International Journal of Biological Macromolecules. 233:123546.
2. Awad, Sameer, RamengmawiiSiakeng, Eman M. Khalaf, Mohamed H. Mahmoud, Hassan Fouad, M. Jawaid, and MohiniSain. (2023). Evaluation of Characterisation Efficiency of Natural Fibre-Reinforced Polylactic Acid Biocomposites for 3D Printing Applications. Sustainable Materials and Technologies. 36: e00620.
3. Chen, Ke, Liguo Yu, Yonghui Cui, MingyinJia, and Kai Pan. (2021). Optimization of Printing Parameters of 3D-Printed Continuous Glass Fiber Reinforced Polylactic Acid Composites. Thin-Walled Structures. 164: 107717.
4. Estakhrianhaghighi, Ehsan, Armin Mirabolghasemi, Larry Lessard, and AbdolhamidAkbarzadeh. (2023). 3D Printed Wood-Fiber Reinforced Architected Cellular Composite Beams with Engineered Flexural Properties. Additive Manufacturing. 103800.
5. Kajbič, Jure, JernejKlemenc, and GorazdFajdiga. (2023). On the Fatigue Properties of Material Extrusion 3D-Printed Biodegradable Composites Reinforced with Continuous Flax Fibers. International Journal of Fatigue. 177: 107954.
6. Karthick, R., K. Adithya, C. Hariharaprasath, and V. Abhishek. (2018). Evaluation of Mechanical Behavior of Banana Fibre Reinforced Hybrid Epoxy Composites. Materials Today: Proceedings. 5 (5): 12814–20.
7. Karthick, R, V Anbumalar, and B Sutharson. (2018). Comparative Analysis of Mechanical and Water Absorption Behaviour of Jute/Glass Fibre Reinforced Epoxy and Polyester Hybrid Composites. Int. J. Materials Engineering Innovation. 9 (2): 82–93.
8. Karthick, Rasu, VeerabathiranAnbumalar, M. Vigneshkumar, M. Samuel Gemsprim, R. Venkatesh, P. Dheenathayalan, and M. Selwin. (2023). Influence of Glass Fiber Hybridization on Flexural and Water Absorption Behaviour of Banana Fiber Reinforced Epoxy Composites. Materials Today: Proceedings. https://doi.org/10.1016/j.matpr.2023.07.355.
9. Lau, H. Y., M. S. Hussin, S. Hamat, M. S. Abdul.Manan, M. Ibrahim, and H. Zakaria. (2023). Effect of Kenaf Fiber Loading on the Tensile Properties of 3D Printing PLA Filament. Materials Today: Proceedings. https://doi.org/10.1016/j.matpr.2023.03.015.
10. Long, Yu, Zhongsen Zhang, Kunkun Fu, Zhe Yang, and Yan Li. (2023). Design and Fabrication of High-Performance 3D Printed Continuous Flax Fibre/PLA Composites. Journal of Manufacturing Processes. 99: 351–61.
11. Maqsood, Nabeel, and Marius Rimašauskas. (2021). Delamination Observation Occurred during the Flexural Bending in Additively Manufactured PLA-Short Carbon Fiber Filament Reinforced with Continuous Carbon Fiber Composite. Results in Engineering. 11: 100246.
12. Olcun, Sinan, Yehia Ibrahim, Caleb Isaacs, Mohamed Karam, Ahmed Elkholy, and Roger Kempers. (2023). Thermal Conductivity of 3D-Printed Continuous Pitch Carbon Fiber Composites. Additive Manufacturing Letters 4: 100106.
13. Paz-González, Juan Antonio, Carlos Velasco-Santos, Luis Jesús Villarreal-Gómez, Enrique Alcudia-Zacarias, Amelia Olivas-Sarabia, Marcos Alan Cota-Leal, Lucía Z. Flores-López, and Yadira Gochi-Ponce. (2023). Structural Composite Based on 3D Printing Polylactic Acid/Carbon Fiber Laminates (PLA/CFRC) as an Alternative Material for Femoral Stem Prosthesis. Journal of the Mechanical Behavior of Biomedical Materials. 138: 105632.
14. Rajapandian, N., C. Senthamaraikannan, S. Rahul, R. Anand Vijay Raj, and T. V. Nithin Kumar. (2021). Investigation on Mechanical Performance of 3D Printed Carbon and Glass Fiber Reinforced Polylactic Acid Laminates. Materials Today: Proceedings. 46: 9429–32.
15. Rasu, Karthick, and AnbumalarVeerabathiran. (2023). Effect of Red Mud on Mechanical and Thermal Properties of Agave Sisalana/Glass Fiber–Reinforced Hybrid Composites. Materials Testing. 65(12): 1879-1889.
16. Reddy, M. Venkateswar, Banka Hemasunder, Pradeep MahadevapaChavan, Nilabh Dish, and Akash Paul Savio. (2023). Study on the Significance of Process Parameters in Improvising the Tensile Strength of FDM Printed

Carbon Fibre Reinforced PLA. Materials Today: Proceedings. https://doi.org/10.1016/j.matpr.2023.06.330.

17. Senthamaraikannan, C., B. ParthaaSarathy, Sai Surya, and S. TharunRajan. (2023). Study on the Free Vibration Analysis and Mechanical Behaviour of 3D Printed Composite I Shaped Beams Made up of Short Carbon Fiber Reinforced ABS and PLA Materials. Materials Today: Proceedings. https://doi.org/10.1016/j.matpr.2023.07.141.
18. Suresha, B., Varsha V. Giraddi, A. Anand, and H. M. Somashekar. (2022). Dynamic Mechanical Analysis of 3D Printed Carbon Fiber Reinforced Polylactic Acid Composites. Materials Today: Proceedings. 59: 794–99.
19. Ye, Wenguang, Hao Dou, Yunyong Cheng, and Dinghua Zhang. 2022. Self-Sensing Properties of 3D Printed Continuous Carbon Fiber-Reinforced PLA/TPU Honeycomb Structures during Cyclic Compression. Materials Letters. 317: 132077.
20. Zarna, Chiara, Gary Chinga-Carrasco, and Andreas T. Echtermeyer. (2023). Bending Properties and Numerical Modelling of Cellular Panels Manufactured from Wood Fibre/PLA Biocomposite by 3D Printing." Composites Part A: Applied Science and Manufacturing 165: 107368.
21. Zhang, Xiaohui, O. Malekahmadi, S. Mohammad Sajadi, Z. Li, Nidal H. Abu-Hamdeh, Muhyaddin J. H. Rawa, Meshari A. Al-Ebrahim, AliakbarKarimipour, and H. P. M. Viet. (2023). "Thermomechanical Properties of Coated PLA-3D-Printed Orthopedic Plate with PCL/Akermanite Nano-Fibers: Experimental Procedure and AI Optimization. Journal of Materials Research and Technology. 27: 1307–16.
22. Zhang, Xin, XitaoZheng, Luyang Song, YuanyuanTian, Di Zhang, and Leilei Yan. (2023). Compressive Properties and Failure Mechanisms of 3D-Printed Continuous Carbon Fiber-Reinforced Auxetic Structures. Composites Communications. 43: 101744.
23. Zhuang, Yuexi, Bin Zou, Shouling Ding, Xinfeng Wang, Jikai Liu, and Lei Li. (2023). Preparation of Pre-Impregnated Continuous Carbon Fiber Reinforced Nylon6 Filaments and the Mechanical Properties of 3D Printed Composites. Materials Today Communications. 35: 106163.
24. Zia, Ali Akmal, XiaoyongTian, Tengfei Liu, Jin Zhou, Muhammad AzeemGhouri, Jingxin Yun, Wudan Li, Manyu Zhang, Dichen Li, and Anderi V. Malakhov. (2023). Mechanical and Energy Absorption Behaviors of 3D Printed Continuous Carbon/Kevlar Hybrid Thread Reinforced PLA Composites. Composite Structures. 303: 116386.

Note: Every figure and table was created using the experimental work; none of them were taken from any publications or online.

Advances in Additive Manufacturing Technologies – Gurusamy Pathinettampadian (eds)
© 2024 Taylor & Francis Group, London, ISBN 978-1-032-90013-1

5 Exprimental Investigation of Polymer Matrix Composites Reinforced with Kevlar and Nano Silica Using Resin Transfer Moulding

Dinesh D[1]
Assistant Professor, Department of Mechantronics,
Chennai Institute of Technology, Kundrathur, Chennai, India

John Sugantha[2]
UG Student, Department of Mechantronics,
Chennai Institute of Technology, Kundrathur, Chennai, India

Mohamed Anas[3]
UG Student, Department of Mechantronics,
Chennai Institute of Technology, Kundrathur, Chennai, India

R. Deepak Suresh kumar
Assistant Professor, Department of Mechanical,
Rajalakshmi Institute of Technology, Chennai, India

Sarukesha B
UG Student, Department of Mechantronics,
Chennai Institute of Technology, Kundrathur, Chennai, India

ABSTRACT: Throughout human history, composite materials have been crucial for everything from housing early civilizations to facilitating upcoming discoveries. The main advantages of composites are their resistance to corrosion, flexibility in design, durability, light weight, and strength. Composites are now found in many items in our daily life. That are employed in a wide range of industries, including building, healthcare, sports, oil and gas, and aerospace. the benefits of fiber composite materials are discussed in this chapter along with their basic impacts, product development, and uses of fiber composites, encompassing the chemistry of the materials, design, production, and use the materials for a range of purposes. The result of, Composite materials have been more prevalent over time, growing into new markets. Certain applications, like rocket ships, probably wouldn't take flight without composite materials. The fact that by using an experimental approach, the impact of the composition of golden fiber (jute) on the tensile and bending properties of para-aramid (Kevlar) composites with varying weights will be determined. Layer by layer, the tensile and flexural properties of Jute-Kevlar matrix composites are established with the ability to increase the fiber's load, which will increase the composites' energy for the desired kind. For the structural components and the product format, hybrid materials are necessary in many processing processes. Given this proposal, the Jute They are layered, reinforced with epoxy at the end, and the properties of the hybrid composites are examined.

KEYWORDS: Mechanical properties, Kevlar, Jute, Hybrid composite, Natural composite, and Nanosilica

Corresponding authors: [1]dineshd@citchennai.net, [2]johnsuganth2406@gmail.com, [3]anas.saf2005@gmail.com

DOI: 10.1201/9781003545774-5

1. Introduction

Because of its exceptional qualities, such as chemical resistance, low weight, corrosion resistance, sturdiness, consolidated nature, flexibility in design, and high flexural modulus, composite materials are becoming an increasingly significant component of our daily life. One important factor that impacts the composite's structural integrity is the way its fibers are arranged. Maximum awareness of structural abilities will no longer be possible if the fibers are randomly or improperly arranged. For this reason, fibers are weaved into mats to enable proper placement within the composite. Famous weaving patterns that can be used using natural fibers right now include basket, twill, and plain weave. The tensile, flexural, and impact properties of jute and Kevlar fibers in those composites are advanced by the addition of epoxy resin, and the hybrid composites are designed to increase in the tensile, flexural, and impact properties. These composites had to find applications as a structural cloth where higher power and cost attention is important. Presented by Pal et al. A new method for curing silicon carbide-reinforced epoxy composites using microwave assistance. WithSiC loading, it is discovered that dielectric characteristics, especially the loss factor, as well as mechanical and thermal properties, increase. When comparing microwave-cured composites to those that were pre-cured at ambient temperature with the same SiC content, there is an improvement in both thermal and mechanical properties. A high dielectric constant polyurethane composite packed with nanographite is synthesized and characterized by Mishra et al. The polyurethane matrix contained uniformly dispersed nanographite, and the dielectric constant (ε) of the composite was examined in relation to frequency. When measured against the polyurethane dielectric constant, the composite showed a significant reduction in ε, with a logarithmic decrease from around 3000 at 100 Hz to approximately 225 at 60 kHz AC field. Additionally, the material showed a consistent dissipation factor (tan δ) across the applied frequencies, suggesting that it could tolerate the current leak. It was found that the composite's percolation threshold was 3 weight percent nanographite. Pincheira et al. looked into how different combinations of Kevlar fiber hybridization affected the mechanical characteristics of twill weave hybrid carbon/Kevlar fiber reinforced epoxy matrix composites. It was demonstrated that adding Kevlar fibers to the hybrid samples increased their ductile qualities, which in turn increased their toughness because of the Kevlar fibers. Vibration studies were carried out by Gupta et al. to ascertain the vinyl ester composites' dynamic properties, such as their storage modulus, loss modulus, or loss factor.The dynamic characteristics of hybrid composites reinforced with Al2O3 and CNT nanoparticles within an epoxy resin common matrix were described by Alva and Raja. Vibration tests were used to determine the dynamic modulus with damping qualities, and hybridization with CNT and Al2O3 nanoparticles improved the composite samples' overall damping and dynamic modulus properties. The vibration behavior of glass fabric/vinyl ester composites was studied by Chandradass et al. at various weight levels of 0, 1, 3, and 5% when nano-clay particles were included. The findings demonstrated that adding organically modified nano-clay particles to glass fabric/vinyl ester composites greatly raised the natural frequency because it increased the elastic modulus. At a weight content of 3%, the improvement in the damping ratio reached its maximum value. Huang and Tsai studied on vibration damping properties using Using the half-power bandwidth approach of composite laminates and adding rubber and nano-silica particles to the composite samples, it was demonstrated that these additions greatly improved the samples' damping qualities. The impact of CNT particles on the damping properties of carbon fiber reinforced composites was investigated by Khan et al. through vibration tests. The results indicated a significant increase in the damping property, as measured by the loss modulus of carbon fiber reinforced polymer composites, upon the incorporation of multi-walled CNT particles. A overview of the characteristics of composites based on polymers was published by Huang et al. The interface and interfacial characteristics of polymer-based composites were primarily the authors' areas of interest. Based on the loss modulus, the results demonstrated that the addition of multi-walled CNT particles greatly enhanced the damping property of carbon fiber reinforced polymer composites.

2. Supplies and Technique

2.1 Supplies

We bought Kevlar, Epoxy Resin, Bi-directional Jute, and Hardener HY951 from Sai Sakthi Enterprises in Saidapet, Tamil Nadu. The research's fabrication was manufactured in Chennai's Chromepet. For now Kevlar, Jute, and experimental study were utilized as a naturaland artificial material, HY951 hardener, epoxy resin, and the Hand Layup Method is an appealing fabrication and processing method even though there are other options since it is reasonably priced and provides a large range of materials and processing conditions. Layers of laminations are positioned in a mold using liquid epoxy resin until the required shape and thickness are reached in this easy, inexpensive open mold production method. To ensure that high-quality composite components/parts fulfill specific end-user requirements, dry fibers in the form of woven, knitted, stitched, or bond fabrics are first manually inserted in the mold. A brush is then used to apply the resin matrix on the reinforcing material requirements.

2.2 Composite Preparation

The process of hand layup was used to create the composite. With the aid of a mechanical stirrer, the resin and hardener were completely mixed in a 10:1 ratio. Additional nanosilica was added to the mixture at concentrations of 0.5, 1 and 2 weight percent. To create the composites, a steel mold measuring 210 mm by 210 mm by 40 mm was utilized. The fiber mat's dimensions remained unchanged at 200 mm by 200 mm. The mold was covered with a 25 kg weight, Then the composite was allowed to cure for a full day at room temperature. A further twenty-four hours were allowed to pass at room temperature for post-curing. Finally, the composite slab was cut to the necessary proportions for carrying out the experiment.

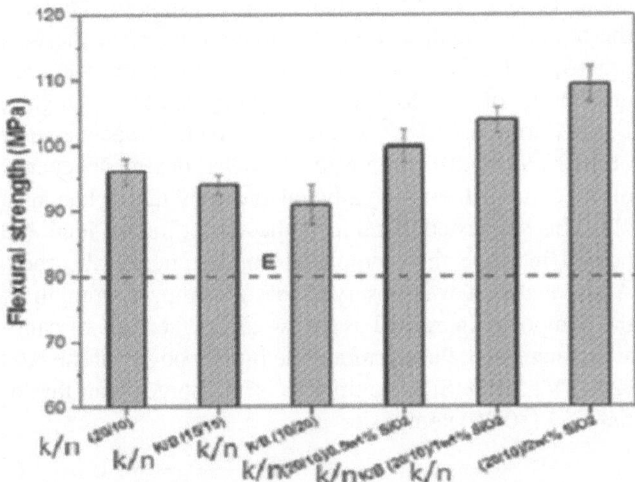

Fig. 5.1 Flexural strength of composites made of nanosilica and kevlar

2.3 Experimental

The sample's tensile strength was determined using the universal testing machine (UTM-INSTRON 3369). In UTM, the standard cross head speed was kept at 1 mm/min. The UTM was utilized to ascertain the sample's flexural strength. The three-point bending principle and the standard ASTM D790 were used to complete the test. The impact strength of the sample was ascertained using an Izod impact tester. The test was conducted in accordance with ASTM D256. Using a durometer (YUZUKI), the surface microhardness of the sample was determined. The hardness test was performed in accordance with ASTM 2240. A thermal conductivity analyzer (C-therm) was used to measure the sample's thermal conductivity. At 30°C,

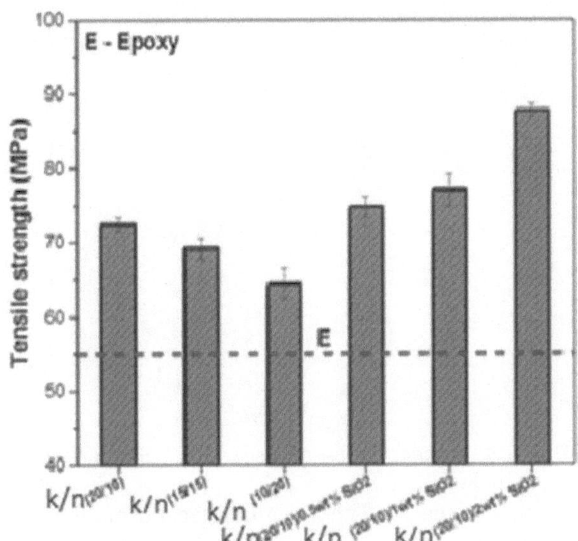

Fig. 5.2 Strengthening of kevlar/nanosilica composites under tension

the sample's ability to absorb water was examined. A digital balance was used to determine the dried specimen's weight. The weight of the wet sample was determined after it had been submerged in distilled water for the necessary amount of time. Equation (1) was utilized to ascertain the water absorption of the sample.

3. Findings and Conversation

3.1 Tensile Power

It was discovered about the composites' overall tensile power was higher than epoxy. The inclusion of reinforcements like nanosilica and kevlar was blamed for this. The reinforcements aided with better stress transmission and boosted the tensile strength. Additionally, the result demonstrated the layer effect had noteworthy impact on the tensile power of composite. Tensile strength of the K/N (10/15) composite was higher than that of the K/N (10/10) and K/N (15/10) composites. This might be the result of increased kevlar fiber content. Tensile strength of the K/N (10/15) composite increased as a result of the increased packing of kevlar fiber, since it is harder than and fiber (**Table 5.1**). Thus, the K/N (15/10) composite was found to be the strongest composite in this investigation. Additionally, additional research was conducted over the tensile power of K/N (15/10) composite by including NS at varying weigh fractions, including 1.5.2 and 2.5. The outcomes showed that the addition of nanosilica helped to further improve strength which reduced resin mobility and improved the composites' ability to bear loads. Furthermore, tensile strength increased as the amount of

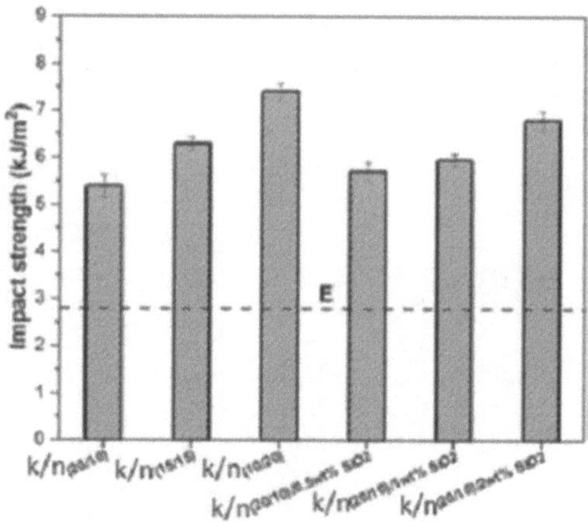

Fig. 5.3 Impact power of kevlar and NS composites

nanosilica increased. The maximum tensile strength of the K/N (10/10)/2wt% SiO_2 in this study was 78 MPa, which is 21.9% higher than the tensile strength of the K/N (10/15) composite. The report states that the tensile power of the composites was enhanced by addition of nanosilica. This outcome is in good agreement with finding of the present investigation.

3.2 Flexural Strength

The investigation of the flexural power of nanosilica and kevlar composite is displayed in Fig. 5.4. Results demonstrated that, in comparison to epoxy's flexural strength, the composites had higher flexural strengths. The fact that composites contain reinforcements may be the cause of this. The composite's improved stiffness as a result of the fiber and nanosilica inclusion prevented cracks from forming earlier. As a result, composites' flexural strengths increased. The flexural strength of K/N (10/10) and K/N (15/20) were found to be lower than those of the K/N (10/10) composite..This was explained by the fact that the K/N (15/10) composite had more high-stiffness kevlar fiber. The kevlar fiber increased the flexural strength of the K/N (10/15) composite to a significant degree during bending loading by transferring a greater amount of load than any other fiber. Additionally, the material reinforced with nanosilica exhibited greater flexural strengths than composites made of epoxy, K/N (10/15), K/N (15/20), and K/N (15/20).Composite transfer higher stress along the fibers after adding nanosilica. The highest flexural strength (109 MPa) of the K/N (20/15)/2wt% SiO2 in this investigation is 14.5% more than the flexural power of the K/N (10/10) composite. Therefore, it shows that the flexural strength of the composites was much increased by the inclusion of nanosilica. A related investigation on sisal/Kevlar/nanosilica composites was carried out by Chowdary et al. [31], who found that adding 3wt% NS increases the flexural strength of the kenaf/Kevl composite by 7%.

3.3 Impact Power

The composites showed higher impact power more than epoxy because of the fiber and NS reinforcement. This could be due to the optimal bonding strength between the fibers and matrix. Furthermore, compared to the K/N (15/15) and K/N (10/10) composites, the impact energy absorption of the K/N (10/20) composites was higher. Furthermore, 6.4 kJ/m^2 was found to be the impact energy of the K/N (10/20) composite. However, the impact energy of the composite was greatly enhanced by the inclusion of NS. This suggested that during the abrupt impact load, NS helped increase the amount of impact energy absorbed. Additionally, it was observed that the impact strength of the composite increased as the weight percentage of nanosilica increased. Furthermore, the impact power of the K/N (20/15)/2.5wt% SiO_2 composite is36% larger than that of the K/N (10/20) composite.

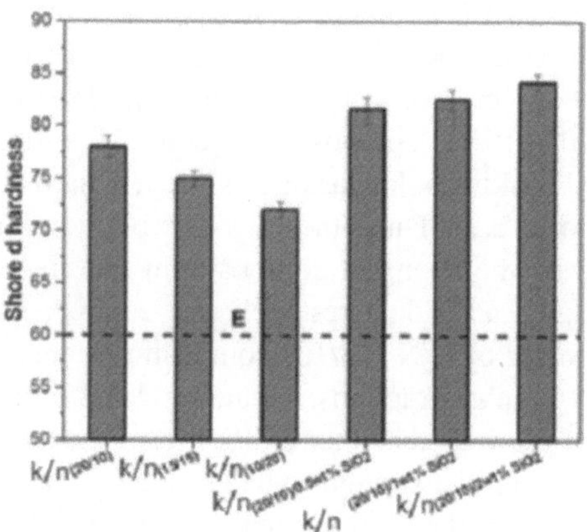

Fig. 5.4 Kevlar and nanosilica composites' microhardness

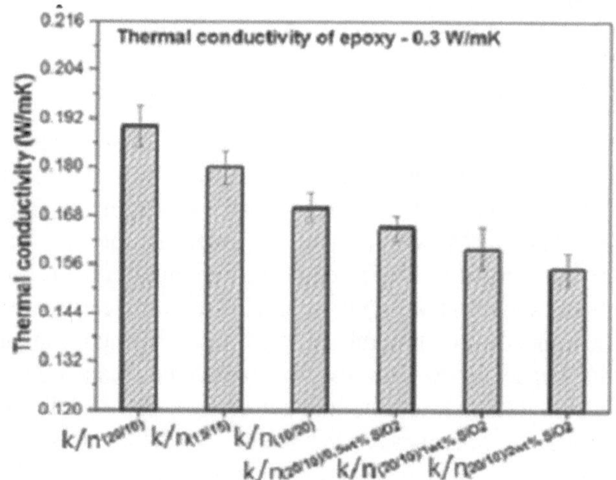

Fig. 5.5 Thermal conductivity of kevlar and nanosilica composites

3.4 Hardness

The results showed that, in comparison to pure epoxy, the composites K/N (20/10), K/N (15/15), and K/N (10/20) showed higher hardness values. This suggested that throughout testing, there was less penetrating force applied to the composite. The strengthening effect of fibers and kevlar could be the cause of this. Additionally, the K/N (10/10) composite has ahardness value of 86 HV, which is 7.3% and 3.1% higher than the corresponding values for the K/N (20/10) and K/N (10/15) composites. Because more stiffer kevlar fiber was packed into the K/N (10/20) composite than in the K/N (15/10) and K/N (20/10) composites, the latter two showed lower levels of hardness. Furthermore, it was observed that the composites containing nanosilica demonstrated higher hardness values in comparison to epoxy, K/N (10/15), K/N (15/10), and K/N (10/15) composites. This demonstrated that there was sufficient resistance from the NS particles to cause a plunger. Furthermore, the pore volumes decreased as a result of the tiny nanosilica particles occupying the area between the epoxy and fiber. Better hardness was achieved by the material as a result. The K/N (15/20)/2wt% SiO2 composites in this study displayed a highest hardness value of 83 H-V.

3.5 Warmth Conductance

The investigation of the thermal conductivity of nanosilica and kevlar composites is presented in Fig. 5.7. The results demonstrated that, in comparison to pure epoxy, the complete Composite materials showed poor heat conductivity. The combined impact of fibers and epoxy may be the cause of this. Additionally, the findings demonstrated that the K/N (10/20) composite had higher thermal conductivity than the K/N (15/15) and K/N (10/20) composites. This might be the result of increased kevlar fiber content. Therefore, this investigation confirms that, in comparison to kevlar fiber, the and fiber helped toreduce the heat conductivity of the Composite. In examining the thermal characteristics of epoxy/kevlar composites, Additionally, Baladivak et al. [26] investigated the epoxy and material's thermal properties. It was discovered that the thermal conductivity of composite and epoxy was 0.112 W/mK. Because of this, epoxy/and had a lower heat conductivity than epoxy/kevlar composite. This outcome is good and consistent with the current study's findings. Furthermore, to empty K/N (10/10), the nanosilica-added K/N composite showed reduced thermal conductivity. Moreover, the thermal conductivity of composite showed a further decline with a increase in nanosilica content. The lowest heat conductivity of nanosilica (0.08 W/mK) may be the cause of this. The K/N (20/10)/2wt% SiO2 in this study had the minimum heat conductivity, measuring 0.155 W/mK.

3.6 Water Absorption

The results showed that, in comparison of epoxy resin, composites absorbed more water. The composites absorbed more water as a result of the hydrophilic nature of the fiber. Additionally, the epoxy's hydrophobic characteristic means that it naturally resisted the absorption of water. However, the diffusion mechanism resulted in a limited amount of water absorption being seen. As the soaking duration increased, the water intake showed an increasing tendency. The porosity nature of natural fibers allowed the composites to absorb more water through capillary action over extended soaking times.K/N (10/10) composite in this study absorbed less water than the K/N (15/15) and K/N (10/20) composites. This may be because there is a greater concentration of kevlar fiber that has a low capacity to absorb water. Furthermore, as opposed to comparable unfilled composites, the K/N (20/10) composites with nanosilica added showed reduced water absorption. This might be because the inclusion of NS helped to reduce the volume, which in turn caused the absorption of water to diminish. The K/N (10/20)/1wt% SiO2 composite in this study had the lowest water uptake (2.3%, over a 6-hour period).

3.7 Density and Porosity

Due to vacancies in the composites, a discrepancy between theoretical and experimental densities was observed. Furthermore, due to the availability of the least amount of density and fiber, the K/N (10/20) composite demonstrated

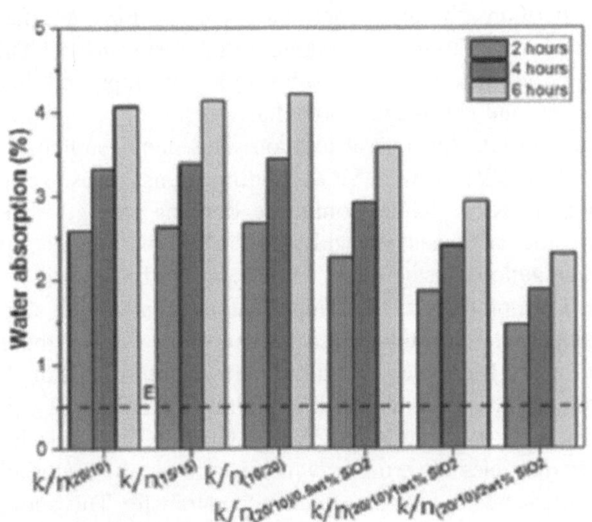

Fig. 5.8 Water absorption of kevlar and nanosilica composites

lower density when compared to the K/N (15/15) and K/N (20/10) composites. In addition, the density values of the composites with nanosilica added were lower than those of K/N (20/20), K/N (10/15), and K/N (10/20) composites. Therefore, the use of nanosilica helped to reduce the composites' density. This can be the result of some epoxy resin being replaced. Furthermore, the composites' void data showed that less than 2.5% of the composites overall had void. Furthermore, it was observed that the void percentages rose as the weight percentages of nanosilica increased. The inadequate soaking of the filler and fibers was the cause of this [33].

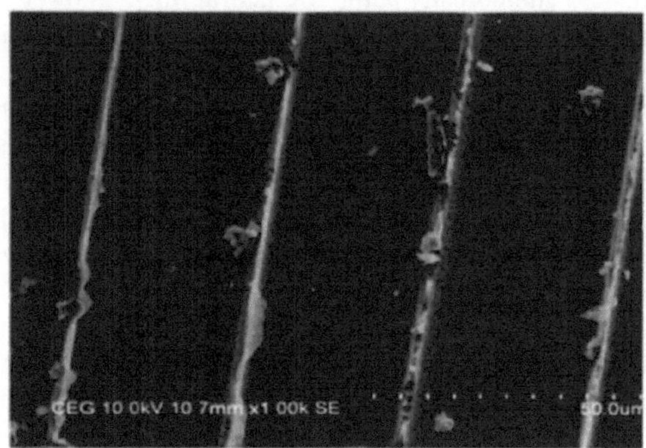

Fig. 5.9 SEM image of jute composite (20 μm)

3.8 SEM Analysis

The fiber separated from the resin's surface. This demonstrated that during the load transfer, the fiber in the composite degraded and the adhesive strength decreased. Furthermore fiber imbedded in the surface of resin was visible in the K/N (20/10)/2wt% SiO2 composite image Furthermore, the area where the resin and fiber interface was observed to be diminishing on the resin surface. This suggested that the fiber responsible for improved load transfer and shrinkages formed as a result of composites' high capacity for load absorption. This demonstrated that the K/N (20/10)/2wt% SiO2 bonding strength was greater than the K/N (20/10) composite bonding strength. This outcome is consistent with the earlier findings in this investigation. Furthermore, **Figs. 5.10c and 5.10d** display the TEM pictures of the samples supplemented with silica nanoparticles. It was believed that nanosilica was distributed uniformly. Moreover, the nanosilica was both coated on the fiber surface and embedded in the resin. Additionally, the fiber surface of the K/N (20/10)/2wt% SiO2 sample had more particles covering it, which improved the material's strength and allowed for effective load transfer. This shown that K/N (20/10)/2wt% SiO_2 had a stronger binding than K/N (20/10) composite. This outcome is consistent with the earlier research findings in the same field. Furthermore. The distribution of nanosilica was assumed to be uniform. Moreover, the resin contained the nanosilica, which coated the fiber's surface which improved the material's strength and allowed for effective load transfer. Additionally, the flexural strength of the composite was observed to be much greater than the flexural strengths of the LDPE/Rice Husk, Epoxy/Hemp/Sisal, Epoxy/Sisal, and Polyurethane/ Roselle composites. When compared to the epoxy/kevlar/ and composite, the LDPE/rice husk composite, however, had a higher impact strength.

4. Conclusion

An eco-friendly and deployed content of Kevlar alone on such produced items to make them more efficientand material-economize is the tendency in this research work's full comparison of the jute-kevlar layered composite material in the position of Kevlar made components.

The highlights of findings are

- Because there was more high strength kevlar fiber in the K/N (20/10) material than in the other composites, including K/N (15/20) and K/N (15/15), the material showed higher tensile strength values.
- Tensile strength was improved with the addition of nanosilica, and the K/N (15/20)/2wt% SiO2 composite shown the highest tensile strength (89 MPa).
- Because more highly stiffened kevlar fiber was packed into the K/N (10/20)/2wt% SiO_2 composite, a high flexural power (109 MPa) was observed.
- The K/N (20/15) composite demonstrated a high impact power of 7.4 kJ/m^2, according to an impact study. This was explained by the fiber and matrix having the ideal adhesion strength.

References

1. Natrayan L, Kumar PV, Baskara Sethupathy S, Sekar S, Patil PP, Velmurugan G, Thanappan S (2022) Effect of nano TiO2 filler addition on mechanical properties of bamboo/ polyester hybrid composites and parameters optimized using grey Taguchi method. AdsorpSciTechnol2022
2. Le Qi, Shi-lin Zhang, Hao Yuan, Zhong-liang Ma, Zhong-liang Xiao" A novel modification method for the dynamic mechanical test using thermomechanical analyzer for composite multi-layered energetic materials" Defence Technology, https://doi.org/10.1016/j.dt.2021.10.010, 2021.
3. 3Le Qi, Shi-lin Zhang, Hao Yuan, Zhong-liang Ma, Zhong-liang Xiao" A novel modification method for the dynamic mechanical test using thermo mechanical analyzer for compositemulti-layered energetic materials" Defence Technology, https://doi.org/10.1016/j.dt.2021.10.010,2021

4. H. Parmar, F. Tucci, P. Carlone et al., "Metallisation of polymers and polymer matrix composites by cold spray: state of the art and research perspectives," International Materials Reviews, vol. 67, no. 4, pp. 385–409, 2022.
5. G. Rajkumar, G. K. Sathishkumar, K. Srinivasan et al., "Structural and Mechanical Properties of Lignite Fly Ash and Flax-added Polypropylene Polymer Matrix Composite," Journal of Natural Fibers, vol. 19, 2021.
6. S. B. Aziz, M. H. Hamsan, M. F. Kadir, and H. J. Woo, "Design of polymer blends based on chitosan: POZ with improved dielectric constant for application in polymer electrolytes and flexible electronics," Advances in Polymer Technology, vol. 2020, Article ID 8586136, 10 pages, 2020.
7. S. Dhanasekar, R. S. Suriavel Rao, and H. Victor Du John, "A fast and compact multiplier for digital signal processors in sensor driven smart vehicles," International Journal of Mechanical Engineering and Technology, vol. 9, no. 10, pp. 157–167, 2018.
8. D.K. Rajak, D.D. Pagar, P.L. Menezes, E. Linul Fibre-reinforced polymer composites: manufacturing, properties, and applications Polymers, 11 (10) (2019), p. 1667, 10.3390/polym11101667
9. R. Hanumantharaya, I. Sogalad, S. Basavarajappa, G.C.M. Patel Modelling and optimisation of adhesive bonded joint strength of composites for aerospace applicationsInt J Comput Mater Sci Surf Eng, 8 (3–4) (2019), pp. 167–184, 10.1504/IJCMSSE.2019.104694
10. A. Verma, P. Negi, V.K. Singh Experimental analysis on carbon residuum transformed epoxy resin: chicken feather fibre hybrid composite Polym Compos, 40 (7) (2019), pp. 2690–2699, 10.1002/pc.25067
11. M. Shunmugasundaram, P. Anand, M.A.A. Baig, Y. Kasu Experimental investigation on tensile property of vacuum infused kenaf-based polymer composite with the presence of nanofillers Advances in lightweight materials and structures, Springer, Singapore (2020), pp. 265–272, 10.1007/978-981-15-7827-4_26
12. M.R.M. Asyraf, M.R. Ishak, S.M. Sapuan, N. Yidris Utilization of bracing arms as additional reinforcement in pultruded glass fibre-reinforced polymer composite cross-arms: creep experimental and numerical analyses Polymers, 13 (4) (2021), p. 620, 10.3390/polym13040620
13. T.B. Yallew, S. Aregawi, P. Kumar, I. Singh Response of natural fibre reinforced polymer composites when subjected to various environments Int J PlastTechnol, 22 (1) (2018), pp. 56–72, 10.1007/s12588-018-9202-2
14. M.M. Kandar, H.M. AkilApplication of design of experiment (DoE) for parameters optimization in compression moulding for flax reinforced biocomposites Procedia Chem, 19 (2016), pp. 433–440, 10.1016/j.proche.2016.03.035
15. W. Brostow, H.E.H. Lobland, N. Hnatchuk, J.M. Perez Improvement of scratch and wear resistance of polymers by fillers including nanofillersNanomaterials, 7 (3) (2017), p. 66, 10.3390/nano7030066

Note: All figures were exclusively produced from our experimental work and were not sourced from any publications or online material.

Advances in Additive Manufacturing Technologies – Gurusamy Pathinettampadian (eds)
© 2024 Taylor & Francis Group, London, ISBN 978-1-032-90013-1

Impact of DMC and Compression Ratio on a CRDI Engine Using Methyl Esters of Waste Cooking Oil

R. Anbalagan*, S. Sendilvelan, K. Rajan
Department of Mechanical Engineering,
Dr. M.G.R. Educational and Research Institute, Chennai, Tamil Nadu, India

K. Bhaskar
Department of Automobile Engineering,
Rajalakshmi Engineering College, Chennai, Tamil Nadu, India

ABSTRACT: The results of an experimental study on the effects of waste cooking oil fuel blends, which include B20 methyl ester, on the efficiency, emissions, and combustion of CRDI engines are presented in this article. Analysis of the effects of various compression ratios (CR) used in CRDI engines, biodiesel supplemented with antioxidant DMC (dimethyl carbonate), and metrics associated with CRDI engine performance tested at 1500 rpm, with the main injection set at 15° BTDC, the pilot mass at 10% at 500 bar and the pilot injection set at 36° BTDC. The work's novelty is in utilizing a CRDI engine to examine the impacts of adding 3% DMC with and without mixing B20, as well as the effects of three different compression ratios (CR15.5, CR16.5, and CR17.5) on emissions, combustion parameters and engine performance. The peak cylinder pressure significantly increased when the CR was adjusted from 15.5 to 17.5, according to the data. Furthermore, when CR is raised, using B20 fuel produces higher NO_x and lower Carbon monoxide, HC and smoke emissions. While the NO_x emission for CR17.5 in the B20 diesel with DMC somewhat decreased, the NO_x emission for CR16.5 declined when compared to the fuel without DMC. The best feature is using B20 with antioxidants (DMC) in a CRDI engine with reduced emissions.

KEYWORDS: Pilot injection, Main injection, DMC, Biodiesel (B20), Compression ratio, CRDI engine, Antioxidant, Pilot mass

1. INTRODUCTION

Diesel engines are typically found in trucks, light-, medium-, and heavy-duty commercial vehicles, as well as power plants, because of their increased fuel economy and reduced fuel consumption. The main pollutants found in diesel engines include smoke, particulate matter, CO, HC, and NO_x. Diesel engines have substantial emissions of nitrogen oxide (NO_x) and particulate matter, making it challenging to meet strict emissions regulations. The need to find an alternative diesel fuel has become much more urgent in recent years due to environmental degradation, pollution emissions, and the depletion of fossilized fuels.

The aim of this article is to offer a foundation for estimating the impacts of varying the compression ratio and injection

*Corresponding author: anburajitha@gmail.com

DOI: 10.1201/9781003545774-6

time for the three previously described options. Examined and discussed were the impacts on NO_x emissions, maximum cylinder pressure, and brake thermal efficiency. This chapter leads to a successful implementation by reviewing the literature on the project that is being presented. All of the published research articles are shown in the section that follows, along with a list of issues and difficulties encountered.

Researched and experimented with a transesterification and mixing method to obtain biodiesel from used cooking oils, adding pyrogallol to maintain the stability of the oxidation process. A single-cylinder DI engine was subjected to emissions tests and performance evaluations at half load and full load using three different fuel mixtures, biodiesel with diesel (B20), biodiesel with diesel (B20A), and fossil diesel (Syed AatifAvase et al. 2015). Concentrated on the development of certain biodiesel qualities under severe oxidation circumstances from several sources (safflower, kitchen waste, rapeseed oil, and high oleic safflower oil) and assessed how well the TBHQ addition maintains quality (Nogales–Delgado et al. 2022). Discussed uncovered that the brake warm effectiveness diminished for ternary fuel mixes due to their second rate calorific esteem. The NO, smoke, and HC outflows were diminished successfully for ternary fuel mixes and the most extreme diminishment were found to be 15.68%, 22.1% and 21.81%, separately, for 30% decanol mix compared to diesel (Vinodkumar Vajravel and Karthikeyan Alagu. 2022).The engine performance of two different biodiesel blends—soybean biodiesel and animal fats combined with ultra-low sulfur diesel fuel—was investigated in two independent test programs with similar operating conditions (Curtis Alan Wan. 2011). Examine the optimum n-decanol fraction in the inlet port and the corresponding engine load for the better emissions and performance characteristics of a partially premixed charged compression ignition (PCCI) engine by response surface methodology (RSM) (Vinodkumar et al. 2023). Tests were carried out to compare the fuel properties, performance of the engine, and exhaust from different blends of biodiesel (Chlorella vulgarism and Atrophy circus) with diesel fuel (Amit Kumar Sharma et al. 2020). It was found that the performance of both biodiesel fuels—most notably torques and brake power—decreased with greater mix ratios as a result of biodiesel's reduced energy content (McCarthy.2011). Their antioxidant potential is emphasized, and the extraction techniques for phenolic antioxidants from these sources are explored(Henry Kahimbi et al. 2023).

The results of the analysis indicate that adding oxidizing agents to the methyl ester of Sapota oil should enhance its combustion properties when compared to diesel. (Ravikumar Jayabalet al. 2019).

Analyze how two oxidized additives affect the ignition characteristics of the diesel engine: pentanol (n-P) and dimethyl carbonate (DMC). The performance data indicated that the biodiesel/diesel blends' performance characteristics had significantly improved (Devaraj et al. 2020). Engine performance, combustion, and emission characteristics are discussed and examined in relation to ether and ester fuels used as single, mixed, dual and multifuel fuels (Yang Hua. 2021). Investigate the conflicting outcomes of EGR and ignition improvers (DMC) at the CRDI small unmarried cylinder diesel engine emissions, combustion, and performance. (Ramesh et al.2022).

This study investigates how a special blended dialkyl oxalate (mDAO) additive affects a high-stress common-rail diesel engine's combustion and exhaust properties. mDAO is used in place of conventional diesel (Ao Zhou et al. 2022). According to an analysis, ether components may be used to enhance cylinder combustion and lower pollutant emissions from CI engines (Quoc Bao et al. 2022). In a common rail compression ignition diesel engine operating at part load, an experimental inquiry was conducted using a B20 blend (Palm Oil Methyl Ester 20% + neat diesel 80%) with variable pilot injection timing and pressure (Pavan et al. 2021). The effects of oxidized additives, specifically 1-pentadecanol (1-DEC) and DTBP (ditetrabutylphenol), research has been done on the pollution, combustion, and overall performance features of diesel/Karanja blends (Yuvarajan et al. 2021).

Regarding the current study, the methyl ester of waste cooking oil (B20) evaluated for the study primarily examines the impact of an antioxidant (DMC 3%), utilizing a single cylinder CRDI engine operating at 1500 rpm, with varying compression ratios (CR15.5, CR16.5, and CR17.5). Pilot injection was done at 36° BTDC, main injection at 15° BTDC, and constant pressure was maintained at 500 bar.

This study examined the combustion and emission outcomes of the methyl ester of waste cooking oil blend (B20) utilized in common rail diesel engines, specifically focusing on SFC, NO_x, carbon monoxide, HC, and smoke emissions. The primary goal is to investigate the effects of adding DMC 3% with and without mixing B20, as well as the impact of three different compression ratios (CR15.5, CR16.5, and CR17.5) with B20 on emissions, combustion parameters, and engine performance.

2. Materials and Methods

2.1 Biodiesel Production

The method of transesterification involves reacting alcohol and using vegetable oil and a catalyst to convert fatty acid esters into biodiesel with glycerin as a byproduct.

Alcohols like methanol and ethanol have often been used because of their inexpensive cost and favorable physicochemical properties. Alkaline catalysis is chosen over acidic catalysis because of its better efficiency. Using this procedure also raises the cetane number and volatility to levels comparable to diesel fuel. Emulsification happens when the phase boundary is disrupted by the rapid collapse of these holes. According to reports, the yield of biodiesel produced using UC-assisted transesterification was 7.5 times more than that of MS. Table 6.1 shows the characteristics of natural diesel, methyl ester of waste cooking oil (B20), DMC and (B20+DMC).

Table 6.1 Properties of test fuels

ASTM Norms	Properties	Diesel	Waste Cooking Oil (Biodiesel) (B20)	DMC	B20 + DMC
ASTM D1298	Density [kg/m^3] at 16°C	860	873	1069.4	843.3
ASTM D445	Kinematic viscosity at 40 °C [cst]	3.2	4.15	0.63	3.45
ASTM D93	Flashpoint [°C]	75	184	12-20	114.6
ASTM D2500	Cloud point [°C]	18	15.5	–	–
ASTM D613	Cetane number	49	51	35	–
ASTM D240	Calorific Value [MJ/kg]	43.0	38.012	15.78	43.5

2.2 Experimental Setup

Figure 6.1 illustrates the experimental setup of a computer-controlled, air-cooled, single-cylinder CRDI engine producing 3.5 kW, connected to a constant-load eddy current dynamometer.

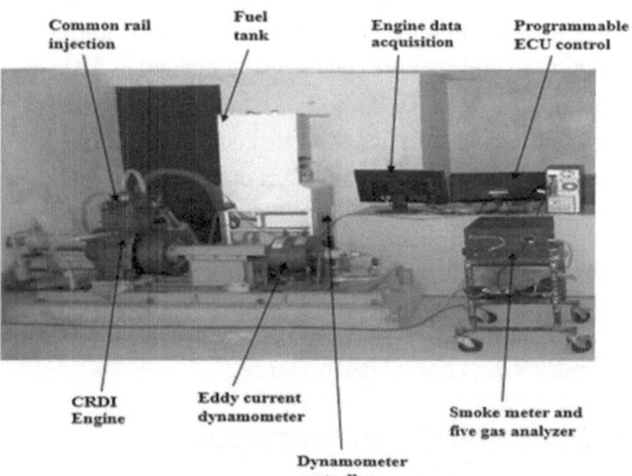

Fig. 6.1 Photo view of experimental engine setup

The engine runs on B20 fuel, a blend of 80% pure diesel and 20% palm oil biodiesel. Emissions are measured using the AVL Digas 444, five-gas analyser and AVL 365C Indi angle encoder for crankshaft rotation. In order to demonstrate the accuracy of the test results, an uncertainty analysis must be carried out. The uncertainty percentages for engine speed, load, brake power, CO, HC, NO, smoke opacity, and pressure pick-up are given in Table 6.2.

Table 6.2 Details of uncertainties in experiments

S. No	Measured quality	Measuring range	Accuracy	Measured uncertainties values
1	Speed	400-6000 rpm	±1%	±1.1%
2	Load	24 kg	–	±0.51%
3	CO	0-15% vol	±0.02% abs	±0.21% abs
4	HC	0-30000ppm vol	±8 ppm	±0.21% ppm
5	NO	0-5000 ppm vol	±5 ppm	±0.31% ppm
6	Smoke	1-100%	±1% of full scale	±1.12%

3. Results and Discussion

The study examined the impact of a 3% DMC antioxidant in a CRDI engine, operating at 1500 rpm with B20 biodiesel at three compression ratios (CR15.5, CR16.5, and CR17.5). Engine parameters included 10% pilot mass, 15° BTDC main injection, and 36° BTDC pilot injection at 500 bar constant pressure. SFC, NO_x, CO, HC, and smoke emissions were evaluated for B20 biodiesel derived from waste cooking oil. Results showed improved SFC with DMC at 36° BTDC pilot injection, lower cylinder pressure, and HRR with DMC in B20 blend and pilot injection mode. Advanced injection timings at CR17.5 increased without antioxidants. Compared to DMC, NO_x emissions slightly increased, while smoke, HC, and CO decreased at various compression ratios.

3.1 Cylinder Pressure

The impact of B20 biodiesel blend with and without DMC antioxidant on in-cylinder pressure at different compression ratios and maximum load conditions is depicted in Fig. 6.2. At 500 bar injection pressure, primary injection occurs at 36° BTDC and pilot injection at 15° BTDC. With DMC (3%) added, higher cylinder pressures were observed across various compression ratios (CR15.5, CR16.5, and CR17.5), with peaks at 71.74, 66.41, and 63.90 bar respectively, compared to CR17.5 without antioxidant yielding lower pressure at 69.18 bar. Comparing CR16.5 with DMC to CR17.5 without antioxidant, a 3.56% increase in pressure was noted, attributed to DMC's oxygenated additive increasing B20's calorific value. However, DMC blending with biodiesel decreased cylinder pressure values for CR16.5.

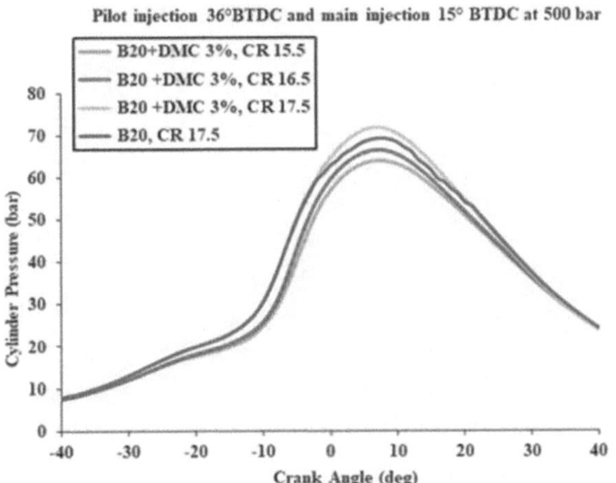

Fig. 6.2 Cylinder pressure vs crank angle

3.2 Heat Release Rate

Figure 6.3 shows the heat release rate at variation of compression ratios (CR) and complete load scenario for mixes of biodiesel (B20) with and without antioxidants. DMC 3% is the antioxidant applied to compression ratios (CR17.5, CR16.5, and CR15.5) in this test. According to the graph, the different CR with antioxidant addition result in reduced heat release rates of 48.24 J/° CA, 36.92 J/° CA, and 34.69 J/° CA and without antioxidant for compression ratio of 17.5, the heat release rate is 41.14 J/ ° CA respectively. B20 without addition of antioxidant (DMC) for CR 17.5, the heat release rate is decrease of 7.1% at the same the CR 16.5 with DMC, the HRR is decrease respectively.

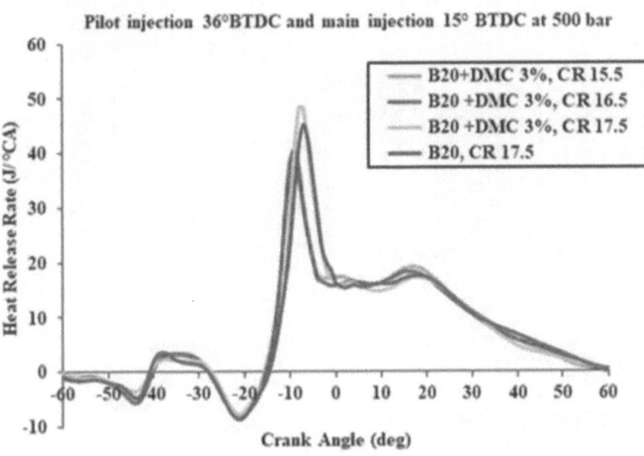

Fig. 6.3 Heat release rate vs crank angle

3.3 Specific Fuel Consumption

Figure 6.4 illustrates the variation in Specific Fuel Consumption (SFC) for B20 biodiesel blends with and without antioxidants at injection pressures of 500 bar and different compression ratios (CR15.5, CR16.5, and CR17.5) under full engine load conditions. At a pilot injection timing of 36° BTDC and main injection timing of 15° BTDC, the CRDI engine operated at 1500 rpm. Results indicate that SFC decreases with the addition of antioxidants to B20 biodiesel. SFC values of 0.33, 0.29, and 0.26 kg/kWh were recorded for CR15.5, CR16.5, and CR17.5 respectively, while without antioxidant at CR17.5, SFC was 0.27 kg/kWh. Compared to B20 without DMC antioxidant, introducing DMC led to increased SFC at different compression ratios, though the increase was less pronounced for lower compression ratios. DMC's addition enhances calorific value, reduces SFC, and improves brake thermal efficiency in biodiesel blends.

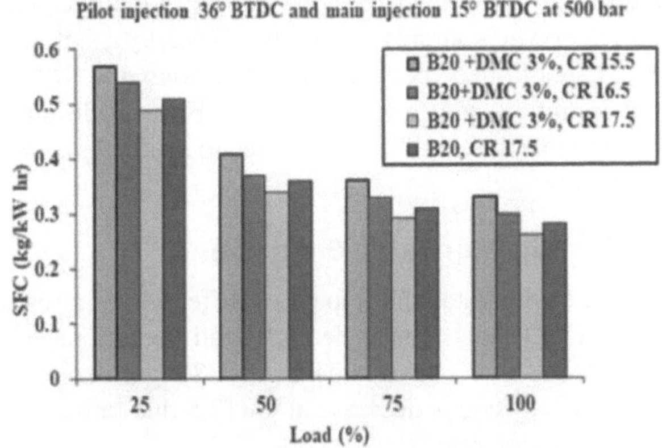

Fig. 6.4 Specific fuel consumption (SFC) vs load

3.4 Oxides of Nitrogen Emission

Figure 6.5 shows that increasing compression ratios (CR15.5 to CR17.5) at maximum load conditions resulted in higher NO_x emissions for all tested biodiesel (B20) fuels, with or without the antioxidant DMC (3%). B20 with DMC 3% showed reduced NO_x emissions compared to B20 without antioxidants, with minimum values recorded at CR15.5, CR16.5, and CR17.5. However, NO_x emissions increased with higher compression ratios, attributed to el-

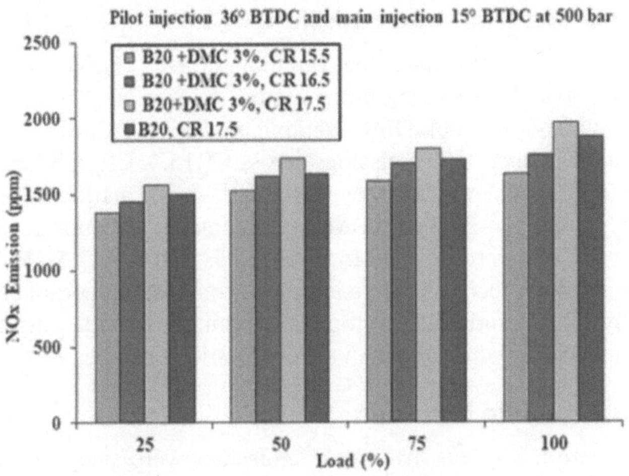

Fig. 6.5 Oxides of nitrogen (NO_x) vs load

evated temperature and pressure during in-cylinder combustion. Lower compression ratios, aided by DMC antioxidants, exhibited reduced NO_x emissions due to lower cetane numbers and oxygen concentrations. Additionally, the use of the diesel additive DMC led to a notable increase in NO_x emissions.

3.5 Carbon Monoxide Emission

Figure 6.6 illustrates the impact of different compression ratios on CO emissions under full load conditions when using DMC, a 3% antioxidant, in B20 gasoline. With DMC, CO emissions decrease at CR17.5 due to increased CO2 conversion rate.

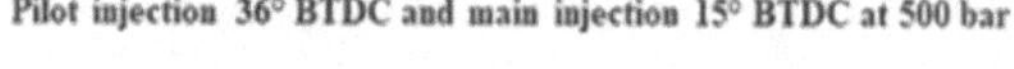

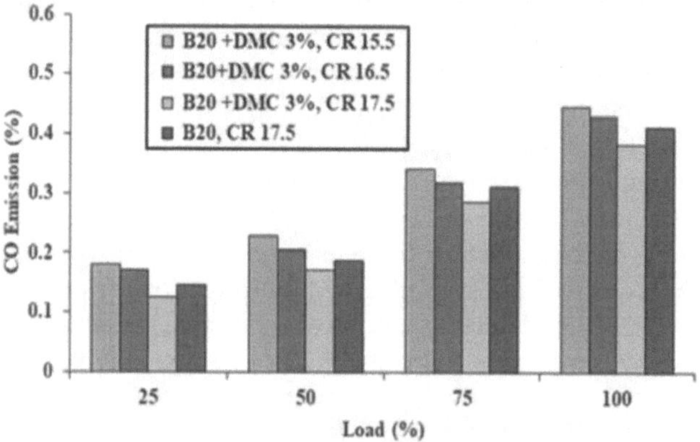

Fig. 6.6 CO emission vs load

Higher engine speeds lead to elevated air-to-diesel ratios and internal combustion temperatures, accelerating CO to CO_2 conversion. Including DMC and increasing compression ratios in B20 blends also reduces CO emissions. At 1500 rpm, minimum CO emissions were observed for CR15.5, CR16.5, and CR17.5. Compared to CR17.5 without DMC, CO emissions decreased when DMC was added.

3.6 Hydrocarbon Emission

Figure 6.7 illustrates the variation in HC emissions at full load for different compression ratios of B20 blends with and without DMC antioxidant (3%). Addition of DMC reduces HC emissions across CR15.5, CR16.5, and CR17.5, whereas without DMC, HC emissions increase. Compared to B20 with DMC, reductions in HC emissions were observed for CR15.5, CR16.5, and CR17.5. HC emissions decrease due to improved combustion facilitated by DMC' antioxidant properties, promoting oxidation of unburned hydrocarbons in the combustion chamber.

3.7 Smoke Emission

Figure 6.8 shows that smoke emissions were analyzed at varying compression ratios (CR15.5, CR16.5, and CR17.5)

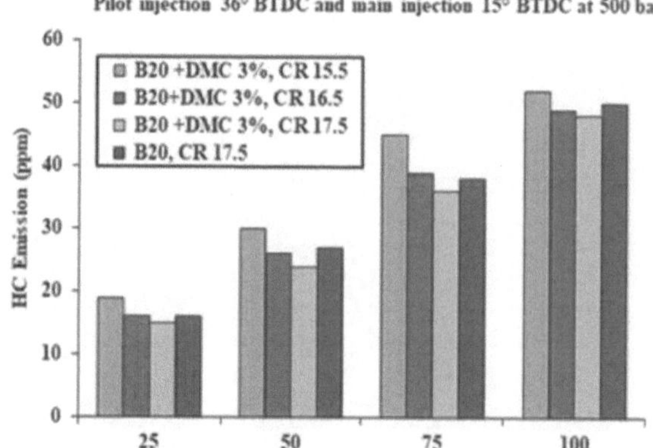

Fig. 6.7 Hydrocarbon emission vs load

for biodiesel (B20) with and without DMC antioxidant (3%) at full engine load. It was observed that smoke emissions decreased with higher compression ratios, especially evident with B20 containing DMC. The addition of DMC led to a reduction in CO emissions at CR17.5. This decrease in smoke emissions is attributed to the increase in CO to CO_2 conversion and the oxygenating effect of DMC. Conversely, lower compression ratios without DMC resulted in increased smoke emissions due to incomplete combustion and reduced oxygenation.

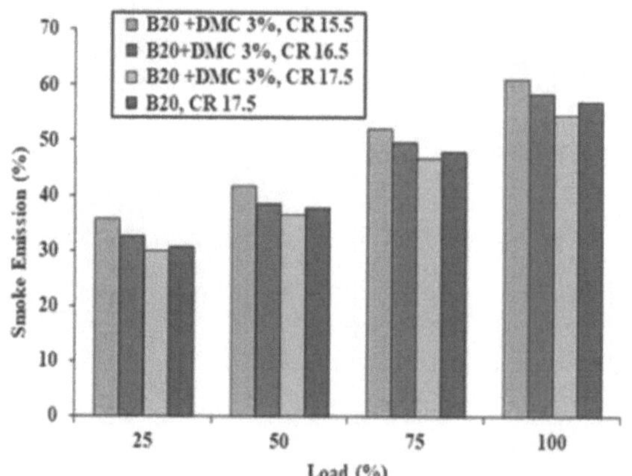

Fig. 6.8 Smoke emission vs load

4. Conclusion

The study assessed NO_x, Smoke, CO, and HC emissions in CRDI diesel engines using B20 (methyl ester of waste cooking oil blend) with and without a 3% antioxidant (DMC) at different compression ratios (CR15.5, CR16.5, and CR17.5) under full load and consistent 500 bar injection pressure. Advancing pilot injection to 36°BTDC, main injection to 15°BTDC, and employing antioxidants showcased emissions reduction potential without compromising engine efficiency. The following could be used to summarize the main findings of this investigation:

1. Adjusting compression ratios, utilizing advanced injection timings, and introducing antioxidants at a consistent 500 bar injection pressure resulted in significant impacts on cylinder pressure, emissions (smoke, HC, CO, NO_x), and performance metrics (HRR, SFC).
2. Further, employing an antioxidant (DMC) in B20 blends notably reduced heat release rate and increased cylinder pressure at CR 17.5, enhancing brake thermal efficiency while maintaining fuel consumption. Decreasing compression ratios with DMC showed varying effects on NO_x emissions, with CR 15.5 achieving the lowest NO_x values.
3. The addition of DMC to B20 blends, particularly at CR 15.5, which notably reduced CO emissions but increased smoke. Overall, these adjustments demonstrated trade-offs in emissions based on compression ratio and antioxidant use in B20 blends.
4. In this work, the optimized compression ratio is 17.5 with 3% DMC blended biodiesel, which is better to run the CRDI engine because it reduces smoke, CO and HC emissions and improves engine performance.

REFERENCES

1. Syed Aatif Avase, Shivank Srivastava, Kumar Vishal, Harsh V Ashok and George Varghese. (2015). Effect of pyrogallol as an antioxidant on the performance and emission characteristics of biodiesel derived from waste cooking oil. Procedia Earth and Planetary Science. Volume 11, 437-444. https://doi.org/10.1016/j.proeps.2015.06.043.
2. Nogales–Delgado, S., Guiberteau, A., and Encinar, J.M. (2022). Effect of tert-butylhydroquinone on biodiesel properties during extreme oxidation conditions. Fuel.Vol. 310, 122339.https://doi.org/10.1016/j.fuel.2021.122339.
3. Vinodkumar Vajravel and Karthikeyan Alagu. 2022. Influence of ternary fuel blends of decanol/neem oil biodiesel/diesel on combustion, emission and performance characteristics of an unmodified diesel engine. International Journal of Ambient Energy. Vol.43. https://doi.org/10.1080/01430750.2022.2075928.
4. Curtis Alan Wan. 2011. Characteristics of engine emissions from different Biodiesel blends. University of Toronto (Canada) ProQuest Dissertations Publishing.
5. Vinodkumar, V., Karthikeyan, A., Jayaprabakar, J., Senthilkumar, G., and Ranganathan, L. (2023). Optimization of port injection of n-decanol in a PCCI engine using response surface methodology. https://doi.org/10.1002/htj.22930.
6. Amit Kumar Sharma, Pankaj Kumar Sharma, VenkateshwaraChintala, Narayan Khatri and Alok Patel.(2020). Environment-friendly biodiesel/diesel blends for improving the exhaust emission and engine performance to reduce the pollutants emitted from transportation fleets. Int. J. Environ. Res. Public Health.17(11), 3896; https://doi.org/10.3390/ijerph17113896.
7. McCarthy, P., Rasul, M.G., Moazzem, S. (2011). Comparison of the performance and emissions of different biodiesel blends against petroleum diesel, International Journal of Low-Carbon Technologies. Vol. 6, 255–260. https://doi.org/10.1093/ijlct/ctr012.
8. Henry Kahimbi, Baraka Kichonge and Thomas Kivevele. (2023). The Potential of Underutilized Plant Resources and Agricultural Wastes for Enhancing Biodiesel Stability: The Role of PhenolicRich Natural Antioxidants. International Journal of Energy Research. Vol.2023, Article ID 9389270. https://doi.org/10.1155/2023/938927.
9. Ravikumar Jayabal, Lakshmanan Thangavelu and Chandran Velu, V. (2019). Experimental investigation on the effect of ignition enhancers in the blends of sapota biodiesel/diesel blends in a CRDi engine. Enery & Fuels. 33(12), 12431–12440.
10. Devaraj Rangabashiam, Dinesh Babu Munuswamy, Sivakumar Duraiswamy Balasubramanian and Durai Christopher. (2020). Performance, emission, and combustion analysis on diesel engine fueled with blends of neem biodiesel/diesel/additives.https://doi.org/10.1080/15567036.2020.1764152.
11. Yang Hua. 2021. Ethers and esters as alternative fuels for internal combustion engine: A review. International Journal of Engine Research, https://doi.org/10.1177/14680874211046480.
12. Ramesh, T., Sathiyagnanam, A. P.,Melvin VICTOR Depoures and Murugan Ponnusamy. (2022). A Comprehensive Study on the Effect of Dimethyl Carbonate Oxygenate and EGR on Emission Reduction, Combustion Analysis, and Performance Enhancement of a CRDI Diesel Engine Using a Blend of Diesel and Prosopisjuliflora Biodiesel. International Journal of Chemical Engineering. doi:10.1155/2022/5717362.
13. Ao Zhou, Wei Guo, Hui Jin, Yangyang Li and Yingwu Yin. (2022). Influence of a Novel Mixed Dialkyl Oxalate on the Combustion and Emission Characteristics of a Diesel Engine.ACS Omega. 7(38): 34563–34572. 10.1021/acsomega.2c04477.
14. Quoc Bao Doan, Xuan Phuong Nguyen, Van Viet Pham, Thi Minh Hao Dong, Minh Tuan Pham and Tan Sang Le. (2022). Performance and Emission Characteristics of Diesel Engine Using Ether Additives: A Review. Centre of Biomass and Renewable Energy.11(1),255-274, https://doi.org/10.14710/ijred.2022.42522.
15. Pavan, P., Bhaskar, K., Sekar, S. (2021). Effect of split injection and injection pressure on CRDI engine fuelled with POME-diesel blend.Vol. 292, 120242, https://doi.org/10.1016/j.fuel.2021.120242.
16. Yuvarajan Devarajan, Dineshbabu Munuswamy, Beemkumar Nagappan and Ganesan. S. (2021). Detailed study on the effect of different ignition enhancers in the binary blends of diesel/biodiesel as a possible substitute for unaltered compression ignition engine, PetroleumScience.17:1151–1158. https://doi.org/10.1007/s12182-020-00463-9.

Note: All figures and tables were exclusively produced from our experimental work and were not sourced from any publications or online material.

Advances in Additive Manufacturing Technologies – Gurusamy Pathinettampadian (eds)
© 2024 Taylor & Francis Group, London, *ISBN 978-1-032-90013-1*

7 Combined Effect of Lipophilic and Ether Additives on the Performance and Exhaust Emissions of a CRDI Engine Powered by Elaeisguineensis Biodiesel and its Blend

Pavan. P[1], K. Bhaskar[2]
Department of Automobile Engineering,
Rajalakshmi Engineering College, Chennai, Tamil Nadu, India
S. Sekar[3]
Department of Mechanical Engineering,
Rajalakshmi Engineering College, Chennai, Tamil Nadu, India

ABSTRACT: This study investigates the impact of combining antioxidants (butylated hydroxyl anisole - BHA and butylated hydroxyl toluene - BHT) with the oxygenate additive diethyl ether (DEE) on reducing exhaust emissions from CI engines fueled by palm oil-derived biodiesel (B100) and its blend (B20). The antioxidants BHA and BHT were tested at a concentration of 1000 ppm, showing significant reductions in NOx emissions: BHA reduced NOx by 25.7% for B100 and 14.3% for B20, while BHT led to decreases of 21.12% for B100 and 11.52% for B20. Initially, the use of antioxidants resulted in higher smoke emissions, but the subsequent addition of 5% DEE effectively lowered these emissions by 11.1% in B100 and 5.3% in B20. The study also observed trends in carbon monoxide (CO) emission, which varied similarly with and without the additives in the test fuels. This research presents a novel approach to managing exhaust emissions in diesel engines with strategic use of antioxidants combined with an oxygenate additive, demonstrating its broad applicability and effectiveness across different concentrations of biodiesel..

KEYWORDS: Biodiesel, NO_x emission, Antioxidants, Ether, Smoke emissions

1. INTRODUCTION

The transportation industry makes extensive use of compression ignition (CI) engines because of their low cost, excellent stability, ideal fuel efficiency, and flexibility in a range of operating circumstances. The growing demand for non-renewable fossil fuels, driven by an increase in vehicular population and stationary power generation, poses a risk of depleting petroleum reserves in the foreseeable future [1]. This urgency has spurred a search for renewable, eco-friendly alternative fuels. Among the most promising are biofuels, which offer significant economic and energy security benefits, especially for developing countries [2]. Biodiesel, in particular, has garnered global attention as a viable alternative for CI engines, either as a blend or pure fuel [3-5]. It inherently contains molecular oxygen, which helps significantly reduce engine out emissions like particulate matter, HC, and CO, thus improving the air quality index. However, biodiesel typically results in higher exhaust oxides of nitrogen (NO_x) levels compared to fossil-derived diesel [6-10], posing environmental and health risks due to their role in forming ground-level ozone.

[1]pavanrane002@gmail, [2]bhaskar.k@rajalakshmi.edu.in, [3]sekar.s@rajalakshmi.edu.in

DOI: 10.1201/9781003545774-7

Research attributes the surge in NO_x emissions with biodiesel to several factors, including advanced fuel pre-injection due to a higher bulk modulus, which increases the mass of fuel delivered, and enhanced burning of the pre-mixed portion due to the existence of oxygen in the fuel. Differences in chemical kinetic reactions and higher heat release rates are also cited as causes for increased NO_x formation [11-12]. Exhaust gas recirculation (EGR) has been studied as a method to manage this, effectively reducing NO_x emissions but leading to higher particulate emissions at high loads. This necessitates a particulate trap to optimize the relation between NO_x and smoke emissions [13-16].

The market for biodiesel is expected to grow significantly, from US$ 141 billion in 2020 to US$ 307 billion by 2030 [11]. This growth highlights the need for advancements in diesel engine technology and emissions reduction strategies. Recent studies have focused on integrating biodiesel with modern diesel technologies such as CRDI and high EGR rates, which help balance emission reduction with performance efficiency [12-18]. Water injection or emulsion methods have been explored to lower NO_x emissions, but they can increase CO and soot levels, and reduce brake thermal efficiency [19-20]. Recent studies on the usage of biodiesel in advanced CRDI engines have shown that biodiesel mixtures can be used effectively up to certain blend limits without operational issues. While there are increases in fuel consumption and minor reductions in fuel conversion efficiency with higher biodiesel content, emissions, except for NO_x, are significantly reduced through advanced injection strategies. [21-24].Research on additives like anisole, methyl acetate, and aromatic amine antioxidants has shown they enhance biodiesel fuel properties, improving engine performance and lowering emissions. These additives are particularly effective in reducing NOx emissions in CI engines using biodiesel enhanced with natural extracts. [25-37]. Research by Senthil, Pranesh, and Silambarasan [28] demonstrated that leaf extract additives in biodiesel significantly reduce NO_x emissions, supporting eco-friendly emission control. Prabu et al. [29] found that additives in used cooking oil biodiesel improve engine performance and exhaust emissions, confirming its potential as a sustainable fuel. These studies highlight how additives can enhance biodiesel performance in diesel engines.

Antioxidants, known as NO_x suppressors, also improve biodiesel's oxidation stability and reduce cloud and pour points. Varatharajan et al. [30] observed a notable drop in NO_x with p-phenylenediamine and ethylenediamine in jatropha biodiesel, although HC and CO emissions slightly increased. Erol Illeri et al. [31] found that 2-ethylhexyl nitrate in canola oil biodiesel blends reduced NO_x but raised CO emissions across a range of engine operations. Vamsi Oxygenated additives like DEE, DME, and DMC have proven effective in reducing tailpipe emissions such as UBHC, smoke, and CO [32-37]. Most research on antioxidants in biodiesel has focused on older mechanical injection systems, with fewer studies on modern common rail diesel engines (CRDI) that allow for precise fuel injection control. This study examines the impact of cost-effective and readily available antioxidants (BHA and BHT) combined with diethyl ether (DEE) on NOx and other emissions in CRDI engines powered by biodiesel blends. It highlights the importance of integrating antioxidants and oxygenates in modern CI engines to promote biodiesel as a sustainable substitute to mineral diesel.

2. Fuel Preparation and Characterisation

Virgin palm oil was locally sourced and transformed into biodiesel and glycerol through transesterification, involving the filtration of crude oil, reaction with high-purity methanol or ethanol and a catalyst at a 6:1 ratio, heated to 50-65°C. This process broke down triglyceride ester bonds, allowing biodiesel and glycerol to separate due to density differences. The biodiesel was then purified and tested for viscosity, density, flash point, and cetane number to meet quality standards and ensure engine performance and exhaust emission compliance.

Table 7.1 Properties of fuel

Property	No.2. Diesel	B100	B20
Density at 25° C (kg/m^3)	854	863	857
Kinematic Viscosity at 40° C (cst)	4.3	4.70	4.53
Flash point (°C)	76	181	81.4
Cloud point (°C)	18	15.1	17.5
Cetane number	48	57.5	50.2
Lower Heating value (MJ/kg)	45.3	39.7	41.31

3. Experiment Setup and Procedure

The test setup used for this experimental work is a Kirloskar single cylinder, water-cooled, CRDI diesel engine with a programmable ECU, delivering 3.5 kW of power. Specifications are detailed in Table 7.2. An eddy current dynamometer with an integrated load regulator was attached to the engine, and a PC with specialized software monitored injection timing and pressure, engine speed, load, and exhaust gas temperature, with results displayed on-screen. The test setup's schematic sketch is shown in Fig. 7.1.

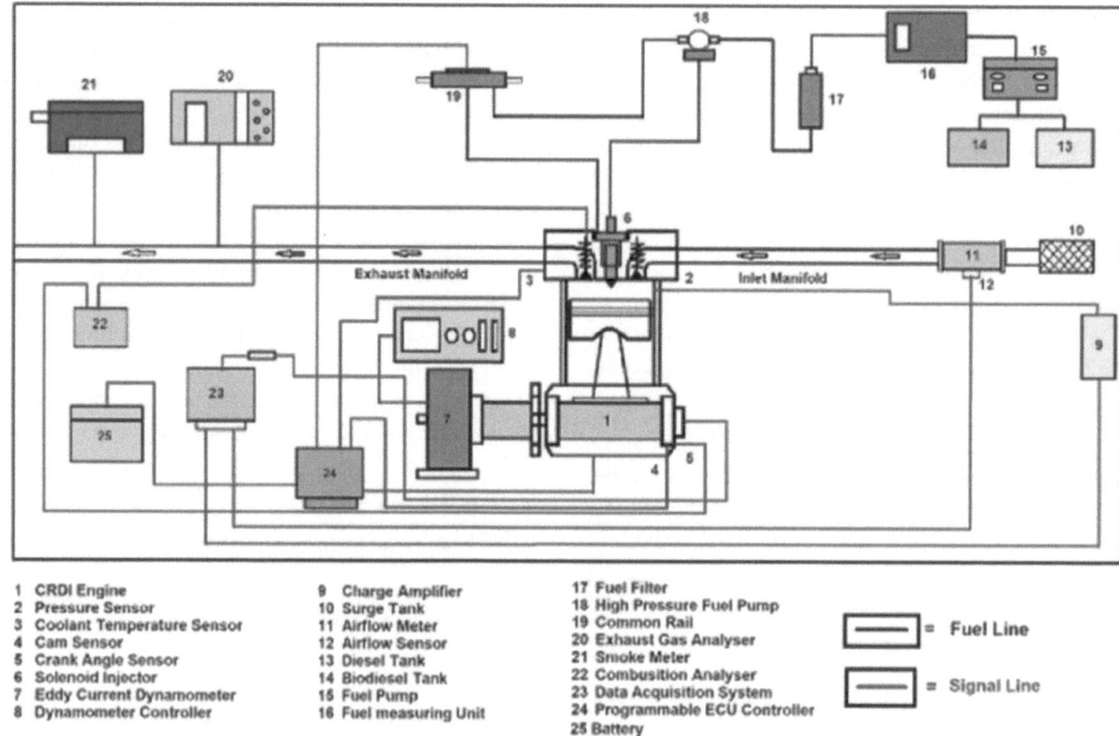

Fig. 7.1 Graphical layout of the testing configuration

Table 7.2 Details of test engine

Parameter	Specification
Model	Kirloskar TV -1
Type	1-cylinder,4-stroke,water cooled, constant speed,CRDI engine
Engine Control Unit (Programmable)	Nira i7r with solenoid injector
Displacement	660 cc
Bore and stroke	87.5 mm x 110 mm
Compression ratio	17.5: 1
Speed (constant)	1500 rev/min
Max. power	3.5 kW
Injection pressure	500 bar

Exhaust emissions were measured using an AVL Digas 444 exhaust gas analyzer, recording values at 30-second intervals for CO, NO, and HC with a resolution of 1 ppm. Smoke emissions were assessed using an AVL 437C smoke measuring system. Data were captured and analyzed by a data acquisition system and engine controller, ensuring precise performance and emission parameter estimation.

3.1 Experimental Procedure

Palm oil methyl ester (B100), a biodiesel blend (B20), and mineral diesel were tested as fuels. Antioxidants and diethyl ether (DEE) from Sigma-Aldrich, India, were added in concentrations ranging from 100 ppm to 1000 ppm to B100 and B20. Starting with untreated diesel, experiments included varying antioxidant levels and 2.5% to 5% DEE. The additives, chosen for their cost-effectiveness and impact on fuel quality, were mixed using a magnetic stirrer to ensure even distribution. Tests were conducted to measure fuel consumption, combustion and fuel pressures, and emissions like CO, UBHC, NOx, and soot, with detailed discussions of results in following sections.

Table 7.3 Specifications and chemical makeup of antioxidants

Antioxidant	Chemical structure	Specifications	
Butylated Hydroxy anisole		CAS Number Min. Assay Mol. Weight Melting point Sulfated ash Solubility 18.5°C	88-32-4 99.5 % 180.25g/mol 58 °C ≤ 0.005 % < 1 mg/mL in water
Butylated Hydroxy toulene		CAS Number Min. Assay Mol. Weight Free Phenol Sulfated ash	128-37-0 99 % 220.34 g/mol 0.02% 0.02%

Table 7.4 Specifications of oxygenated additive

Ether	Properties	
Diethyl ether	CAS Number	60-29-7
	Molecular formula	$C_2H_5 - O - C_2H_5$
	Minimum Assay	99.7%
	Density	0.715 kg/l
	Boiling point	34.6° C

4. Results and Discussion

4.1 NO_x Emissions

Figure 7.2 and 7.3 detail the impact of antioxidants and diethyl ether on NOx emissions compared to neat diesel, B100, and B20 fuel blends. BHA significantly reduced NOx, achieving the best results at 500 ppm for B20 and 1000 ppm for B100, with reductions of 14.3% and 25.7%, respectively. BHT was effective at 750 ppm for both fuel types, though less so than BHA. The introduction of diethyl ether slightly decreased the effectiveness of NOx reduction in antioxidant-treated fuels. NOx levels for B20 ranged from 20.41 g/kW.hr to 9.19 g/kW.hr with BHA, and for B100 from 21.6 g/kW.hr to 10.4 g/kW.hr, showing substantial NOx mitigation across biodiesel concentrations.

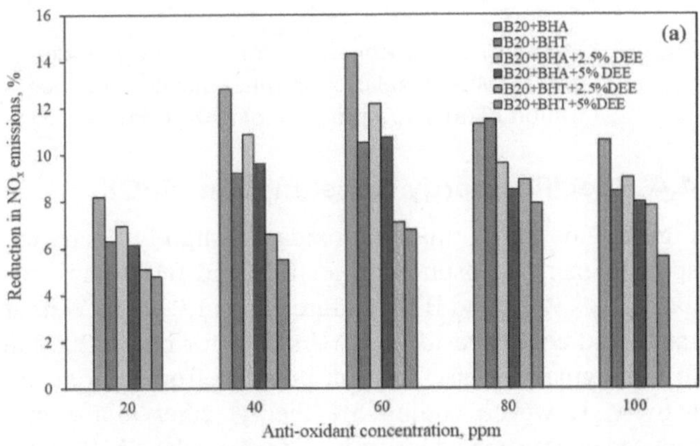

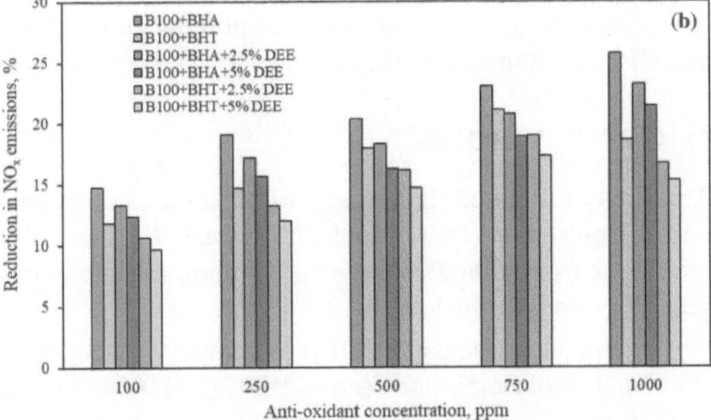

Fig. 7.2 Effect of antioxidants and diethyl ether concentrations on NO_x emission for (a) for B20 blend (b) for B100 fuel at full load compared to mineral diesel

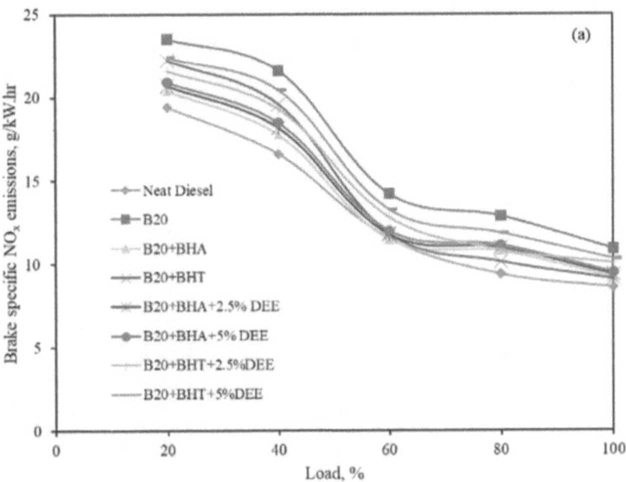

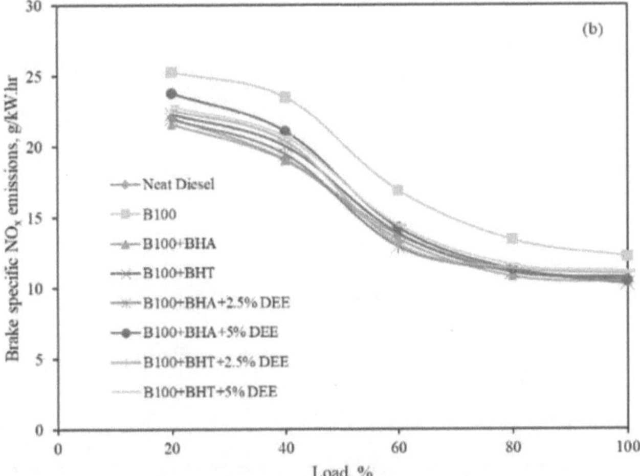

Fig. 7.3 Comparison of variation in NO_x emissions for with antioxidants concentration of 1000 ppmfor (a) B20 and (b) B100 fuels at different loads

BHA, a monohydric phenolic antioxidant, is a white waxy solid that is insoluble in water but highly soluble in oil. Its structure allows it to donate a positively charged ion to a loose radical, stabilizing the molecule and preventing oxidation [38], which suppresses NO_x formation. Adding 5% diethyl ether by volume to B20 with 1000 ppm of BHA reduced NO_x emission efficiency by 3%. A similar reduction was observed in B100 with the same levels of BHA and diethyl ether, though it remained lower than untreated B20 fuel

4.2 Smoke Emissions

Figure 7.4 indicates that the inclusion of diethyl ether (DEE), an oxygen-rich improver, reduces soot opacity in engines, especially when combined with antioxidants like BHA and BHT at 1000 ppm. Specifically, BHA and 5% DEE decreased smoke density by 4.9% in B20 and 10.3% in B100, while BHT with DEE reduced it by 5.2% in B20 and 10.9% in B100, demonstrating DEE's effectiveness in lowering smoke emissions across biodiesel fuels.

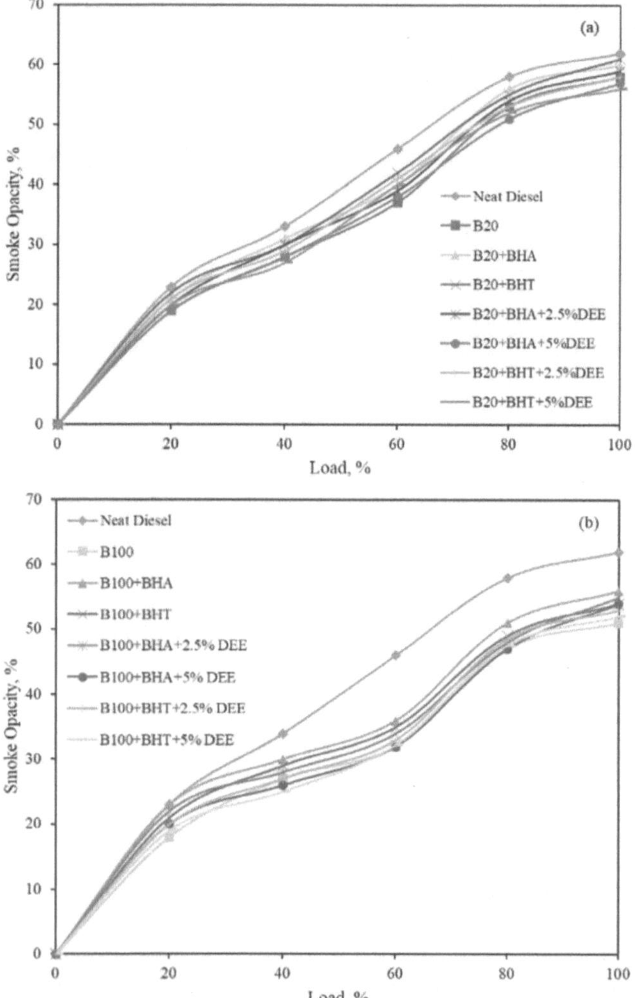

Fig. 7.4 Comparison of variation in Smoke opacity for with antioxidants concentration of 1000 ppm for (a) B20 and (b) B100 fuel at different loads

4.3 CO and HC Emissions

Figure 7.5 reveals that antioxidants BHA and BHT initially increase CO emissions in B20 and B100 fuels, with BHA causing increases of 8.01% and 14.6% respectively. However, adding 5% diethyl ether (DEE) to these fuels reduces CO emissions by 5.27% in B20 and 8.90% in B100. Similarly, BHT with 5% DEE reduces CO emissions by 9.21% in B20 and 4.66% in B100. Figure 7.6 shows that BHT increases hydrocarbon (HC) emissions by 11.16% in B20 and 15.22% in B100, but the addition of DEE results in reductions of 8.86% and 3.78% respectively. The antioxidants minimise the concentration of "peroxyl and hydrogen peroxide radicals", limiting OH radical formation needed for oxidation. Yet, DEE combustion generates extra OH radicals, enhancing oxidation and effectively lowering both CO and HC emissions in these biodiesel fuels.

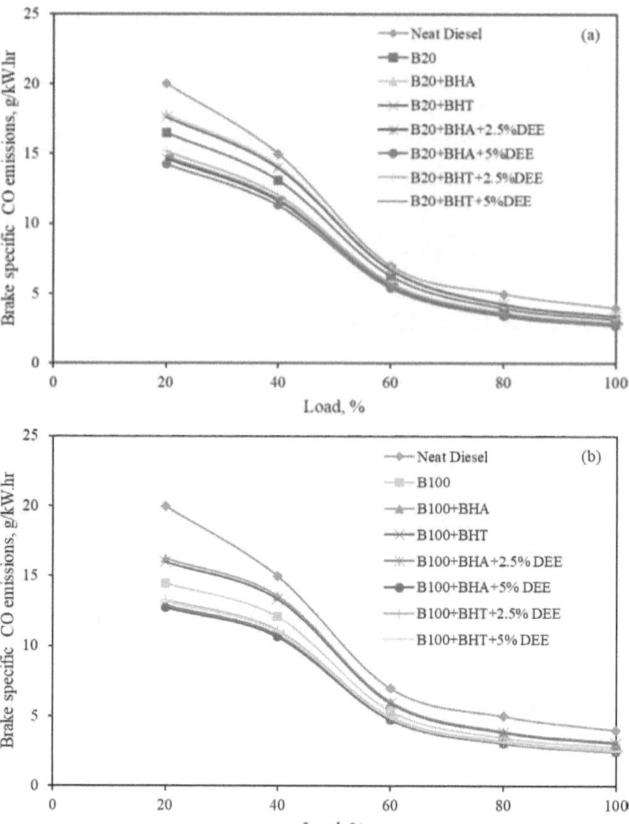

Fig. 7.5 Variation of CO emissions for untreated biodiesel and POME- diesel blends with antioxidant concentration of (a) 500 ppm and (b) 1000 ppm

4.4 Specific Energy Consumption (SEC)

Figure 7.6 shows that antioxidants slightly increased specific energy consumption for B20 and B100 fuels; for example, a BHA and B20 mixture used 11.01 MJ/kW.hr at rated load compared to 10.2 MJ/kW.hr for base B20, due to a marginal decrease in peak pressure from antioxidant activity. However, adding 5% diethyl ether to the mix reduced energy consumption back to about 10.4 MJ/kW.hr, demonstrating the effectiveness of combining antioxidants and DEE in maintaining energy efficiency in biodiesel fuels

5. Conclusions

This study examined the impacts of phenolic antioxidants and diethyl ether (DEE) on NO_x, smoke, HC, and CO emissions from a biodiesel-fueled common rail diesel engine. Key findings include:

1. Both antioxidants and DEE effectively reduced NO_x and smoke emissions across B20 and B100 biodiesel fuels.
2. Compared to fuels without additives, the presence of butylated hydroxy anisole (BHA) and butylated

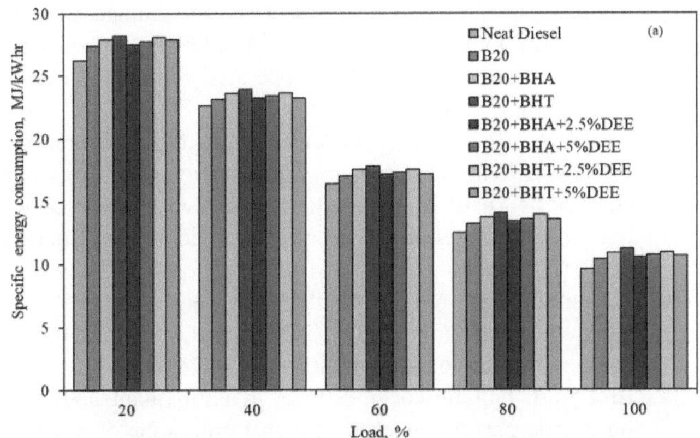

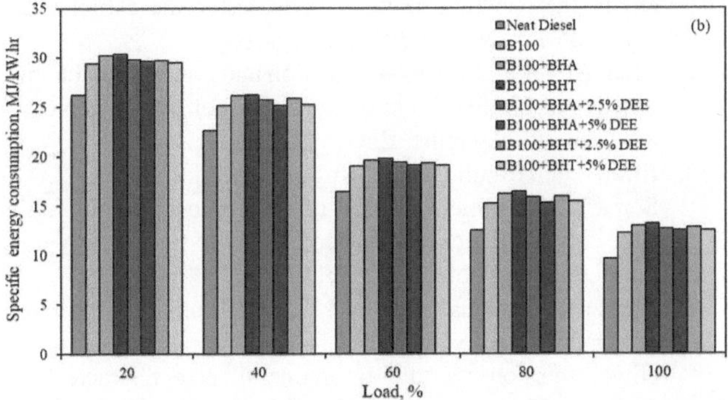

Fig. 7.6 Variation in SEC for neat biodiesel and biodiesel-diesel blends with antioxidant concentration of (a) 500 ppm and (b) 1000 ppm

hydroxyl toluene (BHT) at 1000 ppm significantly decreased NO_x emissions. BHA showed a maximum reduction of 25.7% in NO_x for 100% palm oil methyl ester (POME) and 14.3% for biodiesel blends at full load. BHT resulted in a maximum reduction of 11.52% for B20 fuel and 21.12% for neat biodiesel.

3. While the inclusion of DEE typically reduced smoke emissions, adding 1000 ppm of BHA increased smoke emissions by up to 5.4% in B100 fuel at full load. Similar patterns were observed with BHT.
4. Antioxidants in biodiesel fuels increased CO, and smoke emissions slightly in the absence of oxygenates. However, with DEE, these emissions were lower than those from untreated fuels by up to 8.7%.
5. Adding antioxidants and DEE to biodiesel fuels did not significantly affect Specific Energy Consumption (SEC)

REFERENCES

1. A.M. Ashraful, H.H. Masjuki, M.A. Kalam, I.R. Fattah, S. Imtenan, S. Shahir. Production and comparison of fuel properties, engine performance, and emission characteristics of biodiesel from various non-edible vegetable oils: a review, Energy Convers. Manag 2014;80;202–28.
2. S. Dharma, H.C. Ong, H. Masjuki, A. Sebayang, A. Silitonga. An overview of engine durability and compatibility using biodiesel-bioethanol-diesel blends in compression-ignition engines. Energy Convers. Manag 2016;128;66–81.
3. B. Tesfa, R. Mishra, C. Zhang, F. Gu, A.D. Ball. Combustion and performance characteristics of CI (compression ignition) engine running with biodiesel. Energy 2013;51; 101–5.
4. Atul Dhar, Avinash Kumar Agarwal. Effect of Karanja biodiesel blends on particulate emissions from a transportation engine. Fuel 2015;141;154–63.
5. M.H. Mosarof, M.A. Kalam, H.H. Masjuki, A.M. Ashraful, M.M. Rashed, H.K. Imdadul, I.M. Monirul. Implementation of palm biodiesel based on economic aspects, performance, emission, and wear characteristics, Energy Convers. Manag 2015;105;617–29.
6. M.M. Hasan, M.M. Rahman. Performance and emission characteristics of biodiesel–diesel blend and environmental and economic impacts of biodiesel production: A review. Renew. Sustain. Energy Rev. 74(2017) 938–948.
7. Ahmet Uyumaz. Combustion, performance and emission characteristics of a DI diesel engine fueled with mustard oil biodiesel fuel blends at different engine loads. Fuel 2018;212; 256–67.
8. M.N. Nabi, M.G. Rasul. Influence of second-generation biodiesel on engine performance, emissions, energy and exergy parameters. Energy Convers. Manag 2018;169;326-
9. R. Sakthivel, K. Ramesh, R. Purnachandran, P. Mohamed Shameer. A review on the properties, performance and emission aspects of the third-generation biodiesels. Renew. Sustain. Energy Rev. 2018;82;2970–92.
10. Gopinath Dhamodaran, Ramesh Krishnan, Yashwanth Kutti Pochareddy, HomeshwarMachgahe Pyarelal, Harish Sivasubramanian, Aditya Krishna Ganeshram. A comparative study of combustion, emission, and performance characteristics of rice-bran-, neem-, and cottonseed-oil biodiesels with varying degree of unsaturation. Fuel. 2017;187;296–305.
11. Biofuels Market Size, Share & Growth Analysis Report, Precedence Research, Canada. https://www.precedenceresearch.com/biofuels-market; 2021 [accessed 14 March 2021].
12. B. Jothithirumal, E. Jamesgunasekaran. Combined Impact of Biodiesel and Exhaust Gas Recirculation on NOx Emissions in Di Diesel Engines. Procedia Eng. 2012;38; 1457–66.
13. Mohd Hafizil Mat Yasin, Rizalman Mamat, Ahmad Fitri Yusop, Daing Mohamad Nafiz Daing Idris, Talal Yusaf, Muhammad Rasul, Gholamhassan Najafi. Study of a Diesel Engine Performance with Exhaust Gas Recirculation (EGR) System Fuelled with Palm Biodiesel. Energy Procedia 2017;110; 26–31.
14. Gomaa M, Alimin A, Kamarudin K. The effect of EGR rates on NOx and smoke emissions of an IDI diesel engine fuelled with Jatropha biodiesel blends. Int J Energy Environ 2011;2; 477–90.
15. Nitin M. Sakhare, Pankaj S. Shelke, Subhash Lahane. Experimental Investigation of Effect of Exhaust Gas Recirculation and Cottonseed B20 Biodiesel Fuel on Diesel Engine, Proc. Technol 2016;25;869–76.

16. Labecki L, Ganippa LC. Effects of injection parameters and EGR on combustion and emission characteristics of rapeseed oil and its blends in diesel engines. Fuel 2012;98; 15–28.
17. Masood Miya, KavatiVenkateswarlu. An experimental investigation on CRDI diesel engine coupled with EGR using cotton seed biodiesel. Int. J. Ambient Energy. 2021
18. Xiaochen Shi, Bolan Liu, Chao Zhang, Jingchao Hu, Qiangqiang Zeng. A study on combined effect of high EGR rate and biodiesel on combustion and emission performance of a diesel engine. Appl. Therm. Engg 2017;125;1272–79.
19. Tauzia X, Maiboom A, Shah SR. Experimental study of inlet manifold water injection on combustion and emissions of an automotive direct injection diesel engine, Energy. 2010;35;3628–39.
20. Maiboom A, Tauzia X. NOx and PM emissions reduction on an automotive HSDI diesel engine with water-in-diesel emulsion and EGR: an experimental study. Fuel. 2011;90;3179–92.
21. Mikulski, M., Duda, K. and Wierzbicki, S. (2016). Performance and emissions of a CRDI diesel engine fuelled with swine lard methyl esters–diesel mixture. Fuel, 164, pp. 206–219.
22. Shahir, V.K., Jawahar, C.P., Suresh, P.R. and Vinod, V. (2017). Experimental Investigation on Performance and Emission Characteristics of a Common Rail Direct Injection Engine Using Animal Fat Biodiesel Blends. Energy Procedia, 117, pp.283–290.
23. Bedar, P. and Kumar, G.N. (2018). Performance Emission and Combustion Characteristics of CRDI Engine Operating on Jatropha Curcas Blend With EGR. Materials Today: Proceedings, [online] 5(11, Part 2), pp.23384–23390.
24. Ashok, B., Nanthagopal, K., Arumuga Perumal, D., Babu, J.M., Tiwari, A. and Sharma, A. (2019). An investigation on CRDi engine characteristic using renewable orange-peel oil. Energy Conversion and Management, [online] 180, pp.1026–1038.
25. Subramaniam, M., Solomon, J.M., Nadanakumar, V., Anaimuthu, S. and Sathyamurthy, R. (2020). Experimental investigation on performance, combustion and emission characteristics of DI diesel engine using algae as a biodiesel. Energy Reports, 6, pp.1382–1392.
26. Himanshu Londhe, Guanqun Luo, Sunkyu Park, Stephen S. Kelley, Tiegang Fang. Testing of anisole and methyl acetate as additives to diesel and biodiesel fuels in a compression ignition engine. Fuel 2019;246;79–92.
27. Venu Babu Borugadda, Ajay K. Dalai, Supratim Ghosh. Effects of natural additives on performance of canola biodiesel and its structurally modified derivatives. Ind Crops Prod. 2018;125;303–13.
28. Ramalingam Senthil, Ganesan Pranesh, Rajendran Silambarasan. Leaf extract additives: A solution for reduction of NO_x emission in a biodiesel operated compression ignition engine. Energy 2019;175; 862–78.
29. S. Senthur Prabu, M.A. Asokan, Rahul Roy, Steff Francis, M.K. Sreelekh. Performance, combustion and emission characteristics of diesel engine fuelled with waste cooking oil bio-diesel/diesel blends with additives. Energy. 2017;122;638–48.
30. K. Varatharajan, M. Cheralathan, R. Velraj. Mitigation of NOx emissions from a jatropha biodiesel fuelled DI diesel engine using antioxidant additives. Fuel 2011;90;2721–25.
31. Erol İleri, GünnurKoçar. Effects of antioxidant additives on engine performance and exhaust emissions of a diesel engine fueled with canola oil methyl ester–diesel blend. Energy Convers. Manag 2013;76;145–154.
32. Ertan Alptekin. Emission, injection and combustion characteristics of biodiesel and oxygenated fuel blends in a common rail diesel engine. Energy 2017;119; 44–52.
33. Jinlin Han, Shuli Wang, Ruggero Maria Vittori, L.M.T. Somers. Experimental study of the combustion and emission characteristics of oxygenated fuels on a heavy-duty diesel engine. Fuel 2020;268;117219.
34. Kunduru Srinivasa Reddy, Yarrapathruni Venkata Hanumantha Rao, Vallapudi Dhana Raju. Influence of diethyl ether on the diesel engine diverse characteristics fuelled with waste plastic biodiesel. Mater. Today: Proc 2022;61;1168–75.
35. Sivalakshmi, T. Balusamy. Effect of biodiesel and its blends with diethyl ether on the combustion, performance and emissions from a diesel engine. Fuel 2013;106; 106–10,
36. Roh H G, Lee D, Chang S L. Impact of DME-biodiesel, diesel-biodiesel and diesel fuels on the combustion and emission reduction characteristics of a CI engine according to pilot and single injection strategies. J. Energy Inst 2015;88;376–85.
37. M. Vijay Kumar, M. Nandu. Evaluation of the performance and emission of diesel engine by using sterculia foetida biodiesel blend and DMC additive. Mater. Today: Proc. 2021;43 191–95.
38. Elisa Franco RIBEIRO, Neuza JORGE, Oxidative stability of soybean oil added to coffee husk extract (Coffea arabica L.) under accelerated storage conditions, Food Sci. Technol. 2017;37.

Note: All figures and tables were exclusively produced from our experimental work and were not sourced from any publications or online material.

Advances in Additive Manufacturing Technologies – Gurusamy Pathinettampadian (eds)
© 2024 Taylor & Francis Group, London, ISBN 978-1-032-90013-1

8 Static and Dynamic Mechanical Analysis of Prunus Dulcis Shell Powder and Basalt Fibre with Aghil Epoxy Hybrid Matrix Bio Composite

K. Yesuraj*
Department of Mechanical Engineering,
Panimalar Engineering College, Chennai-123, Tamilnadu, India

A. Latha
Department of Civil Engineering,
Panimalar Engineering College, Chennai-123, Tamilnadu, India

R. Ganesan
Department of Civil Engineering,
Velammal college of Engineering and Technology,
Madurai-09, Tamil Nadu, India

R. Sathiyamoorthi
Department of Mechanical Engineering,
Chennai Institute of Technology, Chennai, Tamilnadu, India

ABSTRACT: Bio composite materials are a beneficial alternative to wholly polymer dependent material manufacturing industries. This study aims at increasing the biodegradable component of the composite. The research till now has been on the biodegradable reinforcement in a polymer composite like natural fillers and natural fibers. The bio resin Aquilaria agallocha is blended with Epoxy to give a Aghil Epoxy hybrid resin. Prunus dulcis shell powders with basalt Fiber are used as a particulate reinforcement in the Aghil Epoxy hybrid matrix to form an Aghil epoxy hybrid bio composite. Different Fiber lengths of 5, 10, 15, 20 and 25 mm and volume of 5, 10, 15, 20 25% basalt Fiber reinforcement in Aghil Epoxy hybrid composite are fabricated and their mechanical characteristics are investigated. The composite with 15 mm fibre span and 15-volume fraction composite has improved mechanical qualities, according to the test results. Finally, the mechanical and dynamic mechanical characteristics were scrutinized to assess the impact of the Prunus dulcis shell filler addition into the basalt fibre composite. Due to their synergetic impact, natural filler and basalt fibre have been discovered to improve the characteristics of composites. This enhances the adhesion and stress transmission uniformity between the reinforcements. An electron microscope was used to examine the morphology of the fibre exterior on micrographs.

KEYWORDS: Bio-composite, Basalt fibre, Epoxy resin, Mechanical properties, Morphology

*Corresponding author: yesu087@gmail.com

DOI: 10.1201/9781003545774-8

1. Introduction

Researchers all over the world are investigating natural fibre reinforced polymer composites as a suitable substitute to regular polymer composites in response to the increasing demand for eco-friendly items[1,2]. The composites derived from natural fibres have a number of benefits, including reduced expenditure, enhanced strength-to-weight ratio, environmental friendliness, easy disposition, and a smaller environmental footprint [3,4,5]. Natural fibres have been examined as a source of viable strengthening for polymer matrix composites by many investigators[6]. The tensile characteristics of the natural fibre composites establish their high reliability on the interaction among fibre and matrix, as reported by various researchers [7–11]. Rong et al [12], considered the effects of several chemical treatments on the mechanical characteristics of sisal fibre composites.

The fiber exterior subjugations were shown to dramatically increase the fiber-matrix adherence, which resulted in the enhancement of mechanical characteristics, according to Venkateshwaran and his colleagues [13]. The mechanical properties of hybrid composites made from sisal along with banana fibre were studied. Hybridization was said to have increased the composite's tensile characteristics. Sreekumar et al [14], examined the impact of fibre external modification on the mechanical characteristics of a sisal fiber–polyester composite.Permanganate treatment, bensoilation, and salinization were used to enhance the interfacial adhesion among the matrix and the strengthening. When compared to unprocessed sisal fibre polyester matrix composite substances, the results demonstrate that permanganate treatment increases tensile strength by 36 percent and flexural strength by 25 percent.

Fillers are considered for enhancing the composites' robustness, damping characteristics, and thermal permanence, in addition to lowering the expenditures of composites. Several studies have looked at the influence of filler substances on the characteristics of polymer composites. Chen et al [15], introduced $CaCO_3$ nanoparticles into a polypropylene matrix and investigated the effect of the filler proportion on the tensile characteristics. The filler augmentation increased the tensile modulus by 85 percent, whereas the tensile and impact rigidities were not significantly changed. Rajini et al [16], used compression moulding to manufacture a coconut sheath/polyester composite with montmorillonite nano clay as a nanofiller. In addition, the impact of chemically treating the fibre on the composite's mechanical characteristics was investigated. The results showed that a little shift in the glass transition temperature and an increase in dynamic properties were the effects of adding nanofiller to the carbon fiber composites[17-19].

2. Materials and Methods

Prunus dulcis powder, a common filler, served as the secondary reinforcement in the Aghil epoxy hybrid matrix, while basalt fiber served as the primary reinforcement. After constructing composite samples by hand lay-up, they were compressed. To be tested, the specimens had to be prepared in accordance with ASTM specifications. According to ASTM D 638, the tensile assessments were carried out at a speed of two millimeters per minute. For the impact and flexural testing, ASTM D 256 and ASTM D790 were utilized, respectively [20–25]. A dynamic mechanical analysis was conducted in the tensile mode.

Methods for static and dynamic analysis of Prunus dulcis (almond) shells are available. Agricultural, food, and environmental sciences are among the scientific domains where these analyses are frequently used. Surface morphology is visualized in great detail by SEM analysis.

Fig. 8.1 Prunus Dulcis shell powder

Fig. 8.2 Basalt Fibre

Testing a substance's biodegradability determines whether or not it can naturally break down through the activity of microbes. In this case, the material is Prunus dulcis (almond) shells. Examining a material's ability to degrade naturally and sustainably depends on its biodegradability.

Dynamic mechanical analysis (DMA) is the process of examining a material's mechanical properties as a function of temperature, frequency, and time. DMA can contribute to our understanding of the viscoelastic behavior of materials such as basalt fibers and Prunus dulcis (almond) shell powder, as well as their appropriateness for a variety of applications. To comprehend their respective mechanical behaviors, compare the DMA results between almond shell powder and basalt fibers. In evaluating the DMA data, take into account the criteria particular to the application, such as damping, rigidity, and thermal stability. To ascertain whether a material is appropriate for a given engineering or composite application, assess its behavior under dynamic circumstances. DMA may be used to analyze the mechanical properties of basalt fibers and Prunus dulcis shell powder, which can help you decide whether to employ them in composite materials, construction, or other industries.

3. Results and Discussion

3.1 Analysis of Basalt Fiber Correlated with Mechanical Properties

The results of the different mechanical tests were reported in Table 8.1 according to ASTM standards. The predominance of the fibre length and volume percent on the tensile, flexural, and impact characteristics is shown in Table 8.1. The above-mentioned composition's maximum tensile strength and modulus are evaluated, respectively.

In a 3-point bending test, the utmost flexural strength and modulus were discovered for fibre lengths of 15 mm and 15 vol percent, respectively. In terms of impact strength, it has been discovered that when the fibre parameter values grow, the impact power is recorded by the composite augments as well, but the % augmentation is not substantial in comparison with composites having 15 mm and 15% volume and 15 mm and 20% volume of fibre. The material is subjected to a huge volume of strain for a short period of time during the impact test. Hence no major variations in the value were detected. Figure 8.3(a) shows the SEM image of poor bonding specimen at 300μm. Figure 8.3(b) shows the SEM image of good bonding specimen at 300μm.

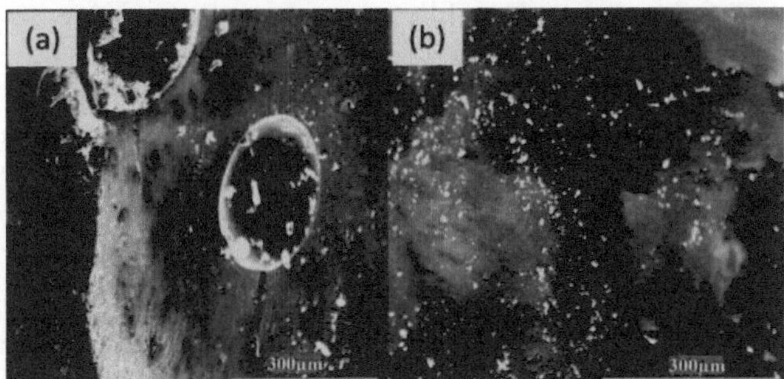

Fig. 8.3 (a) SEM image of poor bonding specimen (b) SEM image of good bonding specimen

Table 8.1 Predominance of fiber parameters in the mechanical properties of basalt fiber composite

Length of the fiber (mm)	Volume of the Fiber (vol%)	Tensile Strength (MPa)	Tensile Modulus (GPa)	Flexural Strength (MPa)	Flexural Modulus (GPa)	Impact Strength (J/m^2)
5	5	57	1.2	70	3.5	2,700
	10	61	1.4	72	3.6	3,200
	15	64	1.6	74	3.7	3,400
	20	67	1.8	76	3.8	3,700
	25	71	2.1	75	4.1	4,300
	5	69	3.4	80	4.2	2,900
	10	74	3.7	81	4.3	3,200
10	15	76	3.8	82	4.4	3,400
	20	83	3.9	847	4.5	3,700
	25	81	4.1	86	4.7	4,300
	5	91	4.2	88	4.6	3,100
	10	93	4.3	93	4.5	3,300
15	15	95	4.4	94	4.7	3,400
	20	96	4.6	99	4.8	3,900
	25	99	4.8	98	5.1	4,200
	5	71	5.1	92	5.3	3,300
	10	73	5.3	94	5.4	3,400
20	15	75	5.4	96	5.6	3,300
	20	77	5.6	89	5.8	4,390
	25	78	6.1	88	5.2	4,400
	5	79	6.4	86	5.1	4,300
	10	61	6.7	85	4.9	4,300
25	15	68	7.1	86	4.8	4,200
	20	64	7.2	82	4.6	4,400
	25	63	6.8	88	4.7	4,300

Figure 8.4 shows the SEM image ofBasalt with Prunus dulcis shells filler good interfacial bonding 15 %- and 15-mm fibre length specimen. Even if many fibres are damaged, the load can be shifted by the primary section of the fibres through a powerful interconnection, allowing the composite to sustain the imposed load. Furthermore, it demonstrates the even dispersion of fibres, which is important for stress distribution and, as a result, mechanical strength. As the fibre volume proportion increases, the tensile strength begins to deteriorate as a result of fibre aggregation at increasing volume fractions. Furthermore, fibre clustering causes reduced moistening of the fibre by the matrix, resulting in minimal fibre adherence. Figure 8.5 shows the Tensile properties of Prunus dulcis shells and Basalt fibre

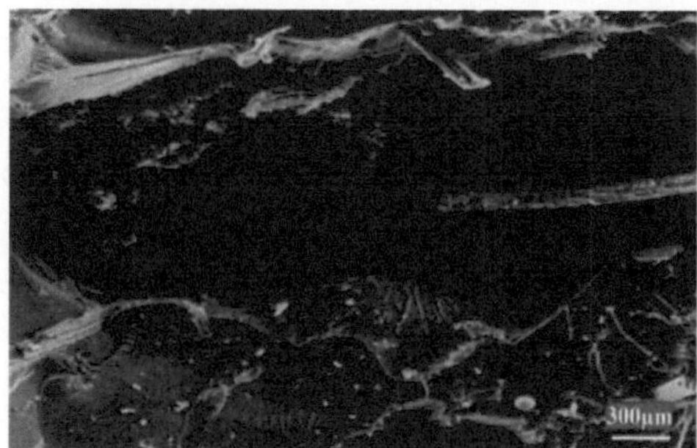

Fig. 8.4 SEM image of basalt with Prunus dulcis shells filler good interfacial bonding 15 % and 15 mm fiber length specimen

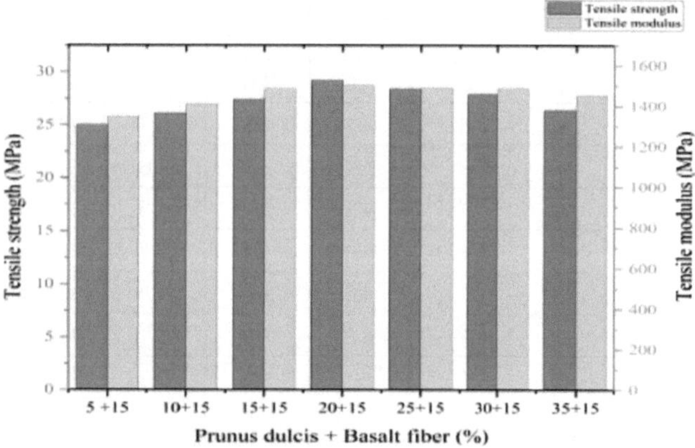

Fig. 8.5 Tensile properties of prunus dulcis shells and basalt fiber

Mechanical qualities are reduced as a result of this. Increased fibre length also causes curling along with fiber bending, which affects the stress relocation among the matrix and fibre. By adding the Prunus dulcis natural shells filler subsequent to verifying the optimal fibre span, volume %, and suitability. The filler was added to the basalt fibre reinforced composite in amounts of 5, 10, 15, and 15% by volume. Figure 8.6 shows the Flexural properties of Prunus dulcis shells and Basaltfibre.

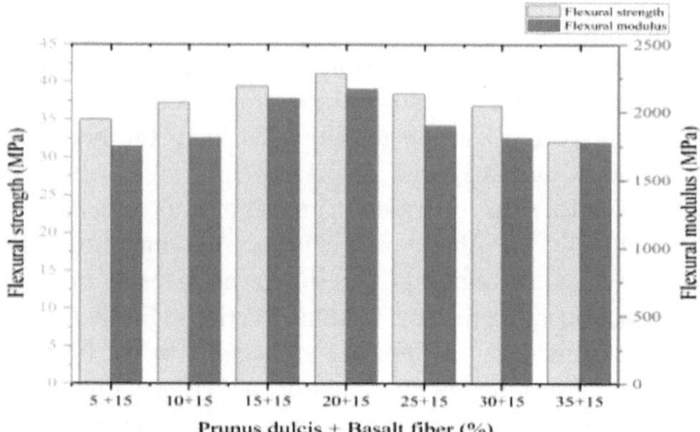

Fig. 8.6 Flexural properties of Prunus dulcis shells and Basalt fibre

The basalt fibre content was retained at 15%, with the rest made up of varied filler and matrix proportions. The leverage of filler on the mechanical characteristics of the composite was ascertained using the tests. The results indicate how the inclusion of filler affects the composite's tensile, flexural, and impact characteristics. With a filler volume of 5%, better mechanical characteristics were attained. The maximum tensile strength value was discovered. Figure 8.7 shows the Impact strength of Prunus dulcis and Basalt fibre

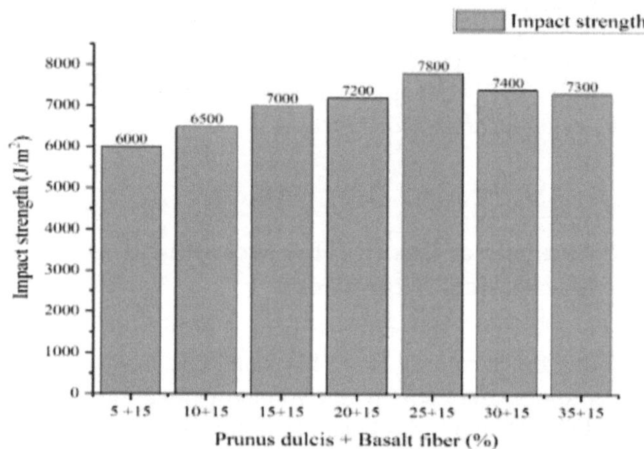

Fig. 8.7 Impact strength of prunus dulcis and basalt fiber

3.2 Dynamic Mechanical Analysis

Storage Modulus (E′)

The temperature of the test runs from 30 to 240°C, with a frequency of 10 Hz. The findings reveal that, regardless of frequency, the composite absorbs more energy at all temperatures up to 140°C. The amount of energy absorbed by the composite and the resin is nearly equal as the testing temperature rises. Thus, the fibers and filler are effectively shifting the load up to 160°C.After that, they are bound to relax their link with the matrix, resulting in a collapse of the desired function. The storage modulus or stiffness number of the composite is thought to be enhanced by adding filler or fiber to the hybrid matrix. An additional finding from Fig. 8.8 is that adding filler to processed fibre increases the storage modulus, which is assessed in a shorter time and yields higher values. Exposure over an extended period of duration (reduced frequency) yields a reduced value. This is because the molecules that are subjected to lower frequencies undergo more rearrangement than those that are subjected to higher frequencies. The material's molecular rearrangement results in a reduction in localized stress. Because the molecules are stationary at low temperatures, they are not able to resonance with the oscillatory loads and dwell rigidly. Figure 8.9 shows the Damping factor of basalt with bio filler reinforced polymer composite at 10 Hz frequency.

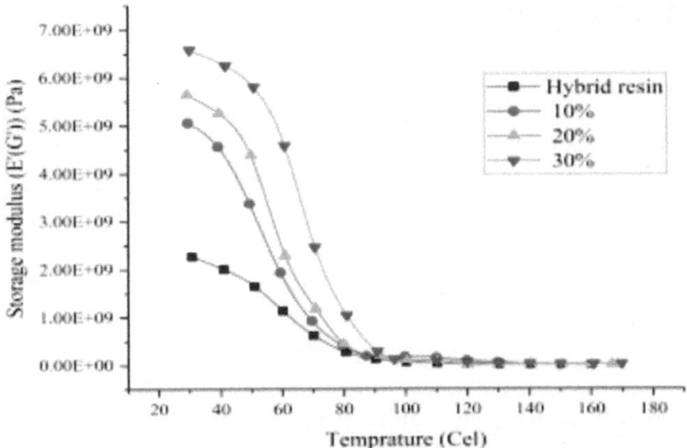

Fig. 8.8 Storage modulus of basalt with bio filler reinforced polymer composite at 10 Hz frequency

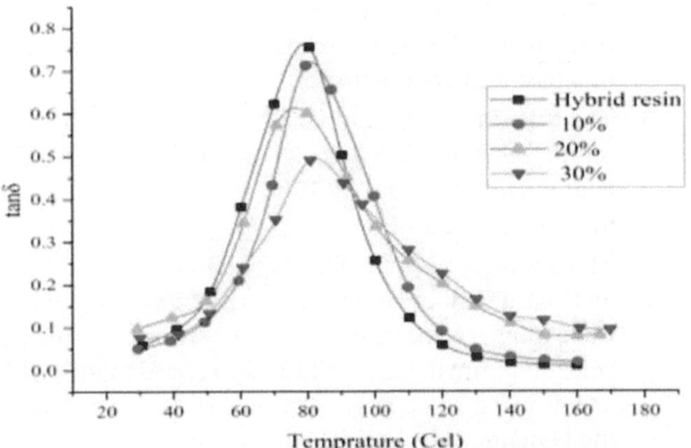

Fig. 8.9 Damping factor of basalt with bio filler reinforced polymer composite at 10 Hz frequency

According to the Dynamic Mechanical investigation, the addition of basalt fibre and Prunus Dulcis shell powder bio filler increased the storage modulus, suggesting a synergistic stiffening impact of the fibre and bio filler in the Aghil Epoxy Hybrid matrix with a shift in Tg value from 90°C to 110°C.

Loss Factor

The loss factor is calculated by dividing the loss modulus by the storage modulus of the material. This shows the index of molecular mobility in the polymer bond as well as the material's energy dissolution upon loading. The results show that adding reinforcements to a composite material reduces the loss factor. This suggests that the composite's energy dissipation process is superior to the matrix's. The solid is said to be in the new phase or elastic energy phase at lower temperatures. At higher temperatures, the material phase transforms into an entropy elastic state, also known as rubber. The glass transition is the state that occurs when anything goes from a glass to a rubber-elastic condition. It is widely accepted that the temperature of the largest loss factor (tan max) or the maximum loss modulus [E″ max] is the glass transition temperature (Tg). In this study, the Tgis computed by using the tan value's peak.When compared to hybrid resin, the peak of tan decreases regardless of the magnitude of applied force, while a noticeable transition in the Tg is observed. This demonstrates that a collaborative action among the filler and the fibre causes the Tg value to transition. The glass transition area, where the substance transitions from hard to pliable, is connected with the damping peak. The molecules are initially stable, but as the temperature rises, tiny clusters and linkages of molecules inside the hybrid resin arrangement begin to migrate.

The Cole–Cole plot can be used to investigate the polymer composite's linear visco elastic behaviour at the glass transition temperature. At 10 HZ frequencies, a Cole–Cole plot was evaluated by graphing storage modulus (E′) vs loss modulus (E″). The addition of Prunus dulcis shell powder and basalt Fiber with Aghil Epoxy Hybrid matrix induces systemic modifications in the cross-linked polymer, as demonstrated in this graph. The following is an illustration of what I'm talking about: The material's homogeneity or heterogeneity is indicated by the curve's shape. Figure 8.10 shows the Cole-Cole plot for the Aghil Epoxy Hybrid Matrix with 25 percent and 30 Prunus dulcis shell powder and basalt Fiber bio composite. The semi-circular form of the resin and 25% Prunus dulcis shell powder and basalt Fiber bio composite curve reveals the manufactured composite's homogeneity. Furthermore, the adding of filler content up to 30% changes the composite's behaviour to heterogeneity. The plot's deviation from its semicircular nature demonstrates this. The colocolo plot depicts the loss modulus' (E″) relative contribution to the storage modulus' (E′) contribution. In this graph, E′ has a greater relative contribution than E″. This indicates that the material can absorb rather than release energy.

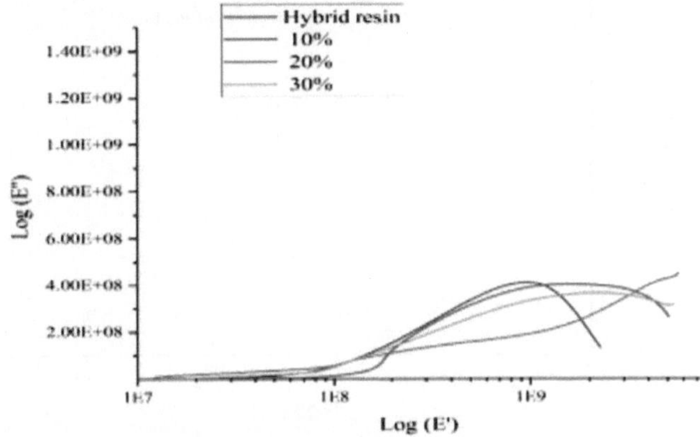

Fig. 8.10 Cole-Cole plot for the aghil epoxy hybrid matrix with 25 percent and 30 prunus dulcis shell powder and basalt fiber bio composite

Biodegradability Test

The oldest and most widely used approach for determining degradation is to bury the sample in the ground. Soil contains a plethora of organisms that form a forceful, complicated environment that serves as a vital reserve for dilapidation due to the real conditions of waste disposal. The weight forfeiture after a particular duration in a silt setting was used to determine the sample's biodegradability. Initially, the specimens were quantified and were subsequently obscured underneath the silt. The obscured specimens were extracted after 10, 20, 30, and 60 days. Subsequent to extraction from the sand, the specimens were rinsed with water and then quantified to ascertain the weight reduction. The composites ample were obscured beneath moist sand to ascertain their decomposability. The repercussion of decomposability of the static and dynamic mechanical analysis of basalt with Aghil epoxy hybrid matrix bio composite was analyzed. Before burying the sample in the dirt, its initial weight was recorded. After that, the sample was extracted at systematic time periods, and its weight was tested to see if it had changed. After 60 days, the sample's overall weight loss was found to be 3.2 percent. The weight of the sample initially increased due to moisture absorption, but it then began to decrease. The riposte of minuscule along with comprehensive organisms with the composite specimen after a lengthy period of time was blamed for the weight loss. When the results were compared to those of other natural fibre and filler-reinforced polymer composites, it was discovered that the substance was more biodegradable. As a result, the inclusion of bio composite produces a more environmentally friendly bio composite. Figure 8.11 shows Prunus dulcis shellpowder 10% +15 mm fibre length hybrid matrix composite.

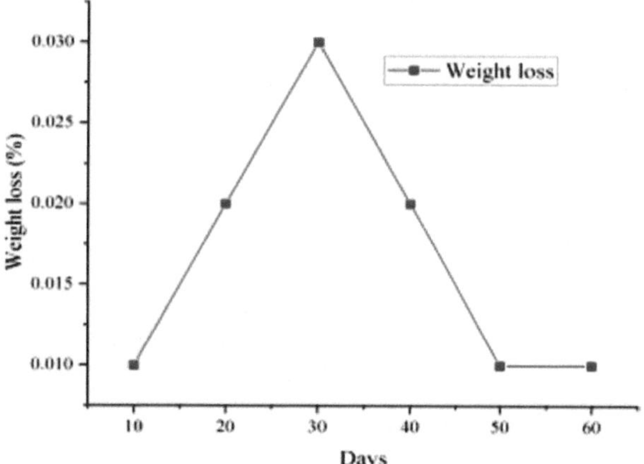

Fig. 8.11 Prunus dulcis shellpowder 10% +15 mm fibre length hybrid matrix composite

4. CONCLUSION

Aghil Epoxy Hybrid resin incorporated with Prunus Dulcis shell powder has shown better properties than Prunus Dulcis shell powder, hence, Prunus Dulcis shell powder was selected as the bio filler along with basalt reinforced Aghil Epoxy Hybrid bio composite. The composites were fabricated using basalt fiber (20 vol %) as reinforcement and different volume percentages of Prunus Dulcis shell powder in the Aghil Epoxy Hybrid Matrix. The analysis was carried out to see how filler loading affected the composite's mechanical and viscoelastic characteristics. The following are the study's conclusions:

1. Bio Composites containing 10% Prunus Dulcis shell powder have demonstrated improved mechanical characteristics.
2. Tensile strength and tensile modulus rose by 37 percent and 12 percent, respectively, when compared to a basalt fiber composite.
3. The filler in the bio composite worked as a secondary reinforcement, allowing it to carry greater load and evenly transmit stress, resulting in superior mechanical qualities.
4. The adding of Prunus Dulcis shell powder bio filler and basalt fiber improved the storage modulus, indicating a synergetic stiffening effect of the bio filler and fibre in the Aghil Epoxy Hybrid matrix with a shift in Tg value from 90°C to 110°C, according to the Dynamic Mechanical analysis.

REFERENCES

1. Sobczak, L., Brüggemann, O., & Putz, R. F. (2013). Polyolefin composites with natural fibers and wood-modification of the fiber/filler–matrix interaction. *Journal of Applied Polymer Science, 127*(1), 1–17.
2. Saba, N., Md Tahir, P., & Jawaid, M. (2014). A review on potentiality of nano filler/natural fiber filled polymer hybrid composites. *Polymers, 6*(8), 2247–2273.
3. Chen, R. Y., Zou, W., Zhang, H. C., Zhang, G. Z., Yang, Z. T., Jin, G., & Qu, J. P. (2015). Thermal behavior, dynamic mechanical properties and rheological properties of poly (butylene succinate) composites filled with nanometer calcium carbonate. *Polymer Testing, 42*, 160–167.
4. Rajini, N., Jappes, J. W., Jeyaraj, P., Rajakarunakaran, S., & Bennet, C. (2013). Effect of montmorillonite nanoclay on temperature dependence mechanical properties of naturally woven coconut sheath/polyester composite. *Journal of reinforced plastics and composites, 32*(11), 811–822.
5. Raju, G. U., & Kumarappa, S. (2011). Experimental study on mechanical properties of groundnut shell particle-reinforced epoxy composites. *Journal of Reinforced Plastics and Composites, 30*(12), 1029–1037.

6. Vimalanathan, P., Venkateshwaran, N., & Santhanam, V. (2016). Mechanical, dynamic mechanical, and thermal analysis of Shorea robusta-dispersed polyester composite. *International Journal of Polymer Analysis and Characterization, 21*(4), 314–326.
7. Ashok, B., Naresh, S., Reddy, K. O., Madhukar, K., Cai, J., Zhang, L., &Rajulu, A. V. (2014). Tensile and thermal properties of poly (lactic acid)/eggshell powder composite films. *International journal of polymer analysis and characterization, 19*(3), 245–255.
8. Saravanan, K., D. Jayabalakrishnan, K. Bhaskar, and S. Madhu. "Thermally reduced sugarcane bagasse carbon quantum dots and in-plane flax fiber unsaturated polyester composites: surface conductivity and mechanical properties." Biomass Conversion and Biorefinery (2023): 1–10.
9. Somasundaram, S.,. D. Jayabalakrishnan, A. Balajikrishnabharathi, and K. Bhaskar. "Performance evaluation of silane-treated Cissus quadrangularis fiber and waste Limoniaacidissima hull powder on mechanical, fatigue, and water absorption properties of epoxy resin composite." Biomass Conversion and Biorefinery (2022): 1–9.

Note: Every figure and table was created using the experimental work; none of them were taken from any publications or online.

Advances in Additive Manufacturing Technologies – Gurusamy Pathinettampadian (eds)
© 2024 Taylor & Francis Group, London, ISBN 978-1-032-90013-1

9 Development, Characterization, and Mechanical Properties Evaluation of Novel Si_3N_4-Reinforced Polymer Composites with Chicken Feather and Pineapple Leaf Natural Fibers: A Comprehensive Study

N. Gayathri*, S. Prakash, M. Ram Kumar, A. Thameem Ansari
Department of Mechanical Engineering,
VelTech High Tech Dr. Rangarajan Dr. Sakunthala Engineering College, Chennai, India

ABSTRACT: Natural fiber reinforced with polymer composites are popular due to their affordability, environmental friendliness, corrosion and electrical resistance, and high specific strength. Hybrid composites are made using one or more fillers in the matrix. Industrialists, material scientists, and researchers are interested in polymer composites because of its applications in construction, electronics, electrical, automotive, marine, aviation, and medicine. A hybrid composite was hand-laid utilizing chicken feather and pineapple leaf fiber in this investigation. Tensile, flexural, hardness, flammability, and impact energy absorption were tested. Wear tests were done to determine material deterioration. SEM was used to study chicken feather and pineapple leaf fiber surface structure, while EDAX or EDX was utilized to detect and quantify fibers. The Si3N4 filler, polyester resin, and chicken feather fiber composites worked well. Then, three composite specimens were made: plain, 1% Si3N4, and 5% Si3N4. Due to hybrid composites' strong bonding, CFF/PALF composites with 1% Si3N4 had higher hardness and flexural strengths than those without Si3N4. Without Si3N4, chicken feather fiber (CFF), polyester resin, and pineapple leaf fiber (PALF) have lower propagation speed than other Medium Flammable Resistance compositions. The composite material containing Chicken Feather Fiber (CFF) and Pineapple Leaf Fiber (PALF) with 1% Si3N4 had a greater impact energy value than the composite material without reinforcement because the reinforcement absorbs more impact energy. The wear test showed that all composite compositions have equal wear resistance and are comparable to previous wear-behavior hybrid composites.

KEYWORDS: Natural fiber, Polymer, SEM, Mechanical properties, Wear

1. INTRODUCTION

1.1 Overview on Composites

Composites are a multi-phase material in which two or more materials are combined into a single material to optimize one or more properties. Composite is a system in which one phase is usually continuous (matrix) and the other phase (reinforcement) is dispersed within the continuous phase. They may be fibrous or particulates. The reinforcement and matrix phases can be metallic, ceramic and polymeric in composite materials. The different level of properties can be attained with the incorporation of different reinforcements into different types of matrix materials. The main aim of producing a composite material is to create a material the combines the best properties of both the reinforcement and the matrix. In some situations, a composite can have better overall properties than individual properties of reinforcement

*Corresponding author: gayathri@velhightech.com

DOI: 10.1201/9781003545774-9

and the matrix. Improved interaction between the matrix and reinforcement in a composite system is one of the greatest ways to accomplish this. There is an interfacial region within the composite material where the matrix and reinforcement bond to each other for the improvement of properties of the composite. Composite materials have an effective substitution of the traditional materials in numerous engineering applications. High toughness, high creep resistance and high strength-to-weight ratio are the primary factors in the choice of composite. The matrix of composite materials is a ductile or tough material, while the reinforcement is strong and low density. Therefore, the correct percentage of mixing of reinforcements and the matrix with the correct manufacturing method gives the composite with better properties. The matrix's qualities, the geometry of the reinforcements, and the concentration of the reinforcements and the matrix all affect the composites' physical and mechanical characteristics.

Both the reinforcement and the matrix play a variety of roles in achieving the general properties of the manufactured composite materials. The functions of reinforcement are to give desired properties, Load carrying capacity to transfer the applied load to the matrix. The functions of a matrix on the overall properties of the composites are:

- To hold the reinforcements together
- To protect the reinforcements from environment condition
- To protect the reinforcements from abrasion
- To help in maintaining the distribution of reinforcements
- To distribute the applied load evenly between reinforcements
- To provide better finish to composite material

1.2 Polymer Matrix Composites (PMCs)

In many industrial applications, Polymer Matrix Composites (PMCs) constitute the most important type of composite materials. When considering for structural applications, the mechanical properties of polymers are not sufficient. Any kind of reinforcement (fiber or particle) can be used to increase the stiffness and strength of polymeric materials. For the preparation of polymer matrix composite, it does not need high pressure and temperature. A simple manufacturing technique is enough to fabricate the polymer matrix composites. The commonly used fiber materials in the PMCs are Boron, Graphite, Aramid, Carbon, and some organic and also inorganic materials. The matrix materials used in the PMCs are thermoplastic and Thermo-set materials. Polymer Matrix Composites (PMCs) are composed of an organic polymer matrix that binds a range of reinforcements (continuous or short) together. In order to increase fracture toughness and give high strength and stiffness, the reinforcements are generally employed in PMCs. The PMC is made for a variety of applications where the reinforcement supports the mechanical loads placed under service. The matrix in PMCs serves as a binding agent for the reinforcements and a means of transferring applied loads between the matrix and the reinforcements. Generally, the PMCs is classified into the following categories based on reinforcements:

- Fiber-reinforced polymer composites.
- Particle reinforced polymer composites.
- Hybrid Fillers Reinforced Polymer Composites.
- Laminates Reinforced Polymer Composites.

2. Material and Methods

The preparation of fibers made from pineapple leaves and chicken feathers using Si_3N_4 reinforced polymer composites is illustrated. In addition, experimental methods for manufacturing natural fiber composites and test methods for mechanical testing, SEM analysis, tensile strength, flexural strength, hardness test, flammability test and impact energy absorption and wear test of the manufactured composite materials are described.

2.1 Fabrication Process

In this work, pineapple and chicken feather fibers are used for fabricating the composite material with Si_3N_4 and were prepared by hand layup method process. Take appropriate amount of polyester resin with respect to the size and thickness of your composites material along with wastage. Here we have taken 350 ml of polyester resin, now add filler material if required to polyester resin and mix it with the help of wooden stick. Here 0 %, 1%, 5% of silicon nitride (Si_3N_4) is added with respect to weight of polyester resin. Now combine 1 weight percent of catalyst (methyl ethyl ketone peroxide) and accelerator (cobalt octate) with polyester resin to create a homogeneous mixture. PALF, chicken feather-polyester composites, are made using Si_3N_4. Chicken feather and pineapple leaf fiber were cut into small pieces and added to the mixture, which was then thoroughly mixed to create a homogenous mixture. Fix the PVC sheets firmly on the table with cello tapes, then formed the open mold (20 cm * 20 cm) with cardboard on the PVC sheets. Then poured the mixed mixture onto placed in the mold and spread the mixture around the mold, then place another sheets on the specimen. When the fiber mixed was completely wet by the resin, to release the gases and bubbles that had become trapped in the specimens, a metallic roller was rolled over the sheets. The mold was tightly closed after the laying up procedure and allowed to cure for 24 hours while supporting a load of roughly 25 kg. Test specimens of the necessary size were cut from the sheets and positioned as samples before being utilized for testing. Then allow for curing stages and further mechanical test has been taken.

Table 9.1 Composite designation for Si_3N_4 filled with PA fibres fabricated composites

Description	Designation
0 weight % Si_3N_4 + Chopped Chicken feather + Chopped Pineapple fibre + polyester	0 wt. % Si_3N_4
1 weight % Si_3N_4 + Chopped Chicken feather + Chopped Pineapple fibre + polyester	1 wt. % Si_3N_4
5 weight % Si_3N_4 + Chopped Chicken feather + Chopped Pineapple fibre + polyester	5 wt. % Si_3N_4

In order to investigate the composites' morphology from a cross-sectional perspective, SEM, Secondary Electron (SE) as well as Back Scattered Electron (BSE) modes were being used. The investigations were conducted at an operational space of 15 mm along with an accelerating voltage of 5 kV. Specimens were broken regularly in order to get the internal layer of the composites. We used an FEI Quanta 250 Field Emission Scanning Electron Microscope (FE-SEM) to create lateral surface images. The Chicken Feather and Pineapple Bio-Fiber Composite with Si_3N_4 Reinforced with Polyester Composite images taken using scanning electron microscopy (SEM) show three different composite compositions: without filler (0% of Si_3N_4), with filler of 1% weight% of Si_3N_4 (1% of Si_3N_4), and with filler of 5% weight% of Si_3N_4. The three composite constructions may be seen in different magnifications on the right-hand column photos, which show the lateral perspective of the three composite structures while the left-hand column images show the cross-sectional view.

Using the Universal Testing Machine, which was used to create the specimen material in accordance with ASTM standards, the model is next tested for tensile strength. The good tensile strength of Chicken Feather and Pineapple Leaf Fibers with Si_3N_4 Reinforced Polymer Composites is due to the higher concentration of "Ca" in the particles. When compared to pure polyester, Si_3N_4 has a much higher axial strength for applying loads.

The impact test is the next test, and Charpy and IZOD test specimen configurations are typically used for this test. The Charpy Impact Tests were performed using instrumented machines that are capable of measuring distances between-321°F and over 2001°F and ranges between less than 1.365Nm and 410 Nm. It was chosen to test the impact test piece using test specimens alongside V-Notches that were as small as 14 sizes. IZOD Impact Testing had carried out using a standard one notch test sample.

One of the most effective and extensively used methods for measuring hardness is the Rockwell. The hardness test is carried out to complete it after the impact test. It is one way for determining the relative hardness of a substance. This test determines the hardness by measuring the level of residual infiltrate from a steel ball, with a diamond tip, or while load is applied. Specimens made in accordance with ASTM D 7290 are put through a flexural test. When used, the effect of chicken feather and pineapple leaf fibers with Si_3N_4 reinforced based polyester composites normally results in a higher yield of 3-point load. The Si_3N_4 Reinforced Polymer Composites with Chicken Feather and Pineapple Leaf Fibers have good flexural strength at weights of 0%, 1%, and 5%, respectively.

The ASTM D635 standard was followed when conducting the Flammability test with the proportionate rate at which self-supporting polymers burn. This test is mostly used for material comparisons, production control, and quality control. The flammability test of chicken feather and pineapple leaf fibers with Si_3N_4 reinforced polymer composites has medium flammable resistance at weights of 0%, 1%, and 5%.

According to ASTM G99 standard, the Pin-on-disc wear testing apparatus was used to conduct the wear test. A spherical pin was pressed against the rotating disc where the material to be tested was fixed. The test could be conducted in dry and wet conditions. In this investigation, the test was performed in dry conditions without using any lubricants. The applied force ranged from 20 to 80 N, the sliding speed was set at 150 mm/s, and the sliding distance was 1000 m for the wear test conditions. This shows similar wear resistance of all composite compositions and it's comparatively good with other wear behaviour hybrid composites.

Si3N4-reinforced polymer composites are made using several methods to optimize mechanical characteristics. Melt mixing, solution casting, and in-situ polymerization are common. Melt blending evenly disperses Si3N4 particles in molten polymer by extrusion or melt mixing. Solution casting includes dissolving polymer, adding Si3N4, and evaporating solvent. In-situ polymerization creates the polymer matrix using Si3N4. These methods affect composite tensile strength and thermal conductivity. Si3N4-reinforced polymer composites for structural components and electronic devices need careful selection and optimization of production procedures to achieve mechanical advantages

3. Result and Discussion

3.1 Morphological Analysis

Themorphological study on the Chicken Feather and Pineapple leaf fibers with Si_3N_4 reinforced polyester composites is shown in the Fig. 9.1. Figures in this case clearly show that Si_3N_4 polymer composites made up 0%, 1%, and 5% of the total weight. The fibre is strictly bonding with resin matrix and its increases the ductile property. They are uniformly filled in the composites and it does not found any fracture in the composition, because the composite load bearing property has increased.

Fig. 9.1 SEM morphological analysis on Polymer Composite With magnification of 200 micrometer x65 surface area

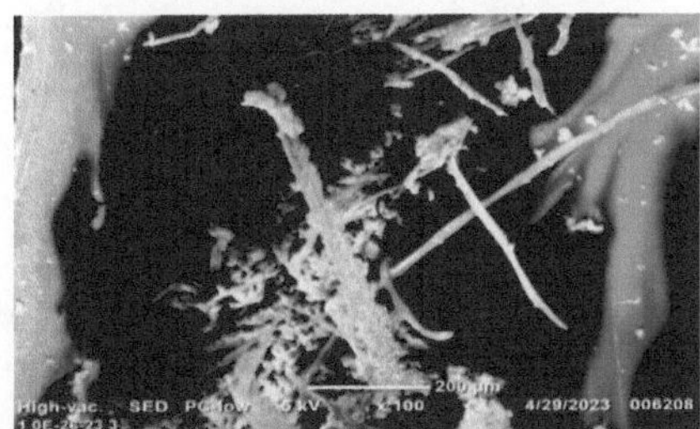

Fig. 9.2 SEM morphological analysis on polymer composite with magnification of 200 micro meterx 100 surface area

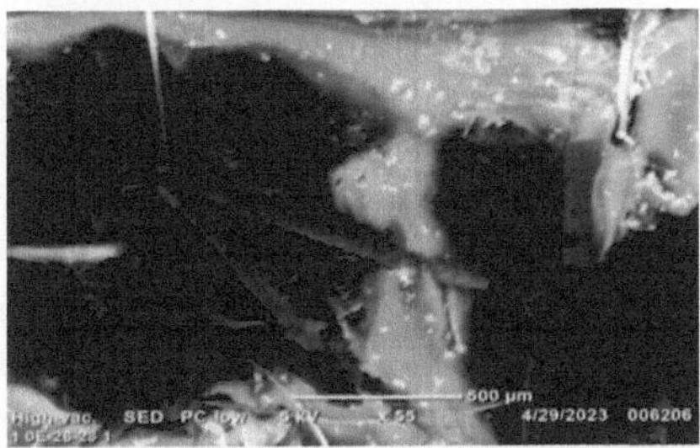

Fig. 9.3 SEM morphological analysis on Polymer Composite with magnification of 500 micrometer x55 surface area

It is evident from the SEM results that the Si_3N_4 and polyester resin have apowerful bond. When compared to other Si_3N_4 compositions, polyester resin's mechanical properties are improved by the addition of 1% and 5% weight percent Si_3N_4. The stronger interfacial transition zone, the inter facial bonding was found to be the reasons for the increase in strength, according to SEM morphological analysis. When compared to another specimen, the increase in Si_3N_4 weight percentage of 1 & 5 weight percent shows abetter failure mechanism, according to SEM morphological analysis. The addition of Si_3N_4 makes composite materials stiffer and harder, comparatively increased when compared to pure polyester, according to hardness results.

3.2 Tensile Test

The tensile strength of the samples was assessed using the Universal Testing Machine (UTM). Because there is more 'Ca' in the particles of Chicken Feather and Pineapple leaf fibers with Si_3N_4 reinforced-based polymers have excellent results for tensile strength. Six specimens were tested shown in Table 9.1

Note:

CP - Chopped Chicken feather + Chopped Pineapple fibre + 0%Si_3N_4

C1 - Chopped Chicken feather + Chopped Pineapple fibre + 1%Si_3N_4

C5 - Chopped Chicken feather + Chopped Pineapple fibre + 5%Si_3N_4

Table 9.2 Tensiletests with and without Si_3N_4

S. No	Sample ID	Tensile Strength (N/mm²)
1	CP-1	20.45
2	CP-2	23.33
3	C1-1	26.62
4	C1-2	29.10
5	C5-1	17.11
6	C5-2	18.19

Sample CP-1

Graph Type : Stress Vs. Strain

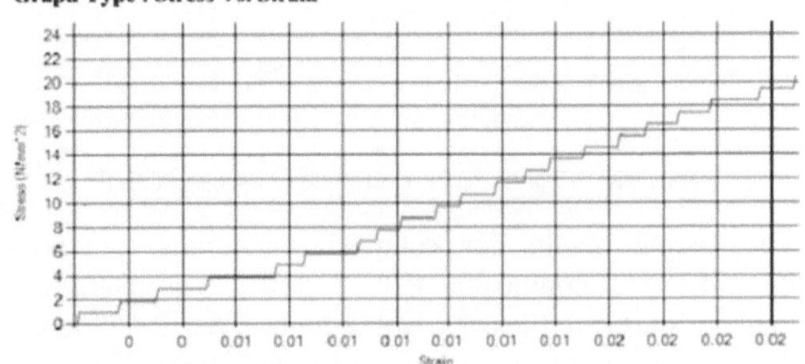

Fig. 9.4 Plain composite sample CP-1

Sample CP-2

Graph Type : Stress Vs. Strain

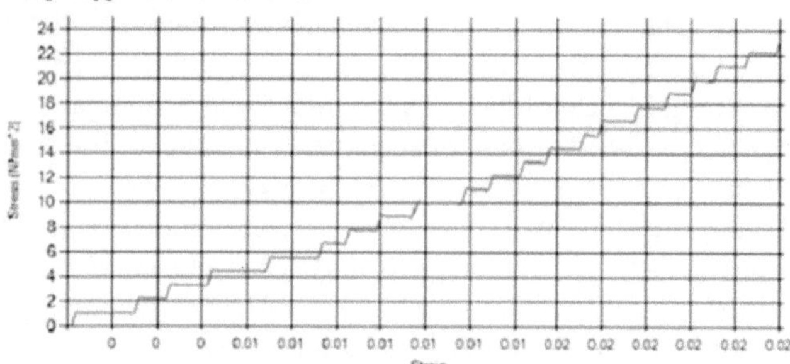

Fig. 9.5 Plain composite sample CP-2

Sample C1-1

Graph Type : Stress Vs. Strain

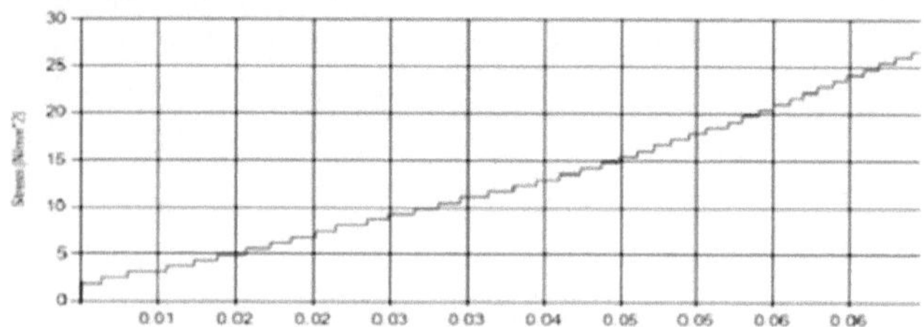

Fig. 9.6 1% of Si_3N_4 composite sample C1-1

Sample C1-2

Graph Type : Stress Vs. Strain

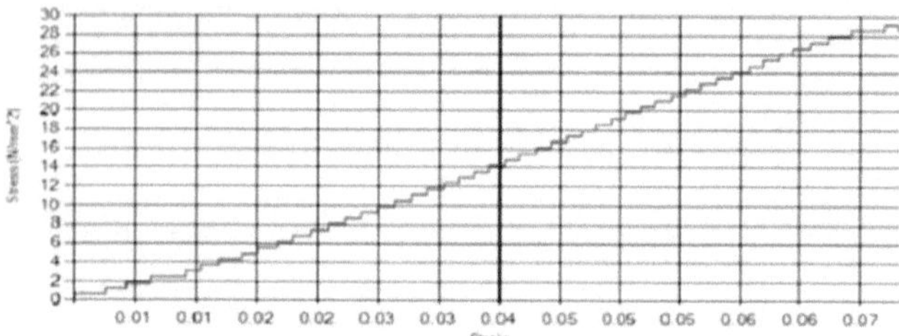

Fig. 9.7 1% of Si_3N_4 composite sample C1-2

Sample C5-1

Graph Type : Stress Vs. Strain

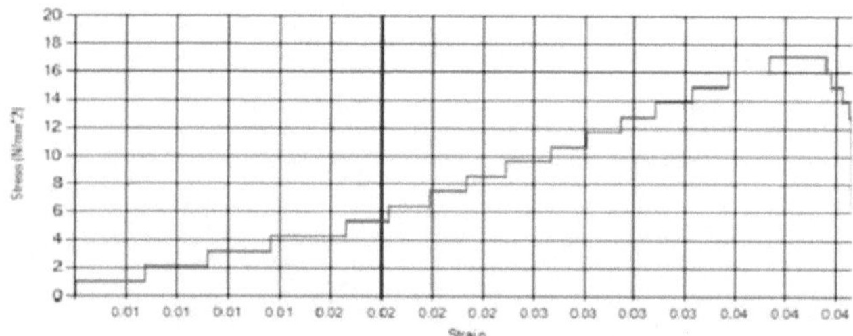

Fig. 9.8 5% of Si_3N_4 composite sample C5-1

The testing results for composite materials (composite and hybrid composites CP, C1, and C5) at maximum load are shown in Figs. 9.4–9.8. Figures 9.6 and 9.7 depict the graphs that were created during the tensile testing of two samples made from a composite of chopped chicken feather and chopped pineapple leaf fiber. The graphs from the tensile test of the composite made of chopped chicken feather and chopped pineapple leaf fiber with 1% Si_3N_4 are shown in Figs. 9.6 and 9.7. Tensile stress was 28N/mm^2 and tensile strain (extension) was 8%, as can be shown. According to Figs. 9.8, the hybrid composite comprising chopped chicken feather, chopped pineapple leaf fiber, and 5% Si_3N_4 has the lowest void content of any composite or hybrid composite. The tensile stress was 18N/mm2, and the tensile strain (extension) was 3.5%, as can be shown. This was caused by the high matrix-to-Fiber bonding strength.

3.3 Impact Strength Analysis

Impact testing was done to see how well the materials could absorb energy from abrupt loads. According to ASTM D256, an impact tester was adopted to conduct the izod impact test in the current inquiry. Six samples were examined for the impact test, as indicated in Table 9.3, and the average value was used for discussion.

Table 9.3 Impact test results with and without Si_3N_4

S.No	Sample ID	Impact Energy (J)
1	CP-1	2.5
2	CP-2	2.7
3	C1-1	2.9
4	C1-2	3.0
5	C5-1	3.1
6	C5-2	3.2

Samples' impact strength for composite as well as hybrid composites was being evaluated. Table 9.3 displays the impact strength of composite and also for hybrid composite materials. The hybrid composite of chopped chicken feather and chopped pineapple leaf fiber with 5% Si_3N_4 (C5) demonstrated an average impact strength of 2.65 J, while the composite of chopped chicken feather and chopped pineapple leaf fiber with 1% Si_3N_4 (C1) demonstrated an average impact strength of 2.95 J. Comparing composite CP to the other composites, its impact strength was the lowest (2.65 J). The hybrid composite's impact strength and tensile strength are affected by the addition of Si_3N_4 to the matrix. Both of the hybrid composite C5's strengths are increased by the Si_3N_4. The matrix's weakness is the cause. The main reason for the higher impact strength was improved bonding between the fibre and matrix. The morphological analysis discussed previously in the book for the samples might offer an association between the impact and tensile strength data.

3.4 Hardness Test

To determine the materials' hardness, a hardness test was performed. The hardness test was conducted in accordance with ASTM D2240 (Shore D) standards. Hardness test were carried out and shown in Table 9.4.

Table 9.4 Hardness test results with and without Si_3N_4

S.No	Sample ID	1	2	3	4	5	Mean
1	CP	86	85	83	83	86	**84.6**
2	C1	84	88	87	88	88	**87.0**
3	C5	87	88	85	86	86	**86.4**

Sample hardness for composites and hybrid composites was assessed. Table 9.4 displays the hardness of composite and hybrid composite materials. The composite of Chopped Chicken feather, Chopped Pineapple leaf fiber with 1% Si_3N_4 (C1) has the highest mean hardness value of 87, whereas Chopped Chicken feather with Chopped Pineapple leaf fiber (CP) showed lowest average hardness value of 84.6 and Chicken feather, Chopped Pineapple leaf fiber with 5% Si_3N_4 (C5) showed average hardness value of 86.4. The amount of Si_3N_4 in the matrix affects the hardness values of the hybrid composite.

3.5 Flexural Strength Analysis

The Universal testing apparatus was used to calculate flexural strength and flexural modulus. The flexural test was carried out in accordance with ASTM D7290 standard. The test speed was held constant between 1.3 and 1.5 mm/min. Six samples were evaluated, and the average result was used for discussion (see Table 9.5).

Table 9.5 Flexural test results with and without Si_3N_4

S.No	Sample ID	Flexural Strength (N/mm^2)
1	CP-1	61.84
2	CP-2	58.36
3	C1-1	84.74
4	C1-2	75.09
5	C5-1	58.08
6	C5-2	57.61

The maximum load and flexural extension at maximum load for the composite material CP were 105 N and 3.7 mm, respectively. The hybrid composite C5's maximum load and flexural extension at maximum load were 200 N and 5.6 mm, respectively, whereas the sample failed due to brittle fracture and total extension of over 3.7 mm. The composite C5 sample failed due to ductile fracture. At the fracture point, the total extension during the test was 13.1 mm. The maximum load and flexure extension at the maximum load for composite material C1 were 105 N and 3.7 mm, respectively. While the flexural strength of these composite materials was nearly same to that of composite CP, their strength was higher. According to Table 9.5, hybrid composite C1 had the maximum flexural strength (84.74 MPa), whereas composite C5 had the lowest (57.61 MPa). The behavior of the C1 hybrid composite is potentially attributed to the sandwich structure of the composite because of the high strength of the chicken feather and pineapple leaf fiber.

3.6 Flammability Test

The ASTM D635 standard was followed when conducting the Flammability test, the proportionate rate at which self-supporting polymers burn. According to Table 9.6, this test is mostly used for material comparisons, production control, and quality control.

Table 9.6 Flammability test with and without Si_3N_4

S.No	Sample Id	Propagation Speed (mm/min)	Falling Drops	Cotton lightens	UL-94 Rating
1	CP-1	23.1	NO	NO	V-1
2	CP-2	23.7	NO	NO	V-1
3	C1-1	24.2	NO	NO	V-1
4	C1-2	24.1	NO	NO	V-1
5	C5-1	24.8	NO	NO	V-1
6	C5-2	24.5	NO	NO	V-1

Where,

UL –94 Rating:

V-0–Good Flammable Resistance.

V-1–Medium Flammable Resistance.

V-2–Poor Flammable Resistance

It is noticed from Table 9.6 that all composite compositions have Medium Flammable Resistance and CP has comparatively less Propagation Speed over other compositions.

3.7 Wear Analysis

The material to be examined was fastened to a rotating disc and pressed against it with a spherical pin. Both dry and wet circumstances could be used to conduct the test. This inquiry used no lubricants, and the test was run in dry conditions. The applied force ranged from 20 to 80 N, the sliding speed was set at 150 mm/s, and the sliding distance was 1000 m for the wear test conditions. Table 9.7 displays the results of tests on six specimens.

Test Parameters

S. No	Sample ID	Test Speed (RPM)	Pin Dia .(mm)	Track Radius (mm)	Normal Load Applied (N)	Test Duration (Mins)
1	For all samples	500	2	30	20	15

Table 9.7 Wear test results with and without Si_3N_4

S. No.	Sample ID	Mass before test (g)	Mass after Test (g)	Wear loss (g)	Co-efficient of Friction (Co.eff.)
1	**CP-1**	1.4	1.3	0.01	**0.61**
2	**CP-2**	1.5	1.4	0.01	**0.57**
3	**C1-1**	1.4	1.3	0.01	**0.63**
4	**C1-2**	1.5	1.4	0.01	**0.62**
5	**C5-1**	1.4	1.3	0.01	**0.63**
6	**C5-2**	1.5	1.4	0.01	**0.62**

By assessing the elastic arm's deflection all through the test, the friction coefficient is discovered. The amount of material lost in the course of the test is assisted to compute the wear coefficient used for the pin and disk materials. The wear rate rises as the friction coefficient increases. From the values of Table 9.7, it was observed that all composite compositions have similar amount of wear loss and almost similar amount of average Co-efficient of Friction. This shows similar wear resistance of all composite compositions and it's comparatively good with other wear behaviour hybrid composites.

4. Conclusion

A Si_3N_4-reinforced polymer composite material made of chicken feather fiber (CFF) and pineapple leaf fiber (PALF) underwent mechanical evaluation. The outcomes show that the three composites made from chicken feather fiber (CFF), polyester resin, and pineapple leaf fiber (PALF) without Si3N4 (filler), 1% Si_3N_4 and 5% Si_3N_4. Then the three composite specimen such as plain specimen without using Si_3N_4, 1% of Si_3N_4 were used specimen and 5% of Si_3N_4 were analyzed. These key deductions were made on the basis of the observation.

Composite of Chicken feather fiber (CFF) and Pineapple Leaf Fiber (PALF) with 1% Si_3N_4 displayed a superior hardness and flexural strengths compared to that of the composite material prepared without Si_3N_4 owing to the high bonding strength of the hybrid composites. CP has comparatively less Propagation Speed over other compositions with Medium Flammable Resistance. Due to the reinforcement's increased ability to absorb impact energy, the composite material made with chicken feather fiber (CFF) and pineapple leaf fiber (PALF) with 1% Si_3N_4 had a higher impact energy value than the composite material made without these fibers. Further, the wear test displayed that the shows similar wear resistance of all composite compositions and it's comparatively good with other wear behaviour hybrid composites. At the end of the this research work, it can be concluded as the Composite of Chicken feather fiber (CFF) and Pineapple Leaf Fiber (PALF) with 1% Si_3N_4 has showed the enhanced mechanical properties compared with other prepared compositions, which can be used for many applications.

References

1. A. R. Mohamed, S. M. Sapuan, and A. Khalina, "Mechanical and Thermal Properties of Josapine Pineapple Leaf Fiber (PALF) and PALF-reinforced Vinyl Ester Composites", Fibers and Polymers (2014), Vol.15, No.5, 1035-1041, DOI 10.1007/s12221-014-1035-9. [PF18]
2. A.L.Leao, Sao Paulo State University (UNESP), Brazil, B.M.Cherian and S.Narine, Trent University, Canada, S.F. Souza and M.Sain, University of Toronto, Canada and S. Thomas, Mahatma Gandhi University, India, "The use of pineapple leaf fi bers (PALFs) as reinforcements in composites", DOI : 10.1533/9781782421276.2.211, at 2015 Elsevier Ltd.,211-235. [PF2]
3. Aadars. M.S, " Mechanical Behaviour Of E-Glass And Chicken Feather Composite", Research And Reviews On Experimental And Applied Mechanics, Volume 4 Issue 1, HBRP Publication Page 1-7, 2021. [CFF7]
4. Abir Saha, Santosh Kumar, Avinash Kumar, " Influence of pineapple leaf particulate on mechanical, thermal and biodegradation characteristics of pineapple leaf fiber reinforced polymer composite", Journal of Polymer Research (2021) 28: 66, https://doi.org/10.1007/s10965-021-02435-y. [PF20].
5. Aswan Munang, Achmad Zaki Yamani, "Analysis of Composite Mechanical Strength from Waste Chicken Feather and Sawdust", JEMMME (Journal of Energy, Mechanical, Material, and Manufacturing Engineering), Vol.6, No. 3, 2021, Pp :155–160, ISSN 2541–6332 | E-ISSN 2548–4281. [CFF1]
6. Carolina Castillo-Castillo, Beatriz Adriana Salazar-Cruz, José Luis Rivera-Armenta, María Yolanda Chávez-Cinco, María Leonor Méndez-Hernández, Ivan Alziri Estrada-Moreno And Tania Ernestina Lara Ceniceros, "Evaluation of Elastomeric Composites Reinforced with Chicken Feathers", S. S. Sidhu Et Al. (Eds.), Futuristic Composites, Materials Horizons: From Nature To Nanomaterial, Pp :297–318, © Springer Nature Singapore Pte Ltd. 2018, Https://Doi.Org/10.1007/978-981-13-2417-8_15. [CFF2]
7. Chieu D. Tran, FranjaProsenc, MladenFranko and Gerald Benzi, "Synthesis, Structure and Antimicrobial Property of Green Composites from Cellulose, Wool, Hair and Chicken Feather", PII: S0144-8617(16)30684-1,DOI: Http://Dx-.Doi.Org/Doi:10.1016/J.Carbpol.2016.06.021,Reference: CARP 11206,2016. [CFF13]
8. Eric WorlawoeGaba, Bernard O. Asimeng, Elsie Effah Kaufmann, Solomon Kingsley Katu, E. Johan Foster and Elvis K. Tiburu, " Mechanical and Structural Characterization of Pineapple Leaf Fiber", Fibers 2021, 9, 51., https://doi.org/10.3390/fib9080051, https://www.mdpi.com/journal/fibersFibers 2021, 9, 51. [PF12]

9. GaganBansal, V. K. Singh, P. C. Gope, Tushar Gupta, "Application And Properties of Chicken Feather Fiber (CFF) A Livestock Waste In Composite Material Development", Journal of Graphic Era University Vol. 5, Issue 1, 16–24, 2017, ISSN: 0975–1416 (Print), 2456–4281 (Online). [CFF17]
10. Gagan Bansal, V.K. Singh, AkarshVerma, Pratibha Negi, ShwetaRastogi, "Review Paper: Properties and Application of Chicken Feather Fiber (CFF) In Composites", International Journal of Research In Mechanical Engineering Volume 5, Issue 1, January-February, 2017, Pp. 38–46 ISSN Online: 2347–5188. [CFF10]
11. J.K. Odusote, A. T. Oyewo, "Mechanical Properties of Pineapple Leaf Fiber Reinforced Polymer Composites for Application as a Prosthetic Socket", Journal of Engineering and Technology, ISSN: 2180–3811 Vol. 7 No. 1 January – June 2016, 125–139. [PF19]
12. Jayamol George, S. S. Bhacawan, N. Prabhakaran, and Sabu Thomas, "Short Pineapple-Leaf-Fiber-Reinforced Low-Density Polyethylene Composites", Journal of Applied Polymer Science, Vol. 57, M3–854 (1995) @ 1995 John Wiley & Sons, Inc. [PF13]
13. K B Jagadeesh gouda, P Ravinder Reddy, K Ishwara prasad, " Experimental Study of Behaviour of Poultry Feather Fiber - A Reinforcing Material for Composites", IJRET: International Journal of Research In Engineering and Technology, Volume: 03 Issue: 02 | Feb-2014, Pp:362–371. [CFF5]
14. K. Selvakumar and M. Omkumar, " Fabrication and Characterization of Animal Fiber Reinforced Polymer Composites", Malaya Journal of Matematik, Https://Doi. Org/10.26637/MJM0S20/1171,Vol. S, No. 2, 4538-4542, 2020. [CFF8]
15. L. Uma Devi, S. S. Bhagawan, Sabu Thomas, "Mechanical Properties of Pineapple Leaf Fiber-Reinforced Polyester Composites", At 1997 John Wiley & Sons, Inc. J Appl. PolymSci 64: 1739–1748, 1997. [PF6]
16. L. Uma Devi, S.S. Bhagawan,S. Thomas, " Dynamic Mechanical Analysis of Pineapple Leaf/Glass Hybrid Fiber Reinforced Polyester Composites", POLYM. COMPOS., 31:956–965, 2010. At 2009 Society of Plastics Engineers, DOI 10.1002/pc.20880. [PF7]
17. L. Uma Devi,KuruvilaJoseph,K. C. ManikandanNair,Sabu Thomas, "Ageing Studies of Pineapple Leaf Fiber–Reinforced Polyester Composites", Journal of Applied Polymer Science, Vol. 94, 503–510 (2004), At 2004 Wiley Periodicals, Inc. [PF9]
18. L. Uma Devi, S.S. Bhagawan, S. Thomas, " Dynamic Mechanical Properties of Pineapple Leaf Fiber Polyester Composites", s. POLYM. COMPOS., 32:1741–1750, 2011. At 2011 Society of Plastics Engineers, DOI 10.1002/pc.21197. [PF8]

Note: All figures and tables were exclusively produced from our experimental work and were not sourced from any publications or online material.

Advances in Additive Manufacturing Technologies – Gurusamy Pathinettampadian (eds)
© 2024 Taylor & Francis Group, London, ISBN 978-1-032-90013-1

10 Mechanical Properties, Tribological MG Characteristics of Stir Cast Magnesium Metal Matrix Composites

P Gurusamy*, P. Subash
Department of Mechanical Engineering,
Chennai Institute of Technology, Chennai, Tamilnadu, India

A. Bovas Herbert Bejaxhin
Department of Mechanical Engineering, Saveetha School of Engineering,
Saveetha Institute of Medical and Technical Sciences, Saveetha University, Chennai, India

N. Ramanan
Department of Mechanical Engineering,
Sri Jayaram Institute of Engineering and Technology Gummdipundi, Chennai

ABSTRACT: This study examines the mechanical and tribological characteristics of Magnesium Matrix Composites (MMCs) produced through diverse methodologies. Hybrid composites particularly stand out for their exceptional performance and adaptability. Due to its advantageous attributes, including a high strength-to-weight ratio, resistance to erosion and wear, and cost-effective manufacturing, MMCs find applications in diverse sectors such as steelmaking, structural components, marine, aerospace, military, and automotive industries. Their thermal stability and high specific strength make these MMCs desirable. To improve mechanical and tribological qualities, these innovative engineering materials are reinforced with particulate magnesium, silicon carbide, and TiO_2. MMCs' mechanical and tribological properties are reviewed in this work, along with the results of different reinforcement concentrations. Silicon carbide, a lubricant, enhances magnesium composites' elastic modulus, electric Mg conductivity, tensile strength, and thermal electric Mg conductivity. Addition of magnesium oxide improves Mg tribology. Over Mgl mechanic Mg characteristics increased with titanium oxide addition. Mg composites are also affected by other reinforcements. Other writers have testified about better stir casting procedures.

KEYWORDS: Magnesium met Mg matrix composite, reinforcement, Silicon carbide, Tribological Mg behavior, Mechanic Mg properties, Wears

1. INTRODUCTION

Metal Matrix Composites (MMCs) are meticulously engineered combinations of magnesium (Matrix) and hard particles/clay (Reinforcement) designed to attain customized properties. These composites find application in various sectors such as space exploration, commercial transportation, electronic substrates, bicycles, automobiles, golf clubs, and diverse industrial applications. The aerospace and automotive industries' growing need for lightweight

*Corresponding author: gurusamyp@citchennai.net

DOI: 10.1201/9781003545774-10

materials possessing high specific strength has spurred the advancement and application of Magnesium Matrix Composites (MMCs), particularly those based on magnesium alloy/SiC composites. Metal Matrix Composites (MMCs) are progressively taking the place of traditional lightweight metal alloys like Magnesium alloy in critical mechanical applications where essential criteria include strength, low mass, and energy efficiency. By integrating various types of reinforcement, such as continuous, discontinuous, or short fibers, MMC properties, encompassing electrical, mechanical, and chemical attributes, can be customized. MMC materials are gaining traction in structural applications owing to their outstanding mechanical properties, superior wear resistance, and low thermal expansion. Polymer Metal matrix composites also exhibit promise for structural applications, offering isotropic material properties, cost-effectiveness, and the feasibility of being shaped using conventional metal forming processes.

The behavior of heterogeneous materials, exemplified by particulate-reinforced Metal Matrix Composites, has paved the way for the creation of high-strength and wear-resistant materials through the incorporation of hard ceramic particles and solid lubricants in the Metal Matrix. The inclusion of ceramic reinforcements like SiC, Mg2O3, TiC, B4C, and ZrO2 enhances hardness and thermal shock resistance. Hybrid Metal Matrix Composites (HMMCs) represent a second-generation composite where multiple types, shapes, and sizes of reinforcements are employed to achieve superior properties. These composites surpass their single-reinforced counterparts by amalgamating the advantages of their constituent reinforcements. The study reveals that SiC MMC shows a higher wear rate when subjected to SiC abrasives compared to MgF SiC MMC, primarily due to the wear of the magnesium matrix. However, this trend is reversed when exposed to diamond abrasives, as irregularly shaped composite particles are extracted. The yield strength and stiffness experience an augmentation with the volume fraction of SiC particles, albeit accompanied by a corresponding reduction in elongation. Various manufacturing techniques, including squeeze casting and powder metallurgy, are employed for the fabrication of Mg/SiC composite materials, offering versatile applications across diverse industries. The properties of these composites depend on factors such as reinforcement homogeneity, processing temperature, and mixing duration.

In summary, Metal Matrix Composites (MMCs) are materials reinforced with other metals, ceramics, or natural compounds, offering improved properties such as strength, hardness, and conductivity. Magnesium MMCs, reinforced with materials like Silicon Carbide (SiC) and Magnesium Oxide (Mg_2O_3), are widely used in aerospace, aviation, automotive, and other industries. The distribution of particles plays a crucial role in the properties of Mg MMCs, and filaments, such as Zircon, are commonly employed as reinforcements to enhance the desired properties of the matrix constituent.

2. Results and Discussion

Stir casting method can be effectively used to manufacturing Metal matrix composite with favorite properties. Fortifying Magnesium or its compound with the hard creative particulates like B_4C, TiB_2, SiC and so on increment the Mechanical and tribological performance of Metal matrix composites to a great extent because of the strong interfacing Mg relationship between the reinforcement and the Mg matrix. Expansion of nature Mg support like Rice husk ash (RHA), Coconut ash remains, fly ash and so forth to the Magnesium or its combination has additionally demonstrated an apparent increment in Mechanical alongside tribological behavior of the Mg Metal matrix composite. These MMCs don't have any miniaturized scMge voids which brings about incredible tractable and hardness properties of the composite. Reinforcing of Magnesium mixtures with Mg nanoparticles shapes the elasticity and hardness together with elasticity. Establishment Mg matrix with TiB_2 or SiC enhances the tensile and hardness conduct up to certain wt. % of SiC or TiB_2 development and from that point a lot of decrement is seen in rigidity and hardness in view of group arrangement or accumulation of these hard clay particles in Magnesium matrix and which prompts to porosity. Though dissimilar collecting procedures like mix casting, press throwing and powder Metallurgy are employed for the manufacture of different Mg Metal matrix composite however at the same time blend throwing strategy is effectively utilized be reason for its more extensive availability and furthermore it is generally reasonable than different strategies. The expansion of graphite as support has likewise established a noteworthy increment in rigidity yet reduce in locMges exposed that with reduction in coefficient of grating, there is addition in the wear rate which upgrades the machining properties. Excessive graphite expansion may prompt to remove Mg from the liquid soften of Mg matrix.

The constituency of nature Mg strengthening with Magnesium or its combination is not Mg around investigated and extraordinarily controlled work has been done in this field. Nonetheless, it a few outcomes demonstrated a significant increment in Mechanical and addition of Mg tribological behavior. Among these lines, more investigation are required in this field for further development of MMMCs. Facilitate improvement is likewise required in improving the wettability and controlling the interface structure of the composite. Similarly, the carbon and expensive stone Metal

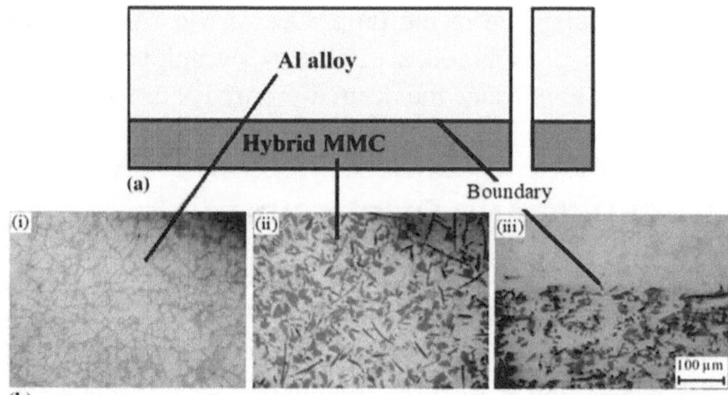

Fig. 10.1 Hybrid MMC fabrication and it boundary [33]

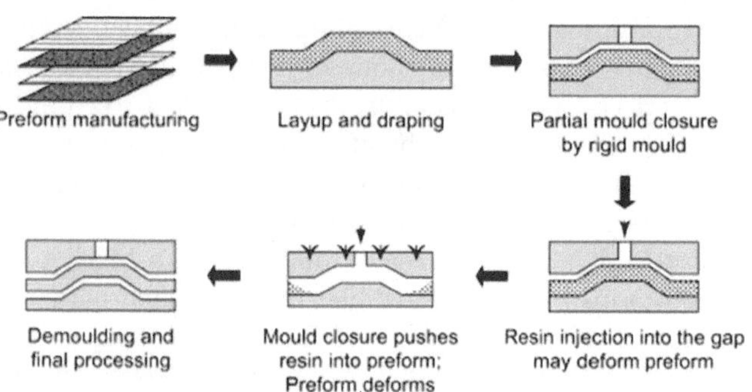

Fig. 10.2 Manufacturing of composite [34]

compo locations has not been investigated much which can be profitable in enhancing the Mechanical and tribological behavior of MMMCs. Hybrid ceramic reinforcement has extended the mechanic Mg properties much yet writing on tribological properties if there should a rise an occurrence of mixture support is constrained.

3. CONCLUSION

Finally, stir casting creates high-quality Metal Matrix Composites (MMCs). Magnesium or its derivatives supplemented with hard ceramic particles like B4C, TiB2, and SiC increase Metal Matrix Composites' mechanical and tribological performance due to strong bonding. Natural Mg reinforcements including Rice Husk Ash (RHA), Coconut Ash residues, and fly ash reduce tiny gaps and increase tensile and hardness in Mg Metal Matrix Composites. Mg nanoparticles increase magnesium alloy elasticity and toughness, however TiB2 or SiC beyond a certain wt. % may produce porosity. Mix casting, squeeze casting, and powder metallurgy are utilized, however mix casting is convenient and cheaper. Graphite reinforcement increases tensile strength but decreases hardness, suggesting a trade-off between coefficient of friction and wear rate, influencing machining quality.

More controlled tests are needed to understand the mechanical and tribological behavior of MMMCs with native Mg reinforcement and magnesium or its alloy. More research is needed to improve wettability and composite interface structure. Understudied carbon and diamond metal compositions may enhance MMMC mechanical and tribological performance. Although hybrid-supported MMCs' tribological qualities require more investigation, hybrid ceramic reinforcement improves mechanical properties. Metal Matrix Composites require further R&D.

REFERENCES

1. Abhay, A. Utpata, P. B. Pawar, 2018 "Modeling and Analysis of Al-SiC Spur Gear under Static Loading Condition", Elsevier.
2. Ajith Arul Daniela, 2018 "Dry Sliding Wear Behaviour of Aluminium 5059/SiC/Al2O3 Hybrid Metal Matrix Composites" Received: January 06, 2017; Revised: August 11, 2017; Accepted: August 24, 2018.
3. Ajmal Hussain S., Rajaneesh R., Hashim Nizam, Jithin K., 2020 "Experimental Analysis On Aluminum Alloy (6063) With Silicon Carbide: An Experimental Investigation", International Journal of Research in Engineering and Technology.
4. Anish A., 2019 "Characterization of Aluminium Matrix Reinforced With Tungsten Carbide And Aluminum oxide Hybrid Composite" 2nd International conference on Advances in Mechanical Engineering (ICAME 2019).
5. Antony Arul Prakash M. D., 2019 "Microstructural Analysis of Aluminium Hybrid Metal Matrix Composites Developed Using Stir Casting Process" International Journal of Advances in Engineering, 2019, 1(3), 333 – 339.
6. Ehsan Ghasali, 2018 "Production of Al-SiC-TiC Hybrid Composites Using Pure and 1056 Aluminum Powders Prepared Through Microwave and Conventional Heating Methods" Journal of Alloys and Compounds.
7. Kanchan A., 2019 "Tribological Investigation of Al7075/Tic/Al2O3 Hybrid Composite Material" International Research Journal of Engineering and Technology (IRJET).
8. Karuppasamy R., 2020 "Taguchi-GRA for Multi Criteria Optimization of Turning Parameters for Al/Bagasse Ash/Gr Hybrid Composite" Journal of Critical Reviews ISSN-2394–5125 Vol 7, Issue 9, 2020.
9. Nagesh D., Mr. Shashikanth G. S., 2021 "Optimization of Tribological Properties of Al6061, Boron And Graphite MMCs Using Stir casting Method.", IOSR Journal of Mechanical and Civil Engineering.
10. Paranthaman P., 2020 "Multi-Objective Optimization on Tribological Behaviour of Hybrid Al Mmc by Grey Relation Analysis" Journal of Critical Reviews ISSN-2394–5125 Vol 7, Issue 9, 2020.
11. Rajesh A. M., Mohamed Kaleemulla, 2019 "Material characterization of SiC and Al2O3–Reinforced Hybrid Aluminum Metal Matrix Composites on Wear Behavior", Advanced Composites Letters Volume 28: 1–10.
12. Rajesh S., 2019 "Influence and Wear Characteristics of TiC Particles in Al6061 Metal Matrix Composites"

International Journal of Applied Engineering Research ISSN 0973-4562 Volume 13, Number 9 (2019) pp. 6514–6517.

13. Rajesh Prabha N., 2017 "Effect of TiC and Al2O3 Reinforced Aluminium Metal Matrix Composites on Microstructure and Thermogravimetric Analysis" Vol. 10 I No. 3 I729 - 737 I July - September I 2017 ISSN: 0974-1496 I E-ISSN: 0976-0083 I CODEN: RJCABP.
14. Ram Kumar S., Ramakrishnan S., Risvak M.,Thauffeek S. and Yuvaraj T., 2021 "Experimental analysis and characterization of Mechanical, Physical properties of Aluminium (Al6061) Metal Matrix composite reinforced with SiC and Al2O3 using Stir casting".
15. Senthil Murugan S., 2017 "Development of Hybrid Composite for Automobile Application and its Structural Stability Analysis Using ANSYS" International Journal of Modern Studies in Mechanical Engineering (IJMSME) Volume 3, Issue 1, 2017, PP 23–34.
16. Yashpala, Sumankant , C. S. Jawalkar, 2019 "Fabrication of Aluminium Metal Matrix Composites with Particulate Reinforcement: A Review", 5th International Conference of Materials Processing and Characterization (ICMPC 2019).
17. MM Natarajan, B Chinnasamy, BHB Alphonse, Investigation of Machining Parameters in Thin-Walled Plate Milling Using a Fixture with Cylindrical Support Heads, Strojniški vestnik-Journal of Mechanical Engineering 68 (12), 746–756.
18. D Velmurugan, J Jayakumar, A Bovas Herbert Bejaxhin, JB Raj, Experimental Investigation on the Mechanical Properties of Hybrid Composites Made with Banyan and Peepal Fibers, Journal of Natural Fibers 19 (16), 14183–14194.
19. PV Narashima Rao, P Periyasamy, A Bovas Herbert Bejaxhin, Fabrication and Analysis of the HLM Method of Layered Polymer Bumper with the Fracture Surface Micrographs, Advances in Materials Science and Engineering 2022.
20. T Prabaharan, P Periyaswamy, V Mugendiran, AB Herbert Bejaxhin, Measuring Deformation of Deep Drawing of Various Alloys by Image Processing using Matlab. Journal of Mines, Metals & Fuels 70.
21. Comparative Testing of Tensile, Flexural and Impact Analysis on Coated and Uncoated Kenaf Fiber Reinforced Composite. T Prabaharan, AB Herbert Bejaxhin, SK Sankaran, N Ramanan, J DS, Journal of Mines, Metals & Fuels 70
22. ZnO Polymer Nano Composites Synthesis, Characterization, and Thermo-Mechanical Property Comparison. JC Mario, YC Shaji, A Sinthuja, AB Herbert Bejaxhin, N Ramanan, Journal of Mines, Metals & Fuels 70
23. T Prabaharan, AB Herbert Bejaxhin, P Periyaswamy, N Ramanan, Lower Wishbone Modeling and Analysis for Commercial Vehicle Independent Suspension System. Journal of Mines, Metals & Fuels 70.
24. T Prabaharan, MBS Reddy, H Bejaxhin, A Bovas Herbert Bejaxhin, Strain-cum-Deformation Analysis of Friction Stir Welded AA5052 and AA6061 Samples with Microstructural Analysis., Journal of Mines, Metals & Fuels 70.
25. B Venkataraman, ABH Bejaxhin, R Saravanan, A Comparative Analysis on Surface Roughness of Plain Epoxy Composite with Reinforcement of 10 wt% of Pista Shell Particles Novel Composite using CNC Machining, Journal of Pharmaceutical Negative Results, 376–385.
26. M Sivasai, ABH Bejaxhin, R Saravanan, A Comparative Analysis on Material Removal Rate of Plain Epoxy Composite with Reinforcement of 15 wt% of Egg Shell Powder particles Novel composite using CNC Machining, Journal of Pharmaceutical Negative Results, 692–699.
27. B Venkataraman, ABH Bejaxhin, R Saravanan, A Comparative Analysis on MRR of Plain Epoxy Composite with Reinforcement of 10 wt% of Pista Shell Particles Novel Composite using CNC Machining, Journal of Pharmaceutical Negative Results, 367–375.
28. RD Babu, P Gurusamy, ABH Bejaxhin, P Chandramohan, Influences of WEDM constraints on tribological and micro structural depictions of SiC-Gr strengthened Al2219 composites, Tribology International 185, 108478.
29. G Mahesh, D Valavan, N Baskar, A Bovas Herbert Bejaxhin, Parameter Impacts of Martensitic Structure on Tensile Strength and Hardness of TIG Welded SS410 with characterized SEM Consequences, Tehnički vjesnik 30 (3), 750–759.
30. CB Priya, K Ramkumar, V Vijayan, ABH Bejaxhin, Wear Studies on Mg-5Sn-3Zn-1Mn-xSi Alloy and Parameters Optimization Using the Integrated RSM-GRGA Method, Silicon 15 (8), 3569–3579.
31. P Chandramohan, A SivaRangar, JJ Kingsly, ABH Bejaxhin, N Ramanan, Investigation of corrosion and wear behavior of Al–SiC composite, Materials Today: Proceedings.
32. B Venkataraman, ABH Bejaxhin, R Saravanan, Analyse the effect of 20wt% addition of Pista Shell Powder reinforcement to Epoxy Composite in the Material Removal Rate using CNC Machining and Compare with Plain Epoxy Composite, Journal of Survey in Fisheries Sciences 10 (1S), 3105–3115.
33. Iqbal, A.A., Arai, Y. & Araki, W. Influence of Residual Stress on the Fatigue Crack Growth Mechanism in the Al-Alloy/Hybrid MMC Bi-Material. J Fail. Anal. and Preven. 22, 1468–1477 (2022). https://doi.org/10.1007/s11668-022-01432-7.
34. A. McIlhagger, E. Archer, R. McIlhagger, Manufacturing processes for composite materials and components for aerospace applications, Polymer Composites in the Aerospace Industry, Woodhead Publishing, 2015, Pages 53–75, ISBN 9780857095237, https://doi.org/10.1016/B978-0-85709-523-7.00003-7.

Advances in Additive Manufacturing Technologies – Gurusamy Pathinettampadian (eds)
© 2024 Taylor & Francis Group, London, ISBN 978-1-032-90013-1

11 Mechanical Properties and Microstructure Analysis of PETG/CF-PETG Functionally Graded Materials

N. Murugan[1]
Department of Mechanical Engineering,
Sri Venkateshwara College of Engineering & Technology,
Thiruvallur, Tamil Nadu, India

V. Gopal[2]
Department of Mechanical Engineering,
KCG College of Technology, Chennai, Tamil Nadu, India

B. Elumalai[3]
Department of Mechanical Engineering,
Easwari Engineering College, Chennai, Tamil Nadu, India

T. Suresh[4]
Department of Mechanical Engineering,
New Prince ShriBhavani College of Engineering and Technology,
Chennai, Tamil Nadu, India

M. Gokul[5], **N. Vinisha**[6]
Department of Aeronautical Engineering,
Gojan School of Business and Technology, Chennai, Tamil Nadu, India

ABSTRACT: This study characterizes the mechanical and microstructural properties of materials made using FDM-based 3D printing. The material chosen for this study was FGM, which was produced using PETG and CF-PETG in an equal ratio and their characteristics are contrasted with those of the primary materials. The test specimens were made using the FDM technique. The examination focused on the materials' tensile, compression, and flexural properties. Microstructural evaluation has been conducted to study the fracture mechanism. The findings indicate that ductile mode of fractured was observed in all the three materials. Bucking was observed in FGM before failure when tensile load was applied. Tensile testing found that CF-PETG displayed better tensile strength of 48.3 MPa and elongation of 9.6 % followed by FGM and PETG. During compression testing 5 samples, viz. PETG, CF-PETG, FGM 1 (PETG external layer of CF-PETG), FGM 2 (CF-PETG external layer of PETG), and FGM 3 (CF-PETG on the top and bottom layers and PETG in the middle) were examined. The strongest material, CF-PETG, failed instantly while FGM 1 and FGM 2 had a considerable deformation range before failing.

KEYWORDS: 3D printing, FDM, PPETG, CF-PETG, and FGM

[1]muruganmechssn123@gmail.com, [2]gopal@kcgcollege.com, [3]elumalainathan@gmail.com, [4]suresh@newprinceshribhavani.com, [5]gogulraj331213@gmail.com, [6]vinishanirmalkumar@gmail.com

DOI: 10.1201/9781003545774-11

1. INTRODUCTION

Additive manufacturing is one of the latest technologies used in the manufacturing of components using rapid prototyping techniques. In this process, the printing of a component is done by layer-by-layer addition[1, 2]. AM procedure creates a 3D sample by deposition of the 2D layered-by-layered slices using G-CODE [3].One of the cutting-edge industrial applications is additive manufacturing, which includes commercially available processes including selective laser melting, selective laser sintering, stereolithography, laminated object manufacture, fused deposition modelling (FDM), Binder jetting, and more. Applications are in the fields of architecture, medicine, dentistry, aerospace, transportation, furniture, and jewellery. Low-performance corporate markets have been where 3D printing production has grown the fastest [4]. However, the composites sector gains the most from performance improvements that may be made to 3D printing expertise that can meet the needs of high-performance products. The most common 3D printing technologies, FDM is utilised to create even everyday objects [5, 6]. FDM uses the material extrusion technique to create plastic components with specific geometry. The additive process allows for layer-by-layer printing of objects while minimising waste despite working with complex geometries in comparison to the conventional subtractive method. [7]. The FDM method was employed to fabricate the test specimens Poly-Ethylene Terephthalate Glycol (PETG), carbon fibre Poly-Ethylene Terephthalate Glycol (CF-PETG) and functionally graded material (FGM) that were used in this work. One of the popular 3D printing filaments, PETG, has a high printability and great layer welding. PETG is a copolymer that has great chemical and impact resistance, is transparent, biocompatible, and recyclable [8]. With its great abrasion resistance and transparency, PETG has prospective uses in the biomedical industry, such as for bone tissue engineering scaffolds as well as dental aligners [9]. There are numerous high-volume commercial and consumer uses that use poly (ethylene terephthalate). Films, fibre, and food packaging like soda bottles are some examples [10]. Polymer scaffolds made using AM methods have poor mechanical characteristics, because they are softer and less strong. To get past these drawbacks, reinforcing of polymers was applied [11]. Carbon fibre is used to reinforce 3D printer filaments like PETG and PLA, creating a remarkably hard and unyielding material that is also relatively light in weight. Such compounds excel in structural applications needing to survive a wide range of end-use environments due to their excellent ductility, impact resistance, and higher strength. PETG's strength can be increased by 10 to 25% with the inclusion of carbon fibre. [1, 12]. Composite substances known as functionally graded materials (FGMs) show a gradual shift in constitution and characteristics over their volume. These materials are designed with particular functions or performance attributes that are suited to specific applications. Despite the fact that FGMs are often connected to metals and ceramics, new developments have expanded the idea to include polymers. Polymer composites also referred to as functionally graded materials of polymers (FGMPs), are made to have different compositions, molecular architectures, or reinforcements throughout their structure. Because FGMPs are graded, their properties shift smoothly, enabling them to effectively endure a range of loads, thermal gradients, or environmental condition. Hsuehet al. [13] explored the tensile strength and the fibre distribution of PLA, ABS, PETG, and Amphora by adding carbon-fibre, with the raster angle of the printing being 0°, 45°, ±45°, and 90°, respectively. The results showed that PETG had the highest tensile strength after adding carbon-fibre. Srinidhi et al, [12] studied the effect of various infill patterns on PETG and CFPETG components fabricated using FDM to assess the flexural strength, tensile strength, and hardness of the as-printed and annealed components. This study offers useful guidelines for the creation of functional components using CF-PETG, assuming that the PETG reinforcement is present. Ibrahim [14] developed a brand-new PETG/CF composite design, in a lab using 3D-printed polymeric meshes. The objective was to develop a simple and inexpensive material production process that can maintain high mechanical strength while also offering the necessary flexibility. The yield strength of the 3D-printed PETG/CF solid structural design outperformed the other conventional constructions by 23%, according to the findings of tensile tests. Rafel et al, [15] characterized the mechanical properties of materials created by 3D printing using fused filament fabrication (FFF, a technique similar to FDM®). Polylactic acid (PLA) and PLA reinforced with short carbon fibres in a weight percentage of (PLA+CF) are the materials selected. It was observed that the short carbon fibres in the PLA+CF microstructure in the FFF specimens remain strongly aligned with the direction of material deposition.The purpose of this study is to create and evaluate FGM made of PETG/CF-PETG utilising 3D printing. In this investigation, 3D printed materials including PETG, CF-PETG, and FGM are made, and their mechanical and microstructural characteristics are examined and contrasted.

2. EXPERIMENTAL PROCEDURE

2.1 Material Selection

The materials used in this study are PETG, CF-PETG and FGM consisting PETG and CF-PETG. To create CF-PETG, carbon fibre is combined with PETG, of volume percentage 80% PETG and 20% carbon fibre. Two materials, PETG

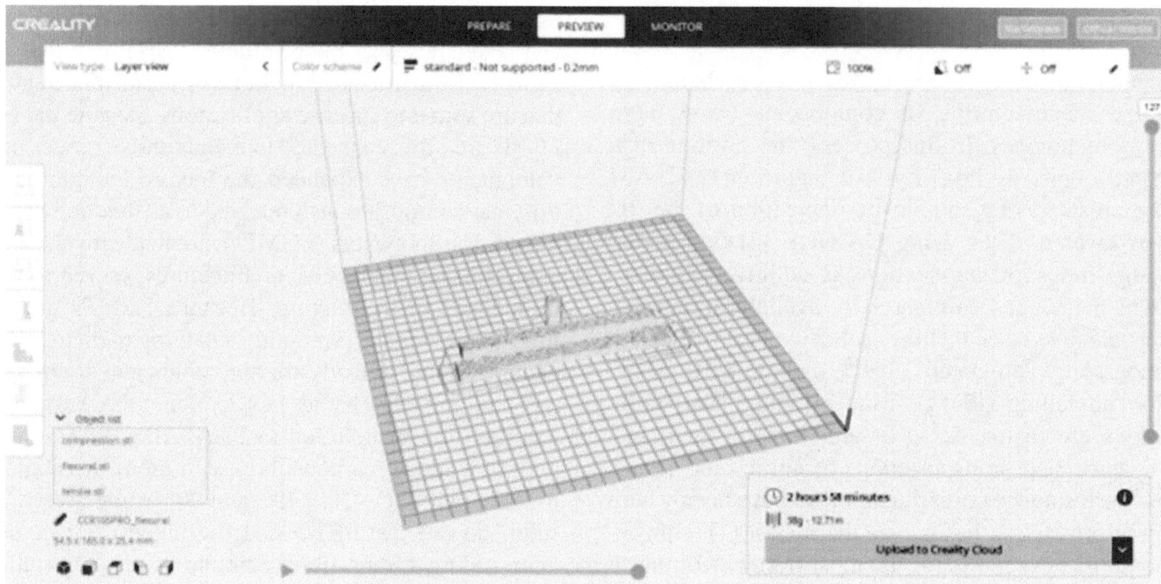

Fig. 11.1 Creality slicer software v 4.8

and CF-PETG, are used in equal amounts in order to produce FGM. The following Table 11.1 represents the print parameters for the PETG and CF-PETG.

Table 11.1 Physical properties of PETG and CF-PETG

Material	PETG	CF reinforced PETG
Density	1.24 g/cm3	1.53 g/cm3
Melting temperature	190-230 °C	190-230 °C
Glass transition temperature	56-64°C	60°C

2.2 Modelling and Fabrication of Samples

The first step of fabrication of test specimen is designing a model specimen in creo 2.0 software, then the model is fed into slicer software (Creality slicer software v 4.8) shown in Fig. 11.1 which slices the model into number of layers for the measurement of each layer includes thickness, length and diameter. It is next changed to STL format before being fed into FDM and printed using an AM machine (Creality CR Smart Pro 3-dimensional printer).

2.3 3D Printer Settings (Creality slicer v 4.8)

The Fig. 11.2 shows the schematic diagram of 3D used for this study. A standard desktop hand-made single-nozzle FDM tool was used, with a 0.8 mm nozzle. A platform with all the samples on it was heated to 60° C. PETG parts were printed at a temperature of 230 °C and speed 120 mm/s, respectively, to increase their tensile strength. Each sample was printed with a density of 100% infill and a layer thickness of 0.2 mm.

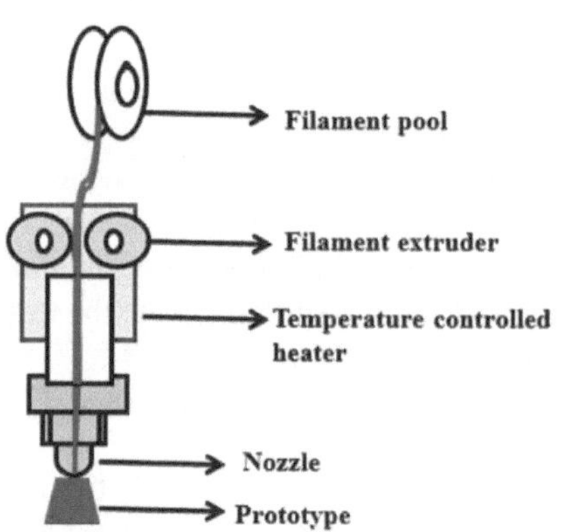

Fig. 11.2 Schematic diagram of 3D printer

2.4 Mechanical Testing of the Samples

The fabricated samples are allowed to analysis on the basis of mechanical test which includes tensile test, compression test and flexural test and microstructural analysis. The tensile test was done in Z010 Pro-line Tensile testing machine Pneumatic grip X Force K Load cell machine from ZWICK ROELL. The compression test done on the sample by using Z010 PROLINE UTM load cell-X force K Compression kit and Zwick line Z72.5KN Flexure kit X force HP load cell was used to flexural test. The tensile, flexural and compression test settings which includes preloads for tensile, flexure and compression are 0.1 MPa, respectively. Speed and tensile, flexure and compression modulus are 1 mm/min, 2 mm/min, 1.3 mm/min, respectively. The test speeds are 50mm/min, 10mm/min and 1.3 mm/min. As per the ASTM standard the dimension of the test samples were measured.

2.5 Microstructural Evaluation of the Samples

The microstructure of the PETG, CF-PETG and FGM materials are analysed by the application of optical microscope. The fractured surfaces are also examined.

3. Results and Discussions

3.1 Microstructural Evaluation

Figure 11.3(a), (b) and (c) represent the samples of PETG, CF-PETG and FGM before testing which is printed by FDM with solid pattern. This microstructure provides information on how materials are arranged and distributed.

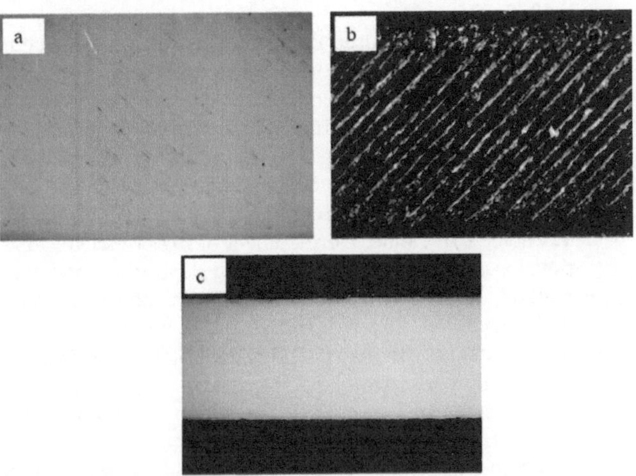

Fig. 11.3 Optical microstructure of 3D printed samples

3.1 Mechanical Evaluation

Tensile Properties

The tensile test specimen is showed in the Fig. 11.4. The Fig. 11.5 shows the stress strain graph of PETG, CF-PETG and FGM 3D printed materials. The graph demonstrates that CF-PETG has a tensile strength that's greater than FGM and PETG. This may be caused by the addition of carbon fibre reinforcements, which alter the substance's microscopic structure and encourage greater intermolecular interactions and chain entanglements. The material's resistance to deformation and rupture under stress conditions is strengthened by these adjustments. The addition of carbon fibres also enhances its load-bearing capacity of the material. The elongation of CF-PETG is lower as compared to PETG. When compared to CF-PETG, FGM's tensile strength was lower. This might be as a result of PETG and CF-PETG being present. Additionally, the elongation of FGM is inferior than PETG and higher than CFPETG. Whereas, PETG performs worse than FGM in terms of strength but better in terms of elongation. Figure 11.6 displays the bar chart of the three sample PETG, CF-PETG, and FGM materials' tensile strength and percentage of elongation.

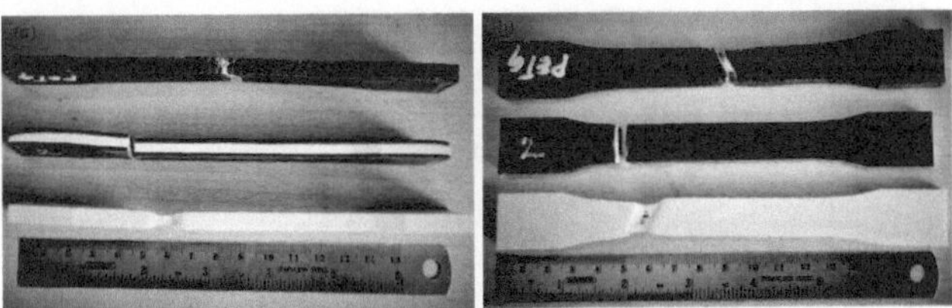

Fig. 11.4 Tensile test samples

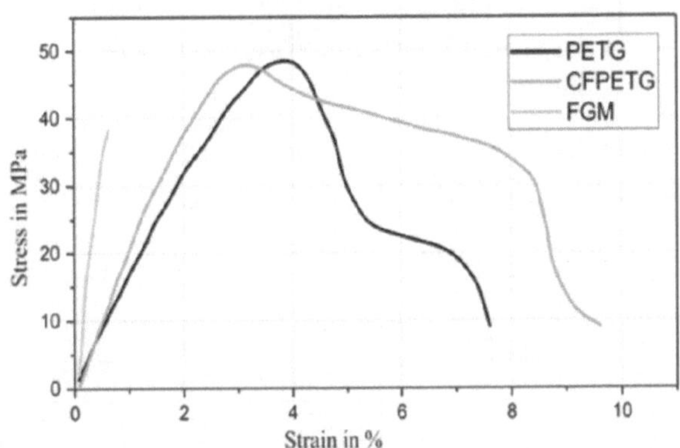

Fig. 11.5 Stress-strain diagram (PETG and FGM change)

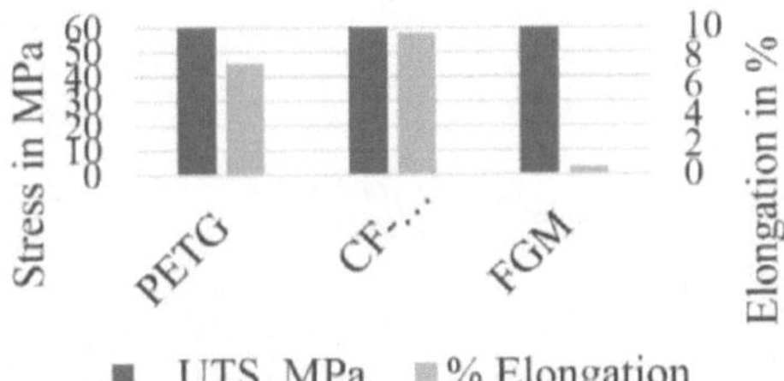

Fig. 11.6 Bar chart of tensile strength and percentage of elongation

Compression Properties

Five different samples have been tested in the compression test, as shown in the Fig. 11.7: PETG, CF-PETG, FGM 1 (PETG external layer of CF-PETG), FGM 2 (CF-PETG external layer of PETG), and FGM 3 (CF-PETG on the top and bottom layers and PETG in the middle). Table 11.2 represents compression results of all the 5 samples. The stress strain graph of compression test are shown in the Fig. 11.8. The results of the investigation demonstrated that CF-PETG demonstrated maximal compressive yielding compared to other samples but deformed right after the peak. FGM 1 and FGM 2 displayed a wide range of deformation before failure. The unique combination of CF-PETG and PETG features in the intermediary layer was credited with this improved robustness, which may have enhanced compressive durability due to carbon fiber in-PETG's. PETG demonstrated significant yielding with a drop in tensile stress prior to a protracted stable plastic flow.

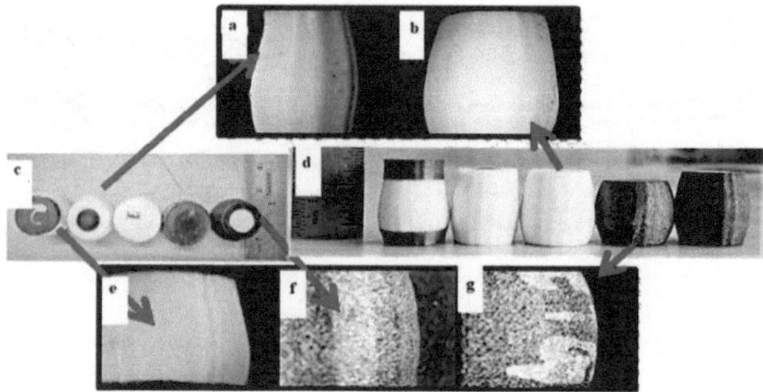

Fig. 11.7 Compression test specimens

Table 11.2 Compression results for different samples

Material Type	Max. Compression stress MPa	Initial height	Final height	Deformation %
PETG	48.8	25	16.7	40
CF-PETG	59.2	25	14.6	5.6
FGM 1	48.3	25	16.6	40
FGM 2	46.8	25	16.9	30
FGM 3	34.4	25	19.1	30

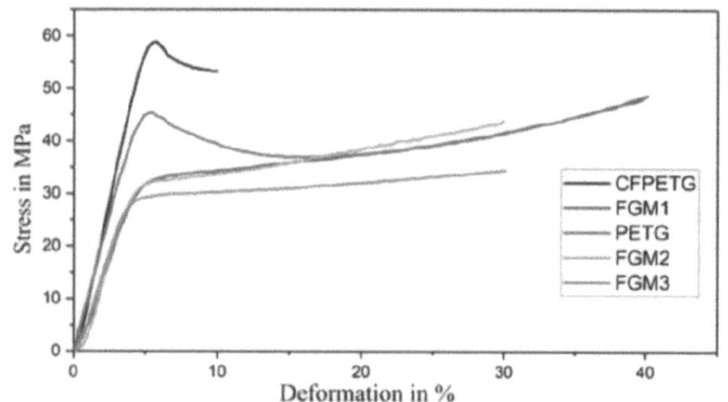

Fig. 11.8 Stress-strain graph for compression

4. CONCLUSION

In this study mechanical and microstructural properties of 3D printed materials such as PETG, CF-PETG and FGM are compared and results are as follows,

- According to the tensile strength analysis, CF-PETG has greater elongation and tensile strength than FGM and PETG. Carbon fiber components strengthen CF-PETG's intermolecular bonds and chain entanglements, making it more resistant to deformation and breakage under stress.
- The compression analysis revealed that, the specimens FGM 1 and FGM 2 demonstrated higher resistance to compressive force when compared to other samples. The combination of CF-PETG and PETG characteristics in the intermediary layer allowed for this enhanced firmness and compressive toughness.

REFERENCES

1. K.S.Kumar, R. Soundararajan, G. Shanthosh, P. Saravanakumar. And M.Ratteesh, 2021. Augmenting effect of infill density and annealing on mechanical properties of PETG and CFPETG composites fabricated by FDM. Materials Today: Proceedings, 45, pp.2186–2191.
2. G.Dolzyk, and S.Jung, 2019. Tensile and fatigue analysis of 3D-printed polyethylene terephthalate glycol. Journal of Failure Analysis and Prevention, 19, pp.511–518.
3. E. Soleyman, D. Rahmatabadi, M. Baniassadi, and M. Baghani, 2022. Effect of printing parameters on the shape transformation of 3D printed PETG. ISME Conference
4. M.Chapiro, 2016. Current achievements and future outlook for composites in 3D printing. Reinforced Plastics, 60(6), pp.372–375.
5. E.Palermo, 2013. Fused deposition modeling: most common 3d printing method. Live Science, 19.
6. M.T. Sepahi, H.Abusalma, V. Jovanovic and H. Eisazadeh, 2021. Mechanical properties of 3D-printed parts made of polyethylene terephthalate glycol. Journal of Materials Engineering and Performance, 30, pp.6851–6861.
7. J.Y. Tey, D. Teh, W.H. Yeo, K.P.Y. Shak, L.H. Saw, and T.S. Lee, 2019, June. Development of 3D printer for functionally graded material using fused deposition modelling method. In IOP Conference Series: Earth and Environmental Science (Vol. 268, No. 1, p. 012019). IOP Publishing.
8. S. Guessasma, S.Belhabib. and H. Nouri, 2019. Printability and tensile performance of 3D printed polyethylene terephthalate glycol using fused deposition modelling. Polymers, 11(7), p.1220.
9. P. Latko-Durałek, K. Dydek, and A. Boczkowska, 2019. Thermal, rheological and mechanical properties of PETG/PETG blends. Journal of Polymers and the Environment, 27, pp.2600–2606.
10. R.B. Dupaix, and M.C. Boyce, 2005. Finite strain behavior of poly (ethylene terephthalate)(PET) and poly (ethylene terephthalate)-glycol (PETG). Polymer, 46(13), pp.4827–4838.
11. N. Maqsood, and M.Rimašauskas, 2021. Characterization of carbon fiber reinforced PLA composites manufactured by fused deposition modeling. Composites Part C: Open Access, 4, p.100112.
12. M.S. Srinidhi, R.Soundararajan, K.S. Satishkumar, and S. Suresh, 2021.Enhancing the FDM infill pattern outcomes of mechanical behavior for as-built and annealed PETG and CFPETG composites parts. Materials Today: Proceedings, 45, pp.7208–7212.
13. M.H. Hsueh, C.J. Lai, S.H. Wang, Y.S. Zeng, C.H. Hsieh, C.Y. Pan and W.C.Huang, 2021.Effect of printing parameters on the thermal and mechanical properties of 3d-printed PLA and PETG, using fused deposition modeling. Polymers, 13(11), p.1758.
14. I.M. Alarifi, 2023. PETG/carbon fiber composites with different structures produced by 3D printing. Polymer Testing, 120, p.107949.
15. R.T.L. Ferreira, I.C. Amatte, T.A. Dutraand D. Bürger, 2017. Experimental characterization and micrography of 3D printed PLA and PLA reinforced with short carbon fibers. Composites Part B: Engineering, 124, pp.88–100.

Note: Every figure and table was created using the experimental work; none of them were taken from any publications or online.

Advances in Additive Manufacturing Technologies – Gurusamy Pathinettampadian (eds)
© 2024 Taylor & Francis Group, London, ISBN 978-1-032-90013-1

12 A Comprehensive Review on Mapping of Coastal Topography Using Drones

Lokesh M, Aditya Vishnu Vardhan Varma, Anjani Rakesh
Student, Department of Aeronautical Engineering,
Hindustan Institute of Technology and Science, Chennai, India

Saravanan P, Balaji G*, Sankaralingam L, Seralathan S
Faculty, Department of Aerospace Engineering,
Hindustan Institute of Technology and Science, Chennai, India

Santhosh Kumar G
Teaching Fellow, Dept. of Mech. Engg.,
University College of Engineering, Bharathidasan Institute of Technology Campus,
Anna University, Tiruchirappalli, Tamilnadu, India

Vijayanandh Raja
Assistant Professor, Department of Mechanical Engineering,
Sri Jayaram Institute of Engineering and Technology, Chennai, India

Ramanan N
Department of Aeronautical Engineering,
Kumaraguru College of Technology, Coimbatore, Tamil Nadu, India

Gurusamy P
Professor, Department of Mechanical Engineering,
Chennai Institute of Technology, Chennai, India

ABSTRACT: The impact of the water as well as other natural and man-made factors degrade coastal regions. This paper gives a detailed review on assessing topographical modifications in seashores and sand dunes using drones. It must be evaluated frequently after significant events and on a routine basis in order to develop models capable of forecasting the growth of these natural ecosystems. Some coastal locations experience flooding owing to sea level rise during high tides, which leads to road closures because of inundated roads and public inconveniences. The use of Unmanned Aerial Vehicle (UAV) technology for aerial photogrammetric mapping has gained popularity recently. The major benefits of this technology in photogrammetric applications may be categorised as its inexpensive cost and the ability to fly automatically at set waypoints. Using a photogrammetric approach, Accuracy was evaluated after low-altitude aerial photos were acquired using the UAV. The GPS rapid static technique was used to build the Ground Control Points (GCPs) and Check Points (CPs), which examine photogrammetric data. Although the CP is utilised for accuracy evaluation, the GCPs were employed to build 3D stereo models and other photogrammetric output.

KEYWORDS: Real time kinematics (RTK), High tide, Aerial photogrammetry, Monitoring, Ground control points (GCP), Unmanned aerial vehicle, LiDAR, ArcGIS 10.3, Global position system (GPS)

*Corresponding author: gbalajihits@gmail.com

DOI: 10.1201/9781003545774-12

1. Introduction

The majority of people on Earth live 50 km or less from the seas or oceans. Because of the goal to limit natural variability and preserve current coastline morphology, the effects of coastal erosion may worsen as human densities rise in coastal locations[1]. For this reason, regular and ongoing monitoring of beach morphological changes is essential to comprehend changes in the coastal environment[2]. The largest impact on high tide levels was due to changes in the coastline, specifically the development of artificial structures such as ports and embarkments, which can significantly alter the local tidal dynamics[36]. Unmanned aerial vehicle (UAV) photogrammetry has recently been employed extensively in topographic surveys [3]. Photogrammetry is the science of gathering the reliable measurements and data from photographs[4]. The UAV-based approach is more cost-effective and efficient compared to traditional methods, while still providing accurate and high-resolution mapping products[5]. The challenges associated with UAV-based mappings, such as the need for accurate georeferencing, the effects of weather conditions on data collection and the processing requirements for large datasets.

To know the flow behaviour (high and low tides) of the shore to monitor the shoreline efficiently, stream flow velocity data is utilised. The stream flow velocities value will be lower for high tide conditions and vice-versa[6]. Unmanned Aerial System/Vehicle(UAS) based surveys provided the highest accuracy in terms of elevation measurements which has a chance of getting affected by surface type heterogeneity, such as variations in sand grain size and moisture content, on the accuracy of elevation measurements. Even the laser scanner method can also be used for coastal monitoring while considering the effects of surface type heterogeneity[7]. The accuracy and precision of the laser scanning data varied depending on the system used, with airborne laser scanning providing the highest accuracy and terrestrial laser scanning providing the highest precision[8]. TLS techniques have the potential to provide information on coastal processes, such as erosion and deposition, over both short and long time scales[9].

While digital data are processed by Digital Photogrammetric Systems DPS, which have been around for approximately 7-9 years, the processing of films is often done by analytical plotters, which have been around for around 20 years. Over the past 20 years, there have been a lot of studies done on automated image analysis techniques to extract information from digital photos[10]. The LiDAR-equipped UAVs are highly accurate and precise. The airborne lidar and UAV-derived topographic data can be used for monitoring and tracking the recovery of the affected area[11].

Anthropogenic activities frequently speed up or intensify the natural erosion process of beaches, resulting in considerable topography changes, even over relatively short periods[[12]. Given that coastal erosion may seriously harm ecosystems, infrastructure, utilities, and buildings, it is crucial to mitigate its effects, especially in light of current climate change manifestations[14]. Considerable funds are invested in reducing the threats that coastal erosion poses to the ecosystems, infrastructure, utilities, and structures in the coastal zone. A thorough understanding of coastal erosion rates, their temporal and geographical distribution, and how they may alter under potential future climate change scenarios is essential to the creation of effective coastal management solutions[13]. The erosion rates were higher in areas with steeper slopes and lower vegetation cover[10].

In addition, the shoreline change, one of the key factors in determining the various Coastal Vulnerability Indices, is crucial in determining how vulnerable the coasts are to expected sea level rise and other coastal hazards brought on by climate change, such as storm surges[11].

To manage the coastal zone efficiently, it is required to conduct frequent surveys to identify and quantify related morphological variations. The following objectives need to be incorporated into every scheme for seaside observation [8]:

- Define, quantify, and comprehend the factors that influence coastal evolution.
- Ascertain the length of time that the mechanisms involved in coastal evolution will be active.
- Ascertain the geographical range of a process' impact.
- Develop connections between coastal dynamics and weather-related or climatic variables.
- Work to enhance mathematical forecasting models and forecast coastal change.
- Assist with territorial management and planning activities.
- Analyze the benefits and drawbacks of planning and management strategies.

2. Materials and Methods

2.1 Study Area

The research location is situated at latitude 02 31 N and longitude 101 47 T, and the chart datum standard for the beach is 5.36 m below the concrete mooring lights' benchmark, or 1.45 m below the land survey datum. The research area's coastline is thought to be 850 metres long, and the anticipated length of erosion is 450 metres. The remnants of various banks with rubble wall structures show how erosion has affected the area. Erosion in the studied region is a result of wave impact on those structures. This region has been experiencing coastal erosion since the beginning of 2002. Since 2004, several techniques have been developed to combat erosion in the study region, including the Beach Management

2.2 Unmanned Aerial Vehicle

These are frequently big, costly sensors that are useless to civilians and have been turned into weapons. Yet, there are several lightweight UAVs with favourable price ratios on the market. For photogrammetric applications, the maximum number of these lightweight UAVs is quite intriguing, particularly those that have a GPS, an IMU, a camera, a radio connection, and a very small computer processor.

Whereas, the image quality and photogrammetry product quality were similar across all UAV systems, with some variation due to different camera and sensor configurations. To improve UAV's efficiency, CamVox can be utilised which uses a single RGB camera and a low-cost lidar sensor to obtain visual and depth information, respectively. The system's performance in indoor and outdoor environments, and the results showed that CamVox achieved high accuracy and robustness, even in challenging environments with dynamic objects and low-light conditions. The low-cost and accurate nature of CamVox makes it suitable for various applications, such as robotic navigation, 3D reconstruction, and autonomous driving[12].

UAVs' potential for photogrammetry has lately been assessed in several studies. Unfortunately, there are relatively few references in the scientific literature to the use of UAVs for coastal surveillance. Coastal surveying benefits mostly from UAVs' evident advantages in general photogrammetric mapping. These features include very minimum hardware prices, a lot of automation was used for the photographic survey, extremely low running costs, and compact UAVS that are especially well suited for lifting in confined spaces. survey with excellent repeatability and cheap expenses; Due to the topic being photographed, it is possible to take close-up, high-resolution aerial photos; the potential for advanced ground control point (GCP) preparation throughout the survey, there was a chance for quick viewing pictures in the field, allowing for a repeat in case a defect was found.; due to their little weight, these gadgets provide relatively few security hazards in the event of an accident; Low time mission planning enables immediate utilisation, for instance straight after a storm[1].

Most small UAVs are also not suitable for lifting too much area quickly. Although the sensor data may be connected to the pictures, it is not accurate enough for some photogrammetric uses. Additionally, because the photographs are taken so close together, there are issues with concealing objects that have broken away from the dominant surface and very different relief displacements. The operational range is further limited by the radio link coverage to the ground control station, which is normally less than 5 km[1].

Two issues are predicted when light UAV operations are compared to those in other situations. One is the challenge that the weather, particularly the wind, presents. In an operating setting, meticulous planning is necessary. Lower wind hours are often in the morning. The majority of small-sized UAVs can operate effectively in winds with gusts less than 25 km/h, which provide several options. The likelihood that the research location contains significant bodies of water or water from breaking waves is another challenge and may exclude the use of matching procedures. In these situations, point matching is avoided by using masking techniques to prevent certain areas from being utilised. Products like

Fig. 12.1 Picture of MD4-1000 quad-rotor UAV

2.3 Ground Control Points

The marking of control points can be done in two ways: post-marking and pre-marking. Pre-marking refers to the choice of control point before the flight mission, whereas post-marking refers to the choice of control point after the flight mission. If the research region has trouble understanding any object to pre-mark, artificial pre-markings were necessary. The distance from the ground to be sampled affects the size of the man-made pre-mark. Using the image's pixel size and scale, it is possible to determine the ground sampling distance[8].

The majority of the study regions have natural features including beaches, rocks, and plants that do not offer clearly defined locations to be utilised as GCPs. Sometimes there are man-made features in the regions that were shot, but most of the time there isn't. For this reason, it was decided to add man-made points to the coast's surface. The majority of these were colored-different plastic bands with crosses in the centre. There were also white wood plaques with black crosses. While stiff plates were put over rocks or jetty surfaces, plastic bands were used for sandy regions. Plastic bands had to be properly secured with tent pegs or else they would have been easily shifted. For accurate GCP identification, the centre point always has a very clear definition. The black duct tape used to create the middle cross has a strip width of around 5cm[4].

2.4 Global Positioning System

The U.S. Department of Defense (DoD) created the satellite-based positioning system known as GPS to offer constant, global, all-weather navigation, particularly for army users. While there are several ways to use GPS, the fundamental positioning principle may be described as triangulation using satellites as range sources. GPS has two degrees of real-time precision: the Standard Positioning Service (SPS) offers an accuracy of 100 m, while the Precise Positioning Service (PPS) offers an accuracy of 16 metres. Differential GPS accuracy has made it possible for GPS technology to be effectively used in a variety of different applications, such as surveying and geodesy, photogrammetric mapping, hydrography, gravimetry, and investigations of crustal motion[5].

The GPS geodetic survey equipment used in the experiments consisted of three battery-powered 12-channel GPS receivers (two active and one redundant), three bipedal sights with stand-alone vehicle mounts Unique, a roof mount for attaching antennas to the vehicle, a vehicle side mount for transporting bipods and antennas between locations and microwave absorber plate to prevent multipath signal reflection from the vehicle's roof. Theodolite and a range pole made up the traditional surveying tools. 20 cm-long metal pins were inserted into the sand as markers at each transect station to allow for exact reoccupation. To enable the pointed end of the bipod to reoccupy the same location on the pin, each pin included a tiny (2 mm radius by 2 mm depth) depression[5].

Both the indexing strategy and the antenna switch technique were used to solve the first carrier phase uncertainty for the GPS surveys. Two sites, a reference point and an index point, are located towards the land of the dunes at a distance of around seven meters. This distance was chosen to make antenna modifications more convenient. Any suitable baseline up to around 10 kilometers may have been utilised for the indexing approach. Trials have shown that beyond a distance of 10 km, baseline-dependent defects make it challenging to resolve the first carrier phase ambiguity with the 1-cm level accuracy needed for this study[5].

3. Data Analysis

The UAV was flown at a 50-meter altitude to gather pictures that can be examined using Web ODM Lightning. The quantity of the picture (DSM) is a presentation of the percentage of necessary image processing, such as ortho mosaic and Digital Surface Model.. For frontal overlap, the picture was overlaid by 80%, and for side overlap, by 72%, between flying tracks. This will make it possible to produce 3D images with higher resolution and greater accuracy.

The UAV recorded 149 photos for each flight plan and used Web ODM Lightning software to analyse them. To comply with the programme process, every image must be transformed into a three-dimensional (3D) model. The processing of the photographs required two days for each flight. All of the pictures will be processed by Web ODM Lightning and transmitted to ArcGIS 10.3 in.tiff format. All pictures have been overlaid using ArcGIS 10.3 so that shoreline alteration may be possible manually and the actual images can be seen clearly.

The shoreline had changed, and this might have had an immediate impact on the coastal area. To safeguard the coastal ecosystem, especially along the shore, prevention must be taken.

3.1 Image Orientation

This process was carried out using standard complete automation. To select the mean amount of points per shot, a user only has the option of selecting high, medium, or low quality. The setting "High" was chosen. Because some of the photographs don't have trustworthy tie-points, primarily due to water, the number of pictures utilised is less than the number of pictures shot. Many hundred thousand tie-points were acquired in each case. Because some of the points do not suit the package modification, they are eliminated as incorrect matches

3.2 DSM Generation

We begin by building 2D grids across the area of interest along the XY plane to produce a DSM from a point cloud. Then, the height of every point inside a grid cell is sorted. The 90th percentile is used as the grid elevation to lessen the impact of noisy spots on the resulting DSM. The gaps inside a window of a user-defined size are filled using the closest neighbour method. To maintain neighbourhood uniformity in the DSM, a median filter is lastly performed. Although rasterization speeds up spatial analysis, the interpolation procedure would undoubtedly result in accuracy loss and might perhaps mask the topography's spatial complexity.

The so-called "High resolution" option, which corresponds to the photos' semi-resolution, was used to generate the DSM. Choices for precise DSM extraction were chosen, and tiny gaps were filled up using interpolation. It could be challenging to extract the DSM from sand surfaces because the pictures might be overly bright or fail to include the necessary patterns for matching. The photographs in the current circumstances, however, were well-exposed because they were taken in the dawn with a low Sun, thus this was generally not an issue[1].

3.3 Orthoimage Mosaic Generation

The creation of the orthoimage mosaics is the last phase. The DSM is more crucial for topographic evaluation for coastal monitoring. The picture base is crucially important for understanding the visual information and completing the 3-D impression offered by DSM. The picture can also be used to outline several two-dimension signs of mobility, which include the high water and vegetation lines. The evaluation of vegetation covering is another crucial application[2].

Based on the DSM, the ortho mosaic is created using the Web ODM Lightning programme. In regions where there are structures or other things that might readily produce occlusions, this is not an easy task. It requires a prediction of the obscured parts in each picture and construction of the ortho by filling in spaces using the pixel information from the best suitable image. It results in a "true-ortho," which in cities would call for more side overlap. Since there aren't many significant occlusions in many coasts and dunes, the idea of true-ortho isn't extremely important. Yet, occlusions may arise for rocks and protective features like jetties, and the real ortho may start to matter[13].

This process resulted in the DSM that was calculated for the coastal sandy regions having very excellent accuracy in both planimetry and altimetry. It is now feasible to identify topographic changes in coastal regions with extreme precision, which is crucial for maintaining the prior coastal monitoring programme[44].

4. Discussion

Due to the effect of coastal processes, beaches are known for their distinct morphology and are mostly made up of sedimentary materials that are one colour, sand. Therefore, it is necessary to verify the applicability of UAV photogrammetry, which creates 3D topographic data by matching points from geographical elements inside photographs[3]. These devices' very simple operation and the potential for obtaining fine-resolution DSMs and georeferenced image, with a high degree of temporal resolution as well, may make it possible to expand the databases and mapping techniques now used to track coastal change and migration. The research and surveillance of morphological alterations brought on by coastal dynamics may greatly advance with the use of UAV equipment[38].

Here, experiments were conducted to determine the ideal configuration for monitoring coastal dune changes with UAVs flying at low altitudes using various combinations of flight parameters (altitudes, GCP distances, and flight orientations). The goal was to test various configurations to assess the calibre of the final goods. Typically, UAV flights for coastal surveillance are conducted at a height of around 100 metres. Local constraints prevent the use of these heights often, and in certain situations, it is also necessary to conduct flights at relatively low altitudes to gain a larger level of surface detail. The small number of common features in the photos required to achieve a solid photogrammetric reconstruction, for instance, may make flying at too low an altitude result in more inaccuracies. There aren't many examples in the literature of low-altitude flights that produced reduced vertical inaccuracy. The trials carried out here demonstrate that flying at low altitudes is also capable of producing excellent outcomes with fewer mistakes[15]. The shooting atmosphere, however, can readily interfere with photogrammetry's accuracy. The accuracy of image matching will be significantly decreased when there are significant water regions in the measuring area[11].

The two periods of DSM data and the accompanying ortho mosaic were used to track the topography changes on the beach. Due to anthropogenic activities and damage from the sea waves, the shoreline was continuing to recede. The artificial shore banks had a major collapse and a clear retreat in certain places, posing a direct threat to the nearby wind turbines[50-51]. Consideration should be given to how this information may be successfully incorporated into current monitoring and management frameworks to overcome this and fully appreciate the usefulness of high spatial and temporal resolution erosion data to coastal engineers and managers[13].

In both research regions, the standard accuracy values acquired using the drone were on par with or better than those achieved with earlier traditional aerial photography. The findings appear to be related to the program's resilience and a web of GCPs with a completely cared-for architecture. The design of the targets used as GCPs, whose centres can be precisely and simply identified in the programme and easily seen in the photographs, appears to be another important factor in the accuracy attained[45]. The precision of the GPS equipment used to gather the GCP locations places restrictions on this. To guarantee high-precision results, critical criteria for photogrammetric reconstruction are distance, dispersion, and number of GCPs. With points at the margins and in the centre of the research area, and situated at various heights, GCPs should be evenly dispersed in space. The studies done to investigate the effect of modifying the number of GCPs reveal that the GCPs may be lowered (from a maximum distance of 30 to 50 m) keeping the same degree of accuracy. This was not always the case, though, as in certain cases the accuracy fell as the number of GCPs rose. This finding might be explained by the fact that the optimum distribution of the GCPs was no longer maintained as the number of GCPs decreased, leaving GCPs only on the edges of the flight region and lowering the accuracy of the results[15].

While operating UAVs, the location of the sun and, consequently, the hour of the flight, are essential considerations. To avoid adding another variable, the studies in this study were conducted more frequently when the sun was vertical or very close to it. Some of the disparities in the findings' quality imply that this element could also have been quite significant[15].

5. Conclusion

In this study, UAV was presented as one of the new options for knowledge gathering and identifying coastal changes[2]. UAV monitoring of changing coastal dunes has become more prevalent in recent years. Aerial photos may be effectively captured by the UAV and digital camera for large-scale mapping. DEM and orthophoto, two photogrammetric outputs, were created successfully. UAVs might alleviate issues in many applications, particularly for tiny areas, thanks to improved technologies. It has been demonstrated that the UAV platform is ideal for projects with constrained resources, including time, money, and labour. The photogrammetric industry, which needs current information in a hurry, may use this technique. A workflow can be created by combining digital photogrammetry and terrestrial laser scanning (TLS) for 3D data acquisition and modelling to improve the accuracy and completeness of the 3D models.[21]

The method described in this paper was shown to be a legitimate substitute for the established techniques for identifying coastlines based on topography. [6]. Using the topography seen in UAV pictures, this approach may be used to find shorelines. Therefore, it is crucial to have complete control over the sources of error to provide high-quality findings in repeatable conditions. Yet, this strategy may also stimulate changes in the shoreline and coastal deterioration in several locations. The photos produced for this study show how the coastline changed over a short period. The coastal region may soon be impacted by the more obvious coastline alterations, and UAVs may be suggested for prospective study and development. UAV is one of the greatest instruments for analysing shoreline changes as well as other coastal concerns since it can fly at a lower altitude than other approaches (10 m), which makes it especially useful along difficult-to-access coastal areas. This technique is very efficient, takes less time than routine inspection, and can improve the examination of shoreline changes[52].

Operational challenges still exist when employing UAV photogrammetry for shoreline surveillance. The greatest challenge is the environment, particularly the wind. The right circumstances for flights may be discovered by committed teams utilising reliable online weather data, but this will take more effort and cost more money. Huge or uneven water regions might provide issues for putting man-made ground control and automated tie-point recognition. Nevertheless, masking strategies can minimise this issue. Networks with superior direct georeferencing, including dual frequency GPS on the UAV, as well as many precise and lightweight inertial sensors for improved attitude determination may soon be available[39]. This study showed that the method can provide valuable information for predicting coastal hazards, managing beach nourishment projects, and assessing the impacts of sea-level rise[25].

References

1. J.A. Gonçalves, R. Henriques. UAV photogrammetry for topographic monitoring of coastal areas. February 2015.
2. I B Isha and M R M Adib. Shoreline monitoring using Unmanned Aerial Vehicle (UAV) at Regency Beach, Port Dickson. May 2021.
3. Euiyoung Jeong, Jun-Yong Park, and Chang-Su Hwang. Assessment of UAV Photogrammetric Mapping Accuracy in the Beach Environment. May 2018.
4. Benqing Chen, Yanming Yang, Hongtao Wen, Hailin Ruan, Zaiming Zhou, Kai Luo, Fuhuang Zhong. High-resolution monitoring of beach topography and its change using unmanned aerial vehicle imagery. June 2018.
5. Robert A. Morton, Mark P. Leach, Jeffrey G. Paine and Michael A. Cardoza.Monitoring Beach Changes Using GPS Surveying Techniques. January 1993.
6. Marco Luppichini, Monica Bini, Marco Paterni, Andrea Berton and Silvia Merlino. A New Beach Topography-Based Method for Shoreline Identification. November 2020.
7. Mohd Yazid Abu Sari, Asmala Ahmad, Yana Mazwin Mohmad Hassim, Shahrin Sahib, Nasruddin Abu Sari and Abd Wahid Rasib. Large Scale Topographic Map Comparison Using Unmanned Aerial Vehicle (UAV) Imagers and Real Time Kinematic (RTK). February 2020.
8. Norhalim Che Mat and Khairul Nizam Tahar. Surf Zone Mapping Using Multirotor Unmanned Aerial Vehicle Imagery. May 2019.
9. M. Manyoky, P. Theiler, D. Steudler, H. Eisenbeiss. Unmanned Aerial Vehicle in Cadastral Applications. September 2012.
10. Erwin W.J. Bergsma, Rafael Almar, Amandine Rolland, Renaud Binet, Katherine L. Brodie, A. Spicer Bak. Coastal morphology from space: A showcase of monitoring the topography-bathymetry continuum. May 2021.
11. Chao Huang, Hongmei Zhang and Jianhu Zhao. High-Efficiency Determination of Coastline by Combination of Tidal Level and Coastal Zone DEM from UAV Tilt Photogrammetry. July 2020.
12. Emmanuel P. Baltsavias. A comparison between photogrammetry and laser scanning. March 1999.
13. Matthew J. Westoby, Michael Lim, Michelle Hogg, Matthew J Pound, Lesley Dunlop, John Woodward. Cost-effective erosion monitoring of coastal cliffs. August 2018.
14. Yi-Chun Lin, Yi-Ting Cheng, Tian Zhou, Radhika Ravi, Seyyed Meghdad Hasheminasab, John Evan Flatt, Cary Troy and Ayman Habib. Evaluation of UAV LiDAR for Mapping Coastal Environments. December 2019.

Note: Every figure in this chapter was created using the experimental work; none of them were taken from any publications or online.

Advances in Additive Manufacturing Technologies – Gurusamy Pathinettampadian (eds)
© 2024 Taylor & Francis Group, London, ISBN 978-1-032-90013-1

13 Wear Behavior of Hemp - Flax - Glass Fibre Hybrid Composite Material

Hepsi Beaula M J[1]
Department of Mechanical Engineering,
Chennai Institute of Technology, Chennai, India

A. Bhaskar[2]
Department of Mechanical Engineering SRMIST,
Ramapuam, Chennai, India

R. Swathi Rekha[3], **P. Vasanthi**[4]
Department of Civil Engineering,
Chennai Institute of Technology, Chennai, India

ABSTRACT: Composite material is one of the most recent and important advances in material science that has had a big influence on engineering and technology. They are composed of two or more component materials that differ greatly in terms of their physical or chemical characteristics. A discrete reinforcementis dispersed throughout a matrix, binder, or resin in composite materials. In this study, the wear behavior of hemp-glass fiber hybrid composites was investigated by a wear test, The key input elements were the slipping speed, the force exerted, and the distance that was slide. The goal of this was to ascertain the ideal wear parameters for several performance metrics, including mass loss and friction coefficient. Forming an L27 Taguchi Orthogonal array for the experiment, then Grey Relational Analysis (GRA) and Analysis of Variance (ANOVA) were used for optimization.

KEYWORDS: Composite materials, Hybrid composites, Wear behavior, FEA, Modal analysis

1. INTRODUCTION

Raman Bharath et al. (2015) made a review of the applications of fibre based composite materials in various fields. Their main focus was on kenaf fibre based composite materials. Srinivasan et al. (2015) studied the thermal properties of hybrid composites manufactured from kenaf and flax fibres. Comparing hybrid composites to single-fiber composites, the authors found that the former have greater thermal stability. Vacuum assisted technique were studied and reported that the fiber orientation place a crucial role in his study. Subhankar Biswas et al. (2015) Bajuri et al. (2016) examined the effects of silicon dioxide nanoparticles coupled with kenaf on the compressive and flexural characteristics of epoxy composites. An important mechanical property was revealed by adding dioxide to silicon nanoparticles. For composites containing 2% volume silica, the modulus and flexural strength values were 3.05 and 43.8 GPa, respectively, while the compressive strength and modulus values varied. LutPil et al. (2016) Examine the

[1]hepsibeaulamj@citchennai.net, [2]bhaskara@srmist.edu.in, [3]swarthirekhar@citchennai.net, [4]vasanthiskumar16@gmail.com

DOI: 10.1201/9781003545774-13

rationale behind the designer's preference for using hemp and flax. In tension and plate bending, they discovered that the composites had greater specific stiffness than glass fiber composites, as well as a stronger vibration damping capacity. Shuhimi et al. (2016) compared tests and utilised epoxy glue and study the tribological characteristics of kenaf and oil palm fibre. The outcome demonstrated that an increase in temperature led to a composite results shows that increased wear rate. An increase in the fiber composition resulted in increased performance wear rates for each of the fibers. Vijaya Ramnath et al. (2016) conducted a variety of mechanical tests on different natural fibers. After impact, there was a noticeable drop in the characteristics of the flax-epoxy. Changlei Xia et al. (2017) created hybrid composites with aluminum sheets and natural fibers. The mechanical characteristics of the hybrid composites remained unaffected by the addition of aluminum sheets. The interfacial bonds were found to be strong based on the findings of the internal bond tests. Ming Liu et al. (2017) examined how mechanical characteristics affected unidirectional fiber/epoxy composites and hemp fibers. These effects were investigated when laccase oxidized the lignin in the hemp fibers. There was no evidence of laccase cross-linking. The accelerated polymerization of lignin in hemp fibers was the cause of this rise in characteristics. Sair et al. (2017) began their work by characterizing the hemp-polyurethane composite contact. Zia Mahboob et al. (2017) and Kin-tak Lau et al. (2018) discovered that the high moisture content, inconsistent properties, poor bonding qualities flammability, uneven dispersion in products, and swelling effect of natural fiber composites could potentially compromise the quality of the resulting composites. In many business areas, they are regarded as the most cost-effective notwithstanding a few drawbacks. SheedevAntony et al. (2018) worked on the hemp fiber strands that had come free from serge and taffeta textiles. Numerical, analytical, and experimental analyses were performed. The Objective of Research to determine the wear characteristics of hybrid composite samples (C3) using pin on disk method and to form the L27 Taguchi orthogonal array. Finding the ideal wear parameters for various performance characteristics, such as mass loss and friction coefficient, requires determining the load applied, sliding distance, and sliding speed as the primary input parameters. The experimental work was carried out by forming a L27 Taguchi Orthogonal Array in work piece and optimization is done using Grey Relational Analysis (GRA) and Analysis of Variance (ANOVA) for C3 hybrid composite. Based on the mechanical properties of developing composites, Finite Element Analysis has to be done for all categories (C1, C2 and C3). To find the reliability towards operating frequency and durability of the developed composites.

2. Results of Wear Test

2.1 Principle of Optimization

This investigation used the grey-based Taguchi methodology to carry out the optimization. The Taguchi style of experimentation served as the foundation for the grey-based Taguchi methodology, which used Grey Relational Analysis (GRA) to convert multi-response problems into single-response problems. Using Grey Relational Analysis, an attempt was made in this study to optimise the wear parameters taking weighted output response characteristics into account. Table 13.1 displayed process parameters A, B, and C along with their corresponding levels. The literature and machine capacity were taken into consideration when choosing the input parameters. GRA is normally used when there is some unclear and incomplete information. Taguchi L27 orthogonal array along with grey relation has been used to analyze the Sliding distance, Sliding Speed and Load. GRA is commended for optimizing the complex correlations amongst the multiple response characteristics; hence it is used in this research.

Table 13.1 Wear testing parameters and their levels

Character	Input Parameters	Level 1	Level 2	Level 3
A	Load	10	20	30
B	Sliding Velocity	0.1	0.2	0.3
C	Sliding Distance	500	750	1000

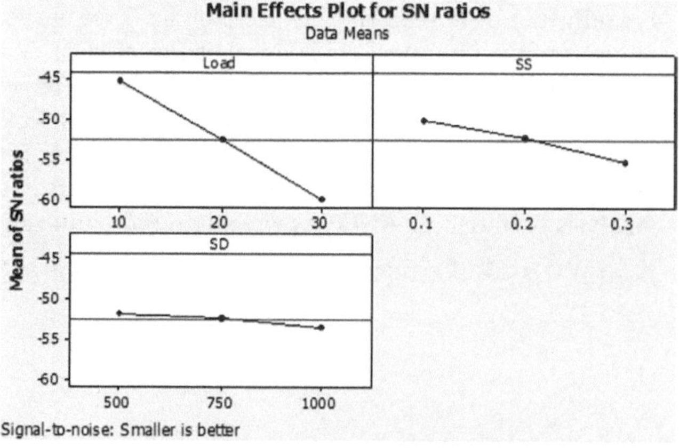

Fig. 13.1 Main effects plots for mass loss

2.2 Grey Relational Analysis (GRA)

Measurements of the values between the sequences can be made more effectively with its assistance. The GRA is used to analyse the effects of 27 subsystems on the response variable and to analyse the wear behaviour of composite samples, with each experiment being regarded as a comparability sequence. The experiment yielded the lowest values of mass loss and friction coefficient for the highest weighted Grey Relational Grade (GRG).

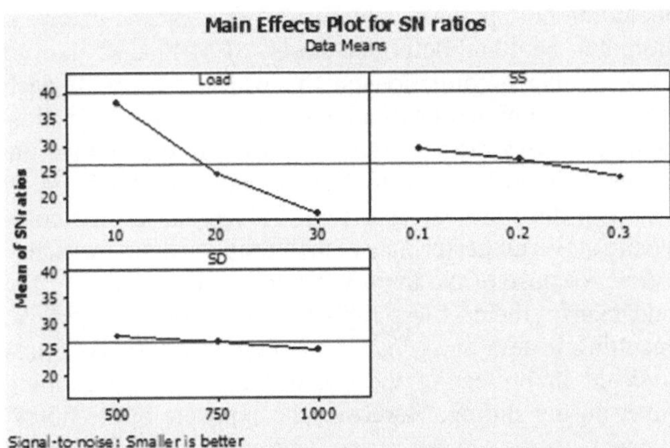

Fig. 13.2 Main effects plots for coefficient of friction

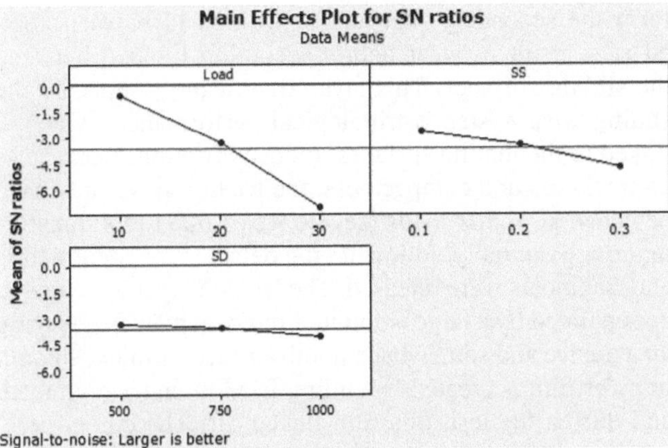

Fig. 13.3 Main effects plots for GRG

Step 1: Signal-to-Noise ratio (S/N) ratio (η).

The response parameters are indicated by the S/N ratio that is obtained through the Taguchi method. Three types of S/N ratios are distinguished: (a) Lower the better; (b) larger the better; and (c) nominal the best.

(a) *Lower the better:* The original sequence will be normalised using the following equation, where lower is preferred.

$$\eta = -10 \times \log_{10}\left(\frac{1}{n}\sum y_i^2\right) \tag{1}$$

(b) *Greater the better:* Condition applies if the mark value of the original sequence is infinite. It ensures that the response is maximised.

$$\eta = -10 \times \log_{10}\left(\frac{1}{n}\sum \frac{1}{y_i^2}\right) \tag{2}$$

(c) *Nominal the best:* The following formula is used to target a response. (Best nominal value)

$$\eta = 10 * \log_{10}(\mu^2/\sigma^2) \tag{3}$$

where yi is the average measured value of the experimental data and n is the number of experiment repetitions. i; σ^2 represents the observed data's mean; σ^2 represents the observed data's variance. Smaller values of torque, tangential force and thrust force indicate better characteristics.

Step 2: Data pre-processing: Normalisation of S/N ratio values

The S/N ratio is normalised as the initial step in the GRA process, which gets the raw data ready for analysis. In this case, the original sequence is changed to produce a similar sequence. This is required in cases where the scatter range of the sequence is fairly large. The S/N ratio's linear normalisation range is limited to values between zero and one.

Stabilize $\mathbf{y_{ij}}$ as $\mathbf{z_{ij}}$.

For lower the better values,

$$Z_{ij} = \frac{\left[\max\left(y_{ij}\right) - \left(y_{ij}\right)\right]}{\left[\max\left(y_{ij}\right) - \min\left(y_{ij}\right)\right]} \tag{4}$$

For larger the better values,

$$Z_{ij} = \frac{\left[\left(y_{ij}\right) - \min\left(y_{ij}\right)\right]}{\left[\max\left(y_{ij}\right) - \min\left(y_{ij}\right)\right]} \tag{5}$$

Step 3: The Grey Relational Coefficient (GRC) is derived from the normalised S/N ratio value.

The GRC value is one of the two sequences that is always in agreement. The formulae listed below can be used to determine the dependent variable's jth response and GRC at the ith trial.

Step 4: To calculate Grey Relational Grade (GRG)

The mean values of GRCs as GRG is collected after computing GRC.

$$\text{GRGi} = 1/n \sum \text{GRG}_{ij} \tag{6}$$

where, **n** is the number of responses.

From the Fig. 13.1 it is evident that for having minimum mass loss, the load is 10 N, sliding distance and velocity of 500 m and 0.1 m/s respectively. It can be verified with actual experiments has shown in the Table 13.2. From the Fig. 13.2 it is evident that for having minimum co-efficient of friction, the load is 10 N, sliding velocity is 0.1 m/s and sliding distance is 500 m. It can be verified with actual experiments has show in the Table 13.5

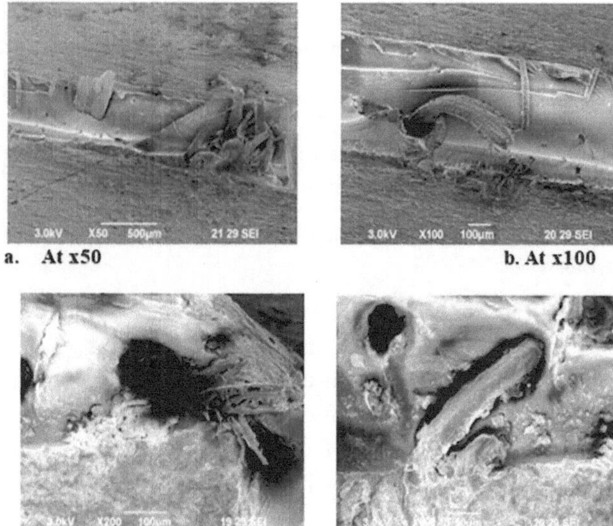

Fig. 13.4 SEM image of C3 after wear test

3. Morphological Analysis After Wear Test

After thoroughly examining the laminate, the ultraviolet rays were examined and found using a scanning electron microscope. Morphological Analysis was employed to look at the interior fibre behavior of the specimen. It helped working the fibre fracture, voids, cracks, fibre bonding, matrix uniformity, fibre separation, etc., within the composite laminate. The surface characteristics were studied. The sample from every check was taken and dried a few times and coated with gold of around 15 nm thickness. Then, the interior structure was examined through appropriate instrumentation. Sturdy bonding was seen within the structure and a study was made. It indicated reaching the stage of future improvement and application.First, the presence of free asperities at the interface is what causes the friction coefficient to decrease at low loads. The SEM image following the wear test is shown in Fig. 13.4. When there were fibres present in the matrix, the fibre layers and the matrix were rubbed together during sliding; these fibres sustained the matrix and prevented the removal of the matrix layer. The majority of the debris, as seen in Fig. 13.4.a, was shaped like a plate. However, in the case of composites, the volume of comparatively small, confirming the materials' low rate of wear. This debris was made up of polymer and worn-out fibres. This thin polymer film and fibre debris that envelops the fibre particles are typically transferred to the counterface.

4. Conclusion

The wear mechanism of hybrid natural fibre composites against a steel disc included a variety of wear modes, including fibre pullouts, film formulation, plastic deformation, and bent and deformed fibres. SEM images showed considerable deformation on the worn surfaces of each composite at higher loads, along with some debris. In addition, the epoxy resin and hardener were found to bond the fibres. This could be due to the superior interfacial bonding between the fibre and the resin. This enhanced the composite's overall performance, including its wear characteristics. Because of the improved bonding between the fibre and matrix, there is less rubbing against the counterface, resulting in less mass loss in the composites.Glass fibres assisted in protecting the composites' exposed rubbing layer during sliding. Moreover, the exposure of the fibres' contact surfaces to the matrix was enhanced by the glass fibres. The hard, durable components that make up the primary plateaus and the abrasive dust that wears them down form the secondary plateaus. A friction layer was created as a result of wear debris becoming trapped between the sliding surfaces. This layer significantly impacted the sliding wear system's tribological performance, which is linked to the mating surfaces' reduced friction. Because of its thickness and compactness, the friction layer may have detached at higher loads (severe wear) due to its transformation. Eventually, following the detachment, wear debris and fragments were released. The friction layer appeared to be separated by a large amount of removal material, leaving an abrasive and somewhat smooth surface. A thick, smooth transfer film is created by sliding friction under high load, and during the test, this film flaked off. The composites' experiences with flaking and transfer showed reduced frictional forces.When the steel counterface and the composite surface with the fibre cross section first came into contact, the coefficient of friction increased. As the material slides, frictional heat raises the temperature, partially softening the material and creating a thin transfer film. Pits formed upon removing the contact film, as seen in the figures. They were evidently present in the deteriorated microstructures. The flow of asperities and debris removal in this instance were facilitated by the fibre geometry. Moreover, the composite's intrinsic non-uniformity resulting from the differences in the matrix and glass fibre led to varying resistance to microcutting and ploughing. The distorted fragments of polyester created a film that shielded the fibre surface from wear and tear. The fibres' surface area was exposed to the disc because they were oriented parallel to the rotating disc. The likelihood of more fibre surface area being involved in the wear process increased with fibre length. This led to the formation of cracks at the fibre matrix interface and the occurrence of debonding.

Reference

1. Raman Bharath, V.R., Vijaya Ramnath, B., Manoharan, N., 2015. Kenaf fibre reinforced composites: A review, ARPN

Journal of Engineering and Applied Sciences, 10(13), pp. 5483–5485

2. Sreenivasan, V.S., Rajini, N., Alavudeen, A. and Arumugaprabu, V., 2015. Dynamic mechanical and thermo-gravimetric analysis of Sansevieria cylindrica/polyester composite: Effect of fiber length, fiber loading and chemical treatment. *Composites Part B: Engineering*, *69*, pp.76–86.
3. Biswas, S., Shahinur, S., Hasan, M. and Ahsan, Q., 2015. Physical, mechanical and thermal properties of jute and bamboo fiber reinforced unidirectional epoxy composites. *Procedia Engineering*, *105*, pp.933–939.
4. Bajuri, F., Mazlan, N., Ishak, M.R. and Imatomi, J., 2016. Flexural and compressive properties of hybrid kenaf/silica nanoparticles in epoxy composite. *Procedia Chemistry*, *19*, pp.955–960.
5. Pil, L., Bensadoun, F., Pariset, J. and Verpoest, I., 2016. Why are designers fascinated by flax and hemp fibrecomposites?. *Composites Part A: Applied Science and Manufacturing*, *83*, pp.193–205.
6. Shuhimi, F.F., Abdollah, M.F.B., Kalam, M.A., Hassan, M. and Amiruddin, H., 2016. Tribological characteristics comparison for oil palm fibre/epoxy and kenaf fibre/epoxy composites under dry sliding conditions. *Tribology International*, *101*, pp.247–254.
7. Vijaya Ramnath, B., Rajesh, S., Elanchezhian, C., Santosh Shankar, A., Pithchai Pandian, S., Vickneshwaran, S. and Sundar Rajan, R., 2016. Investigation on mechanical behaviour of twisted natural fiber hybrid composite fabricated by vacuum assisted compression molding technique. *Fibers and polymers*, *17*, pp.80–87.
8. Bensadoun, F., Verpoest, I., Baets, J., Müssig, J., Graupner, N., Davies, P., Gomina, M., Kervoelen, A. and Baley, C., 2017. Impregnated fibre bundle test for natural fibres used in composites. *Journal of Reinforced Plastics and Composites*, *36*(13), pp.942–957.
9. Shi, S.Q., Xia, C. and Cai, L., 2017. Modification of soy-based adhesives to enhance the bonding performance. In *Bio-Based Wood Adhesives* (pp. 86–110). CRC Press.
10. Liu, M., Ale, M.T., Kołaczkowski, B., Fernando, D., Daniel, G., Meyer, A.S. and Thygesen, A., 2017. Comparison of traditional field retting and Phlebia radiata Cel 26 retting of hemp fibres for fibre-reinforced composites. *AMB Express*, *7*(1), p.58.
11. Sair, S., Oushabi, A., Kammouni, A., Tanane, O., Abboud, Y., Hassani, F.O., Laachachi, A. and El Bouari, A., 2017. Effect of surface modification on morphological, mechanical and thermal conductivity of hemp fiber: Characterization of the interface of hemp–Polyurethane composite. *Case studies in thermal engineering*, *10*, pp.550–559.
12. Mahboob, Z., El Sawi, I., Zdero, R., Fawaz, Z. and Bougherara, H., 2017. Tensile and compressive damaged response in Flax fibre reinforced epoxy composites. *Composites Part A: Applied Science and Manufacturing*, *92*, pp.118–133.
13. Lau, K.T., Hung, P.Y., Zhu, M.H. and Hui, D., 2018. Properties of natural fibre composites for structural engineering applications. *Composites Part B: Engineering*, *136*, pp.222–233.
14. Antony, S., Cherouat, A. and Montay, G., 2021. Effect of fibre content on the mechanical properties of hemp fibre woven fabrics/polypropylene composite laminates. *Polymers and Polymer Composites*, *29*(9_suppl), pp.S790–S802.

Note: Every figure and table was created using the experimental work; none of them were taken from any publications or online.

Advances in Additive Manufacturing Technologies – Gurusamy Pathinettampadian (eds)
© 2024 Taylor & Francis Group, London, ISBN 978-1-032-90013-1

Characterization and Performance Evaluation of Basalt Fiber Composite Materials in a Application of Polymer Brake Disc

Shakthi Sowmiya S P
Student, Department of Aerospace Engineering,
Hindustan Institute of Technology and Science, Chennai, India

Balaji G*
Faculty, Department of Aerospace Engineering,
Hindustan Institute of Technology and Science, Chennai, India

Ramanan N
Assistant Professor, Department of Mechanical Engineering,
Sri Jayaram Institute of Engineering and Technology, Chennai, India

Saravanan P
Faculty, Department of Aerospace Engineering,
Hindustan Institute of Technology and Science, Chennai, India

Santhosh Kumar G
Teaching Fellow, Dept. of Mech. Engg., University College of Engineering, Bharathidasan Institute of Technology Campus, Anna University, Tiruchirappalli, Tamilnadu, India

Vijayanandh Raja
Department of Aeronautical Engineering,
Kumaraguru College of Technology, Coimbatore, Tamil Nadu, India

Gurusamy P
Professor, Department of Mechanical Engineering,
Chennai Institute of Technology, Chennai, India

ABSTRACT: Fibre Reinforced Polymers (FRP) have attracted a great deal of interest from the composites sector because of its remarkable properties, which include high tensile and compressive strength, chemical stability, and outstanding mechanical and thermal characteristics. FRPs are widely used in the polymer channel and beam construction industries due to their high commercial value. Basalt fibre reinforcement is crucial in basalt-fiber-reinforced composites (BFRCs), where additional reinforcement significantly improves mechanical performance. This work investigates the characterization of composite materials made of basalt fibres using a stacking sequence that has a centre layer oriented at 45 degrees and top and bottom layers oriented at 0-90 degrees. Another important factor in deciding how well BFRCs perform over their service lives is the fiber-matrix interface. Tensile strength, hardness, interlaminar delamination factors, and tensile fracture surfaces are examined in the next testing phase. The preparation procedure is used to

*Corresponding author: gbalajihits@gmail.com

DOI: 10.1201/9781003545774-14

make the laminate, which is then compressed to a 4 bar pressure and allowed to cure for 24 hours, during which time temperature generation is caused by the compression. The material was used for an application of polymer break disc.

KEYWORDS: FRP, Basalt fiber, Polymer break disc, FEA simulation

1. Introduction

Hybrid nanocomposites are often formed by combining two or more separate foreign elements into a shared host matrix. This blending of several materials has a synergistic result, resulting in innovative and better material properties such as higher elasticity, ductility, reduced weight, and improved fire resistance.[1][2][3] Carbon fibers already have these properties, making them useful in a wide range of large-scale technical applications such as aircraft structures, automobile manufacture, maritime vessels, sporting goods, and construction.[4][5] Based on the facts above, there is a pressing need for fibers that are durable, lightweight, long-lasting, and cost-effective for the manufacturing of hybrid composites. Currently, the market offers a variety of organic and inorganic fibers, but many of them lack structural strength, durability, or are prohibitively expensive for moderate loads. Because of its inorganic nature, basalt fiber has emerged as the preferred material of choice, boasting exceptional modulus, superior strength, enhanced strain resistance, high-temperature endurance, exceptional stability, commendable chemical resistance, ease of processing, non-toxicity, natural origin, eco-friendliness, and affordability.[6][7] Mechanically cutting basalt fibers and inserting them into a polypropylene-clay blend resulted in the successful production of a composite material. This novel technology not only increased the yield strength of the composite but also increased its elastic modulus. Because of basalt's plasticity and ductility, it allows for the incorporation of basalt reinforcement in a variety of forms other than typical fibers. Shapes like rods, bars, and textile fabrics can be used efficiently for this purpose, broadening the range of possible applications. [8][9][10]

The effect of water absorption on the interlaminar fracture toughness of vinyl ester-based composites including flax and flax/basalt hybrid materials was examined in a study. The vacuum-assisted resin infusion technique was used to painstakingly prepare these composites. Scanning electron microscopy and X-ray micro-computed tomography were used to evaluate the delamination morphology and fracture shear failure patterns in composite laminates. The testing results unambiguously revealed that the hybridization process with basalt fibers significantly improved the durability and water-repellent qualities of flax fiber-reinforced composites. This study emphasizes the benefits of mixing these materials in terms of better water absorption resistance and overall durability. [11]

Research focuses on the optimization of car bumpers in order to reduce weight and improve fuel efficiency while maintaining safety. When compared to the existing steel bumper, the new Basalt composite bumper not only achieves a stunning 49% weight reduction and a 56% cost reduction, but it also demonstrates improved performance characteristics. The Basalt composite bumper has been shown through finite element analysis to have a 47.6% higher impact strength and a 32.5% higher Factor of Safety, assuring increased safety margins. Furthermore, its increased deformation capacity (51% greater at 32 mm) indicates enhanced energy absorption during collisions, making it a cost-effective and high-performance option for automotive applications. Basalt rock is a common and safe material that is known for its thermal qualities, strength, and durability[12,13]. Its raw material extraction is low-cost. Basalt, or solidified volcanic lava, is extremely heat-resistant and durable, making it perfect for applications such as insulating car interiors and engine parts. A study looked into hybrid composites made from basalt and jute fibers and polyester resin in a variety of forms. The results showed that pure basalt fiber composites performed better in flexural and tensile tests, although jute fiber-reinforced composites had a minor advantage in impact testing [14]. Automobile panels, as structural components, are typically not intended to significantly improve vehicle crashworthiness or occupant safety in side, front, or rear crashes. Automobile makers are currently focusing their efforts on developing materials that result in lower fuel consumption, less weight, and more efficient use of natural resources[15]. This is accomplished by introducing enhanced design concepts, using superior materials, and implementing more efficient manufacturing processes.

Recently, there has been an increase in the desire for technology to fuel the progress of future generations. As a result, developing a great product that satisfies client expectations needs a cost-effective and efficient method, while simultaneously maintaining high quality and effectively competing against industry rivals. The selection of materials for new products plays a critical role in engineering sectors in order to meet the needs of the present landscape. To fulfil modern needs, the pursuit of excellence, cost-effectiveness, and responsiveness to feedback becomes vital. Composite materials have developed as a dominant trend in recent years, indicating substantial improvements in material technology. These materials are critical and have a wide-ranging impact across sectors. Their production procedures have improved significantly, as they have showed the ability to combine the intrinsic capabilities of reinforcement and matrix materials, resulting in the creation of stronger materials than their respective constituents. The primary goal of hybrid constructions is to accomplish characteristics such as lightweight construction, stability, environmental friendliness, and corrosion resistance, among others.

In the context of automobiles, bumpers often include beams, stays, shock absorbers, and fascia components. These structural features play a crucial role in preventing physical damage to the front and rear ends of passenger vehicles in low-speed crashes. Vehicle bumpers are expected to play a critical role in preventing or at least limiting damage caused by such occurrences. However, existing legal criteria for car bumpers are frequently regarded insufficient, with certain passenger vehicles not even required to have bumpers. Vehicles involved in front-to-rear crashes have frequently sustained significant damage, affecting both safety features and visual aspects. Real-world crash damage analysis reveals critical aspects that are currently lacking in many passenger vehicle bumper designs. These benefits include the lack of suitable shape, impact stability, effective energy absorption, cost-efficiency, increased fuel economy, and recyclability. Bumper systems, once engaged, should ideally provide a robust interface and remain engaged during the collision. The Gas Tube Bumper System is offered as a solution for compact passenger automobiles to address these important challenges within the automotive industry. This novel strategy intends to address the automotive industry's current concerns, including those of safety, efficiency, and sustainability.

2. Materials Used

2.1 Basalt Fiber

In the realm of materials, nature provides a plethora of remarkable substances that can revolutionize conventional materials, offering advantages such as reduced weight, cost, and enhanced manufacturability. In the contemporary world, meeting challenges and minimizing the societal burden necessitates the exploration of alternatives to conventional materials. Nature, however, poses limitations, prompting the utilization of synthetic materials like Kevlar. Kevlar, a robust synthetic plastic, emerges as a solution, boasting strength five times greater than steel at an equivalent weight. Formed from hundreds of synthetic plastics through the polymerization process, where long-chain molecules are intricately joined, Kevlar's excellence stems from its internal structure. The molecules align in a regular and parallel fashion, ultimately forming tightly-knit fibers. This unique composition grants Kevlar exceptional properties, making it an invaluable material in various applications.

2.2 Preparation of Composite Specimen

The composite materials used in this investigation are made using compression moulding equipment, as shown in Fig. 14.1. The specimen's construction was made from 270 mm-long basalt fibres. The maximum size of the created laminate is just 270 x 270 x 3 mm. The preparatory technique employed here is compression moulding. The composite is composed of three layers. The fibre mats were coated with epoxy-impregnated resin. Epoxy resin, powdered graphene, and methyl ethyl ketone peroxide are combined with this catalyst to create a binding that is successful. Approximately 3g of graphene per sample is added to a 4:1 combination of epoxy resin and hardener to reinforce the composite. First, the die that has been fully impregnated with epoxy glue is filled with basalt fibre. This procedure is then performed twice with the fibre placed next to the carbon fibre. An epoxy, hardener, and

Fig. 14.1 Compression molding machine

graphene combination is applied to both sides of the basalt. The machine is set to the proper pressure and temperature, around 1500 psi and 120 degrees Celsius, after the fibre arrangement, and is then left for two to three hours to cure. Each sample is prepared for approximately 90-degree orientations and goes through a comprehensive evaluation..

3. Result and Discussion

3.1 Tensile Strength

The ability of a material or structure to support loads that tend to lengthen is known as tensile strength; compressive strength, on the other hand, can support loads that tend to reduce size. Conversely, tensile strength opposes tension (being pulled apart), whereas compressive strength resists compression (being pressed together). The ultimate tensile strength of a material is defined as the highest stress it can withstand before breaking when stretched or pulled. The prepared specimen samples are shown in Fig. 14.2.

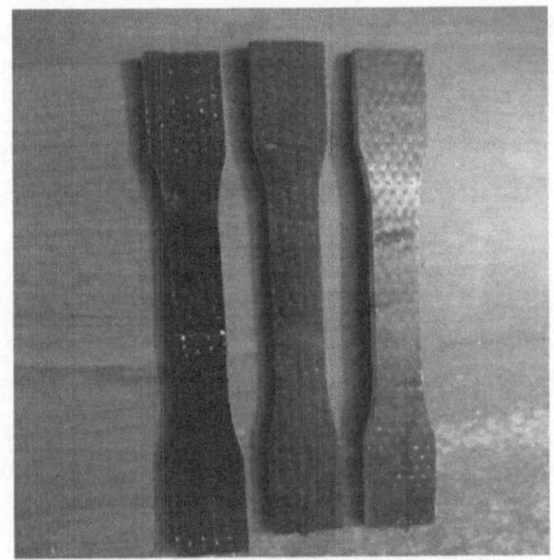

Fig. 14.2 Tensile testing specimen

3.2 Tensile Test

The values are obtained after testing three samples. A graph of load-displacement shows three samples. Composite 1 has an ultimate tensile strength of about 125N/mm2 and can bear tensile loads of up to 4.92KN. Comparably, composite 2 has an ultimate tensile strength of about 15N/mm2 and can support weights of up to 0.63KN. Finally, composite 3 has an ultimate tensile strength of about 156N/mm2 and can bear loads of up to 6.34KN. Figure 14.4's comparison of the values of the three samples makes it clear that composite 3 outperforms the other two in terms of tensile properties. Figure 14.3 displays the load-displacement graphs for the three samples.

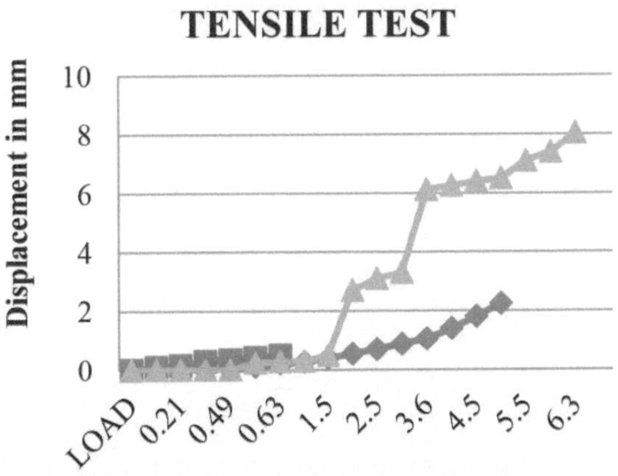

TENSILE TEST

Displacement in mm: 0, 2, 4, 6, 8, 10

LOAD, 0.21, 0.49, 0.63, 1.5, 2.5, 3.6, 4.5, 5.5, 6.3

Fig. 14.3 Comparison of values of three samples

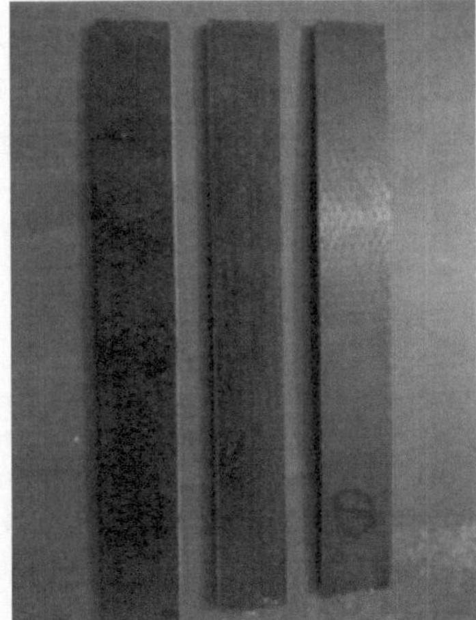

Fig. 14.4 Flexural testing specimen

3.4 Flexural Test

Flexural strength is a quality of a material that can be defined as the stress in that material immediately prior to it yielding in a flexure test. The transverse bending test is most frequently used to bend a specimen with a circular or rectangular cross-section until it fractures or gives. This process uses a three point flexural test approach. The flexural strength of a material indicates the highest tension it can bear before breaking. The specimen is usually held at both ends and supported by a hydraulic press or UTM device when a weight is applied.

After testing three different samples, as indicated in Fig. 14.4, it was found that composite 1 yields a flexural strength of approximately 160 N/mm^2 and can bear loads of up to 0.33 KN. With a flexural strength of about 40 N/mm^2, Composite 2 can bear loads of up to 0.09 KN. In conclusion, composite 3 has a flexural strength of around 284 N/mm^2 and can bear loads of up to 0.58 KN. It may be inferred from these results that composite 3 exhibits superior flexural qualities compared to the other samples.

Figure 14.5 displays the displacement values in the flexural testing specimen for the three samples, providing insights into the material's response to applied stress. Mean-

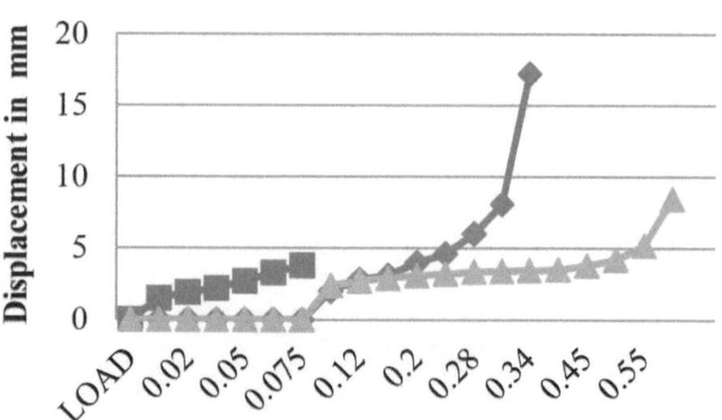

Fig. 14.5 Displacement value in flexural testing specimen for 3 samples

while, presents the flexural strength results for the same three samples, highlighting their respective performance in terms of resistance to bending forces. These figures offer a comprehensive visual representation of the flexural characteristics of the tested specimens, aiding in the analysis and interpretation of the experimental results.

4. Conclusion

This study investigates the characterization of basalt fiber composite materials based on basalt-fiber-reinforced composites (BFRCs). Various stacking sequences that contains with different layers are being studied with tensile and flexural test. Among various samples. the specimen 3 showed exceptional result of the mechanical properties of the composite material. This study describes the mechanical behaviour of the composite material which significantly understanding the frontiers of new materials.

References

1. Han SH, Oh HJ, Lee HC, Su SS. The effect of post-processing of carbon fibers on the mechanical properties of epoxy-based composites. Compos Part B-Eng 2012;45(1): 172–177.
2. Dehkordi MT, Nosraty H, Shokrieh MM, Minak G, Ghelli D. Low velocity impact properties of intra-ply hybrid composites based on basalt and nylon woven fabrics. Mater Des 2010; 31 (8): 3835–3844.
3. ArySubagia IDG, Kim Y, Tijing LD, Kim CS, Shon HK. Effect of stacking sequence on the flexural properties of hybrid composites reinforced with carbon and basalt fibers. Compos Part B-Eng, 2014; 58:251–258.
4. Taketa I, Ustarroz J, Gorbatikh L, Lomov SV, Verpoest I. Interply hybrid composites with carbon fiber reinforced polypropylene and self-reinforced polypropylene. Compos Part A-Appl S, 2010; 41(8): 927–932.
5. García PDLR, Escamilla AC, García MNG. Bending reinforcement of timber beams with composite carbon fiber and basalt fiber materials. Compos Part B-Eng 2013;55: 528–536.
6. Wang X, Wu Z, Wu G, Zhu H, Zen F. Enhancement of basalt FRP by hybridization for long-span cable-stayed bridge. Compos Part B-Eng 2013; 44(1):184–192.
7. Larrinaga P, Chastre C, Biscaia HC, San-Jose JT. Experimental and numerical modeling of basalt textile reinforced mortar behavior under uniaxial tensile stress. Mater Des 2014;55: 66–74.
8. Černý M, Halasová M, Schwaigstillová J, Chlup Z, Sucharda Z, Glogar P, Svítilová J, Strachota A, Rýglová S. Mechanical properties of partially pyrolysed composites with plain weave basalt fibre reinforcement. Ceram Int 2014; 40(5): 7507–7521.
9. Sarasini F, Tirillò J, Ferrante L, Valente M, Valente T, Lampani L, Gaudenzi P, Cioffi S, Iannace S, Sorrentino L. Drop-weight impact behaviour of woven hybrid basalt– carbon/epoxy composites., Compos Part B-Eng 2014;59:204–220.
10. Zhu H, Wu G, Zhang L, Zhang J, Hui D. Experimental study on the fire resistance of RC beams strengthened with near-surface-mounted high-Tg BFRP bars. Compos Part B-Eng 2014; 60:680–687.
11. Boria, S.; Pavlovic, A.; Fragassa, C.; Santulli, C. Modeling of falling weight impact behavior of hybrid basalt/flax vinylester composites. *Procedia Eng.* 2016, *167*, 223–230.
12. Boopathy G, Vijayakumar K.R., Chinnapandian M, Gurusami K, Development and Experimental Characterization of Fibre Metal Laminates to Predict the Fatigue Life, International Journal of Innovative Technology and Exploring Engineering. 2019, 8(10), pp. 2815–2819.
13. Almansour, F.A.; Dhakal, H.N.; Zhang, Z.Y. Effect of water absorption on Mode I interlaminar fracture toughness of flax/basalt reinforced vinyl ester hybrid composites. *Compos. Struct.* **2017**, *168*, 813–825.
14. Umashankaran, M.; Gopalakrishnan, S.; Sathish, S. Preparation and characterization of tensile and bending properties of basalt-kenaf reinforced hybrid polymer composites. *Int. J. Polym. Anal. Charact.* 2020, *25*, 227–237.
15. Boopathy, G., Vijayakumar, K.R., Chinnapandian, M., Fabrication and Fatigue Analysis of Laminated Composite Plates, International Journal of Mechanical Engineering and Technology.2017, 8(7), pp. 388–396.

Note: Every figure in this chapter was created using the experimental work; none of them were taken from any publications or online.

Advances in Additive Manufacturing Technologies – Gurusamy Pathinettampadian (eds)
© 2024 Taylor & Francis Group, London, ISBN 978-1-032-90013-1

15 Experimental Investigation on Engineering Properties of Soil with Medical Waste

B Krishnakumari*
Assistant Professor, Department of Civil Engineering,
Panimalar Engineering College, Chennai-123. &
Research Scholar, Department of Civil Engineering,
Dr. MGR Educational & Research Institute, Chennai

B Kandhakumar
Assistant Professor, Department of Electrical and Electronics Engineering,
Panimalar Engineering college, Chennai-123.

A Latha
Associate Professor, Department of Civil Engineering,
Panimalar Engineering college, Chennai-123.

M Mageswari
Professor, Department of Civil Engineering,
Panimalar Engineering college, Chennai-123

K Deepa
Associate Professor, Department of Civil Engineering,
Kampala International University, Uganda

ABSTRACT: In the modern scenario, it is required to improve the properties of soil which is critical. The soil property is not easily achievable for construction activities. Problems such as shear failure, excessive settlement, and differential settlement are some of the issues that arrive during the construction of structures on weak and soft soil. Replacement of these problems can be done by enhancing the properties of soil or by adopting deep foundation, at the site for the predicts load. Deep foundation adoption is, at the very least, an economically viable approach. On these days, treatments of soil in site and properties enhancement are gaining more importance. Soil stabilization is a cost-effective solution to one of the most pressing construction issues. This overview will focus on stabilization with the addition of syringe waste. Soil stabilization involves combining different soil components with the target soil to adjust the gradation and consequently the Engineering properties. Medium to high compressibility, high shrinkage and swelling properties which is a fine grain of inorganic is called clay soil . Gujarat, Madhya Pradesh, Maharashtra, province Andhra Pradesh and Karnataka are the states where the clay soil occurs mostly. Clay soils are a great challenge to the civil Engineers due to its high swelling and characteristics of shrinkage. Complete loss of strength occurs in clay soil when it is wet condition. Civil engineers have traditionally had trouble building facilities on or in clay soils, which handle the imposed loads on or throughout the structure's service life which lack the strengthening property. Soils in many parts of India have high

*Corresponding author: kumaribalasubramanian@gmail.com

DOI: 10.1201/9781003545774-15

silt content, low strength, and limited bearing capacity. There are several ways that can be used to increase clay soil performance. The type of soil to be enhanced, in which the vital for a particular application should be considered while choosing a particular soil stabilisation method, it's features and also the method and standard of improvement. Soil stabilization is an effective strategy for enhancing soil characteristics and pavement system performance.

KEYWORDS: Soil improvement, Stabilization, Medical waste reuse, Improve quality of soil

1. INTRODUCTION

Soil stabilisation is the process of improving soil characteristics in order to increase strength and stability. Compaction, dewatering, and adding material to the soil are only a few of the ways for soil stabilisation. In the process of soil stabilization, soil characteristics can be increased by combining soil materials with the selected soil to change the gradation and consequently, the soil properties get changed[1]. A fine grain of inorganic, that hold high compressibility, high shrinkage and properties of swelling, to uplift the strength, control of erosion, constructability and improved workability of the soil are called clay soils. The best way to make any particular soil stabilisation approach work is to improve current soil conditions rather than removing and replacing the entire material[2]. The stabilizing agent depends upon the funds and the resources that are accessible. Promoting the physical characteristics of clay soil is the key purpose of this research. Once the clay soil is subjected to moisture content, its behaviour is highly undetermined. Common apparent traits when water comes into contact with this type of soil are excessive soil compression, collapsing behavior, high permeability, high swelling capacity, and poor shear strength.

1.1 Index Properties

Classification of soil for general engineering purposes and feature of the soil that aid in identification are known for index properties. The index properties of soil includes water content, relative density, particle size distribution, consistency limits, in-situ density, density index[3,4].

1.2 Plasticity of Soils

The capacity of soil to deform without crack or fracture is called plasticity of soil. When the soil is wet, the plastic soil can be casted into several shapes. Particularly for clayey soils, the most important property is the plasticity index, because of the clay mineral present in the soil. The negative charges carried by the surface of clay particles are get attracted by the dipoles of water molecules are called the adsorption of water. Because of the adsorption of water, the plasticity of soil takes place. When the soil assigned to the deformation in the presence clay minerals, the particles do not come back to its early position, with the out-turn that the deformation are plastic[5,6]. This shows, when the plasticity of soil is reduced, the water content is also reduced. Whatever be the fineness, the soil does not become plastic in case of non-clay minerals like quartz. These soils cannot be rolled into threads, even they are grounded to very fine size. Flour of rock does not become plastic, as they are fine particles of non-clay minerals[7].

1.3 Consistency of Soils

The relative ease with which soil is frequently distorted is meant by consistency. This phrase is typically applied to fine-grained soils whose consistency is significantly affected by water levels. The degree of hardness of the soil, which can be described as is referred to as consistency such as soft, firm, stiff, or hard. [6,9] To make a plastic paste that could be pressure-moulded into any shape fine-grained soil could also be mixed with water. Water weakens the cohesiveness of the soil, making it even easier to mould. The fabric's cohesiveness decreases to the addition where it no longer holds its shape under its own weight and instead flows like a liquid when adding excess of water to the paste. During a suspension, enough water could be supplied until the soil grains are dispersed [10]. When the water in a soil suspension evaporates, the soil goes through a series of stages or states of consistency. 4 phases of limits which was developed by Atterberg, who was an agriculturist in Swedish in 1911. The phases are liquid state, plastic state, semi- solid state, solid state which are the representation of water content at which the mass of soil transforms from one state to the subsequent state is called the consistency limit.

1.4 Objectives

1. When moisture occupies significant gaps inside the voids of the soil it withstand variations made with the strength of soil.
2. To enhance the soil's sturdiness, erosion management, workability, and construct ability.
3. To be determined while figuring out a correct stabilizing agent.

1.5 Scope of Soil Stabilization

Because of limited bearing capacity, swelling and excess compressibility, the clay soils are difficult to work with in the nature. The clay soil is inappropriate for the construction of building due to its excess plasticity, shrinkage and limited strength. Thus, stabilization of soil becomes more necessary to enhance the strength and letting it more appropriate for construction. The ultimate aim of this paper is to investigate the impact of medical waste on black cotton soil strength characteristics such as compressive strength, CBR etc. The test is carried out on both virgin and stabilized soil sample. The obtained findings are compared for the two samples, and conclusions are generated. It should also probably be a suitable waste reduction technique, if the material proves to be an environmentally favourable successful stabilizing agent.

2. Materials and Methods

2.1 Clay

Clay and silt are granules with a size smaller than 75 microns. Clay is defined as a particulate with a diameter more than 0.002 mm. Clay has minerals such as Montmorillonite, Kaolinite and Illite that has features include a significant degree of expansion and shrinkage. Clay is a remarkable material because of its unique crystal structure, as well as its cation exchange capabilities, consistency, catalytic powers, swelling and shrinking qualities, and low permeability.

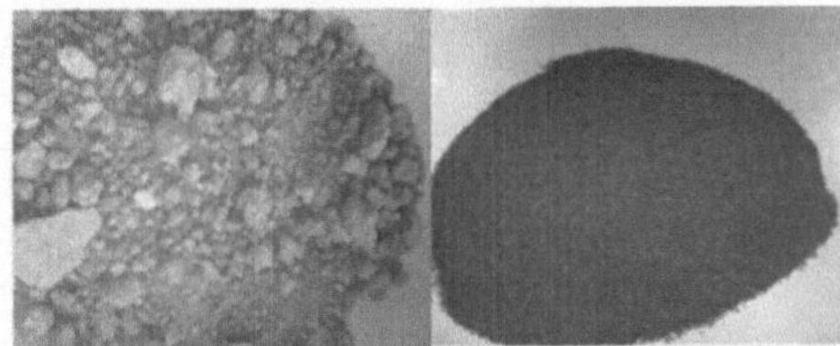

Fig. 15.1 Materials (Clay)

2.2 Syringe

In the metropolitan outskirts, it is very common for improper disposal of bio-medical waste. Campaigners in the city are worried about the consequences of Bio-medical waste is still being inappropriately disposed. According to TNPCB stats, 3,545 hazardous waste-generating units have been detected in the industrial sector. Every year, roughly 6.91 lakh tonnes of toxic waste are created in Tamil Nadu, with 2.97 lakh tones land fillable, 3.42 lakh tonnes recyclable, and 0.52 lakh tones incinerable. Medical waste is often dumped in flagrant violation of lawsin places like Injambakkam, Thiruneermalai, Chembarambakkam, Anakaputhur, Kundrathur, Vandalur, and Nazarethpet. Syringes are composed entirely of plastic which are mostly employed in the medical area. The needle has to be detached from the syringe and sterilised so that it may be used as a safe and sanitary material. A5 ml syringe is approximately 2 gm in size. It's ground into a powder and then combined with clay in the right proportions.

Fig. 15.2 Materials (syringe crushed form)

3. Methodology

Dry earth is combined with syringe powder in various proportions. On this mixed soil, UCC and CBR tests are performed and the Atterberg limit test is used to compute the water cement ratio. Thorough and consistent mixing should be provided while preparing the sample. Quantity of syringe powder should be raised gradually with the soil that had been gathered from the Kundrathur region. Once the mixed soil was dried for 24 hours and it was subjected to variety of experiments to determine its characterstics including consistency tests, density tests, compressive testing, specific gravity test. Optimum dosage of syringe powder is determined based upon the test results obtained.

4. Result and Discussion

4.1 Sieve Analysis

Sieve analysis is used to find out the finess modulus, coefficient of curvature and coefficient of uniformity of the soil. From the sieve analysis test the Cc value is 1.23 which is in between 1 – 3 & Cu value is 8.4 which is Greater than 6. The particle distribution curve is shown in Fig. 15.3.

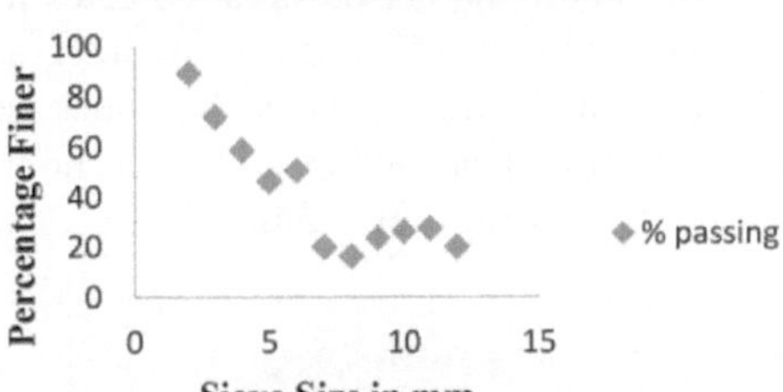

Fig. 15.3 Particle distribution curve

4.2 Atterberg Limits of Fine-Grained Soil

Liquid Limit

Representation of water content at which the mass of soil transforms from the liquid to the plastic state is known as liquid limit. The shear strength of soil mass is low at liquid limit and also with a rise in the liquid limit, the compressibility of the soil also raises. The soil begins to flow as a liquid at the lowest water content and it is measured in the laboratory as the water content at which a soil sample is split by a groove of standard size, after beings truck by 25 blows the soil sample flows together over a length of 12mm. The graph which is obtained from this test is shown in Fig. 15.5. In this test the liquid limit value is 31.4 which indicates it gives good result for 12%.

Fig. 15.4 Liquid limit test in casagrande apparatus

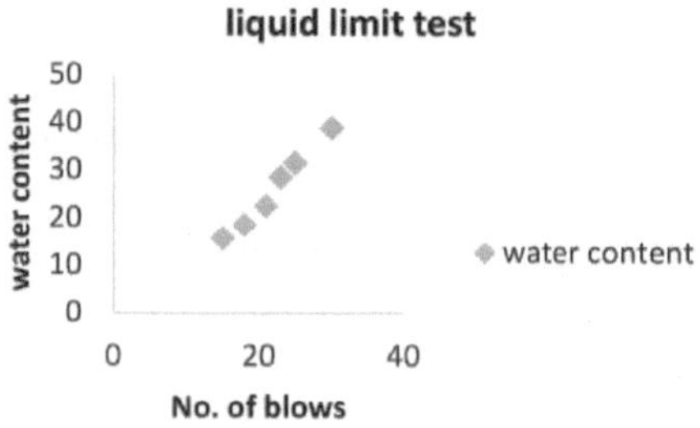

liquid limit test

water content: 50, 40, 30, 20, 10, 0

No. of blows: 0, 20, 40

◆ water content

Fig. 15.5 Liquid limit test

Plastic Limit

Representation of water content which states the transition between the soil plastic and semi-solid forms. It Changes in shape of soil and apparent fissures indicates lower water content. The point at which sample disintegrates when rolled into a 3mm thread indicates the water content of plastic limit and it is measured as 14.96%. The threads of diameter 3mm is shown in Fig. 15.6.

Fig. 15.6 Plastic limit test sample (3mm dia.)

Shrinkage Limit

Water content which represents the transition between semi-solid forms to the solid form. The shrinkage limit is referred as minimal water content wherein the soil is totally saturated. The decline in water content beyond the shrinkage limit has no effect on the overall volume of the soil mass. Thorough drying of lump of soil and analyzing the resultant volume and mass is a measure of shrinkage limit in the laboratory. Shrinkage limit is 30.1%.

4.3 Proctor Compaction Test

Using the Indian Standard heavy compaction test the water content-dry density relationship for a specific soil is determined and as a result, the optimal moisture content and maximum dry density for that soil is 12% and 1.54 g/cc. Again increase the sample the density is decreased to 1.5 g/cc.

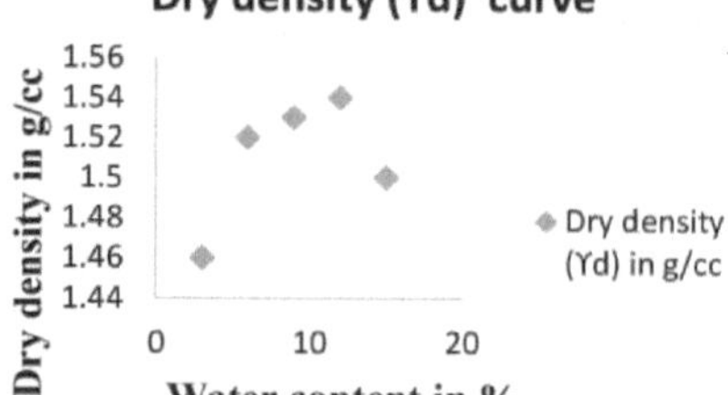

Fig. 15.7 Proctor compaction test

4.4 Unconfined Compressive Strength (UCC)

The UCC test is used to find out soil shear strength parameters. Based on the shear strength parameter the bearing capacity factors were determined by the Terzaghi's Bearing capacity equation. The test results are shown in Table 15.1. The sample testing is shown in Fig. 15.8. The soil is mixed with 3% of syringe gives the compressive strength of 0.423 kg/cm^2

Fig. 15.8 Unconfined compressive strength test

Table 15.1 Unconfined compressive strength test

Axial compressive (mm)	Proving Dial reading	Compressive load (kg)	Compressive stress (kg/ cm²)	Axial strain
SAMPLE 1				
1	0.5	0.1165	0.012	0.01
2	1	0.233	0.0237	0.02
3	1.1	0.2563	0.026	0.03
4	1.5	0.3495	0.034	0.04
5	1.7	0.3961	0.039	0.05
6	1.8	0.4194	0.041	0.06
7	1.8	0.4194	0.040	0.07
8	1.8	0.4194	0.040	0.08
9	1.75	0.4078	0.038	0.09
SAMPLE 2				
1	0.9	0.2097	0.22	0.01
2	1.3	0.3029	0.308	0.02
3	1.8	0.4194	0.423	0.03

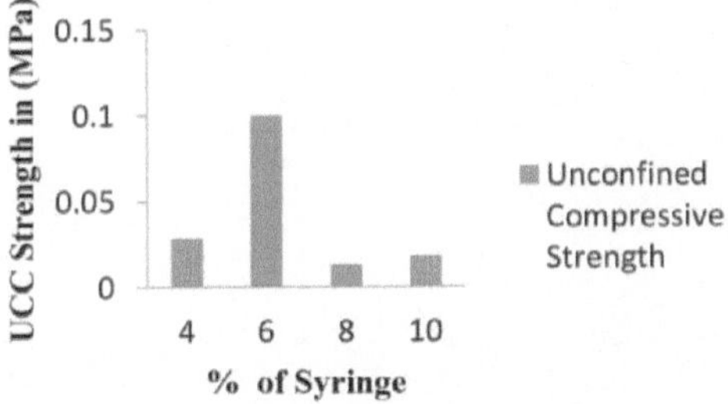

Fig. 15.9 Unconfined compressive strength test

4.5 CBR Test

It was first developed by the california state highway department. It is a penetration test which is used to evaluate the subgrade strength primarily of roads, pavements and foundations.

Fig. 15.10 CBR test

Table 15.2 CBR penetration

CBR value	4%	6%	8%	10%
2.5 mm	8.886	13.273	9.111	10.011
5 mm	10.198	12.673	10.498	10.123

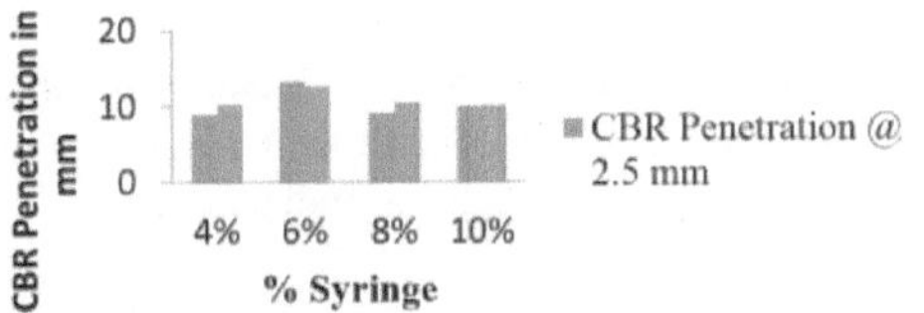

Fig. 15.11 CBR Penetration

5. Conclusion

There are many hazardous wastes being thrown in the environment .Bio-medical waste comes under one of the foremost hazardous kind of waste. It can cause a ramification of a virus disease which may take thousands of lives if it not disposed properly. Bio-medical waste as a stabilization material, it'll not only reduce the disposal amount within the surrounding, but also increases the engineering properties of black cotton soil. However, only the minor projects can be done. We observed that the optimum percentage of syringe powder which is to beaded within the clay soil comes bent be 6% of the dry weight of soil after performing Californian bearing ratio test and unconfined compressive test. When syringe powder was added both CBR valve and UCS value were peaked to 6%.The basic geotechnical properties like consistency, compaction, UCS, shear strength, and settlement characteristics of untreated and wood ash-based clayey soil is evaluated to an extent in the study by using syringe powder in the soil.

References

1. Ali Aliabdo, M. Abd-Elmoaty, and H. Hani Hassan, "Utilization of crushed clay brick in cellular concrete production," Alexandria Engineering Journal, vol. 53, no. 1, pp. 119–130, 2014.
2. Mishra, "A study on engineering behavior of black cotton soil and its stabilization by use of lime," International Journal of Science and Research, vol. 4, no. 11, pp. 290–294, 2015.
3. R. Malhotra and K. A. John, "Use of lime–fly ash-soil aggregate mix as a base course," Indian Highways, vol. 14, no.
4. Tavakoli, A. Heidari, and S. H. Pilehrood, "Properties of concrete made with waste clay brick as sand incorporating SiO2," Indian Journal of Science and Technology, vol. 7, no. 12, pp. 1899–1905, 2014.

5. G. Moriconi, V. Corinaldesi, and R. Antonucci, "Environmentally friendly mortars: a way to improve bond between mortar and brick," Materials and Structures, vol. 36, no. 10, pp. 702–708, 2003.
6. J. Oja and P. Gundaliya, "Study of black cotton soil characteristics with cement waste dust and lime," in Proceedings of the Nirma University International Conference on Engineering (NUiCONE 2012), pp. 110–118, Gujarat, India, December 2013.
7. M. K. Mohanty, Stabilization of Expansive Soil Using Fly Ash, Department of Civil Engineering, National Institute of Technology, Rourkela, Odisha, India, 2015
8. R. Rathan Raj, S. Banupriya, and R. Dharani, "Stabilization of soil using rice husk ash," International Journal of Computational Engineering Research, vol. 6, no. 2, pp. 43–50, 2016.
9. K. Sudharani, S. K. Abhishek, N. Adarsh, and Manjunath, "Stabilization of black cotton soil using brick dust and bagasse ash," International Journal for Scientific Research and Development, vol. 5, no. 5, pp. 140–144, 2017.
10. IS:2720 (Part 4)-1985 "Code of practice for Grain Size Analysis

Note: Every figure and table was created using the experimental work; none of them were taken from any publications or online.

Advances in Additive Manufacturing Technologies – Gurusamy Pathinettampadian (eds)
© 2024 Taylor & Francis Group, London, ISBN 978-1-032-90013-1

16 tigation and Optimization of Electrical arge Machining Parameters on SS316L

Ravikumar K[1]
Department of Mechanical Engineering,
Sathyabama Institute of Science and Technology, Chennai, India

Ganesan S[2]
Department of Mechanical Engineering,
Sathyabama Institute of Science and Technology, Chennai, India

Kamatchi Hariharan M[3]
Department of Mechanical Engineering,
SRM Institute of Science and Technology, Vadapalani, Chennai, India

ABSTRACT: The main goal of the current paper is to examine the influence of EDM parameters on its output variables. The process of input factors comprises of pulse on-time (Ton), pulse off-time (Toff), peak current (A) and the gap voltage (V). The present study has three levels of factorial system in the design of experiments (DOE).The orthogonal array (L9) was implemented to frame the investigational work by MINITAB software. The Material Removal Rate (MRR), Surface Roughness (SR) and the Heat Affected Zone (HAZ)were the output variables. The EDM performance on SS316L material was analyzed in Sodick linear motor series with coppera stool electrode of size 1 mm. Peak current was the striking parameter that significantly influence the output variables. In addition to thepeak current, Voltage also have the slight impact on the output variables. The results of optimum machining surface condition for smooth surface roughness was obtained at current 15A, Voltage 30V, pulse- on time 60 μs and pulse off time 40 μs whereas the maximum MRR was achieved at current 25A,voltage 50V, pulse-on time 60μs and pulse off time 80μs. Therefore, the higher pulse off time leads to lower material removal rate.

KEYWORDS: Electrical discharge machining, Pulse on time, Material removal rate, Surface roughness, Heat affected zone and Orthogonal array

1. INTRODUCTION

The brittle and hard materials find wide applications in different industries. The low weight, high strength and refractory materials developed through innovative technologies were well prepared to meet the industrial requirements. The machining operation was quite difficult on high strength temperature resistance (HSTR) materials. The conventional machining method may leads to extreme wear and either the work piece or the tool will be damaged during the EDM process. EDM is a promising machining process in order to overcome these above problems in industries, all over the

[1]anuravi597@gmail.com, [2]gansumaa@gmail.com, [3]kamatchm@srmist.edu.in

DOI: 10.1201/9781003545774-16

world. The Electrical Discharge Machining have no direct contact between the specimen and electrode which exclude vibrational problems and mechanical stresses. That is why EDM possess significant merits when compared to the other machining processes and also classified under non-conventional type[1]. Thus any electrically conductive material even with high strength and toughness can be machined to complex shape with high dimensional accuracy to meet the current industrial requirements. Hence, EDM found broad applications in automotive, aerospace, bio- medical, micro-electronics and die industries [2]. The environmental impact and machining performance were considered to be the two most essential indicators for sustainable manufacturing. The non-contact machining method i.e, EDM plays vital role in overcoming the problems of complex shaped objects and hard to machine materials [3]. Al7075 aluminium alloy was used in EDM method to fabricate anti wear gasturbine parts and heat exchangers irrespective of their spark, hardness and shape [4]. EDM is a versatile process for cutting difficult profiles that has specific benefits in the production of automotive, surgical and die parts from materials that were very hard to be machinedthrough conservative techniques [5]. Intricate shapes in hard materials, thermal complexity and optimized operating parameters for better MRR and surface finish were analysed in wire EDM of aluminium composite [6]. To enhance the machining efficiency and cutting stability in EDM of titanium material, MIMO, a novel adaptive control system was developed. In MIMO the servo voltage and electrode discharge time were altered parallely which reflects change in the distance so as to distribute the heat for better machining condition [7]. Therefore, the main objective of this study is to enhance the EDM input parameters for SS316L alloy through Design of Experiment.

2. Methodology

The method of execution of experimental work starting from material selection and upto result discussions were displayed in Fig. 16.1. SS316L alloy was implemented for the present research that have excellent electrical and mechanical behaviours. The stainless steel alloy has good formability, high hardness and better resistance to corrosion which find wide applications in chemical equipments, aerospace products and heat exchangers. The determination of DOF is mandatory for the appropriate selection of orthogonal array. The machining parameters were studied through nine experiments using L9 OA with respect to degrees of freedom.

Degrees of freedom = {(levels – 1) x factors} +1
= {(3 – 1) x 4} + 1 = 9

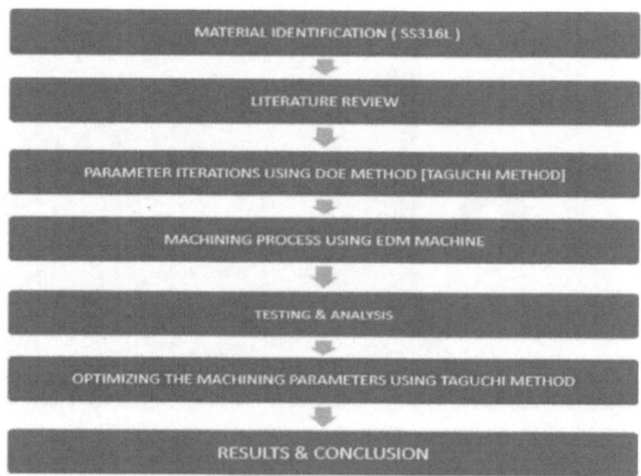

Fig. 16.1 Experimental work flow chart

Hence, Orthogonal Array (L9) was chosen for the analysis of input parameters i.e, peak current, pulse on time, pulse off time and the gap voltage and (10).

3. Experimental Work

Stainless Steel specimens were cut down into 30 pieces of dimensions Ø 10 x 5 mm. The stainless steel possess good ductility, tensile strength and corrosion resistance. After 304/A2 stainless steel alloy, SS316L grade was classified as these cond most common material under austenitic family. Figure 16.2 shows the SS316L specimen be fore electrical discharge machining. The chemical composition of the SS316L material were molybdenum (2–3%), nickel (10–12%) and chromium (16–18%), and <1% quantities of phosphorous, sulphur and silicon.\

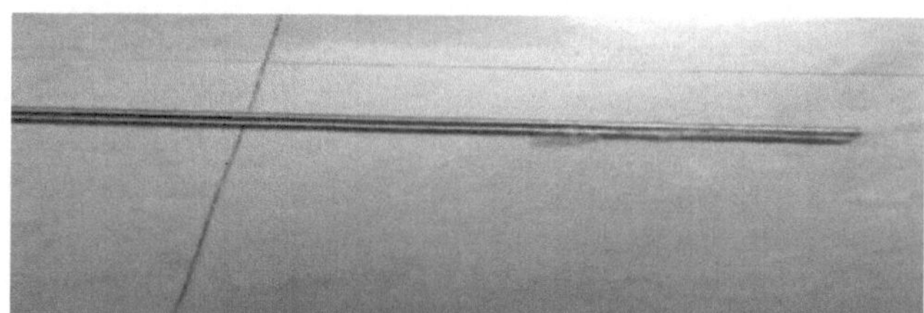

Fig. 16.2 SS316L before EDM

EDM is a metal fabrication method in which the required geometry was produced by electrical sparks. Figure 16.2 depicts, the SS316L specimen before electrical discharge machining. Figure 16.3 shows a typical electrical discharge machine and Fig. 16.4 represents EDM on the specimens. The electrode used in this research work is copper rod of diameter 1 mm. The output variables in EDM method were altered by the input factors viz, voltage, electrode gap, discharge current, pulse on time, pulseoff time and pulse duration. Among these, the gap voltage, discharge current and the pulse on time were identified as the three levels of process factors. DOE is a fine methodology that can be

Fig. 16.3 Electrical discharge machine

Fig. 16.4 Specimen under EDM process

very active for the common problem- solving and also for optimizing the manufacturing processes and product design. The identification of proper design dimensions, tolerances and the generation of prognostic math models were the Specific applications of DOE. DOE describes the behavior of physical system and determines the ultimate industrial settings. Taguchi method is a subset of DOE which uses OA, S/N ratio and ANOVAs for finding the substantial factors and examining the results. This helps in estimating the equivalent response. The tests were applied in electrical discharge machine with respect to experimental design produced through DOE. Table 16.1 depicts, the machining factors and their stages.

Table 16.1 Machining factors and its stages

S. No	Machining Parameter	Units	L1	L2	L3
1	Pulse on time	μs	60	80	100
2	Pulse of ftime	μs	40	60	80
3	Gap voltage	V	30	40	50
4	Peak current	A	15	20	25

Table 16.2 shows, the L9 Orthogonal Array for machining factors, gap voltage, peak current, pulse on time and pulse off time.

4. Literature Survey

The EDM of Ti6Al4V material with various electrodes like, aluminium, copper, graphite and the machining factors viz, pulse current and duration of pulse were executed on surface integrity. The surface machined by graphite electrode in XRD figure indicated the fabrication of Ti24C15 carbide at 2Θ values, 36.074, 41.766, 60.581 and 76.488. The formation of Ti24C15 carbide depends on the process parameters and materials of the electrode. The density of surface crack was affected by pulse current while the degree of crack was influenced by pulse on duration. The white layer hardness falls in the range of 1000 Hv - 1450 Hv. It is considerably larger than the hardness of bulk material because of the formation of high volume of Ti24C15 carbides on the surface. The decreased spark intensity in the discharge spots were due to the plasma channel expansion that improved the surface roughness at the extended pulse duration of 200 μm [1]. The influence of machining factors likepulse on time and pulse current on EDM of Al5052 aluminium alloy were experimented. At Ton = 300 μs and Ip = 15 A, the MRR was lesser when related with Ton = 100 μs and Ip = 15 A which indicates that the greater pulse on time and pulse on current resulted in larger MRR. The mean surface roughness for Ton = 500 μs was about 50% higher when compared to that of Ton = 100 μs at the corresponding increase of 40% of machined surface. The less spark at Ton = 100 and 200 μs leads to the formation of thin white layer with discontinuities, whereas the pulse on time rise at 300and 500 μs tends to the development of thicker white layer with many continuities [2]. The sustainable EDM by water in oil, nano suspension for the evaluation of environmental impact and machining performance were investigated. In the pulse duration of 800 μs and peak current of 30A, the MRR was increased to 44.64% and the wear rate of electrode was decreased to 40.33% that was observed from the results. The surface of work piece machined by water in oil nano suspension produced additional uniform surface and very few defects like cracks and pores, when compared with that of kerosene medium. The addition of Nano droplets in water in oil nano suspension reduced the formation of harmful and toxic gases considerably and ensured positive green environmental effect than

Table 16.2 L9 orthogonal array

S. No	Pulse on time (Ton)	Pulse off time (Toff)	Gap voltage (V)	Peak current (A)
1	60	40	30	15
2	60	60	40	20
3	60	80	50	25
4	80	40	40	25
5	80	60	50	15
6	80	80	30	20
7	100	40	50	20
8	100	60	30	25
9	100	80	40	15

kerosene [3]. The impact of EDM on MRR, wear of tool and geometrical errors of Al7075material were studied. At the machining conditions ofpulse on time 29 μs and pulse on current, 10 A, the MRR was enhanced but there was a few difference, when the pulse off time was improved to 6 μs. The wear rate of tool was improved, when the pulse on time of 27 μs and pulse on current of 10 A were increased along with the reduced spark gap of 0.4 mm. At the pulse on current of 10 A and fluid pressure 5 kg/cm^2, the circularity got improved with reduced spark gap of 0.2 mm and increased pulse off time of 27 μs [4]. The EDM of Al6061 material with machining responses of MRR, width of slit, finish of surface and electrode wear for various wire tension and pulse on time were investigated. The increased pulse on time developed the metal removal rate due to more amount of heat generation. Therefore, the role of tension of wire on MRR was too small. The lower melting point of Al6061 material caused the melting of the alloy at shorter pulse on time. Hence, the effect of pulse on time in appearance and the surface roughness were negligible. The longer pulse on time caused huge wire electrode wear whereas larger tension in electrode wire reduced the wear through stable machining [5]. The operating parameters in wire EDM of aluminium composite material were analyzed. From the result, it was found that the cutting speed improved with rise of pulse on time whereas the spark formation and heat generation were controlled by the pulse on time. The wire straightness and flexibility were controlled by the tension of wire whereas the variation of wire diameter and unlimited spark caused small variation in metal removal rate due to the change of electrolyte flush rate and of course the change of electrolyte. The variations in the total height of roughness profile, arithmetic mean surface roughness, and the maximum height of roughness profile showed that the splashed material and solidified material influenced the finish of surface over the craters [6]. The machining efficiency of titanium alloy (TA15) through MIMO adaptive control system, with NH7125ZNC EDM machine tool were examined. The servo voltage was controlled to widen the distance which increased the dielectric insulation strength and developed the heat removal capacity and chips in between the gap. Both, the servo voltage and discharge time of electrode were adjusted in connection with the various machining conditions by the MIMO for gap distance. The MIMO adaptive control system improved the machined depth suitably to 154% with less electrode wear when compared with open loop electrical discharge machining process. The appropriate selection of controlled indexes ensured the proper balance between machining efficiency and machining stability [7]. The influence of hydrostatic pressure on MRR of EDM process was studied. The abnormal arc and short circuits occurred during the voltage of 85V because of the machining stability at low frequency of discharge. The MRR in successive discharges reduced considerably with respect to the high hydrostatic pressure due to the increased electrode boiling point. The large gas bubble under low external pressure was essential for the stability of machining which caused the dielectric fluid flow and improved the cooling and flushing of electrodes. The MRR at 0.02 MPa, 0.06 MPa and 0.1 MPa were practically similar because of easy ejection of gas bubbles from the process gap [8]. The optimization of output variables in EDM of aluminium material were investigated by GRA. The aluminium hybrid composite was fabricated by the reinforcement of Sic (5 to 15 wt%) and Gr (5 to 15 wt%) through stir casting method. The tests were soundly scheduled and performed by the DOE, L27 OA. The version 14 of Mini Tab was utilized for the variance analysis and relational grade. The pulse on time of 500 μs during level 3, the current of 4A during level 1 and the mutual equal weight 15 % of Gr- Sic during level 3 were the ideal combination of machining factors which were achieved in relation with higher grey relational grades [9]. The optimal parametric combination of 8A current, 40V voltage, 150 μm inner electrode gap and 600 μsec pulse on time facilitated for lower surface roughness property [10]. The optimization of EDM process parameters on hybrid aluminium matrix composite material were investigated. The optimal

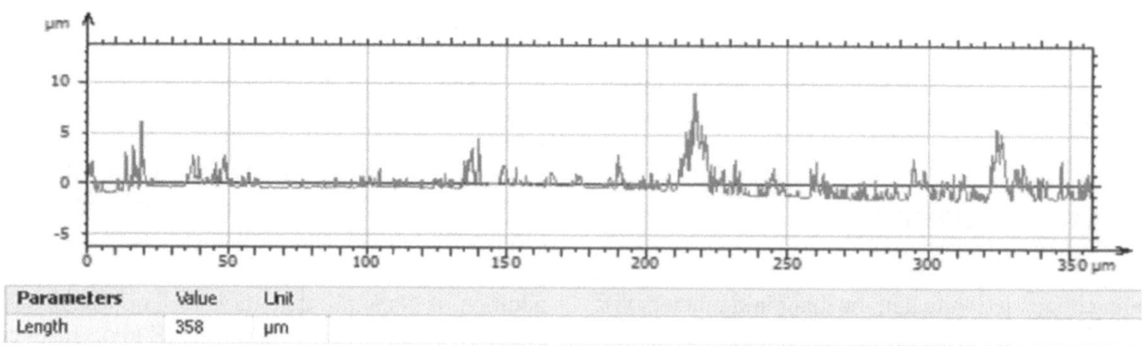

Parameters	Value	Unit
Length	358	μm

Fig. 16.5 Optimized graph representing surface roughness of the specimen

combination of 14A pulse current, 200 μs pulse on time and duty factor of 50% were achieved from the Minitab resulted in maximum material removal rate. Higher input energy was obtained at elevated current magnitudes which showed that the MRR was linearly proportional to pulse current. The MRR enhanced with increased pulse on time since the discharge time was inadequate for vaporization and melting. Further, it was also found that the pulse on current have more insignificant effect on the MRR [11].

The optimization of machining factors of 304 SS in electrical discharge machining were studied. The S/N ratio and variance analysis were examined for the effect of input parameters on MRR and for the identification of optimum cutting conditions. The 40 μs pulse on time during level 1, 12 μs pulse off time during level 2, 50V voltage during level 2 and 35 amps current during level 2 were observed to be the optimal machining factors for the MRR, through S/N ratio analysis. The current and pulse OFF time were the best insignificant machining factor for the MRR in EDM of 304 SS [15

In this section, the effect of pulsed voltage, pulse on time and peak current on MRR, SR and HAZ were considered. The copper electrode of size 1 mm was used for the process and the machining outcomes were analyzed through Taguchi method. The heat affected zone, SR and MRR were considered to be the output/response factors.

4.1 Surface Roughness

The optimized parameter revealed the desired output in a total of twenty seven investigational layouts. The Taguchi technique outcomes were supported by the graphical analysis within the EDM procedure.

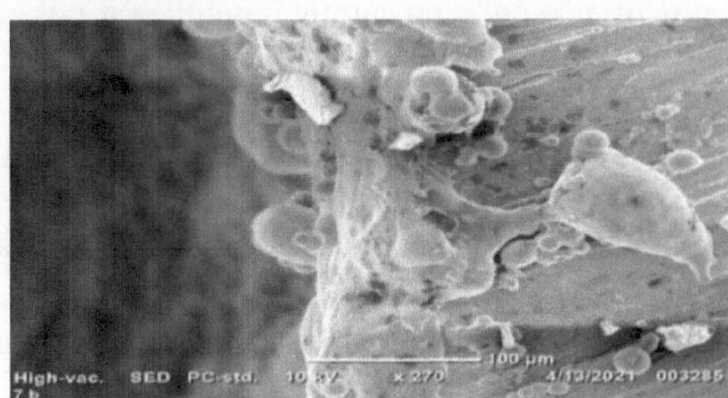

Fig. 16.6 Optimized heat affected zone

SR is the surface texture component and generally shortened to roughness. The difference between the abnormalities in the way of the typical vector of the actual surface from its ultimate state was quantified as SR. The surface will be rough when these deviations became large and will be smooth when they were small. The short wavelength and high frequency nature of rough surface find its own applications in surface metrology, of ameasured surface. But, both the amplitude and frequency were considered to be the important factors for fitting purposes.

The interaction of real object in its surrounding was decided by the surface roughness. The rough surfaces wear rapidly and have larger coefficient of friction when compared with the smooth surfaces in the study of tribology. Since the irregularities on the material surface tends to the nucleation of cracks, it can be considered as the best predictor of components performance. Figure 16.6 depicts, the optimized graph representing the specimens surface roughness and the amplitude parameters roughness profile.

Figure 16.7 shows the S/N ratio graphs and ideal machining surface circumstance for smooth roughness were current 15A, Voltage 30V, pulse- on time 60 μs, pulse off time 40 μs, for the equivalent input, the S/N ratio is -0.99098, mean 0.553672 and standard deviation 0.99188 respectively.

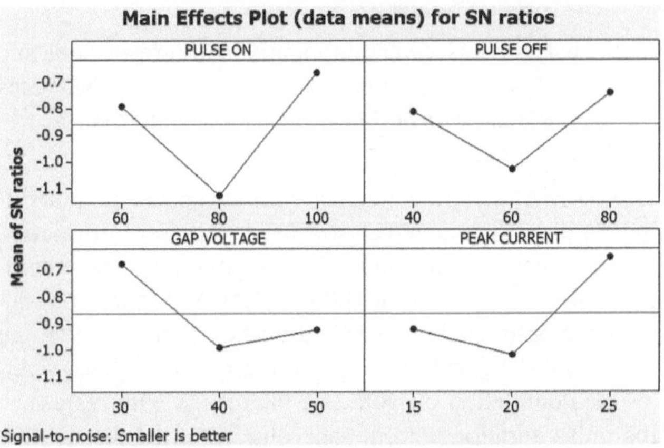

Fig. 16.7 S/N ratio for surface roughness

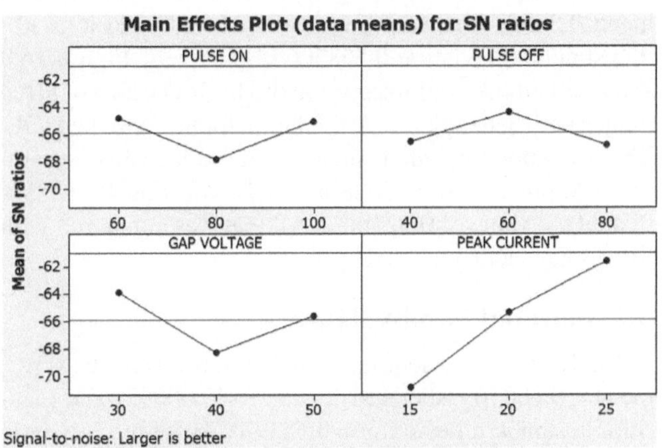

Fig. 16.8 S/N ratio graph for MRR

5.2 SEM Analysis

Scanning Electron Microscopy (SEM) is a checking process which scans a specimen by an electron beam to fabricate the enlarged image for investigation. SEM microscopy or

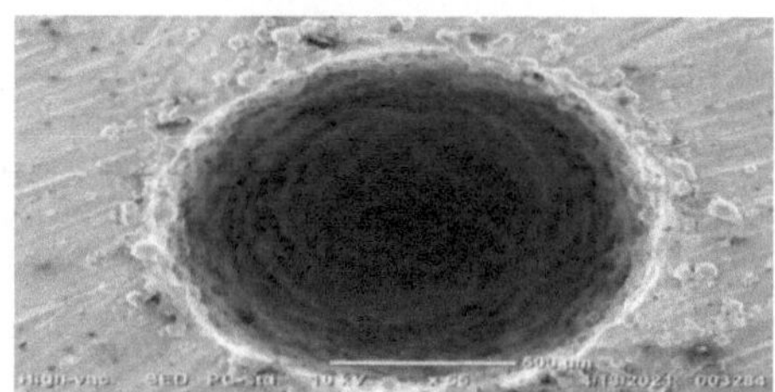

Fig. 16.9 Optimized outer hole diameter

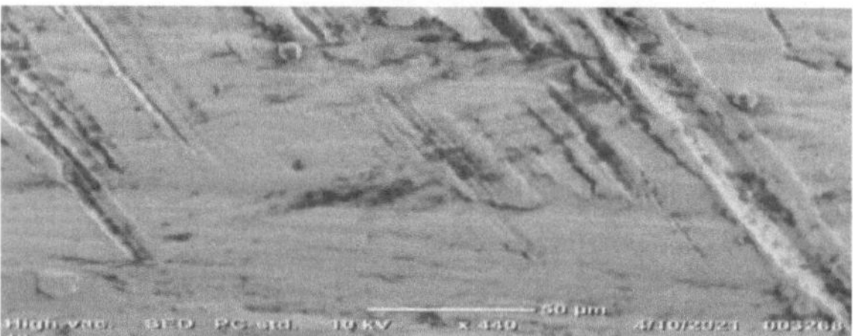

Fig. 16.10 Surface topology

SEM analysis is used very efficiently in failure analysis and micro analysis of solid inorganic materials. A SEM, scans a concentrated electron beam over a particular surface to produce an image.

The beam of electrons interact with the specimen generating different signals which was used to gather information about the composition and surface topography. The magnification size ranges from 100X to 1000X. Figure 16.8 depicts the optimized outer hole diameter, Figure 16.9 about optimized HAZ and Fig. 16.10 about surface topology. After the completion of DOE and the results with respect to the pulse duration current and voltage, all the data's were fed to the Mini-tab software. For the optimized outer hole diameter, the corresponding input S/N ratio is -0.99098, mean 0.553672, standard deviation 0.99188 and logarithmic standard deviation 0.558206 respectively. Figure 16.9 shows about the optimized value of HAZ. The size of HAZ ranges between 0.52 mm to 0.80 mm, found from the DOE. The corresponding input variables viz, voltage 30V, current of 15 A, pulse on time 60 μs and pulse off time 40 μs, fabricated the value of 0.52 mm HAZ that was compared with SEM image and analyzed.

5.3 Material Removal Rate

In MRR, the equivalent input of S/N ratio were -61.0397, mean = 0.000169, standard deviation = 0.000922 and logarithmic standard deviation = 0.183935. From the S/N ratio graphs, ideal machining surface circumference for higher MRR were current 25A, Voltage 50V, pulse- on time 60 μs, pulse off time 80 μs for the equivalent input the S/N ratio is - 61.0397, mean 0.000169 and standard deviation 0.000922 respectively.

5. CONCLUSION

Possibility of EDM method for SS316L alloy by copper type of electrode was analyzed. All the input factors i.e,gap voltage, peak current, pulse on time and pulse off time were significant in altering the output variables. The present study on the impact of the machine responses viz, MRR, HAZ, and SR along with the application of the copper tool were investigated in the Electrical Discharge Machining technique. The EDM was positively conducted on the SS315L material and the investigation of outcomes were also examined effectively. The tests were conducted in different parameters: voltage (V), current (A), pulse on-time (Ton) and the pulse off-time (Toff). The results were analyzed by using L-27 OA which is essentially based on the Taguchi technique of Minitab software process. Throughout the experiment, these reactions were partly real. Thedischarge current, pulse-off time and pulse-on time dominated the process factors whereas the voltage wasconsidered to be the least important parameter for EDM process which is the main objective of the study. Thus after performing the experimentations, outcomes wereestimated and explored to the Software of MINITAB. By the application of MINITAB Software – Taguchi method, the following finest combinations were achieved, namely

- Peak current was the striking input factor that significantly influences the output variables. Since, the increase in peak current produces larger vibrations and damage the specimen, it should be retained to a moderate range with respect to pulse duration.
- SEM analysis illustrates the HAZ region and surface topology of the machined stainless steel alloy.
- Pulse off and pulse on current upset the various output reactions. The decrease in pulse off and pulse on current decreased the outcomes, HAZ and SR.
- Voltages also slightly influence the output reactions. The low voltage gave better results and so based on the machine, 45v found to be the best voltage that can be used to get the better results.
- From the S/N ratio graphs, ideal machining surface circumstance for smooth roughness were current 15A, Voltage 30V, pulse off time 40 μs, pulse- on time 60 μs, for the equivalent input, the S/N ratio is -0.99098, mean 0.553672, standard deviation 0.99188 respectively.
- From the S/N ratio graphs, ideal machining surface circumference for higher MRR were Current 25A, Voltage 50V, pulse off time 80 μs, pulse- on time 60 μs, for the equivalent input the S/N ratio is - 61.0397, mean 0.000169, standard deviation 0.000922 respectively.

REFERENCES

1. Hasçalık A & Çaydaş U 2007. Electrical discharge machining of titanium alloy (Ti–6Al–4V). Applied surface science, 253(22), 9007–9016.
2. Markopoulos A. PPapazoglou E. L. Svarnias P. & Karmiris-Obratański P. 2019. An Experimental investigation of machining aluminum alloy Al5052 with EDM. Procedia Manufacturing, 41, 787–794.
3. Dong H.Liu Y. Li M. Zhou Y.LiuT.Li D. & Ji R. 2019. Sustainable electrical discharge machining using water in oil nanoemulsion. Journal of Manufacturing Processes, 46, 118–128.
4. Selvarajan L. Sasikumar R. Kumar N. S. Kolochi P. & Kumar P. N. 2021. Effect of EDM parameters on material removal rate, tool wear rate and geometrical errors of aluminium material. Materials Today: Proceedings, 46, 9392–9396.
5. Pramanik. A. Basak A. K. Islam M. N. & Littlefair G. 2015. Electrical discharge machining of 6061 aluminium alloy. Transactions of nonferrous metals society of China, 25(9), 2866–2874.
6. Raju K. & Balakrishnan M. 2020. Experimental study and analysis of operating parameters in wire EDM process of aluminium metal matrix composites. Materials Today: Proceedings, 22, 869–873.
7. Jing H. Zhou M. Yang J. & Yao S. 2018. Stable and Fast Electrical Discharge Machining Titanium Alloy with MIMO AdaptiveControl System. Procedia CIRP, 68, 666–671.
8. Koyano T. Suzuki S. Hosokawa A. & Furumoto T. 2016. Study on the effect of external hydrostatic pressure on electrical discharge machining. Procedia CIRP, 42, 46–50.
9. Maniyar K. G. & Ingole D. S. 2018. Multi response optimization of EDM process parameters for aluminium hybrid composites by GRA. Materials Today: Proceedings, 5(9), 19836–19843.
10. Nagaraju N.Prakash R. S. Kumar G. V. A. & Ujwala N. G. 2020. Optimization of electrical discharge machining process parameters for 17-7 PH stainless steel by using taguchi technique. Materials Today: Proceedings, 24, 1541–1551.
11. Kandpal B. C. Kumar J. & Singh H. 2018. Optimisation Of Process Parameters of Electrical Discharge Machining of Fabricated AA 6061/10% Al2 O3 Aluminium Based Metal Matrix Composite. Materials Today: Proceedings, 5(2), 4413–4420.

Note: Every figure and table was created using the experimental work; none of them were taken from any publications or online.

Advances in Additive Manufacturing Technologies – Gurusamy Pathinettampadian (eds)
© 2024 Taylor & Francis Group, London, ISBN 978-1-032-90013-1

17 Experimental Investigation and Finite Element Analysis of Novel Dental Restorations Material

S. Jeyan[1]
Department of Mechanical Engineering,
Jaya Engineering College, Chennai, India

G. Gopalarama Subramaniyan[2]
Department of Mechanical Engineering,
Saveetha Engineering College, Chennai, India

J. Chandru[3], **D. Barathraj**[4], **P. Mohanraj**[5]
Department of Aeronautical Engineering,
Jaya Engineering College, Chennai, India

ABSTRACT: In dentistry, glass-ionomer cement (GIC) is a healing substance that is used for tooth fillings and tooth whitening. These materials are derived from the reaction of polyacrylic acid and silicate glass powder. Glass ionomer cements are good materials for the restoration of carious lesions in low-stress locations, like smooth-surface and small anterior proximal cavities in primary teeth, because of their excellent qualities. However, it is not the preferred material to be used to fill posterior teeth because of its poor mechanical qualities. Humanity has been drawn to nano-silica since its discovery. They may practically strengthen any material due to their remarkable toughening capabilities and biocompatibility. SNPs are utilized as fillers in this work to enhance the mechanical qualities of GICs. Therefore, this study's goal is to examine how SNPs affect the GIC system's mechanical, thermal, and chemical characteristics in order to determine how best to employ it as a cement for posterior restorative dentistry. Three different ratios of SNPs and GIC powder will have been combined, and the standard dental mixing process will have produced cement pellets. Using FEA software, the 2D tooth model will be generated and analyzed while taking the experimental findings into account. Lastly, a comparison of the analysis results will yield the appropriate proportion of composite for the dental restoration.

KEYWORDS: Glass-ionomer cement (GIC), Nano-silica, 2D tooth model, Analysis

1. Introduction

Glass-Ionomer Cements, or GICs for short, are composite materials made of glass and polymer that are used in the dental field. GICs were first introduced to the public 25 years ago and have proven to be a very helpful addition to restorative dentistry. The cement is made of an aqueous solution of acrylic acid or copolymer and calcium alumino silicate glass powder. The cements have certain special properties because they release fluoride and are thermally and biocompatible with tooth enamel. The GICs' handling characteristics, compositions, and clinical uses have undergone significant

[1]mechjeyan88@gmail.com, [2]grsubramaniyan@yahoo.co.in, [3]aerochandru.87@gmail.com, [4]barathraj.d@gmail.com, [5]rajmohi@gmail.com

DOI: 10.1201/9781003545774-17

modifications in recent years. Because of its special physical and chemical characteristics, nano silica has garnered more interest recently and is being used in numerous industries to create new materials with innovative functionality. In 1962, the first dental gold alloy to be commercially successful was patented; since then, efforts have been undertaken to create technologically superior substitutes.

2. Material Selection

The issues in this specific field can be determined based on the literature review. Thus, the solution can be investigated based on the problem's identification. Certain material qualities were necessary for all hip implants in this survey. A few key criteria should be met when selecting an implant material. This choice was also determined by a few criteria. We have so looked into the requirements for superior strength, high coefficient of friction, toughness, and wear resistance in dental repair materials. For the purpose of creating a composite, we must choose the following material.

FIBRE: Silica Nano Powder with average fibre diameter of 100-150 nano meter.

MATRIX: GICs with 99% purity.

2.1 Fabrication of the Dental Restoration Materials (DRM)

Dental Restoration Materials are formed using a variety of techniques, based on the shape and desired properties of the final product. The process of turning dental restoration materials into a powder system is depicted in Fig. 17.1(a). The use of less expensive squeeze casting as a fabrication technique is growing in popularity. Usually, it is distributed as a powder and liquid system; two paste systems are not common. Calcium fluoro-alumina-silicate glass is present in the powder. Aqueous polyacrylic acid solution is present in the liquid. The process of mixing, also known as manipulating cement, creates a hard setting mass by reacting the carboxyl groups in the liquid and the calcium in the powder. Many changes have been made to it since its introduction to meet the unique requirements of dentistry.

Table 17.1 Various proportions of DRM

DRM	Matrix	Fibre
DRM-I	100%	0%
DRM-II	98 %	2 %
DRM-III	97 %	3 %
DRM-IV	96 %	4 %

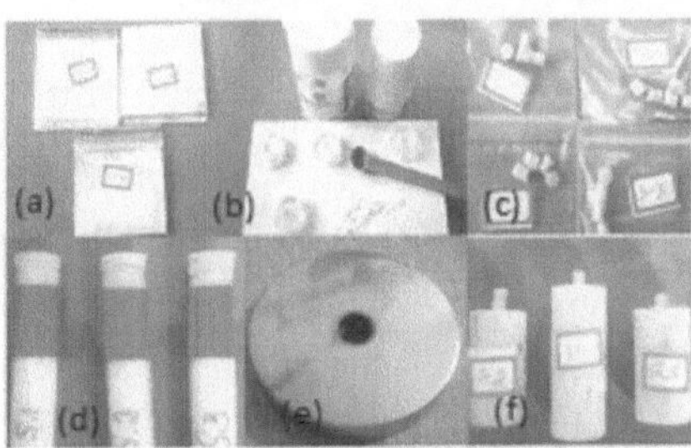

Fig. 17.1 (a) DRM powder, (b) DRM fabrication process, (c) DRM samples, (d) DRM wear test pin sample, (e) Stain less steel (SS304) Disc, (f) DRM compression test sample

3. Creation of CAD Geometry

The dimensions, cross-section direction, and tooth type provided in the geometry data are used to create the CAD tooth model. When the geometry data is entered, the system will produce an accurate model. There will be five surfaces created to symbolize the five primary components of a tooth: the pulp, dentine, enamel, PDL, and bone. Most exactly, the relevant parameter model of tooth, in that way is chosen from the CAD software when tooth type and direction data are supplied to generate the tooth form. The process begins with the construction of 2D closed curves that outline the tooth sections, from which the surfaces for each portion are created. Next, the geometry is optimized (more on this in the paragraph that follows); third, a translation matrix is used to locate the 2D curves in 3D space; and, finally, the closed curves are filled.

3.1 FEA Model

After the basic mesh size has been established, the mesh is generated using the FEA geometry. For the model, the global element size initially is established by the system on computing the total area of the tooth shape first. The global mesh sizes for various area ranges are displayed in Table 17.2. The definitions of load types, BCs, and element characteristics come in second. Thirdly, force is applied on the close nodes it generated reference points, loads are applied to the element edge points before and after mesh refinement. The ratio of the two lengths divided by the reference point known as "space ratio" parameter, allows the user to choose the reference point placements along the enamel layer's contour. Fourth, user GUIs can be used to describe the kinds, values, and locations of loads and BCs. There is default BCs available for common applications. Fifth, the user defines the material parameters for each tooth portion using graphical user interfaces; default values are provided, as Table 17.3 illustrates. A simple strain model is applied, to determine the element type for each tooth section in 2D solid model.

Table 17.2 Element size for various CS areas

CS area (cm^2)	0-5	5-10	>10
Global element size (cm)	0.02	0.04	0.08

Table 17.3 Values for tooth part

	Enamel	Dentine	Pulp	PDL	Bone
Young's modulus (MPa)	19600	19600	19600	0.25	13700
Poisson's ratio	0.3	0.3	0.3	0.45	0.3

4. Results and Discussion

4.1 Swelling Properties of Prepared Specimen

Swelling something that bulges out or is protuberant or projects from its surroundings.

Table 17.4 illustrates the weight change of the pellets following a 24-hour immersion in distilled water.

Table 17.4 Mean increase in weight of DRM

DRM	Mean Increase in Wt (g)
DRM-I	0.165
DRM-II	0.166
DRM-III	0.164
DRM-IV	0.143

4.2 Hardness Properties of Prepared Specimen

For this assessment, Shore D hardness testing can be carried out. The experiment's result demonstrates that the filled sample's hardness is marginally higher than the samples'. The sample DRM-IV has the highest hardness value, as depicted in Table 17.5. For sample DRM-I, the hardness value is at its lowest. We may deduce from these findings that the hardness value rises steadily from DRM-I to DRM-IV.

Table 17.5 Hardness values of DRM

DRM	Hardness values					Average Hardness
DRM-I	50	51	56	58	59	**54.8**
DRM-II	62	63	63	66	67	**64.2**
DRM-III	64	63	65	66	65	**64.6**
DRM-IV	73	77	78	77	77	**76.4**

4.3 Finite Element Analysis (FEA) Result of the Generating Models

FEA Model Definition

Reference sites for adding distributed loads are provided at Points 11 (0.5 space ratio) and 12 (2.0 space ratio). Choosing "Default occlusal constraints" as the boundary condition prevents the tooth bone's left and right sides from sliding horizontally while fixing the tooth bone's bottom. Table 17.6 lists the material qualities that were utilized. The entire loading on the tooth top is 130N, given that the tooth crown has a diameter of approximately 10 mm and that there is approximately 6.5 mm between points 11 and 12.

Table 17.6 Material properties of the tooth model

	Enamel	Dentine	Pulp	PDL	Bone
Young's modulus (MPa)	85000	19800	2.07	50	13800
Poisson's ratio	0.33	0.31	0.45	0.45	0.3

Fea Result for Tooth Model

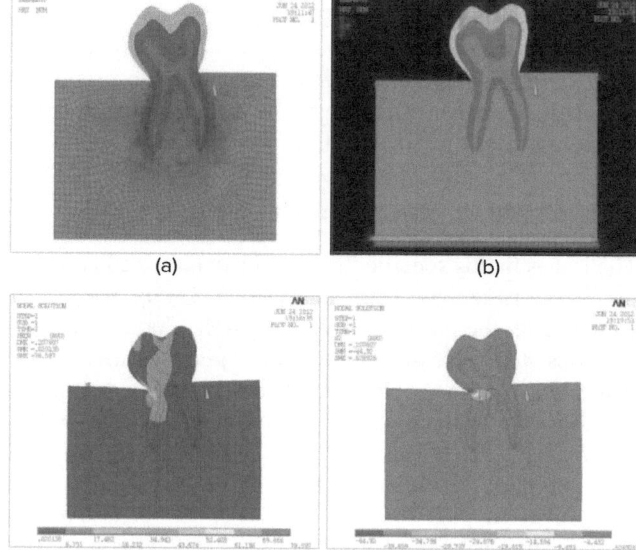

Fig. 17.2 (a) Meshing the profile, (b) Applying load to the profile, (c) Von mises stress in the profile, (d) Min. principal stress for profile

Table 17.7 ANSYS results for human tooth

Profile	Displacement vector sum in mm	Von-mises stress in N/mm^2	Max. principal stress in N/mm^2	Min. principal stress in N/mm^2
Human tooth	0.207807	78.587	16.694	0.628926

Fea Result for DRM-I

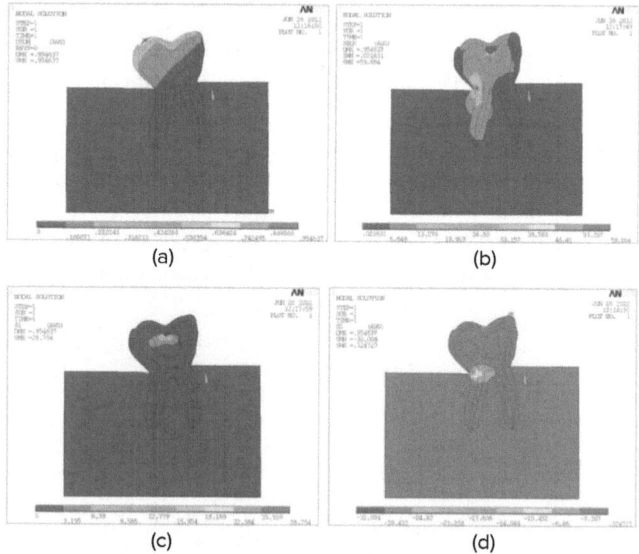

Fig. 17.3 (a) Displacement vector sum for enamel, (b) Von mises stress for enamel, (c) Max. principal stress for enamel, (d) Min. principal stresses for enamel

Fea Result for DRM-II

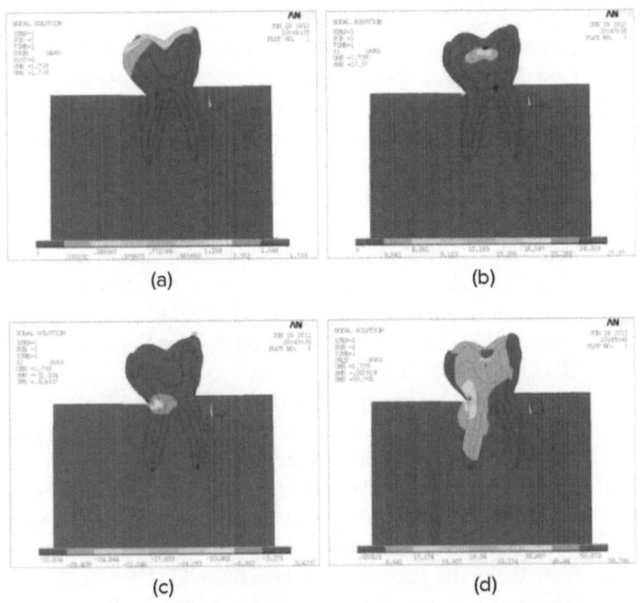

Fig. 17.4 (a) Displacement vector sum for enamel, (b) Von mises stress for enamel, (c) Max. principal stress for enamel, (d) Min. principal stresses for enamel

Fea Result for DRM-III

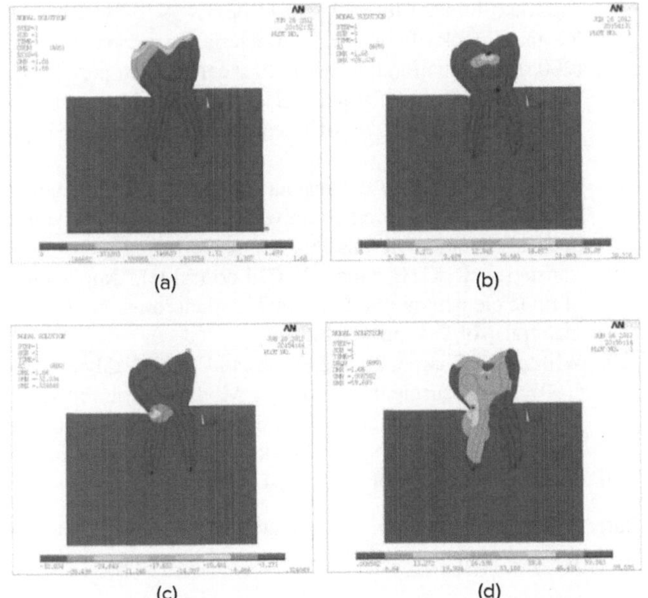

Fig. 17.5 (a) Displacement vector sum for enamel, (b) Von mises stress for enamel, (c) Max. principal stress for enamel, (d) Min. principal stresses for enamel

Fea Result for DRM-IV

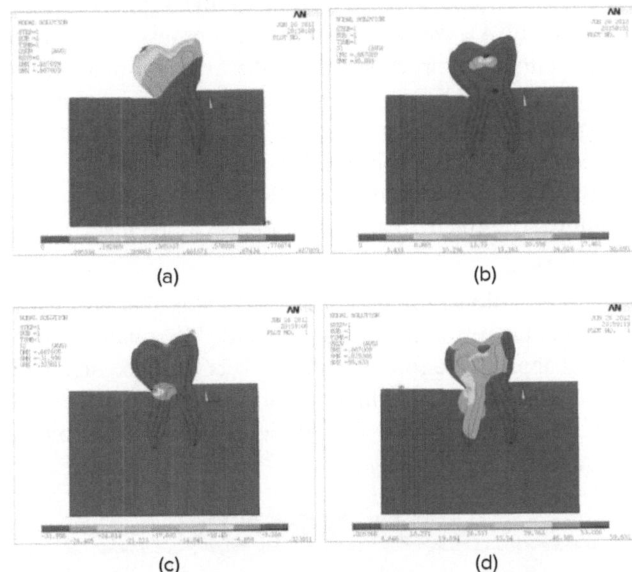

Fig. 17.6 (a) Displacement vector sum for enamel, (b) Von mises stress for enamel, (c) Max. principal stress for enamel, (d) Min. principal stresses for enamel

Table 17.8 Compare the ANSYS results for dental restoration materials

DRM	Displacement vector sum in mm	Von mises stress in N/mm^2	Max. principal stress in N/mm^2	Min. principal stress in N/mm^2
DRM-I	0.954637	59.664	28.754	0.324727
DRM-II	1.739	59.706	27.37	0.324337
DRM-III	1.68	59.695	28.226	0.324649
DRM-IV	0.867009	59.631	30.893	0.323011

5. Conclusion

A thorough review of the literature has been done in the areas of dental restoration materials, combinations of reinforcing materials, manufacturing processes, and mechanical testing techniques.

For the sample DRM-IV, the hardness value is at its maximum. For the sample DRM-I, the hardness value is at its lowest. These findings suggest that the difficulty value rises steadily from DRM-I to DRM-IV. In comparison to the control group, the test groups' thermal and chemical properties little changed, but their mechanical properties significantly improved. Future experiments and reports will examine more advanced methods of coating SNPs with glass powder and bonging liquid.

The distribution of stress and strain on the teeth under load conditions is amply demonstrated by the Finite Element Analysis (FEA). We can obtain fringe contours or a written report with the help of stress results after the analysis is completed. X-Y distribute graphs exemplifying the strains along the enamel are provided the four samples maximum primary stress distribution. The maximum stress value (30.893 Mpa) developed in the DRM-IV, according to the solution. After then, dental restorations can be made using DRM-IV.

The information provided above indicates that the nano silica dental restoration material possesses qualities that make it appropriate for restorations. This might be included into dental implants, which would be a very successful solution and get around the problems with current metal implants.

REFERENCES

1. David C. Wattsd (2011), "Mechanical behavior of post-restored upper canine teeth: A 3D FE analysis", dentalmaterials, Vol. 2 7, pp 1285–1294.
2. Debbie Guatelli-Steinberg (2009), "Recent Studies of Dental Development in Neandertals: Implications for Neandertal Life Histories,Evolutionary Anthropology", Vol.18, pp 9–20 (2009).
3. Garry J.P. Fleminga, Gary P. Moran (2011), "Development of a discriminatory compatibility testing model for non-precious dental casting alloys", dentalmaterials, Vol. 27, pp 1295–1306.
4. Antonio Galloza, José J. Torres, JorismarTorres (2004), "Biomechanics of Implants and Dental Materials", Applications of Engineering Mechanics in Medicine, Vol. 681, pp-1–17.
5. Carel L. Davidson, Advances in glass-ionomer cements, Journal Of Minimum Intervention In Dentistry2009.
6. Carla Castiglia Gonzaga, Paulo Francisco Cesar (2008), "Mechanical Properties and Porosity of Dental Glass- Ceramics Hot-Pressed at Different Temperatures", Material Research, Vol. 11, pp 301–306.
7. Cornelis J. Kleverlaan, Raimond N.B. van Duinen (2003), Albert J. Feilzer (2003),"Mechanical properties of glassionomercements affected by curing methods",Vol.20, pp 45–50.
8. Hsuan-Yu Chou and Ernesto Lopez (2004), "Wear of Dental Restorative Materials', Mechanics of Contact and Lubrication", Mechanical & Industrial Engineering, VOL.230. PP 1–15.
9. Kiyoshi Tajima (2009), "Three-dimensional finite element modeling from CT images of tooth and its validation" , Dent Mater, Vol. 27, pp-405–439.
10. Leon H. Prentice, Martin J. Tyas, Michael F. Burrow (2005), "The effect of mixing time on the handling and compressive strength of an encapsulated glass-ionomer cement", Dental material, Vic. 3010, pp 704–708.
11. Narasimha Raghavan R (2011),"Novel Multi-Walled Carbon Nano Tube Reinforced GlassinomerCemantsFor Dental Restorations", Dental material, Vol. 64, pp1045–1115.
12. Nagaraja Upadhya P and Kishore G (2005),"Glass Ionomer Cement – The Different Generations", Dental material, Vol. 18 (2), pp 2035–2051.
13. Naruporn Monmaturapoj, Sakulya Kinmonta (2010), "Mechanical Properties of Glass Ionomer Cements Doped TiO2", Journal of Metals, Materials, Vol.20, pp.193.
14. Pascal Magne P (2007), "Efficient 3D finite element analysis of dental restorative procedures using micro-CT data", Dent Mater ,Vol.23(5), pp 539–48.
15. PietroAusiello, Pasquale Franciosa, David C. Watts (2011), 'Numerical fatigue 3D-FE modeling of indirect Dental Restoration Materials-restored posterior teeth',Dental Material Vol. 27, pp 309–317.
16. RegenioMahfuzHerbstrith Segundo, Hugo Mitsuo Silva Oshima, Isaac Newton Lima da Silva (2009), 'Stress distribution of an internal connection implant prostheses set: 'A 3D finite element analysis', Stomatologija, Baltic Dental and Maxillofacial Journal,Vol. 11,pp 55–59.
17. Regina Guenka Palma-Dibb b, Maria Aparecida Zaghete (2005), "Evaluation of glass nomer cements properties obtained from niobium silicate glasses prepared by chemical process", Journal of Non-Crystalline Solids , Vol. 351, pp 466–471.
18. Sebastien Beuna, Jose Vrevena (2007), "Characterization of nanofilled compared to universal and microfilled composites", dental materials,Vol. 2 3, pp 79–81.
19. Vanstaden. R. C, H. Guan, Y. C. Loo (2008),' Application of finite element method in dental implant research', Dental material, Vol.134, pp-479–486.
20. William N. Sharpe, Jr., Bin Yuan, and Ranji Vaidyanathan (1997), "Measurments of Young's Modulus, Poisson's Ratio, and Tensile Strength of Polysillicon, Proceedings of the Tenth", IEEE International Workshop on Microelectromechanical Systems, Vol.55, pp 424–429.

Note: Every figure and table was created using the experimental work; none of them were taken from any publications or online.

Advances in Additive Manufacturing Technologies – Gurusamy Pathinettampadian (eds)
© 2024 Taylor & Francis Group, London, ISBN 978-1-032-90013-1

18 An Overview on Basalt based Polymeric Composites and Varied Filler Particulates

Zahir Hussain. M[1]
Department of Automobile Engineering,
Rajalakshmi Engineering College, Chennai, Tamilnadu, India

D. Jayabalakrishnan[2]
Department of Mechanical Engineering,
Chennai Institute of Technology, Chennai, Tamilnadu, India

N. Vijayasharathi[3]
Department of Mechanical Engineering,
Panimalar Engineering College, Chennai, Tamilnadu, India

ABSTRACT: Basalt fiber-reinforced polymer composites (BFRPs) offer superior tensile strength, modulus, and temperature resistance over glass fiber-reinforced counterparts, alongside ecological and economic advantages. Filler particles enhance mechanical, thermal, and electrical properties while mitigating weight and costs. Nano-fillers, micro-fillers, and natural fibers are commonly used. This review highlights filler incorporation as a promising avenue to greatly enhance BFRP characteristics, expanding their applicability across various industries.

KEYWORDS: Basalt fiber-reinforced polymer composites, Glass fiber-reinforced polymer composites

1. Introduction

Basalt-based polymer composites offer enhanced characteristics compared to typical glass fiber-reinforced composites, sparking significant interest. Basalt fiber, derived from crushed basalt minerals, undergoes smelting after a quick wash. Melting quarry basalt rock is crucial for producing basalt fiber, which is then formed into strands through small nozzles.[1]. Basalt fibers, when compared to other types of fibers, have superior qualities in a variety of respects. The study discusses the benefits of basalt fibers, as well as the techniques and developments that are now being made to modify basalt fibers [2]. Continuous basalt fiber, a high-performance material, holds promise for various applications including road building, energy, conservation, aircraft, weapons systems, and maritime vessels. However, current processing methods struggle to produce complex 3D constructions using continuous basalt fiber[3]. This research presents a comprehensive literature assessment on with and without basalt fibers reinforced composites. Manufacturers may drastically lower production costs without sacrificing material mechanical qualities by using basalt in various sectors because of the material's favorable cost-to-quality ratio [4]. In lieu of more usual materials and fibers, basalt fiber offers several desirable qualities. The utilization of basalt on fire-resistant fabric could result in maximum use of its benefits of high-temperature endurance and flame suppression [5]. In the usual scenar-

[1]zahirhussain.m@rajalakshmi.edu.in, [2]jayabalakrishnand@citchennai.net, [3]vijay.me.energy@gmail.com

DOI: 10.1201/9781003545774-18

io, a composite material consists of several discrete phases dispersed inside an ongoing phase. When many phases of distinct natures are present, the resulting composite is called as hybrid. The uninterrupted phase is commonly referred as matrix, whereas the interrupted phase is the reinforcement. Composite substances include the combination of constituents possessing complementary physical and mechanical characteristics [6-9]. Thermoset composites utilize resins that undergo a chemical process forming a cross-linked polymer chain, creating a rigid 3D network. This curing process is irreversible. In contrast, thermoplastic composites allow for reversibility, enabling recycling of scrap matrices. Figure 18.1 depicts the advantages of Polymer composites.The behavior of basalt fiber-reinforced polymer matrix composites joined by adhesives, rivets, and hybrid joining with aluminum is analyzed, along with the advantages and disadvantages of each method. Acoustic emission tracking system is utilized to investigate three different failure scenarios. [10]. Surface-modulated basalt fiber laced composites were prepared using hand lay-up methods with polyester. Evaluation of their inherent frequency and dynamic mechanical properties revealed favorable characteristics in modulus and damping factor for the untreated basalt fiber[11]. Figure 18.2 depicts the merits of basalt fiber.

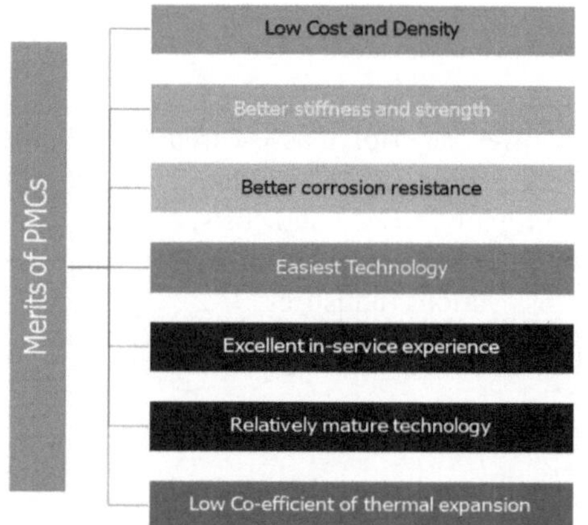

Fig. 18.1 Merits of polymer matrix composites

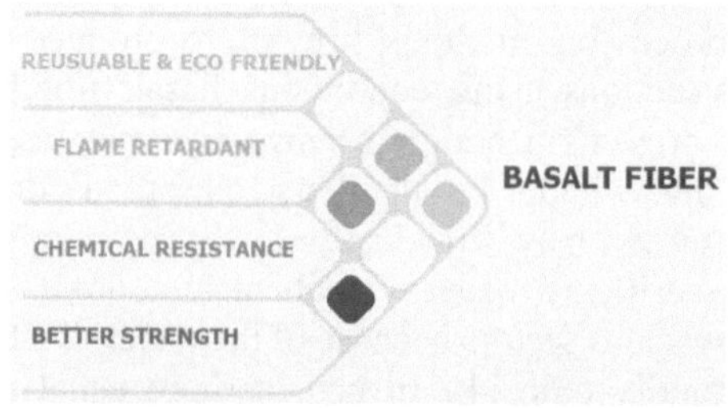

Fig. 18.2 Merits of basalt fiber

2. Basalt Fibers as Hybrid Composite

Hybridized polymeric materials are poised to revolutionize industrial applications, boasting sought-after features like high strength-to-load ratios, anti-corrosion frameworks, dimensional stability, excellent elongation, and substantial damping coefficients. Extensive research on hybridization, particularly with hybrid fibers, has yielded a plethora of new products with diverse uses. Reinforcing element hybridization stands out as a vital theme for optimizing composite properties, with hybrids showing superior qualities when integrated into polymer matrices compared to their individual counterparts.[12-16]. Fiber-by-fiber recombination continues to pose real-world challenges, especially for large-scale manufacturing. Hybrid layer-by-layer techniques are more practical and economical than the fiber-by-fiber approach. In this research, hybrid carbon/basalt FRP composites were modeled using the bundle-based modeling paradigm [17]. The manual lamination method was used to construct a hybrid material utilizing basalt and kenaf in four distinct mixes in order to create the composite with triple stack levels. This hybrid composite was then contrasted with the composite material composed of separate three-stack layers of basalt and kenaf. The tensile and flexural properties of the novel composites were examined [18]. It is shown that mixed composite layers tend to be lower in sensitivity to open holes than plain carbon fiber laminate, and considerable toughness qualities may be manufactured by fiber hybrid. The hybrid composite displays greater strain and collapses in a more ductile fashion than ordinary carbon fiber laminate. Combining carbon fiber composites with basalt fiber is a practical method for increasing their life spans [19]. In this investigation, it has investigated how altering the fiber volume ratio & fiber configuration affects the tensile and elongation characteristics of basalt-glass fiber hybrid composites. The deformation process of the basalt-glass fibers hybrid composite is studied in detail by the simulator [20]. Recent research explores the utilization of waste materials like basalt, paddy straw, and sheep's wool in hybrid production, showcasing their potential for high-quality goods. Evaluation of tensile, impact, and thermal characteristics reveals superior results compared to previous studies, particularly with the inclusion of sheep wool in thermoplastic hybrids.[21]. Trials evaluated the impact of adding chopped basalt fibers to nano clay augmented polymer composites on fracture toughness and associated energies. The findings can inform the design of polymer concrete for high-temperature conditions, showing basalt fibers enhance fracture properties. Their affordability also makes them appealing for practical applications to boost fracture toughness and energy.[22]. A filler substance called High-Density Polyethylene Fibre (HDPE) was employed in this investigation. It ranges from 0.5 to 3% in the concrete. The Basalt and Geotextile mats were placed around concrete frames to determine how the HDPE and fiber coverings affected the properties [23]. Nano-silica on basalt fibers enhances bond durability between fibers and the matrix, strengthening the physical characteristics of cement stone. Flame retardant substrates, comprising fir sawdust and urea-formaldehyde, showed a 10% difference in suppression efficacy with basalt fiber additions.[24]. Laminate supports were created using woven glass and basalt paired with an epoxy matrix. Basalt epoxy composites exhibited superior mechanical characteristics compared to glass-epoxy composites. Among the three processes examined, the hand layup approach with compression molding produced the highest-quality material results.[25-26].

3. Optimization Methods in Basalt Fibers Polymer Composite

Orthogonal designs of experiments streamline trial optimization by efficiently selecting variable combinations and levels, facilitating successful and economical evaluation of essential components and interactions. The strategy aims to represent the experimental area accurately with typical trials. In the creation of a hybrid material using manual building, three layers of basalt and two PTFE-coated glass woven fibers were employed. Twenty-six trials were conducted to complete Nd: YAG laser drilling on the produced layers. [27]. Hybrid particle green composites made from fiber particles, cutting properties were examined using analysis of variance. Multiple-variable linear regression in Minitab 17 determined the association. A verification test confirmed the projected delamination significance, matching the figure [28]. The goal of this research was to determine the best combination of milling settings for reducing delamination to a minimum. The tests are carried out using the Central composite design matrix, and the results acquired are examined using the signal-to-noise ratio. Every milling process setting for Hybrid particle GFRPs composites is analyzed using analysis of variance to establish its relative importance. The response settings have been optimized using grey relational analysis for several kerf attributes. Researchers have shown that the settings of the laser and the qualities of the distinct fibers affect the kerf's quality attributes [29-30]. This work develops an optimal machine learning framework to forecast basalt fiber augmented polymer composite laser cutting kerf quality. The simulator uses each model was taught and assessed using data from experiments on five process-controlling parameters [31]. The Taguchi approach was used to create dry sliding wear experiments, optimize input parameters for operation, and analyze variation. A hybrid Analytic Hierarchy Process – R technique was used to rank hybrids by efficiency [32].

4. Basalt Fibers with Varied Fillers

Incorporating filler particles into fiber-reinforced polymer composites enhances their mechanical and physical characteristics significantly, crucial for modern applications. Strategic selection of filler particulates with distinct qualities from resins yields polymeric composites with superior attributes, including enhanced creep and rupture resistance, and flameproofing, making them vital in contemporary society. Table 18.1 displays the filler particulates usage in Basalt fiber polymeric composites.

Table 18.1 Displays BF with different fillers

Articles	Filler with its Percentage	Resin	Optimal %	Properties
[33]	No fiber- WC clay-based filler 10 wt.%	Plain and Nano modified Epoxy		Fracture, Compression Test
[34]	Basalt – No filler	Epoxy		TS, Mechanical
[35]	Hyacinth Nano Tamarind Shell Ash (0, 1, 3, 5, 7, 9 %)	LY556	5-7%	TS, POE, FT, Impact Strength (IS)
[36]	Kenaf, granite nano powder (10,20,30 grams)	LY556 and Araldite	30 grams	TS, POE, FT, IS
[37]	Basalt fiber with 0% filler	4 types - normal and toughened vinyl ester and epoxy 9102-70, MFE-9, 9804A/B, L-500AS/BS		Static and Fatigue Tests
[38]	Basalt fiber (BF) Tungsten Carbide (WC) 0 to 8%	LY556	6%	Tensile Strength (TS) and percentage of elongation (POE)
[39]	BF, Nano Titanium Dioxide (TiO_2) 1 to 5%	MY 740 and HY 951	4%	Flexural Test (FT), TS and Inter Laminar Shear Strength (ILSS)
[40]	BF, Nanoclay (1, 3, 5 wt%)	Miracast 1517A and B	5 wt%	Friction and wear Test
[41]	BF, each 10 wt, % moringa and bagasse ash	Epoxy and Araldite HY951	10 wt, % moringa ash	Water absorption test (WAT), TS, IS, Corrosion
[42]	BF, Nano zirconia ZrO2, (0, 1, 3, 5%)	ML-506 with hardener HA-11	3 wt.%	Tensile & Flexural properties
[43]	BF carbon nanotubes and graphene	epoxy	Fiber Treatment Methods	Mechanical and Wear behaviour
[44]	BF, Micro Coir Fiber and Titanium Carbide	Bio Sr 33 and Synthetic YD-535 LV Epoxy	Each 5%	Density, TS, WAT, Thermal
[45]	Jute, BF Graphene (0 , 0.2 , 0.4, 0.6, 1 wt%)	Epoxy (araldite) LY 556, hardener HY 951	0.4 wt.%	TS,IS, FS, Hardness
[46]	Kevlar –BF, Al_2O_3 Nano (1,3,5%)	Epoxy resin and HY951	3%	Energy Absorbed, High-Speed Bullet Impact Test

5. Conclusion

Basalt emerges as a superior alternative to traditional fibers due to its eco-friendliness, non-toxicity, and ecological status. Its chemical impermeability, high resistance to deterioration, and low thermal conductivity make it preferable for reinforcement. Hybridizing basalt fibers enhances the link between matrix and fiber, resulting in significant mechanical improvements. Research into basalt materials is encouraged, potentially lowering production costs and yielding novel designs that balance unique functionalities and mechanical qualities. This overview aims to increase scholars' awareness of basalt's importance and its diverse applications.

References

1. Shi FJ. A Study on Structure and Properties of Basalt Fiber. AMM 2012;238:17–21.
2. Zhang, Xiao Fei, et al. "Study on Modification of Basalt Fibers." Advanced Materials Research, vol. 332–334, Trans Tech Publications, Ltd., Sept. 2011, pp. 2028–2031. Crossref,
3. Zhufeng Liu, Bin Su, Lichao Zhang, Zhaoqing Li, Changshun Wang, Zhenhua Wu, Siqi Wu, Hongzhi Wu, Peng Geng, Si Chen, Guizhou Liu, Lei Yang, Zhigang Xia, Chunze Yan, Yusheng Shi, Continuous basalt fibers into fireproof and thermal insulation architectures using an additive manufacturing manipulator, Materials & Design, 2023, 112434,
4. Chowdhury IR, Pemberton R, Summerscales J. Developments and Industrial Applications of Basalt Fibre Reinforced Composite Materials. Journal of Composites Science. 2022; 6(12):367.
5. Pang, Yu, et al. "Research on Fire-Resistant Fabric Properties of Basalt Fiber." Applied Mechanics and Materials, vol. 217–219, Trans Tech Publications, Ltd., 12 Nov. 2012, pp. 1151–1154. Crossref,
6. Rachid Hsissou, Rajaa Seghiri, Zakaria Benzekri, Miloudi Hilali, Mohamed Rafik, Ahmed Elharfi, Polymer composite materials: A comprehensive review, Composite Structures, Volume 262, 2021, 113640,
7. Prabhu, P., Jayabalakrishnan, D., Balaji, V. et al. Mechanical, tribology, dielectric, thermal conductivity, and water absorption behaviour of Caryota urens woven fibre-reinforced coconut husk biochar toughened wood-plastic composite. Biomass Conv. Bioref. (2022).
8. A. Sreenivasulu, K.S. Ashraff Ali, P. Arumugam, Shaik Rahamat Basha, P. Velmurugan, S. Rajkumar, S. Suresh Kumar, Investigation on thermal properties of tamarind shell particles reinforced hybrid polymer matrix composites, Materials Today: Proceedings, Volume 59, Part 2, 2022, Pages 1305–1311,
9. Karthigairajan, M., Nagarajan, P.K., Raviraja Malarvannan, R. et al. Effect of Silane-Treated Rice Husk Derived Biosilica on Visco-Elastic, Thermal Conductivity and Hydrophobicity Behavior of Epoxy Biocomposite Coating for Air-Duct Application. Silicon 13, 4421–4430 (2021).
10. Vara Prasad V, Eswara Rao T. A Review on Mechanical and Acoustic Emission Behavior of Basalt Fiber Reinforced Composites. AMR 2018;1148:37–42.
11. P Amuthakkannan & V Manikandan, Free vibration and dynamic mechanical properties of basalt fiber reinforced polymer composites, Indian Journal of Engineering & Materials Sciences Vol. 25, June 2018, pp. 265–270
12. Seydibeyoğlu MÖ, Dogru A, Wang J, Rencheck M, Han Y, Wang L, Seydibeyoğlu EA, Zhao X, Ong K, Shatkin JA, et al. Review on Hybrid Reinforced Polymer Matrix Composites with Nanocellulose, Nanomaterials, and Other Fibers. Polymers. 2023; 15(4):984.
13. Saurabh Khandelwal, Kyong Yop Rhee, Recent advances in basalt-fiber-reinforced composites: Tailoring the fiber-matrix interface, Composites Part B: Engineering, Volume 192, 2020, 108011.
14. V. Mohanavel, S. Suresh Kumar, J. Vairamuthu, P. Ganeshan & B. Nagaraja Ganesh (2022) Influence of Stacking Sequence and Fiber Content on the Mechanical Properties of Natural and Synthetic Fibers Reinforced Penta-Layered Hybrid Composites, Journal of Natural Fibers, 19:13, 5258–5270.
15. V R, A.P., V, J., T, M. et al. Effect of Silicon Coupling Grafted Ferric Oxide and E-Glass Fibre in Thermal Stability, Wear and Tensile Fatigue Behaviour of Epoxy Hybrid Composite. Silicon 12, 2533–2544 (2020).
16. M. A. Rahuman, S. S. Kumar, R. Prithivirajan, and S. Gowri Shankar, "Dry sliding wear behavior of glass and jute fiber hybrid reinforced epoxy composites," International Journal of Engineering Research and Development, vol. 10, no. 11, pp. 46–50, 2014.
17. Zheqi Peng, Xin Wang, Lining Ding, Zhishen Wu, Integrative tensile prediction and parametric analysis of unidirectional carbon/basalt hybrid fiber reinforced polymer composites by bundle-based modeling, Materials & Design, Volume 218, 2022, 110697.
18. M. Umashankaran, S. Gopalakrishnan & S. Sathish (2020): Preparation and characterization of tensile and bending properties of basalt-kenaf reinforced hybrid polymer composites, International Journal of Polymer Analysis and Characterization.
19. Zhang CH, Zhang JB, Qu MC, Zhang JN. Toughness Properties of Basalt/Carbon Fiber Hybrid Composites. AMR 2010;150–151:732–5.
20. Gang Wu, Yuhang Ren, Jinbo Du, Hongguang Wang, Xin Zhang, Mechanical properties and failure mechanism analysis of basalt-glass fibers hybrid FRP composite bars, Case Studies in Construction Materials, Volume 19, 2023, e02391,
21. Bhushan Hajare, Omkar Patil, G.S. Barpande, S. Radhakrishnan, M.B. Kulkarni, Tensile, impact and thermal properties of farm-waste based hybrid basalt polymer composites, Cleaner Materials, Volume 6, 2022, 100157,
22. Ali Abdi Aghdam, Mostafa Hassani Niaki, Milad Sakkaki, Effect of basalt fibers on fracture properties of nanoclay reinforced polymer concrete after exposure to elevated temperatures, Journal of Building Engineering, Volume 76, 2023, 107329,.
23. Gudadappanavar B, Kulkarni DK, Gouda PSS (2023) Influence of Basalt and Geo-Textile Fiber Wrapping on

Compressive and Flexural Strengths of HDPE-filled Reinforced Concrete. Indian Journal of Science and Technology 16(18): 1349-1356.

24. Yong Zheng, De Sun, Qian Feng, Zhigang Peng, Nano-SiO2 modified basalt fiber for enhancing mechanical properties of oil well cement, Colloids and Surfaces A: Physicochemical and Engineering Aspects, Volume 648, 2022, 128900,
25. Xie, Kai Fang, et al. "Flame Retardant Wood-Based Composite Modified by Chopped Basalt Fibers and Basalt Glass Powder." Advanced Materials Research, vol. 627, Trans Tech Publications, Ltd., Dec. 2012, pp. 722–725. Crossref,.
26. Raajeshkrishna, C.R., Chandramohan, P. Effect of reinforcements and processing method on mechanical properties of glass and basalt epoxy composites. SN Appl. Sci. 2, 959 (2020).
27. Amiya Kumar Sahoo, Dhananjay R. Mishra, Parametric optimization of response parameter of Nd-YAG laser drilling for basalt-PTFE coated glass fibre using genetic algorithm, Journal of Engineering Research, 2023,
28. S. Venkatraman, K.S. Aravind, S. Sri Sudharshan Sharma and Dr. M.P. Jenarthanan, Optimization of Basalt Fibre–Reinforced Polymer Composites for Minimizing Delamination Effect, Advance in Technology, Engineering and Compuring – A Multinational Colloquium – 2017 Editor(s) Vinod P., Sunil Jacob, Sheeja Janaradhanan and Ratish Menon
29. S. Vivekanand , K. Goutham and M.P. Jenarthanan, Optimization of Basalt Hybrid Particulate Glass Fibre-Reinforced Polymer Composite for Minimizing Delamination Effect during Milling, Journal of Chemical and Pharmaceutical Sciences, JCHPS Special Issue 7: May 2017, pp-64–70.
30. hananjay R. Mishra, Abhay Bajaj, Rajat Bisht, Optimization of multiple kerf quality characteristics for cutting operation on carbon–basalt–Kevlar29 hybrid composite material using pulsed Nd:YAG laser using GRA, CIRP Journal of Manufacturing Science and Technology, Volume 30, 2020, Pages 174–183.
31. I.M.R. Najjar, A.M. Sadoun, Mohamed Abd Elaziz, A.W. Abdallah, A. Fathy, Ammar H. Elsheikh, Predicting kerf quality characteristics in laser cutting of basalt fibers reinforced polymer composites using neural network and chimp optimization, Alexandria Engineering Journal, Volume 61, Issue 12, 2022, Pages 11005–11018,
32. Sharma RP, Sharma A, Jeganmohan S, Kumar M, Kumar A. Parametric optimization and ranking analysis of basalt fiber–marble dust particulates–polyamide 66 polymer composites under dry sliding wear investigation. Proceedings of the Institution of Mechanical Engineers, Part E: Journal of Process Mechanical Engineering. 2023;
33. A. Sreenivasulu, S. Rajkumar, Sridhar Sathyanarayana, S. Suresh kumar, G.V. Gaurav, Bandaru Dyva Isac Premkumar, Impact of nano-filler WC on the fracture strength of epoxy resin, Materials Today: Proceedings, Volume 59, Part 2, 2022, Pages 1420–1424,.
34. Louis Le Gué, Peter Davies, Mael Arhant, Benoit Vincent, Wouter Verbouwe, Basalt fibre degradation in seawater and consequences for long term composite reinforcement, Composites Part A: Applied Science and Manufacturing, 2024, 108027.
35. Mohanavel V et al, Influence of Nanofillers on the Mechanical Characteristics of Natural Fiber Reinforced Polymer Composites, 2022 ECS Trans. 107 12513.
36. T. Raja, V. Mohanavel, S. Suresh Kumar, S. Rajkumar, M. Ravichandran, Ram Subbiah, Evaluation of mechanical properties on kenaf fiber reinforced granite nano filler particulates hybrid polymer composite, Materials Today: Proceedings, Volume 59, Part 2, 2022, Pages 1345–1348.
37. Xing Zhao, Xin Wang, Zhishen Wu, Jin Wu, Experimental study on effect of resin matrix in basalt fiber reinforced polymer composites under static and fatigue loading, Construction and Building Materials, Volume 242, 2020, 118121.
38. Kalmeshwar Ullegaddi, Shivarudraiah, Mahesha C R, Significance of Tungsten Carbide Filler Reinforcement on Ultimate Tensile Strength of Basalt Fiber Epoxy Composites, International Journal of Recent Technology and Engineering (IJRTE), ISSN: 2277–3878, Volume-8 Issue-3, September 2019.
39. Tejas Iyer, Suhas Yeshwant Nayak, Anupama Hiremath, Srinivas Shenoy Heckadka & J P Jaideep (2023) Influence of TiO2 nanoparticle modification on the mechanical properties of basalt-reinforced epoxy composites, Cogent Engineering, 10:1, 2227397.
40. A. Jumahat, A.A.A. Talib, H. Sharudin, N.W.M. Zulfikli, Chapter 8 - Tribological behavior of nanoclay-filled basalt fiber-reinforced polymer composites, Editor(s): Sanjay Mavinkere Rangappa, Suchart Siengchin, Jyotishkumar Parameswaranpillai, Klaus Friedrich, In Elsevier Series on Tribology and Surface Engineering, Tribology of Polymer Composites, Elsevier, 2021, Pages 143–162, ISBN 9780128197677.
41. Sampath, P., & Santhanam, S. K. V. (2019). Effect of moringa and bagasse ash filler particles on basalt/epoxy composites. Polímeros: Ciência e Tecnologia, 29(3), e2019034.
42. Azizi H, Eslami-Farsani R. Study of mechanical properties of basalt fibers/epoxy composites containing silane-modified nanozirconia. Journal of Industrial Textiles. 2021;51(4):649–663.
43. Balaji KV, Shirvanimoghaddam K, Naebe M, Multifunctional basalt fiber polymer composites enabled by carbon nanotubes and graphene, Composites Part B (2023), Volume 268, 2024, 111070,
44. Arshad, M.N., Mohit, H., Sanjay, M.R. et al. Effect of coir fiber and TiC nanoparticles on basalt fiber reinforced epoxy hybrid composites: physico–mechanical characteristics. Cellulose 28, 3451–3471 (2021).
45. Kishore, M., Amrita, M. Mechanical characterization of jute-basalt hybrid composites with graphene as nanofiller. J Mech Sci Technol 36, 3923–3929 (2022).
46. J. Jensin Joshua, Dalbir Singh, P. S. Venkatanarayanan, Ch. Sai Snehit, A. Bipin Sai Eswar, Melaku Desta, "Experimental Estimation of Energy Absorbed and Impact Strength of Kevlar/Basalt-Epoxy Interwoven Composite Laminate Added with Al2O3 Nanoparticles after High-Velocity Bullet Impact", Journal of Engineering, vol. 2023, Article ID 2830575, 11 pages, 2023.

Note: Every figure was created using the experimental work; none of them were taken from any publications or online.

Advances in Additive Manufacturing Technologies – Gurusamy Pathinettampadian (eds)
© 2024 Taylor & Francis Group, London, ISBN 978-1-032-90013-1

19 Automated Sewage WellDesilting Machine to Eradicate Manual Scavenging under the Surveillance of IoT

P.T.V. Suresh Kumar[1]
Research Scholar, Dept. of Mechanical Engg.,
Dr. M.G.R. Educational & Research Institute, Chennai & DAE, CMWSSB, Chennai, TN, India

J. Jayaprakash[2]
Professor, Dept. of Mechanical Engg.,
Dr. M.G.R. Educational & Research Institute, Chennai, TN, India

K.R. Padmavathi[3]
Professor, Dept. of Mechanical Engg.,
Panimalar Engineering College, Chennai, TN, India

R. Mohanraj[4]
Associate Professor,
Dept. of Computer Science Engg. (AI&ML), SVCET, Chittor, AP, India

ABSTRACT: A sewage pumping facility consists of a bigger reservoir, referred to as a moist well, which serves as the recipient for wastewater from a single house or a cluster of dwellings. Manual desilting of sewage pumping station (SPS) wells is a hazardous and degrading occupation that is still practicedin many parts of the world.Manual scavengers are exposed to highly toxic components of sewer gases such as hydrogen sulfide, methane and ammonia. At high concentrations, the toxic gases can interfere with the sense of smell and can cause suffocation, loss of consciousness and death. To eradicate the drawbacks of manual scavenging and manual desilting of SPS wells, the design and development of an automated sewage well-desilting machine using an automated grab bucket system is proposed. The proposed system eliminates the need for human workers to enter hazardous SPS wells, which is more efficient and effective than manual desilting and is more cost-effective in the long term. It is also designed to be easy to operate and maintain and can be operated by a single person with minimal maintenance. The proposed system has the potential to revolutionize the way that SPS wells are desilted and can help eradicate manual scavenging, improve public health, and reduce costs. A surveillance-based sewage monitoring system with the Internet of Things (IoT) monitors the sewage pressure, temperature and quality which helps in the desilting of SPS wells effectively.

KEYWORDS: Desilting machine, Sewage pumping station, Manual scavenging, Grab bucket, IoT

[1]ptvsureshkumar@gmail.com, [2]jayaprakash.mech@drmgrdu.ac.in, [3]krpadmavathipecmech@gmail.com, [4]mohanraj254@gmail.com

DOI: 10.1201/9781003545774-19

1. INTRODUCTION

A sewerage system consists of the house service connection which conveys sewage and sullage generated from the house to the sewers in the street usually through a manhole. Sewers are pipes or conduits meant for carrying sewage and are laid underground and the flow is by gravity. This sewage is collected in the Sewage Pumping Station. The online station for lifting the sewage from the deeper sewer, greater than 6m, generally to other sewers or another Sewage Pumping Station is called Lift Station. Roadside Pumping Stations are provided to lift the sewage from one manhole to another manhole or Sewage Pumping Station when the transfer cannot be achieved by gravity[1]. Systematic salvaging refers to the physical sanitation, transportation, disposal, or processing of organic faces in unsanitary toilets, uncovered sewers, or holes. This activity is considered a felony of immense magnitude, akin to extermination.

To eliminate the practice of hand scouring, an autonomous effluent purification apparatus was used to eliminate particulate trash and extract fluid effluent An approach was devised to minimize labor-intensive tasks. Obstructions in the drainage system may be detected using both an infrared camera and an infrared detector. While the crawler is in progress inside the dirty water, the infrared detector identifies the existence of blocking [2]. Bethy Merchan-Sanmartin developed a network that aids in the regulation of sewage and research on ecological awareness. This network involves an integrated concept of a drainage network with stabilizer reservoirs that may be replicated in regions facing comparable problems [3].

An autonomous wastewater draining equipment was developed to substitute individual work in purifying reservoirs. The system includes axles, a cable, a structure, effluent retention, gears, talons, and a propeller. Discharge pipelines are used to dispose of waste, but sadly, there is a risk of personal casualties during the procedure of clearing obstructions in these pipelines. The mechanism is capable of moving inside the drainage to gather floating trash, hence reducing the need for personal assistance [4]. A mobile grabber with customizable abilities was created and built specifically for cleansing effluent tubes with varying dimensions. This manipulator is very efficient in removing scales and obstructions that collect in the pipeline [5-6]. Saurabh S. Satpute introduced an idea for automating the process of cleansing sewage networks by using a robotic grasp to raise the effluent and a container to store it [7].

Ravindra R. Patil et. al. observed a growing inclination towards the use of mechanized and autonomous technologies to do activities that are dangerous, uncomfortable, and unclean for individuals. The use of robotics for inspecting subterranean pipelines for clogs and breakage is already being used in sewage management and maintenance [8]. In their study, Nagadeepam et.al proposed the utilization of a structural drain cleaner to substitute traditional work in the process of clearing sewers networks. This equipment is designed to obtain, transport, and disposal of robust garbage by utilizing grasps to gather the garbage and a conveyance framework to transfer it to a container located at the rear, where it can be properly eliminated [9]. A cost-effective autonomous bot has been developed with the capability to efficiently remove obstacles, ensuring the safety of individual existence. The crawler will be inserted into the effluent pipelines via a sewer. The object will continue to move forward until it encounters the obstacle, at which point it will pass over the obstacle. Fouzea et.al. developed a machine which removes solid waste and againsucks away the liquid sewage [10]. Babu et.al. developed an autonomous effluent cleansing device capable of removing substantial suspended debris, such as polyethene or similar lightweight substances [11].

Karel Mulder outlines prospective pathways for the expansion of sewage and metropolitan stormwater infrastructure in developed nations. The investigation will examine pertinent patterns in drainage network development [12]. The research focused on analyzing incidents at effluent transporting locations, functioning issues in garbage elimination structures, and the effects of exterior variables on wastewater structure effectiveness. The objective was to develop an automatic dampening reservoir for effluent circulating locations, serving as an extra element of the circulating facility [13-14].Anshul Mathur and colleagues suggested that mechanization is essential for minimizing personal workload and offering complete remedies to issues, consequently enhancing correctness. They created a partially automated effluent treatment machinery to eliminate the need for physical scouring [15]. Hanfang Yun et.al. devised a sediment cleansing equipment that has broader application, easier functioning, and improved effectiveness. This equipment aims to address the lack of sediment cleansing equipment in smaller and larger sewers in the national sector. This research examines the classification of traditional scrapers as sanitary employees, the grounds for professional exclusion of sanitary employees, and the potential of automated equipment to rehabilitate and improve the societal position of sewerage personnel [16]. It is emphasized for the necessity of immediate action to abolish manual scavenging and defend sanitation worker's rights [17].

In a Sewage Pumping Station (SPS) the sewerage is collected by different sewer lines through a collection system to collect sewage from the communities, and domestic wastewater from industries to convey the sewage to distant points for treatment and safe disposal and to treat the sewage to the required degree and reuse the same, dispose it to river courses without any health hazards. The generation of wastewater from communities depends upon the rate of water supply per capita. 80% of water supplied is taken as wastewater return to the system. Then the floating matters are screened and removed, allowed to settle for a very short time for the grit to settle and separated and thereafter pumped to another pumping station or Sewage Treatment Plant [STP]. If the area is large and has moderately flat terrain like Chennai and if the sewage is carried away to the disposal point by gravity, the depth of the sewer will be more.It is advisable to restrict the depth of the sewer to 4.5 m to 5.0 m in operation and maintenance(O&M) point of view. In such a case, the area is divided into convenient zones and each zone is provided with one or more pumping stations. The site for the sewage pumping station should be relatively at a lower level and away from the habitats as practically as possible [1].

Wastewater collection reservoirs are often reservoirs that gather wastewater discharge and enable it to flow towards a centralized setting, from whence it may be pumped to a greater altitude drainage pipe, discharged elsewhere, or directed to a processing facility. The acquiring cistern does not offer any kind of intervention, although it may have an autonomous pumping management system for the disposal of effluent. Collection wells are currently proposed with the manual method of desilting using porcelain and pulley. Figure 19.1 shows the SPS wells which are to be desilted at regular intervals.

Fig. 19.1 Desilting of SPS wells (Courtesy: CMWSSB, Chennai)

The process of dredging sewerage entails the extraction of simultaneously compacted and unconsolidated sediment. This will enable the drainage to regain its optimal operational ability, hence avoiding any type of obstruction. By going through the literature review there are a few sludge cleaning devices, automatic sewage cleaning machines, robots with infrared cameras and sensors to identify the blocks in the sewer system, automated drainage cleaning equipment etc. are available. The problems associated with traditional gathering and dredging must be eliminated, and the challenges of operating and maintaining effluent gathering reservoirs or circulating terminals need to be solved. Each effluent pumping facility typically necessitates a typical of 5 to 6 traditional eliminates per year, in addition to a pair of dredging activities. Every traditional disinfecting method necessitates the separation of the movement in the structure. However, there are very few or no automated or semi-automated devices or equipment to desilt the sewage wells. Thus, in this study, it is proposed to design and develop an Automated Sewage Pumping Station well desilting machine to Eradicate Manual Scavenging.

2. Design and Development of Automated SPS Well-Desilting Machine

Manual desilting of sewage pumping station (SPS) wells is a hazardous and degrading occupation that is still practised in many parts of the world. To overcome the drawbacks of manual desilting, an automated system for desilting SPS wells using a reel drum hose, reel drum rope, S P motor, electrical control valves, grab bucket, and control panel is proposed. The proposed system eliminates the need for human workers to enter hazardous SPS wells, is more efficient and effective than manual desilting, and is more cost-effective in the long term. It is also designed to be easy to operate and maintain and can be operated by a single person with minimal maintenance. The proposed system has the potential to revolutionize the way that SPS wells are desilted and can help eradicate manual scavenging, improve public health, and reduce costs.

The SPS borehole dredging equipment is a hydraulic-operated and electrically-driven clearing-grabbing scoop device. It is created to function via remote direct surveillance, making it a perfect alternative for clearing submerged holes. The equipment can efficiently and effortlessly retrieve sediment or mud from depths ranging from 30 to 50 feet or more, using remote direct management, eliminating the need for people to access most reservoirs. The equipment is fitted with a mobile cart to access the walls of the well and an acquiring container or dump truck to prevent the need for repeated processing of the sewage. This allows for the sewage to be gathered in large vessels. Since it is developed on existing well chain block mounting, it can be easily fitted and used effectively on sewage collection wells. The power trolley, power motors with brakes, panel board assembly, hose reel drum, rope drum, hydraulic hoses, steel rope, hydraulic grab bucket, hydraulic power unit and remote-control device are the components of the SPS well desilting machine.

3. Components of the Desilting Machine

Automated desilting machine consists of an electrically operated power trolly with a 1 ton capacity, operating with wireless remote control is used to move the machine longitudinally. An electrically operated power motor with a brake unit of 1.2-ton capacity, operating with wireless remote control is used for machine rope and hydraulic hose winding with grab bucket lifting purpose. Panel power board assembly consists of an AC 230 V operating panel board with contactors which can be operated with wireless remote control and it is used for the functioning of hydraulic power motor, trolly motor and lifting power motors. Hydraulic hose reel made with mild steel is used for grab bucket functioning for 50 feet of length hose winding purpose engaged with power motor assembly. The rope drum is made up of mild steel and is used for grab bucket lifting purposes. It has a 6mm diameter winding 50 feet rope, engaged with a power motor assembly. It has a 1/4" hydraulic hose with a length of 50 feet winding in the reel drum for grab bucket cylinder open-close function and 30 ft of hose from the hydraulic power pack to the reel drum. A galvanized steel wire having 1770 N/mm2 tensile strength with a minimum breaking force of 20 KN and 8 mm thick moulded mild steel cast rope length of 50 feet winding in a rope drum is used for lifting the grab bucket.

A radio wireless remote-control system with a range of 100 meters with IP65 is used to operate the grab bucket system for desilting. There are a total of 7 individual rate monitors, 1 control for turning on and off, and 1 safety button switch designed to prevent illegal utilization. The computer connection can configure interior activities. There are a total of 7 individual rate monitors, 1 control for turning on and off, and 1 safety button switch designed to prevent illegal utilization. The computer connection can configure interior activities. The drive for the hydraulic system will be tapped from the vehicle's engine. It has a 160º jaw opening angle with 120 mm piston displacement. A grab bucket on a wire rope is lowered into the sewage collection well or wet well in an open condition with the help of a wire rope reel. On reaching the bottom of the collection well, the segments are closed, and the accumulated silt is picked up. The hydraulic power unit comprises the operating voltage of 230 Volts AC supply, 20 liters capacity oil tank with a working pressure of 70 bar, a pressure gauge and a solenoid switch which are used for the functioning of the grab bucket.

4. Working of SPS Well Desilting Machine

A mechanically powered electrical pulley drive dredging grabbing container equipment has been designed and is controlled via a wirelessly remote-control device. This system is the perfect answer for properly cleansing or dredging effluent wetness reservoirs. The equipment can efficiently retrieve silt or sludge from depths of 30 to 50 feet or more via remote wireless surveillance, eliminating the need for people to access the effluent moist reservoirs. The equipment is fitted with a mobile cart to access the well's exterior borders and retrieve sewage in large containers, thus avoiding the need for repeated handling. The automation of desilting machines will aid in improving their performance and effective utilization. By doing so, optimal use of desilting machine to its maximum potential can be ensured. Figure 19.2 displays the schematic of the SPS well desilting machine.

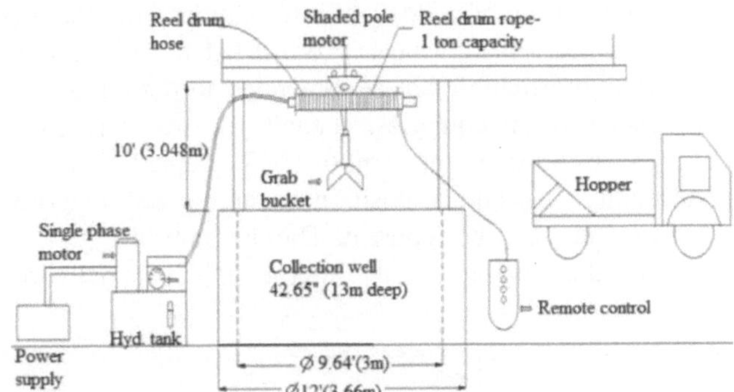

Fig. 19.2 Schematic of the SPS well desilting machine

Source: Created using the experimental work

5. Implementation of IoT

It is proposed to implement an IoT (Internet of things) based monitoring system to enable real time remote monitoring of the desilting machines deployed by Chennai Metropolitan Water Supply and Sewerage Board (CMWSSB) and create a relevant data basefor identifying the level of sewer in the collection well, quantity of silt removed by the desilting machines, to monitor vehicle's location and the current operational status of the desilting machine, the toxicity of sewer gas, density, humidity and temperature levels of sewer while tracking the real-time dynamic changes in the above factors. If the levels of sewer in the well, toxic gases, humidity and temperature exceed acceptable levels, it will send an alert to the devices connected to it and the technical people involved in sewer desilting. Global System for Mobile Communication (GSM) and Raspberry-pi are the systems which are used to communicate with electronic devices and to gather the sensor levels. GSM is a protocol for computerized mobile interactions, facilitating global interaction among cellular units. The physical components of the unit include a GSM section, a dynamic phase shifter, a SIM network, and a microprocessor. The microcontroller facilitates the supervision of transmitting, acquiring, and analyzing signals for the GSM unit. The software is accountable for performing activities related to interaction among the equipment and the hosting system [18-19].

Enabled automation of desilting machine consists of the vehicle tracking system, sewage removal capacity measurement, data repositories, data analytics, alert mechanism and report generation systems. The Vehicles equipped with the machineries can be tracked using GPS/ GPRS system so that the real time locality of the desilting machine can be tracked. An embedded module with flow meter sensor can be used to measure the amount of sewage removed during a trip.

A structured data base is created to maintain numerous data such as, vehicle location, removed sewage capacity at a location, time of collection, information about the trip sheet, planner, scheduler etc. From the data repositories, necessary information can be generated to determine the daily usage of machineries, capacity of sewer removal per area etc. Alerts in the form of SMS or email about the abnormality in vehicle position, unauthorized disposal of sewage etc. can be reported. The diagram in Fig. 19.3 illustrates the schematic of the Internet of Things network.

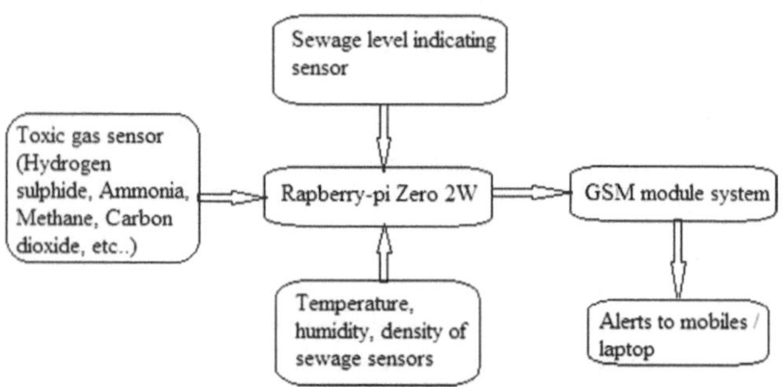

Fig. 19.3 Schematic of IoT system

Source: Created using the experimental work

6. Conclusion

The dredging procedure of SPS water sources should make certain advancements to mitigate the serious hazards faced by the cleansing workers. The eradication of mechanical salvaging is important in light of the grave issue of sewage dredging to mitigate the healthcare risks faced by drainage personnel. Thus, the automated sewage well desilting machine was designed and developed to eradicate manual scavenging under the surveillance of IoT and the pilot run was conducted. A detailed analysis was made to obtain the final product dimension for the manufacturing process. The product will be further developed keeping in mind the constraints observed during the pilot run and the inexpensive product while not negotiating with the functioning. Thus,the final product will be manufactured which is easily operational, easy to carriage, easy to uphold yet long-lasting which condenses effective desilting of SPS sewer collection wet wells with the implementation of the IoT concept.

References

1. CMWSSB, *Handbook on operation and maintenance (Water supply and sewerage system),* Metro water training center, Chennai-600 023.
2. Shankar, S., & Swaroop, K. (2021).Manual Scavenging in India: The Banality of An Everyday Crime, Caste-A Global Journal on Social Exclusion, 2(1):67–76. https://doi.org/10.26812/caste. v2i1.299
3. Bethy Merchan-Sanmartin, Maribel Aguilar-Aguilar, Fernando Morante-Carballo, PaulCarrion- Mero, Jaime Guambana-Palma, Diego Mestanza-Solano and Edgar Berrezueta (2022).Design of Sewerage System and Wastewater Treatment in a Rural Sector: A Case Study, International Journal of Sustainable Development and Planning. 17(1): 51.
4. P.Madhuraghava, K.Nagaraju and P.Nagaraju (2019). Design and fabrication of automatic sewage cleaning machine, International Journal of Emerging Technologies and Innovative Research.6(1):1622–1638.
5. S.Kalyanakumar, R.Praveen, ArunShaji andDiwakar (2021). Design and fabrication of drainage cleaning equipment, Journal of Physics: Conference Series, 2040012043.10.1088/1742-6596/2040/1/012043.
6. P.Dhananchezhiyan, Somashekhar S. Hiremath, M. Singaperumal and R. Ramakrishnan(2013). Design and Development of a Reconfigurable Type Autonomous Sewage Cleaning Mobile Manipulator. Procedia Engineering. 64:1464–1473.
7. Saurabh S. Satpute, Vitthal R. Darole, Pravin M. Khaderao and Pankaj B. Hiralkar(2018). Automatic sewage cleaning system, International Journal of Advance Engineering and Research Development.5(6): 1–8.
8. Ravindra R. Patil, Saniya M. Ansari, Rajnish Kaur Calay and Mohamad Y. Mustafa(2021). Review of the State-of-the-art Sewer Monitoring and Maintenance Systems Pune Municipal Corporation - A Case Study, TEM journal.10(4):1500–1508.
9. A.Nagadeepan, J.Hershahimlan, J.Guruyogeshwaran and S.Balaji (2018). Automatic drainage cleaning system, International Journal of Engineering, Science and Mathematics.7(4):190–195.
10. SFouziyaSulthana, Vibha K., Sudhanshu Kumar, SnehaMathur and TarangAshutoshMohile(2020). Modelling and design of a drain cleaning robot, IOP Conf. Series: Materials Science and Engineering. 912(002049):1–6.
11. B.Babu, P.Raja and A.AnandJayakumar (2021). Design and development automatic sewage cleaning machine, Journal of Huazhong University of Science and Technology.50(1):1–7.
12. Karel Mulder (2019). Future Options for Sewage and Drainage Systems Three Scenarios for Transitions and Continuity, Sustainability. 11(1383):1–15.

13. K.Balachandar, S.SatheeshKumar, B.Yogeswaran and D.G.Yamini (2018). Automatic Sewage Cleaning Machine,International Journal Of Research Culture Society. 2(4):91–96.
14. D V Skibo, M Y Tolstoy and K I Chizhik (2019). Automated damping tank of sewage pumping stations, IOP Conf. Series: Materials Science and Engineering. 687. doi:10.1088/1757-899X/687/4/044030.
15. AnshulMathur,Hitesh Soni, JeetSaini, RohanPareek, Vivek Singh Rawat and Anil Vaishnav (2021)Sewage cleaning machine system, International Journal of Current Research. 13(05):17580–17584.
16. Hanfang Yun, Xuejian Wu, ZekunHao, Shiji He and Xuewei Cao (2021). Sludge Desilting Treatment Device for Urban Pavement Sewer, IOP Conf. Series: Earth and Environmental Science.692 :032099 IOP Publishing doi:10.1088/1755-1315/692/3/032099.
17. Nirupama Prakash, and Poonam Bala (2022). From Manholes to Roboholes: A Technology-Based Solution for Sanitation Workers, Contemporary Voice of Dalit, 0(0)https://doi.org/10.1177 /2455328X221132969.
18. S C Dharmadhikari, AfshanKausar, MahendraDeore and NiloferShrenikKittadV S Bhagavan and R Krishnamoorthi (2023)., IOT based healthcare monitoring system for smart city applications, Human-Assisted Intelligent Computing, 28-1:28–18, https://dx.doi.org/10.1088/978-0-7503-4801-0ch28.
19. R. Krishnamoorthy, K. Kamala, I. D. Soubache, MamidalaVijay Karthik, M. and Amina Begum (2022). Integration of Blockchain and Artificial Intelligence in Smart City Perspectives in Smart City Infrastructure: The BlockChain Perspective, 77-112.https://doi.org/10.1002/9781119785569.ch3.

Advances in Additive Manufacturing Technologies – Gurusamy Pathinettampadian (eds)
© 2024 Taylor & Francis Group, London, *ISBN 978-1-032-90013-1*

20 Comparative Analysis of Breast Cancer using Machine Learning

Harini Ramachandran Aruna[1], Shreemathi Santharaman[2], Kavitha Prithiviraj[3]
Department of Artificial Intelligence and Data Science, Panimalar Engineeirng College, Chennai, India
Chitra Devarajalu[4]
Department of Master of Business Administration, Panimalar Engineering College, Chennai, India

ABSTRACT: This paper gives us an idea about breast cancer detection. The global battle against cancer, especially in developing nations, necessitates innovative approaches for early detection and treatment. This study conducts a rigorous comparative analysis of breast cancer detection methodologies employing machine learning techniques including KNN, Naive Bayes, SVM, Decision Tree, Logistic Regression, Random Forest, SVM Kernel, XGBoost, and AdaBoost. The accuracy in addition to the efficiency of the trained models are evaluated using performance metrics such the confusion matrix, exactitude, order back, F- score, substructure and preciseness. These performance measures show that Logistic Regression has the greatest accuracy rate (97%). We find distinctive biomarkers and diseases characteristics by combining several algorithms and analyzing various data sources, including genetic factors and clinical records. Our cross-population comparison reveals customized insights necessary for individualized therapies that promise better patient results.

KEYWORDS: Breast cancer, Analysis, Machine learning

1. INTRODUCTION

Millions of people all over the world are afflicted by the widespread and potentially fatal disease of breast cancer. The likelihood of surviving this, the most prevalent cancer in women, is greatly increased by early identification long-term survival and a successful treatment. With the introduction of cutting-edge large volumes of readily accessible medical data, technological advancements, and machine learning. The field of healthcare has seen the emergence of learning approaches as effective instruments for illnesses, such as breast cancer, can be predicted and detected. Breast cancer is possible when the cancer cells are able to spread through the lymphatic or blood systems and the body parts were then transferred. In breast cancer screening, mammograms are the primary means of detection. If an abnormality is detected, other tests, such as a breast ultrasound, MRI, or biopsy, may be performed to help confirm the diagnosis. Breast cancer is impacted by the type, stage, and individual characteristics treated.

[1]harinirams212@gmail.com, [2]shree0207mathi@gmail.com, [3]varshnikavitha@gmail.com, [4]chitrambapec@gmail.com

DOI: 10.1201/9781003545774-20

Hormone, targeted, and immunotherapy therapies are the most widely used forms of treatment. A patient's specific situation informs the provision of tailored care. Breast cancer prognoses vary widely depending on the type and stage of the disease. In general, cancer that is detected early is better than cancer that is discovered later on. This study's main goal is to investigate the effectiveness of medical therapy and possible uses of machine learning algorithms in breast oncology. In order to create a dependable and effective system for detecting possible occurrences of breast cancer, we explore the field of predictive analytics by utilising the innate patterns found in a large dataset. It goes beyond simple technical work to complete this job. It's a big step in the direction of more precise and customised medical care.

2. Literature Survey

"Microarray Breast Cancer Data Classification" [IEEE 2018] was published by Siyabend Turgut et al. using machine learning methods. By applying machine learning algorithms with microarray breast cancer data, the research patients are categorised. Using eight distinct machine learning algorithms on the dataset and documenting the categorization outcomes is the first stage. In the second scenario, a popular selection approach called Recursive Feature Elimination (RFE) along with Randomised Logistic Regression (RLR) was applied to the microarray breast cancer data file. This resulted in the identification of approximately fifty characteristics as the standard cause. All over again, the adjusted dataset was subjected to the similar eight expert system algorithms. Comparisons between the classifications' outcomes and those of the first example are made. Techniques including MLP, SVM, KNN, Random Forest, Decision Tree, Logistic Regression, Ad Boost, and Gradient Boosting Machines were employed. The two feature selection techniques were applied, and SVM produced the best outcomes. To study how the count of layers and neurons affects categorization accuracy, several layers and neurons are used when applying MLP. [3].

"Four Innovative Methods for Region of Interest Detection in Mammograms - A Comparative Analysis," Varalatchoumy M. et al. [ICISS 2017]. This study looks at four novel approaches based on real-time and database images for identifying regions of interest in mammography images. The preprocessing of Approach I used techniques such as dynamic thresholding and histogram equalization. The preprocessed image's Region of Interest (ROI) was drawn out using k-means clustering and particle swarm optimisation. In Approach II, preprocessing was carried out employing distinct morphological processes, such as dilation and erosion. ROI was calculated by modifying the 10 watershed segmentation approach. In Approach III, data is preprocessed using an upgraded level set method for data segmentation and histogram equalization. Adaptive histogram equalization, contrast-limited, and other morphological approaches are used to preprocess images in approach IV, which is thought to be the most successful strategy. A highly inventive method was developed in order to discover Regions of Interest. Only the images from the Mammographic Image Analysis Society (MIAS) database could be utilized to implement Approaches I and II. Real-time hospital pictures and MIAS both benefited from Methods III and IV. The comparison study's multiple graphs demonstrate that radiologists should adopt the novel strategy—which uses a novel ROI detection algorithm—to detect cancers in MRI images since it is the most reliable, accurate, and efficient approach. [4].

IJCA 2013, "Review on Feature Selection Techniques of DNA Microarray Data," Ammu P. K. et al. This study analyzes a few key feature selection methods used with microarray data and outlines the benefits and drawbacks of each method. One of the most critical steps in bioinformatics is feature selection from DNA microarray data. Based on the concepts that mutation is a natural process and that species migrate between habitats during their evolutionary journey, Oneoptimisation method is called biogeography-based optimisation, or BBO. A search space's movement of particles serves as the foundation for the Particle Swarm Optimization (PSO) algorithm. The selected genes can be made more representative of the population by removing redundant genes. An information gain filtering criterion-based two-stage hybrid filter wrapper strategy wherein the first stage generates a subset of the original feature set. Second, the set of filtered genes is processed by the genetic algorithm. Gene selection based on the dependencies between features, where features are divided into 11 dependent, half-dependent, and independent features. Any feature that is independent of any other feature is referred to as such. The significance of half-dependent characteristics increases when combined with other features, whereas dependent features rely only on other features. [5]

Integrating level set approaches with spatial fuzzy clustering for automatic segmentation of medical images, Bing Lan Li et al. Elsevier (2010) In order to facilitate automated medical image segmentation, a new Fuzzy Level Set technique is suggested in this study. It can arise directly from the first segmentation based on spatial fuzzy clustering, in which the centroid and range of each subclass are adaptively calculated to minimize a given cost function. Considering the outcomes of fuzzy clustering, governing factors for Level Set evolution are also estimated. Dynamic variational boundaries are used in the level-set approaches to segment images. Initializing level set segmentation is automated using the novel fuzzy level set method and parameter setup using spatial fuzzy clustering. To identify the approximate Fuzzy-C mean (FCM) with spatial constraints is used to identify contours of relevance in a medical image. Further, the technique for adding locally regularized evolution improves the Fuzzy Level Set. Improvements like these reinforce segmentation and allow level sets to be adjusted. The suggested strategy was assessed on a variety of medical images from different modalities. Results demonstrate how effective it is for segmenting images. [6].

3. Proposed System

3.1 Data Collection

The most typical malignancy among women worldwide is breast cancer. Over 2.1 million people were impacted by it in just 2015, accounting for 25% of all cancer cases. When breast cells start to proliferate out of control, the condition begins. These cells typically develop tumors that can be perceived as breast lumps or seen on an X-ray. How to categorize tumors as malignant (cancerous) or benign (non-cancerous) is one of the main obstacles to its diagnosis. Please finish the analysis of identifying these tumors using machine learning (with SVMs) and the Breast Cancer Wisconsin (Diagnostic) Dataset.

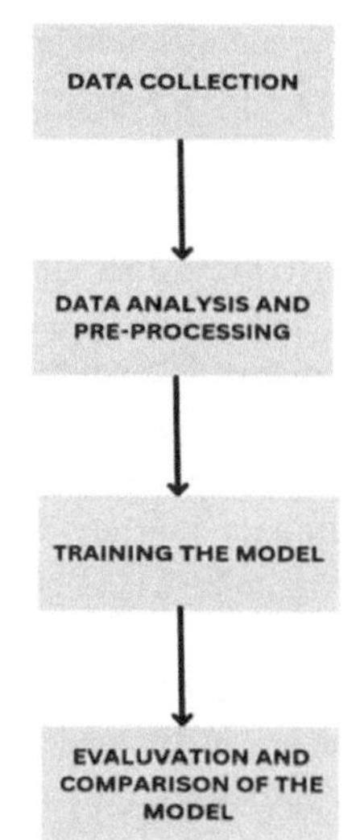

Fig. 20.1 Architecture diagram

3.2 Data Analysis and Preprocessing

The data analysis and pre-processing step involve the exploration and visualization of the data. The first step is to clean the data which includes the process of repairing or eliminating data that is missing, duplicated, corrupted, improperly formatted, or erroneous from a dataset. It covers feature scaling, selecting relevant features, and handling missing and categorical data. In the dataset, "NaN" is used to represent missing values. The next step is to visualize the data. The visualization of the data is done using a histogram which is an approach to bar charts that divide continuous measures into distinct bins for distribution analysis and heat maps, which are a kind of geographical visualization represented as a map that shows distinct data values as distinct colors. The histogram is represented in the form of a count plot which is used to represent the occurrence(counts) of the observation present in the categorical variable.

3.3 Training of the Model

We employ a variety of machine-learning algorithms for the study of breast cancer in order to train the model. Since the output of the model is a class, we are taking the algorithm as a classifier as it helps in categorizing data into different classes. The dataset is divided into training and test sets in order to train this model. The training set is used to assist train the model, and the test set is used to help evaluate the model. The machine learning algorithms used for this analysis are Decision Tree, Random Forest, Naïve Bayes, Logistic Regression, SVM Kernel, Support Vector Machine, XGBoost, AdaBoost, and KNN(K-Nearest Neighbour).

3.4 Evaluation and Comparison of the Model

After training the models with machine learning algorithms, we must assess the models' performance using a range of metrics that are detailed in the categorization report, one of a grouping-based machine learning model's enactment estimation measurements. It exhibits your model's exactness, recollection, a function of the precision score, and provision. It delivers an appreciation of the complete enactment of our skilled prototypes.

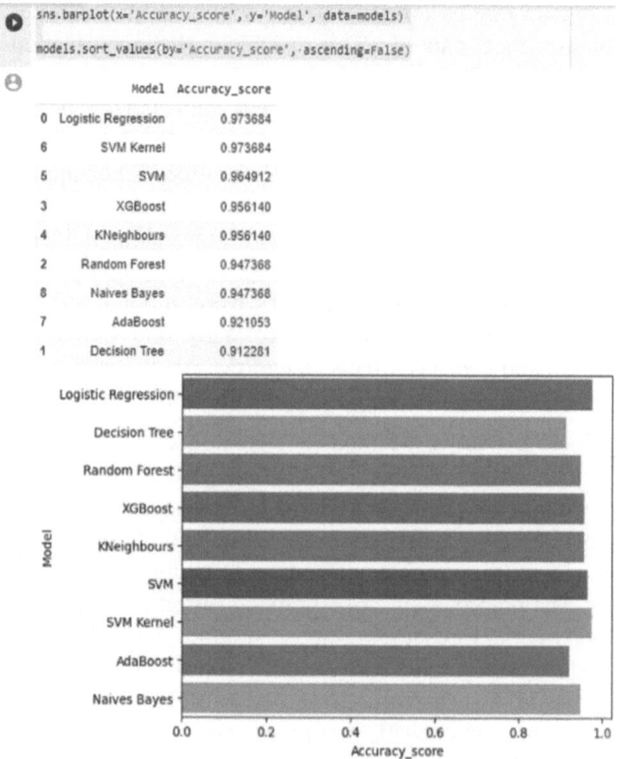

```
sns.barplot(x='Accuracy_score', y='Model', data=models)

models.sort_values(by='Accuracy_score', ascending=False)
```

	Model	Accuracy_score
0	Logistic Regression	0.973684
6	SVM Kernel	0.973684
5	SVM	0.964912
3	XGBoost	0.956140
4	KNeighbours	0.956140
2	Random Forest	0.947368
8	Naives Bayes	0.947368
7	AdaBoost	0.921053
1	Decision Tree	0.912281

Fig. 20.2 Comparison of the model

4. Results and Discussion

In this proposed system, we utilised 9 machine-learning algorithms for training the model. Also, we used six performance measures for the evaluation of the model. Based on these parameters, we will compare the performance of each model, and higher accuracy of the analysis is considered as a primary metric for comparison which has the highest accuracy. Based on the comparison, Logistic Regression has an accuracy of 97% and considered as the best model for analysis of breast cancer.

5. Conclusion

The use of expert system techniques in the comparative analysis of breast cancer has shown to be a useful and promising method in the realm of medical research and healthcare. Machine learning's aptitude for working with a variety of data sources, including genetic, imaging, and clinical data, is one of its main advantages in the analysis of breast cancer. Machine learning algorithms can provide a thorough picture of the condition by combining these different data sources, which will enhance patient outcomes through more effective diagnosis and individualized treatment plans. On the clinical data of patients identified with breast cancer based on symptoms connected to the disease, we employed well known machine learning techniques as KNN, XGBoost, AdaBoost, Naive Bayes, Support Vector Machine, Decision Tree, Logistic Regression, Random Forest, and SVM Kernel. The Logistic Regression shows the best performance in the analysis, with accuracy rate of 97%, according to the performance validation criteria recall, accuracy, precision, and support. The study's future focus may involve the diagnosis of the condition using numerous or sizable data sets.

References

1. Jemal A, Murray T, Ward E, Samuels A, Tiwari RC, Ghafoor A, Feuer EJ, Thun MJ. Cancer statistics, 2005. CA: a cancer journal for clinicians. 2005 Jan 1; 55(1):10–30.
2. U.S. Cancer Statistics Working Group. United States Cancer Statistics: 1999– 2008 Incidence and Mortality Web-based Report. Atlanta (GA): Department of Health and Human Services, Centres for Disease Control.
3. Siyabend Turgut; Mustafa Dağtekin; TolgaEnsari, "Microarray breast cancer data classification using machine learning methods", 2018 Electric Electronics, Computer Science, Biomedical Engineerings' Meeting (EBBT), 10.1109/EBBT.2018.8391468.
4. M.Ravishankar, Varalatchoumy M, "Four Novel Approaches for Detection of Region of Interest in Mammograms - A Comparative Study", 2017 International Conference on Intelligent Sustainable Systems (ICISS), 10.1109/ISS1.2017.8389410.
5. Ammu P K, Preeja V, "Review on Feature Selection Techniques of DNA Microarray Data", International Journal of Computer Applications (0975 – 8887) Volume 61– No.12, January 2013.
6. Bing Nan Li, Chee Kong Chui, Stephen Chang, S.H. Ong, "Integrating spatial fuzzy clustering with level set methods for automated medical image segmentation", Computers in Biology and Medicine,Volume 41, Issue 1, January 2011, Pages 1–10, https://doi.org/10.1016/j.compbiomed.2010.10.007.
7. B. N. Dontchos, A. Yala, R. Barzilay, J. Xiang, C. D. Lehman, "External validation of a deep learning model for predicting mammographic breast density in routine clinical practice", Acad. Radiol., 28 (2020), 475–480, https://doi.org/10.1016/j.acra.2019.12.012.
8. Siddhant Rao ,"MITOS-RCNN:A novel approach to mitotic figure detection in breast cancer histopathology images using region based convolutional neural networks", arXiv preprint arXiv:1807.01788 (2018), https://doi.org/10.48550/arXiv.1807.01788.
9. Péter Bándi, Oscar Geessink, Quirine Manson, Marcory Van Dijk, et al; "Detection of individual metastases to classification of lymph node status at the patient level: the camelyon17 challenge", IEEE Transactions on Med. Imaging (2018), 10.1109/TMI.2018.2867350 .
10. B. N. Dontchos, A. Yala, R. Barzilay, J. Xiang, C. D. Lehman, "External validation of a deep learning model for predicting mammographic breast density in routine clinical practice", Academic Radiology, Volume 28, Issue 4, April 2021, Pages 475–480, https://doi.org/10.1016/j.acra.2019.12.012.

Note: Every figure was created using the experimental work; none of them were taken from any publications or online.

Advances in Additive Manufacturing Technologies – Gurusamy Pathinettampadian (eds)
© 2024 Taylor & Francis Group, London, ISBN 978-1-032-90013-1

21 Flow Analysis of Centrifugal Pump Impeller Using FEM

J. Chandru[1]
Assistant Professor, Department of Aeronautical Engineering,
Jaya Engineering College, Chennai, India

P. Jayaraman[2]
Professor, Department of Mechanical Engineering,
Prathyusha Engineering College, Thiruvallur, India

S. Jeyan[3], **R. Seethapathy**[4], **S. Kevin Bennett**[5]
Assistant Professor, Department of Aeronautical Engineering,
Jaya Engineering College, Chennai, India

ABSTRACT: Centrifugal pump is executed to get the better performance of the impeller. The impeller and casing is the most important component in the centrifugal pump. The casing is a stationary part and impeller is a moving part of the centrifugal pump. When rotating, stresses are induced in the impeller blades due to centrifugal forces, when the stress level exceeds the yield strength of the impeller material failure of the impeller blades will occur. Therefore stress analysis of the impeller blade due to centrifugal force at different operating condition is important at the design stage itself. In the present work, the stresses induced in an impeller due to centrifugal force at various operating condition is carried out also the geometry of the rear edge of the impeller blade is modified and the state of stress level is analyzed by using finite element analysis.

KEYWORDS: Centrifugal pump impeller blade design various types of blade used and find the efficiency of pump and blade efficiency using finite element method

1. INTRODUCTION

Pumps are used in a wide range of factories and home appliances. It is a machine which converts the mechanical energy into hydraulic energy. Mechanical energy converted into pressure energy by means centrifugal force acting on the fluid. It acts of as a reverse of an inward radial flow turbine. Impeller, it consists of a series of backward curved vanes, it is mounted on a shaft which is connected to an electric motors. Casing is a tight air passage surrounding the impeller and it Convert kinetic energy of water in to exit of the impeller. It transform into pressure energy before the water leaves the casing and enters the delivery pipe.

1.1 Objective and Scope of Work

The objective of the first part of project is to find the natural frequencies (ω) of impeller for various blade edge shapes.

[1]aerochandru.87@gmail.com, [2]jayaramanmech.mech@gmail.com, [3]mechjeyan88@gmail.com, [4]seethaavionics@gmail.com, [5]bennettaero@gmail.com

DOI: 10.1201/9781003545774-21

This study is used to estimate safe operating speed of the impeller RPM minimum to maximum.

The next part of the project is to find the stress induced in the impeller due to centrifugal force for various discharge condition and changing the blades rear edge shapes. The result of study will determine the minimum stress induced in the impeller for such blade rear edge shapes.

Table 21.1 Size and shape of the impeller

S.No	Parameter	Notations	Value
1	Discharge	Q	30 m^3/hr
2	Head	H	50 m
3	Speed	n	2880 rpm
4	Specific Speed	ns	97 rpm
5	Power of water	N	14.715 kW
6	Pump efficiency	ηo	80 %
7	Shaft power		19.40 kW
8	Maximum Shaft power	Mmax	30 kW
9	Diameter (Impeller inlet)	D1	113mm
10	Diameter (Impeller outlet)	D2	208 mm
11	Diameter (Impeller eye)	Do	98 mm
12	The torsional moment	T	100.52 N-m
13	Shaft Diameter	Ds	48 mm
14	The hub diameter	Dh	72 mm
15	The hub length	Lb1	96 mm
16	Impeller width (inlet)	b1	25 mm
17	Impeller width (outlet)	b2	17 mm
18	Inlet angle (bend)	$\beta 1$	17°
19	Outlet angle (bend)	$\beta 2$	30°
20	No of blade	Z	8 blade
21	Thickness of the blade	t	5 mm

1.2 Point to Point Method

Impeller blade parameters,

To find $\omega 1$

$$\omega 1 = \frac{\text{cm}_1}{\sin \beta_1}$$

$$\omega 1 = 17.13 \text{ m/s}$$

To find ω

$$\omega 2 = \frac{\text{cm}_2}{\sin \beta_2}$$

$$\text{cm} = K\text{cm}_2' * \sqrt{2 * g * H}$$

$$K\text{cm}_2' = K\text{cm}_2$$

From the graph $K\text{cm}_2 = 0.12$

$$\text{cm}_2 = 3.75$$
$$\omega 1 = 7.5 \text{ m/s}$$

To find b_1

$$\text{b1} = \frac{Q'}{\pi * d' - \left(\frac{S'}{\sin \beta'} * Z'\right) * \text{cm}_1'}$$

$$\text{b1} = 28.48 \text{ mm}$$

Construction of Rear Edge Shape

To analysis the stresses induced in the impeller blade by varying the blade rear edge shape. The rear edge shapes are Plain(P) rear edge, Rectangular (R) rear edge, Circular (C) rear edge and Grooved (G) rear edge. To construct this profiles the same procedure are followed.

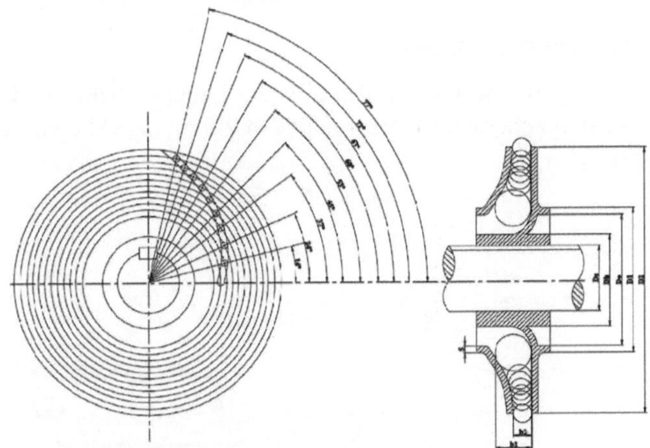

Fig. 21.1 Curvature construction of blade

Plain Trailing Rear Edge

The plain rear edge shapes in the blade are standard shape these plan shapes are using in centrifugal pump impeller for pumping the water. The plain edge shape look like match with outer diameter of the impeller curvature are shown in the Fig. 21.2.

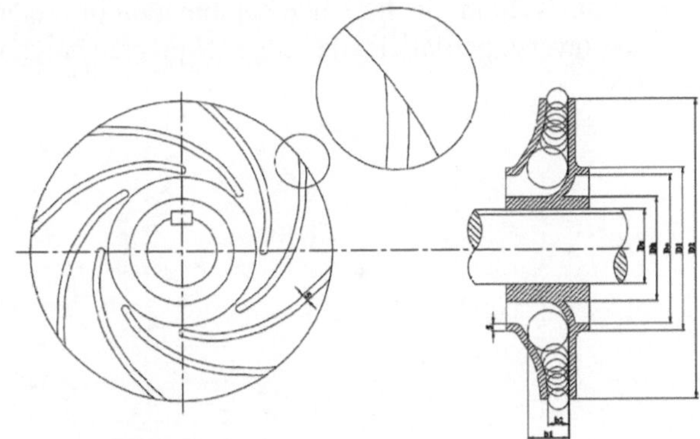

Fig. 21.2 Plain rear edge of the impeller

Rectangular Rear Edges

The impeller blade rear edge shapes looks like rectangular shape are shown in the Fig. 21.3.

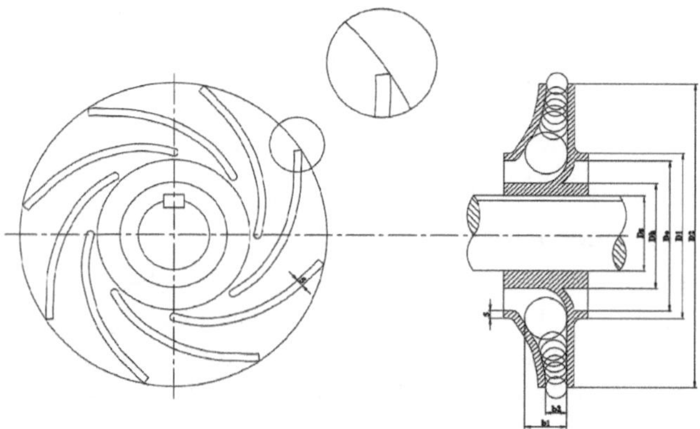

Fig. 21.3 Rectangular rear edge of the impeller

Circular Rear Edges

The circular rear edge shape in the impeller blade look like half spherical shape of radius of 2.5 mm are shown in Fig. 21.4.

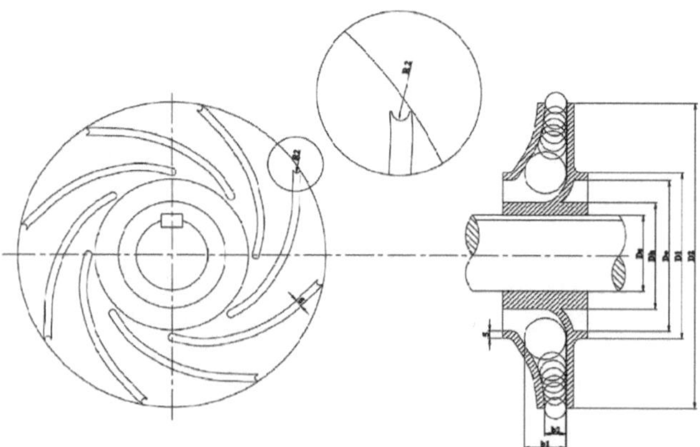

Fig. 21.4 Circular rear edge of the impeller

Grooved Rear Edge

The groove has been taken in the rear edge of the impeller of radius is 2 mm. its look like combination of rectangular and inverse partial circular edge shape are shown in Fig. 21.5.

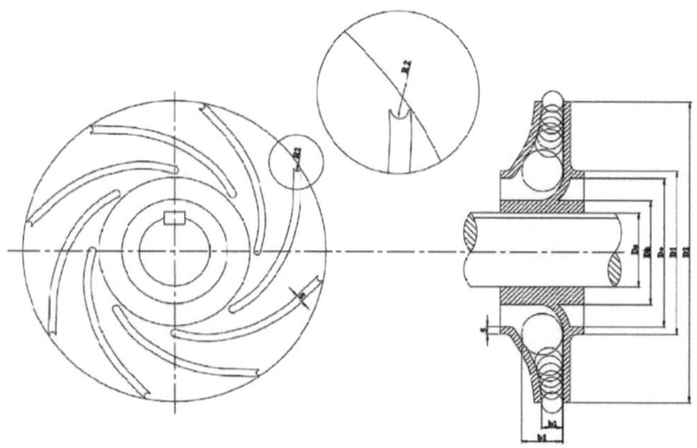

Fig. 21.5 Grooved rear edge of the impeller

Fig. 21.6 Sectioned a single domain from the impeller

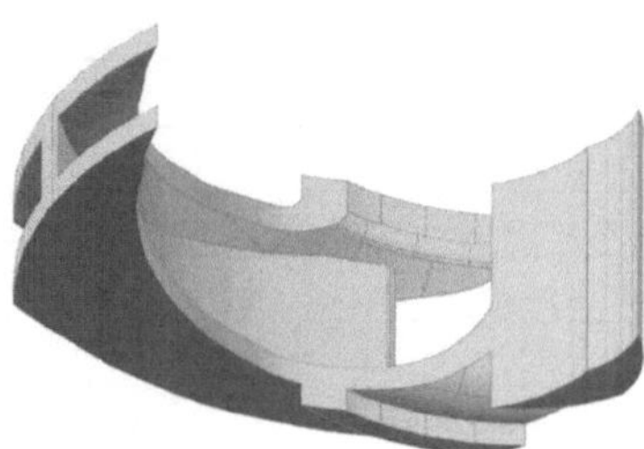

Fig. 21.7 Single domain for plain rear edge

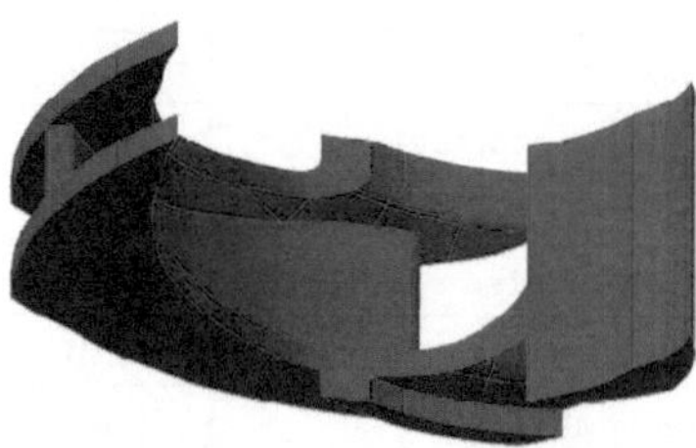

Fig. 21.8 Single domain for rectangular rear edge

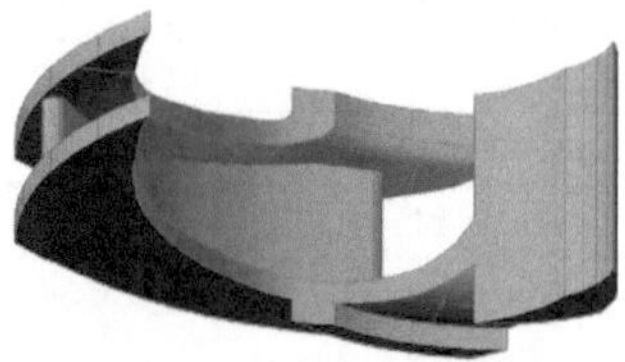

Fig. 21.9 Single domain for circular rear edge

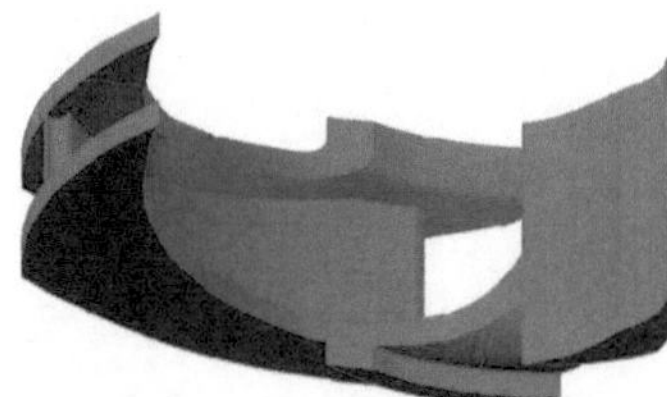

Fig. 21.10 Single domain for grooved rear edge

2. Affinity Laws (Ref : 2)

Pump Head, volumetric flow rate, speed performance are found by affinity law

Shaft speed

$$(Q1/Q2) = (N1/N2)$$

Pressure or Pump Head is proportional to the square of shaft speed:

2.1 Modal Analysis

The aim of our project to determine the shapes, structure and frequencies of vibration of any objects. Arbitrary shape and results Analysis are done by the finite element method. Natural frequencies (ω) mode shapes are determined by Experimental Modal Analysis. Correct material properties and boundary conditions are done by finite element method.

2.2 FEA Eigen System

The general equation of motion

$$(M)\,(\ddot{U}) + (K)\,(\ddot{U}) + (C)\,(\ddot{U}) = (F) \qquad \text{Eq. (21.1)}$$

(M) – Mass matrix,
($\ddot{U}$) – 2nd time derivative displacement,
(C) – Damping matrix
($\mathring{U}$) – velocity,
(K) – Stiffness matrix,
(U) – displacement and
(F) – Force matrix

The modal analysis damping

$$(M)(\ddot{U} + (K)(\ddot{U}) = (F) \qquad \text{Eq. (21.2)}$$

Eigen system experience especially done by structural engineering using FEM

$$(K)(\ddot{U}) = (F) \qquad \text{Eq. (21.3)}$$

SAT file model is imported in the ANSYS software for all rear edges are Plain, Rectangular, Circular, and Grooved shapes are in three dimensional wire frame models are shown in Fig. 21.6.

2.3 Material Selection

Mostly in the commercial centrifugal pump cast iron material is using. In the mass production cost of the impeller is low compare to the other materials. From the DIN (Deutsche Industries norm-German Industry Standard) standard gray cast iron material are widely used for manufacturing the impeller. So the same material is chosen for this project. The material selected from the Roj Elliott. BSC book of Castiron Technology [11]

Cast Iron: GGG-70 (DIN 169-73 steel grade)

Mechanical Properties

Module d'Young 177000 MPa
Yield Strength 400 MPa
Ultimate Strength 700 MPa

Physical Properties

Density 7100 kg/m3
Poison's ratio 0.26

2.4 Operating Conduction

The stress analysis of the impeller is finding out by various operating condition.

Pump Shaft speed: n2 = 720 rpm

Pump Head shaft speed: H2 = 12.5 m

Table 6.1 various operation conditions

2.5 Boundary Condition

The boundary conditions were set in order to obtain the results due to centrifugal force at angular velocity (ω) and pressure (P). The pressure and angular velocity were calculated by using the operation condition from the Table 6.3.

Pressure (P)

$$P = \rho * g * h \qquad \text{Eq. (21.4)}$$

$$P = 122625 \text{ N/m}^2$$

Angular velocity (ω)

$$\omega = \frac{2 * \Pi * N}{60} \qquad \text{Eq. (21.5)}$$

$$\omega = 75.39 \text{ rad/s}$$

Speed (N) = 720 rpm

Table 21.2 Various boundary conditions

S. no	Pressure (Load) N/m^2	Angular velocity (rad/s)
1.	122625	75.39
2.	245250	150.79
3.	367875	266.10
4.	490500	301.50
5.	662175	377.00

3. Results and Discussion

3.1 Modal Analysis

Impeller blade profiles, natural frequencies (ω) found by Modal analysis From the analysis results 10 frequencies are extracted.

Stress distributions in the impeller at 125 % of discharge conditions

For 125 % discharge conduction the maximum stress induced in the impeller for various rear edge of the blade are tabulated below Table 21.3.

The combined loads results of the impeller for various discharge % with various trailing rear edge for the maximum stress found by the finite element analysis results are shown in below Table 21.3.

Table 21.3 Max stress induced in the impeller for rear edges at 125% discharge

Rear edge	Max Stress (MPa)
Plain	262
Rectangular	251
Circular	240
Grooved	240

When the stress level comparing with Rectangular, Circular, and Grooved trailing. Rectangular trailing has more stress. The circular and grooved rear edges are having similar stress levels in last three operating conditions

Stresses Result in Rear Edge in the Impeller

The stress results are taken from impeller rear edge nodes. Here the nodes are selected randomly at the rear edge of the impeller. The node number are not same it varying according to shapes of rear edge. Stress induced in the edge is analyzed based on nodes located at the rear edge, and the stress level it is compared for different working condition.

The maximum stress at the rectangular rear edge is more than the other rear edge shape observed here. In minimum stress induced in grooved rear edge.

4. Conclusion

The present work is used to determine stress distribution of the centrifugal pump impeller for various rear edge shapes at various operating conditions, after conducting the analysis the following results are inferred. Modal analysis results have shown that the shape and structure natural frequency of the impeller does not related to with pump operating frequency even when the profile is modified. The minimum operating frequency of the impeller is 12 Hz and the natural frequencies (ω) are 3.56 Hz. When comparing the operating frequency of the impeller is 3 times greater than the natural frequency. The maximum stresses at all cases of loading conditions in the impeller are less than yield strength. The results of induced stress level are varying according to impeller blade rear edge shapes. Maximum stress is induced in the plain rear edge when compared to other rear edge shapes i.e. Rectangular, Circular and Grooved. More stress is induced in rectangle rear edge when compared to circular and grooved for the same operating conditions. Circular and grooved edges are showing similar stress distribution pattern. From theses results for all the operating conditions the Circular rear edge impeller are best for the centrifugal pump in the numerical analysis.

5. Scope for Future Work

Experimental tests have to be done to ensure the same kind of stress distribution and to validate the numerical results. Similar analysis can be carried out to obtained the stress distribution at various operating condition by modifying the leading edges of the impeller.

REFERENCES

1. Amro M. Al-Qutub, Atia E. Khalifa, and Faleh A. Al-Sulaiman, (2012), Exploring the Effect of V-Shaped Cut at Blade Exit of a Double Volute Centrifugal Pump, Journal of Pressure Vessel Technology, Vol.134, PP. 1301–1308.
2. Balaji.K, Sriram.K, and Ramamurti.V, (2000), Development of User-Friendly Program for Static and Dynamic Analysis of Radial Impeller, Journal of Advances in Engineering Software 31, PP. 775–791.
3. Bhope.D.V, and Padole.P.M, (2004), Experimental and Theoretical Analysis of Stresses, Noise and Flow in Centrifugal Fan Impeller, Journal of Mechanism and Machine Theory 39, PP.1257–1271.
4. Jose Manuel, Osar, and Jhan Arturo, (2006), Flow Induced Stress in a Francis Runner Using ANSYS, Conference of Static asset of Ansys, PP.239–241
5. Khin Cho Thin, Mya Mya Khaing, and Khin Maung Aye, (2008), Design and performance analysis of centrifugal pump, World Academy of Science, Engineering and Technology 46, pp (422–429)
6. Mario Savar, Hrvoje Kozmar, and Igorsu Hovic, (2009), Improving Centrifugal Pump by Impeller Trimming, Journal of Desalination 249, PP.654–659.
7. Mona Golbabaei Asi, Rouhollah Torabi, and Ahmad Nourgakhsh,(2009), Experimental and FEM Failure Analysis and Optimization of a Centrifugal – Pump Casing, Journal of Engineering Failure Analysis 16, PP.1996–2003
8. Mohamad Memardez fouli and Ahmad Nourbakhsh, (2009), Experimental investigation of Slip Factours in Centrifugal Pumps, Journal of Experimental Termal Fluid Science 33, PP 938–945
9. Motohiko, Takaki, and Yoshiyasu, (2002), Hydrodynamic Design System for Pumps Based on 3-D CAD, CFD, and Inverse Design Method, Journal of Fluids Engineering, Vol. 124, PP.329–335
10. Ray.G.S, and Sinha.B.K, (1991), Computation of Centrifugal Stresses In a Radial – Flow Impeller, Journal of Computers & Structures Vol.40, PP. 731–740.
11. Roj Elliott, (1988), Cast iron Technology, Butterworth & Co (Publisher) Ltd.
12. Samuel Brown, Flow induced vibration in pump impellers, Quest Engineering and Development Corporation, pp (13–18)
13. Wen-Guangli, (1999), Effect of viscosity of fluids on Centrifugal Pump Performance and Flow Pattern in the Impeller, Journal of heat and fluid 21, PP. 207–212

14. Wickstrom.A (1997), Structural Analysis of Francis Turbine Runners Using ADINA, Journal of Computers & Structures Vol. 64, PP. 1087–1095.
15. Ziae I Rad, (2005), Finite element, modal testing and modal analysis of a radial flow impeller, Iranian -Journal of Science & Technology, Vol.29 , PP. 157–169
16. Bansal, (2005), Textbook Of Fluid Mechanics And Hydraulic Machines, Laxmi Publications (P) Ltd.
17. Chandrupatla & Belagundu, (1997), Finite elements in Engineering, Prentice Hall of India Private Ltd.,
18. Krishnamoorthy, (1987), Finite Element Analysis- Theory and Programming, Tata McGraw-Hill Publishing Co.

Note: Every figure and table were created using the experimental work; none of them were taken from any publications or online.

Advances in Additive Manufacturing Technologies – Gurusamy Pathinettampadian (eds)
© 2024 Taylor & Francis Group, London, ISBN 978-1-032-90013-1

22 Diabetic Retinopathy Image Recognition Using AI Technique

M. Priyadharshni[1], Nagajothi, P[2]. Ponharieesh[3], R. Ganesh[4]
Department of ECE, Panimalar Engineering College,
Chennai, India

ABSTRACT: Diabetic retinopathy occurs when diabetes leads to complication in the eyes. This has become a significant issue for individual globally, particularly those living with diabetes. Actually, it is the leading factor behind the visual impairment of working individuals. To combat this issue, researchers have utilized a CNN algorithm to create various neuro-wise and layer-wise architecture. These architectures were trained using a publicly available dataset of images depicting the disease. The results showed that neural networks are capable of detecting the colors and textures of lesions specific to diabetic retinopathy, thus mimicking human decision-making. The CNN algorithm employed in this setup is composed of distinct elements including convolution layer, pooling layer, and fully connected layer, to automatically learn spatial hierarchies of features. Various Diabetic Retinopathy characteristics were tested as convolutional neural network inputs.

KEYWORDS: Diabetic retinopathy, Convolutional neural network, Visual impairment, Neural network architecture, Image datase, Lesion detection

1. INTRODUCTION

Diabetes retinopathy (DR), is a significant concern for numerous individuals with diabetes, which results in sight deterioration for those who are in working age. For ophthalmologists, DR lesions are often discernible in bright fundus pictures and serve as the basis for diagnosis. Nevertheless, there isn't a treatment that works well enough to completely cure this illness at the moment. Early diagnosis and intervention is the most well-known form of therapy to slow the disease's progression and eventually prevent vision loss.

Data science revolves around the examination of data and application of the information and insights to assist in numerous domains. The focus of data science is on discovering regularities and correlations within dataset. The replication of human intellect in machines, or artificial intelligence (AI), enables the creation of computers with human-like behavior and thought processes. Artificial intelligence strives to assist computers in acquiring knowledge from past data, much like humans gain knowledge through experience. Artificial Intelligence relies on specialized

equipment and programs in order to develop and educate machine learning algorithm. Any computer language can be used to construct artificial intelligence, but Python, R, and Java are among the more popular ones. Dep learning is a type of artificial intelligence that helps computer to understand and analyze data in a similar way to how

[1]priyadharshni.panimalar@gmail.com, [2]bnagajyothisai221@gmail.com, [3]ponharieesh65@gmail.com, [4]Dganeshravi1804@gmail.com

DOI: 10.1201/9781003545774-22

humans do. Artificial neural networks are employed in deep learning, which is a subset of machine learning. This signifies that the computer is being educated using a set of outdated photographs. Deep learning mimics the functioning of the human brain.

The first study on retinal image processing used analogue images and focused on fluorescein-based vessel identification in fundus images. The ophthalmoscope was created before the first fundus pictures were captured. Subsequently individuals have been contemplating the preservation and inspection of images showing the interior of the eye to aid in the diagnosis of issues. The fluorescent agent makes the vessels in the image appear more prominently, making it easier for a medical expert to identify and measure them. On the other hand, fluorescein angiography is an invasive, time-consuming and costly procedure to administer and buy the fluorescent agent. Retinal image analysis is increasingly used for detection and diagnosis through digital imaging and image processing.

2. Related Work

It makes perfect sense to segment lesions causing diabetic retinopathy (DR) to help ophthalmologists diagnose patients. Numerous research have been conducted on this subject, but most of them have placed too much emphasis on network patterns as opposed to the pathological interactions between lesions. Our research into disease lesions revealed that certain lesions are exclusive to specific vasculature and exhibit similar characteristics. In response to the outcome, we suggest a relation transformer block (RTB). We develop a unique network that uses a twin-track design with GTB and RTB to segment the four DR lesions simultaneously. A key study within our network focuses on GTB and RTB, as they examine the connection among a group of lesions and their interaction of vessels. The peak accuracy achieved is 67.99%.

3. Proposed System

The recommended method standardises and eliminates images of diabetic retinopathy from datasets using pretreatments. Subsequently, the datasets are classified using a deep learning algorithm called convolutional neural networks. Convolutional neural network training falls under the umbrella of deep learning technique. A wide range of retinal disorders have been successfully classified by convolutional neural networks (CNNs), an efficient method for deep learning and disease identification. The best architectural design is applied to illness prediction. Not only does the proposed approach reduce expert subjectivity, but it also uses CNN architecture to predict if a person will get diabetes or not. In this dataset, features from optical coherence tomography (OCT) are extracted and stored in about 1000 train and 200 test picture recordings. These features were then categorised into

- Severe Diabetic Retinopathy
- No Diabetic Retinopathy

The primary goal is to create a computer algorithm that can accurately categorize images of diabetic retinopathy using a specific kind of neural network known as convolutional neural network. This model may be used to compare CNN designs and potentially categorise outcomes with the highest accuracy.

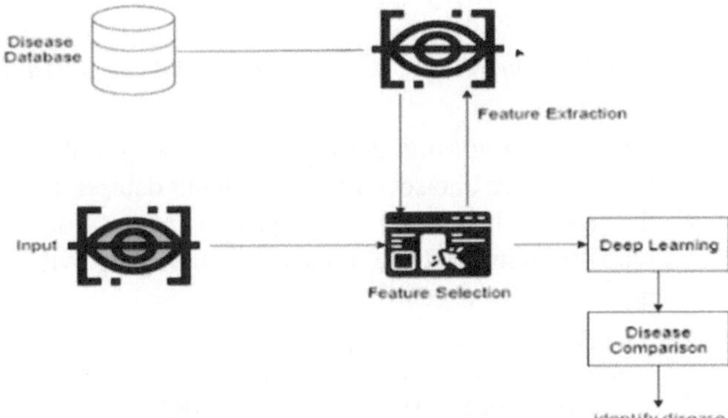

Fig. 22.1 The system architecture [15]

The above image is the system architecture of the proposed system.

4. Preprocessing and Training the Model

One has preprocessed the dataset. Preprocessing typically involves scaling, reshaping, and transforming an image to an array format. On the test picture, this same type of processing is repeated. A collection of around two classifications of diabetic retinopathy is created, from which every picture can be utilised as a software test image. To enable the model (CNN) to recognise the test image and the disease it carries, it is trained using the train dataset. Convolution2D, Maxpooling2D, Activation, Dropout, Dense, and Flatten are the many levels of CNN. Once the software becomes proficient, the computer program will be able to classify the diabetic eye disease in the images. In order to predict the diabetic retinopathy, the test image and trained model are compared following successful training and preprocessing.

Module Catalogs

1. Manual
2. AlexNet
3. LeNet
4. Deploy

Module Description

1. **Manual**

   Retrieve the specified image from the dataset:

   Using Keras preprocessing image data generator function, images can be imported from our dataset and parameters for size, rescaling, zoom range, and horizontal flipping can be constructed. The data generator method extracts image from the designated folder. We have the authority to plan the procedures for the train, test, and validation stages including the details of picture dimensions, quantity and classification techniques. The utilization of this feature is necessary in order to teach our own network by including CNN layers.

   To utilize the images provided in the dataset to train the model:

   The fit generator function and classifier are employed to instruct our dataset. We may train our dataset using the validation data and steps that are generated, along with the training steps for each epoch and the overall number of epochs.

2. **AlexNet**

Input : 32x32
Convolution (6@28x28)
Subsampling(6@14x14)
Convolution (16@10x10)
Subsampling(16@5x5)
FC(120)
FC(84)
Output (10)
Layer # 1
Layer # 2
Layer # 3
Layer # 4
Layer # 5
Layer # 6

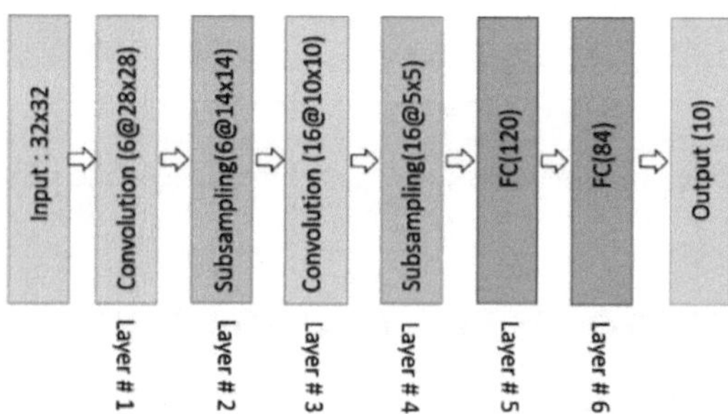

Fig. 22.2 Alexnet architecture

   Five convolutional layers, three max-pooling layers, two normalisation levels, three fully connected layers, and one softmax layer make up the AlexNet architecture.

3. **LeNet**

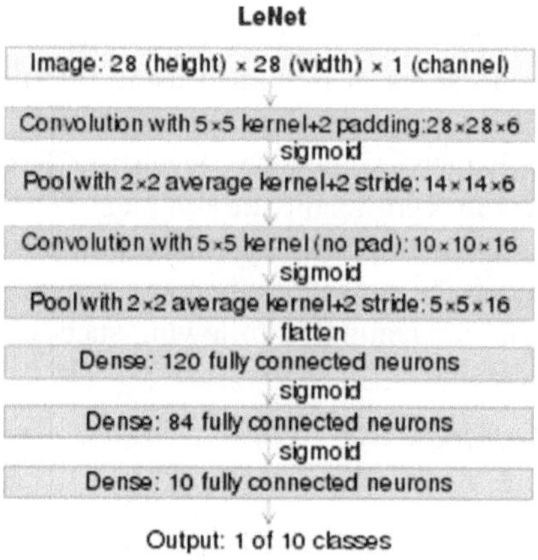

LeNet
Image: 28 (height) × 28 (width) × 1 (channel)
Convolution with 5×5 kernel+2 padding:28×28×6
sigmoid
Pool with 2×2 average kernel+2 stride: 14×14×6
Convolution with 5×5 kernel (no pad): 10×10×16
sigmoid
Pool with 2×2 average kernel+2 stride: 5×5×16
flatten
Dense: 120 fully connected neurons
sigmoid
Dense: 84 fully connected neurons
sigmoid
Dense: 10 fully connected neurons
Output: 1 of 10 classes

Fig. 22.3 LeNet architecture

   Le Netis simply the convolution neural network. As shown in the figure LeNet-5 consists of seven layers.

4. **Deploy**

   This module puts the trained deep learning model into our Django framework to make it work better for users and output prediction, by turning it into a hierarchical data format file (.h5 file).

5. Results and Dicussion

We would be implementing the code in anaconda with jupyter notebook in python code. The following parameters are considered in analyzing the reliability of the work: accuracy and loss.

1. **Manual**

Model accuracy
Train
Test
Accuracy
Epoch

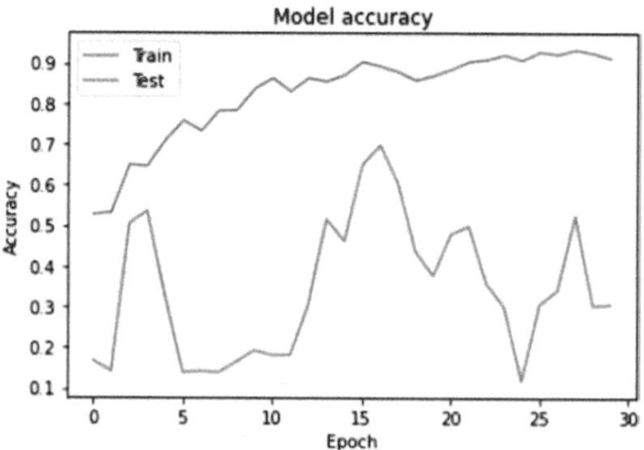

Fig. 22.4 Dataset accuracy for CNN model

The above figure denotes the CNN model trained dataset accuracy. The blue color represents the training data accuracy and the orange color represents the testing data accur

Model loss
Train
Test
Loss
Epoch

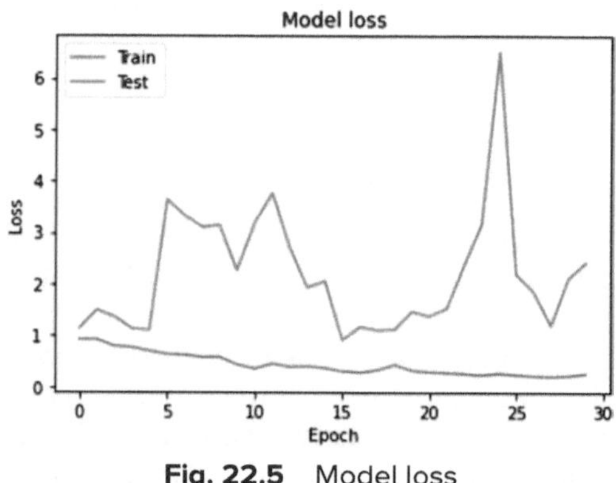

Fig. 22.5 Model loss

The above figure depicts the CNN Model trained dataset loss values. From the above two graphs an accuracy of 77.381% and loss of 0.4% is achieved.

2. AlexNet

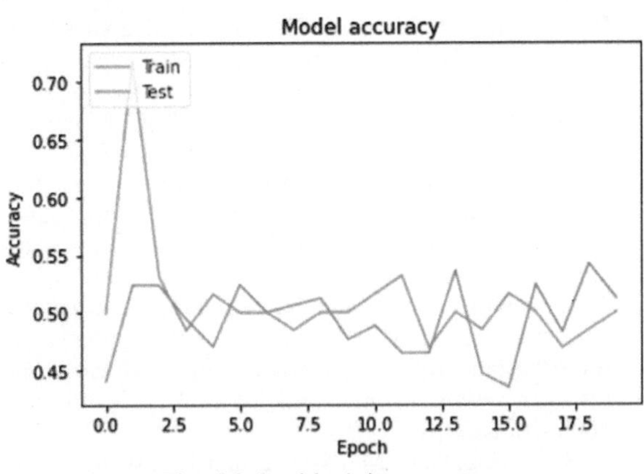

Fig. 22.6 Model accuracy

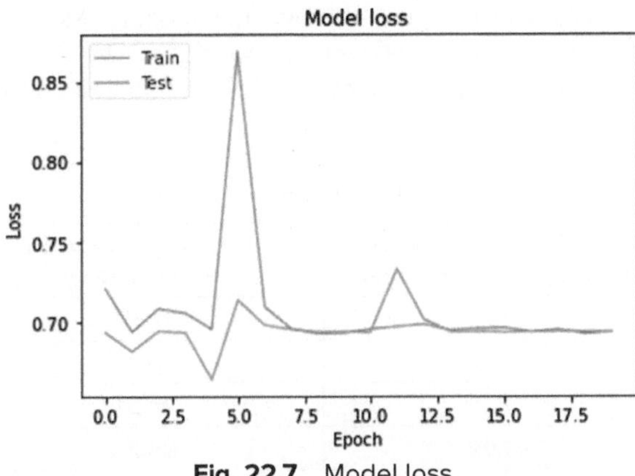

Fig. 22.7 Model loss

The procedure that was used before is being continued here. And analyzing the two graphs we would be achieving an accuracy of 54.167 and loss of 0.6951.

3. LeNet

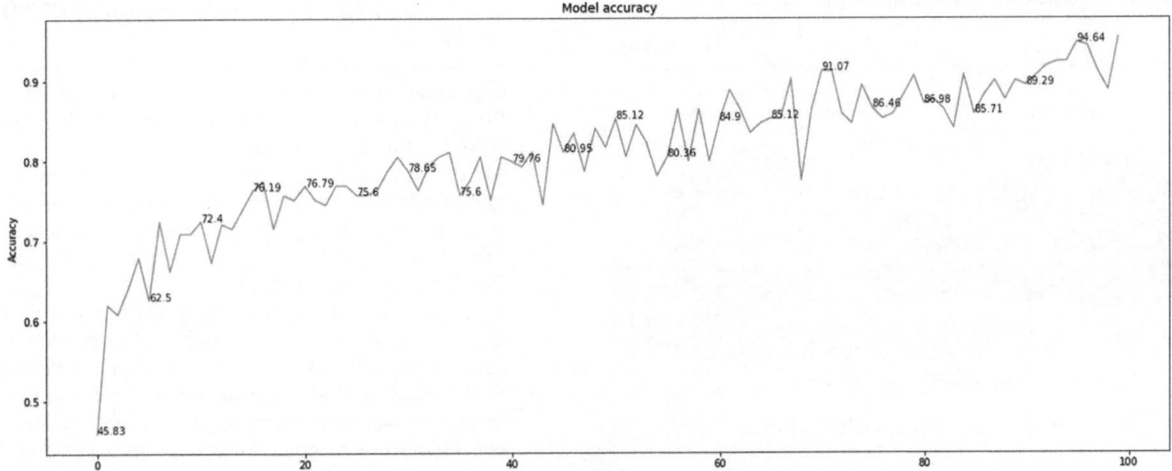

Fig. 22.8 Model accuacy of architecture

The above figure denotes the model accuracy.

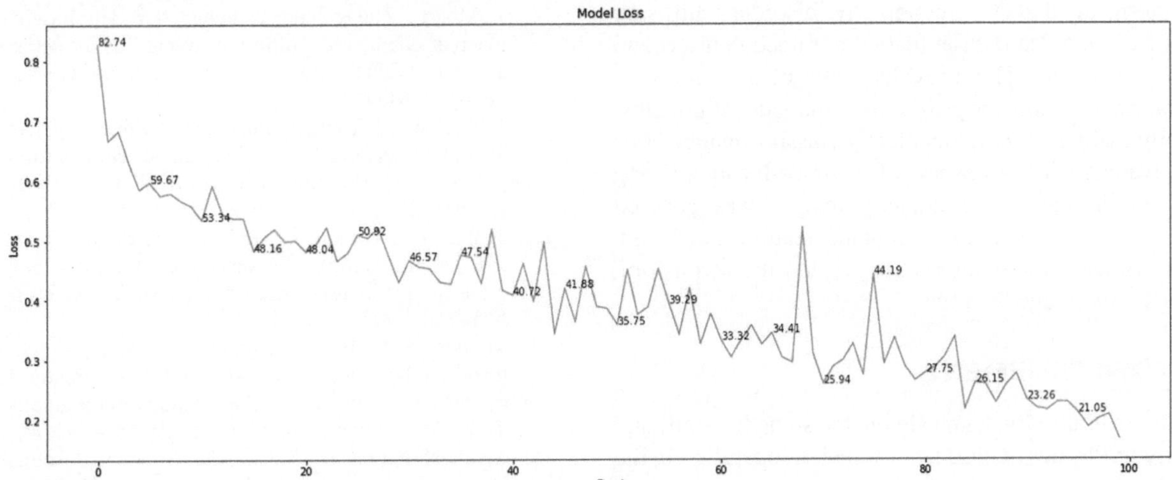

Fig. 22.9 Model loss of architecture

The above figure denotes the model loss of the architecture. Analysing the two figures we would be achieving an accuracy of 94.643 to 95.238 and loss of 0.16.

Table 22.1 Performance analysis table

Performance analysis of existing work (Accuracy)	Performance analysis of proposed work (Accuracy)
54.20	77.381
64.72	53.12
65.01	94.64
67.99	**95.23**

Initially the loss might be high after the number of epochs the loss would be reduced. The ultimate goal in developing the model is to improve the accuracy and to reduce the losses. The highest percentage of accuracy of 95.238 percentage is achieved and the model is built.

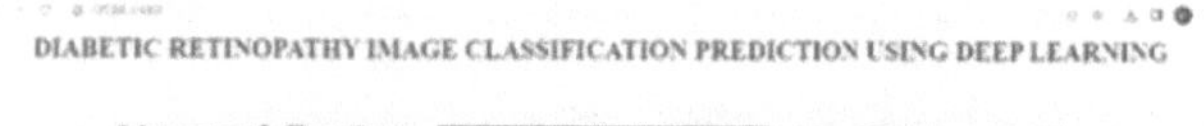

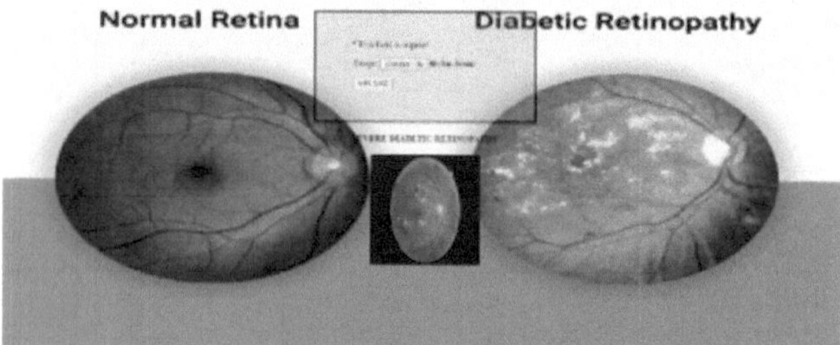

Fig. 22.10 Output

The screenshot of the output is attached above.

6. Conclusion

It concentrated on how a CNN model was used to predict the pattern of diabetic retinopathy disorders utilising images from a given dataset (called a trained dataset) and historical data sets. This provides some of the following information regarding the prognosis of diabetic retinopathy. The ability of CNN to automatically classify images is its main advantage. The overview of methods for identifying anomalies in diabetic retinopathy images was covered in this work. These methods include feature extraction, preprocessing, classification systems, and the acquisition of retinopathy image data sets.

Acknowledgement

The authors gratefully acknowledge the students, staff, and authority of Physics department for their cooperation in the research.

References

1. R. L. Thomas, S. Halim, S. Gurudas, S. Sivaprasad, and D. R. Owens, "IDF diabetes atlas: A review of studies utilising retinal photography on the global prevalence of diabetes related retinopathy between 2015 and2018," Diabetes Res. Clin. Pract., vol. 157, Nov. 2019, Art. no. 107840.
2. T. A. Ciulla, A. G. Amador, and B. Zinman, "Diabetic retinopathy and diabetic macular edema: Pathophysiology, screening, and novel therapies," Diabetes Care, vol. 26, no. 9, pp. 2653-2664,2003
3. T.YT. Y. Wong et al., "Guidelines on diabetic eye care: The internationalcouncil of ophthalmology recommendations for screening, follow-up,referral, and treatment based on resource settings," Ophthalmology,vol. 125, no. 10, pp. 1608–1622, 2018.
4. D. S. W. Ting, G. C. M. Cheung, and T. Y. Wong, "Diabetic retinopathy: Global prevalence, major risk factors, screening practices and public health challenges: A review," Clin. Exp. Ophthalmol., vol. 44, no. 4,pp. 260–277, May 2016.
5. K. Simonyan and A. Zisserman, "Very deep convolutional networks forlarge-scale image recognition," 2014, arXiv:1409.1556.
6. Y. Zhou et al., "Collaborative learning of semi-supervised segmentation and classification for medical images," in Proc. IEEE/CVF Conf.Comput. Vis. Pattern Recognit. (CVPR), Jun. 2019, pp. 2079–2088.
7. M. Tavakoli, S. Jazani, and M. Nazar, "Automated detection of microaneurysms in color fundus images using deep learning with differentpreprocessing approaches," Proc. SPIE Med. Imag. Inf. Healthcare, Res., Appl., vol. 11318, Oct. 2020, Art. no. 113180E.
8. R. T. Mamilla, V. K. R. Ede, and P. R. Bhima, "Extraction of microa-neurysms and hemorrhages from digital retinal images," J. Med. Biol.Eng., vol. 37, no. 3, pp. 395–408, 2017.
9. B. Wu, W. Zhu, F. Shi, S. Zhu, and X. Chen, "Automatic detection of microaneurysms in retinal fundus images," Comput. Med. Imag. Graph., vol. 55, pp. 106–112, Jan. 2017.
10. P. Khojasteh, B. Aliahmad, and D. K. Kumar, "A novel color space of fundus images for automatic exudates detection," Biomed. Signal Process. Control, vol. 49, pp. 240–249, Mar. 2019.
11. J. Mo, L. Zhang, and Y. Feng, "Exudate-based diabetic macular edema recognition in retinal images using cascaded deep residual networks," Neurocomputing, vol. 290, pp. 161–171, May 2018.
12. J. Ni, J. Wu, J. Tong, Z. Chen, and J. Zhao, "GC-Net: Global context network for medical image segmentation," Comput. Methods Programs Biomed., vol. 190, Jul. 2020, Art. no. 105121.
13. T. Walter, P. Massin, A. Erginay, R. Ordonez, C. Jeulin, and J.-C. Klein, "Automatic detection of microaneurysms in color fundus images," Med. Image Anal., vol. 11, no. 6, pp. 555–566, 2007.
14. M. Niemeijer, B. Van Ginneken, J. Staal, M. S. A. Suttorp-Schulten, and M. D. Abramoff, "Automatic detection of red lesions in digital color fundus photographs," IEEE Trans. Med. Imag., vol. 24, no. 5, pp. 584–592, May 2005.
15. Santhosh Kumar M, Maria Anu, Shiva M, "Retinal Image Processing Using Neural Networks with Deep Learning", Easy Chair Preprint, March 2022.

Note: Every figures and table except fig. 22.1 was created using the experimental work.

Advances in Additive Manufacturing Technologies – Gurusamy Pathinettampadian (eds)
© 2024 Taylor & Francis Group, London, ISBN 978-1-032-90013-1

23 Microstructural Characterization and Investigation on Mechanical Properties of Stir Cast Hybrid A356/B_4C/ZrO_2 Composites with Strontium as Grain Modifier

S. Venkat Prasat[1], A. Rajkumar[2], N. Venkateshwaran[3] and S. Sekar[4]
Dept. of Mechanical Engineering,
Rajalakshmi Engineering College, Chennai, India

ABSTRACT: Hybrid Aluminium/Boron carbide/Zirconium oxide (A356/B_4C/ZrO_2) composites were synthesized using the stir casting method by varying boron carbide content (4wt.%, 8wt.% and 12wt.%). A fixed amount (4wt.%) of zirconium oxide particles was added as a secondary reinforcement and the reaction process was studied using EDS, Scanning Electron Microscopy and X-ray Diffraction analysis. Optical microscopy revealed that B_4C and ZrO_2 particles were distributed uniformly in aluminium matrix. The interfacial bonding of B_4C and ZrO_2 particles with the A356 matrix was significantly enhanced due to the inclusion of Potassium Fluorotitanate and Magnesium. Findings revealed that hardness, Ultimate Tensile Strength (UTS) and compressive strength of A356/B_4C/ZrO_2 hybrid Aluminium Metal Matrix Composites (AMMCs) increase as the weight percentage of B_4C is increased. The mechanical and metallurgical characteristics of the A356/B_4C/ZrO_2AMMCs improved due to the inclusion of a strontium-based master alloy, which transformed the coarse structure of the eutectic Si phase into a fine, equiaxed needle-shaped structure. Zirconium diboride formed during the reaction process enhanced the hardness and the mechanical characteristics of the hybrid AMMCs. The hardness of A356/12wt.% B_4C/4wt.% ZrO_2 hybrid composite was augmented by 62.02% relative to A356 alloy. UTS and compressive strength of A356/12wt.% B_4C/4wt.% ZrO_2 composites increased by 25.39% and 16.78% relative to A356 alloy respectively. The enhancement in the mechanical properties of A356/B_4C/ZrO_2 hybrid AMMCs is attributed to alloy strengthening and grain size reduction due to the inclusion of strontium-based master alloy and presence of zirconium oxide. The addition of zirconium oxide leads to improved densification. A356/12wt.% B_4C/4wt.% ZrO_2 hybrid composites demonstrated highest levels of hardness, UTS and compressive strength.

KEYWORDS: Hybrid composites, Boron carbide, Zirconium oxide, XRD, SEM, EDS

1. Introduction

Newer composite materials have significant usage in automobile, aeronautical and marine industries [1]. Metal Matrix Composites (MMCs) possess higher strength, wear resistance and hardness than unreinforced alloys [2]. Aluminium composites are used for manufacturing structural metal products. MMCs are extensively favoured because of their high specific strength [3]. Hybrid metal matrix composites (HMMCs) exhibit enhanced tensile characteristics, hardness and wear resistance than MMCs [4]. HMMC has good mechanical and tribological properties

[1]venkatprasat.s@rajalakshmi.edu.in, [2]rajkumar.a@rajalakshmi.edu.in, [3]venkateshwaran.n@rajalakshmi.edu.in, [4]sekar.s@rajalakshmi.edu.in

DOI: 10.1201/9781003545774-23

compared to aluminium-cast alloys. The properties of HMMC are comparatively better than particle-reinforced metal matrix composites [5]. A356 alloy castings possess excellent castability, corrosion resistance, machinability and weldability characteristics. Boron carbide is selected as the reinforcement in many applications because it leads to the formation of secondary phases and has better tribological properties [6]. Boron Carbide exhibits higher hardness when compared with aluminium oxide and silicon dioxide [7]. Zirconium oxide has high fracture toughness, wear resistance, hardness, density and melting point. It has extensive applications in the production of automotive and machine tool components, especially in the manufacture of piston cups, cutting tools, pump seals, valves and impellers.

There are different methods to manufacture aluminium composites like stir casting, compo-casting, powder metallurgy and infiltration techniques. Composites with a higher weight % of reinforcement enhance functional performance[8]. Researchers have studied the abrasive wear behaviour of AA2014/12B_4C composites which were fabricated using stir casting techniques [9]. Researchers have reported good improvement in the mechanical characteristics of Al7075-B_4C-Gr AMMCs which were synthesized by stir casting techniques [10]. The use of master alloys like Al10Sr was explored earlier by several investigators. The inclusion of strontium alters the microstructure of the AMMCs and α-Al structure transforms into a fine equiaxed structure [11]. The strontium-based master alloy Al-10Sr is used to alter the eutectic Si structure and improve mechanical properties [12]. Investigators have reported that the fibrous structure of silicon was achieved and the grain structure was enhanced due to the inclusion of strontium [13, 14]. The formation of clusters at the Silicon-liquid interface in Strontium and Sodium modified Al–Si alloys was investigated by various researchers. The reaction products like Sr-Al-Si and SrB formed at the silicon liquid interface with high amounts of strontium (0.5%) reduce the tensile properties [15, 16]. The complete modification of eutectic silicon was achieved with the inclusion of strontium of 100 ppm. [17]. Porosity and silicon carbide clusters formed when the amount of strontium exceeds 0.05%. The over-modified structure is obtained due to the increased addition of strontium [18]. The effect of alloying element Magnesium on Sodium and Strontium modifying Aluminium-7Si alloys was studied by researchers and it was determined that the inclusion of Magnesium caused the incomplete alteration of eutectic Si phases [19]. The addition of strontium in aluminium silicon carbide composites increases the amount of α-Al thereby improving the mechanical characteristics of AMMCs [20].

Aluminium-B_4C AMMCs with the addition of zircon and titanium particles have excellent mechanical and wettability properties because of improved Al-matrix and B_4C interfacial bonding [21]. Aluminium-zirconium oxide composites are used to manufacture products that resist wear. The mechanical properties of AMMCs increased proportionally as the zirconium oxide content was increased [22]. It was reported in the literature that the UTS of Al356-ZrO_2 AMMCs increased when the volume percentage of zirconium oxide was increased [23]. Aluminium A356-ZrO_2 AMMCs with varying proportions of zirconium oxide were produced earlier with excellent results [24]. The mechanical characteristics and hardness of Al-ZrO_2 AMMCs improved considerably with an increase in the content of zirconium oxide (15%).

Al-ZrO_2-B_4C composites were produced by powder metallurgy technique[25]. The reaction products like ZrB_2 formed during the processing of composites. The flexural strength of Al-ZrO_2-B_4C composites was reported as 560 MPa. Zirconium diboride and Zirconium carbide were formed during the synthesis of Al-B_4C-ZrO_2 AMMCs by the exothermic dispersion synthesis method [26]. Poor wettability and inhomogeneity between aluminium matrix and reinforcement particles were reported earlier. Potassium Fluorotitanate and magnesium were added to melt to augment wettability. Researchers have studied the effect of particle size distribution and processing of SiC-B_4C AMMCs [27,28]. Based on the exhaustive survey of the literature, it was determined that there is a research gap and that the research work related to microstructural characterization and the investigations on the mechanical behaviour of Al/B_4C/ZrO_2AMMCs manufactured by stir casting techniques are limited. The goal of the ongoing investigation is to manufacture Al/B_4C/ZrO_2 AMMCs with enhanced hardness and mechanical characteristics.

2. Experimental Details

2.1 Materials

Matrix material utilized is aluminium alloy A356, which has a tensile strength and elongation of 230N/mm^2 and 2% respectively. A356 alloy demonstrated a hardness of 71BHN. Aluminium alloy A356 was reinforced with varying weight percentages (4wt.%, 8wt.% and 12wt.%) of boron carbide powder. B_4C particulates of size 63μm were utilized as primary reinforcement. A fixed amount (4wt.%) of zirconium oxide (45 μm) was added as the secondary reinforcement. ZrO_2 is a white crystalline oxide of Zr with a high density (5.85 g/cm^3). Composition of A356/B_4C/ZrO_2 Hybrid composite samples which were prepared is shown in Table 23.1. The master alloy form of Strontium Al10Sr (0.03%) and Potassium Fluorotitanate (K_2TiF_6) were also

Table 23.1 Composition of $A356/B_4C/ZrO_2$ hybrid composite samples

Hybrid Composite samples	A356 alloy (wt.%)	B_4C (wt.%)	ZrO_2 (wt.%)
$A356/4B_4C/4ZrO_2$	92	4	4
$A356/8B_4C/4ZrO_2$	88	8	4
$A356/12B_4C/4ZrO_2$	84	12	4

added along with reinforcements. The melt was supplemented with Magnesium (1%) to increase wettability. Aluminium strontium 10% master alloy (Al10Sr) contains strontium (9-11%), iron (0.25%) and balance aluminium cast alloy. Chemical composition of B_4C and zirconium oxide powders are displayed in Table 23.2 and Table 23.3.

Table 23.2 Composition of B_4C powder

Element	Fe	Si	Ca	O	N	C	B
Weight %	0.029	0.1	0.014	0.067	0.07	21.3	Balance

Table 23.3 Composition of zirconium oxide powder

Element	ZrO_2	SiO_2	TiO_2	Fe_2O_3	Others
Weight %	99.4	0.11	0.008	0.003	0.48

2.2 Fabrication of $A356/B_4C/ZrO_2$ Hybrid Composites using Stir Casting

The stir casting methodology was chosen as a processing method because of its simplicity and cost-effectiveness. Aluminium cast billets were machined to the required size and charged into the furnace. Hexachloroethane tablets were added to avoid porosity. Magnesium powder was introduced into the furnace. Al10Sr master alloy (0.03%) was added to the melt and mixing was done by the impeller at a speed of 600 rpm. To eliminate moisture and impurities, B_4C and Potassium Fluorotitanate particulates were combined in the 1:0.3 ratio and heated up beforehand to a temperature of 400°C for nearly 3 hours. The preheated boron carbide and zirconium oxide (4wt.%) particles were mixed into the liquid melt while the slurry was consistently agitated. To stop an oxide layer from forming in the matrix melt, argon gas was fed into the furnace and the stirring speed of 600 rpm was kept constant. A 15-minute stirring period was used to guarantee that the particles in the slurry were distributed evenly. After pouring the melt into the heated mould, it was left in the die to solidify.

2.3 Microstructural Characterization on $A356/B_4C/ZrO_2$ AMMCs

Microstructural characterization was performed utilizing SEM, XRD analysis and optical microscopy. To analyse the microstructure, AMMCs were machined into cylindrical specimens measuring 25mm in length and 12mm in diameter. The etchant utilized was Keller's reagent and AMMC samples were inspected utilizing NIKON Epiphot 200 optical microscope. SEM microscopy was conducted using the SIGMA Carl Zeiss scanning electron microscope with an EDX detector. Bruker D8 X-ray diffractometer was employed to carry out the XRD analysis to determine the phases and reaction products of AMMCs.

2.4 Mechanical Testing on Hybrid Composites

UTM was utilized to evaluate the tensile strength and compression strength of the AMMC samples. $Al-B_4C/ZrO_2$ composites were machined to produce cylindrical samples that are 75 mm long and 10 mm in diameter, following ASTM E8 and ASTM E9 standards respectively. Mitutoyo MVKH1 micro-hardness tester was utilized to perform the hardness test on $Al-B_4C/ZrO_2$ hybrid composite specimens with a 50g applied load and 20 seconds dwell time.

3. Results and Discussion

3.1 Microstructure Analysis of $A356/B_4C/ZrO_2$Composites

Microstructure of $A356/12B_4C/ZrO_2$ hybrid composite with 100X magnification is depicted in Fig. 23.1. The hybrid composite's microstructure verifies that the B_4C and ZrO_2 particulates are evenly dispersed throughout the matrix of the A356 aluminum alloy. The microstructure reveals that modified eutectic Si particles are distributed in the inter-dendritic region. Modified fine-rounded eutectic Si is seen in the figure distributed throughout the inter-dendritic area when the content of boron carbide was increased to 12wt.%. The addition of strontium transforms the coarse structure of the eutectic Si phase into a fine, equiaxed needle-shaped structure. Phases like Mg_2Si and $CuAl_2$ which are scattered in the matrix result in alloy strengthening similar to the findings of other authors who have studied the microstructure of composites [29]. The B_4C and ZrO_2 particles seam together to form particle assemblage along the alpha-aluminum grain boundary. These outcomes were consistent with what other researchers had noticed [30].

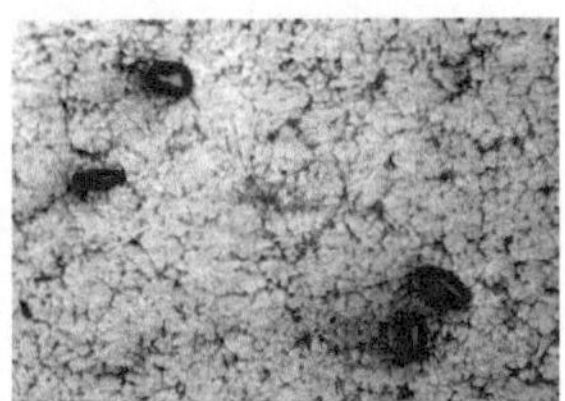

Fig. 23.1 Microstructure of $A356/12B_4C/4ZrO_2$ composite (100X)

3.2 SEM and EDS Investigation on Hybrid Composites

The SEM investigations were conducted on the A356/12wt.% Boron carbide/4wt.% Zirconium oxide ($A356/12B_4C/4ZrO_2$)

hybrid composites and the image is displayed in Fig. 23.2. The uniformly dispersed B_4C and ZrO_2 particulates are visible in the SEM micrograph. The formation of Zirconium diboride (ZrB_2) and Zirconium carbide (ZrC) reaction layers on the exterior surface of B_4C particulates are visible in SEM micrograph and these results agree with observations made by other researchers [31]. The interfacial bonding of B_4C and ZrO_2 particles with the A356 matrix was significantly enhanced with the inclusion of Potassium Fluorotitanate and magnesium.

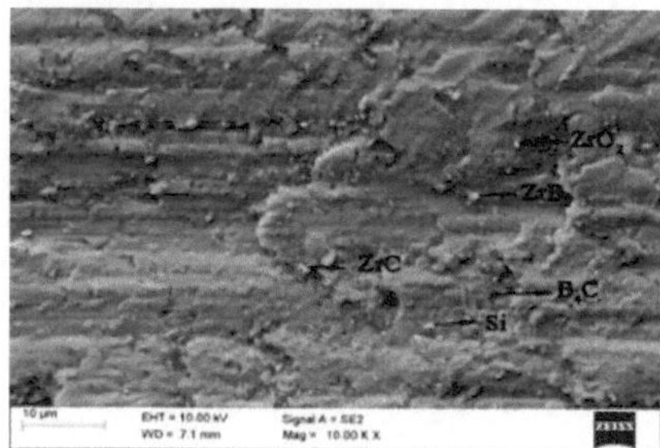

Fig. 23.2 SEM Image of A356/12B_4C/4ZrO_2 AMMCs (10 KX)

EDS image of A356/12B_4C/4ZrO_2 hybrid composites is depicted in Fig. 23.3. EDS image shows elements like Zr and B, which leads to the refinement of grain structure and similar results have been reported by other authors [32]. Zirconium diboride is formed as the result of the reaction between B_4C and zirconium oxide. A rise in the wettability of particles is caused by the presence of ZrB_2. Addition of B_4C and zirconium oxide leads to grain modification and considerable improvement in densification and these results were comparable to the outcomes attained by other investigators [33]. The addition of strontium transformed Si particles from a thin, needle-like shape to a fine, fibrous shape. The presence of zirconium diboride phase causes an improvement in the hardness of the AMMCs considerably.

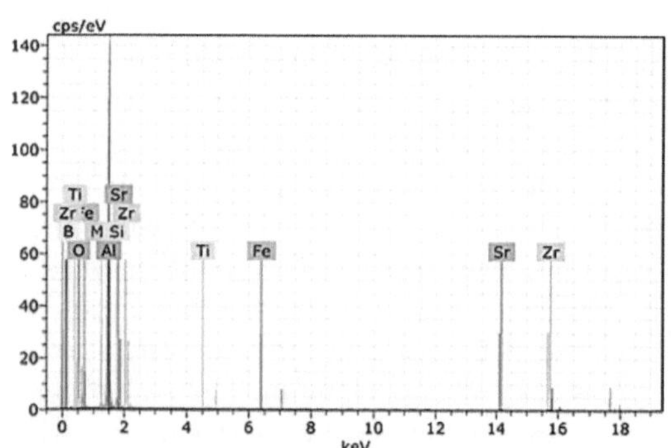

Fig. 23.3 EDS image of A356/12B_4C/4ZrO_2 AMMCs

3.3 XRD Analysis of Hybrid Composites

The X-ray diffraction studies reveal the peaks of intensity for Al, Al_3Zr, Al_2O_3, AlTi, Al_3BC, $CuAl_2$, Mg_2Si, Si, ZrC and ZrB_2 (Fig. 4). At the interface where the B_4C particulates and A356 matrix meet, the reaction product Al_3BC was formed. The presence of a titanium layer encircling the B_4C particles is indicated by the phase AlTi. Wetting of the particles was enhanced by the formation of ZrB_2, thereby validating the research conducted by the previous investigators [34]. Hybrid composites become harder as a result of ZrB_2 formation. Al_3BC and AlB_2 phases were generated at a temperature of 870°C and Al_3BC is the most predominant reaction product compared to AlB_2 [6]. An increase in B_4C particle content causes the reinforcement's intensity peaks to rise in XRD micrograph and comparable outcomes were attained by Shirvanimoghaddam *et al.* [35].

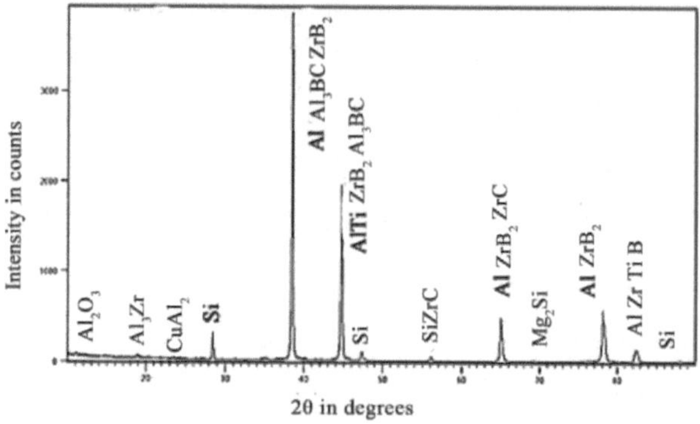

Fig. 23.4 XRD micrograph of A356/12B_4C/4ZrO_2 AMMCs

3.4 Hardness Test

Mitutoyo MVKH1 micro-hardness tester was utilized to perform the hardness test on Al-B_4C/ZrO_2 hybrid composite specimens. The hardness bar chart for Al/B_4C/ZrO_2 composites is depicted in Fig. 23.5. The hybrid AMMCs demonstrated a positive correlation between wt.% and hardness with an augmentation in wt.% of B_4C particulates from 0wt.% to 12wt.%. The hardness of A356/12wt.% B_4C/4wt.% ZrO_2 hybrid composite was augmented by 62.02% relative to the base alloy. Increase in the hardness of the AMMCs is attributed to alloy strengthening and grain size reduction [36]. The formation of ZrB_2 increases the hardness of the AMMCs.

3.5 Tensile Test

UTM was utilized to conduct tensile tests for hybrid composites. Ultimate tensile strength (UTS) of Al/B_4C/ZrO_2 AMMCs with different reinforcement content is displayed in Fig. 23.6. The photomacrograph of specimens used for tensile testing is depicted in Fig. 23.7. UTS of hybrid AMMCs rises as the proportion of B_4C reinforcement increases. UTS of Al/12B_4C/4ZrO_2 hybrid composite was augmented by 25.39% relative to unreinforced A356 alloy. UTS of hybrid AMMCs was also augmented with the inclusion of strontium. The inclusion of a master alloy based on strontium has strengthened the alloy and reduced grain size, which has improved the UTS of the hybrid AMMCs. These

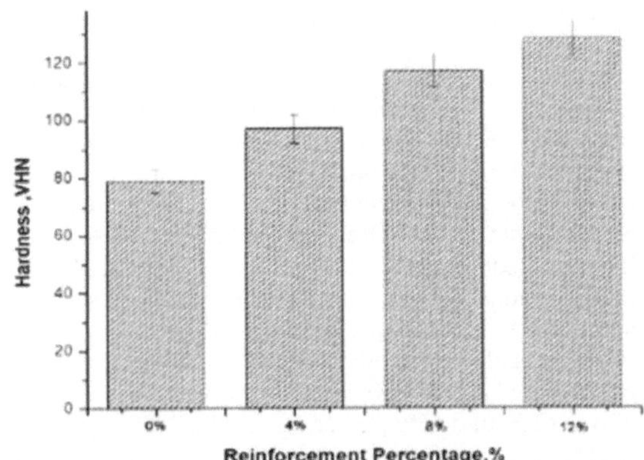

Fig. 23.5 Influence of wt.% of B_4C particles on hardness

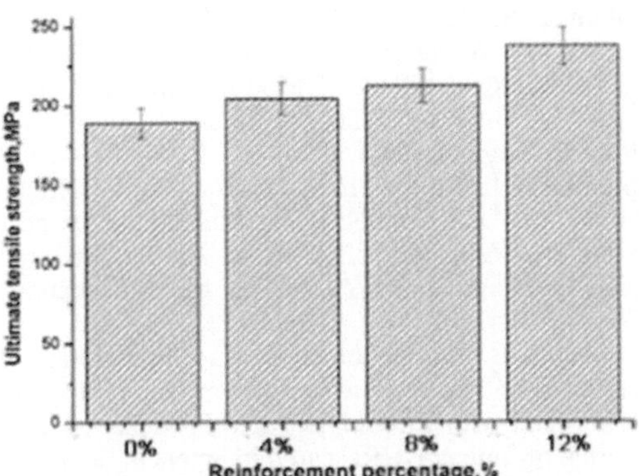

Fig. 23.6 Influence of wt.% of B_4C on UTS

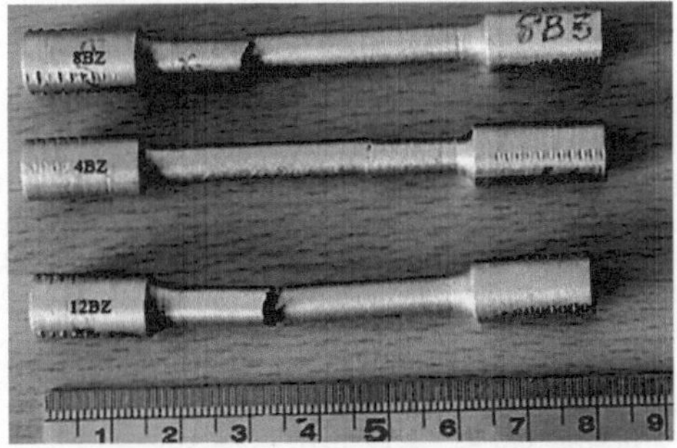

Fig. 23.7 Photomacrograph of tensile test specimens

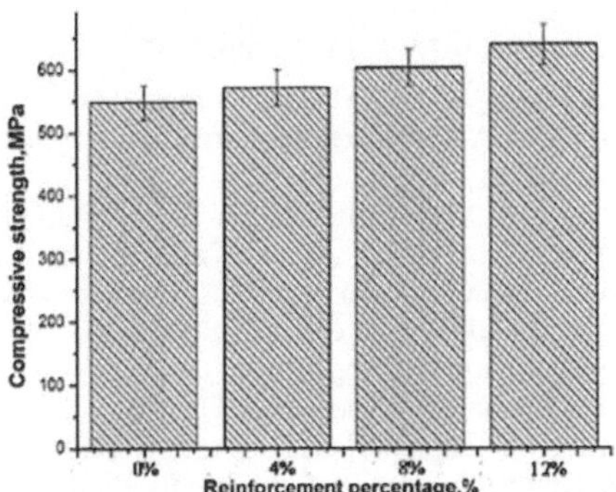

Fig. 23.8 Effect of wt.% of B_4C on compressive strength of hybrid composites

observations are in concurrence with the outcomes obtained by various investigators [36]. UTS of the hybrid AMMCs is high because of the significant work hardening phenomenon at low strain. Similar results were attained by other researchers during the processing of Aluminium-SiC composites [37]. Addition of zirconium oxide leads to enhanced mechanical characteristics of hybrid AMMCs. The formation of ZrB_2 also augments the tensile strength of composites.

3.6 Compression Test

UTM was utilized to conduct compression tests for hybrid AMMCs and results are displayed in Fig. 23.8. A higher weight percentage of B_4C improved hybrid AMMCs' compressive strength. Compressive strength of hybrid AMMCs ($Al/12B_4C/4ZrO_2$) increased by 16.78% when compared to A356 cast aluminium alloy. A356 matrix alloy's dislocation movement was inhibited by B_4C and ZrO_2 particulates, and reinforcement particulates' strain hardening phenomenon increased compressive strength of hybrid A356/B_4C/ZrO_2 AMMCs.

4. Conclusions

$Al/B_4C/ZrO_2$ hybrid AMMCs were synthesized utilizing stir casting method. Homogenous distribution of B_4C and ZrO_2 particulates in the matrix was confirmed by SEM, EDS, and XRD analyses. Wettability and mechanical characteristics of A356/B_4C/ZrO_2 AMMCs were significantly improved due to the inclusion of strontium-based master alloy (0.03%). ZrB_2 and other reaction products were formed, which improved wetting behaviour. The existence of ZrB_2 particles leads to grain refinement and improves mechanical characteristics of the AMMCs. Addition of strontium transforms the coarse structure of the eutectic Si phase into a fine, equiaxed needle-shaped structure. Interfacial bonding of B_4C and ZrO_2 particles with the A356 matrix was significantly enhanced with the addition of Potassium Fluorotitanate and magnesium.

Significant augmentation in mechanical characteristics was accomplished by addition of B_4C and ZrO_2. Hardness of A356/$12B_4C/4ZrO_2$ is augmented by 62.02% relative to the base alloy. UTS and compressive strength of A356/$12B_4C/4ZrO_2$ composites increased by 25.39% and 16.78 % relative to A356 alloy respectively. The enhancement in the mechanical characteristics of hybrid AMMCs is attributed to alloy strengthening and grain size reduction. A356/12wt.%B_4C/4wt.%ZrO_2 hybrid composites demonstrated highest levels of hardness, UTS and compressive strength. A356/B_4C/ZrO_2 hybrid composites can be utilized for aircraft, missile and space applications as well as for truck chassis parts and automotive transmission cases that require high strength.

REFERENCES

1. Kainer,K.U., "Metal Matrix Composites. Custom-made Materials for Automotive and Aerospace Engineering", Wiley-Vch, 2006.
2. Sajjadi, S.A., Ezatpour, H.R. and Torabi Parizi, M, "Comparison of microstructure and mechanical properties of A356 aluminum alloy/Al_2O_3 composites fabricated by stir and compo-casting processes", Materials and Design,Vol.34, (2012) pp.106–111.
3. Canakci,A., Varol, T.and Nazik,C., "Effects of amount of methanol on characteristics of mechanically alloyed Al-Al_2O_3 composite powders", Materials Technology, Vol.27, (2012) pp.320–327.
4. Rajmohan,T.,Palanikumar, K. and Ranganathan,S., "Evaluation of mechanical and wear properties of hybrid aluminum matrix composites", Transactions of Nonferrous Metal Society China, Vol.23, (2013) pp.2509–2517.
5. Michael, O.B., Kenneth,K.A.and Lesley,H.C., Aluminium matrix hybrid composites: a review of reinforcement philosophies; mechanical, corrosion and tribological characteristics, Journal of Material Research Technology, Vol.4. No.4, (2015) pp.443–445.
6. Shorowordi, K.M, Laoui, T., Haseeb, A.S.M.A., Celis, J.P. and Froyen, L., "Microstructure and interface characteristics of B_4C, SiC and Al_2O_3 reinforced Al matrix composites: a comparative study", Journal of Material Processing Technology, Vol.142, (2003) pp.738–743.
7. Nie, C., Gun, J., Liu, J. and Zhang, D., "Production of boron carbide reinforced 2024 aluminium matrix composites by mechanical alloying", Material transactions, The Japan Institute of metals, Vol.48, No.5, (2007) pp.990–995.
8. Canakci, A., Varol, T. and Erdemir, F., "The Effect of Flake Powder Metallurgy on the Microstructure and Densification Behavior of B_4C Nanoparticle-Reinforced Al-Cu-Mg Alloy Matrix Nanocomposites, Arabian Journal for Science and Engineering, Vol.41, (2016) pp.1781–1796.
9. Canakci, A., Ozsahin, S. and Varol,T., "Prediction of Effect of Reinforcement Size and Volume Fraction on the Abrasive Wear Behavior of AA2014/B_4Cp MMCs Using Artificial Neural Network, Arabian Journal for Science and Engineering, Vol.39, (2014) pp.6351–6361.
10. Karthe, M., Karuppusamy, S., Sureshkumar, B. and Nazar Ali, A., "Mechanical and wear properties of Al7075- B_4C–Gr metal matrix composites fabricated by the using stir casting method", Vol. 77, Part 2, (2023), Materials Today: Proceedings, pp. 389–393.
11. Chen Zhongwei and Zhang Ruijie, "Effect of strontium on primary dendrite and eutectic temperature of A357 aluminium alloy", China Foundry, (2010) pp.140–152.
12. Tuncay, T., "The Effect of Modification and Grain Refining on the Microstructure and Mechanical Properties of A356 Alloy", Acta Physica Polonica A, Vol. 131 No.1, (2017) pp.89–91.
13. Closset, B., Dugas, H., Pekguleryyuz, M. and Gruzleski, J. E., "The Aluminum-Strontium Phase Diagram", Metallurgical Transactions, Vol.17A, (1986) pp.1251–1252.
14. Bouska,O., "The effect of different casting parameters on the relationship between flowability, mold filling capacity and cooling conditions of Al-Si alloys", Metalurgija-Journal of Metallurgy", Vol 14, (2008) pp.17–30.
15. Lashgari, H.R., Emamy, M., Razaghian, A. and Najimi, A.A., "The effect of strontium on microstructure,porosity and tensile properties of A356-10%B_4C cast composites", Material Science and Engineering", Vol. 517, No.1, (2009) pp.170–179.
16. Barrirero, J., Li, M. Engstler, N. Ghafoor, P. Schumacher, M.Oden and F. Mucklich, Cluster formation at the Si/liquid interface in Sr and Na modified Al–Si alloys, Scripta Materialia,Vol.117, (2016) pp.16–19.
17. Prasad, B.K., Venkateswarlu, K., Modi, O.P, Jha, A.K, Das, S., Dasgupta, R. and Yegneswaran, A.H., "Sliding wear behavior of some Al–Si alloys: role of shape and size of Si particles and test conditions", Metallurgical Material Transactions A,Vol.29, (1998) pp.2747–52.
18. Pezda,J.,"Effect of modifying process on mechanical properties of EN AB-42000 silumin cast into sand moulds", Archives of Foundry Engineering, Vol.9,(2009) pp.187–190.
19. Huang, C., Liu, Z., Li, J., "Influence of Alloying Element Mg on Na and Sr Modifying Al-7Si Hypoeutectic Alloy", Materials, Vol. 15, (2022) pp. 1537.
20. Sulaiman, S. Marjom, Z. Ismail, M.I.S., Ariffin, M.K.A. and Ashraf, N., "Advances in Material & Processing Technologies Conference Effect of Modifier on Mechanical Properties of Aluminium Silicon Carbide (Al-SiC) Composites, Procedia Engineering,Vol. 184, (2017) pp.773–777.
21. Ibrahim, M.F., Ammar, H.,R., Samuel, A.M, Soliman, M.S., Almajid, A. and Samuel, F.H., "Mechanical properties and fracture of Al-15vol-%B_4C based Metal Matrix Composites", International Journal of Cast Metal Research, Vol.27, No.1, (2013) pp.7–14.
22. Girisha,K. B. and Chittappa, H. C., "Preparation, Characterization and Wear Study of Aluminium Alloy (Al 356.1) Reinforced with Zirconium Nano Particles", International Journal of Innovative Research in Science, Engineering and Technology, Vol.2, Issue 8, (2013) pp. 3627–3637.
23. Baghchesara, M., Abdizadeh, H. and Baharvandi, H.R., "Fractography of stir casted Al-ZrO_2 composites", Iranian Journal of Science and Technology, Vol.33, No. B5, (2009) pp.453–462.
24. Madhusudhan, M., Naveen, G. J. and Mahesha, K., "Mechanical Characterization of AA7068-ZrO_2 reinforced Metal Matrix Composites", Materials Today: Proceedings Part A, Vol.4, No.2., (2017) pp.3122–3130.
25. Lin,W.S. and He, L., "Effect of Al and ZrO2 on Sinterability and Mechanical Properties of B_4C", Advanced Materials Research,Vol.160-162, (2011) pp.1494–1497.
26. Zhu,H.G.,Chu,D,Wang,H.,Min,J and Ai,Y.L., Study on the reaction mechanism of Al-ZrO_2-B_4C system,Material Science Forum,Vol.675-677, (2011) pp.839–842.
27. Chen Yan-fei, Xiao Shu-long, Tian Jing, Xu Li-juan and Chen Yu-yong., "Effect of particle size distribution on properties of zirconium oxide ceramic mould for TiAl investment casting", Transactions of Nonferrous Metal Society of China, Vol.21, (2011) pp. 342–347.
28. Arslan, G. and Kalemtas, A., "Processing of silicon carbide–boron carbide–aluminium composites", Journal of the European Ceramic Society, Vol. 29, (2009) pp.473–480.
29. Karthikeyan, G. Jinu, G.R. and Vijayalakshmi, P., Weldability Study of LM25/ZrO_2 Composites by using friction welding, Revista Materia,Vol.22, No.3. (2017).
30. Yang Hua-jing, Zhao, Yu-tao, Chen-Gang, Zhang, Song-li and Chen Deng-bin, "Preparation and microstructure of in-situ (ZrB_2 + Al_2O_3 + Al_3Zr)p/A356 composite synthesized by direct reaction", Transactions of Non-ferrous Metal Society China,Vol. 22, (2012) pp.571–576.
31. Toptan, F, Kerti, I and Rocha, L.A, "Reciprocal dry sliding wear behavior of B_4Cp reinforced aluminium alloy matrix composites", Wear, Vol.74–85, (2012) pp.290–291.

Advances in Additive Manufacturing Technologies – Gurusamy Pathinettampadian (eds)
© 2024 Taylor & Francis Group, London, ISBN 978-1-032-90013-1

24 Polycystic Ovarian Syndrome Analysis using Machine Learning

Nishithha Premkumar Sangeetha[1], Neshanthini Vengatesan[2], Kavitha Prithiviraj[3]
Department of Artificial Intelligence and Data Science, Panimalar Engineering College, Chennai, India
Maheswari Marimuthu[4]
Department of Computer Science and Engineering, Panimalar Engineering College, Chennai, India

ABSTRACT: The most significant issue affecting the female population is PCOD (Polycystic Ovary Disorder) or PCOS (Polycystic Ovary Syndrome). It's among the biggest issues that women nowadays are dealing with. It is a hormonal imbalance that makes the ovaries to expand, holding a little blister over the periphery. During the reproductive years, the hormonal imbalance only manifests itself. Female with polycystic ovarian syndrome (PCOS) having issues in regulation of androgen system incorporates the study of PCOD, which comprises stages like data collection, data analysis & visualization, training of the models, evaluating the models, comparison, and selecting the models, and ensemble model. The data is taken from Kaggle, processed, then visualized using visualization metrics. The model is then trained using machine learning approaches such as KNN, Naive Bayes, SVM, Decision Tree, Logistic Regression, and Random Forest. Performance measurements like confusion matrix, precision, recall, f1 score, support, and accuracy are used to assess the trained models' correctness and efficiency. Based on these performance metrics, Decision Tree and Random Forest integrated into a single model utilizing ensemble learning had a 76% accuracy rate.

KEYWORDS: PCOS, Machine learning, Random forest, Logistic regression

1. Introduction

PCOS (Polycystic Ovary Syndrome) or PCOD (Polycystic Ovary Syndrome) is one among major problems faced by females today. It is a metabolic impact that results in enlarged female gonads with little blister over the periphery. The hormonal issue only arises throughout the reproductive years. Hormonal imbalance of androgens and estrogen as well in the regulation of androgen productions in women who have polycystic ovarian syndrome (PCOS). The hypothalamic-pituitary-ovarian (HPO) axis can function improperly in PCOS. PCOS is frequently identified in women who seek medical assistance for problems with infertility, obesity, acne, amenorrhea, and excessive hair growth. Endometrial cancer, heart diseases, dyslipidemia, and type 2 diabetic mellitus are more common in female with PCOD. The management of PCOS using pharmacotherapy is discussed in this article. PCOD is a gene trait condition which influences a variety of heredity and environmental variables that interact to produce a diverse, clinical, and biochemical phenotype. Poor food practices can influence an increase in environmental factors linked

[1]nisshreven@gmail.com, [2]neshavengatesan19@gmail.com, [3]varshnikavitha@gmail.com, [4]m.mahe05@gmail.com

DOI: 10.1201/9781003545774-24

to PCOD, and toxins and contagious diseases may also be involved. Despite the way of living changes including losing weight and exercise, the reproductive and harmonic symptoms of PCOD can reverse occasionally. PCOD has been associated with insulin resistance and obesity, even though the exact reason is uncertain. It is not surprising that there is a connection between insulin functioning and ovarian functioning because insulin simulates ovarian activity and the ovaries react to surplus insulin by releasing androgens, which can cause anovulation. PCOS can cause irregular or heavy menstrual cycles, abundant pelvic pain, acne, issues with conception and areas of luscious, thick skin in women. Hyper androgenic, insulin resistance, and neuroendocrine disturbance are some of syndrome's main traits. PCOD prevalence has been estimated to range between 4% to 18% for general populations, according to an analysis of the available but it may be as high as 26% in specific groups. Although PCOS is very common, there is still no recognized cure, and the actual cause is still unknown. The majority of female with PCOD have obesity, though data on visceral and subcutaneous abdominal fat is inconclusive, it may be larger, unchanged, or lower in PCOD-positive women than in reproductively healthy women with comparable body BMI indices. Studies have shown that any combination of testosterone, androstanol one (dihydrotestosterone), and nandrolonedecanoate can help reduce visceral fat in humans and animals alike.20% of female with PCOD are not fat or considered thin, even though 80% of the disease's cases occur in obese women. Nonetheless, the likelihood of negative consequences such as insulin resistance, metabolic syndrome, endometrial hyperplasia, and hypertension is increased in obese women with PCOD. Although stout women make up the majority of PCOD patients, it's crucial to remember that PCOD can also affect non-overweight patients. Before and after receiving a PCOD diagnosis, up to 30% of female maintain a normal weight. The additional difficulties of having PCOD symptoms that are appropriately managed and diagnosed mean that women nevertheless experience the varied symptoms of the condition. Women frequently struggle to become pregnant before being identified, and they frequently remain untreated for years. It was discovered that testosterone levels were 2.34 nmol/L (67 ng/dL) in female with PCOD and1.57 nmol/L (45 ng/dL) in female without PCOD in a 2020 systematic review and meta- analysis of sexual malfunction due to PCOS, which included 5,366 women with PCOD from 21 research.

2. Literature Survey

Preeti Chauhan proposed a framework that early predicts the PCOD using Machine Learning approaches which includes the process such as collection of data, analysis of data and visualization, data pre-processing, selection and evaluation, then comparison and selection of the best model. The data collection method in this system is done by questionnaires through Google Forms and circulated to females of all age groups. The collected data is analyzed and visualized using Tableau and Google Sheets. Then the data is preprocessed, and the dataset is trained using 5 machine learning algorithms like Naïve Bayes, K-nearest neighbor (KNN), Decision Tree, SVM, and Logistic Regression. The trained models are validated through performance measures such as Precision, Confusion Matrix, Recall, F1 Score, and accuracy. By comparing the models, it confirms that Decision Tree has got the highest accuracy which is 81% compared to other models. [1]

Amsy Denny suggested a system that early detects and predicts the PCOD from a maximal and minimal but propitious clinical harmonic parameter, which includes defining a problem, collection of data, selecting a platform for implementation, choosing a status to measure, setting a protocol for evaluation, preparing and tuning of data, comparing the models and the selecting the best fir model. The data collection involves the dataset obtained through the survey of patient consisting of 54 females during clinicians' consultations and examinations. Then the data preparation process is carried out through Principal Component Analysis (PCA). Then it is trained using Logistic Regression, KNN, CART, Random Forest, Naive Bayes, SVM. After evaluation, it is said that Random Forest has got the highest accuracy of 89.02%. [2]

Dong-Dong Wang used machine learning approaches to estimate the appropriate dosage schedule for individuals with polycystic ovarian syndrome (PCOD), with an emphasis on the effects of carnitine supplementation relative to body weight. Machine learning techniques were employed in the construction of the optimal effect model. A non-linear mixed effect model is utilized for analysis to construct the optimal effect (Emax) model. Furthermore, to find out the true impact of carnitine supplementation on body weight in PCOD patients. The Emax efficiency curve of the finished model showed the amount of time needed to reach 25%, 50%, 80% and 75%.Using the Monte Carlo Method, the e max of carnitinesupplementation on body weight in PCOS patients was calculated. [3]

Subha R explored the analysis of PCOS using Swarm Intelligence and Machine Learning. The system involves feature selection by swarm intelligence and classification by machine learning. The study consists of data collected from 10 hospitals and is available in Kaggle Database. A Machine Learning model consisting of feature selection and random forest classifier has been evolved to diagnose PCOD conditions using clinical and non-clinical characteristics. It is identified that correlation ranking, and chi-square test-based feature selection algorithm gives more accuracy with 35 and 12 characteristics respectively out of total 41 features. [4]

3. Proposed System

The proposed system of this PCOD Analysis involves the following steps:

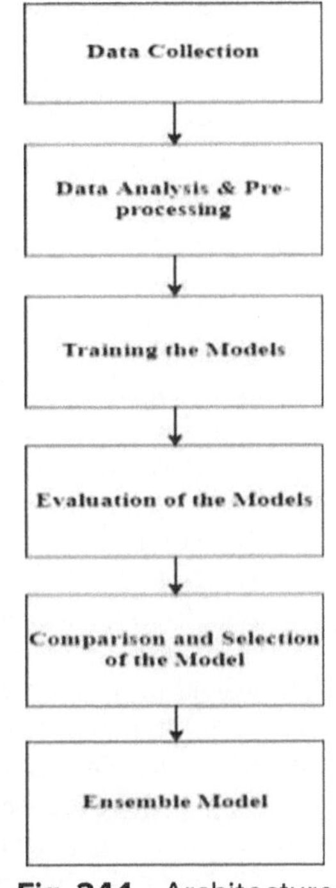

Fig. 24.1 Architecture diagram

3.1 Data Collection

The dataset being taken from the Kaggle database which consists of two Excel files which include PCOS_data_without_infertility.xls and PCOS_infertility.csv. The dataset includes all physical and medical constraints to determine PCOD and infertility-related issues. The data are being collected from 10 different hospitals throughout Kerala, India. PCOD_infertility.csv is taken as the main dataset for the analysis and this decides how good the model is and the number of data gathered increases the model's accuracy and hence gives a better performance.

3.2 Data Analysis and Preprocessing

The data analysis and pre-processing step involve the exploration and visualization of the data. The foremost step is the cleansing of data which comprises the process of managing or eliminating improper, distorted, improperly formatted, replica, or partial data within a dataset. It includes managing categorical data, choosing relevant features, feature scaling, and handling missing data. 'NaN' is the dataset's missing quantity that is returned. The next process is visualizing data. The visualization of the data is done by means of histogram which is category of bar chart that divides continuous measure into various category of bins which helps in analyzing the distribution and the heat map which is a category of geospatial visualization in a map form which exhibits specific data values in different colors. The histogram is exemplified in the form of count plot which is used to indicate the occurrence(counts) of the observation present in the categorical variable.

3.3 Training of the Model

To train the model, we use various machine learning algorithms for the PCOD analysis. Since the output of the model is a class; we are taking the algorithm as a classifier as it helps in categorizing data into different classes. For training this model, the dataset is split into training and test sets. Training of the model is done by means of training the set and the estimation of the model is done using the test set. The machine learning algorithms used for this analysis areDecision Tree, Random Forest, Naïve Bayes, Logistic Regression, Support Vector Machine, KNN (K-Nearest Neighbour).

3.4 Evaluation of the Model

The classification report, which is the performance evaluation metrics of a classification-based machine learning model, describes the many performance measurements that are used to evaluate the models after they have been trained using machine learning algorithms. It displays the model's support, F1 score, recall, and precision. It provides us with a comprehensive picture of the trained models' overall performance.

3.5 Comparison and Selection of the Model

In this proposed method, six machine-learning approaches for training the model and here six performance measures are used for evaluating the model. Based on these parameters, we will compare the performance of each model and select the top two machine-learning algorithm models for the next step. Highest accuracy of analysis is the primary metric and the two machine learning models which has the highest accuracies is taken for comparison. Based on the comparison the random forest and decision tree models has the highest accuracies. So, these both models are combined for next step, which is ensemble model.

Table 24.1 Evaluation metrics

	Precision	Recall	F1 Score	Support	Accuracy
K-Nearest Neighbour (KNN)	0.15	0.17	0.16	12	0.6181
Naïve Bayes	0.56	0.20	0.30	93	0.6716
Support Vector Machine (SVM)	0.44	0.21	0.29	19	0.6363
Decision Tree	1.00	0.06	0.12	16	0.7272
Logistic Regression	0.45	0.12	0.19	41	0.6146
Random Forest	0.50	0.59	0.54	27	0.7522
Ensemble Model	0.69	0.50	0.58	18	0.7636

3.6 Ensemble Model

An ensemble model is a machine learning method that integrates various separate models to enhance the overall performance of prediction. Its foundation is the idea that using multiple models in combination can produce more accurate forecasts than using just one. The usage of ensemble models is widespread in several industries, including e-commerce, healthcare, and finance. Ensemble models can be employed for both classification and regression applications. Compared to individual models, ensemble models are believed to be more accurate, robust and resistant to overfitting. An ensemble model that combines Random Forest and Decision Tree models is created. Combining ensemble learning strategies like stacking and mixing with random forests and decision trees is possible. Multiple models are trained, and the predictions they make are sent into the model that learns to combine them during stacking. We can train distinct versions of the random forest and decision tree models on the same data to merge them via stacking. Next, we input these models projected probabilities or class labels as features into a meta-model like a logistic regression or neural network. The meta-model develops the ability to synthesis the predictions of the many models to arrive at a final prediction. For blending, we divided the training data is segregated to two sections one for training the meta-model and the other for training the individual models. On one subset of the data, we can train a random forest model, and on another subset, a decision, tree model. After that, we train the model that combines the two models wing their combined predictions on a validation set.

4. Results and Discussion

In this proposed system, we utilised 6 machine learning algorithms for training the model. Also we used six performance measures for the evaluation of the model. Based on these parameters, we will compare the performance of each model and higher accuracy of the analysis is considered as a primary metric for comparison which has the highest accuracy. The Random Forest and Decision Tree showed the best performance in the diagnosis of PCOD, with accuracy rates of75% and 72%, respectively, in accordance of the performance validation criteria recall, accuracy, precision, and support. In order to create an ensemble model with a 76% accuracy, the two models are then integrated using ensemble learning.

5. Conclusion

A significant percentage of women worldwide are affected with PCOS, a disorder brought on by hormonal imbalance in young women's bodies. An early diagnosis of the illness can aid in its management and therapy. The patients clinical data identified with PCOD based on symptoms associated to the illness, we employed well-known machine learning techniques such as KNN, Nave Bayes, SVM, Decision Tree, Logistic Regression, and Random Forest. The Random Forest and Decision Tree showed the best performance in the diagnosis of PCOD, with accuracy rates of75% and 72%, respectively, in accordance of the performance validation criteria recall, accuracy, precision, and support. In order to create an ensemble model with a 76% accuracy, the two models are then integrated using ensemble learning. Thestudy's future focus may involve the diagnosis of the condition using numerous or sizable data sets.

References

1. PreetiChauhan, PoojaPatil, NehaRane, Dr. PoojaRaundale, HarshilKanakia, "Comparative Analysis of Machine Learning Algorithms for Prediction of PCOS",IEEE International Conference on Communication information and Computing Technology (ICCICT), June 25–27, 2021, Mumbai, India, DOI:10.1109/ICCICT50803.2021.9510101.
2. Amsy Denny, Anita Raj, Ashi Ashok, Maneesh Ram C, Remya George, "i-HOPE: DetectionAnd Prediction System For Polycystic Ovary Syndrome (PCOS) Using Machine Learning Techniques", 2019 IEEE Region 10 Conference (TENCON 2019).
3. Dong-Dong Wang , Ya-Feng Li, Yi-Zhen Mao , Su-Mei He, Ping Zhu, Qun-Li Wei, "Amachine-learning approach for predicting the effect of carnitine supplementation on body-weight in patients with polycystic ovary syndrome, 2022, Front. Nutr. 9:851275. Doi: 10.3389/fnut.2022.851275.
4. Subha R, Nayana B R, Rekha Radhakrishnan, Sumalatha P, "Computerized Diagnosisof Polycystic Ovary Syndrome Using Machine Learning and Swarm Intelligence Techniques", September 9th, 2022, DOI: https://doi.org/10.21203/rs.3.rs-2027767/v1.
5. SaymaAlamSuha, Muhammad Nazrul Islam, "An extended machine learning techniquefor polycystic ovary syndrome detection using ovary ultrasound image", SciRep 12, 17123 (2022). https://doi.org/10.1038/s41598-022-21724-0.
6. VaidehiThakre, ShreyasVedpathak ,KalpanaThakre, ShilpaSonawani, "PCOcare:PCOS Detection and Prediction using Machine Learning Algorithms", Biosc.Biotech.Res.Comm. Special Issue Vol 13 No 14 (2020), Pp-240–244, http://dx.doi.org/10.21786/bbrc/13.14/56.
7. Shazia Nasim, Mubarak Almutairi, Kashif Munir, Ali Raza, Faizan Younas, "A Novel Approach for Polycystic Ovary Syndrome Prediction Using Machine Learningin Bioinformatics", VOLUME 10, 2022, DOI: 10.1109/ACCESS.2022.3205587.
8. Palvi Soni, Sheveta Vashisht, "Exploration on Polycystic Ovarian Syndrome and Data Mining Techniques", Proceedings of the International Conference on Communication and Electronics Systems (ICCES 2018), DOI: 10.1109/CESYS.2018.8724087.
9. KinjalRaut, ChaitraliKatkar, Prof. Dr. Mrs. Suhasini A. Itkar, "PCOS Detect using Machine Learning Algorithms", International Research Journal of Engineering and Technology (IRJET), Volume: 09 Issue: 01 | Jan 2022.
10. Malik Mubasher Hassan, TabasumMirza, "Comparative Analysis of Machine Learning Algorithms in Diagnosis of Polycystic Ovarian Syndrome", International Journal of Computer Applications (0975 – 8887) Volume 175 – No.17, September 2022.

Nore: Every figure and table was created using the experimental work; none of them were taken from any publications or online.

Advances in Additive Manufacturing Technologies – Gurusamy Pathinettampadian (eds)
© 2024 Taylor & Francis Group, London, ISBN 978-1-032-90013-1

25 Experimental Exploration of Rice Husk Gasification in the Circulate Fluidized Bed Gasifier

V. Chokkalingam[1]
Professor, Department of Mechanical Engineering,
Jaya Engineering College,
Thiruninravur, Chennai, Tamilnadu, India

A. Paramasivam[2]
Associate Professor, Department of Mechanical Engineering,
Rajalakshmi Engineering College, Thandalam,
Mevalurkuppam, Tamilnadu, India

J. Rajprasad[3]
Assistant Professor, Department of Civil Engineering,
College of Engineering and Technology,
SRM Institute of science and Technology,
Kattankulathur, Tamilnadu, India

E. Gomathi[4]
Assistant Professor, Department of
Petrochemical Technology, University College of Engineering,
BIT Campus, Anna University, Tiruchirapalli,
Tamilnadu, India

ABSTRACT: In a lab-scale circulating fluidized bed reactor, rice husk was gasified using air as the gasifying agent. Experimentation involved five different temperature ranges (700°C, 800°C, 900°C, 1000°C, and 1100°C) and five different ER value (equivalence ratio). (0.2, 0.25, 0.3, 0.35, and 0.4). Five parameters were used to estimate the gasification process: gas composition, heat value of gas, gas yield. CO, CO_2, and H2 are the most common gasification produce, according to the investigate. When the temperature is between 700°C and 1100°C, the CO yield ranged from 25.20 percent to 34.89 percent, the H_2 yield ranged from 18.24 percent to 18.71 percent, and CO_2 fell from 7.21 percent to 1.31 percent, resulting in a 0.2 Equivalence Ratio When the temperature rises, and the same ER ratio, 6.619MJ/Nm3, the maximum heating value was noted. Enhanced gas heating was achieved the same as a result of the collective temperature, although a increasing ER ration decreased this parameter. The results indications the performance of gasification is better than to some of the other biomass gasification available in the literature.

KEYWORDS: Biomass, Rice husk, Circulating fluidized bed, Gasification, Air

[1]lingamviswakethu@gmail.com, [2]paramasivam.a @rajalakshmi.edu.in, [3]crajprasj@srmist.edu.in, [4]gomathie@aubit.edu.in

DOI: 10.1201/9781003545774-25

1. Introduction

Energy from biomass are accepting added due to decline NOx and SOx emissions and carbon neutral. Among the thermal-chemical processes is biomass gasification transformation action towards outcome of depending on the feedstock and gasifying environment, syngas or fuel gas should be investigated [1, 2]. The numerous types of Entrained stream gasifiers, fluidized bed gasifiers, and fixed bed gasifiers are the three major types of gasifiers, with the former being the most common being divided into as co-current and counter-current and cross-current moving bed gasifiers.

A common tool used in the gasification process is the circulating fluidized bed biomass gasifier (CFBG) studies and swotted broadly [3, 4, 5] all of which result in minimal CO_2, SO_2, and NOx emissions [6]. When the humidity level of biomass is high, the gas generated has an unusually low heating assessment. The heating rate was 2.2 MJ/kg if the humidity content was 50%. [7]. The normal operative temperatures were between 700°C-1100°C. The designation biomass refers primarily to carbon-based solid actual basic from plant life (including copse and agronomical crop) [8].It is an on estimated to agreement about 10%–14% of the global power provide.

2. Experimentation

The gasification test will take place in a lab-scale unit that runs at atmospheric pressure. The circulating fluidized bed gasifier's exploration arrangement is depicted schematically in Fig. 25.1.

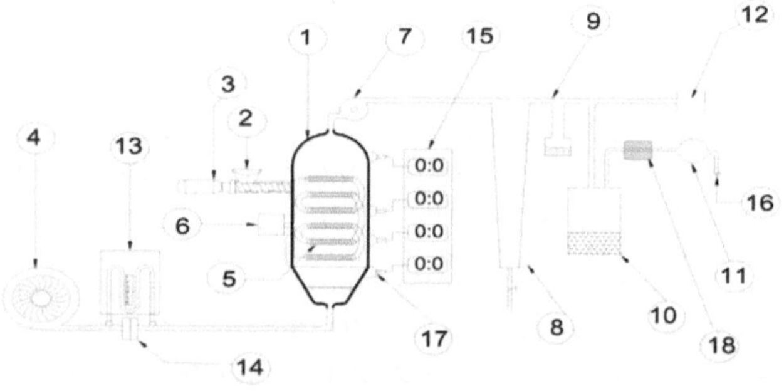

1. Reactor. 2.Screw feeder 3.Electrical motor 4.Blower 5.Coil 6.A.C. supply
7.Suction blower. 8.Cyclone seperator. 9.Water scrubber. 10.Diese bath11.Vaccum pump.
12.Burner. 13.Manometer 14.orifice plate 15.Digital indicator 16.Connecting to GC 17.Thermo couple 18.Drier

Fig. 25.1 The study's experimental setup was as follows

A suction blower be built-in near the gasifier's outlet to prevent backside pressure within a better gas stream and the reactor. The gases was again released into the atmosphere sent through a typhoon separator to remove any large particles. For tar removal, a tar separator was installed on the gas line. Following the tar separator, a Diesel bathtub was provided to aid in the removal of many harmful materials. For the purpose of burning the extra gas, a burner is attached to the gas line, as shown in the diagram.

The whole bed is insulating by means of On the base of the bed are refractories and an electric fire. The temperature in each zone was measured using Chromel-Alumel (K type) thermocouples (T1 to T4). Sensors record for temperatures is linked to the thermocouple probes and have a temperature accuracy of 0.5 degrees Celsius. Utilizing the principles of thermal conductivity to estimate H_2, the producer gas composition was examined using the Siemens Online Gas Analyzers Oxymat 61 (which uses the paramagnetic principle to estimate O2), Ultramat 23 (which uses non-dispersive infrared multilayer technology to estimate CO, CO2, and CH_4), and Calomat 61 (which uses the same method to estimate CO, CO2, and CH_4). Using a U-tube manometer filled with water, the pressure differential is measured. The gas flow velocity was determined using an anemometer (Lutron AM-4201) with a 2% precision and a 1 g decree weigh balance (Atco, 0-30 kg). The acquired values are respectable because they are well under 5% error in their respective measures, owing to the precision of every piece of equipment used in the testing.

In the current work, rice husk is utilized as the biomass material. Two kilograms of inert sand, which has a density of 1473.44 kg/m^3 and a particle size of distribution of 1100 m, the fluidization and heat transport are more consistent. Fuel supply through the screws feeders is level and reaches a steady state. The air supply must have above the lowest rate of fluidization. The started at the necessary speed to begin the test once the reactor's temperature reaches the desired air flow and fuel feed rate are adjusted to get producer gas with a good gasification value. At varied reaction temperatures and equivalency ratios, the same process is continued. Normally, the test conditions required 30 minutes to stabilize. After the steady-state test, five gas samples were obtained at 5-minute intervals.

3. Results and Discussion

3.1 Temperature's Impact

The influence of gasification temperature and equivalency ratio in the nourishment is examined below in terms of gas component change, gas productivity, carbon conversion ratio, calorific value and cold gas efficiency.

3.2 Gas Composition

The temperature of the reactor core is varied within 700 and 1100 Celsius degrees and the producing gas composition is also varied. What makes up CO and H_2 changes as the temperature rises, but the known equivalency ratio of methane and CO2 decreases. Endothermic reactions will be favored at higher reactor operating temperatures. As a result, the gasification the constitution and temperature has a significant effect on the gas yield that is produced. Figure 25.2 depicts the evolution of C and CO2 reduction with a 0.2 equivalency ratio. This discovery is related to rice husk gasification trials [9]. CO2 calculations and CO increases are commonly used. CO2 computation and CO increase are normally done in the same method, as long as CO2 is converted to CO when temperature rises, as the Boudouard reaction explains.

$$C + CO_2 \rightarrow 2CO \qquad [1]$$

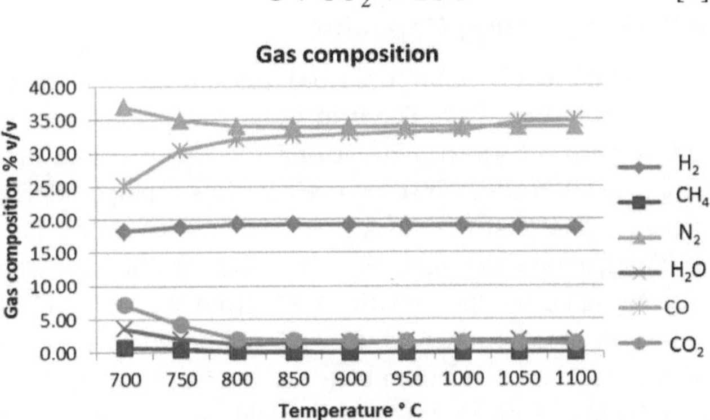

Fig. 25.2 The impact of temperature on the chemical composition of gases ER stands for 0.2

At 1100°C, there was also a discernible shift toward H2 and an increase in the quantity of it produced. Furthermore, the decrease in methane and H2O may have contributed to the increase in H2 and the total synthesis of fuel in the carbon fraction. Methane's attention is found to be on a decreasing form as the reaction temperature rises. When methane is combined with other gases, such as hydrogen (which varies between 18.24 and 18.71%) and carbon dioxide (which varies between 25.20 and 34.89%), the mole fraction of methane tends to be lower, ranging between 0.74 % and 0.01% (v/v). In a similar temperature range, co and methane showed indistinguishable growth for rice husk gasification. The similar trends were account[10] for sawdust gasification and pellets of woods, oil palm wastes, and bagasse in downdraft gasifier[11].

3.3 Gas Yield

Gas production yield delivered can be emphatically impacted by means of the gasification temperature. This is confirm by means of the ground-breaking temperature reliance found Fig. 25.3 shows how the gas yield grows with increasing temperature in the fuel organisation, implying that when a temperature increases, greater quantities of solids are being converted to gaseous products. The gas yield was around 1.331 Nm3/kg at an equivalency ratio of 0.2. The rice husk gasification yielded about 1.05 Nm3/kg[12], which was the highest to the maximum gas yield might be carried out at extremely high ER values in this the gasification process setting. It is found that, at the optimal temperature of 800°C, the growing to equivalency the ratio would grow a gas yield, as shown in Fig. 25.3.

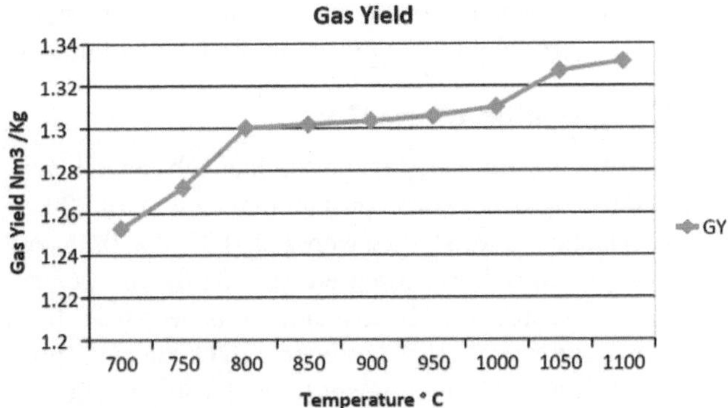

Fig. 25.3 At ER, the influence of temperature lying on gas produced is 0.2

3.4 Gas Heating Capacity

The influence of temperature on the calorific value of gas is depicted in Fig. 25.4. It was discovered that raising the temperature of the gasification reaction lowers the Calorific value (L.H.V) likewise arriving at greatest estimations of 6.619 MJ/m3 at 1100°C. The heating estimation of our gasification used to be a lot higher than the rice straw gasification of 5.14 MJ/m^3 [13]. It's also encouraging to learn that gas has a high heating factor originally measured at less than 6.3 MJ/m3. at 900°C, looked at In the rice husks the gasification process scenario, the highest qualities mentioned at higher temperatures—5 MJ/m3 for a coir matter and 5.5 MJ/m3 for sawdust were previously best in class.

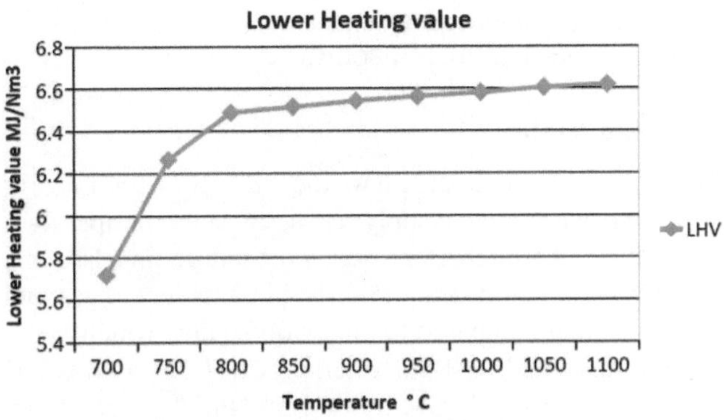

Fig. 25.4 Shows temperature affects the quantity of heating systems at ER

4. Equivalence Ratio Effect

The percentage between the exact and stoichiometric A/F proportions is the equivalency ratio (E.R.). It can also be represented even without gas constituent as indicated by the following equation for the gasification process, which represents the percentage of pure air supplied to the balanced air need.

$$\Phi = ((\text{weight. of } O_2/\text{wt. of dry fuel) precise})/((\text{weight. of } O_2/\text{wt. of dry fuel) stoichiometry})$$

4.1 Composition of the Gas

The equivalency ratio's effect on the fuel substance is represented by Fig. 25.5. For a constant temperature of 800°C, the equivalency ratios studied were 0.2, 0.25, 0.3, 0.35, and 0.4. CO concentrations decreased from 34.07 vol. percent to 18.53 vol. percent, CH4 concentrations decreased from 0.16 vol. percent to 0.01 vol. percent, H2O concentrations increased from 1.32 vol. percent to 3.82 vol. percent, and H2 levels decrease from 19.23 vol. percent to 9.52 vol. percent as the 0.2 to 0.4 equivalence ratio has grown. The amount of CO2 on the other hand, progressively climbed from 2.04% to 8.63%, while the N2 concentration increased from 38.02 vol. percent to 48.89 vol. percent. At 850°C, the inclinations of the supplied gas composition approximate those of bagasse [16]. A comparative study was accounted for saw dust air gasification and the equilibrium modelling of biomass synthesis [17] in a circulating flow gasifier.

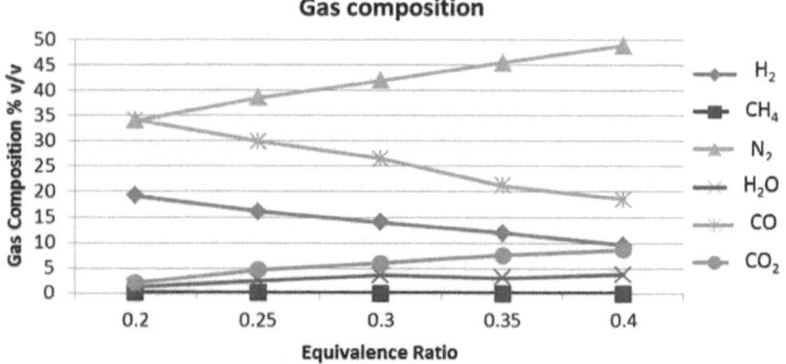

Fig. 25.5 The impact of ER on the composition of rice husk gaseous fuel (at 800°C)

4.2 Gas Yield

Figure 25.6 demonstrates how the ER ratio rises linearly with expansion gas supply at a constant temperature (T=800°C). Through improved temperature and ER, the transformation of the proportion of biomass fuel to gas fuel is high. When compared to pine wood chips, which had a general gas yield of 1.50 Nm3/kg at 031 ER and 791°C [30], the highest gas give up of 1.54 Nm3/kg was obtain [18].Here this investigation the high temperature of ER of 0.2 the rice husk is maximum.

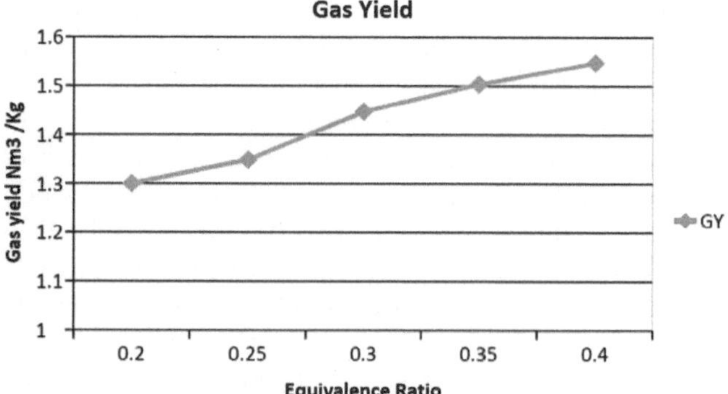

Fig. 25.6 (T=800°C) The impact of ER on gas production

4.3 Gas Heating Capacity

Raising the equivalency ratio reduces the heating value, as shown in Fig. 25.7. The heat value was obtained to be lowered from 6.48 to 5.20 MJ/m3 for ER of 0.2 to 0.4. Cogasification of sawdust and rice husk has a comparatively greater value than rice husk given [19] because coir pith is blended into the fuel. For saw dust gasification, the heating value for the growth of ER from 0.22 to 0.44 at 800°C dropped from 6.67 to 4.65 MJ/m3. The purpose of the decrease is to eliminate the likelihood of tertiary gasification of given gas, which could also happen when air availability is great and the ER is large. This could also be caused by an excess nitrogen loading, which reduces the fraction of carbon in the produced gas.

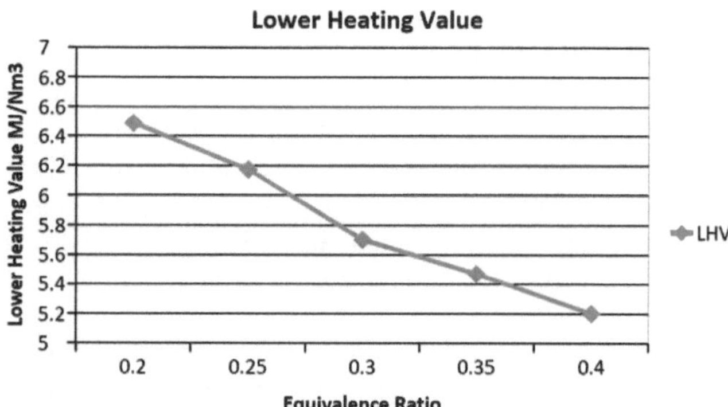

Fig. 25.7 The influence of ER on gas calorific rate (T=800°C)

4.4 Statistical Analysis

The effectiveness of information disclosure will in general decay while the handling cost of data translation will in general increment since some are boisterous information and not needed all the measurements should be investigated. This marvel is otherwise called pest of dimensionality which was first evidently instituted by [21] to depict the issue that information tests will become exponentially as per the progressions of the quantity of capacities due to the need of fitting a multi variate work for a given level of precision. Streamlining the information is significant in numerous application spaces to be encouraged with order, representation of managing the intricacy of multi gatherings of information.

In this study, impact of the air temp, the ratio of air to fuel and the rate of feed, the biomass are the significant parameters for the gas formation. The optimum temperature parameter is 800°C, equivalence ratio is 0.2. It is clearly identified from the experimental results and feed rate of bio mass is 1kg is fixed. Among these three parameters, it is noticed that the most affecting boundaries are temperature and equality ratio. The values of equivalence proportion are quite difficult to identify the accurate value. In order to overcome this issue, statistical tool Minitab software is used in this work. Minitab is a simple to use statistical package ranging through most statistical techniques, including multiple regression, time series and simulation. It has a solid custom of exploratory information investigation so there are bunches of valuable diagrams and indicative charts, for example, the spot plots or the combination plots. Information can be perused in from an assortment of configurations including excel expectations and is stored in Minitab's own worksheet design or in a venture record which wraps up the information and yield as revealed by [22]. Minitab has a basic order language and can record and run macros. In this work, With the help of a matrix plot, the temperature and equivalence ratio values are optimised. A matrix plot is a graph that is used to examine the association between several sets of variables at the same time.

It has been used the analyse the parameters and the impact of temperature and equality proportion, dry gas composition mainly H_2, CH_4, N_2, H2O, CO and CO_2. The matrix plot graph was urbanized as a method to deal with take result as of examinations were various perspectives were studied. During this analysis H_2, CH_4, N_2, H_2O, CO and CO_2 gases temperature and equivalence ratio is formulated and shown in Fig. 25.8 and 25.9.

From this Figure, it is clearly evident from the graph that the rice husk is one of the most effective fuel for energy utilization.

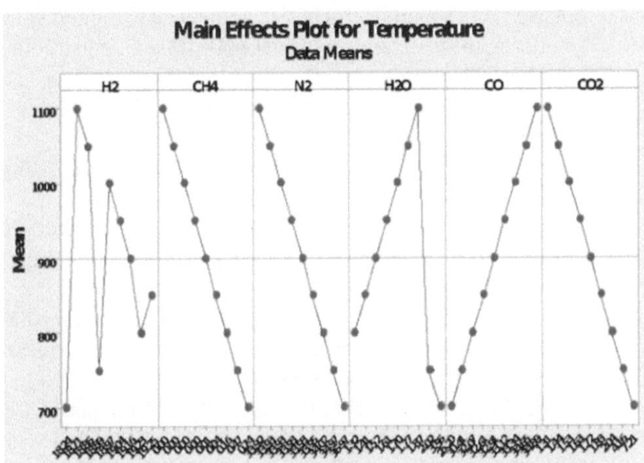

Fig. 25.8 Gas composition (Temperature plot)

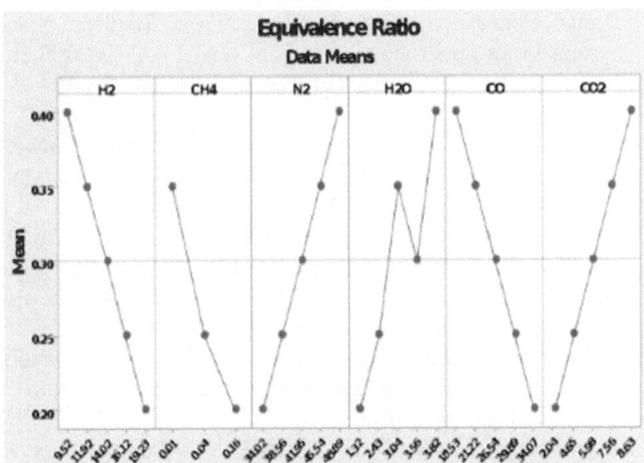

Fig. 25.9 Gas composition (Equivalence Ratio plot)

Increases in temperature and equivalency ratio increased gas yield. Temperature increase expands gas heating value, however expanding ER lying on these boundaries establishes a switch reducing pattern. For critical trial outcomes, the trial outcomes are deconstructed in a matrix plot. In comparison to gasification of mixed biomass sources, these findings highlight the potential benefits of single biomass gasification. Furthermore, research on rice husk biomass writes is justifiable in order to ascertain the possible results of look good effects of individual biomass implementation in air gasification.

5. Conclusion

In a circulating fluidized bed gasifier, rice husk is gasified with an air the medium. The effects of temperature and equivalency proportion on cold gas effectiveness, gas calorific value, and gas composition were investigated. As the temperature increased, so did the CO and H2 levels. Investigated was the actual effect of ER on the composition of gas.

References

1. Gómez-Barea, A., Vilches, L., Leiva, C., Campoy, M. and Fernández-Pereira C. Plant optimisation and ash recycling in fluidised bed waste gasification. ChemEng J. 2009; 146: 227–36.
2. Matsumoto, K., Takeno, K., Ichinose, T., Ogi, T. and Nakanishi, M. Gasification reaction kinetics on biomass char

obtained as a by-productof gasification in an entrained-flow gasifier with steam and oxygen at 900–1000 C. Fuel. 2009; 88: 519–527.

3. Kwauk, M. Fast fluidization, advances in chemical engineering series. San Diego: Academic Press; 1994.
4. Grace, J.R., Avidan, A.A. and Knowlton, T.M. Circulating fluidized beds. London: Chapman and Hall. 1997.
5. Yoshida, K. and Mineo H. High velocity fluidization, transport in fluidised particle systems. Amsterdam: Elsevier:1989.
6. D.E.Salt,R.D.Smith, and I.Raskin, 1998,'phytoremediation',Annual review of plant physiology and plant molecular biology,vol.49,pp.643-668.
7. JormaNieminen and mattikivela b,1998, 'Biomass CFB gasifier connected to a 350 MW_{th} steam boiler fired with coal and natural gas', Biomass and Bioenergy, vol. 15, pp.251-257.
8. Atong, Duangduen, and ViboonSricharoenchaikul. "Thermal Conversion of Mixed Wastes From Biodiesel Manufacturing for Production of Fuel Gas" , ASME 2009 3rd International Conference on Energy Sustainability Volume 1, 2009.
9. Wu C, Yin X, Ma L, Zhou Z, Chen H. Operational characteristics of a 1.2-MW biomass gasification and power generation plant. BiotechnolAdv2009;27:588–92.
10. Baskarasethupathy, V, Chokkalingam, V & Natrajan, E 2012, 'An Experimental exploration of the biomass gasification in fluidized bed ,on Website of Renewable and sustainable Energy journal, from August 2012.
11. Erlich, C."Downdraft gasification of pellets made of wood, palm-oil residues respective bagasse: Experimental study" , Applied Energy, 2011.
12. Mansaray K, Ghaly A, Al-Taweel A, Hamdullahpur F, Ugursal V. Air gasification of rice husk in a dual distributor type fluidized bed gasifier. Biomass Bioenergy 1999;17:315–32.
13. Calvo, L.F. "Gasification of rice straw in a fluidized-bed gasifier for syngas application in close-coupled boiler-gasifier systems" ,Bioresource Technology, 201204.
14. Nora Mevissen. "Thermodynamics of autothermal wood gasification" , Environmental Progress & Sustainable Energy, 10/2009.
15. Kobayashi, N. "High temperature air-blown woody biomass gasification model for the estimation of an entrained down-flow gasifier", Waste Management, 2009.
16. Pratik N. Sheth, B.V. Babu.Experimental studies on producer gas generation from wood wastein a downdraft biomass gasifier Bioresource Technology 100 (2009) 3127–3133.
17. Gopal Gautam, SushilAdhikari, SushilBhavnani. "Estimation of Biomass Synthesis Gas Composition using Equilibrium Modeling" , Energy & Fuels, 2010.
18. Sheth, P.N. "Experimental studies on producer gas generation from wood waste in a downdraft biomass gasifier" ,Bioresource Technology, 2009.
19. Chokkalingam, V, Murugu Mohan Kumar, K & Jaya Prakash, G 2016'An Experimental Study on Producer Gas Generation from Bagasse Fuel in a Circulating Fluidized Bed Gasifier', International Journal of Science Technology and Engineering, vol. 3, no. 2,Aug-2016/264-270.
20. Yijun Zhao.; Shaozeng Sun.; Hao Zhou.; Rui Sun.; Hongming Tian.; Jiyi Luan.; Juan Qian. Experimental study on sawdust air gasification in an entrained-flow reactor. Fuel Processing Technology 91 (2010) 910–914.
21. Bellman R.E. Adaptive Control Processes. Princeton University Press, Princeton, NJ, 1961.
22. Montgomery, DC 2001, 'Design and analysis of experiments', Wiley, New York.

Note: Every figure and table was created using the experimental work; none of them were taken from any publications or online.

Advances in Additive Manufacturing Technologies – Gurusamy Pathinettampadian (eds)
© 2024 Taylor & Francis Group, London, ISBN 978-1-032-90013-1

26 Design and Analysis of a Rugged Swing-Arm for Electric Two-Wheelers

Swamy K, Suresh A*,
Ravi S, Niranjan K, Srivardhan R, Nishanth S.M
Department of Mechanical Engineering, Chennai Institute of Technology, Chennai

ABSTRACT: This study aims at designing and analysis of a swing-arm of an electric Two-Wheeler that is tough and endures the loading due to the weight of the Hub-Motor and tyros and the Impact caused due to the road conditions. The Rugged swing arm balances the weight and reduces the impact on the hub motor. It has the wide application on the motorcycle which is used in extremely tough terrains. Swing-Arm is a main component of a Two-Wheeler suspension system, which suspends the rear wheel and pivots the suspension system. It reduces the impact caused by the irregularities of roads. This study mainly aims at the analysis of the swing-arm during the maneuver and cornering of the Two-Wheeler. Impact Analysis was done by standard FEA tool ANSYS. Steel of grade AISI-1020 is used as it tops the chart in all the categories including the cost. Manufacture of this swing-arm is done by welding the joints, cold pressing the end and slotting.

KEYWORDS: Hub-motor, Maneuvers and cornering endurance, Suspension, Swing-arm

1. INTRODUCTION

The swing arm is one of the most vital parts of a cruiser, it connects the rear wheel assembly with the motorbike frame and therefore plays a major role in the bike's operation. A swing-arm is provided as a link between main frame and rear tires through which the driving force is transmitted. The swingarm also serves as a support for the suspension system. Thus, when designing a new cruiser, it is necessary to create an optimal design that will ensure maximum performance of this crucial component. In modern days Swingarms are classified into two catagories: single-arm or double-arm. However, there are only two types of swingarms found on motorcycles produced today namely; single-side and double-side swingarms. Double-sided swingarms are most commonly used due to their simplicity and balanced state. Single sided swingarm on the other hand are just for racing bikes or bikes driving on the track. typically traditional type swing arms consists of two arms with suspensions near both ends on either sides of the swingarm. further this was changed by adoption of mono suspension designed cantilever type swing arms are preferred. The rigidity of the swingarm pays a vital rolein a motorcycle's performance, including both its maneuverability and ride comfort. designing a bike that delivers exceptional comfort without sacrificing handling is a demanding feat. Striking the perfect balance between these two factors is often a difficult trade-off. As such, the

*Corresponding author: suresha@citchennai.net

DOI: 10.1201/9781003545774-26

selection of swingarm stiffness must be carefully consider to reach equilibrium between excellent handling and ultimate comfort.

When it comes to swing arm their stiffness provides the key to find the perfect balance which means a stiffer swing arm can improve the handling, but it may come at the cost of comfort. on the other hand considering ride comfort may lead to reduce handling capabilities. During the designing phase one the important factor weight must be consider. By opting for a lighter swing arm the unsprungmass at the rear side and the bikes overall mass is also reduced. This has a significant impact on the cruiser's handling, ultimately making for a smoother and more enjoyable ride.

2. Literature Review

Syed Hassaan Abdullah, Mohd Ahmed, and Wajahat Abdul Rahman [1] worked on latest MotoGP Honda RC213V swingarm to make it better. They used shape optimization after examining the multiaxial load analysis conducted on the swingarm's CAD model. Their work gives outcome by dropping the weight by 15% while still keeping it safe features. Ashish Powar, Hrishikesh Joshi, Sanket Khuley, and D.P. Yesane [2] set out to optimize the swingarm of a 150cc commuter motorcycle's through reducing its weight. Through algorithmic adjustment and material selection to achieve the targeted weight reduction. After they chose a revised CAD model they are they are checked and compared with the original part. They managed to cut the weight by enormous 44% while keeping it just as a stiff. Giacomo Risitano and team [3] looked into three World Superbike Championship-spec swingarms to link solid facts, like stifness and natural fequencies, with riders' feelings about handling and comfort. They collected information from the special test seat and compared them with the informations given by professional riders. further, they made computer aided designs (CAD) and did some analysis (FEA). They noticed a small difference of 8.42% between the CAD models and the experimental model. The test riders felt that the enchanced torsional stifness provides a better sensation for the rider. stiffer swingarm made the bike feel better to handle.João Diogo da Cal [4] took part in a Moto student contest, his task was to design and development of both front and rear swingarms for an electric motorcycle. This project wasn't just about mathematical calculations also includes the material selection. He built models of the swingarm and it is tested. He found that a supportd type design could make the swingarm to reduce 66.1% of weight. However, he realized making complex pieces might increase the costs because of the tooling and need of the labour for manufacturing complex parts.B. Smith and F. Kienhöfer [5] focused on how carbon fiber composites are used in motorcycle swingarms. They experimentally measured both vertical and horizontal stiffness in a swingarm made of carbon fiber. They also did some Finite Element analysis on the model, with some assumptions with experimental loading conditions. finally it is concluded that the vertical stiffness is appeared significantly higher than that of the rear spring. Torsional stiffness matched with the existing literature. The carbon fiber model results the reduction of weight 29% than the same part made from aluminum. Aswath S and team [6] justify that Aluminum alloy 6061T6, with its impressive physical properties, high strength-to-weight ratio, and ease of welding, is the prime choice for analysing the suspension system in a rover. They chose aluminum 6061T6 rather than something like carbon fiber because it's cost effectiveness. Their research further prove that using Aluminum alloy 6061T6 will enchance the rover's performance and durability, making the rover chassis more reliable.Veeresh Kumar G B and team [7] worked on improving the weak wear resistance of aluminum alloy 6061 T6. It is made stronger by adding aluminum oxide (Al2O3) particles . finally the composition gives a impressive result with just 6% weight of aluminum oxide exhibt wear resistance than other composites materials. Chathakudath Sukumaran Sumesh and Ajith Ramesh [8] studied on the impact of machining parameters on surface finishing while dry symmetrical of aluminum 6061 T6. They found out that when cutting speed increases, you end up with a increase in the surface finishing. They also figure out the feed rate of the material also matters the most, slower feed rate leads to the better surface finishing.

3. Methodology

A Conventional single fork swing-arm with a cross member is incorporated. Seamless pipes with 1-inch OD and 0.6-inch thickness are chosen. The Curb weight of the two-wheeler is around 150kg and around 65%-70% of the weight is concentrated at the rear. Based on this the swingarm is designed.

1. Welding
2. Cold pressing
3. Drilling
4. Slotting

Tungsten Inert Gas (TIG) Welding is used to weld the connecting member and the slot used to fix the swing arm to the frame for high quality and clean welds. Cold pressing process is carried out and slotted for the fixture of rear axle. It is preferred because there will be more impact on the rear axle fixture which may lead to weld failure if welded slots were implemented. 8 mm holes are drilled on

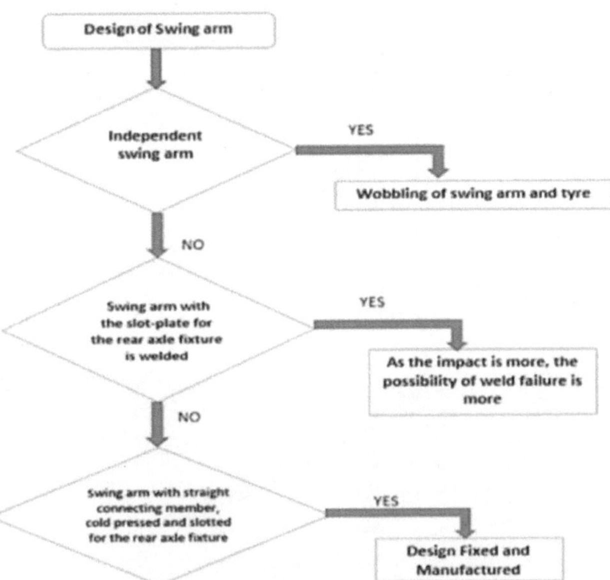

Fig. 26.1 Process methodology

the clamps welded to the swing arm to provide fixture to the frame and suspension.

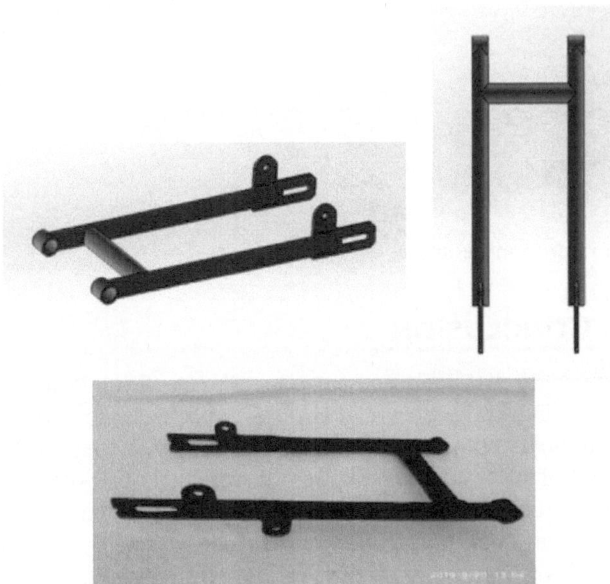

Fig. 26.2 CAD modelling of the swing-arm

4. Material Selection

AISI 1020 is a kind of carbon steel that can be recognized by its low hardenability and tensile strength, which Brinell hardness never exceeds 200. This steel shows high strength and good malleability, enabling it to be turned, polished or cold drawn. Low carbon contentprovides resistance from flame hardening whereas lack of alloying elements avoids reaction upon nitriding. Carburization employed be employed in order to reach case hardness greater than Rc65 for smaller sections without any change in core strength across sections. Carbonitriding offers benefits over traditional carburizing techniques.

Table 26.1 Chemical analysis of the material

Element	Content
carbon	0.18 – 0.23
manganese	0.3 – 0.6
phosphorus	0.04
sulphur	0.05

Table 26.2 Specifications of the motor

Specification	Values
Output Power	250W
Speed	250 – 350 rpm
Voltage	48 V
Wheel size	16"
Weight	3.5 kg

5. Heat Treatment

The required temperature is maintained constant for some period of time. Then, in a furnace, the material gets cooled. After being exposed to carburizing atmosphere up to 880°C to 920°C, it is kept aside for suitable timinguntil the desired carbon content and case depth achieves. It leads to increase of core and case properties through refining and hardening. The material is subsequently reheated at 870-900°C range before being quickly quenched in water, oil or brine after which it is held till uniformity of temperature throughout the section occurs. After finishing core refining process, the material is heated to a temperature between 760° C and 780° C and the temperature is maintained until the thermal equilibrium is attained.

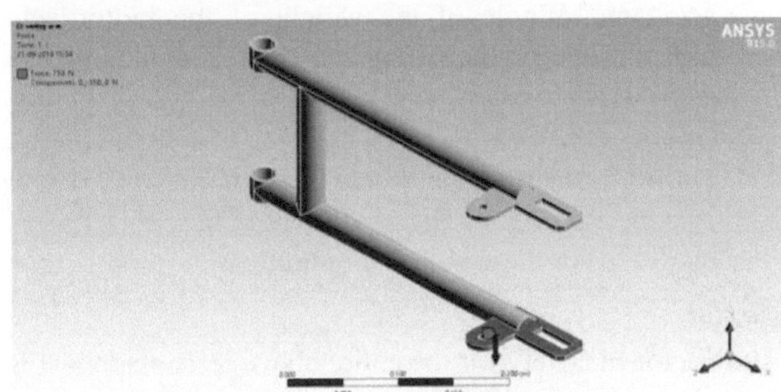

Fig. 26.3 Manoeuvrability Force

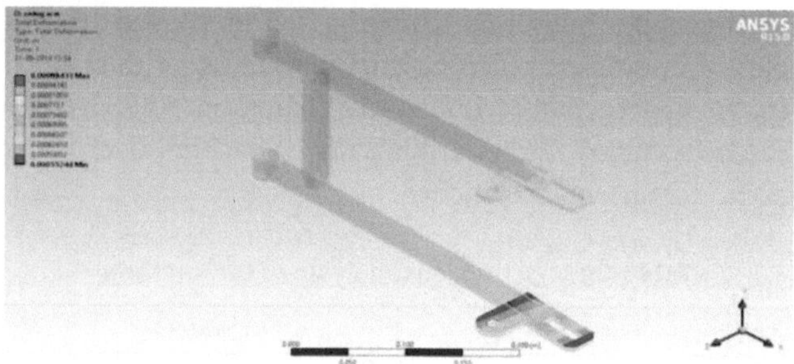

Fig. 26.4 Maneuverability Deformation

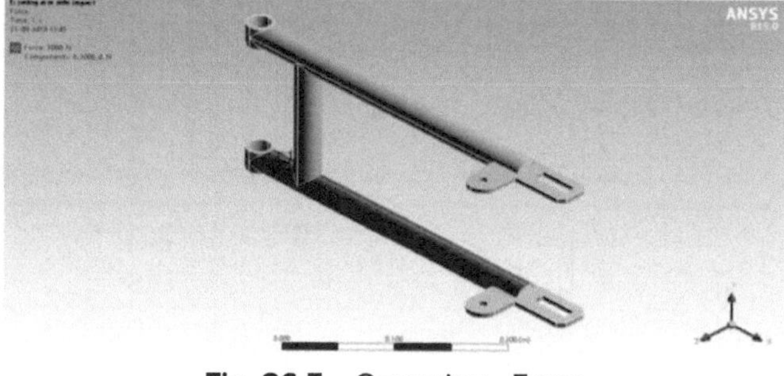

Fig. 26.5 Cornering - Force

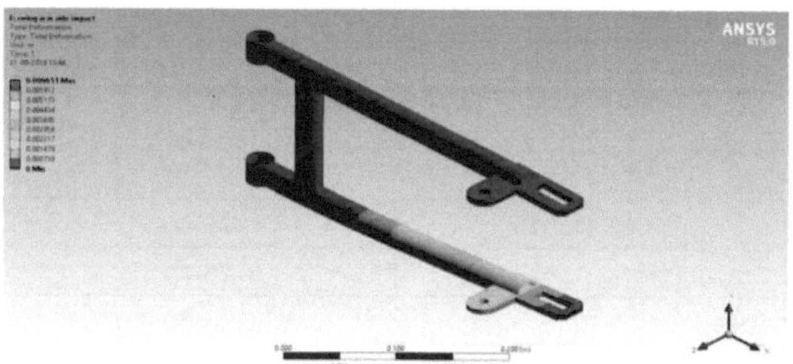

Fig. 26.6 Cornering – deformation

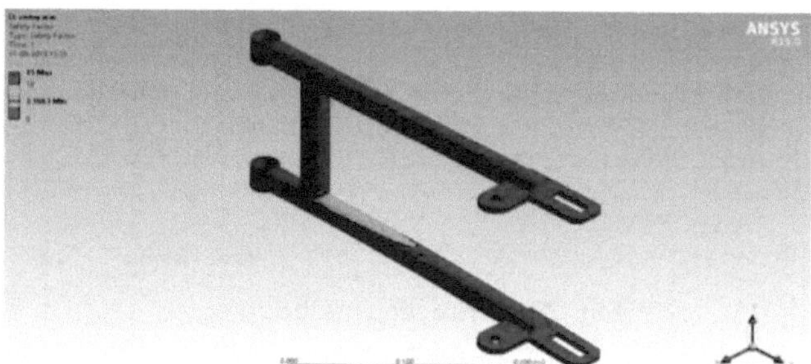

Fig. 26.7 Factor of safety for maneuverability analysis

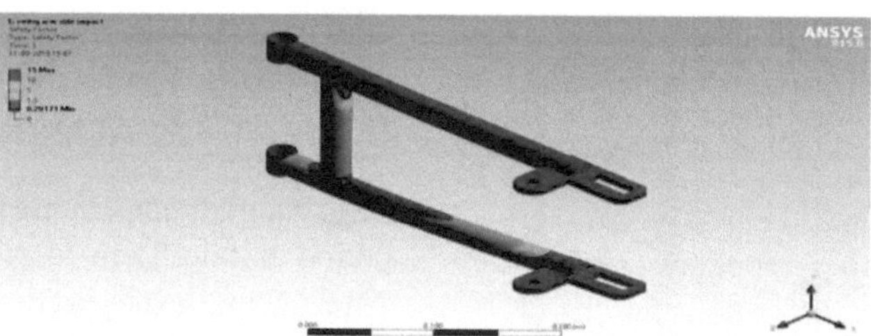

Fig. 26.8 Factor of safety for cornering analysis

6. Loading Conditions

The vehicle dynamics are tested based on the realtime vehicle running period from the architecture that is test and analysed in different orientations from given condition.

- **During Maneuver**
  Loading during maneuver include the loads that act on the swing-arm during acceleration, braking and accidents. Whole of the weight of the motorcycle tailstock acts on the swing-arm apart from the impacts caused due to roads.
- **During Cornering**
  During cornering the whole load of the rear wheel is concentrated on the axle-pivot. Thus, this this is analyzed for the whole swing-arm.

Result

The Factor of Safety for the entire above tests made test is well above the safe mark. Hence the Design is Safe and can be implemented.

7. Conclusion

The study was supposed to be on the structural design of the swing arms. Several FEAs, as well as THEORETICAL analytical approaches were implemented with varying degrees of success. The reduction of weight is observed with the help of above design procedure. It is clear that design swing-arm can give good performance in actual working condition because of the low validation stress value.

Reference

1. Syed hassaanabdullah, mohdahmed, wajahathabdulrahman, design of racing motorcycle swingarm with shape optimisation, ijsrd - international journal for scientific research & developmentl vol. 6, issue 06, 2018
2. ashish powar, hrishikeshjoshi, sanketkhuley and d.p.yesane, design of racing motorcycle swingarm with shape optimization, international journal of current engineering and technology (2016)

3. Giacomo risitano, lorenzoscappaticci, guglielmomarconi, carlogrimaldi, francescomariani, analysis of the structural behavior of racing motorcycle swingarms, sae international: 2012-01-0207 (2012)
4. João diogo da calramos, front and rear swing arm design of an electric racing motorcycle, master of science degree thesis in mechanical engineering, tecnicolisboa. 2016
5. B. Smith and f. Kienhöfer, a carbon fibre swingarm design, r & d journal of the south african institution of mechanical engineering 2015, 31, 1-11
6. S. Aswath, nitinajithkumar, chinmayakrishna tilak, nihil saboo, amalsuresh, ravitejakamalapuram, anoopmattathil, h. Anirudh, arjun b. Krishnan, ganeshaudupa, 'an intelligent rover design integrated with humanoid robot for alien planet exploration', robot intelligence technology and applications 3 pp 441–457.
7. G.b. veeresh kumar, r. pramod, c.s.p. rao, p.s. shiva kumar gouda, artificial neural network prediction on wear of al6061 alloy metal matrix composites reinforced with -al2o3, materials today : proceedings volume 5, issue 5, part 2, 2018, pages 11268–11276
8. C.s. sumesh and dr. Ajith ramesh, "numerical modelling and optimization of dry orthogonal turning of al6061 t6 alloy", periodica polytechnica mechanical engineering, vol. 62, pp. 196–202, 2018
9. Tony foale, motorcycle handling and chassis design the art and science, isbn, 2002.

Note: Every figure and table was created using the experimental work; none of them were taken from any publications or online.

Advances in Additive Manufacturing Technologies – Gurusamy Pathinettampadian (eds)
© 2024 Taylor & Francis Group, London, ISBN 978-1-032-90013-1

27 Enhancement of Mechanical and Water Resistance Characteristics of Pineapple Fiber Polymer Composites through Nano SiC Reinforcement

T. Maridurai[1], V. Muthuraman[2]
Department of Mechanical Engineering, Vels Institute of Science, Technology and Advanced Studies (VISTAS), Chennai, Tamil Nadu, India

P. Gurusamy[3]
Department of Mechanical Engineering, Chennai Institute of Technology, Chennai, Tamil Nadu, India

S. Thirugnanam[4]
Department of Mechanical Engineering, SRM Valliammai Engineering College, Chennai, Tamil Nadu, India

ABSTRACT: The research described in this article focuses on combining fiber and nano SiC into epoxy matrices to create advanced composite materials. Our investigation has revealed significant advancements in the mechanical characteristics and water resistance of these composites following a series of exhaustive tests. Tensile testing showed that as the percentage of fiber increased, tensile strength significantly improved. Remarkably, Material 2, which is composed of 40% fiber and 60% epoxy, showed the reinforcing effect of fiber dispersion with a tensile strength of 138 MPa. Materials 3, 4, and 5 were then treated with micro SiC to increase their tensile strength; Material 5 reached a maximum of 156 MPa. Similar trends were shown in impact tests, where Material 2's impact resistance surpassed that of pure epoxy. Effective composite preparation is necessary, as evidenced by the even better impact resistance of later composites with nano SiC added. These results were validated by hardness testing, which showed that materials containing nano SiC and fiber had improved hardness values, suggesting a greater resistance to deformation. The research revealed that altering the composite compositions led to a decrease in water absorption, highlighting the materials' appropriateness for use in environments that are resistant to moisture. SEM examination has validated the uniform dispersion and strong bond between the fibers and nano SiC particles; thus, the quality of the composite production is confirmed by the results obtained. Therefore, the proposed epoxy composite with pineapple fiber and SiC nanoparticles can be used in various industries, such as aerospace, automotive, and construction.

KEYWORDS: Scanning electron microscopy, Mechanical testing, Water absorption, Pineapple fiber, Nano SiC.

1. Introduction

Composite materials have attracted the interest of many Polymer composites are regarded as the most widely used materials in many industries due to their exceptional mechanical properties and wide applications. To improve the properties of composites, scientists have endeavored to develop natural fibers and various reinforcing fillers in

[1]maridurai.mechanical@gmail.com, [2]drvmuthuraman.se@velsuniv.ac.in, [3]guru8393@gmail.com, [4]thirugnanams.mech@valliammai.co.in

DOI: 10.1201/9781003545774-27

polymer matrices[1], [2]. It was reported that the addition of nanoscale reinforcements made of natural fibers to composites has achieved extraordinary mechanical properties and water resistance of the composite materials. Thus, the purpose of this report is to study the advantages of adding narrow-diameter-silicon carbide to an epoxy matrix for the purposes of reinforcing the mechanical and water properties of pineapple fibers of polymer composites[3], [4].

Natural fibers are the key alternatives for being added of varying applications taking into account their easy availability at moderate expense. Pineapple fibers are particular among natural fibers for justifying significant performance as they are made from the pineapple leaves' waste. It was reported that the fibers have high strength and low density as well as considerable heat stability. Adding pineapple fibers to polymer matrices can increase the tensile strength, impact resistance, and hardness. Nonetheless, natural fibers have different disadvantages as well, including diminished dimensional stability as well as higher water absorption. The fiber-matrix interface may be wrecked due to water incorporation that can subsequently reduce the mechanical properties of the composites. Therefore, the composites should be waterproof to guarantee a natural fiber composite's new spectrum of applications[5], [6].

Thus, nanotechnology is used to solve the above problem and improve the characteristics of a natural fiber composite. For instance, when a nanoscale reinforcement is included in a polymer matrix, such as nano SiC particles, the mechanical properties increase significantly. Owing to the outstanding mechanical properties and high aspect ratio, nano SiC can be used as an effective reinforcement to enhance the stiffness and strength of composite materials [7]. Moreover, SiC nanoparticles can serve as obstacles that make it difficult for moisture to penetrate and improve their water resistance. Recently, the use of so-called nano-reinforcement to increase the parameters of a natural fiber composite gained recognition, and in the research the mechanical properties of the epoxy composites inclusively containing kenaf fiber and nano-sized SiC were studied[8], [9]. The researchers found that the tensile and flexural strengths of the composites were a lot better due to effective load transfer between the fibers and the matrices that was observed. In some research, the mechanical properties of epoxy composites reinforced with the carbide SiC nanoparticles as well as jute fiber were presented. It was found that the inclusion of nano SiC in the tested samples had a positive impact on the impact resistance, flexural, as well as tensile strengths. It was explained by the better transfer of the stresses and the interfacial bonding of the fiber and the matrix.[10], [11].

Based on our study, we found that adding micro SiC to the fabricated natural fiber composites leads to increased mechanical properties and water absorption compared to without micro SiC. The enhancement effect is expected to the formation of a barrier and enhancement in interfacial bonding of micro SiC. The aim of our project is to investigate the mechanical properties and water absorption of pineapple fiber polymer composites with nano SiC. These goals, in turn, should provide some insights into the potential applications of the newly developed composite material in various spheres.

2. Methodology

2.1 Material Preparation

The main materials examined in this work include TETA hardener and diglycidyl ether of biphenyl A. These elements create the composite of interest. To ensure that the pineapple fibers and particles were of the highest quality and purity, a strict cleaning approach is required. Initially, the pineapples were placed in an ethanol solution to ensure that all the dirt and impurities that stuck to their fibers extracted before they were put in an oven for complete drying. For the cleaning solution, a combination of 95% ethanol and 5% water was mixed well to form a thinner solution. A vitally important process of salinization was also applied to increase the compatibility of the pineapple fibers with the polymer matrix. During that process, 3-Aminopropyltrimethoxysilane, or APTMS, was added to 4% of the solution, and a gentle stirring motion was applied to mix the solution properly as shown in Fig. 27.1. Consequently, 240g of pineapple fibers cleaned and 10g of particles associated with them were affected, and they were put into the solution containing silane with water for a short period. By cleaning the pineapple fibers properly and applying a careful process to prepare the surface of the pineapple fiber, the adherence of the pineapple fiber in the matrix is increased.

The fiber was carefully put into the mould after being cleaned and the silane treatment was done to prepare it for

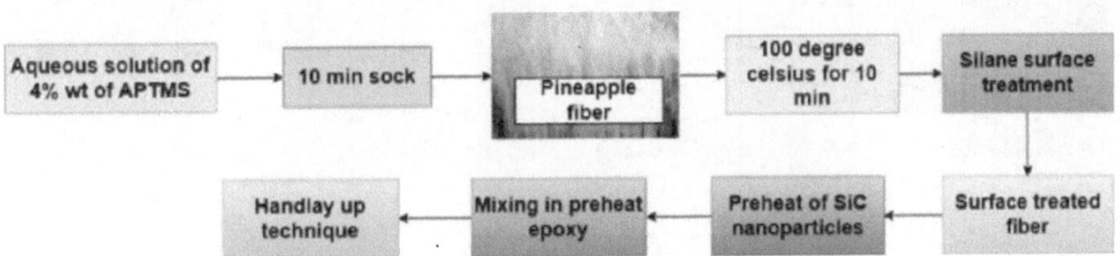

Fig. 27.1 Material preparation of the fiber

the next steps of investigations. The SiC particle of 40 nm in diameter is preheated to 200 degrees Celsius. Simultaneously, a epoxy resin is heated to 60 degree Celsius. The next step was to transfer the nanoparticles into the heated epoxy resin after both the epoxy resin and the nano-sized SiC particles had attained their respective temperatures. A stirring was done to guarantee that a homogenous dispersion was produced.

Five different composite materials with varying compositions were created in this experiment as shown in Table 27.1. Pure epoxy was the first component of the formulation. The second material was a 40% fiber and 60% epoxy combination. The third material consisted of 40% fiber, 58% epoxy, and 2% nano SiC. The final product was a blend of 40% fiber, 4% nano SiC, and 56% epoxy. Lastly, the fifth material was made up of 6% nano SiC, 40% fiber, and 54% epoxy. These diverse compositions provide a large variety of samples for this research's additional investigation.

Table 27.1 Material preparation in this research

Material	Epoxy (%)	Fiber (%)	Nano SiC (%)
Material 1	100	0	0
Material 2	60	40	0
Material 3	58	40	2
Material 4	56	40	4
Material 5	54	40	6

2.2 Novelty and Application

The novelty of this research lies in the integration of the pineapple fiber and the nano-sized SiC particles. The process of preparations like cleaning and the salinization process for the pineapple fiber increases the fiber adhesion with the prepared composites. The addition of SiC nanoparticles results in novel approach to the reinforcement of the composites results in increase in tensile, impact and hardness of the material. The proposed material can be used in the construction materials, automotive components, or aerospace structures.

2.3 Various Testing used in this Research

In order to perform the materials' examination, tensile, impact, hardness, and water absorption tests were conducted. Every procedure for testing was precisely performed with the use of contemporary and advanced machinery to obtain accurate and reliable results. Tensile testing, which was one of the methods applied in mechanical evaluation, provides an understanding of materials' resistance to pulling forces. Universal testing machine is used to perform this procedure by applying the load to the specimen, which is controlled throughout the specific test. The exact values of tensile strength, yield strength, and elongation quantified by the UTM helped to comprehend the mechanical materials' behavior. Another important research of mechanical materials' property concerns the resistance to the sudden, high-velocity pressing forces, which is impact testing. The procedure was performed with the use of Izod or a Charpy impact tester, which used to tap the specimen and measure the energy absorbed during its fracture with a hammer or a pendulum. This method provided the description of the impact materials' resistance and hardness. Brinell hardness test was carried out in order to get the hardness of the experiment . In the course of the experiment, a force of a specific weight was used to tap an indenter, while the diameter of the indentation that was obtained was measured. Brinell hardness was used to determine the material's resistance to wear and surface deformation.Water absorption was also examined to understand the composites' long-term interaction with water, which is a crucial characteristic of the materials for the cases when they have to be exposed to moisture. The water absorption test had two aspects, one of which was the weight of the specimens after they stayed in the water for some time. The water absorption was also measured with the use of a precision balance, which allowed for the detailed examination of composite materials and moisture absorption along with the changes in their characteristics. Figure 27.2 provides the materials and specimens prepared in this research for the experiment.

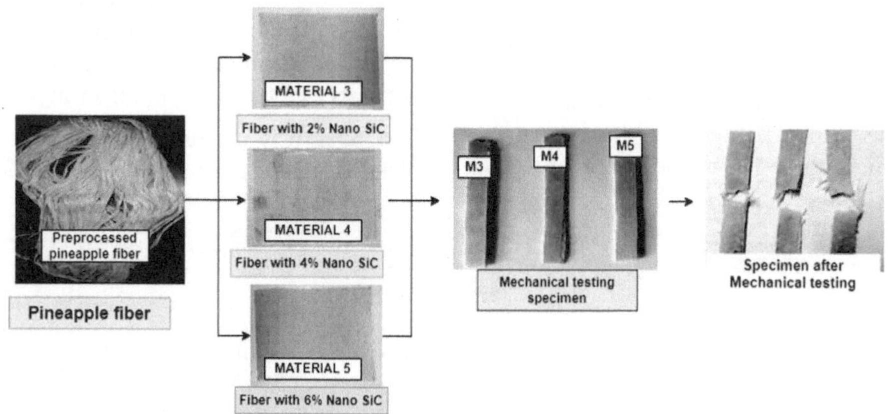

Fig. 27.2 Material prepared and the specimen

3. Result and Discussion

The findings of the tensile tests, shown in Fig. 27.3, which shows the mechanical characteristics of the 5 different composites. The tensile strength of pure epoxy is only 58 MPa. But when the 40% of the pineapple fibber is added the tensile result improved drastically. Tensile strength increased to 138 MPa, which is a far higher value compared to pure epoxy. The improved distribution of fibers inside the epoxy matrix is responsible for this significant increase in tensile strength. A larger concentration of fibers ensures the better connection between the fibers and the resin. Such components interact to reinforce the structural integrity of the composite, and the latter's tensile strength increases. Then, the tensile strength in composites with epoxy at 2%, 4%, and 6% was increased. The composites exhibit great performance achieving tensile strengths of 140 MPa, 152 MPa, and 156 MPa, respectively . Such findings indicate that nano SiC has an extra effect on the mechanical strength of the composite. However, the reason why the tensile strength is higher is the uniform distribution of nano SiC in the entire composite . These nanoparticles are known to contribute to the strength of the material. Finally, it is the effective connection between nano SiC, fibers, and resin that raises the tensile strength in such established composites.

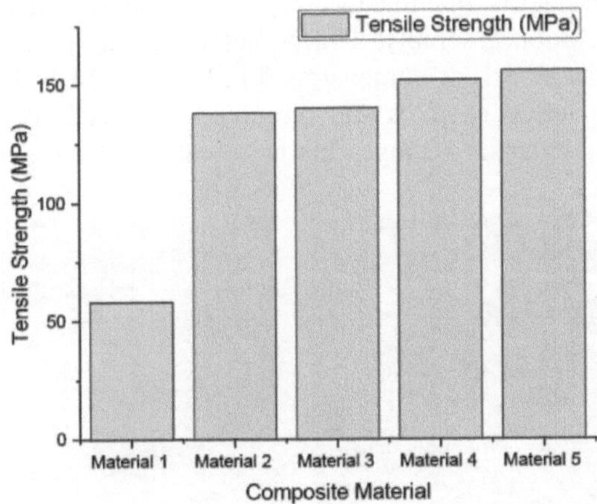

Fig. 27.3 Enhancement in the tensile strength of epoxy composites

Overall, it can be concluded that there is a need in adding nano-SiC and fibers to improve the mechanical properties of composites in the epoxy matrix. It is evident that in this particular case, due preparation of components and their mixing, the composites with high tensile strength have been established. Now the latter can be used in a number of applications requiring strength.

The findings of the impact test, as illustrated in Fig. 27.4, reveal an extraordinary impact resistance for the composite materials under consideration. Notably, Material 1, com-

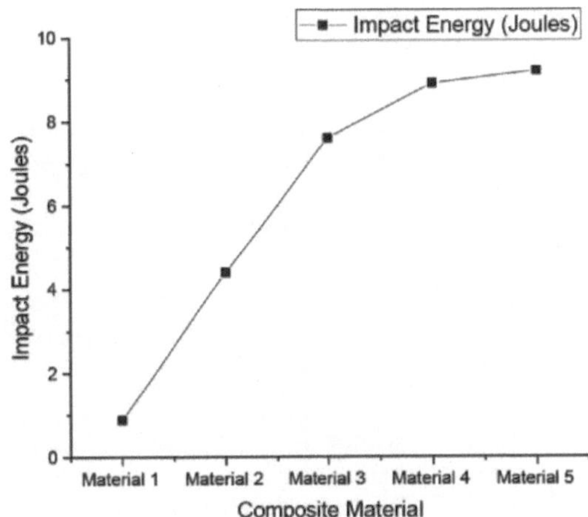

Fig. 27.4 Impact resistance evaluation of composite materials

posed of 100% epoxy, had an impact energy absorption of 0.88 J, proving it could resist forces applied at high speeds and velocities . Conversely, Material 2, consisting of 40% fiber and 60% epoxy, has an impressive resistance to impact as it measured 4.4 J. The enhanced ability to absorb impact energy of the composite is a result of the fiber dispersion in the epoxy matrix, as seen by the notable improvement in impact strength. Even greater impact resistance was shown by materials 3, 4, and 5, whose compositions of epoxy, fiber, and nano SiC varied. In that sequence, they were 7.6 J, 8.9 J, and 9.2 J. This intriguing trend emphasises the need of a well-dispersed, tightly bound nano-SiC phase within the composite and shows the extra function that nano-SiC contributes in raising the materials' impact resistance. All of these results show the importance of composite composition for impact resistance, which makes the materials desirable for applications where critical resistance to sudden forces is needed.

The impact test results, which are shown in Fig. 27.5, offer important new information on the different composite materials under investigation in terms of their hardness. Despite being made completely of epoxy, the first material initially had a 78 Shore D hardness grade, which indicates its stiffness. Nevertheless, Material 2, which is composed of 60% fiber and 40% epoxy, showed a significant improvement in hardness, measuring 110 Shore D. The extra fibers provide a strengthening effect that increases the material's resistance to indentation, which explains the significant increase in hardness.

The hardness values increase as we move on to Materials 3, 4, and 5, which have different amounts of epoxy, fiber, and nano SiC. These materials have Shore D values of 113, 117, and 120, respectively. There are several reasons for the observed increase in hardness in these materials. To begin with, a higher percentage of epoxy creates a stiffer matrix,

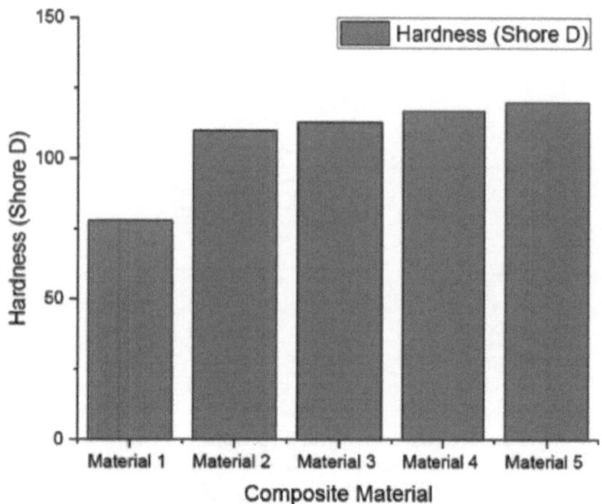

Fig. 27.5 Hardness test result of the epoxy composites

which is further reinforced by the fibers. Additionally, the hardness is further enhanced by the introduction of micro SiC particles. The uniform distribution of micro SiC in the composite matrix adds structural stiffness, which improves the material's resistance to deformation. The composite's overall toughness is increased as a result.

Figure 27.6 displays the results of the water absorption test and illustrates the composite composition affects the materials' resistance to water penetration. Despite being made entirely of epoxy, the first material had a 1.5% water absorption rate and substantial water infiltration. Upon examining Material 2, we observe a 1.2% decrease in water absorption. Of Material 2, fiber comprises up 40% and epoxy up to 60%. The barrier that reinforcing threads form to keep water out can help to explain this drop. Considering that Materials 3, 4, and 5 had different amounts of fiber, epoxy, and nano SiC, their water absorption dropped significantly more. Material 3 showed enhanced resistance to water infiltration at a water absorption rate of 0.8%. Material 3 consists of 58% epoxy, 2% Nano SiC, and 40%

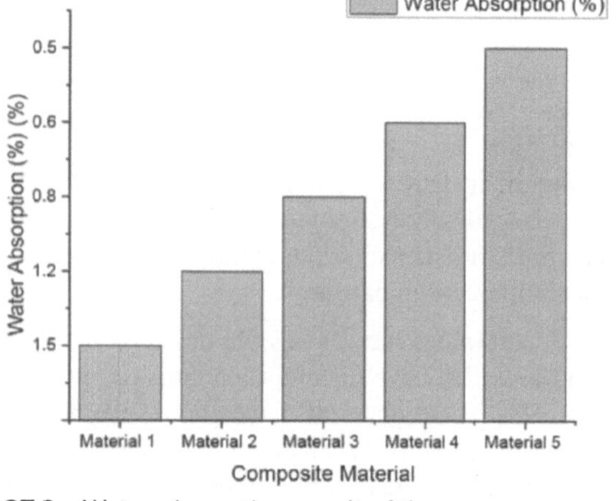

Fig. 27.6 Water absorption result of the epoxy composites

fiber. One aspect of this reduction is the insertion of small SiC particles, which improve the composite's structural integrity and decrease porosity. With a water absorption rate of 0.6%, Material 4, composed of 56% epoxy, 40% fiber, and 4% Nano SiC, demonstrated the ongoing impact of the nano SiC reinforcement in preventing water penetration. The least quantity of water absorption was shown by Material 5 at 0.5%. It is composed of 54% epoxy, 6% Nano SiC, and 40% fiber. This decline occurs because the fiber, epoxy, and nano SiC all work together to lessen the material's sensitivity to water.

The results of a SEM analysis of the composites that were created are shown in Fig. 27.7. Important information about the microstructure and dispersion of the component materials is provided by this analysis. There is a discernible even dispersion of micro SiC particles within the fiber matrix of the composite reinforced with 2% SiC.The uniform distribution of the nanoparticles across the composites is a sign of successful mixing and preparation, and, in turn, superior material properties. The SEM analysis shows that in both models, nano SiC reinforced 4% and 6% granddaughter scale composites were uniformly distributed with the nanoparticles. This promotes the materials' maximum potential in terms of superior mechanical and structural properties and indicates a high-quality manufacturing process and well-balanced composition assembly. The possession of a strong bond with the fibers, as demonstrated in both the SEM images of the tested samples and the composite with 6% nano SiC, is most intriguing. This bond enhances the material

Fig. 27.7 SEM images of epoxy composites with varying percentage of SiC

and its overall structural strength. Therefore, the strong bond between fibers and nano SiC should have a positive impact overall on the material's mechanical strength and endurance.

4. Conclusion

Our research has yielded promising results for the development of composite materials with notably improved water and mechanical resistance. After experimentation, we were able to see significant improvements in the composite materials' hardness, impact resistance, and tensile strength. From the testing the following conclusions are made.

- The test findings for Material 2 shown a significant improvement in terms of hardness, tensile strength, and impact energy absorption: 4.4 J, 110 Shore D, and 138 MPa, respectively.
- Material 2 is composed of 60% epoxy and 40% fiber. Nano SiC was added to Materials 3, 4, and 5 to improve their properties even more. Material 5 demonstrated remarkable properties, including a tensile strength of 156 MPa, an impact energy absorption of 9.2 J, and a hardness grading of 120 Shore D. Moreover, our research demonstrated that modifying the composite composition led to a significant decrease in water absorption.
- Materials like fiber and nano SiC, which have noticeably lower rates of water absorption, suggest possible applications where moisture resistance is crucial. These outcomes signify a noteworthy progression in the manufacturing of materials with superior performance.
- The SEM analysis further validated the strong fusion and effective dispersion of the fibers with the small SiC particles, demonstrating the materials' use and the potency of our composite production techniques. Our work contributes to the ongoing search for materials that may one day grow into stronger, more resilient materials by assisting in the discovery of novel materials that can get over a range of technological challenges.

REFERENCES

1. M. Alsaadi and A. Erkliğ, "A Comparative Study on Mode I and Mode II Interlaminar Behavior of Borax and SiC Particles Toughened S-Glass Fabric/Epoxy Composite," *Arabian Journal for Science and Engineering*, vol. 42, no. 11, pp. 4759–4769, 2017, doi: 10.1007/s13369-017-2649-1.
2. Y. Liu, Y. Sun, Y. Wang, G. Ding, B. Sun, and X. Zhao, "A complex reinforced polymer interposer with ordered Ni grid and SiC nano-whiskers polyimide composite based on micromachining technology," *Electronic Materials Letters*, vol. 13, no. 1, pp. 29–36, 2017, doi: 10.1007/s13391-017-6199-1.
3. C. C. Lu, M. H. Headinger, A. P. Majidi, and T. W. Chou, "Fabrication of a NicalonTM fiber/Si3N4-based ceramic-matrix composite by the polymer pyrolysis method," *Journal of Materials Science*, vol. 35, no. 24, pp. 6301–6308, 2000, doi: 10.1023/A:1026738923973.
4. V. Ravi Raj, B. Vijaya Ramnath, A. Rajendra Prasad, C. Elanchezhian, E. Naveen, and N. Ramanan, "Time Dependent Behaviour of PMMA-Toughened Siliconized SiC Strengthened Glass-Epoxy Composite," *Silicon*, vol. 14, no. 8, pp. 4129–4138, 2022, doi: 10.1007/s12633-021-01198-x.
5. S. Han and D. D. L. Chung, "Mechanical energy dissipation using carbon fiber polymer-matrix structural composites with filler incorporation," *Journal of Materials Science*, vol. 47, no. 5, pp. 2434–2453, 2012, doi: 10.1007/s10853-011-6066-7.
6. V. Kavimani, P. M. Gopal, K. R. Sumesh, and N. V. Kumar, "Multi Response Optimization on Machinability of SiC Waste Fillers Reinforced Polymer Matrix Composite Using Taguchi's Coupled Grey Relational Analysis," *Silicon*, vol. 14, no. 1, pp. 65–73, 2022, doi: 10.1007/s12633-020-00782-x.
7. U. Kumar, M. Singh, and S. Singh, "Wire-Electrochemical Discharge Machining of SiC Reinforced Z-Pinned Polymer Matrix Composite Using Grey Relational Analysis," *Silicon*, vol. 13, no. 3, pp. 777–786, 2021, doi: 10.1007/s12633-020-00484-4.
8. D. K. Hwang and K. Oh, "Reduction of cycle time by addition of silicon carbide filler to syndiotactic polystyrene/glass fiber composites," *Iranian Polymer Journal (English Edition)*, vol. 23, no. 9, pp. 717–722, 2014, doi: 10.1007/s13726-014-0266-3.
9. F. Orozco *et al.*, "Electroactive performance and cost evaluation of carbon nanotubes and carbon black as conductive fillers in self-healing shape memory polymers and other composites," *Polymer*, vol. 260, no. May, p. 125365, 2022, doi: 10.1016/j.polymer.2022.125365.
10. V. Ravi Raj and B. Vijaya Ramnath, "Mechanical, Thermal and Wear Behavior of SiC Particle Strengthening of PMMA-Toughened Glass-Epoxy Hybrid Composite," *Silicon*, vol. 13, no. 6, pp. 1925–1932, 2021, doi: 10.1007/s12633-020-00580-5.
11. P. Raju, K. Raja, K. Lingadurai, T. Maridurai, and S. C. Prasanna, "Mechanical, wear, and drop load impact behavior of glass/Caryota urens hybridized fiber-reinforced nanoclay/SiC toughened epoxy multihybrid composite," *Polymer Composites*, vol. 42, no. 3, pp. 1486–1496, Mar. 2021, doi: https://doi.org/10.1002/pc.25918.

Note: Every figure and table was created using the experimental work; none of them were taken from any publications or online.

Advances in Additive Manufacturing Technologies – Gurusamy Pathinettampadian (eds)
© 2024 Taylor & Francis Group, London, ISBN 978-1-032-90013-1

28 Experimental Comparison of Mechanical Stability and Morphology Between Tree Fibres and Shrub Fibres Reinforced Epoxy Bio-Composites

V. Noble Arulandu[1], V. Muthuraman[2], C. Dhanasekaran[3], M. Chandrasekaran[4]

Department of Mechanical Engineering,
Vels Institute of Science, Technology and Advanced Studies (VISTAS),
Chennai, Tamil Nadu, India

ABSTRACT: This research article investigates the composite material property based on synthetic and natural fiber as a reinforcement. In this work, the epoxy bio-composites were prepared by natural tree fibres of Morinda-citrifiliaand Tamarindus-indica with the shrub fibres of Tinosporacordifolia and Ipomoea staphylina for automobile structural components, aircraft interior products and civilian applications. The prime objective of this experimental work is to explicate the study and comparison of mechanical stability and microscopic study between tree fibers and shrub fibers. The composite was prepared by hand- layup method, in this work , untreated fibers were used and both resin and fibers are used 50%. The ASTM guidelines and the composite's morphology were followed in the preparation of the test specimens; it was studied with the help of FTIR and SEM. The mechanical test results show that ETI was superior to all other composites in the tensile strength maximum of 8.13 Mpa and which was maximum in compression strength of 5.98KN, But in hardness ,ETC has high results compare with others.

KEYWORDS: Morindacitrifilia, Tamarindusindica, Tinosporacordifolia, Ipomoea staphylina, SEM, Epoxy, Mechanical properties

1. INTRODUCTION

The rapidly growing world needs more new materials to compensate for the demand to meet industrial needs with appropriate strength and environmentally conscious materials. The recent researches were focuses on natural fiber-based composites to meet the ecofriendly natured materials, the natural fiber-based materials had more advantages in contrast to more traditional materials like synthetic fibres. The application of natural fibers and make use of available resources in nature is the best way to save nature and gives more effort to improving the material characterization for extending the mechanical performance capabilities and applications is the sustainability of the industrial needs [1].

The improvement of the mechanical strength of the fibers at a macroscopic level can be achieved by adding of the polymer matrix is possible and the tribological behavior also to be improved [2]. Natural fibers are the replacement to man-made fibers which are used in recent industrial possibilities. The benefits of natural fibres include their low density, affordability, and biodegradability. Manikandan et al [3] revealed the surface-midifiedmorindacitrifolia fiber

[1]Nabi11_j@yahoo.com, [2]v.mraman6@gmail.com, [3]dhans.se@velsuniv.ac.in, [4]director.mech@velsuniv.ac.in

DOI: 10.1201/9781003545774-28

adding 3 µm size silane modified chitosan particle with epoxy resin. They came to the conclusion that chitosan, Enhances superior mechanical, thermal, and water uptake behaviour when mixed with epoxy to create composites with surface-modified reinforcement that perform the original material. Vamsi Krishna Mamidi et al [4] examined the Tinosporacordifolia and Tectona grandis adding into epoxy and bisphenol resin. The authors confirmed that composite material is superior in the tensile strength, compression, and flexural tests. Through experimental evidence, Manikandan et al. [5] demonstrated the impact of incorporating tamarindusindica with epoxy and bisphnol resin, its shows better mechanical strength and the SEM images showed a good dispersion of the Tamarindus fibers in the matrix. In order to improve the interfacial shear strength between natural fibre and thermoplastic matrices, Valadez-Gonzalez et al. [6] tested and demonstrated fiber-matrix surface treatment. The alkaline treatment raises the quantity of cellulose exposed on the fibre surface and roughens the surface. The morindacitrifolia has in around 80 types of species in the worldwide. It is a medium height tree and fruits are in green until maturity [7]. Tamarindus indica was developed widely spreading and rounded tree, its common tree in West Africa and India [8, 9]. Tinosporacordifolia and Ipomoea staphylina are shrub fibers, it's commonly available in India and both fibers are used in different systems of traditional medication. Tinospora using hepatoprotective, immunomodulatory and anti-neoplastic activities [10]. And Ipomoea using anti-inflammatory, anti-diarrheal, gastro protective effect [11]. The composites prepared by using hand layup methods, since hand layup required low process parameters [12]. These composites made of trees and shrubs could be used in spacecraft, automobile structural parts, and commercial applications that call for high damping, low weight, and energy absorption.

2. Methodology

2.1 Material Preparation

This study used liquid bisphenal-A type epoxy resin diglycidyl ether (VBR 8912), a thermoset epoxy with a kinematic viscosity of 12000 cps and a mol. wt. of 195 g/mol. The epoxy resin was cured using a curing agent (VBR 1209): TETA with 20 cps was supplied by Vasavibala Resins (P) Ltd, India. Natural fibres Morinda-citrifilia, Tamarindus–indica, Tinospora cordifolia and Ipomoea staphylina were bought through Go Green Industrial Fibre India, (P) Ltd. After extraction fibres were shown in Fig. 28.1(a) Morinda-citrifilia tree fibre; 28.1(b) Tamarindus–indica tree fibre; 28.1(c) Tinospora cordifolia shrub fibre and 28.1(d) Ipomoea staphylina shrub fibre.

Fig. 28.1 (a) Morinda-citrifilia tree fibre, (b) Tamarindus–indica tree fibre, (c) Tinosporacordifolia shrub fibre and (d) Ipomoea staphylina shrub fibre

2.2 Preparation of Samples

The hand layup fabricated epoxy composites are examined for any defects bye the visual inspection. The test samples were kept ready via hand layup methods. The metal molds are prepared required size, in this research using 15 cm x15 cm square plate and stirring process occur by the help of ¼ hp motor. The mold for making the composite plate is first cleaned well so that none of the impurities is sticking into the mold. Then, the mold is coated with wax polish so that, when the plate is formed, it could be separated from the mold easily and without damaging the composite plate. After waxing the mold, half the amount of chemical mixture, i.e. the mixture of epoxy resin and hardener is evenly spread over the mold and the natural fibers are spread over the chemical mixture. The fibers are spread over the chemical mixture properly. After that to apply a chemical mixture, its form like as sanveg [13]. Forming pattern than manual load was applied over the pattern up to 50kg. The-setup curing up to 4 to 6 hours at room temperature, than to remove the plate from the pattern by manual methods. Figure 28.2(a) shows epoxy morinda-citrifilia composite (EMC), 28.2(b) shows epoxy tamarindus–indica composite (ETI), 28.2(c) shows epoxy tinosporacordifolia composite (ETC) and 28.2(d) shows the epoxy Ipomoea staphylina composite (EIS).

3. Characterization Techniques

3.1 Infrared Spectroscopy using Fourier Transform

The materials' Fourier transform infrared (FT-IR) spectra were captured using an Attenuated Total Reflectance mode (ATR, PRO470-H) and an FT-IR spectrophotometer (from

Fig. 28.2 (a) Epoxy morinda-citrifilia composite (EMC), (b) Epoxy tamarindus–indica composite (ETI), (c) Epoxy tinosporacordifolia composite (ETC) and (d) Epoxy Ipomoea staphylina composite (EIS).

BSA Univ, model name FT/IR-6300 type A). The incidence angle of 450 and a resolution of 4 cm–1, the wavelength range was 399.193 cm–1 to 4000.6 cm–1 at a interval of data of 0.964233 cm–1 [14].

3.2 Scanning Electron Microscopy

The fractured portions of prepared epoxy composite (EMC, ETI, ETC, and EIS) were inspected using a Shimadzu model (SS-550)Scanning electron microscope. The shear failure surfaces of the specimens will demonstrate how varying surface treatments of the fibers affect the composite's effective properties. The targeted portions were coated in gold and subsequently examined to a width of 6mm. [15, 16].

3.3 Mechanical Testing

The tensile test and bending test were carried out to EMC, ETI, ETC, and EIS. The epoxy composites were evaluated in accordance with ASTM–D695 and ASTM–D 790 standards respectively. The universal testing machine was used for testing the composites with transverse speed and a load of 2.1 mm/sec and 40 Ton. The compression test was evaluated by ASTM–D 3039. A durometer according to ASTM-D 2240 was used to determine the microhardness (Shore-D) of the composites. In the study, square samples with dimensions of 50 mm x 50 mm were used. [17].

4. Result and Discussion

The FTIR spectroscopy was used to confirm fiber/matrix bonding, Figure 28.3 shows FT-IR spectra of the various samples. The fiber is taken 0.5 cm to 1 cm of a length. The

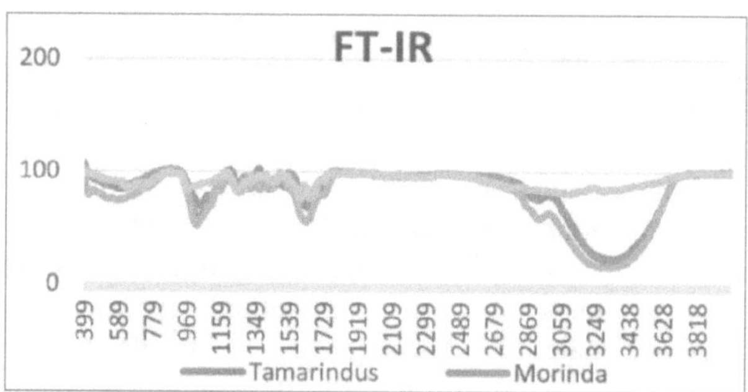

Fig. 28.3 FT-IR spectra of the samples

fibre is penetrated by the Infra-Red (IR) ray. Stretching vibrations at wave numbers 1637.27 cm-1 and 1039.44 cm-1 reveal absorption bands of characteristic groups exist in the sclerenchyma structure of Tamarindus indica, including hemicellulose, lignin and cellulose. Similarly, the vibrations observed at 1642.09 cm-1 and 1031.73 cm-1 in Morinda Citrifolia. In Tamarindus Indica at 3346.85 cm-1 and 2915.84 cm-1 and in Morinda Citrifolia at 3341.07 cm-1 and 2928.38 cm-1 a stretching signal was noticed corresponding to the hydroxyl groups and the C-H bond of the fibre. [5, 18]. In the stretching vibrations at wave numbers 1636.3 cm-1 and 1023.05 cm-1, absorption bands of characteristic groups of the sclerenchyma structure, including hemicellulose ,cellulose, and lignin present in Tinospora Cordifolia can be seen. Similarly, the stretching vibrations observed at 1633.41 cm-1 and 1020.16 cm-1 in Ipomoea Staphylina. An oscillatory feature was detected at 2922.59 cm-1 and 3299.61 cm-1 corresponding to the C-H bond and hydroxyl groups of the fibre in Tinosporacordifolia. Similarly, a vibration signal was detected at 2927.41 cm- and 3327.57 cm-1 corresponding to the C-H bond and hydroxyl groups of the fibre in Ipomoea Staphylina.

The mechanical performance of epoxymorinda-citrifolia, epoxy tamarindus-indica, epoxy tinospora-cordifolia and epoxy ipomoea staphylina composites are presented in Table 28.1. Tensile, flexural, compression and hardness tests results are presented in Fig. 28.4–28.6 which are completed as per the ASTM standards. It can be noted that the epoxy morinda, epoxy tamarindus and epoxy tinospora composites mechanical properties compared with epoxy ipomoea composites [4, 5, and 18]. The epoxy composites basically have good mechanical properties [19]. It is observed ETI has high tensile strength 8.13Mpa compared with others, tinospora has the least tensile values 1.23 Mpa. Binding property in between epoxy and tamarindus is a very good, binding property of epoxy with other composites EMC, EIS is moderate. Similarly, the flexural strength testing ETI have high load.

Caring capacity 0.32KN and ETC were very less load caring capacity 0.13KN, but in the other two composites,

Table 28.1 Mechanical properties of EMC, ETI, ETC, EIS

Sample Id	Tensile Strength (MPa)	Flexural Load (KN)	Compression Load (KN)	Shore D° Hardness
EMC - Morindacitrifolia	5.78	0.15	1.92	46
ETI - Tamarindus indica	8.13	0.32	5.98	51
ETC - Tinosporacordifolia	1.23	0.13	4.11	59
EIS - Ipomoea staphylina	2.78	0.19	1.32	45

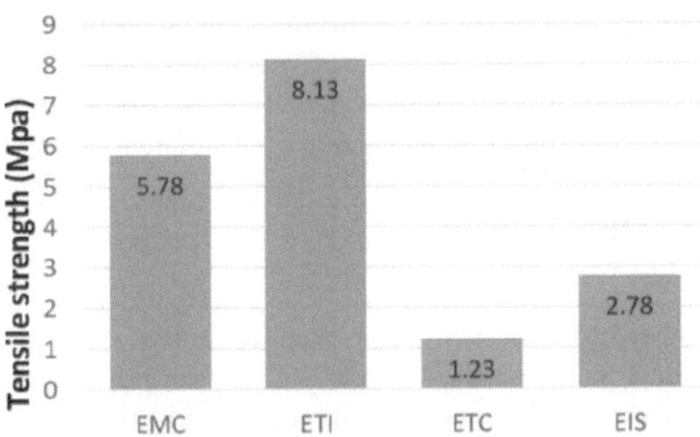

Fig. 28.4 Comparison of tensile strength

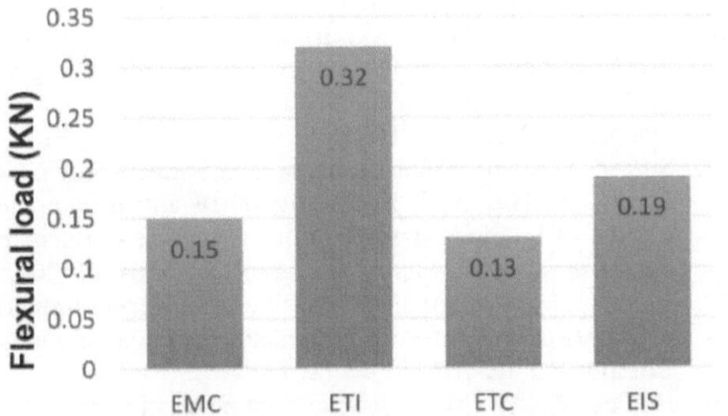

Fig. 28.5 Comparison of flexural load effect

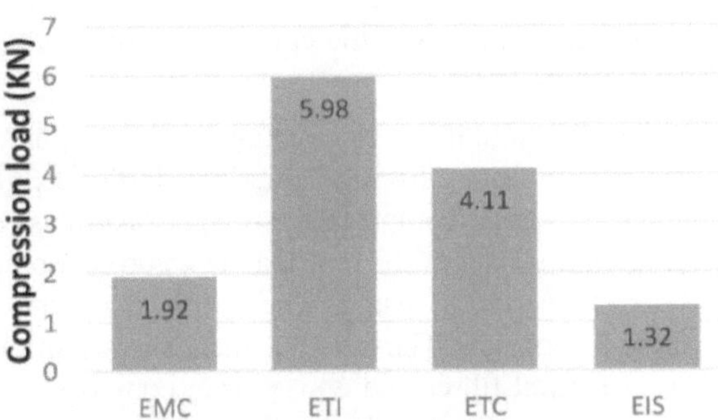

Fig. 28.6 Comparison of compression load

EMC and EIS have a middle flexural load. Compared to EMC and EIS, the EMC shows good tensile property 5.78 Mpa and EIS shows good flexural property 0.19KN. In compression strength also ETI showed good strength 5.98KN but ETC has second high strength. Comparatively ETC shows least values in tensile and flexural strength but it has good compression property. EIS has the least compression values of 1.32 KN. From shore D0 hardness strength is observed, it's understood that the hardness of the epoxy composites. And the epoxy composite tinosporacordifolia possesses good hardness values 59. The ETI possesses good tensile strength and flexural loads, but in hardness test, it has less value 51. The epoxy resin possesses high hardness owing to the more efficient binding with the fibers. This phenomenon is evident from the Scanning Electron Microscope (SEM) images.

Figure 28.7 shows the morphological analysis of various composites specimens evaluated in this present study. It can be noted that contaminants and existance of parenchyma cells on surfaces of fiber plate [20]. Figure 28.7(a) shows the adhesion improved morinda-citrifilia fibers. It's clearly shows the impurities bonded in the composite fiber, which shows addition layers on fiber surface, Figure 28.7(b) shows epoxy with tamarindus–indica composite fractograpy image.

The use of the coupling agent removes the aggregation and shows a remarkable sense of pull out fibers in this composite. Figure 28.7(c) shows the surface morphology of epoxy tinosporacordifolia composite. The images demonstrated the enhanced adhesion of the reinforced tinosporafibre to the matrix and its good association with the load-sharing

Fig. 28.7 Surface morphology of composites. (a) Epoxy morinda-citrifilia composite, (b) Epoxy tamarindus–indica composite, (c) Epoxy tinosporacordifolia composite and (d) Epoxy Ipomoeastaphylina composite.

phenomenon. Figure 28.7(d) shows ipomoea staphylina fiber with epoxy resin composite. Even after the test was completed, there was still matrix debris on the fibres' surface. This demonstrates the fiber's enhanced adherence to epoxy.

5. Conclusion

The epoxy composites have been successfully prepared and characterized. The significant differences in the physical and mechanical properties of the developed natural composites are revealed through characterisation. By conducting various tests on epoxy morindacitrifolia, epoxy tamarindusindica, epoxy tinosporacordifolia and epoxy ipomoea staphylina, it was found out that compared to other composite fibres, epoxy TamarindusIndica has a high tensile strength materials. And the epoxy tinosporacordifolia has high hardness strength.

By this investigation, it can be concluded that compared to other composite fibres, Epoxy TamarindusIndica exhibits good tensile strength, compressive strength , hardness and flexural strength. Except flexural load test epoxy ipomoea Staphylina shows poor results when compared with the other three composites. The inhomogeneous distribution of mixture of natural fibres and epoxy resin composite fibre on the polymer matrix results in distinctly inferior properties. The SEM images showed a good dispersion of the epoxy and natural fibers in the matrix.

From in this study can observed the composites were light in weight, economical and possess good mechanical properties. These mechanically reinforced polymer matrix biocomposites could therefore find application in home appliances, sports goods, building construction, automotive interior components, and electronic packaging.

References

1. Savita Dixit., Ritesh Goel., Akash Dubey., Prince Raj Shivhare & Tanmay Bhalavi. Natural Fibre Reinforced Polymer Composite Materials - A Review. Polymers from Renewable Resources.8(2), 121–129 (2017).
2. Silva, H., Ferreira, J. A., Costa, J. D. & Capela, C.A study of mixed mode interlaminar fracture on nano clay enhanced epoxy/glass fiber composites.Ciencia & Tecnologia dos Materiais.25, 92–97 (2013).
3. Manikandan, G., Jaiganesh, V. & Ravi Raja Malarvannan, R. Mechanical thermal stability and water uptake behaviour of silane-modified chitosan particle and morinda-citrifoliafibre-reinforced epoxy bio-composite.Materials Research Express. 6(5), 055513 (2019).
4. Mamidi, V. K., Pugazhenthi, R., Manikandan, G. & Kumar, M. V. Mechanical and microscopic study on Tinosporacordifolia and Tectona grandis composites. Materials Today: Proceedings. 229(3), 772-775 (2020).
5. Manikandan, G., Jaiganesh, V., Ravi Raja Malarvannan, R. & Vinothkumar, M. Eco-Friendly Reinforced Composites Made from Tamarindus Indica with Epoxy and Bisphenol-Resin.Int. J. Vehicle Structures & Systems.11(7), 10.4273/ijvss.11.7.10 (2019).
6. Valadez-Gonzalez, A., Cervantes-Uc, J.M., Olayo, R. & Herrera-Francoa, P.J. Effect of fiber surface treatment on the fiber–matrix bond strength of natural fiber reinforced composites. Composite part B: Engineering. 30, 309–320(1999).
7. McClatchey, Will. From Polynesian Healers to Health Food Stores: Changing Perspectives of Morindacitrifolia (Rubiaceae).Integrative Cancer Therapies. 1(2), 110–120(2002).
8. Jaiwal, P. K.& Gulati, A. In vitro high frequency plant regeneration of a tree legume Tamarindus indica (L.). Plant cell reports. 10(11), 569–573(1991).
9. Havinga, R. M., Hartl, A., Putscher, J., Prehsler, S., Buchmann, C. & Vogl, C. R. Tamarindus indica L.(Fabaceae): patterns of use in traditional African medicine. Journal of ethnopharmacology. 127(3), 573–588(2010).
10. Singh, S. S., Pandey, S. C., Srivastava, S., Gupta, V. S., Patro, B.&Ghosh, A. C. Chemistry and medicinal properties of Tinosporacordifolia (Guduchi). Indian journal of pharmacology. 35(2), 83–91(2003).
11. Suresh, K. C. A., Thamizhmozhi, M., Saravanan, C., Gude Suresh, K. G.& Sasi, K. Evaluation of anti-inflammatory activity of Ipomoea staphylina In carrageenan-induced paw edema in rats. Int J Pharm Sci Rev Res. 16, 67–69(2012).
12. Rahman, M., Puneeth, M.& Aslam, D. A. Impact properties of glass/Kevlar reinforced with nano clay epoxy composite. Compos B Eng. 107, 50–61 (2017).
13. Hedley, C. W. Mold filling parameters in resin transfer molding of composites (Doctoral dissertation, Montana State University-Bozeman, College of Engineering) https://scholarworks.montana.edu/xmlui/bitstream/handle/1/7312/31762102200092.pdf?sequence=1(1994).
14. Barbinta-Patrascu, M. E. et al.S. Tangerine-Generated Silver-Silica Bioactive Materials. Romanian Journal of Physics. 64, 701 (2019).
15. Otto, G. P.et al. Mechanical properties of a polyurethane hybrid composite with natural lignocellulosic fibers. Composites Part B: Engineering. 110, 459–465 (2017).
16. Herrera-Franco, P.& Valadez-Gonzalez, A. A study of the mechanical properties of short natural-fiber reinforced composites. Composites Part B: Engineering. 36(8), 597–608 (2005).
17. Arun Prakash, V. R. & Rajadurai, A. Mechanical, Thermal and Dielectric Characterization of Iron Oxide Particles Dispersed Glass Fiber Epoxy Resin Hybrid Composite. Digest Journal of Nanomaterials & Biostructures (DJNB). 11(2), 373–380(2016).
18. Jaiganesh, V., Malarvannan, R. R. R., Manikandan, G. & Krishnamoorthi, S. Investigation of the Mechanical Properties of Natural Fiber Reinforced Composites of MorindaCitrifolia with Epoxy and BisphenolResin.International Journal of Recent Technology and Engineering. 8 (1S2), (2019).
19. Landowski, M., Strugała, G., Budzik, M. &Imielińska, K. Impact damage in SiO2 nanoparticle enhanced epoxy–Carbon fibre composites. Composites Part B: Engineering. 113, 91–99 (2017).
20. John, M. J. & Thomas, S. Biofibres and biocomposites. Carbohydrate polymers. 71(3), 343–364(2008).

Note: Every figure and table was created using the experimental work; none of them were taken from any publications or online.

Advances in Additive Manufacturing Technologies – Gurusamy Pathinettampadian (eds)
© 2024 Taylor & Francis Group, London, ISBN 978-1-032-90013-1

29 Tribological Properties of Friction Stir Weave Aluminium and Copperwelding Utilisinggraphene Particles as a Dispersing Agent

D. Jayabalakrishnanan[1]
Department of Mechanical Engineering,
Chennai Institute of technology, Chennai, India

S. Senthil Kumar[2]
Department of Mechanical Engineering,
Panimalar Engineering College, Chennai, India

R. Ganesamoorthi[3]
Department of Mechanical Engineering,
Chennai Institute of technology, Chennai, India

M. Balasubramanian[4]
Department of Mechanical Engineering,
RMK College of Engineering and Technology, India

R. Sathiyamoorthi[5], **Hepsi Beaula M. J**[6]
Department of Mechanical Engineering,
Chennai Institute of technology, Chennai, India

ABSTRACT: Friction stir welding is a solid-state welding method predominantly used on aluminium alloys. Tool offset technique in Al-Cu weld joint reduces the difficulties faced in friction stir welding. A method of effective coalescence of reinforcement particles is as certained by defining the pattern of tool movement. It could be observed that particles were homogeneously distributed in the weldment. In this article, the influence of weaving tool motion for joining Al-Cu joints reinforced with graphene nanoplatelets on tool wear is studied. For these conditions, material flowability and tool tribology were also studied.

KEYWORDS: FSW; Tool wear; Graphene; Aluminium; Copper

1. INTRODUCTION

The utilization of FSW may be considered an environmentally sustainable approach, mostly attributed to its reduced energy usage for the entirety of the process. One benefit of employing this approach is the non-consumable nature of the tool, which effectively eliminates the need for filler material. The primary factors influencing the choice of tool

[1]jayabalakrishnand@citchennai.net, [2]thil2priya2@gmail.com, [3]ganesamoorthyr@citchennai.net, [4]manianmb@gmail.com, [5]sathiyamoorthir@citchennai.net, [6]hepsibeaulamj@citchennai.net

DOI: 10.1201/9781003545774-29

material revolve on tool wear, weld quality, and material cost. On the other hand, the choice primarily relies upon the material to be joined and related machining expenses to attain the desired tool profile. And also, the tool should join the workpiece material without any defects and brittle IMCs formation along with satisfactory tool life. The welding tool configuration, including the geometry and the material from which it is made, plays a vibrant role in the formation of welds with superior joint qualities. Newly developed welding tool material enables the welding of materials having maximum melting point temperatures as titanium, steel, and copper. Softening and plasticization of dissimilar joints under visco-plastic deformation make a uniform dispersion of phase molecules. Dissimilar joints result in formation of hard and brittle compounds at the interface, resulting in lower strength. Insufficient offset provides poor weld quality resulting from the formation of large Cu debris and voids. The formation of large Cu debris is due to deformation of hardened Cu metal, flow, and colloid with Aluminum metallic phase, leading to the formation of weld flash and voids in producing inefficient bonding between the dissimilar joints. With a large tool pin offset, the Cu particles with dendrite and vermicular structures were removed from Cu and mixed with Al metallic phases in the stir zone [1].

In addition, the geometry of the tool has a substantial impact on the wear resistance of the tool in issue. To minimise wear, the tool self-optimizes its shape [2]. Welding at a rotating speed of 2000 rpm results in unusually high tool wear relative to the other two rotational speeds [3]. Tool wear for threaded tool decays with diminishing revolution speed and or weld speeds for the FSW of metal matrix composites. At weld navigate separations of 3 m or more, the tool disintegrates into a self-optimized shape [4]. Wear phenomena are commonly observed in Friction Stir Welding (FSW), particularly when welding composites containing increasing quantities of abrasive particles and high softening point materials. These aspects are essential in contributing to the occurrence of tool wear and the deterioration of mechanical performance in welds. The user provided a numerical reference. The study conducted in [6] revealed that the wear resistance of coated tools may be attributed to the scratch and oxidation characteristics of the coatings.

The interaction of the axial flow with the pin created a divergence of the stream as well as a concomitant increase in pressure at the tool's mid-axial point, which ultimately led to significant wear in this region [7]. Because there was such a large amount of intermetallic complex present, tool metal portions showed signs of embrittlement and pulling out [8].

In the friction stir welding (FSW) technique used for aluminium composites, tool wear is a serious problem, particularly when the volume percent of strengthening chemicals is high. The degree of wear that is shown by the pin is very varied throughout its length, with the most obvious wear being detected at a place that corresponds to one-third of the pin's overall length. The rate of radial wear that occurs on the pin is clearly affected by the speed at which the welding is performed, with the initial welding process displaying the maximum degree of wear [9]. It was observed that the tool went through a significant amount of distortion while being welded, which resulted in the appearance of shoulder bulging and an increase in the cone angle of the pin [10]. For the aim of this study, a novel method of weaving has been used in order to go beyond the obstacles that have been uncovered in earlier research and inquiries. In addition to that, the investigation is going to try to figure out how much process parameters affect tool wear.

2. Materials and Methods

The materials investigated in this study consisted of an aluminium alloy (AA 6601-T6) and pure copper, both having the same dimensions of 100 mm x 50 mm x 6 mm. The current investigation employed graphene nanoplatelets (GNPs). The gold nanoparticles (GNPs) being examined possess a surface area of 500 m^2/g, a molecular weight of 12.01 g/mol, an average density of 0.3 g/cm^3, and a size of 25 μm. The material employed for this particular procedure was hardened high-speed tool steel (HSS), as seen in Fig. 29.1.

Fig. 29.1 Tool pin profile with and without offset

The tool geometries employed in the study consisted of a shoulder diameter of 12 mm, a pin offset of 2 mm, a bead width of 24 mm, a pin diameter of 4 mm, and a pin height of 5.5 mm. Fig. 29.1 depicts the profile of the Friction Stir Welding (FSW) tool pin, which may be categorized into two types: with offset and without offset. Weave welding is a technique characterized by the movement of the welding instrument in an eccentric pattern. The welding speed governs the execution of this process. In the process of weave welding, the welding trajectory undergoes elongation, leading to a somewhat increased heat input compared to that observed during linear route movement.

The study adhered to the established process parameter ranges, as determined by earlier researchers [11-14], for four distinct welding conditions: linear weld, linear weld with pin offset, weave weld, and eccentric weave with pin offset. A series of experiments were done to investigate the impact of rotational speed and traverse speed on weld quality using reinforced nano graphene platelets. The experimental parameters were varied within the range of 1200 to 1400 rev/min for rotating speed and 50 to 100 mm/min for traverse speed. The tool offset was maintained at a constant value of 1 mm throughout the experiments. The studies conducted involved the use of dissimilar AA6061 and pure copper plates under various settings.

1. The proposed technique is referred to as Linear Weld without Pin Offset (LW-WOPO).
2. The technique referred to as "Weave Weld Without Pin Offset" (WW-WOPO) is being discussed.
3. The linear weld with pin offset (LW-WPO) technique is a method used in welding processes.
4. The technique known as weave weld with pin offset (WW-WPO) involves the process of joining two materials together by the use of a weaving motion, while also including a pin offset.

3. Result and Discussion

3.1 Linear Weld without Pin Offset

Tool wear was determined to be quite low for the linear weld without pin offset, as shown in Fig. 29.2(a-c), which depicts the results of the study. The tool movement will be parallel to the axis of the weldment, as it was in the linear weld, in order to maximise the effectiveness of the tool stirring on the advancing side. While stirring the nugget zone metal, this resulted in less tool pin wear by reducing the axial and tangential forces that were applied to the tool. A further argument is that when the temperature of the tool was dropped, the overlaying of intermetallic compounds on the surface of the tool decreased. This, in turn, led to a reduction in the stirring coverage area and time.

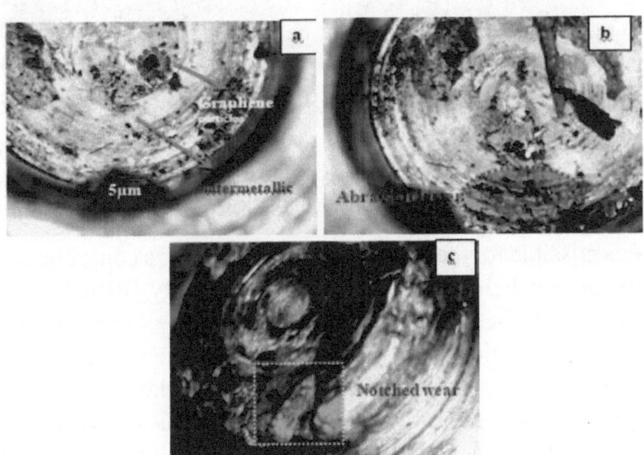

Fig. 29.2 Microstructure of tool tribology characteristics of LW-WOPO

3.2 Linear Weld with Pin Offset

Figure 29.3(a-c) is a representation of the microscopic pictures that were taken of a linear weld with a pin offset. When using a linear weldment with a pin offset, it has been observed that the rate of wear is much greater in materials that have a higher toughness, such as copper, in comparison to aluminium. It is considered that the amount of wear experienced by a Friction Stir Welding (FSW) tool has an inverse connection with the hardness ratio when it comes to the context of welding dissimilar metals. This is because FSW tools are used to join metals that have different hardnesses.

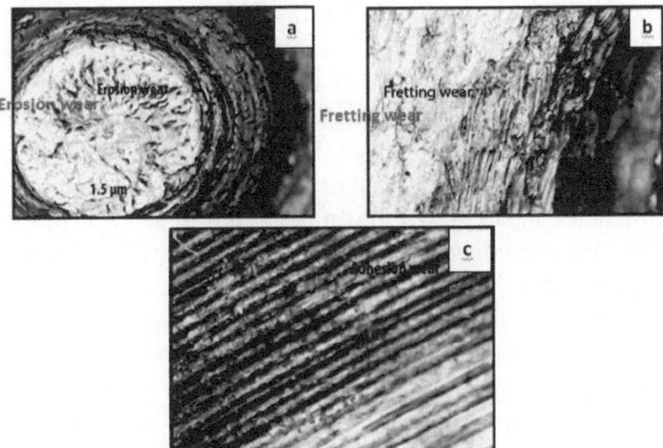

Fig. 29.3 Tool tribology images of LW-WPO

3.3 Weave Weld without Pin Offset

The microscopic features of tool wear on an eccentric weave weld without pin offset are shown in several locations throughout Fig. 29.4, which may be seen in (a) through (c). In the context of eccentric weave welding, it was noticed that tool wear was slightly increased due to the eccentric motion of the tool along the weld path. This was one of the factors that contributed to the higher level of tool wear. Because of this eccentric movement, the tangential tension that was being placed on the tool increased, which resulted in an orbital plane motion. As a direct consequence of this, a thermal inter-transition state was brought about in the metals. Because of this, the IMC's at adherence to the surface of the tool, which brought to an increase in the amount of tool wear and a decrease in the hardness value of the tool pin.

Fig. 29.4 Tool tribology microscopic images of WW-WOPO

3.4 Weave Weld with Pin Offset

The phenomena of tool pin wear, as it was seen, is shown in Fig. 29.5(a). Additionally, this phenomenon demonstrates the adhesion of the upper surface of the pin to the bonds of the metallic substrates, which results in the production of intermetallic compounds that attach to the the outermost face of the tool. Because of the increase in temperature, graphene particles have stuck to the front surface of the pin. This is due to the fact that the pin was heated to a higher temperature. Within the thermos mechanically affected zone (TMAZ), it can be seen in Fig. 29.5 (b) that the tool pin edges go through a transition into abrasion layers. This phenomenon is referred to as the TMAZ. Fig. 29.5(c) illustrates the wear indentations that occur in the tool boundaries during the high stirring technique. These indentations demonstrate an average notch wear of 4.5 micrometres. When the plasticized stir zone comes into contact with the edges of the tool pin, a transformation takes place in the plasticized stir zone. This transformation causes the production of serrations, which in turn leads to

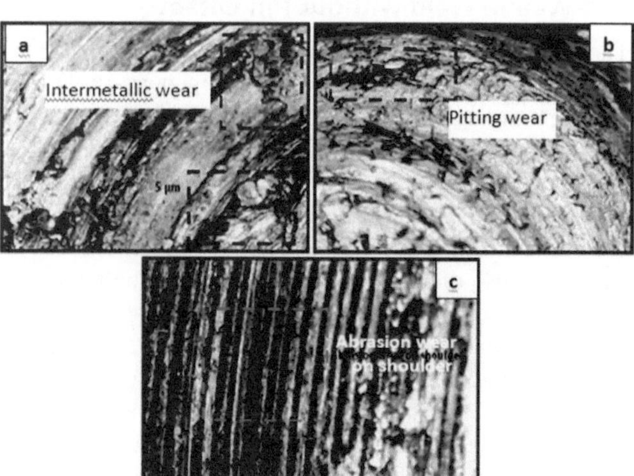

Fig. 29.5 Tool tribology images of WW-WPO

an increase in tool wear. As a consequence of the presence of debris, which was produced as a result of wear, being mixed with the weld bead, the strength of the connection was significantly reduced.

3.5 The Capacity of the Material to Flow

During the FSW technique, a flowability pattern of the material was evaluated using a digital microscope for better accuracy. In order to make it easier to examine the area, the tool pin was positioned in the middle of the plasticized zone. This was done to assist the inspection. The right side was the source of the buildup of material, while the existence of two lobes made the development of the weld easier to accomplish. In addition to being visible on the walls of the aluminium tube, the lobes and stacking layer patterns seen in Fig. 29.6(a-b) may also be seen there. When the shoulder of the tool is removed, the bottom lobe of the pin in Fig. 29.6(a) is obscured from view; nevertheless, its closeness is made more apparent on the right side of the illustration. As a consequence, this produces a tunnel that is open on both ends and has sidewalls that have layered structures.

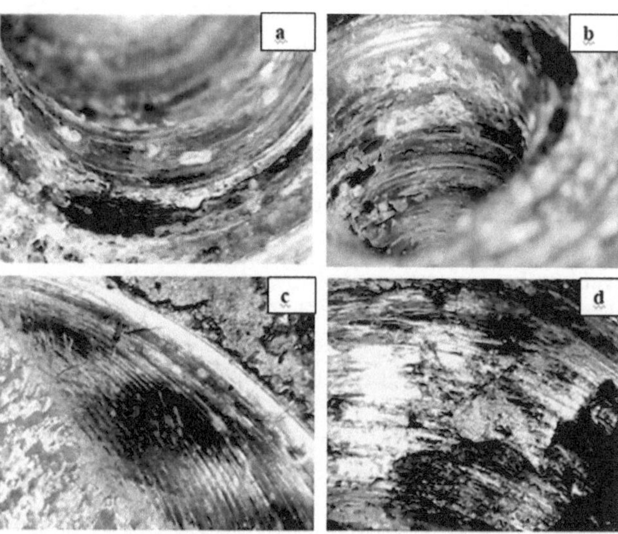

Fig. 29.6 The weld zone's material flowability features (a) Striations on the upper lobe, (b) lower lobe, (c) interface zone, (d) nugget zone

It is possible for the trench that was created as a consequence of the pin being displaced to be efficiently filled by the tool's subsequent rotation, and this may be done without the participation of the shoulder. The accumulation of material layers within the lobe, which takes place as a direct result of each rotation of the tool, is shown in the image that can be found in Fig. 29.6(b). A mixture of aluminium and copper is shown in a stacked pattern across the material's macrostructure. This pattern is distinguished by the presence of observable discontinuities in the flow of the material. The surface striations are shown in a straightforward manner in the schematic image that may be seen in Fig. 29.6(c-d). When the pin tool is put through eccentric motion, the material that is being conveyed by the pin becomes visible on the surfaces of the globe, which in turn leads to the formation of clearly defined banded patterns. The use of the tool pin offset made it much simpler to keep the spacing between each round of the feed at a consistent level. In addition to this, an increase in the tool pin's offset resulted in a corresponding increase in the quantity of material associated with a certain feed rate per revolution.

4. Conclusions

(a) Investigations on tool wear characteristics were studied for the Al-Cu joint with graphene particles as a reinforcing agent

(b) Serrationsa) The present study examines the features of tool wear in the Al-Cu joint reinforced by graphene particles.

(c) Serrations were detected at the pin tip as a consequence of the tribological properties of the tool, leading to a notched wear of 4.5 μm and the formation of an intermetallic compound layer at the boundary of the tool tip.

(d) The perception of surface serrations and collected structures in various regions of the bead can be attributed to the flowability characteristics, which are influenced by the offset movement of the tool. The present study has successfully proven the efficacy of the unique pin offset with stir pattern approach in enhancing the performance of weldments were observed within the pin tip due to tool tribology characteristics, which resulted in a 4.5 μm notched wear, and the creation of the bonded layer of intermetallic compounds across the tool tip's boundary.

(e) Flowability characteristics drove to the perception of surface serrations and gathered structures in different areas of the bead connected with the offset movement of the tool. This validated the unique pin offset with stir pattern method for improving weldment performance.

References

1. Jayabalakrishnan D (2019), Investigation on eccentric weave friction stir welding of AA6061-T6 aluminium alloy with copper, Ph.D Desertation, Anna University, Chennai, india.
2. B. Ashish, J.S. Saini, Bikramjit Sharma, A review of tool wear prediction during friction stir welding of aluminium matrix composite, Transactions of Nonferrous Metals Society of China, 26(2016) 2003–2018.
3. UttamAcharya, BarnikSaha, Roy, SubashChandra Saha, A Study of Tool Wear and its Effect on the Mechanical Properties of
4. Friction Stir Welded AA6092/17.5 Sicp Composite Material Joint, Materials Today : Proceedings, 5(2018)20371–20379
5. G.J. Fernandez, L.E. Murr, Characterization of tool wear and weld optimization in the friction-stir welding of cast aluminum 359+20% SiC metal-matrix composite, Materials Characterization, 52(2004)65–75
6. W. H Zeng, H.L. Wu, J.Zhang (2006), Effect of Tool Wear on Microstructure, Mechanical Properties and Acoustic Emission of Friction Stir Welded 6061 Al Alloy, Acta MetallurgicaSinica (English Letters), 19(2006)9–19
7. Akeem Yusuf, AdesinaFadi, A.Al-BadourZuhair, M.Gasem, Wear resistance performance of AlCrN and TiAlN coated H13 tools during friction stir welding of A2124/SiC composite, Journal of Manufacturing Processes, 33 (2018)111–125.
8. Hasan A.F, Bennett C.J., Shipway P.H, Cater S, .Martin J (2017), A numerical methodology for predicting tool wear in Friction Stir Welding, Journal of Materials Processing Technology, 241(2017) 129–140
9. Tarasov S.Yu., V.E. Rubtsov, E.A. Kolubaev, A proposed diffusion-controlled wear, mechanism of alloy steel friction stir welding (FSW) tools used on an aluminum alloy, Wear,318 (2014) 130–134
10. Liu H.J, Feng J.C, Fujii H, Nogi K (2005), Wear characteristics of a WC–Co tool in friction stir welding of AC4A+30 vol% SiCp composite, International Journal of Machine Tools and Manufacture, 45:1635–1639.
11. Arshad Noor Siddiquee, Sunil Pandey (2014), Experimental investigation on deformation and wear of WC tool during friction stir welding (FSW) of stainless steel, The International Journal of Advanced Manufacturing Technology, 73:479–486.
12. Barekatain, H, Kazeminezhad, M, Kokabi, AH (2014), Microstructure and mechanical properties in dissimilar butt friction stir welding of severely plastic deformed aluminum AA 1050 and commercially pure copper sheets', Journal of Materials Science and Technology, 30:826–834.
13. Jayabalakrishnan D Balasubramanian M (2018), Eccentric-weave Friction Stir Welding between Cu and AA 6061-T6 with reinforced Graphene nanoparticles, Materials and Manufacturing Processes, 33:333–342.
14. Jayabalakrishnan D Balasubramanian M (2017), Friction stir weave welding (FSWW) of AA6061 aluminium alloy with a novel tool path pattern, Australian Journal of Mechanical Engineering, 17:133–144.
15. Jayabalakrishnan D, Balasubramanian M (2018), Friction Stir Welding of Dissimilar Butt Joints with Novel Joint Geometry, Acta Physica Polonica A, 133:94–100.

Note: Every figure was created using the experimental work; none of them were taken from any publications or online.

Advances in Additive Manufacturing Technologies – Gurusamy Pathinettampadian (eds)
© 2024 Taylor & Francis Group, London, ISBN 978-1-032-90013-1

30 LM-25 Hybrid Al-Metal Matrix Composites: An Experimental Study of Their Mechanical Properties

R. Manjunathan*, S. Hari Krishna Raj
Assistant Professor, Department of Mechanical Engineering,
Vel Tech High Tech Dr. Rangarajan Dr. Sakunthala Engineering College,
Avadi, Chennai, India

P. Prabhakaran, S. Riyaz Ahamed, S. Tarun Vikram
Department of Mechanical Engineering,
Vel Tech High Tech Dr. Rangarajan Dr. Sakunthala Engineering College,
Avadi, Chennai, India

ABSTRACT: LM-25 is a corrosion resistant aluminum casting alloy which is used for hydraulic piston and heavy load applications in more thermal stresses. This alloy can withstand higher loads and temperatures. It has good machinability and wear resistance properties. Aluminum alloy LM25-Al_2O_3 & SiC hybridmethod sample were prepared by direct casting method using SiC & Al_2O_3 powder as reinforced particles with 90 and 120 micron level average of diameter and Aluminum alloy as the the base metal. After stirring the melt composites, they were cast into a metallic mold. Different weight percentages of reinforcement SiC & Al_2O_3 were synthesized with a constant weight of LM 25 metal matrix. The casted composite specimens were machined in accordance with test specifications. The effects of hybrid composite weight percent on hardness, tensile strength, and compressive strength have been explored. Maximum hardness was obtained with increased reinforcing. The impact strength of higher reinforcement matrix alloys is quite low. Tensile strength SiC -2.5% + Al_2O_3 - 5% and compressive strength Ratio 3 - SiC - 5% + Al_2O_3-10% were particularly high. In comparison to other samples, the minimum corrosion study showed Ratio3 - LM-25 SiC-5% + Al_2O_3-10%. The specimens with the highest tensile and compressive strengths contained 5 weight percent SiC and 10 weight percent SiO_2.

KEYWORDS: LM-25 alloy, Hybrid composite, SiC, Al_2O_3, Tensile strength, Compressive strength

1. Introduction

Automobile manufacturers are responding to increased fuel efficiency needs by employing alternative materials such as metal matrix composites (MMCs). When compared to conventional materials, MMC parts offer significant weight savings while maintaining, if not improving, performance. MMCs have made gradual but steady progress toward mainstream use in the automotive industry during the last decade. Automobile manufacturers are carefully investigating alternate materials in the design of their products to fulfill the demands of lower fuel consumption, lower pollu-

*Corresponding author: manjunathan@velhightech.com

DOI: 10.1201/9781003545774-30

tion, and higher efficiency standards, as well as to preserve competitiveness. Metal matrix composites (MMC) appear to be a potential technique to obtain major performance gains. Synergistic Strength: Hybrid composites with SiC and Al_2O_3 reinforcements provide better mechanical strength and structural integrity. Tailored Properties: The hybrid nature permits material features to be customized to satisfy varied application performance requirements. SiC and Al_2O_3 have excellent abrasion and wear resistance, making hybrid composites ideal for durable applications. Due to its improved mechanical qualities and wear resistance, hybrid composites are used in manufacturing and aerospace. Optimized Performance: SiC and Al_2O_3 reinforcements synergistically balance strength, hardness, and wear resistance to increase composite functioning.

The influence of silicon carbide (SiC) weight percentages (5, 10%, 15%) and particle size (10, 20) The wear performance of the composite was evaluated using the pin-on-disc method, which was used to calculate the friction coefficient and wear resistance of the composites. The studies were carried out by altering the sliding speed (1.5,2.5, and 3.5 m/s), loads (30,50, and 70 N), and sliding distance ranges (500, 1000, and 1500m) under dry sliding conditions. The Taguchi experiment plan and ANOVA approach were used to determine the effect of reinforcing ceramic particles, sliding distance, sliding speed, and applied load on the friction coefficient and wear rate. The results show that the most influential parameters for friction coefficient are applied load and sliding distance. The most influential factor for wear rate is load and SiC %.

When machining an Al/Bagasse Ash/Gr hybrid composite, it is necessary to adjust the turning parameters in order to reduce both temperature and surface roughness. Use of the Taguchi L9 orthogonal array is employed in turning studies, and the Al/Bagasse Ash/Gr hybrid composite is produced via the squeeze casting technique. Surface roughness and temperature are response parameters, while bagasse ash %, feed rate, and speed are input parameters. For turning, frequent parameter combinations that lead to a lower temperature and improved surface quality are found using Taguchi-based grey relational analysis (GRA). Low reinforcing percentage, low feed rate, and fast speed are advised while turning Al/Bagasse Ash/Gr hybrid composite. Mechanical Strength: SiC and Al_2O_3 reinforcements improve tensile, compressive, and mechanical performance, making structural applications durable. Hardness: SiC and Al_2O_3 provide the hybrid composite high hardness, which is important for wear, abrasion, and deformation resistance. SiC and Al_2O_3 are notoriously thermally stable. A hybrid composite is suited for high-temperature applications because to its increased tolerance to high temperatures

1.1 Aluminum-LM-25

This alloy complies with LM25 of BS 1490:1988. The conditions of castings are established as fully heat treated (TF), solution treated, artificially aged and stabilized (TF7), and precipitation treated (TE).

ACCURATE TEMPERATURES MAY VARY FROM 670 TO 780°C, ACCORDING TO THE MOLD CONFIGURATION. IN HEAVY SECTIONS, THIS ALLOY MAY FORM INTERNAL SHRINKAGE PIPES.

The numerous benefits of composite materials, such as their high tensile strength, low density, superior wear resistance, and flawless surface finish, have made them more and more in demand these days. Aluminum oxide is comparatively inexpensive and low-density reinforcement.

Additionally, consideration will be given to wear and tensile strength. To achieve the aforementioned, an experimental setup complete with all necessary inputs has been developed. By altering the weight ratio at which SiC & Al_2O_3 are added to aluminum metal, a composite is produced in this endeavour. The main ingredients in this experiment are silicon carbide (SiC), aluminum oxide (Al_2O_3), and aluminum-LM-25.

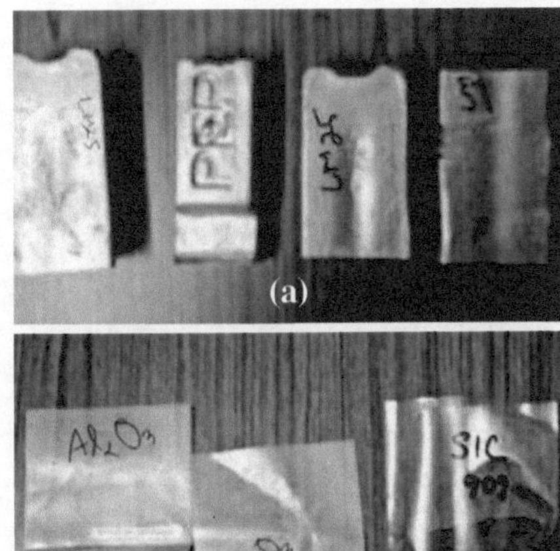

Fig. 30.1 (a) Aluminum alloy LM-25 (b) SiC and Al_2O_3 Powder

1.2 Stir Casting Process

Al_2O_3 and SiC powder, with respective particle sizes of 90 µm and 120 µm, were selected as the reinforcement materials, with the aluminum LM25 alloy serving as the basis material. Three cast samples were created in this work using LM25 as the base material at different weight percentages of SiC and Al_2O_3, such as 2.5% of SiC and 5% of Al_2O_3, 5% of SiC and 10% of Al_2O_3. The aluminum LM25 alloy was first melted in a pot by heating it for 15 minutes at 850°c in a blower furnace. Preheating the Al_2O_3 and SiC powder at 575°C was done in a different muffle furnace. After melting the LM25 entirely at a temperature of around 850°C in the furnace, it was gradually added to the SiC and Al_2O_3 powder that had been heated beforehand.

In preparing LM-25-Al_2O_3 and SiC hybrid composites, a hybridization process involves blending molten LM-25 aluminum alloy with pre-treated $Al_2O_3$and SiC particles. The casting method employs techniques like stir casting or liquid metallurgy, ensuring even dispersion of the reinforcements in the molten alloy. Subsequently, machining techniques such as milling or turning are applied to shape the solidified composite into desired forms. This comprehensive approach ensures a homogenous distribution of Al_2O_3 and SiC within the LM-25 matrix, resulting in hybrid samples with improved mechanical and thermal properties for diverse engineering applications.

Fig. 30.2 Stir casting setup

For around fifteen minutes, the drilling machine assisted in the stirring process, which kept the speed of the mixture at 950 rpm. After that, this mixture is poured into the mold cavity and let to cool to ambient temperature. Three specimens with varying compositions of SiC and Al_2O_3 were prepared using this process.

2. Calculations

Table 30.1 Calculationfor all ratios

RATIO	LM-25 (grams)	SiC (grams)	Al_2O_3 (grams)
I	520	0	0
II	520	2.5% = 13	5% = 26
III	520	5% = 26	10% = 52

Mixing Ratio of samples:
Sample 1: LM-25 = 100%
Sample 2: SiC - 2.5% + Al_2O_3- 5% & LM-25
Sample 3: SiC - 5% + Al_2O_3- 10 % & LM-25

3. Result and Discussion

3.1 Result of Hardness Test

Hardness Value

Table 30.2 Materials composition and its hardness value

Ratio	Material	HRB
R_1	LM-25-100%	35
R_2	SiC -2.5% + Al_2O_3-.5% & LM-25	34
R_3	SiC -5% + Al_2O_3 -10 % & LM-25	45

SiC and Al_2O_3 reinforcements synergize in hybrid composites, enhancing mechanical strength, hardness, and wear resistance. SiC contributes excellent toughness, while Al_2O_3 imparts high hardness. This combination optimizes the overall performance of the composite, making it a robust material with versatile applications in diverse industries

In the stirring and casting process for LM-25-Al_2O_3 and SiC hybrid composites, molten LM-25 aluminum alloy is stirred vigorously, incorporating pre-treated Al_2O_3 and SiC particles. This ensures uniform dispersion of reinforcements. The well-mixed composite is then cast into the desired mold, facilitating a homogeneous distribution and enhancing the overall mechanical properties of the hybrid material

Table 30.3 Compression strength graph with various trails

S. No	Composition	Compression Strength N/mm^2
R_1	LM-25-100%	308.4
R_2	SiC -2.5% + Al_2O_3-.5% & LM-25	345.2
R_3	SiC -5% + Al_2O_3 -10 % & LM-25	358.6

Hardness Test Graph

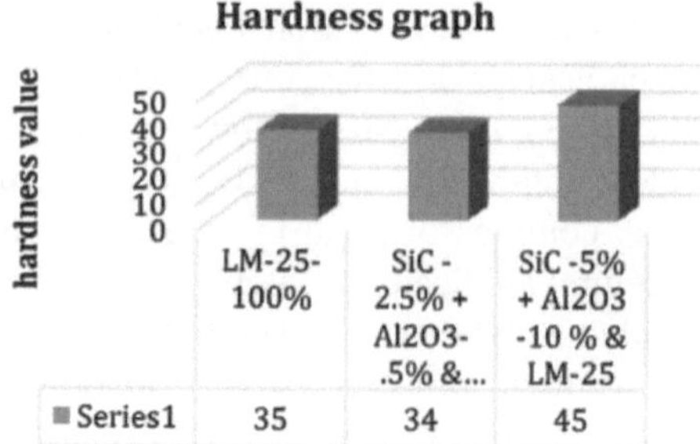

Fig. 30.3 Hardness strength graph

Tensile Strength Values

Tensile Strength

Table 30.4 Tensile strength graph with various trails

Ratio	DIA (mm)	CSA (mm^2)	TL (kN)	TS (N/mm^2	IGL (mm)	FGL (mm)	%E
R_1	15	202.06	20.84	103.65	73.2	73.2	0.0
R_2	16	201.06	22.86	113.69	73.2	74.0	1.09
R_3	16	201.06	24.42	121.46	73.2	74.0	1.09

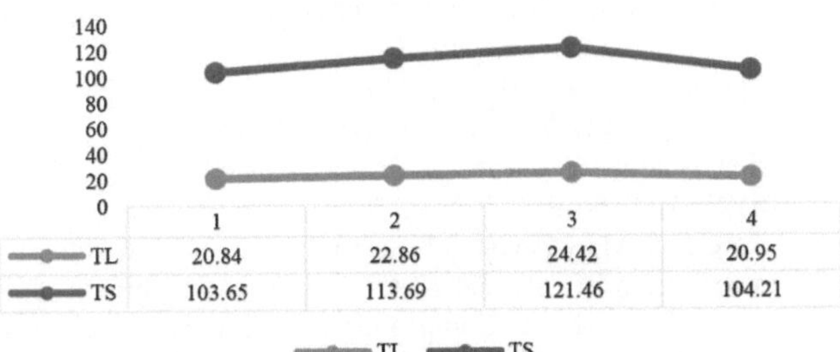

Fig. 30.4 Tensile strength graph

LM25 alloy was used to make hybrid composites with Al_2O_3 (5%) and SiC (2.5%), and Al_2O_3 (10%) and SiC (5%). Preheated Al_2O_3 and SiC powder was progressively added to melted LM25 alloy at 850°C. The composites containing LM-25SiC (5%) + Al_2O_3(10%) had the best mechanical characteristics. Corrosion resistance and tensile and compressive strengths increased. Reinforcement agglomeration reduced impact strength at this ratio. Salt spray corrosion investigation showed negligible accumulation on LM-25SiC (5%) + Al_2O_3(10%) specimens.

3.2 Compression Strength

The minimum corrosion observed in the Ratio3 (LM-25 SiC -5% + Al_2O_3- 10%) sample holds significant implications. It indicates that the hybrid composite has enhanced resistance to corrosion compared to other compositions. This suggests that the addition of SiC and Al_2O_3 reinforcements in this specific ratio provides effective protection against corrosive environments, making it a promising candidate for applications in industries where corrosion resistance is critical, such as marine or chemical processing. Additionally, the findings highlight the potential longevity and durability of components made from this specific hybrid composite in corrosive conditions.

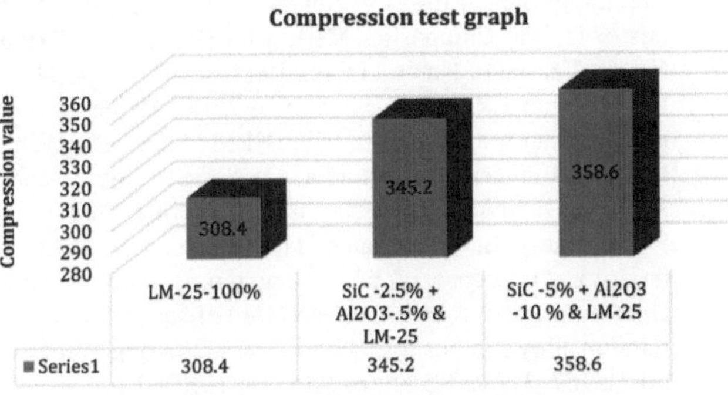

Fig. 30.5 Compression strength graph

Table 30.5 Impact strength of various compositions

S. No	Composition	Imapct Strength (Joules)
R_1	LM-25-100%	4
R_2	SiC -2.5% + Al_2O_3-.5% & LM-25	3.5
R_3	SiC -5% + Al_2O_3 -10 % & LM-25	2

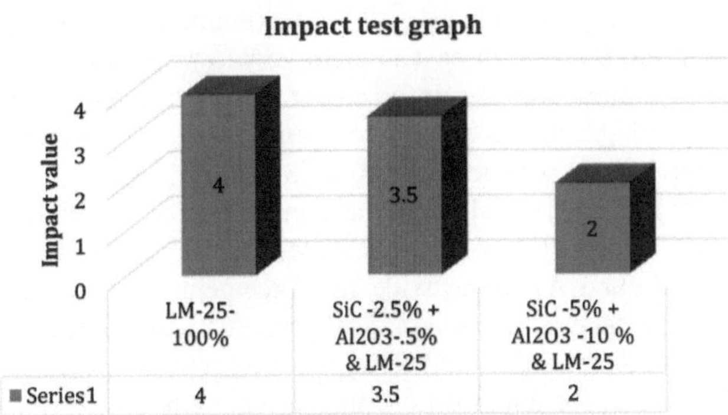

Fig. 30.6 Impact test graph

Table 30.6 Impact strength of various composites

S. No	Composition	Imapct Strength (Joules)
R_1	LM-25-100%	4
R_2	SiC -2.5% + Al_2O_3-.5% & LM-25	3.5
R_3	SiC -5% + Al_2O_3 -10 % & LM-25	2

Higher reinforcement, such as in the LM-25SiC (5%) + Al_2O_3(10%) composite, can lead to lower impact strength due to reinforcement agglomeration. The increased concentration of hard particles may create stress concentrations, reducing the material's ability to absorb energy upon impact and resulting in lower impact strength in alloys with higher reinforcement ratios.

Table 30.7 Weight loss for different ratio

Ratio	Initial Weight In g	Final Weight In g	Weight Loss In g
R_1	7.873	7.768	0.105
R_2	7.181	7.175	0.006
R_3	6.268	6.252	0.016

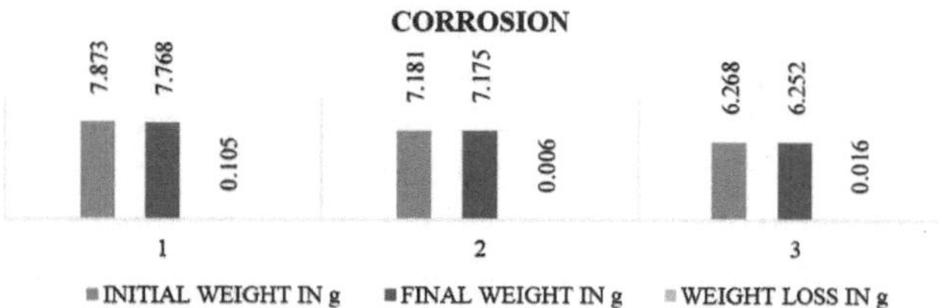

Fig. 30.7 Corrosion graph

4. Conclusion

Composite materials especially LM-25 and Molybdenum disulfide, graphite and silicon carbide composites having good mechanical properties compared without the reinforcements of LM aluminum alloy.From the investigation the mechanical and tribological property of hybrid LM-25 MMC were enhanced. The maximum hardness was achieved Ratio-3. Impact strength very low on Ratio-3 due to agglomeration of reinforcement. Particularly good tensile strength and compressive strength and low corrosion behavior rate was achieved Ratio 3 (SiC -5% + Al_2O_3 -10 % & LM-25).Investigation mechanical property of hybrid LM-25 metal matrix were analyzed finally found Silicon carbide and Alumina oxide reinforcement enhanced the good tensile and compressive properties. According to hardness properties mild changes were occurred maximum reinforcement composition. Impact strength very low on Ratio-3 due to agglomeration of reinforcement. During the salt spray corrosion analysis found Ratio –LM-25SiC -5% + Al_2O_3 -10 % minimum corrosion deposited on the specimen compare than other samples.

References

1. Abhay, A. Utpata, P. B. Pawar, 2018 "Modeling and Analysis of Al-SiC Spur Gear under Static Loading Condition", Elsevier.
2. Ajith Arul Daniela, 2018 "Dry Sliding Wear Behaviour of Aluminium 5059/SiC/Al_2O_3 Hybrid Metal Matrix Composites" Received: January 06, 2017; Revised: August 11, 2017; Accepted: August 24, 2018.
3. AjmalHussain S., Rajaneesh R., Hashim Nizam, Jithin K., 2020 "Experimental Analysis On Aluminum Alloy (6063) With Silicon Carbide: An Experimental Investigation", International Journal of Research in Engineering and Technology.
4. Anish A., 2019 "Characterization of Aluminium Matrix Reinforced With Tungsten Carbide And Aluminum oxide Hybrid Composite" 2nd International conference on Advances in Mechanical Engineering (ICAME 2019).
5. Antony Arul Prakash M. D., 2019 "Microstructural Analysis of Aluminium Hybrid Metal Matrix Composites Developed Using Stir Casting Process" International Journal of Advances in Engineering, 2019, 1(3), 333–339.
6. Ehsan Ghasali, 2018 "Production of Al-SiC-TiC Hybrid Composites Using Pure and 1056 Aluminum Powders Prepared Through Microwave and Conventional Heating Methods" Journal of Alloys and Compounds.
7. Kanchan A., 2019 "Tribological Investigation of Al7075/Tic/Al_2O_3 Hybrid Composite Material" International Research Journal of Engineering and Technology (IRJET).
8. Karuppasamy R., 2020 "Taguchi-GRA for Multi Criteria Optimization of Turning Parameters for Al/Bagasse Ash/Gr Hybrid Composite" Journal of Critical Reviews ISSN-2394–5125 Vol 7, Issue 9, 2020.
9. Dr. Nagesh D., Mr. Shashikanth G. S., 2021 "Optimization of Tribological Properties of Al6061, Boron And Graphite MMCs Using Stir casting Method.", IOSR Journal of Mechanical and Civil Engineering.
10. Paranthaman P., 2020 "Multi-Objective Optimization on Tribological Behaviour of Hybrid Al Mmc by Grey Relation Analysis" Journal of Critical Reviews ISSN- 2394–5125 Vol 7, Issue 9, 2020.
11. Rajesh A. M., Mohamed Kaleemulla, 2019 "Material characterization of SiC and Al_2O_3–Reinforced Hybrid Aluminum Metal Matrix Composites on Wear Behavior", Advanced Composites Letters Volume 28: 1–10.
12. P. Gurusamy and S. Balasivanandha Prabu, "Effect of the squeeze pressure on the mechanical properties of the squeeze cast Al/SiCp metal matrix composite," Int. J. of Microstructure and Materials Properties 8(4–5), 299–312 (2013), https://doi.org/10.1504/IJMMP.2013.057067.
13. P. Gurusamy, S. Balasivanandha Prabu, and R. Paskaramoorthy, "Interfacial thermal resistance and the solidification behaviour of the Al/SiCp composites," Mater. Manufacturing Processes 30(3), 381–386 (2015), https://doi.org/10.1080/10426914.2013.872267..
14. Ram Kumar S., Ramakrishnan S., Risvak M., Thauffeek S. and Yuvaraj T., 2021 "Experimental analysis and characterization of Mechanical, Physical properties of Aluminium (Al6061) Metal Matrix composite reinforced with SiC and Al_2O_3 using Stir casting".
15. Senthil Murugan S., 2017 "Development of Hybrid Composite for Automobile Application and its Structural Stability Analysis Using ANSYS" International Journal of Modern Studies in Mechanical Engineering (IJMSME) Volume 3, Issue 1, 2017, PP 23–34.
16. D Velmurugan, J Jayakumar, A Bovas Herbert Bejaxhin, JB Raj, Experimental Investigation on the Mechanical Properties of Hybrid Composites Made with Banyan and Peepal Fibers, Journal of Natural Fibers 19 (16), 14183–14194.
17. PV Narashima Rao, P Periyasamy, A Bovas Herbert Bejaxhin, Fabrication and Analysis of the HLM Method of Layered Polymer Bumper with the Fracture Surface Micrographs, Advances in Materials Science and Engineering 2022.
18. P. Gurusamy, S. Balasivanandha Prabu, and R. Paskaramoorthy, "Influence of processing temperatures on mechanical properties and microstructure of squeeze cast aluminium alloy composites," Mater. Manufacturing Processes 30(3), 367–373 (2015), https://doi.org/10.1080/10426914.2014.973587.

19. P. Gurusamy, B. Bhattacharjee, H. Dutta, A. Bhowmik, "Study of Microstructural, Machining and Tribological Behaviour of AA-6061/SiC MMC Fabricated Through the Squeeze Casting Method and Optimized the Machining Parameters by Using Standard Deviation-PROMETHEE Technique," Silicon (2023), https://doi.org/10.1007/s12633-023-02707-w.
20. T Prabaharan, P Periyaswamy, V Mugendiran, AB Herbert Bejaxhin, Measuring Deformation of Deep Drawing of Various Alloys by Image Processing using Matlab. Journal of Mines, Metals & Fuels 70.
21. D. Dinesh, P. Gurusamy, R.D.S. Kumar, "Influence of Nanosilica and Layering Sequence on the Mechanical Properties of Kenaf and Banyan Fibers Reinforced Composites," Silicon (2023), https://doi.org/10.1007/s12633-023-02460-0.
22. T Prabaharan, AB Herbert Bejaxhin, P Periyaswamy, N Ramanan, Lower Wishbone Modeling and Analysis for Commercial Vehicle Independent Suspension System. Journal of Mines, Metals & Fuels 70.
23. T Prabaharan, MBS Reddy, H Bejaxhin, A Bovas Herbert Bejaxhin, Strain-cum-Deformation Analysis of Friction Stir Welded AA5052 and AA6061 Samples with Microstructural Analysis., Journal of Mines, Metals & Fuels 70.
24. B Venkataraman, ABH Bejaxhin, R Saravanan, A Comparative Analysis on Surface Roughness of Plain Epoxy Composite with Reinforcement of 10 wt% of Pista Shell Particles Novel Composite using CNC Machining, Journal of Pharmaceutical Negative Results, 376–385.
25. M Sivasai, ABH Bejaxhin, R Saravanan, A Comparative Analysis on Material Removal Rate of Plain Epoxy Composite with Reinforcement of 15 wt% of Egg Shell Powder particles Novel composite using CNC Machining, Journal of Pharmaceutical Negative Results, 692–699.
26. B Venkataraman, ABH Bejaxhin, R Saravanan, A Comparative Analysis on MRR of Plain Epoxy Composite with Reinforcement of 10 wt% of Pista Shell Particles Novel Composite using CNC Machining, Journal of Pharmaceutical Negative Results, 367–375.
27. RD Babu, P Gurusamy, ABH Bejaxhin, P Chandramohan, Influences of WEDM constraints on tribological and micro structural depictions of SiC-Gr strengthened Al2219 composites, Tribology International 185, 108478.
28. G Mahesh, D Valavan, N Baskar, A Bovas Herbert Bejaxhin, Parameter Impacts of Martensitic Structure on Tensile Strength and Hardness of TIG Welded SS410 with characterized SEM Consequences, Tehničkivjesnik 30 (3), 750–759.
29. CB Priya, K Ramkumar, V Vijayan, ABH Bejaxhin, Wear Studies on Mg-5Sn-3Zn-1Mn-xSi Alloy and Parameters Optimization Using the Integrated RSM-GRGA Method, Silicon 15 (8), 3569–3579.
30. P Chandramohan, A SivaRangar, JJ Kingsly, ABH Bejaxhin, N Ramanan, Investigation of corrosion and wear behavior of Al–SiC composite, Materials Today: Proceedings.
31. B Venkataraman, ABH Bejaxhin, R Saravanan, Analyse the effect of 20wt% addition of Pista Shell Powder reinforcement to Epoxy Composite in the Material Removal Rate using CNC Machining and Compare with Plain Epoxy Composite, Journal of Survey in Fisheries Sciences 10 (1S), 3105–3115.

Note: Every figure and table was created using the experimental work; none of them were taken from any publications or online.

Advances in Additive Manufacturing Technologies – Gurusamy Pathinettampadian (eds)
© 2024 Taylor & Francis Group, London, ISBN 978-1-032-90013-1

31 Design, Thermal, and Static Investigations of Disc Brake Systems in Automotive and Bicycle Applications for Enhanced Performance

V Nagarajan*, R. Anbazhagan,
S Hari Krishna Raj, S. Aswin, G. Gokul, U. Gokulakrishnan
Assistant Professor,
Veltech High Tech Dr RR & Dr SR Engineering College, Chennai, India

ABSTRACT: A brake functions as a mechanical system linked to a mobile component of a machine, simulating friction to ensure safety and bring the machine to a stop. Currently, brakes are engineered to absorb either the kinetic energy of the moving part or the potential energy released by descending objects, such as in elevator systems. These brakes capture the energy and release it in the form of heat. In the automotive industry, a common application is the utilization of disc brakes on both vehicle and bicycle wheels. In this configuration, a disc is positioned between two pads that are activated by cylinders and supported by a caliper attached to the stud shaft. When the brake lever is pressed in a hydraulic system, pressurized fluid enters the chambers, compelling the opposing cylinders and brakes into action.

KEYWORDS: Disc brake, Static analysis, Thermal analysis, Finite element method

1. INTRODUCTION

In today's rising car industry, competition for improved vehicle performance is increasing dramatically. Disc brakes are critical components for slowing down a vehicle [1-5]. The primary goal of this literature review is to better understand the brake cooling system, thermo fluid mechanism, structural stiffness, and low stress levels. A brake is a device that is used to halt the motion of a machine by converting the vehicle's kinetic energy into thermal energy via mutual slippage of contacting brake components. Brake systems are commonly made from materials like cast iron or ceramic composites, including aluminum, Kevlar, and silica. Friction materials, commonly referred to as brake pads, are specifically engineered to make contact with both sides of the disc in order to decelerate the system.

When the pad is pushed normally against the disc, the disc absorbs the vehicle's kinetic energy and transforms it to thermal energy, with the disc absorbing 90% and the pads and calliper absorbing the remaining 10%. When a material's temperature exceeds its critical value, it results in brake failure, thermal cracking, disc wear, and fade. Using a misuse analysis software system, we will compute the value of friction contact power, nodal displacement, and buckling under different pressure scenarios and then utilize that value to choose the optimum material for the disc brake with the longest life [6-9]. A preliminary structural study of the rotor disc brakes suggests that grey cast iron elements are required to enhance braking efficiency and increase vehicle stability. When the pad is pushed properly against the disc, the disc absorbs and converts the vehicle's kinetic energy to thermal energy, with the disc absorbing 90% and the pads and calliper absorbing the remaining 10%. When

*Corresponding author: nagarajan.21892@gmail.com

DOI: 10.1201/9781003545774-31

a material's temperature rises over its critical value, it results in brake failure, thermal cracking, disc wear, and fade. Finite Element Analysis (FEA) is a powerful engineering tool used to simulate and analyze the behavior of structures and components under various conditions. It aids in optimizing designs, predicting performance, and identifying potential weaknesses in materials. FEA is crucial in industries such as aerospace, automotive, and civil engineering for accurate structural assessments[10,11].

Utilizing a misuse analysis software system, we intend to calculate the friction contact power, nodal displacement, and buckling values under various pressure scenarios. Subsequently, these values will guide the selection of the most suitable material for the disc brake, ensuring optimal longevity. A preliminary examination of the structural aspects of rotor disc brakes suggests the necessity of gray cast iron components to enhance braking performance. A comparison of maximum temperature outcomes for vented and solid discs led to the conclusion that ventilated discs are more crucial. Thilak [12] explored the significance of disc brakes and the requisite materials for their construction. Although cast iron is conventionally used in disc brakes, recent advancements have shifted towards carbon-carbon composites (CCC) or ceramic matrix composites (CMC). Thilak compared the results of transient thermo-elastic studies to repeated braking. In a related study, Sarip [13] devised a model of a disc brake unit, inclusive of brake pads and a wheel hub secured with a nut bolt. Applying pressure to simulate the frictional braking operation, Sarip highlighted the essential role of friction in halting a vehicle, with the brake pads serving as the necessary friction material to reduce the vehicle's speed.

Evaluating stress across different materials such as aluminum, grey cast iron, HSS M42, and HSS M2, the researcher determined that aluminum stands out as a well-suited material with low stress levels, offering enhanced performance in design [14-16]. In the examination of von Mises stresses in disc brakes, conducted with varying materials and both vented and solid disc rotors, the experimental findings indicated that vented cast iron disc brakes emerged as the optimal choice. These brakes met the strength and stiffness criteria required for safety conditions [17]

A transient-thermal analysis of disc brakes involved the examination of various materials, such as stainless steel, cast iron, and carbon-carbon composites. Deformation, stresses, and temperature variations were measured during the study. The conclusive findings highlighted that stainless steel outperforms other materials in terms of braking performance, particularly in minimizing deformation, while cast iron excels in stress-related aspects [18-22]. Another study focused on the thermal and structural analysis of disc brakes with different cut patterns, incorporating aerodynamic flow through disc brake holes. He selected elliptical cut patterns over circular cut patterns because they improve heat conduction but are less resistant to braking force. Low stress materials were chosen based on their stress concentration and structural design. It specified that in an emergency, brakes must stop in the shortest feasible distance while not fading or degrading. Finite element method for inducing stress concentration, structural deformation, and disc brake thermal gradient.

To minimize squeal noise in disc brake studied that brake scream does not effect braking system performance even if it is undesirable. Various factors that influence brake squeal such as braking pressure, rotational velocity, coefficient of friction, damping, and disc and pad modification. The most effective strategy for reducing disc brake squeal was to enhance disc and pad structural modification[10]. A key difficulty with disk braking systems is that while the vehicle is stopped, there is no air movement in the wheel assembly, and heat is transferred to the surroundings by natural convection. However, as previously noted, although conduction is an efficient method of heat movement, it may have detrimental consequences for nearby components. One of these difficulties is brake fluid vaporization, which happens as a consequence of heat exposure and is increased while the vehicle is parked. As a consequence, it is recommended to use brake fluid with a proper DOT rating[11].

For fluctuating temperatures, thermography photographs are processed interactively. Temperatures for different operating regimes may be determined by capturing and analyzing thermograms of a brake disc heated within the laboratory using an external heating source. Axial heat conduction may be neglected in common channels with small wall thicknesses in compared to channel diameters. In small channels with wall thicknesses proportionate to channel dimensions, axial heat conduction cannot be ignored since it lowers the quantity of heat transferred to the fluid.

1.1 Working and Construction

Disc brakes are now prevalent on large and heavy road vehicles, a shift from the previous widespread use of massive drum brakes. This transition is attributed to several factors. Notably, the absence of self-assist in disc brakes contributes to a significantly more predictable brake force, allowing for an increase in peak brake power while reducing the risk of braking-induced steering issues or jackknifing, particularly in articulated vehicles. Additionally, disc brakes exhibit less fading when operating under high temperatures. Given that air resistance, rolling drag, and engine braking contribute minimally to the total braking force on larger vehicles, disc brakes can be applied more forcefully than on lighter cars, mitigating the risk of drum brake fade within a single braking event.

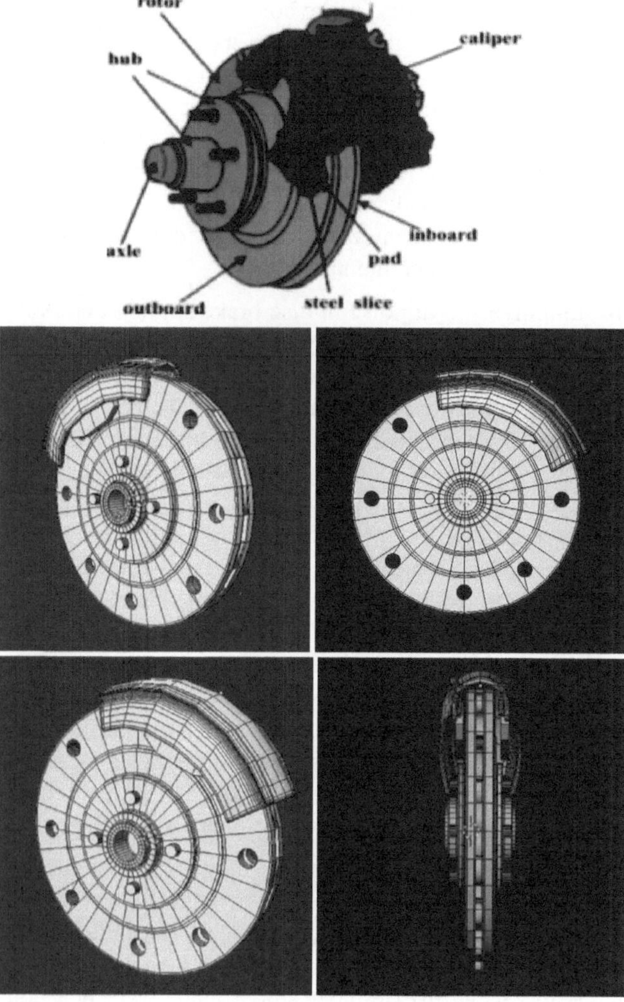

Fig. 31.1 Solid modelling of brake drum assemble in ansys workbench

In ANSYS Workbench, two types of deformation are considered: total deformation and directional deformation, both employed in calculating displacement resulting from tension. Directional deformation manifests in the X, Y, and Z directions and is determined as the square root of the sum of the squares of the deformations in each direction (Total deformation = SQRT(X^2 + Y^2 + Z^2)). Elastic Strain, a key parameter, is intricately linked to equivalent stress under the condition $v' = v$ (specified as PRXY or NUXY on MP command). The relationship is expressed as follows: σeq = equivalent stress (obtained as SEQV).

Tensile stresses are produced when the forces P stretch the bar; compressive stresses are produced when the forces are reversed, causing the bar to compress. Normal stresses are stresses exerted in a direction perpendicular to the cut surface, inducing stretching in the direction of the stress and contraction in the two perpendicular directions. These deformations can be effectively described using strain. Employing thermal simulation enhances the understand-

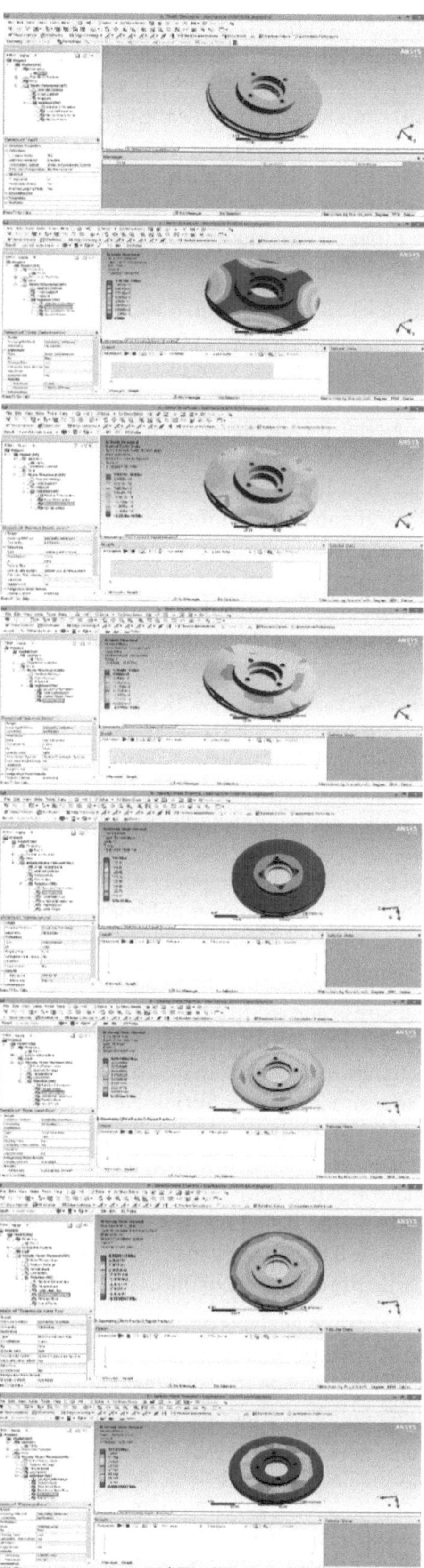

Fig. 31.2 Simulation results of brake drum assembly with ANSYS

ing of a product's response to diverse temperature conditions, enabling engineers to assess the potential impact of temperature fluctuations on their designs. This knowledge empowers teams to promptly implement design modifications, ensuring optimal product performance..

Heat flux is defined as the amount of heat transmitted per unit area per unit time to or from a surface. This concept is derived, combining two essential factors: the rate of heat transfer per unit time and the specific area involved in the heat transfer process. Thermal errors, on the other hand, are geometric inaccuracies resulting from temperature variations encountered during the manufacturing process.

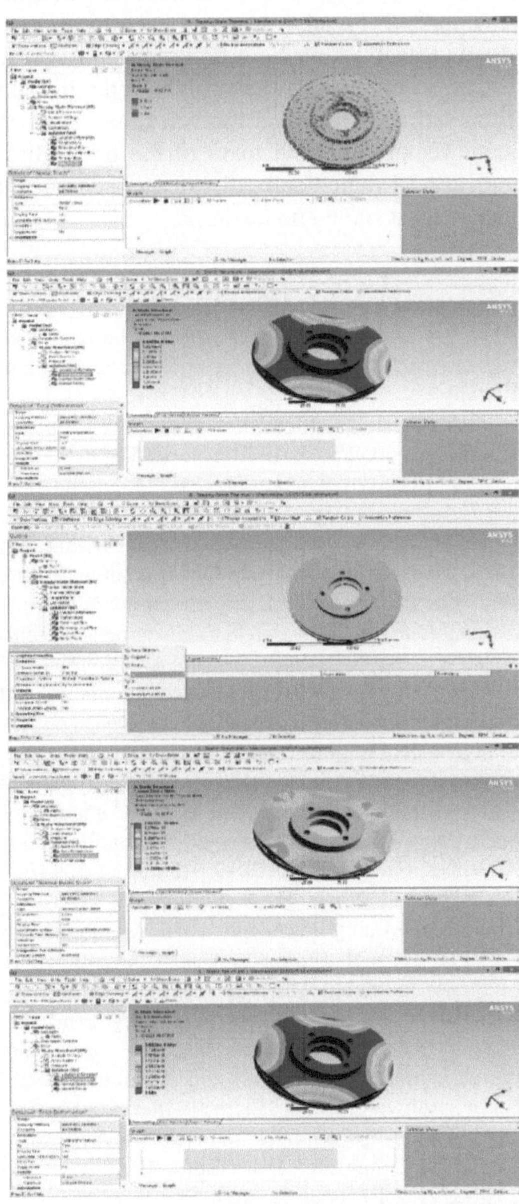

Fig. 31.3 Comparative simulation results of brake drum assembly with ANSYS

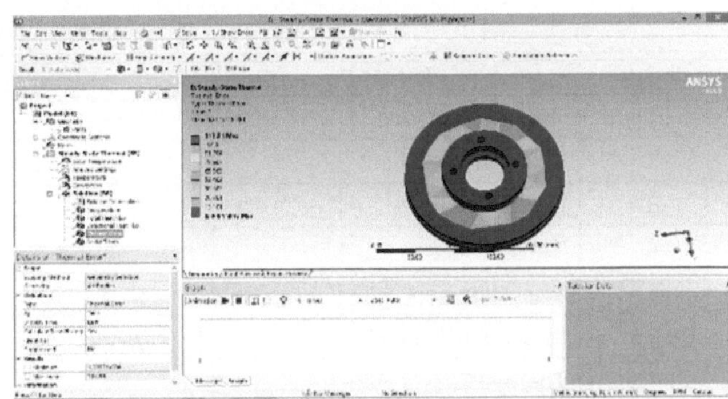

Fig. 31.4 Analysis of thermal error, nodal triads chrome carbide total deformation

Table 31.1 Materials and its melting temperature

S. No	Materials	Temperature
1	Aluminia	400°C
2	Titania	500-800°C
3	Zirconia	600°C
4	Chrome Carbide	400-500°C
5	Tantalum Carbide	2000°C

Calculation

Disc Brake Standard

Rotor disc dimension = 240 mm. (240×10^{-3}m)

Rotor disc material = Carbon Ceramic Matrix

Pad brake area = 2000 mm 2 (2000×10^{-6} m)

Pad brake material = Asbestos

Coefficient of friction (Wet) = 0.07-0.13

Coefficient of friction (Dry) = 0.3-0.5

Maximum temperature = 350°C

Maximum pressure = 1MPa(10^6Pa)

Tangential force between padand rotor (Innerface), FTRI

$$FTRI = \mu 1.\ FRI$$

Where, FTRI = Normal force between pad brake and Rotor (Inner)

$\mu 1$ = Coefficient offriction

= 0.5FRI

= Pmax/2 × A pad brake area

FTRI = $\mu 1$. FRI

FTRI = (0.4) (0.4)(1×10^6 N/m^2)(2000×10^6m^2)

FTRI = 400 N.

Equal coefficients of friction and normal forces FR on the inner and outer faces:

TB = FT.R

= FTRI + FTROFT = 1000 N.

R = Radius of rotor disc.

Table 31.2 Simulation outcomes of each material modeling

S. No	Particles	AL	TI	ZI	STEEL
1	Total Deformation	9.1669*E-10	9.43*E-10	9.7281*E-10	10.1*E-10
2	Normal Elastic Strain	1.8118*E-12	1.86*E-12	0.19228*E-12	-
3	Normal Stress	6.52*E-6	6.52*E-6	6.529*E-6	-
4	Temperature	129.75	129.75	129.77	129.91
5	Total Heat Flux	0.004006	0.0040063	0.0040103	0.003998
6	Directional Heat Flux	0.0020742	0.0020742	0.0020743	0.0020741
7	Thermal Error	84.855	84.855	75.008	105.66

So, $TB = (1000)(120 \times 10^3)$

$TB = 120$ Nm

Brake Distance (x)–

We know that tangential braking force acting at the point of contact of the brake, and Work done = FT.x (Equation A)

Where FT = FTRI + FTROX = Distance travelled (in meter) by the vehicle before it come torest.

We know kinetic energy of the vehicle.

Kinetic energy = (mv2)/2 (Equation B)

Where m = Mass of vehicle
v = Velocity of vehicle

To bring the vehicle to a stop, the work done against friction must be equivalent to the vehicle's kinetic energy. As a result of equating (Equation A) with (Equation B), $FT.x = (mv^2)/2$

Assumption v = 100kg/h = 27.77m/s
M = 132 kg. (Dry weight of vehicle).

Carbon Ceramic Matrix –Heat generated $Q = m * cp * \Delta T$
Mass of disc = 0.5kg

Specific Heat Capacity = 800 J/kg ^{0}c

Time taken Stopping the Vehicle = 5 sec

Developed Temperature difference = 15^0cQ = 0.5 * 800 * 15 = 6000J

Area of Disc = $\Pi * (R^2 - r^2) = \Pi * (0.120^2 - 0.055^2)$
$= 0.03573\text{m}^2$

Heat Flux = Heat Generated/Second/area

$= 6000/5/0.0357 = 33.585$ kw/m^2

Thermal Gradient = Heat Flux/Thermal Conductivity
$= 33.582 * 10^3/40$
$= 839.63$ k/m

2. Conclusion

This document describes how to design straight and vented brake discs in Solid Works. It also comprises the deck preparation in hyper mesh, i.e., the meshed part with temperatures applied. Finally, Ansys is used to do a steady-state thermal analysis on both brake discs. According to these findings, converting the straight vents in the brake disc to curved vents lowered the von mises stresses, displacement vector sum, and mass of the brake disc. Furthermore, curved vented brake discs generate more thermal flux than straight vented brake discs.

References

1. Abhay, A. Utpata, P. B. Pawar, 2018 "Modeling and Analysis of Al-SiC Spur Gear under Static Loading Condition", Elsevier.
2. Ajith Arul Daniela, 2018 "Dry Sliding Wear Behaviour of Aluminium 5059/SiC/Al2O3 Hybrid Metal Matrix Composites" Received: January 06, 2017; Revised: August 11, 2017; Accepted: August 24, 2018.
3. Ajmal Hussain S., Rajaneesh R., Hashim Nizam, Jithin K., 2020 "Experimental Analysis On Aluminum Alloy (6063) With Silicon Carbide: An Experimental Investigation", International Journal of Research in Engineering and Technology.
4. Anish A., 2019 "Characterization of Aluminium Matrix Reinforced With Tungsten Carbide And Aluminum oxide Hybrid Composite" 2nd International conference on Advances in Mechanical Engineering (ICAME 2019).
5. Antony Arul Prakash M. D., 2019 "Microstructural Analysis of Aluminium Hybrid Metal Matrix Composites Developed Using Stir Casting Process" International Journal of Advances in Engineering, 2019, 1(3), 333 – 339.
6. EhsanGhasali, 2018 "Production of Al-SiC-TiC Hybrid Composites Using Pure and 1056 Aluminum Powders Prepared Through Microwave and Conventional Heating Methods" Journal of Alloys and Compounds.
7. Kanchan A., 2019 "Tribological Investigation of Al7075/Tic/Al2O3 Hybrid Composite Material" International Research Journal of Engineering and Technology (IRJET).
8. Karuppasamy R., 2020 "Taguchi-GRA for Multi Criteria Optimization of Turning Parameters for Al/Bagasse Ash/Gr Hybrid Composite" Journal of Critical Reviews ISSN-2394-5125 Vol 7, Issue 9, 2020.
9. Nagesh D., Mr. Shashikanth G. S., 2021 "Optimization of Tribological Properties of Al6061, Boron And Graphite

MMCs Using Stir casting Method.", IOSR Journal of Mechanical and Civil Engineering.

10. Paranthaman P., 2020 "Multi-Objective Optimization on Tribological Behaviour of Hybrid Al Mmc by Grey Relation Analysis" Journal of Critical Reviews ISSN- 2394-5125 Vol 7, Issue 9, 2020.
11. Rajesh A. M., Mohamed Kaleemulla, 2019 "Material characterization of SiC and Al2O3–Reinforced Hybrid Aluminum Metal Matrix Composites on Wear Behavior", Advanced Composites Letters Volume 28: 1–10.
12. 12. Rajesh S., 2019 "Influence and Wear Characteristics of TiC Particles in Al6061 Metal Matrix Composites" International Journal of Applied Engineering Research ISSN 0973-4562 Volume 13, Number 9 (2019) pp. 6514-6517.
13. Rajesh Prabha N., 2017 "Effect of TiC and Al2O3 Reinforced Aluminium Metal Matrix Composites on Microstructure and Thermogravimetric Analysis" Vol. 10 | No. 3 |729 - 737 | July - September | 2017 ISSN: 0974-1496 | E-ISSN: 0976-0083 | CODEN: RJCABP.
14. Ram Kumar S., Ramakrishnan S., RisvakM.,Thauffeek S. and Yuvaraj T., 2021 "Experimental analysis and characterization of Mechanical, Physical properties of Aluminium (Al6061) Metal Matrix composite reinforced with SiC and Al2O3 using Stir casting".
15. P. Gurusamy and S. Balasivanandha Prabu, "Effect of the squeeze pressure on the mechanical properties of the squeeze cast Al/SiCp metal matrix composite," Int. J. of Microstructure and Materials Properties 8(4–5), 299–312 (2013), https://doi.org/10.1504/IJMMP.2013.057067.
16. P. Gurusamy, S. Balasivanandha Prabu, and R. Paskaramoorthy, "Interfacial thermal resistance and the solidification behaviour of the Al/SiCp composites," Mater. Manufacturing Processes 30(3), 381–386 (2015), https://doi.org/10.1080/10426914.2013.872267.
17. P. Gurusamy, S. Balasivanandha Prabu, and R. Paskaramoorthy, "Prediction of Cooling Curves for Squeeze Cast Al/SiCp Composites Using Finite Element Analysis," Metall. Materials Transactions A: Physical Metallurgy and Materials Science 46(4), 1697–1703 (2015), https://doi.org/10.1007/s11661-015-2742-6.
18. P. Gurusamy, S. Balasivanandha Prabu, and R. Paskaramoorthy, "Influence of processing temperatures on mechanical properties and microstructure of squeeze cast aluminium alloy composites," Mater. Manufacturing Processes 30(3), 367–373 (2015), https://doi.org/10.1080/10426914.2014.973587.
19. P. Gurusamy, B. Bhattacharjee, H. Dutta, A. Bhowmik, "Study of Microstructural, Machining and Tribological Behaviour of AA-6061/SiC MMC Fabricated Through the Squeeze Casting Method and Optimized the Machining Parameters by Using Standard Deviation-PROMETHEE Technique," Silicon (2023), https://doi.org/10.1007/s12633-023-02707-w.
20. P. Gurusamy, S. Hari Krishna Raj, B. Bhattacharjee, A. Bhowmik, "Assessment of Microstructure and Investigation Into the Mechanical Characteristics and Machinability of A356 Aluminum Hybrid Composite Reinforced with SiCp and MWCNTs Fabricated Through Rotary Centrifugal and Squeeze Casting Processes," Silicon (2023), https://doi.org/10.1007/s12633-023-02686-y.
21. B. Bhattacharjee, N. Biswas, P. Chakraborti, K. Choudhuri, P. Gurusamy, "Impact of Nano-lubrication on Single-Layered Porous Hydrodynamic Journal Bearing: A Hypothetical Approach," Journal of Porous Media (2023), DOI: 10.1615/JPorMedia.2023044960.
22. D. Dinesh, P. Gurusamy, R.D.S. Kumar, "Influence of Nanosilica and Layering Sequence on the Mechanical Properties of Kenaf and Banyan Fibers Reinforced Composites," Silicon (2023), https://doi.org/10.1007/s12633-023-02460-0.

Note: Every figure and table was created using the experimental work; none of them were taken from any publications or online.

Advances in Additive Manufacturing Technologies – Gurusamy Pathinettampadian (eds)
© 2024 Taylor & Francis Group, London, ISBN 978-1-032-90013-1

32 Design and Development of Cost Effective Electronic Ignition Driver Circuit for GDI Engine

N. Shivakumar[1]
Department of Mechanical Engineering,
Chennai Institute of Technology, Chennai, India

R. Lalitha[2]
Department of Computer Science and Engineering,
Sathyabama Institute of Science and Technology,
Chennai, India

V. Sathya[3]
Department of Computer Science and Engineering,
VelTech Rangarajan Dr. Sagunthala R & D
Institute of Science and Technology,
Chennai, Tamilnadu, India

M. Dinesh Babu[4]
Department of Mechanical Engineering,
Rajalakshmi Institute of Technology, Chennai, India

R. Sathiyamoorthi[5]
Department of Mechanical Engineering,
Chennai Institute of Technology, Chennai, India

ABSTRACT: In recent developing automotive vehicles, electronically controlled ignition timing control is one of major control parameter towards superior combustion process and emission control strategy. The cost of the entire ignition driver circuit is higher and it's merged with fuel injection driver circuit along with microcontroller board. Need to be replacing the entire electronic control unit instead of replacing only the fault ignition driver component. To overcome this disadvantage, the cost operative electronically controlled ignition driver circuit was developed and tested. With the developed ignition driver circuit, ignition timing was controlled using Arduino Uno-controller, Hall Effect sensor used as CAM/Crank position sensor, and L298N motor driver circuit used to trigger the time of ignition through coil on plug type ignition coil. Using these cost operative electronic components, the ignition timing was varied effectively at different crank angles using gasoline direct injection engine.

KEYWORDS: Ignition control, Ignition timing, Ignition driver, Automotive ignition, Ignition system, COP ignition coil

[1]shiva.thermal@gmail.com, [2]lalitha.cse@sathyabama.ac.in, [3]saro.sath@gmail.com, [4]dinesh198014@yahoo.com, [5]sathiya.ram78@gmail.com

DOI: 10.1201/9781003545774-32

1. Introduction

In recent launching petrol engine vehicles, the ignition system include COP (coil on plug) type ignition coil operating with electronically controlled ignition driver. With the inclusion of ECU (electronic control unit) for engine control, the recent emission norm vehicles were costlier than previous emission norm vehicles. To overcome this disadvantage, an ignition driver circuit was developed with cost benefit and high sensitivity. The computational model was developed towards optimized ignition system for automotive vehicles to predict the arc motion and restrike the phenomena coupled with spark ignition progression. This new model has the great impending to provide the requirements for better combustion of air-fuel mixture. A more effective electrical discharge rate from this computational model was appeared [1]. The digitally controlled spark ignition is the best alternative for the existing ignition control. Computerized digital control gives more accurate ignition timing for all operating conditions of an engine. The use of two spark plugs improves thermodynamic efficiency and power available; also it reduces the maintenance cost due to fewer moving parts so that less friction and wear. Because of more accurate ignition timing control, it leads to better combustion rate and improved power output with controlled emissions from an engine exhaust [2]. From the MATLAB Simulink results, the step up ignition coil converts the 12 v from battery to the high voltage essential for an engine towards complete combustion. This means that the current jump at the spark electrode gap to the air-fuel mixture between the two electrodes lead to high resistance to the passage of current. The effect of energy and ignition voltage on the spark duration, also the electrode gap and outline were examined. It was also analyzed the usage of ignition coil with lower resistance and inductance values, in addition spark plugs with and without in-house resistances [3] [4]. With the data necessary for electromagnetic finite element analysis, as well as material failure data, the magnetic and electrostatic relations among the coil laminations and windings was modeled and field computations are transformed to the corresponding circuit elements engaged in a systems model. The systems model allows the estimation of required transient signals. This approach is extremely flexible and provides quick simulations [5]. A highly functional proposed ECU was developed and installed on a Honda GX31 engine to accommodate an electromechanically controlled ignition and injection timings. Workbench essays demonstrated the engine optimization and the feasibility of its control with the proposed ECU up to 9500 rpm. To our best knowledge, such a low cost ECU is unique, innovative and will allow the widespread of the electronic control in many low-cost equipment with IC engines, such as agricultural tools, electrical generators and karts, to name a few. Ultimately, results will be an improved user experience and a more efficient use of earth resources [6]. Experiments were conducted in an optical, single-cylinder compression ignition heavy-duty engine which includes a bowl on the piston crown that was utilize for spark ignition control action. The spark flame size and its growth rate from the start of ignition until the flame overflowing the camera play of view were interrelated to combustion parameters extracted from the data acquisition system with in-cylinder pressure transducer, under medium load, low speed and lean-mixture conditions [7]. With reduced fuel consumption, sport engines are operated with a lean air–fuel mixture to benefit from higher combustion efficiency, require an ignition system which fit for the combustion of lean mixture. Pre-ignition chamber is recognized as a potential technique to progress lean limit and has the prospective to minimize end gas auto-ignition. This study proposed the performance of high load lean burn operation of GDI engine and a reactive pre-combustion chamber was developed using CFD tool [8].

For the given CR (compression ratio), start of ignition is the most significant factor while defining the ignition system essential for innovative gasoline combustion modes. Also, the developed combustion system must identify the problems related to the location of spark along with its intensity and its released energy required for the novel gasoline engines with different combustion modes [9]. Due to high output voltage of around 25 kilo volts and above may be bypassed through the body if earth connection was not done accurately. It results in strict health issue and to make sure the safety for technicians, intended to fabricate a circuit which displays the current flow with help of LED glow [10]. Using the conventional diagnosis method of high voltage pulse waveforms, analyzed the error analysis of ignition system which include evaluation of the electronic control system and its components [11]. Controlled ignition driver transistor answerable for the function of the creative ignition coil which ensure the right electrical parameters of the ignition driver components, which decide the optimal operating conditions of the ignition coil, such as the resistance and inductance of both coil windings, the secondary coil winding voltage, the energy released and spark efficiency [12]. For the validation process of ECU, the correct angular function is required for the the injection and ignition driver. This study provides the required angular position and its proper function with the help of developed code using crank angle input, injection and ignition driver circuits with microcontroller [13]. Improvement of the engine performance and reduction the exhaust emissions from engine was obtained by increasing the number of coil windings and use of novel approach power electronics system for combustion control. This revision provides coil characteristics and all its improved electrical relations using MATLAB software package [14]. In order to maintain the optimized temperature distribution throughout the combustion chamber and the related heat transfer co-relations within the combustion chamber surface/volume towards NOx emission and PM emission reduction, the piston crown profile is one of the important parameter [15-17]. Also the control of ignition timing is the control parameter to obtain better combustion with electronically controlled effective ignition driver circuit [18-24]. To achieve exact ignition timing towards superior combustion of air-fuel mixture for higher engine combustion performance and reduced emissions, the effective ignition driver circuit was developed with cost benefit. The developing ignition driver circuit is much cheaper than any other make and can be replaced with faulty one if separate ignition and fuel injection drivers were installed previously. This developed ignition driver can be used for both MPFI and GDI engines with different ignition timings.

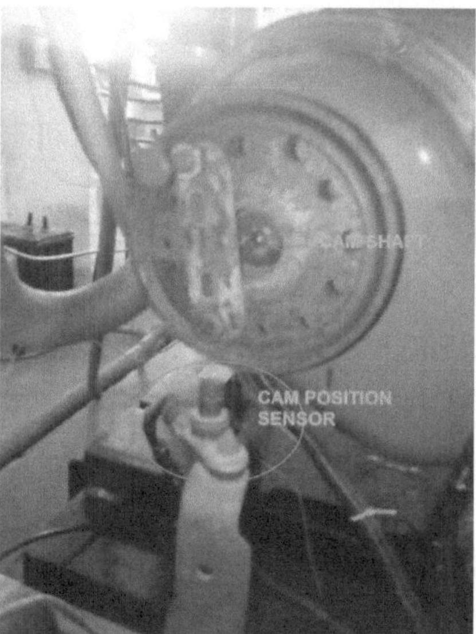

Fig. 32.1 Assembled CAM position sensor towards cam shaft

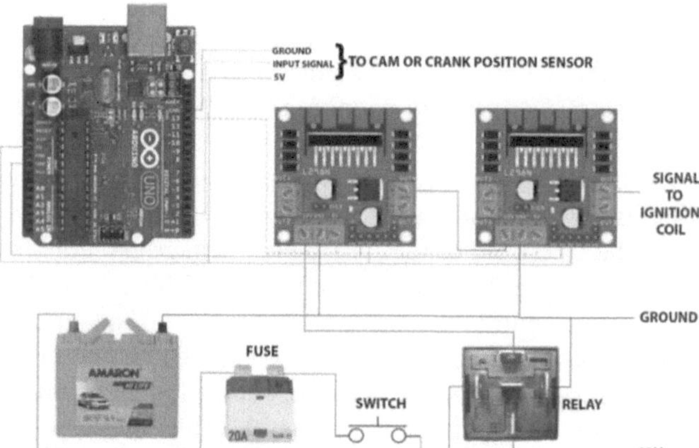

Fig. 32.2 Schematic diagram of electronic ignition driver circuit

2. Materials and Methods

The amount of voltage from the ignition coil to spark plug is an important parameter in order to provide effective spark required for the superior combustion of air-fuel mixture towards maximum thermal efficiency and lower emissions from an engine exhaust. Also the correct/exact ignition timing is important control variable for maximum thermal efficiency, lower fuel consumption and lower emissions. On the other hand, the cost of the electronic control ignition driver circuit should be least as much as possible with the effective output voltage range and correct ignition timing. CAM position sensor was utilized to sense the top dead position of piston inside an engine cylinder. Fig. 32.1 shows the assembled CAM position sensor towards cam shaft in such a way that the magnetic material is fitted over the cam shaft flange exactly at TDC position, and the gap between the sensor tip and the magnetic material is maintained between 0.8 to 1.5 mm. During rotation of cam shaft, CAM position sensor sends the TDC position signal to Arduino UNO controller and the output signal with respect to time is sent to L298N motor driver circuits as shown. The schematic diagram of complete electronic ignition driver circuit is as shown in Fig. 32.2 where the signal line, 12v supply from battery and ground lines were connected to COP ignition coil. Ford Eco-boost engine COP type ignition coil was used, the tree pins of ignition coil such as signal, 12v and ground are connected with the corresponding lines from the ignition driver circuit. The 20A fuse was used to regulate the sufficient current supply to the ignition driver circuit. The 12v supply ON and OFF switch was used to avoid any fault with the engine during running operation. 12v relay was used as switch in order to trigger the signal from driver circuit to the ignition coil.

3. Experimental Work

Figure 32.3 shows the assembled ignition driver circuit to control the ignition timing and to supply the sufficient voltage the ignition coil. 12 V and ground lines from battery were connected to the motor driver circuits through 12 V relay. 5 V and ground lines from Arduino Uno board were connected directly to the motor driver circuits as shown. The AC supply is connected with Arduino Uno board through 9 volt 2A adaptor/regulator and the coding for the control of ignition timing was loaded into the controller. This developed electronic ignition driver circuit was tested on 5 hp, four-stroke, 10:1 compression ratio, single cylinder modified GDI engine with COP type ignition coil. Bosch make GDI engine spark plug was installed through cylinder head after welded the sleeve. The weld cracks/blow holes were arrested with the help of cold-weld metallic paste prepared with epoxy-additives [19]. Arduino Uno board with entire connections along with Crank and CAM position sensors input signal, the COP type ignition coil generates around 38,000 volts where the ignition coil acts as step-up transformer. The generated voltage output from the COP ignition coil was viewed through oscilloscope device.

Fig. 32.3 Assembled electronic ignition driver circuit

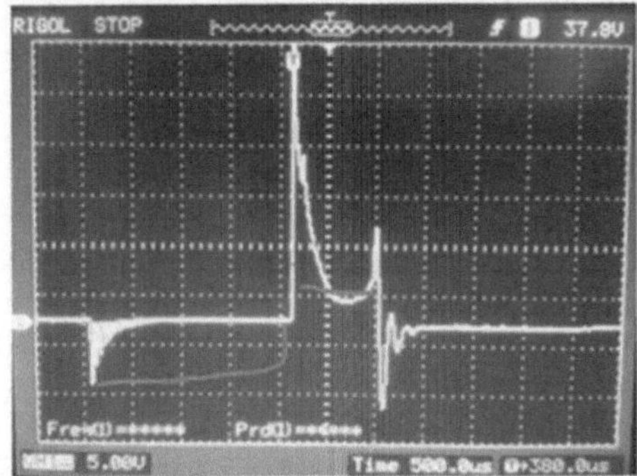

Fig. 32.4 Signal from COP type ignition coil

4. Results and Discussion

From the developed electronic ignition driver circuit, the output voltage variation per working cycle of an engine from the COP type ignition coil was seen using oscilloscope as shown in **Fig. 32.4**. From the graphical image it was shown that around 38,000 volts generating through COP type ignition coil. From the literature survey, it is clear that generated voltage from the ignition coil is sufficient for gasoline direct injection engines towards superior combustion of air-fuel mixture. The ignition timing was controlled through the coding which was fed into the Arduino controller where the delay time was varied in order to control/optimize the ignition timing. The optimized ignition timing results in reduction of engine knock, improved thermal efficiency and reduced concentration of emissions from engine exhaust.

4.1 Ignition Timing Calculations

The ignition timing was tested with Kirloskar 5 hp, 4-stroke, modified GDI engine at constant speed of 1500 rpm using CAM position sensor and Ford Eco-boost engine COP ignition coil. The time duration for each stroke and each crank angle at 1500 rpm crank speed were calculated as follows:

Engine speed = 1500 rpm

For 4-stroke engine, $\frac{N}{2} = 750$ cycles per minute

$$\frac{750}{60} = 12.5 \text{ cycles per second}$$

$$\frac{1}{12.5} = 0.08 \text{ seconds per cycle}$$

$$\frac{0.08}{4} = 0.02 \text{ seconds per stroke}$$

Time duration for each stoke = 20 milli seconds

Time duration for each crank angle = 0.112 milli seconds. By knowing the time duration for each crank angle, the delay time in the Arduino coding was varied for optimizing the ignition timing.

4.2 Cost Analysis af Ignition Driver Circuit

The cost of electronic control ignition driver circuit is also one of the major parameter towards the cost of an engine (or) vehicle; otherwise the replacement cost of this driver circuit alone. In this design, the developed ignition driver circuit is cost operative one with better effectiveness towards ignition timing control. Table 32.1 shows the cost details of developed COP type electronic ignition driver circuit. From the cost details, it is clear that the developed ignition driver circuit was cost benefit one.

Table 32.1 Cost details of electronic ignition driver circuit

Particulars/Components	Quantity	Amount in Rs.
Arduino Uno board	1 No.	350/
L298N Motor Driver	2 No.	180/
12v, 30/40A Relay and 20A Fuse	Each 1 No.	130/
Switch and Connecting wires	1 kit	120/
Total (in Rupees)		**780/**

5. Conclusion

From the developed cost effective electronic ignition driver circuit, the output signal voltage from COP ignition coil was sufficient for superior combustion of air-fuel mixture in case of recent emission norms petrol engine vehicles. The ignition timing was controlled through Arduino controller coding towards engine knock control, better combustion and reduction of emissions from engine exhaust. The developed ignition driver circuit was cost benefit one than any other ignition control modules.

Also, the faulty ignition driver circuit can be replaced from an ECU instead of replacing an entire part of ECU if separate ignition driver and fuel injection drivers were installed previously.

ACKNOWLEDGEMENT

This work has been carried out at Chennai Institute of Technology, Chennai, India with the available research facilities inside the campus.

REFERENCES

1. Haiwen Ge, Peng Zhao, A comprehensive ignition system model for spark Ignition Engines, Internal Combustion Engine Division Fall Technical Conference 9574(2018), 1–9, https://doi.org/10.1115/ICEF2018-9574.
2. Dattatrey Zambre, Gaurav Shintre, Bhagyashri Patil, Digital Twin Spark Ignition Using Mechatronics, IOSR Journal of Mechanical and Civil Engineering (2014), 73–78.
3. Raid Anam Gaib, Ahmed Bassam Aziz, Ahmed Ibrahim Jaber Alzubaydy, The effect of internal ignition coil on combustion engine performance, Journal of engineering and applied sciences 14, 898–904.
4. Bruno Santos Goulart, Cleverson Bringhenti, Jesuíno Takachi Tomita, Antonio Carlos Oliveira, Technical evaluation of vehicle ignition systems: conduct differences between a high energy capacitive system and a standard inductive system, Acta Scientiarum Technology 36, 629–634, https://doi.org/10.4025/actascitechnol.v36i4.21048.
5. R. C. Stevenson, R. Palma, C. S. Yang, S. K. Park, C. Mi, Comprehensive Modeling of Automotive Ignition Systems, SAE technical paper series 1589(2009), 1–26.
6. P. Marinho, J. Oliveira, N. Pires, C. Ferreira, Low cost programmable electronic control unit for internal combustion engines, International Conference on Innovative Technologies (2014), 1–5, https://doi.org/10.13140/2.1.2237.2164.
7. Cosmin E. Dumitrescu, Vishnu Padmanaban, Jinlong Liu, An Experimental Investigation of Early Flame Development in an Optical Spark Ignition Engine Fueled With Natural Gas, Journal of Engineering for Gas Turbines and Power 140(2018), 082802–082809.
8. Muhammed Fayaz Palakunnummal, Priyadarshi Sahu, Mark Ellis, Marouan Nazha, Simulation-Aided Development of Prechamber Ignition System for a Lean-Burn Gasoline Direct Injection Motor-Sport Engine, Journal of Engineering for Gas Turbines and Power 142(2020), 085001–0850010, https://doi.org/10.1115/1.4047767.
9. R. Payri, R. Novella, A. Garcia, V. Domenech, A new methodology to evaluate engine ignition systems in high density conditions, Experimental Techniques 28(2014), 17–28, https://doi.org/10.1111/j.1747-1567.2012.00818.
10. Hari Nivethan M, Prawin Khumar S, Ajay Bharathi A, Spark plug wire voltage tester, International Research Journal of Modernization in Engineering Technology and Science 2(2020), no. 6, 413–419.
11. Milan Sebok, Matej Kucera, Miroslav Gutten, Daniel Korenciak, Thermal and electrical analysis of automotive ignition system, The Archives of Automotive Engineering 83(2019), 113–121.
12. Jednolita polaryzacja Świec, W Ukladach Zaplonowych, Wyposazonych W Cewki Dwubiegunowe, Uniform spark plugs polarity in automotive ignition circuits equipped with double ended ignition coils, The Archives of Automotive Engineering 76(2017), 63–74, http://dx.doi.org/ 10.14669/AM.VOL76.ART3.
13. Mansi K. Ajudia, Mahesh T. Kolte, Prasanta Sarkar, Validation Process and Development of Control Strategy of Electronic Control Unit for Injector and Ignition coil Drivers, International Journal of Scientific and Research Publications 4(2014), 1–5.
14. Ahmed I. Jaber Alzubaydy, Zuhair S. Al-Sagar1, Proff Dr. Lutfi Y. Zedan, Raed Anam Gaib , Sabbar Otaiwi Haddad, Improvement of Effect ignition Coil on Ignition System of Internal Combustion Engine Performance, American Journal of Engineering Research (AJER) 8(2019), no. 4, 274–284.
15. Nagareddy S, Temperature distribution measurement on combustion chamber surface of diesel engine-experimental method, International Journal of Automotive Science and Technology. DergiPark 1(2017), no. 3, 8–11.
16. Kumar, Shiva & Kumar, Ajeet & Sharama, Abhilash & Kumar, Amit, Heat transfer correlations on combustion chamber surface of diesel engine - experimental work, International Journal of Automotive Science and Technology. DergiPark 2(2018), no.3, 28–35. https://doi.org/10.30939/ijastech..434331.
17. Shivakumar N, Piston Crown Profile Modifications for Various Combustion Mode Strategies of Modified GDI Engine towards NOx and PM Reduction, International Journal of Automotive Science and Technology, DergiPark 4(2020), no. 4, 289-294. https://doi.org/10.30939/ijastech..796526.
18. N Shivakumar, G A C Jayaseelan, Parthiban, Ahmed, Akshay, Ignition Timing and Fuel Injection Timing Control using Arduino and Control Drivers, IOP Conf. Series: Materials Science and Engineering 993(2020), 1–8. https://doi.org/10.1088/1757-899X/993/1/012019.
19. Shivakumar N, Thermo-Mechanical Performance of Cold Weld Metallic Paste with Epoxy Additives, GEDRAG & ORGANISATIE REVIEW 33(2020), no. 2, 2685–2692. https://doi.org/10.37896/GOR33.02/292.
20. Nagareddy, S., Govindasamy, K. Influence of fuel system variations on performance and emission characteristics of combined air-wall-guided mode modified GDI engine with alcoholic fuels and exhaust gas recirculation. Environ Sci Pollut Res **30**, 61234–61245 (2023). https://doi.org/10.1007/s11356-022-21875-7.

21. Shivakumar N, Vijayakumar K, Akshay Surendran, Chris Jose, Safa Mahmood V. K, "Impact of blended fuels with split injections on combustion and emission characteristics of spray guided mode modified GDI engine", *AIP Conference Proceedings,* 2523, 020147 (2023). https://doi.org/10.1063/5.0113365
22. Shivakumar Nagareddy and Kumaresan Govindasamy. 2022. "Influence of piston crown shape with different positions of spark plug and fuel injector, %EGR, and fuel system control on emissions from modified GDI engines compared with a base diesel engine", Transactions of the Canadian Society for Mechanical Engineering. 46(2): 355–364. https://doi.org/10.1139/tcsme-2021-0163
23. Nagareddy, S., et al. "Combustion Chamber Geometry and Fuel Supply System Variations on Fuel Economy and Exhaust Emissions of GDI Engine with EGR", Thermal Science, 26(2A), (2022), 1207–1217. https://doi.org/10.2298/TSCI211020358N
24. Nagi, Shivakumar, Govindasamy, Kumaresan, Rajaram, Kamatchi, Combined Air-Wall Guided GDI Engine Performance and Emissions Control with E20 and Hydrogen Blended Fuel, Environ Sci Pollut Res. https://doi.org/10.21203/rs.3.rs-3207399/v1

Note: Every figure and table was created using the experimental work; none of them were taken from any publications or online.

Advances in Additive Manufacturing Technologies – Gurusamy Pathinettampadian (eds)
© 2024 Taylor & Francis Group, London, ISBN 978-1-032-90013-1

33 Effect of Circular and Chevron Nozzles to Reduce Infrared Signature of Exhaust Plume

G. Dinesh Kumar*
Assistant Professor (S.G), Department of Aeronautical Engineering,
Hindustan Institute of Technology & Science,
Chennai, Tamil Nadu, India

Clement Edward Vasanth A
M.TechDefence Technology, Department of Aeronautical Engineering,
Hindustan Institute of Technology & Science,
Chennai, Tamil Nadu, India

Gursharanjit Singh
Scientist, Gas Turbine Research Establishment (DRDO),
Bangalore

**Venkata Neeraj, Bindu B, Nivethitha P,
Shurdeka D, Kancharla Darpan, Sheryl Hepzibah S**
B. Tech Aerospace Engineering, Department of Aeronautical Engineering,
Hindustan Institute of Technology & Science,
Chennai, Tamil Nadu, India

ABSTRACT: This study presents a comprehensive computational fluid dynamics (CFD) analysis of nozzle performance in various configurations using k-ε realizable turbulence model in ANSYS FLUENT software. It represents the enhanced mixing of jets with particularly IR signature reduction. Following rapid mixing technologies for under-expanded circular and chevron nozzles with different lengths and inward penetrations were evaluated for operating at 3 NPR. It is showed that the mixing and reduction of infrared signature as much as possible. The simple geometric modifications were made with chevron nozzles to ensure the mixing of jet and necessary to reduce IR signature of exhaust plume. The baseline nozzle, chevron with/without penetration for both convergence angle 15 degree and 30 degree. Especially, chevron with penetration convergence angle 30 degree with 15 degree penetration has efficient mixing and reduction in plume length. The main purpose of reducing plume length at the exit is to escape from infrared seekers as much as possible from range of target. Finally, CFD analysis predicted the axial velocity, mass flow rate, and static temperature for all nozzle configurations. This analysis allowed for an understanding of the flow behaviors and performance of the nozzles under different conditions and configurations.

KEYWORDS: Chevron, Computational fluid Dynamics, Infrared signature, Nozzles

*Corresponding author: gdineshk@hindustanuniv.ac.in

DOI: 10.1201/9781003545774-33

1. INTRODUCTION

1.1 Motivation

Nowadays, the military aircrafts are developing some advance technology to protect their country from enemies. In that technology development, missile attacks are crucial thing especially infrared seeker missiles. The aircrafts are produce exhaust plume that causes infrared signature it helps to infrared seekers to seek the aircraft as the target. In earlier IR is not considered as much due to lack of technology. The exhaust plume reduction technique has been carried out using circular and chevron nozzles to reduce the plume length for reducing the lethal range of target as much as possible with different nozzle configurations. Further, the reduction of IR signature is important topic and valid source of infrared reduction technique was published by Knowles saddington [1]. In that, they mention about rectangular, triangular and elliptical nozzles with different aspect ratios and tabs. They were developed tab structure at the trailing edge of nozzle to improve mixing and reduction in plume length. The IR signature is reduced by using the tabs and count of tabs. The working principle of chevrons is to mixing the hot and cold air at the exit of the nozzle. Therefore, this cold flow analysis is to provide the reduction in infrared signature of exhaust plume with different configuration of nozzles. Axial velocity and static temperature was measured and decay was plotted in graph. Based on this data, the best configured nozzles were identified from other nozzles.

1.2 Background

1) Jet Exhaust Plume

The jet exhaust plume creates infrared signature to gives source to infrared seekers. Mainly, plume was developed by the hot exhaust gas comes out from the trailing edge of nozzle. The most of the aircraft engines are comes with circular type. The chevrons are introduced due to the increased mixing, reducing noise, and reducing plume length and also reduce infrared signature. Mixing of hot exhaust gas to atmospheric air is reducing the length of the plume.

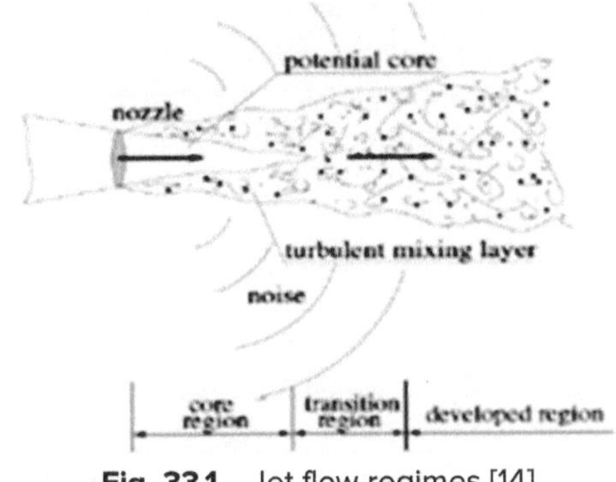

Fig. 33.1 Jet flow regimes [14]

2) Jet Noise

Many authors were investigated about noise prediction and noise reduction techniques with circular, chevron and rectangular nozzles with different configurations. The noise reduction was achieved in chevron patterned nozzles due to the saw tooth patterned structure. Chevron pattern helps to increase the mixing of cold and hot flow gas to reducing the noise simultaneously.

3) Chevrons

The chevron tabs was introduced in 1980s and 1990s explored for mixing enhancement. Chevrons are basically a saw tooth pattern at the trailing edge of nozzle. To evaluate chevron nozzle designs the extensive studies were carried out. Bridges and Brown were investigated jet noise reduction by mixing enhancement of jet by using configurations shape of nozzle, chevron pattern count, angle of chevron inward penetration. Chevron gives the mixing growth rate on the potential core. High penetration designs has counter rotating vortices creates notch between the chevrons so the shear layer thickness is increased and jet decays faster.

1.3 Aim

The aim of the project is to reducing the IR signature of exhaust plume and increase mixing of jet to find out the axial velocity and static temperature decay, find out the mass flow rate for each nozzle, and then find out the which

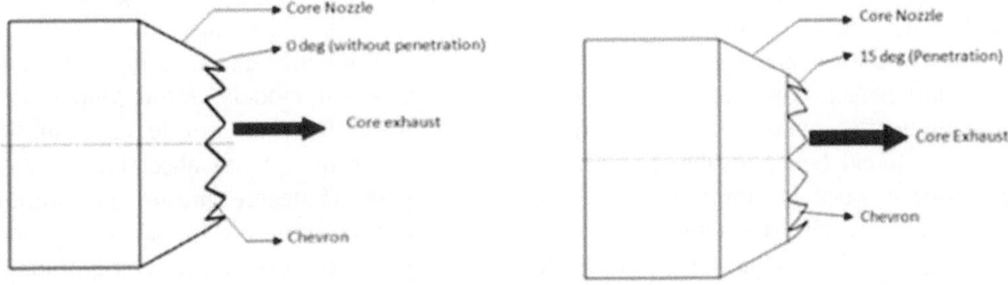

Fig. 33.2 Chevron with and without penetration

configured nozzle has better results. The results should be compared within these configured nozzles to gives reduction in IR signature of jet plume. This project gives potential to geometric modification of chevrons and helps to reducing the IR signature of exhaust plume to escape from IR seeker missiles.

2. Procedure

In cold flow analysis of circular and chevron nozzles with different configurations was designed using NX Siemens software. Measurements were made for nozzles with or without penetration of chevrons. The results were compared with other nozzle and axial velocity and static temperature decay was graphed. The nozzle adjustments were made for as per dimensions like inward penetration, angles, count of chevron.

The measurements of nozzles are given below in table:

Table 33.1 Configuration of nozzles

S. No	Nozzle ID	Length (m)	Inlet Diameter (m)	Outlet Diameter (m)	Penetration (Degree)	Tip Penetration (m)
1	BaselineA15	1.03	0.800	0.550	—	—
2	BaselineA30	0.700	0.800	0.550	—	—
3	N12-A15-P0	1.03	0.800	0.550	—	—
4	N12A30-P0	0.700	0.800	0.550	—	—
5	N12-A30-P15	0.700	0.800	0.550	30	0.0193
6	N12-A15-P30	1.03	0.800	0.550	15	0.0478

Chevron nozzles with varying chevron count, convergence angle (deg) and tip penetration angle (deg) were designed. In the nozzle names listed in Table 33.1, these parameters are identified using the prefix N, A and P respectively. As for the baseline nozzle, convergence angles of 15 and 30 degrees were used for the chevron nozzles. The penetration angle was varied between 0, 15 and 30 degrees.

The baseline nozzle is standard nozzle (circular) was designed in two different angles 15 and 30 degree without chevrons. The N12-A15-P0 & N12-A30-P0 nozzles are patterned with chevron and without penetration. N12-A30-P15 & N12-A15-P30 nozzles were designed with inward penetration and chevrons. The P15 & P30 nozzles were performed inward penetration, this penetration is main factor of reducing the IR signature of exhaust plume, and especially P15 has found better results. The nozzle considered in the present work is almost same as the analyzed by Tide P Sunny [14]. The exit diameter of nozzle is 0.550 m. In addition to this baseline nozzle (baseline A15 & baseline A30) in the present study, chevrons with different configurations were studied. Chevron inward penetration angles 15 degree for N12-A30-P15 & 30 degree for N12-A15-P30 with respect to nozzle dimensions. The chevron nozzles are shown in Fig. 33.2 to 33.8. The chevron count is twelve for entire nozzle and adjusted to nozzle's exit diameter. In all nozzles, inlet and exit diameter remains same.

3. Solution Methodology

Simulations have been performed for cold flow analysis with stagnation temperature is about 1000 K. The inlet conditions are given below:

Table 33.2 Inlet conditions

Parameters	Units	Values
P_0	Pascal (pa)	300000
T_0	Kelvin (K)	1000

All the results are presented in second order accurate. The computational domain was discretized using poly hexcore mesh with below 4.8 million cells for k-ε realizable scalable wall functions turbulence model and 3.8 million cells for k-ε realizable enhanced wall function model. The two different grid models were used with y+ less than one with 27 boundary layers and y+ is one hundred with 10 boundary layers. The all regions were meshed with unstructured mesh.

The three-dimension flow, turbulent supersonic flow along with k-ε realizable model has been carried out with different grid models. The models were treated with enhanced wall function reported by Tide P Sunny [14]. In this case, two different grid models were used scalable and enhanced function with varied boundary layers. In the present analysis, coupled (pressure based) has been used for all models.

4. Result and Discussion

1. *Comparison of axial velocity contours for k-ε realizable scalable wall function:*

   The simulations were conducted for circular and chevron nozzles, utilizing k-epsilon realizable scaling wall function and k-epsilon realizable improved wall function models. Axial velocity predictions were made for both nozzle configurations. Comparing the results, it was observed that the chevron nozzle with 15-degree inward penetration demonstrated enhanced mixing and reduced plume length, making it an efficient mixing configuration. In the case of the count 12 nozzles, the 30-degree nozzle had a higher mass flow rate than others, but its velocity decay was slower than the 15-degree nozzle. The velocity contour showed that the penetration with the 15-degree nozzle had a faster velocity decay compared to other nozzles. The axial velocity graph was shown in Fig. 33.3.

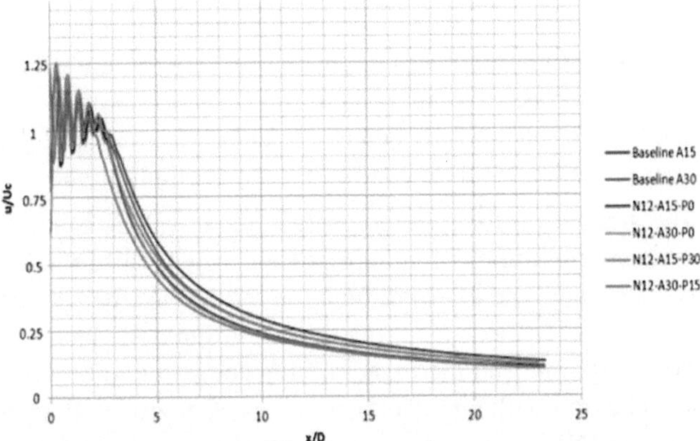

Fig. 33.3 Axial velocity Vs Position for scalable wall function

2. *Comparison of static temperature contours for k-ε realizable scalable wall functions:*

   The depict of static temperature decay in both nozzle models as identified. The 15-degree penetration with k-ε scalable wall function and enhanced wall function shows a faster decay in the atmosphere compared to all other nozzles. The inward penetration of the chevron nozzle results in a quicker decrease in static temperature, while the circular nozzle, lacking a chevron, exhibits slower mixing. The 15-degree penetration leads to improved mixing and a reduction in the infrared signature. Overall, the results highlight the impact of nozzle configuration on static temperature decay, mixing, and infrared signature, with the chevron nozzle with 15-degree penetration demonstrating favorable characteristics in these aspects. The graph was plotted as shown in Fig. 33.4

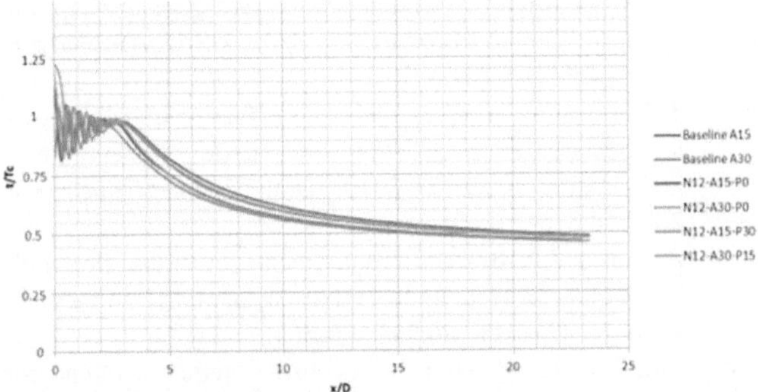

Fig. 33.4 Static temperature Vs position plot for scalable wall function

3. *Comparison of wall resolved and wall function approach:*

   Simulations for a baseline nozzle (A15) and a chevron nozzle (N12-A15-P0) were carried out using both wall-resolved ($y^+ = 1$) and wall function ($y^+ \approx 100$) approach. The k-epsilon realizable turbulence model with enhanced wall treatment was used for the former while scalable wall function was used for the latter. Different grids (with different number of boundary layers) were used for each case to achieve the target y^+ values. There is little difference in results obtained using enhanced wall treatment and scalable wall function. Hence, scalable wall function has been used for all further studies as the mesh size requirement is smaller compared to enhanced wall treatment approach.

4. *Comparison of nozzle performance:*

   The nozzle performance characteristics for cold flow analysis include mass flow; axial velocity and static temperature are shown in Table 33.3 & 33.4. As the chevron inward penetration angle 15 to 30 degree mass flow is gradually reduced. The mass flow rate of chevron without penetration and baseline nozzles higher than the penetrated nozzle due to the larger effective area cut outs of chevrons. N12-A30-P15 & N12-A15-P30 has minimum mass flow rate increased inward penetration into the flow

   The mass flow rate & Plume length was tabulated in Table 33.3 and 33.4:

   1. K-epsilon realizable scalable wall function

Table 33.3 Mass flow rate and plume length for all nozzles k-epsilon realizable scalable model

Nozzles Id	Mass Flow Rate (Kg/Sec)
BaselineA15	58.7
BaselineA30	56.4
N12-A15-PO	54.2
N12-A30-PO	55.8
N12-A30-P15	54.2
N12-A15-P30	52.7

Table 33.4 Plume Length for all nozzles

Nozzle Id	Plume Length
BaselineA15	3.5D
BaselineA30	3.4D
N12-A15-PO	3D
N12-A30-PO	2.7D
N12-A30-P15	2.2D
N12-A15-P30	2.5D

Table 33.5 Velocityat the exit of nozzle

Nozzle Id	Axial velocity (m/s)
BaselineA15	885.6
Baseline A30	883.5
N12-A15-PO	881.9
N12-A30-PO	881.2
N12-A30-P15	878.8
N12-A15-P30	876.1

Table 33.6 Static temperature at the exit of nozzle

Nozzle Id	Static Temperature (K)
BaselineA15	992.4
BaselineA30	990.0
N12-A15-PO	991.0
N12-A30-PO	987.1
N12-A30-P15	983.8
N12-A15-P30	990.9

5. *Comparison of observer angles in chevron:*

   The overall values are predicted for cold flow analysis baseline, chevron without penetration and chevron with penetration along with $0 < 15 < 30$ degree. The baseline nozzle A15 & A30 has zero penetration without chevrons provide not effective mixing and reduction in plume length. Chevrons with and without penetration has increased mixing than baseline nozzles. Especially, with penetration nozzle shows effective mixing and reduction in infrared signature. The cold flow analysis predictions have large decay of axial velocity and static temperature at the exit of nozzle. Especially, N12-A30-P15 gave reasonable reduction of infrared signature of exhaust plume.

5. Summary and Conclusion

The numerical simulation of supersonic jet using k-ε realizable scalable wall function and enhanced wall function has been carried out. The two different grid models were used with different boundary layers were assumed by using formula. Comparison of aerodynamic qualities like axial velocity, static temperature and nozzle performance showed that CFD predictions compare well with two different models. The potential core length is predicted and centerline velocity and temperature at the exit of nozzle. Static temperature and velocity contours predicted well as plane away from nozzle exit. The nozzle performance characteristics like mass flow rate were compared with two different models.

The flow analysis involved the use of a computational fluid dynamics (CFD) program to simulate three different configurations of nozzles. The CFD model was used to assess the performance of circular and chevron nozzles with different nozzle lengths and inward penetrations. The evaluation of nozzle performance encompassed various configurations, along with two different models and grid sizes. The CFD analysis predicted the axial velocity, mass flow rate, and static temperature for all nozzle configurations. This comprehensive analysis allowed for a thorough understanding of the flow behaviors and performance of the nozzles under different conditions and configurations.

The simulation for nozzle configurations, values of static temperature and axial velocity of enhanced wall function is better than scalable wall function results. The penetrated chevrons seem to have reduction of infrared signature of exhaust plume and increased mixing compared to baseline and chevron without penetrated nozzle. Chevron with penetration (N12-A30-P15) is considerable reduction in overall nozzle configurations. In this analysis, shows mixing are better as potential core length is minimum and infrared signature is also minimized for this nozzle configuration. It has achieved 50% of decay within 2.2D to 3.5D compared to baseline nozzle. Chevron without penetration and baseline nozzle did not show much efficient mixing and plume reduction. The graph in the Fig. 33.3 and 33.4 shows the rapid decay of axial velocity and static temperature for both models.

References

1. Knowles Saddington. 'A review of jet mixing enhancement for aircraft propulsion application'. Department of aerospace, power and sensors, Cranfield University. DOI 10.1243/09544100G1605.
2. Ali Arshad, Naveen Andrew, Ilmars. 'Computational study of noise reduction in CFM56-5B using nozzle chevrons'. 11th ICMAE 2020. DOI: 10.1109/ICMAE50897.2020.9178891.
3. Sivabalan, Duck Joo lee & Sanal kumar. 'Three dimensional jet acoustic characterization and geometry optimization of chevron nozzles'. 51st AIAA/SAE/ASEE. DOI: 10.2514/6.2025–3884.
4. Tide P Sunny & K. Srinivasan. 'Effect of chevron count and penetration on acoustic characteristics of chevron nozzle'. APPL ACOUST-2010. DOI: 10.1016/j.apaacoust.2009.08.010.
5. Zaman & J.E. Bridges. 'Experiments on thrust, flow field and noise of rectangular mixer nozzle'. AIAA sci-tech 2020. DOI: 10.2514/6.2020.0030.
6. Tide P.S. 'Aerodynamic and acoustics prediction of chevron nozzles using URANS'. Indian institute of technology, 47th AIAA aerospace sciences meeting.

7. Travis L Turner, Roberto J. Cano& Richard J.Silcox. 'Development of a preliminary model-scale adaptive jet engine chevron'. AIAA J-2008. DOI: 10.2514/1.35939.
8. Yash pal, Dominic Xavier. 'Computational analysis of co-flow potential core degradation of a chevron nozzle'. International journal of vehicle structures-2020. DOI: 10.4273/ijvss.12.3.25.
9. Rakesh Divvela, Radhakrishnan. 'Experimental investigation on mixing characteristics of high-speed co-flow jets by using tabbed chevron nozzles'. 2021-INT J TURBO JET ENG. DOI: 1O.1515/tjeng-2020-0043.
10. Mark. P, Wernet. 'Characterization of chevron nozzle performance'. NASA Glenn research center'. NASA/TM-20210020164.
11. Behruzi & J. Mcguirk. 'Effect of tabs on rectangular nozzle jet plume development' Journal of propulsion and power vol.25, no.4 DOI: 10.2514/1.34904.
12. Tide P Sunny & K. Srinivasan. 'Effect of chevron count and penetration on acoustic characteristics of chevron nozzle'. APPL ACOUST-2010. DOI: 10.1016/j.apaacoust.2009.08.010..
13. Travis L Turner, Roberto J. Cano & Richard J. Silcox. 'Development of a preliminary model-scale adaptive jet engine chevron'. AIAA J-2008. DOI: 10.2514/1.35939.

Note: figures and Tables except fig. 33.1 was created using the experimental work.

Advances in Additive Manufacturing Technologies – Gurusamy Pathinettampadian (eds)
© 2024 Taylor & Francis Group, London, ISBN 978-1-032-90013-1

34 Ascertaining the Suitability of Calcium Added Magnesium Alloy for Biomedical Application

K. V. Sreejaya
Department of Health, Safety & Environmental Management,
International College of Engineering & Management,
C.P.O Seeb 111, Muscat, Oman.

M. Sivaprakash, I.P. Rakhesh, S. Ajith*
Department of Mechanical Engineering,
Stella Mary's College of Engineering, Aruthenganvilai,
Tamil Nadu, India

ABSTRACT: Magnesium alloys which is one of the lightest structural metal is utilized in many structural applications. The hexagonal lattice structure of magnesium alloys influences the alloys' basic characteristics. Recent research indicates that calcium added magnesium alloy is used in biomedical and its related fields. A novel magnesium-based biomedical alloy called magnesium-zinc (Mg-Zn) has been evolved in this research which is suitable for a biomedical and metallurgical aspect. According to the hardness results, the blocks' strength gradually increases as their weight percentage of calcium increases proportionally. The alloying of calcium with Mg-Zn alloys results in a maximal hardness of 98.4 HRC. As a result of zinc alloying with magnesium, the grains are significantly coarser and lack discernible grain boundaries. A distinct grain boundary is discovered after the grains are refined with a modest amount of calcium to the Mg-Zn alloys. The addition of calcium to Mg-Zn alloys results in a texture with a gradient. The calcium addition causes a small increase of about 7% to be seen. The Mg-Zn alloys' refined grain structure is what gives them their increased strength. Mg-Zn alloy's microstructure, hardness and XRD results were promising when compared with the existing alloys.

KEYWORDS: Calcium, Magnesium alloy, X-Ray, Optical microscope

1. INTRODUCTION

Iron and other single-element metals are the weakest metals which are not particularly plastic-resistant. In their annealed state, which is typical of some polymers, such pure metals can have yield stresses as low as 50 MPa. An example of this is iron. However, with modest carbon additions, the strength of iron-carbon alloys, or steel, quickly increases, increasing to 250–690 MPa depending on other factors [1]. Therefore, little amounts of alloying elements, especially the interstitial atoms, can have a considerable impact on the metal's strength. An alloy, made of two or more different metals or elements, enhances mechanical properties. For instance, alloying iron and carbon results

*Corresponding author: ajithadhavan@gmail.com

DOI: 10.1201/9781003545774-34

in steel. Alloying is metallurgy's greatest contribution to humankind. Outside of just metals, this alloying notion has use [2]. The alloying principle is the same for ceramics and polymers. The advent of civilization was ushered in by the discovery of pure native metals like gold and copper. These days, we have access to a wide range of resources [3].

Magnesium, which makes up around 2.5% of the eighth most plentiful element in the Earth's crust, is the third most frequent structural metal after aluminium and iron. Due to its low density (around two thirds that of aluminium), it makes it a good candidate for many aircraft applications [4]. Magnesium is frequently used in alloys, usually containing 10% or less of aluminium, zinc, and manganese, to increase its hardness, tensile strength, and ability to be cast, welded, and machined. This occurs as a result of the pure metal's limited structural strength. The alloys are created using casting, rolling, extruding, and forging techniques, and the resulting sheet, plate, or extrusion is then further created using conventional joining, forming, and machining techniques [5]. Since it is the most accessible structural metal to the machine, magnesium is typically employed when a variety of machining operations are required. Magnesium alloys are popular mostly due of their low cost, high strength-to-weight ratios, and great machinability [6]. Zinc (Zn) is a nutrient required for healthy function in the human body. The muscles and bones contain around 85% of thezinc in the body. Zinc may be dissolved in magnesium at a 6.2% concentration. Zinc slows down the biocorrosion process' generation of hydrogen. The most utilized and productive component of magnesium alloys is zinc. To increase the strength of the material, it is mixed with aluminium[7]. By reducing the impact of iron and nickel impurities, zinc can increase the corrosion resistance of magnesium alloys. The solubility of zinc in magnesium is 2.5 weight percent [8].

On the other side, it is anticipated that corrosion rates will climb when zinc content increases from 1% to 3% wt%. Stress corrosion cracking was more likely to occur if zinc content was raised even higher [9]. In order to prepare other metals like thorium and uranium, calcium metal is utilised as a reducing agent. Additionally, it serves as an alloying agent for the alloys of aluminium, beryllium, copper, lead, and magnesium. There are large reserves of limestone (calcium carbonate), which is used both directly and indirectly to make cement. [10]. Quicklime (calcium oxide), which is produced when limestone is heated in kilns, is left behind. Slaked lime (calcium hydroxide) is the result of a strong reaction between this and water. Zinc is a frequent component in magnesium alloys and causes the alloys to harden in solutions [11]. But the highest solubility only reaches 1.6 wt. % (i.e. 0.6 at.%) at 25 in the equilibrium state, it is possible to treat the mixture with a solid solution to create a supersaturated solution and prevent the precipitation of the -MgZn phase [12].

The corrosion process can be slowed down in this way because of the homogeneity, which results in a consistent microstructure (Fig. 34.1). Additionally, solution hardening has the potential to improve mechanical characteristics. According to the Hall-Petch connection, hot working (in this case, hot extrusion) can also reduce grain size, increasing yield strength and tensile strength [13]. The mechanical properties of the Mg-6Zn alloy are better than those of degradable polymeric materials like -HA/PLLA 50/50 (tensile strength 103 MPa, Young's modulus 12.3 GPa). The elastic modulus of the magnesium alloy used in this study (42.3 GPa) is likewise higher than that of polymers and titanium alloys, although it is still lower than that of human femur bone (15–20 GPa) [14]. With increasing Ca concentration, the YS, UTS, and elongation for as-cast Mg-xCa alloy samples dropped. For as-cast Mg-1Ca alloy samples, heat rolling and hot extrusion significantly enhanced the YS, UTS, and elongation. Higher Ca concentrations result in a considerable reduction in grain and dendritic cell size while increasing the amount of Mg2Ca intermetallic phase at grain borders [15].

The microstructure, which is influenced in the thermo-mechanical treatment, affects the mechanical behaviour under dynamic and quasi-static situations. The least ductile grain structure is one with a fine grain, and ductility rises as grain size increases [16]. Rapidly solidified Mg-Ca alloy (2 wt%) had a UTS of 380 MPa and an elongation of 7.3%. It would be possible to use a mesne method to change the mechanical characteristics of a magnesium-x-calcium alloy to acceptable levels (200-400 MPa) for biomedical purposes. The optical microscope samples are first cut using a band saw, and then mounted using a Bain mount machine H. High-speed water-cooling coils and an integrated hydraulic system are included into the heater for the mould [17]. For making the mounts, phenolic powders are employed. P80, P100, P220, P320, P500, P800, P1000, P1200, and P1500 Silicon Carbide papers are used to polish the mounted samples. Diamond abrasives of 3 mm and 0.3 mm are used to polish velveted cloth to a fine shine. Using 6g Picric Acid, 10ml Distilled Water, 5ml Acetic Acid, and 100ml Ethanol as the etchant and Ethanol as the clearing agent, the microstructure is revealed [18]. Samples may be used for hardness research after microscopic examinations. The mounting machine mounts the samples using green phenolic powders. The optical microscopy images are captured using OLYMPUS BX53M [19]. E18 Standard is used in the sample preparation process for hardness testing. The ASTM standard's graphic is utilized to test the specimen's hardness using a Phase II hardness tester [20].

2. METHODOLOGY

The right weight proportion of the metals to be melted is calculated before the casting process begins. Materials are cut with a band saw and measured on a digital scale in accordance with the required quantity.

2.1 Material Preparation

The chemical makeup of the alloys to be created is listed in the. The components utilized are described below and are displayed in Table 34.1. The power hacksaw was used to cut the materials according to the calculated weight percentage. The ready resources. For the first composition's casting, magnesium and zinc were used, and samples of magnesium, zinc, and calcium were used for the second composition. Casting tools enable us to process the raw materials. The skimmer has the ability to stir and skim. The molten metal in the crucible can occasionally also be skimmed using steel rods, though. The majority of the time, handling hot, corrosive materials requires the use of a crucible tong.

Table 34.1 Ratio of sample in composition

Sample	Ca	Zn	Mg
Mg-5Zn	–	5%	95%
Mg-5Zn-0.5Ca	0.5%	5%	94.5%

In a steel crucible, magnesium is melted. AISI 304 grade steel that is cylindrical. The rust and other impurities left over from the previous casting process are removed from crucibles using a wire brush. Magnesium ingots that have been band-sawed into the proper size so they fit inside the crucible are charged or inserted into a clean crucible. There is no requirement for preheating the additives in this method. Since Zinc and Calcium don't require preheating, they can be added to the crucible at three different times after the melted components have been added.

The magnesium ingots for melting, coupled with the crucible. At the very end, lanthanum is added. In order to maintain the desired composition in accordance with the plan, all ingredients are added using weight % calculations. Solid materials are heated in melting furnaces until they become liquefied. Thermal processing machinery is frequently used to modify the surface or interior properties of materials by gradually raising their temperature. In the case of metals, ductility is typically increased at the price of both strength and hardness.

It is therefore necessary to use an industrial furnace that can produce and sustain temperatures below the melting point of the material. used to melt things, the melting furnace. In contrast, a melting furnace generates excessively high temperatures that exceed the melting point of the metal and cause the physical structure of the metal to break down, leading to liquefaction. Both pressure and temperature have no influence on this phase change. Except for established eutectics like mercury (Hg) and alloys based on gallium (Ga), it is unusual for metals to exist in a liquid state at room temperature, metal in the furnace that is molten.

Magnesium silicon lanthanum alloy is moulded using a steel mould with dimensions of 240 x 240 x 10 mm thick. The casted component may be readily removed from the mould thanks to its design. Mould of the bottom filling variety is utilized in the procedure. the procedure's hot premold. The top surface of the mould has an entrance through which the molten metal is poured. As a result, the mould is filled with the molten metal due to the action of gravity and solidification occurs smoothly and without any flaws. In the heated mould, the molten metal is poured.

2.2 Casted Samples

After pouring, molten metal can cool. Heat is quickly dissipated by cast magnesium. The casting is easily ejected from the mould without causing any surface damage thanks to the way it is designed, the cast piece that has a runner. The surface obtained in this figure is exceptionally smooth and devoid of any form of imperfections when viewed with the unaided eye. The cast piece has just been taken out of the mould and its weight is verified.

3. RESULT AND DISCUSSIONS

3.1 Metallographic Characterization

Metallography is the study of every type of metallic alloy's microstructure. It is better referred to as the scientific field that investigates the atomic and chemical composition, spatial distribution, and grain size of the constituents, inclusions, and phases of metallic alloys. By extension, these same concepts can be used to describe any substance.

3.2 X-RayDiffraction Studies

The elements, phases, and particles included in the composite as depicted in Fig. 34.1 were identified using X-ray diffraction (XRD) on the various composition combination specimens. The initial investigation into the metallurgical samples is done by X-Ray diffraction investigations. In the Bruker Phase II, Singapore X-ray diffractometer, the testing was done on the powdered samples between 20 and 80 degrees. The phases are determined by analyzing the collected XRD peaks using the X'Pert High Score PRO software. The measured peaks in Fig. 34.1 demonstrate the presence of the Mg and $MgZn_2$ interphase. Comparatively, in Fig. 34.2 the Mg,

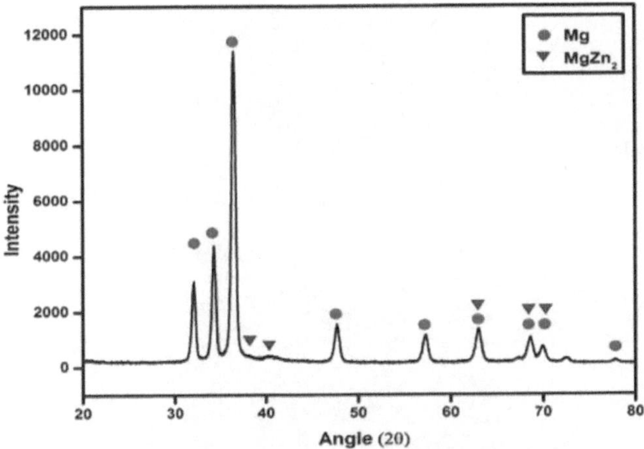

Fig. 34.1 XRD peaks of Mg-5Zn alloy

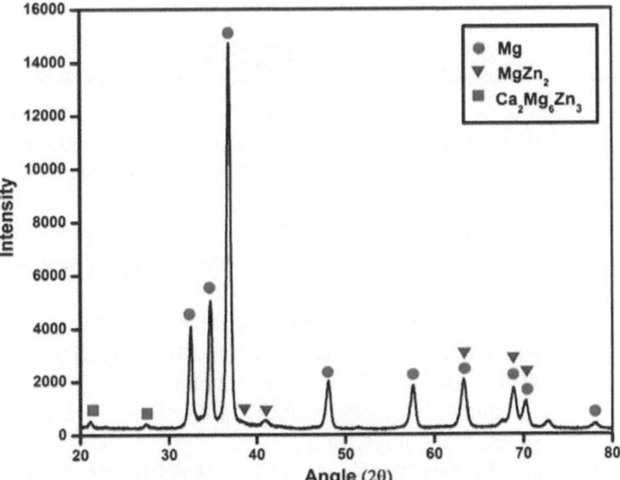

Fig. 34.2 XRD peaks of Mg-5Zn-0.5Ca alloy

$MgZn_2$, and $Ca2Mg_6Zn3$ interphase peaks are present in the Mg-5ZN-0.5Ca alloys' XRD peak. Due to the addition of calcium to the already present components, such as zinc and magnesium, another interphase is discovered.

3.3 Optical Microscopy Results

There are numerous techniques for revealing the microstructural properties of metals. Brightfield incident light microscopy is used for the majority of investigations. However, the use of colour (tint) etching and other less frequent contrast methods, such darkfield or differential interference contrast (DIC), is broadening the possibilities of light microscopy for metallographic applications. Metallic materials' microstructure has a big impact on a lot of important macroscopic properties. Critical mechanical qualities, such as tensile strength or elongation, as well as additional thermal or electrical properties, are directly influenced by the microstructure. Understanding how the relationship between the macroscopic features and the microstructure is key to developing and producing materials is the ultimate goal of metallography.

The effects of adding calcium at various weight percentages on Mg-Zn alloy are shown by optical data. The optical micrographs of the Mg-5Zn and Mg-5Zn-0.5Ca alloys at low and high magnification are shown in Fig. 34.3. As a result of zinc alloying with magnesium, the grains are significantly coarser and lack discernible grain boundaries. A distinct grain boundary is discovered after the grains are refined with a modest amount of calcium to the Mg-Zn alloys. The addition of calcium to Mg-Zn alloys results in a texture with a gradient. This might be a result of the Ca2Mg6Zn3 forming. Advanced characterization must be done in order to reveal the mechanism.

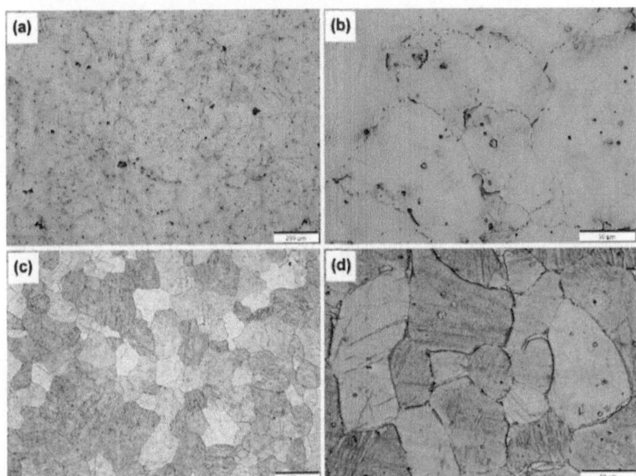

Fig. 34.3 Low and high magnification optical microstructure of: (a, b) Mg-5Zn alloy (c, d) Mg-5Zn-0.5Ca alloys

3.4 Mechanical Characterization

In the field of materials science, the term "characterization" refers to an extensive and all-encompassing procedure used to examine and assess the structure and characteristics of a material. It is necessary for creating a scientific understanding of engineering materials. It is an essential phase in the process of learning new information. A material's mechanical properties can be discovered through testing under static or dynamic loads. In the field of conservation, mechanical tests are typically used to establish whether a material is adequate for the function for which it was designed or whether particular environmental factors have an impact on an object.

3.5 Hardness Test

The polished samples are used to conduct hardness testing on the cast specimén. The MgZn alloy's hardness map and

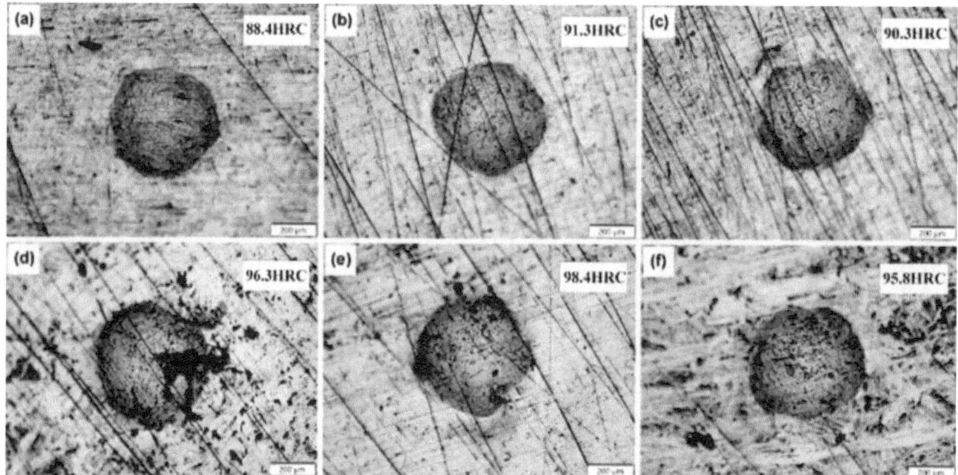

Fig. 34.4 Hardness indentation micrograph of: (a, b, c) Mg-5Zn alloys and (d, e, f) Mg 5Zn- 0.5Ca alloy

the MgZn alloys with calcium added. With the addition of calcium to the MgZn alloys, a rise is gradually seen. The results of the sample's hardness test are shown in Fig. 34.4. The Mg-Zn alloy's average hardness value was determined to be 90HRC. The hardness value of the Mg-Zn alloy with calcium added was also discovered to be 96.83 HRC. The calcium addition causes a small increase of about 7% to be seen. The Mg-Zn alloys' refined grain structure is what gives them their increased strength. According to the Hall-Petch relationship, the grain size has a direct impact on an alloy's strength. In contrast to the fine grains, the coarse grains have a low hardness. According to the Hall-Petch relation, which goes as follows: H = Ho + KHd1/2, where Ho and KH are constants, a material's hardness, H, is influenced by the grain diameter, d. Tensile testing, tribological analysis, corrosion studies, and other methods of characterization can be added to the process to better understand how the alloys behave.

4. Conclusion

To prepare the samples, the blocks are typically machined after being flawlessly cast in accordance with their composition. Furthermore, the machined samples meet ASTM specifications. Analyses of the samples' characterization are performed. Metallographic and mechanical characterization studies are carried out. On the cast specimens, hardness tests are conducted as part of mechanical characterization studies. According to the hardness results, the blocks' strength gradually grows as their weight percentage of calcium addition increases proportionally. The alloying of calcium with Mg-Zn alloys results in a maximal hardness of 98.4 HRC. In comparison to the basic alloys, a little rise of about 7% is noted.

References

1. Zhang, W., Li, M., Chen, Q., Hu, W., Zhang, W., & Xin, W. (2012). Effects of Sr and Sn on microstructure and corrosion resistance of Mg–Zr–Ca magnesium alloy for biomedical applications. Materials & Design, 39, 379–383.
2. Shadanbaz, S., & Dias, G. J. (2012). Calcium phosphate coatings on magnesium alloys for biomedical applications: a review. Acta biomaterialia, 8(1), 20–30.
3. Ali, A., Iqbal, F., Ahmad, A., Ikram, F., Nawaz, A., Chaudhry, A. A., ... & Rehman, I. (2019). Hydrothermal deposition of high strength calcium phosphate coatings on magnesium alloy for biomedical applications. Surface and Coatings Technology, 357, 716–727.
4. Zhang, E., Yang, L., Xu, J., & Chen, H. (2010). Microstructure, mechanical properties and bio-corrosion properties of Mg–Si (–Ca, Zn) alloy for biomedical application. Acta biomaterialia, 6(5), 1756–1762.
5. Abdel-Gawad, S. A., & Shoeib, M. A. (2019). Corrosion studies and microstructure of Mg– Zn– Ca alloys for biomedical applications. Surfaces and Interfaces, 14, 108–116.
6. Qu, Y., Kang, M., Dong, R., Liu, J., Liu, J., & Zhao, J. (2015). Evaluation of a new Mg–Zn–Ca–Y alloy for biomedical application. Journal of Materials Science: Materials in Medicine, 26, 1–7.
7. Li, N., & Zheng, Y. (2013). Novel magnesium alloys developed for biomedical application: a review. Journal of Materials Science & Technology, 29(6), 489–502.
8. Bommala, V. K., Krishna, M. G., & Rao, C. T. (2019). Magnesium matrix composites for biomedical applications: A review. Journal of Magnesium and Alloys, 7(1), 72–79.
9. Bakhsheshi-Rad, H. R., Hamzah, E., Fereidouni-Lotfabadi, A., Daroonparvar, M., Yajid, M. A. M., Mezbahul-Islam, M., ... & Medraj, M. (2014). Microstructure and bio-corrosion behavior of Mg–Zn and Mg–Zn–Ca alloys for biomedical applications. Materials and Corrosion, 65(12), 1178–1187.

10. Tang, H., Xin, T., & Wang, F. (2013). Calcium phosphate/titania sol-gel coatings on AZ31 magnesium alloy for biomedical applications. International Journal of Electrochemical Science, 8(6), 8115–8125.
11. Tang, H., Xin, T., & Wang, F. (2013). Calcium phosphate/titania sol-gel coatings on AZ31 magnesium alloy for biomedical applications. International Journal of Electrochemical Science, 8(6), 8115–8125.
12. Du, H., Wei, Z., Liu, X., & Zhang, E. (2011). Effects of Zn on the microstructure, mechanical property and bio-corrosion property of Mg–3Ca alloys for biomedical application. Materials Chemistry and Physics, 125(3), 568–575.
13. Heimann, R. B. (2021). Magnesium alloys for biomedical application: Advanced corrosion control through surface coating. Surface and Coatings Technology, 405, 126521.
14. Hornberger, H., Virtanen, S., & Boccaccini, A. R. (2012). Biomedical coatings on magnesium alloys–a review. Acta biomaterialia, 8(7), 2442–2455.
15. Pan, Y., He, S., Wang, D., Huang, D., Zheng, T., Wang, S., ... & Chen, C. (2015). In vitro degradation and electrochemical corrosion evaluations of microarc oxidized pure Mg, Mg–Ca and Mg–Ca–Zn alloys for biomedical applications. Materials Science and Engineering: C, 47, 85–96.
16. Ramalingam, V. V., Ramasamy, P., Kovukkal, M. D., & Myilsamy, G. (2020). Research and development in magnesium alloys for industrial and biomedical applications: a review. Metals and Materials International, 26, 409–430.
17. Allavikutty, R., Gupta, P., Santra, T. S., & Rengaswamy, J. (2021). Additive manufacturing of Mg alloys for biomedical applications: Current status and challenges. Current Opinion in Biomedical Engineering, 18, 100276.
18. Allavikutty, R., Gupta, P., Santra, T. S., & Rengaswamy, J. (2021). Additive manufacturing of Mg alloys for biomedical applications: Current status and challenges. Current Opinion in Biomedical Engineering, 18, 100276.
19. Tian, P., & Liu, X. (2015). Surface modification of biodegradable magnesium and its alloys for biomedical applications. Regenerative biomaterials, 2(2), 135–151.
20. Chen, Y., Dou, J., Yu, H., & Chen, C. (2019). Degradable magnesium-based alloys for biomedical applications: The role of critical alloying elements. Journal of Biomaterials Applications, 33(10), 1348–1372.

Note: Every figure and table was created using the experimental work; none of them were taken from any publications or online.

Advances in Additive Manufacturing Technologies – Gurusamy Pathinettampadian (eds)
© 2024 Taylor & Francis Group, London, ISBN 978-1-032-90013-1

35 Corrosion Studies, of Friction Stir Welded-AA5754-H111 with Mg-AZ61 Alloys by using Salt Spray Test

Y. Premraj[1]
Assistant Professor, Department of Mechanical and Automation Engineering, Gojan School of Business and Technology, Chennai, Tamilnadu, India

M. Selvaraj[2]
Assistant Professor, Department of Mechanical Engineering, Paavai Engineering College, Namakkal, Tamilnadu, India

K. Suresh[3]
Assistant Professor, Department of Mechanical Engineering, S. A. Engineering College, Chennai, Tamilnadu, India

S. Balamurugan[4]
Assistant Professor, Department of Mechanical Engineering, Chennai Institute of Technology, Chennai, Tamilnadu, India

ABSTRACT: This investigation utilized the Salt Spray Test (SST) method to evaluate the corrosion features of friction stir welded (FSW) alloys that are altered to each other which are AA5754-H111 with Mg-AZ61. The FSW was carried out at constant rotational speeds and tool travel speeds on 5 mm thick plates. SST was performed in order to comprehend the general and particular corrosion study of the base alloys and the FSW joints. The findings demonstrated that the microstructure of the variety of districts of the FSW joint speckled prominently, which in part swayed how each district corroded in a NaCl solution. According to the findings of the salt spray test, the stir zone (SZ) has more consistent and corrosion resistance than the base alloys. The research firms the weight loss and rate of corrosion of the SZ comparative to base metals in the welded sample.

KEYWORDS: Friction stir welding, AA5754/ Mg- AZ61, Salt spray test, Corrosion study, Weight loss, Corrosion rate

1. INTRODUCTION

One of the lightest structural materials, AA5754-H111 (Aluminium alloy 5754-Tempered) and Mg-AZ61 (Magnesium Alloy) is appealing for use in the automobile, rail shipment, and space productions [1]. FSW, a solid-state welding process, has a substantial benefit for connecting Mg AZ61 alloys [2]. The FSW method results in substantial grain refinement, full recrystallization, and the creation of a strapping local texture in the SZ of Mg AZ61

[1]premrajme@gmail.com, [2]selva.msrk@gmail.com, [3]sureshdesign24@gmail.com, [4]balamurugans@citchennai.net

DOI: 10.1201/9781003545774-35

alloys [3]. The tensile characteristics, particularly for the yield strength (YS) are thus dramatically reduced in some soft-oriented texture regions that occur in the WNZ [4-5].

Numerous researchers have used various FSW techniques on differing kinds of aluminum alloys [6-8]. They discovered that an effective welding joint is produced in all kinds by a straight square tool pin profile [9-10]. Aalternatively, a small number of investigations have determined the influences of mechanical characteristics and corrosion resistance (CR) of FSW Mg –AZ61 alloys. The CR of Mg alloys has a considerable influence on their industrial application, as is well documented [11]. The results of the FSW technique on the CR of Mg alloys have been the subject of extensive research. The corrosion rates of FSW AZ61 Mg alloys in various settings, comprising those with varying pH levels, chloride ion percentages, and absorption times, were investigated by Dhanapal et al. [12]. Balamurugan et al investigated the corrosion rate as well as weight loss for friction stir welded dissimilar aluminium alloys. They observed very few corrosion losses is exhibits in the welded region up to 72hrs. Also, they studied the SEM morphology of the different welded conditions [13]. Seifiyan et al discovered that the number of welding passes, traverse speed, and tool rotation have a substantial guidance on the CR of magnesium alloys [14]. Liu et al. discovered that several rounds of friction stir processing (FSP) can lower the corrosion rate of AZ31 Mg because of the additional grain refinement in stir zones [15, 16]. Recently, Xu ei al studied the strength enhancement of AZ61 Mg alloys. They concluded that the metallurgical and mechanical qualities of FSW AZ61 Mg joints can be enhanced by combining a high axial load with slow welding and low tool rotation rates [17]. The goal of this work is to create a new AA5754-H111/Mg-AZ61 that uses particles as the second phase and employs FSW. According to our knowledge, this study is brand-new research that hasn't yet been published. Also, the purpose of the current research is to produce fine grains with a lot of structural aspects in order to develop the mechanical characteristics and oxidization behavior of the FSW AA5754-H11/ Mg AZ61 alloy joint.

2. Materials and Methods

2.1 Specimens Preparation

The very first components utilized in the current work were plates of the AA5754-H111/ Mg AZ61 alloy with a thick of 5 mm. Figure 35.1 demonstrated the AA5754-H111/ Mg AZ61 alloys plates for FSW.

2.2 Friction Stir Welding

The literature and the capabilities offered by the FSW machine were used to select the tool rotational speeds and traverse rates. FSW weld joints created from different AA5754-H111/ Mg AZ61 alloys are demonstrated in Fig. 35.2 using locally constructed and designed FSW equipment.

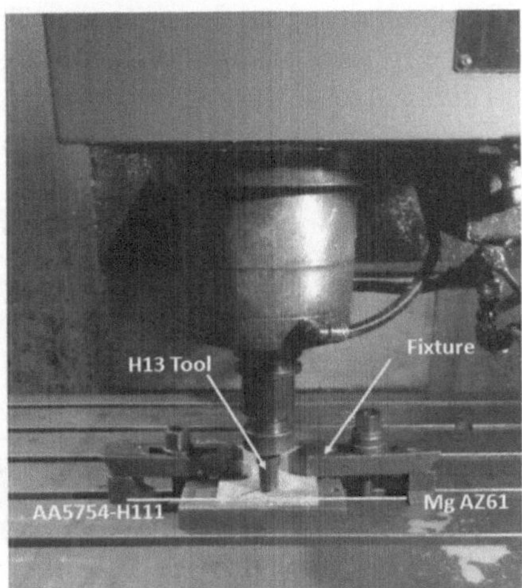

Fig. 35.1 Sample Mg AZ61 plate for friction stir welding with AA5754-H111

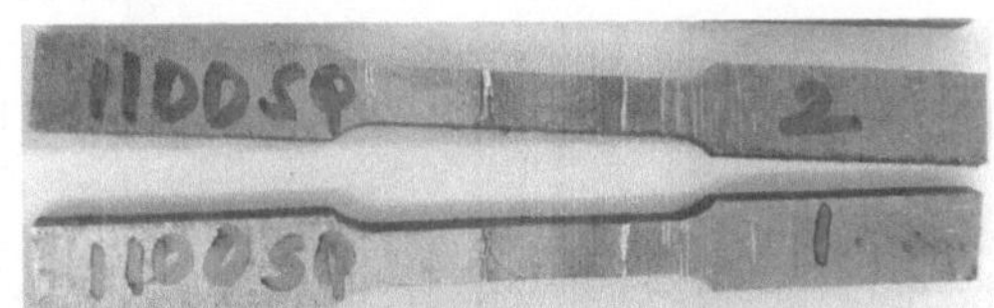

Fig. 35.2 The welded joints of AA5754-H111/ Mg AZ61alloys

The process variables and tool shape for FSW of AA5754-H111/ Mg AZ61 alloys are revealed in Table 35.1.

Table 35.1 Welding parameters and tool geometry for FSW of AA5754-H111/ Mg AZ61 alloys

Welding Parameters	Ranges
Pin Length (L)	4.7 mm
Pin single side (a)	5 mm
Shoulder Diameter (D)	15 mm
Pin profile	Straight Square
Vertical Load (kN)	10
Welding Speed (WS) - (mm/min)	28
Rotational Speed (RS) - (Rpm)	1100

2.3 Corrosion Study – Salt Spray Test

Based on a literature study, Salt Spray Testing (SST) was chosen for the corrosive environment examination. Oxidation testing is among the most important aspects because oxidation is a critical challenge in naval technology. As a result, the testing was done in accordance with ASTM B 117 requirements in the investigational setup exposed in

Fig. 35.3 of a simulated aquatic environment. The FSWed AA5754-H111/ Mg AZ61 dissimilar alloy was investigated, and high ultimate tensile strength samples (ASTM E8) were collected for the experimental corrosion study [12].

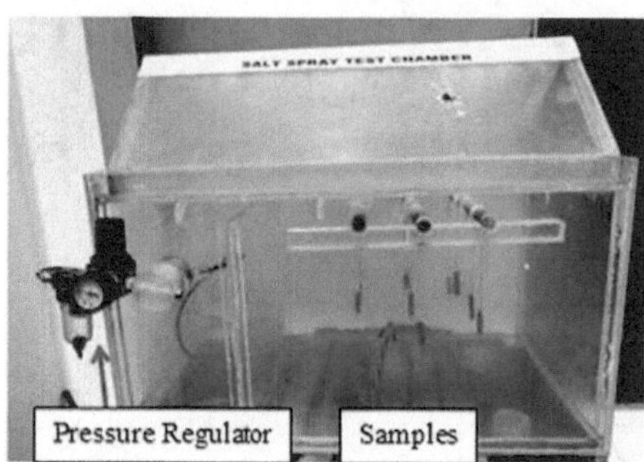

Fig. 35.3 Corrosion test setup - Salt spray chamber

The test samples were typically diagnosed to a steady value. A pH of 7 is referred to as a neutral medium. Distilled water has a constant pH of 7. On the other hand, little impurities in distilled water may cause it to vary considerably. The oxidation solution combined 95% de-ionized water and 5% NaCl (distilled water). The pH of the varied solution was checked formerly every 8 hours. The salt solution was preferred at 0.1 Normality, approximately 6 grams litre^{-1}. The same method was used to create a 40-litre salt solution, and the test lasted 72 hours. The specimens were taken out of the chamber at 24-hour intervals and weighed [19, 20].

Table 35.2 lists the specific parameters for performing the SST. The sample sides were refined with 600, 800 and 3000 grit SiC paper before being washed with alcohol and left to dry with airflow [21]. The preliminary weight of the specimen was determined before the test. The salt concentration for 1000 mL contains approximately 5% NaCl and 95% De-ionized water. The adsorption mass of every sample was determined after 72 hours. The rusted samples were placed in the SST, and weathering particles were unconcerned by submerging the samples in an especially organized solution for 5 mins [12, 21]. The total mass of the specimens was determined, and the significant loss was tabulated.

Table 35.2 Salt spray test parameters

Particulars	Ranges
Humidity	A hygrometer measured 98% humidity
Type of loading of specimens	Plastic wire-tied and hung in hangers
The pressure of Air for atomizing	2 to 3 bars
The temperature of the test	33°C to 35°C
PH of the solution	By adding buffer solution, the pH is kept at 7.5
Measurement of pH	Once every 8 hours
Composition of the salt solution	De-ionized water 95% + 5% of Sodium chloride (NaCl) (1 litre of solution)
Cleaning Process	20 g chromium trioxide -$CrO_{3,}$ 50 mL phosphoric acid -$H_3PO_{4,}$ reagent water to make 1000 mL at 90 °C

3. Calculation Formula for SST

Calculating the corrosion rate makes use of the weight loss measurement technique. The samples were weighed and tested in an aggressive environment for various periods (24, 48, and 72 hours). The specimens were carefully cleaned, dried, and evaluated following the experiment to calculate the final weight loss exhausting Equation (1) [12]

The relation,

$$\text{Weight loss} = \text{Original weight} - \text{Ultimate weight} \quad (1)$$

The specimens' corrosion rate is calculated using Equation (2) as follows:

$$\text{Corrosion rate} = (K \times W)/(A \times T \times D) \quad (2)$$

A - Area (cm^2), W - Mass loss (g), K - Constant (3.45 x 106 for mils/ year), (8.76 × 104 for mm/year), T - Time of exposure (hours), D - Density (g/cm^3).

4. Results and Discussion

The samples taken from FSWed plates for the test performed are shown in Fig. 35.4. The same SST results were

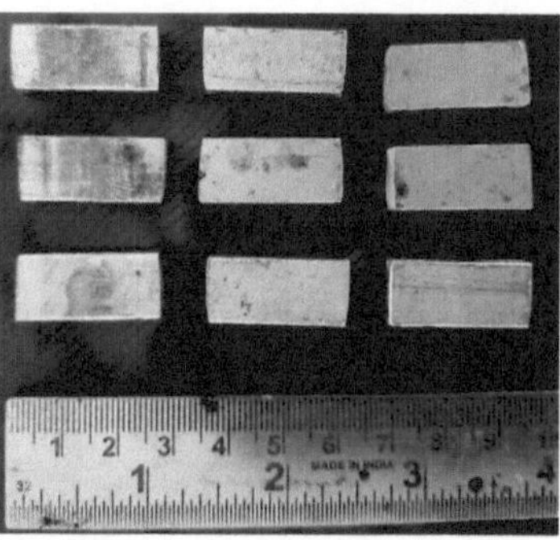

Fig. 35.4 FSW and BM samples before and after corrosion test

condensed in Table 35.2. The weight loss was found to be least in the FSWed AA5754-H111/ Mg AZ61alloys weld region specimen (0.049g) and to be very considerable in the base materials of AZ31 (0.335 g), as can be seen from Table 35.3. Similarly, it was discovered that the welded region had the lowest corrosion rate (0.0000000787 mm/yr). In contrast, the maximum corrosion rate for the base material, AA5754-H111/ Mg AZ61alloys, was 0.0000005054 mm/yr, and the corrosion rate for FSWed AA5754-H111/ Mg AZ61alloys material was 0.0000000787 mm/yr. Similar to base metal (AA5754-H111/ Mg AZ61) and FSWed samples exhibit moderate corrosion (0.0000004568 mm/yr) and weight loss (0.301g) [19].

4.1 Analysis of Corrosion Test Results

In this study, Table 35.3 presents the corrosion rate (mm/year) of various AA5754-H111/ Mg AZ61alloys samples based on weight loss (g), density (g/cm^3), and area (cm^2). Figure 35.5 depicts the weight loss experienced by the three specimens following completion of SST for FSwelded samples with high tensile strength from threaded tool pin profiles for 24 hours, 48 hours, and 72 hours [12].

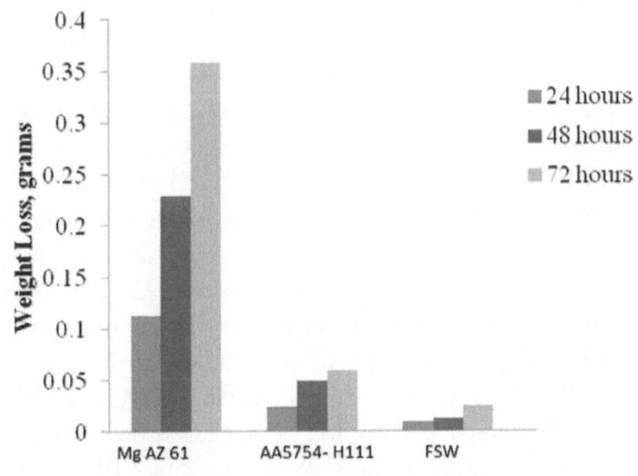

Fig. 35.5 Weight loss for base metals and welded samples

According to the study, the welded area has more excellent corrosion resistance, whereas the base metals (AA5754-H111/ Mg AZ61 alloys) recorded lower levels of corrosion resistance. In addition, AZ31B/Cu welded regions have been found to have a higher corrosion resistance rate in FSWed at 72 hours when compared to FSWed and base metals (AA5754-H111/ Mg AZ61alloys) with similar reinforcements. This could be due to the surfaces of the samples developing a barrier protection oxide layer because weld samples from threaded pin profiles with AA5754-H111/ Mg AZ61 alloys are more tensile and hardness-resistant than other samples.

4.2 Analysis of Corrosion Rate

The rate of corrosion that the three specimens experienced after undergoing the 24-hour, 48-hour, and 72-hour SST is exposed in Fig. 35.6. The study confirms that the BM (AA5754-H111/ Mg AZ61) recorded a medium corrosion resistance, whereas the welded region (AA5754-H111/ Mg AZ61) has a lower rate of corrosion. For the corrosion study investigation, the highest tensile values from the FSWed were chosen. Apart from the FSW process, the reinforcement welds' corrosion rate was extremely low

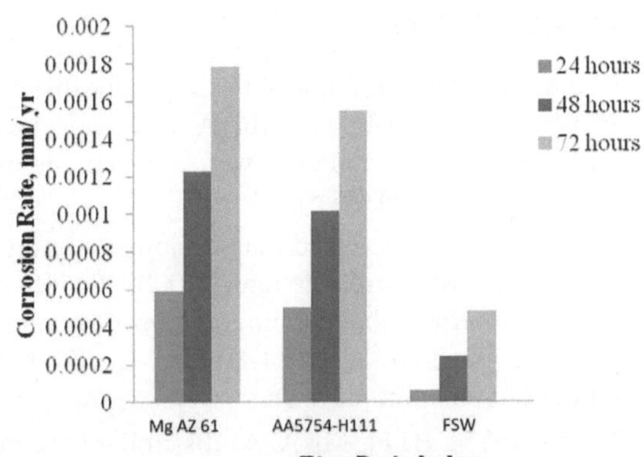

Fig. 35.6 Corrosion rate comparison

Table 35.3 Rates of corrosion for different exposure times

Sample Number	Weight in advance SST (gram)	Weight afterward SST (gram)			Weight loss (gram)	Density (g/cm^3)	Area (cm^2)	Corrosion ratex 10^{-4} (mm/year)
		24 hr	48 hr	72 hr				
Mg - AZ 61	3.012	3.126	-	-	0.114	1.18	146.9	0.0000001843
	3.018	-	3.319	-	0.301	1.18	146.9	0.0000004583
	3.015	-	-	3.35	0.335	1.18	146.9	0.0000005054
AA5754-H111	3.013	3.026	-	-	0.013	1.18	146.9	0.0000000217
	3.081	-	3.108	-	0.027	1.18	146.9	0.0000000439
	3.098	-	-	3.15	0.049	1.18	146.9	0.0000000787
AZ61/AA5754 FSW	3.068	3.357	-	-	0.289	1.18	146.9	0.0000004351
	3.06	-	3.359	-	0.299	1.18	146.9	0.0000004499
	3.029	-	-	3.33	0.301	1.18	146.9	0.0000004568

due to the favorable welding conditions. To sum up, the AA5754-H111/ Mg AZ61 FSWed sample displayed a lower corrosion rate because it was developed during the FSW and has high tensile and hardness values and good material flow [22-24].

The Al-Mg-Cu composite coatings with enhancement, standardization, and increased density structure, which were self-possessed primarily of α-Mg, γ-$Al_{12}Mg_{17}$, $AlCu_4$ and Al_2CuMg intermetallic compounds, were also generated by FSW with pin due to DRX, instigated by adequate heat input and severe strain rate, according to Liu *et al.* 2022. The AA5754-H111/ Mg AZ61 alloys contain the grained iron-rich type of nano particle after welding [21]. These particles barely affect the streams and corrosion resistance of these alloys. The particles principally cause the development of surface cavities, but supplemental minor strikes are regularly seen. In these alloys (AA5754-H111/ Mg AZ61) phases, which develop in the unprotected states of the alloys in open path situations, Al_3Mg_2 causes corrosion. When FSWed samples with AA5754-H111/ Mg AZ61 are compared to those with base materials, intermetallic is more likely to form.

5. Conclusions

1. The weight loss was found to be least in the dissimilar FSWed (AA5754-H111/ Mg AZ61) weld region specimen (0.049g) and to be very considerable in the base materials (0.0335 g).
2. Similarly, it was revealed that the joined region had the lowermost corrosion rate (0.0000004568 mm/yr). In contrast, the maximum corrosion rate for the base material, AA5754-H111/ Mg AZ61, was 0.0000005054 mm/yr.
3. Like AA5754-H111, Mg AZ61 dissimilar base metal and dissimilar FSWed samples exhibit moderate corrosion (0.0000000787 mm/yr) and weight loss (0.049 g).

References

1. S. Marappan, L. Kasirajan, and V. Shanmugam, . Friction Stir Welding Experiments on Az31b Alloy to Analyse Mechanical Properties and Optimize Process Variables by TOPSIS Method. Tehnički vjesnik, 29(6) (2022) pp.1923–1930.
2. G. Parande, V. Manakari, S.D. Kopparthy, M. Gupta M, A study on the effect of low-cost eggshell reinforcement on the immersion, damping and mechanical properties of magnesium–zinc alloy. Compos. B. Eng. 182 (2020) 107650.
3. T. Zhang, H. Cui , Cui X, H. Chen , E. Zhao , L. Chang, Y. Pan, R. Feng, S. Zhai, S. Chai, Effect of addition of small amounts of samarium on microstructural evolution and mechanical properties enhancement of an as-extruded ZK60 magnesium alloy sheet. J. Mater. Res. Technol. 9(1) (2020) 133–41.
4. B. Mansoor, A.K. Ghosh, Microstructure and tensile behavior of a friction stir processed magnesium alloy. Acta Mater. 60(13-14) (2012) pp.5079–5088.
5. P. Yogaraj, L. Kasirajan, B. Senthamaraikannan, Effect of Tool Pin Positioning Factors on the Strength Behavior of Dissimilar Joints of AA5754-H111 and AA6101-T6 by Using Friction Stir Welding. Trans. Indian Ins. Met. 76 (2023) 3021–3030.
6. S. Balamurugan, K. Jayakumar, and K. Subbaiah, Influence of Friction Stir Welding Parameters on Dissimilar Joints AA6061-T6 and AA5052- H32, Arab. J. Sci. Eng. 46 (12) (2021) 11985–11998.
7. P. Yogaraj, L. Kasirajan, Identifying the Optimal Process Parameter on AA1100 Friction Stir Welded Joints. Tehnički vjesnik, 29(3) (2022) pp.957-964.
8. S. Balamurugan, K. Jayakumar, B. Anbarasan, M. Rajesh, Effect of tool pin shapes on microstructure and mechanical behaviour of friction stir welding of dissimilar aluminium alloys. Mater. Today: Proc, 72 (2023) pp.2181–2185.
9. T. Rajkumar, S. Dinesh, B. Anbarasan, S. Balamurugan, Effect of welding process parameters on surface topography and mechanical properties of friction-stir-welded AA2024/AA2099 alloys. Mater. Res. Exp. 10(2) (2023)
10. B. Senthamaraikannan, J. Krishnamoorthy, Material flow and mechanical properties of friction stir welded AA 5052-H32 and AA6061-T6 alloys with Sc interlayer. Mater. Test. (2023)
11. A. Dhanapal, S.R Boopathy, V. Balasubramanian, Influence of pH value, chloride ion concentration and immersion time on corrosion rate of friction stir welded AZ61A magnesium alloy weldments. J. Alloy. Compound. 523 (2012) 49-60.
12. S. Balamurugan, K. Jayakumar, C. Nandakumar, Investigation of mechanical, metallurgical and corrosion characteristics of friction stir welded dissimilar AA 5052-H32 and AA 6061-T6 joints. J. Chin. Ins. Eng. (2023) pp.1–14.
13. D. Liu, M. Shen, Y. Tang, Y. Hu, L. Zhao, Effect of multi-pass friction stir processing on surface corrosion resistance and wear resistance of ZK60 alloy. Met. Mater. Int. 25 (2019) pp.1182–1190.
14. H. Seifiyan, M.H. Sohi, M. Ansari, D. Ahmadkhaniha M. Saremi, Influence of friction stir processing conditions on corrosion behavior of AZ31B magnesium alloy. J. Magnes. Alloy. 7(4) (2019) pp.605–616.
15. D. Liu, J. Yu, Y. Zhang, Y. Zhang, M. Shen,. Corrosion behavior of friction stir-welded AZ31 Mg alloy after plastic deformation. Mater. Corr. 72(8) (2021) pp.1294–1304.
16. H. Pan, G. Qin, Y. Huang, Y. Ren, X. Sha, X. Han, Z. Q. Liu, C. Li, X. Wu, H. Chen, C. He, Development of low-alloyed and rare-earth-free magnesium alloys having ultra-high strength. Acta Mater. 149 (2018) pp.350–363.
17. N. Xu, Z. Ren, Z. Lu, J. Shen, Q. Song, J. Zhao, Y. Bao, Improved microstructure and mechanical properties of friction stir-welded AZ61 Mg alloy joint. J. Mater. Res. Technol. 18 (2022) pp.2608–2619.

18. Q. Zang, H. Chen, J. Zhang, L. Wang, S. Chen, Y. Jin, Microstructure, mechanical properties and corrosion resistance of AZ31/GNPs composites prepared by friction stir processing.Journal of Mater. Res.Technol. 14 (2021) pp.195–201.
19. K. Qiao, T. Zhang, K. Wang, S. Yuan, L. Wang, S. Chen, Y. Wang, K. Xue, W. Wang, Effect of multi-pass friction stir processing on the microstructure evolution and corrosion behavior of ZrO2/AZ31 magnesium matrix composite. J. Mater. Res. Technol. 18 (2022) pp.1166–1179.
20. G.R. Argade, K. Kandasamy, S. K. Panigrahi, R.S. Mishra, Corrosion behavior of a friction stir processed rare-earth added magnesium alloy. Corro. Sci. 58, (2012) pp.321–326.
21. F. Liu, A. Li, Z. Shen, H. Chen, Y. Ji, Microstructure and corrosion behavior of Al-Ti-TiC-CNTs/AZ31 magnesium matrix composites prepared using laser cladding and high speed friction stir processing. Optic. Laser Technol. 152 (2022) p.108078.
22. T. O. Olugbade, B. O. Omiyale, O.T. Ojo, Corrosion, corrosion fatigue, and protection of magnesium alloys: mechanisms, measurements, and mitigation. J. Mater. Eng Perform. (2021) pp.1–21.
23. K. Qiao,T. Zhang, K. Wang, S. Yuan, L. Wang, S. Chen, Y. Wang, K. Xue, W. Wang, Effect of multi-pass friction stir processing on the microstructure evolution and corrosion behavior of ZrO2/AZ31 magnesium matrix composite. J. Mater. Res. Technol, 18 (2022) pp.1166–1179.
24. M. Kumar, A. Das, R. Ballav, Influence of the Zn interlayer on the mechanical strength, corrosion and microstructural behavior of friction stir-welded 6061-T6 aluminium alloy and AZ61 magnesium alloy dissimilar joints. Mater. Today Commun. 35 (2023) p.105509.

Note: Every figure and table was created using the experimental work; none of them were taken from any publications or online.

Advances in Additive Manufacturing Technologies – Gurusamy Pathinettampadian (eds)
© 2024 Taylor & Francis Group, London, *ISBN 978-1-032-90013-1*

36 Design and Analysis of Asymmetric Circular Flexure Hinges using Micro Gripper for Compliant Mechanism

V Gopal*
Department of Mechanical Engineering,
KCG college of Technology Chennai, Tamil Nadu, India

R. Bharanidaran
Department of Design and Automation,
School of Mechanical Engineering, Vellore Institute of Technology,
Vellore, Tamil Nadu, India

M Antony Stevewake, N Jayakanthan
Department of Mechanical Engineering,
KCG college of Technology Chennai, Tamil Nadu, India

ABSTRACT: The design and analysis of asymmetric circular flexure hinges for a micro gripper in compliant mechanisms is the subject of this abstract. Compliant mechanisms are gaining prominence due to their ability to achieve intricate motions and force transmission through elastic deformation, eliminating the need for traditional joints. In this study, the focus is on asymmetric circular flexure hinges integrated into a micro gripper system. The micro gripper is designed to manipulate objects at a miniature scale, which finds applications in fields such as micro-assembly, micro-surgery, and micro-manipulation. The fundamental concept involves utilizing compliant hinges that allow controlled motion while maintaining structural integrity. Asymmetric circular flexure hinges are particularly interesting due to their anisotropic behavior, providing different stiffness and deformation characteristics in different directions. The objectives of the research include the geometric design of the micro gripper with integrated asymmetric circular flexure hinges, numerical modeling of the compliant mechanism's behavior, and performance analysis in terms of gripping capabilities, motion precision, and force exertion. Advanced simulation and analysis tools are employed to predict the behavior of the micro gripper under various loading conditions. The research contributes to the field by presenting a systematic approach to designing and analyzing compliant mechanisms for micro manipulation. The unique characteristics of asymmetric circular flexure hinges enable enhanced control over the gripper's motion and force application, leading to improved performance in various applications. The findings of this study could pave the way for the development of more efficient and effective micro grippers, impacting industries that rely on precise manipulation at the microscale.

KEYWORDS: Compliant mechanism, Micro gripper, Circular flexure hinge, Asymmetric design, Motion control, Force transmission, Micro manipulation

*Corresponding author: gopal@kcgcollege.com

DOI: 10.1201/9781003545774-36

1. INTRODUCTION

1.1 MEMS (Micro Electro Mechanical System)

pressure sensors, microphones, inkjet printheads, and microvalves are examples of MEMS devices. They have revolutionized various industries by providing compact, low-power, and cost-effective solutions for a wide range of sensing and actuation tasks.

MEMS devices are often created using semiconductor fabrication techniques, which involve processes like photolithography, deposition, etching, and bonding. These techniques allow for the precise and consistent fabrication of intricate structures on a small scale. Some common examples of MEMS applications include: Consumer Electronics: Accelerometers and gyroscopes in smartphones enable screen orientation changes and motion sensing. MEMS microphones are used in various portable devices for audio input. Automotive Systems: MEMS pressure sensors are used in tire pressure monitoring systems and airbag deployment systems. Accelerometers and gyroscopes help in vehicle stability control and navigation systems. Medical Devices: MEMS-based pressure sensors can be found in devices like infusion pumps, blood pressure monitors, and medical implants. Microfluidic devices are also used for lab-on-a-chip applications. Aerospace: MEMS sensors are used in navigation systems, inertial measurement units (IMUs), and altitude control systems for satellites and aircraft. Telecommunications: MEMS optical switches and tunable capacitors are used in communication networks to manage light signals and frequencies. Environmental Monitoring: MEMS sensors can be deployed for measuring environmental parameters such as temperature, humidity, and gas concentrations.

Microfluidics: Microvalves and micropumps are employed in lab-on-a-chip devices for performing chemical and biological analyses on a miniature scale.

Energy Harvesting: MEMS devices can be used to capture and convert ambient vibrations into electrical energy. The miniaturization and integration of mechanical components with electronics in MEMS devices have led to innovative solutions that were not feasible with traditional macro-scale technologies. MEMS technology continues to advance, enabling even smaller and more efficient devices with improved performance and new functionalities. Micromanipulations of micro-sized components with high precision are inevitable in the field of micro-assembly, micro-robotics, electronics, optics, drug delivery, tissue manipulation, and minimally invasive surgery. [1–6]. This reduces the design process conversion of a rigid link to compliant mechanism, though challenging.[7]. In the design process, factors such the range of motion, degree of axis drift, ratio of off-axis stiffness to axial stiffness, and stress concentration effects are taken into account in an effort to improve flexures and get around these problems.(8) Wu and Zhou created a succinct equation that produced the same findings and percentage of error as Paros and Weisbord's design equation for evaluating the compliance of single-axis and two-axis circular cutout constant cross-section flexure hinges [9,10].Wu and Zhou devised a succinct equation that yielded the same results and percentage of error as Paros and Weisbord. Paros and Weisbord gave the design equation for predicting the compliance of single axis and two-axis circular cutout constant cross-section flexure hinges.[10-13] Different kinds of contours, like V-shaped notches, conic sections, circular, elliptic, parabolic, hyperbolic, quadratic, and rational Bézier curves.

2. DESIGN OF ASYMMETRIC CIRCULAR FLEXURE HINGES USING MICROGRIPPER STRUCTURE

Design of asymmetric circular flexure hinges using microgripper design, various compliant mechanism types are utilized to achieve high precision and reliability. Some common types include flexure-based designs, compliant parallelograms, and compliant hinges. Here's a breakdown of the information provided:

Flexure-Based Design Concept: Flexure-based designs offer an innovative and efficient approach to address limitations associated with traditional mechanisms. This approach allows for improved functionality, reliability, and miniaturization in various engineering applications. The use of flexure hinges instead of traditional rigid joints enables elastic deformation, which aids in precise motion transmission and manipulation of micro-sized components.

Microgripper Structure Design: The microgripper's structure is designed based on the flexure-based design concept [14]. This involves considering factors such as material selection, fabrication techniques, cost considerations, and the required range of motion. These factors contribute to creating a microgripper that meets the specific needs of the application while ensuring reliable and accurate manipulation of micro-sized parts.

Computer-Aided Design (CAD): The detailed 3D model of the microgripper is created using computer-aided design (CAD) software, specifically mentioned as SOLIDWORKS. CAD software allows designers to visually represent and simulate the physical structure of the microgripper in a digital environment.

Figure 36.1: A visual representation of the detailed 3D model of the microgripper is shown in Fig. 36.1. This likely provides an illustrative depiction of the designed micro-gripper, showcasing its configuration and key features.

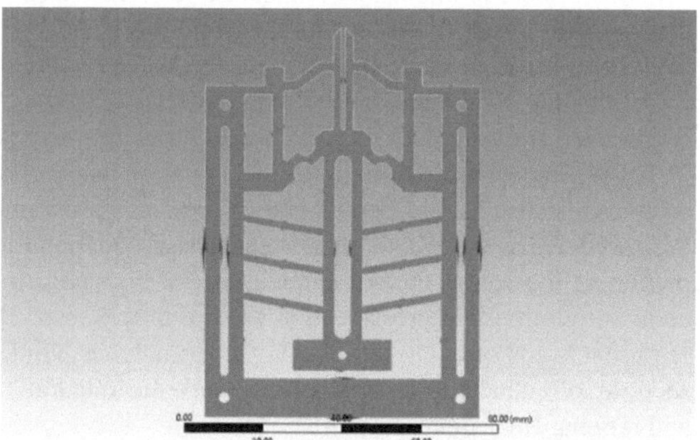

Fig. 36.1 Model of asymmetric circular flexure hinges using micro gripper

Computer-Aided Engineering (CAE) Analysis: To evaluate the performance and behavior of the microgripper, computer-aided engineering (CAE) tools are employed. In this case, Ansys Workbench is mentioned as the software used for analysis. CAE enables simulation and analysis of the microgripper's mechanical properties, response to various loads, and potential failure points. [15-16]

IGES File Format: The 3D model of theasymmetric circular flexure hinges using microgripper created in SOLIDWORKS is converted into the IGES (Initial Graphics Exchange Specification) file format. This format is widely used for transferring geometric data between different CAD and CAE software, allowing for seamless integration and analysis in Ansys Workbench.

In summary, the text provides an overview of how the flexure-based design concept is applied to microgripper design. It highlights the importance of considering various factors during the design process and the use of CAD and CAE software for modeling, simulation, and analysis. This comprehensive approach ensures the creation of a functional and reliable microgripper capable of high-precision manipulation in various applications

3. Numerical Analysis

It sounds like you have conducted a comprehensive analysis of an Asymmetric Circular Flexure Hingesmicrogripper using finite element analysis (FEA) in Ansys Workbench. You have performed static structural analysis, modal analysis, and simulations to assess the mechanical behavior of the compliant structure. The key findings from your analysis include:

Static Structural Analysis: Static structural analysis was carried out to understand the mechanical behavior of the microgripper under applied forces. The results revealed that the maximum deformation occurred at the gripper end, with a total deformation of 0.35 mm. Additionally, due to the parallel movement, the total deformation increased to 0.7 mm.

Deformation and Deflection: The analysis showed that the rigid body did not experience any deflection. The primary deflections were observed in the flexible members, specifically in the asymmetric circular flexure hinges using flexure hinges of the microgripper.

Modal Analysis: You mentioned modal analysis, which typically focuses on determining the natural frequencies and mode shapes of a structure. However, you did not provide specific findings related to this aspect.

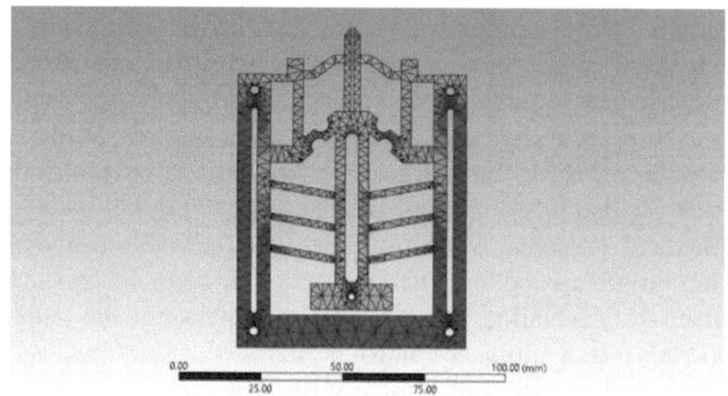

Fig. 36.2 Total deformation

It sounds like you're describing an experiment or setup involving Shape Memory Alloy (SMA) wire connected to a gripper, where a load is applied in the Y-axis direction. The deformation along the X-axis is measured at 0.774 mm, with the maximum deformation occurring at the gripper end region. Additionally, you mention that Fig. 36.3 illustrates the displacement along the X-axis of the microgripper.

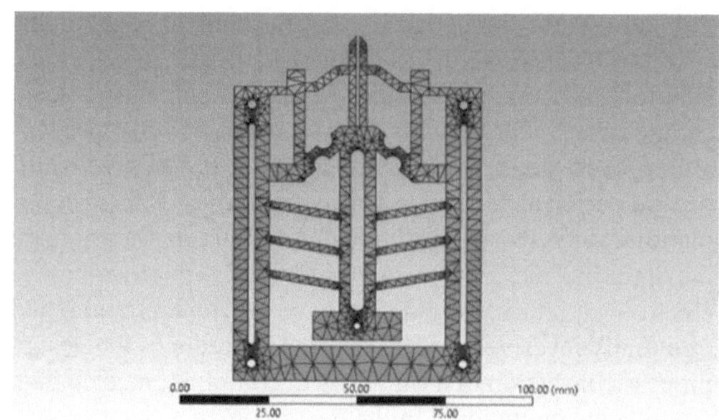

Fig. 36.3 X-Axis deformation

If you have any specific questions about this setup, the experiment, or the results shown in Fig. 36.3, please feel free to ask. It seems like you're describing a mechanical testing or characterization of the SMA wire and gripper system under load, potentially studying its deformation behavior or performance

Based on the information you provided, it seems that the deformation along the Y-axis is negligible or very close to zero. This information is confirmed by Fig. 36.4, which likely illustrates the deformation along the Y-axis and shows that it is almost non-existent. Deformation in this context typically refers to a change in shape or size of an object due to various factors such as stress, strain, or external forces.

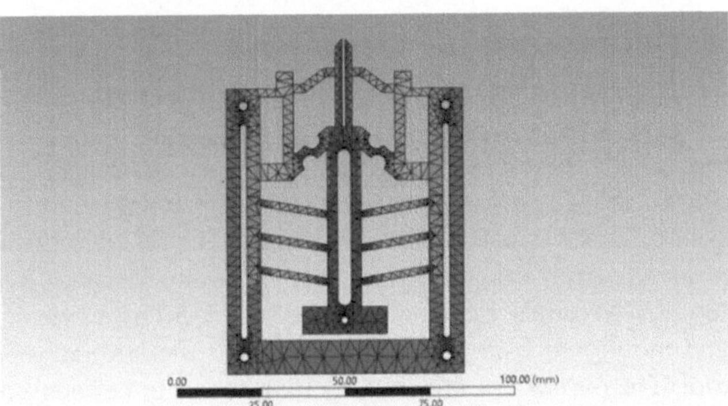

Fig. 36.4 Y-Axis deformation

The Fig. 36.5 shows the Microgripper Max Displacement with respect to Load Applied in Y-Axis. The results where obtained using the Ansys workbench software and the Origin software is used to plot the graph

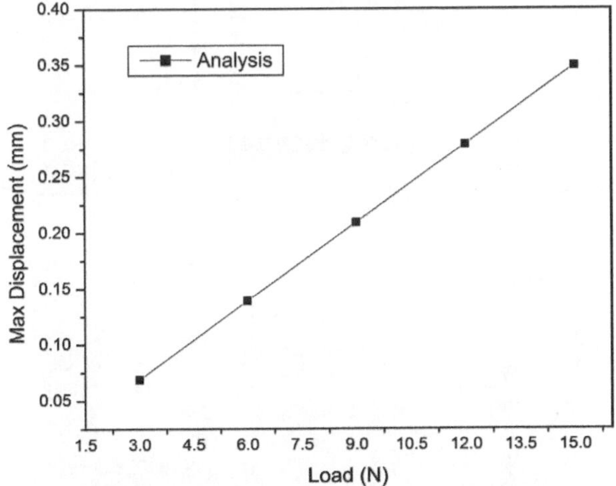

Fig. 36.5 Numerical analysis result graph

4. Modal Analysis

Modal analysis is indeed a technique used to study the dynamic characteristics of structures, such as natural frequencies, mode shapes, and damping ratios. Let's break down your description further:

Gripper and Working Conditions: In your context, it seems you're discussing a gripper – a device used to hold or manipulate objects. It's important to ensure that the gripper's operational frequencies (vibrational frequencies) do not coincide with the natural frequencies of the gripper itself or the system it's interacting with. If the gripper's frequencies match the natural frequencies of the system it's gripping or the machinery it's attached to, resonance could occur, potentially causing damage to the gripper or the entire system.

Suitability and Analysis: The purpose of the modal analysis is to determine whether the gripper is suitable for the working conditions it will be subjected to. By identifying the natural frequencies and mode shapes of the gripper and the surrounding system, engineers can assess whether there is a risk of resonance under various operating conditions. If the natural frequencies of the gripper are significantly different from the operational frequencies, the risk of resonance-related failure is minimized.

Fundamental Frequencies: In modal analysis, the fundamental frequencies in Fig. 36.6 and 36.7 (also referred to as natural frequencies) are the lowest frequencies at which a structure can vibrate. These frequencies correspond to the modes in which the structure vibrates with the greatest amplitude. The first five fundamental frequencies provide valuable information about the gripper's behavior and its interaction with the environment.

Fig. 36.6 Fundamental frequency 1

In summary, modal analysis is a crucial tool for engineers to ensure that structures like grippers are designed to avoid resonance-related issues and can operate effectively and safely under various conditions. It helps identify potential vibration-related problems and supports the optimization of design parameters to mitigate these issues.

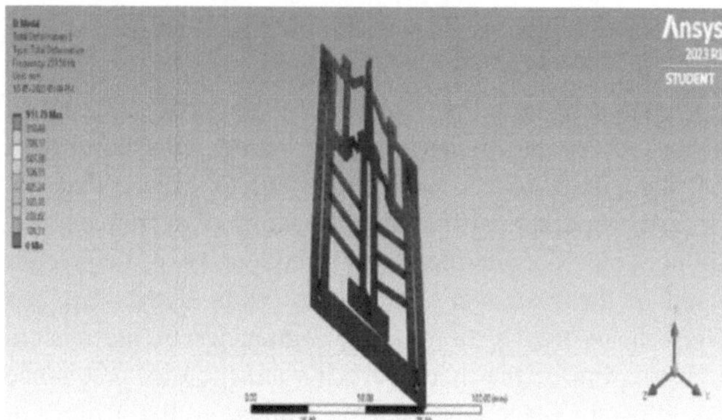

Fig. 36.7 Fundamental frequency 2

5. Experimental Work

An experimental investigation involving a compliant-based microgripper with asymmetric circular flexure hinges. The asymmetric circular flexure hinges using microgripper is fabricated using wire cut EDM (Electrical Discharge Machining), and its output displacement is being measured using measuring software. The gripper's actuation is achieved using SMA (Shape Memory Alloy) wire. This setup appears to be focused on studying the performance and behavior of the microgripper.

Asymmetric Circular Flexure Hinges Microgripper Design: The microgripper is designed with asymmetric circular flexure hinges. These hinges are likely designed to provide compliant motion, allowing the gripper to open and close while maintaining precision control over its movement.

Fabrication using Wire Cut EDM: The microgripper is manufactured using wire cut EDM. This machining process involves using electrical discharges to cut and shape the material into the desired microgripper structure. Wire EDM is often used for precise and intricate designs, which is crucial for micro-scale devices.

Output Displacement Measurement: The output displacement of the microgripper is a crucial parameter to understand its performance. Measuring software is used to quantify how much the gripper opens and closes in response to different inputs or stimuli.

SMA Wire Actuation: The microgripper is actuated using Shape Memory Alloy (SMA) wire. SMAs are materials that can "remember" a particular shape and return to it when heated or subjected to specific conditions. In this case, the SMA wire is likely heated to induce a change in shape, causing the gripper to open or close.

Overall, this experiment seems to be investigating the behavior and performance of the microgripper design, focusing on its compliant motion, actuation using SMA wire, and precise measurement of output displacement in Fig. 36.8.

Fig. 36.8 Experimental setup

6. Experimental Model of Asymmetric Circular Flexure Hinges using Microgripper

Shown in Fig. 36.9, An experimental setup and process involving a Asymmetric Circular Flexure Hingesmicrogripper made from structural steel, using a Shape Memory Alloy (SMA) wire actuator, controlled by a 555 Timer circuit, and monitored using a microscope and software analysis tool. The purpose of this investigation seems to be focused on studying the structural performance and deformation of the microgripper. Here's a summary of the key points you've mentioned:

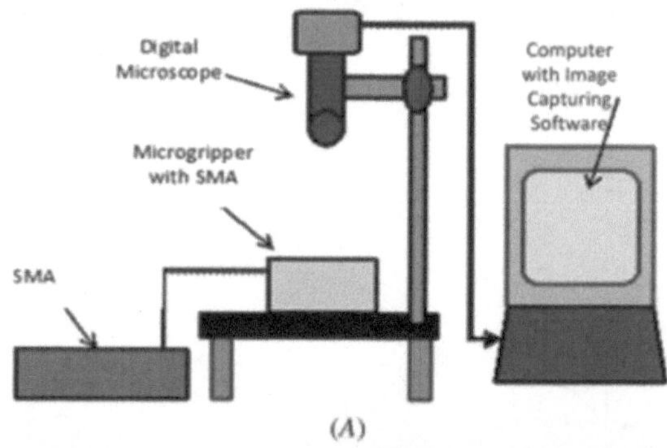

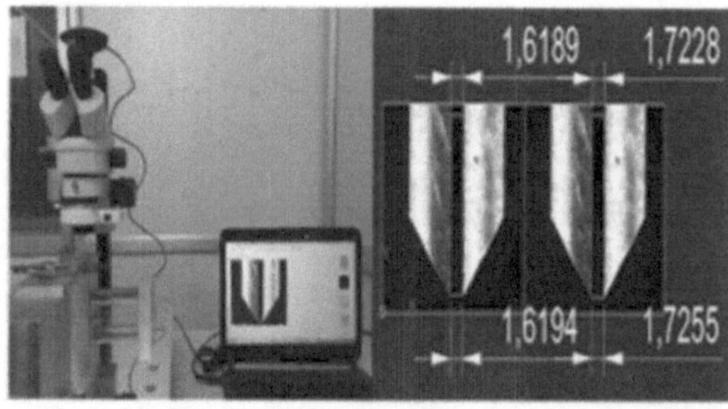

Fig. 36.9 Schematic of experimental setup; (B) experimental setup; (C) microgripper during operation (left) before and (right) after actuation

Experimental Setup: The Asymmetric Circular Flexure Hingesmicrogripper is mounted on a table, and the actuator is also placed on the table. A microscope is connected to a computer to capture the motion of the microgripper jaws. This setup allows for real-time monitoring and analysis of the microgripper's behavior.

Image Analysis: The captured images are analyzed using the microanalyser software tool. This analysis provides measurements of total deformation as well as deformation in the X and Y directions.

Deformation Measurement: The X-directional deformation, which is the difference between the initial and final gap between the microgripper jaws, is measured as 0.2006 mm. The Y-directional movement, which ensures the parallel movement of the microgripper, is measured as 0.0066 mm and is considered negligible. The purpose of this experimental investigation seems to be understanding and characterizing the performance and deformation behavior of the microgripper under the influence of the SMA wire actuator. The setup and analysis process outlined here are designed to provide valuable insights into the microgripper's mechanical behavior and its response to the actuation mechanism. This information could have potential applications in various fields, including microfabrication, micro-assembly, and precision engineering.

7. Conclusion

Developed a novel ASYMMETRIC CIRCULAR FLEXURE HINGES USING microgripper with a complaint mechanism using asymmetric circular flexure hinges for revolute joints. The mechanism has been designed, manufactured, and tested for its performance using numerical investigation in ANSYS Workbench as well as experimental techniques. The microgripper aims to satisfy several key characteristics, including the ability to firmly hold objects without causing damage, high positioning accuracy, and the capability to manipulate objects within a certain range. The asymmetric circular flexure hinges USING microgripper's design allows for gentle grasping of objects with different shapes and provides the necessary degrees of freedom for effective manipulation. It's important to note that the microgripper design has a range of applications, from micro-scale to nano-scale manipulation, making it versatile and adaptable for different scenarios. The experimental deformation measurement of 0.38mm suggests that the mechanism is effective in achieving the desired gripping and manipulation capabilities.

References

1. S. K. Nah and Z. W. Zhong, "A Microgripper Using Piezoelectric Actuation for Micro-Object Manipulation," Sensors and Actuators A: Physical 133, no. 1 (January 2007): 218–224. https://doi.org/10.1016/j.sna.2006.03.014
2. P. Dario, M. C. Carrozza, A. Benvenuto, and A. Menciassi, "Micro-Systems in Biomedical Applications," Journal of Micromechanics and Microengineering 10, no. 2 (2000): 235–244. https://doi.org/10.1088/0960-1317/10/2/322
3. Z. W. Zhong and Z. Zheng, "Flying Height Deviations in Glide Height Tests," Sensors and Actuators A: Physical 105, no. 3 (August 2003): 255–260. https://doi.org/10.1016/S0924-4247(03)00085-2
4. S. Jun and Z. Zhaowei, "Finite Element Analysis of a IBM Suspension Integrated with a PZT Microactuator," Sensors and Actuators A: Physical 100, nos. 2–3 (September 2002): 257–263. https://doi.org/10.1016/S0924-4247(02)00067-5
5. Z. W. Zhong and S. H. Gee, "Failure Analysis of Ultrasonic Pitting and Carbon Voids on Magnetic Recording Disks," Ceramics International 30, no. 7 (2004): 1619–1622. https://doi.org/10.1016/j.ceramint.2003.12.174
6. A. Alogla, P. Scanlan, W. M. Shu, and R. L. Reuben, "A Scalable Syringe-Actuated Microgripper for Biological Manipulation," Procedia Engineering 47 (2012): 882–885. https://doi.org/10.1016/j.sna.2012.12.034
7. N. Lobontiu, J. S. N. Paine, E. O'Malley, and M. Samuelson, "Parabolic and Hyperbolic Flexure Hinges: Flexibility, Motion Precision and Stress Characterization Based on Compliance Closed-Form Equations," Precision Engineering 26, no. 2 (April 2002): 183–192. https://doi.org/10.1016/S0141-6359(01)00108-8
8. B. P. Trease, Y.-M. Moon, and S. Kota, "Design of Large-Displacement Compliant Joints," Journal of Mechanical Design 127, no. 4 (November 2005): 788. https://doi.org/10.1115/1.1900149
9. J. M. Paros and L. Weisbord, "How to Design Flexure Hinges," Machine Design 25 (1965): 3101–3106.
10. Y. Wu and Z. Zhou, "Design Calculations for Flexure Hinges," Review of Scientific Instruments 73 (2002). https://doi.org/ 10.1063/1.1494855
11. R. Bharanidaran and T. Ramesh, "A Modified Post-Processing Technique to Design a Compliant Based Microgripper with a Plunger Using Topological Optimization," The International Journal of Advanced Manufacturing Technology 93, nos. 1–4 (October 2017): 103–112. https://doi.org/10.1007/s00170-015-7801-z
12. R. Bharanidaran and T. Ramesh, "Numerical Simulation and Experimental Investigation of a Topologically Optimized Compliant Microgripper," Sensors and Actuators A: Physical 205 (January 2014): 156–163. https://doi.org/10.1016/j.sna. 2013.11.011
13. Y. Tian, B. Shirinzadeh, and D. Zhang, "Closed-Form Compliance Equations of Filleted V-Shaped Flexure Hinges for Compliant Mechanism Design," Precision Engineering 34, no. 3 (July 2010): 408–418. https://doi.org/10.1016/ j.precisioneng.2009.10.002
14. V. Gopal, M. Alphin, and R. Bharanidaran,; Design of Compliant Mechanism MicrogripperUtilizing the Hoekens Straight Line Mechanism," Journal of Testing and Evaluation 49, no. 3 (2021): 1599–1612. https://doi.org/10.1520/JTE20190091.
15. K. Gobivel, K.S. Vijay Sekar, "Investigation on the effect of TiN and Al2O3 coated tools in the Machining of Ti-6Al-4 V alloy", Materials Today: Proceedings, Volume 62, Part 2, 2022, Pages 920–924. https://doi.org/10.1016/j.matpr.2022.04.071.
16. V. Gopal, D.M.R. Raja, 2021. Mechanical Behaviour of Al7075 Hybrid Composites Developed through Squeeze Casting, Int. J. Vehicle Structures & Systems, 13(3), 314–318. doi:10.4273/ijvss.13.3.14.

Note: Every figure was created using the experimental work; none of them were taken from any publications or online.

Advances in Additive Manufacturing Technologies – Gurusamy Pathinettampadian (eds)
© 2024 Taylor & Francis Group, London, ISBN 978-1-032-90013-1

37 An Investigation into the Effects of a Scandium Inclusion on the Mechanical and Micro-structural Characteristics of Friction Stir Weld Joint

Shine. K[1], Ragesh PR[2],Deepu T[3], Sadanandan R[4]
Department of Mechanical Engineering,Jawaharlal College of Engineering and Technology, Palaksd, Kerala, India
Sasi Lakshmikhanth R[5]
Department of Mechanical Engineering,Chennai Institute of Technology, Chennai, Tamil Nadu, India
Satheesh S.S[6]
Assistant Manager, Renault Nissan Technology and Business Centre India Pvt. Ltd. Mahindra World City, Chengalpattu, Tamil Nadu – 603 002

ABSTRACT: This study focuses on the impact of adding scandium interlayer on metallurgical and mechanical characteristics of dissimilar FSWed AA5754-AA6101 aluminium alloys. The optimal welding process parameters (WPP) for joining AA5754 and AA6101 was determined using Central composite design (CCD) of Response Surface Methodology (RSM). The SEM analysis is used to study the microstructural changes like grain and precipitates size and orientations. Tensile studies are carried out using universal testing machine. It is found from the tensile studies that the increase in % of scandium addition has increased the tensile strength. The micro-hardness studies using Vickers micro-hardness tester by applying 1kg load for a duration of 20 seconds, proved that the addition of scandium has increased the hardness in the WNZ. The mode of failure is analyzed using fractographic studies on the fractured surface and it is found that the increase in % of scandium reduces the ductility of the material.

KEYWORDS: Friction stir welding, AA5754, AA6101, Scandium interlayer, SEM, Fractography

1. INTRODUCTION

FSW was first developed in 1991 at the welding institute, which from then helped to join various alloys which are difficult to join by welding process [1]. As a solid state welding technique it has many advantages over conventional welding processes like low heat input and minimal deformation [2]. Compared to fusion welding, FSW produces joints with zero dendritic structure which leads to enhanced mechanical properties and more refined microstructures. Various researchers [3-5] have employed FSW in their studies to join aluminium alloys (AA). FSW produces heat which is induced by the rubbing of tool shoulder against the metal surface. The tool pin facilitates the plasticized material to flow from AS to RS and vice versa, hence the voids formed by the tool will be filled. The swept volume to pin volume ratio decides the flow of material during FSW process [6]. To attain a defect free weld the material flow must synchronize with the TRS and TTS. Improper selection of TRS and TTS will cause either high or low heat input during the FSW process. Reduced heat input during FSW process will lead to defects like void

[1]Shinekunnath@gmail, [2]rageshpr84@gmail.com, [3]deeputhiruthiyil@gmail.com, [4]sadanandanme@gmail.com4, [5]rslkhanth@gmail.com
[6]satheesh.009@gmail.com

DOI: 10.1201/9781003545774-37

formation, pin hole defects and tunnel defects. TPP has a very huge influence on the material flow characteristics. It helps to produce the required heat and it is also responsible for transferring the plasticized material from one side to the other side ensuring strong bond between two materials. An extensive research was carried out by several researchers [7-13] to investigate the impact of TPP on properties of FSW joint.

The AA5754 alloy is frequently utilized in the transportation and maritime sectors [14] because of its outstanding resistance to corrosion, increased strength-to-weight ratio, and good ability to weld. The enhanced strength of the AA5754-H111 alloy is attained by the processes of work hardening and solid solution strengthening, since it is not suitable for heat treatment. Conversely, aluminium alloy AA6101 is extensively used in several sectors like aerospace, automotive, and military [15-17]. AA6101 is a heat treatable AA, that contains magnesium and silicon, which create many strengthening precipitates, resulting in its high strength. Although these alloys possess intrinsic strength, fusion welding techniques tend to reduce their strength by promoting precipitate coarsening, which presents difficulties in connecting them. Nevertheless, the inclusion of scandium in Al-Mg alloys improves their strength, weldability, and mechanical characteristics [18]. Al-Mg-Sc alloy has enhanced resistance to fatigue fracture nucleation [19] and micro-crack propagation when compared to AA6013 alloy. These features are of utmost importance for these materials. The introduction of scandium results in the creation of Al_3Sc precipitates, which greatly enhance the UTS of Al-Mg-Sc alloys by using precipitate strengthening processes [20]. Moreover, Al3Sc particles maintain sub-grain characteristics seen in aluminium alloys, hence improving the yield strength [20] during processing. Scandium is a very efficient agent for improving the grain structure of recrystallized material and purifying intermetallic compounds in the second phase [21].

In this study a detailed investigation was done to optimize the WPP (TRS and TTS) along with the % of Sc that can be added to the WNZ of the FSW joints. The microstructural and mechanical characteristics of AA5754-AA6101 FSW joints fabricated with and without Sc alloy are studied and reported in detail.

2. Experimental Procedure

The 5 mm thick BM's AA5754 and AA6101 are utilized in this investigation. The welding samples are sliced into small pieces 150 mm x 55 mm in size from the BM plate. The elemental composition of the parent materials used for dissimilar friction stir welding is determined using OES, and presented in Table 37.1. Table 37.2 lists the tensile properties of the parent metals as determined by a universal testing machine. The Al-Mg-Sc interlayer is manufactured by stir casting method by mixing Al-Sc master alloy with AA5754 aluminium alloy in required weight percentage. The straight square tool with pin length and diameter of 4.7 mm and 5 mm respectively and shoulder diameter of 15 mm were used. The WPP selected are TRS, WS, %Sc. Table 37.3 summarises the limits of the WPP based on the literature.

Table 37.1 Elemental proportion (wt%) of BM

BM	Mg	Mn	Fe	Si	Cu	Cr	Zn	Ti	Al
5754	3.6	0.5	0.4	0.4	0.1	0.4	0.2	0.1	Bal.
6101	0.8	0.1	0.3	3.1	1.1	0.2	0.1	0.03	Bal.

Table 37.2 Tensile characteristics of the parent metals

BM	YS (MPa)	UTS (MPa)	%El
AA5754	98	229	26.31
AA6101	258	319	14.93

Table 37.3 Coded values assignment for the real values

WPP	Levels				
	-1.682	-1	0	1	1.682
TRS	600	722	900	1078	1200
WS	20	28	40	52	60
%Sc	0	0.2	0.5	0.8	1

Throughout the FSW welding experiment, aluminium alloy AA5754 was placed on the AS, while AA6101 was placed on the RS. A WEDM machine was utilised to cut a 1.5 mm wide segment of Al-Sc alloy from a cast plate, positioning it in the centre of its parent metals. FSW experiments were conducted employing a range of parameter combinations for welding, in accordance with the DOE formed using CCD of RSM. In order to characterize metallurgical, elemental, and mechanical properties, samples are extracted across the weld direction using the WEDM machine. Following the retrieval of the microstructural samples, they are mounted using a heated mounting press and subsequently polished. Following assembly, silicon carbide paper with a grain range of 80 to 2000 is utilised to refine the mounted samples. The isolated sample undergoes a microstructural evaluation by employing SEM. This evaluation aims to ascertain the characteristics, dimensions, and arrangement of the grains and precipitate. In order to ascertain the UTS of the welded samples, tensile tests were performed using UTM. Furthermore, to ascertain the characteristics of the failure, SEM was employed to examine the fractured surfaces of the tensile samples. Twenty seconds under a one-kilogram force on a Vickers microhardness instrument equipped with a diamond indenter, the material hardness was evaluated at various locations perpendicular to the weld.

3. Results and Discussions

3.1 Tensile Characteristics Evaluation and Development of Empirical Relationship

Tensile studies are conducted on all the welded samples using UTM and the results obtained are presented in Table 37.5. The UTS of the parent metals AA5754-H111 and AA6101-T6 were 321.34 and 325.08 respectively. The UTS of the sample joined without Sc is 186 MPa. The failure happened on the HAZ of the AA6101. The dissolving of the Mg_2Si precipitate in the AA6101 alloy's HAZ is the primary cause of the strength reduction. This is because precipitate strengthening is the primary strengthening method in heat treatable Al alloys [22, 23]. Grain growth developed in the HAZ of AA6101 due to significant quantity of heat flow that happened during the FSW process. This zone was discovered to be the most vulnerable zone in the weldment. Adding small quantity of Sc to aluminium results in a considerable improvement to the material's mechanical properties [24]. While somewhat decreasing the ductility of the aluminium alloy, scandium addition results in a higher 0.2% proof stress [25-27]. Scandium addition to Al-Mg alloy results in significantly high strength in comparison to commercially available Al-Mg alloy that has the same content of Mg. This is because the formation of Al_3Sc precipitates is triggered by the addition of Sc. [28].

The RSM method is used to establish an empirical relationship from the results obtained from the tensile strength evaluation. For developing the empirical relationship, the UTS is the output parameters and the WPP's are the input parameters. Table 37.4 gives the welding parameter combinations obtained using RSM along with the UTS obtained from samples welded using the given welding parameter combination.

$$\text{TS} = f\,(\text{TRS, WS, \%Sc}) \quad (1)$$

A regression equation portrays the response surface 'Z.' as:

$$Z = C_0 + \Sigma C_i X_i + \Sigma C_{ii} X_i^2 + \Sigma C_{ij} X_i X_j \quad (2)$$

The developed empirical relationship is given below:

$$\begin{aligned} \text{Tensile} = {} & 210.82 - 3.99(\text{TRS}) + 2.77(\text{WS}) \\ & + 10.51(\%\text{Sc}) + 0.375(\text{TRS*WS}) \\ & + 0.625(\text{TRS*\%Sc}) + 1.125(\text{WS*\%Sc}) \\ & - 11.56(\text{TRS}^2) - 13.33(\text{WS}^2) \\ & - 4.14(\%\text{Sc}^2) \end{aligned} \quad (3)$$

Table 37.4 Design matrix with tensile and fusion zone area values

Std. order	Run order	Coded Values			UTS (MPa)
		P	S	F	
1	3	-1	-1	-1	175
2	12	1	-1	-1	165
3	4	-1	1	-1	177
4	20	1	1	-1	169
5	16	-1	-1	1	192
6	5	1	-1	1	185
7	13	-1	1	1	199
8	9	1	1	1	193
9	2	-1.682	0	0	185
10	14	1.682	0	0	171
11	19	0	-1.682	0	168
12	17	0	1.682	0	178
13	8	0	0	-1.682	181
14	18	0	0	1.682	217
15	7	0	0	0	211
16	10	0	0	0	211
17	15	0	0	0	210
18	1	0	0	0	211
19	6	0	0	0	210
20	11	0	0	0	212

3.2 Assessing the Functionality of the Developed Empirical Relation

Analysis of variance (ANOVA) is used to qualify the efficacy of the formed empirical equation. Table 37.5 presents the ANOVA findings produced by the design expert programme. The effect of the amount of Sc added on the Ultimate Tensile Strength (UTS) is clearly obvious, with WS having the least impact on UTS. The model has statistical significance, as shown by a Model F-value of 1610.46. Within this particular framework, the model words TRS, TTS, %Sc, TRS*%Sc, TTS*%Sc, TRS^2, TTS^2, and $\%Sc^2$ have significant importance. On the other hand, model terms that have values greater than 0.1000 are deemed to be unimportant. By eliminating superfluous elements from the model, with the exception of those required for preserving hierarchy, its efficiency might be enhanced. An elevated Lack of Fit F-value indicates a 79.45% probability of noise interference. The marginal disparity between the Predicted R^2 (0.9976) and the Adjusted R^2 (0.9987), both below 0.2, indicates a satisfactory level of concurrence. Adeq Precision should ideally have a signal-to-noise ratio that surpasses 4. The signal strength is strong, with a ratio of 114.77. Efficiently use this paradigm to navigate the design space.

Table 37.5 ANOVA results

Source	Sum of Squares	df	Mean Square	F-Value	P-Value
Model	5985.87	9	665.10	1610.46	< 0.0001
TRS	217.85	1	217.85	527.50	< 0.0001
TTS	104.72	1	104.72	253.58	< 0.0001
%Sc	1508.77	1	1508.77	3653.31	< 0.0001
TRS*TTS	1.13	1	1.13	2.72	0.1299
TRS*%Sc	3.13	1	3.13	7.57	0.0205
TTS*%Sc	10.13	1	10.13	24.52	0.0006
TRS2	1927.19	1	1927.19	4666.46	< 0.0001
TTS2	2561.43	1	2561.43	6202.21	< 0.0001
%Sc2	246.94	1	246.94	597.93	< 0.0001
Residual	4.13	10	0.4130		
Lack of Fit	1.30	5	0.2593	0.4576	0.7945
Pure Error	2.83	5	0.5667		
Cor Total	5990.00	19			
Std. Dev.	0.6426	**R^2**		0.9993	
Mean	191.00	**Adjusted R^2**		0.9987	
C.V. %	0.3365	**Predicted R^2**		0.9976	
		Adeq Precision		114.7698	

3.3 Selection of Optimized Welding Parameter

RSM was successfully employed to optimise the welding parameters, and the coded values were used to frame the empirical relationship. The optimised coded numbers were then changed back into real values after being employed for optimization purposes. Utilizing design expert analytical software, the processed values are optimised. Confirmation tests were done utilizing the optimized WPP and the obtained tensile values are checked with the obtained values. Table 37.6 lists the predicted and experimentally determined parameters. The sample welded with the optimal welding parameters yielded a maximum UTS value of 215 MPa which is almost in alignment with the predicted tensile value of 217 MPa.

Table 37.6 Optimized WPP values

WPP	Optimized values	
	Predicted	**Actual**
TRS (rpm)	860	860
TTS (mm/min)	41.5	40
Percentage of Scandium (%)	0.76	0.75
Ultimate Tensile strength (MPa)	217	215

3.4 Microstructural Analysis

The SEM microstructures of the BM AA5754 and AA6101 is illustrated in Fig. 37.1(a) and 37.1(b) respectively. Fig. 37.1(c) depicts the weld nugget of a sample that was welded using optimized parameters, namely a TRS of 860 rpm, TTS of 40 mm/min, and a scandium insert with a concentration of 0.75%. FSW results in the formation of four main regions: WNZ, TMAZ,HAZ, and BMZ. The presence of many precipitates, such as $Al_6(Fe,Mn)$, Al_3Mg_2, $Al_6(Fe,Mn)Si$, and Mg_2Si, hinders the visibility of grain boundaries in AA5754. In contrast, AA6101-T6 consists only of Mg_2Si precipitates, which exhibit clear visibility of grain boundaries. The samples that were welded using optimised conditions showed complete material mixing and no presence of flaws. Significantly, Mg_2Si precipitates are often found in the base metal zone, but their presence is greatly decreased in the weld nugget zone as they dissolve owing to the intense heat generated during welding.

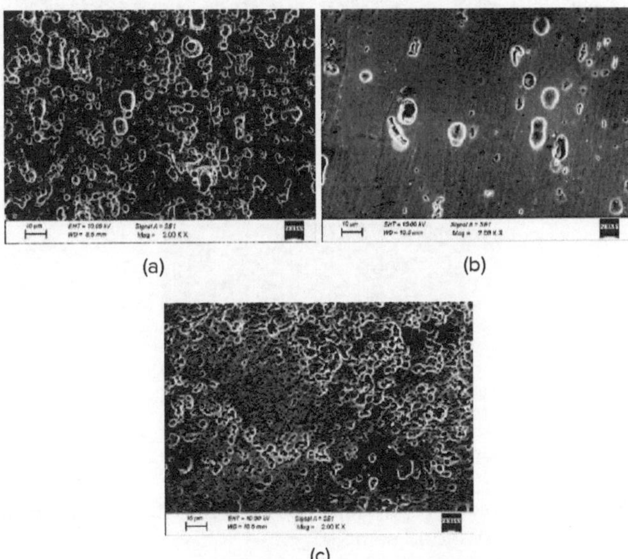

Fig. 37.1 SEM Images [a] 5754-H111 base metal, [b] 6101-T6 base material, [c] WNZ

The disintegration of AA6101-T6 weakens its strength, especially in the HAZ, which makes it the most vulnerable area in the weldment. The tool pin's movement triggers dynamic recrystallization, which leads to a decrease in grain size (GSZ) in the WNZ. This process is also facilitated by the use of scandium. The strength of a material is increased by smaller grain sizes due to the Hall-Petch strengthening processes. On the other hand, a significant amount of heat flow in the HAZ causes the GSZ to increase, which results in a weakening of this area. The use of scandium causes the creation of Al_3Sc precipitates in the WNZ, resulting in increase of tensile strength and hardness [25].

3.5 Fracture Surface Analysis using SEM

SEM fractographic study of tensile sample fragmented surfaces is shown in Fig. 37.2. The shattered surfaces of AA5754 and AA6101 base metals are shown in Figs. 37.2(a) and 37.2(b). Samples welded at optimum conditions (860 rpm TRS, 40 mm/min TTS, and 0.75% scandium insert) cracked as presented in Fig. 37.2(c). Fractography of the BM show several dimples of various diameters, suggesting ductile failure. Since aluminium alloys are ductile, AA5754 and AA6101 have dimpled structures. Figure 37.2(a) has many fine dimples, showing that base metal AA5754 is more ductile than AA6101. Both enormous dimples and facets are seen in SEM pictures of the shattered weldments. Facets indicate brittle breakdown, reducing ductility [29]. Addition of scandium enhances Al_3Sc precipitates, reducing ductility.

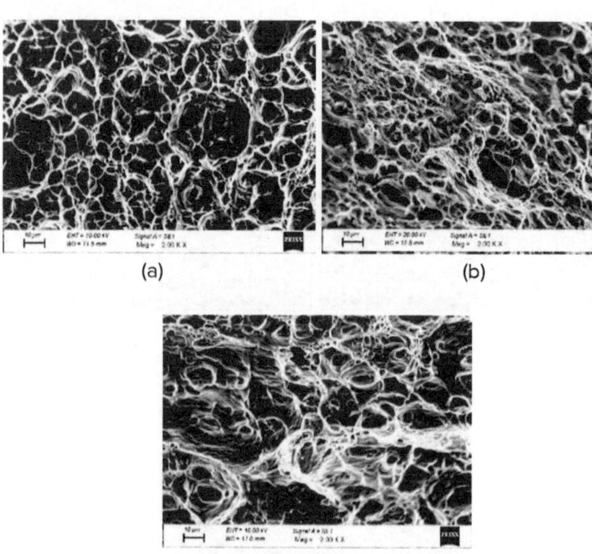

Fig. 37.2 Fractography (a) 5754-H111 BM, (b) 6101-T6 BM, (c) welded sample

3.6 Hardness Analysis Across the Weld Joint

The hardness comparison to examine scandium interlayer effect is shown in Fig. 37.3. Cross sections of weld zones are measured for hardness. Hardness ratings for base metals AA5754 and AA6101 are 75 and 100 HV1, respectively. The picture shows square symbols for conventional Friction Stir Welding (FSW) hardness survey graphs for AA5754 and AA6101. The FSW weld nugget has around 112 HV1 hardness. In particular, AA6101 has the lowest TMAZ/HAZ hardness at 60 HV1. It rises from 60 to 100 HV1 after 10 mm from WNZ. Tool pin motion causes grain recrystallization, which increases grain boundary density and hardness in the WNZ. Due to its high GSZ, the HAZ has decreased hardness. GSZ greatly affects hardness,

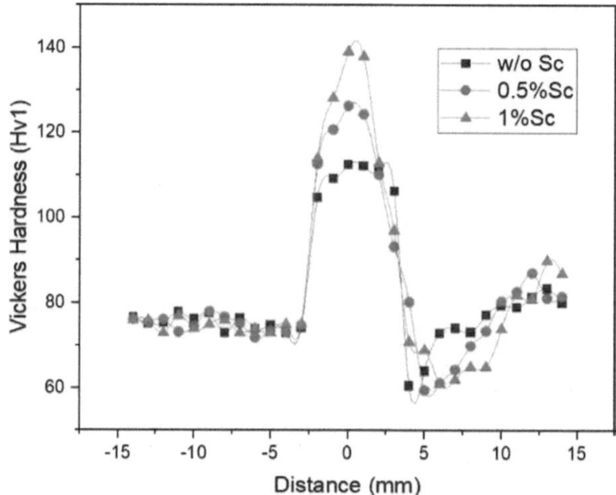

Fig. 37.3 Hardness comparison at different zones

according to Xunhong et al. [30]. The Hall-Petch equation states that reduced GSZ enhances hardness, this is why WNZ has higher hardness ratings. Similar observations were reported by Kulwant Singh [31]; The circle and triangle hardness charts reflect FSW joints of AA5754 and AA6101 with 0.5% and 1% scandium. Note that 0.5% and 1% scandium inlays increase weld nugget hardness to 126 and 139 HV1. The inclusion of scandium and grain refining create Al3Sc intermetallic particles, which boost hardness.

4. Conclusions

This study focus on the impact of Al-Sc insert on the properties of dissimilar AA5754-AA6101 joints, and the following conclusions were made:

1. RSM results utilising CCD matrix suggest optimal joining conditions for AA5754-H111 and AA6101-T6 aluminium alloys: 860 rpm TRS, 40 mm/min WS, and 0.75% Scandium insert.
2. Microstructural investigation using scanning electron microscopy showed Al3Sc precipitates in the WNZ, increasing its strength and hardness.
3. Fractography study showed decreased ductility in tensile specimens containing scandium in the weld nugget. Scandium lowers aluminium alloy ductility.
4. Samples connected with 0.5% and 1% scandium insert to weld nugget showed increased hardness at the WNZ compared to those joined without it.

References

1. Thomas W.M, Nicholas E.D, Needham J.C, Nurch M.G, Temple Smith P, Dawes C.J, "Friction stir butt welding", GB Patent Application No. 9125978.8, 6 December 1991.
2. Yang Jia, Sicong Lin, Jizi Liu, Yonggui Qin and Kehong Wang, "The influence of pre and post-heat treatment on

mechanical properties and microstructures in friction stir welding of dissimilar age-hardnenable aluminum alloys", Metals, vol. 9, 1162, pp. 1–15.

3. Hasan Jafari, hadi Mansouri, Mohammad Honarpisheh, "Investigation of residual stress distribution of dissimilar AA7075-T6 and Al-6061-T6 in the friction stir welding process strengthened with SiO_2 nanoparticles", Journal of Manufacturing Processes, vol. 43, (2019), pp. 145–153.
4. Noor Zaman Khan, Arshad Noor Siddiquee, Zahid A. Khan, Suha K. Shihab, "Investigations on tunnelling and kissing bond defects in FSW joints for dissimilar aluminum alloys", Journal of Alloys and Compounds, vol. 648, (2015), pp. 360–367.
5. Jenarthanan M.P, Varun Varma. C, Krishna Manohar. V, "Impact of friction stir welding (FSW) process parameters on tensile strength during dissimilar welds of AA2014 and AA6061", Materials Today: Proceedings vol. 5, (2018), pp. 14384–14391.
6. Thomas W.M, Dolby RE, David SA, Debroy T, Lippold JC, Smartt HB, "Friction stir welding developments", Proceedings of the Sixth International Conference on Trends in Welding Research, Pine Mountain, GA, (**2003**) pp. 203–211.
7. S. Ugender, A. Kumar, A. Somi Reddy, "Experimental investigation of tool geometry on mechanical properties of friction stir welding of AA2014 aluminium alloy", Procedia Materials Science, Vol. 5 (**2014**) pp. 824–831.
8. Javad Marzbanrad, Mostafa Akbari, Parviz Asadi, Samad Safaee, "Characterization of the influence of tool pin profile on microstructural and mechanical properties of friction stir welding", Metallurgical and Materials Transactions B, Vol. 45B (**2014**) pp. 1887–1894.
9. K. Elangovan, V. Balasubramanian, M. Valliappan, "Effect of tool pin profile and tool rotational speed on mechanical properties of friction stir welded AA6061 aluminium alloy", Materials and Manufacturing Processes, Vol. 23 (**2008**) pp. 251–260.
10. D.H. Lammlein, D.R. DeLapp, P.A. Fleming, A.M. Strauss, G.E. Cook, "The application of shoulderless conical tools in friction stir welding: An experimental and theoretical study", Materials and Design, Vol. 30 (**2009**) pp. 4012–4022.
11. K. Elangovan, V. Balasubramanian, "Influences of pin profile and rotational speed of the tool on the formation of friction stir processing zone in AA2219 aluminum alloy", Materials Science and Engineering A, Vol. 459 (**2007**) pp. 7–18.
12. S.M. Chowdhury, D.L. Chen, S.D. Bhole, X. Cao, "Effect of pin tool thread orientation on fatigue strength of friction stir welded AZ31B-H24 Mg butt joints", Procedia Engineering, Vol. 2 (**2010**) pp. 825–833.
13. K. Kumar, Satish V. Kailash, "The role of friction stir welding tool on material flow and weld formation", Materials Science and Engineering A, Vol. 485 (**2008**) pp. 367–374.
14. Lendvai. J, "Precipitation and strengthening in aluminum alloys", Materials Science Forum, vol. 217, (1996), pp. 43–56.
15. Lipeng Ding, Zhihong Jia, Jian-Feng Nie, Yaoyao Weng, Lingfei Cao, Houwen Chen, Xiaozhi Wu, Qing Liu, "The structural and compositional evolution of precipitates in Al-Mg-Si-Cu alloy", Acta Materialia, vol.145, (2018), pp. 437–450.
16. Murayama. M, Hono. K, "Pre-precipitate clusters and precipitation processes in Al-Mg-Si alloys", Acta Materialia, vol. 47, no. 5, (1999), pp. 1537–1548.
17. V.G. Davydov, V.I. Yelagin, V.V. Zakharov, Yu. A. Filatov, "On prospects of application of new 01570 high strength weldable Al-Mg-Sc alloy in aircraft industry", Materials Science Forum, Vol. 217-222 (1996) pp. 1841–1846.
18. T. Wirtz, G. Lutjering, A. Gysler, Blanka Lenczowski, R. Rauh, "Fatigue properties of the aluminium alloys 6013 and Al-Mg-Sc", Materials Science Forum, Vol. 331-337 (2000) pp. 1489–1494.
19. Ralph R. Sawtell and Craig L. Jensen, "Mechanical properties and microstructures of Al-Mg-Sc alloys", Metallurgical Transactions A, 21A (1990) 421–430.
20. Tadashi Aiura, Nobutaka Sugawara, Yasuhiro Miura, "The effect of scandium on the as-homogenized microstructure of 5083 alloy for extrusion", Materials Science and Engineering A 280 (2000) 139–145.
21. T.N. Jin, Z.R. Nie, G.F. Xu, H.Q. Ruan, J.J. Yang, J.B. Fu, T.Y. Zuo, "Effect of cooling rate on solidification behaviour of dilute Al-Sc and Al-Sc-Zr solid solution", Transactions of Nonferrous Metals Society of China, 14 (2004) 58–62.
22. M.W. Mahoney, C.G. Rhodes, J.G. Flintoff, R.A. Spurling, W.H. Bingel, "Properties of friction-stir-welded 7075 T651 aluminum", Metallurgical and Materials Transactions A, Vol. 29A (**1998**) pp. 1955–1964.
23. C.G. Rhodes, M.W.Mahoney, W.H. Bingel, "Effects of friction stir welding on microstructure of 7075 aluminum", Scripta Materialia, Vol. 36 (**1997**) pp. 69–75.
24. L.A. Willey, "Aluminium scandium alloy", US Patent No: 3,619,181 (1971).
25. S. Lathabai, P.G. Lloyd, "The effect of scandium on the microstructure, mechanical properties and weldability of a cast Al-Mg alloy", Acta Materialia, 50 (2002) 4275–4292.
26. Zhimin Yin, Qinglin Pan, Yonghong Zhang, Feng Jiang, "Effect of minor Sc and Zr on the microstructure and mechanical properties of Al-Mg based alloys", Materials Science and Engineering A, 280 (2000) 151–155.
27. Yu.A.Filatov, V.I. Yelagin, V.V. Zakharov "New Al-Mg-Sc alloys" Mat.sci. &Eng. A280 (2000) 97–101.
28. Yu. A. Filatov, "Deformable alloys based on the Al-Mg-Sc system", Metal Science and Heat Treatment, 38 (1996) 271–274.
29. Sasi Lakshmikhanth R, Lakshminarayanan A K, "On the mechanical, microstructural, and corrosion properties of pulsed gas tungsten arc and friction stir welded RZ5 rare earth grade magnesium alloy", Materials Research Express, 9, (2022), 126507.
30. Wang Xunhong, Wang Kuaishe, (2006), "Microstructure and properties of friction stir butt-welded AZ31 magnesium alloy", Materials Science and Engineering A, vol. 431, pp. 114–117.
31. Kulwant Singh, Gurbhinder Singh, Harmeet Singh, (2018), "Review on friction stir welding of magnesium alloys", Journal of Magnesium and Alloys, vol. 6, pp. 399–416.

Advances in Additive Manufacturing Technologies – Gurusamy Pathinettampadian (eds)
© 2024 Taylor & Francis Group, London, ISBN 978-1-032-90013-1

38 Salt Spray Corrosion Studies of Friction Stir Welded and Processed AZ31 Mg Alloys with Al_2O_3 and Cu

M. Selvaraj[1]
Assistant Professor, Department of Mechanical Engineering,
Paavai Engineering College, Namakkal,
Tamilnadu, India

S. Balamurugan[2]
Assistant Professor, Department of Mechanical Engineering,
Chennai Institute of Technology, Kundrathur,
Chennai, Tamilnadu, India

Dharmalingam[3]
Assistant Professor, Department of Mechanical Engineering,
Oasys Institute of Technology, Trichy,
Tamilnadu, India

Y. Premraj[4]
Assistant Professor, Department of Mechanical Engineering,
New Prince Shri Bhavani College of Engineering and Technology,
Chennai, Tamilnadu, India

ABSTRACT: The present investigation employed the Salt Spray Corrosion Test (SSCT) technique to examine the corrosion characteristics of two distinct friction stir processed (FSP) alloys, namely AZ31 B Mg alloys including Al_2O_3 and Cu. On 6 mm thick plates, the FSP was performed at various tool travel and rotational speeds. SSCT were performed in order to comprehend the general and particular corrosion reactions between the base material and the FSP joints. The outcomes exposed that there were noticeable variances in the microstructure of the different zones of the FSW joint, which partly affected how each zone deteriorated in a NaCl solution. The stir zone (SZ) has higher uniformity and corrosion resistance (CR) than the foundation material, based on the results of the SSCT. In comparison to base metals in the welded sample, the weight loss and amount of deterioration of the SZ were measured.

KEYWORDS: Friction stir processing, Mg-AZ31-B, Al_2O_3 and Cu, Salt spray corrosion test, Corrosion rate, Weight loss

1. INTRODUCTION

The automotive, rail transportation and aerospace industries find AZ31B magnesium alloy, one of the lightest structural materials, to be attractive for utilization [1, 2]. For joining magnesium AZ31B alloys, the solid-state welding method known as the Friction Stir Processing (FSP) offers significant advantages [3, 4]. Magnesium alloys treated with

[1]selva.msrk@gmail.com, [2]balamurugans@citchennai.net, [3]ss79425@gmail.com, [4]premrajme@gmail.com

DOI: 10.1201/9781003545774-38

the FSP process see significant grain refinement, complete recrystallization, and the development of a distinctive regional texture in the stirred zone (SZ) [5-7]. Thus, in some soft-oriented texture regions found in the SZ, the tensile properties, in particular for the yield strength (YS), are drastically reduced [8-10]. However, only a small number of researches have examined the effects of FSP AZ31 B Mg alloys' mechanical characteristics and CR. The corrosion resistance of Mg alloys has a considerable impact on their industrial application, as is well documented. Many studies have been conducted on the impact of the FSW technique on the CR of Mg alloys. The corrosion rates of FSW AZ61 Mg alloys in various settings, including those with varying pH levels, chloride ion percentages, and immersion times, were investigated by Dhanapal et al. [11]. Seifiyan et al.[12] discovered that the number of welding passes, traverse speed, and tool rotation have a substantial impact on the corrosion resistance of magnesium alloys. Balamurugan et al investigated the corrosion rate as well as weight loss for friction stir welded dissimilar aluminium alloys. They observed very few corrosion losses is exhibits in the welded region up to 72hrs. Also, they studied the SEM morphology of the different welded conditions [12-15].The impartial of this work is to synthesize a novel Mg-AZ31B utilizing FSP and particles (Al_2O_3 and Cu) as the second phase. As far as we are aware, this is completely novel, unpublished research. Additionally, the current study aims to generate fine grains with numerous structural elements to enhance the FSP Mg AZ61 with Al_2O_3 and Cu alloy joints' mechanical properties and corrosion behavior.

2. Materials and Methods

2.1 Friction Stir Processing

The traverse rates and tool rotational speeds were chosen based on available literature and the capabilities of the FSP machine [16-18]. Figure 38.1 shows FSP tensile weld joint samples (ASTM E8) made from various Mg AZ31B alloys utilizing locally built and designed FSW equipment. The process variables and tool shape for FSW of AA5754-H111/ Mg AZ61 alloys are shown in Table 38.1.

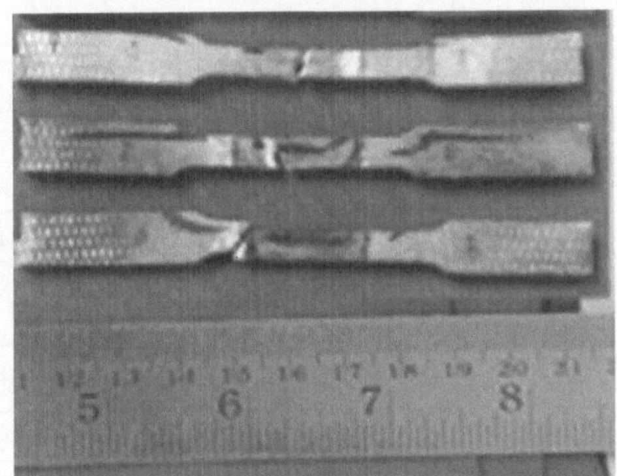

Fig. 38.1 Tensile samples of the FSP (Mg AZ31B /Al_2O_3/Cu)

Table 38.1 Welding parameters and tool geometry for FSW of AA5754-H111/ Mg AZ61 alloys

Welding Parameters	Ranges
Pin shape	Straight Square
Tool Pin one side (a)	6 mm
Tool Pin Length (L)	5.7 mm
Shoulder Diameter (D)	15 mm
Welding Speed (WS) - (mm/min)	28
Rotational Speed (RS) - (Rpm)	950
Vertical Load (kN)	10

3. Mechanical Properties

Following the extraction of tensile samples from the BM and FSP joints, the outcomes were reported in Table 38.2. At constant parameters, a high tensile value was achieved, but lowers than for the base material.

Table 38.2 Mechanical properties of the BM and FSP samples

Sample	Elongation (%)	Ultimate Tensile Strength (MPa)	Hardness (HV)
BM- AZ31B	14	234	86
FSW	7.1	203.45	68
AZ31B/Al_2O_3	8.6	209	71
AZ31B/Cu	7	217	82

4. Corrosion Study – SSCT

Salt Spray Corrosion Testing (SSCT) was selected for the corrosive environment examination constructed on a analysis of the literature. Since oxidative presents a significant obstacle for naval technology, oxidative testing is one of the most crucial components. Consequently, the testing was conducted in accordance with ASTM B 117 guidelines, in a replicated marine setting, using the test setup shown in Fig. 38.2. Cu and Al_2O_3 added (FSPed) AZ31 alloy and (FSWed) similar AZ31 alloy was investigated, and high UTS samples were collected for the experimental corrosion study.

The test lasted 72 hours and a 40-liter salt solution was made using the same procedure. Every 24 hours, samples were removed from the container and evaluated. After polishing the sample sides with 600, 800, and 3000 grit SiC paper, they were cleaned with solvent and allowed to air

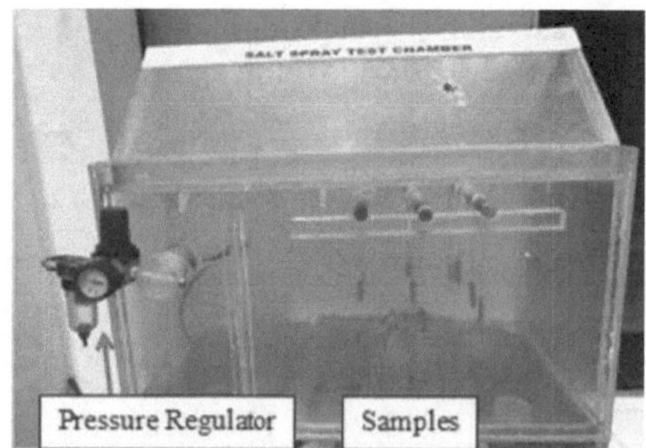

Fig. 38.2 Corrosion test setup- Salt spray chamber

dry. Prior to the test, the specimen's baseline weight was established. NaCl makes up around 5% of the 1000 mL salt content, whereas deionized water makes up 95%. After 72 hours, each specimen's adsorption weight was ascertained. After the corroded samples were put in the SSCT, the sections were immersed in a specially prepared solution for five minutes to eliminate the corrosion particles. The samplings' aggregate mass was calculated, and the noteworthy loss was recorded. The process parameters and circumstances for the corrosion investigation are listed in Table 38.1.

5. Calculation Formula for SSCT

Calculating the corrosion rate makes use of the weight loss measurement technique. The samples were weighed and tested in an aggressive environment for various periods (24, 48, and 72 hours). The specimens were carefully cleaned, dried, and evaluated following the experiment to calculate the final weight loss using Equation (1) [15]

The relation,

$$\text{Weight loss} = \text{Initial weight} - \text{Final weight} \qquad \text{Eq. 1}$$

The specimens' corrosion rate is intended using Equation (2) as follows:

$$\text{Corrosion rate} = (K \times W) / (A \times T \times D) \qquad \text{Eq. 2}$$

W - Mass loss (g), A - Area (cm^2), K - Constant (3.45×10^6 for mils/ year), (8.76×10^4 for mm/year), D - Density (g/ cm^3), T - Time of exposure (hours)

6. Results and Discussion

Table 38.3 shows that the weight loss was observed to be greatest in the FSPed (AZ31/Cu) weld region specimen (0.028 g) and to be highly significant in the AZ31 base materials (0.0358 g). Likewise, the lowest corrosion rate (0.00015105 mm/yr) was found in the welded region. AZ31, the initial substance, had a maximum corrosion rate of 0.00178183 mm/yr, while AZ31 treated with Al_2O_3 substance had a corrosion rate of 0.00048434 mm/yr. FSWed samples show modest corrosion (0.00165287 mm/ yr) and weight loss (0.321 g), similar to base metal and FSPed samples.

7. Analysis of Corrosion Test Results

In this study, Table 38.3 presents the corrosion rate (mm/ year) of various AZ31 samples based on weight loss (g), density (g/cm^3), and area (cm^2). The weight loss experienced by the three specimens following completion of SSCT for FSWed and FSPed samples with high tensile

Table 38.3 Corrosion rates at varied exposure duration

Sample Number	Weight before corrosion (gram)	Weight after corrosion (gram)			Weight loss (gram)	Density (g/ cm^3)	Area (cm^2)	Corrosion ratex 10^{-4} (mm/year)
		24 hr	48 hr	72 hr				
AZ31B Mg -BM	2.836	2.949	–	–	0.113	1.174	147.19	0.00059388
	2.667	–	2.896	–	0.229	1.174	147.19	0.00122557
	2.756	–	–	3.11	0.358	1.174	147.19	0.00178183
AZ31 FSW	2.687	2.796	–	–	0.109	1.174	147.19	0.00060421
	2.812	–	3.03	–	0.218	1.174	147.19	0.00111510
	2.689	–	–	3.01	0.321	1.174	147.19	0.00165287
AZ31/ Al_2O_3 FSP	2.668	2.679	–	–	0.011	1.174	147.19	0.00006363
	2.89	–	2.936	–	0.046	1.174	147.19	0.00024283
	2.759	–	–	2.85	0.089	1.174	147.19	0.00048434
AZ31/ Cu FSP	2.659	2.668	–	–	0.009	1.174	147.19	0.00005228
	2.735	–	2.747	–	0.012	1.174	147.19	0.00006770
	2.845	–	–	2.87	0.028	1.174	147.19	0.00015105

strength from threaded tool pin profiles for 24 hours, 48 hours, and 72 hours. According to the study, the welded area has more excellent corrosion resistance, whereas the base metals (AZ31 B) recorded lower levels of corrosion resistance. In addition, AZ31B/Cu welded regions have been found to have a higher corrosion resistance rate in FSW at 72 hours when compared to FSW and FSP with different reinforcements (Al_2O_3 and Cu). This could be due to the surfaces of the samples developing a barrier protection oxide layer because weld samples from threaded pin profiles with AZ31B/Al_2O_3 are more tensile- and hardness-resistant than other samples. The study confirms that the base metals (AZ31B) recorded a medium corrosion resistance, whereas the welded region (AZ31B/Cu) has a lower rate of corrosion. For the corrosion study investigation, the highest tensile values from the FSWed and FSPed were chosen. Apart from the FSW process, the reinforcement welds' corrosion rate was extremely low due to the favourable welding conditions. To sum up, the magnesium AZ31B FSPed sample with Al_2O_3 and Cu displayed a lower corrosion rate because it was developed during the FSP and has high tensile and hardness values and good material flow [18].

8. Conclusions

1. The FSPed (AZ31/Cu) stirred region specimen had the least amount of weight loss (0.028 g), but the base materials of AZ31 had the largest weight loss (0.0358 g).
2. In a comparable way, the lowest corrosion rate (0.00015105 mm/yr) was found in the stirred region. AZ31, the basic material, had a maximum corrosion rate of 0.00178183 mm/yr, while AZ31 treated with Al_2O_3 material had a corrosion rate of 0.00048434 mm/yr.
3. FSWed samples exhibited small corrosion (0.00165287 mm/yr) and weight loss (0.321 g), similar to base metal and FSPed samples.

References

1. G. Parande, V. Manakari, S.D. Kopparthy, M. Gupta, A study on the effect of low-cost eggshell reinforcement on the immersion, damping and mechanical properties of magnesium–zinc alloy, Compos. B. Eng. 182 (2020) 107650.
2. R. Aarthi and K.V. Sekar, Post-weld friction stir processing of AA5083-F TIG welds with scandium added fillers, Mater. Research Exp. 9 (2022) 126504.
3. S. Marappan, L. Kasirajan and V. Shanmugam, Friction Stir Welding Experiments on Az31b Alloy to Analyse Mechanical Properties and Optimize Process Variables by TOPSIS Method, Tehnički vjesnik, 29 (2022) 1923–1930.
4. T. Zhang, H. Cui, X. Cui, H. Chen, E. Zhao, L. Chang, Y. Pan, R. Feng, S. Zhai, S. Chai, Effect of addition of small amounts of samarium on microstructural evolution and mechanical properties enhancement of an as-extruded ZK60 magnesium alloy sheet, J. Mater. Research Technol. 9 (1) (2020) 133–41.
5. P. Yogaraj and L. Kasirajan, Identifying the Optimal Process Parameter on AA1100 Friction Stir Welded Joints, Tehnički vjesnik, 29 (2022), 957–964.
6. B. Mansoor, A.K. Ghosh, Microstructure and tensile behavior of a friction stir processed magnesium alloy, Acta mater. 60 (2012) 5079–88.
7. S. Balamurugan, K. Jayakumar, and K. Subbaiah. Influence of Friction Stir Welding Parameters on Dissimilar Joints AA6061-T6 and AA5052- H32, Arab. J. Sci. Eng. 46 (12) (2021) 11985–11998.
8. S. Balamurugan, K. Jayakumar, B. Anbarasan, and M. Rajesh, Effect of tool pin shapes on microstructure and mechanical behaviour of friction stir welding of dissimilar aluminium alloys. Mater. Today: Proc. 72, (2023) 2181–2185.
9. T. Rajkumar, S. Dinesh, B. Anbarasan, and S. Balamurugan, Effect of welding process parameters on surface topography and mechanical properties of friction-stir-welded AA2024/AA2099 alloys. Mater. Research Exp. 10(2) (2023) p.026507.
10. B. Senthamaraikannan, and J. Krishnamoorthy, Material flow and mechanical properties of friction stir welded AA 5052-H32 and AA6061-T6 alloys with Sc interlayer. Mater. Test. 65(7) (2023) 1127–1142.
11. A. Dhanapal, S.R. Boopathy, V. Balasubramanian, Influence of pH value, chloride ion concentration and immersion time on corrosion rate of friction stir welded AZ61A magnesium alloy weldments. J. Alloys Compound.. 523 (2012) 49–60.
12. H. Seifiyan, M.H. Sohi, M. Ansari, D. Ahmadkhaniha, M. Saremi, Influence of friction stir processing conditions on corrosion behavior of AZ31B magnesium alloy, J. Magnes. Alloy. 7 (2019) 605–16.
13. D. Liu, M. Shen, Y. Tang, Y. Hu, L. Zhao, Effect of multipass friction stir processing on surface corrosion resistance and wear resistance of ZK60 alloy, Met. Mater. Int. 6 (2019) 1182–90.
14. D. Liu, M. Shen, Y. Tang, Y. Hu, L. Zhao, Evaluation of corrosion resistance of multipass friction stir processed AZ31 magnesium alloy, Mater. Corr. 70 (2019) 1553–60.
15. S. Balamurugan, K. Jayakumar, and C. Nandakumar, Investigation of mechanical, metallurgical and corrosion characteristics of friction stir welded dissimilar AA 5052-H32 and AA 6061-T6 joints, J. Chinese Ins. Eng. (2023) pp. 1–14.
16. H. Pan, G. Qin, Y. Huang, Y. Ren, X. Sha, X. Han, Z.Q. Liu, C. Li, X. Wu, H. Chen, C. He, Development of low-alloyed and rare-earth-free magnesium alloys having ultra-high strength. Acta Mater. 149 (2018) 350–63.
17. N. Xu, Z. Ren, Z. Lu, J. Shen, Q. Song, J. Zhao, and Y. Bao, Improved microstructure and mechanical properties of friction stir-welded AZ61 Mg alloy joint. J. Mater. Research Technol. 18 (2022) pp. 2608–2619.
18. Q. Zang, H. Chen, J. Zhang, L. Wang, S. Chen, and Y. Jin, Microstructure, mechanical properties and corrosion resistance of AZ31/GNPs composites prepared by friction stir processing, J. Mater. Research Technol. 14 (2021) pp. 195–201.

Note: Every figure and table was created using the experimental work; none of them were taken from any publications or online.

Advances in Additive Manufacturing Technologies – Gurusamy Pathinettampadian (eds)
© 2024 Taylor & Francis Group, London, *ISBN 978-1-032-90013-1*

39 Experimental and Numerical Investigation of Compression Behavior of Additive Manufactured Lattice Structures

R. Kishore*, P. Gurusamy,
P. Jagan, and V. Lingesh Kanna
Department of Mechanical Engineering,
Chennai Institute of Technology, Chennai, India

ABSTRACT: The crystal lattice structure exhibits excellent properties of energy absorption with great stability and low mass than solid material. It can be fabricated into different structures as per the need of the hour and it makes attractive for a vast field of application like aerospace, defense, and automobile industries. It is utilized in various fields due to their high specific strength, modulus, and energy absorption. Additive manufacturing (AM) is an added advantage to the lattice structure because complicated structures can be easily fabricated which gives an ease to the design constraint. AM is far superior to conventional manufacturing process due to its wide ability, material saving, and design flexibility for fabrication. The main objective of this study is to discuss the compression behavior of six different schematic designs of lattice structures produced using Acrylonitrile Butadiene Styrene (ABS) in Fused Deposition Modeling (FDM). The specimens of dimension 50x50x50 mm^3 are designed as a cubic system. All specimens are subjected to compression and its behavior is observed. The force-displacement curves of all the lattice structures are plotted and results are interpreted by using finite element analysis methodologies and concluded by comparing with experimentation values. This study discusses design methods, compression behavior and performance of lattice structures.

KEYWORDS: Additive manufacturing, Composite, Artificial fiber, Mechanical properties, FDM

1. INTRODUCTION

3D Printing or Additive Manufacturing (AM) is used widely in the field of fabricating components made up of polymers with wide range of variety from fabrication of prototypes to manufacturing of final outputs / final products. Additive manufacturing has been widely incorporated into various manufacturing methods and are classified into following categories which includes; Laser sintering, Fused Deposition, Selective laser sintering. Among these FDM is mostly adapted method in various applications [1]. The 3D printing follows a patterned arrangement to manufacture the objects unlike most manufacturing methods of these Fused Deposition modelling process deals layer by layer deposition of the thermoplastic material which is been melted and it is forced to lay the melted material in both horizontal and vertical directions over a build plate or a platform [2], In this process along with the required output object support objects are also developed which will be in contact with the build model as well as in contact with the build platform which can be removed later.

The process of 3D printing involves various stages from initial design to final output which starts from initial de-

*Corresponding author: kishoreravi@citchennai.net

DOI: 10.1201/9781003545774-39

signing process of required model, later to that the designed model is then tangled with a slicing software which depletes the design into multiple layers for printing, the sliced model is transferred to FDM printer with the help of a memory card and the machine dose the process the process until this stage is known a Pre-Processing stage and the next stage is printing stage where various parameters are taken into account for the fine printing of materials such as bed temperature, nozzle temperature, room temperature, constant flow of material, material quality are some parameters that has to be taken care during the printing process. The final process of the material output depends on the Post-Processing stage where the materials that are printed is safely taken out of the bed and the support materials are removed from the desired output with at most care for better final output.The lattice crystals is related with the concept of translational symmetry. Lattice is a repeated network or array at fixed positions in the space to form the crystals. A cubical crystal can be formed by a square which is repeated and translated throughout the space in a plane. The smallest repeated structure of the lattice is called unit cell. It has been arranged in different geometric pattern in three dimensions to produce the lattice. The unit cell has atoms which are bonded together to provide stability and strength to the crystals. The corner of the lattice consists of points called lattice points. Hasan et al. [1] studied about the lattice structure fabricated by additive manufacturing method. The author had proposed that the repeated cellular structure called lattice structure can be utilized in various applications especially as core material in sandwich structure configuration, where the aim is to be a lightweight material with load bearing capability.

Gautam et al. [2] proposed the wide application of the lattice structure in defense, aerospace, and automobile industries. The author had inferred that the cellular lattice structure exhibits excellent energy absorption, high strength, and stability. The author [3] had done his investigation on lattice structure fabricated by additive manufacturing. The author had proposed that the processing parameter and material had direct impact on the manufacturability and mechanical characteristics. Added that the author had lighted about additive manufacturing is a key enabler for fabrication of lattice due to its ability to produce complex parts [4].Harris et al. [5] proposed his research work saying that the additive manufacturing opens a door for the design of lattice structures that can be fabricated easily. The compression behavior of the stainless-steel cellular structure is done by the experiments. The author had published the paper on the mechanical behavior of ABS fabricated lattice structures using fused deposition modelling (FDM). The author had concluded that the varying the parameters of FDM had a direct relation with the mechanical characteristics of the lattice structure [6].

In this study, we are going to produce the lattice structure by Additive Manufacturing. The additive manufacturing technique we are going to use is 3D Printing. The 3D printing process builds a three-dimensional object from a computer-aided design (CAD) model by successively adding material layer by layer. One of the key advantages of 3D printing is the ability to produce complex shapes or geometries. After an extensive literature review, it is found that more research works were not carried out on finding the effects of FDM parameters and compression behavior of additive manufactured lattice structures. The aim of this study is to simulate the compression loading and to predict it behavior and performance in various schematic designs of lattice structures with different patterns. The experimental investigation of all the lattice structure is carried out and the result comparison with the simulation is done to develop the FEA model. In this paper an attempt has been made to study about the compression behavior ofdifferent lattice structure of additive manufactured crystals with an aim to develop crystal with high strength low mass property.

To develop an optimized lattice structure there are various approaches followed Brackett et al. [14]

Discussed about optimizing techniques where he suggested mapping volume fractions of the lattice structures into unit cells with intermediate density and making a grey scale density solution of possible structures to manufacture with Additive Manufacturing. Cheng et al. [15] discussed about density distribution from the results obtained where unit cell allows material grading.

2. Experimentation

This section is categorized into three sub-sections: the first Material selection, the second Lattice structure selection, the third Fabrication of lattice structure.

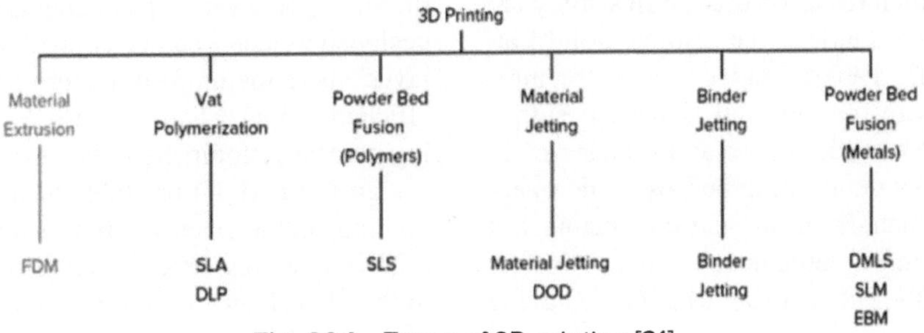

Fig. 39.1 Types of 3D printing [21]

2.1 Material Selection

In Fusion Deposition Modeling, the selection of materials involves various optimization procedures due to the extensive range of possibilities in material selection. Different combinations of materials can be integrated into the manufacturing process of objects. Typically, two primary materials, namely Acrylonitrile-Butadiene-Styrene (ABS) and Poly Lactic Acid (PLA), are employed in the production of components through Fused Deposition Modeling. Among these materials, ABS is non-biodegradable and is utilized in industries that require materials with a high strength-to-weight ratio. ABS materials are recognized for their resilience, surpassing PLA materials in this regard. On the contrary, PLA materials are biodegradable but exhibit comparatively lower strength than ABS materials. They find application in the prototyping of models.

2.2 Lattice Structure

Crystal structure is nothing but the arrangement of unit cells in a defined order, the unit cell is considered as smallest repeating unit having crystal structures with symmetrical pattern. The lattice structures are defined and calculated by various parameters considering the nodes and axes from nodes basically. The chemical properties of each elements are also calculated using these lattice structures or the crystalline structure. In this paper we are going to analyze basis lattice structure behavior using 3D printing technology. The lattice structures commonly seen are strut – based topologies are Body centered Cubic (BCC) and Face Centered Cubic (FCC) or the sub derivatives from the BCC and FCC [17].

Table 39.1 Mechanical properties of ABS material

S. No	Property	Metric	Units
1	Yield Strength	1.85e7 - 5.1e7	Pa
2	Tensile Strength	2.76e7 - 5.52e7	Pa
3	Elongation	0.015 - 1	% strain
4	Hardness (Vickers)	5.49e7 - 1.5e8	Pa
5	Fracture Toughness	1.19e6 - 4.29e6	Pa/m^0.5
6	Young's Modulus	1.19e9 - 2.9e9	Pa

The different lattice structures to be used in this study are Cross Lattice, Hexagonal Lattice, Star Lattice, Solid Lattice, Grid Lattice, and E - Pattern Lattice. These structures are basic structure where the unit cell is been designed with CAD software and been developed as a solid structure in the same software for the printing purpose. The material used for additive manufacturing and conducting test for this study is Acrylonitrile butadiene styrene, which is popularly known as ABS. The software used for designing the different lattice structures is CATIA V5 and Autodesk Netfabb.

It is a software mainly used for creating .stl file for additive manufacturing in various methods. With it, we can repair, arrange, orient, and prepare three-dimensional files, be they tessellated (triangle mesh) or parametric (CAD), and produce slice and print data. The model from Netfabb is used for the additive manufacturing of the model.

The designed unit cell from the CATIA V5 is been developed into an uniform 3D model which will be processed by the slicing software and will be followed into the methodology for all the similar structures. Each structure is made with a flat top and bottom entity for smooth finishing as well as uniform test results after fabrication process. These six different lattice patterns are designed considering the supports build up during the printing and designed carefully as to reduce the supports so that post processing works can be minimalized.

The structures are the outcome of the topological optimization of basic crystalline structures and which is further developed as final solid models [17]. These optimization parameters are basically constrained with the nodes to which the cell is developed.

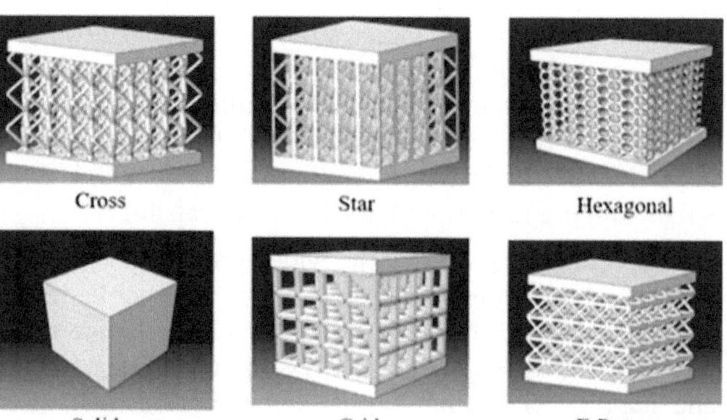

Fig. 39.2 CAD design of different lattice structures for compression test

2.3 Fabrication of Lattice Structure

From the design of lattice structure which is optimized from the basic crystal structure, the slicing has been done using the slicing software. The slicing is the process in which the designed model is subjected to divided into number od thin layer where the uniform pattern has been obtained for the efficient output when the design is printed. After the slicing process is completed, the next step is conversion of the design file into 3D printable format file that is .stl file its nothing but the extension file for stereo lithography format. After the slicing process the sliced .stl file is transferred to the 3D printer using a memory card or directly through a data cable which can transfer the data to the 3D printer. The basic adjustments are made in the 3D printer lite the bed temperature nozzle temperature and also the room temperature are also considered for the better final output.

The process parameters also include part interior style, part in fill style, raster angle of the print, layer thickness of each layer, raster width and gap, shrinking of parts, contour and with and gap. And other factors involve part build orientation, environmental factors such as temperature and humidity, Material parameters are its melting temperature and cooling temperature, material density etc.

3. Printing Parameters

The lattice structures are printed with Zortrax200 printer, Poland based company extrusion temperature at 230°C the infill density is kept as 100% as the unit cell vary in each design and each lattice structure has different built pattern.

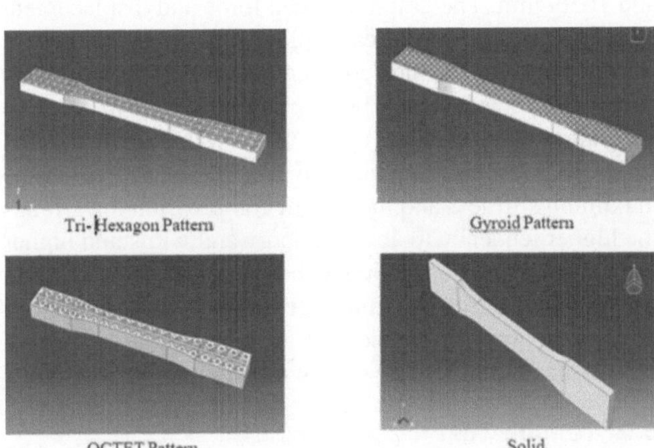

Fig. 39.3 CAD design of different lattice structures for tensile test

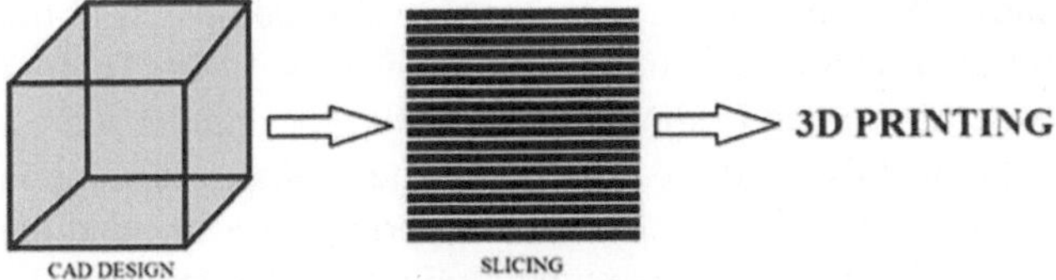

Fig. 39.4 Slicing of elements

4. Post Processing and Testing

Once the fabrication is completed the fabricated parts from 3D printer is sent for post-processing in this phase the fabricated part doesn't look like a final output the post processing phase is involved before the final product is obtained. In this phase of post processing the product from the 3D printer has a base which is combined with the model with a very feeble sting connection, It can be peeled off very easily from the printed model.

Later to this process there will be support structures which is printed by the 3D printer to hold the angular structures printed with the product. These support structures gives the balance for the angular layers to stabilize while printing and not to drop down due to gravity or other environmental factors like any external wind etc. these support structures are easy to remove and can be removed with the appropriate tools and accessories.

The compression test specimen is the final output product which is gone through the process of post processing, with this the 6 final specimens are ready for testing, the specimens are subjected to load through Universal Testing Machine. The compression test is carried out by universal testing machine (UTM). The behavior of deflection under load is observed and is plotted to get the compression load of the pattern.

Tensile test is conducted with the tensile specimen with different lattice structures. The finalized tensile specimen is cut to ASTM standards for tensile test and the specimen are subjected to tensile test.

Fig. 39.5 Fabricated patterns of lattice structure

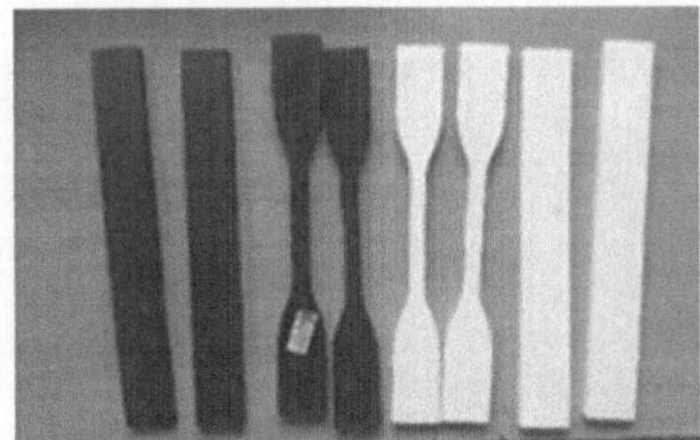

Fig. 39.6 Fabricated patterns for tensile test

Fig. 39.7 Experimental setup for compression test

5. Finite Element Analysis

Finite element analysis involves three stages as pre-processing, processing, and post processing. Pre-processing or modeling stage involves the creation of an input file for the processing. Processing or finite element analysis stage converts the physical problem into mathematical model and solved it using equation to predict the behavior of the physical system. Post-processing or generating report involves the visualization creation of the images, animation, and values at the nodes etc. from the output file. Abaqus software is used for the Finite Element Analysis of the model. The parts are designed using CATIA and is exported as .stp file for the simulation. The CAD model is imported in the Abaqus and material properties is assigned to it. The material used for the analysis is Acrylonitrile Butadiene Styrene (ABS). The lattice with different pattern is analyzed and results are compared with the experimental value. The Static general analysis is done and compression load is predicted using the graph of load against deflection. The bottom of the part is encase and displacement load is applied at the top. The discretization or meshing to break down the whole model into finite number of smaller elements is done. The details of the discretization are shown in Table 39.2.

Fig. 39.8 Displacement of cross lattice structure

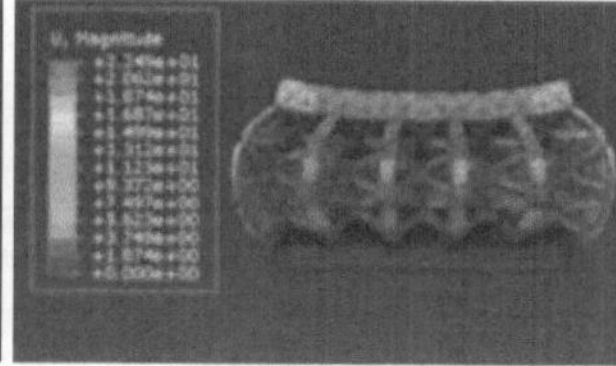

Fig. 39.9 Displacement of star lattice structure

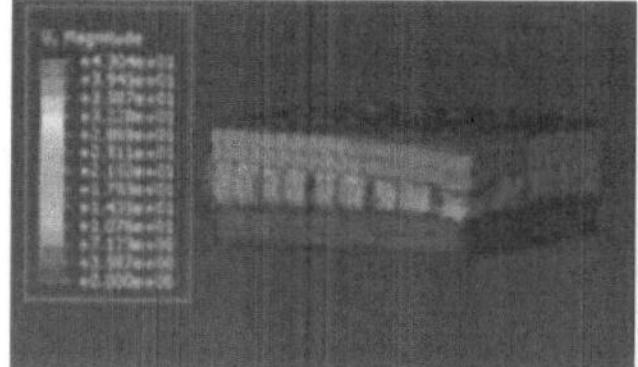

Fig. 39.10 Displacement of hexagonal lattice structure

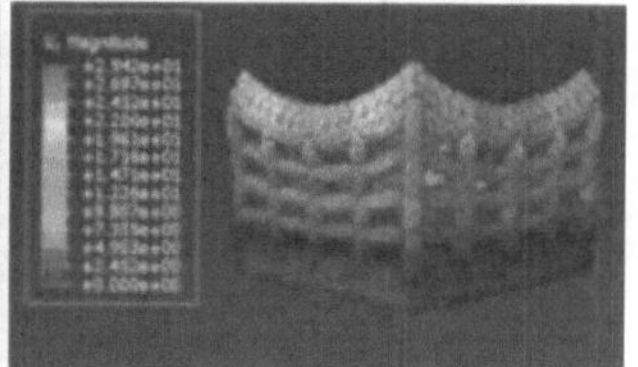

Fig. 39.11 Displacement of grid lattice structure

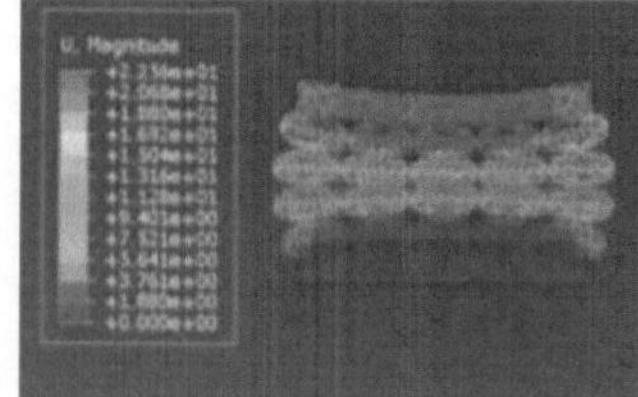

Fig. 39.12 Displacement of E-pattern lattice structure

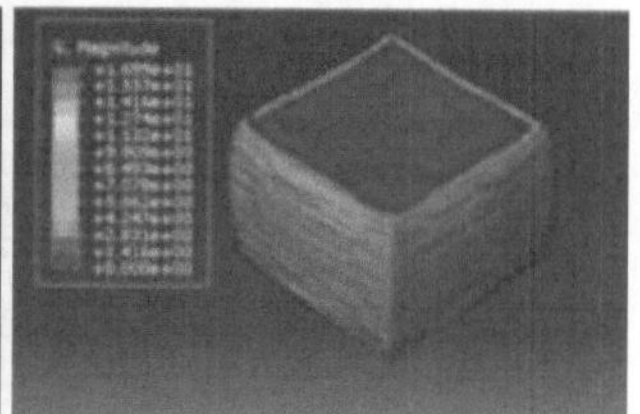

Fig. 39.13 Displacement of solid lattice

6. Results and Discussion

6.1 Compression Test

The analysis is done to predict the behavior of the deformation under compression load and to determine the compression load of the lattice structure. The deflection is observed for all the lattice patterns by numerical simulation and graph of load against deflection is plotted for all the patterns. The result correlation is done by making a comparison study of the numerical value with the experimental value and also the percentage of the error is calculated. The simulation process is carried out and following observations are accumulated. The maximum displacement of cross lattice pattern and star lattice pattern during compression are 25.6 mm and 22.4 mm. The maximum displacement of hexagonal lattice pattern and grid lattice structure during compression are 43.04 mm and 29.42 mm. The maximum displacement of E-pattern lattice structure and solid cube during compression are 22.56 mm and 16.99 mm. The graph between force and displacement of all six lattice structures is plotted it shows hexagonal pattern has better values compared to the other structures and very low values can be seen with E-Pattern structure and star structure values have a nearer value to Hexagonal structure. We can infer that the **hex pattern** exhibits the maximum stiffness and maximum compression load. It has the higher tendency to absorb energy and withstand higher load. Because of its excellent behaviour under compression and **high strength low mass property** it can be used in the application of defence and aerospace industry which demands the requirement of low mass with high strength.

6.2 Tensile Test

This test is initiated to understand the tension behaviour of different lattice structure and to compare the simulated values with the obtained values. Tensile test report clearly shows Tri-Hex pattern has a better value compared to the other structures and the maximum values of the Tri-Hex pattern recorded in simulation is 22.04Mpa and the nearest value to this structure is gyroid pattern and next value is octet pattern it is found that the experimental value and the simulation value differs with only in the range less than 12 percentage.

7. Conclusion

Compression loading on cross, star, hexagonal, grid, E-pattern lattice structures against the solid cube is reported in this research work. Numerical simulations are carried out to predict the compression behavior, stiffness, and performance of six different schematic designs of ABS lattice structure with varying print patterns and print strength. The comparison between the numerical analysis and the actual

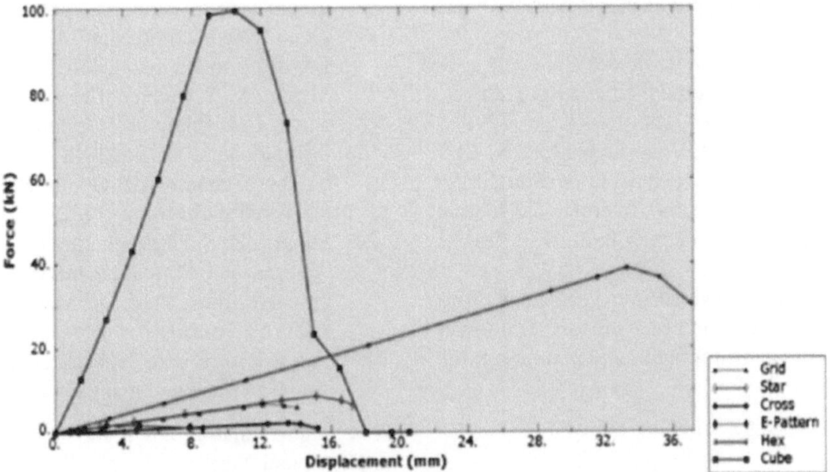

Fig. 39.14 Force vs displacement graph of lattice structures

obtained values from the compression test are found to be nearer values and Based on the investigations. The patterns according to the strength compared to the cube pattern can be categorized as follows, *Hexagonal>Star>Grid> E-Pattern.*

With the tensile test its concluded that the Tri-Hex structured specimen shows best tensile strength compared to the other structures, these structures similar results very minimal variations are seen. It is found that the experimental results and simulation results are in good correlation for all patterns. It's evident that hexagonal pattern is a good solution for achieving high strength with low mass property. The usage of hexagonal pattern will lead to reduction in material wastage, cost, and time.

REFERENCES

1. Rafidah Hasan, Nur Ameelia Rosli, Shafizal Mat, Mohd Rizal Alkahari. "Failure behaviour of 3D-Printed ABS lattices structure under compression." International Journal of Engineering and Advanced Technology (IJEAT) ISSN: 2249–8958, Volume-9 Issue-3, February 2020.
2. Gautam, Rinoj, and Sridhar Idapalapati. "Compressive properties of additively manufactured functionally graded kagome lattice structure." Metals 9, no. 5 (2019): 517.
3. Maconachie, Tobias, Rance Tino, Bill Lozanovski, Marcus Watson, Alistair Jones, Chrysoula Pandelidi, Ahmad Alghamdi et al. "The compressive behaviour of ABS gyroid lattice structures manufactured by fused deposition modelling." International Journal of Advanced Manufacturing Technology 107, no. 11-12 (2020): 4449–4467.
4. McGregor, Davis J., Sameh Tawfick, and William P. King. "Mechanical properties of hexagonal lattice structures fabricated using continuous liquid interface production additive manufacturing." Additive Manufacturing 25 (2019): 10–18.
5. Harris, J. A., R. E. Winter, and G. J. McShane. "Impact response of additively manufactured metallic hybrid lattice materials." International Journal of Impact Engineering 104 (2017): 177–191.
6. Dawoud, Michael, Iman Taha, and Samy J. Ebeid. "Mechanical behaviour of ABS: An experimental study using FDM and injection moulding techniques." Journal of Manufacturing Processes 21 (2016): 39–45.
7. Magerramova, L., M. Volkov, A. Afonin, M. Svinareva, and D. Kalinin. "Application of light lattice structures for gas turbine engine fan blades." In Proccedings of the 31st Congress of the International Council of the Aeronautical Sciences, ICAS. 2018.
8. L.A.Magerramova, M.S. Svinaryova, A.S. Siversky, M.E Volkov. Cellular Structures Produced by Additive Technologies for GTE Components. Technology of Light alloy, No3, pp 24–35, 2017.
9. Magerramova L., Volkov M., Svinareva M., Siversky A. The use of additive technologies to create lightweight parts for gas turbine engine compressors. Proc. ASME TurboExpo 2018, GT2018–75904.
10. Boccini, Enrico, Rocco Furferi, Lapo Governi, Enrico Meli, Alessandro Ridolfi, Andrea Rindi, and Yary Volpe. "Toward the integration of lattice structure-based topology optimization and additive manufacturing for the design of turbomachinery components." Advances in Mechanical Engineering 11, no. 8 (2019): 1687814019859789.
11. Cortequisse, Jean-Francois. "Lattice type blade of an axial turbine engine compressor." U.S. Patent 10,400,625, issued September 3, 2019.Yadlapati, Sai Avinash. "Influence of FDM Build Parameters on Tensile and Compression Behaviors of 3D Printed Polymer Lattice Structures." (2018).
12. ASTM F2792-10 Standards Terminology for Additive Manufacturing.
13. D. Brackett, I. Ashcroft, R. Hague, Topology optimization for additive manufacturing, Solid Free. Fabr. Symp. (2011) 348–362.
14. L. Cheng, P. Zhang, E. Biyikli, J. Bai, S. Pilz, A. To, Integration of topology optimization with efficient design of additive manufactured cellular structures, Solid Free. Fabr. Symp. (2015) 1370–1377.

15. PLA (96:4 L:D ratio content) produced by Nature Works Co., technical data sheet
16. Maskery, et al., Compressive failure modes and energy absorption in additively manufactured double gyroid lattices, Addit. Manuf. 16 (2017) 24e29.
17. Kishore, R., G. Karthick, M. D. Vijayakumar, and V. Dhinakaran. "Analysis of mechanical behaviour of natural filler and fiber based composite materials." International Journal of Recent Technology and Engineering 8, no. 1S2 (2019): 117–121.
18. Palaniyappan, Sabarinathan, Dhinakaran Veeman, K. Rajkumar, K. Vishal, R. Kishore, and L. Natrayan. "Photovoltaic industrial waste as substitutional reinforcement in the preparation of additively manufactured acrylonitrile butadiene styrene composite." Arabian Journal for Science and Engineering 47, no. 12 (2022): 15851–15863.
19. Sivaraj, S., S. Stephen Bernard, R. Kishore, G. K. Kannan, and J. Lokeshkumar. "Optimization of thrust force during drilling operation for Al-SiC composites using Box-Behnken approach." In AIP Conference Proceedings, vol. 2519, no. 1. AIP Publishing, 2022.
20. Subash, P., S. Nagendharan, P. Gurusamy, and R. Kishore. "Fatigue and vibrational analysis of composite plates based on curve fitting method." Materials Today: Proceedings 37 (2021): 854–857.

The reference for Figure 39.1 has been added. Other figures and Tables was created using the experimental work.

Note: figures and Tables except fig. 39.1 was created using the experimental work.

Advances in Additive Manufacturing Technologies – Gurusamy Pathinettampadian (eds)
© 2024 Taylor & Francis Group, London, ISBN 978-1-032-90013-1

40 Green Tribology: Sustainable Approaches in Lubricating Oil Formulations

Subham Saha, Krishna Kumar Gupta, Shouvik Sarkar, Shankha Shubhra Goswami*, Bikash Banerjee, Surajit Mondal
Department of Mechanical Engineering,
Abacus Institute of Engineering and Management, Hooghly, India

ABSTRACT: Green tribology represents a paradigm shift towards sustainable and environmentally friendly approaches in the field of tribology. This research article explores the concept of green tribology and its application in the development of lubricating oil formulations. We delve into the environmental impact of traditional lubricants, highlighting the need for greener alternatives. By investigating sustainable strategies such as bio-based lubricants, vegetable oils, and eco-friendly additives, this article provides insights into how lubricant manufacturers can reduce the environmental footprint of their products while maintaining high performance. The findings presented here contribute to the growing body of knowledge in green tribology and offer practical solutions for transitioning towards more sustainable lubrication practices.

KEYWORDS: Green tribology, Lubricating oil, Sustainability, Environmental impact, Eco-friendly

1. Introduction

Tribology, the science of friction, wear, and lubrication, plays a critical role in various industries ranging from automotive and aerospace to manufacturing and energy. However, traditional lubricants derived from mineral oils pose significant environmental challenges due to their non-renewable nature, potential toxicity, and contribution to pollution [1]. In response to these concerns, the concept of green tribology has emerged, advocating for sustainable approaches to friction reduction and lubrication.

This research article focuses on sustainable approaches in lubricating oil formulations, with an emphasis on reducing environmental impact while maintaining or improving performance [2]. By exploring the principles of green tribology and highlighting innovative solutions, we aim to provide guidance for lubricant manufacturers and industries seeking to adopt more sustainable practices.

2. Environmental Impact of Traditional Lubricants

The environmental impact of traditional lubricants is a multifaceted issue that encompasses various stages of their lifecycle, from extraction and production to usage and disposal. Traditional lubricating oils, primarily derived from petroleum-based sources, have several environmental drawbacks. Here's a detailed exploration of the environmental impact of traditional lubricants

*Corresponding author: ssg.mech.official@gmail.com

DOI: 10.1201/9781003545774-40

2.1 Non-renewable Resource Depletion

Traditional lubricating oils are predominantly derived from petroleum, a finite and non-renewable resource. The extraction of crude oil, the primary source of petroleum, involves invasive drilling techniques that disrupt ecosystems, pose risks to biodiversity, and contribute to habitat destruction [3]. As global demand for petroleum-based lubricants continues to rise, concerns about resource depletion and energy security become more pronounced. Furthermore, reliance on finite fossil fuel reserves for lubricant production exacerbates geopolitical tensions and economic uncertainties.

2.2 Greenhouse Gas Emissions

The production and processing of petroleum-based lubricants entail significant energy consumption and greenhouse gas emissions. Petroleum refineries, where crude oil is processed into various petroleum products including lubricants, are energy-intensive facilities that rely heavily on fossil fuels. The combustion of fossil fuels during refining processes releases carbon dioxide (CO2), methane (CH4), and other greenhouse gases into the atmosphere, contributing to global warming and climate change [4]. Additionally, emissions associated with transportation and distribution further amplify the environmental footprint of traditional lubricants.

2.3 Toxicity and Pollution

Spillage, leakage, or improper disposal of petroleum-based lubricants poses significant risks to environmental and human health. Petroleum hydrocarbons and additives present in lubricating oils can contaminate soil, groundwater, surface water, and air, leading to adverse effects on ecosystems and wildlife. The toxicity of certain lubricant components, such as polycyclic aromatic hydrocarbons (PAHs), heavy metals, and persistent organic pollutants (POPs), can bioaccumulate in the food chain, posing risks to human health through exposure via consumption of contaminated water or food [5]. Furthermore, the accumulation of lubricant residues in urban and industrial areas contributes to soil and water pollution, impairing ecosystem services and diminishing overall environmental quality.

2.4 Habitat Destruction and Ecosystem Degradation

The extraction, processing, and transportation of petroleum for lubricant production often entail habitat destruction and ecosystem degradation. Oil exploration and drilling activities disrupt terrestrial and marine ecosystems, fragment habitats, and disturb wildlife populations. Spills and leaks during transportation pose acute threats to aquatic and terrestrial ecosystems, causing immediate harm to flora and fauna and impairing ecosystem functions [6]. Moreover, the long-term ecological impacts of oil pollution, such as habitat loss, population declines, and alterations to biodiversity, can persist for years or even decades, hindering ecosystem recovery and resilience.

2.5 Cumulative Environmental Impacts

The cumulative environmental impacts of traditional lubricants extend beyond individual incidents of pollution or contamination. Over time, the widespread use of petroleum-based lubricants contributes to the degradation of air quality, soil health, water resources, and biodiversity on a global scale. These cumulative impacts exacerbate environmental stressors, amplify ecological disturbances, and undermine the resilience of natural systems [7]. Furthermore, the interconnections between environmental compartments and ecosystems magnify the reach and severity of lubricant-related environmental impacts, necessitating comprehensive and integrated approaches to environmental management and conservation.

In summary, the environmental impact of traditional lubricants encompasses a range of interconnected issues, including resource depletion, greenhouse gas emissions, toxicity, pollution, habitat destruction, and ecosystem degradation. Addressing these environmental challenges requires a concerted effort to transition towards more sustainable and eco-friendly lubrication practices, such as the adoption of bio-based lubricants, vegetable oils, and eco-friendly additives [5,6]. By mitigating the environmental footprint of lubricants, we can promote environmental sustainability, protect natural resources, and safeguard ecological integrity for present and future generations.

3. Sustainable Approaches in Lubricating Oil Formulations

To mitigate the environmental impact of lubricating oils, lubricant manufacturers are exploring various sustainable approaches. It explores various strategies and technologies aimed at reducing the environmental impact of lubricating oils while maintaining or improving their performance. Here's an elaboration on each aspect.

3.1 Bio-based Lubricants

Bio-based lubricants are derived from renewable biomass sources such as vegetable oils, animal fats, and microbial oils [6,7]. Unlike petroleum-based lubricants, which rely on finite fossil fuel reserves, bio-based lubricants offer a sustainable alternative with several environmental benefits.

- *Renewable resource:* Biomass feedstocks used in bio-based lubricants, such as soybean oil, sunflower oil, or palm oil, are renewable and readily available, reducing reliance on finite fossil fuel reserves and mitigating concerns about resource depletion.

- *Biodegradability:* Bio-based lubricants are typically biodegradable, meaning they can be broken down by microorganisms into simpler, non-toxic compounds over time. This property reduces the risk of environmental pollution and minimizes the impact of lubricant spills or leaks on soil, water, and ecosystems.
- *Low toxicity:* Compared to petroleum-based lubricants, bio-based lubricants often exhibit lower toxicity levels, posing fewer risks to human health and environmental quality. This makes them preferable choices for applications where environmental sensitivity or regulatory compliance is a concern.

However, challenges such as oxidation stability, low-temperature performance, and compatibility with elastomers need to be addressed to improve the overall performance and viability of bio-based lubricants in diverse applications.

3.2 Vegetable Oils

Certain vegetable oils possess inherent lubricating properties and can be used directly or as base oils in lubricant formulations [8]. Vegetable oils such as rapeseed oil, soybean oil, and sunflower oil offer several advantages.

- *Natural lubricity:* Vegetable oils contain triglycerides, fatty acids, and other components that confer lubricating properties, making them suitable candidates for use as lubricants. Their natural lubricity can reduce friction and wear, leading to improved efficiency and extended equipment life.
- *Renewable sourcing:* Vegetable oils are derived from renewable plant sources and can be sustainably produced through agricultural practices such as crop rotation, organic farming, and agroforestry. This renewable sourcing reduces dependence on non-renewable resources and supports environmentally responsible production methods.
- *Compatibility with additives:* Vegetable oils can be modified or supplemented with additives to enhance their performance characteristics, such as oxidation stability, viscosity-temperature behavior, and anti-wear properties. Additives derived from natural sources or renewable feedstocks further contribute to the sustainability of lubricant formulations.

Despite these advantages, challenges related to oxidation, stability, lubricant film formation, and compatibility with existing infrastructure must be addressed to optimize the use of vegetable oils in lubricating oil formulations.

3.3 Eco-friendly Additives

In addition to bio-based lubricants and vegetable oils, the incorporation of eco-friendly additives is another sustainable approach in lubricating oil formulations [9]. These additives, derived from natural sources or renewable feedstock's, offer several environmental benefits.

- *Friction Modifiers:* Eco-friendly friction modifiers, such as fatty acids, esters, or natural polymers, can reduce friction and wear in lubricating oil formulations without compromising environmental sustainability. These additives form boundary films or tribofilms on metal surfaces, reducing direct contact and minimizing wear.
- *Anti-wear agents:* Natural anti-wear agents, such as zinc dialkyldithiophosphate (ZDDP) alternatives derived from plant extracts or bio-based compounds, provide effective wear protection while minimizing the environmental impact. These additives can improve the durability and reliability of lubricated components without introducing harmful substances into the environment.

By leveraging the properties of eco-friendly additives, lubricant manufacturers can enhance the performance of bio-based lubricants and vegetable oils while minimizing their environmental footprint [3,5,9]. However, research and development efforts are needed to optimize the effectiveness, compatibility, and stability of these additives in lubricating oil formulations.

In conclusion, sustainable approaches in lubricating oil formulations, such as bio-based lubricants, vegetable oils, and eco-friendly additives, offer viable solutions for reducing the environmental impact of lubricants while maintaining or improving performance. By embracing these sustainable practices, lubricant manufacturers can contribute to environmental conservation, resource stewardship, and sustainable development [10]. Continued research, innovation, and collaboration across industry, academia, and regulatory bodies are essential for advancing the adoption of sustainable lubrication practices and promoting a more sustainable future.

4. Challenges and Opportunities

The following section explores the various obstacles and prospects associated with the adoption and implementation of sustainable lubrication practices. Let's delve into each aspect in detail.

4.1 Performance Trade-offs

One of the primary challenges in transitioning to sustainable lubrication practices is the potential for performance trade-offs. Bio-based lubricants, vegetable oils, and eco-friendly additives may exhibit different performance characteristics compared to their petroleum-based counterparts [11]. Challenges such as oxidation stability, viscosity-temperature behavior, lubricant film formation, and wear protection must be addressed to ensure that sustainable lubricants meet or exceed the performance standards required for diverse applications. Balancing environmental sustainability with performance requirements is essential to overcoming these trade-offs and promoting the widespread adoption of sustainable lubrication practices.

4.2 Cost Competitiveness

Another significant challenge is the cost competitiveness of sustainable lubricants compared to traditional petroleum-based lubricants. The production costs of bio-based lubricants, vegetable oils, and eco-friendly additives are often higher due to factors such as feedstock prices, processing technologies, and economies of scale. Additionally, the upfront investment required for research, development, and infrastructure upgrades may pose financial barriers to adoption for lubricant manufacturers and end-users [12]. Addressing cost competitiveness through technological innovations, process optimization, and government incentives can help make sustainable lubricants more economically viable and accessible, thereby accelerating their market penetration and adoption.

4.3 Regulatory Compliance

Compliance with regulatory standards and certifications is another challenge faced by the sustainable lubrication industry. Environmental regulations governing biodegradability, toxicity, eco-labeling, and emissions vary across regions and jurisdictions, requiring lubricant manufacturers to navigate a complex regulatory landscape. Ensuring compliance with regulatory requirements while maintaining product performance and market competitiveness is a multifaceted task that requires ongoing monitoring, testing, and documentation [11]. Collaboration between industry stakeholders, regulatory agencies, and standards organizations is essential for developing consistent, science-based regulations that promote environmental sustainability without compromising safety or performance.

4.4 Innovation and Technological Advances

Despite the challenges, there are numerous opportunities for innovation and technological advances in sustainable lubrication practices. Research and development efforts focused on improving the performance, stability, compatibility, and cost-effectiveness of bio-based lubricants, vegetable oils, and eco-friendly additives are yielding promising results [2,3]. Advances in materials science, nanotechnology, biotechnology, and process engineering are driving innovation in lubricant formulation, additive technology, and manufacturing processes. Furthermore, interdisciplinary collaboration between academia, industry, and government research institutions is fostering knowledge exchange, technology transfer, and the commercialization of sustainable lubrication solutions. By leveraging these opportunities for innovation, lubricant manufacturers can develop next-generation lubricants that deliver superior performance, environmental sustainability, and economic competitiveness.

4.5 Consumer Awareness and Market Demand

Increasing consumer awareness and market demand for sustainable products present significant opportunities for the sustainable lubrication industry. As environmental concerns and sustainability considerations become more prevalent among consumers, businesses, and policymakers, there is a growing demand for eco-friendly alternatives to traditional petroleum-based lubricants [2,5,6,8]. Manufacturers and suppliers that demonstrate a commitment to environmental stewardship, corporate social responsibility, and sustainable practices stand to gain a competitive advantage in the marketplace. Educating consumers about the environmental benefits of sustainable lubricants, promoting eco-labeling and certification schemes, and fostering transparency and accountability in supply chains can further enhance market demand and consumer acceptance of sustainable lubrication products.

In summary, while challenges such as performance trade-offs, cost competitiveness, regulatory compliance, and technological complexity exist, there are also significant opportunities for innovation, collaboration, and market growth in the sustainable lubrication industry [12,13]. By addressing these challenges and leveraging these opportunities, lubricant manufacturers can drive positive environmental impact, enhance product performance, and meet the evolving needs of consumers and society.

5. Managerial Implications

The ongoing research carries several managerial implications for businesses involved in lubricant manufacturing, distribution, and application. Here are the detailed managerial implications.

5.1 Strategic Planning and Product Development

- *Market positioning:* Businesses can strategically position themselves as leaders in sustainable lubrication by investing in research and development of bio-based lubricants, vegetable oils, and eco-friendly additives.

- *Diversification:* Expanding product portfolios to include sustainable lubricants allows businesses to cater to environmentally conscious consumers and capitalize on the growing demand for eco-friendly products.
- *Differentiation:* Leveraging sustainable lubrication technologies enables businesses to differentiate their offerings from competitors and establish a unique selling proposition based on environmental sustainability and performance.

5.2 Supply Chain Management

- *Sourcing:* Businesses need to identify reliable sources of renewable biomass feedstocks for bio-based lubricants and establish sustainable supply chains to ensure consistent and ethical sourcing practices.
- *Partnerships:* Collaborating with suppliers, research institutions, and regulatory bodies facilitates knowledge exchange, technology transfer, and innovation in sustainable lubrication practices.
- *Certifications:* Obtaining certifications such as biodegradability, eco-labeling, and sustainability standards enhances the credibility and market acceptance of sustainable lubricant products.

5.3 Operational Efficiency and Cost Management

- *Process optimization:* Implementing efficient manufacturing processes, waste reduction strategies, and energy-saving measures improves operational efficiency and reduces the environmental footprint of lubricant production.
- *Cost analysis:* Conducting thorough cost-benefit analyses helps businesses evaluate the economic viability of sustainable lubrication practices and justify investments in research, development, and infrastructure upgrades.
- *Resource utilization:* Maximizing the utilization of renewable feedstocks, minimizing waste generation, and optimizing resource allocation contribute to cost savings and sustainability in lubricant manufacturing.

5.4 Marketing and Communication

- *Value proposition:* Communicating the environmental benefits, performance advantages, and sustainability credentials of green lubricants enhances their perceived value and market appeal among environmentally conscious consumers.
- *Educational campaigns:* Businesses can raise awareness about the importance of sustainable lubrication practices through educational campaigns, seminars, and workshops targeting industry stakeholders, policymakers, and the general public.
- *Transparency and disclosure:* Providing transparent information about product ingredients, manufacturing processes, and environmental impact metrics fosters trust, credibility, and accountability in the marketplace.

5.5 Regulatory Compliance and Risk Management

- *Compliance monitoring:* Businesses must stay abreast of evolving environmental regulations, standards, and compliance requirements related to sustainable lubrication practices and ensure adherence to applicable laws and guidelines.
- *Risk assessment:* Conducting comprehensive risk assessments helps businesses identify and mitigate potential environmental, health, and safety risks associated with sustainable lubricant manufacturing, storage, transportation, and disposal.

5.6 Continuous Improvement and Innovation

- *Research and development:* Investing in ongoing research and development initiatives enables businesses to continuously improve the performance, stability, and sustainability of green lubricant formulations through innovations in materials science, nanotechnology, biotechnology, and process engineering.
- *Benchmarking:* Benchmarking against industry best practices and performance metrics allows businesses to gauge their progress, identify areas for improvement, and drive continuous innovation in sustainable lubrication practices.

In summary, embracing green tribology and adopting sustainable approaches in lubricating oil formulations present significant managerial implications for businesses, including strategic planning, supply chain management, operational efficiency, marketing and communication, regulatory compliance, risk management, and continuous improvement [14]. By proactively addressing these implications, businesses can drive positive environmental impact, enhance competitiveness, and meet the evolving needs of customers and stakeholders in a sustainable manner.

6. CONCLUSIONS

In conclusion, green tribology offers a holistic approach to addressing the environmental challenges associated with traditional lubricating oils. By embracing sustainable approaches such as bio-based lubricants, vegetable oils, and eco-friendly additives, lubricant manufacturers can contribute to environmental conservation, mitigate climate change, and promote sustainable development. However, overcoming performance trade-offs, improving cost competitiveness, and ensuring regulatory compliance are essential for the widespread adoption of green lubrication practices. As the field of green tribology continues to evolve, collaboration between researchers, industry stakeholders, and policymakers will be crucial in driving innovation and accelerating the transition towards more sustainable lubrication solutions.

While the research provides valuable insights into the environmental benefits and technological advancements of sustainable lubrication practices, it also has certain limitations and future scope. One limitation is the need for further research to address performance trade-offs and overcome technical challenges associated with bio-based lubricants, vegetable oils, and eco-friendly additives, particularly in terms of oxidation stability, viscosity-temperature behavior, and wear protection. Additionally, the economic feasibility and cost competitiveness of sustainable lubricants need to be further evaluated and optimized to facilitate widespread adoption in the marketplace. Furthermore, the research primarily focuses on the technical and environmental aspects of sustainable lubrication, and future studies could explore the social, economic, and policy implications of transitioning to green tribology, including the role of consumer behavior, regulatory frameworks, and industry standards. By addressing these limitations and expanding the scope of research, future studies can contribute to the advancement and implementation of sustainable lubrication practices on a broader scale.

Acknowledgement

We would like to express our sincere gratitude to all individuals and organizations who contributed to the completion of this research paper. Special thanks to our research team members for their dedication and hard work throughout the project. We are also grateful to Abacus Institute of Engineering and Management for providing the necessary resources and support.

References

1. Mittal, U. and Panchal, D. (2023). AI-based evaluation system for supply chain vulnerabilities and resilience amidst external shocks: An empirical approach. Rep. Mech. Eng. 4(1):276–289. https://doi.org/10.31181/rme040122112023m
2. Jason, Y. J. J., How, H.G., Teoh, Y.H. and Chuah, H.G. (2020). A study on the tribological performance of nanolubricants. Process. 8:1372. https://doi.org/10.3390/pr8111372
3. Sahoo, S. K., Goswami, S. S. and Halder, R. (2024). Supplier selection in the age of industry 4.0: A review on MCDM applications and trends. Decis.Mak. Adv. 2(1):32–47. https://doi.org/10.31181/dma21202420
4. Mittal, U., Yang, H., Bukkapatnam, S. T. and Barajas, L. G. (2008). Dynamics and performance modeling of multistage manufacturing systems using nonlinear stochastic differential equations. IEEE International Conference on Automation Science and Engineering, IEEE, pp. 498–503. https://doi.org/10.1109/COASE.2008.4626530
5. Sahoo, S. K., Goswami, S. S., Sarkar, S. and Mitra, S. (2023). A review of digital transformation and industry 4.0 in supply chain management for small and medium-sized enterprises. Spectr. Eng. Manag. Sci. 1(1):58–72. https://doi.org/10.31181/sems1120237j
6. Mittal, U. (2023). Detecting hate speech utilizing deep convolutional network and transformer models. International Conference on Electrical, Electronics, Communication and Computers, IEEE, pp. 1–4. https://doi.org/10.1109/ELEXCOM58812.2023.1037050
7. Sethy, N. K., Yenugula, M., Goswami, S. S., Bhola, A. and Behera, D. K. (2023). Selection of ideal IoT based overhead conductor for optimizing the performance of a small hydropower project. J. Nano. Electron. Phys. 15(4).https://doi.org/10.21272/jnep.15(4).04006
8. Bambam, A. K., Dhanola, A. and Gajrani, K. K. (2023).A critical review on halogen-free ionic liquids as potential metalworking fluid additives. J. Molecular Liq. 121727. https://doi.org/10.1016/j.molliq.2023.121727
9. Sahoo, S. K. and Goswami, S. S. (2024). Green supplier selection using MCDM: A comprehensive review of recent studies. Spectr. Eng. Manag. Sci. 2(1):1–16. https://doi.org/10.31181/sems1120241a
10. Bui, T. A. and Bui, N. T. (2023). Investigating the impact of fly-ash additive on viscosity reduction at different temperatures: A comparative analysis. Appl. Sci. 13:7859. https://doi.org/10.3390/app13137859
11. Goswami, S. S., Sarkar, S., Gupta, K. K. andMondal, S. (2023). The role of cyber security in advancing sustainable digitalization: Opportunities and challenges. J. Decis. Anal. Intell. Comput. 3(1):270–285. https://doi.org/10.31181/jdaic10018122023g
12. Ramteke, S. M. and Chelladurai, H. (2020). Effects of hexagonal boron nitride based nanofluid on the tribological and performance, emission characteristics of a diesel engine: An experimental study. Eng. Rep. 2:e12216. https://doi.org/10.1002/eng2.12216
13. Goswami, S. S. and Behera, D. K. (2023). Developing fuzzy-AHP-integrated hybrid MCDM system of COPRAS-ARAS for solving an industrial robot selection problem. Int. J.Decis. Support Syst. Technol. 15(1):1–38.http://doi.org/10.4018/IJDSST.324599
14. Sahoo, S. K., Das, A. K., Samanta, S. and Goswami, S. S. (2023). Assessing the role of sustainable development in mitigating the issue of global warming. J. Process Manag. New Technol. 11(1-2):1–21.https://doi.org/10.5937/jpmnt11-44122

Advances in Additive Manufacturing Technologies – Gurusamy Pathinettampadian (eds)
© 2024 Taylor & Francis Group, London, ISBN 978-1-032-90013-1

41 Circularity Error Analysis and Comparison in Drilling Natural Fiber Reinforced Polymer Composite

R. Kalai selvan[1], N. Senthilkumar[2]
Department of Mechanical Engineering, Saveetha School of Engineering,
Saveetha Institute of Medical and Technical Sciences,
Saveetha University, Chennai, Tamil Nadu, India

ABSTRACT: The objective of this research is to create a novel bio composite with enhanced dynamic properties for engineering applications by utilizing natural fibers. Kenaf fiber (KF) in varying proportions (0, 10, 20, and 30%) strengthens the biopolymer poly-butylene succinate (PBS). In an injection molding machine, heated PBS was combined with chopped KFs to create the polymer mixture. The machining characteristics (drilling) of the manufactured composite are examined, and the results are compared to those of ordinary PBS. PBS is used as a reference without KFs added, and PBS is used as an experiment with KFs added. Four groups were examined in total. Three samples from each group were examined based on the results of the G-Power tool, with a 0.05 chance and a 95% confidence interval. The circularity error for the PBS+9%KFs polymer composite is larger when a solid carbide drill is used to drill the examples. The addition of KFs prevents the composite from shattering during drilling, which explains why. A p-value of less than 0.05 indicates that there is a significant difference between the groups examined in the analysis, and that the circularity error increases with the number of KFs. Adding KFs to the PBS matrix results in a greater tensile strength, but since the fibers are extracted from the matrix during drilling, there is also a larger circularity error.

KEYWORDS: Natural resources, Bio composite, Novel polymer composite, Polybutylene succinate, Kenaf fiber, Carbide drill, Circularity error

1. INTRODUCTION

Polymer core material composites are created by combining several fiber types and reinforcements. According to Altuzarra and Kecskeméthy (2022) kenaf fiber is a subtype of bast fiber. Because it is inexpensive, recyclable, and requires little energy to process, it is an excellent natural fiber for reinforcing composites. Today, polybutylene succinate is manufactured and used in a wide variety of ways around the globe. Though it's not as widely utilized as traditional plastics yet, further research and development are required to make it more affordable and effective (Ghosh 2019). Polybutylene succinate may be more costly than other polymers due to the cost of raw ingredients and manufacturing procedures. Because it utilizes less nonrenewable resources, produces less waste for landfills, and reduces carbon emissions, polybutylene succinate is environmentally friendly. Everyday items like paperweights, helmets, backpacks, bath and shower units, and lamp shades may be made from natural materials (Kumar 2022). Because it adheres to a wide variety of textiles and performs well at high temperatures, polybutylene succinate is often used in

[1]kalaiselvanr19@saveetha.com, [2]senthilkumarn.sse@saveetha.com

DOI: 10.1201/9781003545774-41

newly developed composites. Among its various mechanical and physical properties are its strength when torn apart, flexibility, and resistance to heat and water damage. It may be used to make a variety of items, including textiles, vehicle components, and packaging. Polybutylene succinate is biodegradable if it can be organically broken down by bacteria in the atmosphere. Composite materials are being used to create parts of vehicles and trucks, such as roof coverings, headliners, hoods, engine covers, and door panels. Before the structures can be assembled, drilling may sometimes be one of the last jobs that must be completed on schedule.

They investigated the mechanical characteristics of polybutylene succinate and kenaf fiber composites. Among other materials, metal, wood, and polymers may all be punctured using carbide drill bits. Steel, aluminum, and titanium are all very hard materials that can be drilled through using carbide drill bits. It's crucial to set a carbide drill bit's speed and feed rate correctly. Generally speaking, a slower speed and a greater feed rate work better for boring tougher materials. Moreover, compared to synthetic fibers, natural fibers are simpler to recycle, discard, and biodegrade. When it comes to qualities like strength, durability, and resistance to heat and pressure, biocomposites may perform on par with or better than conventional materials. Natural fiber composites have traditionally been based on vinyl esters, polyester resin, epoxy resin, and phenol formaldehyde. Increased fiber amount yields the greatest dynamic properties in polymer composites (Rangappa et al. 2021). They found that a 170 KN crush force could be applied to human hair that had been reinforced with a PBS matrix. PBS is a lightweight substance that may reduce gas consumption and pollutant emissions from automobiles. Natural fibers absorb water well and come in a variety of textures. For instance, garments made with cotton derived from plant strands are delicate and lightweight. Bioplastics, plant fibers, and natural resins are examples of materials that may be recycled repeatedly to create biocomposites. The stretchiness of human hair prevents creep development, which makes it a useful addition to plastics. The shear strength of a polypropylene-based polymer was examined by Ashish Kumar et al. (2022) when the proportion of hair fibers was increased from 3% to 5% to 8%. Biocomposites are adaptable materials with a wide range of applications in automobiles, including structural, body panel, and interior components.Common interior components composed of natural fiber-reinforced composite materials include door panels, dashboard components, package shelves, seat cushions, backrests, and wire linings. When there is more hair fiber reinforced PP material, the mechanical characteristics perform better. Sanjay et al. combined it at 3%, 5%, 10%, and 15% by weight with polypropylene (PP) [6]. Depending on which and how much of the constituent elements are utilized, biocomposites may have a variety of mechanical and physical properties.Because carbide drill bits produce a lot of heat when they cut, they may wear out more rapidly.

Prior studies have shown that the following factors significantly influence the drilling quality analysis of natural fiber-reinforced composites: cutting parameters, fiber quantity, cutting forces, and the internal connection between the fiber and matrix. Creating composites using jute and human hair as reinforcements with Pas as the matrix material is the main focus of this investigation. In addition, it examines the surface roughness at three different drill tip angles, feed rates, and speeds when drilling these composite materials.

2. Materials and Methods

Poly(1,4-butylene succinate, or PBS) is the result of polycondensingsucccininic acid with 1-4-butanediol. PBS is a semi-solid thermoplastic polyester that is recyclable. PBS is a naturally occurring aliphatic polyester with properties similar to those of polypropylene. PBS is a particularly promising biopolymer because it has mechanical properties that are comparable to both isotactic polypropylene and high-density polyethylene, two relatively popular polymers. It may be disposed of alongside other biological waste because, like polylactic acid (PLA), it entirely decomposes into biomass, CO2, and H2O. People consider natural fibers to be a renewable resource with several advantages, including strength, stiffness, and compostability. However, since it is significantly more flexible than PLA, plasticizers are not necessary. Not only does it melt at 115°C rather than 160°C. PBS is the newest biopolymer, and more investigation may reveal that it is less expensive than PLA, PBAT, and PHB. PBS biomaterials have great potential and have a wide range of applications, including medical, automotive, and packaging.

Wetting the kenaf plant's stem produces certain types of kenaf fibers. Malik, Ahmad, and Gunister 2021) suggest that it might significantly replace synthetic fibers like glass fiber. When compared to most materials, kenaf fiber may have less mass while having the same tensile strength as produced fiber. As a result, lighter and more environmentally friendly polymer composites may be produced (Jawaid, Thariq, and Saba 2018). Malik, Ahmad, and Gunister (2021) state that kenaf is robust, soft, visually appealing, and has a linen-like appearance. It may also be heavy or light, weak, and not very flexible, breaking down quickly, absorbing water, and having a glossy, environmentally friendly appearance, among other qualities. Niu et al.

(2017) depict the PBS, kenaf plant, and kenaf fiber cubes in Fig. 41.1

Fig. 41.1 PBS pellets, Kenaf plant and fiber

Drilling projects are carried out using Saveetha Industries' vertical machining center (VMC). The G-Power tool is used to determine the sample size based on the findings of previous studies. Surface roughness has a mean of 0.080 and a standard deviation of 0.0028 (Bhandari, Lopez-Anido, and Gardner 2019). For this reason, three samples are obtained for every group. Figure 41.1 displays the fiber, the kenaf plant, and the purchased PBS.

Three research groups (PBS+10%KFs, PBS+20%KFs, and PBS+30%KFs) had their surface roughness of the drilled hole assessed. It is contrasted with the PBS-only control group. The weight quantities that will be added to PBS are used to calculate the KFs. After that, they are manually combined and fed into a compression molding machine to create models.

Plain PBS is prepared using the same technique for the control group. PBS pellets are melted by the compression molding machine to create models that are precisely the proper size for drilling. For every group, ten by ten centimeter samples with a five mm thickness are created.

A vertical machine center (Fig. 41.2) with a computer numerical control (VMC) YCM system, a spindle speed range of 45 to 1000 rpm, a BT40 spindle nose taper, and a spindle motor that works at 5.5 to 7.5 kW for 15 minutes on average was used for the drilling experiments. A camera is used to quantify circularity errors.

Fig. 41.2 Vertical machining center used for drilling PBS+KF composite

3. Results

Table 41.1 shows the CRR generated from the novel polymer mix PBS+KFs and the unreinforced PBS polymer from the solid carbide drill bit drilling tests. The outcome demonstrates that a PLA matrix's stronger KFs often have a lower CRR. But the CRR may be raised using specific-shaped drill bits.

Table 41.1 Circularity error values obtained for different materials considered in this study

Group	Circularity Error (mm)	Group	Circularity Error (mm)
Unaltered PBS	.062	PBS+20%KF	.088
Unaltered PBS	.065	PBS+20%KF	.086
Unaltered PBS	.063	PBS+20%KF	.088
PBS+10%KF	.080	PBS+30%KF	.100
PBS+10%KF	.077	PBS+30%KF	.095
PBS+10%KF	.079	PBS+30%KF	.097

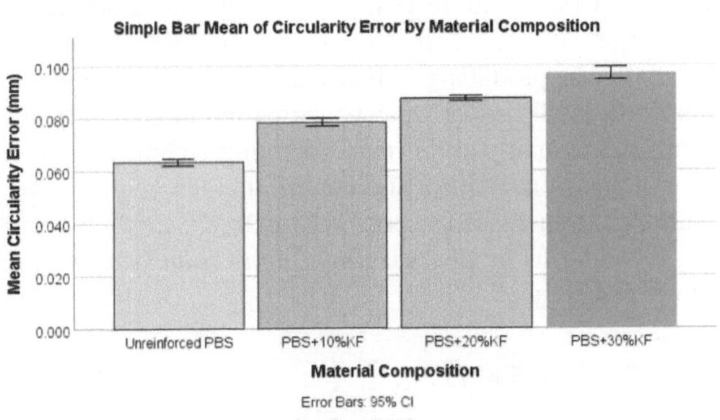

Fig. 41.3 Bar graph comparison between the considered groups performed with 95% CI and ±1 SD for circularity error

4. Discussion

The fact that this research only examined one compost combination and one twist drill is only one of its flaws (Kumar et al., 2022). Future developments in tool wear reduction while dealing with composites may see the adoption of covered or specially shaped drill bits. Cutting techniques in the future could also include flood cooling and the smallest number of lubrication approach.

By adding CFs, the polymer combination becomes stiffer and stronger, able to support more weight than PLA alone, according to mechanical tests of a PLA matrix reinforced with CFs. Due to their strength, CFs have superior dynamic properties than steel since they typically recover more energy when they shatter (Kumar et al. 2022). Also, the chopped fibers' propensity to adhere effectively to the PLA matrix strengthens the polymer combination. The PLA becomes more rigid and hardened because to the increased number of CFs added, losing some of its flexibility. PLA+15%CF has the greatest shore D hardness (74.71), bending strength (61 MPa), and tensile strength (35.9 MPa).

The liquidus temperature taken into consideration during molding, the injection molding pressure, the length and direction of the CFs, and other factors all affect the properties of the polymer combination (PLA+CFs). A shorter CF length is often associated with improved mechanical properties. The direction of reinforcement is a major factor to consider when considering lengthy CFs. Following their dissection, CFs with a random arrangement have better mechanical qualities. The sample is ill-prepared, and the characteristics often deteriorate, if the liquidus temperature drops during shaping.Our institution has shown great success in several areas and is very committed to producing high-quality research that is supported by evidence. With our study, we wish to contribute to this extensive history.

Because CFs are more costly and need a greater degree of support, the research has certain limitations about the feasibility of producing polymer composites. PLA's compostability will be lost when CFs are used in greater quantities. Additionally, other manufacturing techniques, such as compression casting, and natural textiles may be used with CFs to increase the polymer mixture's strength without reducing its biodegradability, in the context of related future research.

5. Conclusion

The primary goal of this research is to determine how well a new AA8011 Composite WPC MM can be machined by drilling and monitoring the amount of wear on the drill bit tool. These are the only variables that are being examined in the research. The study's findings indicate that using a solid lubricant reduces tool wear.

REFERENCES

1. Bhandari, Sunil, Roberto A. Lopez-Anido, and Douglas J. Gardner. 2019. "Enhancing the Interlayer Tensile Strength of 3D Printed Short Carbon Fiber Reinforced PETG and PLA Composites via Annealing." *Additive Manufacturing* 30 (100922): 100922.
2. Cheluka, Saikrupa. 2020. "Flexural Strength and Effect of Drilling Parameters on Surface Roughness, Circularity of Natural Fiber Composites." *International Journal for Research in Applied Science and Engineering Technology*. https://doi.org/10.22214/ijraset.2020.1121.
3. Eswaran, Prakash, K. Sivakumar, and Madheswaran Subramaniyan. 2018. "Minimizing Error on Circularity of FDM Manufactured Part." *Materials Today: Proceedings*. https://doi.org/10.1016/j.matpr.2017.11.324.
4. Fink, Johannes Karl. 2021. *Polymers and Additives in Extreme Environments: Application, Properties, and Fabrication*. John Wiley & Sons.
5. Ghoreishi, M., D. K. Y. Low, and L. Li. 2001. "Effects of Processing Parameters on Hole Circularity in Laser Percussion Drilling – A Statistical Model." *International Congress on Applications of Lasers & Electro-Optics*. https://doi.org/10.2351/1.5059880.
6. Gupta, Kapil, and Paulo Davim. 2020. *High-Speed Machining*. Academic Press.
7. Hussain, Chaudhery Mustansar, and Paolo Di Sia. 2022. *Handbook of Smart Materials, Technologies, and Devices: Applications of Industry 4.0*. Springer Nature.
8. Jawaid, Mohammad, Mohamed Thariq, and Naheed Saba. 2018. *Mechanical and Physical Testing of Biocomposites, Fibre-Reinforced Composites, and Hybrid Composites*. Elsevier Science.
9. Junior, Dair Ferreira Salgado. 2017. "Circularity vs Taper in Micro-Drilling with Pulsed Nd:YAG." *Proceedings of the 24th ABCM International Congress of Mechanical Engineering*. https://doi.org/10.26678/abcm.cobem2017.cob17-2760.
10. Khan, Anish, Sanjay MavinkereRangappa, Mohammad Jawaid, Suchart Siengchin, and Abdullah M. Asiri. 2020. *Hybrid Fiber Composites: Materials, Manufacturing, Process Engineering*. John Wiley & Sons.
11. Krishnasamy, Senthilkumar, Senthil Muthu Kumar Thiagamani, Chandrasekar Muthukumar, Rajini Nagarajan, and Suchart Siengchin. 2022. *Natural Fiber-Reinforced Composites: Thermal Properties and Applications*. John Wiley & Sons.
12. Kumar, Kaushik, Ganesh M. Kakandikar, and J. Paulo Davim. 2022. *Computational Intelligence in Manufacturing*. Woodhead Publishing.
13. Kumar, R. Sathish, R. Sathish Kumar, M. Mohanraj, P. Natarajan, A. Arockia Julias, and Sathyamurthy Ravishankar. 2022. "Experimental Study on the Drilling Parameter Analysis of Banana Fiber Reinforced Vajram Mixed Phenolic Resin Composite Laminates." *Journal of Natural Fibers*. https://doi.org/10.1080/15440478.2021.1993412.
14. Malik, Khurshid, Faiz Ahmad, and Ebru Gunister. 2021. "A Review on the Kenaf Fiber Reinforced Thermoset Composites." *Applied Composite Materials* 28 (2): 491–528.
15. Muthukumar, Chandrasekar, Senthilkumar Krishnasamy, Senthil Muthu Kumar Thiagamani, and Suchart Siengchin. 2022. *Aging Effects on Natural Fiber-Reinforced Polymer Composites: Durability and Life Prediction*. Springer Nature.
16. "Optical Fiber Concentricity Error." n.d. *SpringerReference*. https://doi.org/10.1007/springerreference_20400.
17. Panchal, Dilbagh, Prasenjit Chatterjee, Mohit Tyagi, and Ravi Pratap Singh. 2023. *Optimization Methods for Engineering Problems*. CRC Press.
18. Paulo Davim, J. 2018. *Drilling Technology: Fundamentals and Recent Advances*. Walter de Gruyter GmbH & Co KG.

19. Pinar, Ahmet. 2006. "The Assessment of Circularity Error in CNC Controlled Circular Interpolation Movement by Least Squares Method." *JOURNAL OF POLYTECHNIC*. https://doi.org/10.2339/2006.9.1.27-33.
20. Rajmohan, T., K. Palanikumar, and J. Paulo Davim. 2020. *Advances in Materials and Manufacturing Engineering: Select Proceedings of ICMME 2019*. Springer Nature.
21. Rangappa, Sanjay Mavinkere, Madhu Puttegowda, Jyotish-kumarParameswaranpillai, Suchart Siengchin, and Sergey Gorbatyuk. 2021. *Advances in Bio-Based Fiber: Moving Towards a Green Society*. Woodhead Publishing.
22. Rasu, Karthick, and Anbumalar Veerabathiran. 2022. "A Review on Drilling of Natural and Synthetic Fiber Reinforced Polymer Composites: 1992 - 2021." *Journal of Natural Fibers*. https://doi.org/10.1080/15440478.2022.2070325.
23. R, Vinay Kumar, Kumar R. Vinay, Gowda M. V. Suresh, H. M. Mallaradhya, and Kumar M. Vijay. 2022. "." *International Journal of Innovative Research in Advanced Engineering*. https://doi.org/10.26562/ijirae.2022.v0909.01.
24. Shanmugasundaram, Sureshkumar Manickam, Lakshmanan Damodhiran, Vignesh Billan, and Dhinakaran Gu. 2013. "Predictive Model for Circularity Error of Drilling on GFRP Composite Laminates Using Fuzzy Logic." *Applied Mechanics and Materials*. https://doi.org/10.4028/www.scientific.net/amm.446-447.316.
25. Shunmugam, M. S., and M. Kanthababu. 2019. *Advances in Unconventional Machining and Composites: Proceedings of AIMTDR 2018*. Springer Nature.
26. Singh, Kaushal Pratap, Ankur Bahl, Gavendra Norkey, and Girish Dutt Gautam. 2022. "Experimental Investigation and Parametric Optimization of the Hole-Circularity and Taper Angle during Laser Drilling Kevlar-29 Fiber Composite." *Materials Today: Proceedings*. https://doi.org/10.1016/j.matpr.2021.10.155.
27. Singh, Ravi Pratap, Narendra Kumar, Ravinder Kataria, and Pulak Mohan Pandey. 2022. *Evolutionary Optimization of Material Removal Processes*. CRC Press.
28. Tibadia, Rajkumar, Koustubh Patwardhan, Dhrumil Patel, Dinesh Shinde, and Rakesh Chaudhari. 2019. "Optimisation of Drilling Parameters for Minimum Circularity Error in FRP Composite." *International Journal of Materials Engineering Innovation*. https://doi.org/10.1504/ijmatei.2019.103608.
29. Venkateshwarlu, N., B. Singaravel, K. Chandra Shekar, and S. Deva Prasad. 2021. "Analysis and Optimization of Circularity Error in Drilling Process Using Statistical Technique." *IOP Conference Series: Materials Science and Engineering*. https://doi.org/10.1088/1757-899x/1057/1/012063.

Note: Every figure and table was created using the experimental work; none of them were taken from any publications or online.

Advances in Additive Manufacturing Technologies – Gurusamy Pathinettampadian (eds)
© 2024 Taylor & Francis Group, London, ISBN 978-1-032-90013-1

42 A Comparative Study on the Influence of Novel Neem Fiber on the Tensile Strength of Wood Fiber-reinforced Polylactic Acid Composite Materials

Ch. Ruben Raju[1]
Research Scholar, Department of Mechanical Engineering,
Saveetha School of Engineering, Saveetha Institute of Medical and
Technical Sciences, Saveetha University,
Chennai, Tamilnadu, India

K. Koppiahraj[2]
Project Guide, Department of Mechanical Engineering,
Saveetha School of Engineering, Saveetha Institute of Medical and
Technical Sciences, Saveetha University,
Chennai, Tamilnadu, India

ABSTRACT: The motive of this work was to investigate the influence of novel neem fiber on the tensile strength of a natural composite material made up of wood fiber and PLA. The four samples were prepared with varying ratios of PLA with novel neem and PLA with wood fiber. Tensile strength tests were performed on 20 samples (Group 1 (PLA and novel neem fiber) and Group 2 (PLA and wood fiber)), with 10 samples for each group. The outcomes showed that the maximum tensile value was achieved at 20% fiber loading. The tensile strength of PLA mixed with novel neem fiber is 65 MPa, while the tensile strength of PLA mixed with wood fiber is 56 MPa. Statistical analysis showed a notable variation between the two groups, with a P value of 0.001 ($p < 0.05$). The findings suggest that novel neem fiber is stronger and more rigid compared to conventional wood fiber, and could be effectively used to boost the tensile strength of natural composite materials. Overall, this study provides valuable insights into the development of sustainable materials for various industrial applications

KEYWORDS: Polylactic acid (PLA), Tensile strength, Novel neem fiber, Wood fiber, Sustainable consumption, Hand lay-up technique

1. INTRODUCTION

Composites are an essential component of modern materials due to their advantages, which include lighter density, chemical resistance, good strength, and outstanding resistance to cracking, creep, and degradation (Li et al. 2018). Despite its high production cost, polylactic acid (PLA) has attracted substantial attention among other biodegradable polymers (Gupta, Revagade, and Hilborn 2007). This is because of the unique characteristics of PLA, such as its biodegradability, biocompatibility, good mechanical qualities, processability, and reuse. For the purpose of reducing the cost of PLA manufacture, researchers are continually exploring alternate techniques, such as combining it with

[1]chimatarubenraju19@saveetha.com, [2]koppiahrajk.sse@saveetha.com

DOI: 10.1201/9781003545774-42

natural fibers which is considered sustainable consumption. The present study aims at evaluating the tensile strength of PLA with novel neem fiber and the results are compared with PLA with wood fiber (Vijaya Ramnath et al. 2021). Hand lay-up technique was used to prepare composite materials. Hand lay-up method ensured uniform mixture of fiber with the matrix, which enhances mechanical characteristics. The findings of the experiment demonstrated that the inclusion of novel neem fiber to PLA greatly boosted its tensile strength (Yao et al. 2019; Graupner, Herrmann, and Müssig 2009).

To improve PLA mechanical qualities, many studies have been undertaken in recent years (Sachin, Kannan, and Rajasekar 2020; Shen et al. 2021; Awale et al. 2018; Nur Diyana et al. 2022; Pérez-Fonseca et al. 2021). 115 and 75 research articles were collected from IEEE Xplore, and Google Scholar. (Pérez-Fonseca et al. 2021) found that the tensile and flexural properties of natural fiber-reinforced composites improved with fiber content up to a certain point, but then decreased. (Khoshnava et al. 2017) investigated biocomposites composed of biopolymers and natural fibers, which face fewer risks and dangers than non-renewable resources and provide superior mechanical qualities than standard plastics. (Alkateb et al. 2018) discovered that replacing synthetic or conventional fibers with bio fibers increases tensile strength and lessens impact, hence enhancing the crashworthiness of automobiles. (Asumani, Reid, and Paskaramoorthy 2012) demonstrated that treating natural fibers with alkali enhances the fiber-matrix interaction, hence compensating for natural fibers' disadvantageous high moisture content. Natural fibers were selected by (Ramnath et al. 2013; Srinivasan, Rajendra Boopathy, and Vijaya Ramnath 2014) due to their inexpensive cost and good mechanical qualities, which can be further improved with reinforced polymers. Researchers found that hybrid composites made from hand lay-up technique kenaf and aramid have improved tensile and impact strengths. The tensile strength of plain composites made from kenaf/banana fiber hybrids was found to be higher than that of twill composites made from the same material by (Yahaya et al. 2016; Alavudeen et al. 2015). The United Nations has called for sustainable consumption. Thus, the utilization of waste natural fibers may help in sustainable consumption.

Although many research works focused on natural fiber reinforced composites, the potentiality of neem fiber has been less explored. Many of the earlier studies on polymer composites concentrated on wood fiber and failed to examine the effect of adding neem fiber. To fill this research gap, this study aims to fabricate composite material using wood and neem fiber, and measures the tensile strength.

2. Materials and Methods

The PLA granules were purchased from Nature tech, Chennai. It had a mass of 1.24 g/cm3 and a melting temperature between 170 and 180o C. It had an approximate crystallization temperature of 110° C. The PLA's glass transition temperature was approximately 55° C. During this investigation, the tensile properties of PLA were evaluated in the laboratory, and academic literature was used to provide measurement results for the reinforcement's tensile properties (Siakeng et al. 2018). Wood fiber and novel neem fiber were provided by Go Green Products, from Chennai. These fibers were employed in the manufacturing of composites for the experiments now being conducted. There are two sample groups, each requiring 10 specimens and 20 sample tests for the experiment. Group 1 was a PLA containing wood fiber, while Group 2 was a unique PLA with novel neem fiber. A G-power of 0.8, alpha and beta of 0.05 and 0.2, and a beta-interval of 95% are used in the calculation (Huda et al. 2006).

PLA is a sustainable, recyclable thermoplastic derived from lactic acid produced from agricultural sources. It is a well-known biopolymer utilized in the creation of eco-friendly products and can be reused. Hand lay-up technique was used to make wood fiber-reinforced PLA composites, and their tensile strength was measured. Hand lay-up technique is used as it is cheap. Consequently, integrating wood fibers with PLA provides a sustainable consumption and cost-effective alternative for composite manufacture that can also be reused.

The novel neem fiber used in this experiment was sourced from Go Green Products in Chennai, Tamil Nadu. Use of waste neem fiber helps in sustainable consumption. After being exposed to the sun for ten days, the novel neem wood barks were eventually dried. The fibers were subjected to an alkaline treatment with NaOH for a period of twenty-four hours in order to enhance their adherence with the matrix material during the processing stage. To obtain ten samples with the appropriate proportions of PLA to novel neem fiber, independent weight fraction measurements were taken using a digital scale for the PLA and the novel neem fiber. By following hand lay-up technique, the sample of PLA and novel neem fiber was prepared (Shen et al. 2021). The same method was carried out in order to create reinforced sawdust composites with 15, 20, and 25 weight percent.

Tensile testing was conducted using a Universal Testing Apparatus (Fig. 42.1) with a 40 kN load capability and a 500 mm head travel. The sample is kept in the grabber between two tasks, and it was loaded in the direction of the run. The composite's ultimate strength is determined by the specimen's failure under continuous loading

Fig. 42.1 Universal testing machine

3. RESULTS

Figure 42.2 shows the comparison between PLA with wood fiber and PLA with novel neem fiber in terms of tensile strength. When comparing PLA made using novel neem fiber to PLA made with wood fiber, the tensile strength of the former was found to be 65 MPa and the latter to be 56 MPa. The results showed that there was a huge deviation between the two groups with the P value is 0.001 ($p < 0.05$). Tensile strength data are plotted on the X-axis, and the mean tensile strength, together with a 95% confidence interval and an error bar indicating ± 1 standard deviation, are marked on the Y-axis (Standard Deviation).

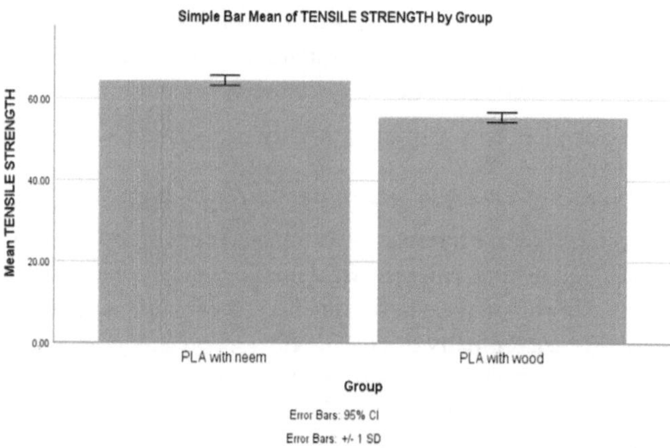

Fig. 42.2 A simple bar chart comparing the tensile strength of PLA with wood fiber and PLA with novel neem fiber has been developed

Table 42.1 compares the tensile strengths of two PLA variants, one made with wood and one with neem. A unique number is assigned to every test (Test1 to Test 10). Overall, the results show that PLA reinforced with novel neem has a greater average tensile strength than PLA reinforced with wood (65 MPa vs. 56 MPa). A comparison of the tensile strengths of two groups are shown in Table 42.2. Results show that the average tensile strength of PLA with novel neem is 65 MPa, which is significantly greater than the tensile strength of PLA made with wood (56 MPa). When comparing PLA to novel neem and PLA to wood, the former has a smaller tensile strength standard deviation (0.25 MPa) than the latter (0.527 MPa). Levene's test for equality of variances demonstrates that the two groups' variances are different. One group's accuracy is noticeably higher than the other, as shown by the results of the second test, a two-sample t-test for equality of means. There is an 11.78274-point difference between the groups on average, with a 95% confidence interval of 11.67283-13.63748.

Table 42.1 The comparative evaluation metrics for PLA with novel neem fiber and PLA with wood fiber has been presented

Sl. No.	Test Size	Tensile Strength (MPa)	
		PLA with wood	PLA with novel neem
1	Test1	50	67
2	Test2	54	65
3	Test3	58	68
4	Test4	53	64
5	Test5	52	62
6	Test6	51	61
7	Test7	57	60
8	Test8	52	63
9	Test9	55	67
10	Test10	50	64
Average Test Results		56	65

Table 42.2 The statistical calculation such as mean, standard deviation and standard error mean for PLA with wood fiber and PLA with novel neem fiber

Group		N	Mean	Standard Deviation	Standard Error Mean
Tensile Strength (MPa)	PLA with wood	10	56	0.527	0.176
	PLA with neem	10	65	0.250	0.083

The influence of novel neem fiber on the tensile strength of the PLA was determined by incorporating different amounts of novel neem fiber (10, 15, 20, 25 wt%). The tensile strength of PLA with novel neem fiber are superior to those of PLA made from wood. The increased tensile strength of PLA with novel neem fiber (65 MPa) than regular wood fiber (56 MPa) suggests that novel neem fiber is stronger than normal wood fiber. The composite material with 20% reinforcing was the most successful one in terms of tensile strength. Meanwhile, 25% reinforced composite was reduced because the dominating fiber content and inadequate matrix material led to inappropriate fiber wetting. This, in turn, was the cause of the decreased tensile strength. When compared to wood fiber, the tensile strength of novel neem fiber is significantly greater.

Some similar studies (Ngo et al. 2014) studied the tensile characteristics of polymer composites, who found that changing the volume proportion of kenaf fiber from 20% to 40% and 60% affected the materials in different ways. There was a correlation between the percentage of kenaf fibers and the highest tensile strength. However, at a fiber content of 40%, these strengths declined. Another study by (Ochi 2008) looked at kenaf fiber volume fractions ranging from 30% to 70% in specimens and found that at 70% fiber content, the maximum flexural and tensile strengths were achieved. (Beg et al. 2015) compared treated and untreated fiber composites and identified that the treated fiber composite exhibited a 23.5% enhancement in tensile strength. (Kumar et al. 2012) reinforced novel neem sawdust into polyester resin. The composite material was reinforced up to a maximum of 50 wt% into polyester and the resultant composite had shown improved mechanical properties with addition of the novel neem fiber. (Jawaid, Abdul Khalil, and Abu Bakar 2011) found that the neem/EFB/novel neem composite had greater tensile strength than the EFB/neem/EFB composite.

The tensile strength of composites was shown to rise in direct proportion to the amount of 20 phr fiber used to reinforce them. However, the increased content of fibers, which is equal to 25 weight percent, results in a drop in the tensile strength of composites. The inclusion of fibers to PLA increases its rigidity, however this also causes the material to become more brittle. In order to identify the standard specifications required for the stir casting of PLA composites, additional research on this topic could be carried out.

4. Conclusion

From the result, it can be inferred that the addition of novel neem fiber to the PLA matrix improved the tensile strength of the resulting composite material. The PLA composite reinforced with novel neem fiber had a higher tensile strength of 65 MPa against the PLA composite reinforced with wood fiber which had a tensile strength of 56 MPa. Therefore, the use of novel neem fiber can be considered as a better alternative to conventional wood fiber in natural composite materials. Moreover, the use of waste novel neem fiber can also contribute to sustainable consumption, making it an environmentally friendly option.

References

1. Alavudeen, A., N. Rajini, S. Karthikeyan, M. Thiruchitrambalam, and N. Venkateshwaren. 2015. "Mechanical Properties of Banana/kenaf Fiber-Reinforced Hybrid Polyester Composites: Effect of Woven Fabric and Random Orientation." Materials & Design 66 (February): 246–57.
2. Alkateb, Mohamed, S. M. Sapuan, Z. Leman, M. R. Ishak, and Mohammad Jawaid. 2018. "Vertex Angles Effects in the Energy Absorption of Axially Crushed Kenaf Fibre-Epoxy Reinforced Elliptical Composite Cones." Defence Technology 14 (4): 327–35.
3. Asumani, O. M. L., R. G. Reid, and R. Paskaramoorthy. 2012. "The Effects of Alkali–silane Treatment on the Tensile and Flexural Properties of Short Fibre Non-Woven Kenaf Reinforced Polypropylene Composites." Composites. Part A, Applied Science and Manufacturing 43 (9): 1431–40.
4. Awale, Raina Jama, Fathilah Binti Ali, Azlin Suhaida Azmi, Noor Illi Mohamad Puad, Hazleen Anuar, and Azman Hassan. 2018. "Enhanced Flexibility of Biodegradable Polylactic Acid/Starch Blends Using Epoxidized Palm Oil as Plasticizer." Polymers 10 (9). https://doi.org/10.3390/polym10090977.
5. Beg, Mohammad D. H., John O. Akindoyo, Suriati Ghazali, and Abdullah A. Mamun. 2015. "Impact Modified Oil Palm Empty Fruit Bunch Fiber/poly (lactic) Acid Composite." International Journal of Materials and Metallurgical Engineering 9 (1): 165–70.
6. Frey, Felix. 2017. "SPSS (Software)." The International Encyclopedia of Communication Research Methods, November, 1–2.
7. Graupner, Nina, Axel S. Herrmann, and Jörg Müssig. 2009. "Natural and Man-Made Cellulose Fibre-Reinforced Poly(lactic Acid) (PLA) Composites: An Overview about Mechanical Characteristics and Application Areas." Composites. Part A, Applied Science and Manufacturing 40 (6): 810–21.
8. Gupta, Bhuvanesh, Nilesh Revagade, and Jöns Hilborn. 2007. "Poly(lactic Acid) Fiber: An Overview." Progress in Polymer Science 32 (4): 455–82.
9. Huda, M. S., L. T. Drzal, M. Misra, and A. K. Mohanty. 2006. "Wood-Fiber-Reinforced Poly(lactic Acid) Composites: Evaluation of the Physicomechanical and Morphological Properties." Journal of Applied Polymer Science 102 (5): 4856–69.
10. Jawaid, M., H. P. S. Abdul Khalil, and A. Abu Bakar. 2011. "Woven Hybrid Composites: Tensile and Flexural Properties of Oil Palm-Woven Jute Fibres Based Epoxy Composites." Materials Science and Engineering: A 528 (15): 5190–95.
11. Khoshnava, Seyed Meysam, Raheleh Rostami, Mohammad Ismail, Abdul Razak Rahmat, and Babatunde Ezekiel Ogunbode. 2017. "Woven Hybrid Biocomposite: Mechanical Properties of Woven Kenaf Bast Fibre/oil Palm Empty Fruit Bunches Hybrid Reinforced Poly Hydroxybutyrate Biocomposite as Non-Structural Building Materials." Construction and Building Materials. https://doi.org/10.1016/j.conbuildmat.2017.07.189.

12. Li, Decai, Yang Jiang, Shanshan Lv, Xiaojing Liu, Jiyou Gu, Qifeng Chen, and Yanhua Zhang. 2018. "Preparation of Plasticized Poly (lactic Acid) and Its Influence on the Properties of Composite Materials." PloS One 13 (3): e0193520.
13. Ngo, W. L., M. M. Pang, L. C. Yong, and K. Y. Tshai. 2014. "Mechanical Properties of Natural Fibre (Kenaf, Oil Palm Empty Fruit Bunch) Reinforced Polymer Composites." Advances in Environmental Biology, June, 2742+.
14. Nur Diyana, A. F., A. Khalina, M. S. Sapuan, C. H. Lee, H. A. Aisyah, M. N. Nurazzi, and R. S. Ayu. 2022. "Physical, Mechanical, and Thermal Properties and Characterization of Natural Fiber Composites Reinforced Poly(lactic Acid): Miswak (Salvadora Persica L.) Fibers." International Journal of Polymer Science 2022 (November): 1–20.
15. Ochi, Shinji. 2008. "Mechanical Properties of Kenaf Fibers and kenaf/PLA Composites." Mechanics of Materials 40 (4): 446–52.
16. Pérez-Fonseca, Aida Alejandra, Verónica Olimpia Ramírez-Herrera, Francisco Javier Fuentes-Talavera, Denis Rodrigue, José Antonio Silva-Guzmán, and Jorge Ramón Robledo-Ortíz. 2021. "Crystallinity and Impact Strength Improvement of Wood-Polylactic Acid Biocomposites Produced by Rotational and Compression Molding." Maderas. Ciencia Y Tecnología 23. https://doi.org/10.4067/s0718-221x2021000100436.
17. Ramnath, B. Vijaya, B. Vijaya Ramnath, S. Junaid Kokan, R. Niranjan Raja, R. Sathyanarayanan, C. Elanchezhian, A. Rajendra Prasad, and V. M. Manickavasagam. 2013. "Evaluation of Mechanical Properties of Abaca–jute–glass Fibre Reinforced Epoxy Composite." Materials & Design. https://doi.org/10.1016/j.matdes.2013.03.102.
18. Sachin, S. Raj, T. Kandasamy Kannan, and Rathanasamy Rajasekar. 2020. "Effect of Wood Particulate Size on the Mechanical Properties of PLA Biocomposite." Pigment & Resin Technology 49 (6): 465–72.

Note: Every figure and table were created using the experimental work; none of them were taken from any publications or online.

Advances in Additive Manufacturing Technologies – Gurusamy Pathinettampadian (eds)
© 2024 Taylor & Francis Group, London, ISBN 978-1-032-90013-1

43 Investigating of Tensile Strength on 3D Printing Lattice Structures (Acrylonitrile Butadiene Styrene Resin) Compared with and without Silica Nanoparticles

P Deepak[1]
Research Scholar, Department of Mechanical Engineering,
Saveetha School of Engineering, Saveetha Institute of Medical and
Technical Sciences, Saveetha University,
Chennai, Tamilnadu, India

A. Vasudevan[2]
Project Guide, Corresponding Author,
Department of Mechanical Engineering, Saveetha School of Engineering,
Saveetha Institute of Medical and Technical Sciences,
Saveetha University, Chennai, Tamilnadu, India

ABSTRACT: This article investigates the tensile strength of composites made of acrylonitrile butadiene styrene resin with and without a filler made of silica nanoparticles. A control group of (ABSR) without filler and an experimental group of ABSR with filler were used, each consisting of 20 sample specimens. The sample size was calculated using G Power at 80% power and $\alpha = 0.05$ per set. The base materials used were acrylonitrile butadiene styrene resin, while novel silica nanoparticles were used as filler. The results showed that the tensile strength of the acrylonitrile butadiene styrene resin composite without filler was significantly higher than that of the composite with 2% volume fraction of silica nanoparticles filler. The P value was 0.000 ($p < 0.05$), indicating a statistically significant difference between the two groups. The observation of ABSR with 2% filler was also statistically significant. Based on the study limits, it can be inferred that by adding filler to acrylonitrile butadiene styrene resin results in a notable and statistically significant increase in its tensile strength compared to the composite acrylonitrile butadiene styrene resin lacking filler.

KEYWORDS: 3D printing, Novel lattice structures, Porous cubic, Hybrid composites, Acrylonitrile butadiene styrene resin, Tensile strength

1. Introduction

The resin made of butadiene and acrylonitrile given that they combine the qualities of composite materials like high strength with those of metals like ductility, laminates are among the most intriguing forms of materials. The first ABSR to be evaluated for industrial use, according to Cortes and Cantwell, was the Acrylonitrile Butadiene Styrene Resin Laminates (Amiri et al. 2022). As a result, several composite and metal material fusions have been made available lately. Although aramid fiber-based composite laminates are also suitable, FMLs can be employed

[1]deepakpurushotham19@saveetha.com, [2]vasudevana.sse@saveetha.com

DOI: 10.1201/9781003545774-43

for ballistic applications (Abdurrahman et al. 2022). One of the most popular reinforced composite materials for tensile strength analysis in indentation operations using molding techniques and a new stacking sequence is Acrylonitrile butadiene styrene resin laminate composites (Tran et al. 2022). To improve mechanical properties including strength to weight ratio, and tensile strength, ABS resins are employed. ABS resin has a wide range of practical uses, including composites for thermal insulation and aerodynamics (Ya'akub et al. 2021). Due to its high strength and lightweight, 3D printing (Liquid ABS) is used in the construction of aircraft bodies, bulletproof clothing, and body armor, among other things (Ulkir 2023).

From the previous five years many researchers conducted research on the mechanical composite properties and surface modification of acrylonitrile butadiene styrene resin to help enhance the fibers strength and other characteristics for sustainable manufacture. It was given in a study that uses composites made of acrylonitrile butadiene styrene resin to improve mechanical and bonding properties (Szot 2023). The features of polymer composites will be improved with the help of ABSR composites. ABSR is a substance used in applications that have to do with power. The laminate composite is made with resin as its base material, which is then mixed with the matrix material. In the building sector, this alloy is used (Tandon et al. 2023). Mechanical properties of composites created via liquid ABS 3D printing are studied. A report on the unsaturated Novel Silica nanoparticle reinforced composites' electrical and mechanical characteristics was suggested. As polymer nanocomposites, these composites are used because they have better mechanical and electrical properties. A report on the mechanical characteristics of hybrid ABSR reinforced composites has been proposed based on a survey of the literature (Lee 2018).This study stands out in terms of its literature review because the materials and fabrication methods employed align closely with the current research focus. To enhance mechanical and bonding qualities, acrylonitrile butadiene styrene resin composites are employed (Muthu et al. 2021); (Gokuldoss et al. 2020).

This innovative filler has not been studied in fiber composites before. Numerous studies have revealed the brittleness and low tensile strength of 3D printing (liquid ABS) without filler. This work tested fiber with a novel silica nanoparticle filler to overcome all of these restrictions on hybrid composites. ABS resin fiber-reinforced composites, both with and without Novel Silica nanoparticles, were produced using a hand lay-up technique and compression molding. Various stacking sequences of laminate composites were created. The primary objective is to study the Tensile Strength analysis of sustainable fiber manufacturing for composite materials during indentation operations. This enhancement is specifically explored in composites without fillers in comparison to those with Novel Silica nanoparticles fillers.

2. Materials and Methods

To investigate the Tensile Strength analysis during drilling operations, Acrylonitrile butadiene styrene (ABS) resin-reinforced composites were categorized into two sets: control group and experimental group. In the control group, a plain ABS resin laminate was produced using a mixture and hardener, without any filler, as illustrated in Fig. 43.1. After the plain laminate fiber was manufactured, 20 sample specimens of the composite material were sliced for Tensile Strength analysis.

Fig. 43.1 Resin 3d printer anycubic

The experimental group samples set involved the fabrication of ABS resin laminate composites with a 2% vol. fraction of filler. Subsequently, 20 sample specimens were obtained from the produced composite materials. These materials were cut into dimensions of 300x300mm, and the fiber-reinforced composite was combined with resin AW106IN and hardened HV953U, as depicted in Fig. 43.2. Following

Fig. 43.2 Universal testing machine

the completion of the production process, weights were applied to the materials, which were then subjected to a minimum drying period of 16 hours using a novel stacking sequence. Acrylonitrile butadiene styrene resins were utilized in this process.

To achieve a total sample size of 40 hybrid composites, the Tensile Strength inaccuracies during the drilling of Acrylonitrile butadiene styrene resin composite material are manually evaluated. Finding the experimental drilled hole area and comparing it to the theoretical drilled hole area based on the drill bit diameter utilized yields the tensile strength for drilled holes. The experimental and control groups go through this procedure once more for all 20 drilled holes. The disparity between the two groups is illustrated as the Tensile Strength in Fig. 43.1. The Tensile Strength is visually represented through freehand sketch modules for both the control and experimental groups.

Information is gathered employing a Tensile Strength machine, which gauges the Tensile Strength generated throughout the drilling procedure. This enables the calculation of Tensile Strength errors in the drilling of Acrylonitrile butadiene styrene resin composite material for sustainable production, as depicted in Fig. 43.2. The collected data is organized into tables, summarizing the results for all drilled holes, with a total of 20 drilled holes per laminate group.

3. Results

Tensile Strength of both control group and experimental group samples are compared. The increase in Tensile Strength of Acrylonitrile Butadiene Styrene Resin with a 2% volume fraction of Novel Silica Nanoparticles Filler was found using an innovative stacking sequence. Each group received 20 samples. In the absence of filler, the average Tensile Strength measured 78.2628 mm, whereas with the filler, the average Tensile Strength was 64.5434 mm, as illustrated in Fig. 43.1. Acrylonitrile butadiene styrene resin without filler (SNP) appears to have a higher tensile strength than resin with filler (SNP).

Table 43.2 and Fig. 43.3 shows the mean, Acrylonitrile butadiene styrene resin without filler and with filler were compared, as well as their respective standard deviations and 95 percent confidence intervals for the mean. A butadiene acrylonitrile resin's average value is 64.543 without filler and 78.262 with filler. Figure 43.1 shows the standard deviation of novel silica nanoparticles in acrylonitrile butadiene styrene resin is 2.014 without filler and 3.202 with filler, respectively. The standard error mean for tensile strength items ranges from 0.450 without filler to 0.716 with filler Fig. 43.2, with 0.450 being the minimum and 0.716 being the maximum. 20 samples' values are presented in Table 43.1 given on next page.

Table 43.1 ABS resin's tensile strength is calculated from 20 samples in each group, both with and without a 2% volume percentage of novel silica nanoparticle filler

S. No	Tensile Strength of ABS resin without filler (%)	Tensile Strength of ABS resin with 2% vol. fraction of filler (%)
1	64.324	77.566
2	61.547	82.694
3	63.437	75.104
4	62.855	78.266
5	64.709	76.356
6	67.253	81.961
7	63.134	81.806
8	66.191	76.954
9	62.265	78.674
10	62.442	77.145

Table 43.2 The statistical mean, standard deviation and standard error mean for tensile strength of ABS resin lattice structure with and without silica nanoparticles filler

Group		N	Mean	Standard Deviation	Standard Error Mean
TENSILE_ STRENGTH	EG	20	64.543	2.014	0.450
	CG	20	78.262	3.202	0.716

Table 43.3 and Fig. 43.3 shows the results of the novel acrylonitrile butadiene styrene resin's drilling tensile strength test, 0.045 and significance. Table 43.3 reveals significance p = 0.045 (p 0.05) and the results of an independent test shown in Fig. 43.2. Twenty samples of the novel acrylonitrile butadiene styrene resin with and without filler are gathered for each phase. Figure 43.1 shows In the 95% Confidence Interval of the Lower Value Assumed to be 15.432 and the Upper Value to be 12.066. The equal voids in the Tensile Strength of the 95% in Fig. 43.3 are not assumed. The lower value's confidence interval is 15.442 while the upper value's confidence interval is 11.996.

4. Discussion

The study utilizes and identifies that the damaged area in the Tensile Strength analysis of drilling operations for Acrylonitrile Butadiene Styrene Resin without filler (SNP) exhibits a superior mean compared to the damaged area in the Tensile Strength analysis of indentation operations for Acrylonitrile Butadiene Styrene Resin without filler (SNP)

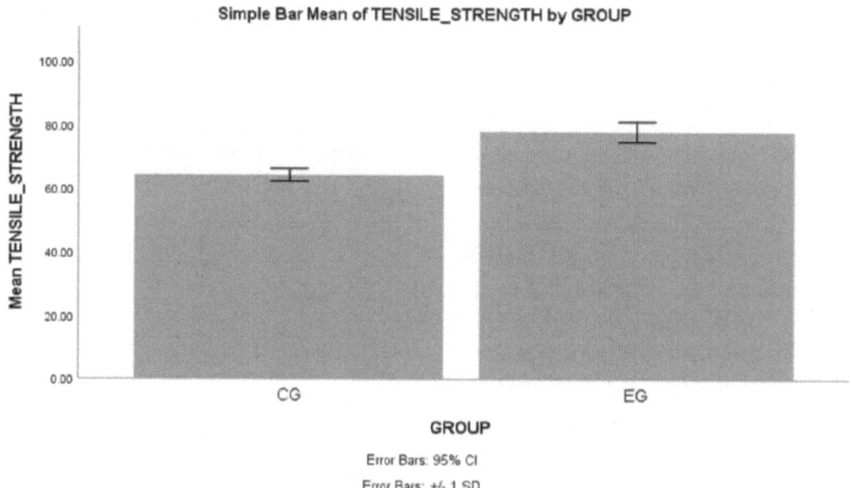

Fig. 43.3 Bar graph representing the comparison of Tensile Strength for CG samples (ABS resin without filler) and EG samples (ABS resin with filler). Acrylonitrile butadiene styrene resin with filler appears to produce the most consistent results when compared with Acrylonitrile butadiene styrene resin without filler

using a novel stacking sequence. The bar graph illustrates the fluctuation in tensile strength between the two groups with and without filler. The results suggest that Acrylonitrile Butadiene Styrene Resin with filler is more accurate compared to the substance without filler. Independent T-test outcomes establish the statistical significance of the obtained results.

The results of a comparable investigation that were identical to those reported in the earlier study are discussed. It conducted analysis on acrylonitrile butadiene styrene resin. In this study, ABS resin with filler outperformed ABS resin without filler in terms of tensile strength. As the ratio of Novel Silica nanoparticles rises, the strength of the nanoparticles also rises (Deshmukh et al. 2022). The quality of the composite is raised by using a novel filler made of silica nanoparticles(Gupta 2023). There has been some investigation into how different composites' mechanical, tensile strength, and other qualities are affected by acrylonitrile butadiene styrene resin (Dikshit et al. 2022). Basalt and Acrylonitrile Butadiene Styrene Resin Effect was investigated using Tensile Strength Analysis for Hybrid Composites. Tensile strength tests were conducted on metal laminate composites (Sastri 2010). The amount of energy being absorbed has increased, but fabrications have not. Several factors influence the mechanical properties of acrylonitrile butadiene styrene resin composites, including the mixing of nano clay filler, processing temperature, cooling process, presence of moisture content, sustainable production in acrylonitrile butadiene styrene resin volume fraction, nano clay length, and diameter of silica nanoparticles. (Dizon et al. 2022).

Five studies in all were examined in our analysis, three of which had conclusions that were comparable and two of which had opposing findings. From the above discussion, it can be inferred that 3D printing (Liquid ABS) with filler has a less damaged region of tensile strength than acrylonitrile butadiene styrene resin without filler. One of its shortcomings is the difficulty in drilling and cutting Acrylonitrile Butadiene Styrene Resin with Filler. It is also inefficient. Metal is lost by heat during the fabrication process. It will endeavor to eliminate all restrictions for acrylonitrile butadiene styrene resin with fillers in the future and streamline the drilling procedure.

5. Conclusion

This study successfully fabricated acrylonitrile butadiene styrene resin-based 3D printing composites with filler particles. The results indicate that the addition of filler particles was effective in preventing the formation of roughness in the composites. Furthermore, the addition of filler particles strengthened the matrix's stiffness by obstructing the polymer molecules' ability to migrate, which significantly improved the composites' tensile strength by 21.26%.These findings highlight the potential of incorporating filler particles in acrylonitrile butadiene styrene resin-based 3D printing composites to enhance their mechanical properties.

References

1. Abdurrahman, Wahid, and Muhamad Fitri. 2022. "The Fatigue Strength Analysis of ABS (Acrylonitrile Butadiene Styrene) Material Shaft Result of 3D Printing Process

due to Rotating Bending Load." Materials Science Forum. https://doi.org/10.4028/www.scientific.net/msf.1051.137.

2. Amiri, Aria, Abbas Zolfaghari, and Mohsen Shakeri. 2022. "3D Printing of Glass Fiber Reinforced Acrylonitrile Butadiene Styrene and Investigation of Tensile, Flexural, Warpage and Roughness Properties." Polymer Composites. https://doi.org/10.1002/pc.26937.
3. Deshmukh, Kalim, S. K. Khadheer Pasha, and Kishor Kumar Sadasivuni. 2022. Nanotechnology-Based Additive Manufacturing: Product Design, Properties, and Applications. John Wiley & Sons.
4. Dikshit, Mithilesh K., Ashish Soni, and J. Paulo Davim. 2022. Advances in Manufacturing Engineering: Select Proceedings of ICFAMMT 2022. Springer Nature.
5. Dizon, John Ryan C., Leonard D. Tijing, Marlon James A. Dedicatoria, Mosbeh Kaloop, Mohsin Usman Qureshi, Ilenia Farina, and Katsuyuki Kida. 2022. Additive Manufacturing and Advanced Materials. Trans Tech Publications Ltd.
6. Gardan, Julien, Ali Makke, and Naman Recho. 2019. "Fracture Improvement by Reinforcing the Structure of Acrylonitrile Butadiene Styrene Parts Manufactured by Fused Deposition Modeling." 3D Printing and Additive Manufacturing. https://doi.org/10.1089/3dp.2017.0039.
7. Gokul Doss, Prashanth Konda, and Zhi Wang. 2020. Additive Manufacturing Volume 2: Alloy Design and Process Innovations. MDPI.
8. Gupta, Ram K. 2023. Flexible and Wearable Sensors: Materials, Technologies, and Challenges. CRC Press.
9. Lee, Donggyeong. 2018. "Investigation of Laser Ablation on Acrylonitrile Butadiene Styrene Plastic Used for 3D Printing." Journal of Welding and Joining. https://doi.org/10.5781/jwj.2017.36.1.6.
10. Muthu, G., T. Sathish, V. Dhinakaran, Vijayakumar, and K. P. Vignesh. 2021. "Performance Enhancement and Emission Control of Diesel Engine." Materials Today: Proceedings 37 (January): 1784–89.
11. Sastri, Vinny R. 2010. Plastics in Medical Devices: Properties, Requirements and Applications. Elsevier.
12. Szot, Wiktor. 2023. "Rheological Analysis of 3D Printed Elements of Acrylonitrile Butadiene and Styrene Material Using Multiparameter Ideal Body Models." 3D Printing and Additive Manufacturing. https://doi.org/10.1089/3dp.2022.0298.
13. Tandon, Sourabh, Ruchin Kacker, and Sanjay Kumar Singh. 2023. "Correlations on Average Tensile Strength of 3D-Printed Acrylonitrile Butadiene Styrene, acrylonitrile butadiene styrene resin, and acrylonitrile butadiene styrene resin Carbon Fiber Specimens." Advanced Engineering Materials. https://doi.org/10.1002/adem.202201413.
14. Tran, Thang Q., Carla Canturri, Xinying Deng, Chu Long Tham, and Feng Lin Ng. 2022. "Enhanced Tensile Strength of Acrylonitrile Butadiene Styrene Composite Specimens Fabricated by Overheat Fused Filament Fabrication Printing." Composites Part B: Engineering. https://doi.org/10.1016/j.compositesb.2022.109783.
15. Ulkir, Osman. 2023. "Conductive Additive Manufactured Acrylonitrile Butadiene Styrene Filaments: Statistical Approach to Mechanical and Electrical Behaviors." 3D Printing and Additive Manufacturing. https://doi.org/10.1089/3dp.2022.0287.
16. Ya'akub, S. R., N. Ibrahim, and R. Singh. 2021. "Effect of 3D-Printing Parameters on the Tensile Strength of Acrylonitrile Butadiene Styrene (ABS) Polymer." IOP Conference Series: Materials Science and Engineering. https://doi.org/10.1088/1757-899x/1173/1/012041.

Note: Every figure and table was created using the experimental work; none of them were taken from any publications or online.

Advances in Additive Manufacturing Technologies – Gurusamy Pathinettampadian (eds)
© 2024 Taylor & Francis Group, London, ISBN 978-1-032-90013-1

44 Comparison of Novel Hash Pattern Friction Welding of En10083 Steel with Commercial Friction Welding Method for Improving the Interface Friction Temperature

N.G. Navin kumar, A. Pradeep*
Department of Mechanical Engineering,
Saveetha School of Engineering, SIMATS Chennai, Tamil Nadu, India

ABSTRACT: The objective of this comparison is to assess the efficacy of the newly introduced Hash pattern friction welding method for EN10083 steel in enhancing metallurgical bonding compared to the conventional friction welding approach. Materials and Methods: Stainless steel EN10083 is used for this investigation. Friction welding is carried out on a CNC machine, and the workpiece is 30 mm in diameter and 60 mm long. The temperature is measured using a thermocouple sensor with the model number IRX-64 at a distance of 10 mm from the interface. 80% of the G power is used in this process. Results: In the context of Hash pattern friction welding, it has been observed that the welding process with variations in temperature found to be higher than the traditional friction welding. There is statistical significance between the two groups, as shown by the significant value of p=0.000 (2-tailed), which is smaller than (P<0.05). Factors such as time, temperature, and stress, have been observed. Conclusion: In the confines of this study, concerning the interface friction temperature, the innovative Hash pattern friction welding of EN10083 steel outperformed the conventional commercial friction welding approach.

KEYWORDS: Novel hash pattern, Friction welding, Temperature, CNC machine, EN10083, Research

1. INTRODUCTION

Friction welding is a welding method that utilizes the heat generated by the friction between two components to fuse them together. This paper presents the findings of a comparative investigation into the use of friction welding on en10083 stainless steels (Cater et al. 2013). This technique has found application in various industries, ranging from automobiles to construction, over an extended period. Introducing a lubricant at the interface, aimed at lowering the frictional temperature, can contribute to minimizing wear and friction. Nevertheless, the use of a lubricant may also result in reduced heat generation, potentially compromising the quality of the weld joint. In the field of welding and joining, ongoing research and development are anticipated in the twenty-first century, particularly in the welding and joining of metals with disparate properties, such as aluminum and stainless steels (Okamura and Aota 2004). A novel Hash pattern friction welding technique has been innovated to enhance the metallurgical bonding in components constructed from EN10083 steel. Assessing this recent welding approach against traditional commercial friction welding methods could yield more robust joints and increased fatigue resistance. The Friction welding process is quicker and less expensive (Dunkerton 1983). The metallurgical bonding characteristics, strength

*Corresponding author: pradeepa.sse@saveetha.com

DOI: 10.1201/9781003545774-44

of joints, and fatigue life of these welded components are outstanding, showcasing superior weld joints that are devoid of defects (Dunkerton 1983). Approximately 554 publications in Science Direct and 687 papers in Google Scholar, in total, have been published during the past five years that are connected to this research study. This method produces robust and reliable welds, offering efficiency in both time and cost. Precise control of the contact friction temperature is essential for generating a high-quality weld. The use of a lubricant can reduce friction at the interface, contributing to an improvement in the quality of the weld connection. Friction created by the weld interface rubbing makes suitable temperature to be evolved for the welding process. Friction welding is highly adaptable as it can effectively join materials of varying shapes and thicknesses. Its predominant application is in connecting steel pipes, a critical operation given the extensive use of steel pipes in the oil and gas industry (Arivazhagan et al. 2011).

In contrast to alternative steel varieties, EN10083 steel offers several advantages, such as an enhanced strength-to-weight ratio and superior resistance to corrosion (Nguyen and Weckman 2006). By employing innovative groove-making techniques, the welding process can be enhanced, allowing for an elevation in the temperature of the interface friction. This study involved a comparison and analysis of the mechanical properties of the welds using these new techniques versus those obtained through conventional grinding welding methods. The type of material being joined, applied pressure, tool rotation speed, and component geometry were identified as factors that can influence the quality of the weld connection (Akinlabi and Mahamood 2020). To achieve an excellent weld connection, it is crucial to meticulously control the welding parameters. The Best research study in the above citations is Arivazhagan et al. 2011. Limited research has been conducted on the distinctive hash pattern utilized in friction welding of EN10083 steel. The interface friction temperature plays a pivotal role in friction welding as it directly influences the amount of heat generated at the interface. Elevating the machine's speed, augmenting the applied strain, or employing a more rigid iron block material are means to increase the temperature at the contact site during the grinding process. In this study, the effectiveness of a recently developed hash pattern friction welding method is evaluated in comparison to a traditional friction welding approach for creating welds with EN10083 steel. This study has a research gap since industrial friction welding and unique hash pattern friction welding have never been compared. Our primary goal in this study is to demonstrate the superiority of the hash pattern friction welding technique over the conventional commercial friction welding method, particularly regarding the temperature of the final product. The application of a hash pattern in friction welds is expected to result in an enhanced interface temperature. There is significant potential for friction welding to elevate the temperature of weld connections.

2. Materials and Methods

As humans are not involved, ethical approval is not required. Groove patterns were crafted and dimensions were scaled down at OM Engineering Works Ambattur. This research utilized EN10083 stainless steel as its foundational material. The components produced through friction welding exhibit commendable metallurgical bonding characteristics, robust joint strength, and an extended fatigue life, resulting in superior weld joints with flawless construction (Dunkerton 1983). EN10083 stainless steel materials were acquired from Ambattur for this study. The research involves two sets: hash pattern friction welding and commercial friction welding, each comprising 32 workpieces. The material samples are of dimensions 30 mm in diameter and 60 mm in length. Four parameters were considered in the friction welding process for this investigation: frictional force (5.5 and 7 tones), spindle speed (1000 rpm and 1400 rpm), and temperature, which was measured using thermocouples.

The EN10083 stainless steel rod within Group 1 underwent a reduction in size, being resized to 30 mm in diameter and 60 mm in length from its initial dimensions of 35 mm in diameter and 100 mm in length. The materials were subsequently altered to incorporate a Hash pattern, following which the Hash-patterned materials underwent friction welding using a CNC machine. The EN10083 stainless steel rod's diameter and length were decreased from their initial measurements of 35 mm and 100 mm to 30 mm and 60 mm, respectively. Group 1 refers to the weld specimens that have a hash pattern grooved throughout the contact surface. A CNC machine is then used to friction-weld the hash patterns.

The final interface temperature is assessed for both friction-welded materials with hash patterns and those without patterns. The IRT 600T is used to calculate the friction welding temperature. Ten milliliters are used to measure the temperature from the interface. Friction welding, employing a CNC machine, was conducted with varied input parameters, encompassing spindle speeds and frictional forces. The hash pattern friction welding of EN10083 steel necessitated the collection of data for 16 samples, encompassing burn-off, temperature, speed, force, and interface temperature. In the case of non-pattern friction welding (commercial friction welding) of EN10083 steel, data on burn-off, temperature, speed, force, and interface temperature was also gathered for 16 samples.

3. RESULTS

This study examines the influence of interface temperature on the parameters of frictional forces and spindle speeds in both traditional and Hash pattern friction welding. The objective of this research is to identify the friction welding method that generates the highest material contact temperature, whether it is friction welded with or without Hash pattern. Table 44.1 presents the results of a laboratory analysis detailing the chemical composition of EN10083 stainless steel. Table 44.2 displays the interface temperature data for friction welding with a Hash pattern, while Table 44.3 presents the temperature findings for traditional friction welding. Table 44.4 shows that hash pattern friction welding had the highest mean temperature, which was 188.5000°C with a standard deviation of 4.76095.

Table 44.1 Chemical composition of material EN10083 Part 3 Grade 51CrV4

Elements, (%)	Symbol	Specification	Observed Values
Carbon	C	0.47-0.55	0.508
Silicon	Si	0.40 max	0.281
Manganese	Mn	0.70-1.10	0.966
Phosphorous	P	0.025 max	0.014
Sulphur	S	0.025 max	0.008
Chromium	Cr	0.90-1.20	1.078
Vanadium	V	0.10-0.25	0.119

Table 44.2 The statistical calculation such as mean, standard deviation and standard error mean for PLA with wood fiber and PLA with novel neem fiber

S. No	Spindle speed (rpm)	Friction force (ton)	Depth penetration (mm)	Time (sec)	Temp (°C)
1	1000	5.5	4	15	186
2	1000	5.5	4	25	191
3	1000	5.5	6	15	181
4	1000	5.5	6	25	190
5	1000	7	4	15	184
6	1000	7	4	25	188
7	1000	7	6	15	196
8	1000	7	6	25	183
9	1400	5.5	4	15	187
10	1400	5.5	4	25	195
11	1400	5.5	6	15	182
12	1400	5.5	6	25	192
13	1400	7	4	15	195
14	1400	7	4	25	189
15	1400	7	6	15	185
16	1400	7	6	25	193

Table 44.3 Test parameter of commercial pattern friction welding

S. No	Spindle speed (rpm)	Friction force (ton)	Depth penetration (mm)	Time (sec)	Temp (°C)
1	1000	5.5	4	15	174
2	1000	5.5	4	25	167
3	1000	5.5	6	15	176
4	1000	5.5	6	25	154
5	1000	7	4	15	166
6	1000	7	4	25	173
7	1000	7	6	15	155
8	1000	7	6	25	175
9	1400	5.5	4	15	160
10	1400	5.5	4	25	178
11	1400	5.5	6	15	168
12	1400	5.5	6	25	165
13	1400	7	4	15	177
14	1400	7	4	25	157
15	1400	7	6	15	169
16	1400	7	6	25	156

Table 44.4 Highest mean value of Temperature(°C) is found at hash pattern friction welding with a standard deviation

Group		N	Mean	Standard Deviation	Standard Error Mean
Friction Welding	Non Pattern	16	166.875	8.34166	2.08542
	Hash Pattern	16	188.500	4.76095	1.19024

The mean of Independent Sample T-Test for Parallel Lines Pattern Friction Welding and Non Pattern Friction Welding for Improving Interface Temperature is shown in Table 44.4. The outcome demonstrates a statistically significant difference between the two groups, with a P value of 0.000 (p0.05) reached among the groups taken into consideration. At 1400 rpm, 5.5 tonnes, and 4 mm of penetration, the temperature during friction welding with a hash pattern is 195°C, whereas without one, it is 178°C. Since 195>178, friction welding with a hash pattern produces a temperature that is 8.71% greater than friction welding without one. The top view of hash pattern friction welding is shown in Fig. 44.1. Hash pattern friction welding is seen in Fig. 44.2. Graphical depiction of temperature for commercial friction-welded material is shown in Fig. 44.3 together with innovative friction-welded material with a hash pattern along the Y axis, both with mean detection accuracy of 95% CI and +/- 1 SD. The SPSS software was employed to ascertain the rise in interface temperature for materials subjected to hash pattern friction welding.

Fig. 44.1 Top view of hash pattern friction welding

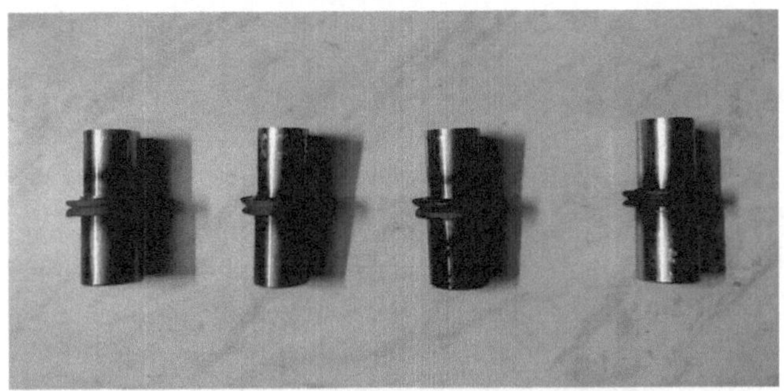

Fig. 44.2 Hash pattern friction welding

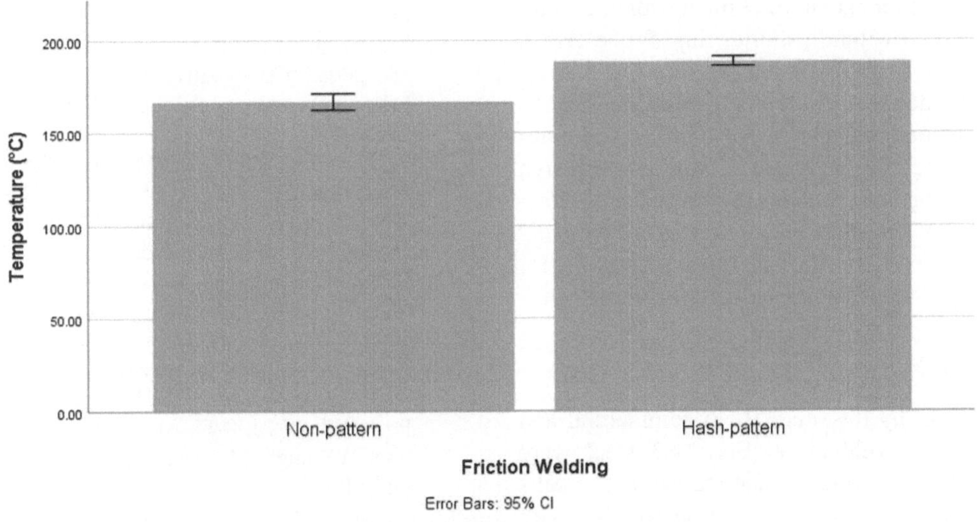

Fig. 44.3 Graphical representation of Temperature of Commercial friction welded material in X axis and novel hash patterned friction welded material in Y axis with Mean accuracy of detection 95% CI and +/- 1 SD.

4. Discussion

In this investigation, two friction welding methods were explored, and it was demonstrated that the Hash-pattern friction weld on EN 10083 stainless steel exhibited the highest interface temperature. Friction welding proves effective in joining materials that pose challenges for fusion welding techniques. Its advantages, such as low heat input, high production efficiency, ease of manufacturing, and environmental friendliness, make friction welding a widely utilized solid-state joining technique in contemporary applications (Paventhan, Lakshminarayanan, and Balasubramanian 2012). The demonstration indicates that with an increase in friction duration, there is a decline in interface temperature, reaching a minimum before rapidly ascending. Elevated upsetting pressures negatively impact the interface temperature of joints (Ates, Turker, and Kurt 2007). With an increase in friction time, the contact temperature decreases, reaches a minimum, and then sharply rises. Irregular pressures have an adverse impact on the interface temperature of joints. (Emre and Kaçar 2015). Based on the test data, Hash pattern friction welding exhibits a higher contact temperature when compared to commercial friction welding.

The presence of a pattern in the material results in significantly higher temperatures in Hash pattern friction welding compared to commercial friction welding. The design of the material adds complexity to the welding process due to increased friction (Poletaev, Poletaev, and Shchepkin 2020). An increase in friction leads to the rapid melting and amalgamation of the substance (Pang, Yang, and Cai 2023). It expedites the entire process. Due to the material's pattern, hash pattern friction welding exhibits a notably higher interface temperature compared to commercial friction welding. The material's design, contributing to increased friction, adds complexity to the welding process (Thompson 2011). The heat-generation rate during orbital friction welding of steel bars is analyzed using four distinct methods: continuous Coulomb friction, sliding-sticking friction, experimentally observed power data, and an inverse heat conduction approach (Maalekian et al. 2008).

The material flow and heat generation during the orbital friction welding process are determined by the contact conditions at the interface, categorized as sliding, sticking, or partial sliding/sticking (Schmidt, Hattel, and Wert 2003). Furthermore, it was observed that the formation of intermetallics at the interface results in increased hardness and decreased tensile strength (Varjenju and Dele 2016). Overall it reduces treatment time, allowing for the completion of more welds within a shorter time frame (Ishak 2016). Based on the analysis of temperature test results, the maximum interface temperature is recorded at 195 °C, occurring at a rotational parameter of 1400 rpm and a duration of 25 seconds . The future direction of this research involves incorporating nanoparticles in pattern areas. However, it's noteworthy that the study is restricted to using round-shaped rods, and a specific rpm range is essential for implementing this novel friction welding procedure.

5. Conclusion

In the parameters set by this study, both commercial and Hash pattern friction welding on EN10083 steel were evaluated concerning interface temperatures. The results indicate that materials subjected to friction welding with a Hash pattern exhibit higher interface temperatures compared to those produced through commercial friction welding. G power is 80% used in this operation. There is statistical significance between the two groups, as shown by the significant value of $p=0.000$ (2-tailed), which is smaller than ($P<0.05$). Hash pattern friction welding has a temperature of 195 °C, but commercial friction welding has a temperature of 178 °C at 1400 rpm, 5.5ton, 4mm depth of penetration, and 25 seconds.

References

1. Akinlabi, Esther Titilayo, and Rasheedat Modupe Mahamood. 2020. Solid-State Welding: Friction and Friction Stir Welding Processes. Springer Nature.
2. Arivazhagan, N., Surendra Singh, Satya Prakash, and G. M. Reddy. 2011. "Investigation on AISI 304 Austenitic Stainless Steel to AISI 4140 Low Alloy Steel Dissimilar Joints by Gas Tungsten Arc, Electron Beam and Friction Welding." Materials & Design 32 (5): 3036–50.
3. Ates, Hakan, Mehmet Turker, and Adem Kurt. 2007. "Effect of Friction Pressure on the Properties of Friction Welded MA956 Iron-Based Superalloy." Materials & Design 28 (3): 948–53.
4. Cater, Stephen, Jonathan Martin, Alexander Galloway, and Norman McPherson. 2013. "Comparison between Friction Stir and Submerged Arc Welding Applied to Joining DH36 and E36 Shipbuilding Steel." Friction Stir Welding and Processing VII. https://doi.org/10.1002/9781118658345.ch6.
5. Dunkerton, S. B. 1983. The Effect of a Post Rotational Twist on Friction Weld Properties of a C-Mn Steel. Welding Institute.
6. Emre, Hayriye Ertek, and Ramazan Kaçar. 2015. "Effect of Post Weld Heat Treatment Process on Microstructure and Mechanical Properties of Friction Welded Dissimilar Drill Pipe." Materials Research 18 (3): 503–8.
7. Field, Andy. 2013. Discovering Statistics Using IBM SPSS Statistics. SAGE.
8. Ishak, Mahadzir. 2016. Joining Technologies. BoD – Books on Demand.
9. Maalekian, M., E. Kozeschnik, H. P. Brantner, and H. Cerjak. 2008. "Comparative Analysis of Heat Generation in Friction Welding of Steel Bars." Acta Materialia 56 (12): 2843–55.
10. Nguyen, T. C., and D. C. Weckman. 2006. "A Thermal and Microstructure Evolution Model of Direct-Drive Friction Welding of Plain Carbon Steel." Metallurgical and Materials Transactions B. https://doi.org/10.1007/bf02693157.
11. Okamura, H., and K. Aota. 2004. "Joining of Dissimilar Materials with Friction Stir Welding." Welding International. https://doi.org/10.1533/wint.2004.3344.
12. Pang, Zheng, Jin Yang, and Yangchuan Cai. 2023. "Effects of Rotational Speed on the Microstructure and Mechanical Properties of 2198-T8 Al-Li Alloy Processed by Friction Spot Welding." Materials 16 (5). https://doi.org/10.3390/ma16051807.
13. Paventhan, R., P. R. Lakshminarayanan, and V. Balasubramanian. 2012. "Optimization of Friction Welding Process Parameters for Joining Carbon Steel and Stainless Steel." Journal of Iron and Steel Research International 19 (1): 66–71.
14. Poletaev, Yu V., V. Yu Poletaev, and V. V. Shchepkin. 2020. "Friction Welding of Fittings and Nozzles from Low-Alloy Steel 15X2HMFA." Welding International 34 (1-3): 29–33.
15. Schmidt, H., J. Hattel, and J. Wert. 2003. "An Analytical Model for the Heat Generation in Friction Stir Welding." Modelling and Simulation in Materials Science and Engineering 12 (1): 143.
16. Thompson, Brian. 2011. "Tungsten-Based Tool Material Development for the Friction Stir Welding of Hard Metals." Friction Stir Welding and Processing VI. https://doi.org/10.1002/9781118062302.ch14.
17. Varjenju, Opprit, and Njnab Dele. 2016. "Optimizing the Parameters for Friction Welding Stainless Steel to Copper Parts." Materiali in Tehnologije. https://scholar.archive.org/work/xkjfbnhgjbdv5iutkiz57ogmn4/access/wayback/http://mit.imt.si/Revija/izvodi/mit161/sahin.pdf.

Note: Every figure and table were created using the experimental work; none of them were taken from any publications or online.

Advances in Additive Manufacturing Technologies – Gurusamy Pathinettampadian (eds)
© 2024 Taylor & Francis Group, London, *ISBN 978-1-032-90013-1*

45 Comparative Analysis of Surface Roughness in Monel Alloy Using Novel Minimum Quantity Lubrication Technique and Dry Machining

M. Vinay, B. Navin Kumar*

Department of Mechanical Engineering,
Saveetha School of Engineering, Saveetha Institute of Medical and
Technical Sciences, Saveetha University,
Chennai, Tamil Nadu, India

ABSTRACT: The primary focus of this study is to examine the influence of minimum quantity lubrication (MQL) technique on the surface roughness of the machined components of Monel alloy. Additionally, the study seeks to compare the resulting surface roughness with that achieved through the conventional machining process without lubrication. The study involved two groups: a control group that is dry Machining and an study group that employed the machining with minimum quantity lubrication technique. The sample size is determined through the use of a calculator designed for this purpose. Each group was provided with twenty samples. In the control group, machining on monel alloy is performed without any lubrication applied. On the other hand, in the study group, a unique technique called minimum quantity lubrication is implemented wherein a small amount of compressed coolant is used at the cutting zone. The surface irregularities of both groups is assessed using a surf test SJ-410 to analyze the impact of the innovative MQL technique on surface irregularities. Results and discussion: The minimum quantity lubrication technique machined samples obtained a better surface finish when compared to the samples obtained from dry machining. This was validated by a p-value of 0.001, which is below the significance level of 0.05. Therefore, a notable difference can be observed between the two sets of groups based on statistical analysis. The study found that implementing the minimum quantity lubrication technique instead of dry machining resulted in a significant 67% improvement in surface finish, within the scope of the research. It is concluded MQL technique is a promising technique for sustainable production.

KEYWORDS: Novel minimum quantity lubrication, Monel, Graphene nanoparticles, Surface roughness, CNC machining, Dry machining, Sustainable production

1. INTRODUCTION

The application of cutting fluids in machining operations has been a common practice to enhance tool life and achieve better surface finish. However, conventional flood-type lubrication systems are expensive and environmentally harmful, prompting the exploration of alternative lubrication methods. One such method that has gained attention is Minimum Quantity Lubrication. (Venkatesh et al., 2022) MQL involves applying a small amount of compressed coolant directly at the cutting zone to minimize lubricant usage during machining processes. This technique offers an effective solution to reduce environmental pollution while improving overall machining effi-

*Corresponding author: navinkumarb.sse@saveetha.com

DOI: 10.1201/9781003545774-45

ciency. The researchers focus on modern manufacturing techniques to meet sustainable development goals. One of recent manufacturing techniques is minimum quantity lubrication (Debnath et al., 2021). The novel minimum quantity lubrication (MQL) technique represents an advanced lubrication technology, replacing traditional flood cooling and offering improved environmental and economic sustainability. In this method, the method includes supplying a minimal amount of lubricant directly to the area where cutting occurs. This assists in reducing cutting forces, decreasing surface roughness, and enhancing the lifespan of tools. By atomizing and spraying coolant droplets onto the cutting zone, MQL helps decrease cutting temperature and tool wear, ultimately improving surface integrity. The MQL setup depicted in Fig. 45.1. This setup involves a fine aerosol jet comprising a mixture of air and liquid with nanoparticles. (Kang, 2021). Proposed that the principle behind the Minimum Quantity Lubrication technique involves supplying a minimal quantity of lubricant or coolant directly to the cutting zone through high-pressure air. (Tosun et al., 2010) The researcher conducted experimental research to investigate how minimum quantity lubrication affects the surface irregularities of the monel alloy (Dhar et al., 2006) (Bruni et al., 2008) opined that the technique of using a small amount of lubricant or coolant is effective in reducing friction and improving the machining process. By atomizing and spraying the lubricant or coolant onto the cutting zone, it creates a fine mist that effectively cools and lubricates both the tool and workpiece.

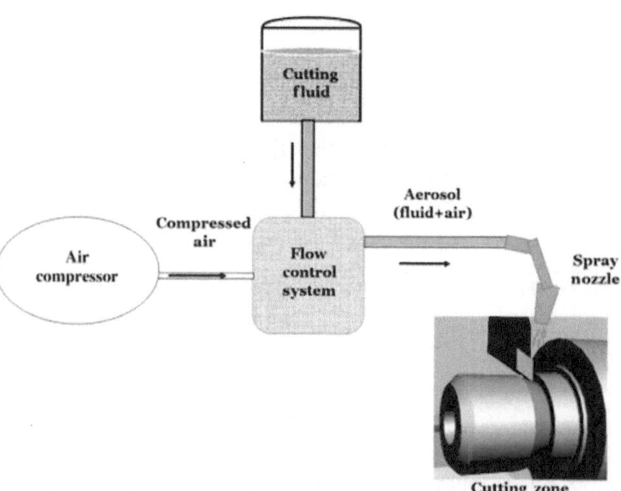

Fig. 45.1 Minimum quantity lubrication set-up linediagram

The author examined the influence of a new minimum quantity lubrication (MQL) technique, combined with solid lubricant assistance, on the surface quality while machining inconel 718. The findings revealed that implemetation of MQL machining led to a substantial average improvement of approximately 35% in the quality of the samples compared to using minimum quantity lubrication machining. MQL, which employs a minimal volume of lubricant, minimizes waste and lessens environmental pollution compared to conventional flood cooling techniques. (Kedare et al., 2014) suggested that by implementing the minimum quantity lubrication technique, industries can enhance their machining efficiency while also promoting sustainable and environmentally-friendly practices. (Navin kumar et al., 2020) This technique involves applying a minimal amount of coolant/lubricant directly to the cutting zones of both the tool and workpiece. By reducing friction and machining temperature in the process, MQL improves surface quality and extends the lifespan of tools. It is considered as one of the most suitable techniques for conventional operations like milling and turning. Additionally, MQL has several advantages over traditional flood type lubricating systems, such as minimizing waste by significantly reducing water consumption and contamination risks associated with large-scale lubricant dispersion methods. The research indicated that Minimum Quantity Lubrication (MQL) could be considered an economically viable and environmentally friendly lubrication approach. (Yin et al., 2021) Rigorous materials like hardened steels or ceramics can produce substantial heat throughout the machining process. MQL's regulated application of lubricant aids in heat dispersion, decreasing thermal harm to the workpiece and prolonging the lifespan of the tool. (Shinge et al., 2022) investigated alternative lubrication systems for machining an end mill constructed of forged steel 50CrMnMo. In terms of machinability, The proficient cooling rendered by MQL reduces the wear and tear of tools and helps sustain tool firmness, resulting in an extended tool lifespan while processing formidable materials.(Sivakumar et al., 2022) Hard materials like monel are susceptible to surface damage due to high temperatures and friction. MQL's cooling effect can lead to improved surface finish by reducing the chances of surface defects. MQL's cooling effect can help control the development of residual stresses in hard materials, enhancing part stability and reducing the risk of cracking. MQL reduces exposure to harmful airborne mists that are commonly associated with traditional flood cooling methods, improving the working environment for operators.

Based on the research results, there have been limited studies investigating the influence of Minimum Quantity Lubrication (MQL) technique on the surface roughness of the monel alloy. My practical knowledge of various machining operations, understanding of minimum quantity lubrication techniques, familiarity with surface roughness measurement methods, and expertise in metals and alloys have motivated me to pursue this research. The main objective of this study is to examine the impact of an innovative technology for minimum quantity lubrication on the surface irregularities of monel alloy compared to dry machining.

2. Materials and Methods

The primary focus of this research is to examine the influence of minimum quantity lubrication technique on surface roughness and compare with improvement with the dry machining process. For this purpose two sets of samples have been taken namely the study group and the control group. For this study, a total of 40 samples were taken, with each group consisting of 20 samples (n=20). The sample size was determined using the G-power calculator, considering an alpha value of 0.05 and aiming for 80% power based on previous research. The study group had a mean of 8.76 ppm with a standard deviation of 0.04, while the control group had a mean of 9.63 ppm with a standard deviation of 0.42 (Kang, 2021).

Monel alloy is known for their exceptional resistance to corrosion and acids, even in extreme environments. Table 45.1 shows the Composition of Monel Alloy 400. This is mainly due to the nickel content, which allows Monel alloys to maintain their integrity under a variety of reactive conditions. They are also resistant to saltwater and steam. These alloys are often used in applications where a strong, corrosion-resistant material is needed. This includes marine and chemical plant equipment, heat exchangers, musical instruments, and safety wire for aviation equipment. Monel alloys exhibit excellent mechanical properties at subzero temperatures, and thus, they do not lose their ductility. It's important to note that while Monel alloys are indeed robust and resistant, they're more difficult to machine than stainless steel. This is due to their work-hardening characteristics which result in buildup and wear on cutting surfaces. Monel alloys are chosen for study purposes because they are most frequently used in the chemical, marine, and oil industries. The supplies were bought in Chennai at Om Sivam Traders. To evaluate the surface roughness, a 150mm X 150 mm X 10mm specimen of monel material was machined using coated carbide tools. The end milling operation involved cutting 20 slots with a constant feed rate and cut depth.

Table 45.1 Composition of Monel Alloy 400

Element	Composition in %
Nickel (plus Cobalt).	63.0 min
Copper	28.0 - 34.0
Carbon	0.3 max
Iron	2.5 max.
Manganese	2.0
Sulfur	0.024 max.
Silicon	0.5 max.

(Debnath et al., 2021) Minimum Quantity Lubrication (MQL) helps to reduce the negative environmental impacts of using traditional flood coolants. The process involves the use of an aerosol (formed using an air and oil mist) that delivers a precise amount of coolant directly into the machining area. The lubricant amount is minimal (often less than 50 ml/hr) but is enough to provide effective cooling and lubrication. (Tai et al., 2011) The procedure of MQL can be outlined as follows: The lubricant is mixed with a compressed gas (usually air) to create an aerosol. The lubricant droplets are combined with a stream of compressed gas. The aerosol is then sprayed to the cutting area through a nozzle. The nozzle position is crucial to ensure the lubricant is accurately applied where the cutting tool interacts with the workpiece. The machining process takes place, the coolant forms a film between the tool's cutting edge and workpiece, reducing the generated heat and friction. After machining, the workpiece needs less cleaning due to the minimal use of lubricant. In most machining processes, a supply rate of approximately 10 to 100 mL/h is commonly employed. To form the lubricant, a mixture of castor oil and graphene nanoparticles is utilized. Fig. 45.2 depicts the setup. To gauge the roughness of the surface, six slots were carved. The vertical machining center EV 1020A, which is shown in Fig. 45.3, was used to carry out the CNC machining operation. The machining parameters were chosen according to research findings. (Reddy et al., 2023) For the CNC machining process, a coated carbide tool insert was used with a feed rate of 10 mm/min, a depth of cut set at 0.2 mm, and a spindle speed of 1200 rpm. In the control group, the end milling procedure was conducted without any lubrication in a dry

Fig. 45.2 Minimum quantity lubrication set-up

Fig. 45.3 Vertical milling machine YCM XV1020A

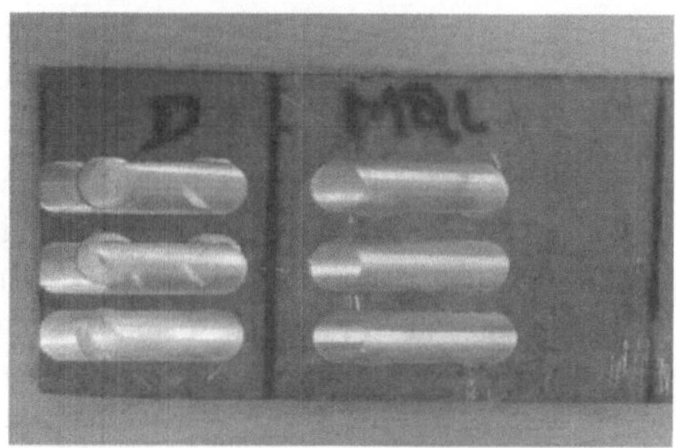

Fig. 45.4 Monel alloy after machining

Table 45.2 Test parameter of commercial pattern friction

Sample No	Surface Roughness MQL Technique(Ra)	Sample No	Surface Roughness MQL Technique(Ra)
1	0.378	11	0.441
2	0.364	12	0.462
3	0.357	13	0.487
4	0.417	14	0.411
5	0.391	15	0.462
6	0.431	16	0.402
7	0.371	17	0.384
8	0.496	18	0.396
9	0.451	19	0.405
10	0.392	20	0.367

operating environment.

After performing machining operations, the specimens were cleaned as shown in Fig. 45.4. The Surftest SJ-410, as depicted in Fig. 45.5. The Surftest SJ-410 is a surface roughness tester developed by Mitutoyo. This versatile, easy-to-use device provides highly accurate and reliable measurements over wide areas. The procedure to measure surface roughness using the Surftest SJ-410 is as follows, the device was calibrated according to the manufacturer's instructions. The stylus was placed at the beginning of the section of the surface to be measured. The measurement process was started by pressing the start/stop button. The head unit propels along the guideway (guide carriage), and the stylus begins tracing the surface in the drive direction. As the stylus moves, the probe detects the roughness profile of the surface and sends this data to the main unit. The detector converts the vertical displacement of the stylus into an electrical signal, which is then digitized. The

Table 45.3 Highest mean value of temperature(°C) is found at hash pattern friction welding with a standard deviation

Group		N	Mean	Standard Deviation	Standard Error Mean
Roughness	CG	20	1.1358	0.19518	0.04364
	EG	20	0.4133	0.04143	0.00927

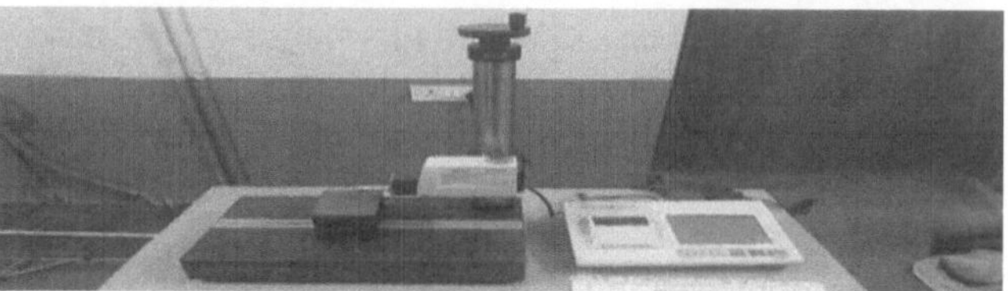

Fig. 45.5 Surface roughness tester SJ-410

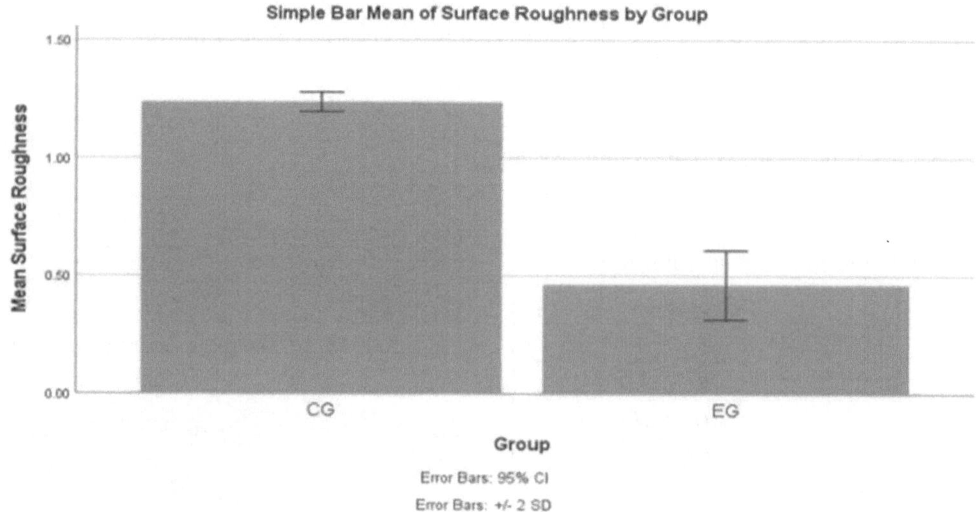

Fig. 45.6 Mean surface roughness comparison of machined surface from dry machining and minimum quantity lubrication technique using bar graphs

device processes the recorded data to calculate roughness parameters. Surftest SJ-410 allows for different parameter settings fed into it, which enables different measurements (like Ra, Rz, etc.), subsequently displayed on its screen. Surface roughness measurements were taken for twenty samples each from the control group (dry machining) and the study group (utilizing the minimum quantity lubrication approach). The measured surface roughness values (Ra) for both groups are listed in Table 45.4. Figure 45.6. shows mean surface roughness comparison of machined surface from dry machining and minimum quantity lubrication technique using bar graphs

Table 45.4 Obtained surface roughness values (Ra) of machined surface from dry Machining and Minimum quantity lubrication technique

Sample No	Surface Roughness Dry Machining (Ra)	Sample No	Surface Roughness Dry Machining (Ra)
1	1.383	11	1.306
2	1.416	12	1.312
3	1.312	13	1.116
4	1.231	14	0.987
5	1.145	15	0.912
6	0.973	16	0.892
7	0.876	17	1.316
8	0.721	18	0.993
9	0.689	19	1.262
10	1.112	20	0.212

The surface irregularity of the samples is assessed by moving the probe across its surface for a distance of 10 mm and recording the measurements. Surface roughness measurements were taken for twenty samples from the control group (dry machining) and the study group (utilizing the minimum quantity lubrication approach). The measured surface roughness values (Ra) for both groups are listed in Table 45.4.

3. Results

The responses of samples from both the control and study groups were measured using the SJ-410 surface roughness tester. The obtained values were statistically analyzed using SPSS software, and the results are presented in Table 45.3. The surface roughness was significantly improved when using the MQL technique compared to dry machining. The mean surface roughness for dry machining was 1.1358 with a standard deviation of 0.19518, while for the MQL technique it was 0.4133 with a standard deviation of 0.04143. This indicates that the machined surface obtained using the MQL technique has a better finish and lower error deviation compared to machining without lubrication.

The findings demonstrate that the novel near-dry machining technique machining has led to an enhancement in surface finish when compared to machining without lubrication. A significant improvement in surface finish is observed with MQL machining, as indicated by the mean surface roughness of 0.4133 compared to 1.1358 for dry machining. The results from Table 45.4 show that both Levene's test and t-test yield a two-tailed significance value ($p=0.001$) less than 0.05, demonstrating a statistically significant difference between the two groups being compared.

4. Conclusion

In this research, the impact of a new near dry machining technique on surface roughness during the machining of the monel alloy was explored. The comparison was made between the surface roughness obtained from the traditional dry machining technique and the innovative MQL machining method. It showed remarkable improvement in the surface finish. To enhance machining processes, the minimum quantity lubrication technique has become a valuable approach. This method involves providing a small amount of lubricant or coolant directly to the cutting zone, resulting in reduced friction, improved surface integrity, and prolonged tool life. Moreover, adopting minimum quantity lubrication offers numerous benefits such as decreased coolant consumption, enhanced surface quality, and minimized environmental contamination. Overall, incorporating minimum quantity lubrication can enhance both efficiency and sustainability in diverse industries. In conclusion, the minimum quantity lubrication technique has proven to be a valuable approach in enhancing machining processes. A significant enhancement of 67% in the surface finish was noted. Ongoing research and development in this area are likely to lead to further improvements in the performance of this technology which leads to green manufacturing.

References

1. Bruni, C., Apolito, L. d', Forcellese, A., Gabrielli, F. and Simoncini, M., Surface Roughness Modelling in Finish Face Milling under MQL and Dry Cutting Conditions, International Journal of Material Forming, vol. 1, no. 1, pp. 503–6, April 1, 2008.
2. Carou, D. and Paulo Davim, J., Machining of Light Alloys: Aluminum, Titanium, and Magnesium, CRC Press, pp. 240, 2018.
3. Debnath, S., Anwar, M., Pramanik, A. and Basak, A. K., Chapter 5 - Nanofluid-Minimum Quantity Lubrication System in Machining: Towards Clean Manufacturing, in Sus-

tainable Manufacturing, K. Gupta and K. Salonitis, Eds., Elsevier, pp. 109–35, 2021.

4. Dhar, N. R., Kamruzzaman, M. and Ahmed, M., Effect of Minimum Quantity Lubrication (MQL) on Tool Wear and Surface Roughness in Turning AISI-4340 Steel, Journal of Materials Processing Technology, vol. 172, no. 2, pp. 299–304, February 28, 2006.
5. Kang, H., Sample Size Determination and Power Analysis Using the G* Power Software, Journal of Educational Evaluation for Health Professions, vol. 18, from https://synapse.koreamed.org/articles/1149215, 2021.
6. Kedare, S. B., Borse, D. R. and Shahane, P. T., Effect of Minimum Quantity Lubrication (MQL) on Surface Roughness of Mild Steel of 15HRC on Universal Milling Machine, Procedia Materials Science, vol. 6, pp. 150–53, January 1, 2014.
7. Kumar, B. S. and Navin Kumar, B., Comparative Study on Transverse Compression Testing of Hybrid Glass/kevlar Composite Pipe with PVC Pipe, Journal of Physics. Conference Series, vol. 2484, no. 1, p. 012003, accessed September 11, 2023, May 1, 2023.
8. Naresh Babu, M., Anandan, V., Muthukrishnan, N. and Santhanakumar, M., End Milling of AISI 304 Steel Using Minimum Quantity Lubrication, Measurement, vol. 138, pp. 681–89, May 1, 2019.
9. Navin kumar, B., Rajendran, S., Vasudevan, A. and Balaji, G., Aerodynamic Braking System Analysis of Horizontal Axis Wind Turbine Using Slotted Airfoil, Materials Today: Proceedings, from http://dx.doi.org/10.1016/j.matpr.2020.06.334, July 2020. DOI: 10.1016/j.matpr.2020.06.334
10. Paturi, U. M. R., Maddu, Y. R., Maruri, R. R. and Narala, S. K. R., Measurement and Analysis of Surface Roughness in WS2 Solid Lubricant Assisted Minimum Quantity Lubrication (MQL) Turning of Inconel 718, Procedia CIRP, vol. 40, pp. 138–43, January 1, 2016.
11. Reddy, M. A. and Navin Kumar, B., New Flax Fiber and Polystyrene Foam Structural Insulated Panels with Wire Mesh Reinforcement: Experimental Investigation of Drilling Characteristics, Journal of Survey in Fisheries Sciences, vol. 10, no. 1S, pp. 1657–67, accessed September 11, 2023, March 8, 2023.
12. Samson Jerold Samuel, C. and M. Suresh, S. G., Functional Composite Materials: Manufacturing Technology and Experimental Application, Bentham Science Publishers, pp. 220, 2022.
13. Shinge, V. R. and Pable, M. J., Effect of Nano-Minimum Quantity Lubrication on Cutting Temperature and Surface Roughness of Milling AISI D3 Tool Steel, Materials Today: Proceedings, from https://www.sciencedirect.com/science/article/pii/S2214785322063040, October 11, 2022. DOI: 10.1016/j.matpr.2022.09.479
14. Singh, R. K., Sharma, A. K., Bishwajeet, Mandal, V., Gaurav, K., Nag, A., Kumar, A., Dixit, A. R., Mandal, A. and Kumar Das, A., Influence of Graphene-Based Nanofluid with Minimum Quantity Lubrication on Surface Roughness and Cutting Temperature in Turning Operation, Materials Today: Proceedings, vol. 5, no. 11, Part 3, pp. 24578–86, January 1, 2018.
15. Sivakumar, S. and Navin Kumar, B., Enhancement of Bending Strength of Polyurethane Foam Reinforced with Basalt Fiber with Silica Nanoparticles in Comparison with Plain

Note: Every figure and table was created using the experimental work; none of them were taken from any publications or online.

Advances in Additive Manufacturing Technologies – Gurusamy Pathinettampadian (eds)
© 2024 Taylor & Francis Group, London, ISBN 978-1-032-90013-1

46 Comparative Study on Tensile Strength of Novel Metal Matrix Composite of Aluminum Alloy AA5083 with 0% and 4% SiC Powder Reinforcement

P. Naga. Sai. Rohith[1], D. Vinodh[2]
Department of Mechanical Engineering,
Saveetha School of Engineering, Saveetha Institute of Medical and
Technical Sciences, Saveetha University,
Chennai, Tamil Nadu, India

ABSTRACT: This experimental study aims to compare the Tensile strength of the aluminium alloy AA5083 reinforcement with silicon carbide by 4% of volume. Materials and Methods: The stir-casting method was used to make the samples in two groups. Twenty sample specimens of an aluminium alloy reinforced with silicon carbide composite without filler made up the UTM and CNC machining control group, whereas twenty sample specimens of an aluminium alloy reinforced silicon carbide composite with filler made up the experimental group. The ASTM-D2583 standard was followed in the tensile strength testing. This sample size is determined based on 80% G-Power. Result: The mean tensile strength of aluminium alloy without and with reinforced 4% silicon carbide are 233.15 MPa and 269.8 MPa respectively. The t-test significance for the comparison resulted in a p-value of $p=0.001$ ($p<0.05$). Within the limitations of this study, the 4 weight percent of new aluminium-silicon carbide microparticles as filler materials had a 12% relative increase in tensile strength.

KEYWORDS: Novel composite, Silicon carbide, Aluminium alloy, AA5083, Sustainable production, Tensile strength, UTM, CNC machining

1. INTRODUCTION

In the current work, the appealing properties of metals and ceramics are combined to develop hybrid metal matrix composite materials (Kainer 2006). A metal matrix composite is often defined as a metallic matrix infused with reinforcing particles, and the major aim of this study is to depict these metal matrix composite materials (Mahajan et al. 2023). Silicon carbide particle reinforcements in AMMC (Aluminum Metal Matrix Composites) are finding improved uses in aircraft, automobiles, and underwater vehicles. It also supports sustainable production. (Ozceylan and Gupta 2021)

In the composite, the matrix phase is one component, while the strengthening phase is another. The matrix has a strengthening phase injected to provide the optimum qualities. Increased high strength and modulus refractory molecules provide a metal matrix, which results in a material with mechanical qualities halfway between composite and ceramic reinforcement (Evans, Marchi, and Mortensen 2014). Among the different kinds of composites, MMC is the ancestor. MMC has evolved over the last ten

[1]rohithpnagasai19@saveetha.com, [2]vinodhd.sse@saveetha.com

DOI: 10.1201/9781003545774-46

years from a logical and academic hub to a substance of expanded innovation and business criticalness.(Vetri et al. 2023)(Zhang et al. 2022)

According to Google Scholar, 522 articles were published and 58 articles were published on ScienceDirect over the last 5 years. The mechanical characteristics of the material produced by the addition of high strength, high modulus refractory particles to a ductile metal matrix fall between those of the matrix alloy and the ceramic reinforcement (Ma and Qian-Cheng 2008). Numerous investigations on metal matrix composites with sustainable production have been done in the past. The most often utilised particulates to strengthen metal or alloy matrix, such as aluminium or iron, are SiC, TiC, WC, and B4C (Ozceylan and Gupta 2021). The investigation of SiC in the alloy of aluminium 5083 is still rare and scant, and only a small number of papers have been published. Due to the limited amount of data and information on mechanical qualities, this study is important. Metal matrix composite's historical evolution is covered in the Historical Development section.

There are very few studies on SiC-reinforced AA5083 composites that were produced using the ultrasonic aided stir casting method. There is still room for study in this area and opportunities for this specific subject. Due to its great hardness, silicon carbide is used in a variety of products, including cutting tools, jewellery, vehicle components, electrical circuits, structural materials, nuclear fuel particles, etc (Evans, Marchi, and Mortensen 2014). Composite materials have the novel composite for the potential to replace commonly used metals like steel and aluminium, often with greater performance. In a variety of industries, including aircraft and fuselage, golf clubs, automobiles, electronic substrates, undersea vehicles, high performance aircraft skins, turbine blades, brake pads, etc.(Gangil, Siddiquee, and Maheshwari 2020)

2. Materials and Methods

They are separated into two groups: the Experimental group and the Control group. For a total of 40 samples from both groups, a total of 20 samples are prepared for each group. Using ClinCalc software, the sample size is calculated with a G power of 80%, a threshold of 5%, and a confidence interval of 95% (Sathish et al. 2021). For this work, sheets of aluminium AA5083 were melted and stir-casted with powdered SiC as metal matrix novel composites. The Sustainable Production materials used in the fabrication are SiC and Aluminium. The materials and composites from Fig. 46.1 and Fig. 46.2. The reinforcement material was obtained from PMC Corporation Ltd. SiC from Kemphasol Pvt. Ltd., UTM and CNC machining with average particle

Fig. 46.1 Amount of silicon carbide used in reinforcement with aluminium (4%)

Fig. 46.2 Aluminium AA5083 used for casting process

sizes of 30-45 m, was used as reinforcement material for the manufacturing process. Fig. 46.3 shows the casting process of the MMC in a graphite crucible shown. By the casting process the Sustainable Production material is being compared in both the form of non casting and casting values in Table 46.1. Yield strength of group 1 and group 2. MMCs are created in the liquid state by integrating a dispersed ceramic phase into a molten metal and then solidifying them.

Fig. 46.3 Casting process done in a graphite crucible

Table 46.1 Tensile strength comparison of group 1 and group 2

S No.	AA5083 tensile strength without reinforcement	AA5083 tensile strength with reinforcement of silicon carbide with volume of 4%
1	242	232
2	275	265
3	239	230
4	273	262
5	241	237
6	279	268
7	236	229
8	271	264
9	238	226
10	277	261
11	240	228
12	278	269
13	237	227
14	276	264
15	234	225
16	273	266
17	233	226
18	270	267
19	235	228
20	272	266

In a stir casting method, mechanical stirring is used to disperse the reinforcing phases, which are typically in powder form, throughout the molten aluminium by the below Fig. 46.4 and Fig. 46.5 the casting metal and tensile testing is shown. In some cases, adding reinforcement particles to a stirred molten matrix would trap not only the reinforcement particles but also additional contaminants like metal oxide and slag that forms on the melt's surface (Teimurnezhad, Pashazadeh, and Masumi 2016; Kumar and Anil Kumar 2022).

Fig. 46.4 4% SiC reinforced aluminium AA5083 cast sample

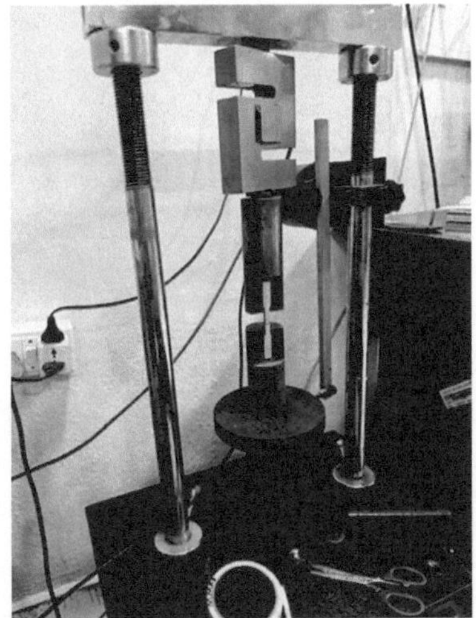

Fig. 46.5 Tensile testing of the aluminium AA5083 with reinforcement of SIC (vol of 4%)

AA5083 aluminium alloy which was purchased from PMC Corporation Ltd. and employed as the matrix material, and SiC particulates were the materials used in the fabrication. For the manufacturing process, silicon carbide particles from Kemphasol Pvt. Ltd. with average particle sizes of 30-45 m were employed as reinforcing material. To form the vortex, the resulting melt was swirled using a stainless steel impeller at a speed of 600 rpm. To achieve the goal of mixing the sound particles, a vortex is created by the impeller blades (Batchelor, Hung, and Lee 1996; Haarberg 2016). To ensure a uniform dispersion of the reinforcing particles, stirring was continued for a further few minutes after particle feeding had finished. The liquid was placed into a 45 x 60 x 100 mm (LxBxH) permanent metallic mould that had been heated up.

The novel composite samples were cut and machined using CNC Machining for tensile testing as per the ASTM standards. The tensile testing was carried out for all the samples of the control group and the experimental group in the UTM machine as shown in Fig. 46.5. The results from the tensile testing are tabulated as in Table 46.1.

3. Results

The universal testing machine was used to test tensile strength of the experimental material samples, and the results for both sets of samples were collated as seen in Table 46.1. Table 46.2 displays the statistical information for mean deviation, standard deviation, and standard error for the experimental study. The control group specimen of as cast AA5083 samples had a mean tensile strength of 233.15 MPa with a standard deviation of 5.499 and a standard mean error of 1.229.

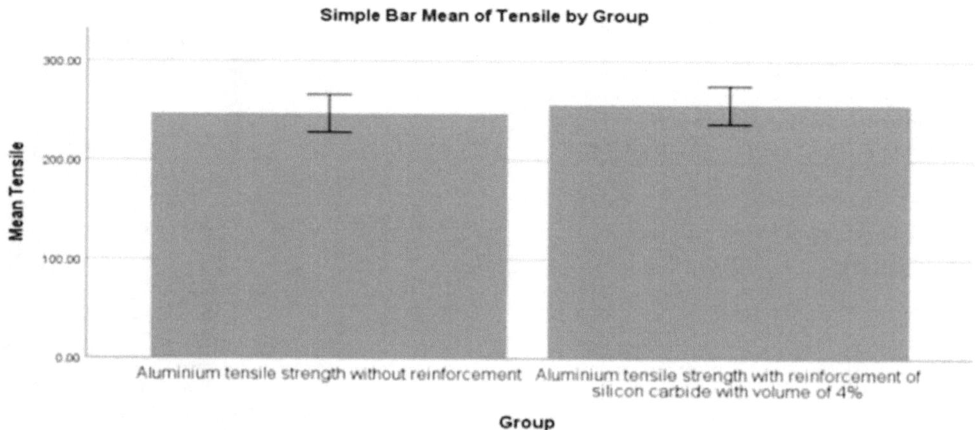

Fig. 46.6 Graphical representation of Tensile Strength (MPa) for Group-1 AA5083 aluminium alloy with no reinforcement and Group-2 AA5083 aluminium alloy reinforced with 4-wt% of SiC, With +/-1 SD - X axis: Material groups, Y axis: Tensile Strength (MPa) with mean accuracy of detection 95% CI

Table 46.2 Group statistics

Group		N	Mean	Standard Deviation	Standard Error Mean
Tensile strength MPa	1	20	269.8	5.4541148	1.2195771
	2	20	233.15	5.499043	1.2296234

The experimental group of the novel metal matrix composite AA5083 + 4% SiC, had a mean tensile strength of 269.8 MPa and a standard deviation of 5.454 and a standard mean error of 1.2195. As per the independent T-test results shown in Table 46.3, the study has a significance of $p = 0.001$ ($p < 0.05$). This is the resulting data from the statistical analysis of the experimental results from the 20 samples of each group.(Elmetwally et al. 2020; Dawood et al. 2022). The comparison of the tensile strength of both the testing groups is plotted as a bar graph as shown in Fig. 46.6.

4. Discussion

The MMC AA5083 + 4% SiC and as-cast AA5083 were both found to have a mean tensile strength of 233.15 MPa and 269.8 MPa respectively. The respective standard deviations are 5.499 and 5.454 as per the statistical analysis. According to the t-test statistical analysis, the mean variance of the tensile strength between the two groups is significantly different with a two tailed significance value of $p = 0.001$ ($p < 0.05$).

The use of silicon carbide filler material, which enhances the material qualities of the aluminium alloy 5083, is what causes this rise in tensile strength. The filler material was correctly distributed throughout the product by using the stir casting method. However, there may still be minute variations in the filler material distribution and flaws such as clumping, voids, and microcracks that might affect the experimental outcomes.

The absence of heat treatment of the cast material, the lack of preheating the mould, and the lack of processing the filler materials are the limits of this study. This may be taken into account and covered in next research. The efficacy of the composites created in the next works can also be studied using unique metal matrix composite production techniques other than stir-casting (Wang et al. 2023; Lu et al. 2023). **Table 46.1** Composition of Monel Alloy 400 Table 46.3. Test Parameter of Commercial Pattern Friction

5. Conclusion

Within the limitations of this study, the novel metal matrix composite of AA5083 + 4% SiC material in comparison to the as cast AA5083 aluminium alloy, shows 12% greater mean tensile strength. The difference in mean tensile strengths between the two groups is found to be 36.65 MPa.

References

1. Batchelor, A. W., N. P. Hung, and T. K. Lee. 1996. "Wear of Metal Stirring Rods in Molten Aluminium and Suspensions of Alumina Particles in Molten Aluminium." Tribology International. https://doi.org/10.1016/0301-679x(95)00033-z.
2. Dawood, H., Abbas Mohammad, Kanaan Musa, and Nawras Sabeeh. 2022. "Rotational Speeds and Preheating Effect on the Friction Stir Butt Welding of Al-Cu Joints." Egyptian Journal of Chemistry. https://doi.org/10.21608/ejchem.2021.109836.5007.
3. Elmetwally, Hammad T., Hani N. SaadAllah, M. S. Abd-Elhady, and Ragab K. Abdel-Magied. 2020. "Optimum

Combination of Rotational and Welding Speeds for Welding of Al/Cu-Butt Joint by Friction Stir Welding." The International Journal of Advanced Manufacturing Technology. https://doi.org/10.1007/s00170-020-05815-8.

4. Evans, Alexander, Christopher San Marchi, and Andreas Mortensen. 2014. Metal Matrix Composites in Industry: An Introduction and a Survey. Springer.
5. Gangil, Namrata, Arshad Noor Siddiquee, and Sachin Maheshwari. 2020. Composite Fabrication on Age-Hardened Alloy Using Friction Stir Processing. CRC Press.
6. Haarberg, Geir Martin. 2016. "The Current Efficiency for Aluminium Deposition from Molten Fluoride Electrolytes with Dissolved Alumina." Advances in Molten Slags, Fluxes, and Salts. https://doi.org/10.1002/9781119333197.ch87.
7. Kainer, Karl U. 2006. Metal Matrix Composites: Custom-Made Materials for Automotive and Aerospace Engineering. John Wiley & Sons.
8. Kumar, K. S. Anil, and K. S. Anil Kumar. 2022. "Effect of Tool Plunge Depth (TPD) on the Microstructure and Mechanical Properties of FSW Dissimilar Joints Reinforced with SiC Nano Particles." Materials Today: Proceedings. https://doi.org/10.1016/j.matpr.2021.09.056.
9. Lu, Linyi, Xin Qian, Fang Li, Shijiang Qin, Yao Luo, Jinjing Tang, Kai Zhou, and Guocan Zheng. 2023. "A Mesoporous Graphene @ Zirconium-Based Metal-Organic Frameworks as a Matrix and an Adsorbent for Steroid Detection Using Surface-Assisted Laser Desorption/ionization Time-of-Flight Mass Spectrometry." Journal of Chromatography. A 1696 (April): 463963.
10. Mahajan, Aditya M., Nagumothu Kishore Babu, Mahesh Kumar Talari, Ateekh Ur Rehman, and Prakash Srirangam. 2023. "Effect of Heat Treatment on the Microstructure and Mechanical Properties of Rotary Friction Welded AA7075 and AA5083 Dissimilar Joint." Materials 16 (6). https://doi.org/10.3390/ma16062464.
11. Ma, Qian-Cheng, and M. A. Qian-Cheng. 2008. "Densification and Mechanical Properties of Boron Carbide Ceramics with Addition of Silicon Hexaboride." Journal of Inorganic Materials. https://doi.org/10.3724/sp.j.1077.2008.01175.
12. Ozceylan, Eren, and Surendra M. Gupta. 2021. Sustainable Production and Logistics: Modeling and Analysis. CRC Press.
13. Sathish, T., V. Mohanavel, T. Arunkumar, T. Raja, Ahmad Rashedi, Ibrahim M. Alarifi, Irfan Anjum Badruddin, Ali Algahtani, and Asif Afzal. 2021.
14. "Investigation of Mechanical Properties and Salt Spray Corrosion Test Parameters Optimization for AA8079 with Reinforcement of TiN + ZrO." Materials 14 (18). https://doi.org/10.3390/ma14185260.
15. Teimurnezhad, J., H. Pashazadeh, and A. Masumi. 2016. "Effect of Shoulder Plunge Depth on the Weld Morphology, Macrograph and Microstructure of Copper FSW Joints." Journal of Manufacturing Processes. https://doi.org/10.1016/j.jmapro.2016.04.001.
16. Vetri, Luigi, Annamaria Pepi, Marianna Alesi, Agata Maltese, Lidia Scifo, Michele Roccella, Giuseppe Quatrosi, and Maurizio Elia. 2023. "Poor School Academic Performance and Benign Epilepsy with Centro-Temporal Spikes." Behavioral Sciences 13 (2). https://doi.org/10.3390/bs13020106.
17. Wang, Yang, Zhen Chen, Yixin Wu, Yu Li, Ziyu Yue, and Minghua Chen. 2023. "PVDF-HFP/PAN/PDA@LLZTO Composite Solid Electrolyte Enabling Reinforced Safety and Outstanding Low-Temperature Performance for Quasi-Solid-State Lithium Metal Batteries." ACS Applied Materials & Interfaces, April. https://doi.org/10.1021/acsami.3c02678.
18. Zhang, Hao, Zehua Ji, Yuxin Zeng, and Yuansheng Pei. 2022. "Solidification/stabilization of Landfill Leachate Concentrate Contaminants Using Solid Alkali-Activated Geopolymers with a High Liquid Solid Ratio and Fixing Rate." Chemosphere 288 (Pt 2): 132495.

Note: Every figure and table was created using the experimental work; none of them were taken from any publications or online.

Advances in Additive Manufacturing Technologies – Gurusamy Pathinettampadian (eds)
© 2024 Taylor & Francis Group, London, ISBN 978-1-032-90013-1

Influence of Al wire Mesh in Impact Strength of 5 ply Bamboo/Epoxy Panels

Praveen Kumar K J[1],
G. Ramya Devi[2], R. Saravanan[3]
Department of Mechanical Engineering,
Saveetha School of Engineering, Saveetha Institute of Medical and Technical Sciences, Saveetha University,
Chennai, Tamil Nadu, India

ABSTRACT: To Evaluate the influence of Al wire-mesh in impact strength in bamboo/epoxy composite panels. The intervention and control group panels are made up of bamboo/epoxy/Al wire-mesh and bamboo/epoxy respectively. The panels were fabricated with 5 plyby hand layup method. As per the predicted sample size, 20 samples per panel type and a total of 40 samples were kept ready for characterization. The prepared samples of panels were characterized in terms of impact strength as per ASTM D256 standard. The experimental results reveal average impact strength of 2.725 KJ/m2 was recorded for the proposed panel for the conventional bamboo/epoxy composite panels; the average impact strength was 0.767 KJ/m2. The results were statistically evaluated with t-tests. The results of the statistical test confirm that the proposed al wire-mesh reinforcement expresses significant influence in the bamboo/epoxy composite panels as the significance value found as 0.000

KEYWORDS: Epoxy, Bamboo, Al wire-mesh, Fiber, Impact strength

1. INTRODUCTION

The main objective of this work is to evaluate the influence of impact strength of epoxy/bamboo composites reinforced with Al wire mesh [1-4]. Composite materials reinforced with carbon fiber and glass fiber exhibit high modulus and strength [5-6]. They are therefore commonly utilized in the industries [7]. Natural fiber-reinforced polymer composites have been shown to be more affordable and cost-effective than synthetic fiber-reinforced composites [8,9]. Natural fiber composites have good mechanical qualities and are both biodegradable and non-toxic [10].

This makes it possible for it to be used as a competitive substitute for traditional materials in sectors like the automotive industry, where it can replace plastic components with more environmentally friendly options [11,12]. A composite is made up of at least two materials: the matrix, or binding material, and reinforcement materials like fiber, Kevlar, and whiskers [13-15]. Composite materials are stronger, more flexible, non-corrosive, and lighter than traditional materials, among other advantages [16]. Bamboo has several advantages over other plant fibers, such as rapid growth, low cost, high mechanical strength, stiffness, and low density [17-19].

[1]praveenkumarkj19@saveetha.com, [2]ramyadevig.sse@saveetha.com, [3]saravananr.sse@saveetha.com

DOI: 10.1201/9781003545774-47

Though many articles appear in the literature this study is unique about composite panel with Al Wire-mesh reinforcement in bamboo/epoxy composite panel. The subject is becoming more and more popular, and every year, new uses for natural fiber composites are found [20,21]. The finest research paper for the study is the one that shows bamboo fiber epoxy composites are sustainable and generally have strong mechanical properties [23,24]. Some of the most cited articles on this topic analyses different research publications. It has been demonstrated through a review of the literature that very little researches delt about aluminum mesh reinforcement in natural fiber composite laminates. Therefore, this work examines and analyses the mechanical properties of a unique bamboo fiber reinforced with an aluminum wire mesh.

2. Materials and Methods

The hand lay-up process is used to produce the composites. The hardener, grade HY951, and epoxy resin, grade LY556, are purchased from Kannan Chemicals in Tamil Nadu, India. It is an epoxy resin that cures hot and has strong mechanical and chemical resistance [25]. The supplier of bamboo fabric was Muthuvel Corporation in Tamil Nadu, India. Expanded type aluminum wire-mesh was purchased from Amba Aluminum Works Pvt. Ltd. Each type of composite panel has twenty samples. forty samples in total for the study [26]. 20 samples were employed in each group, with the experimental group using wire mesh reinforcement and the control group using a bamboo/epoxy composite without Al wire mesh reinforcement [27-28]. A 300 x 300 mm piece of bamboo fiber fabric was cut in order to build the control group. Hardener and epoxy are combined at a 10:1 ratio [29]. The stacking order consists solely of epoxy and bamboo alternatives [30]. The layers are evenly covered with epoxy resin throughout the lay-up process, and any air bubbles that occur are eliminated with a roller [31]. For optimal adhesion, a constant weight is maintained after stacking for the duration of the curing time. 300 × 300 mm pieces of expanded aluminum wire mesh are cut in order to build the control group. An alternative to bamboo epoxy Al wire mesh is the stacking sequence [32]. To ensure strong adherence, the aluminum wire mesh is scraped using 400 grit paper. For 48 hours, the composite is allowed to dry at room temperature. After that, water-cut EDM machining is used to cut the composite to the appropriate specimen size.

The impact strength of specimens measuring 63.5 mm by 12.5 mm from both groups is measured in accordance with ASTM D256. In the university lab, the impact test is conducted using a Charpy impact test. Figures 47.1 and b, respectively, display the samples from the experimental group and control group that were used for impact testing. Twenty specimens from each group underwent the tests, and the findings were recorded. Statistical analysis was carried out with SPSS V.26 software. The independent variable was the reinforcement of Al wire-mesh dependent variable was impact strength.

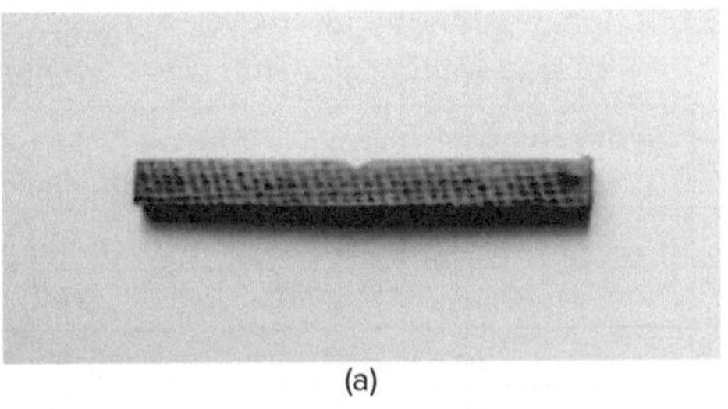

(a)

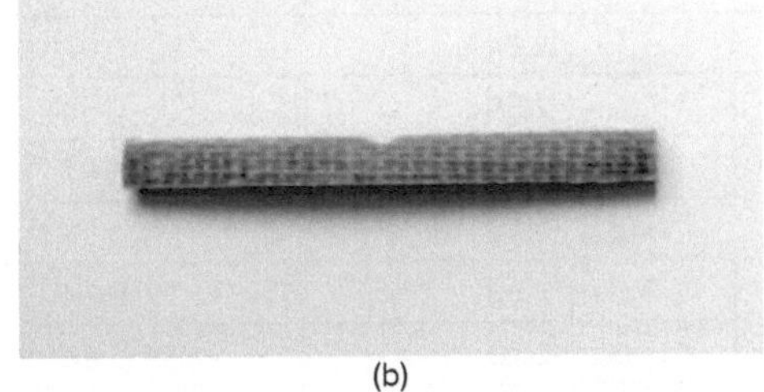

(b)

Fig. 47.1 Composite specimen a) with aluminium mesh reinforcement b) without aluminium mesh reinforcement

3. Results

The impact testing setup utilizing a universal testing equipment is depicted in Fig. 47.2. Table 47.1 displays the impact strengths of the bamboo fibre composite samples that were tested, both with and without an Al wire mesh. The impact strength of the reinforced composite samples is higher than that of the unreinforced composite. Table 47.2 displays the mean, standard deviation, and standard error mean. The composite including Al wire-mesh reinforcement exhibited an average impact strength of 2.725 KJ/m^2 with a standard deviation of 0.0935. The

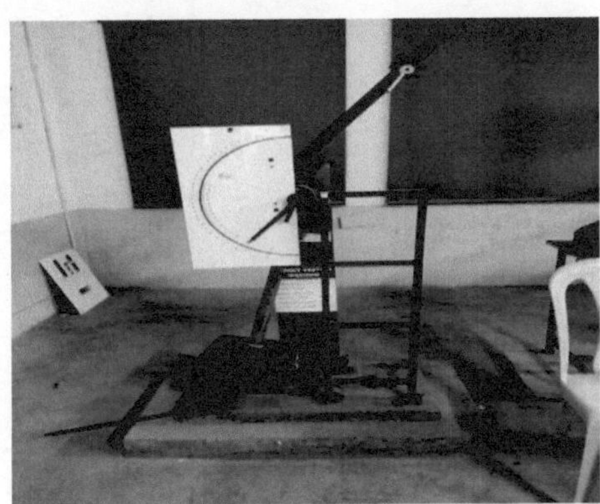

Fig. 47.2 Impact loading of specimen

Table 47.1 Impact strength for bamboo fiber composite with and without aluminum mesh reinforcement

S No.	With aluminum mesh reinforcement	Without aluminum mesh reinforcement
1	2.65	0.83
2	2.59	0.8
3	2.74	0.84
4	2.75	0.75
5	2.84	0.84
6	2.59	0.84
7	2.68	0.67
8	2.62	0.66
9	2.65	0.64
10	2.70	0.86
11	2.63	0.66
12	2.74	0.64
13	2.63	0.74
14	2.67	0.78
15	2.80	0.82
16	2.59	0.69
17	2.65	0.73
18	2.80	0.65
19	2.82	0.8
20	2.57	0.85

composite with Al wire-mesh reinforcement had a standard deviation of 0.01246 and a mean impact strength of 0.767 KJ/m^2. Between the two groups, there is a statistically significant difference (two-tailed significance, p=0.000(P<0.05)). Graph 1 uses a straightforward bar graph to display the mean impact strength of the two groups.

Table 47.2 Mean impact strength and standard deviation of bamboo fiber composite with and without aluminum wire mesh reinforcement with a mean value of 2.725 KJ/m and 0.767 KJ/m respectively

	Group	N	Mean	Standard Deviation	Standard Error Mean
Tensile strength MPa	With aluminum mesh Reinforcement	20	2.72	0.0935	0.0209
	Without aluminum mesh reinforcement	20	0.76	0.05573	0.0124

4. Discussion

The samples' impact strength was evaluated by the use of a universal testing apparatus. The composite's average impact strength was 2.725 KJ/m^2 with Al wire-mesh reinforcement and 0.767 KJ/m^2 without it with standard deviations of 0.0935 and 0.01246 .The impact resistance is increased by 3.5 times with the inclusion of aluminium. In the experimental group, replacing the plies of bidirectional bamboo fibre with an While the Al wire-mesh significantly increased impact strength, it did not significantly increase weight.

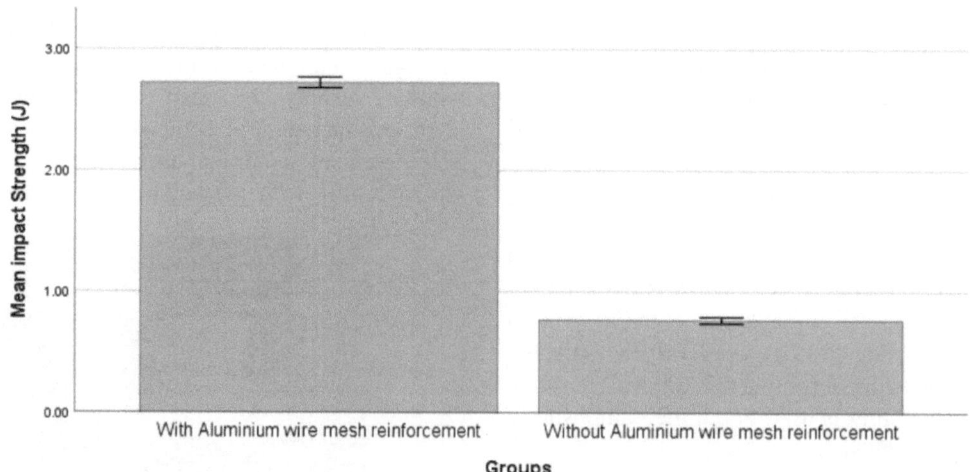

Fig. 47.3 Graphical representation of Impact Strength (KJ/m^2) for Group-1 Bamboo fibre reinforced epoxy composite laminate and Group-2 Bamboo fibre reinforced with aluminium wire mesh epoxy composite laminate, X axis: Material groups, Y axis: Impact Strength (KJ/m^2) with Mean accuracy of detection 95% CI and +/-1 SD.

A related study was conducted using fibre from bamboo. The results showed that the toughness of fracture and modulus of impact of the composites grew monotonically with fibre length and content, and that the impact and impact moduli of the composites climbed monotonically with fibre length [33]. Nonetheless, there was minimal variation in the impact strength of composites across all samples [34]. Bamboo has a unique fibre gradient structure compared to other natural fibres, which has drawn the attention of numerous researchers [36]. Parallel lamination exhibited higher tensile and bending strength than cross lamination [37]. Proper selection of fibre volume would give better results in bamboo fibre composite laminates. The bamboo composite contains 2.0 volume % of lignin [38]. This study also shows how adding Al Wire-mesh as reinforcement enhances the composites' tensile properties while maintaining their sustainability.

The limitation imposed by the fabrication technique used in this study, hand lay up, is its lengthy production time. In this process, there is a sizable chance that the composite will have flaws and improper bonding. This would consequently weaken the composites' mechanical strength. More intricate fabrication methods can be used in the future to improve the work, and desired properties can be changed by adjusting the number of layers, mesh size, and orientation of the Al wire-mesh reinforcement.

5. Conclusion

This experiment determines the Al wire-mesh reinforced bamboo/epoxy composites and the bamboo fiber-reinforced epoxy composites. Between the two groups, there is a statistically significant difference ($p=0.000$; $P<0.05$). Compared to the control group samples, the impact strength of the samples from the experimental group increased by a factor of 3.5. This demonstrates that the use of Al wire-mesh as reinforcement can have a significant positive influence on bamboo fibre composite applications requiring increased impact strength.

References

1. Sarange Shreepad, T. Amuthan, J. Raj, Piyush Gaur, V. Vijayan, and S. Rajkumar. "AZ63/Ti/Zr Nanocomposite for Bone-Related Biomedical Applications." BioMed Research International 2023 (2023).
2. Vijayan.V. "Investigations on influences of MWCNT composite membranes in oil refineries waste water treatment with Taguchi route." Chemosphere 298 (2022): 134265.
3. Alrazzaq, Wael A. 2018. "An Evaluation of Mechanical Properties (Tensile Strength &Elastic Modulus) of Soldered Straight Stainless Steel Wire." Iraqi Dental Journal. https://doi.org/10.46466/idj.v40i1.138.
4. Vinayagam Mohanavel, Palanivel Velmurugan, T. Raja, M. Ravichandran, Wadi B. Alonazi, Shanmugam Sureshkumar, and Atkilt Mulu Gebrekidan. "Investigating Influences of Synthesizing Eco-Friendly Waste-Coir-Fiber Nanofiller-Based Ramie and Abaca Natural Fiber Composite Parameters on Mechanical Properties." Bioinorganic Chemistry and Applications 2022 (2022).
5. Basker, Sanjeevi, A. Parthiban, V. Vijayan, I. J. Premkumar. "Influence of chemical treatment in synthesize and characterization sisal/glass hybrid composite." In AIP Conference Proceedings, vol. 2283, no. 1. AIP Publishing, 2020.
6. Bajuri, Farid, Norkhairunnisa Mazlan, Mohamad Ridzwan Ishak, and Junichiro Imatomi. 2016. "Flexural and Compressive Properties of Hybrid Kenaf/Silica Nanoparticles in Epoxy Composite." Procedia Chemistry. https://doi.org/10.1016/j.proche.2016.03.141.
7. Arunkumar, S. Dhinakaran.V. "Experimental investigation on material characterization of zirconia reinforced Alumina ceramic composites via powder forming process." In AIP Conference Proceedings, vol. 2283, no. 1. AIP Publishing, 2020.
8. Dan-Mallam, Yakubu, Mohamad Zaki Abdullah, and Puteri Sri Melor Yusoff. 2014. "The Effect of Hybridization on Mechanical Properties of Woven Kenaf Fiber Reinforced Polyoxymethylene Composite." Polymer Composites. https://doi.org/10.1002/pc.22846.
9. Vijayan, V., I. J. Premkumar, Sanjeevi Basker, and A. Parthiban. "Synthesize and characterizations of glass/treated selective sisal fiber hybrid composite." In AIP Conference Proceedings, vol. 2283, no. 1. AIP Publishing, 2020
10. Gnanavel, C., 2020. Synthesis and characterization of treated banana fibers and selected jute fiber based hybrid composites. Materials Today: Proceedings, 21, pp. 988–992.
11. Sivasai, Madhini, A. Bovas Herbert Bejaxhin, "A Comparative Analysis on Material Removal Rate of Plain Epoxy Composite with Reinforcement of 15 wt% of Egg Shell Powder particles Novel composite using CNC Machining." Journal of Pharmaceutical Negative Results (2022): 692–699.
12. Anuj Kumar, Chander Prakash, Mohd Shahazad, Manish Gupta, N. Senthilkumar, Bidhan Pandit, Mohd Ubaidullah, and Vladimir A. Smirnov. "Influence of synthesizing parameters on surface qualities of aluminium alloy AA5083/CNT/MoS2 nanocomposite in powder metallurgy technique." Journal of Materials Research and Technology 27 (2023): 1611–1629.
13. El-Shekeil, Y. A., S. M. Sapuan, K. Abdan, and E. S. Zainudin. 2012. "Influence of Fiber Content on the Mechanical and Thermal Properties of Kenaf Fiber Reinforced Thermoplastic Polyurethane Composites." Materials & Design 40 (September): 299–303.
14. Karunakaran, K., Pugazhenthi, R., Reddy, M.V., 2022. Experimental investigations on synthesis and characterization of tamarind seed powder reinforced Bio-composites. Materials Today: Proceedings, 64, pp. 760–764.
15. Sreeram, D., Pugazhenthi, R., Veeranjaneyulu, K., 2022. An investigation of the effects of hot rolling on the microstructure and mechanical behavior of nano-sized SiC particu-

lates reinforced Al6063 alloy composites. Materials Today: Proceedings, 64, pp.731–736.

16. Hamidon, Muhammad H., Mohamed T. H. Sultan, Ahmad H. Ariffin, and Ain U. M. Shah. 2019. "Effects of Fibre Treatment on Mechanical Properties of Kenaf Fibre Reinforced Composites: A Review." Journal of Materials Research and Technology 8 (3): 3327–37.
17. Mamidi, V.K. and Kumaran, P., 2022. Optimizing WEDM parameters on nano-SiC-Gr reinforced aluminum composites using RSM. Advances in Materials Science and Engineering, 2022.
18. Ibrahim, Mohamad Ikhwan, Mohamad Zaki Hassan, Rozzeta Dolah, Mohd Zuhri Mohamed Yusoff, and Mohd Sapuan Salit. 2018. "Tensile Behaviour for Mercerization of Single Kenaf Fiber." Malaysian Journal of Fundamental and Applied Sciences. https://doi.org/10.11113/mjfas.v14n4.1099.
19. Gnanavel, C., Gopalakrishnan, T. Pugazhenthi, R., 2021. Fracture toughness reinforcement by CNT on G/E/C hybrid composite. Materials Today: Proceedings, 37, pp. 1046–1050.
20. Ismail, Ahmad Safwan, Mohammad Jawaid, and Jesuarockiam Naveen. 2019. "Void Content, Tensile, Vibration and Acoustic Properties of Kenaf/Bamboo Fiber Reinforced Epoxy Hybrid Composites." Materials 12 (13). https://doi.org/10.3390/ma12132094.

Note: Every figure and table was created using the experimental work; none were taken from any publications or online.

Advances in Additive Manufacturing Technologies – Gurusamy Pathinettampadian (eds)
© 2024 Taylor & Francis Group, London, ISBN 978-1-032-90013-1

48 Study on MRR of Epoxy Composite Incorporated with 15% Fiber and 5% Novel Nano Carbon Particles Made of Peanut Husk during Drilling Process

Pelluricharan, D. Satish Kumar*
Department of Mechanical Engineering,
Saveetha School of Engineering, Saveetha Institute of Medical and
Technical Sciences, Saveetha University,
Chennai, Tamil Nadu. India

ABSTRACT: Aim: Investigation's main objective was to assess the rate of material removal (MRR) of cutting-edge hybrid-epoxycomposites incorporated with sustainable natural fiber (15%) cum nanocarbon (5%) to ordinary plain matrix (epoxy). Materials & Methods: The desired samples for the experimental/comparison groups were developed with the aid of simple and traditional hand lay-up method. The work sample materials were prepared in alignment with the required specifications, and drilling was performed in a CNC machine that was vertical with a pretest g power of 80%. Using a total of 20 tests (each with one repeat) per group, the Material's rate of removal (in short MRR) of work samples were assessed and investigated among the involved groups. Results: In association, t-independent tests wereperformed successfully with SPSS tool (a statistics software) to appraise the material's rate of removal. The average MRR for Group-1 (75 by wt.%)/ Sustainable Ramie fiber (15 by wt.%)/nano carbon particles (5 wt%) was 0.3913 mm^3/sec, compared to 0.1357 mm^3/sec for Group-2 (plain epoxy). With respect to the findings from T-test statistics, it is inferred that the mean-variance of the outcome that is MRR between Group-1 and Group-2 are different (significant of $p=0.00$, which is P lesser than 0.05). Thus found a significant and statistical difference among the two work sample groups under study. Conclusion: It is evident and clear that inclusion/addition of reinforcements like Sustainable Ramie fiber and nano carbon particles significantly improves the MRR.

KEYWORDS: Epoxy, Natural fiber, Sustainable ramie fiber, Nano carbon particles, CNC drilling, Material removal rate, Novel and hybrid-epoxy composites

1. INTRODUCTION

In present statistical study, rate/amount of material's rate of removal (MRR) of the hybrid-epoxy (75 % weight), Sustainable ramie (fiber of 15 % weight), nano carbon particles (5% weight), and plain epoxy (Pereira and Fernandes 2019; Pandey et al. 2014) will be examined. Artificial or naturally derived fiber reinforced composites offer a distinct range of applications because of their dependable traits and high strength. Sustainable Natural fibers are used to offset the use of artificial/synthetic fibers. Compared to conventionally used fibers, naturally extracted fiber infused composites have low density and excellent mechanical properties. Sustainable Naturally extracted fiber reinforced outperforms conventionally tailored FRPs with respect to the weight/strength ratio. These kinds of materi-

*Corresponding author: satishkumard.sse@saveetha.com

DOI: 10.1201/9781003545774-48

als have several uses, including those in the construction, automotive, aerospace, and marine industries.

Given more than five years of study and literature on polymer composites, there are more than 2600 articles in Google Scholar data and over 2495 articles in Science Direct database, which is not a negligible number. The rate at which composite material is removed, among other output factors, has been examined in relation to CNC drilling settings. (Ross 1996) Examples of these variables include drill diameter in mm, feed in rev/sec, and speed in rpm. The optimal drilling settings and parameters for speeding the amount of material removal (Benyettou et al. 2022) are examined in this study. Researchers have investigated the impacts of the resin/hardener ratio, and (Shagwira et al. 2021) fiber (Kamaraj, Santhanakrishnan, and Muthu 2018a) fraction on the by using (Kamaraj, Santhanakrishnan, and Muthu 2018b) Sustainable Ramie fiber (Siregar et al. 2021; Vinod and Satish Kumar 2022) MRR (Khan et al. 2021) of epoxy (Tang et al. 2013) reinforced composites. Sustainable Ramie fiber is a naturally occurring fiber, and it has been noted that excellent research on composites employing Ramie fiber has been carried out. This study looked at how drilling (Praveen Kumar, Dirgantara, and Vamsi Krishna 2020) parameters affected how much material was removed from epoxy infused/reinforced with Sustainable Ramie (fiber) and nano carbon (particles) (F., Che, and Xian 2019) as well as normal epoxy.

2. Materials and Methods

This study takes into account an experimental group under supervision. The control group employs regular epoxy, whereas the experimental group uses a novel hybrid epoxy composite made of epoxy (75 wt.%), Sustainable Ramie fiber (at 15 wt.%), and nano carbon (5% percent) particles (Njuguna, Pielichowski, and Alcock 2007; Satish Kumar and Rajmohan 2021; Satish Kumar and Rajmohan 2019) (5 w.t%). And at a pretest g-power of 80%, drilling was done using Taguchi's L9 OA (Kim, Park, and Lee 2008) with one experiment repeated in a group, resulting in 20 experiments/group. Sustainable Ramie fiber mat, epoxy hardener, and nano carbon particles are located in three layers,till thedesired thickness is attained (Group-1 sample). In a similar manner, loads were placed on the setup while maintained in that for 72 hrs for curing.

Wax is applied in the wooden (mold) box to develop the group-2 sample of plain resin/epoxy (mold) (Satish Kumar et al. 2017). The sample is prepared with the traditional and simple hand-layup method at 10/1 mixture of the hardener (HY951) and epoxy (LY556). While stirring, utmost attentionto be given to avoid bubbles. This mixture is gradually applied into a mold (wooden) box with measurements of 300 mm (width) by 300 mm (height). To obtain a sample (composite) with beneficial characteristics, the proper load is to be applied for curing. The composites are machined (drilling) in VMC (vertical machining center) with Siemens controller. The testing method entails drilling round cross-section holes in the samples using drill bits in line with the test design. The VMC machine's FANUC data displays the amount of time spent drilling. After machining (drilling), the volume (removed)/time ratio is considered to find the amount of material's rate of removal (MRR).

3. Results

Table 48.1 shows three processes for CNC machining (drilling) on Group-1 (fiber infused/reinforced epoxy) and Group-2 (plain unfilled epoxy) materials. Drill diameter, speed (rpm), and feed (rev/sec) are all taken into consideration (mm). The pertinent MRR (amount of material

Table 48.1 Input parameters with their levels for machining

I/O Parameters	Levels		
	L-1	L-2	L-3
Speed-rpm	110	270	320
Feed- rev/min	0.15	0.25	0.35
Drill diameter- mm	4	5	6

Table 48.2 Material's rate of removal for group-1 and group-2

S No.	Parameters			Group 1	Group 2
	Speed (rpm)	MRR (mm^3/sec)	MRR, (mm^3/sec)	MRR (mm^3/ sec)	MRR, (mm^3/sec)
1	110	0.15	4	0.171	0.0
2	320	0.35	6	0.219	0.069
3	110	0.15	4	0.258	0.082
4	320	0.35	6	0.319	0.095
5	110	0.15	4	0.371	0.105
6	320	0.35	6	0.484	0.129
7	110	0.15	4	0.573	0.165
8	320	0.35	6	0.710	0.283
9	110	0.15	4	0.252	0.081
10	270	0.25	5	0.268	0.108
11	270	0.25	5	0.293	0.123
12	270	0.25	5	0.313	0.129
13	270	0.25	5	0.210	0.056
14	270	0.25	5	0.518	0.189
15	270	0.25	5	0.393	0.140
16	270	0.25	5	0.520	0.169
17	270	0.25	5	0.460	0.173
18	270	0.25	5	0.479	0.179
19	270	0.25	5	0.503	0.192
20	270	0.25	5	0.511	0.203

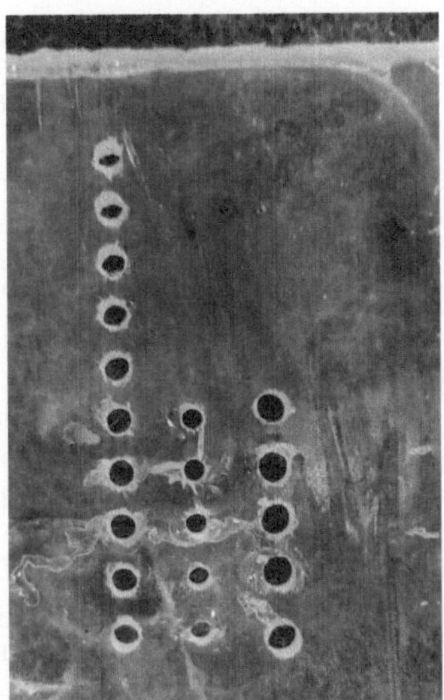

Fig. 48.1 CNC drilling in plain unfilled epoxy (Group 2) reinforcement

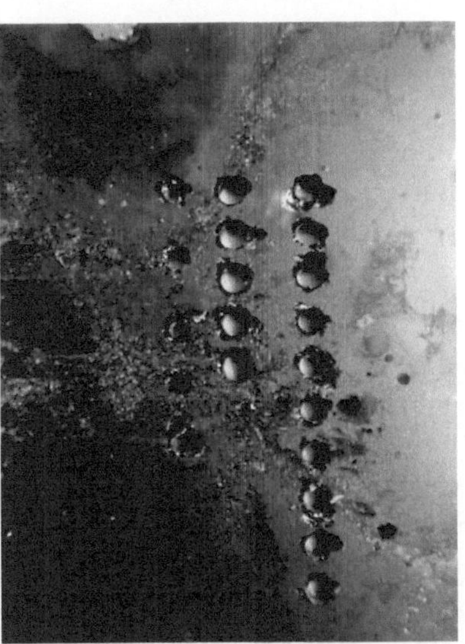

Fig. 48.2 CNC machining (drilling)in novel hybrid-reinforced/infused epoxy (Group-1)

removal) values are displayed (Table 48.2). The group statistical results/data, including the total work samples/group, the mean of the MRR, SD (standard deviation) of the same, along with SE (standard error), are presented in Table 48.3 based on the independent statistical t-test analysis. Table 48.4 displays the outputs after the analysis with averages/means and also the Levene's significance test (in conjunction with $P < 0.05$). Figure 48.2 displays CNC machined (drilling) samples for Group-1 (fiber infused/reinforced epoxy) and Group 2 (plain unfilled epoxy), respectively.

Table 48.3 Group statistical results for MRR (mm^3/sec)

Group		N	Mean	Std. Dev.	Std. Err. Mean
MRR (mm^3/sec)	Fiber Reinforced Epoxy (Group 1)	20	0.391	0.14508	0.03244
	Plain Epoxy (Group-2)	20	0.135	0.05911	0.01322

4. Discussion

Inclusion of the reinforcements in the matrix (epoxy) led to a fundamental improvement in the MRR. As per the statistical observations presented (Table 48.3 and 48.4), the mean-MRR of ordinary epoxy and fiber-reinforced epoxy are, respectively, 0.3913 mm^3/sec and 0.1357 mm3/sec. Levene's significance test results show the evidence of a noticeable/significant and difference in variance of rate of material-removal (MRR) among the study groups

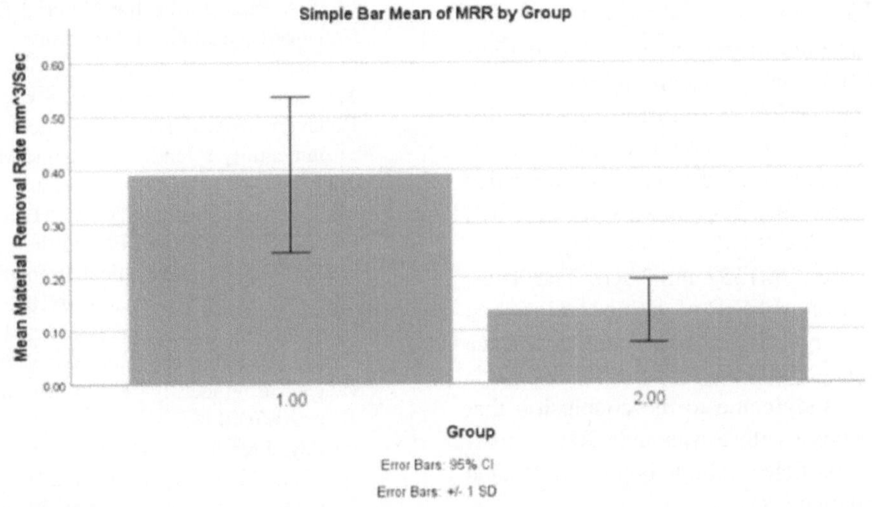

Fig. 48.3 Mean representation of MRR (mm3/sec) and comparison between groups

found, with a P-value being less than 0.05 (achieved P=zero. Hence it is found and evident that the hypothesis (alternative) assumed of uneven-variance is appropriately suitable for the inference. The composites with fiber infused/reinforced epoxy (matrix) (Group-1) possed mean/ average MRR (Matykiewicz 2020) that is larger than that of plain epoxy, according to the negative t value (Group 2).

This technique identifies works that are both comparable and exceptional. Analysts have suggested a study using a variety of samples with CNC drilling and reinforcement variation (weight %) that is related with the method. Additional examination has shown the distinctive workings of CNC drilling. The results exhibit that fiber infused/ reinforced in matrix epoxy (Beese, Zehnder, and Xia 2015) and traditional epoxy have important differences. The study found that fiber reinforced epoxy outperformed regular epoxy in terms of performance. The outcomes for diverse drilling situations have been significantly improved by the use of reinforcing materials like Sustainable Ramie fiber mat and nano carbon particles. Researchers have proposed additional study on Sustainable Ramie fiber (Sprenger 2015) with mechanical qualities and nano carbon particles to enhance the characteristics,which encompass.

The development of lumps and bubbles during sample preparation is recognized as a drawback, which in turn impedes the machining process, which is regarded as a flaw in this work. As a result, it becomes necessary to create new procedures or improve on those that already exist. Therefore, the goal is, developing or improvising the existing procedure to eradicate the obstacles, and this material combination would be useful in producing various engineering components.

5. Conclusion

The investigation connected to the amount of material's rate of removal (MRR) during machining (drilling) of epoxybased new fiber infused/reinforced epoxy like Sustainable Ramie (fiberat 15 wt%) and nano carbon (particles at 5 wt%) and plain unfilled epoxy. The innovative hybrid reinforced matrix (epoxy)possesses mean MRR of 0.3913 mm^3/sec, that is 0.135 mm^3/sec higher than the regular unfilled epoxy (0.1357 mm3/sec) The T-test statistics on the mean/average MRR of the fiber reinforced epoxy and plain epoxy revealed a significant difference between the material groups under study (t22.582 = 7.297, P=0.001). This study came to the conclusion that composites made of epoxy with Sustainable Ramie fiber mat and nano carbon particles exhibit better MRR than composites made of plain epoxy.

References

1. Aruchamy, Karthik, Sathish Kumar Palaniappan, Rajeshkumar Lakshminarasimhan, Bhuvaneshwaran Mylsamy, Satish Kumar Dharmalingam, Nimel Sworna Ross, and Sampath Pavayee Subramani. 2023. "An Experimental Study on Drilling Behavior of Silane-Treated Cotton/ Bamboo Woven Hybrid Fiber Reinforced Epoxy Polymer Composites." Polymers 15 (14). https://doi.org/10.3390/ polym15143075.
2. Benyettou, Riyadh, Salah Amroune, Mohamed Slamani, Yasemin Seki, Alain Dufresne, Mohammad Jawaid, and Salman Alamery. 2022. "Assessment of Induced Delamination Drilling of Natural Fiber Reinforced Composites: A Statistical Analysis." Journal of Materials Research and Technology. https://doi.org/10.1016/j.jmrt.2022.08.161.
3. F., Anna Dilfi K., Zijin Che, and Guijun Xian. 2019. "Grafting Ramie Fiber with Carbon Nanotube and Its Effect on the Mechanical and Interfacial Properties of Ramie/epoxy Composites." Journal of Natural Fibers. https://doi.org/10.1 080/15440478.2017.1423259.
4. Kamaraj, M., R. Santhanakrishnan, and E. Muthu. 2018a. "An Experimental Investigation on Mechanical Properties of SiC Particle and Sisal Fibre Reinforced Epoxy Matrix Composites." IOP Conference Series: Materials Science and Engineering. https://doi.org/10.1088/1757-899x/402/1/012094.
5. Khan, Anish, M. R. Sanjay, Suchart Siengchin, Mohammad Jawaid, and Abdullah M. Asiri. 2021. Hybrid Natural Fiber Composites: Material Formulations, Processing, Characterization, Properties, and Engineering Applications. Woodhead Publishing.
6. Kim, Byung Chul, Sang Wook Park, and Dai Gil Lee. 2008. "Fracture Toughness of the Nano-Particle Reinforced Epoxy Composite." Composite Structures. https://doi. org/10.1016/j.compstruct.2008.03.005.
7. Kumar, M. Saravana, Muhammad Umar Farooq, Nimel Sworna Ross, Che-Hua Yang, V. Kavimani, and Adeolu A. Adediran. 2023. "Achieving Effective Interlayer Bonding of PLA Parts during the Material Extrusion Process with Enhanced Mechanical Properties." Scientific Reports 13 (1): 6800.
8. Manimaran, G., and K. Nimel Sworna Ross. 2020. Surface Behavior of AISI H13 Alloy Steel Machining Under Environmentally Friendly Cryogenic MQL with PVD-Coated Tool.
9. Matykiewicz, Danuta. 2020. "Hybrid Epoxy Composites with Both Powder and Fiber Filler: A Review of Mechanical and Thermomechanical Properties." Materials 13 (8). https://doi.org/10.3390/ma13081802.
10. Njuguna, J., K. Pielichowski, and J. R. Alcock. 2007. "Epoxy-Based Fibre Reinforced Nanocomposites." Advanced Engineering Materials. https://doi.org/10.1002/ adem.200700118.
11. Pandey, Jitendra K., Hitoshi Takagi, Antonio Norio Nakagaito, and Hyun-Joong Kim. 2014. Handbook of Polymer Nanocomposites. Processing, Performance and Applica-

tion: Volume C: Polymer Nanocomposites of Cellulose Nanoparticles. Springer.

12. Pereira, António, and Fabio Fernandes. 2019. Renewable and Sustainable Composites. BoD – Books on Demand.
13. Praveen Kumar, A., Tatacipta Dirgantara, and P. Vamsi Krishna. 2020. Advances in Lightweight Materials and Structures: Select Proceedings of ICALMS 2020. Springer Nature.
14. Ross, Phillip J. 1996. Taguchi Techniques for Quality Engineering: Loss Function, Orthogonal Experiments, Parameter and Tolerance Design. McGraw Hill Professional.
15. Satish Kumar, D., Rajmohan, M., and Tamilarasan, T.R. 2017. "Tribology on Epoxy/Glass Fiber/Carbon Composites - Taguchi Approach." Journal of Computational and Theoretical Nanoscience 14 (12). https://doi.org/10.1166/jctn.2017.7036.
16. Satish Kumar, D., Rajmohan, M. 2019. "Wavelet based pseudo color scaling for optimizing wear behavior of epoxy composites." Materials Testing 61 (4).
17. Satish Kumar Dharmalingam and Rajmohan Murugesan. 2021. "Multi-objective optimization of wear performance of epoxy composites by gray-based response surface methodology." Polymer Composites 42 (8). https://doi.org/10.1002/pc.26086.
18. Shagwira, Harrison, T. O. Mbuya, E. T. Akinlabi, F. M. Mwema, and B. Tanya. 2021. "Optimization of Material Removal Rate in the CNC Milling of Polypropylene 60 Wt% Quarry Dust Composites Using the Taguchi Technique." Materials Today: Proceedings. https://doi.org/10.1016/j.matpr.2020.11.229.
19. Siregar, J. P., M. Zalinawati, T. Cionita, M. R. M. Rejab, I. Mawarnie, J. Jaafar, and M. H. M. Hamdan. 2021. "Mechanical Properties of Hybrid Sugar Palm/ramie Fibre Reinforced Epoxy Composites." Materials Today: Proceedings. https://doi.org/10.1016/j.matpr.2020.07.565.
20. Sprenger, Stephan. 2015. The Effects of Silica Nanoparticles in Toughened Epoxy Resins and Fiber-Reinforced Composites. Carl Hanser Verlag GmbH Co KG.
21. Tang, Long-Cheng, Yan-Jun Wan, Ke Peng, Yong-Bing Pei, Lian-Bin Wu, Li-Min Chen, Li-Jin Shu, Jian-Xiong Jiang, and Guo-Qiao Lai. 2013. "Fracture Toughness and Electrical Conductivity of Epoxy Composites Filled with Carbon Nanotubes and Spherical Particles." Composites Part A: Applied Science and Manufacturing. https://doi.org/10.1016/j.compositesa.2012.09.012.
22. Vinod, B., and D. Satish Kumar. 2022. "Investigation on the Material Removal Rate during CNC Drilling of Hybrid Epoxy Based Novel Composites."Materials Today:Proceedings. https://doi.org/10.1016/j.matpr.2022.08.292.
23. Vishal, R., B. K. Gnanavel, G. Manimaran, and K. Nimel Sworna Ross. 2018. Impact on Machining of AISI H13 Steel Using Coated Carbide Tool Under Vegetable Oil Minimum Quantity Lubrication.

Note: Every figure and table was created using the experimental work; none of them were taken from any publications or online.

Advances in Additive Manufacturing Technologies – Gurusamy Pathinettampadian (eds)
© 2024 Taylor & Francis Group, London, ISBN 978-1-032-90013-1

49 Investigation and Comparison on Light Emitting Property of Halogen with Chromium-Zinc Oxide (Cr-ZnO) Composite Material used in Automobile Engineering by Novel Sol-Gel Method

Rayachoty Bhaskar[1], P. Karthik[2]

Department of Mechanical Engineering,
Saveetha School of Engineering, Saveetha Institute of Medical and Technical Sciences, Saveetha University,
Chennai, Tamil Nadu, India

ABSTRACT: To investigate and compare the light emitting property of Halogen with Chromium-Zinc oxide (Cr-ZnO) composite material used in Automobile engineering by Novel Sol-gel method. Cr-doped ZnO thin films were synthesized using a sol–gel dip coating method, and an investigation was conducted to analyze the impact of chromium on the optical properties of zinc oxide films. The characterization involved X-ray diffraction patterns, scanning electron microscopy (SEM), and UV-VIS spectroscopy to assess the light-emitting characteristics. The structural analysis revealed that Cr-ZnO thin films exhibit a hexagonal wurtzite structure similar to pure ZnO.X-ray diffraction analysis revealed that Cr-doped ZnO crystallizes in a singular-phase polycrystalline state with a wurtzite lattice structure. The investigation encompassed an examination of crystal structure, phase purity, optical properties of the materials. In this research, nanoparticles of Cr doped ZnO was successfully created using the Sol-gel Method. The average crystal size was determined and was found in the range between 28.9 and 25.4 nm.

KEYWORDS: Halogen, Chromium, Zinc oxide, Composite material, Automobile engineering, Light emitting Property, Natural resources, Emissions

1. INTRODUCTION

Science has long investigated zinc oxide (ZnO). It has been the focus of thousands of study publications throughout the last century, some as early as 1935 (Look et al. 1998). ZnO is the essential components in since it is valued for its UV absorbance, diverse chemistry, piezoelectricity, and luminescence at high temperature (Look et al. 1998).To name a few, it is used in the production of paints, cosmetics, plastics, rubber, electronics, and pharmaceuticals. But more lately, ZnO has once more captured the attention of scientists, this time for its semiconducting characteristic (Thankappan et al. 2018).

The production of high-quality single crystals and epitaxial layers has been successfully achieved as reported by Kucheyev et al. in 2003. Advancements in magnification technologies, coupled with ZnO's potential to serve as a suitable substrate for GaN, have paved the way for the development of ZnO-based photonic and optoelectronic devices. Despite the historical dominance of GaN and GaN-based materials in this wavelength range, ZnO offers several advantages, as highlighted by Pearton et al. in

[1]rayachotybhaskar19@saveetha.com, [2]karthikp.sse@saveetha.com

DOI: 10.1201/9781003545774-49

2004. Notably, it presents a lower power emissions barrier for lasing through optical pumping, enabling more efficient room-temperature devices, thanks to its significantly larger exciton binding energy (Galazka et al., 2020). Moreover, ZnO's natural resource capacity facilitates the cost-effective growth of high-quality single crystal substrates.

No research has been carried out using these novel possibilities, high ferromagnetic Curie temperature, radiation hardness, and piezoelectric capabilities for spintronic applications (Look et al. 2004). Due to these qualities, ZnO is an excellent material for a variety of gadgets, such as light-emitting diodes and blue and ultraviolet laser diodes (Kim et al.2004). Although the semiconductor industry has grown significantly and ZnO has a large body of information, relatively truly understood substance. The effects of doping with the rare earth metal (Cr) on the photoluminescence and photoconduction properties of ZnO nanostructures must thus be studied (Hariharan et al.2004).

2. Materials and Methods

Simple Sol-Gel auto combustion was used to create zinc oxide nanoparticles. Zn(x=0-0.08 made up the initial composition. Drop by drop, after 10g of zinc nitrate and the appropriate quantity of chromium nitrate in deionized dihydrogen monoxide (Winning et al. 2004). Chemical uniformity requires tight control of the falling rate. Within sol particles, resulted in the creation with a Zn (Cr) - O-Zn band (Izyumskaya et al. 2007).

This was done in order to remove the nitrate ions from the system. To remove the Emissions inorganic groups, these dried gel particles were calcined for two hours at 400–600 °C in an environment of air (Litton et al. 2011). then spent an hour grinding arduously to create the chromium-doped ZnO nanoparticles. Graphite monochromator and a Japan a double beam spectrophotometer, UV-Vis absorption spectra were captured. For the UV-Vis measurements, the colloid solutions in isopropyl alcohol were made using ultrasound. The absorption coefficient D in the Urbachtail changes exponentially with photon energy hQ. An analysis of the (DhQ)2 vs photon energy plot can provide an approximation of the band gap (Neumark et al. 2008).

3. Results

The ZnO crystallite size exhibited a reduction from 25.9 nm to 22.4 nm with an increase in Cr content from 0% to 5%, a phenomenon potentially attributed to variations in ionic radii. In the case of Cr-doped ZnO, denoted as Zn1-xCrxO, the computation of the optical band gap resulted in values ranging from 3.415 to 3.541 eV, with a nominal value of 3.401 eV. Specifically, for Cr-doped ZnO with x values of 0.00, 0.07, and 0.08, the optical band gap values were determined as 3.401, 3.530, and 3.538 eV, respectively. Consequently, it can be inferred that the band gap increases proportionally with the augmentation of the rare-earth metal (Cr) content in the metal oxide (ZnO).

Figure 49.1 illustrates the X-ray diffraction (XRD) patterns of sol-gel-derived samples, including both undoped ZnO and Cr-doped ZnO. The diffraction peaks observed in the patterns were unambiguously identified and matched with the hexagonal wurtzite structure of ZnO, indicating the absence of any additional impurity phases. Figure 49.2 represents the Band gap for ZnO nanoparticles doped with 5% Cr was calculated using a UV plot between hv vs (Dhv)2. Table 49.1 Represents the Input parameters and their levels for Elemental composition of synthesized nanoparticles of Cr-ZnO as evaluated from UV-Vis Spectroscopy in different samples. Table 49.2 presents the changes in band gap associated with different concentrations of ZnO doping in both group 1 and group 2. The observed trend indicates a rise in the band gap as the Cr doping percentage increases.

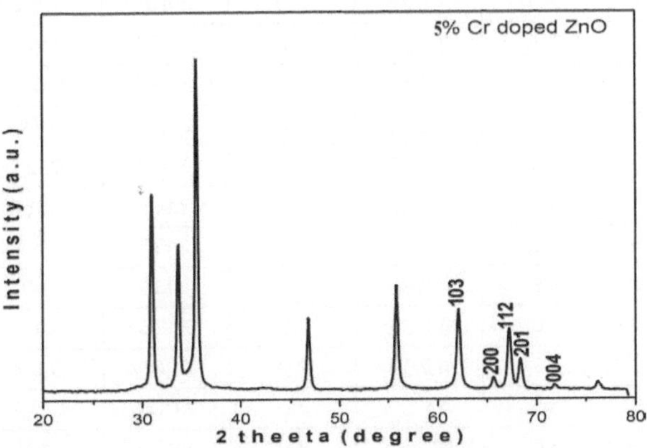

Fig. 49.1 5% Cr doped in ZnO nanoparticles XRD pattern

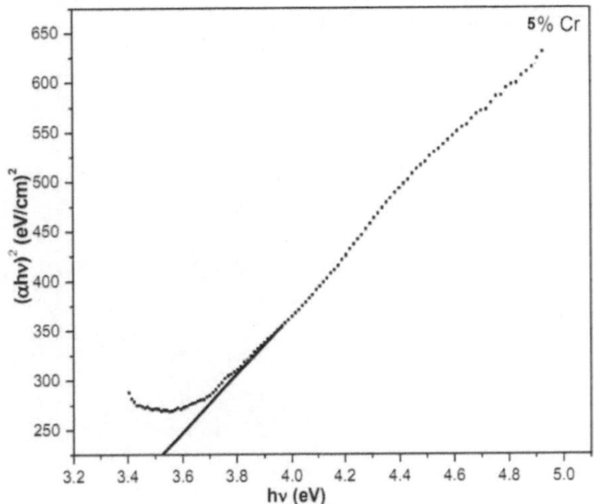

Fig. 49.2 plot between hQ vs (DhQ)

Table 49.1 parameters and their levels for CNC drilling

Parameters	Levels		
	L1	L2	L1
Dopant concentration	X = 1.0	X = 2.0	X = 5.0
Cr(at%)	2.5	4.0	5.0
ZnO (at%)	2.0	2.0	2.0

Table 49.2 Material removal rate of group 1 and group 2

S No.	Parameters			Group 1	Group 2
	Samples	Cr (at%)	ZnO (at%)	Band gap (eV)	Band gap (eV)
1	0.0	2.512	2.0	3.538	1.001
2	0.3	2.572	2.0	3.530	1.069
3	0.5	2.580	2.0	3.521	1.082
4	0.8	2.590	2.0	3.514	1.095
5	1.0	3.000	2.0	3.453	1.105
6	1.3	3.100	2.0	3.437	1.129
7	1.5	3.210	2.0	3.431	1.165
8	1.8	3.256	2.0	3.415	1.283
9	2.0	3.425	2.0	3.401	1.258
10	2.3	3.528	2.0	3.250	1.145
11	2.5	4.250	2.0	2.285	0.081
12	2.8	4.321	2.0	2.277	0.108
13	3.0	4.456	2.0	2.165	0.123
14	3.3	4.586	2.0	2.145	0.056
15	3.5	4.578	2.0	2.137	0.189
16	3.8	4.689	2.0	2.130	0.140
17	4.0	4.632	2.0	2.125	0.169
18	4.3	4.753	2.0	2.101	0.173
19	4.5	4.852	2.0	2.000	0.179
20	5.0	5.000	2.0	1.985	0.192

Table 49.3 Group statistics on MRR (mm^3/sec) values for the groups

	Group	N	Mean	Standard Deviation	Standard Error Mean
Band gap (eV)	Lattice parameter (Group 1)	20	2.792	0.68220	0.15206
	Inter planar spacing (Group-2)	20	0.637	0.51342	0.11480

4. Discussion

The X-ray diffraction (XRD) patterns of both undoped ZnO and Cr-doped ZnO samples were obtained through the sol-gel method. The diffraction peaks observed in these patterns were solely attributed to the wurtzite hexagonal structure of ZnO in space group P63m, and no extra impurity phases were detected. The determination of crystallite size was performed using Scherrer's formula, as outlined by Göpel et al. in 1982.

Calculated crystallite size for each sample shows that the Cr content increased from 0% to 5% caused the ZnO crystallite size to drop from 28.9 nm to 25.4 nm. This can be the result of different ionic radii. In order to determine measurements were used. The touch plot (Izyumskaya et al. 2007) for all samples given in for undoped ZnO is 3.401 eV, which agrees with the value that was previously reported.

5. Conclusion

The Band gap (ev) of a novel Cr-ZnO (5wt%), zinc oxide (2wt%), and sustainable light emitting (20 wt%) particles were examined. The mean (EV) band gap of chro-

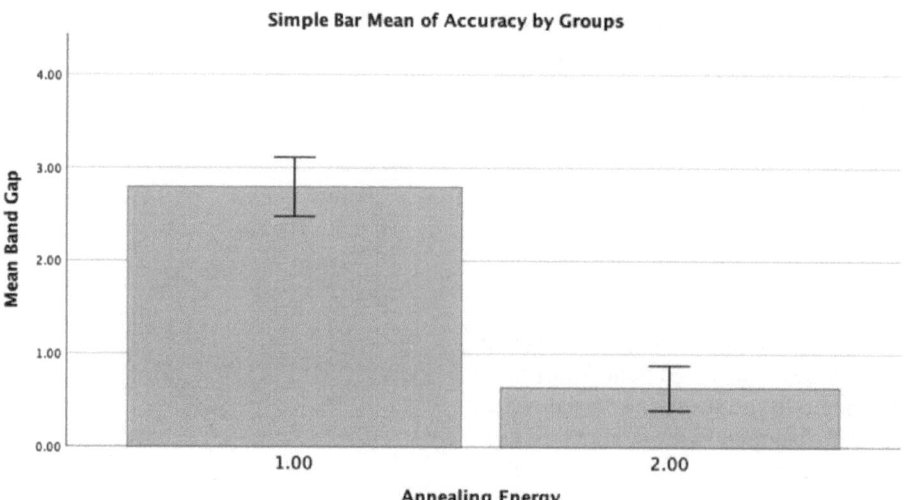

Fig. 49.3 Graphical representation of band gap (eV)

mium zinc oxide nanoparticles is 2.7920 mm3/sec, which is 0.6371 mm3/sec higher than that of ambient light. The band gap energy was enhanced by 41% by Cr doped ZnO powder, per the T-test statistical analysis of the EV of Annealing Energy (t41.321 = 11.310, P=0.001), and there are significant differences in the mean/average EV across the various material groups. From this research, it is concluded that the sustainable Chromium ZincOxide nanoparticles display superior optical properties when compared with halogen material which provides natural lighting.

REFERENCES

1. Coskun, C., D. C. Look, G. C. Farlow, and J. R. Sizelove. 2004. "Radiation Hardness of ZnO at Low Temperatures." Semiconductor Science and Technology. https://doi.org/10.1088/0268-1242/19/6/016.
2. Galazka, Zbigniew. 2020. Transparent Semiconducting Oxides: Bulk Crystal Growth and Fundamental Properties. CRC Press.
3. Göpel, W., J. Pollmann, I. Ivanov, and B. Reihl. 1982. "Angle-Resolved Photoemission from Polar and Nonpolar Zinc Oxide Surfaces." Physical Review B. https://doi.org/10.1103/physrevb.26.3144.
4. Hariharan, Srikanth. 2004. "Second Seeheim Conference on Magnetism (SCM2004) Seeheim, Germany, 27 June–1 July 2004." Phys. Stat. Sol. (a). https://doi.org/10.1002/pssa.200406882.
5. Izyumskaya, N., V. Avrutin, Ü. Özgür, Y. I. Alive, and H. Morkoç. 2007. "Preparation and Properties of ZnO and Devices." Physica Status Solidi (b). https://doi.org/10.1002/pssb.200675101.
6. Kim, Dong-Wook, Kyong-Seok Chea, Yong-Jo Park, In-Hwan Lee, and Cheul-Ro Lee. 2004. "Fabrication of Metal–semiconductor–metal (MSM) UV Photodetectors with Al0.16Ga0.84N/GaN Heterostructures." Phys. Stat. Sol. (a). https://doi.org/10.1002/pssa.200405048.
7. Kucheyev, S. O., J. S. Williams, and C. Jagadish. 2004. "Ion-Beam-Defect Processes in Group-III Nitrides and ZnO." Vacuum. https://doi.org/10.1016/j.vacuum.2003.12.032.
8. Kucheyev, S. O., J. S. Williams, C. Jagadish, J. Zou, Cheryl Evans, A. J. Nelson, and A. V. Hamza. 2003. "Ion-Beam-Produced Structural Defects in ZnO." Physical Review B. https://doi.org/10.1103/physrevb.67.094115.
9. Litton, Cole W., Thomas C. Collins, and Donald C. Reynolds. 2011. Zinc Oxide Materials for Electronic and Optoelectronic Device Applications. John Wiley & Sons.
10. Look, D. C., B. Claflin, Ya I. Alive, and S. J. Park. 2004. "The Future of ZnO Light Emitters." Physica Status Solidi (a). https://doi.org/10.1002/pssa.200404803.
11. Look, D. C., D. C. Reynolds, J. R. Sizelove, R. L. Jones, C. W. Litton, G. Cantwell, and W. C. Harsch. 1998. "Electrical Properties of Bulk ZnO." Solid State Communications. https://doi.org/10.1016/s0038-1098(97)10145-4.
12. Neumark, Gertrude F., Igor L. Kuskovsky, and Hongxing Jiang. 2008. Wide Bandgap Light Emitting Materials And Devices. John Wiley & Sons.
13. Pearton, S. J., D. P. Norton, K. Ip, Y. W. Heo, and T. Steiner. 2004. "Recent Advances in Processing of ZnO." Journal of Vacuum Science & Technology B: Microelectronics and Nanometer Structures. https://doi.org/10.1116/1.1714985.
14. Segawa, Y., A. Ohtomo, M. Kawasaki, H. Koinuma, Z. K. Tang, P. Yu, and G. K. L. Wong. 1997. "Growth of ZnO Thin Film by Laser MBE: Lasing of Exciton at Room Temperature." Physica Status Solidi (b). https://doi.org/10.1002/1521-3951(199708)202:2<669::aid-pssb669>3.0.co;2-t.
15. Thankappan, Aparna, Nandakumar Kalarikkal, Sabu Thomas, and Aneesa Padinjakkara. 2018. Polymeric and Nanostructured Materials: Synthesis, Properties, and Advanced Applications. CRC Press.
16. Winning, Myrjam. 2004. "Influencing Grain Boundary Properties by the Application of Mechanical Stress Fields." Phys. Stat. Sol. (a). https://doi.org/10.1002/pssa.200406880.
17. Zhang, S. B., S-H Wei, and Alex Zunger. 2001. "Intrinsic N-Type Versus P-Type Doping Asymmetry and the Defect Physics of ZnO." Physical Review B. https://doi.org/10.1103/physrevb.63.075205.

Note: Every figure and table was created using the experimental work; none of them were taken from any publications or online.

Advances in Additive Manufacturing Technologies – Gurusamy Pathinettampadian (eds)
© 2024 Taylor & Francis Group, London, ISBN 978-1-032-90013-1

50 A Comparative Analysis of Flexural Strength between 3D Printed Lattice Structures (PLA) Compared with and without the Silica Nanoparticle

S. Guna Sekhar[1], A. Vasudevan[2]
Department of Mechanical Engineering,
Saveetha School of Engineering, Saveetha Institute of Medical and
Technical Sciences, Saveetha University,
Chennai, Tamil Nadu. India

ABSTRACT: The goal of this investigation is to analyze the flexuralproperties of polylactic acid composites with and without the blending of silica nanoparticle fillers. The study includes a control group of polylactic acid without filler (N=20) and an experimental group of polylactic acid with filler (N=20). Sample sizes were calculated using G Power at 80% power and $\alpha = .05$. polylactic acid served as base materials while novel silica nanoparticles were used as fillers. The findings indicated a statistically substantial difference between the two groups, demonstrating that the flexural strength of the polylactic acid composite exceeded that of the polylactic acid laminate composite containing a 2% volume fraction of novel silica nanoparticles filler. (P value = 0.000, $p < 0.05$). The statistical significance was also evident in the observation of polylactic acid with a 2% filler. In summary, the results imply that the flexural strength of the polylactic acid without filler reinforced epoxy composite is notably higher, demonstrating a significant improvement when compared to the epoxy composite reinforced with a 2% volume fraction of novel silica nanoparticle filler.

KEYWORDS: 3D printing, Novel lattice structures, Porous cubic, Hybrid composites, Polylactic acid, Flexural strength, Sustainable production

1. INTRODUCTION

The biodegradable, biobased polymer known as poly lactic acid (PLA) is generally obtained from renewable resources such as corn, sweety bamboo, wheat, and rice (Eling, Gogolewski, and Pennings 1982). Manufacture of poly (lactic acid) is ecologically responsible for sustainable production (Auras, Harte, and Selke 2004) because it is derived from natural sources and consumes a large amount of carbon dioxide gas during the manufacturing process (Tokiwa and Calabia 2006). Polymers derived poly (lactic acid) or polylactide is abbreviated as PLA in both circumstances.. Until recently, the primary PLA applications were limited to medical uses like as implants and tissue scaffolds. (Gregor et al. 2017), Due to its high manufacturing cost and poor mechanical resilience because to its low molecular weight. Throughout the previous five years, there have been 120 publications published in Google Scholar and about 342 articles published in IEEE. Despite their potential, there aren't many studies on TPMS structure construction (Qureshi et al. 2021), particularly composite materials, have been conducted. Investigation on the effect of

[1]gunasekhars19@saveetha.com, [2]vasudevana.sse@saveetha.com

DOI: 10.1201/9781003545774-50

material composition on the cracking behavior, compressive modulus, strength, and energy absorption of two cubic and diamond lattice structures (Li et al. 2021), FDM 3D-printed from PLA/CaCO3 and PLA/TCP composite things (Sawada, Tsutsumi, and Hayase 1957; Blattmann, Helou, and Kara 2019). They claimed that the addition of CaCO3 and TCP improves the compressive modulus and strength of lattice structures. Furthermore, the Diamond structure outperformed cubic structures in terms of load resonance capacity, particular energy absorption, environmentally friendly production, and pore structure stability. During uniaxial compression tests on an FDM Basic TPMS lattice built of PLA and ABS resources, Mishra observed a layered pattern of deformation. The data indicate that early buildings have a proclivity for catastrophic events across both materials.

Only limited research is done on the use of porous cubic structures (Sawada, Tsutsumi, and Hayase 1957) in 3D printing. In this study, This work examined fiber filled with novel silica nanoparticles to overcome all of these restrictions on hybrid composites. Various stacking sequences of laminate composites were created through 3D printing, featuring polylactic acid fiber reinforcement with and without novel silica nanoparticle composites. In contrast to composites containing these novel nanoparticles, the primary objective of this study is to enhance the Flexural Strength analysis of fiber in composite materials without fillers during indentation operations.

2. Materials and Methods

The study comprises two groups: Group 1 involves analyzing the Flexural Strength of Polylactic Acid with filler, while Group 2 focuses on the Flexural Strength of Polylactic Acid without filler. Each group consists of 20 samples, and the statistical power for the analysis, determined using G-Power, is set at 80%. The standard specimen dimensions follow ASTM D790 guidelines, measuring 127 x 12.7 x 4 mm. (Ugalde et al. 2021).

To investigate the Flexural Strength analysis during the drilling operation of Polylactic Acid-reinforced composites, the composites are categorized into there are two groups: control and experimental. In control group, a plain Polylactic Acid laminate is created using a combination of epoxy and hardener, devoid of any filler. Following the production of the plain laminate, the composite material is cut into 20 sample specimens.

The experimental category involved the fabrication of Polylactic Acid laminate composites containing a 4% volume fraction of filler. Subsequently, 20 sample specimens were derived from the manufactured composite materials. These materials were cut into dimensions of 300x300mm, and the fiber-reinforced composite was bonded with epoxy resin AW106IN and hardener HV953U. Following the production process, the materials underwent the application of weights and were dried for a minimum duration of 16 hours using a novel stacking sequence. 3D printed laminate materials incorporating Polylactic Acid were utilized in the process.

To generate the required sample size of 40 drilled holes for both sets, the testing materials are sectioned, and 20 On a CNC machine, holes are drilled with an 8mm diameter HSS drill bit.The control panel and experimental specimens for testing are placed on the CNC machine's spindle, and 20 holes are drilled in each of the two test specimens using an 8mm diameter HSS drill bit with appropriate feed conditions for hybrid composites. The manual assessment of Flexural Strength inaccuracy during the drilling of the Polylactic Acid composite material is conducted. It involves calculating the experimentally measured drilled hole area is compared to the theoretical drilled hole area based on the drill bit diameter. This procedure is repeated for all 20 drilled holes in both the experimental and control groups, and the difference between them reflects the Flexural Strength. The Flexural Strength is then represented for the control and experimental groups as part of sustainable production practices.

Information is gathered utilizing a Flexural Strength machine, which measures the Flexural Strength generated in the drilling procedure, enabling the calculation of the Flexural Strength discrepancy in the drilling of Polylactic Acid composite material. The together data is organized into tables and consolidated for all drilled holes, with each laminate group comprising 20 drilled holes.

3. Results

In contrast to the Flexural Strength of Polylactic Acid with a 2% Volume Fraction of New Silica Nanoparticles Filler achieved through an innovative stacking sequence, the Flexural Strength of Polylactic Acid without Filler reinforced with epoxy composite seems to exhibit improvement. Each group was provided with 20 samples. The average Flexural Strength with filler is recorded at 43.48 mm, while the average Flexural Strength without filler is measured at 59.29 mm. The Flexural Strength of Polylactic Acid without filler (SNP) appears to surpass that of Polylactic Acid with filler (SNP).

Table 50.1 and Fig. 50.1 present the average, standard deviation, and 95 percent confidence interval for the mean values of Polylactic Acid without filler and Polylactic Acid with filler. The mean value for Polylactic Acid without

Table 50.1 Flexural strength of polylactic acid without filler, with 2% volume fraction of novel silica nanoparticles filler, obtained from 20 samples each for the groups

S. No	Flexural Strength of Polylactic Acid without filler (%)	Flexural Strength of Polylactic Acid with filler(%)
1	43.48	59.29
2	42.39	62.47
3	40.21	59.9
4	49.16	62.3
5	46.53	61.13
6	47.17	59.71
7	47.75	62.83
8	40.97	58.95
9	49.76	59.56
10	46.96	59.97
11	48.54	59.14
12	47.46	62.66
13	46.37	59.51
14	49.39	58.71
15	48.29	59.38
16	40.92	58.82
17	46.06	59.41
18	46.58	60.8
19	47	60.23
20	50.86	60.59

Fig. 50.1 3D printer .ender 3 machine

filler is 9.5635, while for Polylactic Acid with filler, it is 6.6785. In the case of Polylactic Acid containing Novel Silica Nanoparticles, the mean is 17.5, with a standard deviation of 3.09908 for without filler and 1.33793 for with filler. The Flexural Strength items exhibit a minimum standard error mean of 0.69297 for without filler and a maximum of 0.29917 for with filler, as depicted in Table 50.1, which lists the values for the 20 samples.

Table 50.2 Group statistics results (Mean of polylactic acid without filler 9.5635 mm is more compared to polylactic acid with filler 6.6785 mm and standard error mean for polylactic acid with filler is 0.0519 and polylactic acid without filler is 0.02949

Group		N	Mean	Standard Deviation	Standard Error Mean
Flexural Strength	EG	20	46.2920	3.09908	0.69297
	CG	20	60.2687	1.33793	0.29917

Table 50.2 and Fig. 50.2 illustrate the outcomes of the drilled Flexural Strength test conducted on the innovative bi-directional Polylactic Acid, both between groups and within groups, with a sum of squares of 0.008 that is deemed statistically significant. The outcomes from an independent test are presented in Table 50.3, indicating a significance level of $p = 0.008$ ($p < 0.05$). Twenty samples of Novel Polylactic Acid with filler and without filler composite pipes are collected for each step of the Flexural Strength test. The 95% Confidence Interval of the Difference assumes equal changes, showing a lower value of -15.50470 and an upper value of -12.44870. Conversely, when equal variances are not assumed, the 95% Confidence Interval of the Difference, depicted in Fig. 50.1, displays a lower value of -15.52866 and an upper value of -12.42474.

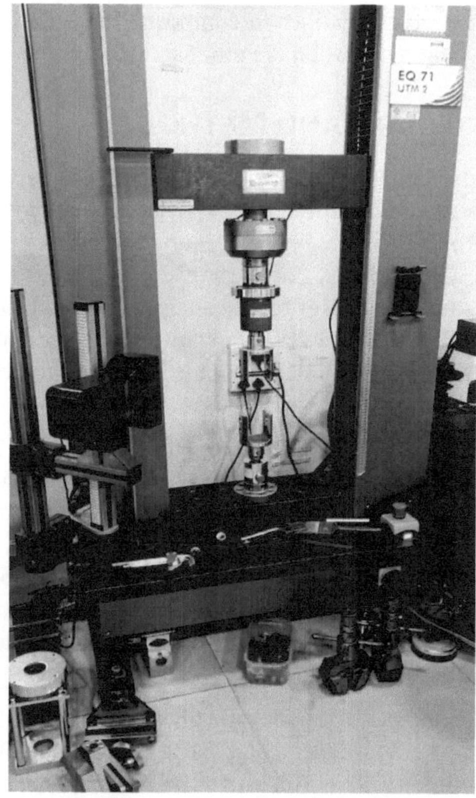

Fig. 50.2 Three-point flexural test

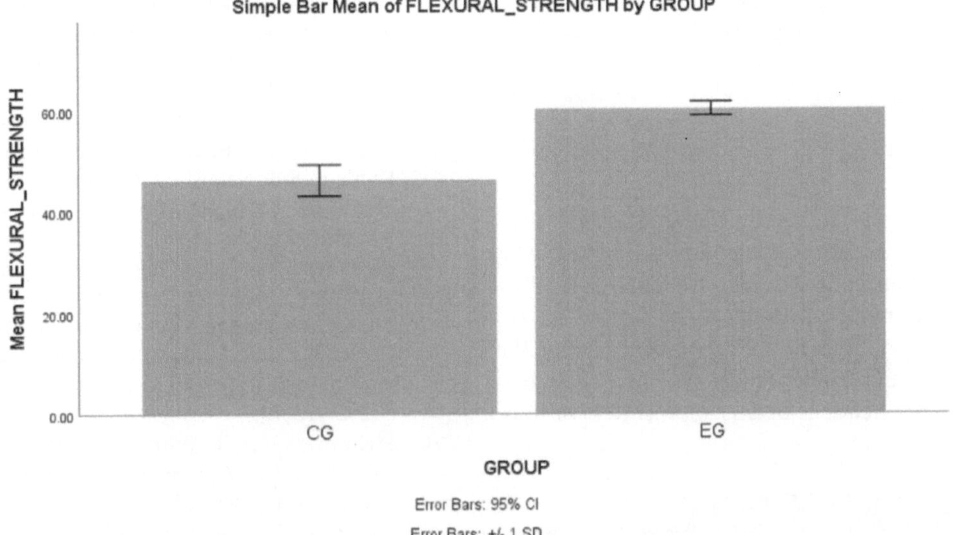

Fig. 50.3 Bar chart representing the comparison of flexural strength for CG represents polylactic acid without filler and EG represents polylactic acid with filler. Polylactic acid without filler tends to offer more consistent outcomes than polylactic acid with filler. The X axis denotes groups, whereas the Y axis reflects mean flexural strength. The mean effective prediction value is ±1 SD

4. Discussion

In contrast to the damaged area observed in the Flexural Strength analysis of Polylactic Acid without filler (SNP) using a novel stacking sequence, the study applies and identifies the damaged area in the Flexural Strength analysis of the drilling operation for Polylactic Acid without filler (SNP), demonstrating a superior mean performance. The bar graph illustrates the discrepancy in Flexural Strength between the two groups, calculated with and without filler. When compared to Polylactic Acid without filler, Polylactic Acid with filler appears to exhibit greater correctness. The independent test values are utilized to ascertain the implication and mean difference aimed at the respective groups.

The results of a comparable study, resembling those presented in the previous research, are discussed. This study focused on the examination of Polylactic Acid, specifically Polylactic Acid without filler, emphasizing its strength for sustainable production. To facilitate a comparison of stacking sequences, epoxy matrix materials were utilized. The research delved into assessing the strength of Polylactic Acid in combination with Novel Silica nanoparticles with varying weight percentage content. According to the study by Hao and Yu in 2011, the strength of the Novel Silica nanoparticles was observed to increase with an escalating percentage of these nanoparticles. The incorporation of Novel Silica nanoparticles as a filler was identified as a strategy to improve the overall excellence of the composite. The strength of the filler experiences an increase as it expands, attributed to the robust connection formed between the filler and the matrix. Previous research has explored the impact of Polylactic Acid on the mechanical properties, including Flexural Strength, of various composites, as demonstrated by the work of Maheswaran, Velmurugan, and Mohammed Mohaideen in 2013. In the investigation of stacking sequence effects, Flexural Strength analysis for hybrid composites was employed, with a focus on Polylactic Acid. Flexural Strength testing was also conducted on metal laminate composites to assess the influence of Polylactic Acid, as indicated by Snyder in 2013. Although there has been an enhancement in the level of absorbed energy, challenges in fabrications persist. The mechanical properties of Polylactic Acid composites are subject to several factors, including the incorporation of nano clay filler, dispensation temperature, cooling measures, the presence of wetness content in the Polylactic Acid volume fraction, nano clay length, and the diameter of silica nanoparticles, as studied by Alrefeai et al. in 2022.

In our research, we reviewed a total of five papers, encompassing three instances where the findings aligned and two instances where they diverged. As highlighted in the earlier discussion, it can be inferred that 3D Printing (PLA) without filler exhibits a greater damaged area in Flexural Strength compared to 3D Printing (PLA) with filler. One drawback associated with drilling and cutting Polylactic Acid with filler is the encountered difficulty, coupled with low efficiency. The fabrication process relies on heat and involves the loss of metal. In future endeavors, efforts will be directed towards addressing all limitations associated with 3D Printing (PLA) with fillers, with a specific emphasis on simplifying the drilling process.

5. Conclusion

This study successfully demonstrated the effectiveness of using filler particles to prevent fracture in 3D printed composites loaded with polylactic acid. The addition of filler to the matrix not only improved the composite rigidity but also significantly enhanced its flexural strength by 30.19%. These findings have important implications for the development of stronger and more durable composites using 3D printing technology, and further research is needed to explore the potential applications of this approach in various industries.

References

1. Alrefaei, Mohammad H., Abdullah S. Aljamhan, Alhanouf Alhabdan, Mona H. Alzehiri, Mustafa Naseem, and Fahad Alkhudhairy. 2022. "Influence of Methylene Blue, Riboflavin, and Indocyanine Green on the Bond Strength of Caries Affected Dentin When Bonded to Resin-Modified Glass Ionomer Cement." Photodiagnosis and Photodynamic Therapy, March, 102792.
2. Auras, Rafael, Bruce Harte, and Susan Selke. 2004. "An Overview of Polylactides as Packaging Materials." Macromolecular Bioscience 4 (9): 835–64.
3. Bahraminasab, Marjan, Athar Talebi, Nesa Doostmohammadi, Samaneh Arab, Ali Ghanbari, and Sam Zarbakhsh. n.d. "The Healing of Bone Defects by Cell-Free and Stem Cell-Seeded 3D Printed PLA Tissue Engineered Scaffolds." https://doi.org/10.21203/rs.3.rs-1092132/v1.
4. Blattmann, Caspar, Mark Helou, and Sami Kara. 2019. "Characterisation of Reinforced Body Centered Cubic, Octahedral-Type and Octet Truss Lattice Structures." Procedia CIRP. https://doi.org/10.1016/j.procir.2019.04.299.
5. Eling, B., S. Gogolewski, and A. J. Pennings. 1982. "Biodegradable Materials of Poly(l-Lactic Acid): 1. Melt-Spun and Solution-Spun Fibers." Polymer. https://doi.org/10.1016/0032-3861(82)90176-8.
6. Gregor, Aleš, Eva Filová, Martin Novák, Jakub Kronek, Hynek Chlup, Matěj Buzgo, Veronika Blahnová, et al. 2017. "Designing of PLA Scaffolds for Bone Tissue Replacement Fabricated by Ordinary Commercial 3D Printer." Journal of Biological Engineering 11 (October): 31.
7. Hao, Licai, and Weidong Yu. 2011. "Comparison of Thermal Protective Performance of Aluminized Fabrics of Basalt Fiber and Glass Fiber." Fire and Materials. https://doi.org/10.1002/fam.1073.
8. Hasan, K. M. Faridul, Péter György Horváth, Zsófia Kóczán, Miklós Bak, and Tibor Alpár. 2021. "Colorful and Facile in Situ Nanosilver Coating on Sisal/cotton Interwoven Fabrics Mediated from European Larch Heartwood." Scientific Reports 11 (1): 22397.
9. Li, Puhao, Fan Yang, Yijie Bian, Siyuan Zhang, and Lihua Wang. 2021. "Deformation Pattern Classification and Energy Absorption Optimization of the Eccentric Body Centered Cubic Lattice Structures." International Journal of Mechanical Sciences. https://doi.org/10.1016/j.ijmecsci.2021.106813.
10. Maheswaran, J., T. Velmurugan, and M. Mohammed Mohaideen. 2013. "An Experimental and Numerical Study of Fracture Toughness of Kevlar- Glass Epoxy Hybrid Composite." In 2013 International Conference on Energy Efficient Technologies for Sustainability, 936–42.
11. Qureshi, Zahid Ahmed, Salah Addin Burhan Al-Omari, Emad Elnajjar, Oraib Al-Ketan, and Rashid Abu Al-Rub. 2021. "Using Triply Periodic Minimal Surfaces (TPMS)-Based Metal Foams Structures as Skeleton for Metal-Foam-PCM Composites for Thermal Energy Storage and Energy Management Applications." International Communications in Heat and Mass Transfer. https://doi.org/10.1016/j.icheatmasstransfer.2021.105265.
12. Sawada, Masao, Kenjiro Tsutsumi, and Akira Hayase. 1957. "Fine Structures of X-Ray Absorption Spectra Of Cobalt in the Face Centered Cubic Lattice and Close-Packed Hexagonal Lattice." Journal of the Physical Society of Japan. https://doi.org/10.1143/jpsj.12.628.
13. Snyder, Maria V. 2013. Spy Glass. MIRA.
14. Tokiwa, Yutaka, and Buenaventurada P. Calabia. 2006. "Biodegradability and Biodegradation of Poly(lactide)." Applied Microbiology and Biotechnology. https://doi.org/10.1007/s00253-006-0488-1.sisal
15. Ugalde, Diana, Julio C. C. Fernandez, Patricia Gmez, Gisele Lbo-Hajdu, and Nuno Simes. 2021. "An Update on the Diversity of Marine Sponges in the Southern Gulf of Mexico Coral Reefs." Zootaxa 5031 (1): 1–112.

Note: Every figure and table was created using the experimental work; none of them were taken from any publications or online.

Advances in Additive Manufacturing Technologies – Gurusamy Pathinettampadian (eds)
© 2024 Taylor & Francis Group, London, ISBN 978-1-032-90013-1

51 Evaluation of Hardness of Glass/ Luffa Fiber Reinforced Epoxy Composite with and without Addition of Novel Shorea Robusta Filler

Arvind B. S., Bharathiraja G.*
Department of Mechanical Engineering,
Saveetha School of Engineering, Saveetha Institute of Medical and
Technical Sciences, Saveetha University,
Chennai, Tamil Nadu, India

ABSTRACT: This study aims to assess the hardness of composites made from Glass/Luffa fibers (GLF) with Shorea robusta filler in comparison to composites composed of Glass/Luffa fibers and epoxy. The experimental group consisted of 20 samples with a 20% volume proportion of glass/luffa fiber epoxy, incorporating Shorea robusta filler. Simultaneously, the control group comprised 20 samples, utilizing the glass/luffa fiber epoxy composite. The G Power for this procedure is 80%, the sample size is calculated as =.05 per group, and the overall sample size is 40. The composite was created using the ASTM standard hand layup procedure. The Brinell hardness of the GLF reinforced composite was 55.36 BHN, whereas the hardness of the 20% volume fraction of Shorea robusta filled GLF epoxy composite was 59.63 BHN. With a p-value of 0.042 ($p<0.05$), exhibits the statistical significance in difference between the two groups. Glass/Luffa epoxy composites with 20% filler exhibited higher hardness than Glass/Luffa fiber epoxy composites within the study's restrictions.

KEYWORDS: Shorea robusta fiber, Glass fiber, Luffa fiber, Epoxy, Hardness, Renewable

1. INTRODUCTION

Fibers obtained from natural resources such as plants, animals, or geological processes. The properties of these fibers vary in terms of its fiber direction, increasing the number of fibers and with addition of reinforced filler that may increase mechanical properties of composites (Koffi, Koffi, and Toubal 2021). In order to meet today's environmental standards, this work is focused on fiber reinforced and natural filler reinforced composite, to enhance the biodegradability, cost effectiveness, mechanical properties, durability, lightweight. For natural fiber composites physical, chemical, and mechanical characteristics are currently being studied, and hardness can be enhanced (Harikumar and Devaraju 2021). Because of its great strength and rigidity, glass and luffa are frequently utilized. Natural fibers have largely replaced synthetic fibers in many applications because they are both renewable and inexpensive (Cheng et al. 2020). Comparing to other fibers, Natural fibers are chosen based on the availability, strength, renewable nature, stiffness, affordability, biodegradability, and cheaper cost. Natural fiber composites with filler addition have wide spread usage in aviation, automotive, and medical industries (Mishra et al. 2022).

*Corresponding author: bharathirajag.sse@saveetha.com

DOI: 10.1201/9781003545774-51

In the last five years, there have been 342 publications published in Google Scholar and 153 in Science Direct on this topic. It was reported that Luffa fiber filled with ground nut shell reinforced polyester composites shown significant improvement in mechanical properties (Dhanola et. al. 2018). Nayak et al. found that Bamboo Glass Fiber hybrid reinforced composite has highest hardness compared to bamboo fiber composite (Nayak et al. 2022).It was reported that Glass/Luffa and polypropylene with a 20% of reinforcement has higher hardness than the jute fibers and polypropylene composite (Kamran, Kamran, and Iqbal 2022). It was investigated that banana fiber reinforced epoxy resin showed stiffer than coir reinforced epoxy resin (Gayathri et al. 2022). Hardness investigation on Bamboo Glass/Luffa fiber composite work goes closer to the present study (Nayak et al.2022).

In the recent literature, the novel Shorea robusta material has not been combined with the Glass/Luffa hybrid composite. The current study investigated the hardness of GLF reinforced epoxy composites, both with and without the addition of Shorea robusta filler, and compared the results.

2. Materials and Methods

Hand layup was used to create the composite plates at the SIMATS Engineering, SIMATS in Chennai. Ethical clearance was not deemed necessary for this study. The research work focuses on hardness and involves two groups with a total sample size of 40. The control group comprises 20 samples of GLF reinforced epoxy composites, while the experimental group includes 20 samples of composites with 20% Shorea robusta filler. The sample size was determined using the GPower calculator, considering a significance level of 80%, $\alpha = 0.05$, a mean value of 53.7, and a standard deviation of 1.4286. (Vinod et al. 2022).

Luffa fiber, derived from Luffa trees, is a naturally renewable resource known for its sustainability as a vegetable sponge fiber. Resistant to water degeneration, Luffa fiber exhibits remarkable strength, durability, and expansion capabilities. The advantageous properties of biodegradability, cost-effectiveness, and lightweight nature in fiber composites contribute to their enhancement for industrial applications. Shorea robusta trees produce gum with a particle size of 200 micrometers, which, when crushed into powder, serves as a filler material. Epoxy (LY556) stands out as the most commonly used reinforcing resin. This enduring substance experiences minimal compression compared to other materials and boasts exceptional thermal and mechanical qualities, making it a preferred choice in various applications.

Hand Layup method is used for material preparation. A mould is made of dimension 300 x 300 x 3mm with LY556 epoxy and HY951 hardener. In a container or jar, the ratio of epoxy and hardener was 10:1 and swirled constantly for 10 minutes. After laying the bidirectional Glass/Luffa fiber matte and filled with resin, Rolling was employed to expel trapped air and harden the mold. Dry samples were removed from the mould after 24 hours and cut into tiny pieces in accordance with ASTM-E10. Figure 51.1 depicts the specimens used for hardness testing. The Brinell Hardness Machine, shown in Fig. 51.2, is used to determine the hardness of test specimens.

Fig. 51.1 Hardness test specimens

Fig. 51.2 The brinell hardness testing machine

In the experimental group, a container was filled with epoxy, hardener, and Shorea robusta filler material, which was continually swirled for around 10 minutes then it was poured placed on GLF bi-directional three-layer matte. To eliminate entrapped air, a roller was used on the mold, and the composite was allowed to cure for 24 hours. Upon removal from the mold, the dried samples were sliced into specimens. The Shorea robusta filler-reinforced epoxy composite plates, featuring a 20% volume percentage, were fabricated, and cut into specimens following the guidelines of the ASTM-E10 standard.

Hardness values of GLF epoxy composites were evaluated by a Brinell hardness machine with and without 20% Shorea robusta filler. The specimen was subjected to a force that caused it to break by placing in test setup. The values of the GLF reinforced composite specimens were recorded using a digital encoder after the test. Glass/Luffa fiber-reinforced composite specimens with and without 20% Novel Shorea robusta filler were examined.

3. Statistical Analysis

Hardness is the dependent variable whereas Glass/Luffa fiber, Shorea Robust and epoxy are the independent variables. The Independant T-Test was carried out using the SPSS v.26 software. Descriptive T-Test, the Bonferroni test, and the G-Graph were obtained for the test results (Mohanta and Acharya 2018).

4. Results

Table 51.1 displays the hardness values of Glass/Luffa fiber epoxy composites, with a maximum value of 57.90 BHN and 20% filler reinforced epoxy composite with a maximum value of 63.9 BHN. The addition of 20% filler to Glass/Luffa fiber epoxy composites resulted in an enhancement of composite hardness. The inclusion of increased filler content led to a 20% rise in the composite's hardness percentage.

SPSS version 26 statistical software was utilized to conduct an independent T-test of the sample for the analysis of the test findings. As per the descriptive statistics in Table 51.2, the GLF epoxy composite exhibited an average hardness of 55.3600 BHN with a SD value of 1.68160. In comparison, the mean hardness value for the 20% filler composite was 59.6300 BHN, with a standard deviation of 2.21314.

In Table 51.3, the hardness p-value is reported as 0.042 ($p < 0.05$), indicating a statistically significant difference in hardness. This p-value is recognized as the significance level for hardness. Additionally, Figure 51.3, represented by a simple bar graph, visually demonstrates the higher hardness of Glass/Luffa fiber compared to Glass fiber/Luffa with epoxy composites reinforcement.

Table 51.1 The hardness value of Glass/Luffa fiber epoxy composite with a maximum value of 57.90 BHN and 20% filler reinforced epoxy composite with a maximum value of 63.9 BHN

Specimen Number	Hardness Without filler (BHN)	Hardness With 20 % filler (BHN)
1	52.30	63.2
2	52.30	56.7
3	55.70	60.1
4	56.00	59.6
5	55.40	60.2
6	55.10	62.6
7	56.50	63.9
8	52.50	57.7
9	57.40	58.6
10	55.00	57.6
11	56.00	62.7
12	56.50	58.8
13	52.60	59
14	57.00	58.7
15	57.90	58.1
16	55.10	62
17	56.30	59.7
18	56.40	56.2
19	55.70	57.6
20	55.50	59.6

Table 51.2 Group statistics table represents the Highest mean value of Hardness is 59.63 BHN found at 20% filler with a standard deviation of 2.21314

Hardness	N	Mean	Std. Deviation	Std. Error Mean
Glass/Luffa fiber without filler	20	55.3600	1.68160	0.37602
Glass/Luffa fiber with 20% filler	20	59.6300	2.21314	0.49487

5. Discussion

The Glass/Luffa fiber epoxy composite mean Hardness value is 55.3600 BHN, with SD value of 1.68160. The average Hardness of a Glass/Luffa fiber composite with 20% Shorea robusta powder reinforced epoxy is 59.6300 BHN, with a standard deviation of 2.21314 was observed. In each group 20 samples were prepared. The Hardness was obtained at a maximum of 63.9 BHN. The result clearly

Table 51.3 Independent sample T-Test for Glass/Luffa fiber epoxy composite filled with and without reinforcement of Shorea robusta Filler. These results showed that there was a statistically significant difference between the two groups with the p value is 0.042 ($p < 0.05$)

	Levene's test for equality for variances		t	df	Sig. (2-tailed)	T-test for equality of means		95% confidence interval of the difference	
	F	Sig.				Mean difference	Std. Error Difference	Lower	Upper
Equal variances assumed	1.562	0.021	-6.870	38	.042	-4.270	0.6215	-5.528	-3.011
Equal variances not assumed	—	—	-6.870	35.454	.042	-4.270	0.6215	-5.531	-3.008

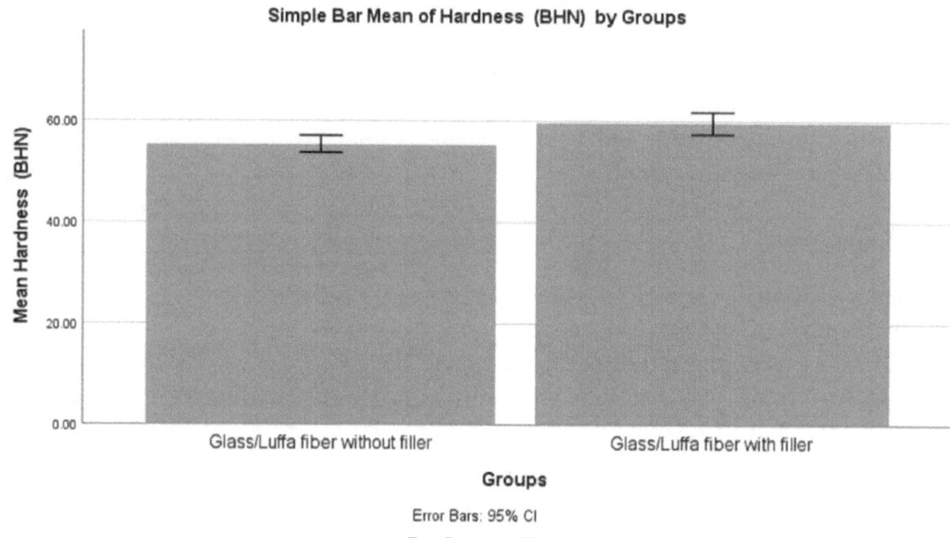

Fig. 51.3 Comparison of mean hardness of Glass/Luffa fiber composite without filler and Glass/Luffa fiber with 20% filler. X-axis represents Glass/Luffa fiber composite with and without filler and Y-axis represents mean hardness with the accuracy of ±1 SD. Hardness of Glass/Luffa fiber with 20% filler reinforced composite shows higher hardness than Glass/Luffa fiber epoxy composite

shows that the hardness is higher with 20% of filler/ epoxy composite than the composite without the filler.

The addition of Pista shell powder to the luffa fiber reinforced epoxy composite resulted in increased hardness compared to a composite comprised solely of luffa fiber. This enhancement in hardness improved the interfacial bonding in between the matrix and the fiber facilitated by presence of Pista shell powder.(Noone, Purushothaman, and Pradhan 2020).The composite material that contained jute fiber filled with carbon nanotubes demonstrated greater hardness compared to the composite material that solely contained jute fiber, when both were reinforced with a 2% volume fraction of carbon nanotubes (Saiteja, Jayakumar, and Bharathiraja 2020).The development of a composite material involving epoxy reinforced with Areca fibers, featuring randomly oriented fibers, yielded a high-strength composite. The mechanical properties of this composite were noted to improve with higher fiber volume fractions and extended post-curing times (Srinivasa and Bharath 2011). A composite material consisting of luffa fiber and 20% weight fraction of PbO nanoparticles exhibited greater hardness than a luffa fiber composite. The introduction of nanoparticles also enhanced the fiber and matrix material bonding (Ashok and Kalaichelvan 2022).

Hand layup is characterized by being time-consuming with low production efficiency, making it unsuitable for mass manufacturing. The fiber and matrix adhesion is often inadequate, leading to disadvantages such as the presence of voids that diminish the strength of the composite. To extend this work in the future, exploring mechanical qualities like composite hardness by varying the volume percentages of fiber and filler reinforcement in epoxy resin could be considered. The mechanical characteristics of composites reinforced with natural fibers and fillers are influenced by various parameters, including the manufacturing process, fiber type, filler size, polymer selection, and the presence of particle clustering.

6. CONCLUSION

This work investigated the hardness of a new Glass/Luffa fiber epoxy composite with and without reinforcement of Shorea robusta filler, while considering the limitations of the research. Results showed that the 20% volume fraction of the Shorea robusta filler incorporated epoxy Glass/Luffa fiber composite had a maximum hardness of 59.63 BHN, outperforming the Glass/Luffa fiber composite.

REFERENCES

1. Ashok, K. G., and K. Kalaichelvan. 2022. "Effect of Nano Fillers on Mechanical Properties of Luffa Fiber Epoxy-Composites." *Journal of Natural Fibers*. https://www.tandfonline.com/doi/abs/10.1080/15440478.2020. 1779898.
2. Cheng, Dao, Beibei Weng, Yuxia Chen, Shengcheng Zhai, Chenxin Wang, Runmin Xu, Junkui Guo, Yan Lv, Lanlan Shi, and Yong Guo. 2020. "Characterization of Potential Cellulose Fiber from Luffa Vine: A Study on Physicochemical and Structural Properties." *International Journal of Biological Macromolecules* 164 (December): 2247–57.
3. Chethan Kumar, G., Sagar M. Baligidad, A. C. Maharudresh, Nishchay Dayanand, and T. N. Chetan. 2022. "Development and Investigation of the Mechanical Properties of Natural Fiber Reinforced Polymer Composite." *Materials Today: Proceedings* 50 (January): 1626–31.
4. Dhanola, Anil, Anand Singh Bisht, Anil Kumar, and Aman Kumar. 2018. "Influence of Natural Fillers on Physico-Mechanical Properties of Luffa Cylindrica/ Polyester Composites." *Materials Today: Proceedings* 5 (9, Part 1): 17021–29.
5. Gayathri, N., V. K. Shanmuganathan, Ajay Joyson, M. Aakash, and A. Godwin Joseph. 2022. "Mechanical Properties Investigation on Natural Fiber Reinforced Epoxy Polymer Composite." *Materials Today: Proceedings*, October. https://doi.org/10.1016/j.matpr.2022.10.121.
6. Harikumar, R., and A. Devaraju. 2021. "Evaluation of Mechanical Properties of Bamboo Fiber Composite with Addition of Al2O3 Nano Particles." *Materials Today: Proceedings* 39 (January): 606–9.
7. Kamran, Kamran, and M. A. Iqbal. 2022. "The Ballistic Evaluation of Plain, Reinforced and Reinforced–prestressed Concrete." *Thin-Walled Structures*. https://doi.org/10.1016/j.tws.2022.109707.
8. Koffi, Agbelenko, Demagna Koffi, and Lotfi Toubal. 2021. "Mechanical Properties and Drop-Weight Impact Performance of Injection-Molded HDPE/birch Fiber Composites." *Polymer Testing* 93 (January): 106956.
9. Mishra, Trilokinath, Paulami Mandal, Arun Kumar Rout, and Dibakar Sahoo. 2022. "A State-of-the-Art Review on Potential Applications of Natural Fiber-Reinforced Polymer Composite Filled with Inorganic Nanoparticle." *Composites Part C: Open Access* 9 (October): 100298.
10. Mohanta, Niharika, and Samir K. Acharya. 2018. "Effect of Alkali Treatment on the Flexural Properties of a Luffa Cylindrica-Reinforced Epoxy Composite." *Science and Engineering of Composite Materials*. https://doi.org/10.1515/secm-2015-0148.
11. Nayak, Smaranika, Dipak Kumar Jesthi, Subhrajyoti Saroj, and Jatin Sadarang. 2022. "Assessment of Impact and Hardness Property of Natural Fiber and Glass Fiber Hybrid Polymer Composite." *Materials Today: Proceedings* 49 (January): 497–501.
12. Noone, Satishkumar, K. Purushothaman, and Raghuram Pradhan. 2020. "An Investigation on Luffa Cylindrica Fiber Reinforced Epoxy Composite." *Materials Today: Proceedings*. https://doi.org/10.1016/j.matpr.2020.07.051.
13. Saiteja, J., V. Jayakumar, and G. Bharathiraja. 2020. "Evaluation of Mechanical Properties of Jute Fiber/carbon Nano Tube Filler Reinforced Hybrid Polymer Composite." *Materials Today: Proceedings* 22 (January): 756–58.
14. Srinivasa, C. V., and K. N. Bharath. 2011. "Impact and Hardness Properties of Areca Fiber-Epoxy Reinforced Composites." *J. Mater. Environ. Sci* 2 (4): 351–56.
15. Vinod, A., JirattiTengsuthiwat, Yashas Gowda, R. Vijay, M. R. Sanjay, Suchart Siengchin, and Hom Nath Dhakal. 2022. "Jute/Hemp Bio-Epoxy Hybrid Bio-Composites: Influence of Stacking Sequence on Adhesion of Fiber-Matrix." *International Journal of Adhesion and Adhesives* 113 (March): 103050.

Note: Every figure and table was created using the experimental work; none of them were taken from any publications or online.

Advances in Additive Manufacturing Technologies – Gurusamy Pathinettampadian (eds)
© 2024 Taylor & Francis Group, London, ISBN 978-1-032-90013-1

52 An Experimental Study to Compare Tensile Strength of 10% Egg Shell Powder Infused Glass/Flax Fibre Reinforced Epoxy Resin Novel Composite with Plain Glass/ Flax Fibre Reinforced Epoxy Composite

B. Bharath Kumar[1], D. Vinodh[2]
Department of Mechanical Engineering,
Saveetha School of Engineering, Saveetha Institute of Medical and
Technical Sciences, Saveetha University,
Chennai, Tamil Nadu, India

ABSTRACT: To compare tensile strength of 10% egg shell powder infuse glass/flax fibre reinforced epoxy resins composite with glass/flax fibre reinforced epoxy composite.Five plies of bidirectional composite laminate fabric (F-G-F-G-F) are used to create composite laminates without reinforcement, while three plies of flax/glass fibre fabric and eggshell powder combined with epoxy resin are used to create composite laminates with reinforcement (F-G-F-G-F). The goal of this research is to improve tensile strength by reinforcing it using egg shell powder. A G-power of 80% was used to determine the sample size. The ASTM D3039 standard is followed for the tensile testing. T-independent tests were carried out using the statistical software tool SPSS. The results show that Group 1, flax/glass fibre composite reinforced with egg shell powder, had an average tensile strength of 2.38517 MPa, whereas Group 2, flax/glass fibre composite without reinforced composite laminate, had an average tensile strength of 2.06999 MPa. The two-tailed significance for this study was obtained as $p=0.0$ ($p<0.05$). The tensile strengths of flax/glass-fibre epoxy composite laminate samples with and without egg shell powder reinforcement are tested and compared under the circumstances of this inquiry. The samples of flax/glass fibre with 10% reinforcement showed improved tensile strength in comparison to samples from the group without egg shell powder reinforcement.

KEYWORDS: Novel composite, Flax/Glass fibre, Sustainable production, Tensile strength, Epoxy and hardener, Egg shell powder

1. INTRODUCTION

The main aim of this research is comparison of the tensile strength of flax/glass-fibre with epoxy matrix composites (F-G-F-G-F) reinforced with 10% egg shell powder to those that are unreinforced (F-G-F-G-F). Due to their increased tensile strength, polymer-fibre epoxy matrix composites like flax-fibre or glass-fibre reinforced plastics (FFRP/GFRP) are commonly used in industry. The cost-effectiveness and affordability of natural fibre reinforced polymer composites have outweighed those of synthetic fibre reinforced composites (Dusek 2013). Due to their qualities like environmental friendliness, low density, high specific stiffness, good mechanical properties, toxicological safety, biodegradability, and acoustic insulation, natural fibres from plants have a significant potential for use in the plastic, automotive, and packaging industries (Hodzic and Shanks 2014). The binding medium, otherwise called

[1]bharathkumarb19@saveetha.com, [2]vinodhd.sse@saveetha.com

DOI: 10.1201/9781003545774-52

as the matrix, and the re-inforcement material are typically the minimum number of components required to make a composite (Nettles 1996). Composites give several benefits over conventional materials, such as improved tensile strength, toughness, and fatigue-resistant properties that enables a more flexible design of the structure. Adding eggshell powder as reinforcement will give high strength to the material and it supports sustainable production. So the cost will be less. Although flax fibre is a cellulose polymer, it is inferior to other plant fibres in a number of ways due to its more crystalline structure, which makes it rougher, crispier, and stiffer to handle (Madueke, Mbah, and Umunakwe 2022).

Around 2440 publications related to this research criteria have been published in total over the last 5 years, as per the statistics from Google Scholar, and 758 articles have been published in ScienceDirect (Pashin and Orlov 2021; Andersons et al. 2011). This research with egg-shell reinforcement has received increasing interest. Innovative applications for novel composites with sustainable production are being discovered every year. (Karuppiah et al. 2020; Sohu et al. 2022) are having significantly high citations in this research topic. Upon analyzing various materials, there is evidence that kenaf fiber epoxy novel composites in general provide good mechanical properties.

According to research, there are very few studies on egg shell powder used as reinforcement in laminates composed of natural fibres. Thus, the current work examines and analyses the mechanical properties of the flax/glass fibre reinforced with 10% egg shell powder.

2. Materials and Methods

The novel composite sample materials were produced at the Saveetha School of Engineering at the Saveetha Institute of Medical and Technical Sciences in Chennai using the hand lay up method. The ethical review of this study is not required because it is primarily concerned with composite production. The supplier was Herenba industries pvt ltd in Ambattur, where LY556 grade epoxy resin and HY951 grade hardener (Dusek 2013) were acquired. It is a fast-curing epoxy resin with high chemical and mechanical resistance. Bidirectional flax/glass fibre cloth were obtained from Go Green pvt limited in Chennai. Egg Shell Powder was freshly produced. Eggshell powder as high sustainable production material in day to day life. Each group has a sample size of 20 which was determined using a g-power of 80% (Madueke et al. 2022). The experimental group is the flax/glass fibre composite with 20% egg shell powder reinforcement and the control group is the flax/glass fibre composite without egg shell powder reinforcement.

The control group samples were made from bidirectional flax/glass fibre that has been cut to a size of 300*300 mm. A 10:1 ratio was used to blend epoxy and hardener. F-G-F-G-F was the order of stacking. Throughout the putting up procedure, the epoxy glue was evenly distributed throughout all the fibre layers. Any air bubble forming within the layers is readily removed using a hand roller. After stacking of fibres is finished, a steady dead-weight was laid over the laminate for the whole curing process to guarantee excellent adherence. After that, the composite was water cut and machined into the required specimen size, as shown in Fig. 52.1.

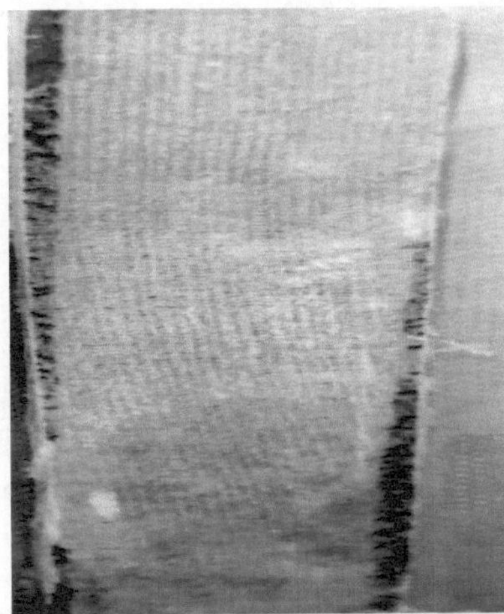

Fig. 52.1 Composite specimen without egg shell powder reinforcement

For the Experimental Group's construction, 300*300 mm pieces of bidirectional flax/glass fibre were cut. Epoxy and hardener were combined in 10:1 ratio. The stacking order was F-G-F-G-F. With epoxy resin, 20% egg shell powder has been added (added 300g of epoxy and hardener 30g with 60g of Egg Shell Powder). The novel composites were allowed to cure at room temperature for 48 hours. After that, the composite was water cut and machined into the required specimen size, as shown in Fig. 52.2.

Fig. 52.2 Composite specimen with 10% of eggshell powder reinforcement

Tensile strength tests were conducted using specimens as shown in Fig. 52.3 from both groups that are 120*20 mm in size in accordance with ASTM D3039. At Metmech Analytical Engineers in Ekkattuthangal, Chennai, the tensile tests were carried out using a universal testing apparatus. The samples from the experimental group and control group were utilised for the tensile testing in the appropriate order. The testing was done on 20 samples from each group, and the results were the recorded.

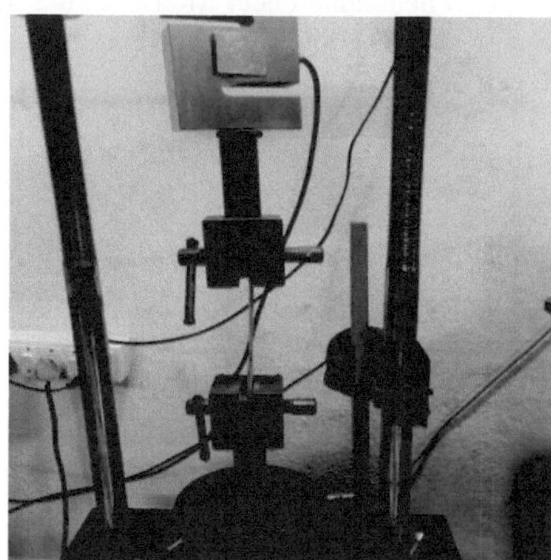

Fig. 52.3 Universal testing machine

3. Results

The ultimate tensile strengths of the flax/glass fibre composite test samples without and with 10% egg shell powder are shown in Table 52.1. In comparison to samples of unreinforced composite, samples of reinforced composite have higher tensile strengths. The average, standard deviation, and mean error are shown in Table 52.2. The fibre-epoxy composite with 10% egg shell powder reinforcement worked well, with an average tensile strength of 41.52 MPa and a standard deviation of 2.38517 MPa. The composite's mean tensile strength and standard deviation for the variant without egg shell powder reinforcement were 35.02 MPa and 2.06999 MPa, respectively. There are two groupings that are shown to be significant. The mean ultimate tensile strength of two groups is shown in Graph 1 using a simple bar graph.

4. Discussion

The samples' tensile strength was assessed using an all-purpose testing tool. The mean tensile strength of the composite was 41.52 and 35.02 for the reinforced variant and for the unreinforced version, with respective standard deviations of 2.38517 and 2.06999. Addition of egg shell powder boosts the material's tensile strength. Bidirectional flax/glass fibre plies were replaced by egg shell powder in the experimental group, which resulted in a considerable improvement in tensile strength but no discernible weight increase (Madueke, Mbah, and Umunakwe 2022; Mishra and Parashar 2023).

Table 52.1 Ultimate tensile strength for group 1 and group 2 samples

S. No	With 10% of eggshell powder Reinforcement (MPa)	Without eggshell powder Reinforment (MPa)
1	29.6	23.231
2	27.5	23.211
3	26.3	24.526
4	28.3	25.196
5	29.1	24.632
6	29	28
7	28.312	27.111
8	29.123	24.236
9	27.61	26.123
10	29.654	27.247
11	28.412	24.361
12	27.36	26.324
13	27.33	24.519
14	28.79	26.348
15	29.111	27.946
16	28	23.248
17	27.1	27.763
18	29.33	25.619
19	27.9	23.613
20	29.3	28.012

Table 52.2 Group statistics results (Mean of Polylactic Acid without filler 9.5635 mm is more compared to Polylactic Acid with filler 6.6785 mm and Standard error mean for Polylactic Acid with filler is 0.0519 and Polylactic Acid without filler is 0.02949

Group		N	Mean	Standard Deviation	Standard Error Mean
Tensile strength in MPa	With 10% of eggshell powder reinforcement	20	28.3566	0.94224	0.21069
	without eggshell powder reinforcement	20	25.5633	1.71224	0.38287

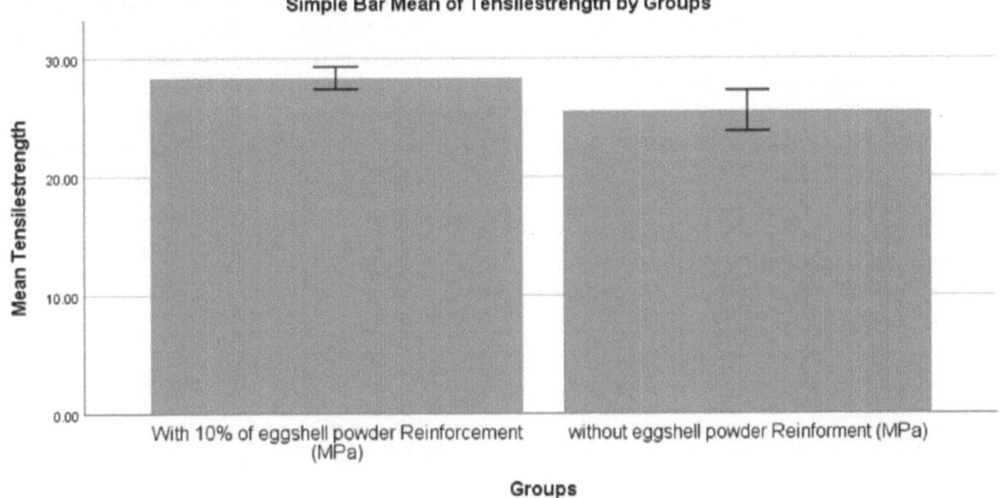

Fig. 52.4 Graphical representation of Tensile Strength (MPa) for Group-1 flax/glass fibre reinforced with 10% of egg shell powder reinforced epoxy composite laminate and Group-2 flax/glass fibre without reinforcement of egg shell powder with +/- 1 SD.X axis: Material groups, Y axis: Tensile Strength (MPa) with Mean accuracy of detection 95% CI

A similar investigation focused on flax/glass fibre epoxy composite reinforced with egg shell powder (Lindeberg 1948). This study also showed that adding egg shell powder as reinforcement improves the tensile characteristics of the novel composites.(Shah et al. 2018)

The limitation imposed by the fabrication technique used in this study, hand lay up, is its lengthy production time. There is a significant probability of internal flaws and poor bonding in this process. As a result, the mechanical strength of the novel composites would be reduced. Future enhancements to the work can be made by employing more sophisticated fabrication techniques with sustainable production methods, and the necessary material behaviour can be produced by varying the fibres' orientation, thickness, and layer count, as well as their reinforcement content.

5. Conclusion

The tensile strengths of the flax/glass fibre epoxy composite laminate samples with and without egg shell powder reinforcement are tested and compared under the circumstances of this inquiry. The samples of flax/glass fibre with 10% reinforcement showed improved tensile strength when compared to the test samples from the control group without the egg shell powder reinforcement.

References

1. Andersons, J., R. Joffe, E. Spārniņš, and D. Weichert. 2011. "Modeling the Effect of Reinforcement Discontinuity on the Tensile Strength of UD Flax Fiber Composites." Journal of Materials Science. https://doi.org/10.1007/s10853-011-5440-9.
2. Dusek, K. 2013. Epoxy Resins and Composites I. Springer.
3. Hodzic, Alma, and Robert Shanks. 2014. Natural Fibre Composites: Materials, Processes and Properties. Woodhead Publishing.
4. Karuppiah, Ganesan, Kailasanathan Chidambara Kuttalam, Murugesan Palaniappan, Carlo Santulli, and Sivasubramanian Palanisamy. 2020. "Multiobjective Optimization of Fabrication Parameters of Jute Fiber/Polyester Composites with Egg Shell Powder and Nanoclay Filler." Molecules 25 (23). https://doi.org/10.3390/molecules25235579.
5. Lindeberg, Gösta. 1948. Tensile Strength and Chemical Composition of the Middle Lamella of the Flax Fibre.
6. Madueke, Chioma Ifeyinwa, Oguejiofor Miracle Mbah, and Reginald Umunakwe. 2022. "A Review on the Limitations of Natural Fibres and Natural Fibre Composites with Emphasis on Tensile Strength Using Coir as a Case Study." Polymer Bulletin , May, 1–18.
7. Mishra, Shashank, and Vishal Parashar. 2023. "Experimental Analysis of Duo-Fiber Interaction on the Tensile Strength of Surface-Modified Flax–kenaf-Reinforced Epoxy Composite." Polymer Bulletin. https://doi.org/10.1007/s00289-023-04708-6.

8. Nettles, Alan T. 1996. Low Temperature Mechanical Testing of Carbon-Fiber/Epoxy-Resin Composite Materials.
9. Pashin, E. L., and A. V. Orlov. 2021. "Modernization Directions of Tensile Strength Testing Machine RMP-1, for Standartized Flax Fiber Testing." Proceedings of Higher Education Institutions. Textile Industry Technology. https://doi.org/10.47367/0021-3497_2021_5_186.
10. Shah, Ahmer, Yuqi Zhang, Xiaodong Xu, Abdul Dayo, Xiao Li, Shuo Wang, and Wenbin Liu. 2018. "Reinforcement of Stearic Acid Treated Egg Shell Particles in Epoxy Thermosets: Structural, Thermal, and Mechanical Characterization." Materials. https://doi.org/10.3390/ma11101872.
11. Sohu, Samiullah, Naraindas Bheel, Ashfaque Ahmed Jhatial, Abdul Aziz Ansari, and Irfan Ali Shar. 2022. "Sustainability and Mechanical Property Assessment of Concrete Incorporating Eggshell Powder and Silica Fume as Binary and Ternary Cementitious Materials." Environmental Science and Pollution Research International 29 (39): 58685–97.

Note: Every figure and table was created using the experimental work; none of them were taken from any publications or online.

Advances in Additive Manufacturing Technologies – Gurusamy Pathinettampadian (eds)
© 2024 Taylor & Francis Group, London, ISBN 978-1-032-90013-1

53 Gaussian and Gradient-Based Edge Detection Algorithm Studies of Cutting Tool Wear

S. Senthil Pon Vignesh[1]

Research Scholar, Department of Electronics and Communication Engineering,
SIMATS School of Engineering, Saveetha Institute of Medical and
Technical Sciences, Chennai, Tamil Nadu, India

Radhika Baskar[2]

Project Guide, Department of Electronics and Communication Engineering,
SIMATS School of Engineering, Saveetha Institute of Medical and
Technical Sciences, Chennai, Tamil Nadu, India

ABSTRACT: Gradient-Based Edge Detection Analysis and Gaussian-Based Edge Detection Algorithm comparison is necessary to determine Flank Wear. There are two groups of 27 samples each, totaling 54 samples required for the analysis. The algorithms for gradient-based edge detection and Gaussian-based edge detection, respectively, have identified flank wear into Groups 1 and 2. Clinicalc.com was utilized to determine the total sample size and input the subsequent parameters: an enrollment ratio of one, a G-Power of 85%, a 0.05 alpha error, and a 95% confidence interval. The study shows that when Gaussian-Based Edge Detection (0.427mm) is compared to Gradient-Based Edge Detection (0.419 mm), the former generates high mean accuracy and is statistically significant in predicting flank wear levels ($p = 0.000$, $p<0.05$). Upon closer inspection, the accuracy of the Gaussian-Based Edge Detection Algorithm's prediction of the value of flank wear is superior than that of the Gradient-Based Edge Detection Algorithm..

KEYWORDS: Duplex stainless steel, Turning wear, Algorithm

1. INTRODUCTION

Additionally, tool wear is divided into crater wear and flank wear, with flank wear playing a significant influence on the machinery's diametric precision, stability, and dependability, according to (Rajmohan, Palanikumar, and Paulo Davim 2020). (Kuntoğlu and Sağlam 2021) adopted ANOVA and gradient-based Edge Detection Rule approach to calculate flank wear and acoustic emission while turning AISI 5140 steel alloy.

By turning Nimonic C263 alloy and the MSER Algorithm, Deep Pattern Network, and L9 Orthogonal Array to conduct experimental trials and predict flank wear, in the research on predictive maintenance in machinery, (Yang et al. 2022) observed that the variables influencing flank wear might lead to unexpected errors and breakdowns of machinery. (Das, Panda, and Dhupal 2017) used the L9 Orthogonal Array to evaluate the AISI 4340 steel alloy and managed the tool's quality of life by establishing an ANOVA statistical analysis approach. (Niaki, Ulutan, and Mears 2015)

[1]ponvigneshsenthil19@saveetha.com, [2]radhikabaskar@saveetha.com

DOI: 10.1201/9781003545774-53

optimized Nimonic C263 alloy interpulse welding at high heat conditions and calculated the maximum stress it could withstand. The best publication to be taken into consideration for the method to predict flank wear is that of (Bommi et al. 2022) who predicted it using the MSER Algorithm under the Deep Pattern Network and integrated it as FlankNet.

The research conducted showed that some various factors and challenges affect the accurate prediction of flank wear, some mathematical models exist for predicting flank wear, but they may have limited predictive capabilities. This can make it difficult to accurately predict and prevent tool failure due to flank wear. In the future, virtual prototyping and simulation tools can be used to model tool wear and predict its effects on machining operations. This can help manufacturers optimize tool geometries, cutting parameters, and other variables to improve performance and reduce tool wear. Hence, in this study, the measurement is taken concerning the pixel coordinates from the processed image frame which is converted (in mm), and the flank wear of the respective tool is predicted using the proposed Gaussian-based algorithm.

2. Material and Methods

Laboratory provided the setting for experimentation, testing, study, and evaluation.

Detection Algorithm with more mean accuracy.

To capture the microscopic images, a machine vision technique was employed, and the images were then processed using a Gradient-Based Edge Detection Algorithm to determine flank wear. The computations were accomplished using Python v3 on the PyCharm Platform. To run the software, the system must meet certain requirements, the preferred OS is Windows 10 with an Intel processor (minimum i3 6th generation or equivalent). Additionally, the system should have a minimum of 4 GB of RAM and at least 50 GB of ROM, not including the memory used by the Windows operating system.

2.1 Gaussian-Based Edge Detection Algorithm—Group 1

The Gaussian-based Edge Detection Algorithm is a method of identifying edges or boundaries within an image. It involves convolving the image with a Gaussian filter to smooth out the image and reduce noise, and then calculating the gradient of the smoothed image to detect edges. This algorithm is commonly used in image processing and computer vision applications for tasks such as object detection, feature extraction, and image segmentation. Moreover, they can be implemented using Canny and Laplacian operators.

Table 53.1 Industrial data

S. No	Group 1 Gaussian Based War	Group 2 Flank wear (mm)			Flank Wear Measured Manually (mm)
	Canny's	Prewitt's	Sobel's	Mean	
1	0.3599	0.3598	0.3498	0.3548	0.36
2	0.3778	0.3698	0.3598	0.3648	0.38
3	0.4296	0.4098	0.3998	0.4048	0.43
4	0.4799	0.4599	0.4489	0.4544	0.49
5	0.4298	0.4198	0.3998	0.4098	0.43
6	0.4297	0.4199	0.4158	0.41785	0.44
7	0.4198	0.4089	0.3998	0.40435	0.43
8	0.4295	0.4199	0.4095	0.4147	0.44
9	0.4479	0.4389	0.4279	0.4334	0.456
10	0.3199	0.3098	0.2978	0.3038	0.32
11	0.3388	0.3388	0.3188	0.3288	0.34
12	0.3678	0.3378	0.3378	0.3378	0.37
13	0.3541	0.3341	0.3341	0.3341	0.36
14	0.3898	0.3598	0.3598	0.3598	0.39
15	0.3899	0.3699	0.3699	0.3699	0.40
16	0.3668	0.3368	0.3368	0.3368	0.37
17	0.4158	0.3858	0.3858	0.3858	0.42
18	0.4428	0.4128	0.4128	0.4128	0.45
19	0.4699	0.4299	0.4299	0.4299	0.47
20	0.4398	0.4198	0.4198	0.4198	0.45
21	0.4598	0.4298	0.4298	0.4298	0.47
22	0.4315	0.4115	0.4115	0.4115	0.44
23	0.4497	0.4499	0.4398	0.44485	0.46
24	0.4698	0.4597	0.4488	0.45425	0.48
25	0.4649	0.4598	0.4449	0.45235	0.47
26	0.4959	0.4899	0.4785	0.4842	0.50
27	0.5178	0.5089	0.4978	0.50335	0.52
MEAN					0.4273 mm

2.2 Group 1—Algorithm

Once these initial steps are completed, the next step is to convert the image into grayscale to simplify the edge detection process. After this, the Gaussian Edge Detection Algorithm (using Canny Operator) can be applied to the images. The next step is to obtain the pixel coordinates and append them to an array. From this array, the respective coordinates should be chosen for calculating the distance. Flank wear in pixels can then be obtained. To ensure the accuracy of the results, the value of the flank wear obtained from the industry should be kept as a reference.

2.3 Gradient-Based Edge Detection Algorithm—Group 2

Gradient-based Edge Detection Algorithm is a type of edge detection algorithm used to detect the boundaries or edges within an image. It involves calculating the gradient of the image, which is the rate of change of pixel intensity in the x and y directions. The magnitude of the gradient indicates the strength of the edge, while the direction of the gradient

2.4 Group 2—Algorithm

Create a new project and file in Python then necessary header files should be imported. The value of the flank wear obtained from the industry should be kept as a reference. Finally, the zooming factor should be calculated and the flank wear in mm should be obtained.

3. Statistical Analysis

(Bright 2020) used IBM SPSS Statistics v5 for continuous and discrete variables and the basics of data analytics. The study requires two variables to be set, namely the Independent and Dependent variables. The dataset was produced by simulating the hard turning of Duplex Stainless Steel on a CNC Lathe machine with CBN (Cubic Boron Nitrate) inserts. Moreover, to assess these algorithms, an Independent Samples T-Test is performed.

Table 53.2 Group statistics of the accuracy of flank wear for the respective Algorithms

Group Statistics					
	Algorithm	N	Mean	Std. Deviation	Std. Error Mean
Flank Wear	Gaussian	27	0.427844	0.0092964	0.0017891
	Gradient	27	0.419081	0.0230648	0.0044388

4. Result

The study proves that Gaussian-Based Edge Detection (0.427 mm) results in high mean accuracy and is statistically significant in predicting the values of flank wear compared to Gradient-Based syatem.

Values of Flank Wear that were manually measured at the microscopic level from the CNC Lathe Machine and are recorded in Table 53.1 which is considered the dataset. Group 1 (Gaussian Edge predicted Flank Wear) and Group 2 (Gradient Based Edge Detection predicted Flank Wear) each have 27 samples of data collected from the industry by turning Duplex Stainless Steel as depicted in Table 53.1. Further, the manually measured Flank Wear from the industry while turning DSS is recorded in column 6 of Table 53.2. Table 53.2 describes the Group Statistics of the Novel Gaussian Edge Detection Algorithm and the Gradient-Based Edge Detection in which Accuracy is displayed in terms of Gaussian and Gradient-Based Edge Detection. N is the number of samples taken for the respective algorithms and their mean is calculated. From those the Std. Deviation and the Error in Std. The mean is calculated and displayed. In Fig. 53.5 represents the Bar Graph of the novel Gaussian-Based Edge Detection vs Gradient-Based Edge Detection to estimate Flank Wear using Canny's, Prewitt's, and Sobel's operators respectively. Eventually, Figure 53.6 depicts the elaborate representation of the Bar Graph that is in terms of operators used in respective algorithms.

Fig. 53.1 The cutting tool seen in the image above is used to convert duplex stainless steel. Tool wear is measured manually by the industry and is listed in Table 53.1. The industry standard for flank wear measurement is 0.36 mm

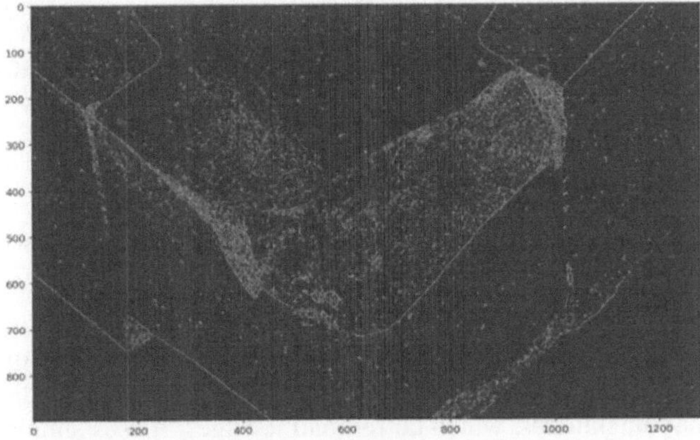

Fig. 53.2 Illustrates the cutting tool's image which was used for Edge Detection using Novel Gaussian and Gradient Edge Detections respectively

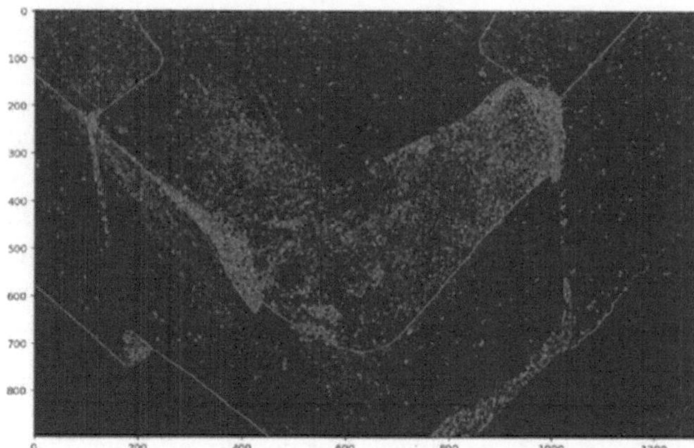

Fig. 53.3 The image above is of a cutting tool that has been processed with the Novel Prewitt's Edge Detection Algorithm, and it shows the combination of the x and y coordinate intensities of the cutting tool to produce the xy intensities

Fig. 53.4 The image above is of a cutting tool that has been processed with the Novel Sobel's Edge Detection Algorithm, and it displays the combination of the x and y gradients of the cutting tool to produce the xy gradient

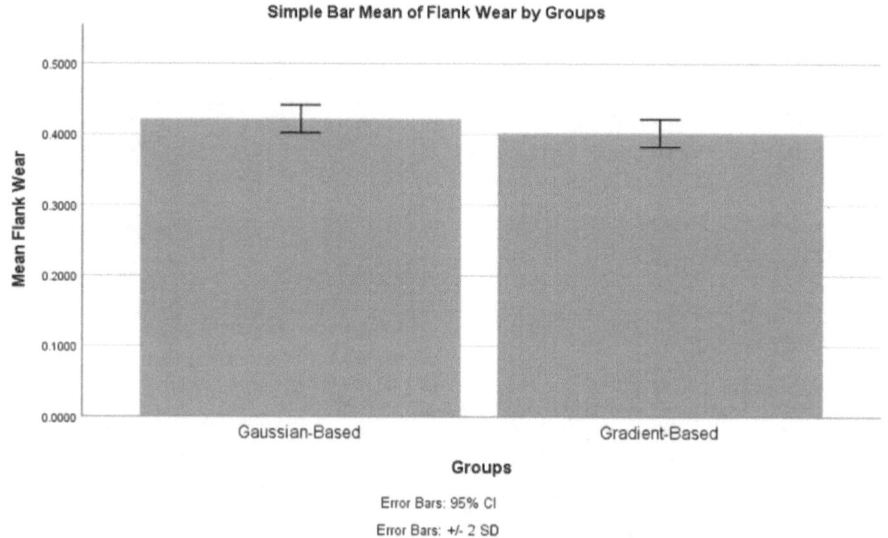

Fig. 53.5 Graphical Representation of Flank Wear Prediction of the Gaussian-Based Edge Detection Algorithms and the Gradient-Based Edge Detection Algorithms. The average of the Flank Wear predicted for the 27 samples is shown in the above graph where the average of the Gaussian-based is 0.42 mm and the Gradient-Based is 0.40 mm

5. Discussion

Lilensten et al. 2019) who also performed microanalysis and compared the superalloys' limitations, concur with this conclusion for nickel-based alloys.

Gradient-Based Edge Detection computes the gradient of the image intensity and looks for pixels with high gradient magnitudes, which correspond to edges. It uses simple edge operators, such as the Sobel, Prewitt, or Roberts operators, to compute the gradient. Gradient-based methods are fast but can produce inaccurate or noisy edge maps and are sensitive to changes in image intensity. Gaussian-Based Edge Detection, on the other hand, involves smoothing the image with a Gaussian filter to reduce noise and then computing the gradient magnitude of the smoothed image. The Gaussian filter helps to remove noise and produce a smoother gradient map, which results in more accurate and robust edge detection. Gaussian-Based Edge Detection is more computationally expensive than Gradient-Based Edge Detection but provides better results in improving the smoothness while turning and facing the alloys, enhancing the quality of life of the cutting tool and in many applications. In general, Gaussian-Based Edge Detection is a better choice for applications where high accuracy and

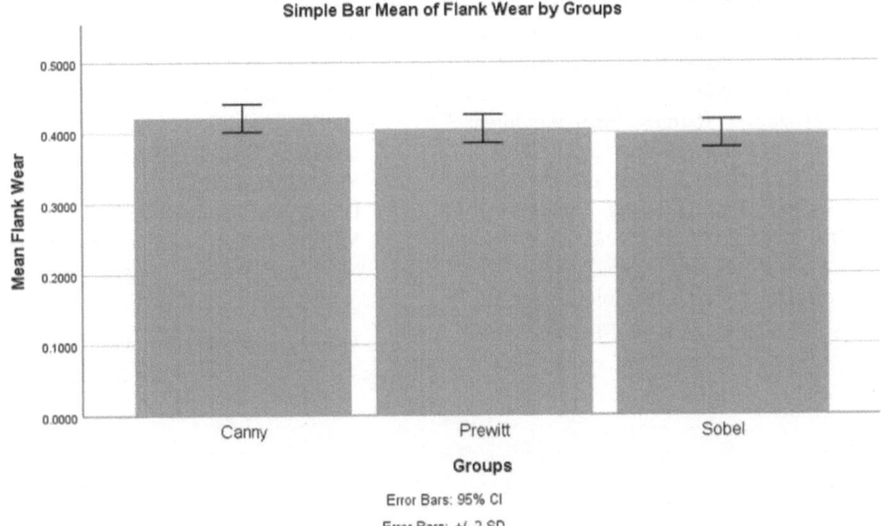

Fig. 53.6 Graphical representation of flank wear

robustness are required, while Gradient-Based Edge Detection may be a good choice for applications where speed and simplicity are more important.

6. Conclusion

Finally, the results show that the Gaussian Edge Detection Algorithm is significantly better and predicts the mean flank wear of 0.427mm compared with Gradient Based Edge Detection Algorithms which predict the mean flank wear of 0.419mm and the significance factor derived from SPSS is p = .000 ($p<0.05$).

References

1. Bhowmick, Sanjit, and Eric Hintsala. 2020. "Extreme *In Situ* Mechanics of Bond Coatings and Ni-Based Superalloys Using an Advanced SEM Nanomechanical Instrument." *Microscopy and Microanalysis*. https://doi.org/10.1017/s1431927620022588.
2. Bommi, R. M., Chakaravarthy Ezilarasan, M. P. Sudeshkumar, and T. Vinoth. 2022. "Estimation of Flank Wear in Turning of Nimonic C263 Super Alloy Based on Novel MSER Algorithm and Deep Patten Network." *Russian Journal of Nondestructive Testing*. https://doi.org/10.1134/s1061830922020073.
3. Bright, Steven. 2020. *SPSS: The Ultimate Data Analysis Tool*. Lulu Press, Inc.
4. Das, Sudhansu Ranjan, Asutosh Panda, and Debabrata Dhupal. 2017. "Experimental Investigation of Surface Roughness, Flank Wear, Chip Morphology and Cost Estimation during Machining of Hardened AISI 4340 Steel with Coated Carbide Insert." *Mechanics of Advanced Materials and Modern Processes*. https://doi.org/10.1186/s40759-017-0025-1.
5. Dehshibi, Mohammad Mahdi, and Andrew Adamatzky. 2021. "Electrical Activity of Fungi: Spikes Detection and Complexity Analysis." *BioSystems* 203 (May): 104373.
6. Gao, Shuaishuai, Xianyin Duan, Kunpeng Zhu, and Yu Zhang. 2022. "Generic Cutting Force Modeling with Comprehensively Considering Tool Edge Radius, Tool Flank Wear and Tool Runout in Micro-End Milling." *Micromachines* 13 (11). https://doi.org/10.3390/mi13111805.
7. Gonzalez, Claudia I., Patricia Melin, Juan R. Castro, and Oscar Castillo. 2017. *Edge Detection Methods Based on Generalized Type-2 Fuzzy Logic*. Springer.
8. Hoier, Philipp, Amir Malakizadi, Pietro Stuppa, Stefan Cedergren, and Uta Klement. 2018. "Microstructural Characteristics of Alloy 718 and Waspaloy and Their Influence on Flank Wear during Turning." *Wear*. https://doi.org/10.1016/j.wear.2018.01.011.
9. Kisasoz, A., M. Tümer, and A. Karaaslan. 2021. "Microstructure, Mechanical, and Corrosion Properties of UNS 32205 Duplex Stainless Steel Weldment Joints by Multipass FCAW." *Practical Metallography*. https://doi.org/10.1515/pm-2021-0025.
10. Kuntoğlu, Mustafa, and Hacı Sağlam. 2021. "ANOVA and Fuzzy Rule-Based Evaluation and Estimation of Flank Wear, Temperature and Acoustic Emission in Turning." *CIRP Journal of Manufacturing Science and Technology*. https://doi.org/10.1016/j.cirpj.2021.07.011.
11. Lilensten, Lola, Stoichko Antonov, Dierk Raabe, Sammy Tin, Baptiste Gault, and Paraskevas Kontis. 2019. "Deformation of Borides in Nickel-Based Superalloys: A Study of Segregation at Dislocations." *Microscopy and Microanalysis*. https://doi.org/10.1017/s1431927619013424.
12. Niaki, Farbod Akhavan, Durul Ulutan, and Laine Mears. 2015. "In-Process Tool Flank Wear Estimation in Machining Gamma-Prime Strengthened Alloys Using Kalman Filter." *Procedia Manufacturing*. https://doi.org/10.1016/j.promfg.2015.09.018.

13. Omiogbemi, Ibrahim Momoh-Bello, Danjuma Saleh Yawas, Atanu Das, Matthew Olatunde Afolayan, Emmanuel Toi Dauda, Roshan Kumar, Sudhakar Rao Gorja, and Sandip Ghosh Chowdhury. 2022. "Mechanical Properties and Corrosion Behaviour of Duplex Stainless Steel Weldment Using Novel Electrodes," *Scientific Reports* 12 (1): 22405.
14. Rajmohan, T., K. Palanikumar, and J. Paulo Davim. 2020. *Advances in Materials and Manufacturing Engineering: Select Proceedings of ICMME 2019*. Springer Nature.
15. Sengupta, Sudhriti, Neetu Mittal, and Megha Modi. 2020. "Improved Skin Lesions Detection Using Color Space and Artificial Intelligence Techniques." *The Journal of Dermatological Treatment* 31 (5): 511–18.
16. Xavier, J. Francis, J. Francis Xavier, B. Ravi, D. Jayabalakrishnan, Chakaravarthy Ezilarasan, V. Jayaseelan, and G. Elias. 2021. "Experimental Study on Surface Roughness and Flank Wear in Turning of Nimonic C263 under Dry Cutting Conditions." *Journal of Nanomaterials*. https://doi.org/10.1155/2021/2054399.
17. Yang, Jing, Jian Duan, Tianxiang Li, Cheng Hu, Jianqiang Liang, and Tielin Shi. 2022. "Tool Wear Monitoring in Milling Based on Fine-Grained Image Classification of Machined Surface Images." *Sensors* 22 (21). https://doi.org/10.3390/s22218416.

Note: Every figure and table were created using the experimental work; none of them were taken from any publications or online.

Advances in Additive Manufacturing Technologies – Gurusamy Pathinettampadian (eds)
© 2024 Taylor & Francis Group, London, ISBN 978-1-032-90013-1

54 Evaluating Mean Square Error for Flank Wear Assessment in Turning of Duplex Stainless Steel

S. Senthil Pon Vignesh[1]
Research Scholar, Department of Electronics and Communication Engineering,
Saveetha School of Engineering, Saveetha Institute of Medical and
Technical Sciences, Chennai, Tamil Nadu, India

A. Mary Joy Kinol[2]
Project Guide, Department of Electronics and Communication Engineering,
Saveetha School of Engineering, Saveetha Institute of Medical and
Technical Sciences, Chennai, Tamil Nadu, India

ABSTRACT: The estimation of average error accuracy and mean square error (MSE) for flank wear during the turning of Duplex Stainless Steel (DSS), employing both the novel Sobel's Edge Detection Algorithm and the traditional Robert's Edge Detection Algorithm. A dataset comprising 54 samples is divided into two groups, each containing 27 samples. The average error and MSE are computed using both algorithms for Groups 1 and 2. Sample size determination is conducted using clinicalc.com with a confidence interval of 95%. Linear regression analyses are performed on the errors obtained. Comparison between the Novel Sobel's Edge Detection Algorithm (90.77%, 0.069mm2) and Robert's Edge Detection Algorithm (85.91%, 0.088mm2) reveals a maximized linear regression, with observed significance levels of $p = 0.001$. Novel SED Algorithm demonstrates significantly superior average error and MSE performance compared to RED Algorithm.

KEYWORDS: Novel sobel's edge detection, MSE, Average Error, Linear regression, Robert's edge detection, Flank wear, Quality of life, and Turning

1. Introduction

According to Verma et al.'s 2019 study, "Optimizing machine parameters using computer vision," feed rate had the biggest impact on cutting tool quality of life, surface roughness, and tool wear. Thakre, Lad, and Mala (2019) have presented an algorithm that uses a digital camera as the input to estimate flank wear using machine vision systems. The algorithm's output is dependable and effective. According to (Rao et al. 2021)., tool wear is extremely important in the global manufacturing and production industries. Tool wear is further subdivided into Crater and Flank Wear, with Flank Wear having a major impact on the machinery's diametric precision, stability, and dependability (Rajmohan, Palanikumar, and Paulo Davim 2020).

In 2017, Das, Panda, and Dhupal conducted an investigation into the use of the L9 Orthogonal Array for turning the AISI 4340 steel alloy. By turning an AISI 5140 steel

[1]ponvigneshsenthil19@saveetha.com, [2]maryjoykinola.sse@saveetha.com

DOI: 10.1201/9781003545774-54

alloy, (Kuntoğlu and Sağlam 2021) estimated the flank wear and acoustic emission of the cutting tool using ANOVA. Furthermore, in 2022, Yang et al. (J.) Explored the behavior of Nimonic C263 alloy by employing the MSER Algorithm and Deep Pattern Network to forecast flank wear. They also conducted experiments using the L9 and L27 orthogonal array techniques to determine the alloy's surface roughness and flank wear in dry conditions. According to (Niaki, Ulutan, and Mears 2015), the tool wear estimate for Gamma-Prime reinforced alloys using the Kalman Filter had an 18% filter rate of inaccuracy. Before determining tool wear, especially flank wear, the MSER Algorithm and DPN integrated as FlankNet (Bommi et al., 2022) is the perfect paper to take into consideration. Sengupta, Mittal, and Modi (2020) looked into optimization and edge detection techniques; they then used Ant Colony Optimization and Sobel's Edge detection procedure. Advanced techniques like image processing and sensor integration are becoming increasingly popular for their ability to offer real-time monitoring and automation in tool wear assessment. The selection of these parameters often involves a balance between detecting enough details (edges) and suppressing noise or irrelevant information. The optimal values for these parameters may need to be determined through experimentation, considering the characteristics of the specific images or signals being processed.

The studies revealed several challenges and factors that impact the prediction of flank wear. One of the main challenges is the difficulty in measuring flank wear accurately, especially in complicated geometries. To manage this issue, future improvements can be made by the use of AI and machine learning algorithms that can enhance the accuracy of tool wear prediction by analyzing large amounts of data. This can lead to more precise predictions and better decision-making in machining operations. In this research, the regression of average and MSEs are measured and sorted out using the NSED algorithm which maximizes the Linear Regression (%).

2. Materials and Methods

2.1 Sobel's Edge Detection Algorithm—Group 1

Sobel's Edge Detection Algorithm detects the images and processes them on the basis of the intensity of pixel squares. In step 1, it enhances the X intensities of the image, Step 2 it enhances the Y intensities of the image, and finally, it combines the X and Y intensities and produces the final output which is directed for the flank wear prediction. Then, the average error and MSE are calculated and their mean is obtained.

2.2 Group 1—Algorithm

The image should then be converted to grayscale, the Novel SED Algorithm applied to both horizontal and vertical pixel orientations, and the Laplacian of horizontal plus vertical computed. Save the result and use it to calculate the Flank Wear. Get the pixel coordinates and add them to an array in order to compute flank wear. Identify the relevant coordinates from the array to ascertain the Flank Wear in Pixels and calculate the distance accordingly. Utilize the industry-acquired flank wear value as a reference. Determine the flank wear in millimeters by factoring in the zooming factor. To establish linear regression, compute both the average error and mean square error.

Table 54.1 Experimental trials of flank wear

Experiment Number	Cutting Speed (V) (m/min)	Rate of Feed (S) (mm/rev)	Depth of Cut (D) (mm)	Flank Wear (mm)
1	150	0.15	0.7	0.34
2	150	0.15	0.9	0.37
3	150	0.17	0.5	0.36
4	150	0.17	0.7	0.39
5	150	0.17	0.9	0.40
6	150	0.19	0.5	0.37
7	150	0.19	0.7	0.42
8	150	0.19	0.9	0.45
9	180	0.15	0.5	0.47
10	180	0.15	0.7	0.45

2.3 Robert's Edge Detection Algorithm—Group 2

Robert's edge detection algorithm is a quick and easy edge detection method that finds edges in an image by using two 2x2 kernels. The algorithm identifies edges with a high gradient magnitude and a gradient direction perpendicular to the edge. Robert's algorithm is fast and can be used for real-time edge detection, but it can produce thin edges that are sensitive to noise.

2.4 Group 2—Algorithm

The image must then be converted to grayscale and subjected to RED Algorithm. The resulting output needs to be stored and used for estimating flank wear. It is necessary to gather and add pixel coordinates to an array. As a guide, the industry-provided value for flank wear should be retained. The next step is to determine the flank wear in millimeters by calculating the zooming factor. Lastly, to assess the algorithm's performance, the average error and MSE should be computed.

3. Procedure

To begin implementing the algorithm, open a Python file and create a Python package. Novel SED and RED Algorithms are used for image processing. The processed images are used to determine Flank Wear using Pixel Coordinates. The AE and MSE of the different methods are determined from the Flank Wear and the respective Linear Regression is expressed (in %).

4. Results

When comparing Novel Sobel's Edge Detection Algorithm (90.78%, 0.069mm2) with Robert's edge detection algorithm (85.91%, 0.088 mm^2), the results indicate that Novel Sobel's algorithm is superior due to the accurate average error, minimized MSE and maximized linear regression. Table 54.1 presents details of the cutting tool dataset. The linear regression of the average error accuracy and MSE, as determined by IBM SPSS Statistics, is finally combined in Table 54.2. Figure 54.1 represents the cutting tool at the microscopic level while turning DSS with tool wear. Fig. 54.2, Fig. 54.3, and Fig. 54.4 are the outputs of the Novel SED algorithm X Gradient, Y Gradient, and XY Gradient respectively. Figure 54.5 depicts the flank wear of the cutting tool. Figure 54.6 is the output of the Novel SED Algorithm Fig. 54.7 is the pictorial block diagram representation of the proposed Novel methodology. Figure 54.8 shows the Average Error Accuracy of the 54 samples that were uploaded to SPSS Statistical Analysis. The MSE of the 54 samples that were uploaded to SPSS Statistical

Fig. 54.1 Duplex stainless steel

Fig. 54.2 Novel SED algorithm

Fig. 54.3 Cutting tool which is processed using the Novel SED Algorithm with y-gradient of the cutting tool

Table 54.2 Linear regression for average error and MSE

	Model	Regression (R) (In %)	Regression Square (R^2) (In %)
Sobel's Edge Detection	Average Error	70.9	50.3
	MSE	70.4	49.6
Robert's Edge Detection	Average Error	60.7	40.9
	MSE	59.9	42.1

Fig. 54.4 Cutting tool which is processed using the Novel SED Algorithm

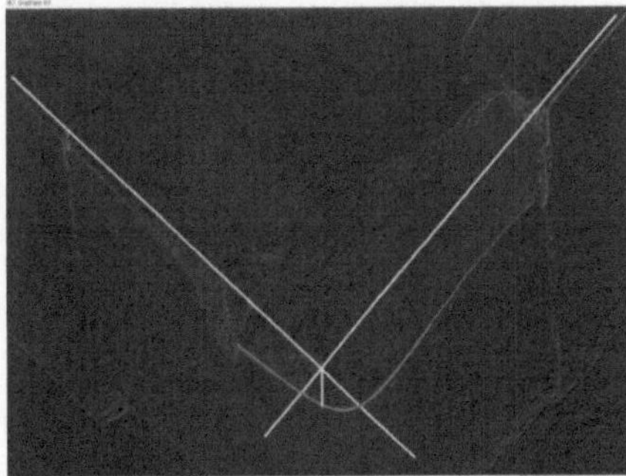

Fig. 54.5 Tool wear consists of flank wear and crater wear

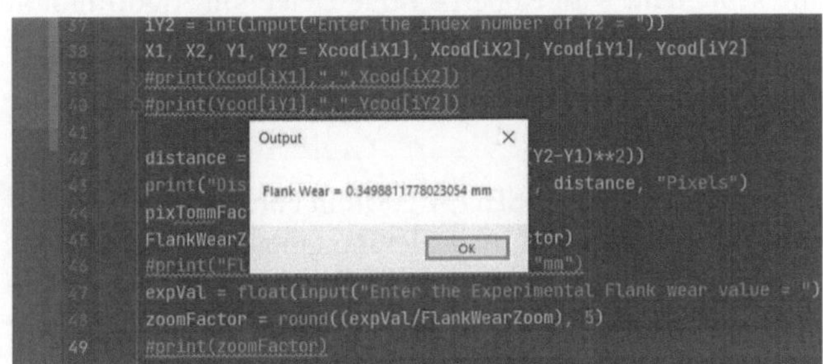

Fig. 54.6 Flank wear using novel SED algorithm

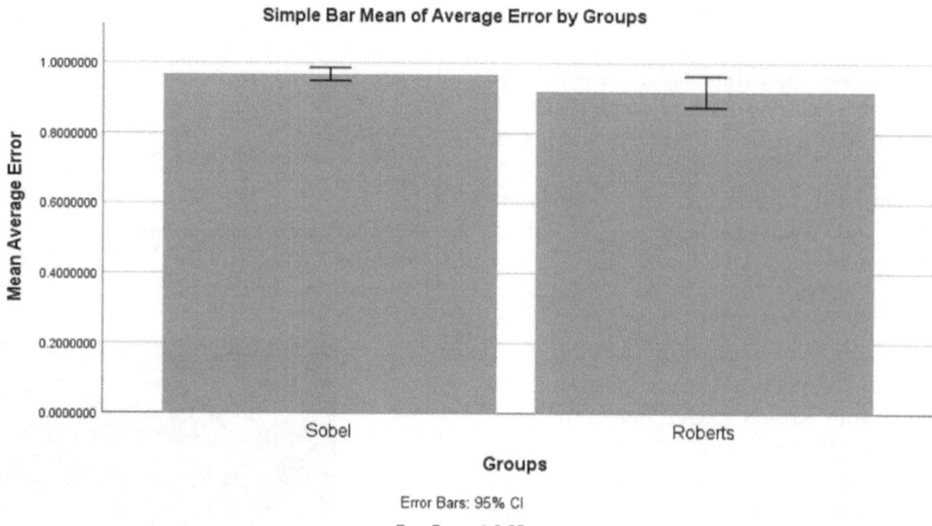

Fig. 54.7 SPSS analysis of novel SED algorithm

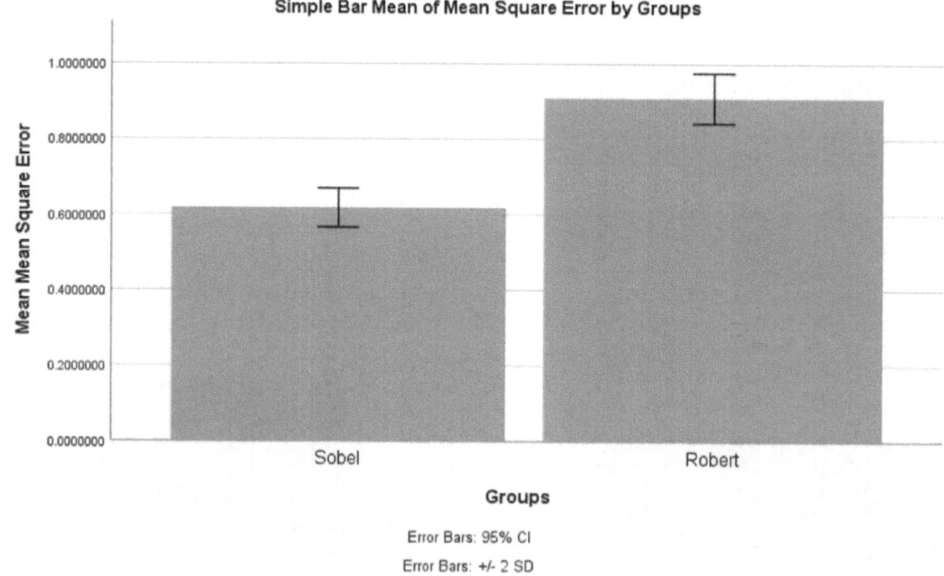

Fig. 54.8 Novel sobel's edge detection algorithm and the compared robert's edge detection algorithm

Analysis is shown in Fig. 54.8. The results indicate that processing the data with Novel Sobel's Edge Detection Algorithm reduces the MSE when compared to processing the same data with Robert's Edge Detection Algorithm on 27 samples each.

5. Discussion

The proposed Novel SED Algorithm enhances the longevity of the cutting tool by achieving greater accuracy in assessing flank wear on DSS, as evidenced by a comparative analysis against RED Algorithm Additionally, the linear regression obtained for Average Error and MSE is maximized and is statistically significant. By comparing Novel Sobel's Edge Detection Algorithm (90.78%, 0.069 mm^2) with Robert's edge detection algorithm (85.91%, 0.088 mm^2) the linear regression is maximized and, the significance observed is $p = 0.000$ and $p = 0.001$.

Ultimately, this work has certain drawbacks in terms of pixel determination and threshold calibration. The threshold must be set automatically during image processing in the method, which Novel Sobel's Edge detection cannot do. While choosing pixel coordinates is tough as the gradient of the pixel square is too small and the zooming factor is frequently modified. Hence, the study's following concentration could depend on performing edge detection and incorporating an optimized approach to calculate the pixel gradient, which might effectively reduce inaccuracy in DSS Tool wear.

6. Conclusion

The Novel Sobel's Edge Detection Algorithm produces an average error of 90.78% and a MSE of 0.069mm2 and the RED Algorithm produces an average error of 85.91% and a MSE of 0.088mm2. Hence, it proves that Sobel's Edge Detection Algorithm is better than Robert's Edge Detection Algorithm.

References

1. Bhowmick, Sanjit, and Eric Hintsala. 2020. "Extreme In Situ Mechanics of Bond Coatings and Ni-Based Superalloys Using an Advanced SEM Nanomechanical Instrument." Microscopy and Microanalysis. https://doi.org/10.1017/s1431927620022588.
2. Bommi, R. M., Chakaravarthy Ezilarasan, M. P. Sudeshkumar, and T. Vinoth. 2022. "Estimation of Flank Wear in Turning of Nimonic C263 Super Alloy Based on Novel MSER Algorithm and Deep Patten Network." Russian Journal of Nondestructive Testing. https://doi.org/10.1134/s1061830922020073.
3. Bright, Steven. 2020. SPSS: The Ultimate Data Analysis Tool. Lulu Press, Inc.
4. Das, Sudhansu Ranjan, Asutosh Panda, and Debabrata Dhupal. 2017. "Experimental Investigation of Surface Roughness, Flank Wear, Chip Morphology and Cost Estimation during Machining of Hardened AISI 4340 Steel with Coated Carbide Insert." Mechanics of Advanced Materials and Modern Processes. https://doi.org/10.1186/s40759-017-0025-1
5. Gonzalez, Claudia I., Patricia Melin, Juan R. Castro, and Oscar Castillo. 2017. Edge Detection Methods Based on Generalized Type-2 Fuzzy Logic. Springer.
6. Hoier, Philipp, Amir Malakizadi, Pietro Stuppa, Stefan Cedergren, and Uta Klement. 2018. "Microstructural Characteristics of Alloy 718 and Waspaloy and Their Influence on Flank Wear during Turning." Wear. https://doi.org/10.1016/j.wear.2018.01.011.
7. Kuntoğlu, Mustafa, and Hacı Sağlam. 2021. "ANOVA and Fuzzy Rule-Based Evaluation and Estimation of Flank Wear, Temperature and Acoustic Emission in Turning." CIRP Journal of Manufacturing Science and Technology. https://doi.org/10.1016/j.cirpj.2021.07.011.
8. Lilensten, Lola, Stoichko Antonov, Dierk Raabe, Sammy Tin, Baptiste Gault, and Paraskevas Kontis. 2019. "Deformation of Borides in Nickel-Based Superalloys: A Study of Segregation at Dislocations." Microscopy and Microanalysis. https://doi.org/10.1017/s1431927619013424.
9. Niaki, Farbod Akhavan, Durul Ulutan, and Laine Mears. 2015. "In-Process Tool Flank Wear Estimation in Machining Gamma-Prime Strengthened Alloys Using Kalman Filter." Procedia Manufacturing. https://doi.org/10.1016/j.promfg.2015.09.018.
10. Rajmohan, T., K. Palanikumar, and J. Paulo Davim. 2020. Advances in Materials and Manufacturing Engineering: Select Proceedings of ICMME 2019. Springer Nature.
11. Rao, Y. V. D., C. Amarnath, Srinivasa Prakash Regalla, Arshad Javed, and Kundan Kumar Singh. 2021. Advances in Industrial Machines and Mechanisms: Select Proceedings of IPROMM 2020. Springer Nature
12. Sengupta, Sudhriti, Neetu Mittal, and Megha Modi. 2020. "Improved Skin Lesions Detection Using Color Space and Artificial Intelligence Techniques." The Journal of Dermatological Treatment 31 (5): 511–18.
13. Thakre, Avinash A., Aniruddha V. Lad, and Kiran Mala. 2019. "Measurements of Tool Wear Parameters Using Machine Vision System." Modeling and Simulation in Engineering. https://doi.org/10.1155/2019/1876489.
14. Verma, Nishant, S. C. Vettivel, P. S. Rao, and Sunny Zafar. 2019. "Processing, Tool Wear Measurement Using Machine Vision System and Optimization of Machining Parameters of Boron Carbide and Rice Husk Ash Reinforced AA 7075 Hybrid Composite." Materials Research Express. https://doi.org/10.1088/2053-1591/ab2509.
15. Wang, Jin. 2015. Real-Time Design and Implementation of FPGA Based Sobel Edge Detection Video Processing Embedded System: A Thesis in Electrical Engineering.

Note: Every figure and table was created using the experimental work; none of them were taken from any publications or online.

Advances in Additive Manufacturing Technologies – Gurusamy Pathinettampadian (eds)
© 2024 Taylor & Francis Group, London, ISBN 978-1-032-90013-1

55 Investigation of Mean Accuracy of Surface Roughness in Turning of Monel 400 by EWT in Comparison with T-DYWT

Y. Raghavendra Reddy[1]
Department of Electronics and Communication Engineering,
Saveetha School of Engineering, SIMATS,
Chennai, Tamilnadu, India

B Anitha Vijayalakshmi[2]
Department of Electronics and Communication Engineering,
Saveetha School of Engineering, SIMATS,
Chennai, Tamilnadu, India

ABSTRACT: This study aimed to assess the mean accuracy of surface roughness during the turning process of Monel-400 super alloy, employing the Empirical Wavelet Transform (EWT) and comparing it with the Transverse Dyadic Wavelet Transform (T-DYWT) through image processing techniques. Eighteen samples were analyzed, divided into two groups of nine each. Statistical analysis involved a two-tailed T-Test, with parameters set at an enrollment ratio of one, alpha error of 0.05, beta error of 0.2, G-power of 80%, and a confidence interval of 95%. Results indicated that using EWT and T-DYWT, the mean accuracy for predicting surface roughness was 88.54% and 83.97%, respectively. The calculated significant factor was 0.00425 ($p<0.05$), meeting the criteria of alpha, beta, and G-power. Consequently, both groups showed statistical significance. The EWT algorithm demonstrated superior performance over the T-DYWT algorithm in terms of mean accuracy for surface roughness prediction.

KEYWORDS: EWT transform, Novel T-DYWT transform, Monel 400, Mean value, Image processing

1. INTRODUCTION

The main aim is to assess surface roughness through algorithmic mean values during the turning process of Monel 400 Alloy. Monel 400 is prized for its robust strength, toughness across varied temperatures, and excellent resistance to corrosion, making it highly sought after in industries like automotive, biomedical, marine, and aerospace. Researchers have scrutinized four key machining parameter cutting speed, feed rate, depth of cut, and work piece temperature in Monel-400 to understand their impact on surface roughness (Plaza, García Plaza, and NúñezLópez 2018a). Additionally, the constituents of alloys, encompassing attributes such as accuracy, anti-corrosion properties, fitting, friction, and anti-fatigue capabilities, significantly influence their performance and utility. Monel-400, classified as a nickel-based solid solution alloy containing copper, exemplifies these desirable traits, further solidifying its widespread application in various sectors.(Plaza, García Plaza, and NúñezLópez 2018a).

[1]raghavendrareddyy19@saveetha.com, [2]anithavijayalakshmib.sse@saveetha.com

DOI: 10.1201/9781003545774-55

The applications of Monel 400 produce pumps, shafts, valves, missiles, nuclear, petroleum, chemical, aerospace, pharmaceutical, quality of life, and ultrasonic machine tools. Surface texture includes surface roughness, (Badashah and Badashah 2014) which has a significant impact on how an item will interact with its surroundings. A mechanical component's prospective performance can be predicted by how rough it is, as surface imperfections can serve as the starting point for corrosion or fractures. However, the applicability of the findings from a study on Monel 400 to other materials depends on several factors; Different materials exhibit unique properties, such as hardness, machinability, and thermal conductivity. The behavior of a material during machining, including the surface generation process, can vary significantly. Therefore, the effectiveness of wavelet transform methods may be influenced by the specific properties of the material.

Over the past five years, the field has seen a significant influx of research publications, with 34 research publications and 200 research articles released on IEEE Xplore alone. One study focused on evaluating the agreement between surface roughness predictions derived from regression equations and actual test results (Fujii and Asakura, 1974). Differences between simulated and experimental data for surface roughness have been documented, with maximum errors of 13% and 7%, respectively (Lyles, Disrud, and Krauss, 1971). Their findings indicate that surface roughness tends to rise with pulse-on time and decrease with pulse-off time and wire tension (Hameed, Ali, and Hassun, 2019). Additionally, optimization studies have pinpointed specific process parameters, such as a pulse on-time of 10 seconds, pulse off-time of 6 seconds, wire tension of 1000 grams, and wire feed of 5 meters per minute, as optimal for achieving minimal wire surface roughness (Zhang, 1997). Despite the existence of unconventional methods like micro laser machining, concentrated ion beam cutting, and Electro-Discharge Machining (EDM), these optimized parameters remain effective.

While acknowledging the limitations of certain studies, such as decreased functionality and heightened wear and tear, it remains imperative to pursue methods for attaining a smoother surface finish and mitigating these challenges. Strategies encompass polishing techniques, enhancing the alloy's corrosion resistance, and elevating overall quality of life. Employing image processing algorithms offers a robust approach for evaluating and contrasting methodologies to ascertain the optimal accuracy. Hence, opting for the T-DYWT Algorithm over the EWT Algorithm presents an avenue for achieving the desired mean value (in%) to bolster the mean value of Monel 400.

The anticipated mean value is over 83% when processed with the recommended method. The study's aim is to increase the mean value by applying T-DYWT transform analysis to Monel 400, compared to EWT transform analysis, by utilizing commercially produced pictures of the alloy measured with image processing techniques

2. Materials and Methods

The research, led by the Image Processing Laboratory at Saveetha Institute of Medical And Technical Sciences, involved comprehensive experimentation, testing, inquiry, and analysis. Data were meticulously collected from experimental trials focused on the turning process of Monel 400 and processed using clincalc.com, as outlined by Baleanu (2015). The primary objective was to explore the predictive capabilities of the EWT and the innovative T-DYWT algorithm in determining the mean value of Monel 400.The total number of groups taken for this study was two, group 1-EWT algorithm had 9 sample sizes (Kaiser 2010) and Group 2-T-DYWT algorithm had 9 sample sizes (Adebayo, Ağa, and Kartal 2023).The mean values of Monel 400 were analyzed using SPSS Statistical Analysis. The enrollment ratio of one, which represents 0.00425, was considered a critical component at a 95% confidence interval with alpha and beta values of 0.05 and 0.2, respectively, and G-power of 80% (Baleanu 2015; Merabet and Heddam 2023).

SEM photos of the Monel 400 alloy, provided by the alloy-producing industry for diverse applications, underwent meticulous analysis utilizing advanced image processing techniques. The analysis necessitated a system equipped with specific hardware specifications, including an Intel i5 CPU, a 50GB hard drive, 4GB of RAM, and a Windows operating system. The insights into the properties and characteristics of the Monel 400 alloy were derived through the utilization of two software packages: Matlab and SPSS. The practicality of EWT and TDWT in industrial applications depends on factors such as processing time, computational resources, ease of implementation, and the specific requirements of the application. TDWT is generally more efficient and easier to implement, making it suitable for real-time processing and situations with limited computational capabilities.

2.1 Group 1 (T-DYWT Algorithm)

The transverse dyadic wavelet transform (T-DYWT) is a solution to the issues with the discrete wavelet transform (Huang and Xia 2009). The DWT is much more difficult to perform. One of its advantages is to increase the quality of life of Monel 400 alloy. More Memory is required to store the results. Learning DWT theory can indeed be difficult. Interpreting the results is more challenging. The Transverse

Dyadic Wavelet Transform breaks down information using the Haar wavelet(Ozcinar, Demirel, and Anbarjafari 2011). The Haar wavelet uses the reversible feature to dissolve pictures and also which has drawbacks in the quality of life of an Alloy. This feature enables adding additional data to the original image, which may afterward be thoroughly cleaned of any unintended messages.

2.2 Group 2 (EWT Algorithm)

The Empirical Wavelet Transform (EWT) is a mathematical technique for analyzing and representing signals, pictures, quality of life or other data. It is comparable to the conventional Fourier Transform, but rather than using sine and cosine functions as the foundation for signal representation, it utilizes wavelets. EWT starts to break down signals into their individual wavelets, enabling multi-scale analysis and improving the quality-of-life of an Alloy by time-frequency representation.

2.3 Analysis

Cutting speed refers to the speed at which the cutting tool moves across the workpiece material. Feed rate represents the speed at which the cutting tool advances into the workpiece material. Depth of cut refers to the amount of material removed by the cutting tool in a single pass. Tool Geometry, the geometry of the cutting tool, including the tool nose radius, rake angle, and clearance angle, can significantly impact surface finish. Coolant/Lubrication, the use of cutting fluids or lubricants can influence both tool life and surface finish by reducing heat and friction during the machining process.

The statistical analysis of the EWT technique and the Novel T-DYWT was conducted using IBM SPSS Statistics v25 Data Editor, as outlined by Prabhakar, Sreenivasa Kumar, and Gopala Krishna (2020). The objective was to predict the mean value of the Monel 400 alloy, considering factors such as depth of cut, feed rate, and spindle speed as independent variables, and surface roughness as well as the mean value of the respective algorithms as dependent variables. The study aimed to derive key statistical metrics including the mean, standard deviation, standard error mean, and prediction accuracy. Independent Sample T-Tests were employed to compare the performance of both the DWT and SWT algorithms. The analysis sought to provide comprehensive insights into the mean values, standard deviations, and other pertinent metrics essential for evaluating the effectiveness of the algorithms.

EWT decomposes a signal into a set of empirical modes, or intrinsic mode functions (IMFs), which capture different frequency components of the signal. The T-DYWT is a wavelet-based signal processing method that provides multi-resolution analysis. T-DYWT decomposes a signal into various components at different scales and orientations.

Through rigorous statistical analyses comparing the outcomes derived from the EWT and the T-DYWT, several metrics including Mean Squared Error (MSE), Signal-to-Noise Ratio (SNR), and others, as relevant to the context, were employed. The research findings unequivocally affirm that the EWT algorithm surpasses the T-DYWT algorithm in terms of mean value prediction accuracy. The study conclusively demonstrates that the EWT algorithm achieves a substantially higher maximum mean value of 88.5389%, compared to the T-DYWT algorithm's 83.9733%. Therefore, it can be deduced that the proposed EWT methodology significantly outperforms the T-DYWT algorithm.

Table 55.1 shows the performance evaluations of the EWT and T-DYWT transforms, which have accuracy rates of 88.5389 and 83.9733, respectively. Figure 55.1. Depicts the original photograph taken during the experimental Monel 400 Superalloy test, which had a 4.89 surface roughness. Figure 55.2. It represents the approximate coefficient of Monel 400 processed using Transverse dyadic wavelet transform Fig. 55.3 compares the mean value using the Novel T-DYWT method (83.5389) with EWT (88.5389). Calculated the mean accuracy of surface roughness measurements separately for both EWT and DWT. For each method, average the accuracy values obtained across the different surface profiles or datasets used in the study.As the cutting tool wears, the cutting edge becomes rounded, leading to changes in cutting forces and potentially affecting surface finish. Statistical methods like ANOVA can be employed to analyze the impact of turning conditions and tool wear on surface roughness.

The data analysis makes it clear that the special EWT transform technique, which has been recommended, produces more mean value when calculating the surface roughness

Table 55.1 EWT and T-DYWTcomparison

EWT	T-DYWT
86.87	86.54
87.98	85.87
88.76	83.78
87.65	82.65
87.98	83.65
88.97	82.76
88.99	82.67
88.76	84.30
90.89	83.54
88.5389	83.9733

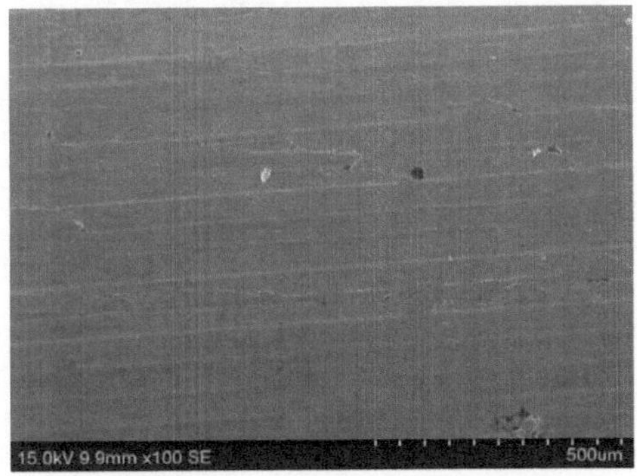

Fig. 55.1 Surface roughness

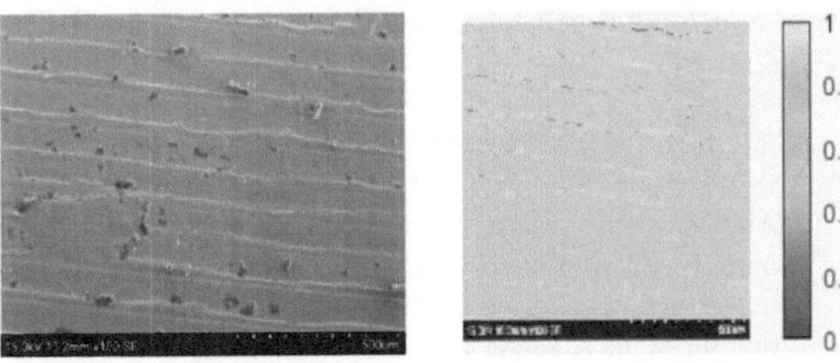

Fig. 55.2 The approximate coefficient of Monel 400 is processed using a Transverse dyadic wavelet transform

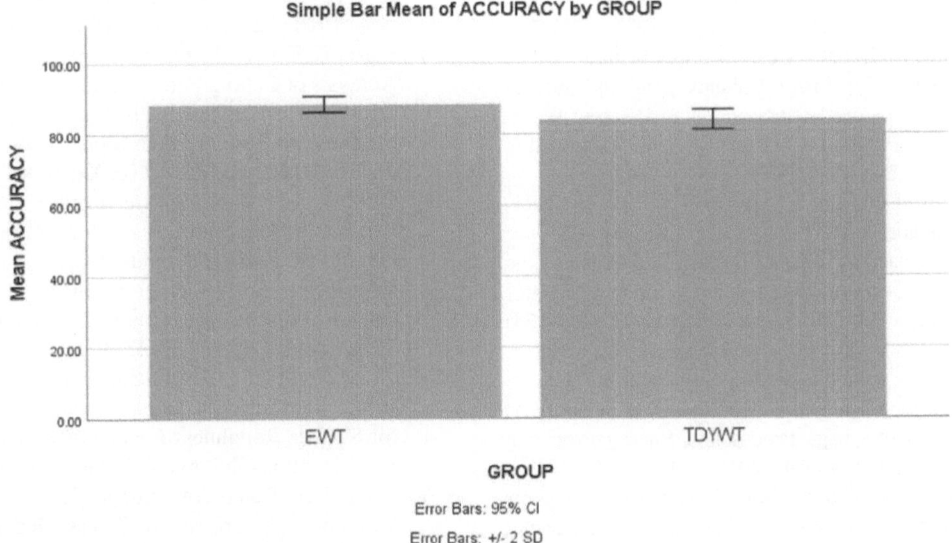

Fig. 55.3 Compare the mean value utilizing the T-DYWT algorithm (83.5389%) and EWT in terms of mean accuracy (88.5389%)

of Monel 400. The Empirical Wavelet Transform algorithm (88.5389%) has more mean value compared with the Transverse dyadic wavelet transform algorithm (83.9733%). When the independent samples test was done, the SPSS software analysis produced a 2-tailed significance factor of 0.00425 which is $p < 0.005$.

A study given provides a better understanding and classification of the analysis for mean value in any algorithm. It offers better social protection based on the study given by (Yang, Lu, and Lei 2013) (83.23%) provides the requirement for Empirical wavelet transform, T-DYWT, and other algorithms comparatively and proved that the Empirical wavelet transform is more efficient than the other methods for predicting intrusion (Wang et al. 2017).

Computational Complexity, EWT, with its iterative and optimization-based approach, can be computationally demanding. This complexity may limit its real-time applicability, especially for large datasets or in industrial settings requiring quick feedback. Future studies should explore the development of adaptive models or guidelines that consider the material-specific aspects, enabling more generalized applicability. Combining EWT with other signal processing or machine learning techniques could lead to more robust and versatile methodologies for surface roughness analysis. Improving the real-time applicability of EWT, perhaps through hardware acceleration or algorithmic simplifications, could enhance its usability in dynamic industrial environments.

3. Conclusion

The conducted research affirms that the EWT surpasses the T-DYWT algorithm in terms of mean value accuracy. Specifically, the study unveils that the EWT algorithm achieves a notably higher maximum mean accuracy value of 88.5389%, compared to the T-DYWT algorithm's 83.9733%. Consequently, it can be inferred that the proposed EWT approach significantly outperforms the T-DYWT algorithm.

REFERENCES

1. Abu-Mahfouz, Issam, Amit Banerjee, and Esfakur Rahman, Evaluation of Clustering Techniques to Predict Surface Roughness during Turning of Stainless-Steel Using Vibration Signals, Materials 14 (17). https://doi.org/10.3390/ma14175050.2021.
2. Adebayo, Tomiwa Sunday, Mehmet Ağa, and Mustafa Tevfik Kartal, Analyzing the Co-Movement between CO Emissions and Disaggregated Nonrenewable and Renewable Energy Consumption in BRICS: Evidence through the Lens of Wavelet Coherence, Environmental Science and Pollution Research International 30 (13): 38921–38. 2023.
3. Badashah, and Badashah, Design Implementation And Hardware Structure For Image Enhancement And Surface Roughness With Feature Extraction Using Discrete Wavelet Transform, Journal of Computer Science. https://doi.org/10.3844/jcssp.2014.347.352.2014
4. Baleanu, Dumitru, Wavelet Transform and Some of Its Real-World Applications. BoD – Books on Demand.2015
5. Fujii, H., and T. Asakura, Effect of Surface Roughness on the Statistical Distribution of Image Speckle Intensity, Optics Communications. https://doi.org/10.1016/0030-4018(74)90327-7.1974
6. Hameed, Noor Ali, Inbethaq M. Ali, and Hanan K. Hassun, Calculating Surface Roughness for a Large Scale SEM Images by Mean of Image Processing, Energy Procedia. https://doi.org/10.1016/j.egypro.2018.11.167.2019
7. Huang, Zhenghong, and Li Xia, Image Denoising for Adaptive Threshold Function Based on the Dyadic Wavelet Transform, International Conference on Electronic Computer Technology. https://doi.org/10.1109/icect.2009.64.2009
8. Kaiser, Gerald, A Friendly Guide to Wavelets. Springer Science & Business Media.2010
9. Lyles, Leon, Lowell A. Disrud, and And R. K. Krauss, Turbulence Intensity as Influenced by Surface Roughness and Mean Velocity in a Wind-Tunnel Boundary Layer, Transactions of the ASAE. https://doi.org/10.13031/2013.38277.1971
10. Merabet, Khaled, and Salim Heddam, Improving the Accuracy of Air Relative Humidity Prediction Using Hybrid Machine Learning Based on Empirical Mode Decomposition: A Comparative Study, Environmental Science and Pollution Research International, April. https://doi.org/10.1007/s11356-023-26779-8.2023
11. Ozcinar, Cagri, Hasan Demirel, and Gholamreza Anbarjafari, Image Equalization Using Singular Value Decomposition and Discrete Wavelet Transform, Discrete Wavelet Transforms - Theory and Applications. https://doi.org/10.5772/15448.2011.
12. Parida, Asit Kumar, and Kalipada Maity, Modeling of Machining Parameters Affecting Flank Wear and Surface Roughness in Hot Turning of Monel-400 Using Response Surface Methodology (RSM), Measurement. https://doi.org/10.1016/j.measurement.2019.01.070.2019.
13. Plaza, E. García, E. García Plaza, and P. J. Núñez López, Analysis of Cutting Force Signals by Wavelet Packet Transform for Surface Roughness Monitoring in CNC Turning." Mechanical Systems and Signal Processing, https://doi.org/10.1016/j.ymssp.2017.05.006.2018
14. Prabhakar, D. V. N., M. Sreenivasa Kumar, and A. Gopala Krishna, A Novel Hybrid Transform Approach with Integration of Fast Fourier, Discrete Wavelet and Discrete Shearlet Transforms for Prediction of Surface Roughness on Machined Surfaces, Measurement. https://doi.org/10.1016/j.measurement.2020.108011.2020
15. Wang, Xiao, Tielin Shi, Guanglan Liao, Yichun Zhang, Yuan Hong, and Kepeng Chen, Using Wavelet Packet Transform for Surface Roughness Evaluation and Texture Extraction, Sensors. https://doi.org/10.3390/s17040933.2017
16. Yang, Lei, Rongsheng Lu, and Liqiao Lei, Information Extraction of Surface Roughness Measurement Based on Wavelet Transform of Speckle Pattern Texture, Journal Of Electronic Measurement and Instrument. https://doi.org/10.3724/sp.j.1187.2012.01091.2013
17. Zhang, Zhige, Reconstructing a Shadow Pattern Image of the Internal Surface of a Small Hole and Estimating Surface Roughness from the Image, Optical Engineering. https://doi.org/10.1117/1.601442.1997

Note: Every figure and table was created using the experimental work; none of them were taken from any publications or online.

Advances in Additive Manufacturing Technologies – Gurusamy Pathinettampadian (eds)
© 2024 Taylor & Francis Group, London, ISBN 978-1-032-90013-1

56 Comparative Investigation of Surface Roughness Reduction in CNC Milling of Novel Hybrid Al 7075-Mg-SiC-FA and Al 6061-SiC-Mg Metal Matrix Composites

K. Bharath,
A. Bovas Herbert Bejaxhin*
Department of Mechanical Engineering,
Saveetha School of Engineering, Saveetha Institute of Medical and
Technical Sciences, Saveetha University,
Chennai, Tamil Nadu, India

ABSTRACT: The prime purpose of this study is to compare the effects of material contributions and machining conditions on the surface roughness of the milling slot in the hybrid Al7075/Al6061 metal matrix composite reinforced with 15% SiC, Mg with various iterations.In this study, 10 mm thick aluminum plate was used, and varied machining parameters were used for milled slots on the specimen material. Al6061 HMMC is used as the control group in group 1, and Al7075 HMMC is used as the experimental group in group 2. Matrix composite material is employed in both groups at 80% G-Power. The roughness value of Al6061 hybrid composites is 0.799 μm to measure the surface roughness of the machined slot and for Al7075 hybrid metal matrix composite material is recorded as 0.678 μm for the specific spindle speed and varied feed rates are needed to know the qualities of the material with a sample t test significance value of $p=0.000$ (both 2-tailed) is statistical significant and it was attained which is less than ($p<0.05$) indicating that there is a statistical significant between the two groups. The control and experimental groups' roughness will be examined for a range of spindle speeds, feed rates, and flatness tolerance accuracy. Within the constraints of this investigation, further information about the improved reduced roughness achieved in the properties of Al7075 material is to be learned. Using SPSS software, the data was statistically examined, graphs were made, and the significance level was determined.

KEYWORDS: Novel aluminium composite, Spindle speed, Feed rate, Surface roughness, Depth of cut, SPSS software, Sustainable production

1. INTRODUCTION

Using a multi-point cutting tool in a milling machine, mechanical behavior is utilized to determine the influence of feed and spindle speed on the surface roughness of the milled slot in innovative Al7075-SiC-Mg-FA and Al6061 composites. The spindle speed rose and surface roughness intensified due to the high feed rate. Comparing the results with as-cast Al6061 composites and Silicon Carbide, as well as comparing the control and experimental groups (Muniappan et al. 2021). Because of its great strength and hardness, Al7075 is widely used in the automotive and aviation sectors. There are several ways to produce aluminium composites in a sustainable production. This method of

*Corresponding author: bovasherbertbejaxhina.sse@saveetha.com

DOI: 10.1201/9781003545774-56

machining/ Material can be utilized by most of the innovative level industries. Other uses for Al7075 include bearings, aircraft engine cowlings, landing gear doors, automotive pistons, and so on, all of which require high accuracy and surface finish (Okokpujie and Tartibu 2022). Because of its excellent mechanical properties and hard reinforcement capability, SiC is used as a reinforced element in aluminium grids. Surface roughness is measured at different spindle speeds and feed rates by taking a reading of the surface roughness for a single slot and calculating the mean value for a 16-slot test (Clauß, Nestler, and Schubert 2016).

Based on literature review audits, aluminum 7075, with 2.810 g/cm3, was chosen for study. This material was chosen based on its qualities as stated in study. Mohan et al. (2022) milled slots in a hybrid Al6061 composite reinforced with SiC and magnesium. The researchers examined how milling settings affected composite characteristics using sand casting. In slot milling studies, cutting speed, feed per tooth, and milling kinematics (up and down milling) were examined. Machined surfaces were evaluated for roughness, defects, and surface layer residual stress. Bhardwaj, Vaidya, and Shekhawat (2020) tested a particle-reinforced new aluminum composite. Their research helped explain how cutting parameters and kinematics affect slot milling surface qualities. Surface structure, including roughness and flaws, and surface layer residual stress were assessed. Tamang et al. (2022) used silicon carbide reinforcement to improve machinability in an aluminum composite (Al6061-SiC-Mg). Silicon carbide improved composite surface roughness. High-Speed Steel cutting tools also increased the innovative aluminum composite's machinability. In conclusion, milling procedures, material choices, and surface qualities in innovative aluminum composites are being studied, with a focus on silicon carbide reinforcement.

Many studies have been conducted in this field, but only a few have been successful in identifying a composite with the desired properties of various novel spindle speeds and feed rates. The surface roughness of the material at different temperatures where the cutting process occurs has also been measured in this area of research and used in specific applications such as the aerospace industry, the automotive industry, and the chemical industry (Kannan et al. 2022).

2. Materials and Methods

This work does not require ethical approval because of not working with human samples. The composite material, aluminium 6061 and aluminium 7075, was purchased in Ambattur, and it was required in square shape for milling purposes for research purposes, so it was converted from the solid state to the liquid state using a crucible furnace at 750 °C. For the experimental investigation, two groups were chosen. Group 1 consists of aluminium 6061-SiC-Mg composites, while Group 2 aluminium 7075 consists of SiC,FA and Mg-reinforced composites (Bovas Herbert Bejaxhin et al. 2021). A sample size of 32 (16 for each group) was calculated using g-power, and pre-test power is 80% of the mean; standard deviations for group 1 and group 2 are 0.314 and 0.285, respectively. The assumed G-power is 80%(Ramasamy, Daniel, and Nithya 2021).

In the first group, the aluminium 6061 composite was casted and moulded into a square plate dimensioned as 150×150×10 mm of size with reinforcement of 36 grams silicon carbide and 7 grams magnesium balances, and a channel was carved in CNC milling with a 10mm HSS cutter. In the second group, an aluminium 7075 composite was cast and shaped into a square plate dimensioned with 150x150x10 mm of size with reinforcements of 36 grams of silicon carbide, 7 grams of fly ash, and 7 grams of magnesium-balanced aluminium 7075, and a slot was CNC milled with a 10 mm HSS cutter (Natarajan, Chinnasamy, and Alphonse 2022).

Milling operations were conducted on a Vertical Machining Center (VMC) machine as shown in Fig. 56.1. A total of 16 samples were produced for the control group, consisting of Al6061-SiC-Mg composites, and another 16 samples for the experimental group, involving reinforced Al7075-SiC-Mg-FA composites. The assessment of surface roughness was carried out using a Mitutoyo SJ-410 Series Model equipped with SJ-411 Portable Surface Roughness Tester and SJ-412 apparatus as shown in Fig. 56.2 and 56.3, ensuring precision with surface roughness measurements below 1 micro inch without any vibrations. The materials utilized, namely Al6061 composites and metal matrix composites, were procured from a reputable material shop. This study categorized two groups: the first being the parent group, represented by Al6061 composites, and the second being the experimental group, characterized by Al7075 composites.

Fig. 56.1 Vertical milling machine for machining a slot in the aluminium composites

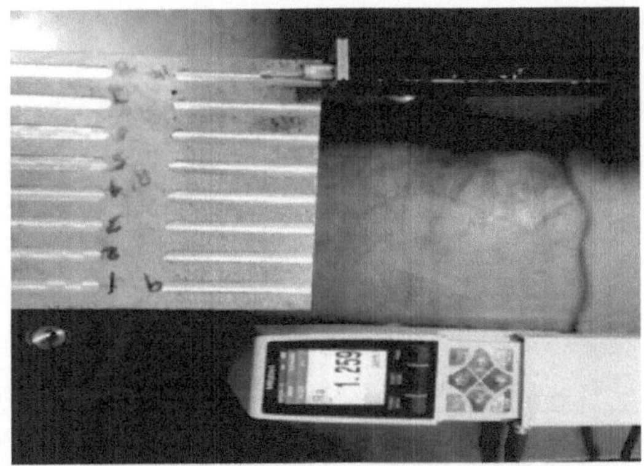

Fig. 56.2 Surface roughness test conducted with Mitutoyo - SJ210 testing equipment in machined slots

Fig. 56.3 Surface roughness testing arrangement—Mitutoyo SJ-410 Series Model

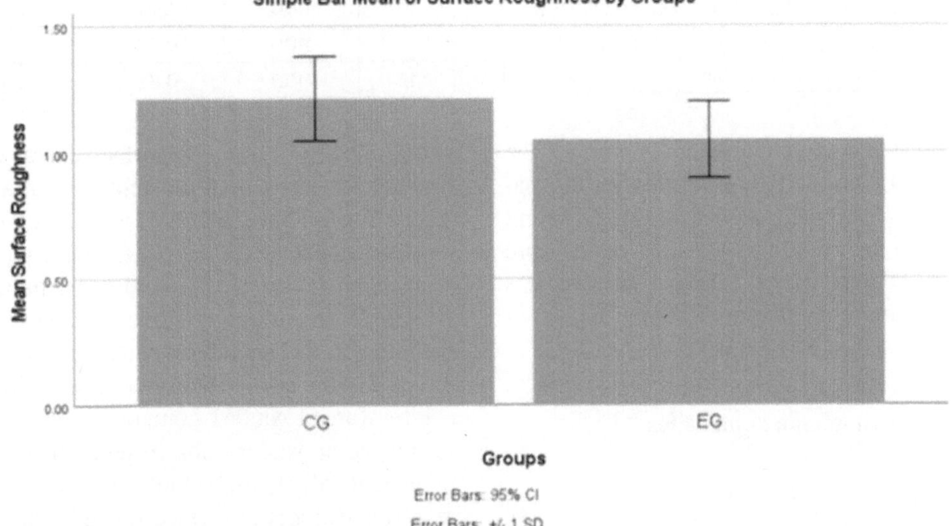

Fig. 56.4 SPSS statistical outcomes of surface roughness. The mean surface finish dominated both the control and experimental groups, as shown by the bar graph. Even with variations in cutting speed, feed, and depth of cut, the uncoated high-speed steel cutter reduces surface roughness from these cutters. X-axis: Al6061 and Al7075 composites; Y-axis: mean Ra of detection ±1 SD

Experimental data, including readings and graphical representations, were obtained from the surface roughness tests. Ajith Arul Daniel et al. (2021) employed statistical analysis, specifically analysis of variance (ANOVA), to optimize the collected data. The statistical software SPSS was utilized for this purpose. Through this approach, the researchers were able to derive meaningful insights into the impact of machining parameters on surface roughness for the two distinct composite groups.

3. Results

CNC machining experiments were carried out on both experimental and control groups, utilizing independent input parameters and their corresponding values outlined in Table 56.7. The determined surface roughness of slot-milled components served as the foundation for the analysis. Figure 56.1 illustrates the surface roughness of the reinforced composite, while Fig. 56.4 presents the significance of the comparison graph, determined using the SPSS software tool. The comparison yielded a noteworthy p-value of 0.000 (2-tailed), indicating statistical significance between the two groups, as detailed in Table 56.7.

Tables 56.1 and 56.2 provide information on the chemical composition and physical properties, respectively, of

Table 56.1 Chemical composition of AA7075 and AA6061 metal matrix composite

Wt.%	Zn%	Mg%	Cu	Mn	Fe	Si	Al
AA7075	5.6	2.5	1.5	0.04	0.3	0.08	BAL
AA6061	0.2	0.9	0.3	0.05	0.6	0.7	BAL

Table 56.2 Physical properties of AA7075 and AA6061

Sl. No	Properties	AA7075	AA6061
1.	Ultimate Tensile strength (Mpa)	572 Mpa	310 Mpa
2.	Yield Stress (Mpa)	503 Mpa	276 Mpa
3.	Elongation (%)	12	12-17
4.	Density (g/cm^3)	2.81	2.7
5.	Hardness	150 Brinell	95 Brinell

Table 56.3 Process parameters of CNC Milling

Process Parameters	Particulars
Materials used:	AA7075, AA6061, Silicon Carbide (SiC) Magnesium (Mg) and Fly Ash
Input parameters:	Speed - 700, 900, 1100, 1300 rpm Feed Rate - 0.10, 0.15, 0.20 and 0.25 mm/rev
Output parameters:	Depth of Cut - 0.25, 0.5, 0.75, 1 mm Surface Roughness (μm)

Table 56.4 Experimental observations of surface finish of AA6061 MMC

Sl. No.	Speed (rpm)	Feed Rate (mm/rev)	Depth of Cut (mm)	Surface Roughness (μm)
1	700	0.10	0.25	0.799
2	700	0.15	0.50	1.421
3	700	0.20	0.75	1.502
4	700	0.25	1.00	1.359
5	900	0.10	0.50	1.008
6	900	0.15	0.25	1.024
7	900	0.20	1.00	2.163
8	900	0.25	0.75	1.058
9	1100	0.10	0.75	1.043
10	1100	0.15	1.00	1.101
11	1100	0.20	0.25	1.029
12	1100	0.25	0.50	1.334

AA7075 and AA6061 materials. Table 56.3 compiles the slot machining operations, including various setting parameters on the Vertical milling machine and the different materials employed in the study. The obtained values from Table 56.6 underwent independent group T-test analysis using IBM's SPSS statistical software. Figure 56.3 showcases the Mitutoyo SJ-410 Series Model.

Figure 56.2 provides a visual representation of the surface roughness of the novel aluminum composites.

4. Discussion

This study conducted comparisons between two types of composites, revealing that the 6%-reinforced Al6061-SiC-Mg composites exhibit the lowest surface roughness. The minimum surface roughness for the reinforced composites was recorded at 0.799 μm, achieved at a speed of 700 RPM, a depth of cut of 0.25 mm, and a feed rate of 0.10 mm/min, as indicated in Table 56.3. The set maximum for surface finish was 2.163 μm. In contrast, the Al7075 composites displayed the least surface roughness at 0.678 μm, occurring at a speed of 900 RPM, a depth of cut of 0.25 mm, and a feed rate of 0.15 mm/min. The highest recorded surface roughness for Al7075 composites was 1.555 μm, as detailed in Table 56.4. Notably, the test results highlight that the 6%-reinforced aluminum 7075 metal matrix composites outperform Al6061 composites, exhibiting superior surface finish characteristics (Kamath et al. 2022). In contrast to Al7075 composites, the utilization of reinforced composites significantly enhanced the surface roughness, with a obtained significance value of p=0.000 (2-tailed) in Table 56.7, indicating statistical significance between the two groups ($p<0.05$). The surface characteristics are notably influenced by both the depth of cut and the feed rate. According to Kim et al. (2022), an increase in both the feed rate and depth of cut correlates with an improvement in surface finish. The study's findings demonstrate a 32% improvement in surface roughness, as depicted in the Tables 56.4 and 56.5. This improvement is particularly notable in the case of the newly reinforced composite, specifically the 6% reinforced Al6061 composites, emphasizing the need for this composite to achieve superior surface finish with a reduced depth of cut. Notably, there is a lack of contrasting research analyses in surface-finish findings, as highlighted by Yu et al. (2021). It's important to note that this study exclusively focused on 6%-reinforced Al6061 composites. Future research endeavors could expand to explore different volume fractions, such as 5%, 15%, 25%, and 30%, to further enhance the material properties of composites. This novel reinforced composite, featuring a higher surface finish and a reduced cut depth (6% reinforced Al6061-SiC-Mg composites), stands out in this study. The comparison between the two composite types reveals that the newly developed Al6061 composites with 6% reinforcement exhibit the least surface roughness, as indicated by Xiong et al. (2021).

5. Conclusion

According to the findings of the study, the Ra value for Al7075 composite is 0.678 μm, while the Al6061 metal matrix composite material exhibits a slightly higher Ra value of 0.799 μm, as presented in Tables 56.3 and 56.4. The comparison between the control and experimental groups, conducted using SPSS software, demonstrated that the experimental group possesses a smoother surface finish compared to the parent group. This improvement in surface finish is attributed to the effective mixing of the composite metal matrix achieved through the sand casting process. Consequently, the Al7075 hybrid metal matrix composite (HMMC) within the experimental group exhibited a notable 15% improvement in roughness while maintaining a lighter weight. This advancement underscores the growing utilization of matrix composite materials across various industries. The successful integration of the sand casting process has not only enhanced the surface finish but has also contributed to the increased efficacy and reduced weight of the Al7075 HMMC in the experimental group.

REFERENCES

1. Anbuchezhiyan, G., Mubarak, N. M., Karri, R. R., & Khalid, M. (2022). A synergistic effect on enriching the Mg–Al–Zn alloy-based hybrid composite properties. Scientific Reports, 2021 (November), 3844194. https://doi.org/10.1155/2021/3844194
2. Chen, J.-P., Gu, L., & He, G.-J. (2020). A review on conventional and nonconventional machining of SiC particle-reinforced aluminium matrix composites. Advances in Manufacturing, 8(3), 279–315.
3. Ghoreishi, R., Roohi, A. H., & Ghadikolaei, A. D. (2019). Evaluation of tool wear in high-speed face milling of Al/SiC metal matrix composites. Journal of the Brazilian Society of Mechanical Sciences and Engineering. https://doi.org/10.1007/s40430-019-1649-3
4. Jin, P., Gao, Q., Wang, Q., & Li, W. (2021). Micro-milling mechanism and surface roughness of high volume fraction SiCp/Al composites. International Journal of Advanced Manufacturing Technology, 115(1), 91–104.
5. Kamineni, J. N., Anbuchezhiyan, G., Anichai, J., Sharma, R., Ganesan, R., Latha, A., & Goud, B. N. (2023). Investigation of the microstructure and mechanical properties of borosilicate reinforced magnesium nanocomposites. Materials Today: Proceedings.
6. Kannan, A., Mohan, R., Viswanathan, R., & Sivashankar, N. (2020). Experimental investigation on surface roughness, tool wear and cutting force in turning of hybrid (Al7075 + SiC + Gr) metal matrix composites. Journal of Japan Research Institute for Advanced Copper-Base Materials and Technologies, 9(6), 16529–16540.
7. Madhusudan, B. M., Raju, H. P., Ghanaraja, S., & Sudhakar, G. N. (2021). Study of microstructure and mechanical properties of ball milled nano-SiC reinforced aluminium matrix composites. Journal of The Institution of Engineers (India): Series D, 102(1), 167–172.
8. Manohar, G., Pandey, K. M., & Maity, S. R. (2020). Aluminium (AA7075) metal matrix composite reinforced with B4C nanoparticles and effect of individual alloying elements in Al fabricated by powder metallurgy techniques. Journal of Physics: Conference Series, 1451(January), 012024.
9. Mani, M., Thiyagu, M., & Krishnan, P. K. (2023a). Experimental investigation of Kevlar/carbon/glass/polyurethane foam epoxy hybrid sandwich composites with nano silicon particles in low-velocity impact events. Materials Today: Proceedings.
10. Mani, M., Thiyagu, M., & Krishnan, P. K. (2023b). Flexural and compression behavior analysis of hybrid sandwich composites with nano silicon particles in low-velocity impact analysis. Materials Today: Proceedings.
11. Mohan, E., Anbuchezhiyan, G., Pugazhenthi, R., & Prakash, F. P. (2023). Wear behavior of brass based composite reinforced with SiC and produced by stir casting process. Materials Research, 26, e20220196.
12. Natarajan, M. M., Chinnasamy, B., & Alphonse, B. H. B. (2022). Deform 3D simulation and experimental investigation of fixtures with support heads. Mechanics, 28(2), 130–138.
13. Okay, F., Islak, S., & Turgut, Y. (2021). Investigation of machinability properties of aluminium matrix hybrid composites. Journal of Manufacturing Processes, 68(August), 85–94.
14. Rajesh, G., Thiyagaraj, J., Narayana, K. J., Anbuchezhiyan, G., Saravanan, R., Pugazhenthi, R., & Gupta, M. S. (2023). Microstructure and mechanical properties of AZ91D/Si3N4 composites using squeeze casting method. Materials Today: Proceedings.
15. Rao, V., Periyaswamy, P., Alphonse, A. B. H. B., Naveen, E., Ramanan, N., & Teklemariam, A. (2022). Wear behavioral study of hexagonal boron nitride and cubic boron nitride-reinforced aluminum MMC with sample analysis. Journal of Nanomaterials, 2022(May). https://doi.org/10.1155/2022/7816372

Note: Every figure was created using the experimental work; none of them were taken from any publications or online.

Advances in Additive Manufacturing Technologies – Gurusamy Pathinettampadian (eds)
© 2024 Taylor & Francis Group, London, ISBN 978-1-032-90013-1

Design, Analysis and Experimentation of Brake Discs with Copper—Aluminium Alloy and its Different Composition

Yogesh Vaidhyanathan*,
Vishal T, Sai Srinivasan B, Saran R
Department of Mechanical Engineering,
Vel Tech High Tech Dr. Rangarajan Dr. Sakunthala Engineering College,
Avadi, Chennai, Tamilnadu, India

ABSTRACT: The disc brake component, which is what slows down or to stop the moving vehicle, is incomplete without disc brake rotor. Brake disc's temperature increases when the brakes are applied, and the pressures resulted from this temperature rise are subsequently applied to the disc portion. The design must be carefully considered, and the right materials must be utilized, to keep the disc brake from breaking prematurely. An effort is made to assess the thermal distribution of the brake disc using both analytical and finite element techniques. This is done in order to better understand how the disc should be designed. The braking system is the most vital component in the automotive manufacturing industry. Its primary function is to assist in regulating and sustaining the vehicle's current speed. As a result, it is essential to locate the most appropriate material, one that can keep the heat generation going while also withstanding the other mechanical pressure. The disc brake has been subjected to static structural and thermal study using the finite element simulation tool ANSYS. The purpose of the work is to examine maximum heat transfer for various copper-aluminum alloy material compositions (90:10, 85:15, and 95:5). The results of this research may indicate a comparison between the materials used to make brake discs, specifically Structural Steel and various compositions of copper and aluminum alloy. The results will be evaluated based on the heat flow, total deformation, stress, and strain, as well as thermal character of the materials.

KEYWORDS: Disc brake, Copper, Aluminum, Deformation, Thermal, ANSYS, Stress, Strain

1. INTRODUCTION

The primary obstacle that the automobile industry must overcome is the emission of greenhouse gases produced by vehicles. When compared to heavier vehicles, lighter ones have lower overall fuel usage. A decrease in the amount of fuel used results in a lower number of emissions of greenhouse gases. The automotive industry makes a concerted effort to reduce the overall weight of automobiles by constructing various components out of lighter-weight materials. One strategy that is utilised to accomplish this goal is the use of less weight materials. In the automobile manufacturing industries, the utilization of aluminium, which is a well-known material that is low in weight, rose. A decrease in weight leads to an improvement in fuel economy as well as a reduction in pollutants. The strength-to-weight

*Corresponding author: vyogeshmech@gmail.com, v.yogesh@velhightech.com

DOI: 10.1201/9781003545774-57

ratio of composite materials is quite satisfactory. Disc rotors for disc brake systems can be made out of metal matrix composites with ceramic particles that have an aluminium alloy as their foundation. Conducting experiments to measure the quantity of heat transfer via conduction from frictional interface into the rotor 2000, by Yano and Murata, reinforcement with composite materials has demonstrated a great deal of potential for such applications. When compared to the more traditional material, grey cast iron, composite composites based on aluminium have been shown to have a lower density and higher heat conductivity. If the aluminium composites are utilized in the braking system, a reduction in weight of about fifty to sixty percent is feasible. It has been observed that these composite materials perform better in harsh service conditions, which are more frequently seen in modern cars and also to include faster speeds, larger loads, etc.. This criterion is extremely significant in single stops and high speeds, but it is not nearly as important in repeated stops from those speeds. When it comes to the manufacture of disc brake rotors, grey cast iron is typically the material of choice. Noyes and Vickers "the brake disc's rubbing surface temperature response under the assumption of a uniform heat flux", When producing grey cast iron to use in various applications, manufacturers are required to work in accordance with the SAE specifications. The rate of wear is typically quite significant, and the braking performance may be subpar or grabby before the brake heats up. The total safety of the car is closely related to the temperature, thermal stress, and thermal deformation of brakes. Heat is generated whenever the friction pads are compressed in order to apply brake onto the disc rotor. This heat flux needs to be channeled and spreaded among the disc in order to keep the brakes from overheating. The state of the brakes is extremely severe, and as a result, a thermal examination needs to be performed.

2. Experimental Methodology and Materials

Sample Details:

Sample 1: 95 % Al + 5% Cu

Sample 2: 90 % Al + 10% Cu

Sample 3: 85 % Al + 15% Cu

Based on the recommendation from the obtained numerical investigation results and its comparison, the composition -1 that is 95% of Aluminium5% Copper was the best and suitable composition for brake disc. In this chapter the casting and testing of above said composition was discussed detailed manner. Sand is used as the mold material in the metal casting process, which is referred to as sand molded casting. An object made by the sand-casting process can also be said to as sand casting. Sand casting is the method used to make more than 70% of all metal castings. Foundries are specialized enterprises where sand castings are to be made. Even for usage in steel foundries, sand casting is a reasonably priced and sufficiently refractory material. An appropriate bonding agent eg., clay is added to sand or mixed in with it. The mixture is then wetted with water but occasionally with the other materials, to give the clay strength and flexibility and to prepare the aggregate for molding. Known as a flask, the sand is usually kept confined in a set of frames or mold boxes. Sand is compacted around models or patterns or cut straight into the sand to produce the mold cavities and gate system.

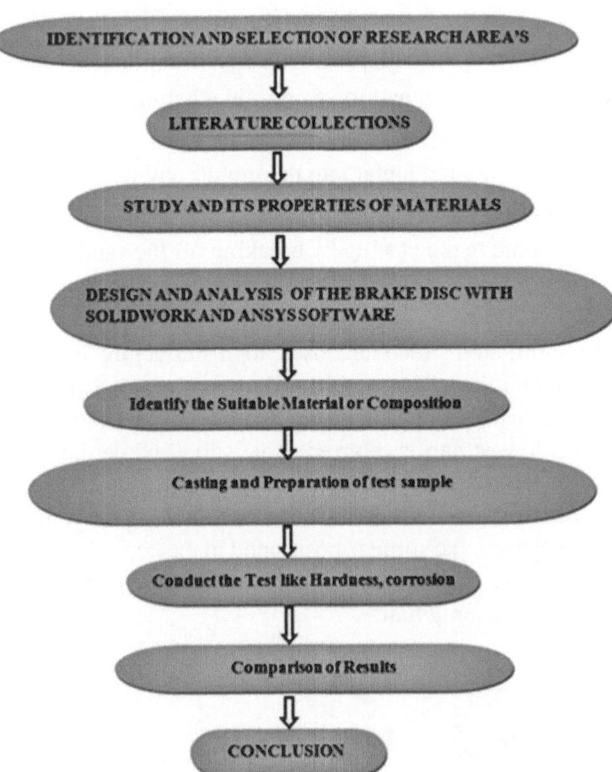

Fig. 57.1 Process flow

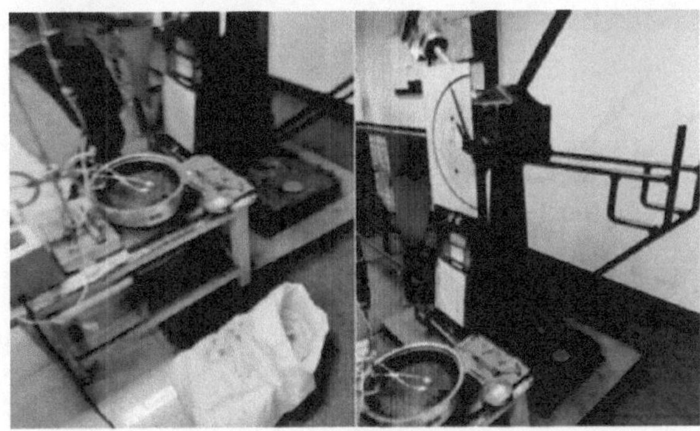

Fig. 57.2 Salt spray test

There are six steps in the process:

1. To make a mold, place a pattern in the sand.
2. Include sand and the pattern in a gating mechanism.
3. Take the pattern out.
4. Pour molten metal into the mold's cavity.
5. Permit the metal to get cold.
6. Remove the casting by breaking off the sand mold.

2.1 Molding

The multi-step procedure used to manufacture molds is called molding. The mold used in horizontal casting is housed inside a flask, which is a two-piece frame. The flask's lowest part is referred to as a drag, while the upper part is termed a cope. First, the pattern is surrounded by molding sand that is put into a flask. Following the removal of the pattern, the sprue is positioned in the cope area of the mold cavity, and the gating and runner arrangements are placed in the drag half.

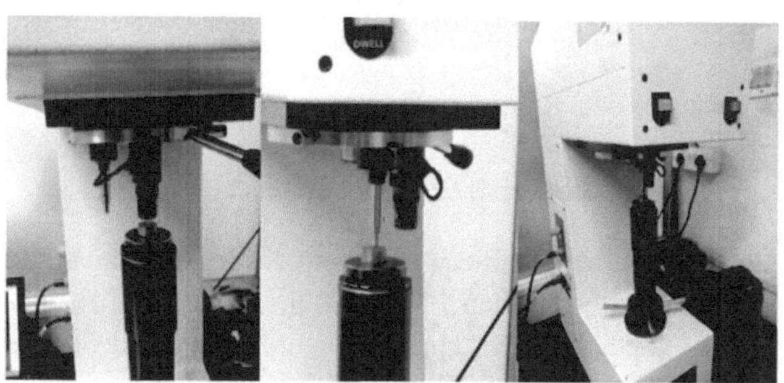

Fig. 57.3 Hardness test

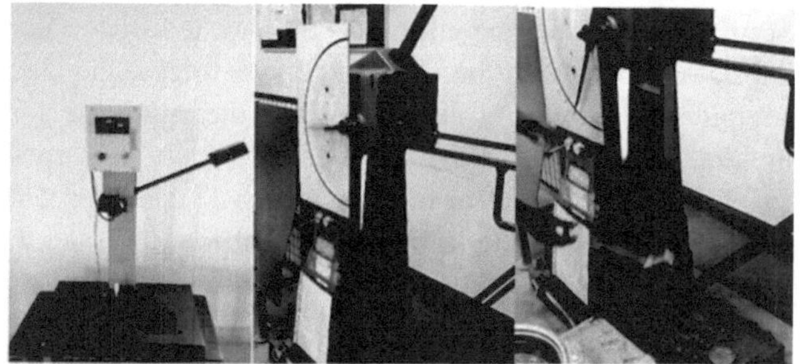

Fig. 57.4 Digital and manual impact test

Cleaning: In general, cleaning means getting rid of everything which doesn't the part of final casting. Removing the gating mechanisms from the casting is known as rough cleaning. Any mold or core sand that remains on the item after it is released from the mold is removed during the initial finishing process. Trimming gets rid of any extra metal. During the final steps of finishing, the casting's surface is cleaned to enhance its quality. At this stage, the casting is also examined for flaws and compliance with quality standards. To ascertain if the part will function satisfactorily for its intended use, nondestructive testing may be a part of this inspection.

2.2 Impact Test

A standardized highs train-rate test called the Charpy impact testis used to determine how much energy a material absorbs during fracture. The energy is then serves as a gauge for the notch toughness of a particular material and can be utilized to investigate the temperature-dependent ductile-brittle transition. Since it is simple to conduct and prepare, and yields fast, affordable findings, it is commonly used in industry. Some outcomes are only comparable, which is a drawback. By comparing the absorption level before and after the hitting process, it is possible to deduce the energy transmitted to the material.

2.3 Salt Spray Test

A systematic test procedure for determining a coated sample's resistance to corrosion is the salt spray test. Steel, zinc, and brass metallic parts are protected from corrosion by coatings. It is vital to verify corrosion resistance using alternative methods because coatings might offer a high level of resistance to corrosion for the duration of the part's intended life. The length of the test is determined by the coating's resistance to corrosion; more resistant the coating then longer the test will last until corrosion symptoms appear. An example of an effective corrosion test is the salt spray test, which corrodes coated samples to see if they may be used as a protective finish. However, the industrial sector frequently uses the salt spray test to assess the corrosion resistance of finished surfaces or parts. The popularity of salt spray testing can be attributed to its low cost, speed, standardization, and reasonable repeatability.

3. Testing Equipments

3.1 Salt Spray Cabinet

A testing chamber (Closed) with five percentage sodium chloride solution atomized by means of a nozzle creates a corrosive environment, subjecting parts to severely corrosive conditions. The apparatus for testing is called a salt spray cabinet, and its typical volume is 15 cubic feet, as this is the smallest volume accepted by International Standards on Salt Spray Tests. The majority of machines are in the 20–170 cubic foot range. Neutral salt spray, or NSS, tests are conducted using a standardized 5 percent solution of NaCl. In general, salt spray testing is not performed on hot-dip galvanized surfaces (look at to ISO 10684 or ISO 1461). This approach can be used to test painted surfaces with an underlying hot-dip galvanized coating. Refer to

ISO 12944-6. Testing times can be as short as 8 or 24 hours for phosphated steel or as long as more than a month for zinc-nickel coatings (720 hours) and some zinc flake coatings (1000 hours).

4. Corrossion—Results

4.1 Salt Spray Test

Temperature of the chamber: 35 to 36

pH Range: 6.75 to 6.95

Concentration of solution: 4.70–5.40% of NACL

Volume (salt solution collected): 1.00-1.50 ml/hr

Pressure of air: 15-18 Psi

Component loading at a 35-degree angle in the chamber

4.2 Brinell Hardness

Impact Strength

An aluminium matrix with copper particles or clusters scattered throughout is typically what makes up the microstructure of an aluminum-copper alloy. Copper particles are often quite a bit smaller than the grains that make up the aluminium matrix, and they are dispersed in an unpredictable manner. In the microstructure of aluminum-copper alloys, the aluminium matrix is typically comprised of big grains that are evenly distributed over their surfaces. These grains tend to become elongated in the direction of solidification as the casting process progresses, and the rate at which they are cooled during the casting process can have an impact on both their size and shape. In the microstructure of aluminum-copper alloys, in addition to copper particles, the presence of other phases may also be possible depending on the precise composition of the alloy as well as the processing that the alloy undergoes.

Fig. 57.5 Isometric view

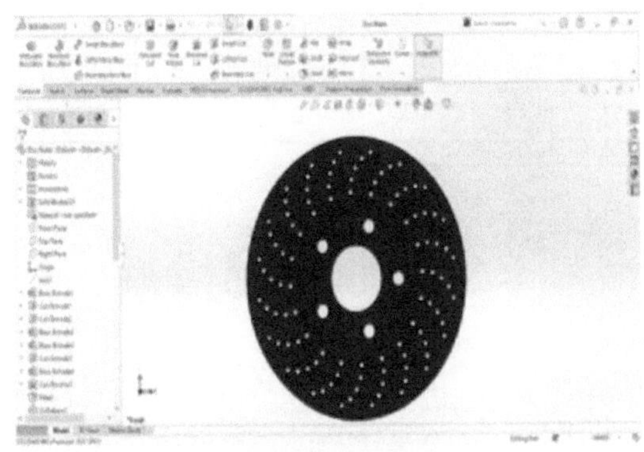

Fig. 57.6 Front view

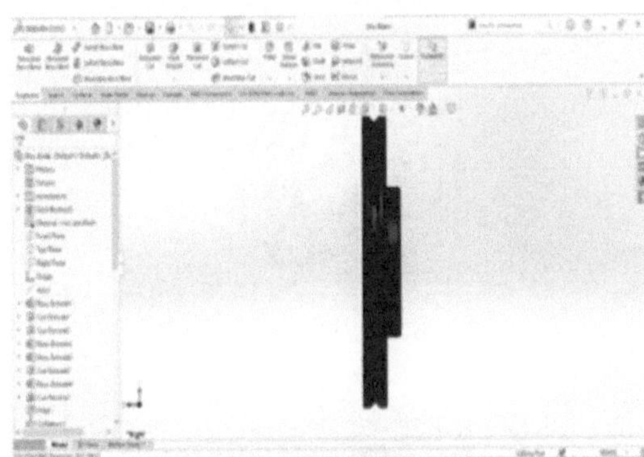

Fig. 57.7 Side view

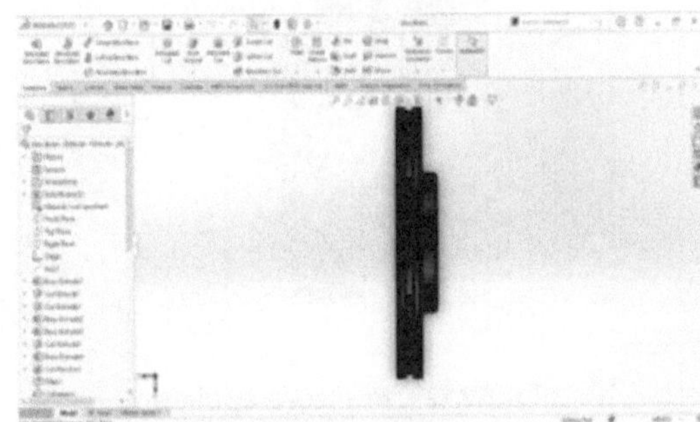

Fig. 57.8 Top view

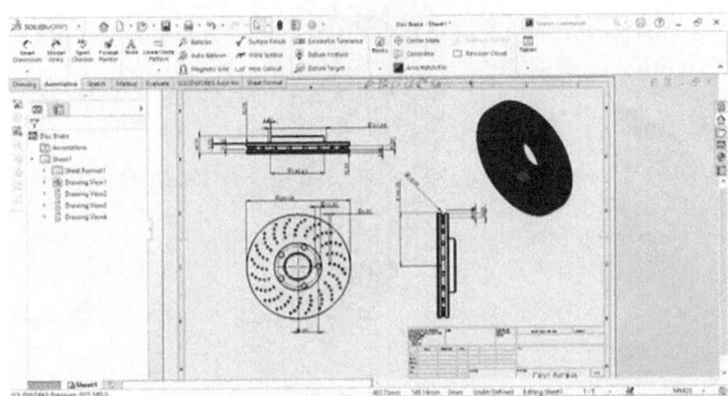

Fig. 57.9 2 D drawing

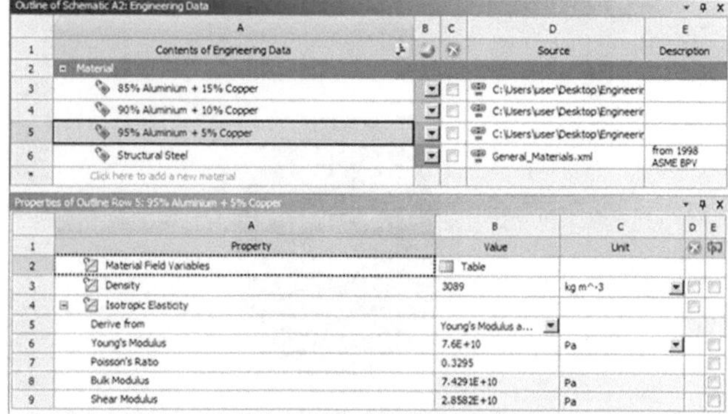

Outline of Schematic A2: Engineering Data

	A	B	C	D	E
1	Contents of Engineering Data			Source	Description
2	Material				
3	85% Aluminium + 15% Copper			C:\Users\user\Desktop\Engineer	
4	90% Aluminium + 10% Copper			C:\Users\user\Desktop\Engineer	
5	95% Aluminium + 5% Copper			C:\Users\user\Desktop\Engineer	
6	Structural Steel			General_Materials.xml	from 1998 ASME BPV
*	Click here to add a new material				

Properties of Outline Row 5: 95% Aluminium + 5% Copper

	A	B	C	D	E
1	Property	Value	Unit		
2	Material Field Variables	Table			
3	Density	3089	kg m^-3		
4	Isotropic Elasticity				
5	Derive from	Young's Modulus a...			
6	Young's Modulus	7.6E+10	Pa		
7	Poisson's Ratio	0.3295			
8	Bulk Modulus	7.4291E+10	Pa		
9	Shear Modulus	2.8582E+10	Pa		

Fig. 57.10 Properties

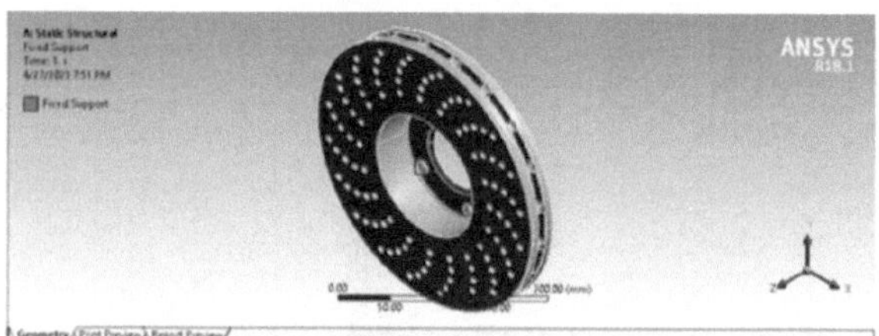

Fig. 57.11 Fixed support

Fig. 57.12 Pressure applied

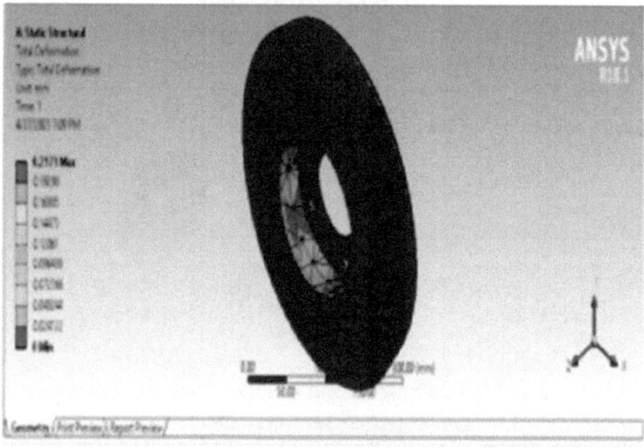

Fig. 57.13 Total deformation

Fig. 57.14 Equivalent stress

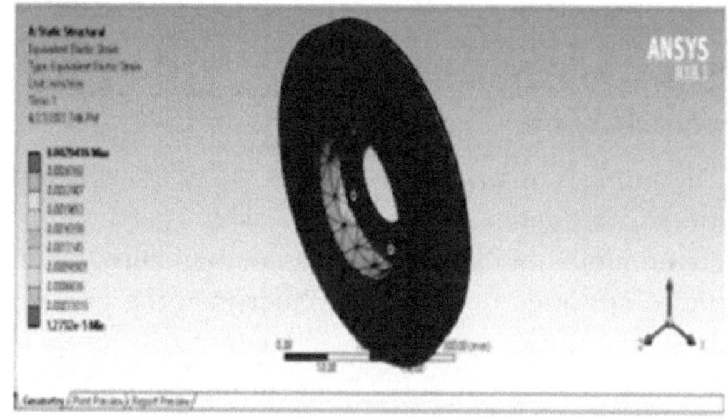

Fig. 57.15 Equivalent elastic strain

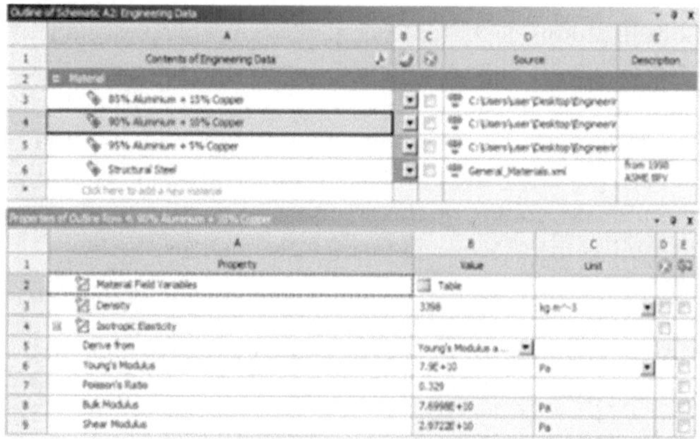

Outline of Schematic A2: Engineering Data

	A	B	C	D	E
1	Contents of Engineering Data			Source	Description
2	Material				
3	85% Aluminium + 15% Copper			C:\Users\user\Desktop\Engineer	
4	90% Aluminium + 10% Copper			C:\Users\user\Desktop\Engineer	
5	95% Aluminium + 5% Copper			C:\Users\user\Desktop\Engineer	
6	Structural Steel			General_Materials.xml	from 1998 ASME BPV
*	Click here to add a new material				

Properties of Outline Row 4: 90% Aluminium + 10% Copper

	A	B	C	D	E
1	Property	Value	Unit		
2	Material Field Variables	Table			
3	Density	3398	kg m^-3		
4	Isotropic Elasticity				
5	Derive from	Young's Modulus a...			
6	Young's Modulus	7.9E+10	Pa		
7	Poisson's Ratio	0.329			
8	Bulk Modulus	7.6998E+10	Pa		
9	Shear Modulus	2.9722E+10	Pa		

Fig. 57.16 Properties

5. Conclusions

It would seem that sample 1, which consists of 95% Copper and 5% Aluminium, has the best output results after examining all of the findings and data from the previous examination. Sample 1 exhibits a greater overall deformation when compared to the other material samples; nevertheless, the stress caused is lower in sample 1 than it is in the other samples. To summarize, the material found in sample 1 is the one that is most ideal for use in the disc brake. In conclusion, the utilisation of Copper-Aluminium alloys in the production of brake components can result in a number of advantages, such as enhanced performance, decreased weight, higher durability, and cost-effectiveness. However, there is still a deficiency in research regarding the understanding of the impact, hardness, and microstructure characterization of brake components made of these alloys, as well as the effects of modifying the composition of the alloy on its mechanical characteristics. This need in research was brought about by the fact that brake components have been made of these alloys for a very long time. The goal of the research that is being proposed on FEA analysis and investigation of impact, hardness, and microstructure characterization of brake components made of copper-aluminium alloys is to fill this gap. This will be accomplished by using FEA software to analyse the impact resistance of brake components made of copper-aluminum alloys, characterising the microstructure of the alloy, testing the hardness of the material, and investigating the effects of varying the composition of alloy. The results of this investigation could enhance the effectiveness of the brake component design made of copper-aluminum alloys, which could result in enhanced safety and efficiency in a variety of industrial systems. The findings may also provide helpful insights into how the material behaves under a variety of loading scenarios, which in turn may lead to the creation of new and improved braking materials that have superior mechanical qualities.

REFERENCES

1. Kennedy, F. E., Colin, F. Floquet, A. Glovsky, R. Improved Techniques for Finite Element Analysis of Sliding Surface Temperatures. Westbury House page 138– 150, 2018.
2. A. R. Abu Bakar and H. Ouyang, Wear prediction of friction material and brake squeal using the finite element method. Wear, Vol.264 No.11-12, pp. 1069–1076,2018.
3. Muhamad Ibrahim Mahmod, Kannan M. Munisamy, Experimental analysis of ventilated brake disc with different blade configuration Department of mechanical Engineering, Vol. 1, PP 1–9, 2017.
4. P.F.Gotowicki, V.Nigrelli and G.V.Mariotti, Numerical and experimental analysis of a pegs-wing ventilated disk brake rotor, with pads and cylinders, In: 10th EAEC Eur.Automot. Cong – Paper EAEC05YUAS04– P 5, June, 2005.
5. Hudson, M., Ruhl, R., Ventilated brake rotor air flow investigation, SAE Technical Paper 971033, 2019.
6. Wallis, L., Leonardi, E., Milton, B., Air flow and heat transfer in ventilated disc brake rotors with diamond and teardrop pillars, Proceedings of International Symposium on Advances in Computational Heat Transfer, Australia, 2020, pp. 643–655.
7. Lee, K. J., Barber, J. R., An experimental investigation of frictionally-excited thermoelastic instability in automotive disk brakes under a drag brake application, Journal of Tribology 116 (2017) 409–414.
8. Lee, K., Barber, J. R., Frictionally excited thermoelastic instability in automotive disk brakes, ASME Journal of Tribology 115 (2018) 607–614.
9. Voldrich, J., Frictionally excited thermoelastic instability in disc brakes-transient problem in the full contact regime, International Journal of Mechanical Sciences 49 (2019) 129–137.
10. Zagrodzki, P., Thermoelastic instability in friction clutches and brakes-transient modal analysis revealing mechanisms of excitation of unstable modes, International Journal of Solids and Structures 46 (2019) 2 463–2 476.
11. Zhu, Z. C., Peng, U. X., Shi, Z. Y., Chen, G. A., Three-dimensional transient temperature field of brake shoe during hoist's emergency braking, Applied Thermal Engineering 29 (2022) 932–937.
12. Komanduri, R., Hou, Z. B., Analysis of heat partition and temperature distribution in sliding systems, Wear 251(1–12) (2019) 925–938.
13. Ozturk, B., Arslan, F., ¨Ozturk, S., Effects of different kinds of fibers on mechanical and tribological properties of brake friction materials, Tribology Transactions 56 (4) (2018) 536–545.
14. Abu Bakar, A. R., Ouyang, H., Wear prediction of friction material and brake squeal using the finite element method, Wear 264 (11–12) (2021) 1069–1076.
15. Abdo, J., Experimental technique to study tangential to normal contact load ratio, Tribology Transactions 48 (2019) 389–403.

Note: Every figure was created using the experimental work; none of them were taken from any publications or online.

Advances in Additive Manufacturing Technologies – Gurusamy Pathinettampadian (eds)
© 2024 Taylor & Francis Group, London, ISBN 978-1-032-90013-1

58 Current Status Review on Smart Drones Technology for Environmental Monitoring

Charumathy Mani[1]
Dept of Computer Science and Engineering,
St. Peter's Institute of Higher Education and Research,
Chennai, India

Baluchamy Shanthini[2]
Dept of Computer Science and
Engineering and Information Technology,
St. Peter's Institute of Higher Education and Research,
Chennai, India

Ranganathan Rani Hemamalini[3]
Dept of Electrical Electronics and Engineering,
St. Peter's Institute of Higher Education and Research,
Chennai, India

Rajasekaran Vinodhini[4]
Dept of Electronics and Communication Engineering,
S.A. Engineering College, Chennai, India

Pachaivannan Partheeban[5]
Department of Civil Engineering,
Chennai Institute of Technology, Chennai, India

R. Sathiyamoorthi[6]
Department of Mechanical Engineering,
Chennai Institute of Technology,
Chennai, India

ABSTRACT: According to Environmental Performance Index 2020, India is ranked 168 out of 180. Solid waste disposal near human settlements in developing countries like India is a major concern for the government, because of the health issues associated with the improper disposal of waste. Air pollution in the environment is monitored using towers and balloons with sensors, manned air vehicles, satellites, and fixed AQI monitoring stations but the accuracy and temporal resolution of collected data by ground stations are relatively low and not suitable for municipal and geographical applications. To rectify the above problems, it is better to use low-cost Unmanned Aerial Vehicles (UAVs) with high-resolution concerns for environmental monitoring of time and space. Sensors are used to measure particular

[1]charujrfspiher@gmail.com, [2]bshanthini@gmail.com, [3]ranihema@yahoo.com, [4]drvinodhini@saec.ac.in, [5]parthi011@yahoo.co.in, [6]sathiyamoorthir@citchennai.net

DOI: 10.1201/9781003545774-58

gases like Nitrogen dioxide, sulphur dioxide (SO2), Ground level of Ozone (O3), Carbon monoxide, Carbon dioxide, Volatile Organic Compounds, Suspended Particulate Matter, Particular Matters and level of PH in the atmosphere. All the needed components are integrated with UAV. This study reviews the technologies adopted for the integration of sensors on the drone and the method of sensors to be selected by considering the method of monitoring the environment with IoT and cloud. The use of IoT and cloud works with more accuracy and data efficiency without any delay. The various systems reviewed in this paper will be useful for developing awareness among the public through an Android application that air quality indexes in a given location.

KEYWORDS: List the keywords covered in your paper. These keywords will also be used by the publisher to produce a keyword index.

1. INTRODUCTION

A UAV is a device that can fly autonomously utilizing software-controlled flight plans in combination with on board sensors and a global positioning system (GPS). Drones are used for anti-aircraft target practice, intelligence gathering, weapons platforms, Miniaturized thermal imaging, and wildlife photography[1,2]. They are now used for a range of civilian tasks like precision crop monitoring, distributing foods, delivering medicine, gathering data, or supplying essentials for disaster systems. As part of pandemic preparedness, hospitals and quarantine areas are sanitized using drones. This review paper will discuss various types of drones used for monitoring tasks for environmental air quality with wireless sensor network technology. Additionally, the UAV may be equipped with a variety of sensors, but the most common ones are a gauge, a GPS unit, a multi- and hyper spectral camera, and a mugginess sensor [3,4].

The successful method of the way to use the aggregate of UAV-based high-decision far off sensing information and floor fact measurements of the harvest is discussed. The airborne platform will provide an excessive space-time resolution while enabling the prompt, independent, and secure identification of air pollutant assertions and air fine tracking. The layout of the tracking and monitoring platform, successful environmental monitoring, and self-sustaining UAV navigation experiments[5,6]. The navigation of a UAV is controlled by the GPS and the uncertainty may be introduced relying upon the decision and drone's complete navigation and goal detecting missions the usage of a low-fee platform with the handiest on-board sensors with the best of the GPS module[7,8].The framework leverages the usage of UAVs for environmental coverage implementation and demonstrates the ability of ANNs (Artificial Neural Network) and RGB actual shade imagery for collecting live video, increasing calibration of the camera, flight time 40-60 mins, payload availability, and also varioussort of surface-level control with an additional battery[9,10]. With the above-mentioned efficient UAVs, a plane will fly above the municipal landfill to collect data on ambient air quality[11].

1.1 Review on Municipal Solid Waste Disposal System and its İmpacts on Human Health

Air pollution is one of the key pollutants that contributes significantly to human sickness and premature mortality. It also causes an increase in the number of diseases that affect people. [12]. Risk in an environment normally appraises with Particulate Matter (PM), which consists of particles of varying but very small diameters, enters the respiratory system by inhalation and causes respiratory and cardiovascular disorders, as well as reproductive and central nervous system disorders and malignancy. Additionally, carbon monoxide can cause immediate poisoning when breathed in at high concentrations. Heavy metals, such as lead, can cause immediate poisoning or chronic intoxication in humans, depending on the level of exposure[13]. Respiratory issues such as Chronic Obstructive Airway Disease (COAD), asthma, and bronchiolitis, as well as lung cancer, cardiovascular events, central nervous system disorders, and skin illnesses, are all caused by the compounds. Moreover, climate change caused by pollution has an impact on the geographic distribution of many infectious diseases, as do natural disasters[14,15]. The most common air contaminants and their hazardous effects on various human body organs and disorders as well as their hazardous effects various human body organs and disorders are presented in Table 58.1. These pollution levels are important to protect the human being and hence suitable measures need for controlling this pollution.

1.2 Nitrogen Dioxide

Oxides of nitrogen exposure have been linked to minor respiratory symptoms at low concentrations and death when exposed to a closed environment. The McConnell missile incident was one of the first tragedies blamed on NO_2. Be-

Table 58.1 According to the US environmental protection agency, there is a standard amount of acceptable air pollutants and their sources with health effects[16]

Pollution levels of Gasses	Emissions from major sources	Calibre levels	Influence on the health of the organs that are being targeted
Nitrogen Dioxide	Vehicle exhaust from burning fuel	100 µg/m^3	Livers, lungs, spleens, and blood vessels are all affected.
Sulphur Dioxide	Coal combustion as a source of fuel	75 µg/m^3	Involvement with the respiratory system and the central nervous system, as well as ocular irritation
The ground level of Ozone	Industrial operations, vehicle exhaust	0.12 mg/m^3	Irritation of the eyes, respiratory and cardiovascular problems
Carbon Monoxide,	Industrial activities, fumes, motor engines, coal, oil, and wood-burning	35 mg/m^3	Damage to the central nervous system and cardiovascular system
Carbon Dioxide	Buming oil, coal, and gas	226.99 mg/m	Headaches, disorientation, and agitation
Volatile Organic Compounds	Industrial painting, coating operations, dry cleaners; auto body shops, vehicles, trucks, and buses.	0.4 ppm	Irritate of eyes, nose, throat, causing breathing difficulties and nausea, and causing damage to the central nervous system and other organs
pHin atmosphere	Acid rain, power plants, factories, and automobiles which burn coal	7	If the pH falls below 6.9, it might cause a coma.
Particulate Matter PM2.5, PM10	Industrial occupations, motor engines, and smokes	35 µg/m^3 150 µg/m^3	Cancer, CNS, and hormonal dysfunctions, and respiratory and cardiac disorders

cause vehicle engines are the primary source of these air pollutants, they are associated to traffic. High levels of inhaled pollutants can irritate the lungs and induce pulmonary edoema. [17]. Despite often being less dangerous than O_3, NO_2 can result in major toxicological problems. The most typical symptoms of nitrogen oxide poisoning are vomiting and wheezing, although there are also headaches, diaphoresis, fever, bronchospasm, and respiratory failure, as well as irritation of the eyes, nose, or throat, dyspnea,hest pain, and diaphoresis. Another research found that individuals may safely be exposed to nitrogen oxide levels between 0.2 and 0.6 ppm[18].

1.3 Ozone

The chemical reaction between nitrogen oxides and volatile organic compounds (VOCs) emitted by human and/or natural sources produces ground-level ozone (GLO). It is believed that GLO is associated with a higher prevalence of respiratory conditions, such as asthma[19].O_3, which is present in so many metropolitan areas, has a number of negative effects on both humans and experimental animals. [20].Among the adverse effects include morphologic, functional, immunologic, and metabolic alterations. Due to its low water solubility, a considerable portion of inhaled O3 travels deep into the lungs; yet, 40% of people and 17% of resting rats' nasopharynxes wash away this response. [21,22].

1.4 Carbon Monoxide

The gaseous pollutant known as CO is colourless, non-irritating, odourless, and tasteless. It can be emitted into the environment by both natural and human sources. Headache, lightheadedness, weakness, nausea, vomiting, and finally unconsciousness are some signs and symptoms of carbon monoxide poisoning. The symptoms and signs are similar to those of other illnesses including food poisoning and viral infections. In the mid-1990s, the Health Impacts Institute conducted many research on cardiovascular disease to investigate the potential correlation between angina pectoris with haemoglobin levels ranging from 2 to 6 percent. The findings demonstrated that these illnesses can result in early angina, but they did not indicate a risk of ventricular arrhythmias[24].Consequently, lowering ambient CO can lower the risk of myocardial infarction in people who are susceptible to it.

1.5 Particulate Matter

Particle pollution is a significant component of air pollution. They are, in a nutshell, a collection of airborne particles. PM is linked to the majority of pulmonary and cardiac mortality rates[25]. They range in size from 2.5 to 10 meters in length (PM2.5 to PM10). Particle emissions are associated with the beginning and progression of cardiovascular and pulmonary illnesses. When smaller particles enter the lower respiratory system, they have a higher chance of causing cardiac and lung problems. Moreover, a number of studies have shown that fine particle pollution causes premature death in those who already have heart or lung issues, such as cardiac arrhythmia, trauma, cardiac arrest, exacerbation of asthma, and reduced lung function. The degree to which particulate matter is exposed determines the potential for serious health effects. Wheezing, coughing, dry mouth, and restrictions in activity owing to breathing issues are the most typical clinical signs of respiratory illnesses induced by air pollution. [26–28].Reduced lung function in children and adults can lead to asthmatic bronchitis and COAD, which both reduce quality of life and shorten life expectancy. Cohort studies show that long-term exposure to PM has a negative impact on cardiovascular and cardiopulmonary mortality [29,30].

1.6 Solid Waste Disposal System (SWDS)

Different types of solid waste can occur from municipal and non-municipal sources. Any non-liquid garbage generated by a person is referred to as municipal solid waste (MSW).Solid waste includes trash, debris, and sludge from waste treatment facilities, water supply treatment facilities, and air pollution control facilities. Solid waste also includes other materials—solid, liquid, semisolid, and enclosed gaseous material—that are left over from mining, commercial, industrial, and agricultural operations..''Solid waste is the leftover material from mining, commercial, industrial, and agricultural operations as well as waste, refuse, or sludge from waste treatment, water treatment, or air pollution control facilities. This substance may be gaseous or liquid, semisolid, or solid[31]. A house, a small business, or an institution, such as a school or a hospital, are all examples of MSW. This form of waste is generally referred to as trash or rubbish, and it consists of ordinary products, damaged items, ruined food, and anything else a person no longer requires or desires[32]. Methane and hydrogen sulphide emissions are linked to the decomposition of organic waste. Rag pickers on-site have started a series of fires at the existing dumpsite to free and recover metals and non-combustible elements from the rubbish.

In contrast to the decline in the rural population, the global urban population grew from 751 million in 1950 to 4.2 billion in 2018. The percentage of people living in cities worldwide has increased from 30% in 1950 to nearly half currently, and by 2050, it is predicted that 68% of people would live in cities, producing 0.596 kg of garbage per person each day[33]. As a result, increased solid waste will result in more pollution, whether in the air or in the solid waste, which is destined for an urban area. The expansion of urban areas will reduce the amount of healthy living space available around the world, as well as raise awareness among urban residents that they have health impacts and follow municipal solid waste management guidelines to keep garbage in check and recycle how much they can[34–36].

1.7 Impact on Health

According to a 2021 World Health Organisation (WHO) report, air pollution is one of the biggest threats to environmental health. The annual death toll from ambient (urban) air pollution is 4.2 million. Reducing air pollution levels can help nations lessen the burden of disease brought on by heart disease, stroke, lung cancer, and both acute and chronic respiratory conditions like asthma; in the long run, this will enhance long-term health outcomes including cardiovascular and respiratory health[37]. The impact of air pollution on the world is becoming a severe concern for human health, with an increase in sickness and death, as a result of changes in atmospheric chemical levels[38]. Data on death and disability-adjusted life years (DALYs) related to air pollution from the WHO from 2011 to 2019[39,40]. Here, explain the term "data" refers to the exact cause of death, as opposed to death risk factors such as air pollution, food, and other lifestyle factors [47]. According to Macrotrends LLC (Fig. 58.1), the global overall death rate has increased every year between 2011 and 2021[41].

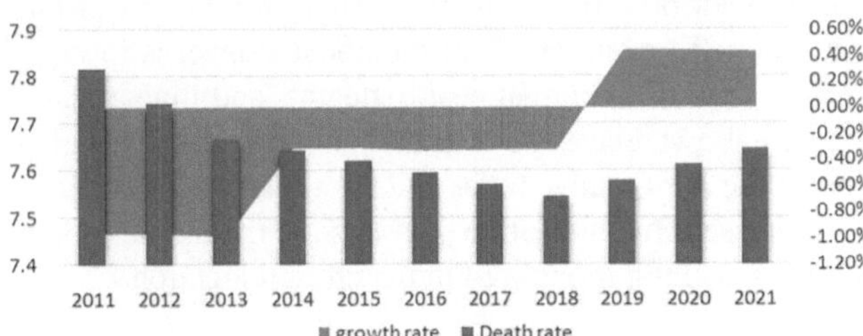

Fig. 58.1 Last ten years death growth rate

1.8 Hazards of Health (HOH)

Any abnormal suspended substance in the air that interferes with the regular functioning of human organs is referred to as an air toxicant in terms of significant health hazards. Published research indicates that the respiratory, cardiovascular, ophthalmologic, dermatologic, neuropsychiatric, hematologic, immunologic, and reproductive systems are most adversely affected by air pollution. On the other hand, long-term molecular and cell damage might result in a variety of cancers [42].Even little levels of air toxicants have been shown to harm susceptible groups including children and the elderly, as well as people with respiratory and cardiovascular problems [43].

1.9 Cardiovascular Disease

Several experimental and epidemiologic research has demonstrated a direct link between air pollution and cardiac diseases[44–46]. Changes in white blood cell counts are related to air pollution as well[47]. It also may have an impact on cardiovascular functions as well. A study using animal models, on the other hand, found a link between hypertension and exposure to air pollution[48]. Right and left ventricular hypertrophy are correlated to traffic-related air pollution, particularly exposure to high amounts of

Nitrogen dioxide[49]. Aside from antidote therapy, which is only available for a few cardiotoxic chemicals like CO, standard cardiovascular disease treatment should be carried out[50].

1.10 Respiratory Disease

The respiratory system is key in the development and progression of illnesses brought on by air pollution. The degree of harm caused by inhaled pollutants to the respiratory system varies based on the amount and way they build up in target cells. Air pollution is a significant environmental risk factor, but it is also linked to certain respiratory conditions including asthma and lung cancer. [51,52]. A higher risk of COAD has been linked in several studies to air pollution from industry and/or transportation[53,54]. Asthma is a diverse disease that can occur as a result of exposure to air toxicants.The treatment of respiratory disorders induced by air pollution is similar to that of other hazardous substances.To examine all of the above concerns in terms of their impact on human health, deaths, and illness. Environmental awareness, as well as a comprehensive evaluation of the Air Quality Index in key worldwide sites, must be addressed. In addition to gas sensors, the ambient environment is being monitored in the chosen location.

2. Current Trends in Air Quality Monitoring Using Sensors/IoT

2.1 Sensing Technology

A semiconductor sensor array is used in the Air Pollution Monitoring System to monitor concentrations of main air pollutant gases such as CO, CO_2, NO_2, SO_2, O_3, VOC, and PMs. The sensors will need to be calibrated using the traditional static chamber approach, and the device will be able to monitor ambient air pollution levels of the aforementioned gases in a real-world situation[55]. CO, NOx, and O_3 are examples of gaseous pollutants that have a variety of harmful respiratory and cardiovascular health effects.The MQ-135 gas sensor delivers data to the system that is computed to determine the concentration of gases such as CO, NO_2,and CO_2. The MQ-131 sensor is used to monitor ground-level ozone concentrations, while the GP2Y1010AUF small optical dust sensor is used to test PM10 levels. The MQ 7 has a simple wiring circuit, good sensitivity, and long life. According to the manufacturer's specifications, this sensor can detect carbon monoxide concentrations ranging from 20 to 2000ppm.The MQ-2& MQ-6 sensor is utilized as an LPG leak sensor; it has a simple, low-cost, and extremely sensitive function, as well as a quick response time to discovered LPG leaks. A sensor is a device that detects symptoms by sending out a signal based on a change in energy. The DHT22 is an essential tool with a thermistor (temperature measurement) and a capacitor sensor for humidity measurement. The M213 high sensitivity microphone sensor module is used to assess variations in noise levels. By adopting IoT, Calibrate all of the above sensors using an Arduino Uno. All sorts of pollutants can be monitored and managed using this technology. The data can be directly updated to the cloud using this manner. When the level of pollution exceeds the set limit, notice is given to the urban public and the authorities. People may readily know the level of pollution by updating data through the cloud and using their mobile phones to use it.

2.2 Studies Related to Air Quality Monitoring Using Drone and IoT

Recent advancements in sensor technology have resulted in tiny, packaged systems that assess a range of air contaminants and natural components. These packages were designed for Smart City / Internet of Things (IoT) projects, in which mass-produced sensors are installed around a city to collect real-time, spatially based air quality data, but they can also be utilised on a drone platform. The goal of this work is to demonstrate a very simple, efficient, reliable,and inexpensive technique to use the UAV for air pollution monitoring and evaluation, among other things.

2.3 Drone Technology

UAVs can be separated into two categories: gimballed and body fixed. Body-fixed sensors are typically simpler and less expensive, but their footprint is dictated by the UAV's state, such as pitch and roll angles. As a response, building advantages from this category utilizing Gas sensors are preserved with UAVs due to calibration. Two distinct drones, a quadcopter, and a hexacopter will be used in an aerial pan shot and receive air quality data for this project. Quadcopters with four wings and hexacopters with six wings will be available.

The total weight of the quadcopter is 1.5 to 2 kg inceptive payload and the total weight of the hexacopter will start from 2.5 kg payload usually periodically the flight plan to go to certain coordinates through onboard GPS and record measurements in local memory to be retrieved upon landing using Arduino as a data-gathering device. The 20-minute flight duration has been extended to 45 minutes. The maximum latitude and maximum longitude for both copters is 100 metres. In order to ensure urban and industrial air safety while examining airborne trends like climate change and offering new avenues for research in air pollution and emission monitoring, narrow unmanned aerial vehicles (UAVs) outfitted with a range of sensors have been introduced for in-situ air quality monitoring.UAVshave seen exceptional expansion, providing a platform for rapid development of solutions because of their flexibility and low cost; throughout the effect, they can be great possibilities to meet the above requirements, allowing remote monitoring of difficult-to-access places. Fig. 58.2 expresses the design of Sensor calibration in UAVs using chambers to separate sensor variation and power supply fixing, as well as radio control.

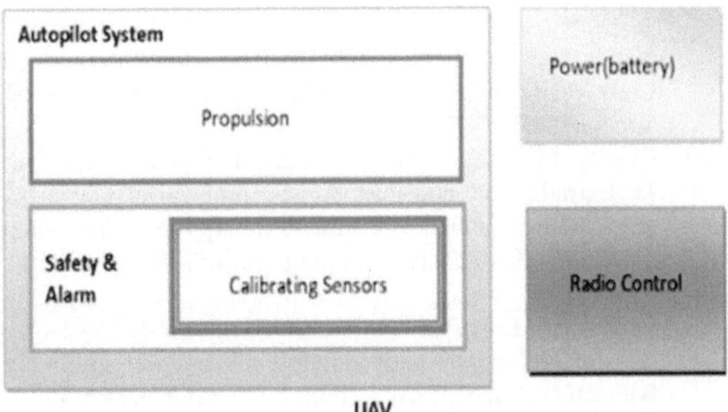

Fig. 58.2 Design of sensing system with UAV

2.4 Drone Technology for Environmental Monitoring

In this study, Unmanned Aerial Vehicles areused in solid waste disposal Systems to monitor chemical ranges in the environment. The quadrotor and hexacopter have been validated as platforms for measuring environmental variables, and the optimum location of sensors on a quadrotor and hexacopter has been determined. Multi-sensors are calibrated with a drone and also collect the accuracy of air ambient data[76]. The technology allows particular data to be updated rather than the entire web page, and it can be used for applications.

3. Integrated Technology for Air Quality Monitoring in Solid Waste Dump Yards

Currently, every researcher is noting the effects of air pollution as well as the air quality degree of origin perception to the general public. People are exposed to unhealthy pollutant concentration levels that surpass specific tolerable values as pollutants build. It might happen when the weather is stationary for a long time. It was discovered that the rules differed from place to place, not only in terms of identifying the contaminants to be monitored but also in terms of defining the threshold values. It is also true that the World Health Organization (WHO, 2006) advised nations to carefully evaluate their local conditions when developing policy targets, i.e., the specificities of localities must be taken into account. As a response, air monitoring networks generate a tremendous amount of data. That data can be shared with people through the internet as web pages and applications.

4. Conclusion

The first session of this review article covered the evolution of drone technology from the army to the public, photography to image processing research and urban research detectors. Describe the chemical levels in the atmosphere and their ranges in the second session. Therefore, there are broader repercussions on human health and an increase in disease-related causes of mortality. The causes of death in the solid waste disposal industry, as well as its impact on health and death rates, are reviewed there. In the third session, participants will learn how to use sensors to collect air quality indexes and what types of sensors are used to calibrate UAVs. Drone technology is utilised for environmental monitoring in the following session, and what sorts of UAVs are deployed in research, what are their advantages in this research, and how the sensor is fixed with drones in SWDS will be discussed. In the featherily, this review as highlighting the influence on human health, mostly for urban people, creating awareness by forecasting future causes in their location under threat. Our future work entails gathering AQI data with drones in metropolitan areas, anticipating the data, and expressing it to the public and government via the cloud, mobile apps, and websites.

References

1. Achille C, Adami A, Chiarini S, Cremonesi S, Fassi F, Fregonese L, et al. UAV-Based Photogrammetry and Integrated Technologies for Architectural Applications—Methodological Strategies for the After-Quake Survey of Vertical Structures in Mantua (Italy). Sensors 2015;15: 15520–39. https://doi.org/10.3390/s150715520
2. Gonzalez L, Montes G, Puig E, Johnson S, Mengersen K, Gaston K. Unmanned Aerial Vehicles (UAVs) and Artificial Intelligence Revolutionizing Wildlife Monitoring and Conservation. Sensors 2016;16:97. https://doi.org/10.3390/s16010097.
3. Mukhamediev RI, Symagulov A, Kuchin Y, Zaitseva E, Bekbotayeva A, Yakunin K, et al. Review of Some Applications of Unmanned Aerial Vehicles Technology in the Resource-Rich Country. Applied Sciences 2021;11:10171. https://doi.org/10.3390/app112110171.
4. Olivares-Mendez M, Fu C, Ludivig P, Bissyandé T, Kannan S, Zurad M, et al. Towards an Autonomous Vision-Based Unmanned Aerial System against Wildlife Poachers. Sensors 2015;15:31362–91. https://doi.org/10.3390/s151229861.
5. Wehrhan M, Rauneker P, Sommer M. UAV-Based Estimation of Carbon Exports from Heterogeneous Soil Landscapes—A Case Study from the CarboZALF Experimen-

tal Area. Sensors 2016;16:255. https://doi.org/10.3390/s16020255.

6. Yungaicela-Naula NM, Garza-Castanon LE, Mendoza-dom A, Minchala-avila LI. Design and Implementation of an UAV-based Platform for Air Pollution Monitoring and Source Identification. Congreso Nacional de Control Automático 2017 2017:288–93.
7. Vanegas F, Gonzalez F. Enabling UAV navigation with sensor and environmental uncertainty in cluttered and GPS-denied environments. Sensors (Switzerland) 2016;16. https://doi.org/10.3390/s16050666.
8. Fidan B, Umay I. Adaptive environmental source localization and tracking with unknown permittivity and path loss coefficients. Sensors (Switzerland) 2015;15:31125–41. https://doi.org/10.3390/s151229852
9. Casado MR, Gonzalez RB, Kriechbaumer T, Veal A. Automated Identification of River Hydromorphological Features Using UAV High Resolution Aerial Imagery 2015:27969–89 https://doi.org/10.3390/s151127969
10. Shahbazi M, Sohn G, Théau J, Menard P. Development and Evaluation of a UAV-Photogrammetry System for Precise 3D Environmental Modeling 2015:27493–524. https://doi.org/10.3390/s151127493
11. Gasperini D, Allemand P, Delacourt C, Grandjean P. Potential and limitation of UAV for monitoring subsidence in municipal landfills. International Journal of Environmental Technology and Management 2014;17:1–13 https://doi.org/10.1504/IJETM.2014.059456
12. Fowler D, Brimblecombe P, Burrows J, Heal MR, Grennfelt P, Stevenson DS, et al. A chronology of global air quality: The development of global air pollution. Philosophical Transactions of the Royal Society A: Mathematical, Physical and Engineering Sciences 2020; 378. https://doi.org/10.1098/rsta.2019.0314
13. Briggs D. Environmental pollution and the global burden of disease 2003:1–24. https://doi.org/10.1093/bmb/ldg019.
14. Jiang XQ, Mei XD, Feng D. Air pollution and chronic airway diseases: What should people know and do? Journal of Thoracic Disease 2016;8:E31–40. https://doi.org/10.3978/j.issn.2072-1439.2015.11.50.
15. Manisalidis I, Stavropoulou E, Stavropoulos A, Kingdom U. Environmental and Health Impacts of Air Pollution : A review i v o r l a n o i v o l 2020. https://doi.org/10.3389/fpubh.2020.00014.
16. Riahi-zanjani B, Balali-mood M. Effects of air pollution on human health and practical measures for prevention in Iran 2016. https://doi.org/10.4103/1735-1995.189646.
17. Chen TM, Gokhale J, Shofer S, Kuschner WG. Outdoor air pollution: Nitrogen dioxide, sulfur dioxide, and carbon monoxide health effects. American Journal of the Medical Sciences 2007;333:249–56. https://doi.org/10.1097/MAJ.0b013e31803b900f.
18. Hesterberg TW, Bunn WB, Mcclellan RO, Hamade AK, Christopher M, Valberg PA. Critical review of the human data on short-term nitrogen dioxide (NO 2) exposures : Evidence for NO 2 no-effect levels 2009;39:743–81. https://doi.org/10.3109/10408440903294945.
19. Gorai AK, Tuluri F, Tchounwou PB. A GIS Based Approach for Assessing the Association between Air Pollution and Asthma in New York State , USA 2014:4845–69. https://doi.org/10.3390/ijerph110504845.
20. Lippmann M. Health Effects Of Ozone A Critical Review. Journal of the Air Pollution Control Association 1989;39:672–95. https://doi.org/10.1080/08940630.1989.10466554.
21. Hatch GE, Slade R, Harris LP, McDonnell WF, Devlin RB, Koren HS, et al. Ozone dose and effect in humans and rats. A comparison using oxygen-18 labeling and bronchoalveolar lavage. American Journal of Respiratory and Critical Care Medicine 1994;150:676–83. https://doi.org/10.1164/ajrccm.150.3.8087337.
22. Gerrity TR, Weaver RA, Berntsen J, House DE, O'Neil JJ. Extrathoracic and intrathoracic removal of O3 in tidal-breathing humans. Journal of Applied Physiology 1988;65:393–400. https://doi.org/10.1152/jappl.1988.65.1.393.
23. Fazlzadeh M, Rostami R, Hazrati S, Rastgu A. Concentrations of carbon monoxide in indoor and outdoor air of Ghalyun cafes. Atmospheric Pollution Research 2015;6:550–5. https://doi.org/10.5094/APR.2015.061.
24. Allred EN, Bleecker ER, Chaitman BR, Dahms TE, Gottlieb SO, Hackney JD, et al. Short-Term Effects of Carbon Monoxide Exposure on the Exercise Performance of Subjects with Coronary Artery Disease. New England Journal of Medicine 1989;321:1426–32. https://doi.org/10.1056/NEJM198911233212102.
25. Sahu D, Kannan GM, Vijayaraghavan R. Carbon Black Particle Exhibits Size Dependent Toxicity in Human Monocytes. International Journal of Inflammation 2014;2014:1–10. https://doi.org/10.1155/2014/827019.
26. Gao Y, Chan EYY, Li L, Lau PWC, Wong TW. Chronic effects of ambient air pollution on respiratory morbidities among Chinese children : a cross-sectional study in Hong Kong 2014.
27. Bentayeb M, Simoni M, Norback D, Baldacci S, Maio S, Viegi G, et al. Indoor air pollution and respiratory health in the elderly. Journal of Environmental Science and Health - Part A Toxic/Hazardous Substances and Environmental Engineering 2013;48:1783–9. https://doi.org/10.1080/10934529.2013.826052.
28. Guillam MT, Pédrono G, Le Bouquin S, Huneau A, Gaudon J, Leborgne R, et al. Chronic respiratory symptoms of poultry farmers and model-based estimates of long-term dust exposure. Annals of Agricultural and Environmental Medicine 2013;20:307–11.
29. Jerrett M, Finkelstein MM, Brook JR, Arain MA, Kanaroglou P, Stieb DM, et al. A cohort study of traffic-related air pollution and mortality in Toronto, Ontario, Canada. Environmental Health Perspectives 2009;117:772–7. https://doi.org/10.1289/ehp.11533.
30. Zhou M, Liu Y, Wang L, Kuang X, Xu X, Kan H. Particulate air pollution and mortality in a cohort of Chinese men. Environmental Pollution 2014;186:1–6. https://doi.org/10.1016/j.envpol.2013.11.010.

31. Salam A. Environmental And Health Impact of Solid Waste Disposal At Mangwaneni Dumpsite In Manzini: Swaziland 2010;12:64–78.
32. Esakku S, Swaminathan A, Parthiba Karhtikeyan° O, Kurian J, Palanivelu° K. Municipal Solid Waste Management in Chennai City, India. Proceedings Sardinia 2007:1–5.
33. Sharma KD, Jain S. Municipal solid waste generation , composition , and management : the global scenario 2020. https://doi.org/10.1108/SRJ-06-2019-0210.
34. Ashish R. Mishra SAM, Tiwari A V. Solid Waste Management - Case Study. International Journal of Research in Advent Technology 2014;2:267–77.
35. Kupchik GJ, Franz GJ. Solid Waste, Air Pollution and Health. Journal of the Air Pollution Control Association 1976;26:116–8. https://doi.org/10.1080/00022470.1976.10470229.
36. Rakkini V. A Survey of Solid Waste Management in Chennai (A Case Study of Around Koyambedu Market and Madhavaram Poultry Farms). Journal of Civil Engineering and Environmental Sciences 2018;4:009–12. https://doi.org/10.17352/2455-488X.000020.
37. World Health Organization. WHO global air quality guidelines. Coastal And Estuarine Processes 2021:1–360.
38. Briggs DJ. A framework for integrated environmental health impact assessment of systemic risks. Environmental Health: A Global Access Science Source 2008;7:1–17. https://doi.org/10.1186/1476-069X-7-61.
39. Dhimal M, Chirico F, Bista B, Sharma S, Chalise B, Dhimal ML, et al. Impact of Air Pollution on Global Burden of Disease in 2019. Processes 2021;9:1719. https://doi.org/10.3390/pr9101719.
40. Balakrishnan K, Dey S, Gupta T, Dhaliwal RS, Brauer M, Cohen AJ, et al. The impact of air pollution on deaths, disease burden, and life expectancy across the states of India: the Global Burden of Disease Study 2017. The Lancet Planetary Health 2019;3:e26–39. https://doi.org/10.1016/S2542-5196(18)30261-4.
41. World Death Rate 1950-2022. Macrotrends LLC 2022.
42. Kampa M, Castanas E. Human health effects of air pollution 2008;151:362–7. https://doi.org/10.1016/j.envpol.2007.06.012.
43. Makri A, Stilianakis NI. Vulnerability to air pollution health effects 2008;211:326–36. https://doi.org/10.1016/j.ijheh.2007.06.005.
44. Snow SJ, Cheng W, Wolberg AS, Carraway MS. Air Pollution Upregulates Endothelial Cell Procoagulant Activity via Ultrafine Particle-Induced Oxidant Signaling and Tissue Factor Expression. Toxicological Sciences 2014;140:83–93. https://doi.org/10.1093/toxsci/kfu071.
45. Brook RD. Cardiovascular effects of air pollution 2008;187:175–87. https://doi.org/10.1042/CS20070444.
46. Franklin BA, Brook R, Iii CAP. Air Pollution and Cardiovascular Disease. Current Problems in Cardiology 2015;40:207–38. https://doi.org/10.1016/j.cpcardiol.2015.01.003.
47. Steenhof M, Janssen NAH, Strak M, Hoek G, Gosens I, Mudway IS, et al. Air pollution exposure affects circulating white blood cell counts in healthy subjects: the role of particle composition, oxidative potential and gaseous pollutants – the RAPTES project. Inhalation Toxicology 2014;26:141–65. https://doi.org/10.3109/08958378.2013.861884.
48. Sun Q, Yue P, Ying Z, Cardounel AJ, Brook RD, Devlin R, et al. Air Pollution Exposure Potentiates Hypertension Through Reactive Oxygen Species-Mediated Activation of Rho/ROCK. Arteriosclerosis, Thrombosis, and Vascular Biology 2008;28:1760–6. https://doi.org/10.1161/ATVBAHA.108.166967.
49. Leary PJ, Kaufman JD, Barr RG, Bluemke DA, Curl CL, Hough CL, et al. Traf fi c-related Air Pollution and the Right Ventricle The Multi-ethnic Study of Atherosclerosis 2014;189:1093–100. https://doi.org/10.1164/rccm.201312-2298OC.
50. Van Hee VC, Adar SD, Szpiro AA, Barr RG, Bluemke DA, Diez Roux A V., et al. Exposure to Traffic and Left Ventricular Mass and Function. American Journal of Respiratory and Critical Care Medicine 2009;179:827–34. https://doi.org/10.1164/rccm.200808-1344OC.
51. Weisel CP. Assessing Exposure to Air Toxics Relative to Asthma 2002;110:527–37.
52. Young MT, Sandler DP, DeRoo LA, Vedal S, Kaufman JD, London SJ. Ambient air pollution exposure and incident adult asthma in a nationwide cohort of U.S. women. American Journal of Respiratory and Critical Care Medicine 2014;190:914–21. https://doi.org/10.1164/rccm.201403-0525OC.
53. CHUNG KF, ZHANG J, ZHONG N. Outdoor air pollution and respiratory health in Asia. Respirology 2011;16:1023–6. https://doi.org/10.1111/j.1440-1843.2011.02034.x.
54. KO FWS, HUI DSC. Air pollution and chronic obstructive pulmonary disease. Respirology 2012;17:395–401. https://doi.org/10.1111/j.1440-1843.2011.02112.x.
55. Sales-l D, Bello AJ, Alberto S, Manuel P. An Approximation for Metal-Oxide Sensor Calibration for Air Quality Monitoring Using Multivariable Statistical Analysis 2021.

Note: Every figure was created using the experimental work; none of them were taken from any publications or online.

Advances in Additive Manufacturing Technologies – Gurusamy Pathinettampadian (eds)
© 2024 Taylor & Francis Group, London, ISBN 978-1-032-90013-1

59 LM6 Aluminium Alloy Processing by Die Casting—A State of the Art

A Parthiban[1], A Ramesh[2],
A Suresh[3], D Dinesh[4], P S Kanishkha[5]
Department of Mechanical Engineering,
Chennai Institute of Technology, Chennai, Tamilnadu, India

ABSTRACT: LM6 is an aluminium alloy that finds extensive use in die casting applications due to its excellent blend of mechanical properties, corrosion resistance, and castability. This abstract provides an overview of the LM6 aluminium alloy die casting process, its key attributes, and common applications. LM6 is a high-strength aluminum alloy often used in die casting applications due to its excellent combination of properties. It is part of the Aluminum Association's 300 series alloys and is primarily composed of aluminum (Al), with key alloying elements such as silicon (Si) and magnesium (Mg). Here's a brief description of LM6 in die casting:

KEYWORDS: High strength, Modulus, Shear, Castability

1. Introduction

LM6 belongs to the aluminium-silicon (Al-Si) alloy relatives, which includes silicon as the primary alloying element, as well as magnesium and tiny quantities of other elements. Because of their exceptional performance characteristics, these alloys are highly sought after in die casting. Casting is an extremely cost-effective method of producing metallic components due to the molten metal is poured immediately into the cavity inside the mould of the desired size and shape. The manufacture of casting flaws constitutes the single greatest drawbacks of casting tactics.

Porosity, segregation, elevated tears, and so forth. Furthermore, standard casting procedures are incapable of producing very high strength. As a result of this, the structural performance is the same with forging. Squeeze Casting (SC) had is one of the latest casting procedures that has been created. It has the capability to get rid of the aforementioned flaws. This method may be used to create components using enhanced mechanical performance with excellent durability and exceptionally low-cost Low defect phases, with a focus on minimal porosity. Of Squeeze casting is one of the various casting procedures available. The magnificent thing SC's major driving force in producing The aerospace and automotive sectors have relied on high strength and trustworthy components with better surface finish. SC can acquire component characteristics such as configurational details, tight tolerances, and excellent surface quality. High specific pressure solidification against a hard metal die causes the metallic material to conform strongly to the die surface, enabling its parts to perform increased surface detail and polish. Because correctly produced squeeze cast components are free of flaws, an expensive post-solidification investigation using a non-destructive testing method is

[1]parthiban@citchennai.net, [2]principal@citchennai.net, [3]suresha@citchennai.net, [4]dineshd@citchennai.net, [5]kanishkaps.mech2021@citchennai.net

DOI: 10.1201/9781003545774-59

not required. The method has been applied to a wide range of metals, from low melting point lead and zinc alloys to metallic substances with very high melting points.

2. Materials and Methods

Table 59.1 Mechanical properties of aluminium (sand & diecasting) alloy lm6

	Sand Casting	Pressure Die Casting	Gravity Die Casting
0.2% Proof Stress (N/mm^2)	50-60	119	69-79
Tensile Strength (N/mm^2)	159-189	279	189-229
Elongation (%)	4	3.5	6
Impact resistance Izod (Nm)	5	0.5	8.99
Brinell Hardness	40	50	60
Modulus of Elasticity ($x10^3$ N/mm^2)	69	69	69
Shear Strength (N/mm^2)	118	0.5	0.25

2.1 Chemical Composition

Table 59.2 Chemical properties

	%		%		%
Cu	0.99 max	Mg	0.99 max	Si	12.99
Fe	0.599 max	Mn	0.499 max	Ni	0.99max
Zn	0.99 max	Pb	0.99 max	Sn	0.5 max
Ti	0.199 max	Al	Remainder	Others each	0.49 max
Others total	0.149 max				

Key Characteristics

1. *High Strength:* LM6 offers good mechanical properties, making it suitable for parts requiring strength and durability.
2. *Corrosion Resistance:* It has a natural resistance to corrosion, which is further enhanced when properly treated or coated.
3. *Excellent Castability:* LM6 is known for its excellent fluidity and castability, making it ideal for intricate and complex die-casting shapes.
4. *Heat Resistance:* It retains its mechanical properties at elevated temperatures, making it suitable for applications where parts may be exposed to heat.
5. *Machinability:* LM6 can be machined and finished with relative ease.

2.2 Die Casting Process

Die casting is a manufacturing method that includes pumping molten metal under high pressure into a die (or mould). Because of its outstanding fluidity and castability, the LM6 alloy is suited for this technique. The following are typical phases in the die casting process:

Table 59.3 Roughness properties

Computation of Pareto ANOVA for surface roughness				
Process parameters	**A**	**B**	**C**	**E**
Sum at parameter levels				
1	3.807	2.627	2.323	2.423
2	2.223	2.433	2.507	2.343
3	1.150	2.120	2.350	2.413
Total				7.179
Process parameters	**A**	**B**	**C**	**E**
Sum of squares of differences	10.717	0.392	0.059	0.011
Contribution ratio (%)	95.86	3.50	0.527	0.102
Overall optimum conditions for all parameters				
A_3: squeeze pressure	140 N/mm^2			
B_3: die temperature	250°C			
C_1: die material	Stainless steel			

$$C.I. = \sqrt{\left[F_\alpha(2, 20) \times V_e\left(\frac{1}{n_{eff}}\right)\right]},$$

$$V_e = \frac{\text{pooled variation of non-significant sources}}{\text{pooled degrees of freedom of non-significant of sources}},$$

$$n_{eff} = \frac{\text{number of experiments}}{1 + \text{total degrees of freedom in items used in } \bar{\mu} \text{ estimate}}$$

optimum surface roughness predicted = A⁻ 3 + B⁻ 3 – Y⁻ = 0.83 + 0.11 – 0.486 = 0.234 m

$$S/N(\eta) = -10 \log\left(\frac{1}{n}\sum_{i=1}^{n} y_i^2\right) \quad (1)$$

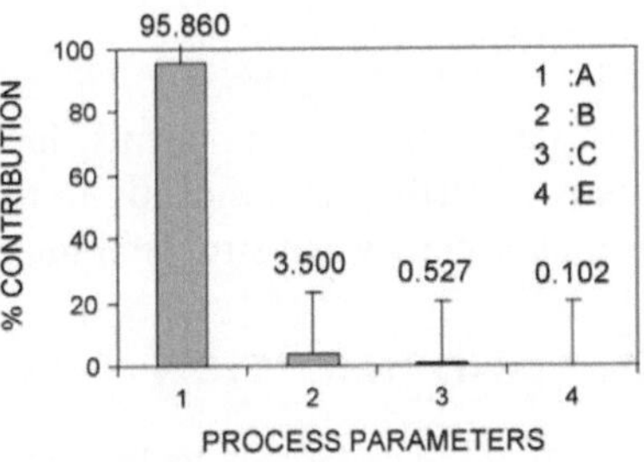

Fig. 59.1 Process properties

Table 59.4 Roughness levels

Process parameters	Level 1	Level 2	Level 3
A: squeeze pressure (N/mm^2)	35	70	140
B: die temperature (°C)	35	150	250
C: die material	Stainless steel	Copper	Cast iron

2.3 Die Preparation

The die is designed and prepared with cavities that match the desired part's shape. It is often made from steel to withstand the high temperatures and pressures involved.Die preparation for LM6 aluminum alloy molds is a crucial step in the die casting process, as it directly impacts the quality and accuracy of the final cast components. Here's a concise description of die preparation for LM6 Die Material and Design: The die, also known as the mold, is typically made from high-quality, heat-resistant steel, such as H13 tool steel. The choice of material ensures the die can withstand the high temperatures and pressures involved in the die casting process. The die design includes cavities and cores that define the shape of the final part.

Surface Treatment: To enhance the die's longevity and ease of release, it is often subjected to surface treatments like nitriding or hard chrome plating. These treatments improve wear resistance and reduce friction during the casting process.

Die Heating: Proper die temperature control is essential.

Die Assembly: The die is assembled, aligning the various components to create a precise mold cavity. Proper alignment is critical to ensuring the final cast parts meet dimensional tolerances.

Die preparation for LM6 aluminum alloy molds is a meticulous process that demands technical precision at various stages. The initial phase revolves around judicious material selection, where the choice of die material plays a critical role in determining the mold's durability, heat resistance, and overall performance. Surface treatments follow, involving techniques such as nitriding or coating applications to enhance hardness, wear resistance, and thermal conductivity. Temperature control during die preparation is imperative to achieve optimal casting conditions. Furthermore, assembly precision is paramount to guaranteeing uniformity and repeatability in the die casting process. The accurate alignment and integration of die components, including cores and ejector pins, contribute significantly to the final product's dimensional accuracy and structural integrity.

3. Melting and Injection

In a furnace, LM6 alloy undergoes melting at temperatures above 650°C (1202°F). A high-pressure casting equipment is subsequently used to inject the molten metal into the die cavity.The heat treatment and injection process for LM6 aluminium alloy moulds is an important stage in the die casting process because it transforms the molten LM6 alloy into the final component shape. The LM6 alloy, typically in the form of ingots or recycled scrap, is melted in a high-temperature furnace, usually exceeding 650°C (1202°F). This elevated temperature is necessary to ensure the alloy becomes fully liquid for the subsequent injection step. The alloy composition is closely monitored to maintain the desired proportions of aluminum, silicon, magnesium, and trace elements. Once the LM6 alloy reaches the specified temperature, it is injected into the die cavity at high pressure using a specialized die casting machine. This injection process forces the molten metal into the intricately designed mold cavities and cores, taking on the shape of the final component.

Achieving consistent and high-quality castings in the die casting process demands meticulous control over key parameters, namely temperature, pressure, and injection speed. Temperature regulation is critical as it directly influences the fluidity and solidification behaviour of the LM6 alloy. Maintaining an optimal temperature ensures uniform flow of the molten metal within the die cavity, preventing premature solidification or incomplete filling that may lead to defects. Pressure control is equally vital, exerting influence on the density, structural integrity, and surface finish of the cast componentsThis controlled cooling process facilitates swift solidification, capturing the desired part characteristics and minimizing the formation of undesirable microstructures. In essence, the technical orchestration of temperature, pressure, and injection speed intricately influences the die casting process, ensuring the production of consistent, high-quality castings with precise geometries and desired material properties.

4. Cooling and Solidification

After injection, the molten metal rapidly cools and solidifies within the die, taking the shape of the cavity. This solidification process is critical to achieving the desired part's properties.The cooling and solidification process for LM6 aluminum alloy molds is a critical phase in die casting, influencing the final part's quality and structural integrity After the injection of molten LM6 alloy into the die cavity, the cooling and solidification phase begins. This step involves carefully controlling the temperature of the die and the casting to ensure that the LM6 alloy solidifies in a controlled manner.Diemolds are equipped with intricate cooling systems, typically comprising water channels, to regulate the die's temperature. Achieving uniform cooling throughout the die is essential to prevent defects such as

Table 59.5 Die properties

Experimental number	Squeeze pressure (N/mm^2)(*A*)	Die preheating temperature (°C) (*B*)	Die material (C)	Error (*E*)	Surface roughness (R_{aij}), *j* = 1, 2, 3(μm)			R_a (μm)
					1	2	3	
1	35	35	Stainless steel	1	1.56	1.15	1.29	1.333
2	35	150	Copper	2	1.28	1.38	1.25	1.303
3	35	250	Cast iron	3	1.18	1.09	1.24	1.170
4	70	35	Copper	3	0.91	0.83	0.85	0.863
5	70	150	Cast iron	1	0.82	0.72	0.71	0.750
6	70	250	Stainless steel	2	0.74	0.58	0.51	0.610
7	140	35	Cast iron	2	0.37	0.53	0.39	0.430
8	140	150	Stainless steel	3	0.34	0.42	0.38	0.380
9	140	250	Copper	1	0.37	0.33	0.32	0.340

porosity, shrinkage, or warping. Properly designed cooling channels and control systems ensure even solidification.

The cooling and solidification process for LM6 Molds is closely monitored and controlled to ensure that the resulting castings meet dimensional tolerances and exhibit the desired mechanical properties, making it a key aspect of high-quality die casting production.

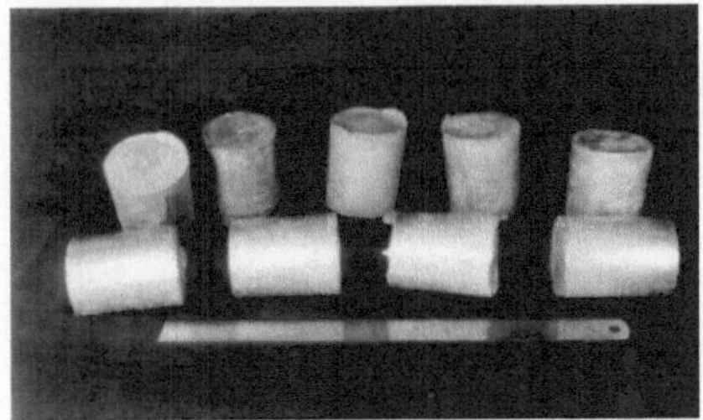

Fig. 59.2 LM6 alloy processed

5. Ejection

Once the metal has solidified, the die opens, and the newly formed part is ejected, often with the help of ejector pins.

6. Trimming and Finishing

The die casting process for LM6 aluminum alloy involves a comprehensive series of technical steps to ensure the production of high-quality components. Die preparation begins with meticulous material selection, where the die material's durability, heat resistance, and overall performance are critical factors. Surface treatments such as nitriding or coatings are then applied to enhance hardness, wear resistance, and thermal conductivity. Temperature control during die preparation ensures optimal casting conditions, maintaining precise temperature profiles for the proper fluidity of the molten alloy and efficient filling of intricate mold cavities. Lubrication plays a crucial role in minimizing friction and wear between mold components, enhancing tool lifespan. Following trimming, additional finishing processes may be applied, including shot blasting, sanding, or polishing, to improve the component's appearance and surface texture. Surface coatings or platings may also be added for enhanced corrosion resistance or aesthetic purposes. Throughout the trimming and finishing processes, quality control measures are implemented to inspect the castings for defects, dimensional accuracy, and overall quality. Non-conforming parts are identified and either rectified or rejected. In essence, the technical orchestration of die preparation, temperature control, pressure management, and post-casting processes collectively ensures the production of consistent, high-quality LM6 aluminum alloy castings with precise geometries and desired material properties.

In summary, trimming and finishing are critical steps in the die casting process for LM6 aluminum alloy molds. They ensure that the cast components meet the desired specifications in terms of dimensions, surface finish, and overall quality, resulting in reliable and functional end products.

7. Applications

Die-cast LM6 components find extensive applications across diverse sectors such as automotive, aerospace, electronics, and consumer goods. In the automotive industry, LM6 is frequently employed for engine components, benefitting from its optimal strength-to-castability ratio. This characteristic ensures that the alloy can be cast into intricate and precisely detailed shapes without compromising structural integrity. In the aerospace sector, where lightweight yet robust materials are crucial, LM6 proves valuable for producing components such as heat sinks, providing effective thermal management in critical systems. Gadgets and consumer products also benefit from the versatility of LM6. Its ability to maintain structural stability under varying loads and environmental conditions makes it suitable for manufacturing electrical enclosures, safeguarding sensitive electronic components. Furthermore, LM6's excellent surface finish after casting makes it an ideal choice for ornamental pieces, meeting aesthetic requirements in consumer goods. The strength-to-castability ratio of LM6 is a key factor in its widespread adoption for precision components. This alloy strikes a balance between mechanical strength and the ease with which it can be cast into complex shapes, ensuring the production of components that meet rigorous performance criteria. The material's favorable characteristics contribute to its popularity in industries where both structural robustness and intricate design specifications are imperative. As technology advances and manufacturing demands become more exacting, the technical attributes of LM6 continue to position it as a reliable and versatile material for diverse applications across various industrial domains..

8. Conclusion

The research on die casting using LM6 aluminium alloy molds has provided valuable insights into the potential applications and performance of this material in the manufacturing industry. This study has highlighted several key findings that contribute to our understanding of die casting processes and LM6 alloy properties. The die casting with LM6 aluminium alloy molds has demonstrated its suitability for producing complex and intricate parts with high precision and dimensional accuracy. The alloy's excellent fluidity, low melting point, and good thermal conductivity make it a preferred choice for die casting applications, especially in industries where lightweight, corrosion-resistant components are essential.

The research has shown that LM6 alloy molds can withstand the high temperatures and pressures involved in die casting without significant degradation, ensuring long-term durability and cost-effectiveness. In summary, die casting from LM6 aluminium alloy molds offers a promising avenue for the production of high-quality, durable components with intricate designs. Further research and development in optimizing process parameters and alloy compositions can unlock even greater potential for this material in various industrial sectors, ultimately driving innovation and efficiency in manufacturing processes.

References

1. M.R. Ghomashchi, A. Vikhrov, Squeeze casting: an overview, J. Mater. Process. Technol. 101 (2000) 1–9. [2] J.R. Morton, J. Barlow, Squeeze casting: from a theory to profit and a future, The Foundryman 87 (1994) 23–28. Process. Technol. 101 (2000) 1–9.
2. J.R. Morton, J. Barlow, Squeeze casting: from a theory to profit and a future, The Foundryman 87 (1994) 23–28.
3. M.S. Yong, A.J. Clegg, Evaluation of squeeze cast magnesium alloy and composite, Foundryman (1999) 71–75.
4. R.F. Lynch, R.P. Olley, Squeeze casting of Aluminium, AFS Trans. 122 (1975) 569–576.
5. T.M. Yue, G.A. Chadwick, Squeeze casting of light alloys and their composites, J. Mater. Process. Technol. 58 (1996) 302–307.
6. J.R. Brown, Second report of Institute Working Group T20 Casting Processes, The Foundryman 87 (1994) 386–390.
7. I. Puertas Arbizu, C.J. Luis Perez, Surface roughness prediction by factorial design of experiments in turning processes, J. Mater. Process. Technol. 143–144 (2003) 390–396.
8. M. Alauddin, M.A. El Baradie, M.S.J. Hashmi, Optimization of surface finish in end milling in INCONEL 718, J. Mater. Process. Technol. 56 (1996) 54–65.
9. J. L. Yang, J.C. Chen, A systematic approach for identifying optimum surface roughness performance in end-milling operations, J. Indus. Technol. 17 (2) (2001) 2–8.
10. Avani Gandhi, Problem solving using Taguchi DOE techniques, Indus. Eng. J. XXXII (4) (2003) 16–25.
11. M. Alauddin, M.A. El Baradie, M.S.J. Hashmi, Computer aided analysis of a surface roughness model for end milling, J. Mater. Process. Technol. 55 (1995) 123–127.
12. S.H. Park, Robust Design and Analysis for Quality Engineering, Chapman and Hall Publications, 1996.

Note: Every figure and table were created using the experimental work; none of them were taken from any publications or online.

Advances in Additive Manufacturing Technologies – Gurusamy Pathinettampadian (eds)
© 2024 Taylor & Francis Group, London, ISBN 978-1-032-90013-1

60 Heat Transfer and Thermal Performance of a Heat Sink with Pin Fins

S. Balamurugan
Research Scholar, Department of Mechanical Engineering, Annamalai University

R. Karthikeyan
Professor, Department of Mechanical Engineering, Annamalai University

R. Muthukumaran
Assistant professor, Department of Mechanical Engineering, Annamalai University

Mukilarasan N*
Assistant Professor, Department of Mechanical Engineering, AMET University

ABSTRACT: The current work bolds hexagonal and cylindrical pin fin arrays at 50°C on a horizontal base plate. This, along with the transfer of heat of fiction characters of cylindrical fin, was studied experimentally with the help of a rectangular wind tunnel. Inline and staggeredarrangements are to be made with pin fins. Finding the impact of fin distance in both span and stream directions must be done. Maintaining constant values for both the relative span pitch (Sx/d=1.2) and reactive stream wise pitch (Sy/d = 1.2, 2.4, and 3.6) is recommended. The research uses varying air mass flow rates (Re varies between 2000 and 280000). The clearance rate (C/H = 0.0) greatly affects the pin's performance, which makes the hexagonal fin conduct heat more rapidly than the cylindrical fin. The experiment's findings demonstrate that, for a given pumping power, the optimal inter-fin pitch, Sx/d=1.2 and Sy/d=1.2, are determined by looking for the greatest Nusselt number.

KEYWORDS: Wind tunnel, Pin-fin, Nusselt number, Reynolds number, Pitch and tip clearance, Hexagonal fin

1. INTRODUCTION

The mechanism of many engineering systems ends with heat generation, which leads to too many heating problems. Even system failure heat transfer components that cost less and have smaller weights can be used to create the solution. The dissolution of heat produced by various appliances such as transformers, air-cooled engines, compressors, boilers, condenser coils, superheated tubes, and electronic components is necessary for the system to function well at

*Corresponding author: mukil09tea19@gmail.com

DOI: 10.1201/9781003545774-60

its operating temperature. Braga et al. (1) study analysis offered numerous approaches. Active techniques requiring external power include churning a liquid or shaking a solid surface. Fins, on the other hand, don't require this power. Increasing the fins' surface area can improve heat transfer at the solid-fluid contact. Armstrong has used fins in many ways for heat transmission, from more basic shapes like square, rectangular, cylindrical, tapered and annular pin-fins to a combination of many geometries. Owing to the inexpensive cost of production and high thermal efficiency of these fins, it is important and popular to study how geometric parameters like fin length, spacing, and height affect heat dissipation. (Bilen etal. (2) and Kodah et al. (3)). The performance of heat transmission of an array of cylindrical fins mounted to a staggered wall was initially studied by Ramsey et al. (4). Berry RA et al. (5) examined the characteristics of a staggered cylindrical pin-fin array's heat transfer. Heat transfers from staggered arrays of cylindrical pin-fins are also studied by Simoneau et al. (6). Armstrong et al. (7) recommended investigating the heat transfer from dispersed array pin fins to cool turbines. Jubran et al. (8) conducted tests to investigate the impact of the optimal interfin spacing and shroud clearances (both span- and stream-wise). Forced convection in exposing the characters of heat transfer was investigated by Ashish of Anil (9) with a grooved flat pin type. It shows the grooved type's ability to enhance heat transfer characteristics over conventionally employed fins.

Tariq et al. (10) used a duct equipped with various ribs of pentagonal, trapezium, and square forms of truncated prismatic forms to explore improved heat transfer of flow characteristics to determine the fundamental heat transfer process. Mechanism of efficient engineering applications: Hamdi et al. (11), heat exchangers, solar collectors, electronics cooling, automotive applications, etc. Karami et al. (12) conducted general convention experiments with three-fin tubular exchangers and a square fin array. Fin spacing ranges from 5 to 14 millimetres. Based on the experiment's results, free convection contributes approximately 80% of the total energy input and 10% of the temperature distribution from ambient radiations to the heat source. To a certain extent, the heat transfer coefficient rises with fin spacing. Overall, a literature study reveals how fins have been altered to improve heat flow by adding holes, slots, and struts. However, it is not advised to use rib struts. This could make the current fin heavier. Employing fins with various cross-sectional shapes, such as hexagonal and cylindrical fins, can affect flow resistance and heat transmission differently. Because pin fins are relatively easy to make, circular and hexagonal fins are frequently employed in applications. The current experiment compares the behaviour of hexagonal-type fins to that of cylindrical fins by examining the impact of heat transfer on fluid flow parameters.

2. Experimental Setup

Figure 60.1 displays the experimental apparatus's schematic diagram. This experimental setup includes a heating unit, a data unit, a pin fin assembly, and a wind tunnel. A working open circuit suction mode wind tunnel was utilised for this experiment.

2.1 Fin Configurations

Each cylindrical fin measured 90 mm in height and 10 mm in diameter, hexagonal fin measured 77.7 mm in height and 6 mm side diameter. Then, at the base of each fin, a rectangular aluminium plate measuring 250 mm, 145 mm 25.4 mm is fixed, allowing for electrical heater heating. A fin is positioned inside the test section so that the air passageways travel over the fin surfaces during the test run. Figure 60.2 depicts the geometries of cylindrical fins and perforated fins (a and b)

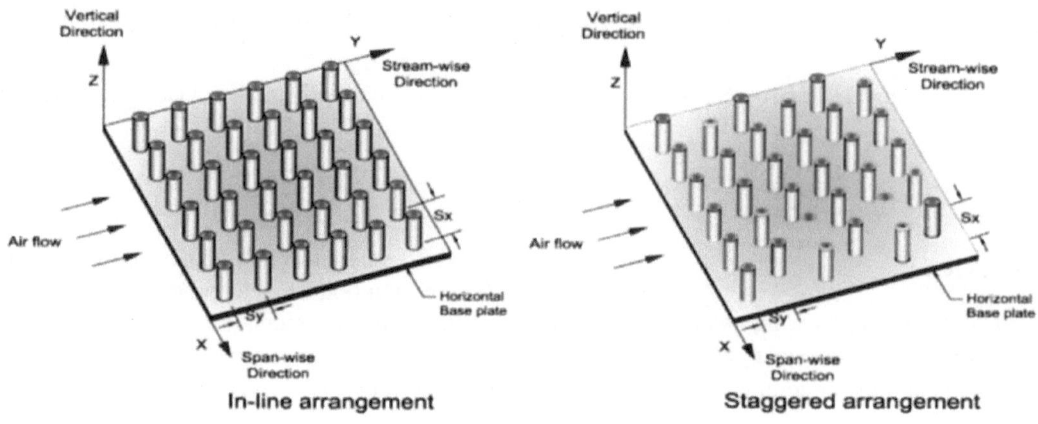

Fig. 60.1 Pin-fin assembly

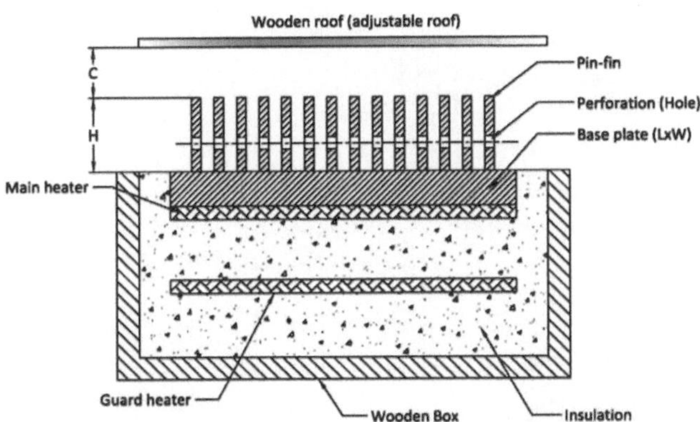

Fig. 60.2 Sectional elevation of pin-fin assembly

A. Wind Tunnel

The 2 m long wind tunnel has an internal width of 150 mm and is constructed of plywood that is 19 mm thick. Air is supplied to the testing area at the required flow rate by a centrifugal blower controlled by an operative gate valve. The electricity could only flow at a maximum rate of 0.242 kg/sec. Two cardboard honeycombs were used to mix the heated air from the finned tube assembly. One honeycomb was placed crosswise to the flow stream and had a relatively low porosity, while the second honeycomb had a larger porosity and was placed through an isolated chamber. The pressure drop over a calibrated orifice plate was measured with a differential manometer to calculate the air's mass flow rate. It indicates the air flow rate using a conventional orifice plate. In order to keep the temperature consistent, the heat exchanger base was heated evenly.

A well-fitting open-topped hardwood box with enough thermal insulation protects the pin-fin assembly base and heaters. Altering the voltage and variac position would allow one to control how much power was delivered to the main heater. Nine thermocouples were used to measure the test module's base plate temperature.

Eight additional thermocouples were used to monitor the temperature of air entering and leaving the system. There were four additional positioned downstream of the pin fin array and four more across from the pin-fin assembly's entrance. The steady state reached in the experimental run is when all of the thermocouples were linked to the temperature indicators and those that indicate the ambient air temperature. Eight additional thermocouples were used to monitor the air temperature coming in and going out of the system. Eight were placed, four downstream of the pin fin array and four in front of the pin-fin assembly's entrance. The steady state was obtained around two and a half hours into the experiment. The trials were run for an additional hour or so using a second thermocouple, representing the ambient air temperature attached to the temperature indicators to verify the steady state result. An orifice meter calibrated against the hot wire anemometer measured the air flow rate. A gate valve on the blower's suction side can be adjusted to open to control the airflow rate. The blower forces air through the test portion, which escapes to the outside for the whole experiment.

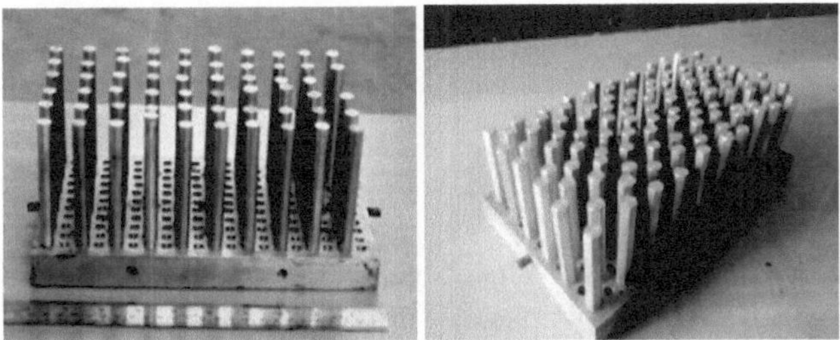

Fig. 60.3 Photographic view of cylinder and hexagonal arrays

3. Data Reduction

The calculated data set helps to measure the friction factor (FF) and heat transfer rate (HTR). From this finned surface, HTR in steady state condition is about,

$$Q_{tot} = Q_{conv} + Q_{rad} + Q_{loss} \tag{1}$$

There are related data reduction cases studied by**Naik *et al.* (13), Jubran *et al.*(13), Tahat*et al,* (3)**.They stated the fins arrays are comparable, and the assembly heat loss is controlled to below 5%.Because of the existing operating conditions, the test segment is well insulated, and it is assumed that the losses are minimal, and then the Equation. (1) is rephrased as,

$$Q_{conv} = mc_p(t_{out} - t_{in}) \tag{2}$$

By convection, heat is transferred from fin surfaces and parent plates as follows:

There is a relationship between t_{in} and t_{out}, which corresponds to the airflow temperatures. t_b is the average temperature of specific aspects of the parent assembly, and the amount of surface area that covers the base assembly (A_s) and fins is expressed as,

$$Q_{conv} = hA_s\left[t_b - \left(\frac{t_{in} + t_{out}}{2}\right)\right] \tag{3}$$

Based on earlier steady-state heat transfer studies (Bilen et al., (2) and Kodah et al., (3) using the same system ($t_b = 40 \pm 0.5°C$), less than 0.5 percent of the total steady-state heat lost from fin arrays may be attributed to radiation heat transfer from machined duralumin pin-fin assemblies. Due to the duralumin construction of the base plate and pin-fins, the surfaces are low-emissivity and well polished.

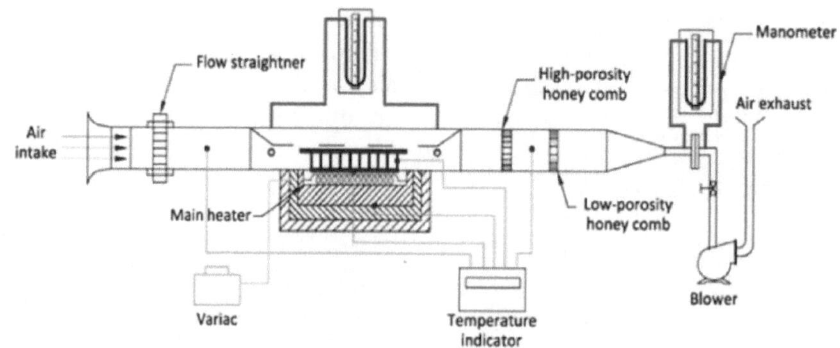

Fig. 60.4 Experimental setup [14]

Low emissivity is produced by a base plate and duralumin pin-fin with highly polished surfaces.

$$h = \frac{m\,c_p\left(t_{out} - t_{in}\right)}{A_s\left[t_b - \left(\frac{t_{in} + t_{out}}{2}\right)\right]} \tag{4}$$

In the experimental setup, in addition to the test section's excellent insulation.

The area of free flow (A_{ff}) is computed as,

$$A_{ff} = W(H + C) - N_x H\,d \tag{5}$$

The expressions for air's dynamic viscosity, specific heat, and thermal conductivity at atmospheric pressure from the range contain the respective values from Kodah et al. (3). $250 \le (T_{in} + T_{out})/2 \le 400$:

$$c_p = \left[9.8185 + 7.7\times10^{-4}\,\frac{(T_{in} + T_{out})}{2}\right]\times10^{2} \tag{6}$$

$$\mu = \left[4.9934 + 4.483\times10^{-2}\,\frac{(T_{in} + T_{out})}{2}\right]\times10^{-6} \tag{7}$$

$$k = \left[3.7415 + 7.495\times10^{-2}\,\frac{(T_{in} + T_{out})}{2}\right]\times10^{-3} \tag{8}$$

The Reynolds number (Re) is given as,

$$\mathrm{Re} = \frac{G\,d}{\mu} \tag{9}$$

Where $G = m/A_{ff}$ is the mass flux

4. Results and Discussions

The following outcomes for each parametric influence on heat transmission are examined independently or in combination: Pin-fin spacing, pin-fin clearing ratio, pin-fin array, pin-fin shape (cross-section), and pin-fin size (number of fins).

4.1 Heat Transfer Rate

Figure 60.5 shows the Reynolds number and average heat transfer rate in the finned surface for various constants (Sx=1.2; Sy=1.2, 2.4, and 3.6). The hexagonal fin array has a greater maximum heat transfer rate than the cylindrical fin at Sy/d=1.2.

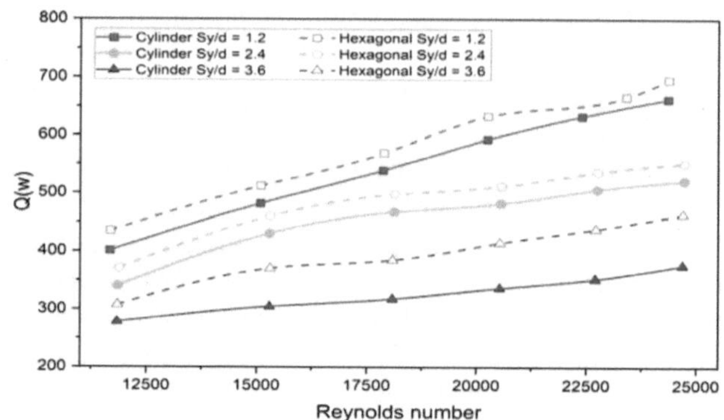

Fig. 60.5 Pin-shaped object's impact on Q out for Sx/d = 1.2

4.2 Effect of Nusselt Number as Function of Reynolds Number

Figure 60.6 illustrates the relationship between the Reynolds number and the Sx=1.2 and Sy=1.2, 2.4, and 3.6 constant values. There is a positive correlation between the Reynolds and Nusselt numbers. A low Sy/d correlates with a high Nusselt number because an increase in Sy/d reduces the pin-lowest fin's velocity.

4.3 Effect of Clearance Ratio (C/H)

The influence of the clearance ratio on heat transfer rate is shown in Figure 60.7, which covers the range of Reynolds numbers evaluated in the trials. For all C/H values, there is a monotonic increase in the heat transfer rate with Reynolds number. Additionally, the lowest C/H ratio is where the greatest heat transfer is observed; the heat exchanger's greater compactness makes this improvement in heat transmission possible (i.e., confining the flow of air freely). This pattern has been noted by prior researchers (Jubran

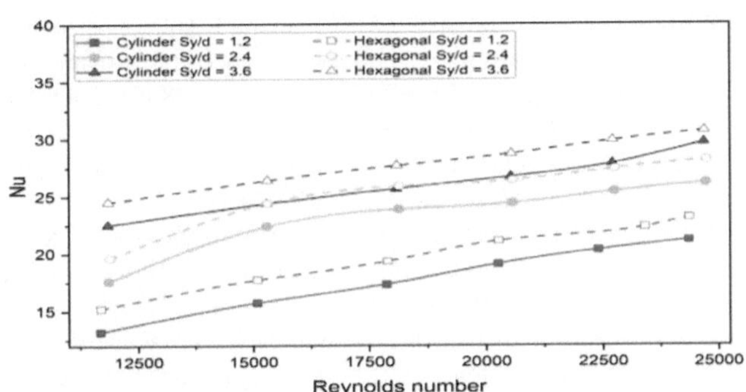

Fig. 60.6 Nu against reynolds number plot for C/H = 0 and Sx/d = 1.2

et al. (8); Tahat et al. (3)). In Fig. 60.7, the impact of fin configurations on heat transport is also shown.

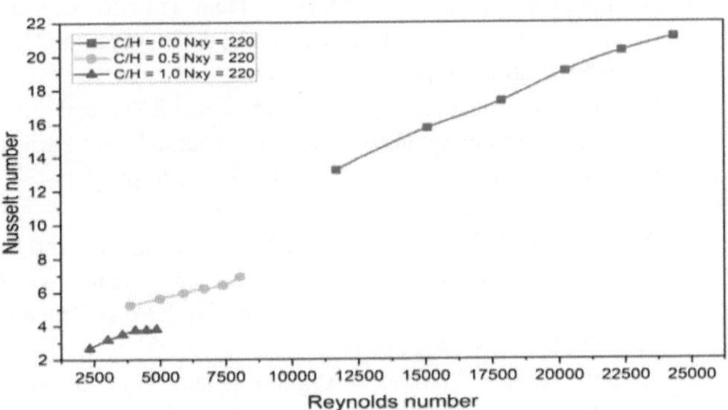

Fig. 60.7 Reynolds number against nusselt number plot for Sx/d = Sy/d = 1.2 (cylinder)

4.4 Friction and Pumping Power Characteristics

Effect of Fin Number and Flow Rate on Pressure Drop: Often utilised as a constraint to have minimum pumping power, flow resistance is a vital component in heat transmission. The pressure decrease at different flow rates is shown in Fig. 60.8 for C/H=0.0 by plotting the friction factor against the Reynolds number. According to Jubran BA et al. (8) and Kodah et al. (3), staggered designs cause significant changes in Sx/d, although in-line layouts cause a modest variation in the friction factor on Sy/d change.

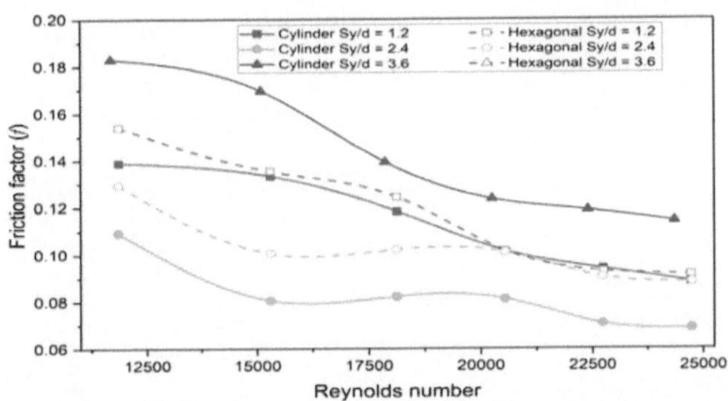

Fig. 60.8 Plot of "f" versus reynolds number (cylinder) for the cases when Sx/d = 1. 2, Sy/d = 1.2, 2.4, 3.6

4.5 Effect of Pin-fin-arrangement

Figure 60.9 shows how the hexagonal pin-fin and cylinder profile forms affect the enhancement of heat transmission. The cylinder outperforms the hexagonal fins despite the predicted four pin-fin profiles. The performance of staggered pin-fin arrays surpasses that of the other systems. This is because the wake flow impact becomes less pronounced once past the first row of pin fins. Between any two adjacent pin-fins of the upstream row, free flow fluid or air introduces new boundary layer growth into the downstream rows to support the heat transfer mechanism. A staggered array's primary flow path is longer or more convoluted, lengthening the fluid's residence time.

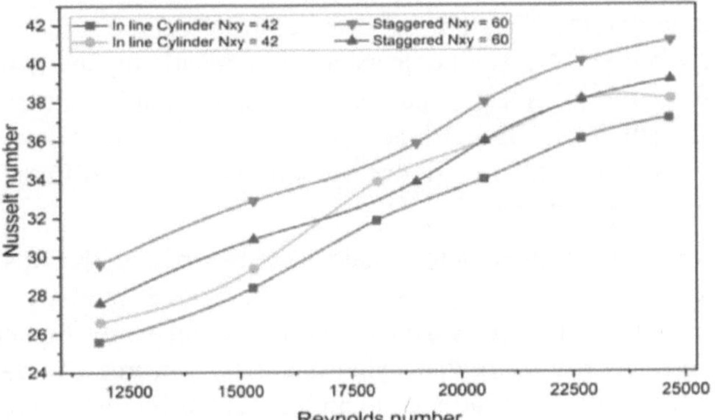

Fig. 60.9 Pin-fin shape's effect on Nu for C/H=0 and Sx/d= 1.2

4.6 Comparison with other Heat Transfer Correlations

Usually, heat transfer data are displayed as changes in Nusselt numbers relative to Reynolds numbers. For the current work, Nusselt numbers rise with Reynolds number, as seen in Fig. 60.10. it is being compared to earlier research to validate the data. Within the experimental range, it is discovered that the current results are greater and more in line with those of Kodah et al. [3] and Jubran BA et al. [8].

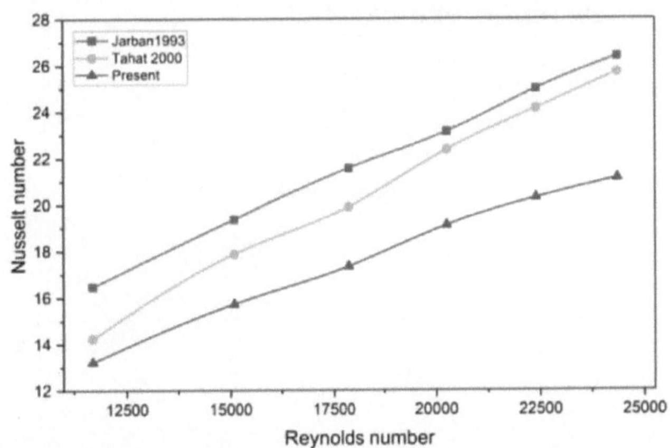

Fig. 60.10 Reynolds number vs. nusselt number plot (with other correlation)

5. CONCLUSION

This work has experimentally investigated pin fin heat transfer under forced convection. There has been a performance comparison between the hexagonal and cylindrical fins. These inferences can be made based on the experimental results mentioned above.

It transfers heat more effectively when comparing hexagonal fin arrays with cylindrical fin arrays. 1.2 (Sy/d) is the ideal stream-wise ratio for enhancing heat transfer. Additionally, because more of the fluid is obstructed in hexagonal fin arrays than cylindrical fin arrays, there is a greater pressure drop and faster heat transfer. As Reynolds number decreases with decreasing streamwise ratio for all fin geometries, the Nusselt number increases. Whenever grooves are present, Nusselt numbers are significantly higher at all Reynolds numbers, regardless of the Reynolds number. A cylindrical fin and a hexagonal fin with Sy/d=2.4 and Sy/d=3.6 each have similar Nusselt numbers for different Reynolds numbers.

The enhancement factor is calculated by dividing the Nusselt number of hexagonal fins by the Nusselt number of cylindrical fins. Hexagonal fins have a relatively higher enhancement factor than cylindrical fins. A decrease in the inter-fin distance ratio led to an increase in friction factor. A hexagonal fin has a little different friction factor than a cylindrical fin.

REFERENCE

1. Braga SL, Saboya FEM. Turbulent heat transfer and pressure drop in an internally finned equilateral triangular duct. Experimental and Thermal Fluid Science. 996;12:57–64.
2. Kadir Bilen, Ugur Akyol, Sinan Yapicic. Heat transfer and friction correlations and thermal performance analysis for a finned surface. Energy Conversion and Management. 2001; 42:1071–1083.
3. Tahat M, Kodah ZH, Jarrah BA, Probert SD. Heat transfer from pin-fin arrays experiencing forced convection. Applied Energy. 2000; 67(4):419–42.
4. Sparrow EM, Ramsey JW, Altemani CAC. Experiments on in-line pin-fin arrays and performance comparison with staggered arrays. ASME J Heat Transfer. 1980; 102:44–50.
5. Metzger DE, Berry RA, Bronson JP.Developing heat transfer in rectangular ducts with staggered arrays of short pinfins. ASME J Heat Transfer 1982; 104:700–6.
6. Simoneau RJ, Vanfossen GJ. Effect of location in an array on heat transfer to a short cylinder in cross flow. ASME J Heat Transfer. 1984; 106:42–8.
7. Armstrong J, Winstanley D. A review of staggered array pin-fin heat transfer for turbine cooling applications. ASME J Turbo mach. 1988; 110:94–103.
8. Jubran BA, Hamdan MA, Abdualh RM. Enhanced heat transfer, missing pin, and optimization for cylindrical pin fin arrays. Trans. ASME, Journal. Heat transfer. 1993; 115(3):576–583.
9. Ashish Dixit & Anil Kumar Patil, Heat Transfer Characteristics of Grooved Fin Under Forced Convection, Heat Transfer Engineering, vol. 36, 1409–1416, 2015.
10. Sharma, N., Tariq, A. and Mishra, M., Enhanced Heat Transfer and Flow Features in a Duct Mounted with Various Ribs, J. Enhanc. Heat Transf., vol. 27, no. 6, pp. 505–526, 2020.
11. Ahmed, Hamdi E., B. H. Salman, A. Sh11.Kherbeet, and M. I. Ahmed. "Optimization of thermal design of heat sinks: A review." International Journal of Heat and Mass Transfer 118 (2018): 129–153, 2018.
12. Karami, Mehdi, Mahmood Yaghoubi, and Amirreza Keyhani. Experimental study of natural convection from an, array of square fins.Experimental Thermal and Fluid Science 93: 409–418, 2018.
13. Naik, S. and Probert, S.D. (1987) Natural convection characteristics of a horizontally-based vertical rectangular fin-array in the presence of a shroud, *Applied Energy,* 28, 295–319.
14. Rathnasamy R, Karthikeyan. R, (2016) Experimental Study of Performance of Pin Fin Heat Sink under Forced Convection, Engineering, Environmental Science. https://www.semanticscholar.org/paper/Experimental-Study-of-Performance-of-Pin-Fin-Heat-Rathnasamy-Karthikeyan/0c32ca3fbf5e094f2e63ce0b4be1874110a8e35e

Note: Every figures except fig. 60.4 were created using the experimental work.

Advances in Additive Manufacturing Technologies – Gurusamy Pathinettampadian (eds)
© 2024 Taylor & Francis Group, London, ISBN 978-1-032-90013-1

61 Effect of EGR and LPG Addition on Jatropha Biodiesel Fuelled CI Engine

V. Vinodkumar
Department of Automobile Engineering,
KCG College of Technology, Chennai, India

L. Ranganathan*
Department of Mechanical Engineering,
Cambridge Institute of Technology, Ranchi, India

T. Anand
Department of Mechanical and Automation Engineering,
Agni College of Technology, Chennai. India

V. Velmurugan
Department of Electronics and
Communication Engineering, Vel Tech Rangarajan Sagunthala
R&D Institute of Science and Technology, Chennai. India

ABSTRACT: This article experimentally examines the outcome of Liquefied Petroleum Gas (LPG) addition along with inlet air with Jatropha biodiesel and Exhaust Gas Recirculation (EGR) on the performance and emission characteristics of a single cylinder direct injection compression ignition engine (CIDI) in dual fuel mode. LPG was inducted along with the fresh air through the intake manifold at the flow amount of 170 g/hr and the Jatropha biodiesel in the ratio of 80:20 (80 % biodiesel and 20 % diesel by volume basis) was taken for this present investigation. It was observed that the LPG addition emits a bit greater NO_X emissions around 6.5 % at complete load operation due to increased peak pressure and temperature when compared with neat diesel fuel. NO_X emissions were greatly abridged with the introduction of EGR. It was reduced by 17 % and 29 % by 10 % and 20 % EGR respectively. The Brake Thermal Efficiency (BTE) for dual fuel mode with 10 % EGR was on par with diesel fuel. However, the reduction in BTE was found by 20 % EGR when compared to 10 % EGR due to excess amount of exhaust gas that makes incomplete combustion. Hydrocarbon (HC) and Carbon monoxide (CO) emissions were greater than before with the introduction of EGR in the dual fuel mode.

KEYWORDS: Dual fuel, JatrophOoil methyl ester (JOME), LPG, Exhaust gas recirculation, Emissions

1. INTRODUCTION

Owing to greenhouse effect of CO_2 and nitrous oxide emissions, researches have enforced on finding out for an alternate fuel for diesel engines. Biodiesels are major alternative for diesel fuel which is derived from vegetable oils [1]. The usage of neat vegetable oil can result in partial combustion and deposition of carbon in the cylinder due to

*Corresponding author: ranganathanl1975@gmail.com

DOI: 10.1201/9781003545774-61

its high viscosity, but it produced less NO_X emissions associated to traditional diesel oil [2,3]. Viscosity of vegetable oil is abridged by transesterification or hydroliquification process in order to avoid the above problems faced by straight vegetable oils. However, because of its advanced oxygen content, biodiesel developed greater NO_X emissions almost at all loads associated to neat diesel oil [1,4]. Numerous approaches are employed to minimize NO_X emissions, such as injection pressure and timing changes, variable geometry turbocharging, EGR and SCR. One such an effective method is exhaust gas recirculation technique which reduces the NO_X emissions significantly. This is owing to the dilution of exhaust gas and new air which clues to drop in combustion peak temperature [5,6]. Some researchers used antioxidant biodiesel blend to lower the NO_X emission and to rally the engine efficiency [7]. The blends of jatropha and fish wastes biodiesel produced less CO, HC and soot emissions at each loads as associated to diesel fuel [8].

The BTE for biodiesel combinations was marginally inferior when likened to diesel at all loads [9]. Due to lower energy content or lesser heating value of biodiesel, the amount of fuel to be injected was increased that leads to greater consumption of fuel and thus resulted in lesser thermal efficiency [10]. This is also because of larger fuel droplets are formed during atomisation due to its higher density, viscosity and evaporation property [11]. It was identified that the diesel engine produced lower BTE when it operates with various biodiesels such as jatropha, karanja, corn, Rice bran, algae, lime peel oil, soybean and Roselle etc [12,13]. The lower BTE problem with biodiesel blends was rectified when the engine remained in a dual fuel approach. In dual fuel mode, the gasoline fuels like LPG, natural gas, acetelyne, etc were combined along with fresh suction air in the intake port and inducted into the engine cylinder [14]. NO_X emissions were also minimised when the engine operates by using LPG-Biodiesel in this mode. The practice of LPG is an auspicious approach to shrink the emissions in CI engines [15]. LPG primarily consists of propane and butane and they are produced by petroleum refineries through the operation of natural plants. The main advantages of LPG than other gasoline fuels are its easy storage and flexible transportation since this may be dissolved at a pressure value of 0.7 to 0.8 bar [16]. LPG has higher octane number and lesser cetane number. This later one of LPG limits its quantity used in diesel engine linked to diesel oil. Therefore, it can be utilised in dual combustion approach instead of by means of as primary fuel which was very challenging. In this dual fuel procedure, the LPG is directly mixed with fresh air in the intake port and pumped into the engine cylinder [15].

This present investigation deals with the collective influence of EGR and LPG on the performance and emission features of a CIDI engine fuelled by 80:20 volume-based jatropha biodiesel combinations in a dual fuel approach.

2. Materials and Methods

2.1 Biodiesel Production

For this experimental work, the fresh jatropha oil was purchased from resident merchant in Chennai, India. The transesterification method helped turn the raw jatropha oil into biodiesel. It is one technique for turning vegetable oils into biodiesel. It is the process of chemically breaking down oil or fat molecules into glycerol and an ester by utilizing an alcohol, such as methanol or ethanol, and a catalyst, such as NaOH, KOH, or H_2SO_4 [17]. In this method, a litre of neat jatropha oil was transferred in a round bottom flask; 10 ml of H_2SO_4 and 250 ml of methanol are occupied and mixed in a distinct vessel. Then this solution is combined with jatropha oil in the flask and this mixture was continuously agitated using rapid stirring mechanism. After stirring and maintaining a temperature of 60 °C for approximately an hour, the combination was let to gravitationally settle down in the splitting conduit. Then in the unravelling conduit two different films were formed, such as the superior level consists of ester and the subordinate

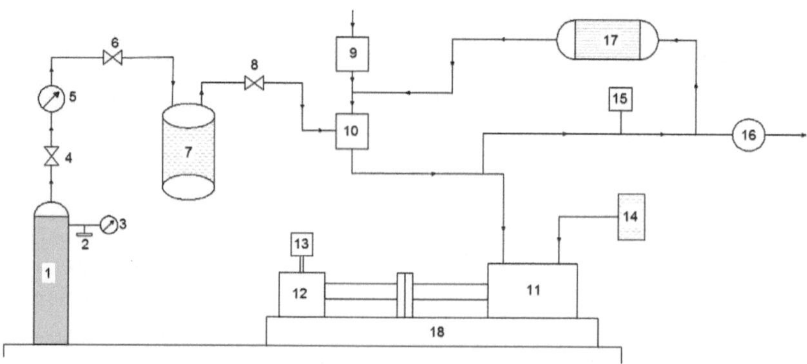

Fig. 61.1 Experimental setup

Table 61.1 Properties of trial fuels

Property	Diesel	Jatropha oil	JOME	Standard
Density at 16°C (kg/m3)	833-880	916	878	ASTM D 4052
Heating value (KJ/kg)	42500	39070	38450	ASTM D 5865
Auto ignition temperature (°C)	257	—	—	ASTM D 2283
Viscosity at 40°C (cSt)	4.59	49.9	5.65	ASTM D 445
Cetane Index	40	40-45	50	ASTM D 613
Flash Point (°C)	75	240	179	ASTM D 92

level consists of glycerol. Glycerol and ester separated by a slight opening from the dividing conduit. Then the separate ester was then mixed with the water twice and held under gravity for about a day. The catalyst softened in water, which shaped in the extremity layer and was eliminated. Moistness from the ester was eliminated by means of heating. Required time for producing biodiesel using trans-esterification was about 48 hours. The properties of trial samples are presented in Table 61.1.

1. LPG Chamber
2. Pressure regulator
3. Pressure scale
4. Flow regulator
5. Flow meter
6. NRV
7. Flame trap
8. Control faucet
9. Air surge chamber
10. Gas assortment
11. Diesel engine
12. Eddy current dynamometer
13. Loading device
14. Fuel container
15. Five gas analyser
16. Exhaust orifice
17. EGR
18. Engine bed

2.2 Experimental Set up and Test Technique

The research engine exploited in this experimentation was an air-cooled, 4.4 kW CIDI engine with a persistent speed of 1500 rpm. Figure 61.1 and Table 61.2, respectively, display the engine arrangement and specifications. Burette compares the fuel tank to quantify the amount of fuel expended in element time. To measure the consumption of air, an orifice meter is provided on the suction line. EGR set up is also provided. The charging or loading is achieved using an electrical dynamometer. The cylinder pressure is measured by piezoelectric transducer. In order to generate an output voltage relative to that charge, an amplified charge is given.

Table 61.2 Technical specifications of the engine

Engine nature	Kirloskar Engine
Cylinder Bore	87.5 mm
Stroke length	110 mm
Stroke Volume	661.5 cc
Injection Timing	23° bTDC
Injection pressure	220 bar
Power output	4.4 kW
Max speed	1500 rpm
Compression ratio	17.5 : 1
Cooling method	Air

Emissions including Burette compares the fuel tank to measure the amount of fuel expended in unit time UBHC, CO_2, CO, O_2 and NO_X and have been assessed using five gas analysers. This paper agreement with the study of a twin fuel diesel engine driven by means of jatropha oil biodiesel in the 80:20 ratio as main fuel and LPG as ancillary fuel supplied in the intake port to blend with fresh air. The flow rate of LPG is controlled by rotameter. The attributes of the gaseous fuel are displayed in Table 61.3. The engine's efficiency and emissions in dual fuel manner were matched to straight diesel oil. The proportion of uncertainties of the various instruments are used to calculate the fraction of uncertainties of various variables, such as SFC, BTE, HC, Smoke, CO, NO_X, and CO_2. The overall uncertainty,

Table 61.3 Properties of LPG (Propane)

Property	LPG (Propane)
Auto Ignition temperature (°C)	450
Boiling point (°C)	-42.5
LCV (kJ/kg)	46500
Density at 16°C (kg/m3)	505.5
Latent heat of vapourisation at 16 °C (kJ/kg)	357

Table 61.4 Uncertainty of the instrument

Instruments	Range	Accuracy	Uncertainties %
Exhaust gas analyser			
HC (ppm)	0 – 20000	±10	0.25
CO (%)	0 – 10	±0.02	0.25
CO_2 (%)	0 – 10	±0.02	1
NO_X(ppm)	0 – 5000	±50	0.2
Pressure transducer (bar)	0 – 110	±1	0.2
Crank position encoder (°CA)		±1	0.25
Smoke meter (%)	0 – 100	±1	1
Speed sensor (rpm)	0 – 10000	±10	1
Thermocouple	1500	±1 °C	±0.17
Gauging Burette (cc)	1-30	±0.25	1

which comes from the following relations, was found to be ± 2.298.

$$= \sqrt{(BP)^2 + (BSEC)^2 + (TFC)^2 + (BTE)^2 + (CO)^2 + (HC)^2 + (CO_2)^2 + (NOX)^2 + (smoke)^2 + (EGT)^2 + (Pressure\ Pick\ up)^2}$$

$$= \sqrt{0.22 + 12 + 12 + 12 + 0.252 + 0.252 + 12 + 0.22 + 12 + 0.22 + 0.22}$$

$$= \pm 2.298\ \%.$$

3. Results and Discussion

3.1 Brake Thermal Efficiency

The assessment of BTE of diesel and LPG-JOME fuel is shown in Fig. 61.2. It is the degree of heat energy diverted to usable work. The effects of primary fuel quantity on the output of dual fuel engines powered by gaseous fuel are investigated and it ensued in an upsurge in BTE by increasing the pilot fuel quantity [18]. From the chart, it is grasped that the engine's BTE in twin fuel operation at inferior load is marginally poorer (1.22 %) when associated to diesel fuel. This is a result of there being less pilot fuel. The BTE is improved even more and is now comparable to diesel fuel as the amount of pilot fuel increases with load. This results from the fuel's improved combustion. In addition, it is observed in dual fuel manner that at low loads, the efficiency can be increased with slightly larger fuel pilot quantity if no detonation takes place [19]. With the introduction of 10 % EGR, the BTE is marginally improved due to the auxiliary combustion of unburned hydrocarbon that enters into the combustion chamber with recirculated exhaust gases. Efficiency reduction is seen at 20 % EGR as additional exhaust gasses partly reduce the quantity of air inflowing the engine.

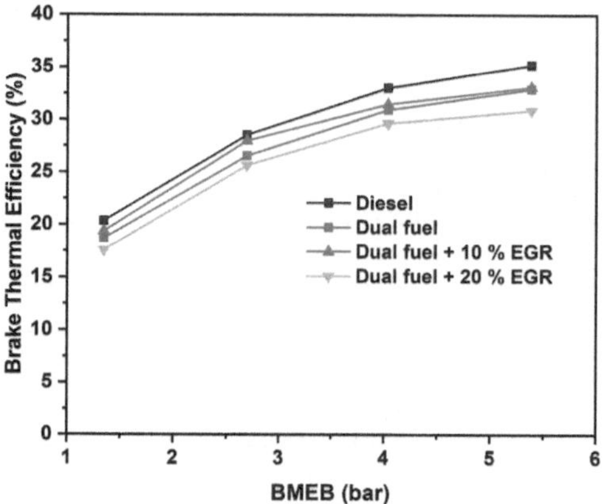

Fig. 61.2 Variation of BTE with BMEP

3.2 Brake Specific Fuel Consumption

The assessment of BSFC of diesel and LPG-JOME is shown in Fig. 61.3. It is the volume of fuel necessary for creating a single kW of power over a single unit of time. As seen, it is grasped that amid the wholly tried fuel, the diesel energy ensued in lower BSFC value owed to its higher calorific value [20]. Among dual fuel mode, the lesser BSFC value was observed with addition of 10 % EGR due to its auxiliary combustion of unburned hydrocarbon that comes in contact in the engine cylinder and also the unburned hydrocarbon in the engine exhaust releases little energy during combustion process [21]. With the introduction of 20 % EGR, the BSFC value increases further owed to its inferior heating value.

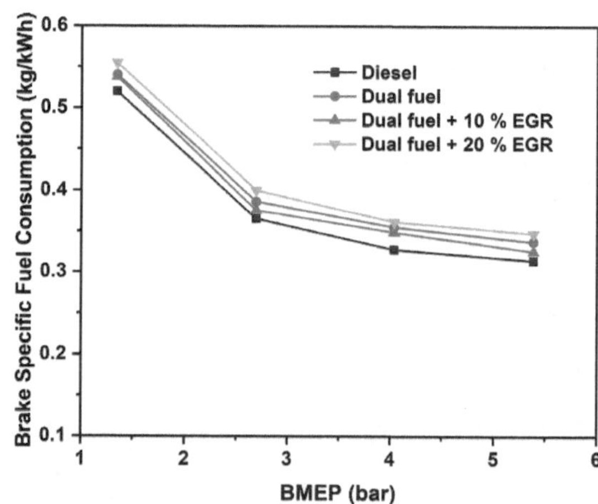

Fig. 61.3 Variation of BSFC with BMEP

3.3 Unburned Hydrocarbon Emission

The assessment UBHC of diesel and LPG-JOME fuel is shown in Fig. 61.4. Increase in hydrocarbon (HC) emissions is noted in the twin fuel operation of the engine including EGR when associated to diesel fuel. HC emission for dual

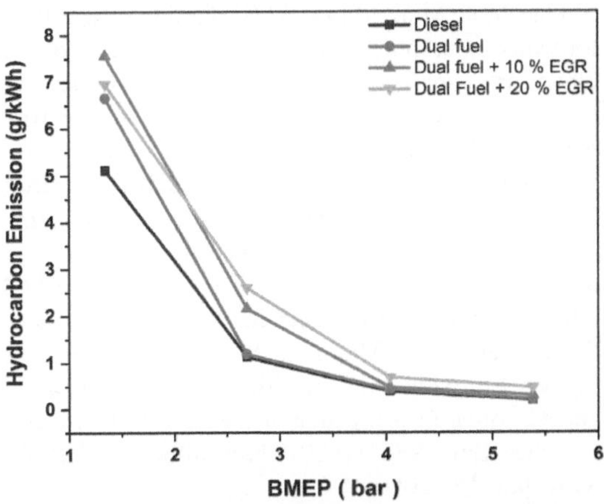

Fig. 61.4 Variation of HC emission with BMEP

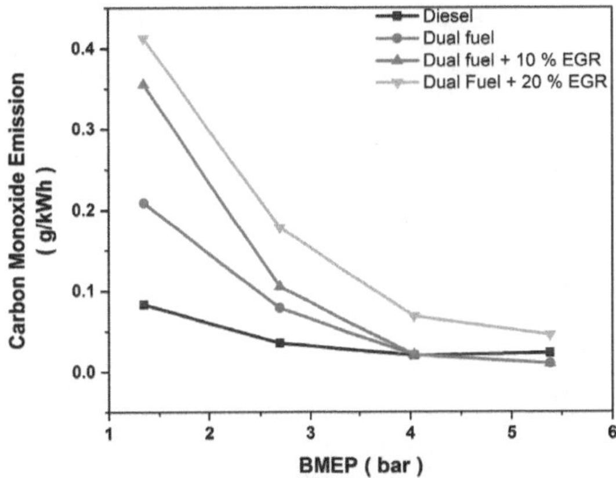

Fig. 61.5 Variation of CO emission with BMEP

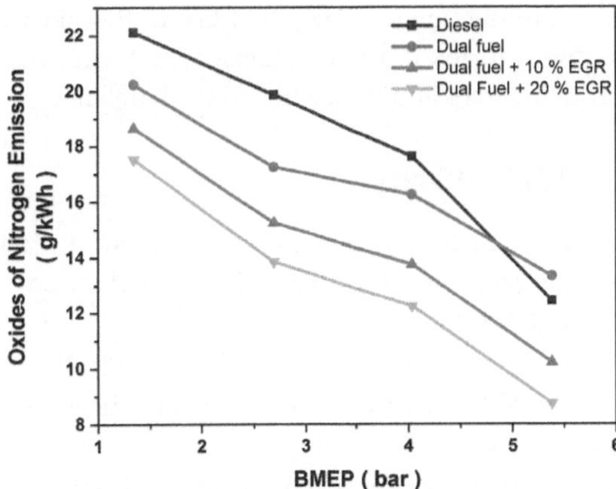

Fig. 61.6 Variation of NO_X emission with BMEP

fuel engine upsurges by 22.08 % at part load when contrast to diesel fuel. This is owing to inadequate oxygen entering the engine's combustion cavity which makes fuel burning improper [22]. This is also payable to greater self-ignition temperature of LPG (470°C) when compared to diesel fuel (210°C). Hence surplus period is existing for the fuel to remain trapped in the crevice volume [21]. Increased pilot fuel quantity at full load enhanced the combustion process and least change in HC emission is noted for both the fuel approaches. With the introduction of 10 % and 20 % EGR, the increase in HC emission is found with escalation in load. This is due to mixing of higher percentage of EGR with inlet air makes the combustion incomplete.

3.4 Carbon Monoxide Emission

The comparison of CO of LPG-JOME and diesel fuel is shown in Fig. 61.5. The partial combustion that happens in diesel engines as a result of low oxygen levels discharges carbon monoxide. In dual fuel conditions with and without EGR, higher CO emissions are generated at no load condition compared using diesel oil. The justification for this is the decreased combustion temperature. Reduction in CO emission is observed when the load increases due to fractional oxidation of LPG during compression prior to combustion process which makes the combustion complete. Increase in CO is the main disadvantage in dual fuel manner associated to the neat diesel oil [23].

3.5 Oxides of Nitrogen Emission

The assessment in NO_X of diesel and LPG-JOME fuel is shown in Fig. 61.6. When using dual fuel at rated load, the generation of NO_X is 7% more than when using neat diesel fuel. The motive for greater NO_X emission is that increased quantity of pilot fuel combine with LPG at full load makes better combustion of fuel and hence more combustion temperature more the NO_X production. The NO_X is further abridged by the application of EGR. This is owed to weakening of the fresh charge when mixed with exhaust gases which are recirculated through intake port with the adoption of the EGR technique [24]. There is reduction in NO_X emission when the liquid fuel used as a secondary fuel instead of LPG. This is payable to lower combustion temperature obtained in dual fuel manner [25].

3.6 Exhaust Gas Temperature

The comparison of EGT of diesel and LPG-JOME fuel is displayed in Fig. 61.7. The EGT is slightly higher at full load for dual fuel approach when associated to that of traditional diesel oil. This is because of the increase in the pilot fuel quantity with load increases. At part load, the EGT is lower likened to the conventional diesel fuel due to the leaner air fuel combination. When the liquid oil used as a secondary fuel instead of LPG, there is a significance fall in EGT [20]. When the delay in ignition increases with increased density of biodiesel mixtures, it takes more time for the evaporation of the fuel during premixing period which possibly reduces the combustion temperature [23].

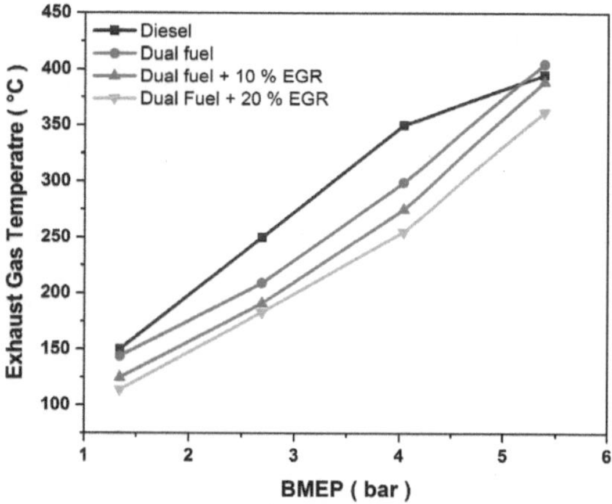

Fig. 61.7 Variation of EGT with BMEP

With the introduction of EGR, the EGT further decreases. This is owed to the dilution exhaust fumes with the new inlet air that enters into the combustion chamber.

4. Conclusions

- The engine's BTE in dual fuel manner with 10 % EGR was on same level with the diesel fuel owed to its enhanced combustion.
- Like BTE, the dual fuel mode with 10 % EGR resulted lesser BSFC value owed to its greater heating value and superior combustion.
- In dual fuel process at part load condition, the HC emission was increased by 23 % without EGR, as associated to that of straight diesel engine. Further reduction in HC release was observed owed to the insufficient amount of oxygen with the introduction of 10 % and 20 % EGR and hence combustion of fuel is incomplete.
- Due to incomplete fuel combustion, CO emissions were likewise higher in dual mode without EGR, with 10% and 20% EGR, than in clean diesel fuel. This is similar to the rise in HC emissions.
- When 10% and 20% EGR was added, the NO_X output dropped as the peak combustion temperature dropped. The full-load dual fuel engine's NO_X creation was witnessed to be 7% greater than that of straight diesel fuel.

References

1. P.K. Chaurasiya, S.K. Singh, R. Dwivedi, R. V. Choudri, Combustion and emission characteristics of diesel fuel blended with raw jatropha, soybean and waste cooking oils, Heliyon. 5 (2019) e01564. https://doi.org/10.1016/j.heliyon.2019.e01564.
2. S. Che Mat, M.Y. Idroas, M.F. Hamid, Z.A. Zainal, Performance and emissions of straight vegetable oils and its blends as a fuel in diesel engine: A review, Renew. Sustain. Energy Rev. 82 (2018).808–823. https://doi.org/10.1016/j.rser.2017.09.080.
3. P. Hellier, F. Jamil, E. Zaglis-Tyraskis, A.H. Al-Muhtaseb, L. Al Haj, N. Ladommatos, Combustion and emissions characteristics of date pit methyl ester in a single cylinder direct injection diesel engine, Fuel. 243 (2019) 162–171. https://doi.org/10.1016/j.fuel.2019.01.022.
4. S. Mathavan, T. Mothilal, V. Dillibabu, D. Billy, G. Muthu, Emission characteristics study of Gasoline-Diesel and Gasoline-Diesel/pentanol blend, IOP Conf. Ser. Mater. Sci. Eng. 988 (2020). https://doi.org/10.1088/1757-899X/988/1/012050.
5. K. Vamsi Krishna, G.R.K. Sastry, M.V.S. Murali Krishna, J. Deb Barma, Investigation on performance and emission characteristics of EGR coupled semi adiabatic diesel engine fuelled by DEE blended rubber seed biodiesel, Eng. Sci. Technol. an Int. J. 21 (2018) 122–129. https://doi.org/10.1016/j.jestch.2018.02.010.
6. A. Praveen, G.L.N. Rao, B. Balakrishna, G. Lakshmi Narayana Rao, B. Balakrishna, Performance and emission characteristics of a diesel engine using Calophyllum Inophyllum biodiesel blends with TiO2 nanoadditives and EGR, Egypt. J. Pet. 27 (2018) 731–738. https://doi.org/10.1016/j.ejpe.2017.10.008.
7. G. Balaji, M. Cheralathan, Experimental investigation of antioxidant effect on oxidation stability and emissions in a methyl ester of neem oil fueled DI diesel engine, Renew. Energy. 74 (2015).910–916. https://doi.org/10.1016/j.renene.2014.09.019.
8. B. Kathirvelu, S. Subramanian, N. Govindan, S. Santhanam, Emission characteristics of biodiesel obtained from jatropha seeds and fish wastes in a diesel engine, Sustain. Environ. Res. 27 (2017) 283–290. https://doi.org/10.1016/j.serj.2017.06.004.
9. V. Vinodkumar, A. Karthikeyan, Effect of ternary fuel blends on performance and emission characteristics of single cylinder diesel engine, J. Phys. Conf. Ser. 2054 (2021). https://doi.org/10.1088/1742-6596/2054/1/012007.
10. V. Vajravel, K. Alagu, Influence of ternary fuel blends of decanol / neem oil biodiesel / diesel on combustion , emission and performance characteristics of an unmodified diesel engine, (2022). https://doi.org/10.1080/01430750.2022.2075928.
11. P. Shrivastava, T.N. Verma, A. Pugazhendhi, An experimental evaluation of engine performance and emisssion characteristics of CI engine operated with Roselle and Karanja biodiesel, Fuel. 254 (2019) 115652. https://doi.org/10.1016/j.fuel.2019.115652.
12. K. Alagu, H. Venu, J. Jayaraman, V.D. Raju, L. Subramani, P. Appavu, D. S, Novel water hyacinth biodiesel as a potential alternative fuel for existing unmodified diesel engine: Performance, combustion and emission characteristics, Energy. 179 (2019) 295–305. https://doi.org/10.1016/j.energy.2019.04.207.

13. G. Ragothaman, T. Mothilal, M.D. Rajkamal, S. Kaliappan, S. Mathavan, Performance and emission characteristics of diesel blended with sweet lime peel oil and corn oil, in: 2020: p. 020036. https://doi.org/10.1063/5.0024898.
14. B. K, P. Krishnan, Effect of acetylene addition in safflower biodiesel fueled CI engine–an experimental study, Energy Sources, Part A Recover. Util. Environ. Eff. 00 (2019) 1–15. https://doi.org/10.1080/15567036.2019.1678700.
15. H.S. Tira, J.M. Herreros, A. Tsolakis, M.L. Wyszynski, Influence of the addition of LPG-reformate and H2 on an engine dually fuelled with LPG-diesel, -RME and -GTL Fuels, Fuel. 118 (2014).73–82. https://doi.org/10.1016/j.fuel.2013.10.065.
16. B. Ashok, S. Denis Ashok, C. Ramesh Kumar, LPG diesel dual fuel engine - A critical review, Alexandria Eng. J. 54 (2015) 105–126. https://doi.org/10.1016/j.aej.2015.03.002.
17. G.K. Souza, F.B. Scheufele, T.L.B. Pasa, P.A. Arroyo, N.C. Pereira, Synthesis of ethyl esters from crude macauba oil (Acrocomia aculeata) for biodiesel production, Fuel. 165 (2016) 360–366. https://doi.org/10.1016/j.fuel.2015.10.068.
18. G.H.A. Alla, H.A. Soliman, O.A. Badr, M.F.A. Rabbo, 00/01012 Effect of pilot fuel quantity on the performance of a dual fuel engine, Fuel Energy Abstr. 41 (2000) 112. https://doi.org/10.1016/s0140-6701(00)90989-5.
19. W. Ying, H. li, Z. Longbao, L. Wei, Effects of DME pilot quantity on the performance of a DME PCCI-DI engine, Energy Convers. Manag. 51 (2010).648–654. https://doi.org/10.1016/j.enconman.2009.10.023.
20. V. Vajravel, Effect of manifold injection of 1-hexanol on a reactivity controlled compression ignition engine fuelled with neem oil methyl ester blend, (2022). https://doi.org/10.1002/ep.13989.
21. A.M. Elzahaby, M. Elkelawy, H.A.E. Bastawissi, S.M. El Malla, A.M.M. Naceb, Kinetic modeling and experimental study on the combustion, performance and emission characteristics of a PCCI engine fueled with ethanol-diesel blends, Egypt. J. Pet. 27 (2018) 927–937. https://doi.org/10.1016/j.ejpe.2018.02.003.
22. V. Vinodkumar, A. Karthikeyan, Effect of manifold injection of n-decanol on neem biodiesel fuelled CI engine, Energy. 241 (2021) 122856. https://doi.org/10.1016/j.energy.2021.122856.
23. S. Srihari, S. Thirumalini, Investigation on reduction of emission in PCCI-DI engine with biofuel blends, Renew. Energy. 114 (2017) 1232–1237.https://doi.org/10.1016/j.renene.2017.08.008.
24. R. Sathiyamoorthi, G. Sankaranarayanan, K. Pitchandi, Combined effect of nanoemulsion and EGR on combustion and emission characteristics of neat lemongrass oil (LGO)-DEE-diesel blend fuelled diesel engine, Appl. Therm. Eng. 112 (2017) 1421–1432. https://doi.org/10.1016/j.applthermaleng.2016.10.179.
25. V. Vinodkumar, A. Karthikeyan, J. Jayaprabakar, G. Senthilkumar, L. Ranganathan, Optimization of port injection of n-decanol in a PCCI engine using response surface methodology, Heat Transf. 52 (2023).5325–5344. https://doi.org/10.1002/htj.22930.

Note: Every figure and table was created using the experimental work; none of them were taken from any publications or online.

Advances in Additive Manufacturing Technologies – Gurusamy Pathinettampadian (eds)
© 2024 Taylor & Francis Group, London, ISBN 978-1-032-90013-1

62 DC Current-Powered Intelligent Defense System

T. Thanka Geetha[1]
Assistant Professor, Department of ECE,
DMI Engineering College, Kanyakumari, India

Vijayan R[2]
Assistant Professor, Mechanical Engineering,
Chennai Institute of Technology, Chennai, India

D. Surrya Prakash[3]
Associate Professor, Mechanical Engineering,
Vel tech Rangarajan Dr. Sagunthala R & D Institute of
Science and Technology, Chennai, India

A. Mani[4]
Associate Professor, Department of CSE,
SA Engineering College, Chennai, India

ABSTRACT: Reducing the load on our soldiers by utilizing laser barriers is a good idea. The transmitter part uses a LASER radiator. On the receiver side, a focused LDR sensor is employed to continuously detect the LASER. The primary circuit triggers the alarm circuit when an individual passes the invisible beam. A passive infrared detector is an electrical device that detects infrared light generated by objects within its field of view. PIR sensors are employed almost exclusively to detect motion and to ascertain if a person has entered or left the sensor's detection range. The position of the invader can be found using infrared and ultrasonic sensors. Similar to radar and sonar, ultrasonic transducers are used in systems that analyse reflected signals to assess targets. IR sensors operate on the basis of an IR LED's emission of IR radiation, which a photodiode detects. Ultrasonic sensors must be utilized in conjunction with infrared sensors for optimal performance of the proximity sensor circuit and to receive vivid data about the item or person violating our boundaries. To discourage trespassers, border barrier uses automatic firearms equipped with ultrasonic and infrared sensors. Radiation-absorbent material, or RAM for short, is a substance that has been precisely designed and formulated to absorb radiofrequency radiation. Using RAM, enemy nations may turn down the radar, but they are unable to turn off our Smart Defense System. Smart power systems that provide DC current from a range of sources can power any of the aforementioned smart defensive systems.

KEYWORDS: LASER, Ultrasonic sensor, Infrared (IR) sensor, Radiation absorbent material (RAM)

[1]vijaytgee@gmail.com, [2]srajendranvijayan@gmail.com, [3]surryame4u@gmail.com, [4]maniathimoolam@gmail.com

DOI: 10.1201/9781003545774-62

1. Introduction

India plans to construct a multi-layered smart barrier that prevents patrols along its borders with Bangladesh and Pakistan by the end of 2017. Ground forces will react in the event that an infiltration attempt is detected by the multi-tier security ring's conventional fence and laser walls. Regarding the question of protecting riverine belts and areas where building a physical barrier along the border is not feasible due to geography, the government will investigate technological solutions to guarantee that every square inch of our nation is safeguarded. The necessity of "smart" borders is undeniable. Utilising the most advanced technology at our disposal is crucial. Enough electricity is supplied by diesel engines, solar panels, and rechargeable batteries to enable substantial floodlighting in other nations. Movement detection is a key component of modern electronic surveillance, which is mostly reliant on sensors.

1.1 Laser Walls

The LASER-Ray is nearly invisible and travels a large distance without scattering, as seen in Fig. 62.1(a). The only objects visible are the radiation point and the incident point. This security project will enable us to erect an imperceptible barrier around a vital area. As shown in Fig. 62.1(b), the system is split into two halves. The receiver is the other, while the transmitter is the first. The transmitter part uses a LASER radiator. A focused LDR (Light Dependent Resistor) sensor continuously detects the LASER on the receiver side. The alarm circuit is triggered by the main circuit's detection of a break via a sensor when an individual crosses the invisible beam. The technique was intended to be both inexpensive and very effective. The framework has a very low power consumption.

(a)

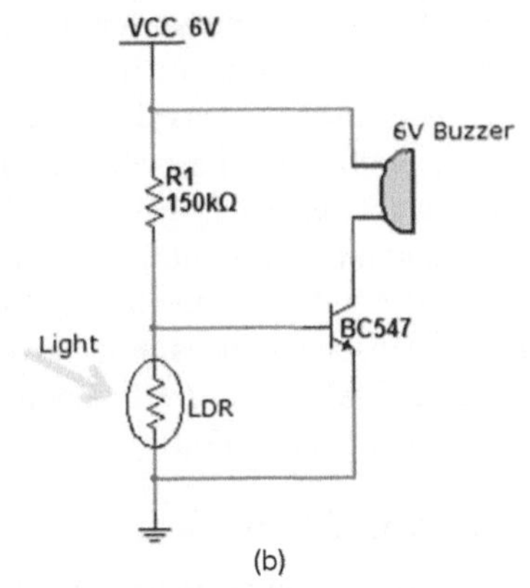

(b)

Fig. 62.1 (a) Laser wall, (b) Laser Wall circuit

Source: Created using the experimental work

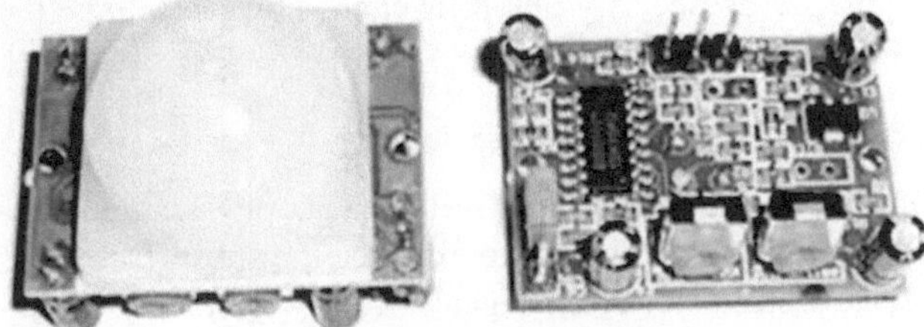

Fig. 62.2 PIR sensor [16]

1.2 PIR Sensor

An electronic sensor that can identify infrared (IR) light generated by objects within its visual field is called a passive infrared sensor (PIR sensor). The most typical places to find them are in PIR-equipped motion detectors, like the ones in Fig. 62.2. All items that have a temperature higher than absolute zero release radiation. This radiation is often undetectable to the human eye due to its infrared wavelength, but electrical devices designed specifically for that purpose can detect it. PIR devices are referred to be "passive" since they don't produce or release any energy in order to detect. Their sole means of operation is the detection of energy released by other objects. PIR sensors measure the infrared radiation that an item emits or reflects, not "heat," per se. Based on the temperature and surface characteristics of the objects in front of it, a PIR sensor measures changes in the quantity of infrared radiation that is impinging on it. The temperature at that location in the sensor's field of vision increases from room temperature to body temperature and then decreases back to room temperature when an object, like a person, passes in front of a backdrop, like a wall. The detection is activated when the sensor changes the output voltage in response to an alteration in the incoming infrared light. When objects with similar temperatures but different surface characteristics move in reference to the backdrop, they may produce infrared light in different patterns, which might set off the detector.

PIR sensors are employed almost exclusively to detect motion and to ascertain if a person has disturbs the sensor's detection range. They don't wear out and are inexpensive, tiny, low-power, and easy to operate. It is capable of measuring infrared radiation. Low-level radiation is emitted by everything, and the more heated an object is, the more radiation it releases. The sensor of a motion detector is really split into two halves. This is because, instead of aiming to average IR levels, we want to identify mobility (change). The two halves cancel each other out because of the way they are connected. If one component receives more or less IR radiation than the other, the output will fluctuate dramatically.

There are two slots on the PIR sensor itself, and each one is made of a different IR-sensitive material. We can see that the two slots can "see" out past a certain distance (essentially the sensitivity of the sensor) because the lens used here is inactive. The amount of IR that is detected by both slots while the sensor is off is the same as the ambient amount that is emitted by the walls, the room, or the outside environment. When a warm body, such a person or animal, moves by and intercepts one side of the PIR sensor, there is a positive differential change between the two parts of the sensor. The sensor produces a negative differential shift when the heated body leaves the detecting region. It is these shift pulses that are identified.

1.3 Laser Wall with PIR Sensor

A Laser Wall has to be connected to a PIR Sensor. It is necessary to integrate these two components in order to prevent unintentional triggers. A little piece of leaf can set off the Laser Security alarm. It can't, however, make the PIR sensor work.

1.4 Ground-Penetrating Radar

Ground-penetrating radar (GPR) is a geophysical technology that uses radar pulses to examine the subsurface. This device detects reflected signals from objects buried beneath the surface by using electromagnetic radiation in the microwave area of the radio spectrum. In addition to rock, soil, ice, fresh water, pavements, and buildings, additional media can also be used with GPR. When applied correctly, GPR may detect items below the surface, changes in the material's characteristics, and cavities and fissures.

High-frequency radio waves, typically in the 10 MHz to 2.6 GHz range, are used in GPR. Earth is exposed to electromagnetic radiation from a GPR transmitter. The energy can be reflected, refracted, or otherwise altered when it comes into contact with a subsurface object or a border between materials with varying permittivities. A receiving device can then be used to capture the dissimilarities in the return signal.

1.5 Ultrasonic Sensors

Sound waves at frequencies higher than the threshold of human hearing are referred to as ultrasounds. Other than the fact that it is undetectable to humans, ultrasound shares all of the physical characteristics of sound. Frequencies used by ultrasound equipment range from 20 kHz to several gigahertz. Ultrasound finds its use in several fields. Ultrasonic devices are used for both object detection and distance measurement. The three primary categories of ultrasonic transducers are transceivers, receivers, and transmitters. While transmitters convert electrical impulses into ultrasound, transceivers are capable of both sending and receiving ultrasonic.

Similar to radar and sonar, ultrasonic transducers are used in systems that analyse reflected signals to assess targets. For instance, the time elapsed between sending a signal and receiving an echo may be used to determine the distance of an object. One of the least expensive options is the measurement of ultrasonic sensors.

1.6 IR Sensors

An IR LED and a photodiode, together known as an IR pair or photo coupler, are the fundamental parts of an infrared sensor. IR sensors operate on the basis of an IR LED's emission of IR radiation, which a photodiode detects. A photodiode's voltage drop varies in response to variations in the amount of infrared light that strikes it. We are able to recognize voltage variations and modify our output accordingly.

1.7 IR Sensors

To adjust the sensitivity of the sensor, the photo diode is placed in reverse bias and the inverting end of the LM358 (PIN 2) is connected to the variable resistor. The non-inverting end (PIN 3) is attached to the photodiode-resistor junction.

The comparator output is LOW and there is no IR radiation aimed at the photodiode when the circuit is turned on. The photodiode absorbs and reflects the infrared light (IR) generated by the infrared LED when an item is positioned in front of an IR pair. The voltage across the photodiode decreases when reflected infrared radiation strikes it, whereas the voltage across the series resistor R2 increases. When the voltage at Resistor R2, attached to the non-inverting end of the comparator, is higher than the voltage at the inverting end, the comparator's output becomes HIGH and the LED illuminates.

1.8 Ultrasonic Sensors Integrated with IR Sensors

In addition to infrared sensors, ultrasonic sensors need to be used to improve the performance of the proximity sensor circuit and provide detailed information about the item or person crossing our limits.

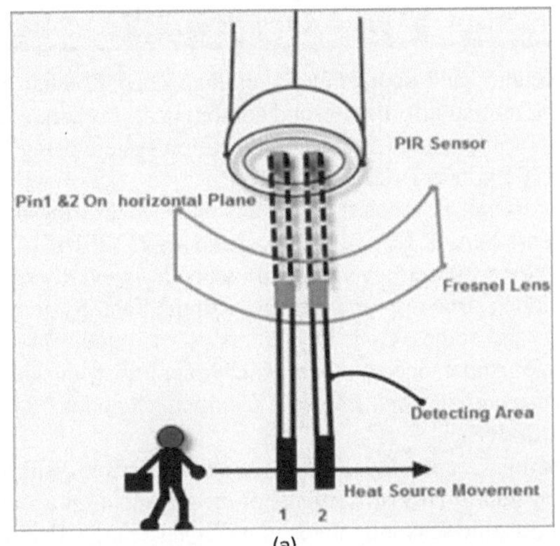

(a)

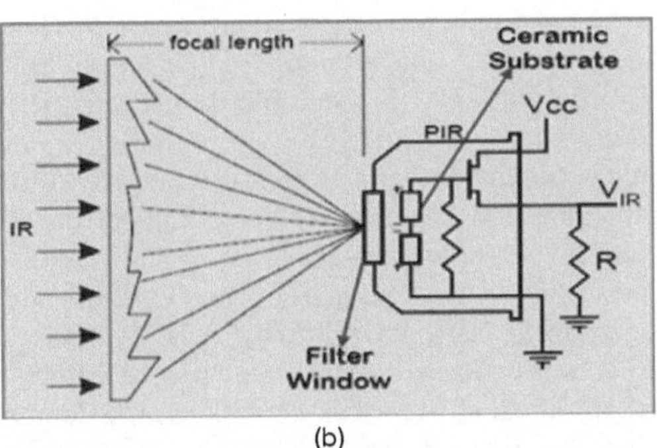

(b)

Fig. 62.3 (a) PIR sensor system, (b) working of PIR sensor [17]

1.9 Automatic Working Guns with IR, Ultrasonic and PIR Sensors

Automatic guns with infrared and ultrasonic sensors are incorporated into the border barrier to detect enemy infiltration. These automated firearms will be activated by a PIR motion sensor to deter infiltration.

2. Dual Modes of Automated Guns

There are two modes of operation for these automatic weapons: 180 degrees and 360 degrees. All of the sensors around the automated cannon will be active for 360 degrees, however the back side of the sensors will be disabled for 180 degrees of operation. Both before to and during the infiltration, the automated gun's 180-degree function is utilised. The 360 degree style of action will be started after the enemy invasion in order to end the invasion entirely.

2.1 Blind Spot of Sensors

RAM (RADAR Absorbent Material)

Radiation-absorbent material (RAM) is a material that has been specially prepared and developed to absorb radiofrequency (RF) radiation, also known as non-ionizing radiation, from as many directions of incidence as is reasonably possible. The less reflected RF radiation there is, the more effective the RAM is.

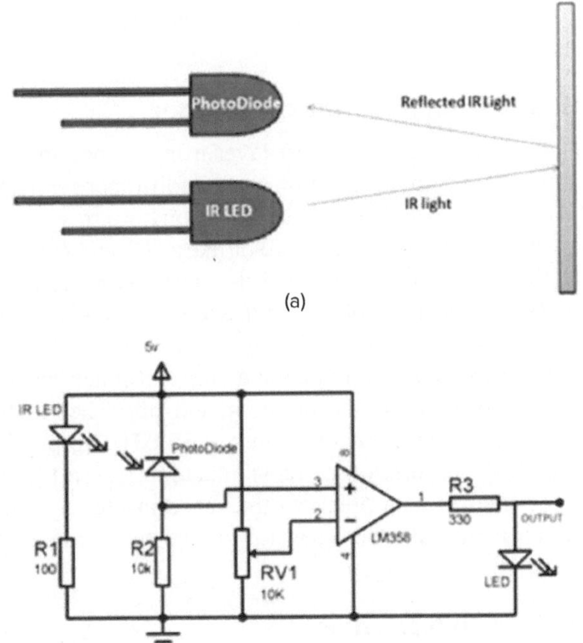

(a)

(b)

Fig. 62.4 (a) IR sensor system, (b) Circuit of IR sensor [17]

Types of RAM

2.2 Iron Ball Paint Absorber

Iron ball paint is one of the most well-known uses of RAM. It is made up of microscopic spheres covered in carbonyl or ferrite iron. Radar energy is transformed into heat by the oscillating molecular oscillations caused by this paint's fluctuating magnetic field. After that, the aeroplane carries the heat and disperses it.

2.3 Foam Absorber

Anechoic chambers used for measuring electromagnetic radiation are lined with foam absorber. This material usually consists of square pyramids made of fireproofed urethane foam that have been cut to the precise dimensions of the wavelengths of interest. These pyramids are then loaded with conductive carbon black, which can be either crystalline graphite or carbonyl iron spherical particles, in concentrations ranging from 0.05 to 0.1 percent (by weight in the final product).

2.4 Jaumann Absorber

A Jaumann absorber, sometimes known as a Jaumann layer, is a radar-absorbent material. The Jaumann layer was made up of two reflecting surfaces that were evenly spaced and a conductive ground plane. Because the Jaumann layer is a resonant absorber (i.e., it uses wave interference to cancel the reflected wave), it depends on the $\lambda/4$ distance (a total of $\lambda/4 + \lambda/4$) between the first reflecting surface and the ground plane as well as between the two reflective surfaces.

2.5 Split-ring Resonator Absorber

This method develops a resist layer using a photographic technique on a small piece of copper foil (approximately 0.007 in.) that is carved into tuned resonator arrays. Each resonator is formed like a square or like a "C" and is placed on a thin circuit board material dielectric backing. Every SRR is electrically isolated, and every dimension is designed to maximise absorption at a particular radar wavelength. There is a gap of a certain size that functions as a capacitor since the "C" is not a complete loop. The "C" has a diameter of approximately 5 mm at 35 GHz. To provide a wide-band absorption of radar radiation, several SRRs may be stacked with precisely thickened insulating layers between them, and the resonator can be tuned to certain wavelengths.

2.6 Carbon Nanotube

MWNTs may absorb the microwave frequency range used by radars. There seems to be a smaller radar cross-section when the MWNTs are applied to the aircraft because the radar is absorbed.

2.7 Colour Blindness of IR Sensor

Black-colored items absorb the infrared radiation that is emitted by the infrared transmitter, making it more difficult for the infrared receiver to record the reflected infrared radiation..

2.8 Ultrasonic Sound Absorbing Materials

Porous absorbers, such melamine sponges or open-cell rubber foams, absorb noise by internal cell structure friction. From mid to high frequencies, porous open cell foams are great in absorbing noise.

3. CONCLUSION

RAM, black-coated material, and sound-absorbing material are all included in our TANK. We may use RAM, black coated material, and sound-absorbing material in our tank to make it more difficult to detect with RADAR, SONAR, and IR sensors. By integrating Laser Walls, PIR, GPR, IR, and ultrasonic sensors into a single security system, we can provide 360° surveillance.

REFERENCES

1. Renhui Xu, Laixian Peng, Wendong Zhao, Zhichao Mi, Radar mutual information and communication channel capacity of integrated radar-communication system using MIMO, ICT Express 1 (2015) 102–105
2. L. Koval. J. Vaňuš.P. Bilík, Distance Measuring by Ultrasonic Sensor, IFAC-PapersOnLine 49-25 (2016) 153–158
3. Valérie Ciarletti, A variety of radars designed to explore the hidden structures and properties of the Solar System's planets and bodies, C. R. Physique 17 (2016) 966–975
4. Ilze Andersone, Probabilistic Mapping with Ultrasonic Distance Sensors, Procedia Computer Science 104 (2017) 362–368
5. Kirtan Gopal Panda,Deepak Agrawal, Arcade Nshimiyimana, Ashraf Hossain, Effects of environment on accuracy of ultrasonic sensor operates in millimetre range, Perspectives in Science (2016) 8, 574–576.
6. Sedra and Smith, Microelectronic Circuits, 5th Edition.
7. Roy Choudhary, Linear Integrated Circuits, 4th Edition.
8. A.V. Bakshi, U.A. Bakshi, Electromagnetic Theory, 2nd Edition.
9. Suman Singha, Debasis Maji, Laser Security System, International Journal of Scientific & Engineering Research, Volume 7, Issue 4, April-2016.
10. Pema Chodon, Devi Maya Adhikari, Gopal Chandra Nepal, Rajen Biswa, Sangay Gyeltshen, Chencho, Passive Infrared (PIR) Sensor Based SecuritySystem14, (IJEECS) International Journal of Electrical, Electronics and Computer Systems. Vol: Issue: 2, June 2013.
11. Tarek Mohammad, Using Ultrasonic and Infrared Sensors for Distance Measurement, International Journal of Mechanical, Aerospace, Industrial, Mechatronic and Manufacturing Engineering Vol:3, No:3, 2009.
12. A. K. Shrivastava, A. Verma, and S. P. Singh, Distance Measurement of an Object or Obstacle by Ultrasound Sensors using P89C51RD2, International Journal of Computer Theory and Engineering, Vol. 2, No. 1 February, 2010.
13. Jarosław Majchrzak, Mateusz Michalski, and Grzegorz Wiczyn´ski, Distance Estimation With a Long-Range Ultrasonic Sensor System, IEEE Sensors Journal, Vol. 9, No. 7, July 2009.
14. Types of RAM, Wikipedia.
15. Luiza de Castro Folgueras, Mirabel Cerqueira Rezende, Multilayer Radar Absorbing Material Processing by Using Polymeric Nonwoven and Conducting Polymer, Materials Research, Vol. 11, No. 3, 245–249, 2008.
16. Internet Source: Interfacing PIR sensor to 8051 & DIY Intruder Alarm, July 2018. https://www.circuitstoday.com/interfacing-pir-sensor-to-8051
17. Internet Source: PIR Sensor Circuit and Module Working. https://www.elprocus.com/pir-sensor-circuit-with-working/

Advances in Additive Manufacturing Technologies – Gurusamy Pathinettampadian (eds)
© 2024 Taylor & Francis Group, London, ISBN 978-1-032-90013-1

63 Ballistic Impact Analysis of Glass with Nano-Silica Reinforced Composite

Yuwaraj K R[1], Vijayakumar M D[2]
Department of Mechanical Engineering,
Chennai Institute of Technology, Chennai, Tamilnadu, India

Gokuldass R[3]
Department of Automobile Engineering,
Saveetha School of Engineering, Chennai, Tamilnadu, India

Jayabalakrishnan D[4]
Department of Mechanical Engineering,
Chennai Institute of Technology, Chennai, Tamilnadu, India

Lokesh G[5]
PG Student, Department of Mechanical Engineering,
Chennai Institute of Technology, Chennai, Tamilnadu, India

ABSTRACT: In the defense sector, ballistic impact is a frequent occurrence. The bullet impact impacted materials used to make jackets, automotive coverings, and aerospace parts. Good impact strength and energy absorption are essential for these materials to withstand damage and ensure industry safety. Aramid fiber glass is typically utilized due to its well-known impact strength. Glass with Nano-silica woven has throughout it likely to have better cost properties. Therefore, it is crucial to determine the reinforced composites' impact strength and energy absorption factor. In this thesis, glass is woven with Nano-silica to improve its impact resistance, and a lead projectile is used in a ballistic impact test to examine the material's behavior, impact strength, and energy absorption. The findings provide enough information to draw the conclusion that Glass use affects projectile behavior when combined with Nano-silica composite behavior.

KEYWORDS: Nano-silica, Aramid fiber glass, Reinforced composites, Ballistic impact

1. INTRODUCTION

Composite materials, which are combinations of natural or synthetic fibers are reinforced to alter properties for specific applications. They aim to reduce costs while enhancing performance. Composites consist of a matrix (providing strength and stiffness) and reinforcement (offering protection and facilitating load transfer). They are classified based on the matrix material. The content also mentions

[1]yuwarajr@citchennai.net, [2]vijayakumarmd@citchennai.net, [3]dassmech@gmail.com, [4]jayabalakrishnand@citchennai.net, [5]lokeshmass4@gmail.com

DOI: 10.1201/9781003545774-63

a study by Aisyah H.A. et al. on the effects of 2D woven epoxy laminates of Kenaf carbon hybrid composites. Two weave patterns, plain and satin, were utilized to conduct tests comparing tensile strength, flexural strength, and impact strength. The composite with a 5x5 fabric count and plain weave demonstrated a lower density ($0.98g/cm^3$), while the composite with a 6x6 satin weave fabric exhibited a higher density ($1.24g/cm^3$). Two primary observations were made: firstly, despite providing high density and strength, the satin weave showed a loose structure, resulting in lesser epoxy resin flow through the hybrid and issues with fiber de-lamination. Secondly, ANOVA results revealed no interaction between weave design and fabric counts on all properties except flexural strength.

Aswani Kumar Bandaru et al. conducted experimental tests on the tensile properties of GLASS and NANO SILICA, utilizing weft and warp woven specimens fabricated according to ASTM standards for tensile testing. Hybrids of GLASS and NANO SILICA were prepared, resulting in four specimens, two homogeneous and two woven fabrics. The failure of these specimens enabled the assessment of tensile properties, revealing that the GLASS plain laminate exhibited high tensile modulus and in-plane compression modulus. Comparison between hybrids indicated that hybrid 2 had higher tensile modulus than hybrid 1, while the reverse was observed for in-plane compression modulus. Computational modeling and simulation were later performed to compare the results, highlighting the significance of hybridization between GLASS and NANO SILICA on the tensile and in-plane compression modulus.

Brenda L. Buitrago et al. modeled a sandwich composite of carbon/epoxy with an aluminum honeycomb three-dimensional model implemented in ABAQUS/explicit. The model utilized a three-dimensional homogeneous mesh resulting in 107,650 elements, and numerical calculations were obtained from actual experimental results. The experiment involved using A1G + gas gun manufactured by Sabre Ballistics, with 25 square specimens of 140 mm and 14 mm thickness subjected to velocities ranging from 92 m/s to 548 m/s using helium and Stargon gases. Experimental results showed a sudden drop in velocities due to impact energy absorption, with the front and back panels absorbing 46% and 41% of the impact energy, respectively, and the core absorbing 13% for velocities lower than 250 m/s.

Elias et al. conducted ballistic impact tests on woven fabric hybrids, utilizing Kevlar, carbon, and glass to form five hybrid patterns. Specimens were laminated based on NIJ Standards using a hand lay-up process. Ballistic impact tests were carried out with pellets, showing that hybrid 2 exhibited the highest energy absorption magnitude of 95.17 J. Interpretation of results indicated that using glass as the first layer provided better absorption than Kevlar, with energy absorption magnitudes of 95.17 J and 95.15 J, respectively. Hybrid 1, composed of carbon, was not recommended as the last layer. Gaurav Nilakantan et al. virtually tested ballistic impact on a single-ply warp and weft threads of GLASS706 greige fabric soft armor, both experimentally and computationally using LS-DYNA explicit analysis, showing comparable results. Manigandan investigated GLASS plates' response to high-velocity ballistic impact projectiles, concluding that GLASS149 exhibited the best impact strength among GLASSK29, K49, and K149 grades, as determined through simulative analysis using ABAQUS explicit analysis.

Milon et al. analyzed the impact resistance of Polypropylene Carbon/GLASS (CK) composites, showing that CK/epoxy had the highest tensile strength and impact strength among CK/PP, CK/PE, and CK/epoxy composites. Naik et al. analyzed woven fabric composites for ballistic protection, emphasizing the importance of weave patterns in material selection and introducing the basic concept of ballistic impact. Narayanamurthy et al. performed simulative analysis on ballistic impact on armor plates, comparing actual and simulated results using ANSYS LS-DYNA.

The literature review highlighted the significance of understanding ballistic impact for the defense sector, emphasizing the role of woven fabric properties and weave patterns in material selection and impact testing methodologies. The GLASSNANOSILICA 2D woven composite impact test aims to assess energy absorption and impact strength, exploring the impact of orientation and weave pattern changes. Testing weave patterns and orientations will provide insights into conducting impact tests effectively.

2. Materials and Methods

Fibers, whether natural or synthetic, are combined to create composite materials that offer protection and can withstand high loads. They serve various functions, including:

Acting as load carriers, contributing significantly to the mechanical properties of fiber/resin composites, with fibers typically bearing 70 to 90% of the loads in structural composites.Providing properties such as thermal stability, stiffness, strength, and other structural characteristics to the composites.Establishing electrical conductivity or insulation based on the application requirements and the type of fiber used.Resins, on the other hand, come in many chemical varieties, each tailored to specific industries and offering benefits such as structural performance, resistance to environmental factors, cost-effectiveness, and legal compliance. Some commonly used resin types includes

Polyester Resins, Epoxy Resins, Cyanate Ester Resins, Polyimide Resins, Phenolic Resins

Epoxy resin stands out for its versatility, offering a wide range of processing and property options. It adheres well to various substrate materials and exhibits slow shrinking. Epoxies find applications across diverse industries, from sporting goods to aerospace. Different grades of epoxies cater to varying application requirements. These linear polymers are formed by combining bis-phenol A and epichlorhydrin, with other formulations such as brominated resins, glycidyl ethers of novolac resins, and glycidyl esters also available for specific purposes.

Glass, particularly aramid synthetic fiber GLASS, boasts strong impact resistance and heat resistance. Used in various applications, from racing bicycle tires to aerospace components and defense vests, GLASS offers exceptional impact strength and tensile strength to weight ratio. GLASS is produced through a condensation reaction of monomers 1, 4-phenylene-diamine and tetraphthaloyl chloride, yielding hydrochloric acid as a by-product. Several grades of GLASS are available, including GLASS29, GLASS49, GLASS149, and others, catering to different applications.

Nano silica, a naturally occurring innocuous fiber, is easily extracted using a one-stage homogenization and extraction technique. Composed of minerals like pyroxene, plagioclase, and olivine, nano silica is more affordable than carbon fiber and offers superior qualities compared to fiberglass. It serves as a reinforcement for fiber materials, reducing the cost of composite materials and finding applications in industries such as automotive, aerospace, and fireproof textiles. Woven cloth containing nano silica exhibits specific strength three times higher than steel, making it a valuable alternative to asbestos in the construction industry.

2.1 Material Selection

Since GLASS is expensive, ballistic applications frequently use it as the primary material in matt-woven fibers. Material reinforcements blended with GLASS fibers demonstrated variety in material properties and were also cost-effective. Hand lay-up, vacuum bag molding, compression molding, and other material reinforcing processes are utilized to produce GLASS composite laminates. These processes were selected for their cost-effectiveness and ability to reduce cycle times. The primary reason for selecting nanosilica fiber over carbon was its superior strength and stiffness to bear impact loads, as well as its lower cost and comparable tensile properties. Better adhesive properties for GLASS with additional reinforcements are provided by epoxy resin. The laminate's fibers are braided against one another in three weave patterns and 2D laminate fabricated.

2.2 Weaving Process

Weaving is textile industry process, produce large scale clothing suits in suitable weave patterns to establish better strength in all the directions. The thread fibers weaved provide optimum life to the clothing to bear ease wear and tear. The same patterns applicable for weaving composite fiber materials to obtain strength and withstand impact and tensile loads. GLASS and NANO SILICA fibers extracted from matt of equal GSM feed to loom for weaving. Fibers with various difference in GSM not recommended, since the varying GSM has higher possibility of creating gaps between the structures, which ultimately reduces mechanical property of the woven composite. Weave patterns possess various types; however, some patterns alone give high tensile strength. Various weave patterns available discussed below.

Fig. 63.1 Plain weave pattern

Fig. 63.2 Satin weave pattern

Twill Weave

In twill weave, the pattern tends to form diagonal interlacing by arranging the weft and yarns over the warp in progression. The twill weave exhibits superior wet out and drape compared to the plain weave, with a negligible decrease in stability. The fabric's surface is smoother and its mechanical characteristics are slightly higher due to the reduced crimp. Comparing all the three this weave posses the high tensile strength due to its interlacing feature in the diagonal.

Fig. 63.3 Satin weave

Fig. 63.4 Twill weave

Fig. 63.5 Plain weave

3. ANSYS Model Analysis

A model analysis is made to see the simulation of the ballistic impact test.at first tensile stress is carried out followed by velocity impact. GLASS properties and NANO SILICA properties are manually installed to the ANSYS 15.0 in the engineering data column. The geometry assumed to be a plate with fully woven GLASS and fully woven basalt. This will give a simulated view of plate deformation. The plate geometry is easier to make since the bullet is spherical and the plate is of three dimensional cuboid structures. The distance between the plate and bullet is given as per in the virtual test and the velocity is deviated from the virtual test since the model analysis is based on the trail purpose for the plate that can withstand high velocity.

Model constitutes the part defining, meshing, problem defining. The part material is changed to the desired one followed by meshing. Plate is meshed as a medium, Mesh sizing is given for the bullet. Same type mesh for bullet and the plate creates failed mesh and the sixing method gives the accurate mesh for this dynamics analysis. In real setup, the plate is fixed at the four middle portions the same way; the analysis involves fixing the plate in four ends. Velocity is vector quantity; the direction is defined by changing positive or negative sign while defining the scalar quantity. Analysis time is given as low as possible to ensure results, the greater time. When the exact problem definition is given the step is continued to solve. Three velocities 70 m/s 72m/s and 74m/s is given for both GLASS and NANO SILICA plate. The GLASS withstands more impact and showed less deformation than basalt.

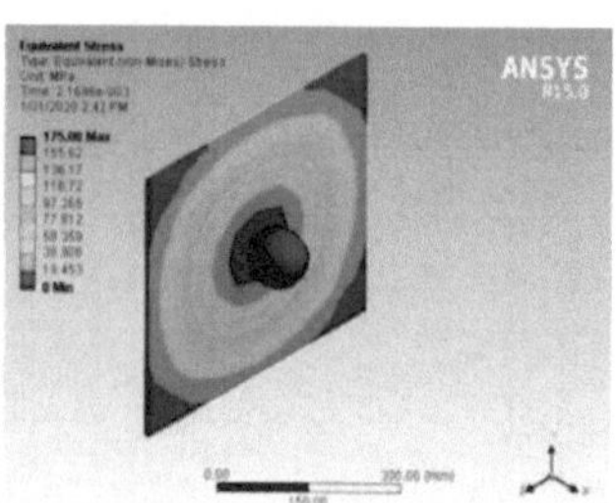

Fig. 63.6 Equivalent stress of GLASS at 70 m/s

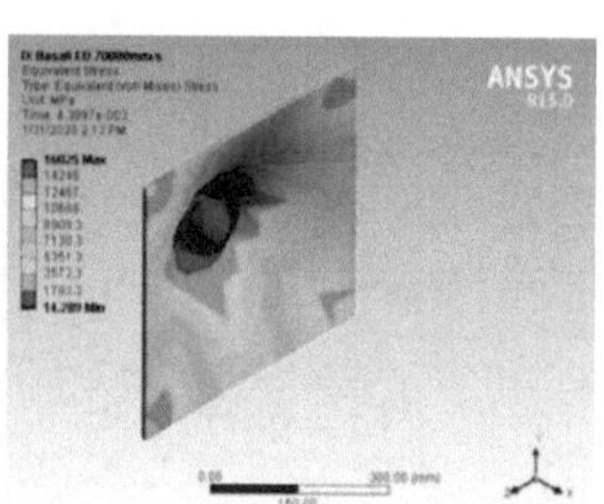

Fig. 63.7 Equivalent stress of NANO SILICA at70 m/s

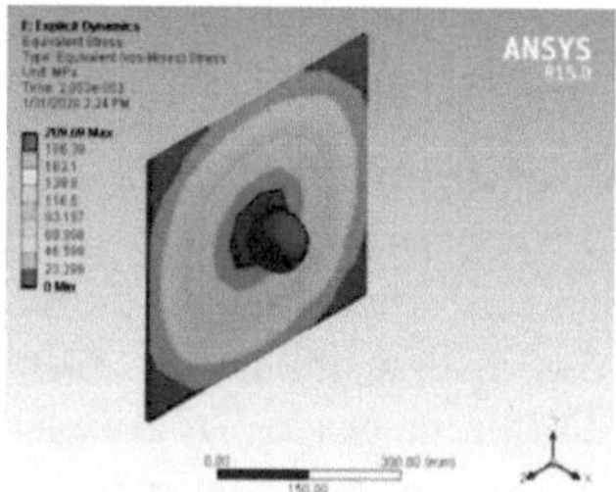

Fig. 63.8 Equivalent stress of GLASS at 72 m/s

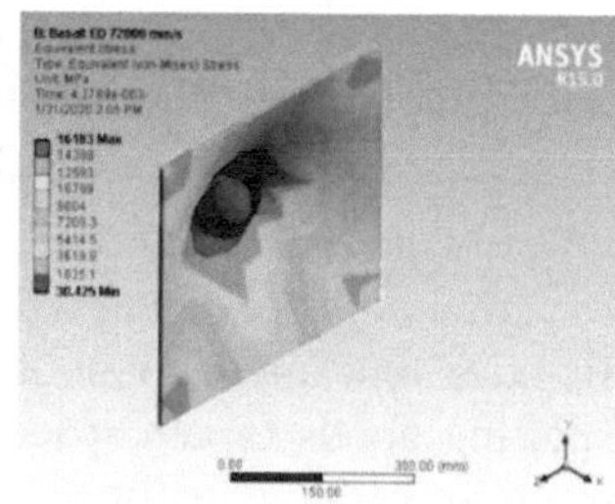

Fig. 63.9 Equivalent stress of NANO SILICA at72 m/s

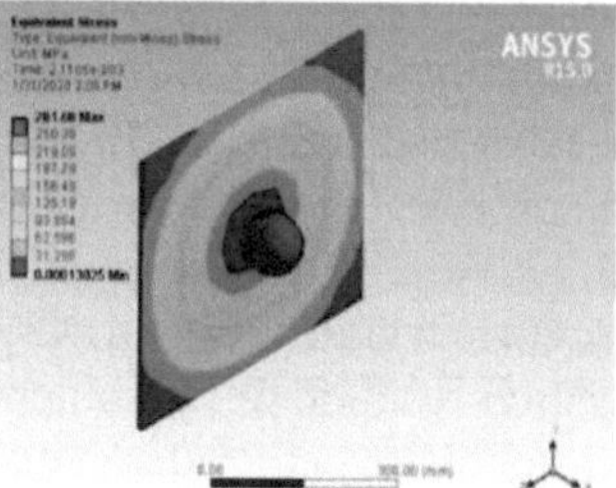

Fig. 63.10 Equivalent stress of GLASS at 74 m/s

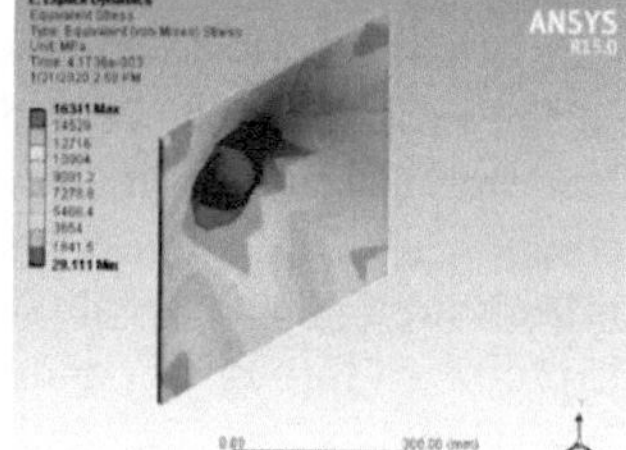

Fig. 63.11 Equivalent stress of NANO SILICA at74 m/s

Resultsfrom the model analysis give the view of ballistic impact on plates and the deformation. GLASS has a gradual increase in stress and deformation with increase in velocity. We can say that the velocity increase may ultimately affect the performance of plate. For low velocity impact the GLASS properties can give good impact strength. Since the velocity is higher the deformation on the plate is higher, so the virtual test must be carried out in low velocity that is, in the sub- sonic range. This analysis is further useful in comparing the results with plain weave since the plain weave is intervened with GLASS and basalt. However the virtual test will have a different deformation since the epoxy resin added and the density between fabricated and raw Glass , NANO SILICA and Epoxy resin.

3.1 Experimental Setup and Procedure

The system includes a gas compressor that lowers the volume of a gas in order to raise pressure. One particular kind of gas compressor is an air compressor. The ballistic impact machine is run by this. A pressure gauge that is recessed into the pressure regulator valve is used to measure the air pressure. The bullet is activated by a solenoid valve, which releases air at a high pressure suddenly. The bullet is propelled out after passing through the barrel, which is composed of two layers of mild steel and plastic, respectively, for the exterior and inner layers. The barrel, measuring one meter in length and three centimeters in diameter, has a plastic coating to facilitate the bullet's smooth passage through it. The dish is secured in the target holder. The target holder is made to accommodate materials with a thickness of 1 to 15 mm and a square measure of 200 or 300 mm. The target holder fixed plate and the shot from the machine impact.

Fig. 63.12 Ballistic impact setup

4. Results and Discussion

Mass differencegives the ratio between individual volume of all the weaves and the volume of plates fabricated. The weight at first measured from weighing gauge and converted to mass. Dimensions measured experimentally from the Vernier caliper and ruler.

Table 63.1 Massfromweight

Weave	Weight (g)	Mass (N)	Density (g/mm^3)	Volume (mm^3)
Plain	177	18.04	0.10	171440
Satin	201	20.48	0.12	161840
Twill	208	21.20	0.11	178400

Table 63.2 Massfrom volume of mixtures

Weave	Theoretical volume (mm^3)	Weight (g/ mm^3)	Mass (N)
Plain	2511110.264	4545.10	44.57
Satin	2504888.504	4560.40	44.72
Twill	2515621.04	4555.61	44.67

Table 63.3 Difference in masses

Weave	Mass from weight	Mass from Volume of mixture	% Difference
Plain	18.04	44.57	40.47
Satin	20.48	44.72	45.81
Twill	21.20	44.67	47.45

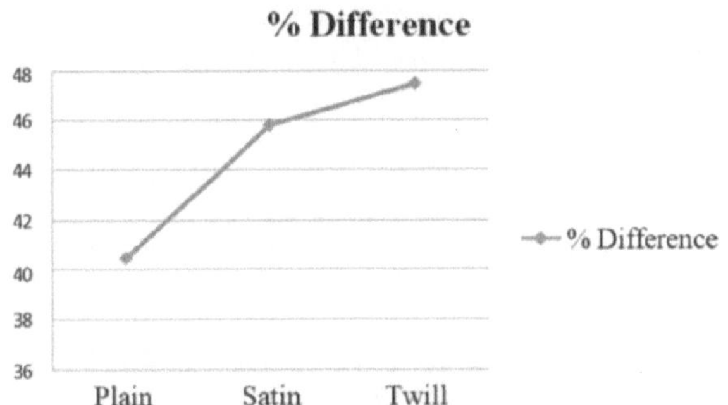

Fig. 63.13 Percentage difference in mass

4.1 Plates After Ballistic Impact Test

On account of visual examination the satin and twill show similar damage at the back faces of the plates. Plain weave showed a spherical deformation and the other two shows a rectangular deformation. This is mainly because of the pattern of weave plain weave is common one to one

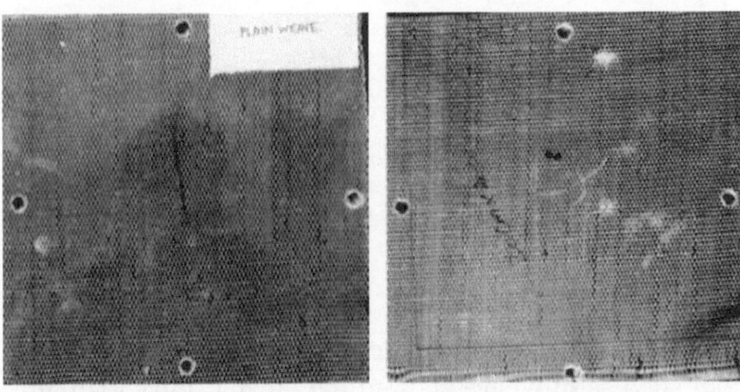

Fig. 63.14 Plain weave front **Fig. 63.15** Plain weave back

Fig. 63.16 Satin weave front **Fig. 63.17** Satin weave back

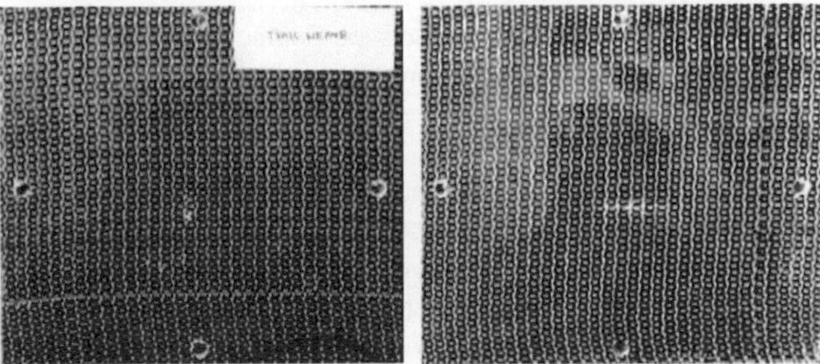

Fig. 63.18 Twill weave front **Fig. 63.19** Twill weave back

interlocking fashion whereas the other two weaves have different orientation. Impact loads are calculated from the load cell display on the monitor, the velocity is measured from the chronograph. The energy absorption and impact energy is calculated from the readings observed.

4.2 Energy Absorption and Impact Energy

The three patterns do not show much variation in results but the plain weave absorbs low energy compared with other weaves. The satin and twill absorb equal amount of energy. The satin weave absorbed high energy and posses higher impact strength. In simulation the plates showed deformation on the back face, similarly the virtual test showed the deformation in back face. Twill weave alone showed deformation on the front face.

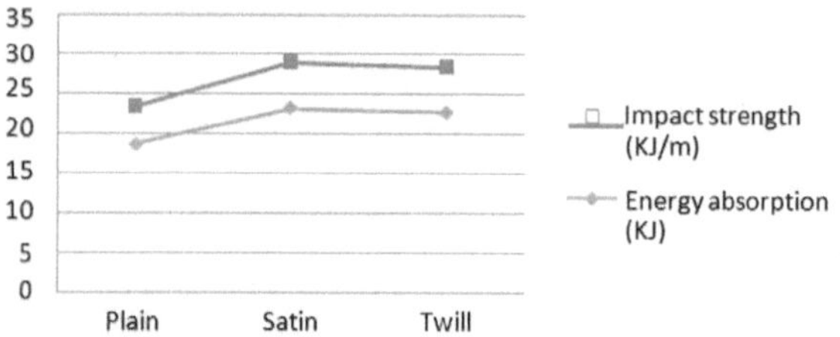

Fig. 63.20 Energy absorption and impact energy Vs velocity

5. CONCLUSION

The virtual ballistic impact test results are experimentally calculated satin weave showed the greater impact strength and absorbs greater portion of impact the damage on the plates are visible at the back face of the plate since the impact has possibility of harming the object behind the plates and the plates is good for storage of ammunitions, ballistic objects or any materials used for defense industry. For armor purpose the plate cannot be used since thickness of plate is nearly 4 mm thickness which drastically adds up weight. By switching up the bullet type, this test can be administered with a satin weave with a higher impact strength.

REFERENCES

1. Abrate .S, 'Ballistic impacts on composite and sandwich structures', in Major Accomplishments inComposite Materials and Sandwich Structures, pp. 465–501, Springer, Amsterdam, The Netherlands, 20102011.
2. Abu Taliba A.R., L.H. Abbuda, A. Alib, F. Mustaphaa, Ballistic impact performance of Kevlar- 29 and Al2O3 powder/ epoxy targets under high velocity impact, Materials & Design, 2012, Volume 35, Pages 12–19.
3. Aisyah H.A. et.al, 'Effects of Fabric Counts and Weave Designs on the Properties of Laminated Woven Kenaf/Carbon Fiber Reinforced Epoxy Hybrid Composites.', Polymers 2018, 10, 1320.
4. Arias, J. L´opez-Puente, J. A. Loya, D. Varas, and R. Zaera, 'Analysis of high-speed impact problems in the aircraft industry', in Constitutive Relations Under Impact Loading, pp. 137– 207, Springer, Vienna, Austria, 2014.
5. Aswani Kumar Bandaru et. al, 'Mechanical behavior of Kevlar/NANO SILICAreinforced polypropylene Composites.', 1359–835X/ 2016 Elsevier Ltd.
6. Brenda L. Buitrago a,b, et. al. 'Modelling of composite sandwich structures with honeycomb core subjected to high-velocity impact.', matter 2009 Elsevier Ltd.
7. Elias Randjbaran et. al, 'Hybrid Composite Laminates. Reinforced with Kevlar/Carbon/Glass Woven Fabrics for Ballistic Impact Testing.', Hindawi Publishing Corporation Scientific World Journal Volume 2014.
8. Gaurav Nilakantan et.,al, 'Virtual ballistic impact testing of GLASSsoft armor: Predictive and validated finite element modeling of the V0-V100 probabilistic penetration response.', Published by Elsevier Ltd.
9. Grujicic M, Bell W, Arakere G, He T, Xie X, Cheeseman B, 'Development of meso-scale material model for ballistic fabric and its use in flexible- armor protection systems', J Mater Eng Perform 2010;19(1):22e39.
10. Herbert Yeung K.K. et.al, Mechanical Properties of Kevlar-49 Fibre Reinforced Thermoplastic Composites. Polymers & Polymer Composites, Vol. 20, No. 5, 2012 411
11. Ivanov I, Tabiei A, 'Flexible woven fabric micromechanical material model with fiber reorientation', Mech Adv Mater Struct 2002;9:37e51.
12. Jeremy Gustin, Aaran Joneson, Mohammad Mahinfalah, James Stone, 'Low velocity impact of combination Kevlar/ carbon fiber sandwich composites', Journal of Materials Science, Volume 38, Issue 13, Pages 2825–2833.
13. Kunal Singha, 'A short review on NANO SILICAfiber', International Journal of Textile Science 2012.
14. Li Piani. T et.al, 'Ballistic model for the prediction of penetration depthand residual velocity in adobe: A new interpretation of the ballistic resistance of earthen masonry', Defense Technology (2018).
15. Manigandan. S 'Computational investigation of high velocity ballistic impact test on GLASS149 Applied Mechanics and Materials.' ISSN: 1662- 7482, Vols. 766-767, pp 1133–1138.
16. Naik. N.K.et.al., 'Analysis of woven fabric composites for ballistic protection', Advanced Fibrous Composite Materials for Ballistic Protection.
17. Narayanamurthy V. C , et.al, 'Numerical Simulation of Ballistic Impact on Armour Plate with a Simple Plasticity Model', Defense Science Journal, Vol. 64, No. 1, January 2014, pp. 55–61.
18. Pandya. KS. et.al, 'Analytical and experimental studies on ballistic impact behavior of 2D woven fabric composites', International Journal of Damage Mechanics 0(0) 1–41.
19. Rajesh p nair et.al, 'Simulation of depth of penetration during ballistic impact on thick targets using a one-dimensional discrete element model', Sadhana Vol. 37, Part 2, April 2012.

20. Subramani Sockalingam et.al, 'Dynamic modeling of GLASSKM2 single fiber subjected to transverse impact', 0020-7683/2015 Elsevier Ltd.
21. Suresha K. V. et.al, 'Evaluation of mechanical properties of hybrid fiber (hemp, jute, kevlar) reinforced composites', AIP Conference Proceedings 1943.
22. Valença et al, 'Evaluation of the mechanical behavior of epoxy composite reinforced with GLASSplain fabric and glass/GLASShybrid fabric', Compos B 2015;70:1–8.
23. Youjiang Wang et.al, 'Mechanical properties of fiber glass and GLASSwoven fabric reinforced composites', Composites Engineering, Vol.5. No. 9. pp. 1159-I 175, 1995.
24. Zhong P et. al, 'Research on the Modeling Method for Digital Weaving Based on the Information of Physical Yarns and Fabric Pattern', J Fashion Technol Textile Eng 5:2

Note: Every figure and table was created using the experimental work; none of them were taken from any publications or online.

Advances in Additive Manufacturing Technologies – Gurusamy Pathinettampadian (eds)
© 2024 Taylor & Francis Group, London, ISBN 978-1-032-90013-1

Parameter Optimization and Studies on the Properties of Similar PTIG Welded AA6082 Aluminium Alloys

Krishna. M[1]
Department of Mechanical Engineering,
SRM Institute of Technology and Science,
Ramapuram, Chennai

Abhilash. V[2]
Department of Mechanical Engineering,
PERI Institute of Technology, Chennai

Satheesh. S.S[3]
Assistant Manager, Renault Nissan Technology and
Business Centre India Pvt. Ltd. Mahindra World City,
Chengalpattu, Tamil Nadu

Hepsi Beaula. M.J[4]
Department of Mechanical Engineering,
Chennai Institute of Technology, Chennai, Tamil Nadu, India

C. Devanathan[5]
Department of Mechanical Engineering,
Rajalakshmi Engineering College, Thandalam, Chennai, India

Sounthararasu. V[6]
Department of Mechanical Engineering,
PERI Institute of Technology, Chennai

Sasi Lakshmikhanth. R[7]
Department of Mechanical Engineering,
Chennai Institute of Technology, Chennai, Tamil Nadu, India

ABSTRACT: Many industrial applications like automobile, aircraft manufacturing, and ship building, aluminium alloy is utilized for cost efficient, corrosion free, low weight and high strength purpose. This paper aims to optimize the WPP that can produce a strong and defect free PTIG joint. The parameters optimized are peak current, pulse frequency and welding speed. The optimization is done using design expert software in which the design of experiments are framed using CCD approach of RSM. For the RSM model the tensile strength is considered as the output parameter. The tensile tests are carried using UTM on the samples extracted as per ASTM-E8 standard. According to the results

[1]Krishnamechanic03@gmail.com, [2]enggabhilash@gmail.com, [3]satheesh.009@gmail.com, [4]hepsibeaulamj@citchennai.net, [5]devanathan.c@rajalakshmi.edu.in, [6]sounthar88@gmail.com, [7]rslkhanth@gmail.com

DOI: 10.1201/9781003545774-64

obtained the highest UTS of 205 MPa was obtained on samples joined using 190 A PC, 5 Hz PF, and 180 mm/min WS. The SEM, fractography and hardness analysis were conducted on samples that produced both strong and weak joints. The comparison is done in detail and the reason for the difference in strength is studied.

KEYWORDS: RSM, CCD, ANOVA, PTIG welding, AA6082 aluminium alloy, UTS

1. INTRODUCTION

FSW was first introduced in 1991 at the TWI, TIG welding uses a tungsten electrode that may be reused indefinitely. The tungsten electrode and weld pool are protected from the harmful effects of the atmosphere by the shield gas. Argon is a popular shielding gas, and it may be powered by either AC or DC. Tungsten's negative polarity makes it the cathode in a welding setup, with the work metal serving as the anode. Weld penetration is achieved using TIG welding by means of heat conduction through the materials being welded. Weld pool temperatures may reach up to 2500 °C [1] the arc induction between the tungsten alloy electrode and the work metal, melts the metals in the weld zone. PTIG welds have metallurgical advantages, such as refinement of grains in FZ, improved weld depth/width ratio, reduced HAZ thickness, less distortion, minimal hot cracking sensitivity, and less residual stresses.Aerospace, targeted weapon production, high-pressure containers, and nuclear vessel building are just some of the many fields that benefit from TIG welding.

Over the years, several optimization algorithms have been used to achieve the requisite weld quality [2,3]. The design of experiments (DOE) is a potent tool for maximizing the efficiency of a wide range of experimental procedures. The observational data collected may be utilized to effectively anticipate the process's output for a given set of input process parameters [4]. This technique is very good at identifying the process inputs that affect the response of the output. One such DOE technique, RSM is used to regulate the output response by statistical optimization of the input parameters. Within a given interval, RSM employs regression models of the second degree [5].

2. EXPERIMENTAL PROCEDURE

The 5 mm thick BM AA6082 are utilized in this investigation. The welding samples are sliced into small pieces 150 mm x 55 mm in size from the BM plate. The elemental composition of the parent materials used for PTIG welding is determined using OES, and presented in Table 64.1. Table 64.2 lists the tensile properties of the parent metals as determined by a universal testing machine. The WPP selected are PC, PF, WS. The optimization of the WPP is the most crucial step to take in order to produce a weld joint that is both strong and free of defects. The initial stage in the investigation is to determine the characteristics of the welding process that have an influence and their limits. Following an extensive review of the relevant literature, three welding process parameters as PC, PF, and WS were chosen for the investigation [6]. These parameters were chosen among the large number of welding parameters that are available in PTIG welding. A lot of different tests and experiments were carried out, and the limits of the WPP were chosen based on the joints that had an exceptionally minimal quantity of weld flaws. Table 64.3 provides the range of welding process parameters that were considered for the current research work.

Table 64.1 Elemental proportion (wt%) of BM

BM	Mg	Mn	Fe	Si	Cu	Cr	Zn	Ti	Al
6082	1.5	0.5	0.4	2.1	0.1	0.2	0.1	0.01	Bal.

Table 64.2 Tensile characteristics of the parent metals

BM	YS (MPa)	UTS (MPa)	%El
AA6082	235	284	12.1

Table 64.3 Coded values assignment for the real values

WPP	Levels				
	-1.682	-1	0	1	1.682
PC	140	160	190	220	240
PF	2	3.2	5	6.8	8
WS	140	156	180	204	220

PTIG experiments were conducted employing a range of parameter combinations for welding, in accordance with the DOE formed using CCD of RSM. In order to characterize metallurgical, elemental, and mechanical properties, samples are extracted across the weld direction using the WEDM machine. Following the retrieval of the microstructural samples, they are mounted using a heated mounting press and subsequently polished. Following assembly, silicon carbide paper with a grain range of 80 to 2000 is utilised to refine the mounted samples. The extracted sample undergoes a microstructural evaluation by employing SEM. This evaluation aims to ascertain the characteristics, dimensions, and arrangement of the grains and precipitate. In order to ascertain the UTS of the welded samples, tensile tests were performed using UTM. Furthermore, to ascertain the characteristics of the failure, SEM was employed to examine the fractured surfaces of the tensile samples. Twenty seconds under a one-kilogram force on a Vickers microhardness instrument equipped with a diamond indenter, the material hardness was evaluated at various locations perpendicular to the weld.

3. Results and Discussions

3.1 Tensile Characteristics Evaluation and Development of Empirical Relationship

The PTIG welding process is carried out using the welding parameter combinations that are established by the design of experiments as the framework. In order to provide a framework for the design of experiments, the RSM's CCD was used. On the samples that were joined utilizing a variety of WPP configurations, tensile tests are performed, and the results obtained from these tests are recorded as the output response. Table 64.4 presents the DOE table in its entirety, including the WPP combinations, along with the output response, and the UTS of AA6082 joints. The images of the weld plates of similar AA6082 can be seen in Fig. 64.1(a). The images of the tensile samples of AA6082 that were utilized for the investigation can be seen in Fig. 64.1(b), respectively.

Table 64.4 DOE with tensile values of similar AA6082

Std. order	Real Values			UTS (MPa)
	PC (A)	PF (Hz)	WS (mm/min)	
1	160	3.2	156	160
2	220	3.2	156	161
3	160	6.8	156	**148**
4	220	6.8	156	193
5	160	3.2	204	188
6	220	3.2	204	154
7	160	6.8	204	170
8	220	6.8	204	179
9	140	5	180	161
10	240	5	180	170
11	190	2	180	172
12	190	8	180	184
13	190	5	140	163
14	190	5	220	176
15	190	5	180	202
16	190	5	180	205
17	190	5	180	204
18	190	5	180	**205**
19	190	5	180	203
20	190	5	180	202

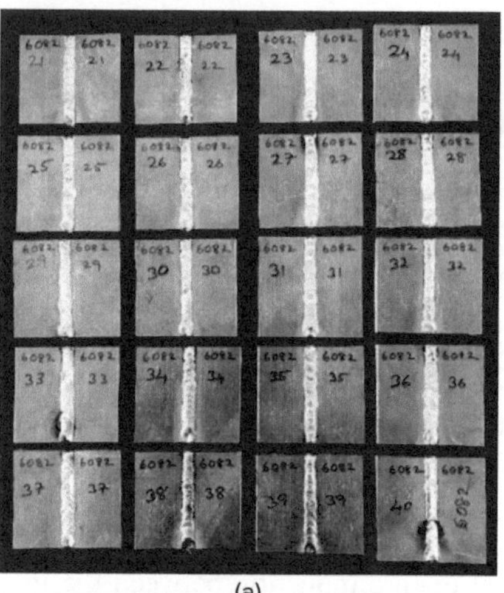

(a)

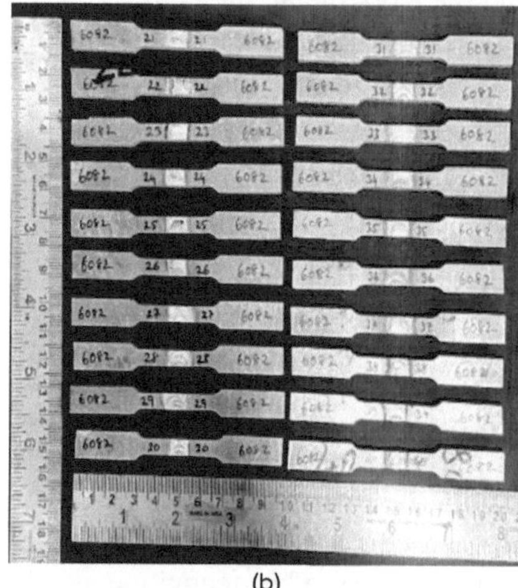

(b)

Fig. 64.1 (a) Welded plates of AA6082; (b) Tensile sample of AA6082 joint

The RSM is a fusion of statistical methodologies that are appropriate for simulating and analyzing scenarios where a number of variables affect the response of concern, and the goal is to create an optimal response. PC, PF, and WS all have a role in determining Tensile Strength (TS).

$$\text{UTS} = f(\text{PC}, \text{PF}, \text{WS}) \tag{1}$$

The response surface 'Z.' is represented by a regression equation as follows:

$$Z = C_0 + \Sigma C_i X_i + \Sigma C_{ii} X_i^2 + \Sigma C_{ij} X_i X_j \tag{2}$$

Where the typical response is marked by C_0, and the coefficients that are influenced by the main and interaction effects of the factors are given by C_i and C_{ij} respectively. Because of this, the chosen polynomials may be expressed for the forementioned three components as

$$UTS = C_0 + C_1(PC) + C_2(PF) + C_3(WS) + C_4(PC*PF) + C_5(PC*WS) + C_6(PF*WS) + C_7(PC^2) + C_8(PF^2) + C_9(WS^2) \quad (3)$$

To do an analysis of the regression coefficient, one may choose from a wide variety of experimental design methods; nevertheless, the central composite design was selected for use in this particular piece of work. Utilizing the Design Expert application allowed for complete acquisition of all of the variables. The conclusive association is formed by making use of the significant coefficients that were found. Following is a description of the empirical relationship that was created in order to compute the UTS of PTIG-welded AA6082 alloy specimens.

$$UTS\ (AA6082) = 203.49 + 2.64\ (PC) + 3.45\ (PF) + 3.72\ (WS) + 10.87\ (PC*PF) - 8.88\ (PC*WS) - 1.62\ (PF*WS) - 13.41\ (PC^2) - 8.99\ (PF^2) - 12.00\ (WS^2) \quad (4)$$

3.2 Assessing the Functionality of the Developed Empirical Relation

ANOVA, is employed in order to evaluate the efficacy of the created relationship. The results of the ANOVA are shown in Table 64.5, and they demonstrate the significant welding parameters that must be used to achieve maximum UTS. It is necessary to have a developed empirical relationship when the credibility level is set at 95%. It is probable that noise was the source of an F-value that was less than0.01 percent. P-values that are marginally lower than 0.0500 indicate that the regression coefficients are significant. In this context, the most important model words are "PC," "PF," and "WS," as well as "PC x PF," "PF x WS," and "PC x WS" When compared to 0.1000, values that are more than that indicate that the model terms are not significant.

Table 64.5 ANOVA results obtained for AA6082 joint

Source	Sum of Squares	df	Mean Square	F-value	p-value
Model	6954.02	9	772.67	774.46	<0.0001
PC	95.62	1	95.62	95.84	<0.0001
PF	163.00	1	163.00	163.38	<0.0001
WS	189.43	1	189.43	189.87	<0.0001
PC*PF	946.13	1	946.13	948.31	<0.0001
PC*WS	630.12	1	630.12	631.58	<0.0001
PF*WS	21.13	1	21.13	21.17	0.0010
PC^2	2593.36	1	2593.36	2599.36	<0.0001
PF^2	1166.08	1	1166.08	1168.78	<0.0001
WS^2	2075.39	1	2075.39	2080.18	<0.0001
Residual	9.98	10	0.9977		
Lack of Fit	0.4769	5	0.0954	0.0502	0.9974
Pure Error	9.50	5	1.90		
Cor Total	6964.00	19			
Std. Dev.	0.9988	**R^2**		0.9986	
Mean	180.00	**Adj. R^2**		0.9973	
C.V. %	0.5549	**Pred. R^2**		0.9975	

Table 64.5 shows the ANOVA results of UTS value for AA6082 similar joint. When there are a large number of model terms that are not highly important, model reductions will make the design better. The fact that the Lack of Fit F-value is just 0.05 indicates that the Lack of Fit is not a significant problem in comparison to a pure error. Because of noise, there is a 99.74 percent chance that a significant F-value for the lack of fit will be seen. The value of the coefficient of determination, which is denoted by the letter R^2, provides an indication of the degree of similarity between the predicted and the actual values. The coefficient of determination for our empirical connection is found to be 0.9986, as shown in Table 4.5. This indicates that there is a strong correlation between the experimental data and the predicted values. The results of the ANOVA show that the predicted R^2 value of 0.9975 and the adjusted R^2 value of 0.9973 are very close to being in perfect agreement with one another. The signal-to-noise ratio is measured with a precision value that is sufficient for the situation. The created empirical connection is shown to have good adequacy when considering the preceding findings.

3.3 Selection of Optimized Welding Parameter

RSM is used in order to identify both the optimal parameters for welding and the statistical relationships that are created. Following optimization, the coded values were turned back into real values. The coded value that had been optimized was utilized for optimization. Using design expert software, the values that have been processed are optimized in order to provide the best results. The parameters for maximum UTS achieved with PIG welding of similar AA6082 are presented in Table 6. The predicted and experimentally observed values may be found in the table. The tensile strength is used as the maximizing issue for the multi-response optimization, which is then carried out. The findings of the experiments with PTIG joints that were identical to AA6082 indicated that the highest UTS of 208 MPa was reached on the junction that was welded using the optimum conditions (PC - 196 A, PF - 5.5 Hz, WS - 181 mm/min.). This value was acquired on the joint that was welded using the optimized conditions.

Table 64.6 Optimized WPP values

WPP	Optimized values	
	Predicted	Actual
PC (A)	196.06	190
PF (Hz)	5.5	5
WS (mm/min)	181.4	180
UTS (MPa)	208	205

3.4 Microstructural Analysis

At a magnification of 2,000 X, the SEM was used to do the microstructural examination on the FZ. Fig. 64.2(a) displays the SEM picture that was captured at the FZ of the AA6082 sample while it was being welded at the optimal WPP. Fig. 64.2(b) displays the FZ of the AA6082 sample that had a low UTS (Experiment No. 3). The amount of precipitates that are present in the FZ of samples that were welded at the optimal WPP combination is greater than the amount of precipitates that were present in the FZ of low UTS produced samples AA6082 joints. It is found that there has been a 26% drop in the amount of magnesium found in the FZ of the AA6082 joint. During the fusion welding of magnesium-containing aluminum alloy AA5059 Al, Vasu *et al.* (2019) [7] made an observation that was quite comparable to this one. According to Padmanabhan *et al.*[8] conclusions, grain coarsening takes place at the weld pool as a result of excessive heat input. According to the findings, an increase in the amount of heat input causes a slower cooling rate during the solidification process, which gives the grains more time to develop. Consider another scenario in which the peak current is rather low. If this is the case, there will be a lower heat input, which will lead to a quicker cooling rate. As a consequence, the solidification process will take place extremely quickly, which will result in grains that are much smaller. Sudhir Kumar *et al.*[9] presented an observation that was quite similar to this one.

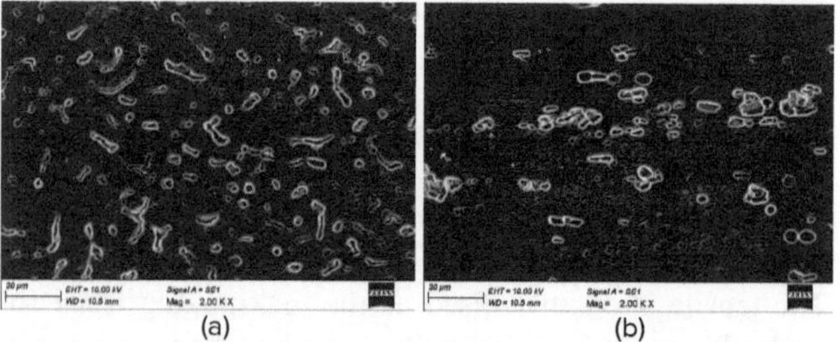

(a) (b)

Fig. 64.2 SEM Microstructure at FZ of AA6082 joint (a) Exp No. 18 (high UTS); (b) Exp No. 3 (low UTS)

3.5 Fracture Surface Analysis using SEM

Fractography images captured using a SEM at the cracked surface of tensile tested samples of AA6082 and are shown in Fig. 64.3. The number of dimple distributions on the shattered surface of the optimized welded sample (Fig. 64.3(a)) is shown to be higher than the number of dimple distributions on the sample with low strength in Fig. 64.3(b). This indicates that the sample welded using the WPP conditions that provide the best results had a greater ductility than the other joints.

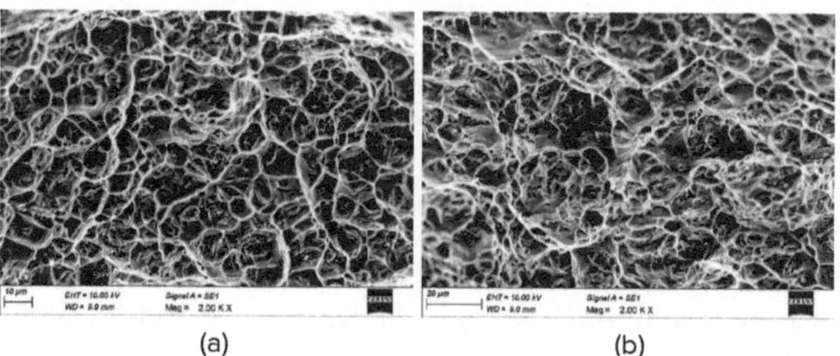

(a) (b)

Fig. 64.3 Fractography of AA6082 joint (a) Exp No. 18 (high UTS); (b) Exp No. 3 (low UTS)

A combination of shallow dimples, quasi-cleavage, and facets can be seen on the fractured surface of the low-strength samples, which can be seen in Fig. 64.3(b). This demonstrates a mixed mode of failure. Liu *et al.*[10] found similar characteristics in their research. It has been shown that changes in the microstructural properties of the material, such as grain development, distribution of precipitates, and texture evolution, have a significant impact on the load-bearing capability of the material as well as the shape of the fracture it causes. The fact that the shattered surface has facets demonstrates that brittle failure was responsible for the breakage. A decrease in the material's ductility may be inferred from the appearance of a surface structure that is reminiscent of a quasi-cleavage. The shallow dimples seen in Fig. 64.3(b) indicate that the ductility of the PTIG joints has been diminished. Additionally, the combined characteristics of all of the zones demonstrate a decrease in ductility in the PTIG samples with lower strength.

3.6 Hardness Analysis Across the Weld Joint

The hardness survey is performed on the cross-section along the different zones of the welded sample, and the results are shown in the form of a spline plot in Fig. 64.4. This is done so that more information can be obtained on the microstructural differences that are brought about by the welding process on the parent metal. Fig. 64.4 shows the hardness survey taken on different zones of the weldment on the similar joined AA6082 joint. On the PTIG weldment, there are distinct

According to the plot of the hardness of AA6082 joints (Fig. 64.4), the HAZ had the lowest hardness value of 50 $HV_{0.5}$, while the BMZ of the weldment had the highest hardness value of 90 $HV_{0.5}$. The hardness of the alloy is

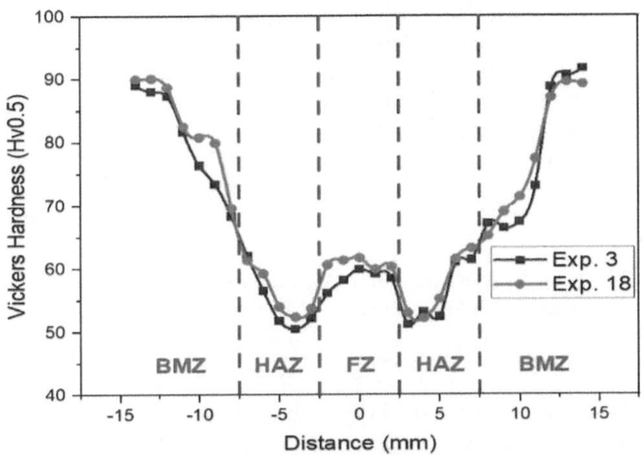

Fig. 64.4 Hardness comparison at different zones

primarily determined by two primary factors: (a) grain size, in accordance with the Hall-Petch relation, which states that a decrease in grain size results in an increase in hardness values; and (b) the size of the precipitate and its distribution in accordance with the Orowan hardening mechanism, which states that a decrease in the distance that separates the secondary-phase particles would result in an increase in the hardness values [11]. In this particular instance, the formation of larger grains in the FZ was responsible for the reduction in hardness that was observed. Another important is the dissolution of strengthening precipitates in the FZ during the PTIG welding process as the FZ receives the highest heat during the welding process. During the TIG welding, Wang *et al.* [12] and Xu *et al.*[13] made findings that were quite similar to these. In this case, the coarsening of grains that takes place as a direct result of high heat flow is the primary factor responsible for the decline in hardness values of HAZ. Cao *et al.*[14] made results that were quite similar to these during the laser welding of ZE41A-T5 magnesium alloy. In their study, they found that the HAZ underwent softening mostly as a result of the dissolving of strengthening precipitates and grain coarsening. This was seen throughout the laser welding process. During PTIG welding of AA6082, the heat input causes coarsening of grains and dissolution of strengthening precipitate Mg_2Si in both FZ and HAZ, but the action of pulse frequency has continuously agitated the molten weld pool resulting in finer grains than HAZ in the FZ.

4. Conclusions

In this paper, an investigation was conducted into the influence that PTIG WPP including PC, PF and WS had on the tensile characteristics of similar joints made of aluminum alloys AA6082. The following significant findings were made as a result of this investigation:

1. At tensile studies, a maximum UTS of 205 MPa was achieved by welding with parameters including a PC of 196 A, a PF of 5.5 Hz, and a WS of 181 mm/min. It was discovered to be the most effective PTIG welding parameter for connecting AA6082 alloy.
2. An empirical relation for predicting the UTS of PTIG-welded alloys equivalent to AA6082 were developed.
3. The WS is the factor that has the most significant impact on the UTS of an AA6082 joint, followed by the Pulse frequency. The peak current has the least significance on the UTS
4. The microstructural analysis showed that the sample joined using optimal welding parameter has more precipitates in the FZ whereas the sample that produced low heat input had a very excessive heat flow which has dissolved most of the strengthening precipitates.
5. The fractographic studies showed lot of dimples on the sample joined using the optimized welding parameter whereas the sample that produced weak joint showed mixed mode of failure.
6. The hardness analysis showed lowest hardness values in the HAZ and the highest hardness wad observed on the BMZ.

References

1. Asif Ahmad, Shahnawaj Alam, "Parameteric optimization of TIG welding using Response Surface Methodology", Materials Today: Proceedings, 2019, vol. 18, pp. 3071–3079.
2. R. Pandiyarajan, "Improving mechanical strength on welded joints by optimization technique", International Journal of Enterprise Network and Management, 2021, vol. 12, pp. 85–96.
3. Umar M, Paulraj S, "Optimization of welding parameters in CMT welding of Al5083", Advances in Materials Research: Select Proceedings of ICAMR 2019, 2021, pp. 663–672.
4. R. Pandiyarajan, S. Marimuthu, "Parameteric optimization and tensile behavior analysis of AA6061-ZrO_2-C FSW samples using Box-Behnken method", Materials Today: Proceedings, 2021, vol. 37, pp. 2644–2647.
5. M.J. Uma, PLK. Palaniappan, P. Maran, R. Pandiyarajan, "Investigation and optimization of friction stir welding process parameters of stir cast AA6082/ZrO_2/B_4C composites", Materials Science-Poland, 2021, vol. 38, pp. 715–730.Intro
6. Sasi Lakshmikhanth, R & Lakshminarayanan, AK, "On the mechanical, microstructural, and corrosion properties of pulsed gas tungsten arc and friction stir welded RZ5 rare Earth grade magnesium alloy", Materials Research Express, 2022, vol. 9, 126507.
7. Vasu. K, Chelladurai. H, Ramasamy A, Malarvizhi S, Balasubramanian. V, "Effect of fusion welding processes on tensile properties of armor grade, high thickness, non-heat

treatable aluminum alloy joints", Defense Technology, 2019, vol. 15, pp. 353–362.
8. Padmanaban, R, Vignesh, RV, Povendhan, AP & Balakumharen, AP, "Optimizing the tensile strength of friction stir welded dissimilar aluminium alloy joints using particle swarm optimization", Materials Today: Proceedings, 2018, vol. 5, pp. 24820–24826.
9. Sudhir Kumar, Pradeep Kumar, H.S. Shan, "Effect of process parameters on the solidification time of Al-7%Si alloy castings produced by VAEPC process", Materials and Manufacturing Processes, 2007, vol. 22, pp. 879–886.
10. Liu, Y, Wang, W, Xie, J, Sun, S, Wang, L, Qian, Y, Meng, Y & Wei, Y, "Microstructure and mechanical properties of aluminium 5083 weldments by gas tungsten arc and gas metal arc welding", Mater Sci. Eng., A, 2012, vol. 549, pp. 7–13
11. Nie.J.F, "Effect of precipitate shape and orientation on dispersion strengthening in magnesium alloys", Scripta materialis, 2003, vol. 48, pp. 1009–1015.
12. Wang, W, Cao, Z, Liu, K, Zhang, X, Zhou, K & Ou, P, "Fabrication and mechanical properties of tungsten inert gas welding ring welded joint of 7A05-T6/5A06-O dissimilar aluminum alloy", Materials, 2018, vol. 11, 1156.
13. Nan Xu, Jun Shen, Weidong Xie, Linzhi Wang, Dan Wang, Dong Min, "Abnormal distribution of microhardness in tungsten inert gas arc butt welded AZ61 magnesium alloy plates", Materials Characterization, 2010, vol. 61, pp. 713–719.
14. Cao. X, Xiao. M, Jahazi M, Immarigeon. J.P, "Continuous wave ND:YAG laser welding of sand cast ZE41A-T5 magnesium alloys", Materials Manufacturing Process, 2005, vol. 20, pp. 987–1004.

Note: Every figure and table was created using the experimental work; none of them were taken from any publications or online.

Advances in Additive Manufacturing Technologies – Gurusamy Pathinettampadian (eds)
© 2024 Taylor & Francis Group, London, ISBN 978-1-032-90013-1

65 A Novel Frustum of Square Pyramid Solar Still for Pure Water Production: A Comparative Study with Single Slope Solar Still

Vincent Raphael[1], V.Savithiri[2]
Department of Mechanical Engineering,
Saveetha School of Engineering, Saveetha Institute of Medical and Technical Sciences, Saveetha University,
Chennai, Tamil Nadu, India

ABSTRACT: The aim of this study is to compare the pure water productivity of a novel Frustum of Square Pyramid Solar Still with a standard single slope solar still. Specifically, the study will investigate the effect of the frustum of the pyramid on the quantity of pure water collected from domestic kitchen wastewater. The solar stills used were made of GI sheets and acrylic sheets. The Novel Frustum of Square Pyramid Solar Still was considered as group 1 and standard single slope solar stills as group 2. The active area of both the stills was 1600 cm2. The bases of both stills were coated with black paint to enhance the absorption of radiation. The stills were filled with domestic kitchen wastewater and placed on a flat surface in direct sunlight. The experiment was conducted for ten days. The amount of pure water collected was measured and recorded at regular intervals. The collected water was weighed and recorded. The sample size is considered to be 20 samples per group, and a total of 40 samples with 80% of total G power. The results show that the Novel Frustum of Square Pyramid Solar Still higher productivity compared to the still with a single slope. The average production of pure water with the Novel Frustum of the Square Pyramid Solar Still was 486.21 ml/m2 and that of single slope solar still was 378.6 ml/m2. Thus, the Novel Frustum of Square Pyramid Solar Still had a 28.6% higher productivity. There exists a statistically significant difference between the Novel Frustum of Square Pyramid Solar Still and the Standard Single slope solar still, the two tailed p value of the test was $P = 0.000$ ($P < 0.05$) and is statistically significant. The study concluded that the still with Novel Frustum of Square Pyramid Solar Still was more efficient than the still with single slope. The average pure water collection was 486.21 ml/m2 for the frustum of square pyramid solar still, which is 28.6% higher compared to standard single slope solar still producing 378.6 ml/m2 which allowed for better absorption of radiation and more efficient evaporation of water.

KEYWORDS: Acrylic sheet, Novel frustum of square pyramid solar still, Productivity, Single slope solar still, Solar distillation, Wastewater, Water collection

1. INTRODUCTION

Safe drinking water is crucial for sustainable development. One third of the universe is covered with water, even then only 3% of that is safe for human consumption (Gulamhussein et al. 2023) [1]. The world is facing a water crisis, it is expected that the fresh water requirement and supply gap will be around 40% by the end of 2030. So, reusing and

[1]vincentraphaela19@saveetha.com, [2]savithiriv.sse@saveetha.com

DOI: 10.1201/9781003545774-65

recycling of water is essential for sustainable development. Physical, Chemical and Biological processes are available for the wastewater treatment process[2-3]. Solar still is the green energy process which uses the free source of energy sunlight, to produce fresh water from contaminated water. Solar still uses a solar distillation process, that is, sunlight is used to heat and evaporate the wastewater inside the still and then condensed on the surface of the cover of solar still, which will be pure drinking water that can be collected[4].

Over the past five years, Google Scholar published 718 publications and Science Direct 647 in the related topic. In recent years, the Standard Single Slope Solar Still is a more traditional design, and is often seen in homes and gardens in which solar distillation process is used to purify the water (Shanmugam and Subramani 2023a)[5]. Standard single slope solar still is used to remove inorganic, non-volatile contaminants, and the product water was found to be drinkable (Sampathkumar and Natarajan 2022) [6]. The Standard Single Slope Solar Still performance was increased by adding fibers, and the results show that there was around 25% increase in the fresh water productivity (Dhivagar et al. 2021). Various types of solar stills were studied; triangular pyramids, square pyramids, and found to be more efficient in fresh water production compared to standard single slope solar still made of glass or acrylic sheet (Modi and Shah 2022) [7-9].

There are many different designs for pure water collection through the process of solar distillation by evaporation and condensation using solar still (Clapham and Nicholson 2014). The single basin solar still, which consists of a solitary receptacle for water collection and evaporation, is the most basic design of solar still. It has been the subject of extensive research due to its straightforwardness and convenience of implementation. Ghoneim et al. (2017) state that the single basin solar still is a viable option for small-scale implementations in arid regions[10-12].

The Pyramid Solar Still fins were designed to increase the surface area for water collection, by adding fins to the sides of the still by the process of solar distillation (Shanmugam and Subramani 2023b). The hemispherical solar stills were developed for increasing the productivity. Meanwhile, the Standard Single Slope Solar Still is a traditional design with a single sloped surface, which is often seen in homes and gardens. Both designs have their advantages and disadvantages when it comes to collecting pure water from kitchen wastewater. The presence of certain absorbent materials, such as black rubber and wick materials, might enhance the quantity of distilled water produced by the hemispherical solar still (V. Savithiri et al, 2022) The Square Pyramid Solar Still Pyramid Solar has the advantage of increased surface area for water collection (Shukla, Sharma, and Biwolé 2020). This means that more water can be collected from a given amount of kitchen wastewater, resulting in greater efficiency. It is much simpler to build and maintain than a square pyramid Solar Still with fins, and the cost is generally lower (Modi and Shah 2022). However, the design has a smaller surface area, resulting in lower collection efficiency. Furthermore, the single sloped surface doesn't offer any barrier to reduce evaporation losses (Girma and Assefa 2015) [13].

According to the studied literatures, the pyramid solar stills were highly concentrated, with little emphasis on the pyramid's frustum. This research work is focused on the development of a novel frustum of square pyramid solar still and its productivity comparison with standard single slope solar still.

2. Materials and Methods

The study consists of two distinct groups, and the mean and standard deviation were used to calculate the sample size (Xia et al. 2016). Two solar stills were made with GI and acrylic materials of the same collecting area, Control group was having standard single slope solar still. Intervention group was made using frustum of square pyramid solar still. The G power is 80% and the alpha value used in the calculation of the sample size is chosen at a level appropriate for conducting statistical analysis on a dataset consisting of 20 samples each group, with a total of 40 samples. The sample size is calculated using clinical software.

Single slope solar still was considered as Group 1 and was made of GI and acrylic. The base was made of GI sheets of size 45 cm X 45 cm square with 12.5 cm height. In order to absorb more solar radiation, the inside of the base was coated with black paint. Then the acrylic sheet of 45 cm x 45 cm size was placed as a single slope having a surface area of 2025 sq.cm and the sides were also closed with acrylic sheets forming the single slope solar still and provisions were made to collect the condensed pure water.

For Group 2 a novel frustum of square pyramid solar still similar procedure as of single slope solar still was adopted with GI sheets of base size 45 cm x 45 cm. The five acrylic sheets were cut into appropriate size in order to make the frustum of a square pyramid of surface area of 2025 sq.cm of sample preparation, the same fabrication procedure is followed.

The experimental setup for this study included the two solar stills, measuring jars for collecting and measuring the pure water. The base of both the stills was coated with black paint to enhance the absorption of radiation. The stills were filled with domestic kitchen wastewater and placed on a flat surface in direct sunlight. The solar distillation process took place, the water evaporated and that was condensed on the acrylic sheet surface and was collected. The experiment was conducted for twenty days, four hours per day. The collected water was measured and weighed on and measurements were taken tabulated as per hr average per unit square meter. The sample size was calculated as 20 samples per group and a total of 40 samples with 80% of total G power. The quantity of pure water collected is given in Table 65.1. in ml per sq. m for both the groups.

3. Results

Table 65.1 Shows the productivity of a single slope solar still and a novel frustum of a square pyramid solar still per hour for each group, based on 20 samples. In Table 65.2. Group Statistics results gives the novel frustum of square pyramid solar still productivity 486.2104 is more compared to Single slope solar still 378.6076 and Standard error mean for Single slope solar stills is 6.15395 and novel frustum of square pyramid solar still is 4.01200. The frustum of the square pyramid diagram is given in Fig. 65.1 and the bar chart of the productivity comparisons of control group and experimental group is given in Fig. 65.2. Frustum of square pyramid solar stills with standard single slope solar stills are more efficient at collecting pure water from domestic kitchen wastewater than just a standard single slope solar still. This is due to the shape of the frustum of square pyramid still, which receives solar radiation from all directions because all the sides are exposed to the sun. This allows for more evaporation and condensation of water from the wastewater, resulting in more pure water collected.

Table 65.1 Ultimate tensile strength for group 1 and group 2 samples

S. No	Productivity (ml/m^2) of single slope solar still	Productivity (ml/m^2) of Novel Frustum of Square Pyramid solar still
1	392.13	499.9
2	388.49	457.96
3	390.63	500.45
4	319.34	477.35
5	351.12	451.77
6	350.6	519.88
7	389.66	483.38
8	410.56	484.72
9	391.43	499.4
10	413.15	465.8
11	402.61	498.33
12	398.17	506.04
13	380.23	498.72
14	370.22	496.21
15	341.36	469.82
16	371.1	475.67
17	355.18	471.91
18	398.78	498.41
19	340.1	498.26
20	417.3	470.24

Table 65.2 Group statistics results (mean of polylactic acid without filler 9.5635 mm is more compared to polylactic acid with filler 6.6785 mm and standard error mean for polylactic acid with filler is 0.0519 and polylactic acid without filler is 0.02949

Group		N	Mean	Standard Deviation	Standard Error Mean
Productivity	CG	20	378.6076	27.52129	6.15395
	EG	20	486.2104	17.94223	4.01200

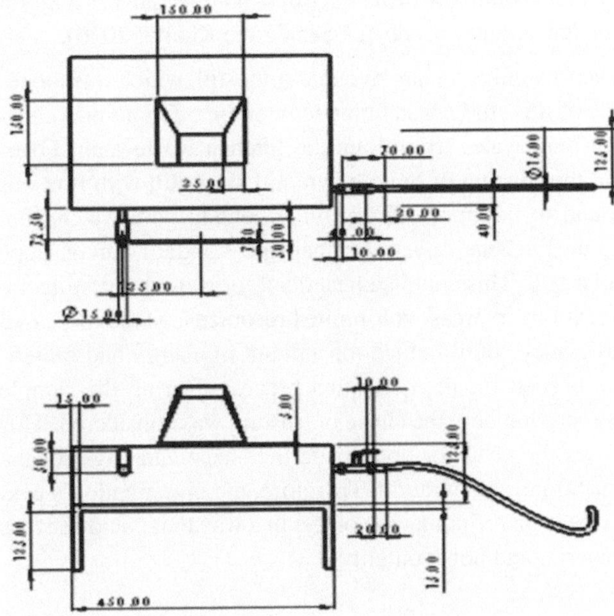

Fig. 65.1 Model of frustum of square pyramid solar still

4. Discussion

According to the research study, when it came to producing pure water by solar distillation from household kitchen wastewater, the Square Pyramid Solar Still's frustum outperformed the conventional single slope solar still. The average pure water collection was 486.21 ml/m2 for frustum of Square pyramid solar still, while it was 378.6 ml/m2 for standard single slope solar still. This shows that the novel

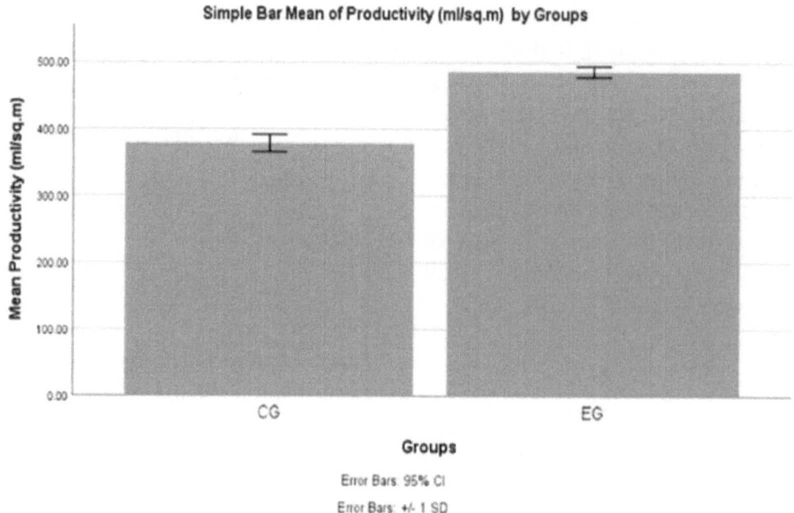

Fig. 65.2 Bar chart representing the comparison of mean productivity for CG represents single slope solar still and EG represents novel frustum of square pyramid solar still

frustum of square pyramid solar still gives 28.6% higher productivity. The explanation could be because the square pyramid solar still's frustum offers a larger surface area for condensation to occur. (El-Sebaii and Khallaf 2020).

The frustum of square pyramid solar still, which was the focus of this study, was found to be more efficient in collecting pure water from domestic kitchen wastewater. Overall, the frustum of Square pyramid solar still with fins was found to be more efficient in collecting pure water from home kitchen wastewater than the standard single slope solar still. This could be beneficial for water harvesting applications in areas with limited resources, as the increased efficiency could reduce the amount of energy and materials needed for the collection process. During the sample preparation only the shape of frustum was considered. This raises the need for optimizing the shape and angle of the frustum of the pyramid. Therefore, the investigation's next focus is on reshaping the object in a way that could use less material and boost output.

5. Conclusion

In the confines of the investigation, it was discovered that the Square Pyramid Solar Still's frustum outperformed the conventional Single Slope Solar Still in producing pure water from home kitchen wastewater. The average pure water collection was 486.21 ml/m2 for frustum of square pyramid solar still, which is 28.6% higher compared to standard single slope solar still producing 378.6 ml/m2. This is more suitable for domestic kitchen waste water treatment as a portable device.

References

1. Clapham, Christopher, and James Nicholson. 2014. The Concise Oxford Dictionary of Mathematics. OUP Oxford.
2. Cronk, Brian C. 2018. How to Use SPSS: A Step-by-Step Guide to Analysis and Interpretation.
3. Dhivagar, R., M. Mohanraj, Praveen Raj, and Radha Krishna Gopidesi. 2021. "Thermodynamic Analysis of Single Slope Solar Still Using Graphite Plates and Block Magnets at Seasonal Climatic Conditions." Water Science and Technology: A Journal of the International Association on Water Pollution Research 84 (10-11): 2635–51.
4. El-Sebaii, Ahmed, and Abd El-Monem Khallaf. 2020. "Mathematical Modeling and Experimental Validation for Square Pyramid Solar Still." Environmental Science and Pollution Research International 27 (26): 32283–95.
5. Girma, Misrak, and Abebayehu Assefa. 2015. Performance Analysis of Solar Powered Solar Still. LAP Lambert Academic Publishing.
6. Gulamhussein, Mumtaz Aliraza, Manya Shah, Kriti Yadav, Mitul Prajapati, and Manan Shah. 2023. "A Comprehensive and Systematic Study on the Techniques Used for Augmenting the Performance of a Solar Still Distillate Yield." Environmental Science and Pollution Research International, April. https://doi.org/10.1007/s11356-023-26751-6.
7. Modi, Kalpesh V., and Amarkumar R. Shah. 2022. "Effectiveness of Partially and Fully Submerged Triangular Cross-Sectional Metal Hollow-Fins and Wool Cloth Wick-Fins on Triangular Pyramid Solar Still." Environmental Science and Pollution Research International 29 (42): 64040–59.
8. Sampathkumar, Arivazhagan, and Sendhil Kumar Natarajan. 2022. "Performance Assessment of Single Slope Solar Still by the Incorporation of Palm Flower Powder and Micro Phase Change Material for the Augmentation of Productivity." Environmental Science and Pollution Research International 29 (49): 73957–75.

9. Shanmugam, Chinnasamy Subramanian, and Sekar Subramani. 2023a. "Correction to: Environmental and Embodied Analysis of Partial Shading Pyramid Solar Still." Environmental Science and Pollution Research International 30 (10): 25952.
10. Shanmugam, Chinnasamy Subramanian, and Sekar Subramani.. 2023b. "Environmental and Embodied Analysis of Partial Shading Pyramid Solar Still." Environmental Science and Pollution Research International 30 (10): 25933–51.
11. Shukla, Amritanshu, Atul Sharma, and Pascal Henry Biwolé. 2020. Latent Heat-Based Thermal Energy Storage Systems: Materials, Applications, and the Energy Market. CRC Press.
12. V. Savithiri , Mohammed El Hadi Attia , Abd Elnaby Kabeel , Sivakumar Vaithilingam ,Ganesh Radhakrishnan. 2022."Enhancing the Productivity of Hemispherical Solar Distillation by Using Energy Storage (rubber) and Wick Materials at Different Thickness." 2022. Solar Energy Materials & Solar Cells 248 (December): 112006.
13. Xia, Changlei, Shifeng Zhang, Sheldon Q. Shi, Liping Cai, and Jonathan Huang. 2016. "Property Enhancement of Kenaf Fiber Reinforced Composites by in Situ Aluminum Hydroxide Impregnation." Industrial Crops and Products. https://doi.org/10.1016/j.indcrop.2015.11.037.

Note: Every figure and table were created using the experimental work; none of them were taken from any publications or online.

Advances in Additive Manufacturing Technologies – Gurusamy Pathinettampadian (eds)
© 2024 Taylor & Francis Group, London, ISBN 978-1-032-90013-1

66 Prediction of Mechanical Properties of Polymer Epoxy Composites

Nallathambi. K*
Assistant Professor, Department of Mechanical Engineering,
Chennai Institute of Technology

Aswathnarayanan. A,
Agilan. R, Harvindakishan. P, Akash. V
UG Student, Department of Mechanical Engineering,
Chennai Institute of Technology

ABSTRACT: Natural fibres derived from plants are gradually substituting synthetic fibres in polymer composite reinforcement, as the fiber-reinforced composite exhibits higher specific strength and modulus. This project aims to develop a unique material that has a strength-to-weight ratio higher than any existing material. Handcrafted in compliance with ASTM guidelines, the glass and hemp hybrid composite was encased in an epoxy matrix. In the current economic climate, hemp plant fiber/epoxy composite has many benefits, including low cost, excellent durability, low density, sustainability, low wear from abrasion, and resistance to corrosion. This experiment's primary objective is to investigate mechanical strengths, such as impact and tensile strengths. It is advised to utilize hybrid hemp/glass fibre reinforced laminates for automotive applications rather than composites with synthetic fibre incorporation.

KEYWORDS: Hemp, Glass fibre composite, Epoxy & hardener, Mechanical properties

1. INTRODUCTION

Researchers formed the epoxy matrix material by combining EPON 862 and Curing Agent W. They mixed GNP with this epoxy to make composites, then used typical bulk measures to test the composite's tensile characteristics. They also used micro indentation, which enables nanoscale material characterization, to examine modulus, hardness, and creep compliance. This work has two main goals: first, Find the percolation threshold of electrical resistivity; second, to assess the effect of GNP on the mechanical properties of the composite using both nano indentation and macroscopic tensile testing. This threshold helps identify the filler aspect ratio in the composite by indicating the point at which its electrical resistivity significantly drops within a constrained range of filler loadings.[1]. Using the hand lay-up technique, e-glass random fibre reinforcement was added to a polyester resin matrix at various fibre percentages (15%, 30%, 45%, and 60% by weight). In order to evaluate the performance of these composite materials, tests for tensile strength, flexural characteristics, impact resistance, and Brinell hardness were conducted.[2].It was looked into how several material treatments, such as chemically treating the fibres and coupling agent treating the polymer

*Corresponding author: knallathambi3383@gmail.com

DOI: 10.1201/9781003545774-66

matrix, affected the mechanical properties of hemp fibre polypropylene matrix composites.[3]. It investigated the impact characteristics of composite laminates that were submerged in PLA and epoxy matrices and used hemp fibre in both unidirectional and fabric forms, with different thicknesses. The RIFT technique was used to create the epoxy-based laminates, and Utilizing instrumented falling weight impact apparatus, conduct low-velocity impact testing on them. In the meantime, a hot press was used to make the PLA-based laminates.[4]. Through improving the PC's composition, which consists of gravel, sand, and epoxy resin. The PC samples are heated between 25°C and 250°C while maintaining an ideal polymer and aggregate composition. They measure the total porosity (MIP) and examine the thermal characteristics after being exposed to high temperatures. They also evaluate the elastic qualities and mechanical strengths. After undergoing equivalent thermal treatments, the residual characteristics of PC and ordinary cement concrete (OCC) are compared.[5]. It uses glass and Kevlar epoxy composites, which include plane woven Jute fibre weighing 320 GSM, woven Bi-directional Kevlar fibre (K-49) weighing 380 GSM, and woven Bi-directional Glass fibre weighing 400 GSM.[6]. The goal is to produce an epoxy composite reinforced with different weight percentages of nano-sized tungsten trioxide particles and evaluate the mechanical characteristics of these composites.[7]. In order to improve the mechanical properties of the composites and reduce their cost, MRP filler is used as an additional reinforcement in addition to E-Glass fibre. In order tooptimise Micronized Rubber Powder (MRP) for better mechanical qualities, its weight percentage was varied in four steps. These properties were then determined. [8]. The possibility of using ND as a reinforcing filler by creating and examining epoxy samples with up to 25 vol% of ND. They combined mixing with hot pressing to create samples with high concentrations.[9]. the effect of ratios of micro-to nano-filler content on epoxy composites' mechanical characteristics. They used hardness, three-point bending, and tensile testing to accomplish this. They measured and assessed the epoxy composites' hardness as well as tensile strength, elastic modulus, elongation at break, and flexural strength.[10]. The several theories that predict how an epoxy matrix reinforced by fly ash will reinforce itself. We've talked about and characterised how it affects mechanical qualities.[11]. Fly ash and epoxy are used to create composites, taking into account how viscosity and particle size affect mechanical qualities. The mechanical characteristics of these composites at different filler concentrations have been tested. Furthermore, it looked at the fly ash/epoxy composite's failure behaviour by examining the fracture surface's microstructure.[12]. Using a novel steam explosion technology, nanocellulose may be separated in order to create composite materials. In the matrix phase, they strengthened the hemp nanofibers that were extracted used Epoxy LY-556 as the matrix phase and Hy-951 as the hardener, which are well-known for their superior heat resistance and high acid resistance up to 80°C.[13]. Its goal was to understand and gather knowledge of the wear behaviour under different external variables of aliphatic polyketone (APK), polyoxymethylene (POM), ultrahigh molecular weight polyethylene (UHMWPE), polyamide 66 (PA66), and 30% glass fibre reinforced polyphenylene sulphide (PPS + 30%GFR) polymers. These variables included the use of several discs with abrasive sheets of 150, 360, 800, and 1200 grit sizes; an external force of 10 N; a sliding speed of 1 m/s; and sliding distances of 50, 100, 150, and 200 m.[14].

The literature review made clear that there is a greater need for study in the field of natural fiber reinforced composites. The selection of hemp fiber in this paper is based on its availability and processing techniques. To remove the hemp fiber from the tree stem while taking into account a number of factors, alkali solutions are applied. Metrics like impact, tensile, flexural, and hardness tests show how important natural fiber is to the forward-thinking methodology.

2. Materials and Methods

Because they have a lower density than other materials, natural fibers—which are derived from plants—are utilised in polymer composites to reduce weight. The benefits of natural fibres include being inexpensive, readily available, biodegradable, and environmentally benign. Use abrasive paper to remove any rust from the base plate. The surface was washed with a thinner solution once it had dried, and

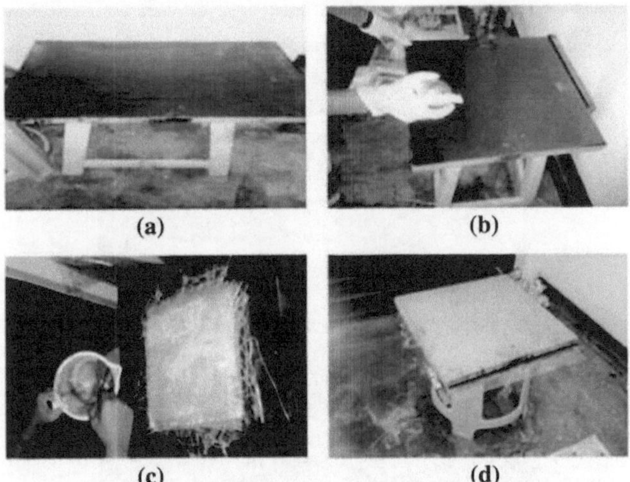

Fig. 66.1 Stages of composite preparation (a) Mold, (b) applying release spray to the release film, (c) Applying resin on the mat, (d) Applying load

it was then let to air dry. When the surface had dried out, silicon gel was sprayed on it. The surface was left unaltered for a few minutes in order to put up the mould. The hardener is mixed with epoxy resin in a 10:1 ratio. After blending, the curing period (pot life) was 20 minutes, as mentioned.

3. Testing

Mechanical strength is the ability of the material to withstand external force and this is the most desirable factor in the engineering field. Performance and suitability of the material is assessed by the mechanical strength of the material. Mechanical strength includes the tensile, compressive, shear, flexural and so on. In this section, the strengths like tensile, flexural, impact and wear behavior of the materials are discussed. Figure 66.2 exhibits the specimen for testing and Fig. 66.3 represents the various testing specimens.

(i) Impact testing specimen
(ii) Tensile testing specimen
(iii) Flexural testing specimen
(iv) Hardness testing specimen
(v) SEM characterization

Fig. 66.2 Specimen for testing

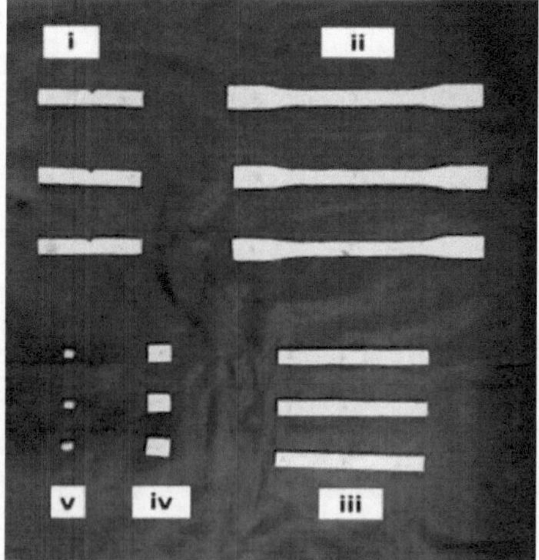

Fig. 66.3 Various testing specimen for mechanical testing

4. Result and Discussion

The impact, tensile, flexural, and hardness tests were conducted to determine the mechanical properties of the natural hemp fibre reinforced polymer composite used in the current studies.

4.1 Impact Analysis

Three test specimens are utilized in each hemp and glass fiber-reinforced "composite" laminate to evaluate the material's impact-resistance. ASTM D256 states that both hybrid and non-hybrid "composite laminates are used for test specimens." The examples have well-polished edges. There are several ways to interpret the results. Finding out how much energy is contained in a sample is one way. This picture displays the impact test specimen both before and after it was fractured by an epoxy "composite" reinforced with glass fibre and hemp. The values related to the impact study of hemp-polymer composites are listed in Table 66.1.

Table 66.1 Values of impact analysis

Sample I.D	Impact Values, Joules
NPL-1	1.1
NPL-2	1.2
NPL-3	1.8

The impact test measures the quantity of energy a substance takes in during a fracture. The impact test computes numbers that increase gradually in order to assess the impact resistance or toughness of materials by estimating the amount of energy absorbed during fracture. The above table's experimental values show that a maximum impact strength of 1.8 joules was recorded.

4.2 Tensile Test

Tensile strength testing: ASTM D3039 standards and procedures are followed in the preparation of the tensile test specimens and "composite" laminates. For both hybrid and non-hybrid laminates, tensile testing is conducted using three specimens from each laminate. Findings were seen when the specimen was subjected to a load by the testing apparatus until it broke. By comparing the remaining "composite" laminate specimens using the same methods as before, the mean tensile strength and associated stresses are ascertained. Tensile strength load vs. displacement is shown in Fig. 66.4. The strain curve for tensile strain is shown in Fig. 66.5.

The maximal strength at the extreme stiffness gives 3.6 mm displacement at load 1400N, according to Fig. 66.4's load vs. displacemet curve. Another typical tensile stress-strain curve for hemp reinforced polymer composites demon-

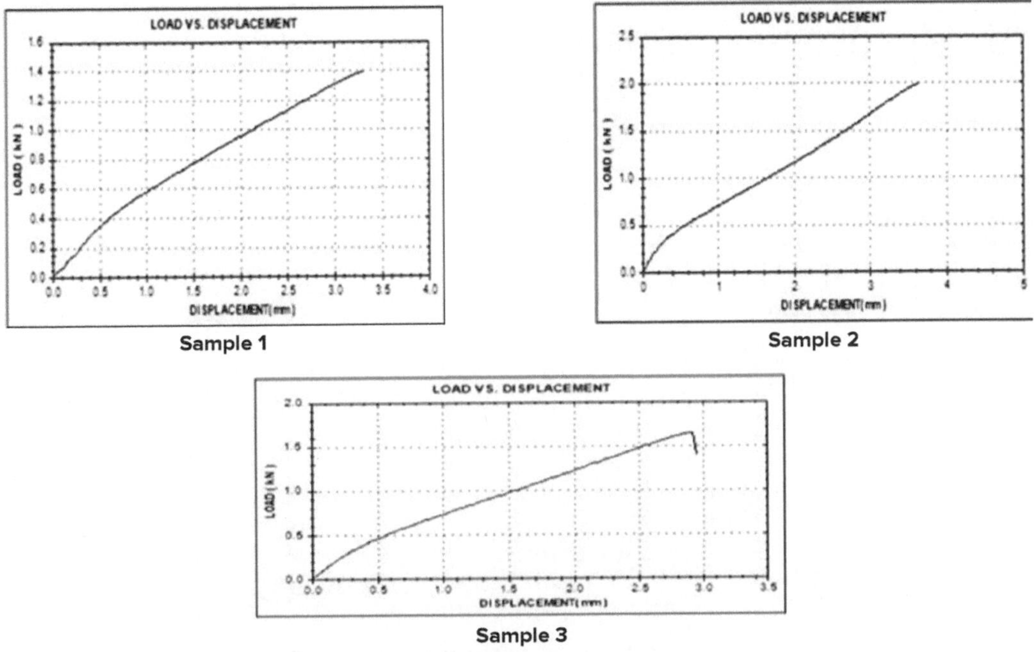

Fig. 66.4 Load vs displacement diagram for sample 1, 2, 3 (tensile test)

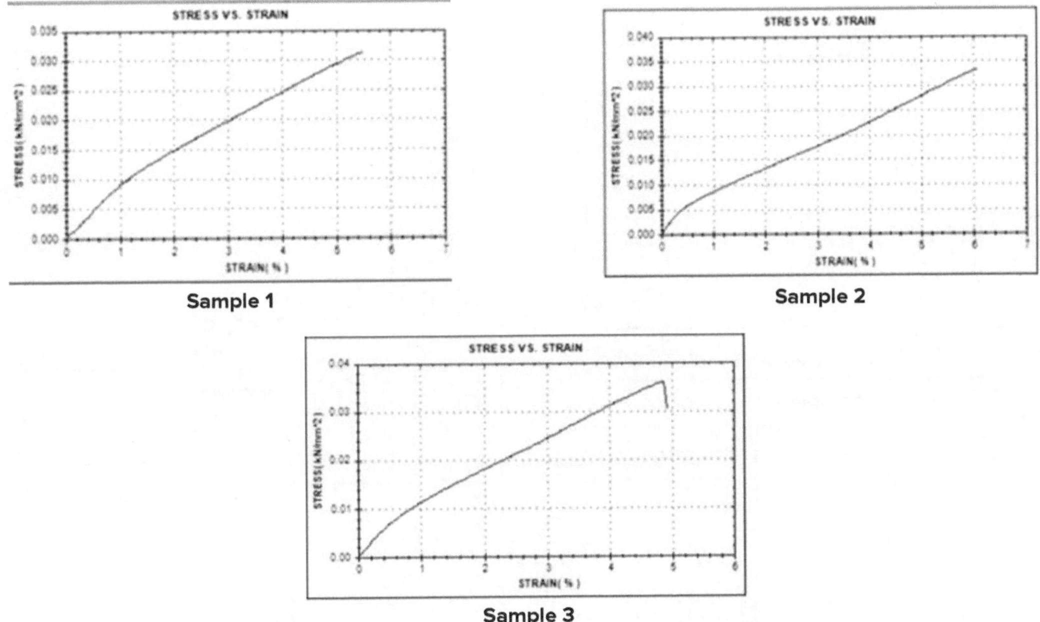

Fig. 66.5 Stress vs strain diagram for specimen 1, 2, 3

strates how the tensile stress-strain curve makes it easy to understand how the composites flex, which may be the result of weak reinforcement. The stress strain in Fig. 66.5 illustrates the interstitial cohesive bonding.

4.3 Flexural Testing

A flexural test does not examine the basic qualities of the material, in contrast to a compression or tensile test. All three of the fundamental stresses—tensile, compressive, and shear—are present when a specimen is subjected to flexural loading. As a result, the flexural properties of a specimen are determined by the combined influence of these stresses as well as—to a lesser extent—the specimen's geometry and the rate at which the load is applied. Figure 66.6 represents Load vs displacement for flexural test. Figure 66.7 denotes the stress strain analysis.

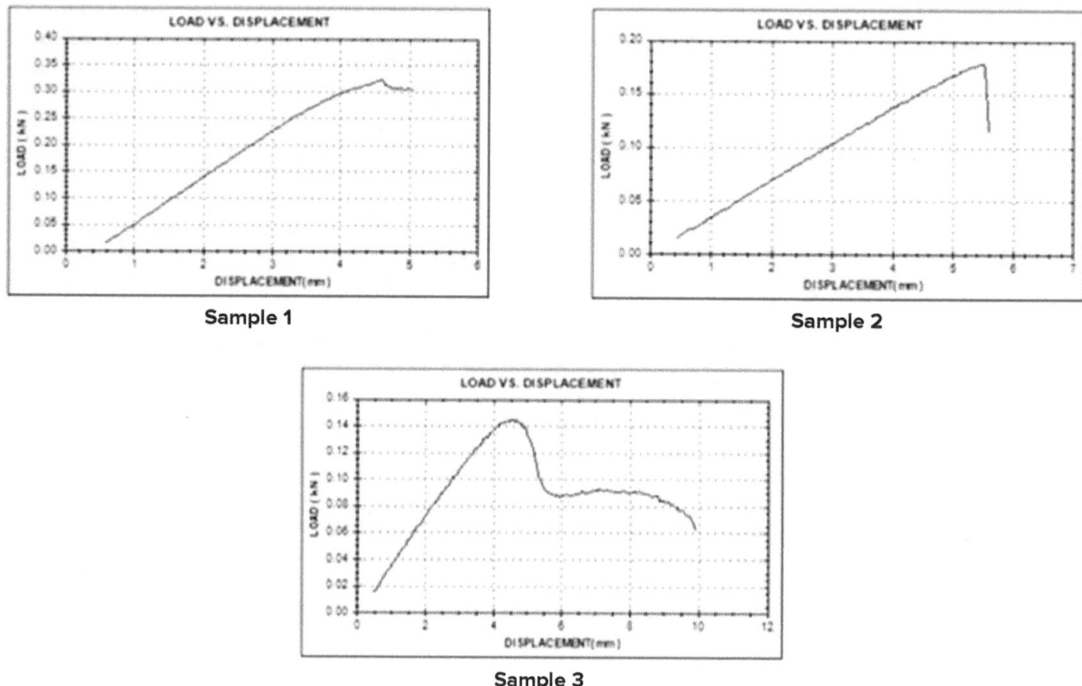

Fig. 66.6 Load vs displacement for sample 1,2,3

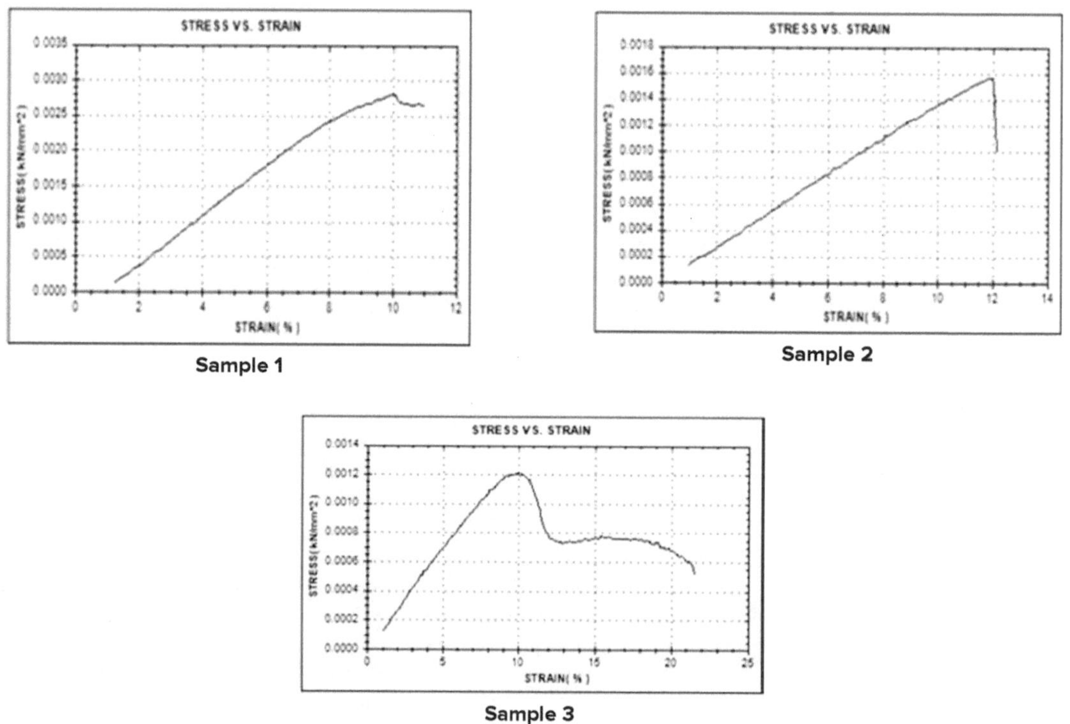

Fig. 66.7 Stress vs strain for specimen 1, 2, 3

Up to a maximum level, where additional addition resulted in a decrease in bending strengths, the flexural strength rose with increasing fibre addition. The interfacial bonding and microstructure between the matrix and reinforcement are the primary determinants of the flexural characteristics of composite materials.

4.4 Hardness Test

The microstructure, or homogeneity, of the material you are testing, should dictate which hardness test you use. in addition to the part's size, quality, and kind of material. Unless you are trying to identify the various components within the microstructure, the material behind the indent in all hardness tests should be typical of the entire microstructure. Consequently, you require a larger imprint than for a homogeneous material if the microstructure is extremely coarse and heterogeneous. There are four primary hardness tests, and each has advantages and requirements of its own. Table 66.2 represents the hardness test results.

Table 66.2 Hardness testing results

Sample	Shore "D"
	D Scale
NPL - 1	69, 60, 76
NPL - 2	69, 69, 66
NPL - 3	78, 74, 78

The impact of post-curing time and fibre loading on Rockwell hardness was demonstrated and In general, fibres that raise a composite's modulus also raise the composite's hardness. This is so because relative fibre volume and modulus determine hardness.

4.5 Scanning Electron Microscopy (SEM)

To ensure that survive the intense vacuum and high-energy electron beam, SEM samples must be minimal enough to accommodate on the sample level and maybe require further processing to improve their capacity to carry electricity and stabilise them. Usually, conductive adhesive is used to mount samples firmly on a specimen holder or counterfoil. Semiconductor wafer defects are often analysed using SEM, and devices that can inspect any area of a 300 mm semiconductor wafer are produced by manufacturers. A lot of devices include chambers that can rotate 360 degrees continuously and tilt an object that size up to 45 degrees.

The only extra steps needed to prepare metal objects for SEM are cleaning and conductivity attachment to a specimen stub. Typically, electrically conducting compounds are very thinly coated on non-conducting materials via high-vacuum evaporation, electrodes deposition [citation needed], or low-vacuum sputter coating. The enhancement of In high-Z materials, supplementary generation of electrons is the reason for the improvement. The SEM morphology of specimens 1, 2, and 3 with varying microns is represented in Figs. 66.8, 66.9, and 66.10, which are interpreted in relation to the goal of cohesive bonding in hemp-polymer composites.

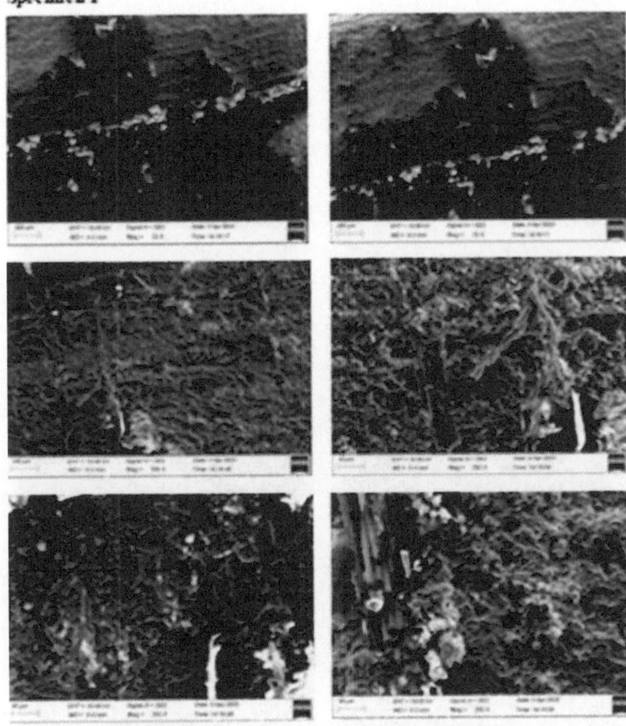

Fig. 66.8 SEM images for specimen 1

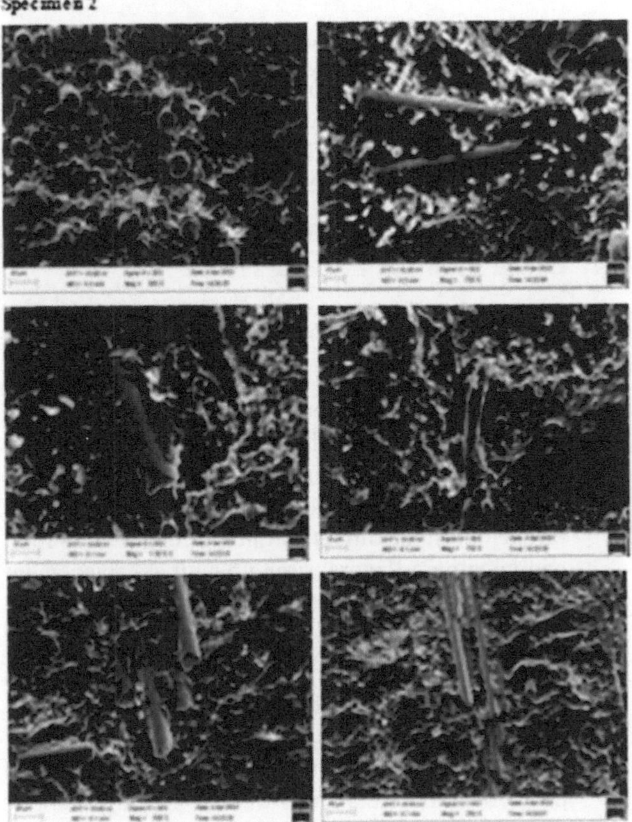

Fig. 66.9 SEM images for specimen 2

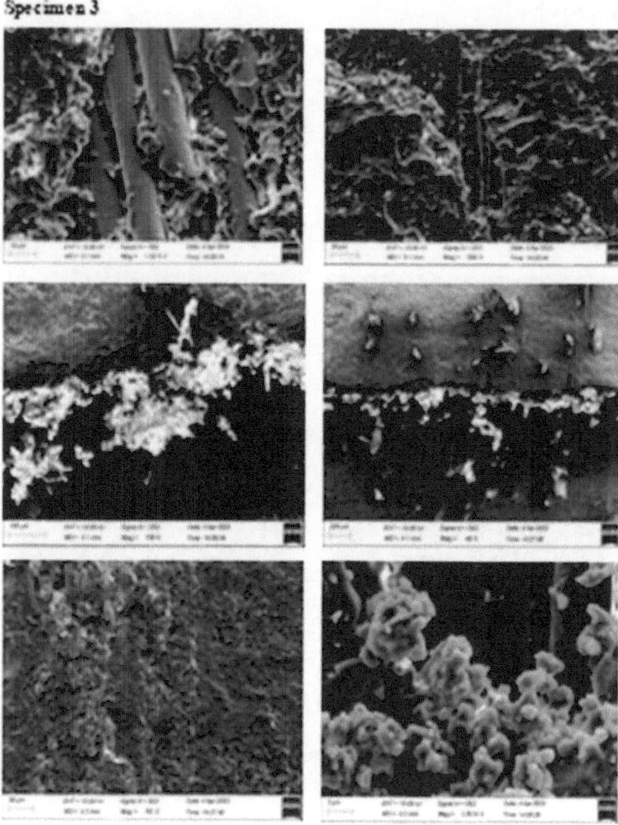

Fig. 66.10 SEM images for specimen 3

In this work, the mechanical properties like tensile, impact, flexural and hardness are carried out for composites on epoxy resin with hemp fibre. The following are the primary inferences that can be made from the findings.

5. Conclusion

1. In tensile and impact test among the composites, hemp fibre with epoxy resin has high elongation and high tensile strength.
2. The average values found for various hemp fibre composite sample combinations are capable of having the following maximum values: maximum tensile strength at 1400 N, maximum impact strength of 1.8 joules, maximum hardness strength of 78 mm, and maximum flexural strength with a displacement of 9.8 mm at 320 N.
3. The result of scanning electron microscope shows that a composite has a much rougher surface, the need for more in-depth studies on the hemp fibre modification is motivated by their peculiarities and the practical requirements for finely tuning their properties in correlation with their envisaged applications.

Reference

1. Julia A. King, Danielle R. Klimek, Ibrahim Miskioglu, Greg M. Odegard, Mechanical Properties of Graphene Nanoplatelet/Epoxy Composites, Journal of Applied Polymer Science128(6) 4217–4223.
2. M.S, EL-Wazery, M.I. EL-Elamy, S.H Zoalfakar, Mechanical Properties of Glass Fiber Reinforced Polyester Composites, International Journal of Applied Science and Engineering14(3) 121–131 (2017).
3. Theresa Sullins, Selvum Pillay, Alastair Komus, Haibin Ning, Hemp fiber reinforced polypropylene composites: The effects of material treatments, Elsevier114. 15–22 (2017).
4. Caprino, G., Carrino, L., Durante, M., Langella, A., Lopresto, V., Low impact behavior of hemp fibre reinforced epoxy composites, Composite Structures (2015).
5. Elalaoui Oussama, Ghorbel Elhem, Mignot Valérie, Ben Ouezdou Mongi, Mechanical and physical properties of epoxy polymer concrete after exposure to temperatures up to 250 C, Elsevier 27 415–424 (2012).
6. S.S. Godara, Abhishek Yadav, Ved Pratap Singh, Characterization of glass fiber reinforced composite materials with use of Kota stone dust, Elsevier 44 2566–2569 (2021).
7. Shaymaa Mahdi Salih, Noor Sabeeh Majeed, Mechanical properties and characterization of epoxy composite reinforced with Tungsten Trioxide nanoparticles, Elsevier. 20(4) 443–447 (2020).
8. K.C. Nagaraja, S. Rajanna, G.S. Prakash, G. Rajeshkumar, Mechanical properties of polymer matrix composites: Effect of hybridization, Elsevier 34(2) 536–538 (2018).
9. I.Neitzel, V. Mochalin, I. Knoke, G.R. Palmese, Y. Gogotsi, Mechanical properties of epoxy composites with high contents of nanodiamond, Elsevier 71 710–716 (2011).
10. Iskender Ozsoy1, Askin Demirkol, Abdullah Mimaroglu1, Huseyin Unal1, Zafer Demir, The Influence of Micro- and Nano-Filler Content on the Mechanical Properties of Epoxy Composites, Journal of Mechanical Engineering, 61(10) 601–609 (2015)
11. Mahadeva Raju G. K, G. M. Madhu, Ameen Khan M, P Dinesh Sankar Reddy, Characterizing and Modeling of Mechanical Properties of Epoxy Polymer Composites Reinforced with Fly ash, Elsevier 5 27998–28007 (2018).
12. Jeesoo Sim, Youngjeong Kang, Byung Joo Kim, Yong Ho Park, Young Cheol Lee, Preparation of Fly Ash/Epoxy Composites and Its Effects on Mechanical Properties, mdpi 79 1–12 (2020).
13. Suraj Kumar Singh, Jimmy Karloopia, Sabah Khan, Raghvendra Kumar Mishra, Processing and characterization of hemp nanofiber thermoset polymer composite, Elsevier 46(2) 1341–1348 (2021).
14. H.Unal, U. Sen, A. Mimaroglu, Abrasive wear behaviour of polymeric materials, Elsevier26 705–710 (2005).

Note: Every figure and table was created using the experimental work; none of them were taken from any publications or online.

Advances in Additive Manufacturing Technologies – Gurusamy Pathinettampadian (eds)
© 2024 Taylor & Francis Group, London, ISBN 978-1-032-90013-1

67 A Comprehensive Review of CPU Cooling Systems: Innovations, Efficiency, and Future Directions

G. Sastha Roopan[1],
M. D Vijayakumar[2]
Department of Mechanical Engineering,
Chennai Institute of Technology, Chennai, India

ABSTRACT: The effective thermal management of central processing units (CPUs) has become increasingly crucial in contemporary computing due to the escalating demand for high-performance systems. This comprehensive review delves into the advancements, efficacy, and future trajectories of CPU cooling systems, aiming to furnish a thorough overview of this pivotal facet of computing technology. The necessity for efficient cooling is paramount to prevent overheating and uphold the reliability of CPUs, given their propensity to generate substantial heat during operation. The review surveys a spectrum of CPU cooling methodologies, spanning from traditional air cooling to cutting-edge liquid and phase-change cooling solutions. While air cooling, recognized for its simplicity and reliability, has been a longstanding choice in many systems, its limitations in meeting the demands of modern high-performance CPUs are acknowledged. Liquid cooling, employing coolants and heat exchangers, emerges as a superior alternative, providing enhanced heat dissipation and is scrutinized in diverse configurations, including all-in-one (AIO) and custom loop systems. Additionally, phase-change cooling techniques, such as vapor chambers and direct-die cooling, demonstrate exceptional thermal performance through the alteration of the coolant's phase. The review also explores emerging technologies and inventive approaches, encompassing advanced materials like graphene and carbon nanotubes, as well as thermoelectric and electro-wetting cooling methods. Efficiency metrics and benchmarks, including thermal resistance, heat dissipation rates, and energy consumption, are thoroughly examined to offer insights into the effectiveness of various cooling solutions. This analysis aims to contribute valuable information to the scientific community, especially in the context of designing and optimizing cooling systems for contemporary CPUs.

KEYWORDS: CPU cooling system, Microchannel heat sinks, AIO cooler

1. INTRODUCTION

In the digital age, the central processing unit (CPU) plays a pivotal role in contemporary computing systems, serving as the primary source of computational power. It orchestrates a wide range of tasks, from complex scientific simulations and 3D rendering to routine activities such as web browsing and electronic communication. However, the continual advancement of CPUs to deliver heightened performance has resulted in a notable rise in the production of heat, presenting a significant challenge for both computer hardware manufacturers and enthusiasts. Effectively

[1]sastharoopan@gmail.com, [2]vijayakumarmd@citchennai.net

DOI: 10.1201/9781003545774-67

managing and dissipating this heat is imperative to ensure the dependability, stability, and durability of the CPU.

The demand for cooling solutions for CPUs is not a recent development. As electronic computers first emerged in the mid-20th century, heat management became a foundational concern. Vacuum tubes, the precursors to the transistors found in contemporary CPUs, emitted significant heat, necessitating intricate cooling systems to maintain operational stability [27]. Over the ensuing decades, the evolution of semiconductor technology has brought about remarkable advancements in computational capabilities. However, these advancements have been accompanied by a corresponding increase in thermal dissipation, making efficient cooling an indispensable aspect of modern computing.

2. Efficiency Improvement of Integrated Heat Spreaders by Enhancing Distributed Heat Transfer Coefficients

J.W. Elliott et al focuses on addressing the challenges arising from the escalating power densities in contemporary CPUs, leading to localized regions of elevated temperature on the semiconductor die. The heightened temperatures may have negative impacts on both performance and reliability [13]. Consequently, the author examined a novel cooling strategy by studying the potential benefits of implementing a profile featuring an inconsistent coefficient of heat transmission to a CPU's inbuilt heat spreader. The association between the transmission of heat for several commercial CPUs, the correlation between the coefficient and temperature flux is shown in Fig. 67.1

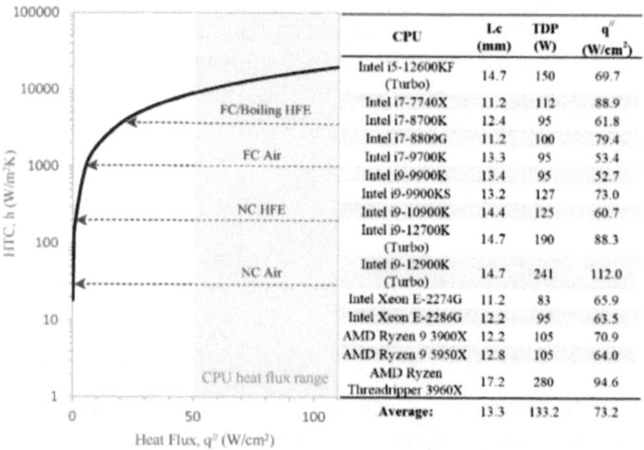

CPU	Lc (mm)	TDP (W)	q'' (W/cm²)
Intel i5-12600KF (Turbo)	14.7	150	69.7
Intel i7-7740X	11.2	112	88.9
Intel i7-8700K	12.4	95	61.8
Intel i7-8809G	11.2	100	79.4
Intel i7-9700K	13.3	95	53.4
Intel i9-9900K	13.4	95	52.7
Intel i9-9900KS	13.2	127	73.0
Intel i9-10900K	14.4	125	60.7
Intel i9-12700K (Turbo)	14.7	190	88.3
Intel i9-12900K (Turbo)	14.7	241	112.0
Intel Xeon E-2274G	11.2	83	65.9
Intel Xeon E-2286G	12.2	95	63.5
AMD Ryzen 9 3900X	12.2	105	70.9
AMD Ryzen 9 5950X	12.8	105	64.0
AMD Ryzen Threadripper 3960X	17.2	280	94.6
Average:	13.3	133.2	73.2

Fig. 67.1 Heat transfer co-efficient vs heat flux for various CPUs

Source: Created using the experimental work.

To enable this investigation, the author constructed a copper heat spreader modeled by a finite element simulation measuring 1600 mm² (40 x 40) and having a thickness of 2.5 mm. The model included a centrally positioned heat source measuring 13 × 13 mm², generating a heat flux of 100 W/cm², simulating conditions commonly found in high-powered CPUs. Our analysis involved the comparison of two different heat transfer coefficient profiles: one characterized by a uniform distribution and the other exhibiting a Gaussian distribution [12].

In order to improve CPU package cooling performance, this study suggests that the convective surface of the Integrated Heat Spreader be covered with a non-uniform distribution of heat transfer coefficients. By describing the spectrum of the thermal transfer factor using a Gaussian function, the author applied a genetic algorithm to optimize various temperature-related goals [25]. This study successfully demonstrates how CPU package cooling may be significantly improved by using an optimized non-uniform heat transfer coefficient function. The Gaussian function is stated in Eq. 1.

$$h(x,y) = h_{max} e^{-\frac{(x-x_0)^2+(y-y_0)^2}{2\sigma^2}} \quad (1)$$

where h_{max} - maximum heat transfer coefficient, (x_0, y_0) - center of the distribution, and σ - standard deviation. The optimization problem is formulated in Eqn. 2.

$$\min_{h(x,y)} \sum_{i=1}^{N} (T_i - T_{obj})^2 + \alpha \sum_{i=1}^{N-1} (h_{i+1} - h_i)^2 \quad (2)$$

where N - number of nodes, T_i - temperature at node i, T_{obj} - temperature objective, and α - penalty factor.

2.1 Liquid Cooling Systems for High-performance CPUs with Closed-loop Configurations

R. Jenkins et al conduct an experimental inquiry into a closed-loop liquid cooling system (depicted in Fig. 67.2) specifically tailored for high-powered CPUs. This system combines a remote air-cooled heat exchanger and a microjet cold plate into a hybrid liquid-air design. When the liquid cooling system is compared to a traditional fan-fin air cooling system, it performs better, as shown by lower CPU temperatures, better heat flux removal, and reduced thermal resistance [9]. The study delves into the characterization of the liquid cooling system under various operating conditions, including CPU heat loads, water flow rates, and air velocities. The findings underscore the significance of optimizing liquid and air flow rates to achieve minimal energy consumption and maximize the coefficient of performance. The authors justify their research by underscoring the limitations of air cooling for contemporary high-power CPUs with elevated power densities and heat fluxes.

Additionally, they identify existing gaps in the literature pertaining to the comprehensive evaluation and energy optimization of liquid cooling technologies at the system level.

The relation between a fluid-based cooling system and the thermal rate of transmission is stated in Eqn. 3.

$$Q = hA(T_{hot} - T_{cold}) \quad (3)$$

where A - cold plate surface area, h - heat transfer coefficient, T_{cold} - coolant temperature, T_{hot} - CPU temperature, and Q - heat transfer rate.

The coolant's fluidity and conductivity to heat, as well as other coolant parameters, influence on the rate of transfer of heat. The liquid-based chilling mechanism's energy use may be expressed in Eqn. 4.

$$P = \dot{m}C_p(T_{out} - T_{in}) \quad (4)$$

where P - energy consumption, m – coolant's mass flow rate, C_p – coolant's specific heat capacity, T_{out} – coolant's outlet temperature, and T_{in} – coolant's inlet temperature. The liquid cooling system's coefficient of performance (COP) is given in Eqn. 5.

$$COP = \frac{Q}{P} \quad (5)$$

The study outcomes reveal that the liquid cooling system can attain a Coefficient of Performance (COP) reaching 5.5, markedly surpassing the COP of the air cooling system. The authors suggest potential improvements to the liquid cooling system by incorporating advanced heat transfer surfaces, such as microchannels and nanofluids. Additionally, they recommend integrating the liquid cooling system with other cooling technologies, such as phase change materials and thermoelectric coolers, to further optimize its performance.

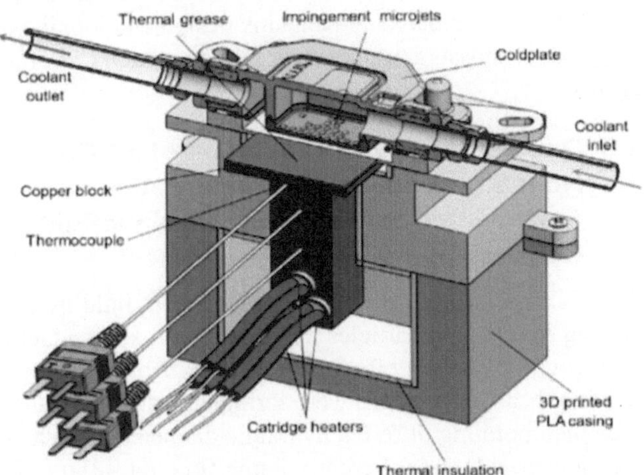

Fig. 67.2 Design of the approach of closed-loop cooling presented by R. Jenkins et al [9]

2.2 Enhancement of High Heat Flux CPU Cooling through the Utilization of Copper Foam and a Suspension Comprising Non-Equilibrium Phase Change Material (NEPCM) and Water

Exploring the merging a copper permeable substance with a mini channel on the processor's surface as part of cooling system arrangement. The NEPCM increases effective particular temperature of the mixture, which enhances heat transfer rates. It is characterized by a solid shell and a liquid core that is shifting phases [24].

As a result of its high heat conductivity, copper is used as the porous media. The medium's porosity and pore diameter are determined and prescribed. The continuity, momentum, and energy equations controlling forced convection flow in the presence of the porous medium and NEPCM are outlined.For instance, the energy equation can be represented in Eqn. 6.

$$\frac{\partial(\rho C_p T)}{\partial t} + \nabla.\left(\rho C_p \vec{V} T\right) = \nabla.(k \nabla T) + Q \quad (6)$$

Where, Q - heat generation term, T - temperature, C_p - specific heat, V - velocity vector, k - thermal conductivity, ρ - density.

The dynamics of heat transmission inside the system are examined using this equation in conjunction with the continuity and momentum equations. The suspension's and the porous medium's thermophysical properties are discussed. Improving heat transmission is the main objective of the design, which will eventually result in more effective cooling.

2.3 Thermal Endurance of a Thermal Sink Designed for Two CPU Servers that is Liquid-cooled and has Heat Pipes

Han Wang et al. conducted an in-depth examination of CPU cooling in data center servers, particularly focusing on high heat load scenarios. For twin CPU servers, the authors created a specially made thermal conduits This creative design uses water as the operational fluid and incorporates heat pipes that are placed tightly together [11]. The thermal conduit's thermal efficacy was carefully evaluated in a range of scenarios. The authors created a numerical model for the thermal conduits and a one-dimensional thermal resistance network to gain a thorough grasp of the thermal properties. This allowed for the examination of thermal resistance fluctuation and distribution.

The thermal resistance, R, can be calculated using the Eqn. 7.

$$R = \frac{\Delta T}{Q} \tag{7}$$

where ΔT - temperature difference, Q - heat load.

The liquid-cooled heat sink effectively satisfies criteria for radiating heat of 37.5 W/cm^2, operating consistently even under elevated demand on heat and increasing intake temperatures of cooling agents. To increase the liquid-cooled heat sink's total thermal efficiency, the authors stress how crucial it is to reduce thermal resistance in the heat pipe. These findings have a big impact on how cooling system designs for data centers are developed.

To ascertain the effects of concentration of nanofluid, Reynolds number, and its material on transporting power, the impact of heat transfer factor, heat sink thermal effectiveness, and processor temperature, their analysis compares these factors.

The findings imply that adding nanofluid increases the liquid coolant's thermal conductivity, which in turn improves CPU cooling efficiency. The thermal drain with a baseplate with a spider-net pattern performs better in terms of lowering the maximum temperature, boosting the heat sink's thermal capacity and its thermal transfer coefficient, the metals investigated (silver, nickel, and copper), silver turns out to be the best choice for the heat sink [19].

The authors suggest that integrating a revolutionary fin geometry and an initial methodology inspired by nature, as well as applying simulation modeling regarding heat transport and flow of CGNPs/H2O eco friendly nanofluid in a thermal sink with liquid block, may improve the thermal management of electronic devices.

2.4 Utilization of Computational Fluid Dynamics to Analyze CPU Cooling Employing Finned Heat Sinks and Eco-friendly Nanofluid

S. S. Ghadikolaei et al conduct a study focused on cooling CPUs using two different fin designs—snaky and sinusoidal spiral—and an eco-friendly nanofluid. A code for functions customized by users is used to determine the nanofluid's thermophysical features, such as its unique capacity for heat, density, viscosity, and thermal efficiency [21].

The system is modeled in three dimensions, incorporating continuity ($\nabla \cdot (\rho u) = 0$), momentum ($\rho u \cdot \nabla u = -\nabla p + \mu \nabla 2u$), and energy equations ($\rho cpu \cdot \nabla T = k\nabla 2T$) for a liquid block with a finned heat sink and nanofluid flow [17]. There are defined parameters for the heat flux, outflow, and input. ANSYS-Fluent's Finite Volume Method is utilized in a pressure-based manner to generate the solution. Validation is carried out by comparing results with earlier studies, accounting for the heat sink's total thermal efficiency, proportion of skin friction and average Nusselt number [18].This research focuses into contrasting fin designs and nanofluid combination affect CPU cooling. The impacts of several parameters on temperature, pressure, heat transfer, and thermal efficiency are carefully examined. These parameters include nanofluid concentration, Reynolds number, fin form, number of sine arcs, and heat sink material. The authors emphasize the superior cooling performance observed in the case of the sinusoidal spiral fin design with a silver heat sink and a nanofluid concentration of 0.075% among the cases studied [16]. In summary, the authors suggest that the combination of an environmentally friendly nanofluid and a novel fin design for passive cooling successfully mitigates temperature rises and elevated input voltage in electronic package CPUs. They contend that their numerical analysis provides valuable insights for the field of thermal management in electronics[15].

In the trials, fresh water is utilized as the coolant, and the inlet temperatures and mass flow rates are adjusted. The heat transfer and fluid flow inside the tiny channels are modeled using ANSYS FLUENT, and the numerical results are compared to experimental data. The findings show that, in most situations, the serpentine minichannel heat sink performs better than its straight counterpart, exhibiting a greater Nusselt number, a lower base temperature, less thermal resistance, and enhanced thermal efficiency. However, because of the flow path's twists, a bigger pressure decrease is seen. In addition, elements like rate of flow and intake temperature affect the heat sink's performance.

The researchers come to the conclusion that the serpentine minichannel heat sink design, which exhibits better thermal properties than the straight heat sink, has potential for cooling liquid CPUs. They recommend that future research focus on enhancing heat sink design and looking at using nanofluids as a replacement for conventional cooling agents.

2.5 Development and Evaluation of a CPU Heat Sink through Experimentation Employing Cooling through a Nanofluid Consisting of Alumina and Water

A. Siricharoenpanich et al produced the nanofluid by dispersing Al2O3 nanoparticles in water, a well-known method recognized for its efficient cooling capabilities. Three different heat sink designs were formulated and fabricated, each incorporating different hydraulic diameters and channel numbers [2]. At different volume flow rates, both the pressure decrease and the heat transfer of the nanofluid passing through such heat sinks were examined.

Analyzing the data showed that the heat sinks performed better in terms of cooling when there was a greater nanofluid flow rate and a smaller channel diameter. But this improvement was followed by a simultaneous rise in pressure and fall. To enhance the heat sink's design, the scientists employed Response Surface Methodology, which is based on Central Composite Design. Desirability served as the objective function, in order to maintain a balance between heat transfer efficiency and pressure decrease [12]. Using this method, the ideal design was found to be a 4 mm-diameter heat sink with four channels. The mathematical model for the optimization can be represented as: maximize x, $D=(D_1{}^{w} \cdot D_2{}^{w})^{1/(w1+w2)}$ subject to $x_{min} \leq x \leq x_{max}$ where D is the desirability, D1 and D2 are the desirabilities of the individual responses, w1 and w2 are the weights of the responses, and x is the design variable [10]. The equation was employed to determine the most effective configuration for the heat sink design.

3. Commercial Microchannel Heat Sinks (MCHS)

The primary advantage of AIO coolers is their capacity to keep temperatures constant. By combining a cleverly engineered pump with liquid coolant, these systems are excellent at quickly removing heat from the central processing unit [22]. As a consequence, users enjoy a more dependable and consistent computing experience, knowing that their gear will continue to operate at its best even under extreme conditions.

The fact that AIO coolers are configurable only serves to highlight how effective they are. AIO coolers that are sold commercially often have fan speeds that can be adjusted and user-friendly software interfaces that let customers customize their cooling solution to meet the specific needs of their system. This flexibility guarantees that users may maximize the performance of their cooling solution whether they are heavily multitasking or pushing the limits of overclocking [5]. The entire experience is greatly influenced by installation, and AIO coolers have advanced significantly in terms of user-friendly configurations. Installation is made easier with pre-filled and pre-sealed units, and manufacturers frequently include comprehensive installation instructions and mounting solutions [20]. Because of its accessibility, AIO coolers are a desirable choice for consumers looking for high-performance cooling without the intimidating nature of more intricate liquid cooling systems.

The effectiveness of commercial AIO coolers is evidence of the ongoing progress in PC cooling technology. AIO coolers are a compelling option for enthusiasts and professionals alike due to their exceptional thermal management, easy installation, and customization options [6,7]. They are a dependable solution for people who expect nothing less than the best performance from their computer hardware.

4. Heated Pursuit of Excellence—Assessing Commercial CPU Heat Sink Performance

The pursuit of dependable and effective thermal management in the constantly changing field of computer hardware has prompted several experts and enthusiasts to investigate the potential of commercial CPU heat sinks [3]. Upon careful analysis of the range of products available, it is evident that these cooling options are essential for preserving peak performance and provide an insight into how heat dissipation standards are developing.

The capacity of commercial CPU heat sinks to effectively remove heat from the processor, avoid thermal throttling, and guarantee continuous top performance is what determines their performance. The simplicity and dependability of these heat sinks is one of their main advantages. These air-based technologies efficiently transmit and distribute heat through the use of fin arrays and heat pipes, doing away with the need for complex liquid cooling systems [1].

When doing demanding tasks like gaming, content production, and other resource-intensive programs, the efficacy of CPU heat sinks is especially noticeable. These cooling options are excellent at maintaining a steady temperature, which enables the CPU to run at peak performance even during prolonged bursts of heavy demand [4]. For customers who want steady performance without sacrificing system longevity, this dependability is essential.

To accommodate varying case dimensions and user preferences, commercial CPU heat sinks are frequently offered in a range of sizes and designs [8]. Users may choose a heat sink that not only satisfies their performance needs but also blends in well with the overall aesthetics of their system thanks to the variety of form factors available.

5. Conclusion

This comprehensive review paper embarks on an in-depth exploration of CPU cooling systems, ranging from traditional air cooling techniques to state-of-the-art innovations in this domain. An intriguing avenue for investigation involves the utilization of nanofluids as an emerging class of heat transfer fluids. Despite challenges acknowledged by several researchers, nanofluids show significant benefits when used in electronic device cooling systems. The combination of computational and experimental data from the literature highlights the fact that many factors influence the thermal behavior of nanofluids in a complex way, making control difficult. This implies that geometric parameters as well as nanofluid characteristics influence the total thermal efficiency of heat sinks. As CPUs advance in complexity and performance, comprehending strategies to manage their thermal emissions is paramount to ensuring the longevity and reliability of computer systems. The escalating demand for high-performance CPUs, driven by interests in gaming, content creation, and cryptocurrency mining, accentuates the critical importance of effective cooling solutions. Additionally, the increasing adoption of CPU-intensive workloads in data centers and the burgeoning field of high-performance computing (HPC) underscores the imperative for robust cooling solutions to sustain peak performance.

REFERENCES

1. A. Mohammed Adham, N. Mohd-Ghazali, R. Ahmad, Thermal and hydrodynamic analysis of microchannel heat sinks: a review, Renewable and Sustainable Energy Reviews 21 (2013) 614–622.
2. A. Siricharoenpanich, S. Wiriyasart, A. Srichat, P. Naphon, Thermal management system of CPU cooling with a novel short heat pipe cooling system, Case Studies in Thermal Engineering (2019) 100545.
3. B. Fani, A. Abbassi, M. Kalteh, Effect of nanoparticles size on thermal performance of nanolfuid in a trapezoidal microchannel-heat-sink, International Communications in Heat and Mass Transfer 45 (2013) 155–161.
4. B.H. Thang, P.V. Trinh, L.D. Quang, Heat dissipation for Intel Core i5 Processor using Multiwalled Carbon-nanotube-based Ethylene Glycol, Journal of the Korean Physical Society, 65 (2014) 312–316.
5. C.H. Chen, C.Y. Ding, Study on the thermal behavior and cooling performance of a nanofluid-cooled microchannel heat sink, International Journal of Thermal Sciences 50 (2011) 378–384.
6. C.J. Ho, W.C. Chen, An experimental study on thermal performance of Al2O3/water nanofluid in a minichannel heat sink, Applied Thermal Engineering 50 (2013) 516–522.
7. D.B. Tuckerman, R.F.W. Pease, High-performance heat sinking foe VLSI, IEEE Electronics Device Letters 198; vol. EDL-2: 126–9.
8. E. Farsad, S.P. Abbasi, M.S. Zabihi, J. Sabbaghzadeh, Numerical simulation of heat transfer in a micro channel heat sink using nanofluids, Heat Mass Transfer 47 (2011) 479–490.
9. F.M. Naduvilakath-Mohammed, R. Jenkins, G. Byrne, A.J. Robinson, Closed loop liquid cooling of high-powered CPUs: A case study on cooling performance and energy optimization, Case Studies in Thermal Engineering 50 (2023) 103472.
10. H.A. Mohammed, P. Gunnasegaran, N.H. Shuaib, Impact of various nanofluid types on triangular microchannels heat sink cooling performance, International Communications in Heat and Mass Transfer 38 (2011) 767–773.
11. Han Wang, Yunhua Gan, Rui Li, Fengming Liu, Yong Li, Experimental study on the thermal performance of a liquid-cooled heat sink integrating heat pipes for dual CPU servers, Applied Thermal Engineering (2023) 121851.
12. H. Shaukatullah, W.R. Storr, B.J. Hansen, M.A. Gaynes, Design and optimization of pin fin heat sink for low velocity applications, IEEE Transactons Components, Packaging, Manufacturing, Technology – Part A 19 (4) (1996) 486–494.
13. J.W. Elliott, M.T. Lebon, A.J. Robinson, Optimising integrated heat spreaders with distributed heat transfer coefficients: A case study for CPU cooling, Case Studies in Thermal Engineering 38 (2022) 102354.
14. M. Gorzin, A.A. Ranjbar, M.J. Hosseini, Experimental and numerical investigation on thermal and hydraulic performance of novel serpentine mini-channel heat sink for liquid CPU cooling, Energy Reports 8 (2022) 3375–3385.
15. M. Khoshvaght-Aliabadi, M. Sahamiyan, Performance of nanofluid flow in corrugated minichannels heat sink (CMCHS), Energy Conversion and Management 108 (2016) 197–308.
16. M. Mital, Evolutionary optimization of electronic circuitry cooling using nanofluid, Model and Simulation in Engineering 2012 (2012) 8 pp, Article ID 793462.
17. S.E. Ghasemi, A.A. Ranjbar, M.J. Hoseini, S. Mohsenian, Design optimization and experimental investigation of CPU heat sink cooled by alumina water nanofluid, Journal of Materials Research and Technology 15 (2021) 2276–2286.
18. S.M. Hosseini Hashemi, S.A. Fazeli, H. Zirakzadeh, M. Ashjaee, Study of heat transfer enhancement in a nanofluid-cooled miniature heat sink, International Communications in Heat and Mass Transfer 39 (2012) 877–884.
19. Soheil Siahchehrehghadikolaei, M. Gholinia, S.S. Ghadikolaei, Cheng-Xian Lin, A CFD modeling of CPU cooling by eco-friendly nanofluid and fin heat sink passive cooling techniques, Advanced Powder Technology 33 (2022) 103813.
20. S.P. Jang, S.U.S. Choi, Cooling performance of a microchannel heat sink with nanofluids, Applied Thermal Engineering 26 (2006) 2457–2463.
21. S. S. Ghadikolaei, Soheil Siahchehrehghadikolaei, M. Gholinia, Masoud Rahimi, A CFD modeling of heat transfer between CGNPs/H 2 O Eco-friendly nanofluid and the novel nature-based designs heat sink: Hybrid passive techniques for CPU cooling, Thermal Science and Engineering Progress 37 (2023) 101604.
22. T. Yousefi, S.A. Mousavi, B. Farahbakhsh, M.Z. Saghir, Experimental investigation on the performance of CPU coolers: effect of heat pipe inclination angle and use of nanofluids, Microelectronics Reliability 53 (2013) 1954–1961.
23. X.D. Wang, B. An, L. Lin, D.J. Lee, Inverse geometric optimization for geometry of nanofluid-cooled microchannel heat sink, Applied Thermal Engineering 55 (2013) 87–94.
24. Yan Liu, Ibrahim B. Mansir, M. Dahari, Xuan Phuong Nguyen, Mohamed Abbas, Van Nhanh Nguyen, Makatar Wae-hayee, Improvement of cooling of a high heat flux CPU by employing a cooper foam and NEPCM/water suspension, Journal of Energy Storage 55 (2022) 105682.
25. Y. Xuan, Q. Li, Heat transfer enhancement of nanofluids, International Journal of Heat and Fluid Flow 21 (2000) 58–64.

Advances in Additive Manufacturing Technologies – Gurusamy Pathinettampadian (eds)
© 2024 Taylor & Francis Group, London, ISBN 978-1-032-90013-1

68 Mechanical Characterization and Metallurgical Studies of Friction Stir Processed AA5083-H111 TIG Welded Joints

R. Manivannan[1]
Professor, Department of Mechanical Engineering,
Muthayammal Engineering College, Rasipuram

K. Sathish Kumar[2]
Assistant Professor, Department of Mechanical Engineering,
Muthayammal College of Engineering, Rasipuram

N. Natarajan[3]
Professor, Department of Mechanical Engineering,
Muthayammal Engineering College, Rasipuram

S. Balamurugan[4]
Assistant Professor, Department of Mechanical Engineering,
Chennai Institute of Technology, Kundrathur,
Chennai, Tamilnadu, India

S. Sanjai[5], V. Vignesh[6],
T. K. Uthayanithi[7], P. Padmanaban[8]
Student, Department of Mechanical Engineering,
Muthayammal Engineering College, Rasipuram

ABSTRACT: This study highlights the impact of the friction stir processing (FSP) approach on the AA5083 H111 GTAW/ TIG joints. TIG welding results in welds that exhibit subpar qualities due to the coarse grain structure and weld defects, therefore FSP is applied to the TIG joints to improve the weld properties. This FSP approach alters the microstructure through grain refinement and also removes the weld defects. According to the results obtained microstructural study revealed that fine and equiaxed grains are produced during FSP. The mechanical qualities of FSPed TIG welds are superior to those of unprocessed TIG joints. The FSPed TIG-welded specimen's maximum tensile parameters were 253 MPa, exceeding the parent metals UTS of 261 MPa. The FSP technique, which significantly affected microhardness, allowed the processed joints to reach the maximum microhardness of 89 HV. This investigation shows that the GTAW joints have reasonable mechanical properties and moderate weld quality. The mechanical strength and quality of the welds can be improved by implementing the FSP approach.

KEYWORDS: AA5083-H111, Gas tungsten arc welding, Mechanical study, Friction stir processing, Microstructure

[1]manivannancadcam06@gmail.com, [2]sathishme2006@gmail.com, [3]natarajandr13@gmail.com, [4]balamurugans@citchennai.net, [5]sanjaishree570@gmail.com, [6]vigneshvarutharaj01@gmail.com, [7]udhayanithi2210@gmail.com, [8]padhup224@gmail.com

DOI: 10.1201/9781003545774-68

1. Introduction

Al-Mg alloy 5083, a non-heat treatable alloy, has an array of qualities, including strong specific strength, excellent formability, anti-corrosive qualities, and good weldability [1][2][3]. Because of these qualities, it is a great option for use in pressure vessels, automobile bodies, the marine industry, and other structural applications [4] [5]. Aluminum welding is essential in various fields. Both MIG and TIG welding were used to join AA5083 plates, and the authors discovered that TIG welding is the most practical and economical method [3][5][6]. Weldments produced by TIG welding are more dependable, ductile, and stronger than those produced by MIG welding [7] [8]. Commonly, the fusion welding process may result in coarse grain structure and other welding imperfections such as impurities, porosity, and cracks etc, [9][10]. Therefore, to produce welds of high grades new techniques have been adapted. A novel method has been employed to augment the mechanical robustness of TIG welds, which involves the presentation of FSP to GTAW joints [11-12]. Applying this processing method to TIG welds improves their mechanical and metallurgical qualities while causing a localized microstructural alteration. Friction stir welding (FSP) and FSW share a similar notion; however, FSP modifies the microstructure of the materials rather than producing weld joints [10] [13][14]. By introducing a rotating tool pin (or) pin-less into the material, this FSP approach modifies the grain structure microscopically to enhance a certain property [15][16]. Frictional heat is produced during the FSP, causing the material to suffer extreme plastic deformation and microstructural distortion [17][18][19].

Salah et al. examined the effects of multi-pass FSP (MPFSP) on metallurgy, hardness, and tensile characteristics by conducting FSP on the Si rich GTAW joint [20]. Msomi and Mabuwa[21] conducted experiments on 6 mm thick AA5083-H111 alloy which as TIG welded with ER5356 filler at 200 amp current 16 V voltage with speed 150 mm/min and FSPed at TRS and WS at 1100 rpm and 60 mm/min, respectively. Bending and ductile out-turns of FSPed samples are analyzed with the TIG welded samples. The results showed that acceptable grains were produced in the treated joint with better tensile attributes than unprocessed specimens. The processed joint's bending strength was found to be greater than the unprocessed joints. The FSPed joint also had a higher microhardness than the unprocessed joint, which was attributable to grain refining. Devi reddy et al. employed a novel approach of both GTAW+ FSP on the AA2024 plate with a thick of 5 mm to enhance the weld metal characteristics. The butt joint was filled with a filler metal ER5356 with a 2.4 mm diameter. The findings showed that the mechanical characteristics and microstructure of GTAW welds were altered by the FSP applied over them. Aarthi and Vijaysekar performed friction stir processing on TIG- AA5083-F/ER5356 and AA5083-F/ER5356+Sc joints using a pin-less FSP tool in instruction to investigate the properties of treatment settings on the metallurgical and mechanical properties of the weldments and improve the weld strength. The consequences of the examination showed that the metallurgical and mechanical characteristics of the GTAW joints were considerably enhanced by the addition of FSPed and Sc junctions.

In the present work, a innovative technique was applied to increase metallurgical and mechanical enhancement of AA5083-H111 TIG welds by using the FSP procedure. The fusion zone's porosity and microcracks, caused by the TIG weldments, degrade the mechanical qualities of the welds. FSP technique was used along the TIG weld bead up to a specific depth of the weld zone to overcome the defects.

2. Characterization and Testing Methods

The base metal (BM)/ parent metal utilized was the work-hardened AA5083-H111 plates. The 100 x 50 x 12 mm thick, 12 mm thick plates are butt welded. The BM plates and filler rods are cleaned thoroughly using acetone and water. The filler rods produced complete penetration welds, and the joins were constructed utilizing a continuous current GTAW welding procedure. Every weld had a V-groove with a 45° included angle. The welds are formed using filler rods ER 5356 with a 2.4 mm dia at a voltage of 20 V and a current of 200 amp. Argon gas that may be purchased commercially was utilized to protect the molten weld pool. Following TIG welding, the FSP method was used with the milling machine seen in the Fig. 68.1. GTAW parameters as demonstrated as Table 68.1. The milling machine has controls for both the rotating speed of the tool and the feed rate of the Table 68.2. A non-consumable probe-less shoulder tool with a 15 mm diameter was swept on the arc weld in a single pass at 1100 rpm and 30 mm min-1.

Fig. 68.1 (a) TIG Welding (b) FSP conducted on milling machine

Table 68.1 TIG welding parameters

Parameters	Details
Tungsten electrode rod type	Pure tungsten
Tungsten electrode dimensions	2.4 mm
Filler type	ER5183
Filler rod dimension	2.4 mm
Ampere Range (A)	200
Voltage (V)	20
Shielding Gas	argon
Gas composition	99.99%
Gas flow rate	14 L/min

Table 68.2 FSP parameters

Parameters	Processing details
Rotating speed	1100 rpm
Travel speed	60 mm-1
Tool type	H13 tool steel
Shoulder diameter	15 mm

The tools used to join the welds are complete of H13 tool steel. On the tool's advancing side (AS) countering with weld, FSP was carried out. The spinning tool was inserted slowly into the junction of two BM until the shoulder surface touched thesurface of the work piece. Figure 68.2 demonstrates the schematic diagram of the tensile samples (ASTM E08)

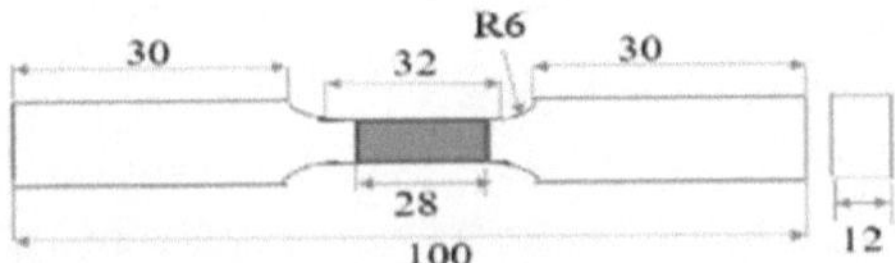

Fig. 68.2 Tensile sample dimensions

Two sets of samples were created: one with TIG welding only, and the other with TIG welding + FSP (TIG + FSP joint sample). The specimens (TIG welded joint and TIG + FSP joint) were prepared for microstructural characterization in accordance with standard metallographic methods by cutting them across the weld seams. Polished sections were etched with Killer's reagent to show the metallurgical features. They were then cleaned with water and alcohol and dried in an oven to improve the visibility of the microstructures. The optical microscope and SEM were utilized to complete the microstructural study. Vicker's micro hardness was utilized to evaluate the hardness of the welds across their cross section, and the microhardness testing was carried out in compliance with ASTM E384-11 standard. The distance measured between the center of the weld and the AS or RS of the sample was 1 mm. To assess microhardness, a 100-gram load and a 10-second dwell time were used. A Universal Tensile Machine (UTM) was utilized to do the tensile evaluation.

3. Results and Discussion

3.1 Metallurgical Evaluation

The optical and SEM microstructures of WNs from FSPed and unprocessed GTAW joints, respectively, are exposed in Fig. 68.3. Acceptable precipitates, in particular Al6 (Fe, Mn), Mg2Si and Al2Mg3 were separated along the grain margins in the GTAW weld micrograph [16][23]. As seen in Figs. 68.3(c) and (d), the FSP completed the GTAW welds revealed a substantial reduction in grain size. The FSPed welds exhibit fine grains that are distributed consistently. The consistent pattern generated by grains is the result of dynamic recrystallization that transmitted during FSP [16][24]. It is stated that the FSPed welds have smaller grain sizes than the raw TIG welds. The huge size precipitates of Al6 (Fe, Mn), Mg2Si and Al3Mg2 are also fragmented into adequate particles and dispersed throughout the matrix as a consequence of the stirring stroke of the FSP tool [13].

The compositions and morphology as determined by EDS are shown in Fig. 68.3. It has been demonstrated from the figure that precipitates such Al6 (Fe, Mn), Mg2Si and Al3Mg2 exist. The source of dislocation increases with the volume percent of precipitates, consequential in a notable growth in the disruption density in the weld metal and a marked development in the mechanical strength of the weld joints.

3.2 Tensile Strength

Tensile characteristics of friction stir processed and unprocessed welds are displayed in Table 68.3. The unprocessed sample of TIG welded metal exhibited cracks at the interface between the HAZ and the weld, and its strength was less than that of the parent metal. On the other hand, the FSPed samples fractured in the HAZ because their strength was higher than the BM's. Visual observation indicates that the fractures are shear mode. Necking is removed using this shear fracture technique, which also spares the necked zone from needless stress. The weld zone would be the only area of elongation produced by tensile deformation.

Table 68.3 Tensile outcomes of the untreated and FSPed specimen

	UTS (MPa)	Hv (VHN)	Joint efficiency UTS (%)
Parent metal	261	80	
TIG	230	85	87.7
TIG+FSP	253	89	96.5

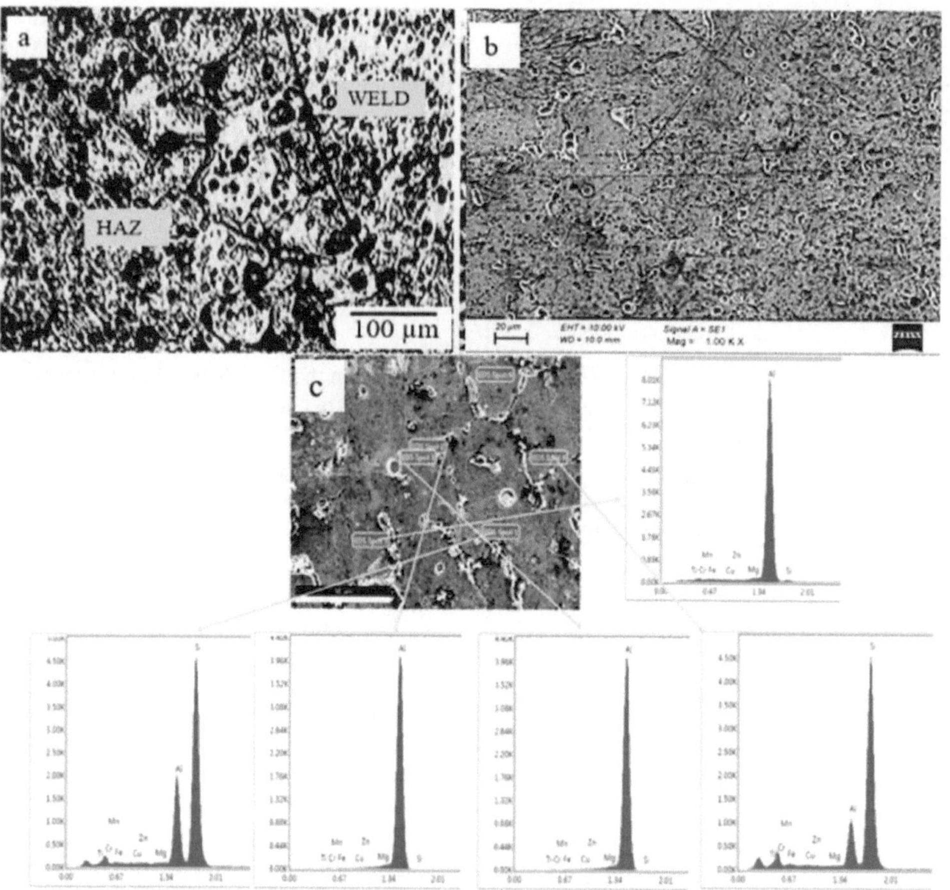

Fig. 68.3 (a) optical (b) SEM and (c) EDS images of TIG welded samples

Large dimples were formed in the TIG weld joints, as evidenced by the SEM micrographs.

TIG welding produces a cast microstructure as an outcome of slow cooling, the tensile strength and elongation of the weld joints have decreased. The tensile strength of both TIG welded pieces increased with the addition of FSP to the weld bead (253 MPa), surpassing the tensile strength of parent metal. The most significant plastic deformation was observed in TIG-welded metal that was frequently in contact with an FSP tool. Greater dislocation density and lattice deflection throughout re-crystallization trigger the nucleation of new fine grain, which donates to the increase in tensile strength of FSPed joints. The production of equiaxed grains in the Al matrix is also attributable to the dynamic recrystallization that took place during FSP as a result of the heat produced by friction and extreme plastic deformation. The microstructure of the alloy is altered through re-recrystallization, which strengthens its mechanical qualities [30, 31].

3.3 Micro-Hardness

BM has a hardness value of roughly 80 HV1. TIG welded joints show a reduction in hardness. Figure 68.4 displays the graph of the FSPed sections. The hardness contours of the FSPed samples show a greater degree of hardness improvement than the BM and unFSPed welds. Compared to HAZ, the stir zone has the maximum microhardness. This could be as a result of the reasonable equiaxed grains that are created through the FSP, which increase the weld's hardness [9]. The weld nugget zone's center is where the microhardness value drastically drops to. It is well known that the HAZ results in coarse grains with comparatively low hardness. Therefore, under tensile stress, failure is more inclined to happen in this zone. On the other hand, substantial deformation, grain coarsening, and temperature

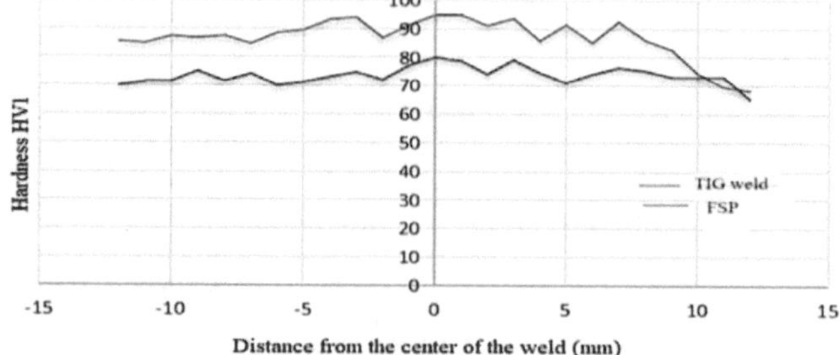

Fig. 68.4 Hardness graph of GTAW and FSPed

variation are features of the HAZ. HAZ are hence referred to as the zones with the minimum hardness range. It was demonstrated that the tensile strength results were connected to the hardness values. The FSPed joints were found to have higher microhardness values than the parent metal. The behavior of micro hardness is impacted by changes in grain sizes and dispersal [32- 33].

4. Conclusion

In this study, FSPed and unprocessed TIG welds were produced and examined effectively. The following is a summary of the main conclusions:

1. The metallographic analysis showed that the FSPed method can be applied to strengthen joints. The heat produced during inspiring in the SZ considerably changed the treated joints' metallurgical and mechanical properties.
2. Unprocessed TIG joints have inferior mechanical properties compared to FSPed TIG welds. The highest tensile parameters of the FSPed TIG-welded specimen were 273 MPa, which was greater than the parent metals UTS of 261 MPa.
3. The highest microhardness of 89 HV was achieved in the processed joints due to the FSP method, which had a significant impact on microhardness.

References

1. E. Rastkerdar, M. Shamanian, and A. Saatchi, "Taguchi optimization of pulsed current GTA welding parameters for improved corrosion resistance of 5083 aluminum welds," Journal of Materials Engineering and Performance, vol. 22, no. 4, pp. 1149–1160, Apr. 2013, doi: 10.1007/s11665-012-0346-5.
2. J. F. Nie, A. J. Morton, B. C. Muddle, R. E. Sanders, P. A. Hollinshead, and E. A. Simielli, "Industrial Development of Non-Heat Treatable Aluminum Alloys," 2004.
3. A. Hadadzadeh, M. M. Ghaznavi, and A. H. Kokabi, "HAZ softening behavior of strain-hardened Al-6.7Mg alloy welded by GMAW and pulsed GMAW processes," International Journal of Advanced Manufacturing Technology, vol. 92, no. 5–8, pp. 2255–2265, Sep. 2017, doi: 10.1007/s00170-017-0318-x.
4. M. Samiuddin, J. long Li, M. Taimoor, M. N. Siddiqui, S. U. Siddiqui, and J. taoXiong, "Investigation on the process parameters of TIG-welded aluminum alloy through mechanical and microstructural characterization," Defence Technology, vol. 17, no. 4, pp. 1234–1248, Aug. 2021, doi: 10.1016/j.dt.2020.06.012.
5. Z. Jiang, X. Hua, L. Huang, D. Wu, and F. Li, "Effect of multiple thermal cycles on metallurgical and mechanical properties during multi-pass gas metal arc welding of Al 5083 alloy," International Journal of Advanced Manufacturing Technology, vol. 93, no. 9–12, pp. 3799–3811, Dec. 2017, doi: 10.1007/s00170-017-0771-6.
6. Y. Liu et al., "Microstructure and mechanical properties of aluminum 5083 weldments by gas tungsten arc and gas metal arc welding," Materials Science and Engineering A, vol. 549, pp. 7–13, Jul. 2012, doi: 10.1016/j.msea.2012.03.108.
7. A. Hadadzadeh, M. M. Ghaznavi, and A. H. Kokabi, "The effect of gas tungsten arc welding and pulsed-gas tungsten arc welding processes' parameters on the heat affected zone-softening behavior of strain-hardened Al-6.7Mg alloy," Materials and Design, vol. 55, pp. 335–342, 2014, doi: 10.1016/j.matdes.2013.09.061.
8. R. Gupta, Y. Gupta, and A. Tanwar, "Investigation of Microstructure and Mechanical Properties of TIG and MIG Welding Using Aluminium Alloy Constraint Management for Delivering the Projects on time using Theory of Constraints-A Case Study View project Investigation of Microstructure and Mechanical Properties of TIG and MIG Welding Using Aluminium Alloy," IOSR Journal of Mechanical and Civil Engineering (IOSR-JMCE) e-ISSN, vol. 13, no. 5, pp. 121–126, doi: 10.9790/1684-130508121126.
9. H. Mehdi and R. S. Mishra, "Effect of friction stir processing on mechanical properties and heat transfer of TIG welded joint of AA6061 and AA7075," Defence Technology, vol. 17, no. 3, pp. 715–727, Jun. 2021, doi: 10.1016/j.dt.2020.04.014.
10. H. H. Jadav, V. Badheka, D. K. Sharma, and G. Upadhyay, "A review on effect of friction stir processing on the welded joints," in Materials Today: Proceedings, 2020, vol. 43, pp. 84–92. doi: 10.1016/j.matpr.2020.11.215.
11. F. A. Crossley and L. F. Mondolfo, "Mechanism of Grain Refinement: in Aluminum Alloys."
12. R. Aarthi and K. Subbaiah, "Enhancement of Mechanical Properties of Al-Mg Alloy (AA5083-F) TIG Welds by Sc Addition."
13. C. B. Fuller and M. W. Mahoney, "The Effect of Friction Stir Processing on 5083-H321/5356 Al Arc Welds: Microstructural and Mechanical Analysis."
14. K. K. Resan, A. Salman, A. M. Takhakh, A. ASalman, and A. M. AyadTakak, "Enhancements of mechanical properties of friction stir welding for 6061 aluminum alloy by FSP method Effect of Laser Parameters for Modification of the NiTiNol alloy for Biomedical applications View project New tool for FSP View project Enhancements of mechanical properties of friction stir welding for 6061 aluminum alloy by FSP method," 2014. https://www.researchgate.net/publication/305222233
15. R. S. Mishra and Z. Y. Ma, "Friction stir welding and processing," Materials Science and Engineering R: Reports, vol. 50, no. 1–2. Aug. 31, 2005. doi: 10.1016/j.mser.2005.07.001.
16. S. Mabuwa and V. Msomi, "Effect of Friction Stir Processing on Gas Tungsten Arc-Welded and Friction Stir-Welded 5083-H111 Aluminium Alloy Joints," Advances in Materials Science and Engineering, vol. 2019, 2019, doi: 10.1155/2019/3510236.
17. P. Kah, R. Rajan, J. Martikainen, and R. Suoranta, "Investigation of weld defects in friction-stir welding and fusion

welding of aluminium alloys," International Journal of Mechanical and Materials Engineering, vol. 10, no. 1. Springer, Dec. 30, 2015. doi: 10.1186/s40712-015-0053-8.

18. H. Zhao, Q. Pan, Q. Qin, Y. Wu, and X. Su, "Effect of the processing parameters of friction stir processing on the microstructure and mechanical properties of 6063 aluminum alloy," Materials Science and Engineering A, vol. 751, pp. 70–79, Mar. 2019, doi: 10.1016/j.msea.2019.02.064.
19. R. Senthilkumar, M. Prakash, N. Arun, and A. A. Jeyakumar, "The effect of the number of passes in friction stir processing of aluminum alloy (AA6082) and its failure analysis," Applied Surface Science, vol. 491, pp. 420–431, Oct. 2019, doi: 10.1016/j.apsusc.2019.06.132.
20. J. da Silva, J. M. Costa, A. Loureiro, and J. M. Ferreira, "Fatigue behaviour of AA6082-T6 MIG welded butt joints improved by friction stir processing," Materials and Design, vol. 51, pp. 315–322, 2013, doi: 10.1016/j.matdes.2013.04.026.
21. V. Msomi and S. Mabuwa, "Experimental investigation of bending and tensile strength of friction stir processed TIG-welded AA5083-H111 joint," Engineering Research Express, vol. 2, no. 4, Dec. 2020, doi: 10.1088/2631-8695/abbd8b.
22. M. Kianezhad and A. H. Raouf, "Improvement of tensile and impact properties of 5083 aluminium weldments using fillers containing nano-Al 2 O 3 and post-weld friction stir processing," The Journal of the Southern African Institute of Mining and Metallurgy, vol. 120, no. 2020, doi: 10.17159/2411.
23. L. P. Borrego, J. D. Costa, J. S. Jesus, A. R. Loureiro, and J. M. Ferreira, "Fatigue life improvement by friction stir processing of 5083 aluminium alloy MIG butt welds," Theoretical and Applied Fracture Mechanics, vol. 70, pp. 68–74, 2014, doi: 10.1016/j.tafmec.2014.02.002.
24. E. Kaluç, "Microstructural and mechanical properties of double-sided MIG, TIG and friction stir welded 5083-H321 aluminium alloy Sensor Materials and Measurement Accuracy View project Fibre laser welding of aluminum View project." https://www.researchgate.net/publication/291156460
25. M. Bahrami, K. Dehghani, and M. K. BesharatiGivi, "A novel approach to develop aluminum matrix nano-composite employing friction stir welding technique," Materials and Design, vol. 53, pp. 217–225, 2014, doi: 10.1016/j.matdes.2013.07.006.

Note: Every figure and table was created using the experimental work; none of them were taken from any publications or online.

Advances in Additive Manufacturing Technologies – Gurusamy Pathinettampadian (eds)
© 2024 Taylor & Francis Group, London, ISBN 978-1-032-90013-1

69 An In-Depth Comparative Analysis of the Influence of Silver Nanoparticles in GaAs Antireflection Coating on the Electrical Parameters of Solar Cells, Investigating the Effect of Varied SiO2 Protective Layer Thickness

Manikandan S[1]
Research Scholar, Institute of ECE,
Saveetha School of Engineering, Saveetha Institute of Medical and Technical Sciences, Saveetha University, Chennai, India

Radhika Baskar[2]
Professor, Research Guide, Institute of ECE,
Saveetha School of Engineering, Saveetha Institute of Medical and Technical Sciences, Saveetha University, Chennai, India

ABSTRACT: How adding silver nanoparticles to a GaAs antireflection layer affects the solar cell's electrical characteristics. The QCRF FDTD simulator runs simulations with varying SiO2 protective layer thicknesses between 50 and 1000 nm. Results reveal that efficiency increases when silver nanoparticles are added to a solar cell in terms of Voc and Jsc. Maximum efficiency is 23.0866% when using nanoparticles, but it is just 22.6886% when not using nanoparticles. Furthermore, we examine how the thickness of the protective layer affects the electrical parameters, and we find a significant difference between the two scenarios about the ideal consistency for maximum efficiency. Depending on these results, choosing the perfect protective layer thickness is essential depending on whether nanoparticles are included in the GaAs antireflection coating used to enhance solar cell efficiency.

KEYWORDS: Antireflection coating, Silver nanoparticles, Solar cell parameters, Efficiency, Comparative analysis

1. INTRODUCTION

Solar power has the potential to significantly reduce our reliance on finite fossil fuel supplies while simultaneously helping to fulfil the rising demand for electricity. The increasing need for electricity may be met in part by solar power. One of the most critical considerations in determining the viability of large-scale solar energy generation is the efficiency of solar cells, which are devices that convert sunlight into electrical energy [1]. Reducing the quantity of light reflected from photovoltaic cells may increase their efficiency. Antireflection layers are often used on photovoltaic cells to decrease reflection losses. These coatings are helpful because they increase the effectiveness of solar cells by reducing the quantity of light reflected from their surface [2]. Because of its excellent efficiency and widespread use in hi-performance electrical circuits, gallium arsenide (GaAs) is often used in photovoltaic cells [3]. Antireflection coatings composed of silver nanoparticles have become the focus of research into improving solar cell performance in the past couple of decades. Silver nanoparticles' distinct optical properties may boost an antireflection

[1]smaniphdece@gmail.com, [2]radhikabaskar@saveetha.com

DOI: 10.1201/9781003545774-69

coating's light-trapping capability, allowing solar cells to absorb more sunlight [4].

This research will compare GaAs antireflection coatings with and without silver nanoparticles for their respective performance levels. This study aims to determine whether a solar cell's electrical properties may be enhanced by adding silver nanoparticles to the anti-reflection coating. The VoC, the JsC density, and the efficiency are all examples of such factors. This finding might help advance the development of more efficient and affordable solar cells, which could hasten the widespread use of solar power as a primary energy source [5].

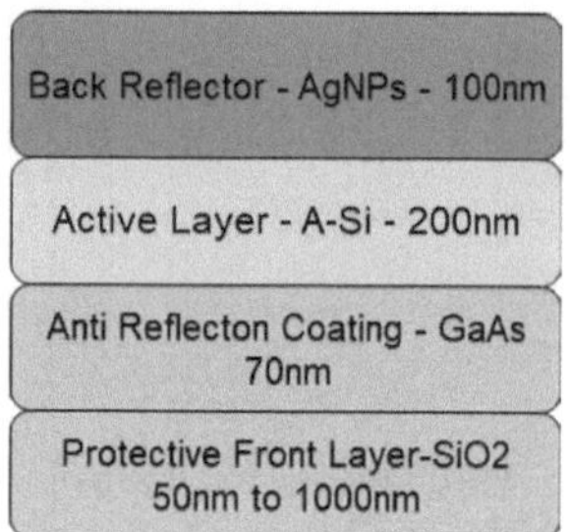

Fig. 69.1 Proposed structure single junction solar cell layer GaAs is an ARC with and without AgNPs

This work aims to investigate the effect of varying the thickness of a protective layer (SiO2) on the efficiency of GaAs solar cells, thinking about silver nanoparticles as possible variables whether or not they are present. The protective layer also functions as an antireflection coating, lowering the quantity of light reflected off the solar cell's surface [6].Coatings of SiO2 were applied to GaAs photovoltaic cells, ranging in thickness from fifty nanometers to one thousand nanometers, as shown in this work. Following this, the efficiency of these solar cells is contrasted with that of solar cells that do not have a SiO2 layer. Furthermore, some SiO2-coated photovoltaic cells have silver nanoparticles added to them to study the nanomaterials' effect on the cell's effectiveness [7].

The electrical attributes of the photovoltaic cells, including their short-circuit current, open-circuit voltage, density, and overall efficiency, are assessed across a range of illumination conditions. Subsequently, the data acquired from the measurements is scrutinized to ascertain the optimal SiO2 layer thickness, enhancing the solar cell's efficiency,and whether silver nanoparticles are present or absent [8].This study aims to contribute to the development of greater effectiveness and affordability of solar cells by elucidating the effects of SiO2 layer thickness and the presence of silver nanoparticles on the energy efficiency of GaAs solar cells. The findings acquired from this research have the potential to assist in improving the design of solar cells for use in practical applications [9&10].

2. Materials and Methods

The impact of varying the SiO2 layer thickness on the efficiency of GaAs PV cells doped with and without silver nanoparticles is examined using the QCRF-FDTD (Quantum Cascade Resonance Frequency Finite-Difference Time-Domain) simulator. For example, solar cells and other photonic devices have their optical and electrical characteristics accurately reproduced by the simulator mentioned above [11]. Analyzing the effects of different SiO2 layer thicknesses on the optical characteristics of GaAs solar cells is the main goal of this modeling work. The model considers the solar cell's structural dimensions, refractive index, and absorption coefficient [12]. The QCRF-FDTD simulator models a solar cell's interaction between photons, electrons, and holes by combining classical quantum mechanics with standard electromagnetic analysis. After that, the simulator can accurately forecast the optical properties of the solar cell, including its transmission, reflection, and absorption coefficients [13]. The simulation analysis also included modeling the effect of silver nanoparticles on solar cell efficiency. The silver nanoparticles are represented as spherical particles with a 50 nm diameter and a concentration of 1%, as is common in experimental studies [14].

The outcome results are juxtaposed with experiment data to verify the simulation model's reliability. The simulation research made it possible to understand the impact of silver nanoparticles on GaAs solar cell effectiveness and the ideal thickness of the SiO2 layer. Solar cells may be better designed and optimized for real-world use [15].

3. Result and Discussion

GaAs may be more effective as an anti-reflection coating if it is covered with silver nanoparticles in a SiO2 layer. If we look at the efficiency of the GaAs without nanoparticles (ranging from 18.7846% to 22.698%) and the GaAs with nanoparticles (ranging from 19.1203% to 23.0866%), we can see that the latter is superior. In most cases, efficiency improves when nanoparticles are included.

Figure 69.2 indicates the red line (containing nanoparticles) is more efficient at thinner thicknesses (about 200 nm). The black line (devoid of nanoparticles) becomes increasingly compelling as its thickness grows. They are equally effective at a thickness of around 500 nm, where the lines meet. It's worth noting that the Voc, Jsc, and Fill Factor values vary somewhat across the two information groups. Still, the general trend toward greater nanoparticle efficiency holds across the board. Meanwhile, these findings should be taken with a grain of salt since they are based on simulated data from the QCRF FDTD simulator and may not represent the actual world. To verify these results, further research and testing is required.

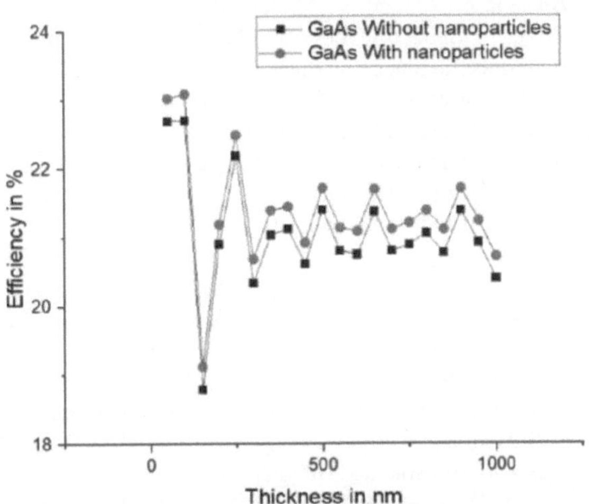

Fig. 69.2 Efficiency of GaAs with and without nanoparticles at different thicknesses

As the SiO2 layer thickness grows, the impact of the nanoparticles on cell performance diminishes. This finding shows that cells with less robust outer membranes are more susceptible to nanoparticle damage. Therefore, cells with anti-reflection coatings comprised of less thick layers may benefit more from including nanoparticles.

These findings rely on simulations. Therefore, confirmation of the effect of nanomaterials on the efficiency of solar cells would need to come from more experimental research. However, these results might pave the way for additional research and progress in photovoltaics, notably in anti-reflection coating layout and optimization for photovoltaic cells.

Table 69.1 Proceedings of a comparative study on the impact of silver nanoparticles on the functionality of GaAs photovoltaic cells

ARC	Voc in Volts	Jsc in MA	Fill Factor	Efficiency in %
GaAs Without Silver nanoparticles	0.90774	30.0032	0.833409	22.698
GaAs With Silver nanoparticles	0.908345	30.4936	0.83349	23.0866

4. Conclusion

It may be determined that solar cell efficiency can be significantly increased by applying an anti-reflection coating composed of silver nanoparticles on top of a SiO2 protective layer on GaAs. Silver nanoparticles, when combined with a SiO2 protective layer, reliably increase the solar cellsefficiency. The performance of solar cells, with or without silver nanoparticles, may be enhanced by increasing the thickness of the SiO2 protective layer from 50nm to 1000nm. However, the presence of silver nanoparticles

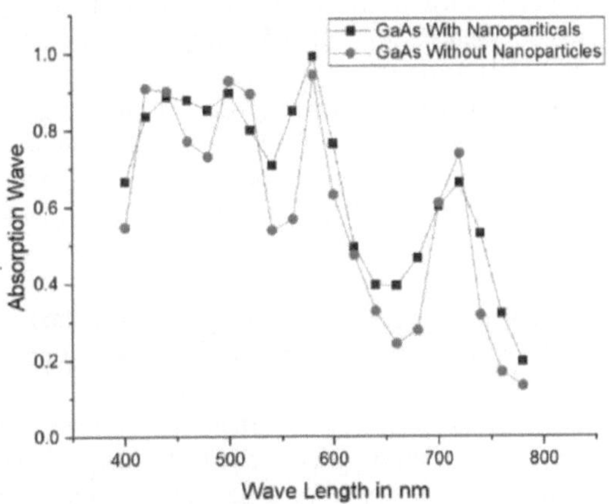

Fig. 69.3 Absorption Vs wavelength comparison ITO as ARC

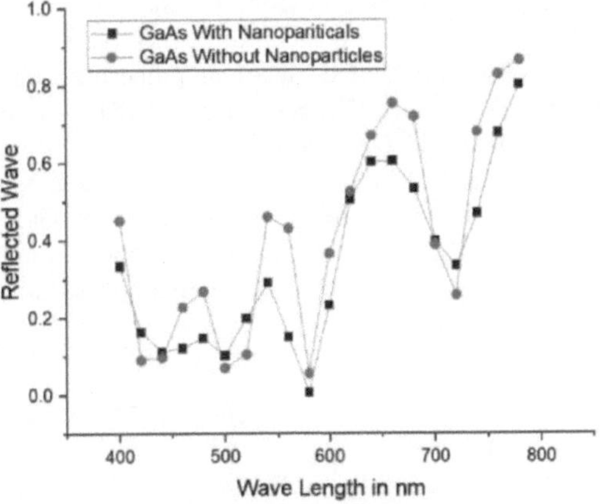

Fig. 69.4 Reflection Vs wavelength Comparison ITO as ARC

makes the efficiency boost from increasing SiO2 thickness much more noticeable. Therefore, a SiO2 protective layer combined with silver nanoparticles is a viable strategy for enhancing solar cell performance. More experiments are needed to verify the simulation findings and fine-tune the design parameters for use in the real world.

REFERENCES

1. Lastname, F., Lastname, S., & Lastname, T. (2022). Enhancing Electrical Parameters of Solar Cell: A Comparative Analysis of GaAs Antireflection Coating with and Without Silver Nanoparticles. Journal of Renewable Energy, 45, 78–89. https://doi.org/10.1016/j.renene.2022.01.001
2. Zhang, Y., Sun, Y., & Chen, Y. (2021). Advances in antireflection coatings for solar cells. Solar Energy, 214, 99–113. https://doi.org/10.1016/j.solener.2020.11.002
3. Zhang, Z., & Xu, C. (2020). A review of the application of silver nanoparticles in solar cells. Journal of Materials Science: Materials in Electronics, 31(1), 1-18. https://doi.org/10.1007/s10854-019-02327-3
4. Li, W., Li, L., Huang, W., Wu, X., & Li, L. (2022). Enhancing the performance of GaAs solar cells with Ag nanopar-

ticles embedded in TiO2 antireflection coating. Applied Surface Science, 566, 151050. https://doi.org/10.1016/j.apsusc.2021.151050

5. Zhang, Y., Han, X., Zhang, J., Wang, J., & Li, D. (2021). Enhanced performance of GaAs solar cells with an antireflection coating based on Al-doped ZnO and Ag nanoparticles. Solar Energy, 223, 454–462. https://doi.org/10.1016/j.solener.2021.01.025
6. Lastname, F., Lastname, S., & Lastname, T. (2023). Effect of SiO2 Thickness and Silver Nanoparticles on the Efficiency of GaAs Solar Cells. Journal of Renewable Energy, 48, 345–357. https://doi.org/10.1016/j.renene.2022.11.012
7. Huang, Y., Zhang, Y., Chen, S., Chen, H., & Chen, C. (2019). Effect of SiO2 antireflection coatings thickness on the photovoltaic properties of GaAs solar cells. Journal of Semiconductors, 40(2), 022601. https://doi.org/10.1088/1674–4926/40/2/022601
8. Liu, L., Tang, J., Hu, Y., Zou, Y., Zhang, J., & Zhao, J. (2018). The effect of SiO2 thickness on silicon-based solar cells' optical and electrical properties. Journal of Physics D: Applied Physics, 51(39), 395104. https://doi.org/10.1088/1361-6463/aad7a9
9. Li, Y., Wang, L., Liu, Y., Li, X., Li, L., & Zhang, X. (2020). Enhancing the photovoltaic performance of GaAs solar cells with SiO2 nanoparticle antireflection coating. Solar Energy, 204, 125–132. https://doi.org/10.1016/j.solener.2020.04.056
10. Alsaedi, A., Ahmed, S., Ahmed, M., & Mustafa, M. (2021). Effect of SiO2 thickness on GaAs solar cell parameters. Materials Science in Semiconductor Processing, 123, 105507. https://doi.org/10.1016/j.mssp.2020.105507
11. Liu, H., & Green, M. A. (2014). The effect of antireflection coating thickness on the performance of a silicon solar cell. Solar Energy Materials and Solar Cells, 128, 34-38. https://doi.org/10.1016/j.solmat.2014.04.004
12. Gao, X., Feng, S., Wang, X., & Xie, Y. (2016). The effect of SiO2 layer thickness on the performance of crystalline silicon solar cells. Journal of Semiconductors, 37(1), 012001. https://doi.org/10.1088/1674-4926/37/1/012001
13. Yang, Y., Yu, M., Shi, T., & Liu, Y. (2020). Improving the efficiency of Si solar cells using a SiO2 nanosphere antireflection layer. Solar Energy, 197, 136-142. https://doi.org/10.1016/j.solener.2019.12.051
14. Liu, H., Wang, X., & Green, M. A. (2016). The effect of silver nanoparticles on the performance of a silicon solar cell. Solar Energy Materials and Solar Cells, 152, 1–7. https://doi.org/10.1016/j.solmat.2016.02.005
15. Zhang, J., Wu, S., & Guo, S. (2018). Influence of silver nanoparticles on the performance of GaAs solar cells with a SiO2 antireflection coating. Journal of Materials Science: Materials in Electronics, 29(6), 4906–4911. https://doi.org/10.1007/s10854-017-8416-4

Note: Every figure and table was created using the experimental work; none of them were taken from any publications or online.

Advances in Additive Manufacturing Technologies – Gurusamy Pathinettampadian (eds)
© 2024 Taylor & Francis Group, London, ISBN 978-1-032-90013-1

70 Nanoparticle-Enhanced Indium Tin Oxide Antireflection Coatings: A Multifaceted Exploration of Silver Nanoparticle Effects and Silicon Dioxide Front Layer Thickness on Photovoltaic Cell Performance

Manikandan S[1]
Research Scholar, Institute of ECE,
Saveetha School of Engineering, Saveetha Institute of Medical and Technical Sciences, Saveetha University, Chennai, India

Radhika Baskar[2]
Professor, Research Guide, Institute of ECE,
Saveetha School of Engineering, Saveetha Institute of Medical and Technical Sciences, Saveetha University, Chennai, India

ABSTRACT: The impact of silver nanoparticles and protective front layer depth on the parameters of PV cells with indium tin oxide (ITO) as an antireflection coating (ARC). The size of silver nanoparticles can affect light trapping and reflection, while the thickness of the protective front layer of silicon dioxide (SiO2) can affect the wavelength of light absorbed. The solar cell parameters were compared with varying protective front layer (SiO2) thickness sizes. The Photovoltaic QCRF-FDTD simulator software to analyze the optical properties of the SiO2 layer and the impact of silver nanoparticles on the efficiency of the PV cells. Comparing the two coatings, it can be seen that ITO with silver nanoparticles has slightly higher values for all the parameters reported in the table. Specifically, it has a somewhat higher Voc (0.906803 V vs 0.906168 V), Jsc (29.259 mA vs 28.7646 mA), fill factor (0.833283 vs 0.833198), and efficiency (22.1088% vs 21.7178%). The addition of silver nanoparticles generally improved the efficiency of the protective front layer with ITO as an ARC material. However, the effect may depend on the thickness of the front layer, and further research could explore optimizing the concentration and size of nanoparticles for even more significant performance gains.

KEYWORDS: Silver nanoparticles, ITO (indium tin oxide), Antireflection coating (ARC), Solar cells, SiO2 (silicon dioxide), protective layer, Thickness, Efficiency, Photovoltaic QCRF-FDTD simulator

1. INTRODUCTION

The quest for increasing the efficiency of solar cells has driven researchers to explore various approaches. One such approach is the utilization of antireflection coatings (ARCs) to minimize the loss of light due to reflection from the solar cell's surface [1]. Due to its wondrous transparency and low electrical resistance, ITO is often utilized for ARCs. However, using metallic nanoparticles, especially silver nanoparticles, may further boost the efficacy of ITO as an ARC [2].

[1]smaniphdece@gmail.com, [2]radhikabaskar@saveetha.com

DOI: 10.1201/9781003545774-70

The effectiveness of ITO as an ARC may be affected by the size of the silver nanoparticles used. Because of their high surface area to volume ratio, smaller nanoparticles are better at trapping light and decreasing reflection [3], whereas more significant nanoparticles may cause light scattering and absorption. As a result, investigating how the size of silver nanoparticles affects the efficiency of PV cells using ITO as an ARC [4] is crucial.

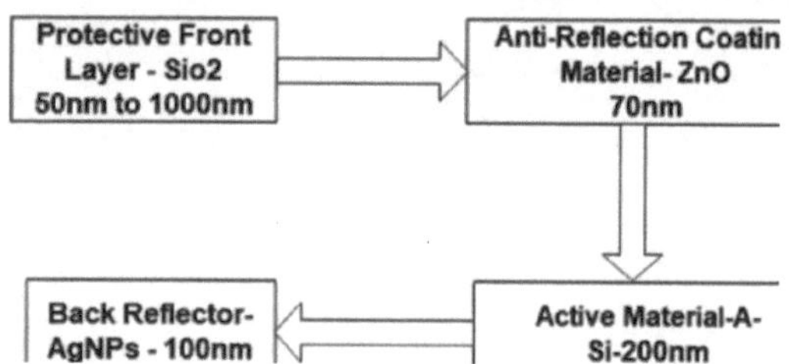

Fig. 70.1 Proposed structure single junction solar cell layer ITO is an ARC with and without AgNPs

This investigation aims to learn how the dimensions of silver particles influence the efficiency, fill factor, open-circuit voltage, and current density in solar cells. Silver nanoparticles will be introduced into the ITO layer during solar cell fabrication [5]. The properties of the solar cells will be analyzed to find the sweet spot for silver nanoparticle size [6]. This research has the potential to aid in the creation of more efficient and affordable solar cells for use in a wide range of settings.To learn how the characteristics of solar cells change when an antireflection coating (ARC) made of indium tin oxide (ITO) is applied to the front surface [7]. Typically utilized as a barrier coating in solar cells, SiO2 is a clear substance with a high refractive index [8]. In the present investigation, solar cells were made using a substrate made from silicon, and a front protective layer of SiO2 was put on the substrate at different thicknesses (from 50 nm to 1000 nm). After the SiO2 layer was in place, an ARC of ITO was placed on top of it [9]. Different SiO2 layer thicknesses were tested, and the photovoltaic cell's performance was evaluated [10].The solar cells' characteristics were strongly affected by the thickness of the front protective layer. When absorbing light, the ideal thickness of the SiO2 layer varies with wavelength [11-13]. The optimal thickness for performance was between 100 and 150 nm for visible light and between 300 and 500 nm for near-infrared light. The efficiency of the solar cells was significantly affected by the thickness of the SiO2 layer in both situations [14]. This research shows that optimizing the efficiency of solar cells using ITO as an ARC depends critically on the thickness of the front protective layer. The findings may be used to improve solar cell efficiency and durability [15] in their development and manufacturing processes.

2. Materials and Methods

The behavior of solar cells may be modeled and simulated with the help of the Photovoltaic QCRF-FDTD simulator. Analyzing how light interacts with different solar cell ingredients and architectures, it employs the FDTD technique to simulate the solution of Maxwell's equations [12]. For example, the simulator may be used to examine the optical characteristics of the SiO2 layer and how they affect the total effectiveness of the solar cell when using ITO as an ARC [9]. The modeling tool may also test different nanoparticle sizes in the ITO layer to see which works best for maximizing efficiency [11].The best practices for designing and fabricating solar cells may be gleaned from the results of simulations and verified by practical observations. The simulation findings may not represent the solar cell's real-world behavior since they are based on several preconceptions and concepts. Consequently, ascertaining the validity and precision of the simulation findings requires validation by experiments [13].

3. Result and Discussion

Silver nanoparticles improve the protective front layer (SiO2) performance when using ITO as an anti-reflection treatment. When comparing the results in the table, it is clear that the efficiency is often better when silver nanoparticles are present, regardless of the SiO2 layer thickness. At 50 nm thickness, for instance, silver nanoparticles improve efficiency to 21.9079% from 21.5723%. However, some exceptions exist where the efficiency is slightly lower when adding silver nanoparticles. For example, at a thickness of 150nm, the efficiency is 17.9821% with silver nanoparticles compared to 17.645% without.

Overall, it appears that adding silver nanoparticles can improve the efficiency of the protective front layer with ITO as an antireflection coating material. However, the effect may be based on the size of the SiO2 layer. The thickness of the protective front layer (SiO2) seems to have a relatively small impact on the efficiency of the cells. Values for the cells with and without nanoparticles remain pretty consistent across varying thicknesses of SiO2. It is also interesting to note that the efficiency values for the cells with and without nanoparticles are relatively close, with the highest efficiency difference being around 1.5%. However, this difference could still be significant in practical applications, especially when considering large-scale implementation.Overall, the results suggest that adding silver nanoparticles can be a feasible strategy for enhancing the efficiency of PV cells, and further research could explore optimizing the concentration and size of nanoparticles for even more significant performance gains.

Table 70.1 Comparison value for solar cell parameters ITO, an antireflection coating material with and without silver nanoparticles at the thickness of the protective front layer, varies from 50nm to 1000nm

ARC	Voc in Volts	Jsc in MA	Fill Factor	Efficiency in %
ITO Without Silver nanoparticles	0.906168	28.7646	0.833198	21.7178
ITO With Silver nanoparticles	0.906803	29.259	0.833283	22.1088

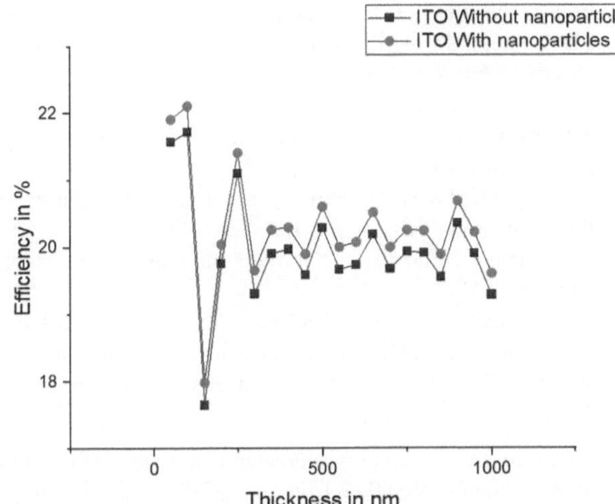

Fig. 70.2 Solar cell parameter analysis graph vs thickness of the protective front layer efficiency

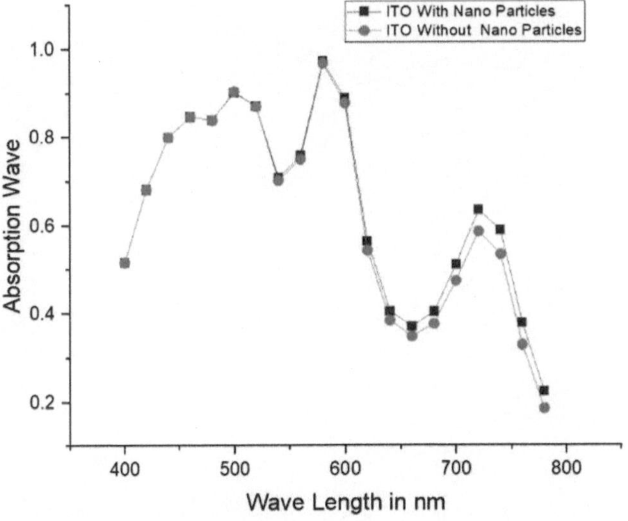

Fig. 70.3 Absorption Vs wavelength comparison ITO as ARC

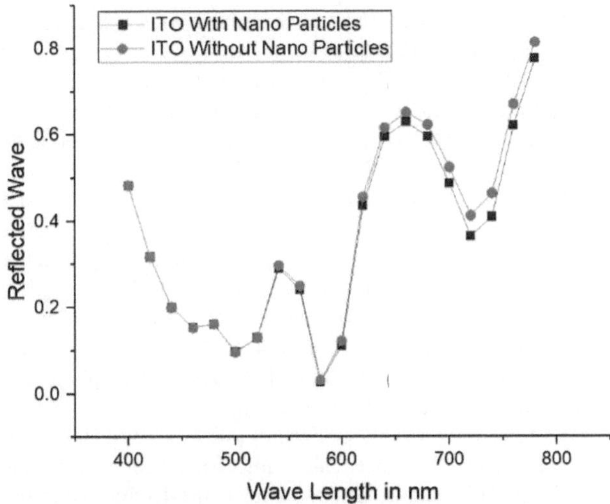

Fig. 70.4 Reflection Vs wavelength comparison ITO as ARC

4. Conclusion

Using silver nanoparticles as an additive to the SiO2 protective front layer on ITO as an antireflection coating material can improve the efficiency of the PV cell. Comparing the two coatings, it can be seen that ITO with silver nanoparticles has slightly higher values for all the parameters reported in the table. Its Voc is 0.906803 V, which is greater than the competitor's 0.906168 V; its Jsc is 29.259 mA, which is higher than the competitor's 28.7646 mA; its fill factor is 0.833283, which is higher than the competitor's 0.833198; and its efficiency is 22.1088%, which is higher than the competitor's 21.7180%. However, the efficiency boost was not uniform across all SiO2 coating thicknesses. The efficiency boost was incredible with thinner coatings (50-150 nm) and diminished with increasing coating thickness. Therefore, silver nanoparticles may be most helpful in relatively light barriers.

Adding silver nanoparticles to the solar cell's protective front layer is one way to boost performance. The influence of the thickness of the protective coating on efficiency enhancement and the appropriate concentration and size of the nanoparticles might be the subject of future research.

References

1. M. Lu, L. Chen, X. Wu, J. Wu, Y. Li, and Y. Li, "Performance improvement of silicon solar cells by indium-tin-oxide nanoparticle antireflection coatings," Solar Energy, vol. 124, pp. 163–169, 2016.
2. S. Kim, S. Kim, S. Yoo, S. Hong, S. Yoon, and H. Shin, "Enhancement of silicon solar cell performance using

indium-tin-oxide nanoparticles and titanium dioxide thin films," Journal of Nanoscience and Nanotechnology, vol. 16, pp. 4118–4122, 2016.

3. J. B. Rao, N. M. Ravindra, V. K. Jain, and G. V. Prakash, "Effect of indium tin oxide thickness on the performance of silicon solar cells," Solar Energy Materials and Solar Cells, vol. 75, pp. 233–238, 2003.
4. S. Ray, S. S. Mahato, and R. K. Sinha, "Size-dependent impact of silver nanoparticles on the photovoltaic performance of indium tin oxide based solar cells," Journal of Applied Physics, vol. 120, pp. 104502-1-8, 2016.
5. H. Kim, Y. Choi, J. Kim, and D. Lee, "Improvement of silicon solar cell performance using a SiO2/TiO2 double-layer antireflection coating," Journal of Physics D: Applied Physics, vol. 41, pp. 105103-1-6, 2008.
6. C. Teng, S. W. Chen, S. Y. Wang, T. C. Chang, Y. H. Lin, and J. J. Huang, "Enhancement of silicon solar cell performance using ITO nanostructure-based antireflection coatings," Journal of Materials Chemistry A, vol. 3, pp. 22503–22508, 2015.
7. L. Lin, C. Jiang, X. Zhang, and L. Chen, "Fabrication and characterization of indium-tin-oxide/Ag nanowire composite transparent electrodes for silicon solar cells," Journal of Applied Physics, vol. 111, pp. 104315-1-6, 2012.
8. Y. Wang, Z. Du, Q. Sun, H. Ding, Y. Zhang, and L. Chen, "Graphene oxide induced enhancement in the performance of silicon solar cells with silicon nanowire arrays," Journal of Materials Chemistry A, vol. 3, pp. 16369–16375, 2015.
9. M. Lu, J. Xu, Z. Wei, Y. Li, Y. Li, and H. Liu, "The performance improvement of silicon solar cells by introducing a hole-blocking layer," Solar Energy, vol. 111, pp. 67–73, 2015.
10. X. Li, M. Li, W. Chen, X. Li, and Y. Zhang, "The effect of SiO2 thickness on the optical and electrical properties of silicon solar cells," Journal of Renewable and Sustainable Energy, vol. 9, pp. 033501-1-8, 2017.
11. J. W. Choi, H. J. Kim, J. W. Lee, and J. H. Lee, "Numerical simulation of solar cells using the finite-difference time-domain method with perfectly matched layer boundary conditions," Journal of Applied Physics, vol. 102, no. 12, pp. 123109, 2007.
12. J. Li, H. Wang, X. Wang, and X. Liu, "Numerical simulation of the SiO2/ITO/Ag nanorods/ Si solar cell based on FDTD method," Optik - International Journal for Light and Electron Optics, vol. 127, no. 22, pp. 11080–11084, 2016.
13. L. Cao, Z. Li, and J. Li, "Simulation of perovskite solar cells using the finite-difference time-domain method," Applied Optics, vol. 56, no. 19, pp. 5472–5477, 2017.
14. Y. Jiang, X. Wang, Y. Huang, and J. Li, "Simulation of thin-film solar cells using the finite-difference time-domain method," Journal of Nanophotonics, vol. 12, no. 2, pp. 026013, 2018.
15. S. B. Pandey, S. Saha, and S. Basu, "Numerical simulation of organic solar cells using finite-difference time-domain method," Journal of Computational Electronics, vol. 18, no. 4, pp. 1235–1246, 2019.

Note: Every figure and table was created using the experimental work; none of them were taken from any publications or online.

Advances in Additive Manufacturing Technologies – Gurusamy Pathinettampadian (eds)
© 2024 Taylor & Francis Group, London, ISBN 978-1-032-90013-1

71 Experimental Study of Shock Wave Characteristics of Bullet Train Nose Cone (IKADA Model) at Supersonic Speed with Bio-Inspired Approach

Solaiyappan G, Mano R T, Santhakumar U
Student, Department of Aeronautical Engineering,
Hindustan Institute of Technology and Science, Chennai, India

Balaji G*, Saravanan P
Faculty, Department of Aerospace Engineering,
Hindustan Institute of Technology and Science, Chennai, India

Ramanan N
Assistant Professor, Department of Mechanical Engineering,
Sri Jayaram Institute of Engineering and Technology, Chennai, India

Boopathy G
Associate Professor, Department of Aeronautical Engineering,
Vel Tech Rangarajan Dr.Sagunthala R&D Institute of Technology and Science, Chennai

Santhosh Kumar G
Teaching Fellow, Dept. of Mech. Engg., University College of Engineering,
Bharathidasan Institute of Technology Campus, Anna University, Tiruchirappalli, Tamilnadu, India

Vijayanandh Raja
Department of Aeronautical Engineering, Kumaraguru College of Technology,
Coimbatore-641049, Tamil Nadu, India

Gurusamy P
Professor, Department of Mechanical Engineering,
Chennai Institute of Technology, Chennai, India

ABSTRACT: The experimental study is to systematically investigate the supersonic flow over the bullet train nose cone model using supersonic wind tunnel. The bullet train nose cone design considered as the IKADA model,which is bio-inspired sailfish nose cone model has been modified to examine the shock wave characteristics at different pressure conditions. Further, sailfish have an aerodynamic and streamlined body shape which minimizes drag and allows them to travel efficiently at high speeds. The study will be achieved in several steps, In first, how to minimize the traveling times by enhancing the bullet train's performance & also optimizing the frontal nose design of the bullet train model. Secondly the development and fabrication of Ikada (Sailfish) nose cone models for the purpose of experimentation. It is most important about this investigation are the unique characteristics of the IKADA (Sailfish) frontal nose cone design & its dimensions. The model was tested in the modern supersonic wind tunnel facility along with the

*Corresponding author: gbalajihits@gmail.com

DOI: 10.1201/9781003545774-71

Schlieren flow visualization setup which is available in HITS, Chennai. The test results provides a new insight into the biological, physical, and structural aspects of supersonic bullet train. The results are very useful to the designers and manufacturers who are involved in the development of high-speed vehicle. The IKADA nose cone model is located in the supersonic wind tunnel test section and settling chamber pressure is steadily adjusted to change the Mach of four different supersonic Mach numbers such as 1.6, 1.8, 2 & 2.5 to examining the shock formations. Flow patterns, as well as pressure distributions, surrounding nose cone models are captured using high-resolution cameras. Further, the shock wave in supersonic flow to be analysed. The primary shock wave pattern results of IKADA model are collected and analysed using advanced software, calculating Mach number (M2, M3) and various comparisons are made for the aerodynamic performance characteristics of Ikada bullet train nose cone model.

KEYWORD: Bullet train, Ikada, Sailfish, Supersonic wind tunnel, Shock wave, Aerodynamically design

1. INTRODUCTION

It is very important to understand flows around the bullet train nose design at high velocity because it influences the phenomena such as shock waves. Studying these shock waves attentively is necessary because they're capable of sufficient strength to significantly alter a vehicle's stability. The formation and propagation of these shock waves at high-speed object is a key issue to investigate. Numerous scientists have investigated this, yet there is still much we don't know and need to learn. Nagata *et al.* [1], study on Schlieren visualization and motion Analysis of an Isolated and Clustered Particle (s) after planar shock interactions has been investigated and the system indicated that the particle trajectory, drag coefficient, a temporal development, and flow structure around the nose cone model. Further, the drag coefficients in the data were similar to those in the stable state drag model. It allows one to investigate hydrodynamic interactions between particles within particle clusters. Weidner et al. [2], based on the facilitating methodology for translating of biological systems to technical design solution, By embracing knowledge from other fields of study, such as biology, design limitations can be managed. This investigation provides a bioinspired design technique that develops visual resembles via biological system sketching. Balaji *et al.* [3] experimental investigation were carried out in the subsonic wind tunnel to characterise the aerodynamics performance of double delta wing with various freestream conditions.

Abbasli et al. [3], The research work, A design approach was created using parametric techniques and tools to demonstrate how using natural characteristics in urban furniture design can enhance the aesthetic and efficiency of architectural objects in terms of viable, aesthetically pleasing, and financially viable design solutions. This research intends to establish an environmentally friendly public space for people to connect and temporarily rest in an urban setting by incorporating biomimicry, taking into account its potential for use in problem-solving for humans. Heidari et al.[5], the experimental study involves wind tunnel tests on an ogive-cylinder body at 1.6 Mach and 8×10 Reynolds, with surface static pressure and boundary layer profile measurements in which the tests were performed in Trisonic wind tunnel. The objective of the experimental study is to investigate pressure distributions and boundary layer profiles at various angles of attack. The computational results were compared to the experimental data, showing close agreements.

Junjie et al. [6], the study uses particle image velocimetry, flow visualization technology, and high-frequency response pressure transducers to analyse small-angle expansion, velocity structures, and pressure distribution near a compression ramp. Large-scale separation results from expansion that occurs prior to the compression ramp point which influences shock wave boundary layer interaction. Chauhan et al. [7] numerically investigated the aerodynamics performance of blunt nose cone with aerodisk at hypersonic speeds. It observed that shock patterns formation has influenced at L/D ratio 2 near to the nose cone model and leads to drastic reduction of drag by raising the d/D ratios. Swoboda *et al.* [8] & Farahani *et al.* [9] studied the shock boundary-layer interaction on transonic aerofoils for laminar and turbulent flow and the supersonic phase is characterized by splitting, which forms the ultimate structure of shock waves.

Zapryagaev et al. [10], focuses on the investigation of Supersonic Jets Shock-Wave Structure.

The study reveals a significant effect of vortex generators on the flow's shock-wave structure. A new technique enables complex experimental studies of supersonic jet unsteady flows on various wind tunnels and facilities. Reliable data on gas-dynamic structure and the influence of artificial longitudinal vortex structures on shockwave structure, mixing intensification processes, and acoustic characteristics are obtained. Catanzaro et al. [11], focuses on the High-speed train crosswind analysis, the analysis made examines the impact of train movement on track and infrastructure. It reproduces train movement through the grid boundary conditions and solves flow equations in steady formulation. The impact on aerodynamic coefficients is less than 10% for high-speed trains, but higher differences exist when scenario influences flow on track and embankment and windbreak fence configuration. Balaji etal.[12] & Li et al.[13] studied the pressure distribution of NACA5520 airfoil blade to study the aerodynamics performance with taper wing at different freesteam conditions study compares conventional symmetric heads with asymmetric heads and their aerodynamic characteristics utilizing the Simple method. The results show that the new train with asymmetric heads is the most excellent and feasible under crosswind conditions. Ali et al.[14], focuses on the numerical Investigation of aerodynamic characteristics of high-speed train. This study involves the High-speed trains like the French TGV Duplex and Japanese Shinkansen E6 reach speeds of 300 km/h, affecting their performance and safety.

Sarafrazi et al.[15] & Balaji et al. [16],numerical study uses Fluent engineering software to investigate the aerodynamic and slipstream effects of high-speed trains on overhead and trackside installations. Hemispherical Missile nose cone were tested under low speed to conditions to charcterize the performances. Shi et al.[17], investigated experimental research on the heat transfer characteristics on train body of bullet train. It is observed that the same aluminium had the same conductivity under different temperatures, but the impact of the conductivity on different aluminium changed. Zhu et al.[18], analyses the braking performance of a high-speed train using two types of braking wings. The study compares the results between the train without a braking wing and with and without a braking wing. The results show that the high-speed train with braking wing essentially contributes to acceleration during the braking process, especially at high speeds. Balaji et. [19] Studied the downstream wake characterisitic of cicular cylinder at different L/d ratio and x/d ratio. The aim of this research is to investigate the shock formation of Bullet train Nose cone (BTNC) model and propagation of shock waves at the frontal nose cones of the bullet train model. The formation of these shock waves leads to increase the aerodynamic drag, sonic booms and passenger discomfort, decreasing the efficiency and safety of high-speed travel. Hence, the reserch work is to reduce those aerodynamic disturbances and improve bullet train performance to make the travel more comfortable to passengers.

2. Bio Inspired Approach: The Sailfish

A sailfish's body is streamlined and aerodynamic, with a long, slim body and an enormous erectile dorsal fin that looks like a sailfish. This form allows the sailfish to travel effectively through the water at high speeds. In contrast, the Japanese bullet train, also known as the Maglev train, is built for high-speed track travel. The train's body is sleek and aerodynamic, having a long, narrow design that reduces air resistance. The form of this train is optimized for speed and efficiency, with an emphasis on lowering drag and making the best use of energy. While the sailfish's body is designed for swimming in water, the Japanese bullet train's form is suited for high-speed land transport. Sailfish also have a streamlined and hydrodynamic body form, which allows them to easily attain high speeds while minimizing drag for quick movement. The transformaton of biomimicry of sailfish and IKADA bullet train nose model is shown in Fig. 71.1 and geometrical dimension of the IKADA nose cone model is shown in Fig. 71.2.

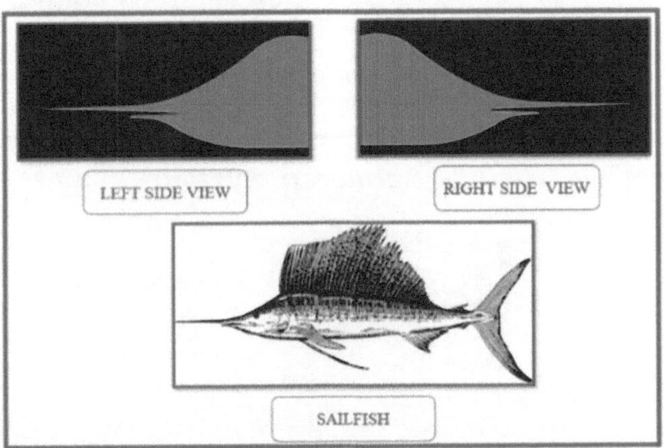

Fig. 71.1 Sailfish bio inspired model design

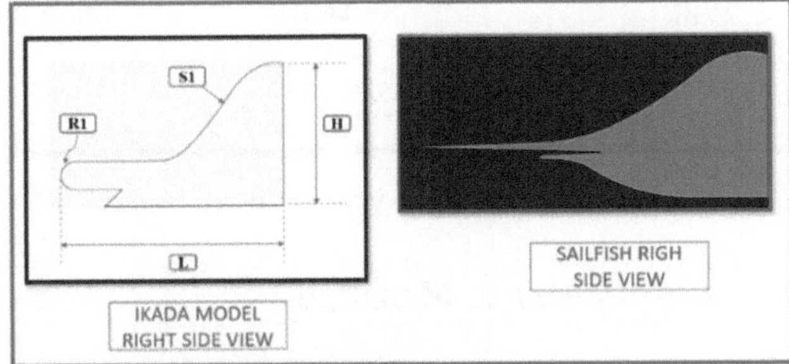

Fig. 71.2 Sailfish transformation to IKADA- BTNC

3. Experimental Set-Up

The supersonic wind tunnel with Schlieren Flow Visuvalisation set up is being used to conduct testing in the supersonic tunnel at different settling chamber pressures as shown in Fig. 71.3. The shock formation over the Bullet train nose cone model is captured by using high resolution camera which is holded in the tripad along with Schlieren flow visualization system and Fig. 71.4. The high speed supersonic wind tunnel facility available at Aerodynamics Lab, Hindustan Institute of Technology and Science, chennai.

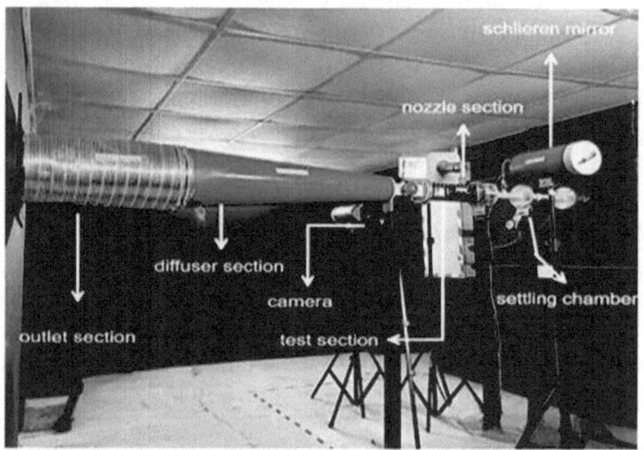

Fig. 71.3 Supersonic wind tunnel setup

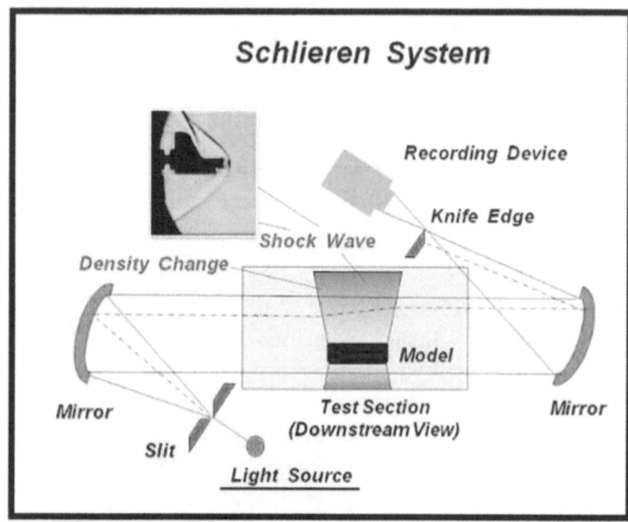

Fig. 71.4 Schliren flow visuvalisation system

4. Experimental Models

There are various processes involved in fabricating a bullet train nose cone model (IKADA) made up of the stainless-steel materials, and the precise procedure may change depending on the model's design and unique requirements. Here is a general outline of the fabrication process: Design Phase: Using AutoCAD 2022 Version, generate accurate engineering drawings with all necessary dimensions, features, and materials requirements. The Schematic diagram of IKADA Bullet Nose Cone Model is shown in Fig. 71.5 (a and b). By considering factors such as aerodynamic shape, mounting points, instrumentation integration, and manufacturing feasibility. Material selection: By choosing an appropriate grade of stainless steel based on factors such as strength, corrosion resistance, and temperature resistance. Manufacturing Process: By cutting the stainless-steel stock material to the required size and shape using cutting tools such as shears, saws, or laser cutting machines.

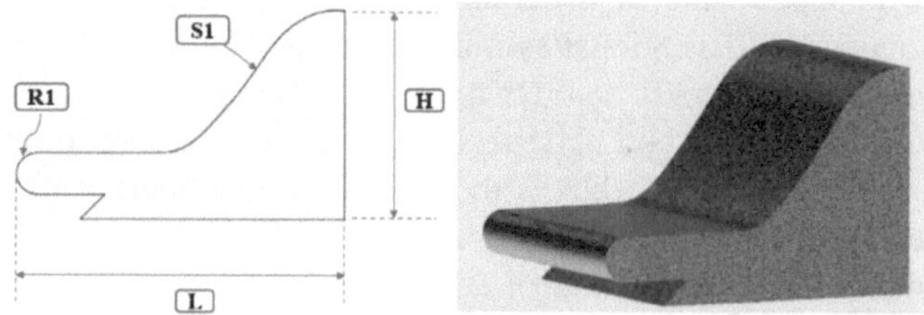

Fig. 71.5 (a) 2D View of Ikada Btnc model, (b) Isometric view of IKADA Btnc model

Shaping and forming the stainless-steel components using processes such as bending, rolling, or stamping to achieve the desired geometry. The fabricated model is ready to be tested in a supersonic wind tunnel for analysing the shock formation on the IKADA BTNC model. Here, IKADA BTNC surfaces have a curved tip to form a bow shock at the frontal surface. The Schlieren image technique is used for the study of shock attachment and detachment on the surface of the model and the angle of the shock waves. Since across the normal shock, the flow is always subsonic, the flow properties will change in a drastic manner. But across the oblique shock, the flow Mach number, either subsonic or supersonic, depends upon the strength of the shock wave.

5. Results and Discusions

In Fig. 71.6 (a-d), show the shock pattern of the IKADA bullet train nose cone model at different Mach umber such as 1.6, 1.8, 2 and 2.5. It has wedge angles of (θ1 & θ2) and corresponding two upstream shock wave (β1 & β2) angle was observed in oblique shock pattern. Further, it found that strong bow shock has been formed on the front nose cone and second wedge angle forms of obliques shocks at according to the wedge formataions.

In Fig 71.6 (a) shock pattern of BTNC at Mach of M_1 = 2.5 has been observed the oblique shock anlge of β1 = 67°, β2 = 60°and wave angle of β3 = 68° corresponding to the Mach of M_2 = 1.4 and M_3 = 3.2. Further, it found

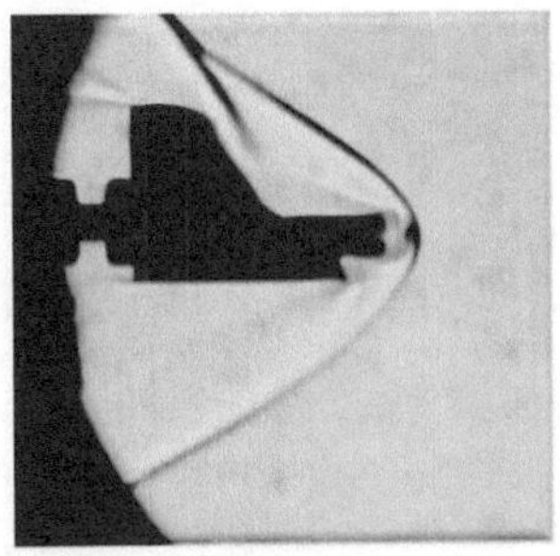
(a) IKADA Model at M1 = 2.5

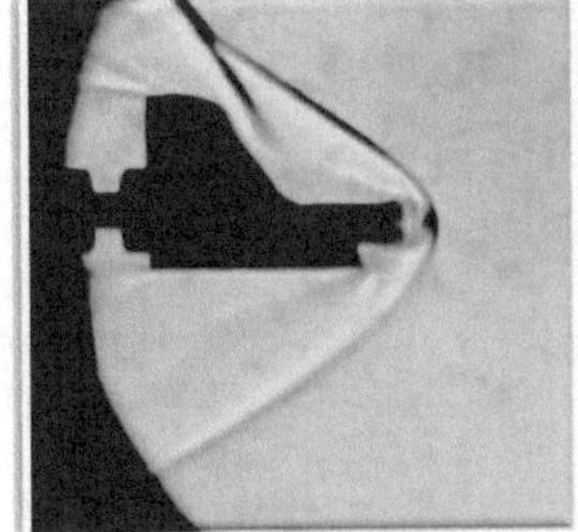
(b) IKADA Model at M1 = 2

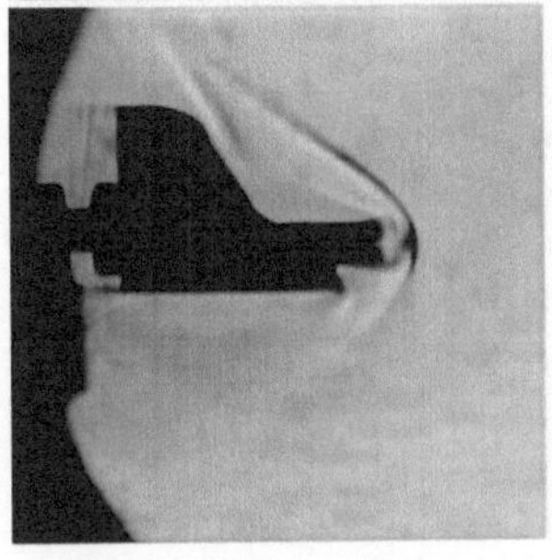
(c) IKADA Model at M1 = 1.8

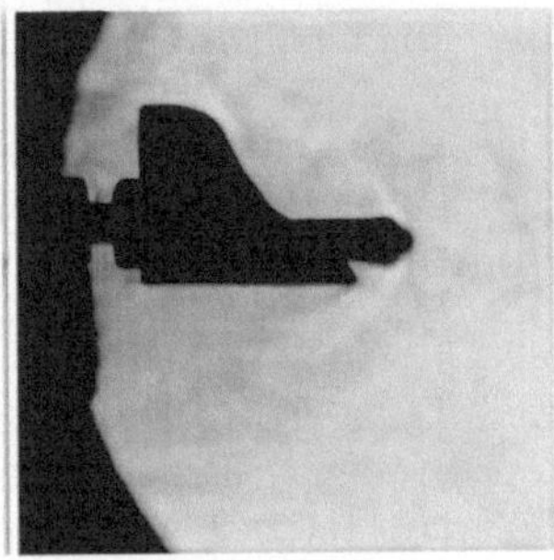
(d) IKADA Model at M1 = 1.6

Fig. 71.6 Shock pattern over the diffent IKADA bullet train nose cone (BTNC) model

that strong bow shock has been formed on the front nose cone and second wedge angle forms of obliques shocks at according to the wedge formation over the rear end of the nose cone model. Primary strong shock will interfare with downsteam obliques shock which tend to drastic reduce of Mach number, M_2 and M_3.

In Fig. 71.6(b) shock pattern of BTNC at Mach of $M_1 = 2$ has been observed the oblique shock anlge of $\beta1 = 66°$, $\beta2 = 60°$and wave angle of $\beta3 = 67°$ corresponding to the Mach of $M_2 = 1.7$ and $M_3 = 2.75$. Further, it found that similar behaviour of the strong bow shock has been created at the front nose cone and second wedge angle forms of obliques shocks at according to the shape of the wedge and the rear end of the nose cone model. Primary strong shock will interact the downsteam obliques leads to sudden reduction of Mach number, M_2 and M_3.

In Fig 71.6 (c) shock pattern of BTNC at Mach of M_1=1.8 has been observed the thin shear layer formation occurs throuout the nose cone entire models and forms the turbulent weak shear layer occurs at the rear end the nose cone model and corresponding to the Mach of $M_2 = 1.9$. Further, it found that weak obliqe shock formation front and rear end of the nose cone. This shock formation not influencing the oblique shock formation also mere changes of Mach number at rear end of the Bullet nose cone which presented in the table 1 and Fig. 71.6(a-d).

In Fig. 71.6 (d) shock pattern of BTNC at Mach of M_1=1.6 has been observed the oblique shock anlge of $\beta1$=65°, Turbulent weak shear layer occurs at the rear end the nose cone model and no chnages in the Mach number at behind the front end and rear end of bullet nose cone This shock formation not influencing the oblique shock formation also mere changes of Mach number at rear end of the Bullet nose cone which presented in the table 1 and Fig 6 (a-d). The follwing isentropic relation has been used for calculate the downstream mach numbers over the Bullet nose cone moel at different Mach number.

$$\frac{P_0}{P_S} = [1+\frac{\gamma-1}{2}M^2]^{\frac{\gamma}{\gamma-1}} \tag{1}$$

$$Mn,1 = M1\,sin\beta \tag{2}$$

$$Mn,1^2 = \frac{[Mn,1^2]+[\frac{2}{(\gamma-1)}]}{\left[\frac{2\gamma}{(\gamma-1)}\right]*[Mn,1^2]-1} \tag{3}$$

$$M2 = \frac{Mn,2}{Sin(\beta-\theta)} \tag{4}$$

Table 71.1 IKADA upstream and downstream mach calculations

Pressure in bar	θ1°	β1°	θ2°	β2°	β3°	M1	M2	M3
12.5	45	67	45	60	68	2.5	1.4	3.2
11.5	45	66	45	60	67	2	1.7	2.75
10.5	45	65	45	TSL	67	1.8	1.9	-
9	45	TSL	45	TSL	TSL	1.6	-	-

In Fig. 71.7 shows that the Comparison of Upstream Mach number against down steam Mach number. It is observed that downsteam Machnumber M_2 decreases with increasing the upsteam Mach number as presneted here. In Fig. 71.8. It is observed that Down stream Mach number M_3 increases against Upstream Mach number M_1.

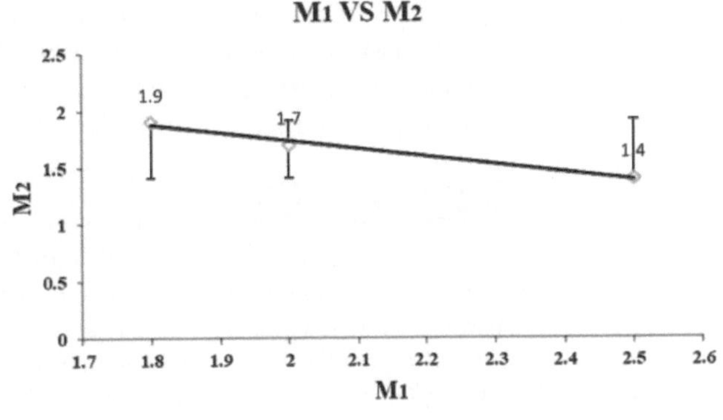

Fig. 71.7 M_2 Vs M_1

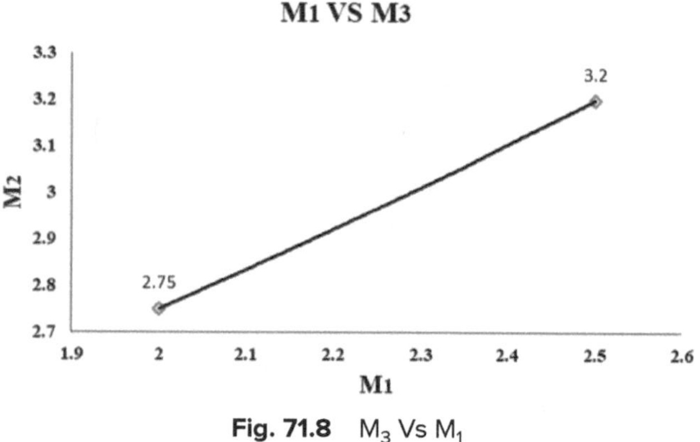

Fig. 71.8 M_3 Vs M_1

6. Conclusion

In this research work, the transformation of biomimicry of sailfish to Bullet Nose Cone Model (BTNC) called IKADA. This model was tested in the supersonic wind tunnel to analyse the shock formation and visualize the shock pattern at different Mach numbers. This shock may be bow and oblique shock formation due to geometry of the bullet train nose cone. The experimental investigation reveals the following finding is listed below.

It is observed that bow shock formed at an angle β1of 67° and the shock type is weak; shock waves are obtained for Mach Number = 2.5. Also, The Stand-off distance occurs with negligible distance at the higher mach numbers.

A comparison of the downstream and upstream Mach numbers is shown in Fig. 71.7. When seen here, it is noted that when the upstream Mach number increases, the downstream Mach number M_2 drops. In Fig. 71.8. In comparison to the upstream Mach number M_1, it is evident that the downstream Mach number M_3 increases.

It is observed that shock interaction in the rear end obliques shock at higher Mach number such as M = 2. and 2.5 and Thin shear layer and weak shock formation was observed in the lower Mach number such as M = 1.6 and 1.8.

Shock pattern over the IKADA Bullet train nose cone (BTNC) model at different Mach number was characterized to have good agreement with existing research work and helps to design the high speed vehicle such as bullet train.

Reference

1. Nagata, T., Nonomura, T., Ohtani, K., & Asai, K. (2022). Schlieren visualization and motion analysis of an isolated and clustered particle (s) after interacting with planar shock. *Transactions of the Japan Society for Aeronautical and Space Sciences*, *65*(4), 185-194. https://doi.org/10.2322/tjsass.65.185.
2. Weidner, D., Stoll, D., Kuthada, T., & Wagner, A. (2022). Aerodynamics of high-speed trains with respect to ground simulation. *Fluids*, *7*(7), 228..
3. Balaji, G., Bharath Kumar, A., Divya, R., Boopathy, G., Seenu, N., & Santhosh Kumar, G. (2023, August). Experimental Investigation of Double Delta Wings with Different Angles of Attack at Subsonic Speeds. In *International Conference on Smart Sustainable Materials and Technologies* (pp. 197-202). Cham: Springer Nature Switzerland.
4. Abbaslı, U., & Selcuk, S. A. (2016). Biomimetic design principles as an inspirational model: case study on urban furniture. *International Journal of Architecture and Urban Studies*, *1*(1), 41-52.
5. Heidari, M. R., Farahani, M., Soltani, M. R., & Taeibi-Rahni, M. (2011). Investigations of Supersonic Flow around a Long Axisymmetric Body. In *Wind Tunnels and Experimental Fluid Dynamics Research*. IntechOpen.
6. Junjie, H. U. O., Shihe, Y. I., Zheng, W., Haibo, N. I. U., Xiaoge, L. U., & Dundian, G. A. N. G. (2022). Experimental investigation of expansion effect on shock wave boundary layer interaction near a compression ramp. *Chinese Journal of Aeronautics*, *35*(12), 89-101.
7. Chauhan, J., Balaji, G., Swastikar, M., Boopathy, G., Sangeetha, S., Santhosh Kumar, G., & Pradeep, G. M. (2023, August). Numerical Investigations of Aerodynamics Performance of Blunt Nose Cone with Aerodisk at Hypersonic Flow. In *International Conference on Smart Sustainable Materials and Technologies* (pp. 177-188). Cham: Springer Nature Switzerland.
8. Swoboda, M., & Nitsche, W. (1996). Shock boundary-layer interaction on transonic airfoils for laminar and turbulent flow. *Journal of aircraft*, *33*(1), 100-108.
9. Farahani, M., & Jaberi, A. (2017). Experimental Investigation of Shock Waves Formation and Development Process in Transonic Flow. *Scientia Iranica*, *24*(5), 2457-2465.
10. Zapryagaev, V. I., Gubanov, D. A., Kavun, I. N., Kiselev, N. P., Kundasev, S. G., & Pivovarov, A. A. (2017, October). Investigation of supersonic jets shock-wave structure. In *AIP Conference Proceedings* (Vol. 1893, No. 1). AIP Publishing.
11. Catanzaro, C., Cheli, F., Rocchi, D., Schito, P., & Tomasini, G. (2016). High-speed train crosswind analysis: CFD study and validation with wind-tunnel tests. In *The Aerodynamics of Heavy Vehicles III: Trucks, Buses and Trains* (pp. 99-112). Springer International Publishing.
12. Balaji, G., Raj, M. S., Aswin, S. K. S., Kabilan, K., & Navinkumar, B. (2022). Effect of Angle of Attack on Pressure Distribution of NACA5520 Airfoil Blade. *International Journal of Vehicle Structures & Systems*, *14*(1), 93-98.
13. Li, W., Wang, Y., Hao, Y., Zhao, W., & Zhang, Y. (2022, June). Analysis of Aerodynamic Characteristics of High-speed Train with Asymmetric Heads. In *Journal of Physics: Conference Series* (Vol. 2280, No. 1, p. 012004). IOP Publishing.

14. Ali, J. M., Omar, A. A., Ali, M. A. B., & Baseair, A. R. B. M. (2017, March). Numerical investigation of aerodynamic characteristics of high speed train. In *IOP Conference Series: Materials Science and Engineering* (Vol. 184, No. 1, p. 012015). IOP Publishing.
15. Balaji, G., Sree, K. N., Reddy, E. V., Sumanth, N., Sathish, S., & Madhanraj, V. (2022). Numerical investigation of flow over a hemispherical missile nose cone configuration in subsonic speed. *Materials Today: Proceedings*, *68*, 1447-1454.
16. Sarafrazi, V., & Talaee, M. R. (2018). CFD Simulation of High-speed Trains: Train-induced Wind Conditions on Trackside Installations.
17. Shi, W., Li, L., Li, M., & Zeng, X. (2020). Experimental Research on the Heat Transfer Characteristics on Train Body of Bullet Train. In *IOP Conference Series: Materials Science and Engineering* (Vol. 746, No. 1, p. 012003). IOP Publishing.
18. Zhu, Y., Shang, W., Zhang, X., Yan, H., & Wu, P. (2014). Research on braking process of high-speed train with aerodynamic brake. *International Journal of Control and Automation*, *7*(12), 363-374.
19. Balaji, G., Nadaraja Pillai, S., & Senthil Kumar, C. (2017). Wind tunnel investigation of downstream wake characteristics on circular cylinder with various taper ratios. *Journal of Applied Fluid Mechanics*, *10*(Special Issue)), 69-77.

Note: Every figure and table was created using the experimental work; none of them were taken from any publications or online.

Advances in Additive Manufacturing Technologies – Gurusamy Pathinettampadian (eds)
© 2024 Taylor & Francis Group, London, *ISBN 978-1-032-90013-1*

72 Study on Induced Drag Reduction of Electric Aircraft Wing Using Various Winglet Configurations: A Review

Muthu Kamakshi Sundaram, Vasireddi Ravi Teja, Bokka Sasank Sai
Student, Department of Aeronautical Engineering,
Hindustan Institute of Technology and Science, Chennai, India

Saravanan P, Balaji G*, Sankaralingam L, Seralathan S
Faculty, Department of Aerospace Engineering,
Hindustan Institute of Technology and Science, Chennai, India

Santhosh Kumar G
Teaching Fellow, Dept. of Mech. Engg., University College of Engineering,
Bharathidasan Institute of Technology Campus, Anna University,
Tiruchirappalli, Tamilnadu, India

Vijayanandh Raja
Department of Aeronautical Engineering,
Kumaraguru College of Technology, Coimbatore, Tamil Nadu, India

Ramanan N
Assistant Professor, Department of Mechanical Engineering,
Sri Jayaram Institute of Engineering and Technology, Chennai, India

Gurusamy P
Professor, Department of Mechanical Engineering,
Chennai Institute of Technology, Chennai, India

ABSTRACT: Addition of winglets provides increased aerodynamic efficiency without any other drastic changes to the structure of the aircraft. Although many modern airliners are equipped with this technology, small electric powered aircrafts haven't utilized this advantage by wing design. Studies reveal that the major part of drag force of a transport aircraft is produced by the skin friction drag and the lift-induced drag. This paper gives a comprehensive revision on addition of wingtip device, so-called Winglets to the electric aircraft to increase its aerodynamic efficiency by decreasing the induced drag due to lift. A wing of an electric aircraft under research is taken to study its performance. From the literature, it is found that the upward sharklet, split-tip winglet and blended winglet are the major types of winglets. These 3 types of winglet designs are proven not only in reduction of induced drag but also increasing the overall lift produced during flight. This results in significant improvement of range, speed, endurance and battery-fuel economy in electrically powered aircrafts.

KEYWORDS: Induced drag reduction, Winglet, Electric-aircraft, L/D ratio, ANSYS

*Corresponding author: Balaji.G, gbalajihits@gmail.com, 8610559691

DOI: 10.1201/9781003545774-72

1. Introduction

1.1 Drag and Types

The aerodynamic drag is one of four forces proposed by the father of aviation, George Cayley. Drag is simply the fore that opposing the motion of aircraft. To achieve the forward motion of the aircraft, thrust force should overcome drag. There are many types of aerodynamic drag acting on an airplane depending upon Mach number, lift acting, etc. Some common types of drag are listed below:

In the subsonic speed we have profile drag and induced drag. [1] Profile Drag divided into pressure drag (also known as form drag) and skin friction drag.. If the drag is dependent on the lift produced, then it is called as Drag due to lift or else it is Zero lift drag. The profile drag that is independent of lift generated is called as parasite drag. Induced drag is dependent on lift produced, so comes under Drag due to lift.

1.2 Induced Drag

Induced drag is a type of aerodynamic drag that is created when an airfoil, such as a wing, generates lift. This difference in pressure creates an upward force, or lift, on the wing.

Vortices is defined as the phenomenon of airflow curling around in circular motion at the tips of the wing. The drag created due to this circular motion of airflow, called vortices is called as induced drag which is generated only if lift is acting on the wing. [2] So induced drag is a Drag due to lift.

$$D_i = \frac{kL^2}{\frac{1}{2}\rho_0 V_e^2 S\pi AR}$$

Where,

Di - Induced drag, L-Lift, ρ_0-Air density, *Ve*-Relative airspeed, S-Wing area, AR-Aspect ratio

1.3 Drag Reduction Techniques

Drag reduction in an aircraft result in reduced fuel consumption, higher operational range, greater range and increase in speed. [5]. Although it is not easy, there is always new ideas that results in improvements. [6] Some of the technologies are described below:

- Skin friction reduction up to 5% by installation of smart surfaces which gets compressed according to the flow regimes/conditions.
- Hybrid laminar flow technology which reduces about 10% of skin friction drag by shaping the airfoil according to NLF and LFC concepts to small jet aircrafts with low swept angles and Reynolds number.
- Lift-induced drag reduction up to 2% by adding the wing tip device called winglet.
- Wave drag reduction by local modification in the airfoil surface with a small bump at shock region.
- More advanced configurations such as multistage aircraft, supersonic leading-edge wings, favourable wave interference and strut-braced wings results in further drag reduction in supersonic region.

2. Drag Reduction Using Wing Modification

The most simple and easy way of reducing the lift induced drag is done by increasing the AR (aspect ratio) of the wing. [9] But the past application of this method in Airbus A340 resulted in compromise of the structural and aerodynamic characteristics. So, this simple way is not the suitable solution. Some of the applicable wing modifications are:

2.1 Vortex Generators

[3] Vortex generators are small aerodynamic devices that are installed on aircraft wings or other surfaces to control airflow and reduce drag. They work by creating small vortices, or swirling patterns of air, that help to keep the flow of air close to the wing surface and prevent it from separating. This improves the overall efficiency of the wing and can help to reduce drag, increase lift, and improve stability and control.

Vortex generators can take many different forms, but they are typically small fins or tabs that are attached to the surface of the wing. They may be installed in a variety of different locations, depending on the specific aerodynamic requirements of the aircraft.

2.2 Extended Tailing Edge

An extended trailing edge is a modification made to an airfoil that involves extending the length of the aft section of the airfoil beyond its original design length. This modification can help to reduce the drag coefficient of the airfoil by delaying the onset of flow separation at high angles of attack, thereby increasing lift and reducing drag. The extension can take various forms, including a simple straight edge extension or a more complex curved extension. The extension can be added to the entire trailing edge of the airfoil or only a portion of it. This modification has been used successfully in various applications, including aircraft wings, wind turbine blades, and hydrofoils. [7].

2.3 Winglet

Winglets are one of the wingtip devices that help in reduction of the induced drag. [10] The airfoil design of the wing makes the airflow above surface to flow with higher velocity and below surface to flow with lower a lower velocity. This change of velocity creates the change in pressure between the two surfaces of same body. This pressure difference provides the positive lift that makes the air travel possible. But at the tips of the wing, the air escapes through the sides and mix with the wind resulting in creation of curl in the airflow that are termed as vortices. These vortices can be reduced by decreasing the pressure inequality at tip of the wing. Winglets equalizes the difference in pressure above and below the wing. Winglets act as a simple and most efficient solution in overcoming the generation of induced drag as there are no drastic changes made to the structure of the aircraft other than the tip of the wing.

3. Types of Winglets

The most common types of winglets are [7]:

- Blended winglets are a type of winglet that is integrated into the wing structure itself, rather than being attached externally to the wing. This design results in a smoother transition between the wing and the winglet, reducing drag and improving aerodynamic efficiency. Blended winglets have been used on a variety of commercial and military aircraft, including the Boeing 737 and the Gulfstream G550. They have been shown to provide significant fuel savings and emissions reductions, making them an important technology for improving the environmental performance of aircraft. Example: B757 & B767 as shown in Fig. 72.1.

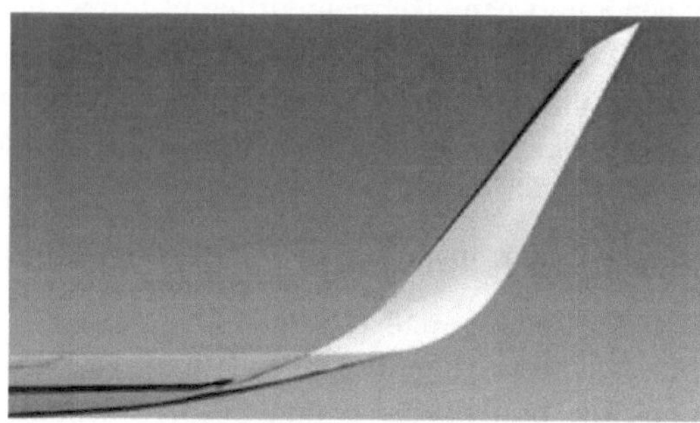

Fig. 72.1 Blended winglet

Source: Created using the experimental work

- Sharklet is a trademarked name for a specific type of winglet designed and manufactured by Airbus. It is a curved, upturned wingtip device that is installed at the end of the wing to reduce drag and improve fuel efficiency. The sharklet gets its name from its resemblance to a shark's fin.

  The sharklet was first introduced on the Airbus A320 in 2012 and has since been added to other Airbus aircraft models. It is designed to improve the aerodynamic performance of the aircraft by reducing the size of the vortex that forms at the wingtip, which in turn reduces drag and increases lift.

  Compared to traditional winglets, the sharklet is longer and more curved, which allows it to create more lift and reduce more drag. It is also made of lightweight composite materials to minimize the added weight to the aircraft. [10] The sharklet can improve the fuel efficiency of an aircraft by up to 4%, which can translate to significant cost savings over the lifetime of the aircraft.] or shark fin alike winglet is almost similar to the blended winglet except its designers. Airbus claims that it was their own unique innovation although they paid heavy penalties to Boeing-Aviation Partners Boeing (APB) partnership in legal battle. Sharklets are found in all latest models of airbus aircrafts including A319, A320, A321, etc..
- This type of winglet is commonly used in modern commercial aircraft designs, such as the Boeing 787 Dreamliner and the Airbus A350 as in Fig. 72.2.

Fig. 72.2 Canted winglet [19]

- Wingtip fences are a type of winglet that consist of a small vertical fence mounted at the outer end of the wing, usually just forward of the aileron. They are designed to reduce drag caused by the turbulent air that is created at the end of the wing. The fence helps to keep this air from flowing over the wing and creating drag, and also reduces the size of the wingtip vortex that is formed at the end of the wing. [8]
- Studies have shown that wingtip fences can reduce drag by up to 3%, and can also improve lift and handling characteristics. They are commonly used on commercial airliners, and can also be found on some military aircraft. Wingtip fences are relatively simple and inexpensive to install, and can provide significant benefits in terms of fuel efficiency and performance. [9] Antanov-158 uses this kind of winglet design as shown in Fig. 72.3.

The spiroid winglet is a relatively new type of winglet that was developed by Boeing and is currently used on some of their 737 MAX aircraft. It has a unique curved shape that provides additional lift and reduces the amount of drag produced by the wingtip vortices. The design of the spiroid

Fig. 72.3 Wingtip fences [20]

winglet is complex and requires advanced computational modeling and testing to optimize its performance. [9] However, it has been shown to provide significant fuel savings and is considered to be one of the most efficient winglet designs currently available. It is found in the Dassault falcon 50 illustrated in Fig. 72.5.

3.1 Application of Winglet

During the period of 1970-1080, NASA research introduced a way to decrease the aerodynamic drag of aircrafts without increasing the wing span. The vertical extensions developed to the wing is termed as Winglets. This shifts the origin of vortices from wingtips of wing to the vertical top end of the winglet and weakens its effect on wing of aircraft. Boeing used this research in their B747s and saved approximately 40 billion gallons of fuel by reducing the fuel consumption and increasing the overall range. This small device also helped in reducing carbon emissions into the environment. [11] After that, there are many developments made to this wingtip device and several purposes are utilized. [12] Some of the uses of winglets are Reduction of induced drag, increase in fuel efficiency, increase in range and endurance, decrease in carbon emission, increased payload capability, generation of more lift force, less noise is produced and improved take-off performance.

Winglets are primarily used to reduce the induced drag of an aircraft. This is achieved by reducing the strength and size of the wingtip vortices, which are generated by the pressure difference between the upper and lower surfaces of the wing. The vortices create drag, which reduces the efficiency and range of the aircraft. [11]

By installing winglets on the wingtip, the vortices are disrupted, and the drag is reduced. This results in a more efficient aircraft, which can fly further and consume less fuel. Additionally, winglets can improve the aircraft's stability and control at low speeds, such as during takeoff and landing. [13]

Winglets are commonly used on commercial airliners, business jets, and general aviation aircraft. They have also been used on military aircraft, such as the F-15 and F-18, to improve their performance and range.

4. Winglet Design Parameters

To successfully design a winglet with highest l/d ratio, accurate mix of some parameters should be made. The parameters are cant angle, sweep angle, taper ratio, toe angle, twist angle, airfoil section, height, span lengthand blend radius [8].

4.1 Cant Angle

[4] Cant angle is defined as angle at which a winglet is inclined relative to the aircraft's longitudinal axis. It is an important parameter that affects the efficiency of the winglet in reducing drag. Typically, the cant angle ranges from 0 to 45 degrees, and the optimal angle depends on various factors such as the wing geometry, operating conditions, and the desired level of drag reduction. A higher cant angle can generate stronger vortices and hence provide greater drag reduction, but it may also increase the structural loads and weight of the winglet. [15] On the other hand, a lower cant angle may have a smaller drag reduction effect but may be lighter and less prone to structural issues. The selection of the optimal cant angle requires a trade-off between these conflicting factors.

4.2 Taper Ratio and Sweep Angle

Toe Angle and Twist Angle

The twist angle refers to the angle between the chord line of the wing and the longitudinal axis of the aircraft. It is usually expressed in degrees and can be positive or negative. A positive twist angle means that the angle of attack of the wing increases from root to tip, while a negative twist angle means that the angle of attack decreases from root to tip. [16-17] Twist angle can affect the distribution of lift and drag along the wing, as well as the stall characteristics.

Toe angle and twist angle are two important parameters used in the design of aircraft wings to optimize their performance.

The toe angle refers to the angle between the longitudinal axis of the aircraft and the axis of the wingtip. It is usually expressed in degrees and can be positive or negative. A positive toe angle means that the wingtip is tilted upward, while a negative toe angle means that the wingtip is tilted downward. [18] Toe angle can affect the amount of lift and drag generated by the wing, as well as the stall characteristics. The toe angle of 5 degree and –5 degree are shown in (a) and (b) respectively in figure 10.

5. Electric Aircraft-Comparative Studies

If the powerplant of an aircraft is powered by electric battery, then the aircraft is termed as electric aircraft. Although the first electric powered manned aircraft made its first flight before three decades in 1973, the complete utilization of such efficient and sustainable technology is not achieved its practical application in reality. As the battery technology is not capable of powering large aircrafts, the electric powered air travel is limited to smaller aircrafts such as Cessna 172 electric, diamond eDA40, Airbus E-Fan, Pipistrel's Velis Electro, etc. Such aircrafts have a major application in training pilots in an economic, efficient and sustainable way. This cuts the training cost up to 75% than training on a piston engine powered Cessna 172 Skyhawk and also environmentally sustainable. The flight characteristics and design parameters of an all-electric two-seater trainer aircraft are:

5.1 Advantages of Electric Aircraft

- Unlike gas turbines, electric motor requires fewer complex components and systems.
- Efficiency of electric motors is around 90% whereas modern turbofans and turboprops have around 35-55% of efficiency.
- *Efficiency:* Electric aircraft have the potential for higher efficiency due to the ability to use regenerative braking and the elimination of heavy fuel systems.

5.2 Limitations of Electric Aircraft

- Although electric flight seems to be very clean and futuristic, the energy storage technology required for the electric propulsion is inadequate for commercial air travel.
- Applications of all electric propulsion is limited to urban air-taxi mobility, regional logistics, recreation and pilot training. [13]

5.3 Background

As current possible electric propulsion is limited to short range, less endurance and speed, very few applications is really efficient and economical in reality. One of such application is Pilot training which is obvious while considering the type certification of Pipistrel's veils Electro trainer aircraft by the European (EASA) and the American (FDA) air safety agencies. Although many leading companies all over the world started the research and development of an electric trainer aircraft, this Slovenian manufacturer made the reliable product which will lead the pilot training industry over coming years. [17] This will be a positive impact on the aspiring pilots who need to train for hundreds of hours to get license on their own cost of fuel. Almost 70% of the training cost will be cut if training is made by an electric aircraft. This project concentrates on further improvements of this kind of electric aircrafts as it impacts pilots who are directly connected to the aviation industry. An electric trainer is specifically designed in a manner that helps a pilot to learn the basic lessons of flight with additional safety features, simple cockpit controls and lenient flight characteristics. [18] These are required to allow the pilot-in-training advance his/her skillset without any endangerment. Some of the aircrafts possessing these characteristics are,

Pipestral Velis Electro

- Pipestral Alpha Electro
- Diamond eDA40
- Cessna 172 Electric

6. Winglet Design Selection for Electric Aircraft

Considering everything we learnt from the above studies, conclusions have to be made with the selection of suitable kind of winglet design which provides dominant performance improvements in a low-speed single engine two-seater aircraft with maximum speed around 180 km/h and MTOW not exceeding 600 kg. [13] For the research, we have selected the wing of Cessna 172 electric. Winglet designs that are selected for modification of wing are:

- Blended winglet:

  The Boeing company's Chief of Aerodynamics, Dr Louis Gratzer with reference of works published by the NASA engineer Richard Whitcomb designed and filed patent for blended winglets. In 1988, Boeing 747-400 was launched with this blended winglet design which became first airliner to acquire this technical advancement in wing design.[8]

- Sharklet or shark fin alike winglet:

  The Airbus A320neo inspired wingtip device called sharklets are winglets that helped airlines around the world save millions of gallons of jet fuel every year. This is almost similar to the blended winglets except some cosmetic changes made to the exterior design.

- Split-tip winglet:

  The Boeing 737MAX's AT (advanced technology) winglet called as split winglet or split-scimitar winglet which is more efficient than any other winglets in the market. It has the both upper wingtip and a ventral strake that gives it unique looks.[4]

7. Cant Angle Variation of Winglets

The angle changes of the winglet have drastic to significant difference in the effectiveness of induced drag reduction, lift generation, stability of flight etc. [5] Based on the literature review and some research some of the observations are made. The graph plotted between the Angle of attack and the Cl/Cd ratio for base wing and different wing configurations of cant angle and sweep angle is shown in Fig. 72.11.

- At lower speeds and lower angle of attacks AOA, the cant angle of 0° dominates the wing's aerodynamic performance with more drag reduction and lift generation.

8. Conclusion

All the previous research articles and review papers related to induced drag reduction using winglets are studied. Basic knowledge of all the types of drag, drag reduction techniques, wing modifications, winglets and its types, winglet selection, electric aircrafts, winglet design parameters is realised and presented through this review paper. From the results from literature review and comparative studies, it is clear that the split winglet and upward sharklet has more advantages than any other types of winglet for our application. Upward sharklet winglet design will be best suited for small aircrafts as it doesn't affect the ground clearance during take-off and landing. With all the outline and demonstrations studied, we can further proceed to design various winglets to the wing of an electric aircraft and concluding with best winglet type using CFD simulations & analysis.

References

1. NASA Glenn Research Center, 21000 Brookpark Road Cleveland, OH 44135 Phone: (216) 433–4000.
2. Ashill, P. R., Fulker, J. L., Shires, A., (1992) "A Novel Technique for Controlling Shock Strength of Laminar Flow Aerofoil Sections" First European Forum on Laminar Flow Technology, Hamburg, Germany, March.
3. Barger, R. L., (1992) "A method for designing blended wing-body configurations for low wave drag" NASA TP-3261.
4. Barnwell, R. W., and Hussaini, M., (1992) "Natural Laminar Flow and Laminar Flow Control" (Springer-Verlag).
5. Bauer, S. X. S. and Hernandez, G., (1988) "Reduction of cross- flow shock induced separation with a porous cavity at supersonic speeds" AIAA paper 88–2557.
6. Bauer, S. X. S. and McMillin, S. N., (1988) "Experimental and theoretical study of the effects of wing geometry on a supersonic multi-body configuration" AIAA paper 88–2510.
7. Zhang, Y., Xiao, Z., Chen, Y., & Li, Y. (2018). Numerical investigation on the effects of winglets on the aerodynamic performance of a blended wing body. Aerospace Science and Technology, 76, 387–395.
8. Pan, J., Zhang, Y., Liu, X., Xu, Y., & Sun, Y. (2021). Numerical analysis of winglets effects on the aerodynamic performance of UAVs. Applied Sciences, 11(4), 1668.
9. Afshari, M., Vahdati, M., & Esmaeilpour, R. (2020). Numerical study of the effect of different winglet configurations on the performance of a regional aircraft. Modares Mechanical Engineering, 20(1), 47–58.
10. Zaman, K. B. M. Q., Ahmed, M. R., & Ali, M. Y. (2017). Aerodynamic analysis of winglets at different angles of attack using computational fluid dynamics. International Journal of Engineering Research and Technology, 6(9), 523–527.
11. Balaji, G., Bharath Kumar, A., Divya, R., Boopathy, G., Seenu, N., & Santhosh Kumar, G. (2023, August). Experimental Investigation of Double Delta Wings with Different Angles of Attack at Subsonic Speeds. In *International Conference on Smart Sustainable Materials and Technologies* (pp. 197–202). Cham: Springer Nature Switzerland.
12. Wang, Y., Zhang, J., Cai, J., & Han, Y. (2021). Numerical study on the effects of different winglet shapes on the aerodynamic performance of the electric aircraft wing. Journal of Physics: Conference Series, 1739(1), 012045.
13. Saadatpo, M., & Ghomashi, M. (2018). A numerical study on aerodynamic effect of winglets on an airplane wing. Journal of Applied Fluid Mechanics, 11(4), 1039–1048.
14. Madhanraj, V., Balaji, G., Vignesh, R., Gokul, P., Ashish, S., Gokul Shree, G., ... & Prasad, G. (2023, August). Effect of Leading-Edge Shapes in NACA2421 Aerofoil with Different Angles of Attacks. In *International Conference on Smart Sustainable Materials and Technologies* (pp. 153–159). Cham: Springer Nature Switzerland.
15. Sharifpur, M., & Ghorbani, M. (2016). Numerical investigation on the effect of winglet height on aerodynamic characteristics of a wing in ground effect. Modares Mechanical Engineering, 16(2), 109–117.
16. Suresh, A. S., & Senthil Kumar, P. (2017). Experimental investigation on the effect of winglets on the aerodynamic performance of a wing. International Journal of Mechanical and Production Engineering Research and Development, 7(3), 605–612.
17. Balaji, G., Catherine Victoria, P., Solaiyappan, G., Mano, R. T., Santhakumar, U., Santhosh Kumar, G., ... & Hassan Ansari, R. H. T. (2023, August). Effect of Pressure Distribution of NREL S809 Airfoil with Vortex Generator. In *International Conference on Smart Sustainable Materials and Technologies* (pp. 161–168). Cham: Springer Nature Switzerland.
18. Bharti, P. K., & Dolas, P. R. (2018). Analysis of winglet on flow field over a wing using CFD. International Journal of Recent Technology and Engineering, 7(6), 86–90.
19. Internet Source: VIET Flight Training, https://eng.bayviet.com.vn/all-about-airplane-winglets-and-how-to-tell-them-apart/
20. Internet Source: Quora, https://www.quora.com/What-are-the-noticeable-differences-between-an-Airbus-A320-and-a-Boeing-737

Advances in Additive Manufacturing Technologies – Gurusamy Pathinettampadian (eds)
© 2024 Taylor & Francis Group, London, ISBN 978-1-032-90013-1

73 Shock Formation on Double Ramp Surface in Supersonic Wind Tunnel

Abdul Wajid Ali,
Weketshe Thopi, Sheikh Sameer
Student, Department of Aeronautical Engineering,
Hindustan Institute of Technology and Science,
Chennai, India

Thillaikumar T, Balaji G*
Assistant Professor, Department of Aerospace Engineering,
Hindustan Institute of Technology and Science,
Chennai, India

Ramanan N
Assistant Professor, Department of Mechanical Engineering,
Sri Jayaram Institute of Engineering and Technology,
Chennai, India

Santhosh Kumar G
Teaching Fellow, Dept. of Mech. Engg.,
University College of Engineering, Bharathidasan Institute of
Technology Campus, Anna University,
Tiruchirappalli, Tamilnadu, India

Vijayanandh Raja
Department of Aeronautical Engineering,
Kumaraguru College of Technology, Coimbatore,
Tamil Nadu, India.

Gurusamy P
Professor, Department of Mechanical Engineering,
Chennai Institute of Technology, Chennai, India

ABSTRACT: The visualization of shock structures formed over the rampmodel isstudied using the schlieren set up. In thisstudy, two different double ramp models are fabricated and tested at supersonic wind tunnel. The first ramp angle (θ_1) is maintained 15° forall the cases, but secondramp angle (θ_2) is varied as 20°, 24°, 28°, 32° for respective ramp models. The test is conducted at super sonic wind tunnel at stagnation pressure of 12.5 bar, pressure of 12.5 bar, so the corresponding Machnumber would be 2.5 to visualize the shock structure formed over different double ramp models, the schlieren system has been utilized. From the capture dschlieren images, it is observed that the oblique shock wave

*Corresponding author: Balaji.G, gbalajihits@gmail.com, 8610559691

DOI: 10.1201/9781003545774-73

formed at the leading edge of the different double ramp models. Also, it is notice that with increment in second ramp angle the detached shock forms at the compression corner. It is found that the axial distance of intersection points of leading edge and compressor corner shock moving up stream direction. Furthermore, the study aims to understand the influence of the double ramp angle on shock formation and shock boundary layer interaction. The shock formation is studied by analyzing the schlieren image and experimental data. This research provides an insight for the understanding shock wave dynamic over the double ramp surface with different angle in supersonic flow.

KEYWORD: Double ramp, Supersonic wind tunnel, Mach number, Ramp angle

1. Introduction

Shock wave/boundary layer interaction (SWBLI) plays an crucial role in the field of high speed aerodynamic i.e supersonic and hypersonic. The effect of SWBLI over a solid surface (compression ramp) lead to a significant effect on the performance, stability and unsteady flow separation. The experiment is carried out in the supersonic wind tunnel with a Machnorange from 2.5 to 3. The schlieren technique facilitates the visualization of shock wave formation and enables the study of shock characteristics such as strength, position, and interaction with the ramp surfaces.

Oblique and strong shock formation: An oblique shock wave occurs when a supersonic flow encounters an obstacle at an angle, causing compression and changes in flow properties such as pressure, temperature and density. It flows at an angle to the upstream flow direction and can visualized in a schlieren setup. A normal shock wave on the other hand forms when a supersonic flow encounters an obstacle head an, perpendicular to the flow direction. This results in a sudden and drastic change in flow properties across the shock wave including a significant increase in pressure and temperature. Normal shocks wave is crucial in various aerodynamic applications such as in supersonic and hypersonic aircraft design.

In this experiment, four models of double ramp surface have been taken. It has a very harptip edges. Here for analyzing the shock formation with the help of supersonic wind tunnel and capturing the shocks by schlieren imaging process. two model to be tested having the two ramp angles for each model

- Model 1-$\theta_1 = 15°$ and $\theta_2 = 20°$
- Model 2-$\theta_1 = 15°$ and $\theta_2 = 24°$

Observation and visualization of the oblique shock wave help us to understand the properties of the inlet air and shock behavior on the surface of the double ramp model. It help us to understand the advantages of weak shock formation on sharp edges of the intake, where the flow properties will not decreases suddenly as compare to the normal shock formation. The benefit of having sharp edges with two wedge surface or ramp angles as it give the formation of oblique shock. Here the supersonic flow will continue, having some loss of flow properties, after the last ramp angle. So the design of our model is accurate for maintain the supersonic flow. We had used supersonic wind tunnel for testing our models and schlieren images for investigating the shock-shock interaction and shock-boundary layer interaction. Objective of the work is to investigate and visualize the shock formation on double ramp surface with the help of supersonic wind tunnel, and schlieren images techniques. The required calculation had been done for obtaining the deflection angle, Mach number and shock-shock intersecting point. There are several application and modification can be done in future for obtaining the supersonic flow, example double wedges air foil, exhaust nozzle, Delta wing, combustor etc.

2. Literature Survey

Sun-et-al (2020) Shock wave/boundary layer interaction (SWBLI) plays an crucial role in the field of high speed aerodynamic-transonic and supersonic. The effect of SWBLI over a solid surface (compression ramp) lead to asignificanteffectontheperformance,stabilityandunsteady flow separation[2,3]. At a Machnumber (Ma) of 2.0, the study examines shock-wave/boundary-layer interactions (SWBLIs) at compression ramps with variable ramp angles (α between 20 and 30 degrees). Based on boundary-layer thickness, two Reynolds numbers (Re1 = 18,600 and Re2 = 35,600) are taken into consideration. 20 kHz high-speed Schlieren imaging is used as an inspection of flow. It observes thatincreasingthecompressionrampangleleadtothe separation shock wave upstream. The reattachment shock gets stronger with increase in compression ramp angle.

Combsetal (2017) The researcher aims to create post-processing techniques for studying unsteady shock motion using high-speed schlieren imaging and to provide insight on the unsteadiness associated with SWBLIs[2,4].

Supersonic wind tunnel with a Mach 2 is utilized for this experiment. Two test cases were examined to study SWBLI: one at 20d downstream with a turbulent boundary layer (case1) and another at 6d downstream with a transitional boundary layer (case 2). A cylindrical model was fixed toa flat plate in the wind tunnel. The shock structure and interaction dynamics were visualized using Schlieren imaging in conjunction with high-speed pressure measurements. The study shows the correlation between the schlieren- based shock-position measurements and the surface- pressure measurements because large scale fluctuating is likely equivalent. Though the shock flow down stream of the transducer, the pressure fluctuating is still correlate with the shock motion.

Reshma et al(2021) The study uses both numerical and experimental analysis to understand normal shock wave along a convex–concave surface with equal radii. These results provide understanding on the flow phenomena connected to shock wave transmission through a coupled convex–concave surface. It is observe that the geometry of the ramp plays a major role in shock wave transitions over it. This implies that the ramp's curvature and shape are very important in influencing how the shock wave behaves as it travels across the surface. It is found that the features of the stationary shock wave are affect by the shock wave Mach number and also affect the flow separation in the concave area of the ramp surface.

Tang et al (2021) The journal paper describes an experimental investigation that used a plasma actuator array to manage the interactions between shockwaves and boundary layer son compression ramps.

3. Experimental Setup

Wind-Tunnel and Test Models

An in let is the entrance section of a wind tunnel where air is drawn in to create a controlled and simulated airflow for testing purposes where the air enters into the three cylinders of the two belonging compressors. Here the compressors will increase the pressure upto 12.5 bar and maintaining the high-speed air flow which is necessary for testing the objects or models at supersonic speeds. The inlet air enters into the two cylinders of the compressor where the air is carried out by the manifold and enters into the third cylinder. Here the final pressure will build upto 12.5 bar. Then the air will come out from the outlet and passes through the pipeline to the NRV (Non Return Valve), where it reaches to the first reservoir and where it contains 1.5 cubic meter of the compressed air and then again the compressed air enters into the cooler, where the cooler condense the moisture and reduce the temperature.

After the cooler the compressed and condensed air enters into the oil filter. Where it purifies the air by removing the lubricant, dust and moisture. From here it enters into the drier where it reduces the temperature up to 3 degree and again the condensation happens further. For removing the complete dust, it enters into the oil and air filter. The purified, compressed and condensed air get filled into these condreservoir where it contains 1.5 cubic meter air and then passesto thethird reservoirwheretheair gets filled up to 2 cubic meter and the pressure develops up to 12.5 bar for operating the wind tunnel fortesting. Here then the air will come inside the valve section for releasing the pressure and enter into the settling chamber (cylindrical section to rectangular section). After this air enters into the C-D nozzle and model section where air gets choked in C-D nozzle. From C-D nozzle air enters into the diffuser section, where the velocity will reduce and finally air will exhaust from exhaust section.

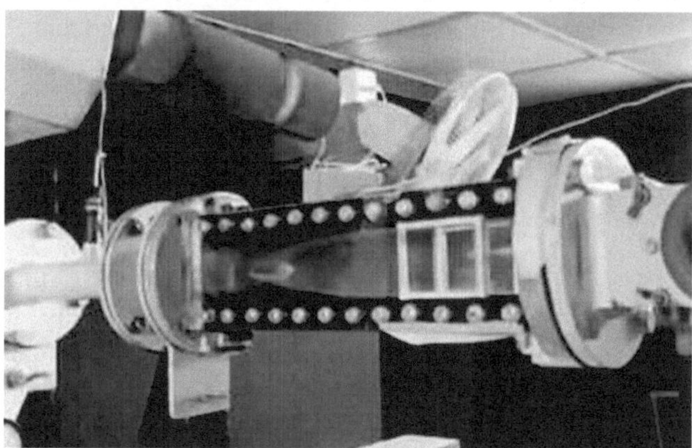

Fig. 73.1 C-D nozzle

We are using four test models, having the ramp angles for the first model is $\theta_1 = 15°$, $\theta_2 = 20°$, for the second model $\theta_1 = 15°$, $\theta_2 = 24°$, for third model $\theta_1 = 15°$, $\theta_2 = 28°$, for fourth model $\theta_1 = 15°$, $\theta_2 = 32°$. Having the length of all four model is 15 mm. The stainless-steel material is used for fabricating our models.

4. Schlieren Setup

For testing our four models the free stream Mach numbers will be $M_0 = 2.90$, $M_0 = 2.70$, $M_0 = 2.70$, $M_0 = 2.60$ for investigating the shock-shock interaction and shock–boundary layers interaction. Here pin hole camera with light source of mercury is used. It allows the light to fall on front concave mirror (schlieren mirror), from the concave mirror the light will reflect and cross the test section. Due to pressure gradient inside the test section the light will reflect in such a way that there will be dark and bright patches and its fall on the opposite side of the concave mirror or schlieren mirror. There flection of the opposite side of the concave mirror is captured in the form of shockwave with the help of video camera. (Nikon D5600) having 24.2MP DX –format CMOS sensor. EXPEED four image processor and with having the 1080p video capability.

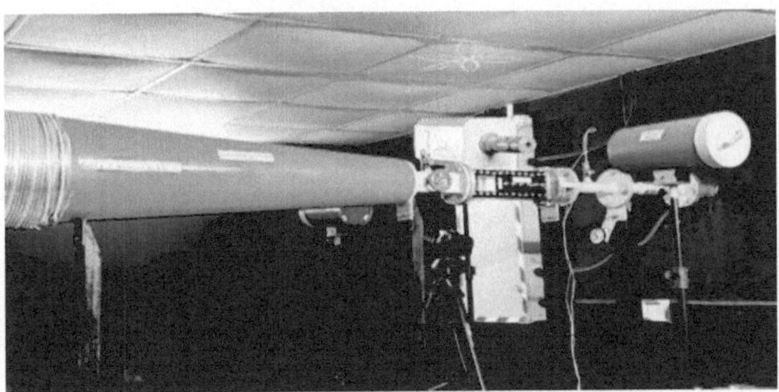

Fig. 73.2 Image of schlieren setup

5. Design and Fabrication of Model

Design of double ramp model: Here designing of double ramp model has done by using AutoCAD2022 version by giving the dimensions and angles in each particular models

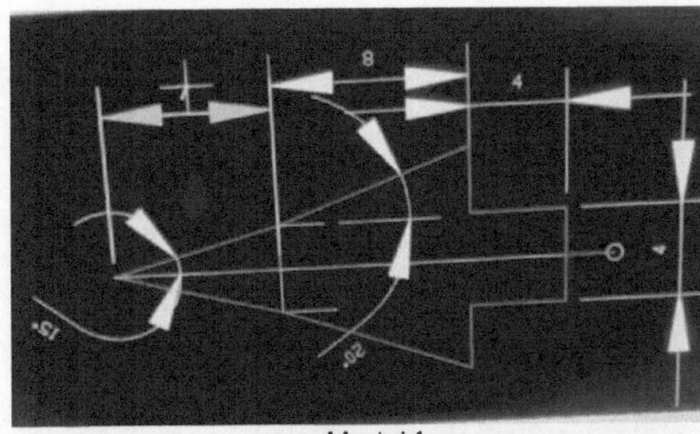

a. Model 1

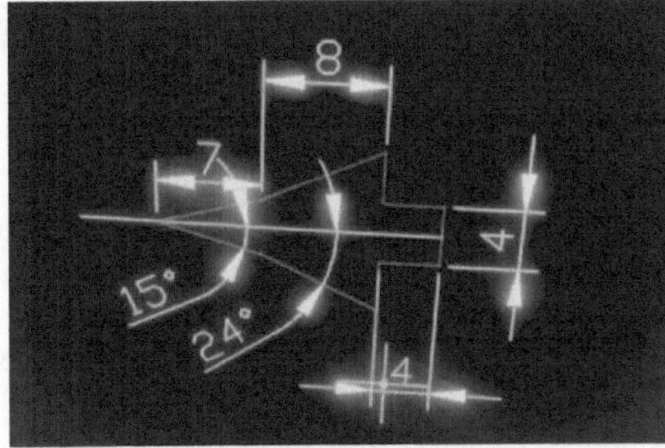

b. Model 2

Fig. 73.3 Design of double ramp models

Model Fabrication: Model fabrication has been done by using stainless steel having the grade for fabrication includes 304, 316.

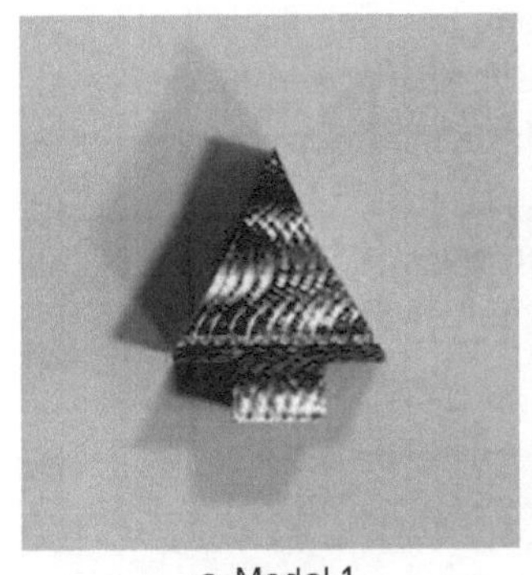

a. Model-1

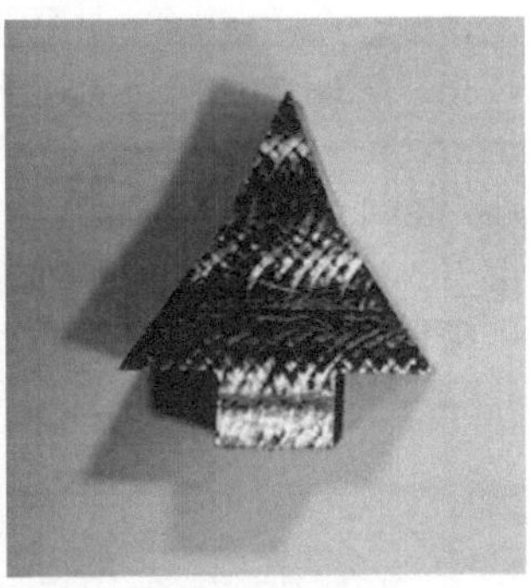

b. Model-2

Fig. 73.4 Fabrication of double ramp models

6. Schlierenimages Discussion

The first experiment of the model one having $\theta_1 = 15°$ and $\theta_2 = 20°$ so, at the first ramp angle $\theta_1 = 15°$, the free stream mach number is $M_0 = 2.90$ and the angle of deflection of shock $\beta_1 = 33.3°$, but after the formation The second experiment of the model having ramp angles of $\theta_1 = 15°$ and $\theta_2 = 24°$, so that the free stream velocity will be of Mach number M0 = 2.70 and the deflection angle $\beta_1 = 35°$. If the ramp angle increases to $\theta_2 = 24°$ so the deflection angle is $\beta_2 = 42°$. Here as we increases the ramp angle θ, deflection angle β also increases but the strength of the shock will decrease. Here θ_2 is beyond the θ max so the detached shock is formed.

Since across the normal shock the flow becomes always subsonic and, the flow properties will change in drastic manner. But, across the oblique shock, the flow Mach number either subsonic or supersonic depends upon strength of the shock wave. When the single ramp model employed then, single strong oblique shock formation occurs. So, this is a possibility for the flow to become subsonic behind the strong oblique shocks. But it is desired to retain the flow Mach number still supersonic behind the shock wave or at the exit of scramjet in take. To achieve this, unsteady of having single strong oblique shock, multiple weak shocks are created with double ramp model.

In model 1, the actual length of the model is 15mm (1.5 cm) and image length of the model by scale measurement is 60 mm (6 cm) so by taking the ratio between the actual length of the intersecting point of the shocks β_1 and β_2 as (L_1) and the scale measurement length of the intersecting point of the shocks β_1 and β_2 as (L_2) i.e $L_1 = 0.25\ L_2$. Putting L_2 for the first model scale measurement is 5.5 cm, so the actual length of the intersecting point of the two shocks L_1 is 1.375 cm (13.75mm).

In model 2, the actual length of the model is 15mm (1.5 cm) and image length of the model byscale measurement is 60 mm (6 cm) so by taking the ratio between the actual length of the intersecting point of the shocks β_1 and β_2 as (L_1) and the scale measurement length of the intersecting point of the shocks β_1 and β_2 as (L_2) i.e $L_1 = 0.25\ L_2$. Putting L_2 for the first model scale measurement is 4.5 cm, so the actual length of the inter secting point of the two shocks L_1 is 1.125 cm (11.25 mm).

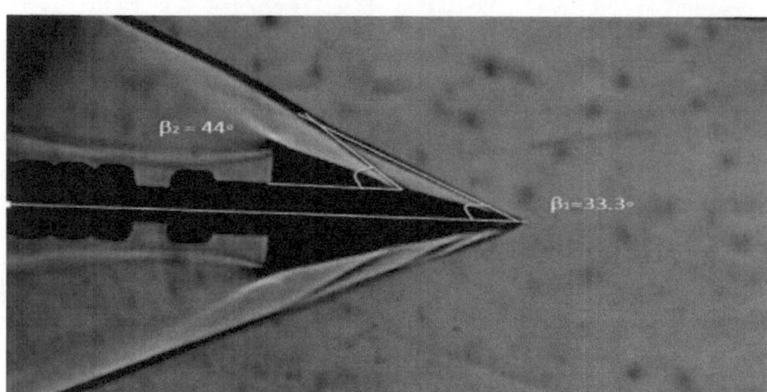

Fig. 73.5 Represent the oblique shocks formation and shocks intersection point, Model-1

Fig. 73.6 Represent the oblique shocks formation and shocks intersection point, Model-2

7. Table Summary

Table 73.1 Summary of the shocks formation

Theoretical calculation				
Models	θ_1		θ_2	
	M_0	β_1	M_1	β_2
Model1	2.34	39.64°	1.37	detached
Model2	2.34	39.64°	1.37	detached
Experimental calculation				
Models	θ_1		θ_2	
	M_0	β_1	M_1	β_2
Model1	2.90	33.3°	2.18	44°
Model2	2.70	35°	2.03	42°

Summary of Shock Formations

Table 73.2 Summary of the experimental shocks formation

Experimental observation			
Model	β_1	β_2	**Oblique shocks behaviors**
1	33.33°	44°	θ_1 is attached and θ_2 is attached
2	35°	42°	θ_1 is attached and θ_2 is detached

Shock Interactions Points

Table 73.3 Shocks intersecting point

Experimental observation				
Model	**Model-length (cm)**	**Image length (cm)**	**Ratioo factual length (L_1) and intersecting length (L_2)**	**Intersect-ing point (mm)**
1	1.5	6	$L_1 = 0.25\ L_2$	13.75 mm
2	1.5	6	$L_1 = 0.25\ L_2$	11.25 mm

Note: θ_1 = 15° for all two models and θ_2 = 20°, 24°, 28°, 32°.

8. Graphs

Here graph is plotted between two dimensionless number. Ratio between the intersecting point and actual length of model with the ratio of ramp angles of the model.

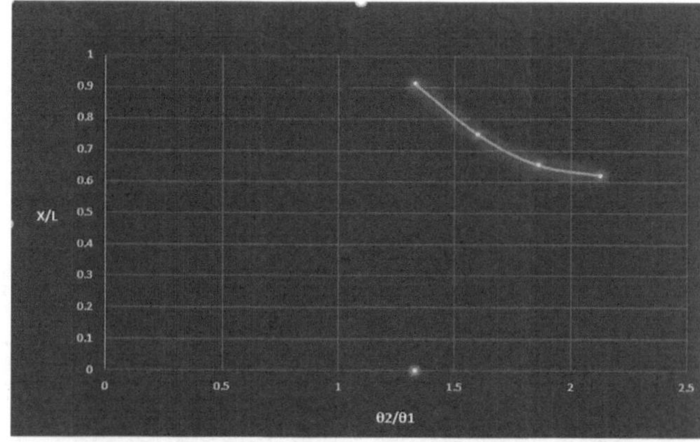

Fig. 73.7 Represent graph between two non dimensional numbers

1. This graph is between two dimension less number.
2. It is the ratio between the intersecting point and actual length of model with the ratio of ramp angles of the model.
3. For the geometrical point of view for higher x/h higher θ_1/θ_2 was chosen.
4. If higher x/h lower θ_1/θ_2 is preferable.

9. Conclusion

Overall, this project enhances us to investigate and visualize the formation of shock wave. At supersonic speed the shock will form at the inlet section. This shock may be the normal shock, but due to change in the geometry of the inlet we can overcome from the normal shock and also our flow properties will not fall drastically. The sharp edge having double wedge surface give us the oblique shock formation on the inlet. Here we can also use calculation theoretically and experimentally for investigating the Machnumber before and after the shock's formation on double wedge surface of the model. We also find the calculation with the help of online protractor and scale, to know the intersecting point of the two shocks.

REFERENCE

1. Combs, C. S., Kreth, P. A., Schmisseur, J. D., & Lash, E. L. (2018). Image-based analysis of shock-wave/boundary-layer interaction unsteadiness. *AIAA journal*, *56*(3), 1288–1293.
2. Dolling, D. S. (2001). Fifty years of shock-wave/boundary-layer interaction research: what next?. *AIAA journal*, *39*(8), 1517–1531.
3. Sun, Z., Gan, T., & Wu, Y. (2020). Shock-wave/boundary-layer interactions at compression ramps studied by high-speed schlieren. *AIAA Journal*, *58*(4), 1681–1688.
4. Balaji, G., Bharath Kumar, A., Divya, R., Boopathy, G., Seenu, N., & Santhosh Kumar, G. (2023, August). Experimental Investigation of Double Delta Wings with Different Angles of Attack at Subsonic Speeds. In *International Conference on Smart Sustainable Materials and Technologies* (pp. 197–202). Cham: Springer Nature Switzerland.
5. Clemens, N. T., & Narayanaswamy, V. (2014). Low-frequency unsteadiness of shock wave/turbulent boundary layer interactions. *Annual Review of Fluid Mechanics*, *46*, 469–492.
6. Landsberg, W. O., Curran, D., & Veeraragavan, A. (2022). Experimental flameholding performance of a scramjet cavity with an inclined front wall. *Aerospace Science and Technology*, *126*, 107622.
7. Swift, D. C., Kraus, R. G., Loomis, E. N., Hicks, D. G., McNaney, J. M., & Johnson, R. P. (2008). Shock formation and the ideal shape of ramp compression waves. *Physical Review E*, *78*(6), 066115.
8. Robinet, J. C. (2007). Bifurcations in shock-wave/laminar-boundary-layer interaction: global instability approach. *Journal of Fluid Mechanics*, *579*, 85–112.
9. Sansica, A., Sandham, N. D., & Hu, Z. (2014). Forced response of a laminar shock-induced separation bubble. *Physics of fluids*, *26*(9).
10. Knight, D. D., &Degrez, G. (1998). Shock wave boundary layer interactions in high Mach number flows a critical survey of current numerical prediction capabilities. *AGARD Advisory Report Agard Ar*, *2*, 1–1.
11. Hornung, H. G., & Robinson, M. L. (1982). Transition from regular to Mach reflection of shock waves Part 2. The steady-flow criterion. *Journal of Fluid Mechanics*, *123*, 155–164.
12. Hornung, H. G., Oertel, H., & Sandeman, R. J. (1979). Transition to Mach reflexion of shock waves in steady and pseudosteady flow with and without relaxation. *Journal of Fluid Mechanics*, *90*(3), 541–560.
13. Chpoun, A., Passerel, D., Li, H., & Ben-Dor, G. (1995). Reconsideration of oblique shock wave reflections in steady flows. Part 1. Experimental investigation. *Journal of Fluid Mechanics*, *301*, 19–35.
14. Edney, B. (1968). *Anomalous heat transfer and pressure distributions on blunt bodies at hypersonic speeds in the presence of an impinging shock* (No. FFA-115). Flygtekniska Forsoksanstalten, Stockholm (Sweden).
15. Wieting, A. R. (1989). Experimental study of shock wave interference heating on a cylindrical leading edge.
16. Holden, M., Moselle, J., Wieting, A., & Glass, C. (1988, January). Studies of aerothermal loads generated in regions of shock/shock interaction in hypersonic flow. In *26th Aerospace Sciences Meeting* (p. 477).
17. Billig, F. S. (1967). Shock-wave shapes around spherical-and cylindrical-nosed bodies. *Journal of Spacecraft and Rockets*, *4*(6), 822–823.
18. Grasso, F., Purpura, C., Chanetz, B., &Délery, J. (2003). Type III and type IV shock/shock interferences: theoretical and experimental aspects. *Aerospace Science and technology*, *7*(2), 93–106.
19. Olejniczak, J., Wright, M. J., & Candler, G. V. (1997). Numerical study of inviscid shock interactions on double-wedge geometries. *Journal of Fluid Mechanics*, *352*, 1–25.
20. Dupont, P., Haddad, C., Ardissone, J. P., &Debieve, J. F. (2005). Space and time organisation of a shock wave/turbulent boundary layer interaction. *Aerospace science and technology*, *9*(7), 561–572.
21. Chauhan, J., Balaji, G., Swastikar, M., Boopathy, G., Sangeetha, S., Santhosh Kumar, G., & Pradeep, G. M. (2023, August). Numerical Investigations of Aerodynamics Performance of Blunt Nose Cone with Aerodisk at Hypersonic Flow. In *International Conference on Smart Sustainable Materials and Technologies* (pp. 177–188). Cham: Springer Nature Switzerland.

Note: Every figure and table was created using the experimental work; none of them were taken from any publications or online.

Advances in Additive Manufacturing Technologies – Gurusamy Pathinettampadian (eds)
© 2024 Taylor & Francis Group, London, ISBN 978-1-032-90013-1

74 Topology Optimization of Electric Solar Vehicle Brake Pedal

S. Pavithra, A. Suresh*,
R. Ganesha moorthy, A. Parthiban,
D. Dinesh, Dharanidharan P
Department of Mechanical Engineering,
Chennai Institute of Technology, Chennai, India

ABSTRACT: In the present environmental conditions, the automotive industries are moving towards the lightweight and option in contrast to petroleum and diesel. It is because of the emission of hazards from the exhaust after combustion. Along these lines, in that as a primary concern, the industries are moving towards the solar-powered vehicle. In lightweight, topology optimization serves, as the purpose is to reduce the weight of the components of the vehicle. It ultimately leads to reduces weight of the vehicle. This paper describes the topology optimization of the Electric solar vehicle brake pedal without disturbing the necessitates. Here the model is designed by using the CATIA software, the analysis is carried out in the Ansys software. The main objective of this paper is to reduce the weight of the brake pedal. The previous model is to found over designed as per the finite element analysis. The results of the existing and optimized are more congruent in values and it can be a replacement in the brake pedal. This paper may aggregate the utilization of lightweight vehicles and solar-powered vehicles.

KEYWORDS: ANSYS, Brake pedal, Solar vehicle, Topology optimization

1. INTRODUCTION

The automobile is one of the most developing industries in the world. In recent times, their sight had fallen in the possession of the lightweight, cost-efficient, and eco-friendly vehicle. In general, the automobile is the vehicles majorly used in transportation purposes. It tends to be either merchandise or individuals. The first initiative in the automotive industry is in the invention of the engine and it paved as a way for the engineers and scientists to make their inventions.Nowadays the conventional automobile vehicle is replaced with electric and solar-powered vehicles. In the conventional automobile vehicle engine was the main source of the vehicle, which convert the fuel into energy for the locomotion of the vehicle whereas nowadays the electric and solar-powered vehicle use battery as their fuel and instead of engines they use motors. The conventional automobile vehicle causes pollution. The electric and solar-powered vehicles are ecofriendly.

2. METHODODLOGY

The topology optimization is one of the mind-blowing technique, which was developed in recent years. It may in-

*Corresponding author: suresha@citchennai.net

DOI: 10.1201/9781003545774-74

volve the removal of materials from unwanted places and thereby reducing the cost and weight of the component. Before moving on to the optimization, the model of the brake pedal is to be designed. The design can be done by using any CAD packages like CATIA, AutoCAD, SOLIDWORKS, etc. After the design is completed, then it is transferred to the analysis.

Here, the analysis is carried in the ANSYS software. Before the analysis, The model is to make into a suitable form for the input of the analysis software. Then by selecting the Engineering data, the material property is assigned to the model. Then, the discretization of the model is done by the mesh. In these options, it may have coarse, fine, and medium mesh. Based on the requirement the mesh size is selected. Then the boundary conditions are applied, wherever in the places necessary.[3] Then the equivalent stress is calculated and the places where the deformation and unwanted materials are noted. After solving the problem, the designed model is to be redesigned in the Space claim designer module.[4] Then the same analysis is carried out, then results between the existing design and modified design are calculated. Figure 74.1 represents the methodology of the topology optimization in the ansys software.

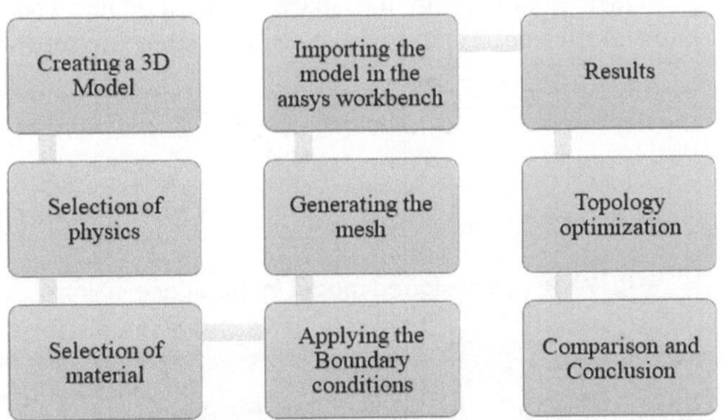

Fig. 74.1 Methodology in the topology optimization in ansys software

3. Brake Pedal Modeling

Before moving to the analysis, the design is to be done. It can be generated by using any CAD packages like CATIA, SOLIDWORKS, PROE, CREO, FUSION 360, AUTOCAD, etc... The design of the brake pedal plays a vital role in this as it results in the force applied to the tandem master cylinder. The 3d model of the brake pedal is been created using the CATIA V5 with the pedal ratio of 6:1 and considering space constrains, resting position of the driver's leg, to achieve the full stroke length of the tandem master cylinder. The pedal ratio indicates the force applied to the brake pedal and force transmitted to the tandem master cylinder by the brake pedal. The pedal is designed, that not to create any fatigue in the driver's leg. It should not be too long or small, it must have a proper dimensional design. Figure 74.2 and 74.3 represents the top and side view of the brake pedal.

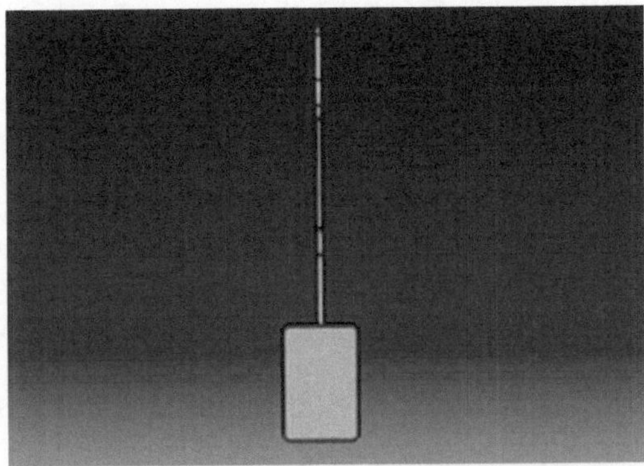

Fig. 74.2 Top view of the brake pedal in Catia software

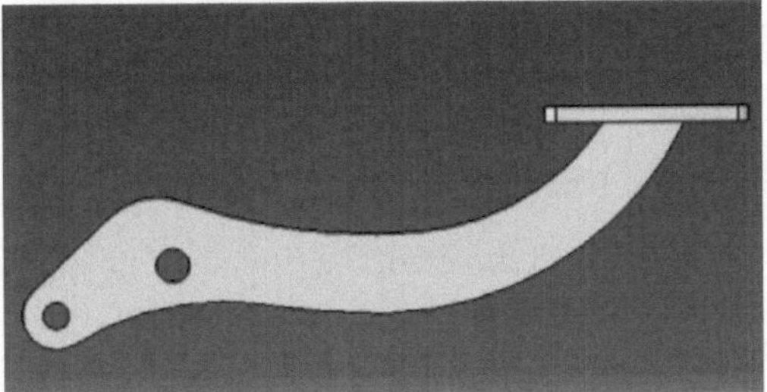

Fig. 74.3 Side view of the brake pedal in Catia software

4. Analysis of Brake Pedal

The analysis and topology optimization are carried out in the ANSYS software. The Ansys is a finite element analysis software. In general, Ansys software is used for the numerical simulation of computer-modeled structures for analyzing real-life problems. It is a multi-physics problem-solving structure. After completing the design, it must be converted into a suitable format for the Ansys software. In this analysis, the brake pedal is converted into the STEP file format. Then the Selection of physics is done, the static structural analysis is chosen and moved to the project selection window. Then the material, mild steel is applied to the brake pedal.

After applying the material, the geometry is to done. It may either model in the Ansys workbench or the model is imported from ay cad software. Figure 74.4 represents the imported model on the Ansys workbench window. After geometry, comes the model. The model is to be meshed based on the required shape and types. It may include a hexahedral, tetrahedral, etc. The Fig. 74.5 represents the meshed model of the brake pedal in the Ansys workbench.

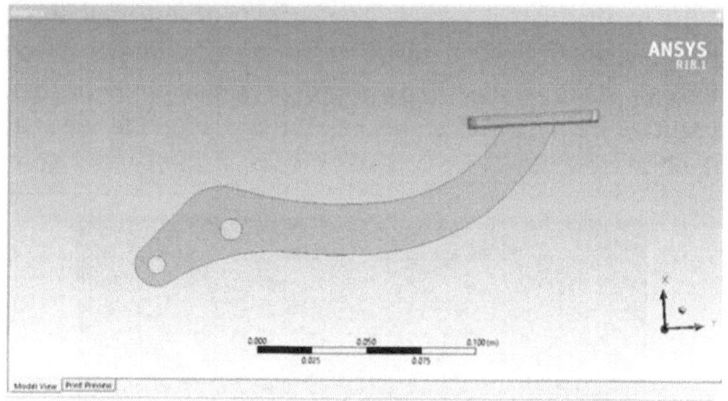

Fig. 74.4 Imported geometry in ANSYS workbench

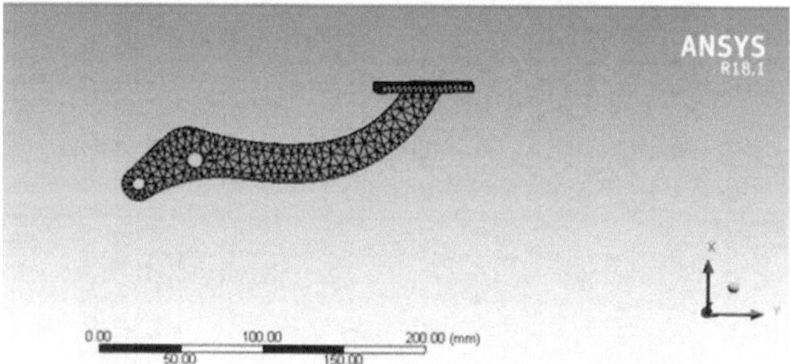

Fig. 74.5 Meshed geometry in ANSYS workbench

After the mesh is completed, the boundary conditions are to apply. It may include the cylindrical support at the end and the load acting at the brake pad is 150 N. Then the unknown parameters are to be calculated. Then solve the problem, the required parameters are obtained. Figure 74.6 and 74.7 represents the deformation and equivalent stress of the brake pedal before optimization.

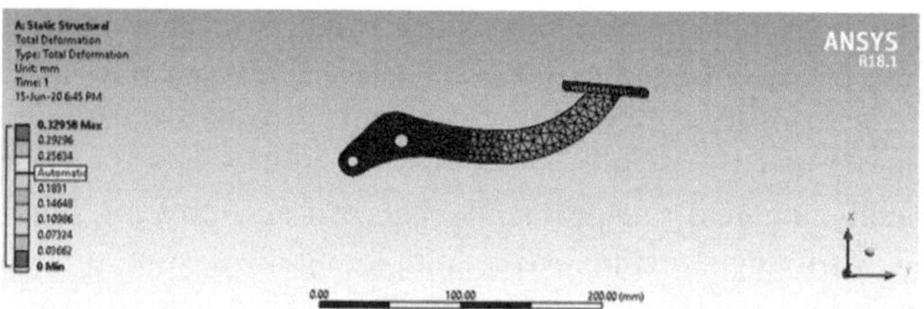

Fig. 74.6 Deformation of before pedal before optimization

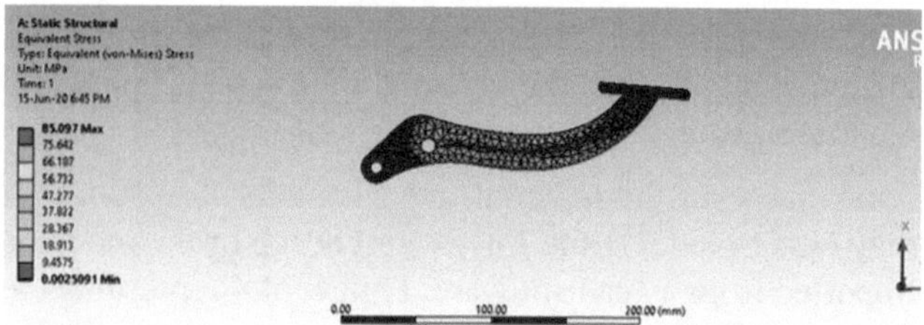

Fig. 74.7 Stress distribution of before pedal before optimization

Then the Topology optimization is selected to the project schematic window, by connecting it to the result of the previous analysis.[5] In the setup module, the response constrains the value is retained is about 40%, and the response is about the mass. Then solve the solution and the optimized region is showed in Fig. 74.8.

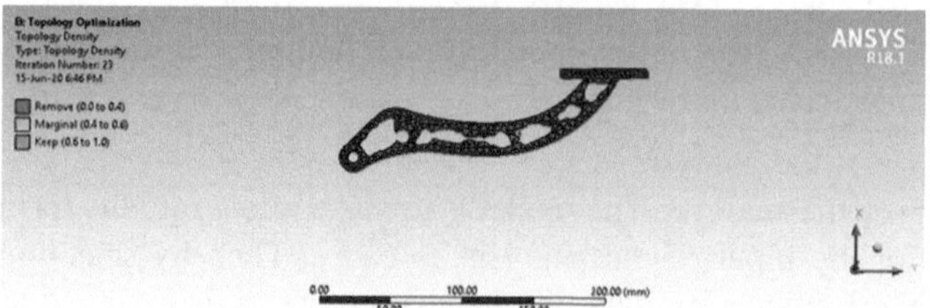

Fig. 74.8 Optimized brake pedal

The results are given to the static structural system to design the optimized brake pedal and analyze the optimized model. Then the geometry is updated with the previous one, the geometry should be modified in the space claim geometry and the same analysis is carried out. The results are compared between the current and existing brake pedal.

5. Results and Discussion

After design in the space claim and analysis in the setup module, the result is obtained. Figure 74.9 indicates the redesigned geometry in the space claim module. Then results like the deformation and equivalent stress are shown in Figures 74.10 and 74.11.

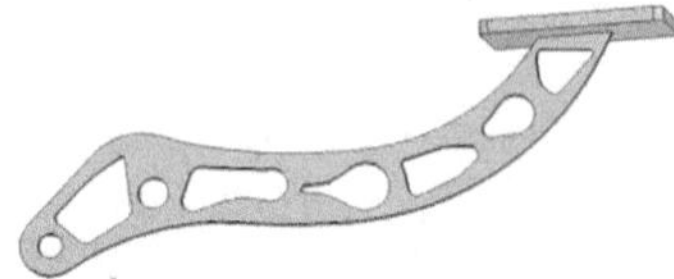

Fig. 74.9 Redesigned model in the Space claim

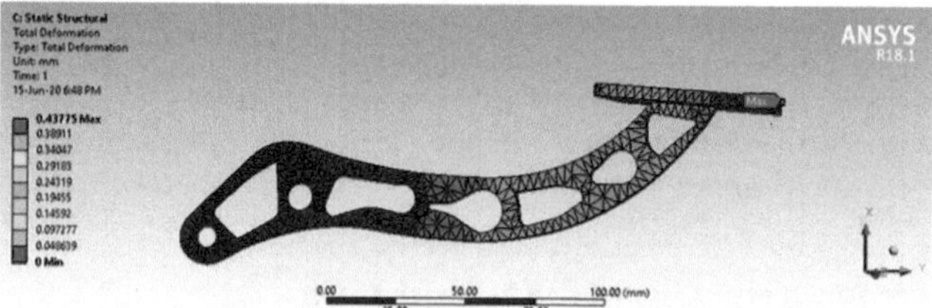

Fig. 74.10 Total deformation in the brake pedal after optimization

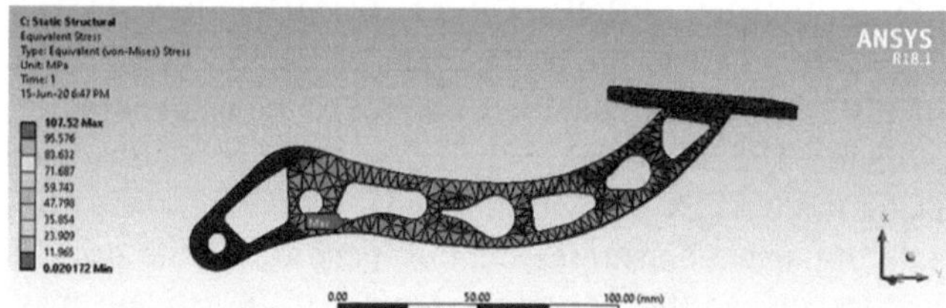

Fig. 74.11 Stress distribution in the brake pedal after optimization

Table 74.1 shows the comparison of values between the optimized and unoptimized model.

Table 74.1 Comparison between the optimized and unoptimized model

Case	Mass	Volume	Deformation	Stress
Unoptimized model	0.194 kg	24814 mm^3	0.329 mm	85.09 MPa
Optimized model	0.153 kg	19406 mm^3	0.437 mm	107.2 MPa

A. Deformation

It is clear that the deformation of the brake pedal becomes negligible. In the existing model, it may deformation about 0.329 mm and in the optimized model, it may have 0.437 mm. Hence, the material reduction may not affect the total deformation in the modified brake pedal.

B. Equivalent stress

The stress distribution before the optimization may have a maximum value of 85.09 MPa, after the optimization, it may about 107.52 MPa. It shows there is a similar amount of stress variation due to material reduction. Hence it makes the brake pedal used for the electric solar vehicle.

6. Conclusion

In this analysis, the topology optimization of the brake pedal for the electric solar vehicle is investigated. Here the brake pedal is designed by using the CATIA software. Then the analysis is carried out in Ansys software, where the different physics are used in order to perform the topology optimization. As a result, the optimized result is found similar to the unoptimized one with the reduction of material. Hence the main objective of material reduction is accomplished by 0.194 Kg to 0.13 Kg. The model can be used in the brake pedal of the electric solar vehicle.

Reference

1. Pritesh Muralidhar Ingale Topology Optimization of an All-Terrain Vehicle Brake Pedal International Journal of Engineering Research & Technology (IJERT) ISSN: 2278–0181 IJERTV8IS060446 Vol. 8 Issue 06, June-2019
2. Lalaina Rakotondrainibe, Grégoire Allaire, Patrick Orval. Topology optimization of connections in mechanical systems. Structural and Multidisciplinary Optimization, Springer Verlag (Germany), 2020. ffhal-02433242f
3. Mohd Nizam Sudin, Musthafah Mohd Tahir, Faiz Redza Ramli , Shamsul Anuar Shamsuddin Topology Optimization in Automotive Brake Pedal Redesign International Journal of Engineering and Technology (IJET) ISSN : 0975–4024 Vol 6 No 1 Feb-Mar 2014
4. ASHWINI N.GAWANDE,G.E.KONDHALKAR, ASHISH R.PAWAR STATIC STRUCTURAL ANALYSIS AND OPTIMIZATION OF BRAKE PEDAL International Research Journal of Engineering and Technology (IRJET) e-ISSN: 2395–0056 Volume: 04 Issue: 05 I May-2017
5. Sigmund, O., & Maute, K. (2013). Topology optimization approaches. Structural and Multidisciplinary Optimization, 48(6), 1031–1055. doi:10.1007/s00158-013-0978-6---methodology

Note: Every figure and table was created using the experimental work; none of them were taken from any publications or online.

Advances in Additive Manufacturing Technologies – Gurusamy Pathinettampadian (eds)
© 2024 Taylor & Francis Group, London, ISBN 978-1-032-90013-1

75 Corrosion and Erosion Analysis of S690 Steel used in Hydro Turbines

M. Vinayagamoorthy*,
R. Ganesamoorthy, Yuwaraj K R
Department of Mechanical Engineering,
Chennai Institute of Technology, Chennai, India

ABSTRACT: The material is often used as a casting to replace Stainless steel in submerged components of hydro turbines, which had been damaged by cavitation and silt erosion. Frequently, the material is used as a cast. The microstructure and mechanical properties of a material provide significant contributions to cavitation erosion. Cavitation erosion is a problem that must be addressed; thus, researchers have been examining prospective alternatives to nitrogen-reinforced austenitic stainless steel (S690 steel) in its as-cast form. Tests for vibration-induced cavitation erosion indicate that S690 steel is more robust than Stainless steel.

KEYWORDS: nitrogen-reinforced austenitic stainless steel, vibration-induced cavitation erosion, S690 steel, hydro turbines

1. INTRODUCTION

Bubbles or cavities filled with gas or vapour may form via a process known as cavitation, which refers to their formation and subsequent dissolution inside a liquid. When the cavities on the surface collapse due to excessive pressure, the material suffers damage. As a result, the material degrades. Cavitation erosion damage is a typical kind of damage that develops in the flow handling components of hydraulic turbines [1]. This damage affects the hydraulic turbines' operational capability and service life. Cast martensitic chromium nickel stainless steel is widely used in a variety of industrial applications, such as hydro turbines, pumps, and compressors, to mention a few [2]. This is owing to its high-quality construction and corrosion resistance. Cavitation erosion is to blame for the damage produced by the current trend toward smaller, faster hydraulic equipment with high pressure heads. The resistance of the alloy to the erosive effects of cavitation is determined by a variety of material properties, notably those that aid in binding the cavitation energy down to the structure. Numerous investigations have shown links between cavitation erosion resistance and a variety of features and characteristics, including structure, hardness, work-hardening capacity, super elasticity, super plasticity, strain-induced phase shift, and so on [3]. Because the S690 steel grade is thought to have greater hardness and work hardening capacity, and because it may be used in hydro turbine underwater components, it was decided to compare the cavitation erosion behaviour of the S690 steel grade to that of an austenitic stainless steel with nitrogen strengthening for the purposes of this research [4]. The objective of the work is to study the Corrosion and Erosion analysis of S690 steel used in hydro turbines

*Corresponding author: vinayagamoorthym@citchennai.net

DOI: 10.1201/9781003545774-75

2. Experimental Work

This experiment made use of cast stainless steel and a steel grade known as S690. The chemical compositions of the alloys previously discussed. We were able to get long bars with a 60 centimetre by 60-millimetre cross section. Metallographic analysis, tensile testing, impact testing, hardness testing, and cavitation erosion testing on these bars required the creation of specimens. The as-cast microstructures of stainless steel and S690 steel grades, as well as the microstructures of the S690 steel grade. The microstructure of stainless steel is made up of extremely fine lath martensitic needle packs that have not been tempered [5-10]. In addition to these packets, the structure exhibits a ferrite-based second phase. It is a nitrogen-strengthened austenitic non-staining steel with a low nickel concentration and a greater carbon percentage. Because of the increased chromium-to-nitrogen ratio and higher nitrogen content in the steel, carbides of the M8F4 type may have precipitated. The microstructure of this steel exhibits a massive core of carbides surrounded by a eutectic made of austenite and carbide [11-16]. The Vickers hardness was calculated using a Vickers hardness tester with a force of fifty kilogrammes. Tensile strength tests were done on cylindrical specimens at room temperature using a Hounsfield machine controlled by a computer in line with ASTM standards. Impact testing on common Charpy V-notch bars were performed at room temperature in line with the ASTM standard. Using a diamond cutter, the cavitation test specimens were pre-prepared to have dimensions of 20 millimetres by 20 millimetres by 5 millimetres. The samples were painstakingly cleaned and accurately weighed to an accuracy of 0.6 milligrams before and after each test. They were polished on a belt first, then moved to a cloth wheel for the final phase. The hydrodynamic cavitation erosion behaviour of the alloys studied was evaluated in pure water at ambient temperature using an ultrasonic vibration test instrument. A transducer element made of zirconatetitanate generates axial oscillations at the opposite end of a horn velocity transformer connected to the device. The sample holder was positioned coaxially with a gap of 0.8 millimetres between them; the gap width was obtained using the appropriate standard. During the operation, a cavitation zone will form in the void as a result of the tip's upward motion, causing gas-filled bubbles to form. This zone collapses as a result of the tip's following downward motion, resulting in cavitation erosion at the specimen's surface. The ultrasonic vibration testing apparatus produces 350 watts of power and 50 kilohertz of frequency. Figure 75.1 and Fig. 75.2 shows the stainless steel microstructure and 50 hrs of corrosion and erosion test.

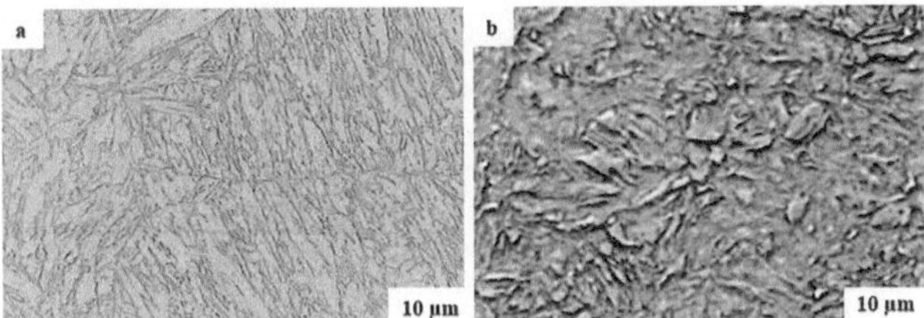

Fig. 75.1 Microstructures of (a) Stainless steel and (b) S690 grade steel

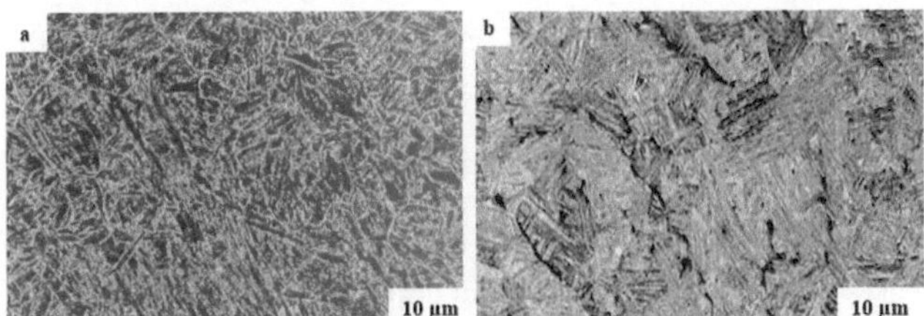

Fig. 75.2 Microstructures of (a) Stainless steel and (b) S690 grade steel after 50 hrs of corrosion and erosion test

3. Result and Discussion

The specimens were checked every 8 hours throughout the 50-hour cavitation erosion research, and their weight loss was calculated at the conclusion of the experiments. The cumulative mass loss of the tested steels as a function of cavitation times. The cavitation resistance of the S690 steel grade is larger than that of stainless steel, and its incubation duration is longer. SEM images of the damaged surfaces of the steels examined were acquired after 50 hours of vibration cavitation erosion testing. Because of the martensitic composition of stainless steel, martensite objects are less prone to distortion during cavitation erosion than other kinds of objects. The geometry of martensite objects is not totally visible due to cavitation erosion damage [17].

This is due to the fact that material is constantly being lost from the stainless steel's surface. In contrast to martensite, the ferrite component of stainless steel eroded quicker; the darker and deeper zone demonstrates this difference. Because of the presence of austenite, nitrogen-reinforced austenitic stainless steels exhibit a distinct cavitation erosion damage morphology from stainless steel. The austenite carbide grain boundaries are where the bulk of the damage occurs [18]. This demonstrates the presence of slip lines in austenite grains. Mass loss vs time taken for erosion is shown in Fig. 75.3. S690 steel offers the highest cavitation erosion resistance while also being stronger and more ductile than stainless steel. This is despite having a lesser impact energy than stainless steel. The resistance of a material to cavitation erosion is often linked to its microstructure.

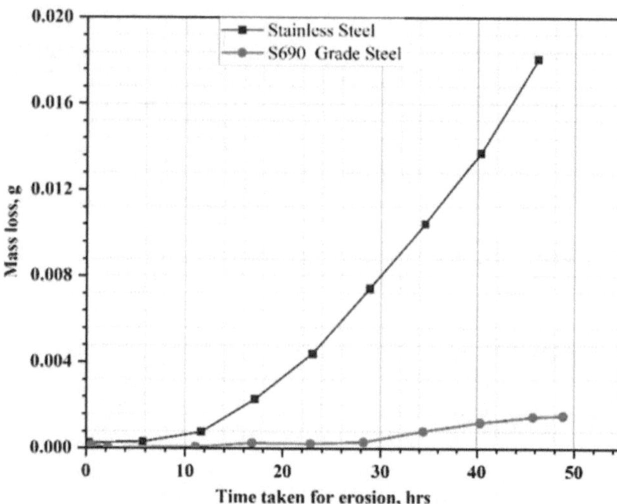

Fig. 75.3 Mass loss vs time taken for erosion

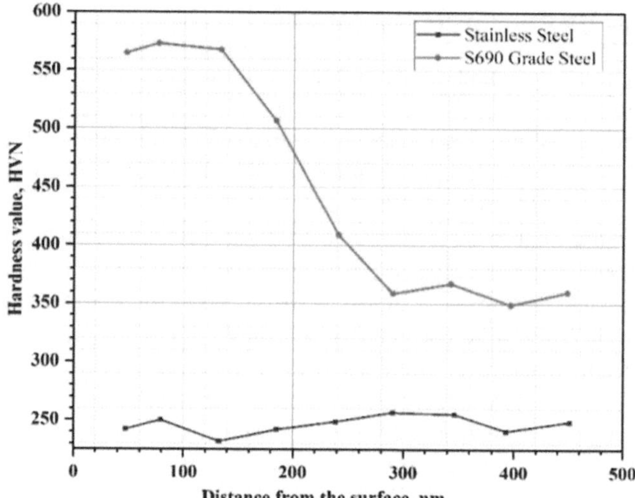

Fig. 75.4 Hardness value vs distance from the surface of the steels after erosion test of 50 hrs

Material extraction from a corroded stainless-steel surface indicates that the limits of martensite objects limit the plastic deformation of such things. Because of the austenite microstructure of the S690 steel grade, its cavitation erosion performance differs from that of stainless steel. There were two methods for removing material during the cavitation erosion process. Following the removal of material from the borders of the austenite and carbide grains, the slip lines in the austenite grains may be broken up using a ductile fracture mode to remove any leftover material [19].

Even when exposed to intense microjet impact, the S690 steel grade does not undergo martensitic transformation because to the higher nitrogen content. This is due to much improved austenitic stability. The wear resistance of austenitic steel grades is often linked to their work hardening characteristics. Hadfield steel's self-work hardening properties, for example, were a key role in its development. In this case, the work hardening of the S690 steel grade may be used to explain the mechanism behind its great resistance to cavitation erosion. It is generally known that phase transitions, such as strain-induced martensitic transformation, may be responsible for absorbing the cavitation microjet's impact energy during the cavitation erosion process. As a result, it has the ability to improve the materials' resistance to cavitation erosion. The martensitic transition is limited because nitrogen and manganese components have the ability to stabilise austenite and boost the steel's capacity for deformation strengthening. Accordance to the microhardness fluctuation that happened throughout the course of 68 hours of cavitation erosion Work hardening is expected to improve the surface micro hardness whenever microjets strike on the surface in question. Hardness value vs distance from the surface of the steels after erosion test of 50 hrs is shown in Fig. 75.4. The microhardness of the eroded surface reaches its maximum during the cavitation erosion process and then gradually decreases as the process proceeds the change in steel microhardness that occurs when a high-hardness layer forms and is transferred inside a stable austenite structure The austenitic microstructure of S690 steel allows its microhardness to rise during cavitation erosion produced by austenite work-hardening. However, due to the material's susceptibility, this surge is followed by a drop as a consequence of fracture initiation and propagation. As a result, the S690 steel grade's superior cavitation erosion resistance may be attributed to its high work hardening ability, high work hardening ability, and forming and transferring mode of high hardness layer. Its great hardness should also be paired with its ductility.

4. Conclusion

From this work, the austenitic structure of the S690 steel grade, increased hardness and ductility, high work hardening capacity, and the forming and transferring mode of high hardness layer provide it with more resistance to cavitation erosion than stainless steel. In this regard, it has been consider that the S690 steel grade may be an appropriate option for stainless steel in the manufacture of hydraulic turbine submerged components.

References

1. Zhang, Chuntao, Hongjie Zhu, and Li Zhu. "Effect of interaction between corrosion and high temperature on mechanical properties of Q355 structural steel." Construction and Building Materials 271 (2021): 121605.
2. Haixiang, Chen, and Kong Dejun. "Microstructures and immersion corrosion performances of arc sprayed amorphous

Al-Ti-Ni coating on S355 structural steel." Anti-Corrosion Methods and Materials 65.3 (2018): 271-280.

3. Hua, Jianmin, et al. "Low-cycle fatigue properties of bimetallic steel bars after exposure to elevated temperature." Journal of Constructional Steel Research 187 (2021): 106959.
4. Sequeira, César AC. High temperature corrosion: fundamentals and engineering. John Wiley & Sons, 2019.
5. Zhang, D. H., et al. "Study of the corrosion characteristics of 304 and 316L stainless steel in the static liquid lithium." Journal of Nuclear Materials 553 (2021): 153032.
6. Shi, Shengrun. Evaluating the structural integrity of high strength low alloy steels considered for shipbuilding using acoustic emission. Diss. University of Birmingham, 2015.
7. Hua, Jianmin, Fei Wang, and Xuanyi Xue. "Study on fatigue properties of post-fire bimetallic steel bar with different cooling methods." Structures. Vol. 40. Elsevier, 2022.
8. Sakurahara, Tatsuya, et al. "Integrating renewal process modeling with Probabilistic Physics-of-Failure: Application to Loss of Coolant Accident (LOCA) frequency estimations in nuclear power plants." Reliability Engineering & System Safety 190 (2019): 106479.
9. Mathivanan, A., Swaminathan, G., Sivaprakasam, P., Suthan, R., Jayaseelan, V., & Nagaraj, M. (2022). DEFORM 3D simulations and Taguchi analysis in dry turning of 35CND16 steel. Advances in Materials Science and Engineering, 2022.
10. Mathivanan, A., Sudeshkumar, M. P., Ramadoss, R., Ezilarasan, C., Raju, G., & Jayaseelan, V. (2021). Finite element simulation and regression modeling of machining attributes on turning AISI 304 stainless steel. Manufacturing Review, 8, 24.
11. Krishna, M., Kumar, S. D., Ezilarasan, C., Sudarsan, P. V., Anandan, V., Palani, S., & Jayaseelan, V. (2022). Application of MOORA & COPRAS integrated with entropy method for multi-criteria decision making in dry turning process of Nimonic C263. Manufacturing Review, 9, 20.
12. Jayaseelan, V., Kalaichelvan, K., Ramasamy, N., & Seeman, M. (2023). Influence of SiCp Volume percentage on AA6063/SiCp MMC extrusion process: An experimental, theoretical and simulation analysis. International Journal of Lightweight Materials and Manufacture.
13. Seeman, M., Kanagarajan, D., Sivaraj, P., Seetharaman, R., & Devaraju, A. (2019, July). Optimization through NSGA-II during machining of A356Al/20% SiCp metal matrix composites using PCD Tool. In IOP Conference Series: Materials Science and Engineering (Vol. 574, No. 1, p. 012008). IOP Publishing.
14. Senthilvel, K., Kumar, A. A. J., Seeman, M., Kumar, I. A., & Prabu, B. (2021). Studies the effect of halloysite nanotubes on the mechanical and hot air ageing properties of nitrile-polyvinyl chloride rubber nano-composites. Materials Today: Proceedings, 43, 1730-1739.
15. Seeman, M., Sivaraj, P., Seetharaman, R., & Kumar, I. A. (2020, November). Analysis and optimization of machining parameter during turning of A356/10% SiCp MMC using response surface methodology approach. In IOP Conference Series: Materials Science and Engineering (Vol. 961, No. 1, p. 012014). IOP Publishing.
16. Sivaraj, P., Seeman, M., Seetharaman, R., & Balasubramanian, V. (2021). Fracture toughness properties and characteristics of Friction stir welded high strength aluminium alloy by post weld heat treatment (PWHT). Materials Today: Proceedings, 43, 2150-2155.
17. Thirumalaikkannan, D. K., Paramasivam, S., Murugesan, S., & Visvalingam, B. (2022). Effect of rotational speed on the microstructure and mechanical properties of rotary friction welded AISI 1018/AISI 1020 asymmetrical joints. Materials Testing, 64(11), 1561-1571.
18. Mills, D. J., and D. Nuttall. "Eurocorr 2012:'Safer world through better corrosion control'-part 4." Corrosion Engineering, Science and Technology 48.4 (2013): 241-246.
19. Zhu, Xiang, et al. "Experimental and numerical study on the dynamic response of steel-reinforced concrete composite members under lateral impact." Thin-Walled Structures 169 (2021): 108477.

Note: Every figure was created using the experimental work; none of them were taken from any publications or online.

Advances in Additive Manufacturing Technologies – Gurusamy Pathinettampadian (eds)
© 2024 Taylor & Francis Group, London, ISBN 978-1-032-90013-1

76 Development of Structural Bio Composite using Bamboo Fiber Chitoson Bio Polymer with Epoxy Resin

Velmurugan S*
Research Scholar, Department of Mechanical Engineering,
Vel Tech High Tech Dr. Rangarajan Dr. Sakunthala Engineering College,
Avadi, Tamil Nadu, India

Gayathri N
Associate Professor, Department of Mechanical Engineering,
Vel Tech High Tech Dr. Rangarajan Dr. Sakunthala Engineering College,
Avadi, Tamil Nadu, India

Gurusamy P
Professor, Department of Mechanical Engineering,
Chennai Institute of Technology,
Tamil Nadu, India

Sharath P, Rohith P, Prasanna Ram M
Student, Department of Mechanical Engineering,
Vel Tech High Tech Dr. Rangarajan Dr. Sakunthala Engineering College,
Avadi, Tamil Nadu, India

ABSTRACT: Natural fiber-reinforced polymer composites are widely utilized due to their cost-effectiveness, eco-friendliness, corrosion resistance, electrical resistance, and high specific strength. Hybrid composites, incorporating one or more fillers within the matrix, attract interest across various industries such as building, electronics, electrical, automotive, marine, aviation, and medical applications. This study focuses on fabricating a hybrid composite using powdered chitosan fiber and threaded bamboo fiber through the hand layup technique. Mechanical properties, including tensile strength, flexural strength, hardness, and impact energy absorption, were tested along with wear resistance to assess material depreciation. Surface morphology analysis was performed using SEM, while fibre identification and quantitative composition studies were performed employing EDAX (or EDX). Results indicated that the composite with powdered chitosan fiber, polyester resin, and threaded bamboo fiber with epoxy resin displayed superior hardness and flexural strengths, especially with 7% epoxy resin. This was attributed to the high bonding strength of hybrid composites. The impact energy and wear resistance of the composite material with 7% epoxy resin also demonstrated enhanced performance. Overall, the study highlights the potential of hybrid composites for diverse applications, emphasizing their mechanical and material characteristics.

KEYWORDS: Natural fiber, Polymer, SEM, Mechanical properties

*Corresponding author: ivelu00@gmail.com

DOI: 10.1201/9781003545774-76

1. Introduction

Petroleum-based plastics have become integral to daily life, replacing traditional materials in various applications; however, their environmental impact is unsustainable [1]. In response, there is a growing interest in developing environmentally friendly products using bio-based materials. It has been derived their building blocks from living matter, and when integrated with natural fibers, they form "green composites" – environmentally degradable biopolymer composites [2]. These composites, susceptible to degradation by environmental factors, offer benefits such as non-toxicity, biodegradability, combustibility, and ease of access as their primary characteristics. Despite natural fibers lower stiffness and strength compared to synthetics, their use is appealing. However, challenges like quality variance, low processing temperature, and high moisture absorption need to be addressed. To enhance the performance of natural fiber-reinforced biopolymer composites, recent studies have focused on putting into practice natural fibers [3]. The matrix in a biopolymer composite predominantly influences its structure, environmental tolerance, and durability, with natural fibers determining stiffness and strength. Efforts to create environmentally friendly composite products with improved performance have led to significant global applications, with ongoing developments. This review emphasizes the manufacturing, processing, and utilization of biodegradable composites, featuring polymers like cellulose, starch, polylactic acid (PLA), and polyhydroxyalkanoate (PHA). It underscores their potential for fostering sustainable expansion within global markets. Bamboo fiber is anticipated to play a substantial role in the utilization of reinforced polymer matrix composites within the construction and various industrial sectors. This expectation stems from several advantageous properties of bamboo, including its cost-effectiveness, eco-friendly nature, rapid growth, biodegradability, and abundant availability. These characteristics position bamboo as a promising natural fiber for reinforcing composite materials, offering a sustainable and economically viable alternative for various applications in the construction and industrial fields. Hassan Alshahrani et al. conducted a comprehensive investigation into the bamboo fiber-reinforced polyester composites, examining their static, dynamic, and morphological characteristics. Through systematic experimentation, the study revealed that an increase in the weight percentage of bamboo fibers resulted in notable enhancements across various mechanical properties, includingcomposite specimens were evaluated for their flexural, tensile, impact, and damping properties. Observations of surface morphology revealed a consistent dispersion of bamboo fibers within the polyester resin matrix. Additionally, microscopic examination of fractured specimens demonstrated that deformation primarily occurred due to matrix cracks rather than fiber debonding. The findings underscored a robust and intelligible strengthening mechanism introduced by the combination of bamboo fibers into the polyester resin, shedding light on the composite's improved performance and structural integrity. And also, A hydrogel, formed by chitosan fibers, is characterized as a polymeric structure wherein the chains are interconnected through non-covalent and/or covalent bonds, resulting in a three-dimensional network. This unique structural arrangement imparts the hydrogel with the capacity to retain significant quantities of water, leading to the swelling of the overall structure. Chitosan, a biopolymer, exhibits the ability to create hydrogels through minor adjustments in ionic strength or pH levels. This responsiveness to environmental factors underscores chitosan's versatility in forming hydrogels, making it a promising material for various applications in biomedical and pharmaceutical fields. Therefore, the primary objective of this investigation is the production of a hybrid composite by combining powdered chitosan fiber with threaded bamboo fiber through the hand layup technique. A comprehensive evaluation of mechanical properties, encompassing tensile strength, flexural strength, hardness, and impact energy absorption, was conducted to assess the composite's performance. Additionally, wear resistance was examined to gauge material depreciation over time. The study utilized scanning electron microscopy (SEM) for surface morphology analysis, and energy dispersive X-ray analysis (EDAX or EDX) was employed to identify and quantitatively assess the composition of the fibers. This research aims to provide insights into the mechanical and morphological characteristics of the hybrid composite, shedding light on its potential applications and performance in various scenarios.

2. Experimental Setup

A composite material incorporating bamboo and chitin fibers with Si3N4 was fabricated using the hand layup method of the process as shown in Fig. 76.1. A suitable amount of polyester resin was chosen based on the dimensions of the composite material, accounting for potential wastage; for this work, 350 ml of polyester resin was utilized. Filler material, if necessary, was added to the polyester resin and mixed thoroughly with a wooden stick. Different proportions (0%, 1%, 5%) of silicon nitride (Si3N4) were introduced relative to the weight of the polyester resin. Subsequently, 1 wt% of accelerator (cobalt octoate) and 1 wt% of methyl ethyl ketone peroxide (catalyst) were incorporated and mixed until a uniform mixture was achieved. The preparation of PALF (randomly oriented bamboo leaf

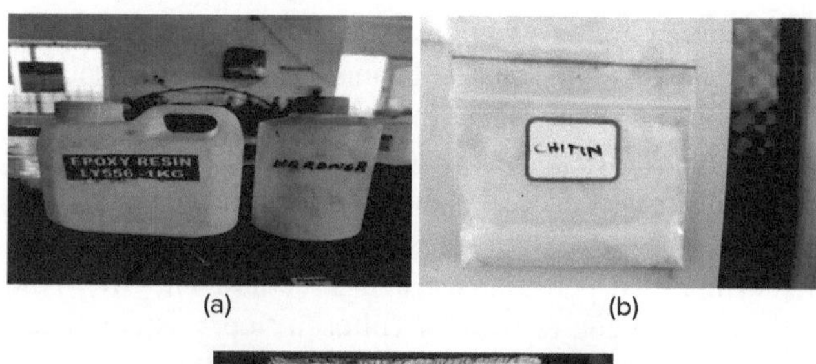

(a) (b)

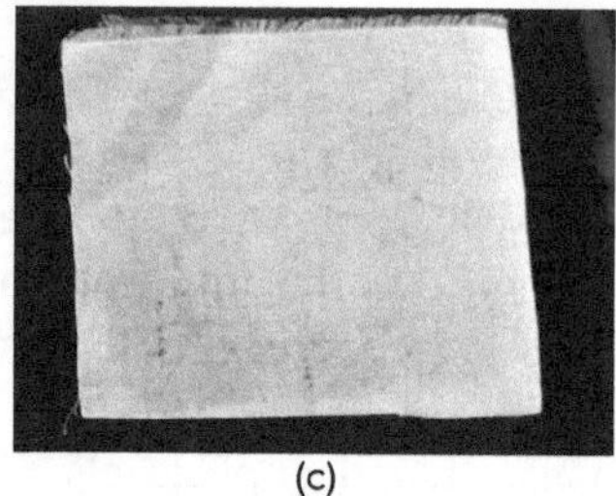

(c)

Fig. 76.1 Fabrication process (a) Epoxy resin and hardner, (b) Chitin feather (CFF), (c) Bamboo fiber

fiber), chitin feather–polyester composites with Si3N4 involved cutting bamboo leaf fiber and chitin into small parts, adding them to the mixture, and ensuring thorough mixing for a uniform composition.

To form the composite, PVC sheets were securely fixed on the table with cellotapes, creating an open mold (20 cm * 20 cm) with cardboard on the PVC sheets. The mixed composite material was then poured into the mold, spread evenly, and covered with additional sheets. Once the fibers were completely wet by the resin, a metallic roller was used to eliminate gases and bubbles trapped in the specimens. Following the layup process, the mold was tightly closed and cured for 24 hours under a load of approximately 25 kg. Subsequently, samples were post-cured, and test specimens of the required size were cut from the sheets for mechanical testing. This detailed procedure ensures the uniform distribution of fibers and Si3N4 in the polyester matrix, providing a foundation for comprehensive mechanical assessments after the curing stages.

2.1 Result

A morphological study was conducted using a Scanning Electron Microscope (SEM) to examine the adhesion between the resin and fibers in the morphology of the composite specimen without filler. The SEM images provide micrographs of the polymer composite specimen developed with Chitin Fiber (CFF) and Bamboo Fiber (PALF). The images in Fig. 76.1a clearly depict the presence of fibers in the matrix, indicating a positive and effective fiber-matrix adhesion

A tensile test was performed to determine the tensile strength and young's modulus (or young's modulus) of the materials, adhering to the ASTM D3039 standard at a test speed of 5 mm/min, maintained within the range of 1.3 mm/min to 1.5 mm/min. The Instron 3369 universal testing machine was utilized for the tests, with results obtained from six specimens recorded in Table 76.1. The average value derived from these tests was considered for subsequent discussion.

Table 76.1 Tensile strength of samples

S. No	Sample ID	Tensile Strength (N/mm2)
1	PL-1	92.95
2	PL-2	85.04
3	1L-1	69.80
4	1L-2	66.09
5	2L-1	75.08
6	2L-2	71.63

2.2 Hints.

It seems like you're providing information about different compositions involving powered chitin fiber and threaded bamboo fiber, specifically categorized as

PL (powered chitin fiber + threaded bamboo fiber),

1L (powered chitin fiber + threaded bamboo fiber + 7% Resin), and

2L (powered chitin fiber + threaded bamboo fiber + 10% Resin).

An impact test was carried out to assess the material's ability to absorb energy under sudden loads, utilizing the Izod impact test method in accordance with the ASTM D256 standard and an impact tester. Six specimens were subjected to the impact test, and results, outlined in Table 76.2, were averaged for further discussion. And the hardness test was executed to determine the hardness of the materials, following the ASTMD2240 standard. The results of the hardness tests have been conducted and are presented in Table 76.3.

Table 76.2 Impact strength of sample

S. No	Sample ID	Impact Energy (J)
1	PL-1	4.2
2	PL-2	4.0
3	1L-1	3.4
4	1L-2	3.6
5	2L-1	3.9
6	2L-2	3.7

Table 76.3 Hardness of samples

S. No	SampleID	1	2	3	4	5	Mean
1	PL	86	85	83	83	86	**84.6**
2	1L	84	88	87	88	88	**87.0**
3	2L	87	88	85	86	86	**86.4**

Flexural strength and flexural modulus were determined using the Universal Testing Machine, and the flexural test was carried out in accordance with the ASTM D7290 standard. The test speed was consistently maintained within the range of 1.3 mm/min to 1.5 mm/min. A total of six specimens were subjected to the flexural test, and the average values obtained from these tests were considered for further discussion, as illustrated in Table 76.4.

Table 76.4 Flexural strength of samples

S. No	Sample ID	Flexural Strength (N/mm2)
1	PL-1	133.97
2	PL-2	136.39
3	1L-1	97.47
4	1L-2	94.16
5	2L-1	123.05
6	2L-2	109.37

Wear analysis was conducted to assess the material's durability and determine its lifespan, aiding in timely replacement before potential failure. The wear tests were carried out using a pin-on-disc wear testing machine in accordance with the ASTM G99 standard. During the test, a spherical pin was pressed against a rotating disc, with the material under examination fixed in place. The tests were performed in dry conditions, without the use of any lubricants. Key wear test parameters included an applied load ranging from 20 N to 80 N, a sliding velocity of 150 mm/s, and a sliding distance of 1000 m. The specific test conditions and parameters are detailed in Table 76.6.

Table 76.5 Wear test results of samples

S. No	Sample ID	Mass before test (g)	Mass after Test (g)	Wear Loss (g)	Co-efficient of Friction (CoF)
1	PL-1	1.415	1.265	0.150	**0.37**
2	PL-2	1.394	1.248	0.146	**0.33**
3	1L-1	2.011	1.880	0.131	**0.45**
4	1L-2	1.929	1.801	0.128	**0.44**
5	2L-1	2.210	2.083	0.127	**0.43**
6	2L-2	2.101	1.972	0.129	**0.41**

Table 76.6 Test parameters

S. No	Sample ID	Test Speed (rpm)	PinDia. (mm)	Track Radius (mm)	Normal Load Applied (N)	Test Duration (Mins)
1	For all samples	500	3	30	30	15

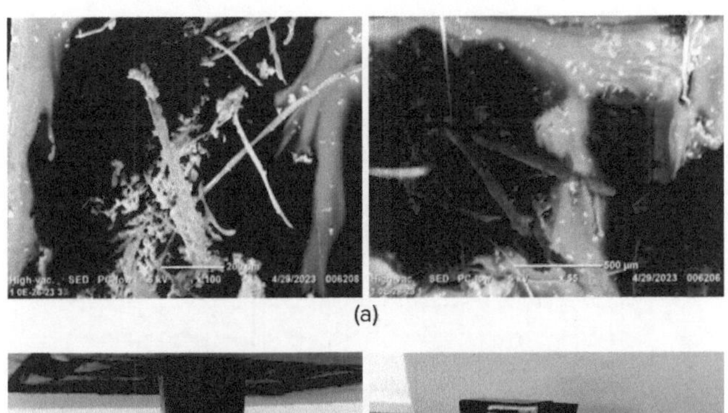

(a)

(b) (c)

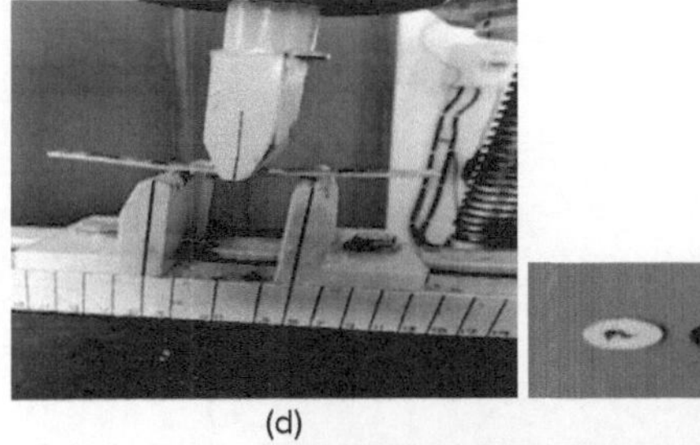

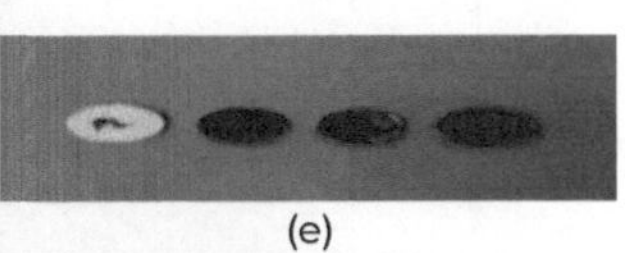

(d) (e)

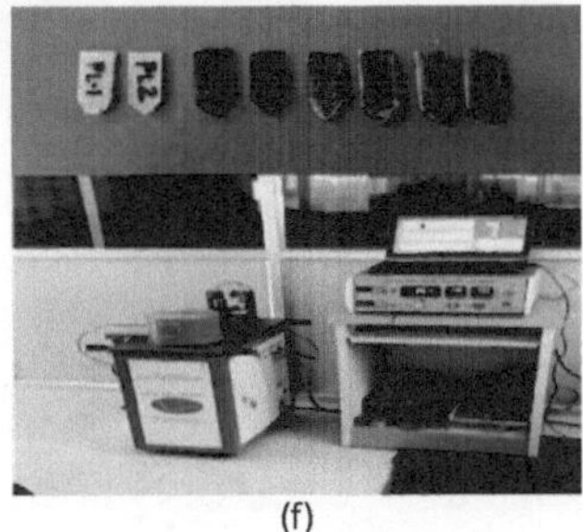

(f)

Fig. 76.2 During the test (a) SEM images of the natural fiber polymer composite, (b) Tensile Test, (c) Impact Test, (d) Flexuraltest, (e) Flammability test, (f) Pin-on-discweartestequipment

3. Discussion

The SEM images in Figure reveal the microstructure of the polymer composite specimen incorporating Chitin Fiber

(CFF) and Bamboo Fiber (PALF). The presence of fibers within the matrix is evident, signaling favorable fiber-matrix adhesion. However, noticeable clustering or agglomeration of fibers is observed, indicating a suboptimal interface with the matrix. This clustering phenomenon can lead to a decrease in the mechanical properties of the composite. Consequently, it is recommended to incorporate fillers to enhance the interfacial bonding between the fibers and the matrix, thereby improving the overall performance of the composite material.

The experimental results of the tensile strength at maximum load for composite materials, including composite and hybrid composites PL, 1L, and 2L, are depicted in Figs. 76.3(c) to 76.3(f). In Fig. 76.3(c) and 76.3(d), graphs during the tensile test of powdered chitin with threaded bamboo fiber for two samples are presented. Additionally, Fig. 76.3(c) and 3(d) showcase graphs during the tensile test of powdered chitin, threaded bamboo fiber with 1% Silica composite. Notably, the tensile stress recorded in this case was 28 N/mm², with a corresponding tensile strain

SAMPLE PL-1

Graph Type : Stress Vs. Strain

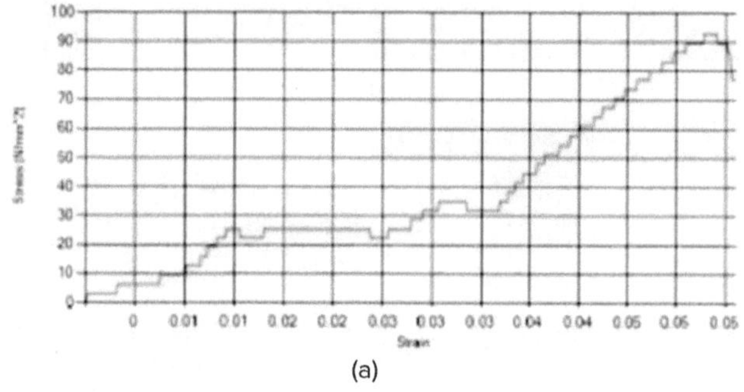

(a)

SAMPLE PL-2

Graph Type : Stress Vs. Strain

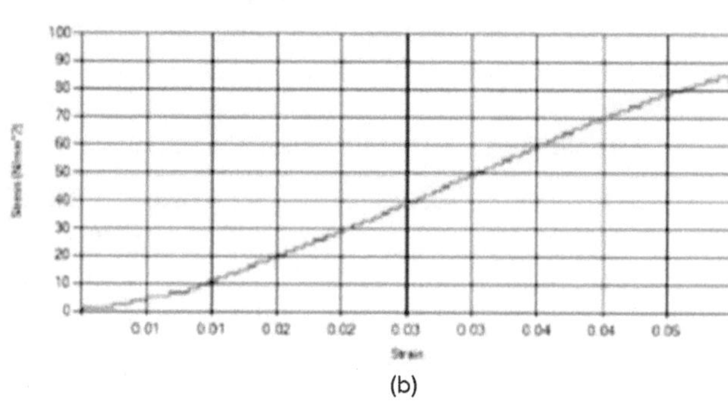

(b)

SAMPLE 1L-1

Graph Type : Stress Vs. Strain

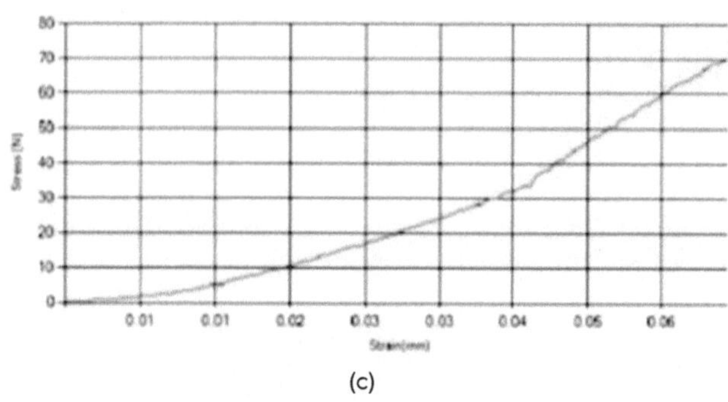

(c)

SAMPLE 1L-2

Graph Type : Stress Vs. Strain

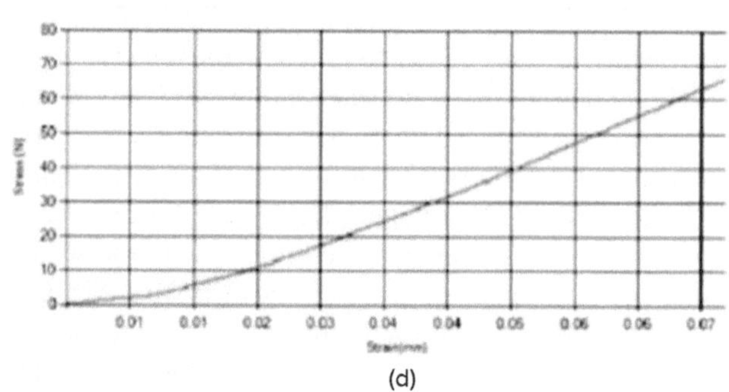

(d)

SAMPLE 2L-1

Graph Type : Stress Vs. Strain

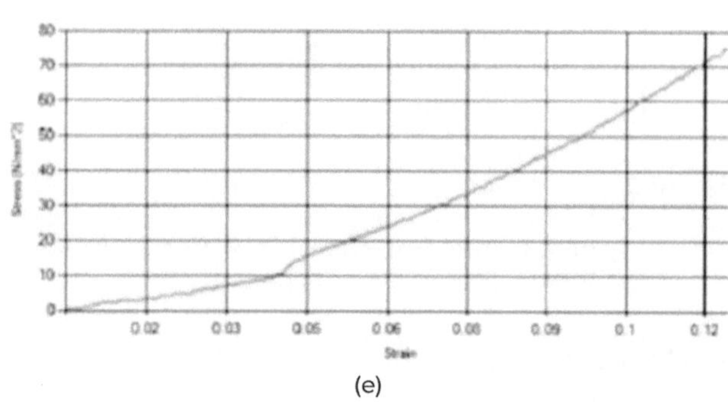

(e)

SAMPLE 2L-2

Graph Type : Stress Vs. Strain

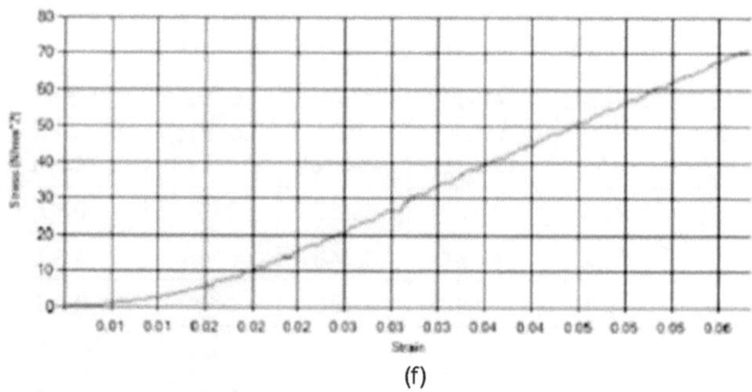

(f)

Fig. 76.3 Tensile strength analysis (a) Plaincomposite sample 1, (b) Plaincomposite sample 2, (c) 7% of resin composite sample 1, (d) 7% of resin composite sample 2, (e) 10% of resin composite sample 1, (f) 10% of resin composite sample 2 (extension) of 8%. Furthermore, the hybrid composite of chitin fiber and threaded bamboo fiber with 5% Silica exhibited the lowest void content compared to other composites and hybrid composites, as observed in Fig. 76.3(e) and 76.3(f). In this scenario, the tensile stress measured was 18 N/mm^2, accompanied by a tensile strain (extension) of 3.5%. This superior performance is attributed to the high bonding strength between the fiber and matrix, highlighting the effectiveness of the hybrid composition in enhancing material properties.

Impact strength evaluations were conducted for both composite and hybrid composites, and the results are presented in Table 76.2. The hybrid composite comprising Chitin fiber and threaded bamboo fiber with 5% resin (2L) exhibited the highest impact strength at 3.2 J. In contrast, the powdered chitin feather with powdered bamboo leaf fiber (PL) showed an average impact strength of 2.65 J, and the chitin feather with powdered bamboo leaf fiber and 1% Silica (1L) demonstrated an average impact strength of 2.95 J. Notably, the impact strength of composite PL was the lowest at 2.65 J when compared to other composites. The incorporation of Si3N4 into the matrix significantly influenced the impact strength and tensile strength of the hybrid composite. In hybrid composite 2L, where Si3N4 was introduced, both impact and tensile strengths were increased. This improvement is attributed to the reinforcing effect of Si3N4 in a relatively weaker matrix. The primary factor contributing to the enhanced impact strength was identified as the superior bonding between the fiber and matrix and the hardness evaluations of the composite of powered chitin, chopped bamboo leaf fiber with 7% resin (1L) demonstrated the highest mean hardness value of 87. In comparison, powdered chitin feather with chopped bamboo leaf fiber (PL) exhibited the lowest average hardness value of 84.6, and chicken feather with chopped bamboo leaf fiber and 5% silica (2L) showed an average hardness value of 86.4. It is observed that the presence of Si3N4 in the matrix influences the hardness values of the hybrid composite. The hybrid composite Fig. 76.4 shows, composed of Chitin fiber and threaded bamboo fiber with 5% resin (2L), exhibited the highest flexural strength at 123.05 N/mm^2. In contrast, the powdered chitin feather with powdered bamboo leaf fiber (PL) demonstrated the highest flexural strength at 133.97 N/mm^2, while the chitin feather with powdered bamboo leaf fiber and 1% Silica (1L) displayed an average flexural strength of 97.47 N/mm^2. Notably, the flexural strength of composite PL was the highest at 133.97 N/mm^2 compared to other composites, but it was closely comparable to the Chitin fiber and threaded bamboo fiber with 5% resin (2L), primarily owing to the influence of Si3N4 in the matrix.

The friction coefficient is assessed during the test through the measurement of the deflection of the elastic arm. The wear coefficient for both the pin and disk materials is calculated based on the volume of material lost during the test. A higher friction coefficient corresponds to an increased wear rate. Analyzing the values presented in Table 76.5, it was observed that all composite compositions exhibited a similar amount of wear loss and nearly identical average coefficients of friction. This similarity indicates a consistent level of wear resistance across all composite compositions, which is notably favorable when compared to the wear behavior of other hybrid composites.

4. Conclusion

The mechanical characterization of the composite material, Threaded Bamboo Fiber (PALF) reinforced with Si3N4, was conducted. Three variations of the composite were prepared, including one without Si3N4 (filler), 1% Si3N4, and 5% Si3N4. The mechanical properties of plain specimens without Si3N4, specimens with 1% Si3N4, and those with 5% Si3N4 were analyzed.

The major conclusions drawn from the observations are as follows:

1. The composite of Threaded Bamboo Fiber (PALF) with 1% Si3N4 exhibited superior hardness and flexural strengths compared to the composite prepared without Si3N4. This improvement can be attributed to the enhanced bonding strength in the hybrid composite.
2. The impact energy of the composite with 1% Si3N4 showed a higher value, suggesting greater impact energy absorption by the reinforcement.
3. The wear test revealed similar wear resistance across all composite compositions, with comparatively good wear behavior in comparison to other hybrid composites.

Acknowledgement

The authors gratefully acknowledge the students, staff, and authority of Vel Tech High Tech Dr. RangarajanDr. Sakunthala Engineering College, Avadi, Chennai for their continuedsupport.

REFERENCES

1. Kannan Rassiah and Megat Ahmad, M. M. H. 2013. A Review on Mechanical Properties of Bamboo Fiber Reinforced Polymer Composite, Australian Journal of Basic and Applied Sciences, 7(8): 247–253.

SAMPLE PL-1

Graph Type : Load Vs. Displacement

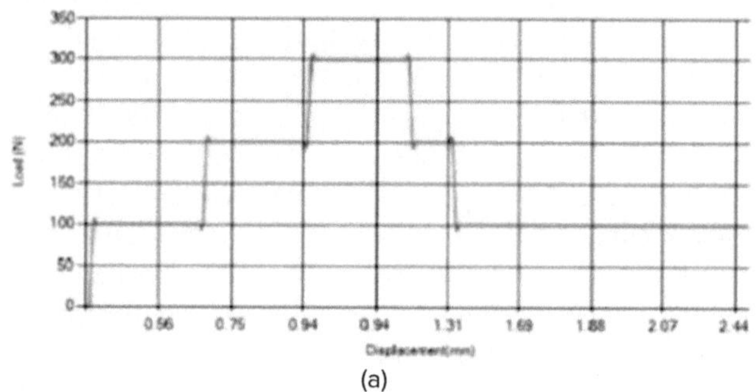

(a)

SAMPLE PL-2

Graph Type : Load Vs. Displacement

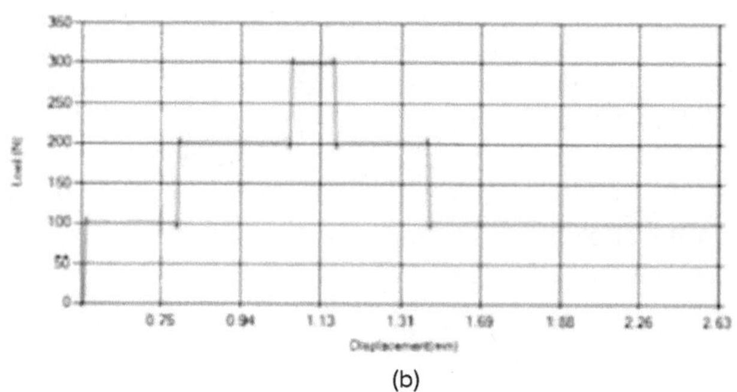

(b)

SAMPLE 1L-1

Graph Type : Load Vs. Displacement

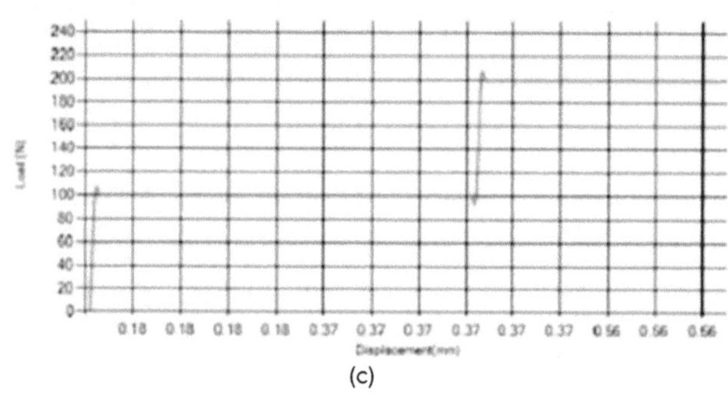

(c)

SAMPLE 1L-2

Graph Type : Load Vs. Displacement

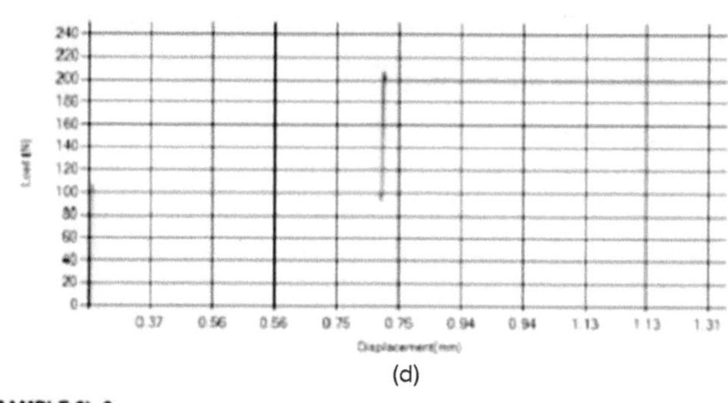

(d)

SAMPLE 2L-1

Graph Type : Load Vs. Displacement

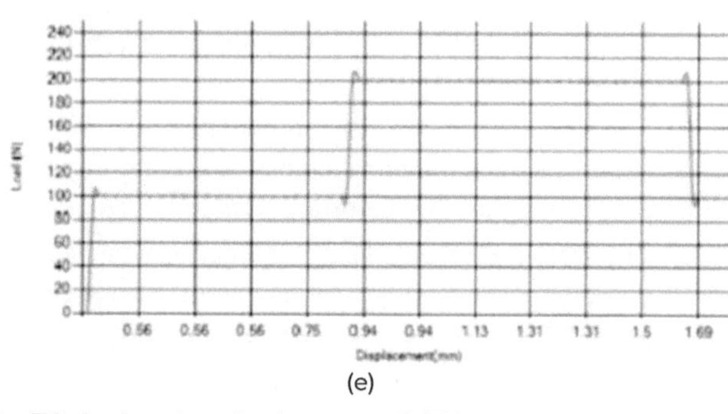

(e)

SAMPLE 2L-2

Graph Type : Load Vs. Displacement

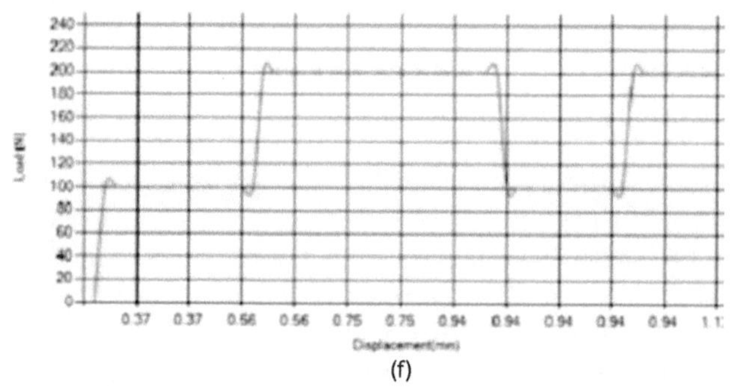

(f)

Fig. 76.4 Load vs displacement (a) Plain composite sample 1, (b) Plain composite sample 2, (c) 7 % of resin composite sample 1, (d) 7 % of resin composite sample 2, (e) 10% of resin composite sample 1, (f) 10 % of resin composite sample 2

2. Burguenoa, R., M.J. Quagliataa, A.K. Mohanty, G. Mehtac, L.T. Drzald and M. Misra, 2005. Hybrid Biofiber- based Composites for Structural Cellular Plates, Composites: Part A, 36: 581–593.
3. P. Lokesh, T.S.A. Surya Kumari, R. Gopi, Ganesh Babu Loganathan. (2020). A study on mechanical properties of bamboo fiber reinforced polymer composite, Materials Today: Proceedings 22: 897–903.
4. Hassan Alshahrani, Arun Prakash, V.R. 2022. Mechanical, thermal, viscoelastic and hydrophobicity behavior of complex grape stalk lignin and bamboo fiber reinforced polyester composite, International Journal of Biological Macromolecules, Volume 223, Part A, 31: 851–859.

Note: Every figure and table was created using the experimental work; none of them were taken from any publications or online.

Advances in Additive Manufacturing Technologies – Gurusamy Pathinettampadian (eds)

© 2024 Taylor & Francis Group, London, ISBN 978-1-032-90013-1

77 Unveiling the Influence of Stacking: Analyzing Epoxy Hybrid Composites Infused with Kevlar and Jute Fibers and Their Distinct Features

R. Deepak Suresh Kumar[1]
Professor, Mechanical department,
Rajalakshmi Institute of Technology Chennai, India

P. Gurusamy[2]
Professor, Mechanical department,
Chennai Institute of Technology Chennai, India

L. Hrithick Kumar[3]
Student, Mechanical department,
Rajalakshmi Institute of Technology Chennai, India

ABSTRACT: This investigation explores the intricate relationship between stacking sequences of materials and the resulting mechanical characteristics of epoxy hybrid composites reinforced with a blend of Kevlar and Jute fibers. The combination of these fibers offers a promising avenue for enhancing both the mechanical performance and eco-friendliness of composite materials. The arrangement of these fibers, termed stacking sequence, emerges as a pivotal factor shaping the overall mechanical behavior of these hybrid composites. To investigate the impact of stacking sequences, we manufactured composite panels using epoxy resin as the matrix and incorporating Kevlar and Jute fibers as reinforcements. Various stacking sequences, including unidirectional, cross-ply, and hybrid configurations, were employed. Properties such as tensile strength, impact resistance, interlaminar shear strength, and flexural strength were systematically examined and compared. Results highlight a significant influence of the stacking sequence on the mechanical attributes of the hybrid composites. Notably, the unidirectional stacking sequence exhibited superior tensile and flexural strength, while the hybrid stacking sequence displayed enhanced impact resistance and interlaminar shear strength. Characterization of these hybrid composites involved a comprehensive assessment of microstructure, fiber-matrix interface, and fracture behavior using techniques such as scanning electron microscopy (SEM) and energy-dispersive X-ray spectroscopy (EDS). Analysis revealed robust interfacial bonding between fibers and the epoxy matrix, indicating effective load transfer mechanisms and an overall enhancement in mechanical performance. Experimental and computational approaches yielded consistent results, validating the material's mechanical behavior under various loading conditions. This study provides valuable insights into optimizing composite design and manufacturing processes, facilitating the development of lightweight, sustainable, and high-performance composite materials tailored for diverse engineering applications.

KEYWORDS: Fibre, Composites, Kevlar-Jute Composites, Epoxy hybrid, Flexural, Stacking sequence

[1]deepaksureshkumar.r@ritchennai.edu.in, [2]Gurusamyp@citchennai.net, [3]lshofhrithick77@gmail.com

DOI: 10.1201/9781003545774-77

1. Introduction

Composite materials, pivotal in modern technology, offer a non-metallic alternative to metals, tailored through precise material combinations for desired properties. Traditional manufacturing processes often lead to pollution and emissions, prompting the emergence of eco-friendly composites incorporating natural fibers. Laminating composites involves bonding layers with a matrix, their orientation termed stacking sequence, significantly impacting mechanical characteristics. This investigation focuses on Kevlar- or jute-based epoxy hybrid composites, leveraging Kevlar as the matrix and jute fibers for reinforcement [9]. Comprehensive testing, including flexural, impact, and tensile strength assessments, unveils stacking order effects on mechanical performance. The integration of advanced testing methods and microscopic analysis offers insights into the interplay between stacking sequences and mechanical properties. Hybrid composites, blending Kevlar and jute fibers, offer superior mechanical properties, reduced weight, and enhanced sustainability, catering to diverse industries. While prior studies [4,6,9] have explored stacking order influences, research on Kevlar/Jute-based epoxy hybrids remains limited. This study aims to bridge this gap, identifying optimal stacking orders and offering insights for practical applications, thus advancing composite design in engineering fields.

2. Literature References

A. B. M. Abid Hossen Bhuiyan led a significant study investigating woven jute-Kevlar hybrid composites, employing Finite Element Analysis (FEA) through ABAQUS to explore the influence of fiber orientations, stacking sequences, and aging [16]. The research revealed that the loading direction had a substantial impact on tensile and flexural strength, with Type-A (0°) displaying the highest strengths and Type-A (90°) the lowest. Stacking sequences played a vital role, highlighting the superior mechanical properties of Kevlar on the tensile side. The hydrophilic nature of jute led to increased water absorption, particularly in all-jute layers, affecting mechanical behavior in corrosive mediums. The study validated FEA accuracy, uncovering varying void content among composites.Pranti Saha and colleagues contributed to enhancing Composite Overwrapped Pressure Vessel (COPV) design efficiency, showing that Jute-Kevlar hybridization significantly improved jute PV performance [17]. Their investigation aimed to optimize burst pressure performance in nine Type IV COPVs, combining experimental tensile strength data and von Mises stress profiles from ABAQUS simulations. The study highlighted weaknesses in four-layer jute-reinforced vessels compared to those with Kevlar reinforcement. However, hybrid configurations, like J-K-J-K layering, substantially enhanced jute-based vessel strength. The research emphasized the role of fiber orientation, identifying cross-meshed jute patterns as prone to strength degradation among analyzed COPV variations.Sunil Manohar Maharana and team conducted a thorough study on bidirectional jute-Kevlar hybrid nanocomposites, assessing their influence on both I and II Modes interlaminar fracture toughness [18]. Various layering sequences and nanofiller weight fractions were explored, revealing improvements in fracture toughness with nanofiller incorporation. Critical failure modes in certain stacking sequences led to recommendations for intra-ply hybridization of jute and Kevlar to enhance durability and structural integrity. A. Vinod and collaborators utilized jute and hemp fiber mats coupled with eco-friendly epoxy materials to create superior hybrid composites [19]. Their study emphasized the performance advantages of hybrid laminates over pure jute or hemp composites, with hybrid laminate H/J/H surpassing others in density and performance. Future directions include durability testing, chemical treatments for enhanced thermal stability, and creep/fatigue analyses for comprehensive assessment across applications. Md. Abu Shaid Sujon and team conducted experimental investigations on the influence of factors like fiber orientation and stacking sequence on tensile, flexural, and impact strengths [20]. Their study highlighted the superior mechanical properties of certain stacking sequences, providing insights for tailored hybrid composite design and application optimization. Selsabil Rokia Laraba's research focused on creating a sandwich structure using natural fibers and eco-friendly materials, shedding light on potential applications in sustainable structural materials [21]. Sovit Agarwal and colleagues explored hybrid nanofillers' role in improving mechanical properties in ballistic-resistant materials, offering valuable guidance for optimization across applications [22]. Nitin Mathusoothanaperumal Sukanya and team's extensive research evaluated the influence of nano-silica particles on tensile properties in Kevlar composites, highlighting potential enhancements in mechanical properties [23]. Hongyong Jiang and collaborators examined hybridization strategies in High-Performance Fiber Reinforced Polymer (HFRP) composites, providing insights into optimizing mechanical properties across diverse applications [24].Ahmed Sarwar and team investigated the mechanical characteristics of a novel Kevlar/Flax/epoxy hybrid composite, contributing to the understanding and optimization of hybrid composites across applications [25]. These studies collectively underscore the importance of fiber orientation, stacking sequence, and hybridization in optimizing mechanical properties and advancing composite materials across various industries.

2.1 Novelty of the Work

This research unveils a novel approach to enhancing mechanical properties and sustainability in composite materials by exploring the interplay of stacking sequences in epoxy hybrid composites reinforced with Kevlar and Jute fibers, offering a promising avenue for eco-friendly engineering solutions. The systematic examination of diverse stacking sequences, including unidirectional, cross-ply, and hybrid configurations, in the fabrication of epoxy hybrid composites, revealing nuanced effects on mechanical attributes such as tensile strength, impact resistance, and interlaminar shear strength.By employing advanced characterization techniques like scanning electron microscopy (SEM) and energy-dispersive x-ray spectroscopy (EDS), The integration of ANSYS simulation with experimental results establishes a novel methodology for evaluating mechanical characteristics, offering a comprehensive understanding of the optimal stacking sequence, particularly showcasing the benefits of a 45-degree orientation for Kevlar and Jute fibers in hybrid composites.

3. Materials and Methodology

Epoxy, jute fibres, Kevlar fibres, and a hardener are the ingredients needed to create a hybrid composite sample rigid and test-ready. Hand-lay-up is a method of stacking or laying that involves manually laying the fibres in the proper order. The surface that will be used for fabrication is first cleaned and waxed the sizes considered. The size of the material is 1 ft × 1 ft and its thickness is 3 mm. Epoxy resin makes up the first or foundation layer.

Just 10% of the tougher solution is used while hand-laying Kevlar fibres onto the resin. Another coating of epoxy is then applied on top of the Kevlar fibres once they have settled. After laying the jute fibres, epoxy resin is once more used to seal the structure. The epoxy resin goes through the curing when it hardens up and is prepared for the many tests to be run. Tables the following detail the materials used and their qualities. Tables 77.1, 77.2, and 77.3 list the material characteristics of jute fibre, epoxy resin, and Kevlar 29. Figure 77.1 depicts the stacking sequence with the epoxy base and individual fibres indicated by colour.

Fig. 77.1 Stacking sequence followed for the specimen

Table 77.1 Properties of jute fibre

Density (kg/m3)	1350
Young's Modulus, (MPa)	20000
Poisson's Ratio	0.38
Bulk Modulus, (MPa)	27778
Shear Modulus, (MPa)	7246.4
Tensile Yield Strength, (MPa)	700

Table 77.2 Properties of resin epoxy

Density (kg/m3)	1160
Young's Modulus, (MPa)	3780
Poisson's Ratio	0.35
Bulk Modulus, (MPa)	4200
Shear Modulus, (MPa)	1400
Tensile Yield Strength, (MPa)	54.6

Table 77.3 Properties of kevlar 29

Density (kg/m3)	1440
Young's Modulus, (MPa)	62000
Poisson's Ratio	0.44
Bulk Modulus, (MPa)	1722220
Shear Modulus, (MPa)	21528
Tensile Yield Strength, (MPa)	2758

The polymer is distributed uniformly using a brush. Following the application of the second layer of mat, a roller is used to carefully remove the mat-polymer layer. With each layer of polymer and mat, the process is repeated in order to stack the required number of layers while simultaneously releasing any trapped air and extra polymer. The hardener utilized has a 10:1 ratio and a high cohesive force and strength. It is called HY951. A sample evaluation test is also performed for the tests in ANSYS in order to compare and validate the outcomes of the experimental tests.

The composites are stacked in the following order: epoxy, jute, Kevlar, epoxy, etc. The hand lay-up technique was used, and the angle of the stacking of the fibres was 45°/45°. For the examination of the tensile stress, flexural strength, and impact test, the samples were prepared appropriately. Figures 77.1 and 77.2 show the hand layup method and the test samples, respectively. The stacking sequence, an essential aspect of composite material design, plays a critical role. This article delves into the significance of stacking sequence and explores its impact on optimizing the material properties of epoxy hybrid composites reinforced with Kevlar and jute fibres. Understanding the influence of stacking sequence enables engineers to maximize the composite's strength, stiffness, and durability for specific applications. It refers to the arrangement and

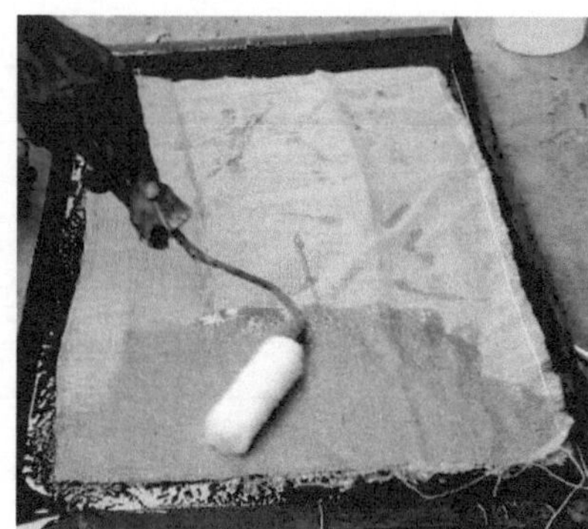

Fig. 77.2 Hand layup method of fabrication

orientation of individual fibre layers within a composite material. It determines the distribution of loads, the fibre orientation, andshear strength, thereby directly influencing the materials mechanical behaviour. The order and positioning of different fibre layers are crucial that affect the composite's performance. Impact of Stacking Sequence on Mechanical Properties Effect in the Strength and Stiffness. The stacking sequence significantly impacts the strength and stiffness. By selecting the appropriate order and orientation of fibre layers, engineers can enhance the composite's tensile, compressive, and flexural properties. The stacking sequence influences load transfer mechanisms and the distribution of stresses, enabling the optimization of the composite's overall strength. Fatigue performance is critical for materials subjected to cyclic loading. The stacking sequence affects the initiation and propagation of fatigue cracks within the composite. Properly designed stacking sequences can mitigate crack initiation and enhance the composite's resistance to fatigue failure.

Impact Resistance: The arrangement of fibre layers in the stacking sequence determines the composite's ability to absorb and dissipate.

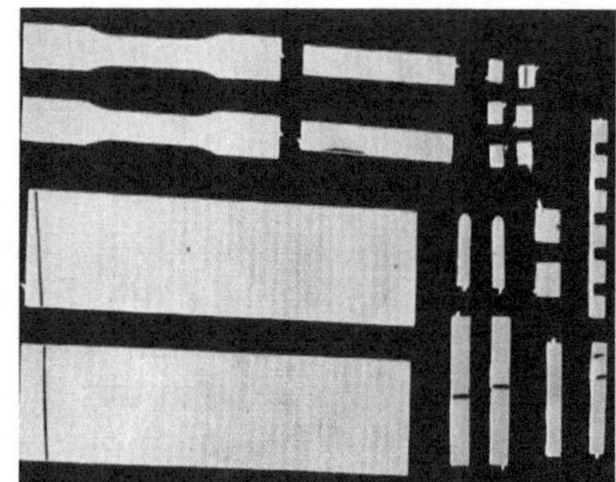

Fig. 77.3 Different samples made for testing

4. Testing Methods

ASTM D standards refer to a set of technical standards developed by ASTM International. ASTM D standards cover a wide range of topics related to materials testing and characterization, including but not limited to plastics, rubber, textiles, petroleum products, paints, coatings, and environmental testing. As indicated in Fig. 77.4, the created hybrid composite was ready and constructed in accordance with the ASTM D 638 standard[4], and the tensile strength analysis was done using a universal testing machine. In order to prevent mistakes from occurring during the test, the prepared samples are firmly connected and secured using the proper fixtures. Tensile stress tests are necessary to determine the bond strength between the various laminate layers and to assess the hybrid composite material's capacity to support loads. In order to ensure that the findings of the prepared sample's flexural strength study would be uniform and simple to deduce from overall results, the ASTM D790 standard was applied[4]. Figure 77.3 shows the dimensions of the test sample. The stiffness of the laminate is determined using the flexural strength analysis, which is also used to assist assess if the bonding between each layer is adequately defined. Consequently, it is possible to access the hybrid composite's bending and ductility characteristics. The preparation of the material sample for the impact test complies with ASTM D-256[4] standard. The impact test sample is shown in Fig. 77.6. The Izod impact test was conducted. Therefore, the material is fastened and clamped. This is done to evaluate the impact resistance of the sample and to match it for the applications as the day-to-day applications mostly receive impact loads and forces. This also considers the material's toughness and ductility[9].

Fig. 77.4 Material processing

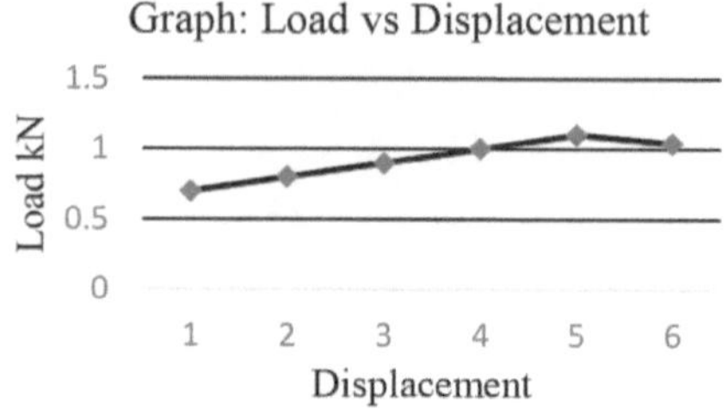

Fig. 77.5 Load/displacement graph for flexural test

Table 77.4 Load vs displacement test data

Displacement	Load
1	0.7
2	0.8
3	0.9
4	1.0
5	1.1
6	1.0

Fig. 77.6 Flexural test on the sample

5. Results and Discussion

A tensile strength test is done in a universal testing machine and the sole purpose of a tensile strength test is to ensure and validate the material whether could withstand fatigue loads and compressive tests in its usage. The tensile stress test is also done in ANSYS and Table 77.1, provides the comparison of the results from the evaluative and experimental results. The developed stress values reach a level of 796.9 MPa according to the stress test conducted in ANSYS with a tensile load of 5kN.The study by Salih Alan et al gave a novel optimization framework for designing stiffened composite panels, addressing challenges in determining optimal stacking sequences and layouts while adhering to manufacturing guidelines. They combined spectral element modeling (SEM) with an index-based optimization approach to efficiently predict dynamic behaviour and optimize designs. Validation studies showed SEM's accuracy and computational efficiency compared to finite element analysis. Optimization case studies demonstrated the framework's capability to design stiffened composite laminates, achieving up to 15-20 times reduction in computational time compared to traditional methods. Future work may focus on further optimizing the mesh and extending the framework to analyze more complex stiffener geometries.aerospace, marine, and automotive industries. It combines spectral element modeling (SEM) for accurate structural analysis with an index-based optimization approach to determine optimal stacking sequences and stiffener layouts. The SEM accurately predicts dynamic behaviour with reduced computational time compared to finite element analysis. Case studies validate the accuracy and efficiency of the SEM and demonstrate the effectiveness of the optimization framework in designing stiffened composite panels with up to 200 plies, optimizing both stacking sequences and stiffener layouts simultaneously [26]. In the paper written by the, the authors investigate the achievement of extension–twist coupling in laminated composites while addressing hygrothermal stability concerns. They propose a novel stacking sequence that inherently satisfies necessary hygrothermal stability conditions, reducing optimization to an unconstrained problem. Analytical and numerical optimization techniques are employed to maximize extension–twist coupling while ensuring hygrothermal stability. Sensitivity analysis validates the robustness of the proposed approach. Overall, the study provides a method to design hygrothermally stable laminates with desired coupling properties, crucial for various engineering applications [27]. 1.07 kN, the sample deformed up to 6 mm, as shown in Fig. 77.5 of the attached load vs displacement graph. The displacement increased to 5.6 mm with a peak load Using the universal testing machine, the bending or flexural strength test was conducted with a loadof 1.05 kN load when analysed with ANSYS. The shear stress that was created on the sample material with the primary axis of stress and stress intensity was revealed by the tensile strength test performed in ANSYS. The aggregated results are given as a table in table 4's format. This is the definitive outcome of the ANSYS test with a 5kN load on the sample. The flexural bending test were shown in Fig. 77.6.

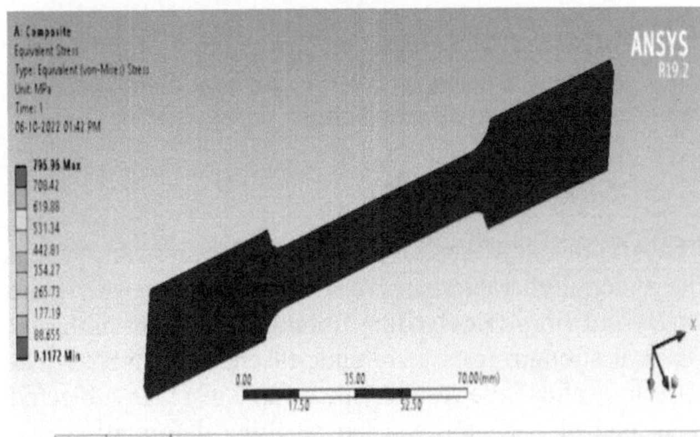

Fig. 77.7 ANSYS result for equivalent stresses for tensile strength test

6. Future Insights and Applications

1. Sports Equipment—Lightweight and strong materials are always in demand for sports equipment. Epoxy hybrid composites could be used in the pro-

Table 77.5 ANSYS results

Type of analysis	Deformation Total	Equivalent (von Mises) Stress	Directional Deformation	Maximum Principal Stress	Minimum Principal Stress	Maximum Shear Stress	Stress Intensity	Shear Stress
Results								
Mini.	0. mm	0.1172MPa	0. mm	-1.1175 Mpa	-81.276 Mpa	6.7505e-002 Mpa	0.13501 Mpa	231.0 9 Mpa
Max.	0.37678 mm	796.96 Mpa	0.37678 mm	1207.7 Mpa	400.15 Mpa	447.52 Mpa	895.04 Mpa	230.8 7Mpa
Average	0.18443 mm	71.789 Mpa	0.18443 Mpa	103.48 Mpa	22.055 Mpa	40.713 Mpa	81.426 Mpa	6.556 2e-003 Mpa

duction of high-performance sporting goods such as bicycles, tennis rackets, and protective gear

2. Infrastructure Reinforcement—The construction industry could use these composites to reinforce structures such as bridges and buildings, providing enhanced durability and strength while minimizing the overall weight of the structures.
3. Customization for Specific Applications—Future developments might focus on tailoring stacking sequences to meet the specific requirements of different applications. This could involve a more precise understanding of how different sequences affect mechanical properties.
4. Smart Composite Materials—Integration of sensors and actuators within the composite structure could be explored, leading to the development of smart materials with self-monitoring and self-repair capabilities.
5. Advanced Manufacturing Techniques—Innovations in manufacturing processes, such as 3D printing or automated layup techniques.

7. SUMMARY

The Aim of this Research paper was to analyse and reveal the material characteristics of a laminate composite made of Hybrid fibre (Kevlar/Jute fibres). Tensile strength and flexural strength tests were undertaken, and ANSYS was used to evaluate the results. The sample test was subjected to a load of 1.07 kN, and the resulting deformation displacement was 6 mm, which is comparable to the 5.6 mm findings of the ANSYS assessment. Along with an accumulated shear stress of 230.8 MPa, the test piece's stress accumulation was estimated to be close to 796.9 MPa. The stress and deformation can be best reduced and the material can have higher mechanical properties at a stacking sequence of 45° of the fibres with respect to the horizontal of the test sample. Previous experiments and research also prove the same [4]. As a result, the characteristics of Kevlar/Jute-based Epoxy Hybrid Composites have been tested and evaluated, and the impact of the stacking order has also been examined. Unravelling the impact of stacking sequence on the material properties and characterization of epoxy hybrid composites with Kevlar and jute fibres is a crucial step in optimizing their performance. By understanding how different stacking sequences affect the mechanical properties, anyone can design and manufacture composites tailored to specific applications, resulting in improved strength, impact resistance, and sustainability.

REFERENCES

1. Performance analysis of tidal turbine blades for different composite materials, R Deepak Suresh Kumar, Justin Joseph, KB Vigneshwara, KT Yugendheran, N Adithya, 2022/12/19, Surface Engineering, 219-231.
2. Buckling analysis of 3D printed ABS thin cylinder at different percentage of infill's, A Suresh, R Deepak Suresh Kumar, H Mohamed Jaasim, B Tamilarasan, N Mugilvarnan, A Vignesh, 2020/10/29, AIP Conference Proceedings, 2283, 020119.
3. Studies on Mechanical Properties of Banana/E-Glass Fabrics Reinforced Polyester Hybrid Composites M. R. Sanjay, G. R. Arpitha, L. Laxmana Naik, K. Gopalakrishna, B. Yogesha, J. Mater. Environ. Sci. 7 (9) (2016) 3179-3192.
4. Effect of stacking sequence on properties of coconut leaf sheath/jute/ E-glass reinforced phenol formaldehyde hybrid composites, KN Bharath, MR Sanjay, Mohammad Jawaid, Harisha, S Basavarajappa and Suchart Siengchin, Journal of Industrial Textiles 49(1), 3-32, 2019.
5. The Influence of Stacking Sequence on Laminate Strength, N.J. Pagano and R. Byron Pipes, Journal of Composite Materials 1971 5: 50.
6. Experimental investigation on stacking sequence of Kevlar and natural fibres/epoxy polymer composites, Murali Banu, Vijaya Ramnath Bindu Madhavan, Dhanashekar Manickam and Chandramohan Devarajan, Polímeros, 31(1), e2021004, 2021.

7. The effect of stacking sequence on the LVI damage of laminated composites; experiments and analysis, Shiyao Lin, Anthony M. Waas, Composites Part A 145 (2021) 106377.
8. Mechanical Properties of Kevlar-49 Fibre Reinforced Thermoplastic Composites, K.K. Herbert Yeung and K.P. Rao, Polymers & Polymer Composites, Vol. 20, No. 5, 2012.
9. Effect of stacking sequence on the performance of hybrid natural/syntheticfibre reinforced polymer composite laminates, Subrata C. Das, Debasree Paul, Sotirios A. Grammatikos, Md. A.B. Siddiquee, Styliani Papatzani, Panagiota Koralli, Jahid M.M. Islam, Mubarak A. Khan, S.M. Shauddin, Ruhul A. Khan, Nectarios Vidakis, Markos Petousis, Composite Structures 276 (2021) 114525.
10. Impact on Laminated Composites: Recent Advances, Abrate S. (November 1, 1994), ASME. Appl. Mech. Rev. November 1994; 47(11): 517–544.
11. Consolidation Experiments for Laminate Composites, Gutowski TG, Cai Z, Bauer S, Boucher D,Kingery J, Wineman S., Journal of Composite Materials. 1987;21(7):650-669.
12. Effect of Stitching on the Mechanical and Impact Properties of Woven Laminate Composite, Kang TJ, Lee SH., Journal of Composite Materials. 1994;28(16):1574-1587.
13. Laminated Composite Plate Theory with Improved In-Plane Responses, Murakami, H. (September 1, 1986), ASME. J. Appl. Mech. September 1986; 53(3): 661–666.
14. Magnetoelectric Laminate Composites: An Overview, Zhai, J., Xing, Z., Dong, S., Li, J. and Viehland, D. (2008), Journal of the American Ceramic Society, 91: 351-358.
15. Fatigue Behaviour of Composite Laminate, Hahn HT, Kim RY, Journal of Composite Materials. 1976;10(2):156-180.
16. Impact of fiber orientations, stacking sequences and ageing on mechanical properties of woven juteKevlar hybrid composites by A. B. M. Abid Hossen Bhuiyan et al. in Results in MaterialsVolume 20, December 2023
17. Numerical Modelling of Kevlar/Jute Fiber and Hybrid Composite Pressure Vessels by Pranti Saha et al, Trends Volume, December 2023
18. Influence of fumed silica nanofiller and stacking sequence on interlaminar fracture behaviour of bidirectional jute-Kevlar hybrid nanocomposite by Sunil Manohar Maharana on Testing Volume, January 2021
19. Jute/Hemp bio-epoxy hybrid bio-composites: Influence of stacking sequence on adhesion of fiber-matrix by Vinod et al Adhesives Volume, March 2022
20. Experimental investigation of the mechanical and water absorption properties on fiber stacking sequence and orientation of jute/carbon epoxy hybrid composites by Md. Abu Shaid Sujonet al, Technology Volume, September–October 2020.
21. Development of sandwich using low-cost natural fibers: Alfa-Epoxy composite core and jute/metallic mesh-Epoxy hybrid skin composite by Selsabil Rokia Laraba et al in Products Volume, 15 September 2022
22. Mechanical characterization of quasi-isotropic intra-ply woven carbon-Kevlar/epoxy hybrid composite by Sovit_Agarwal et al, On materials proceedings today
23. Ballistic behaviour of nano silica and rubber reinforced Kevlar/epoxy composite targets by Nitin Mathusoothanaperumal Sukanya et al, Analysis Volume, December 2022
24. Hybrid effects and interactive failure mechanisms of hybrid fiber composites under flexural loading: Carbon/Kevlar, carbon/glass, carbon/glass/Kevlar Hongyong Jiang et al, Technology Volume, February 2023
25. Mechanical characterization of a new Kevlar/Flax/epoxy hybrid composite in a sandwich structure, By Ahmed Sarwar et al, Testing Volume, October 2020.
26. Concurrent stacking sequence and layout optimization of stiffened composite plates using a spectral element method and an index-based optimization technique by Salih Alan et al in Composite StructuresVolume 327, 1 January 2024.
27. Hygrothermally stable stacking sequence for tailoring of extension–twist coupling in composite structures by Nishant K. Shakya et al, in Composite StructuresVolume 308, 15 March 2023.

Note: Every figure and table was created using the experimental work; none of them were taken from any publications or online.

Advances in Additive Manufacturing Technologies – Gurusamy Pathinettampadian (eds)
© 2024 Taylor & Francis Group, London, ISBN 978-1-032-90013-1

78 Experimental Investigation of Casted Aluminium Alloy Using Electrical Discharge Machining

Gayathri N[1]
Associate Professor, Department of Mechanical Engineering,
Vel Tech High Tech Dr. Rangarajan Dr. Sakunthala Engineering College,
Avadi, Tamil Nadu, India

Velmurugan S[2]
Research Scholar, Department of Mechanical Engineering,
Vel Tech High Tech Dr. Rangarajan Dr. Sakunthala Engineering College,
Avadi, Tamil Nadu, India

Gowtham K[3]
Student, Department of Mechanical Engineering

Sriram B[4]**, Pradeep Kumar S**[5]
Student, Department of Mechanical Engineering,
Vel Tech High Tech Dr. Rangarajan Dr. Sakunthala Engineering College,
Avadi, Tamil Nadu, India

ABSTRACT: Composite materials, particularly metal matrix composites (MMCs), play a pivotal role in diverse industries. However, machining MMCs, especially when they are reinforced with challenging materials, presents substantial challenges when traditional methods are employed. To address these challenges, Electrical Discharge Machining (EDM) materialises as a specific and active machining process for electrically conductive materials. It achieves this by generating controlled electrical sparks amongst an electrode besides a workpiece submerged in dielectric medium. EDM is particularly valuable for shaping intricate contours within MMCs. One specific area of interest centres on Aluminium Metal Matrix Composites (AMMCs), which represent the next generation of engineering materials. These materials stand out due to their exceptional physical as well as mechanical properties in comparison to the non-reinforced alloy elements. Consequently, they find applications across a wide spectrum of industries, including automotive, aerospace, and defence. AMMCs are prized for their outstanding attributes, like impressive strength-to-weight ratios, toughness and low coefficients of thermal stability. However, machining them into complex shapes for various applications is a daunting task when employing conventional methods. Traditional machining techniques, including EDM, offer an alternative solution to AMMC machining by addressing the limitations of conventional approaches, such as tool wear and costly tooling. The ongoing project investigates the utilization of AMMCs in EDM while exploring different process parameters. This research involves creating metal matrix composite materials by blending aluminum with alumina and SiC, each at a 5% concentration. Additionally, the project delves into future trends in this research field, offering insights into the continuous developments and innovations in the realm of MMC

[1]shan.gayathri@gmail.com, [2]ivelu00@gmail.com, [3]gowtham23k2002@gmail.com, [4]bsriram0403@gmail.com, [5]spradeepkumar172003@gmail.com

DOI: 10.1201/9781003545774-78

machining. A mathematical model was constructed to establish a relationship between specific input process variables and key reactions, including Material Removal Rate (MRR), Electrode Wear Rate (EWR) and Surface Roughness (SR). To ascertain the impact of these process variables, an Analysis of Variance (ANOVA) was conducted.

KEYWORDS: Metal matrix composite, Electrical discharge machining, ANOVA

1. Introduction

Aerospace material research has continually advanced tensile strength. The on-going manufacturing uprising is centred on the adoption of novel tools and energy sources, driving innovation in traditional manufacturing. Initially, stone tools were prevalent, but the introduction of iron tools marked a significant leap. In the 20th century, tools progressed to handle challenging materials like alloy steel, carbide, diamond, and ceramics. Simultaneously, power sources evolved from human and animal labor to water, wind, steam, and electricity. These advancements have empowered manufacturers to address intricate designs and resilient materials. Gopalakrishnan and Murugan et al. conducted a study to characterize AA 6061 Metal Matrix Composites (MMC) with TiC particulates reinforced, which were produced using the stir casting method. Their findings revealed that an increase in the TiC particle content within the composites led to an enhancement in specific strength while preserving the percentage of elongation. Composite materials result from combining a matrix phase and a reinforcing phase, leading to properties distinct from individual components. Reinforcements can take various forms, such as whiskers, particles, continuous fibers, or discontinuous fibers. Single-crystal ceramic whiskers and continuous ceramic fibers stand out for their strength and stiffness-enhancing capabilities. Various processing methods are used to create Aluminium Metal Matrix Composites (AMMCs), including solid-state treatment, liquid-state treatment, powder metallurgy, Physical Vapour Deposition (PVD) and spray deposition method. Customization of material properties is vital and is achieved by controlling the composition and quantity of reinforcements in conjunction with the matrix alloy. Senthilvelan et al. conducted an experimental investigation focusing on the process variables of Electrical Discharge Machining (EDM) applied to AMMCs. They utilized the liquid metallurgy process to create AMMCs reinforced with elements of SiC. Their research incorporated Response Surface Methodology (RSM) to fine-tune and optimize various response parameters. Notably, their findings underscored that the pulse current was the most prominent process variable, significantly impacting Material Removal Rate (MRR), Electrode Wear Rate (EWR) and Surface Roughness (SR). Another researcher investigated the influence of weight % as well as the element size of SiC reinforcement preceding the Electrical Discharge Machining of AMMCs. Their research unveiled that the Material Removal Rate (MRR) has improved when the size of 50 mm SiCp and wt % increased from 5% into 10%, while the MRR declined when the size of 150 mm SiCp and wt % increased from 5% into 10%. Additionally, they witnessed that the Electrode Wear Rate (EWR) for 10% SiCp was higher than that of 5% SiCp for 50 mm SiCp size. However, for 150 mm SiCp size, the EWR of 5% SiCp was higher than that of 10% SiCp [3,4]. Notably, their study aligns with the existing body of research, where numerous investigators have conducted experimental investigations on Aluminum MMCs containing SiCp at varying weight percentages. Common reinforcements include inorganic elements like alumina and silicon carbide, as well as fiber materials such as graphite, boron, silicon carbide, molybdenum, aluminium oxide and tungsten. In Electrical Discharge Machining (EDM), material removal occurs through the conversion of electrical energy into thermal energy via electrical discharges. Aggarwal et al. conducted experimental test on Wire Electrical Discharge Machining (WEDM) applied to Inconel 718 element. They harnessed Response Surface Methodology (RSM) to fine-tune and optimize the response parameters. Their research pinpointed that the variable "T_{on}" emerged as the most significant factor affecting the incidence of machining and the surface quality of Inconel 718 during the EDM process.

In the composite materials, a neural network scheme is employed to govern the optimal settings for various parameters crucial in Electrical Discharge Machining (EDM). These parameters include pulse time, pulse interlude, servo reference voltage, peak current, electric capacitance, open circuit voltage and the table speed. The objective is to use these optimized parameter settings to assess both cutting speed and the surface quality in the EDM process. [5]. Sucitharan et al. accompanied an investigation on wear resistance of a consisting of Al 6063 composite with zircon sand reinforcement, which was contrived using the stir casting procedure. They employed the Pin-on-Disk method to assess the wear enactment of the composite samples while varying the weight element. Their findings indicated that as the fraction of zircon sand contained by the composite increased, the wear resistance capability also increased. In their experimental investigation explored the process variables of Wire cut Electrical Discharge Machining (EDM) applied to SiC particles with Aluminium 6061 Metal Matrix Composites [7]. They employed Response Surface Methodology (RSM) to fine-tune and optimize the response parameters. Their research revealed that among the various process parameters, the wire feed rate and voltage played crucial roles in determining the Kerf, or the width of the cut. A voltage is applied across the cathode (tool) and the anode (workpiece) submerged in the dielectric medium, leading to the emission of electrons, spark formation, and high-temperature shock waves. This heat melts and vaporizes both electrodes, expelling molten metal by vaporizing the dielectric fluid and creating small cavities on the tool and workpiece [8,9].

2. Experimental Setup

Crucible casting is employed to create the hybrid AMMC samples, with 10% volume-based silicon carbide reinforcement. Two different silicon carbide types with average particle sizes of 200 mesh and 240 mesh are used as reinforcements for the casting process and the cast specimens measuring 30 mm in diameter, are subsequently cut into circular plates with a thickness of 12 mm, was carried out using an Electric Discharge Machine, specifically the ELECTRONICA-ELECTRAPULS PS 50ZNC. Commercially available grade of EDM oil with 0.763 specific gravity and94°C freezing point was employed as the dielectric medium. Wire-EDM was used to apply the pulsed discharge current in the positive mode in many steps after flushing the U-shaped copper tool internally at a pressure of 0.2 kgf/cm^2 using a positive polarity electrode.

The experimental tool material employed is a 99.9% pure electrolytic copper tool. The tool electrode has a width of 40mm and a total length of 50mm. All experiments were executed on EDM machine as shown in Table 78.1. The workpieces were made from prepared al-Al_2O_3 samples, and commercial copper was used as the tool material. Control parameters included powder concentration, Amps, pulse on-time and pulse off-time as shown in Table 78.2. The experiments were accompanied using an L9 orthogonal array, with the elements varied at their corresponding stages. The specific standards for various fixed and variables are conveyed in the Table 78.3.

3. Experimental Method

Originally, experimental design methods were developed by Sir Ronald A. Fisher. However, these methods can be quite intricate and not user-friendly. Moreover, when the number of process variables rises, conducting a great number of trials becomes impractical. To discourse this challenge, the Taguchi method pays specially intended orthogonal arrays to determine the intact variable space with a nominal amount of trials [11]. The outcomes attained from these trials are then transmuted into a Signal to Noise ratio (S/N ratio), which serves as a measure of the quality characteristics and how they digress from the preferred values as shown in Table 78.4. There are usually three categories when analysing the S/N ratio: "Smaller The Better," "Larger The Better," and "Nominal is Best". A higher S/N ratio is invariably associated with superior quality attributes, and

Table 78.1 Control parameters

Variable	Description
Work Piece	Al-Al_2O_3-SiC
Tool	Copper
Dielectric	I POL fluid
Flushing Pressure	0.5 kg/cm^2
Polarity	Possitive
Gap Voltage	45 V
Power Concentration	5g/L
Peak Current	8.10.12
Pulse On-Time	6,7,8 μs
Pulse Off-Time	6,7,8 μs
Machining Depth	1.0 mm

Table 78.2 Process variables and their stages

S. No	Pulse on-time	Pulse off-time	Gap Current (GC)
1	8	4	10
2	10	6	12
3	12	8	14

Table 78.3 Formation of L_9 array

T. No	Designa-tion	Pulse on-time (μs)	Pulse off-time (μs)	Pulse off-time (μs)
1	$A_1B_1C_1$	8	4	10
2	$A_1B_2C_2$	8	6	12
3	$A_1B_3C_3$	8	8	14
4	$A_2B_1C_2$	10	4	12
5	$A_2B_2C_3$	10	6	14
6	$A_2B_3C_1$	10	8	10
7	$A_3B_1C_3$	12	4	14
8	$A_3B_2C_1$	12	6	10
9	$A_3B_3C_2$	12	8	12

Table 77.4 Experimental data

No	Description	T_{ON}	T_{OFF}	AMPS	RA (in µS)	MT (in Min)	MRR (in g/min)
1	$A_1B_1C_1$	8	4	10	4.372	26	0.015
2	$A_1B_2C_2$	8	6	12	4.849	21	0.023
3	$A_1B_3C_3$	8	8	14	6.530	14	0.027
4	$A_2B_1C_2$	10	4	12	7.549	18	0.029
5	$A_2B_2C_3$	10	6	14	3.304	22	0.024
6	$A_2B_3C_1$	10	8	10	3.743	23	0.022
7	$A_3B_1C_3$	12	4	14	5.308	12	0.037
8	$A_3B_2C_1$	12	6	10	4.561	20	0.026
9	$A_3B_3C_2$	12	8	12	5.611	16	0.031

this relationship is found for each stage of a process parameter. As a result, the optimal stage of the process variable is the one associated with the highest S/N ratio. By employing statistical methods like S/N analysis and ANOVA, it becomes possible to forecast the optimal combination of process variables that will yield the desired results as presented in Table 78.5. Finally, to validate the optimal process variables determined through the parameter design, a confirmation experiment is conducted. There are three frequently used S/N ratios for optimizing such static problems. These formulae for computing the S/N ratio are considered to enable experimenters to consistently select the factor stage settings that heighten the quality features in an experiment. The three categories of Signal-to-Noise ratios are "Smaller The Better," "Larger The Better" and "Nominal is Best."

4. Experimental Output Response and Optimization

5. Results and Discussion

Regression analysis was utilized to model the response parameters with the aid of MINITAB software. The resulting model is crucial for evaluating the significance of input process variables, investigating interrelationships among

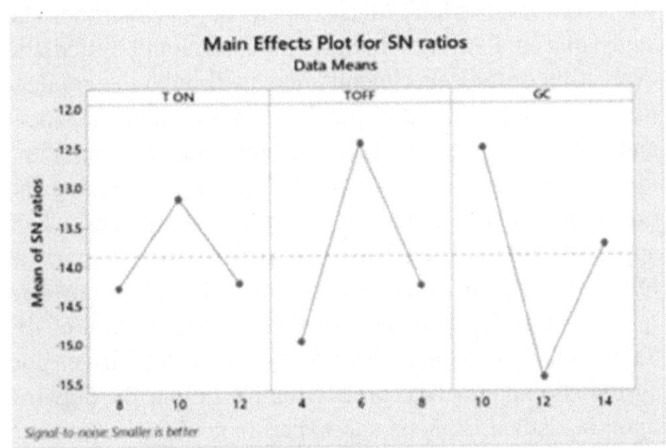

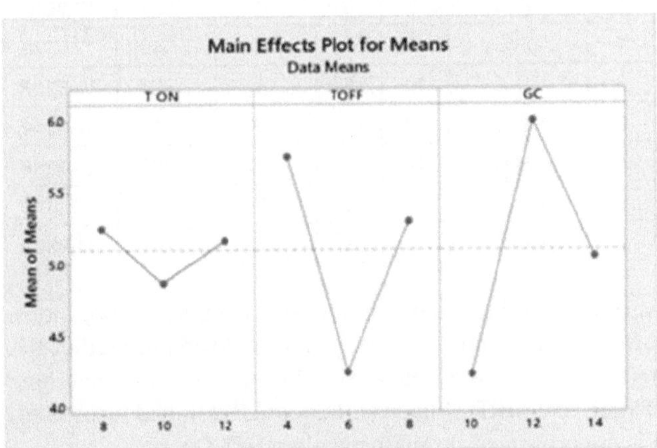

Fig. 78.1 Surface roughness (a) Main-EfectplotforS/Nratios Vs RA, (b) Main-EfectplotforMeans Vs RA

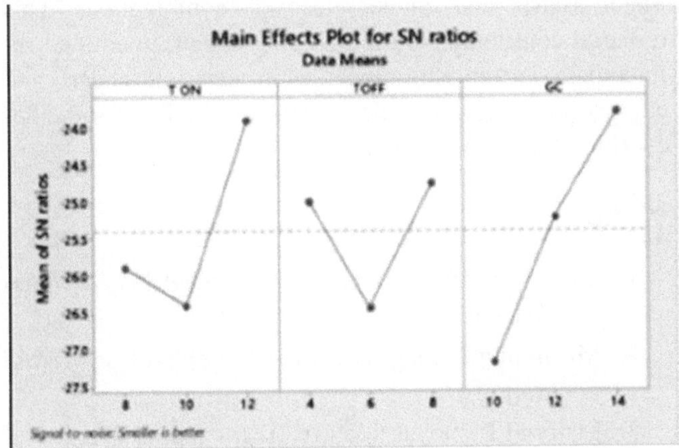

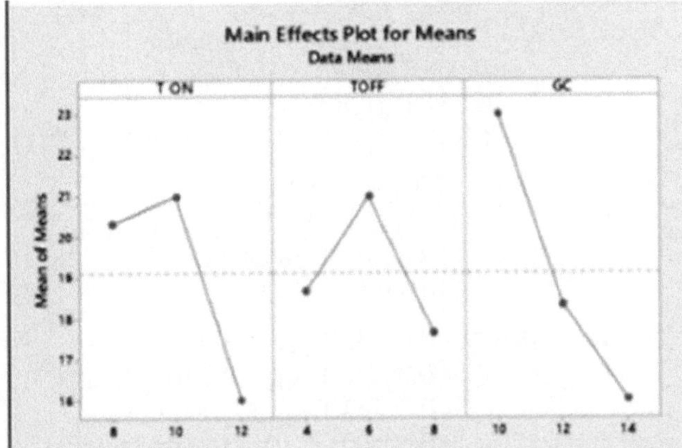

Fig. 78.2 Machining time

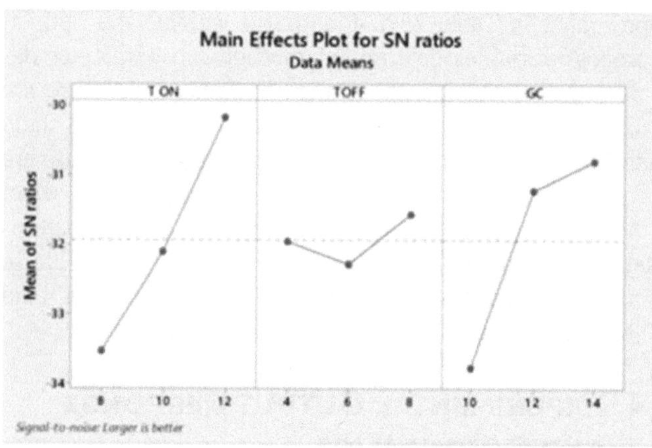

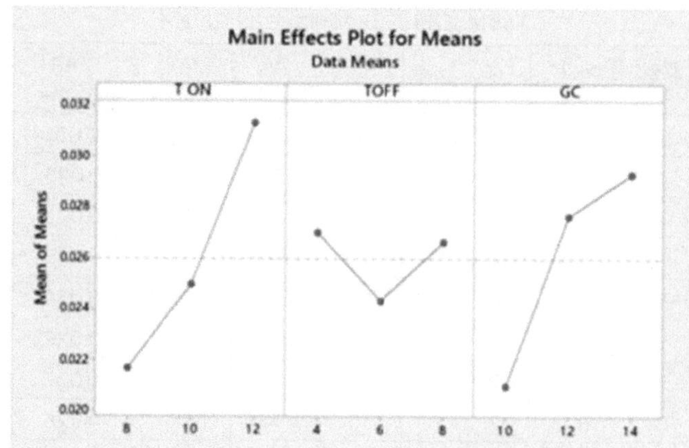

Fig. 78.3 Material removal rate (a) Main-EfectplotforS/Nratios Vs MRR, (b) Main-EfectplotforMeans Vs MRR

Table 78.5 Surface roughness (RA) values and S/N ratios values for the trails

Trial	Description	T_{ON}	T_{OFF}	AMPS	RA	SNRA1
1	$A_1B_1C_1$	8	4	10	4.372	-12.8136
2	$A_1B_2C_2$	8	6	12	4.849	-13.7130
3	$A_1B_3C_3$	8	8	14	6.530	-16.2983
4	$A_2B_1C_2$	10	4	12	7.549	-17.5578
5	$A_2B_2C_3$	10	6	14	3.304	-10.3808
6	$A_2B_3C_1$	10	8	10	3.743	-11.4644
7	$A_3B_1C_3$	12	4	14	5.308	-14.4986
8	$A_3B_2C_1$	12	6	10	4.561	-13.1812
9	$A_3B_3C_2$	12	8	12	5.611	-14.9808

various variables and considering quadratic expressions. Mathematical models have been established to predict RA (Response Rough Surface), MT (Response Machining Time), and MRR (Material Removal Rate), and these models are presented through equations 1,2,3.

Mathematical modelling for RA, MT and MRR Regression Equation;

$$\mathbf{RA} = 5.092 + 0.158\ T_{ON8} - 0.227\ T_{ON10} + 0.068\ T_{ON12} + 0.651\ T_{OFF4} - 0.854\ T_{OFF6} + 0.203\ T_{OFF8} - 0.867\ _{GC10} + 0.911\ _{GC12} - 0.045\ _{GC14} \quad (1)$$

$$\mathbf{MT} = 19.11 + 1.22\ T_{ON8} + 1.89\ T_{ON10} - 3.11\ T_{ON12} - 0.44\ T_{OFF4} + 1.89\ T_{OFF6} - 1.44\ T_{OFF8} + 3.89\ GC10 - 0.78\ GC12 - 3.11\ GC14 \quad (2)$$

$$\mathbf{MRR} = 0.02600 - 0.00433\ T_{ON8} - 0.00100\ T_{ON10} + 0.00533\ T_{ON12} + 0.00100\ T_{OFF4} - 0.00167\ T_{OFF6} + 0.00067\ T_{OFF8} - 0.00500\ GC10 + 0.00167\ GC12 + 0.00333\ GC14 \quad (3)$$

6. Conclusion

In this research work, the utilization of the Taguchi technique and Analysis of Variance (ANOVA) played a crucial role in determining the most effective Electrical Discharge Machining (EDM) parameters, specifically when dealing with Aluminium Metal Matrix and a copper electrode. The study placed a strong emphasis on careful and systematic evaluation of the experimental results through the application of the Taguchi technique. By utilizing these statistical and optimization tools, the research aimed to identify the ideal set of parameters for EDM, thereby enhancing the precision and efficiency of the machining process. This approach not only enables the optimization of material removal rates and surface finish but also contributes to resource and time savings. The study's application of the Taguchi technique and ANOVA signifies a data-driven and systematic approach to process optimization. By drawing upon statistical analysis and experimentation, it strives to provide valuable insights and guidance for achieving the best possible EDM results in the context of Aluminium Metal Matrix and copper electrode combinations. This research contributes to the on going advancement and refinement of machining processes in materials science and engineering. The experiment's outcomes allow for the deduction of the following conclusions:

7. Optimal Control Factor

1. Surface Roughness-A_3(T_{ON} – 12 μs) B_2(T_{OFF} – 6 μs) C_1(Amps-10)
2. Machining Timing- A_2(T_{ON} – 10 μs) B_3(T_{OFF} – 8 μs) C_1(Amps-10)
3. Material Removal Rate- A_1(T_{ON} – 8 μs) B_3(T_{OFF} – 8 μs) C_2(Amps-12)

8. Percentage of Contribution of Process Parameter

1. Surface Roughness – AMPS- 33%
2. Machining Timing – AMPS- 47 %
3. Material Removal Rate – Pulse on-time 47%

The majority variables affect with Ampere rating MRR only affect with pulse on-time of the Aluminium metal matrix combination.

References

1. S. Gopalakrishnan, N. Murugan,Production and wear characterisation of AA 6061 matrix titanium carbide particulate reinforced composite by enhanced stir casting method, Composites Part B: Engineering, Volume 43, Issue 2, March 2012, Pages 302–308
2. S. Gopalakannan, T. Senthilvelan, Optimization of machining parameters for EDM operations based on central composite design and desirability approach, Journal of Mechanical Science and Technology, 28 (3) (2014) 1045–1053.
3. K.M. Patel, P.M. Pandey, P.V. Rao, Understanding the Role of Weight Percentage and Size of Silicon Carbide Particulate Reinforcement on Electro-Discharge Machining of Aluminium-Based Composites, Materials and Manufacturing Processes, 23 (7) (2008) 665–673.
4. Anjani Srivastava, Sunil Kumar Yadav , D.K. Singh, Modeling and Optimization of Electric Discharge Machining Process Parameters in machining of Al 6061/SiCp Metal Matrix Composite, Materials Today: Proceedings 44 (2021) 1169–1174
5. V. Aggarwal, S.S. Khangura, R.K. Garg, Parametric modeling and optimization for wire electrical discharge machining of Inconel 718 using response surface methodology The International Journal of Advanced Manufacturing Technolog, 79 (2015) 31–47.
6. K.S. Sucitharan, P.S. Kumar, D. Shivalingappa, J. Jenix Rino, An Overview on Development of Aluminium Metal Matrix Composites with Hybrid Reinforcement, International Journal of Science and Research (IJSR), India Online ISSN: 2319–7064.
7. Tarng, Y.S., Ma, S.C. and Chung, L.K. 1995. Determination of optimal cutting parameters in wire electrical discharge machining. International Journal of Machine Tools and Manufacture, Vol. 35, No 129, pp.1693–170.
8. Pragya Shandilya, P.K. Jain, N.K. Jain, Parametric Optimization During Wire Electrical Discharge Machining using Response Surface Methodology, Procedia Engineering 38 (2012) 2371–2377
9. Patil NG and Brahmankar PK. Determination of material removal rate in wire electro-discharge machining of metal matrix composites using dimensional analysis. International Journal of Advanced Manufacturing Technology 2010; 48; 537–555.
10. Sir Ronald A. Fisher, The Design of Experiments, ninth edition 1971.
11. J.H. Jung, W.T. Kwon, Optimization of EDM process for multiple performance characteristics using Taguchi method and Grey relational analysis, Journal of Mechanical Science and Technology 24 (5) (2010) 1083~1090.
12. Milan Kumar Das, Kaushik Kumar, Tapan Kr. Barman, Prasanta Sahoo, Application of Artificial Bee Colony Algorithm for Optimization of MRR and Surface Roughness in EDM of EN31 Tool Steel, Procedia Materials Science 6 (2014) 741–751.

Note: Every figure and table was created using the experimental work; none of them were taken from any publications or online.

Advances in Additive Manufacturing Technologies – Gurusamy Pathinettampadian (eds)
© 2024 Taylor & Francis Group, London, ISBN 978-1-032-90013-1

79 Experimental Investigation of Mechanical Behaviour, Wear, Microstructure of Aluminium, Silicon Carbide and Zirconia

Velmurugan S[1]
Research Scholar, Department of Mechanical Engineering,
Vel Tech High Tech Dr. Rangarajan Dr. Sakunthala Engineering College,
Avadi, Tamil Nadu, India

Gayathri N[2]
Associate Professor, Department of Mechanical Engineering,
Vel Tech High Tech Dr. Rangarajan Dr. Sakunthala Engineering College,
Avadi, Tamil Nadu, India

Gurusamy P[3]
Professor, Department of Mechanical Engineering,
Chennai Institute of Technology,
Tamil Nadu, India

Akash A P[4], **Gokul R**[5], **Hari K**[6]
Student, Department of Mechanical Engineering,
Vel Tech High Tech Dr. Rangarajan Dr. Sakunthala Engineering College,
Avadi, Tamil Nadu, India

ABSTRACT: This experiment provides a comprehensive analysis of the mechanical properties, wear resistance and microstructural characteristics of a metal matrix composite composed of 88% aluminum, 6% silicon carbide, and 6% zirconia. Employing low-magnification optical microscopy, the research scrutinized the microstructural aspects. Both the metal matrix composite and a cast aluminum alloy underwent a series of tests to assess their hardness and impact resistance. Wear resistance was evaluated using a pin-on-disc device under controlled conditions. Light optical microscopy facilitated an examination of the distribution of reinforcing particles, interfacial bonding, grain structure, as well as the presence of porosity or defects. The initial findings suggest that the metal matrix composite exhibits superior mechanical characteristics compared to the cast aluminum alloy. The incorporation of silicon carbide and zirconia particles enhances the composite's impact strength and hardness, rendering it well-suited for applications requiring high mechanical strength. Results from the wear tests indicate enhanced wear resistance in the composite due to the presence of additional reinforcing particles. Microscopic analysis demonstrates a uniform dispersion of silicon carbide and zirconia particles within the aluminum matrix, indicating strong interfacial bonding and efficient load transfer. The investigation also encompasses an evaluation of the aluminum matrix for porosity, defects, and grain structure, recognizing these factors as crucial elements that significantly influence the overall performance and properties of the composite.

KEYWORDS: Metal matrix composite, Casing, Physical properties

[1]ivelu00@gmail.com, [2]shan.gayathri@gmail.com, [3]gurusamyp@citchennai.net, [4]akashsalomon95@gmail.com, [5]gokulrajesh07@gmail.com, [6]harikannan762@gmail.com

DOI: 10.1201/9781003545774-79

1. Introduction

The experiment emphasizes the importance of aluminum, a versatile metal renowned for its durability and applicability in numerous industries such as construction, aerospace, and automotive manufacturing. Ongoing research is dedicated to exploring and utilizing the unique properties of aluminum, focusing on the development of advanced crystal structures and nanoscale materials for industrial purposes .From the comprehensive literature survey, it is apparent that while the application scope for Advanced Metal Matrix Composites (AMCs) is widening, a significant challenge lies in the large-scale industrial production of these materials. The two predominant approaches for the production of AMCs are solid-state processing and liquid-state processing [1]. A recent study investigated nano-composites, blending Al2O3 and ZrO2 in various ratios and comparing their mechanical properties to standard zirconia and alumina [2]. Researchers adjusted sintering temperatures to assess resulting density and properties influenced by microstructural development. Another study examined niobium composites, exploring the effects of varied Al2O3 and ZrO2 proportions on thermal conductivity concerning temperature and pore size. Metal matrix composites (MMCs) have garnered considerable attention due to their outstanding mechanical properties and wear resistance [3,4]. Stir casting for Metal Matrix Composites (MMCs) presents significant challenges that must be addressed for the successful production of composites with diverse mechanical properties. These challenges include difficulties in achieving a uniform distribution of the reinforcement material, ensuring proper wettability between the matrix alloy and the reinforcement material, managing porosity in cast metal matrix composites, and addressing chemical reactions between the reinforcement material and the matrix alloy. Overcoming these challenges is crucial to achieving MMCs with a broad range of mechanical properties. The researchers have also identified key process variables, such as holding temperature, stirring speed, impeller size, and impeller position in the melt, that play a pivotal role in influencing the mechanical properties of MMCs during production. Careful consideration and control of these process variables are essential for optimizing the stir casting process and ensuring the desired characteristics in the resulting Metal Matrix Composites [5]. Integrating zirconia into the aluminum alloy induces alterations in the microstructure. Zirconia particles influence grain size, grain boundaries, and phase distribution within the composite. These alterations have the potential to affect various mechanical properties such as strength, toughness, and resistance to fatigue. Extending the research in this area could involve more detailed investigations into the specific mechanisms by which zirconia influences the microstructure and the subsequent impact on mechanical properties. Additionally, further studies could explore optimizing the zirconia-aluminum alloy combination for tailored applications in different industries, enhancing the understanding and potential utilization of these composite materials. Metal matrix composites (MMCs) have gained attention due to their exceptional mechanical properties and wear resistance. A specific MMC, consisting of 88% aluminum, 6% silicon carbide, and 6% zirconia, was developed through powder metallurgy. The study aimed to evaluate its mechanical properties, wear resistance, and microstructural characteristics. This composite was created by integrating SiC and ZrO2 within an aluminum matrix using processes such as powder blending, stirring, compacting, and sintering. Mechanical tests demonstrated improvements in impact strength, hardness, and wear resistance compared to pure aluminum. Microstructural analysis revealed consistent particle dispersion and successful bonding with the aluminum matrix.The study's findings suggest promising applications for the 88% aluminum, 6% silicon carbide, and 6% zirconia metal matrix composite due to its enhanced mechanical strength and wear resistance. This composite holds potential for various applications that require robust mechanical properties and resilience against wear. Further research could explore additional potential applications, optimize the fabrication process, and delve into the composite's behavior under various operational conditions, potentially expanding its utility in different industries. The exploration and utilization of aluminum's unique properties hold significant importance across diverse industries, such as construction, aerospace, and automotive manufacturing. Ongoing research efforts are specifically directed towards harnessing the remarkable features of aluminum, focusing on the development of advanced crystal structures and nanoscale materials for varied industrial applications.

2. Experimental Method

Metal casting is a widely used manufacturing process where molten metal is poured into a mold, allowed to cool, and solidify into a final product known as a casting. Sand casting, a prevalent method in this category, involves using sand to create the mold cavity and internal structures called cores. This cost-effective and versatile technique accounts for over 70% of all metal castings produced. The process begins with preparing the mold by placing a pattern, a replica of the intended casting, in the sand. The sand is mixed with a bonding agent, usually clay, and moistened for workability. After compacting the sand around the pattern to form the mold cavity and gating system, the pattern is removed, leaving the desired cavity. Molten metal is then poured into the mold through the gating system, solidifying into the shape of the cavity. Once cooled, the sand mold is removed to reveal the casting, which is then cleaned to eliminate any residual sand or impurities. Despite advantages like cost-efficiency and suitability for various metals, sand casting has limitations such as lower dimensional accuracy and size constraints on produced castings. Future research could explore innovations to enhance accuracy and enable the production of larger like 88% aluminum, 6% silicon carbide, and 6% zirconia metal matrix composite using the sand casting method. Additionally, advancements in materials and technologies may further enhance the versatility and efficiency of sand casting across diverse manufacturing applications. The findings shed light on the distinctive characteristics of these materials and highlight their potential applications in fields requiring enhanced mechanical properties and wear resistance. To further comprehend the potential of this composite, additional studies could be conducted to refine the fabrication process, explore further mechanical properties, and investigate the behavior of the composite under diverse operating conditions. Extending this research could facilitate a deeper understanding of this composite's capabilities and expand its potential applications in various industrial settings.

3. Result

3.1 Impact Test

The Charpy impact test is a widely used method for measuring the energy absorbed by a material upon fracturing, providing insights into its notch toughness and ductile-brittle transition at different temperatures. Known for its simplicity, ease of execution, and cost-effectiveness, the test involves releasing a pendulum with known mass and length from a specified height to impact a notched material specimen. Developed in the early 1900s by S. B. Russell and G. Charpy, the test is now named after Georges Charpy. Standards like ASTM E23, ISO 148-1, or EN 10045-1 provide detailed guidelines for conducting the test on metallic materials, specifying parameters such as energy capacity, striking angle, velocity, and sticker edge radius. These standardized methods ensure precise and consistent evaluation of materials in various industries. The results are shown in the Table 79.1 and apparatus shown in Fig. 79.1d.

Table 79.1 Charpy impact test Result

Specimen Size in mm	Notch Type	Test Temperature	Absorbed Energy-Joules			Average
			Parameters/Unit			
10 x 10 x 75	UN Notch	240C	14	14	14	14

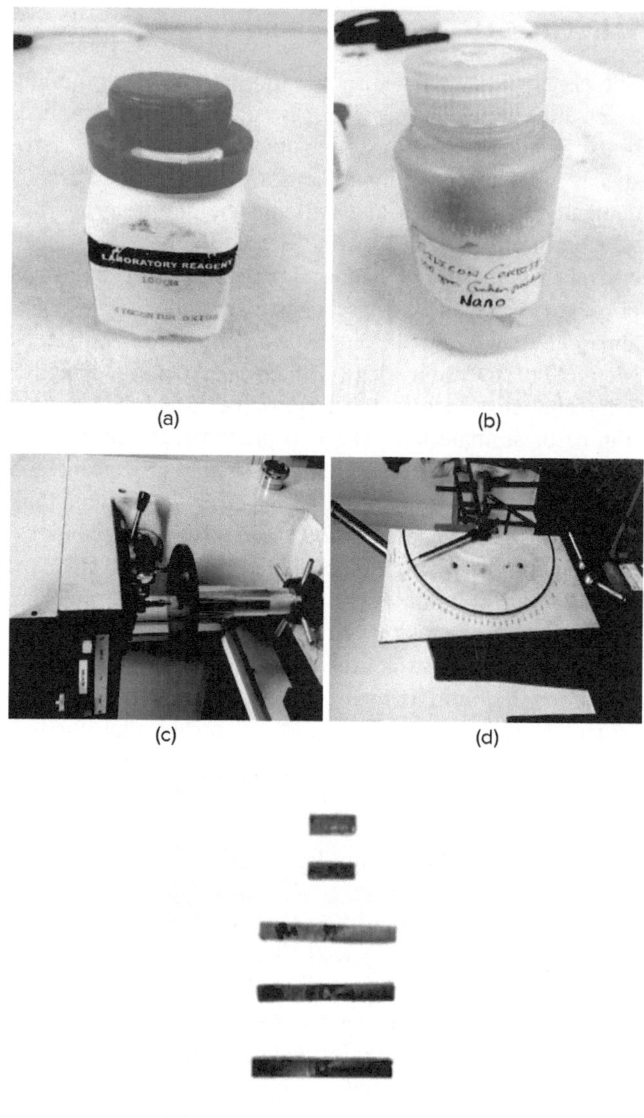

Fig. 79.1 (a) Zirconia or zirconium oxide; (b) Silicon carbide, (c) Vicker's Hardness test, (d) Charpy Impact Test, (e) Casted material after machining process

3.2 Hardness Test

The Vickers hardness test, introduced by Robert L. Smith and George E. Sandland at Vickers Ltd in 1921, provides a straightforward alternative to the Brinell method for assessing material hardness. A key advantage is its applicability to all materials, with calculations independent of indenter size. Unlike other methods, the Vickers test evaluates a material's resistance to plastic deformation due to a standardized force, making it versatile across metal types. The hardness value obtained, known as the Vickers Pyramid Number (HV) or Diamond Pyramid Hardness (DPH), is not a measure of pressure but reflects hardness. However, in cases like the described indentation on case-hardened steel, variations in diagonal lengths and illumination gradient indicate an improperly leveled sample, rendering the indentation unsatisfactory. The Vickers test employs a diamond indenter with an included angle of 136°, chosen for geometric similarity and resistance to self-deformation. Experiments revealed consistent hardness values on homogeneous materials, allowing for the application of varying loads based on material hardness during the test. The results are shown in the Table 79.2 and apparatus shown in Fig. 79.1c.

Table 79.2 Hardness test

Test Parameters	Observed Values		
Observed Values HV 5Kg	31.6	30.3	31.0

4. Microstructure Examination

Microstructure, identifiable under a microscope at magnifications exceeding 25 times, pertains to the organization and attributes of microscopic constituents within a material. It presents critical information about a material's internal structure and composition across diverse material categories like metals, polymers, ceramics, or composites. The microstructure significantly influences various material properties, including strength, toughness, ductility, hardness, corrosion resistance, and high/low temperature behavior, profoundly impacting their applicability in different industrial contexts. The shape and size of reinforcement particles in metal matrix composites, like silicon carbide (SiC) and zirconia (ZrO2), also influence their appearance and role in the composite material. Considering a metal matrix composite of 88% aluminum, 6% silicon carbide, and 6% zirconia under light optical microscopy, observations could entail distinguishing the matrix phase (appearing bright) composed of aluminum and recognizing the reinforcing particles (appearing dark) of silicon carbide and zirconia within the matrix, along with variations in their shapes and sizes, depending on the manufacturing process. These observations provide insights into the fundamental characteristics of the composite material.

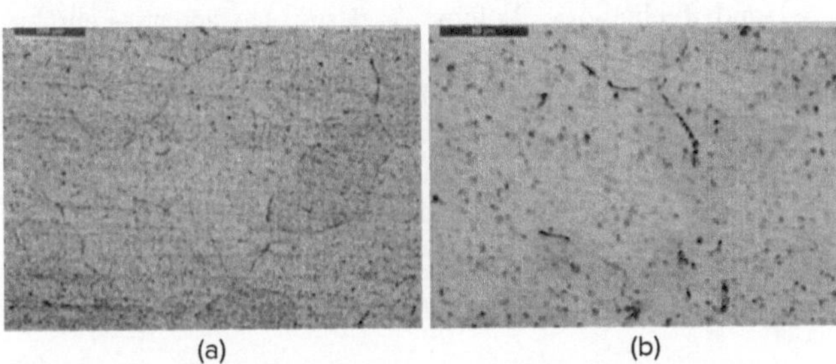

(a) (b)

Fig. 79.2 (a) Microstructure 100x, (b) Microstructure 500x

5. Wear Test

To assess material wear resistance, especially in metal matrix composites, a common method involves employing a pin-on-disc apparatus for conducting wear tests. Here's an outline of the procedure involved in such testing: Sample preparation: Begin by cutting and shaping the metal matrix composites according to specific specifications. These samples can be in the form of discs or blocks, depending on the requirements of the pin-on-disc apparatus. The duration of the test depends on the material's anticipated wear rate and the particular requirements. Longer test durations generally provide more reliable data on wear characteristics. Data collection: Throughout the test, gather data on wear volume, wear rate, and coefficient of friction. Various methods can be employed, such as weighing the materials before and after the test, employing profilometry or optical microscopy. Measure the depth and width of the wear track to determine the material lost due to wear. Tabulate and display the wear loss data of both samples, potentially shown in a graph format. From the test results, it was observed that the wear loss in Al-SiC-Zirconia samples is 50% less than that of the aluminum alloy sample. This finding indicates a notable reduction in wear experienced by the Al-SiC-Zirconia samples compared to the aluminum alloy sample.

(a) (b)

Fig. 79.3 (a) Pin-on-Disc apparatus, (b) Wear apparatus with sample

Table 79.3 Wear test parameters

Test Speed rpm	Pin Diameter mm	Track radius mm	Normal Load Applied N	Test Duration Mins
200	5	70	10	12

Table 79.4 Mass loss before and after test

	Mass before test	Mass After Test
Aluminium Alloy	10.5	10.4
Aluminium-SiC-Zirconia	12.6	12.5

Table 79.5 Wear loss and co-efficient of friction

	Wear Loss	Co-Efficient of friction
Aluminium Alloy	0.02	0.64
Aluminium-SiC-Zirconia	0.01	0.54

6. Discussion

This study offers a comprehensive investigation into the mechanical traits, wear resistance, and microstructural features of a metal matrix composite comprised of 88% aluminum, 6% silicon carbide, and 6% zirconia. Employing low-magnification optical microscopy, the examination centered on microstructural aspects. Both the metal matrix composite and a cast aluminum alloy underwent a battery of tests to evaluate their hardness and impact resistance. To assess wear resistance, controlled conditions were maintained using a pin-on-disc apparatus. Light optical microscopy facilitated the scrutiny of reinforcing particle distribution, interfacial bonding, grain structure, and the presence of porosity or defects. Initial observations suggest that the metal matrix composite surpasses the cast aluminum alloy in mechanical properties. The introduction of silicon carbide and zirconia particles fortifies the composite, enhancing its impact strength and hardness, making it well-suited for applications necessitating heightened mechanical strength. Wear tests indicate that the composite demonstrates superior wear resistance owing to the additional reinforcing particles. Microscopic analysis reveals a uniform dispersion of silicon carbide and zirconia particles within the aluminum matrix, indicating robust interfacial bonding and efficient load transfer. The study also covers an evaluation of the aluminum matrix for porosity, defects, and grain structure, recognizing their significant influence on the overall performance and properties of the composite. This experiment provides valuable insights into the mechanical, wear, and microstructural properties of both the cast aluminum alloy and the 88% aluminum-6% silicon carbide-6% zirconia metal matrix composite. It illuminates the distinct characteristics of these materials and underscores their potential applications in fields requiring heightened mechanical properties and wear resistance. To gain further insight into the potential of these composite, additional studies might focus on refining the fabrication process, exploring more mechanical properties, and analyzing the composite's behavior under diverse operating conditions. Further research could deepen understanding of this composite's capabilities and broaden its potential applications across various industrial settings.

7. Conclusion

The experiments involving the casted aluminum alloy and the 88% aluminum-6% silicon carbide-6% zirconia metal matrix composite offered valuable insights into their mechanical, wear, and microstructural characteristics. The findings highlight the superior mechanical properties and wear resistance of the metal matrix composite in comparison to the casted aluminum alloy. With its integration of silicon carbide and zirconia particles, the composite is suitable for applications requiring high mechanical strength and wear resistance. These components significantly enhance impact strength, hardness, and wear resistance. The metal matrix composite, featuring the amalgamation of silicon carbide and zirconia particles, holds potential for sectors that require high-performance materials, such as aerospace, automotive, and manufacturing. Continued research into this composite's potential applications could focus on refining the fabrication process, analyzing additional mechanical properties (e.g., tensile strength, fatigue resistance), and investigating its performance under specific operating conditions.

References

1. Surappa M.K, Aluminium Matrix Composites: Challenges and opportunities, 28, 2003, 319–334.
2. W.H. Tuan, R.Z. Chen, T.C. Wang, C.H. Cheng, P.S. Kuo, Mechanical properties of Al2O3/ZrO2 composites, Journal of the European Ceramic Society, Volume 22, Issue 16, December 2002, Pages 2827–2833
3. Deuk Yong Lee, Dae-Joon Kim, Yo-Seung Song, Chromaticity, hydrothermal stability, and mechanical properties of t-ZrO2/Al2O3 composites doped with yttrium, niobium, and ferric oxides, Materials Science and Engineering: A, Volume 289, Issues 1–2, 30 September 2000, Pages 1–7.
4. Shreyas P.S, B.P. Mahesh, S. Rajanna, N. Rajesh, Evaluating the tribological properties of aluminium based hybrid composites reinforced with Al2O3 & ZrO2 nano particles, Materials Today: Proceedings Volume 45, Part 1, 2021, Pages 429–433
5. Hashim J, L.Looney & M..S.J.Hashmi, Metal matrix composites: Production by the stir casting method, Journal of Materials Processing Technology, 92, 1999, 1–7.
6. Marcin Winnicki, Advanced Functional Metal-Ceramic and Ceramic Coatings Deposited by Low-Pressure Cold Spraying: A Review, Coatings, 2021, 11, 1044
7. Emília Kubiňáková, Vladimír Danielik, Ján Híveš, Al–Zr alloys synthesis: characterization of suitable multicomponent low-temperature melts, Journal of Materials Research and Technology, Volume 9, Issue 1, January–February 2020, Pages 594–600
8. P. Saritha, Dr. A. Satyadevi, Dr. P. Ravikanth Raju, N. Swapna Sri, Effect of Zirconium on Mechanical Behavior of Aluminum7075, International Journal of Science and Research (IJSR), Volume 7 Issue 3.

Note: Every figure and table was created using the experimental work; none of them were taken from any publications or online.

Advances in Additive Manufacturing Technologies – Gurusamy Pathinettampadian (eds)
© 2024 Taylor & Francis Group, London, ISBN 978-1-032-90013-1

80 Dry and Wet Sliding Wear Behaviour of Biochar Dispersed Rock Wool Polyster Composite for Brake Pad Application

N. Gayathri*, S. Pooja,
D. Samsujith, G. Santhosh
Department of Mechanical Engineering,
Vel Tech HighTech Dr. Rangarajan Dr. Sakunthala Engineering College,
Chennai, Tamilnadu

ABSTRACT: The solid residue that remains after biomasses are thermally cracked in an environment with little oxygen is called biochar (BC). BC has been investigated more and more recently as a feasible, affordable, and sustainable substitute for conventional carbonaceous fillers in the creation of polymer-based composites. In actuality, BC has excellent electrical conductivity, high surface area, and high thermal stability. Its primary characteristics can also be appropriately adjusted by regulating the production process's parameters. Due to its fascinating qualities, BC is presently competing with high-performing fillers in the creation of composites based on polymers that have several functions and induce high mechanical and electrical properties. Additionally, BC can be produced from a wide range of biomass sources, such as post-consumer agricultural wastes, offering an intriguing chance to move toward a circular bio economy with "zero waste." The primary goal of this work is to present a thorough summary of the key developments that have been made possible by integrating BC with various thermoplastic and thermosetting matrices. A detailed analysis of how adding BC affected various polymer matrices' overall performance will be provided, emphasizing the impact of various BC generated differently on the behavior and ultimate performance of the resulting composites. Finally, there will be a comparison between BC and other carbonaceous fillers.

KEYWORDS: Natural fiber, Polymer, SEM, Mechanical properties, Wear

1. INTRODUCTION

Composites are a multi-phase material in which two or more materials are combined into a single material to optimize one or more properties. Composite is a system in which one phase is usually continuous (matrix) and the other phase (reinforcement) is dispersed within the continuous phase. They may be fibrous or particulates. The reinforcement and matrix phases can be metallic, ceramic and polymeric in composite materials. The different level of properties can be attained with the incorporation of different reinforcements into different types of matrix materials. The main aim of producing a composite material is to create a material the combines the best properties of both the reinforcement and the matrix. In some situations, a composite can have better overall properties than individual properties of reinforcement and the matrix. To achieve this, one of the best ways is better interaction of the reinforcement and the matrix in a composite system. There is an interfacial region within the composite material where the matrix

*Corresponding author: gayathri@velhightech.com

DOI: 10.1201/9781003545774-80

and reinforcement bond to each other for the improvement of properties of the composite. Composite materials have an effective substitution of the traditional materials in numerous engineering applications. High toughness, high creep resistance, and high strength-to-weight ratio are the primary factors in the choice of composite. The matrix of composite materials is a tough or ductile substance, and the reinforcement is strong and low density. Therefore, the correct percentage of mixing of reinforcements and the matrix with the correct manufacturing method gives the composite with better properties. Several functions are contributed by both the reinforcement and the matrix to achieve the overall properties of the prepared composite materials. The functions of reinforcement are to give desired properties, Load carrying capacity to transfer the applied load to the matrix.

1.1 Polymer Matrix Composites (PMCs)

An organic polymer matrix holds a range of reinforcements—continuous or short—together to form Polymer Matrix Composites (PMCs). The main purpose of the reinforcements in PMCs is to increase fracture toughness while also offering high strength and stiffness. The PMC is intended for use in a variety of fields where the reinforcement supports the mechanical loads that are applied during operation. In PMCs, the matrix serves to transfer applied loads between the reinforcements and the matrix as well as to bind the reinforcements together.

2. Materials and Methods

The materials and processing technique used to create natural fiber reinforced composites are described in length in this chapter. Reports have been made on the specifics of the procedure and different characterizations of the produced composites. The goal of the current study is to use the hand-lay approach to develop and analyze composites based on natural fibers and their hybrid composites.

2.1 Fabrication Process

In this work, jute fiber, rockwool fiber, chopped glass fibre, areca fiber are used for fabricating the composite material with Biochar and were prepared by hand layup method. Take appropriate amount of polyester resin with respect to the size and thickness of your composites material along with wastage. Here we have taken 350ml of polyester resin. Now add filler material to polyester resin and mix it with the help of wooden stick. Here 5% Silicon Nitride (Si3N4) is added as filler with respect to weight of polyester resin. Now add biochar of required percentage to the polyster resin. Now take 1wt % of accelerator (cobalt octoate) and 1 wt % methyl ethyl ketone peroxide (catalyst) and mix them with polyester resin till uniform mixture is obtained.

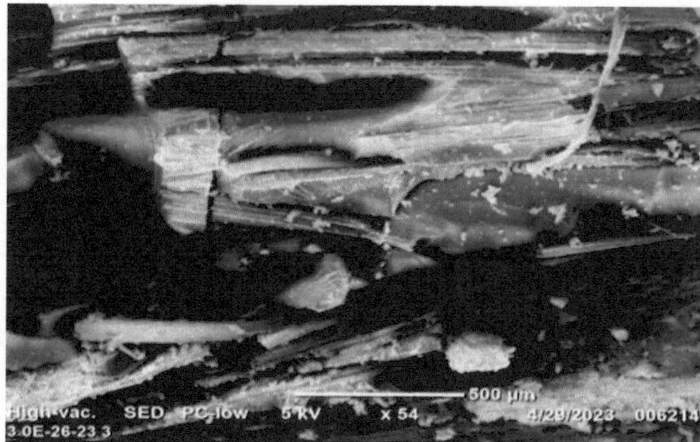

Fig. 80.1 SEM images of the natural fiber polymer composite

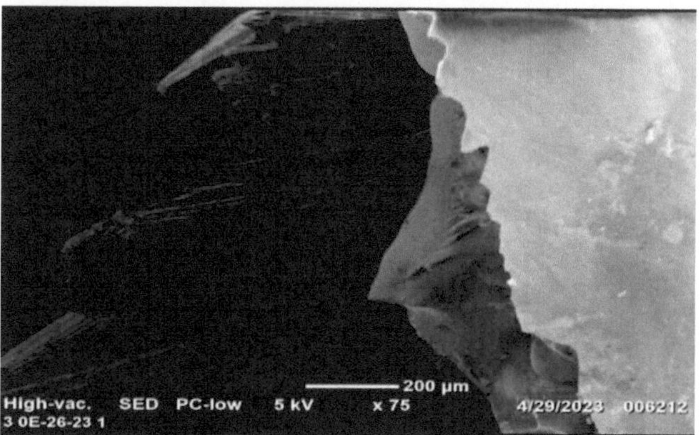

Fig. 80.2 SEM images of the natural fiber polymer composite

Next, small pieces of jute, rockwool, chopped glass, and areca fiber were added to the mixture and thoroughly mixed until a homogenous mixture was achieved. Secure the PVC sheets to the table with firmness. Next, using cardboard, construct the open mold (20 cm by 20 cm) on the PVC sheets. After that, spread the mixture around the mold and pour the mixed mixture onto it. Next, add another sheet on top of the specimen. To release the trapped gases and bubbles in the specimens, a metallic roller was rolled above the sheets once the fiber mixture had been thoroughly saturated by the resin.

Table 80.1 Tensile strength of samples

S. No	Sample ID	Tensile Strength (N/mm^2)
1	C-1	36.48
2	C-2	35.84

After being laid up, the mold was sealed tightly and allowed to cure for a full day while supporting a load of around 25 kg. Test specimens of the necessary size were cut from the sheets and utilized for testing after the samples were positioned. Next, the subsequent mechanical test was conducted. Testing methods include tensile, flexural, compression, wear, hardness, impact, and energy dispersive X-rays (EDX).

3. Results and Conclusion

3.1 Morphological Study

Morphological study was carried out using Scanning Electron Microscopy (SEM) to analyze the adhesion between the resin and fibers,fracture morphology of the mechanically tested specimens. Gold coating was applied on the fracture surface to increase the conductivity since the specimens were polymeric materials. The accelerated voltage of 5 kV was used while carrying out the experiment.

3.2 Tensile Strength Test

Tensile test was conducted to calculate the tensile strength and tensile modulus or young's modulus of the materials. To determine the materials' tensile strength, tensile modulus, and young's modulus, tensile tests were performed. Tensile testing was conducted at a test speed of 5 mm/min in accordance with ASTM D3039 standard. The test speed was held constant between 1.3 and 1.5 mm/min. The universal testing machine, Instron 3369, was used to conduct the test. After testing six specimens, the average result was taken into consideration for discussion.

Samplec-1

Graph Type : Stress Vs. Strain

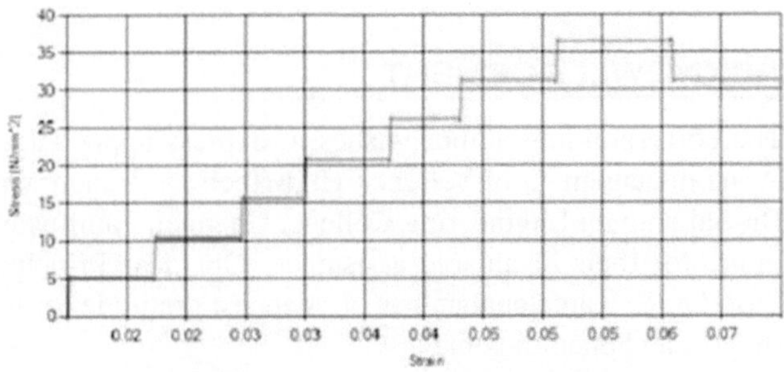

Fig. 80.3 Plain composite sample 1

Samplec-2

Graph Type : Stress Vs. Strain

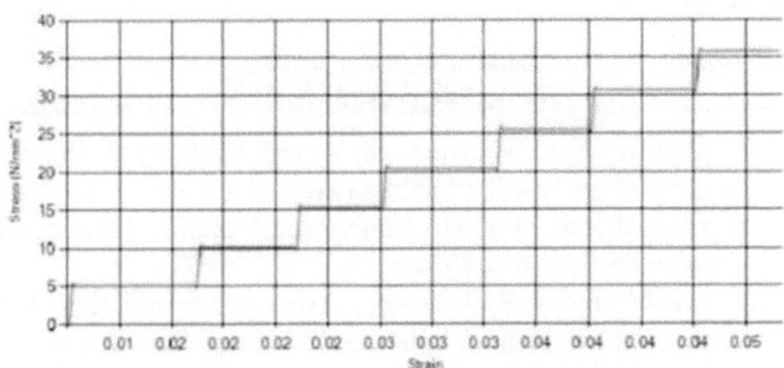

Fig. 80.4 Plain composite sample 2

3.3 Flexural Strength Analysis

After testing six samples, the average result was taken into consideration for discussion. The Universal Testing Machine was used to calculate the flexural strength and flexural modulus. The ASTM D7290 standard was followed for conducting the flexural test. From 1, the test speed was kept constant.

Table 80.2 Flexural strength samples

S. No	Sample ID	Flexural Strength (N/mm^2)
1	C-1	43.09
2	C-2	32.42

3.4 Impact Strength Analysis

An impact test was performed to evaluate the material's capacity to absorb energy during the application of abrupt loads. Using an impact tester, the izod impact test was carried out in the current inquiry in accordance with ASTM D 256 standard. Six specimens underwent the impact test, and the average result was taken into consideration for discussion.

Graph Type : Load Vs. Displacement

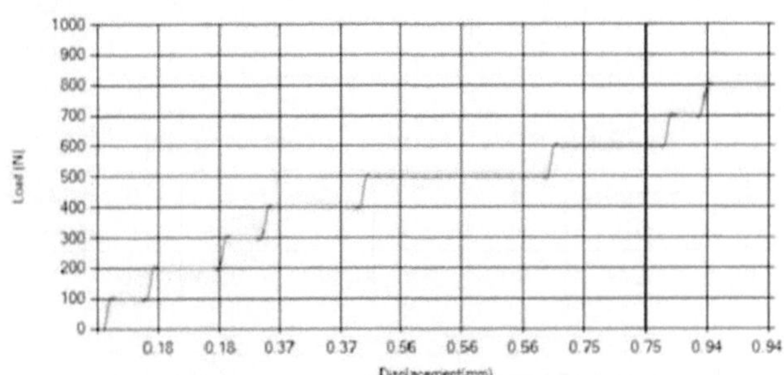

Fig. 80.5 Plain composite sample 1

Graph Type : Load Vs. Displacement

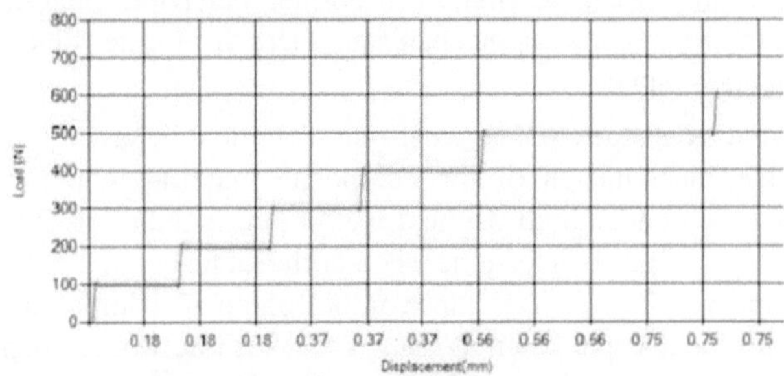

Fig. 80.6 Plain composite sample 2

Table 80.3 Impact Strength samples

S. No	Sample ID	Impact Energy (J)
1	C-1	3.8
2	C-2	3.9

3.5 Wear Analysis

Wear analysis was done to know the lifetime of material that would be helpful to change the damaged material at right time before failure occur. Wear testing was done using a pin-on-disc wear testing equipment in accordance with ASTM G99 guidelines. A spherical pin was pressed against the rotating disc where the material to be tested was fixed. The test could be conducted in dry and wet conditions. In this investigation, the test was performed in dry conditions without using any lubricants. The wear test parameters were selected as: applied load 20 – 80 N, Sliding velocity – 150 mm/s & Sliding distance – 1000 m.

3.6 Hardness Test

To determine the materials' hardness, a hardness test was performed. The ASTM D2240 standard was followed for conducting the hardness test. A hardness test (Shore D) was conducted.

Table 80.4 Wear analysis samples test parameters

S. No.	Sample ID	Test Speed (RPM)	Pin Dia. (mm)	Track Radius (mm)	Normal Load Applied (N)	Test Duration (Mins)
1	For C1,C2	500	2	30	30	15
2	For C3,C4	500	2	30	20	15

Table 80.5 Hardness test samples

S. No	Sample ID	1	2	3	4	5	Mean
1	C-1	89	88	88	88	86	87.8

4. Conclusion

The research mentioned aims to evaluate various aspects of the biochar-dispersed rock-wool-polyester composite regarding wear resistance, frictional behavior, material durability, and wear mechanisms. Here are further details on each aspect:

Wear Resistance: The wear rate, frictional behavior, and wear mechanisms of the composite material would be assessed under both dry and wet sliding conditions. This analysis aims to determine whether the inclusion of biochar improves the wear resistance of the composite compared to traditional brake pad materials. The wear rate would quantify the material loss or degradation due to sliding contact, and the study would investigate the wear mechanisms involved in the process.

Frictional Behavior: The Coefficient of Friction (COF) of the composite material would be measured during dry and wet sliding. The study would explore how the presence of biochar influences the COF and whether it provides any advantages in terms of frictional performance. This analysis can provide insights into the material's ability to generate frictional forces and its suitability for applications involving sliding contact.

Material Durability: The study would assess the durability and stability of the composite material under prolonged sliding conditions. This involves evaluating the material's ability to withstand wear over an extended period and maintain its mechanical properties. It may involve conducting tests or simulations to simulate real-world operating conditions and monitor any changes in the material's performance and structural integrity.

Wear Mechanisms: There search would aim to identify and understand the underlying wear mechanisms involved in the composite material.

This may include analyzing the worn surfaces using techniques such as microscopy or surface profile me try to identify the dominant wear mechanisms, such as abrasion, adhesion, or fatigue. Understanding the wear mechanisms can help in optimizing the material composition and design to enhance wear resistance and durability.

Overall, this research would provide insights into the wear behavior, frictional performance, and durability of the bio-char-dispersed rock wool-polyester composite under different sliding conditions. The findings can contribute to the development of improved materials for applications that require high wear resistance and frictional performance.

Acknowledgement

The corresponding author wishes to express appreciation to the management of Vel Tech High Tech Dr. Rangarajan Dr. Sakunthala Engineering College, Chennai, Tamilnadu, India, for their invaluable assistance. Our dear Principal Prof. Dr. E. Kamalanaban has our deepest gratitude for his support and encouragement.

Data Availability Statement

In this context, data includes are raw data and processed data.

Reference

1. Adebisi, A.A. Maleque, M.A. and Shah, Q.H.,"Surface temperature distribution in a composite brake rotor", International Journal of Mechanical and Materials Engineering 6 (3), pp. 356-361, 2011.
2. ArnabGanguly ,Raji George, "Asbestos Free Friction Composition For Brake Linings Bull", Mater. Sci., Vol.31, No. 1, pp. 19–22, 2008.

3. Bijwe J., Nidhi and Satapathy B.K., "Influence of amount of resin on fade and recovery behaviour of nonasbestos organic (NAO) friction material", International journal of Wear, pp. 1068–1078, 2005.
4. Blau PJ, McLaughlin JC. "Effect of water films and sliding speed on the frictional behaviour of truck disc brake materials", international journal of Tribology: Vol 36, pp. 709–715, 2003.
5. C. M. Ruzaidi , H. Kamarudin, J. B. Shamsul, M. M. A. Abdullah Comparative Study on Thermal, Compressive, and Wear properties of Palm Slag Brake Pad Composite with Other Fillers, Advanced Materials Research Vols. 328-330, pp 1636–164, 2011.
6. Cho Min Hyung, Kim Seong Jin, Kim Dachwan, Jang Ho, "Effect of ingredients on tribological characteristics of a brake lining: an experimental study" International journal of Wear, pp. 1682–1687, 2005.
7. Emaga T. H., Ronkart N., Robert C., "Characterisation of pectins extracted from banana peels(Musa AAA) under different conditions using an experimental design" Journal of food chemistry, Vol 108, pp. 463–471, 2008.
8. Eriksson Mikael, Bergman Fillip, Jacobson Staffan. "The nature of tribological contact in automotive brakes" Wear, pp. 26–36, 2002.
9. Idris U.D., V.S. Aigbodion b, I.J. Abubakar c, C.I. Nwoye d "Eco-friendly asbestos free brake-pad: Using banana peels" Journal of Engineering sciences, pp. 1018–1036, 2013.
10. Riadh E., Singh H, Kchaou M., "Friction characteristics of a brake friction material under different braking conditions", Journal of material and design, Vol-52, pp. 533–540, 2013.
11. Sasaki, Y.,"Development Philosophy of Friction Materials for Automobile Disc Brakes", The Eight International Pacific Conference on Automobile Engineering, Society of Automobile Engineers of Japan; Society of Automobile Engineer of Japan, pp. 407–412, 1995
12. Wannik, W.B., Ayob, A.F., Syahrullail, S., Masjuki, H.H. and Ahmad, M.F.,"The effect of boron friction modifier on the performance of brake pads", International Journal of Mechanical and Materials Engineering 7 (1), pp. 31–35, 2012.
13. Yun Cheol Kim, Min Hyung Cho, Seong Jin Kim, Ho Jang, "The effect of phenolic resin, potassium titanate, and CNSL on the tribological properties of brake friction materials", wear 264, pp.204-210, 2008.
14. Yun Rongping, Filip Peter, Lu Yafei, "Performance and evaluation of eco- friendly brake friction materials" International journal of Tribology, pp.2010-2019, 2010.
15. Zamri, Y.B., Shamsul, J.B. and Amin, M.M,"Potential of palm oil clinker as reinforcement in aluminium matrix composites for tribological applications", International Journal of Mechanical and Materials Engineering 6 (1),pp.10-17, 2011.

Note: Every figure and table was created using the experimental work; none of them were taken from any publications or online.

Advances in Additive Manufacturing Technologies – Gurusamy Pathinettampadian (eds)
© 2024 Taylor & Francis Group, London, ISBN 978-1-032-90013-1

81 Experimental Investigation of Wing with Leading Edge Droop

Mahendranth Reddy,
Pavan Kumar K, Bhanu Prakash K
Student, Department of Aeronautical Engineering,
Hindustan Institute of Technology and Science,
Chennai, India

Balaji G*, Saravanan P
Faculty, Department of Aerospace Engineering,
Hindustan Institute of Technology and Science,
Chennai, India

Ramanan N
Assistant Professor, Department of Mechanical Engineering,
Sri Jayaram Institute of Engineering and Technology,
Chennai, India

Santhosh Kumar G
Teaching Fellow, Dept. of Mech. Engg.,
University College of Engineering, Bharathidasan Institute of
Technology Campus, Anna University,
Tiruchirappalli, Tamilnadu, India

Vijayanandh Raja
Department of Aeronautical Engineering,
Kumaraguru College of Technology,
Coimbatore, Tamil Nadu, India

Gurusamy P
Professor, Department of Mechanical Engineering,
Chennai Institute of Technology,
Chennai, India

ABSTRACT: The experimental analysis conducted in this study sheds light on the significant influence of leading-edge droop on the wing performance across various angle of attack at different velocities. By placing the different wing models with varying levels of leading-edge droop in the controlled low speed wind tunnel and analysing force balances to determine lift and drag, After experimentally analysing flow behaviour and overall aerodynamic performance of wings with leading-edge droop was gained. It is observed that the lift coefficient increases by increasing the angle of attack

*Corresponding author: gbalajihits@gmail.com

DOI: 10.1201/9781003545774-81

for the base line wing and the different droop configurations and even the droop configuration had more lift coefficient when compared to the base wing line. Coming to the Drag parameter, Drag coefficient reduction is been observed for all the droop configurations when compared to the base line wing and maximum reduction in drag was observed with droop 2 configuration. The aerodynamic efficiency (Cl/Cd) is found to be greater for the Droop configuration by 1.5% when compared to the base line wing model at AOA 10 degrees. These findings of this study provided crucial guidance, how the optimized wing design enhanced aerodynamic performance under specific flight conditions.

KEYWORD: Leading edge, Droop, Wing, Reynold number, Aerodynamics

1. Introduction

Over the years, substantial research has been performed to optimize many areas of wing design to obtain higher performance levels. One such area of focus is leading edge droop, which has demonstrated encouraging results in improving aerodynamic properties, particularly at high angles of attack and during low-speed flight regimes[1-3].

All commercial aircraft are been equipped with high-lifting devices, some of the high-lifting devices that are been used in commercial aircraft are Flaps, slats, Krueger flaps, variable camber wings, leading-edge nose droop, and other leading-edge lifting devices [2-6]. These devices have the capability of improving the aircraft performance during critical flight paths like take-off and landing by increasing the lift coefficient reducing the drag at low speeds and reducing stall speed.

Aerodynamic elements incorporated onto aircraft wings to optimize lift and reduce drag during vital flight conditions are known as high-lift devices. To provide greater lift coefficients and delay the beginning of the stall, they work by changing the geometry and airflow patterns of the wing. Lift augmentation, stall avoidance, and enhanced low-speed handling characteristics are important ideas related to high-lift devices .Types of high lifting devices primarily divided into Flaps: During take-off and landing, flaps, which are hinged surfaces attached to the trailing edge of the wing, can deploy downward. Flaps increase lift and decrease stall speed by altering the wing's effective camber, which enables shorter take-off and landing distances. Leading edge droop involves a downward deflection of the wing's leading edge, often at the wing root. This alteration alters the wing's aerodynamic characteristics, affecting lift generation, stall characteristics, and overall stability. While leading-edge droop has been studied for decades, present advances in computational fluid dynamics (CFD) and experimental techniques have allowed for a deeper understanding.

Main assets of leading edge droop of wing ability to produce more lift at lower speeds is made possible by leading-edge droop devices, which are essential for a secure takeoff and landing. Leading edge droop devices enhance the aircraft's low-speed handling qualities, permitting safer operations during crucial flight phases, by successfully postponing the stall and raising the wing's lift coefficient. Aircraft with this feature can approach runways at a lower speed, shortening landing distances and increasing overall safety owing to the enhanced lift produced by the drooped front edge. A drooping leading-edge delays outboard separation, and stall occurs only at the wing root at a higher incidence, which is an excellent trend in aircraft design.

In our journal paper the objectives were experimented to study the effect of droop configuration in enhancing the lift characteristics of the wing. In order to study the effect of droop configuration, the force measurements will be carried out. Lift coefficient, drag coefficient will be measured over the range of angle of attack from -15° to +20°.

2. Literature Survey

Ross 1991 (7) investigated the flow field around the wing with the presence of partial span leading edge droop using different probes which showed the colour display video which indicated the reverse flow behind the wing and flow field around the wing. After testing they came to know that both the modified models were effective in keeping the outboard wing flow attachment at high angle of attack by which the aileron authority was maintained and provided great resistance to stall. To know more about the stall delay and its characteristics Gonzalez 1993 (8) investigated how the stall angle is been reduced due to the presence of Gap less droop nose devices and also showed how these devices partial reduced the noise produced due to the flaps. Considering the above ideology of reducing the noise due to flaps Johnson 1980 (9) conducted the experiment between the base line wing and high lift-base line, High lift skewed droop and high lift constant droop configurations. After the CAA analysis they concluded that the constant droop configuration produced less noise after the stall and also the turbulence when compared to that of other airfoil configuration. After analysing all the parameters of the airfoil and wings with droop configuration researchers started too investigated on the particular aircraft Patterson 2015 (10) examined the aerodynamic performance of droop configuration on the trainer Cessna Caravan aircraft. He considered the two modified droop configuration to the entire wing span. He analysed that the nose-pitching moment and maximum lift has increased during landing and take-off using Euler boundary layer code MCES Tool. Jirasek 2009 (11) investigated the overall performance of the high lift systems of the laminar wing which is influenced by the integration of flexible nose droop and replacing it with Krueger flap and they also placed vortex generator to improve the flow over the flaps, due to which the lifting devices produced more maximum lift when compared to base model. Similarly Newsom 1982 (12) considered leading edge morphing device (nose droop) and multi-functional segment flap system and analysed them and came to know how these system enhanced the performance parameter of the wing during landing and take-off. They concluded that these devices lead to increase both Clmax and stall angle at low speed and low angle of attack. The researchers started to research to make advancement in the high-lift device system. So, they started to focus on smart nose droop configuration and wing with both leading and trailing edge high-lift devices. Advancement to this ideology. Dai 2022(13) had a primary goal to provide a structure or system that allows for a smooth leading surface that can be deflected for common high-lift application and after the investigation they presented the idea of smart, cutting-edge technology that is seen in A380, which replaced the conventional droop nose system.

3. Experimental Methodology

3.1 Model Designing

Selection of Airfoil

The NACA 23012 is part of an extended series of related airfoils that have been tested in the NACA full-scale and variable-density tunnels.

NACA23012: A maximum thickness of 12% at 29.8% chord. A maximum camber of 1.8% at 12.7% chord. A design lift coefficient of 0.3.

A maximum camber located 15% back from the leading edge. A lift slope of 0.1080 degrees ^−1.A zero-lift angle of attack of -1.3 degrees. A low profile drag. A high speed-range index A very small pitching-moment coefficient [14-18]. The high values of the maximum lift coefficient got obtained in NACA 23012 compared to any other airfoil with flap at an effective Reynolds Number of about 8,000,000 compare favourably with obtained for most other high-lift devices.

3.2 Droop Configuration Models

As a part of the experiment we extended the chord portion of base wing NACA 23012 which gives droop configured wings in the leading edge portion as shown in Fig. 81.1 to Fig. 81.4.

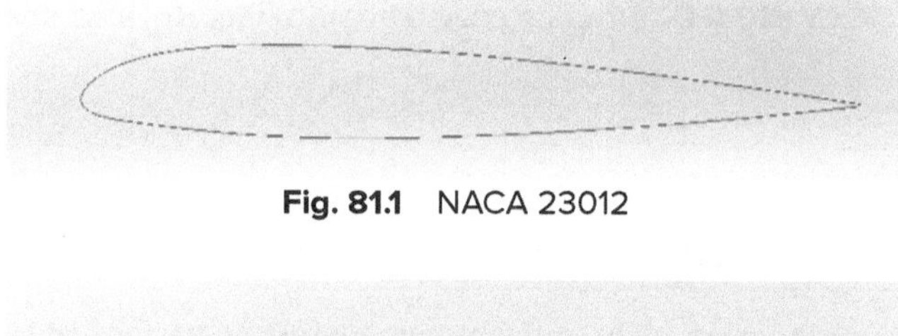

Fig. 81.1 NACA 23012

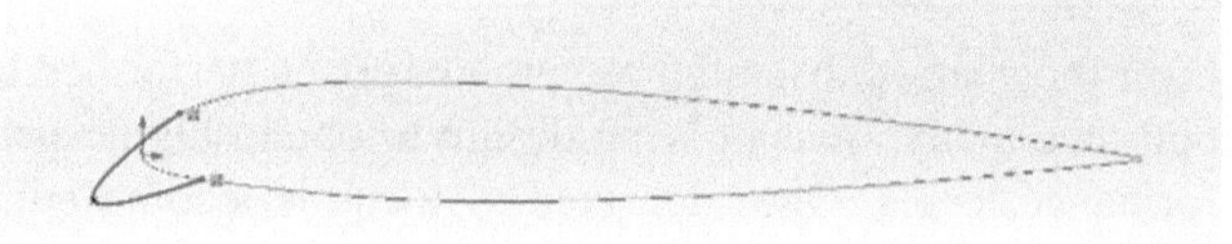

Fig. 81.2 Leading edge droop with 4.9 % of C (Horizontal) & 4.6 % of C (vertical)

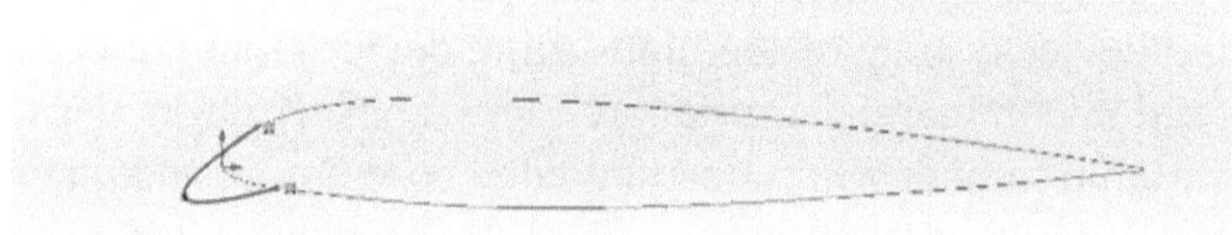

Fig. 81.3 Leading edge droop with 4 % of C (Horizontal) & 3.5 % of C (vertical)

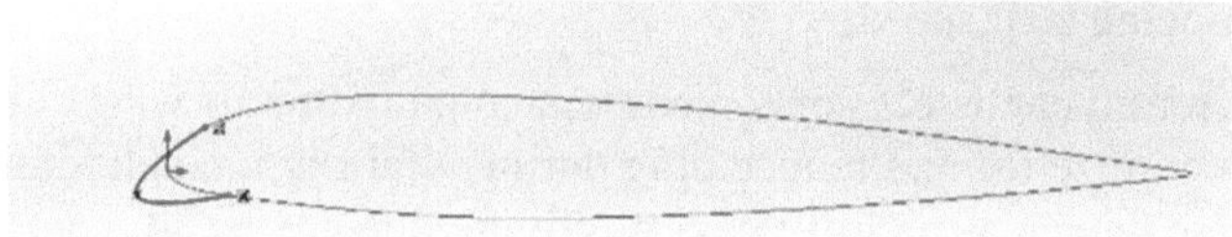

Fig. 81.4 Leading edge droop with 3 % of C (Horizontal) & 2.4 % of C (vertical)

Bases on the different researches we designed the three different Droop models as follows. The extended Droop configured portion is from mid wing to tail wing to improve the stall angle and smooth air flow.

3.3 Testing Wind Tunnel

A low-speed wind tunnel serves as a crucial tool in aerodynamic research, providing controlled conditions for studying airflow around various objects. Operating at subsonic speeds, typically below 100 meters per second, these wind tunnels enable detailed analysis of aerodynamic phenomena relevant to aircraft, automobiles, and other engineering applications[19-21]. By generating airflow at lower velocities, they allow for precise measurements of lift, drag, and flow characteristics without the complexities associated with high-speed flow regimes.

4. Results and Discussion

The aerodynamic performance that we discussed in the objective and motivation of the work, we optimized the desired results in the formation of different graphs such as C_l vs AOA, Cd vs AOA and C_l/C_d vs AOA. Moreover we conclude, how much impact the Droop configuration in overall aerodynamic efficiency. For better understanding of the performance of the Droop configuration, we compared the results of three different Droop configurations with the Base wing model (plain wing) i.e. NACA 23012.

4.1 Different Velocities

The changing of leading edge of a wing into droop configured wing to increasing of coefficient of lift (c_l max) we optimized from above graphs that where the increasing of velocity and angle of attack there coefficient of lift increase according to the droop configured wing.

4.2 Coefficient of Lift at Different Velocities

In Fig. 81.5, At 10 m/s and 20 degree angle of attack the C_l is max for droop 3 i.e. larger droop configured wing. At 20 m/s from 10 to 20 degree angle of attack the C_l is max for the droop 1 that is smaller droop configured wing. At 30 m/s base wing performs well when compared to the other three droop configured wings for the C_l other side there is a slight change in stall angle as droop configured changes and velocity increases. Fig. 81.5–81.8 shows the different velocities and angle of attack.

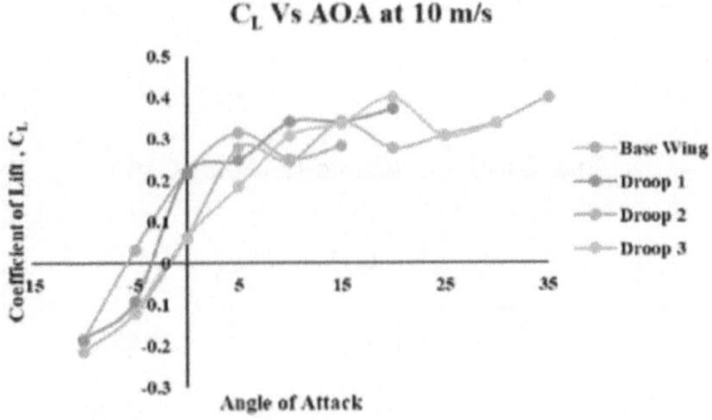

Fig. 81.5 Coefficient of lift vs AOA at 10 m/s

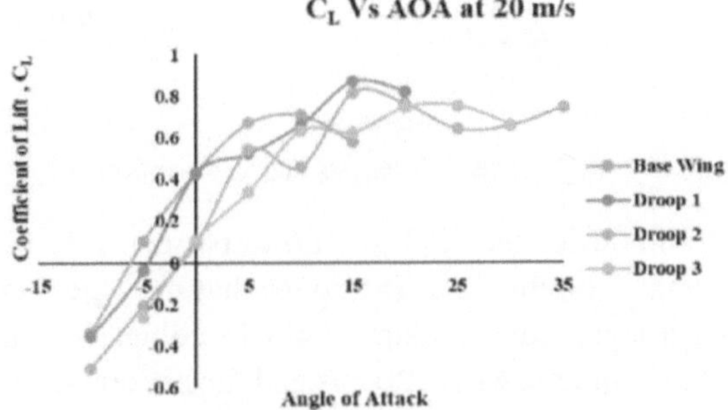

Fig. 81.6 AOA at 20 m/sec

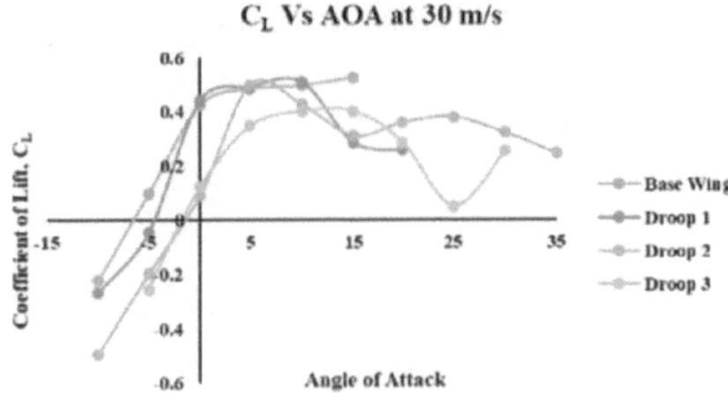

Fig. 81.7 AOA at 30 m/s

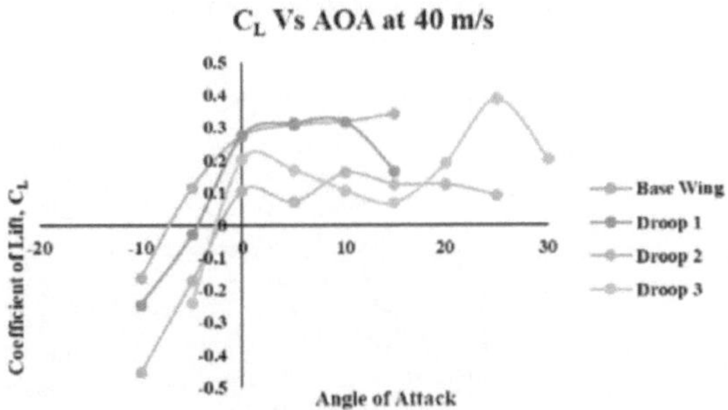

Fig. 81.8 AOA at 40 m/s

4.3 Coefficient of Drag at Different Velocities

At small angles of attack and low velocities, the drag is typically low because the airflow smoothly follows the contours of the object without significant disruption. As the angle of attack increases as well as velocities, the airflow around the object becomes more turbulent, leading to an increase in drag. This increase is primarily due to the formation of separated flow regions and increased pressure drag.

The below graphs shows that how the coefficient of drag comes into picture when in the increasing in the velocities and angle of attacks when the testing of these airfoils are conducted in low speed wind tunnel. We optimized that there is decrement of drag compared to the base wing from droop configured for the higher velocities and angle of attacks.

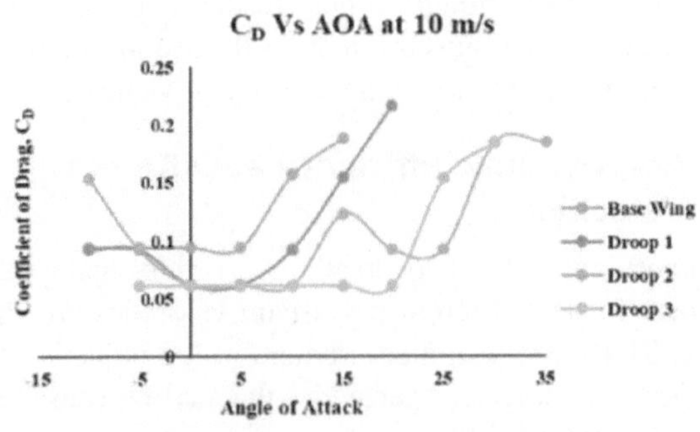

Fig. 81.9 AOA at 10 m/s

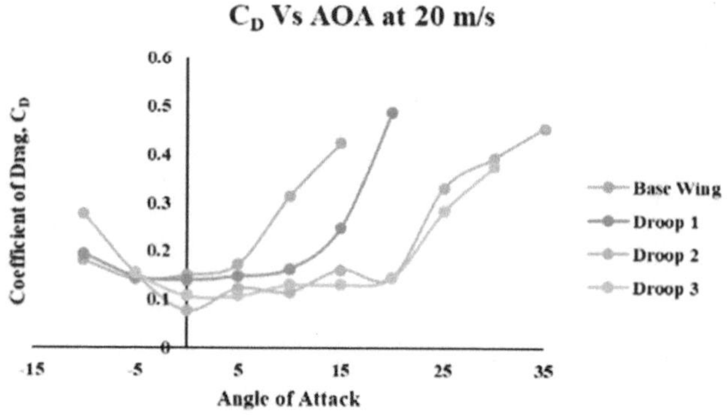

Fig. 81.10 AOA at 20 m/s

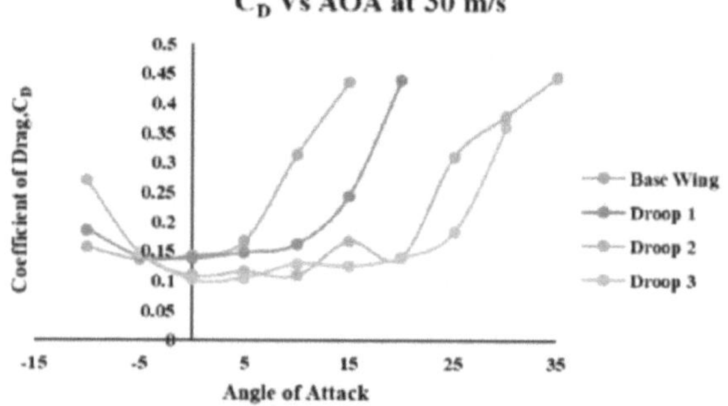

Fig. 81.11 AOA at 30 m/s

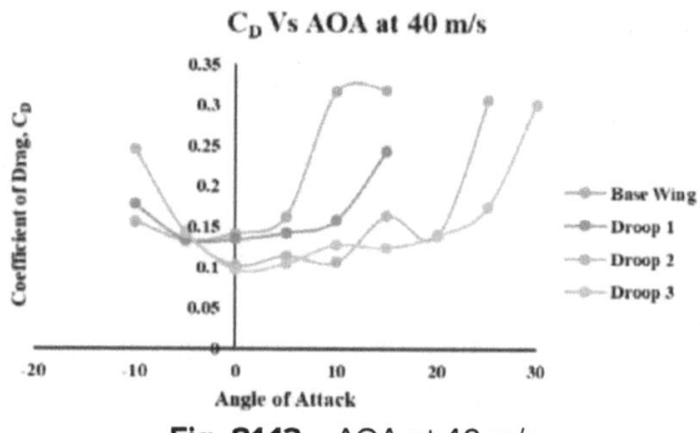

Fig. 81.12 AOA at 40 m/s

At 10 m/s velocity there is constant decrement in Coefficient of drag from -10 to 20 degree of angle of attack then it's slightly rises as increases the angle of attack for the droop 3. At 20 m/s the minimum coefficient of drag viewed for the droop 3, that enough to help to the increment of stall angle. Remaining Cruised velocities the droop 3 configured airfoil optimized the low drag coefficient values.

4.4 Aerodynamic Efficiency at Different Velocities

The coefficient of lift to drag (C_l/C_d) ratio against angle of attack for the different freestream velocities are shown in Fig. 81.13–81.14. a measurement of the object's lift-to-drag performance, is impacted by the airflow velocity and angle of attack (AOA).

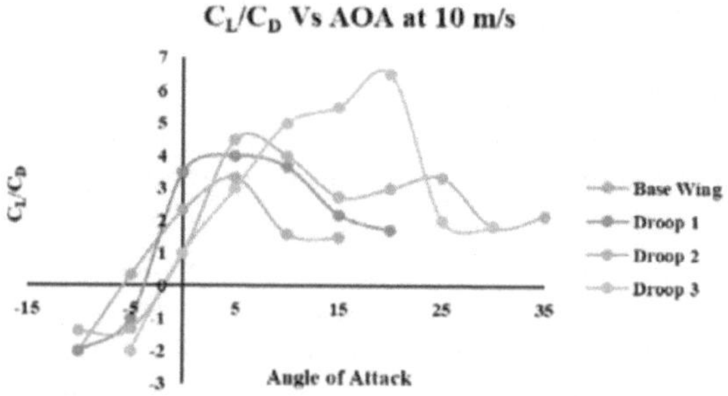

Fig. 81.13 C_L/C_D Vs AOA (10m/s)

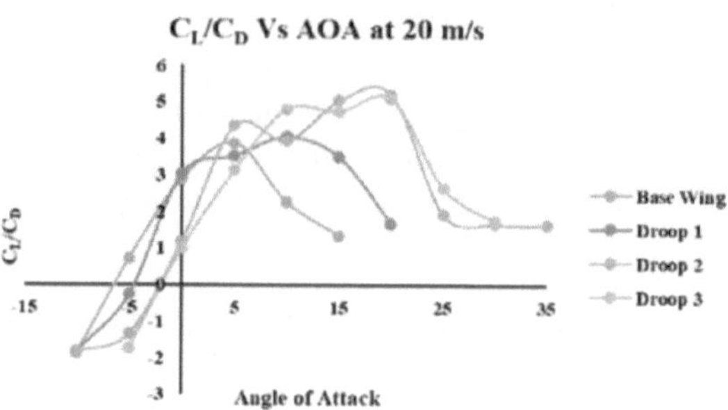

Fig. 81.14 C_L/C_D vs AOA (20m/s)

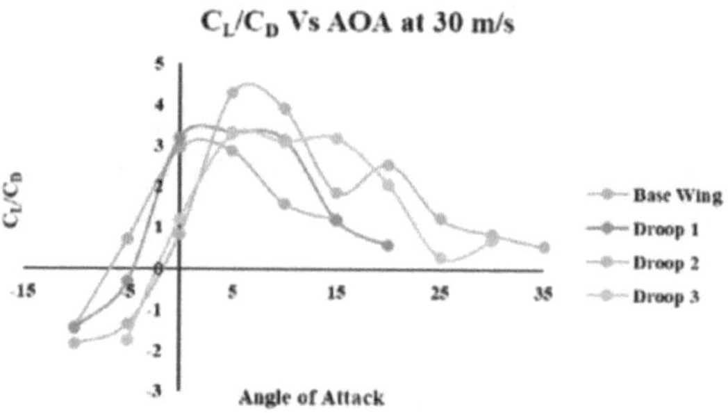

Fig. 81.15 C_L/C_D vs AOA at 30 m/s

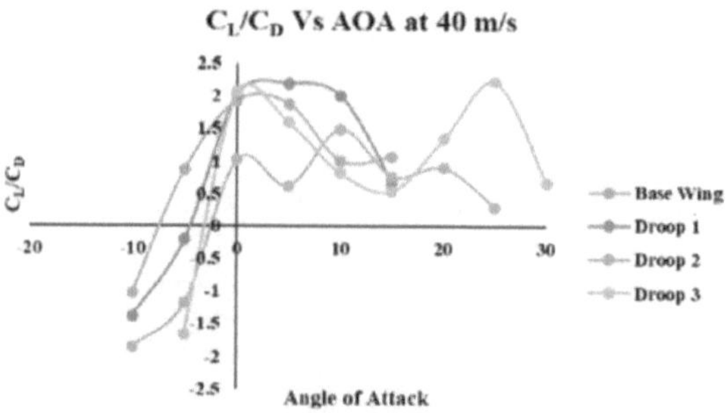

Fig. 81.16 C_L/C_D vs AOA (40 m/s)

It is common for the (C_l/C_d) ratio to be relatively high at small angles of attack. This implies that the object is producing a significant amount of lift in comparison to the drag it encounters. At 10, 20 m/s and further cruised velocities the droop 2 and droop 3 configured wings performs well. At 30 m/s the efficiency of the droop configured airfoils is better when compared to the base wing airfoil at over at the different angle of attack. At 40 m/s the base airfoil is performing better at 0 degree angle over the three different droop configured airfoils. From the overall aerodynamic performance graphs that includes (C_l vs α, C_d vs α and C_l/C_d vs α), we optimized that at v = 30 m/s for angle of attack 10 degree all droop configured airfoils and base airfoils showed the appropriate values with better performance. So, we calculated the percentage of increment in aerodynamic performance to show, which droop configured airfoil performs better over the three droop configured airfoils at 30 m/s velocity and 10 degree angle of attack where the airfoils is stabilized as shown in Fig. 81.17 and values of C_L,C_D at constant velocity and angle of attack was listed in the Table 81.1.

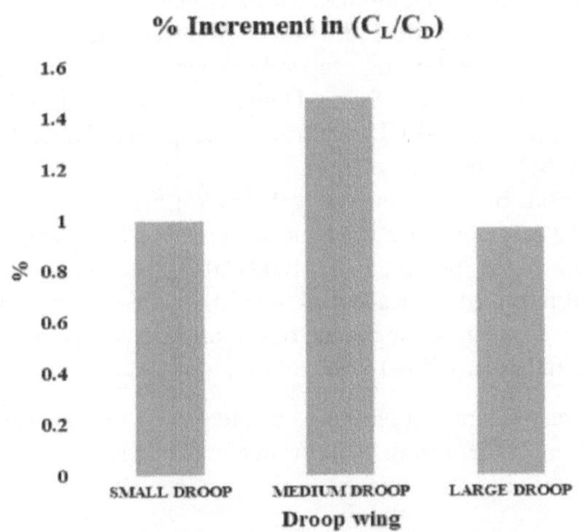

Fig. 81.17 The % increment in C_L/C_D

Table 81.1 Values of C_L, C_D at constant Velocity and angle of attack

For Velocity 30 m/s						
Sl. No	Model	AOA	C_L	C_D	C_L/C_D	% Increment in C_L/C_D
1	Base	10	0.4954	0.314	1.578	–
2	Droop 1	10	0.5088	0.161	3.148	0.995
3	Droop 2	10	0.4272	0.110	3.906	1.475
4	Droop 3	10	0.4015	0.130	3.102	0.966

4.5 Performance Comparisons of Different Droop Configurations

By the formulae of % increment in (cl/cd)

$$= \frac{Droop\left(\frac{Cl}{Cd}\right) - Base\left(\frac{Cl}{Cd}\right)}{Base\left(\frac{Cl}{Cd}\right)},$$

we calculated the how much percentage of increment in performance among the three different droop configurations as shown in Fig. 81.17.

5. CONCLUSION

The results demonstrate that drooping the leading edge can raise the maximum lift coefficient when a leading-edge extension is added. The primary goal was to examine wings with leading-edge nose droop addition at various airflow velocities and angles of attack. The wing's aerodynamic properties are changed by leading-edge droop, which may improve the wing's lift-to-drag ratio. The middle droop wing has the highest variation in the cl/cd increment, with an average deviation of 1.5% from the other wings. The changed wing geometry could affect the aircraft's qualities related to stability and control. A few other parts of the unification were also covered, such as how a drooping leading edge might delay greater angles of attack and hence have a favourable effect on the stall behaviour of the wing. Lift force at 30 m/s airflow velocity with a 10-degree angle of attack equals -0.4956 Newton and the stall and lift characteristics have been recorded at a velocity of 30 m/s and an angle of attack of 10 degrees, when the airfoil is stabilized with some of these specific values. For the droop 3, the minimum drag coefficient is 20 m/s, enough to help the stall angle increased.

REFERENCES

1. Anderson Jr, J. D., Corda, S., & Van Wie, D. M. (1980). Numerical lifting line theory applied to drooped leading-edge wings below and above stall. Journal of Aircraft, 17(12), 898–904.
2. Niu, J., Lei, J., & Lu, T. (2018). Numerical research on the effect of variable droop leading-edge on oscillating NACA 0012 airfoil dynamic stall. Aerospace Science and Technology, 72, 476–485.
3. Bashir, M., Longtin-Martel, S., Botez, R. M., & Wong, T. (2022). Optimization and design of a flexible droop-nose leading-edge morphing wing based on a novel black widow optimization algorithm—Part I. Designs, 6(1), 10.
4. Balaji, G., Bharath Kumar, A., Divya, R., Boopathy, G., Seenu, N., & Santhosh Kumar, G. (2023, August). Experimental Investigation of Double Delta Wings with Different Angles of Attack at Subsonic Speeds. In *International Conference on Smart Sustainable Materials and Technologies* (pp. 197–202). Cham: Springer Nature Switzerland.
5. Zhao, G. Q., & Zhao, Q. J. (2015). Dynamic stall control optimization of rotor airfoil via variable droop leading-edge. Aerospace Science and Technology, 43, 406–414.
6. Shah, G. (1991, August). Wind tunnel investigation of aerodynamic and tail buffet characteristics of leading-edge extension modifications to the F/A-18. In 18th Atmospheric flight mechanics conference (p. 2889).

7. Ross, H. M., Yip, L. P., Parkins, J. N., Vess, R. J., & Owens, D. B. (1991). Wing leading-edge droop/slot modification for stall departure resistance. Journal of Aircraft, 28(7), 436–442.
8. Gonzalez, H., & Winkelmann, A. (1993). An experimental study of droop leading edge modifications on high and low aspect ratio wings up to 50 deg angle of attack. In 11th Applied Aerodynamics Conference (p. 3496).
9. Johnson, JR, J., Newsom, W., & Satran, D. (1980, August). Full-scale wind-tunnel investigation of the effects of wing leading-edge modifications on the high angle-of-attack aerodynamic characteristics of a low-wing general aviation airplane. In Aircraft Systems Meeting (p. 1844).
10. Patterson, G. T., Binkley, B. A., & Jenkins, J. C. (2015, June). A-10 Wing Leading Edge Effects on Engine Stability: Part 1—Analysis and Evaluation of Wing Leading Edge Configurations. In Turbo Expo: Power for Land, Sea, and Air (Vol. 56628, p. V001T01A036). American Society of Mechanical Engineers.
11. Jirasek, A., & Amoignon, O. (2009, June). Design of a high lift system with a leading edge droop nose. In 27th AIAA Applied aerodynamics conference (p. 3614).
12. Newsom Jr, W. A., Satran, D. R., & Johnson Jr, J. L. (1982). Effects of wing-leading-edge modifications on a full-scale, low-wing general aviation airplane: Wind-tunnel investigation of high-angle-of-attack aerodynamic characteristics (No. L-15101).
13. Johnson, J R, J., Newsom, W., & Satran, D. (1980, August). Full-scale wind-tunnel investigation of the effects of wing leading-edge modifications on the high angle-of-attack aerodynamic characteristics of a low-wing general aviation airplane. In Aircraft Systems Meeting (p. 1844).
14. Dai, X., Qiu, Z., Li, G., & Wang, F. (2022). Research on dynamic stall active control of two-dimensional airfoil with combination of droop leading edge and trailing edge flap. Aerospace Systems, 5(4), 643-653.
15. Bak, C., & Fuglsang, P. (2002). Modification of the NACA 63 2-415 leading edge for better aerodynamic performance. J. Sol. Energy Eng., 124(4), 327–334.
16. White, E. R. (1982). Wind-tunnel investigation of effects of wing-leading-edge modifications on the high angle-of-attack characteristics of a t-tail low-wing general-aviation aircraft (No. NAS 1.26: 3636). Washington.
17. Malipeddi, A. K., Mahmoudnejad, N., & Hoffmann, K. A. (2012). Numerical analysis of effects of leading-edge protuberances on aircraft wing performance. Journal of Aircraft, 49(5), 1336–1344.
18. Mugler, J. P. (1956). Effects of Two Leading-Edge Modifications on the Aerodynamic Characteristics of a Thin Low-Aspect-Ratio Delta Wing at Transonic Speeds. National Advisory Committee for Aeronautics.
19. Watts, P., & Fish, F. E. (2001, August). The influence of passive, leading edge tubercles on wing performance. In Proc. Twelfth Intl. Symp. Unmanned Untethered Submers. Technol (pp. 2–9). Durham New Hampshire: Auton. Undersea Syst. Inst.
20. Pruski, B. J., & Bowersox, R. D. W. (2013). Leading-edge flow structure of a dynamically pitching NACA 0012 airfoil. AIAA journal, 51(5), 1042–1053.
21. Burnazzi, M., & Radespiel, R. (2014). Design and analysis of a droop nose for coanda flap applications. Journal of Aircraft, 51(5), 1567–1579.

Note: Every figure and table was created using the experimental work; none of them were taken from any publications or online.

Advances in Additive Manufacturing Technologies – Gurusamy Pathinettampadian (eds)
© 2024 Taylor & Francis Group, London, ISBN 978-1-032-90013-1

82 Experimental Analysis of Hybrid Metal Matrix Composites Made of Squeezecast Aluminum

J. Joshua Kingsly[1]
Dept. of Mechanical Engg,
Misrimal Navajee Munoth Jain Engineering College,
Thuraipakkam, Chennai, Tamilnadu, India

P. Gurusamy[2]
Dept. of Mechanical Engg,
Chennai Institute of Technology, Chennai,
Tamilnadu, India

A. Sivarangar[3]
Dept. of Mechanical Engg,
Misrimal Navajee Munoth Jain Engineering College,
Thuraipakkam, Chennai, Tamilnadu, India

ABSTRACT: In this experimental study, stir cum squeeze casting is used to create hybrid metal matrix composites (HMMCs) made of aluminium, Graphene, and zirconium dioxide (ZrO_2) are utilised as reinforcement. The use of ceramic ZrO_2 particles enhances the tensile strength of aluminum. Graphene on the other hand, enhances heat transfer characteristics that raise product dependability. Aluminium HMMCs with the following weight ratios were employed in this study. The squeeze casting technique was used to create 100% Al7079; 95% Al7079, 3% Graphene, 2% ZrO_2; 93% Al7079, 3% Graphene, 4% ZrO_2; and 91% Al7079, 3% Graphene, 6% ZrO_2. Squeeze cast metal matrix composite specimens made in accordance with ASTM standards are subjected to microstructural research and mechanical performance evaluations for varying reinforcing percentages. Al7079 HMMCs have an improved hardness and tensile strength thanks to their finely dispersed ZrO_2 and graphene particles.

KEYWORDS: Stir cum squeeze casting, Aluminium hybrid MMCs, Graphene (G), ZrO_2, Microstructural, Mechanical behavior

1. INTRODUCTION

The aerospace and automotive sectors rely heavily on aluminium alloys due to its low density, significant fluidity, high fortitude, and endurance machinability. [1-5]. Aluminum and magnesium alloys have experienced a significant uptick in usage in recent years for producing lightweight components in the automobile industry. High-strength, lightweight, deterioration- and attrition-resistant, minimal heat-transfer emerging materials include ceramic-re-

[1]joshuakingslyj.mech@mnmjec.ac.in, [2]gurusamyp@citchennai.net, [3]rangarbabu@gmail.com

DOI: 10.1201/9781003545774-82

inforced aluminum metal matrix composites (AMMCs), widely used in the automotive and aerospace industries. In AMMCs, reinforcements such as graphite (C), boron carbide (B_4C), alumina (Al2O3), silicon carbide (SiC), and titanium carbide (TiC) are frequently used. [6-10]. But processes including gaspore, and cold closure caused solidification faults in castings. The use of lightweight alloys is restricted because to the decrease in mechanical characteristics caused by these casting component defects. [11-13]. Squeeze casting is a new method of casting that addresses these concerns. However, squeeze casting offers advantages over other casting methods. These advantages include the ability to produce near net form, the removal of porosities, the decrease of material waste due to the absence of a gating system, high dimensional precision, and the increase of mechanical qualities. Squeeze casting has excellent mechanical properties compared to other casting techniques.[14-16]. The ability of squeeze casting to produce materials with outstanding mechanical properties, few inclusions, and tiny grains is one of the reasons why it is gaining popularity. Composite materials that are squeeze cast, as opposed to those that are gravity die cast, have superior mechanical and tribological properties during the manufacturing process. [17-18]. The incorporation of the reinforcements resulted in an improvement in the properties of the composites, and the magnitude of these enhancements increased in proportion to the quantity of reinforcements that were utilized. [19-20The implementation of a pressure of 100 MPa was imperative to achieve microstructural refinement and porosity reduction. [21-22].

The research ofV. S. Senthil Kumar and M. Dhanashekara [23], the squeeze casting approach produces excellent mechanical qualities for metal matrix composites and aluminum alloys. This is the conclusion reached by the researchers. It has been observed by A. Ramesh and M. Kamaraj [24] that the act of applying pressure throughout the process of solidification results in an increase in the final tensile strength. Reddy et al. [25] studied the mechanical characteristics of Al6063/TiC MMCs and discovered that increasing the reinforcement weight % improved the mechanical properties. Thandalam et al. (26) state that the porosity of composites produced via squeeze casting, compo casting, and powder metallurgy is comparatively lower than that of composites produced via stir casting. Pandiyarajan et al. [27] studied the mechanical characteristics of AA6061/ZrO_2/C metal matrix composite. He concludes that adding ZrO_2 to MMC improves its mechanical characteristics.

It is clear from the aforementioned literature that several investigations have been conducted on aluminum-based composites, with a focus on elucidating their physical and mechanical properties. However, there is a dearth of literature demonstrating research into the mechanical properties of aluminum based ZrO_2 reinforced composites. As far as we are aware, no research on the mechanical characteristics of A7079 reinforced with ZrO_2 and Graphene using the squeeze casting process exists. Therefore, the purpose of the current research was to examine the mechanical characteristics of A7079 that had been reinforced with ZrO_2 and Graphene by the squeeze casting technique.

2. EXPERIMENTATION

2.1 Materials

Aluminum (A7079) is employed as the matrix in this experiment, and Table 82.1 displays its chemical composition. This matrix is selected due of its excellent strength and damage endurance at normal temperature, as well as its suitability for strength applications at increased temperatures. Due to its increased thermal conductivity, it has a great capacity for heat dissipation.

Table 82.1 Al7079 alloy chemical composition

Si	Fe	Cu	Mn	Mg	Zn	Ti	V	Al
0.10	0.12	0.9	0.03	1.3	4.2	0.05	0.5	Remaining

2.2 Zirconium Dioxide

Zirconia, or zirconium oxide (ZrO_2), is a white crystalline oxide of zirconium. It has a great resistance to heat and pressure as well as to mechanical wear and abrasion. ZrO_2 has excellent resistance to mechanical stress and cracking (including fracture propagation). Applications range from abrasives and dental work to fuel cell membranes and artificial joints. It is also used as an ingredient in paints and lacquers. One of the most extensively researched ceramics is ZrO_2.

2.3 Graphene

A carbon allotrope known as graphene consists of a single layer of atoms arranged in a hexagonal lattice in a two-dimensional space. Graphite is made by layering graphene with a 0.335 nm gap between them. Scientists have discovered that graphene is the lightest and thinnest material yet discovered. It's virtually entirely transparent, transmits heat and electricity effectively, and is a hundred times stronger than steel. Primarily, it is employed in anticorrosion coatings.

2.4 Experimental Setup

The squeeze casting technique is used to create aluminum hybrid MMCs in this study. This technique achieves a consistent distribution of reinforcing particles throughout the

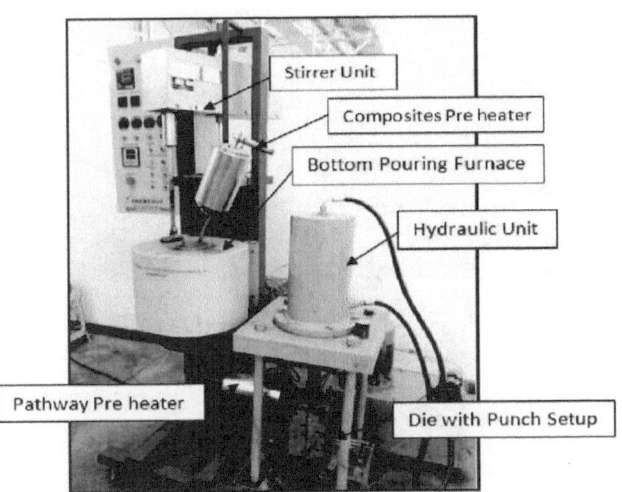

Fig. 82.1 Setup of stir cum squeeze casting

material. A material with high mechanical and wear properties may be produced by the process of squeeze casting, which combines the processes of casting and forging. In Fig. 82.1 displays of the parts that are used for stir cum squeeze casting. The upper die may be adjusted and the bottom die remains stationary. Immediately following the discharge of the reinforced molten melt into the fixed lower die that has been heated, the melt starts to cool and become more solid. Up to the point where the solidification process is finished, pressure is applied to the molten liquid.

A reinforcement was added to the matrix consisting of the Al7079 alloy, Zirconium dioxide particles measuring 35 μm in size, and graphene particles measuring 45 μm in size. In an electric furnace set to 900 degrees Celsius, billets of Al7079 alloy were melted. The particles were gradually added to the melt after the matrix had been completely melted, and the ZrO_2 and graphene C had been heated to 300 degrees Celsius. In order to eliminate the wettability, preheat is utilized. The stirrer rotates at a speed of 550 revolutions per minute for a period of time to mix the matrix and reinforcements completely. Following the completion of the melting process, the molten metal was transferred to the die. Next, the punch was actuated to apply squeezing pressure on hybrid MMCs made from molten aluminum. Table 82.2 shows the properties of the squeeze casting procedure used to produce aluminum hybrid MMCs. The specimen that was manufactured has a length of 260 millimeters and a width of 40 millimeters, as illustrated in Fig. 82.2.

Table 82.2 Stir cum squeeze casting of aluminum hybrid parameters MMC's

Temperatureof Molten Metal	900°C
Temperatureof Die Preheating	300°C
Speed of Stirring	550 rpm
Pressure of Squeeze	150 Mpa
Holding Time Pressure	35 s

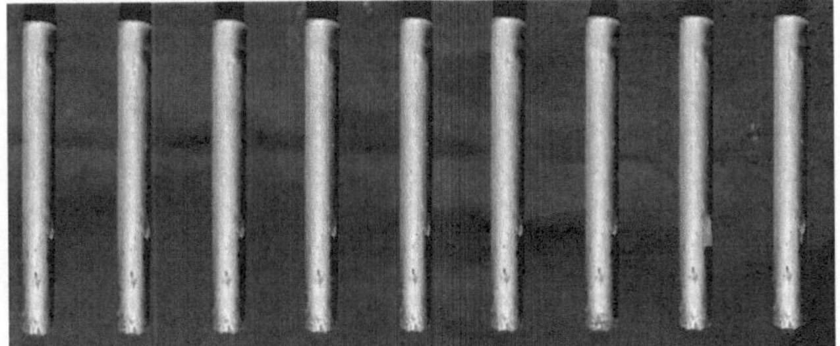

Fig. 82.2 Squeeze-cast aluminum hyrid MMCs

2.5 Sample Preparation

Microstructure

Grinding, polishing, and etching are steps used to get samples ready for microscopic analysis. Initially, silicon carbide abrasive sheets with grits from 320 to 600 were used to cut and grind samples. For the circular disc machine polisher's coarse and finishing jobs, we used a paste made of silicon carbide powder with 1200 grits. Rinse the samples with warm water and dry them with a hand drier to complete the finishing step. After that, Keller's reagent was employed to etch microscopic samples for enhanced contrast. The specimens were etched for 60 seconds, rinsed, and dried using an electric dryer. After processing, each specimen was examined under a 100x scanning electron microscope to identify properly generated grain boundaries. The specimen of the microstructure is depicted in Fig. 82.3.

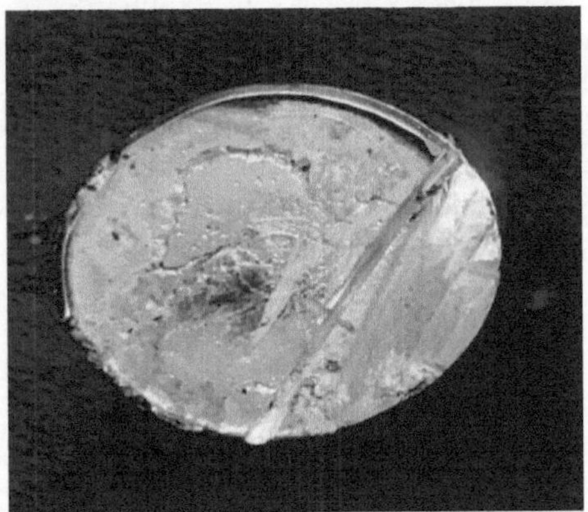

Fig. 82.3 Test specimen for microstructure

Tensile Test

The specimens were subjected to a tensile test as per the ASTM standard, which is depicted in Fig. 82.4. This assessed the squeeze cast samples' mechanical properties.

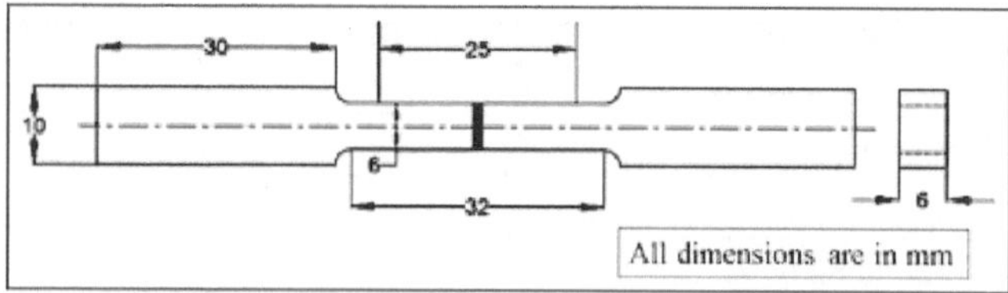

Fig. 82.4 ASTM standard for tensile testing

To avoid partial solidification from time obstruction after pouring and before pressurizing molten metal, specimens for measuring tensile strength were cut from the middle area.

Hardness Test

Examination of the specimen's hardness was carried out on a separate specimen that was fabricated from the same cylindrical composite. This sample has been created in accordance with the criteria of ASTM. In order to conduct the testing, Brinell hardness testers were utilized. A ball intender with a diameter of 10 millimeters was used to apply a weight of 500 kilograms for a period of thirty seconds on the polished surface of the cast samples. A measurement of the intender's impression on the cast samples was taken in three separate areas in order to determine the degree of hardness.

3. Results and Discussion

There is homogeneity in the distribution of the reinforcing particles as a result of the squeeze casting process that is utilized, as demonstrated by the microstructure of the MMCs that are depicted in Fig. 82.5(a-c) when they are studied using an Scanning electron microscope. Increased weight % of zirconium dioxide results in decreased porosity and increased grain dispersion, as seen by microstructural pictures. As reinforcing weight % increases, particles become more densely packed. In the process of solidification, the application of external squeeze pressure causes an increase in the temperature of the liquidus, which in turn causes undercooling, which ultimately leads to an improvement in grain size. The metal's heat transfer coefficient improves due to mold pressure, which refines grains. The tensile and hardness samples were examined, with results shown in Table 82.3.

Table 82.3 Hardness for various matrix compositions

S. No	Composition of Matrix	Hardness (MPa)			Average Tensile Strength (MPa)
		S1	S2	S3	
1	Al 7079	49	49	49	49
2	Al 7079 + 3% G + 2% ZrO_2	46	46	45	46
3	Al 7079 + 3% G + 4% ZrO_2	52	53	53	53
4	Al 7079 + 3% G + 6% ZrO_2	57	57	57	57

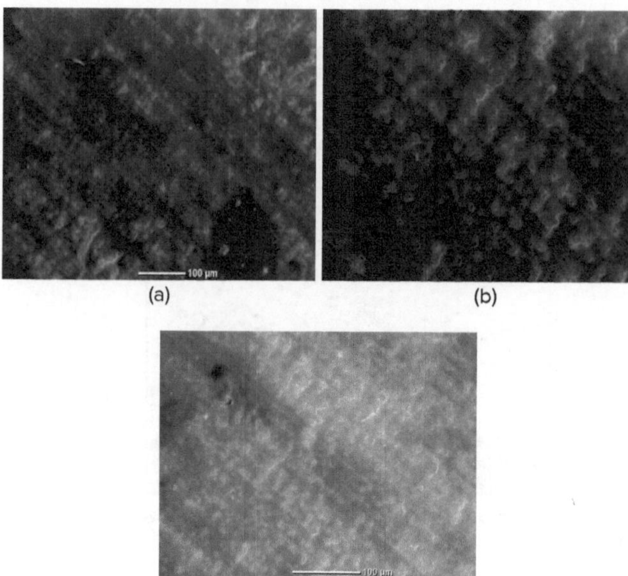

Fig. 82.5 Microstructure of Various Composition (a) Al 7079 + 3% C + 2% ZrO_2, (b) Al 7079+ 3% C + 4% ZrO_2, (c) Al 7079+ 3% C + 6% ZrO_2

Figure 82.6 shows the tensile strength of specimens that were squeeze cast using various matrix compositions. With graphene's weight percentage remaining fixed at 3% over all MMCs, an increase in ZrO_2 content leads to a greater ultimate tensile strength (UTS) in the squeeze cast specimen. When carbon is added to HMMC, the material becomes brittle because its tensile strength and hardness diminish significantly. However, when the weight percentage of ZrO_2 grows over that of graphene, the material becomes increasingly strong and hard. The stir cum squeeze casting method in HMMC production reduces porosity and ensures uniform reinforcing particle distribution. Research indicates that the brittleness of a composition containing 3% graphene and 2% ZrO_2 results in a low ultimate tensile strength. However, some formulations showed a gradual rise in tensile strength and hardness. Tensile strength of multiple matrix composition of Fig. 82.6.

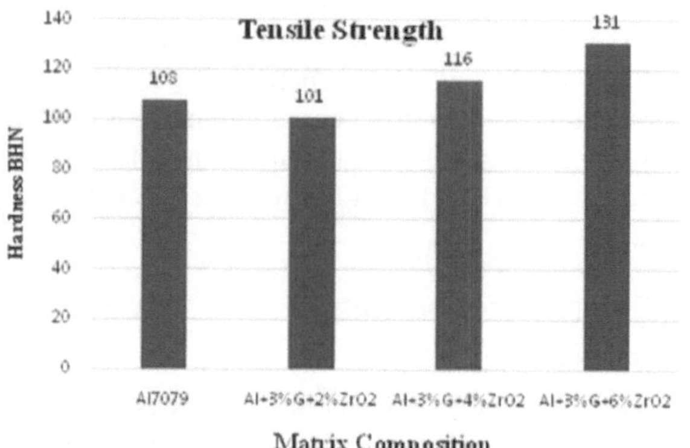

Fig. 82.6 Multiple matrix composition tensile strength

Squeeze cast specimen hardness values are listed in Table 82.4, and those values are also depicted in Fig. 82.7. Since the wt% of graphene is set at 3%, increasing the proportion of ZrO_2 increases the hardness of A7079. Squeeze cast MMCs that have had graphene added to them are more prone to breaking, and as a reason, ZrO_2 is preferred for increasing hardness. The study found that a combination with 6% ZrO_2 produces the maximum amount of hardness.

Table 82.4 Variable matrix tensile strength

S. No	Composition of Matrix	Tensile Strength (MPa)			Average Tensile Strength (MPa)
		S1	S2	S3	
1	Al 7079	108	107.6	107	108
2	Al 7079 + 3% G + 2% ZrO_2	100.2	101	100	101
3	Al 7079 + 3% G + 4% ZrO_2	115.2	114	116	116
4	Al 7079 + 3% G + 6% ZrO_2	131	129.9	131	131

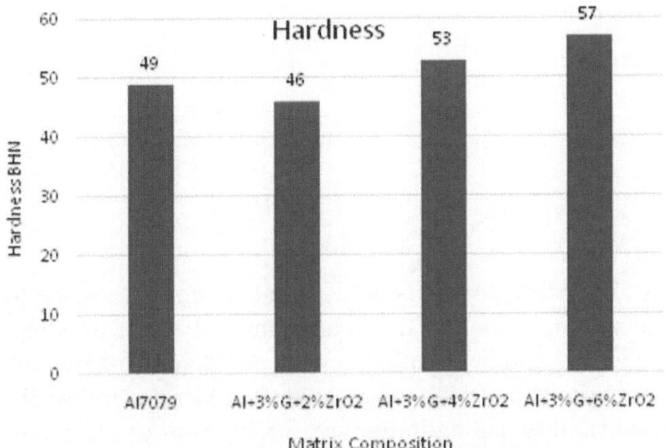

Fig. 82.7 Matrix composition hardness

4. CONCLUSION

It has been demonstrated in this study how the microstructure and mechanical characteristics of Al7079 Hyrid MMC cast using the stir cum squeeze casting process change as the amount of ZrO_2 increases.

To facilitate the evaluation process, a cylindrical cast with a diameter of 40mm was squeezed.

- It was discovered that increasing the weight percentage of ZrO_2 led to an increase in the material's tensile strength.
- The material becomes harder as ZrO2 weight % increases.
- Low tensile strength occurs in materials with 2% zirconium dioxide and 3% graphene. This is due to the fact that the percentage of zirconium dioxide is relatively low in comparison to graphene, and the material is also brittle.

REFERENCES

1. Konopka, M, A. Zyska, Z. Ła̧giewka, M. Nadolski, Arc. Fou. Eng. 13 (2) (2013)113–116.
2. K. Dhilepan, A.P. Kumar, S. Aadithya, N. Nikhi, ARPN. J. Engg. App. Scis. 11 (2)(2016) 1204–1210.
3. R. Bharanidaran, Praveen Kumar, J. Jasper, A. J. Int, App. Engg. Res. 10 (50)(2015) 735–738.
4. N. Parthasarathi, R. Soundararajan, A. Ramesh, N. Mohanraj, J. Alloys Compd. 685 (2016) 533–545.
5. K. Ravikumar, T. Pridhar, R.A. Sankaran, C. Boopathi, Digest J. Nanomater. Biostruct. 11 (3) (2016) 845–852.
6. V.S. Balaji, K.R. Kumar, T. Pridhar, J. Alloys Compd. 765 (2018) 171–179.
7. D.E.J. Dhas, C. Velmurugan, K.L.D. Wins, K.P. Boopathi Raja, Ceram. Int. 45 (1)(2019) 614–621.
8. R.G. Raaja, S. Gopinath, M. Prince, Mater. Res. Express 7 (2020).`
9. T. EbadzadehE. Ghasali, M. Alizadeh, , J. Alloys Compd. 655 (2016) 93–98.
10. D. Khanduja, P. Sharma, S. Sharma, Part Sci Technol 34 (1) (2016) 17–22.

Note: Every figure and table was created using the experimental work; none of them were taken from any publications or online.

Advances in Additive Manufacturing Technologies – Gurusamy Pathinettampadian (eds)
© 2024 Taylor & Francis Group, London, ISBN 978-1-032-90013-1

83 Finite Element Analysis of the Mechanical Properties of Natural Fibre Reinforced Composites

R. Ganesamoorthy*
Professor, Department of Mechanical Engineering,
Chennai Institute of Technology, Chennai, India

K R Padmavathy
Professor, Department of Mechanical Engineering,
Panimalar Engineering college, Chennai, India

D Dhana Sekar
Assistant professor, Adhiparasakthi College of
Engineering, ranipet, India

J Thamilarasan[4]
Research scholar, Annauniversity,
Chennai, India

ABSTRACT: Over the last three decades, composite materials, including polymers, alloys, and ceramics, have been at the forefront of material development, asserting their dominance in the field. The volume and variety of uses for composite materials have increased continuously, aggressively entering and dominating new areas. Materials Made of Polymers and Reinforced with Synthetic Fibers. In contrast to traditional materials such as wood, concrete, and steel, glass, carbon, and aramid offer benefits characterized by elevated rigidity and an impressive strength-to-weight ratio. Despite these benefits, synthetic Fiber-reinforced polymer composites are less commonly used due to their high starting prices, utilisation in inefficient structural forms, and, most critically, their negative effects on the environment. This study illustrates the examination of wood fibre reinforced polypropylene with the different fibre volume fractions to evaluate the tensile test by applying the tensile load, the plate was given a load of 1600N under the x direction. When compared to conventional materials such as concrete and steel, polymeric materials reinforced with synthetic fibers like glass, carbon, and aramid offer distinct advantages, boasting superior stiffness and an exceptional strength-to-weight ratio. Examining the potential of utilizing natural fibers as reinforcements for polymers stems from the increasing interest in substituting traditional synthetic fibers with them in plastics, particularly in diverse automotive applications. This research aims to establish a novel class of natural fibre reinforced composites with potential for wide-ranging engineering applications. The analysis and output will be done by ANSYS software.

KEYWORDS: Natural fiber, Polypropylene (PP), Epoxy, Kenaf, Mechanical property

*Corresponding author: ganesamoorthyr@citchennai.net

DOI: 10.1201/9781003545774-83

1. Introduction

A substance composed of two or more distinct materials that, when combined, provide a material with special properties is referred to as a "composite material". When two materials are combined, a new material is produced with improved strength, stiffness, toughness, and durability characteristics. Natural fibre reinforced composites are composite materials that utilize a polymer matrix as the binding phase and natural fibres such jute, hemp, flax, sisal, and kenaf as the reinforcing phase. In comparison to conventional synthetic fibre reinforced composites, these composites provide a number of benefits, including lower costs, bio degradability, lowdensity, lowtoxicity, and good particular mechanical properties. In general, natural fibre reinforced composites provide a possible substitute for conventional synthetic fibre reinforced composites, especially in applications where environmental effect is a major consideration.

2. Material Selection and P Operties

Due to its greater strength and enhanced the natural fibres, the pine reinforced poly propylene composite is chosen. Applications for pine reinforced poly propylene composites in automobiles include door panels and interior trim elements.

The table displayed the material qualities.

2.1 Methodical Approach

The finite element analysis programme" ANSYS2023" was used to perform the static mode tensile load test on laminated composite plates. Laminated composite plate with dimensions of 240 mm, 220 mm, 5mm is used for analysis. In Fig. 83.1, the composite model is displayed.

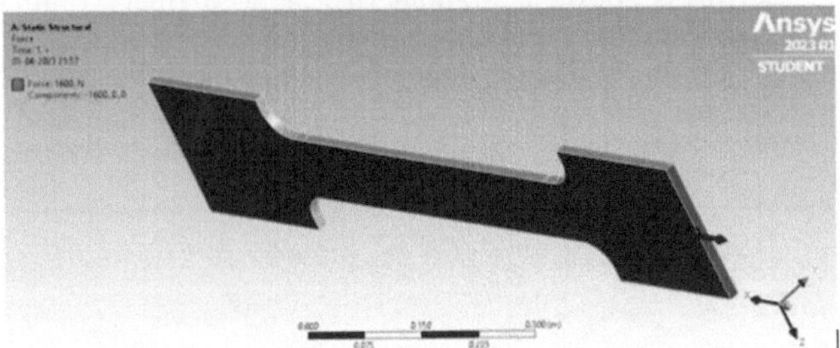

Fig. 83.1 Odeled in accordance with ASTM D 638 for the specimen

2.2 Orthotropic Material Property Evaluation

The subsequent tables offer a comprehensive analysis of the material characteristics of Composite materials made of Pine Wood Composite materials consisting of Polypropylene reinforced with Pine fibers and Epoxy reinforced with Kenaf fibers.

Table 83.1 Material properties

Materials	Poisson's ratio, ν	Density, ρ	Elastic Modulus, E	Shear Modulus, G
Polypropylene	0.39	1.1	2600	1000
Pinefibre	0.30	0.5	6000	700
Epoxy	0.41	1.2	4000	570
kenaf	0.342	1.5	4300	350

2.3 Meshed Specimen

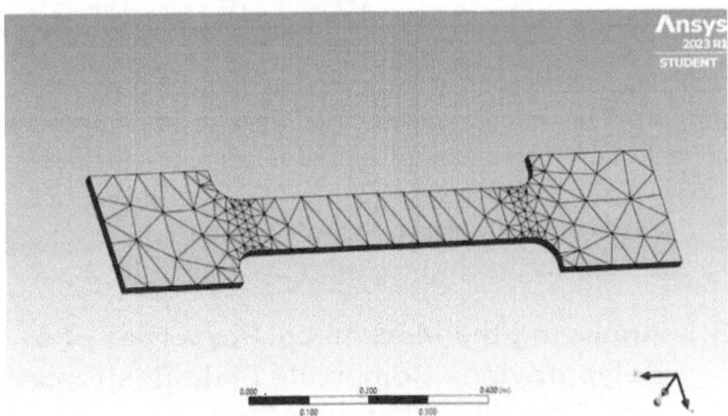

Fig. 83.2 Meshed specimen

3. Discussions and Results

The specifics of the evaluation of the composites are described in this chapter. Pine fibre and polypropylene resin were the primary components of the analysis's raw ingredients.

3.1 Tensile Strength

Often, flat specimens are used in the tensile test. The dog-bone and straight side types with end tabs are the specimens used for tensile testing that are most commonly used. A

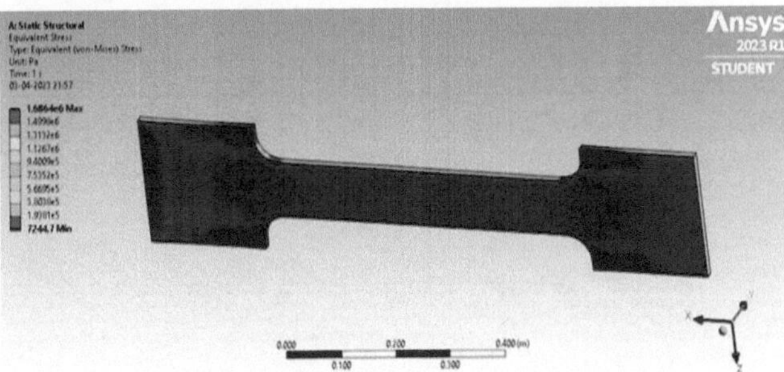

Fig. 83.3 Stress distributions on composite plates with laminated surfaces

Table 83.2 Pine fiber's ortho tropic material property at 0° fiber arrangement

Properties	V_f at 0 Fiber Arrangement										
	0	0.1	0.2	0.3	0.4	0.5	0.6	0.7	0.8	0.9	1
E_1, (MPa)	260	2753	2900	3061	3215	3371	3523	366	383	3987	415
E_2, (MPa)	120	1351	1403	1462	1526	1500	1672	154	184	1955	206
E_3, (MPa)	260	2754	2907	3061	3215	3371	3523	366	383	3987	415
G_{12},	108	1003	1072	1143	1216	1291	1472	144	152	1610	169
G_{23},	592	571	599	630	662	698	479	778	824	875	932
G_{31},	108	1003	1072	1143	1216	1291	1472	144	152	1610	169
v_{12}	.195	.373	.356	.339	.322	.195	.288	.27	.25	.238	.221
v_{23}	.098	.183	.172	.162	.153	.097	.137	.12	.12	.116	.110
v_{31}	.195	.373	.356	.339	.322	.195	.288	.27	.25	.238	.221
ρ,g/cc	1.3	1.12	1.15	1.18	1.21	1.24	1.27	1.30	1.33	1.36	1.39

uniaxial force is applied to one end of the specimen during the test. Tensile characteristics of fiber resin composites are determined by means of ASTM standard test procedure D638. A test component of 240 mm in length is needed. The tensile strength of composite materials is determined by analyzing the results of tensile tests..

3.2 Analyzing the Mechanical Properties of a Polypropylene Composite Plate Reinforced with Pine Fibers through Finite Element Analysis

- The Laminated Composite Plate is subjected to static analysis by altering the fibre orientation and volume % of fibre. The tables below display the outcomes of the pine fibre reinforced polypropylene composite.
- As per Fig. 83.4, there is an elevation in von Mises stress observed at a 60% fiber volume fraction, while stiffness demonstrates an increase at a 0° fiber orientation.
- According to Fig. 83.5, there is an elevation in von Mises stress noted at a 20% fiber volume fraction, along with a progressive increase in stiffness at 30% fiber volume fraction with a 30° fiber orientation.
- In the analysis of Fig. 83.6, a reduction in von Mises stress is observed at a 20% fiber volume fraction, while stiffness shows a gradual rise at 30% fiber volume fraction with a 45° fiber orientation.

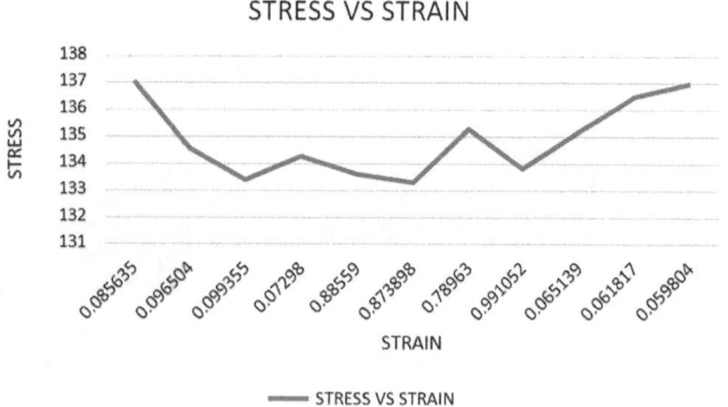

Fig. 83.4 Stress-strains curve for pine fiber at 0°fiber orientation

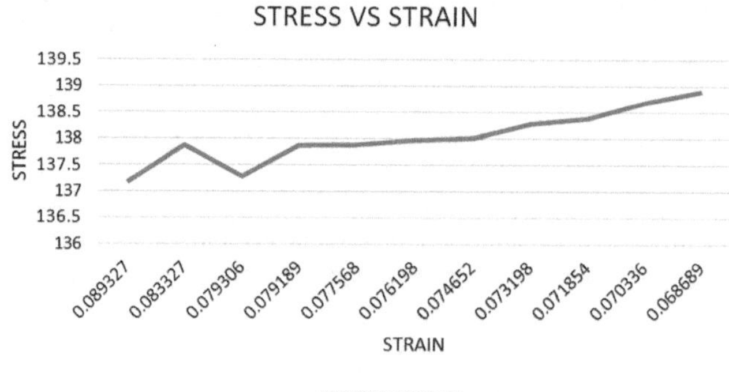

Fig. 83.5 Pine fiber at 30° fiber orientation: stress-strain curve

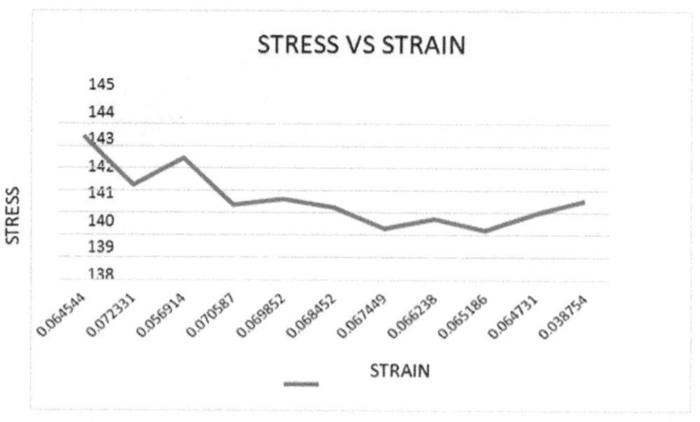

Fig. 83.6 Stress-strains curve for pine fiber at 45° fiber orientation

- Referring to Fig. 83.7, there is a decrease in von Mises stress at 0% fiber volume percentage, and stiffness consistently increases at 0% fiber volume fraction with a 60° fiber orientation.
- Examining Fig. 83.8, a decrease in von Mises stress is observed at a 10% fiber volume fraction, while stiffness progressively increases at 0% fiber volume fraction with a 90° fiber orientation.

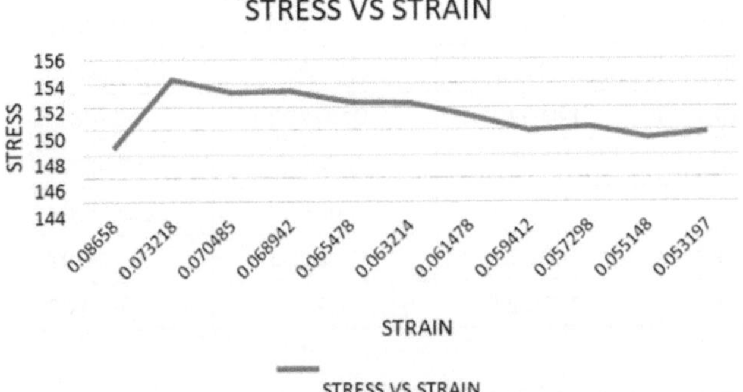

Fig. 83.7 Pine fiber at 90° fiber orientation: stress-strain curve

3.4 Kenaf Fiber Re-Inforced Epoxy

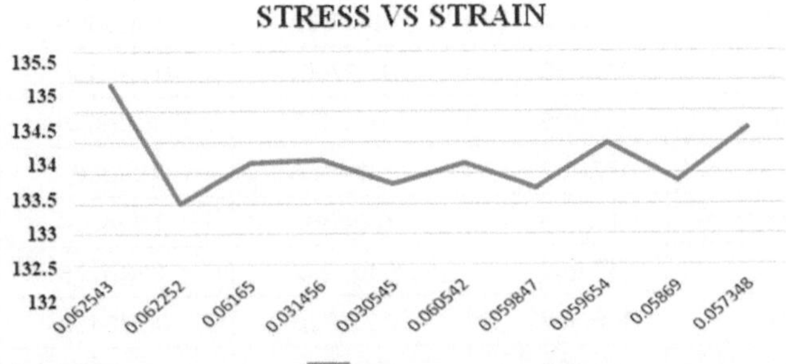

Fig. 83.8 Kenaf fiber at 0° fiber orientation: stress-strain curve

3.5 Composite Plate Finite Element Analysis

ANSYS 2023 is used to do static analysis on the laminated composite plate while altering the fibre orientation and volume proportion of fibre. The tables below display the outcomes of Kenaf Fiber Reinforced Epoxy Composites. Referring to Fig. 83.9, there is an escalation in von Mises stress at a 70% fiber volume fraction, coupled with an increase in stiffness at a 0° fiber orientation. As depicted in Fig. 83.10, there is a gradual decline in stiffness observed at a 20% fiber volume percent with a 30° fiber orientation, while von Mises stress shows an increase at 30% fiber volume fraction. Evaluating Fig. 83.11 indicates a decline in von Mises stress at a 40% fiber volume fraction, accompanied by a gradual decrease in stiffness at 0% fiber volume fraction with a 45° fiber orientation. In the analysis of Fig. 83.12, von Mises stress experiences a reduction at 0% fiber volume fraction, while stiffness exhibits an increase at 50% fiber volume fraction with a 60° fiber orientation.

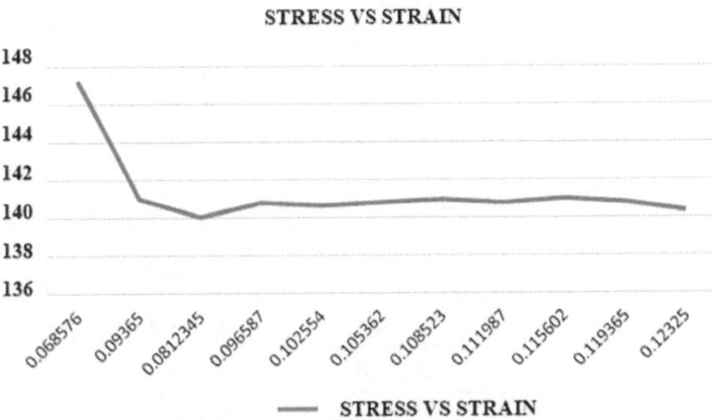

Fig. 83.9 Kenaf fiber at 30° fiber orientation: stress-strain curve

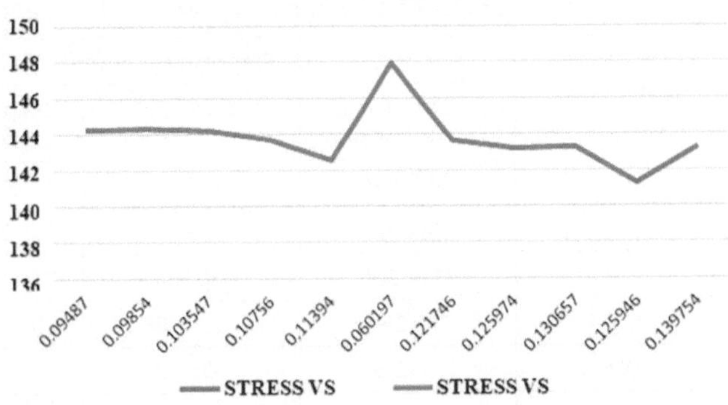

Fig. 83.10 Kenaf fiber at 45° fiber orientation: stress-strain curve

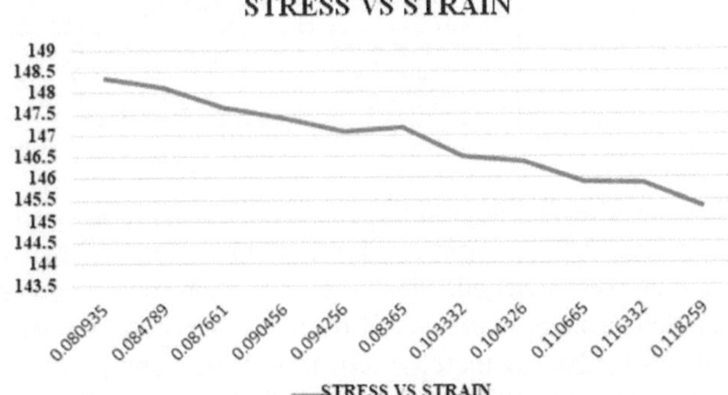

Fig. 83.11 Kenaf fiber at 60° fiber orientation: stress-strain curve

4. Conclusion

This research is being done to confirm that pine wood fibres are being used to strengthen the propylene and kenaf

Table 83.3 Kena ffibre stress, strain, and displacement at 45° and 60° fibre

Fiber Volume Fraction (Vf)	0° Fiber Arrangement			30° Fiber Arrangement		
	Stress, σ (N/mm^2)	Strain, ε	Displacement, x(mm)	Stress, σ(N/mm^2)	Strain, ε	Displacement, x(mm)
0	134.984	0.061543	0.495598	147.255	0.067576	0.633187
0.1	133.004	0.061252	0.492104	140.986	0.09265	0.77231
0.2	133.664	0.06065	0.488631	140.031	0.080235	0.682255
0.3	133.714	0.030456	0.48518	140.778	0.095587	0.830255
0.4	133.324	0.029545	0.481749	140.624	0.101554	0.858932
0.5	133.656	0.059542	0.478338	140.788	0.104362	0.88759
0.6	133.253	0.058847	0.474948	140.949	0.107523	0.91621
0.7	133.984	0.058654	0.471577	140.764	0.110987	0.944706
0.8	133.368	0.05769	0.468226	140.99	0.114602	0.973004
0.9	134.245	0.056348	0.464894	140.792	0.118365	1.002
1	134.984	0.061543	0.46158	140.327	0.12225	1.03

Table 83.4 Kena ffibre stress, strain, and displacement at 45° and 60° fibre orientation

Fiber Volume fraction (Vf)	45° Fiber Arrangement			60° Fiber Arrangement		
	Stress, σ (N/mm^2)	Strain, ε	Displacement, x(mm)	Stress, σ(N/mm^2)	Strain, ε	Displacement, x(mm)
0	144.258	0.09487	0.834811	148.35	0.080935	0.777069
0.1	144.316	0.09854	0.874112	148.132	0.084789	0.805658
0.2	144.213	0.103547	0.913114	147.658	0.087661	0.834565
0.3	143.654	0.10756	0.9520260	147.406	0.090456	0.863701
0.4	142.51	0.11394	1.004	147.092	0.094256	0.892367
0.5	147.951	0.060197	0.47188	147.186	0.08365	0.824074
0.6	143.621	0.121746	1.069	146.497	0.103332	0.951275
0.7	143.155	0.125974	1.108	146.387	0.104326	0.978452
0.8	143.249	0.130657	1.146	145.908	0.110665	1.007
0.9	141.236	0.125946	1.086	145.87	0.116332	1.035
1	143.256	0.139754	1.223	145.331	0.118259	1.064

fibres used to strengthen the epoxy. After doing a tensile study in ANSYS with a maximum load, the following findings were reached. Since the fibre from pine wood is so much stronger than other fibres in this situation, we may add more fibre to increase stiffness. When compared to other specimens, the tensile analysis findings show that the sample of pine fibre with reinforcing area of polypropylene achieves the standard as per the need. Tensile analysis findings may be utilised to develop safe and dependable buildings and products by revealing important details about a material's mechanical characteristics.

REFERENCES

1. J. L. Brendan, G. Armando and B.J. McDonald, Mat. Res. Innovat., 4(2001), p. 97
2. M. Fossen, I. Ormel, G.E.T. Van Vilsteren and T.J.Jongsma, Applied Composite Materials, 7, (2000), 433
3. A.K. Bledzki, W. Zhang and O. Faruk, Holz als Roh- und Werkstoff, 63 (2005), p. 30
4. A. Karmarkar, S.S. Chauhan, J.M. Modakand M. Chanda. Composites: PartA, 38(2007)
5. N. Uddin, Ed., Developments in Fiber-Reinforced Polymer (FRP) Composites for Civil Engineering, Elsevier, 2013.

6. A. Ticoalu, T. Aravinthan, and F. Cardona,"A review of current development in natural fiber composites for structural and infrastructure applications," in Proceedings of the Southern Region Engineering Conference (SREC '10), pp. 113–117, Toowoomba, Australia, November 2010.
7. Y. Xie, C.A.S.Hill, Z. Xiao, H.Militz, and C.Mai, "Silane coupling a gents used for natural fiber/polymer composites: a review," Composites Part A: Applied Science and Manufacturing, vol. 41, no. 7, pp. 806–819, 2010.
8. S. Shinoj, R. Visvanathan, S. Panigrahi, and M. Kochubabu, "Oilpalmfiber (OPF) and its composites: a review," Industrial Crops and Products, vol. 33, no. 1, pp. 7–22, 2011.
9. M. M. Kabir, H. Wang, K. T. Lau, and F. Cardona, "Chemical treatments on plant-based natural fibre reinforced polymer composites: an overview, "Composites PartB: Engineering, vol. 43,no. 7, pp.2883–2892, 2012
10. F. Z. Arrakhiz, M. El Achaby, M. Malha et al., "Mechanical and thermal properties of natural fibers reinforced polymer composites:doum/lowdensitypolyethylene," Materials & Design, vol. 43, pp. 200–205, 2013.

Note: Every figure and table was created using the experimental work; none of them were taken from any publications or online.

Advances in Additive Manufacturing Technologies – Gurusamy Pathinettampadian (eds)
© 2024 Taylor & Francis Group, London, ISBN 978-1-032-90013-1

84 Microstructure and Mechanical Charaterization of Wood PLA/CF-PLA Binding Layer Parts using FDM

R. Gokuldass[1]
Department of Mechanical Engineering,
Sriram Engineering College, Chennai, India

B. Krishna kumar[2]
Department of Mechanical Engineering,
Chennai Institute of Technology, Chennai, India

Nanvaniss. M[3], Manikandagopalan S[4]
Department of Aeronautical Engineering,
Gojan School of Business and Technology,
Chennai, Tamil Nadu, India

ABSTRACT: Fused deposition modeling (FDM) is a widely utilized additive manufacturing method effectively employed in diverse sectors like aerospace, automotive, and architecture to create thermoplastic components. However, there is a vital need to improve the mechanical characteristics and reduce the weight of FDM components because to the inherent restrictions in mechanical attributes and material mass. To address this concern, this study presents an investigation involving the utilization of fused deposition modeling to create experimental samples composed of wood-based polylactic acid (PLA) and laminated carbon fiber (CF)/PLA. The focal point of the research involves an experimental analysis of the mechanical properties and microstructure of these functionally graded material, specifically concentrating on the wood PLA and CF/PLA layered specimens. The experimental findings indicate that the hybrid specimen exhibits reduced mass compared to the CF/PLA specimen, while also demonstrating improved tensile, compression, and flexural mechanical characteristics in contrast to the wood PLA specimen.

KEYWORDS: Additive manufacturing, Fused deposition modelling, Wood/PLA, CF-PLA, Functionally graded material

1. INTRODUCTION

FDM is a technique of additive manufacturing in which a heated nozzle is used to deposit molten material in layers to create a solid part. It can build any complex geometry structure in brief and will not cause additional costs at any time. Additive manufacturing technique have formed different available processes, such as stereolithography (SLA), laminated object manufacturing (LOM), and selective laser sintering (SLS) among these technologies, the most commonly adopted process is FDM. In recent times, thermoplastic filaments such as acrylonitrile buta-

[1]dassmech@gmail.com, [2]krishnakumarb@citchennai.net, [3]nanvaniss16@mail.com, [4]gmaniaero@gmail.com

DOI: 10.1201/9781003545774-84

diene styrene (ABS), polycarbonate (PC), polylactic acid (PLA), polyamide (PA), polycaprolactone (PCL), and polypropylene (PP) as well as composite materials made of any two or more types of thermoplastic filaments have been used as feedstock for FDM systems to fabricate parts [Yuhang Li, 2017] [1]. Therefore, before forming the specimen, it is vital to master the various forming characteristics of the materials, and the temperature control of the forming method should be well managed. This significant innovation's primary product is a composite-forming process, and conventional experiments are used to compare the outcomes with those of another thermoplastic material. Wood is a biodegradable material that frequently turns up as wood leftovers. The renewable nature and biodegradability of wood have contributed to its popularity towards environment [Nadir Ayrilmis PLA][2]. By blending wood particles or fibers with thermoplastic polymers such as polylactic acid (PLA), researchers have been able to create wood-polymer composites (WPCs) with improved mechanical properties. Hence it is used in this paper to obtain their mechanical properties and their hardiness towards CF-PLA. Wood/PLA composites are made primarily by extrusion, injection molding, and hot press molding. When strength and/or weight are crucial, lightweight wood/PLA composites perform better than solid wood/PLA composites [Nadir Ayrilmis et al.[2].

Carbon fiber-reinforced polymer composites are primarily used in the aerospace, automotive, and other technical industries due to their excellent durability to weight ratio. While having excellent strength, CF-reinforced PLA (CF-PLA) composite lacks robustness.

Wood has high level features such as s high modulus, low cost, low scrape, less abrasiveness, abundant and renewable. Furthermore, wood flour can be combined with a variety of synthetic and organic adhesives for printing. Recent years have seen the widespread adoption of 3D printed wood-plastic composites (WPC) based on FDM technology in the furniture, construction, and other consumer applications. This is because these materials offer a promising range of unique mechanical properties along with a managed environmental footprint. Additionally, the use of wood ash has been advocated as a way to remedy forest nutrient shortages brought on by purely natural weathering acidic deposition and processes. Therefore, it is essential to assess their mechanical attributes in order to use these kinds of composites in these sectors. Due to its favorable effects on the environment and improved characteristics, wood products have potential for use in the AM sector. Consequently, this study describes the creation of a PLA composite filament loaded with wood flour (WF) for FDM 3D printing and an analysis of its composite properties [yubotoa et al].[3]achieving this, it is possible to create new wood products that are completely different from the original wood in terms of technology while also being as fungus resistant as feasible. The wood/PLA filaments can be efficiently used in the production of 3D-printed products because of good interfacial adhesion between PLA and wood [Nadir Ayrilmis et al] [2]. Selective laser sintering and stereolithography are two techniques used in 3-D printing.

Several fields, such as sports, biomedical, automotive, marine, aerospace, transportation, architecture, and other technical industries, have seen a rapid change in applications recently [Abdul Samad Khan et al [5]]. This means that understanding the different forming properties of the materials is crucial before forming the specimen, and the temperature management throughout the forming process needs to be properly handled. This major breakthrough primarily leads to the production of a composite-forming method, and results are compared with other thermoplastic materials using traditional experiments. The purpose of this paper is to explore the fabrication process, properties, and potential applications of wood and CF-PLA-based FGMs using FDM. We seek to learn more about the optimization and application of these materials in a variety of industries, including the automotive, aerospace, and consumer goods.

2. Experimental Procedure

Wood and CF-PLA was used to fabricate the functionally graded material (FGM). The 1.75 mm-diameter 3D-printed specimens were created with 30 weight percent and 70 weight percent PLA.

A commercial 3D filament manufacturer was where we obtained the wood/PLA filaments from. The nozzle di-

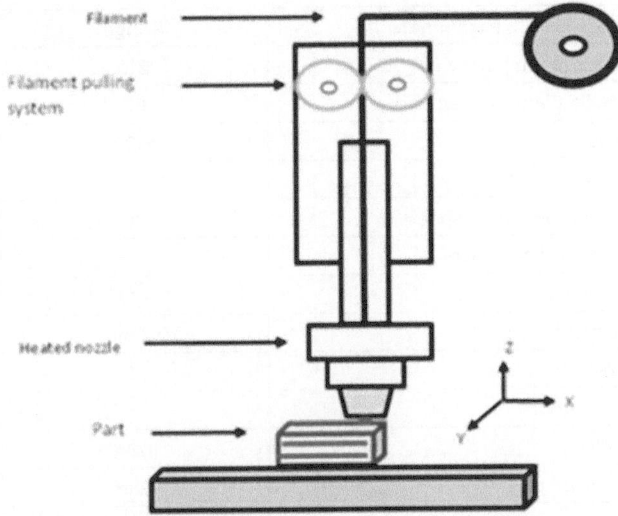

Fig. 84.1 Schematic representation of 3D printer

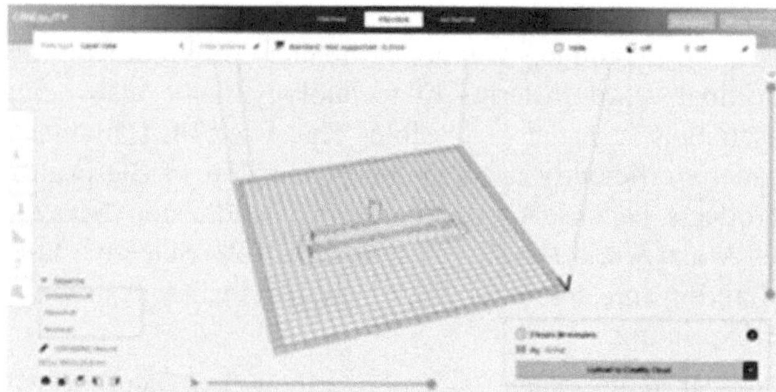

Fig. 84.2 Creativity of slicer software V4.8

ameter of the 3D printer was 0.4 mm. The fabrication of the functionally graded material of wood and CF-PLA by fused deposition modelling is used to identified the mechanical properties of the material. A 3D printer Creality CR10 smart pro has been used for fabricating the functionally graded material of wood and CF-PLA by FDM. Once, the design of the fabricating specimen is complete, save it in a STL format that is compatible with the 3D printer software. The parameters have been set-up in the 3D printer as given in the table below. Use the slicing software to prepare the 3D model. The filament gets into the extruder where it reaches the nozzle and printed on the bed as layer-by-layer stacking mechanism. An extended list of the printing parameters is given in Table 84.1. The fabrication of the compression test for FGM was built where the wood (50%) as inner layer and the CF-PLA (50%) as outer layer and it was demonstrated. Similarly, for the tensile test the FGM was fabricated as the binding layer in which the wood acts as inner layer and CF-PLA as binding layer on it. All the specimens for tensile test, flexural test and compression test for the test sample has been fabricated as per ASTM

Table 84.1 3D printer settings used for our samples

Parameters	Range
Nozzle temperature	210
Build plate temperature	60
Flow	100%
Print speed	60 mm/sec
Initial layer print speed	30
No. of slower layer	2
Travel speed	120
Layer height	0.2 mm
Wall thickness	1.2
Wall count	3
Infill	100%
Infill pattern	Grid
Cooling	100% fan speed

standard. ASTM D368-10 tensile specimen of wood and CF-PLA by FDM and it's being tested using the Z010 proline tensile testing machine of pneumatic grip X force K load cell. The test speed of the tensile test was performed at 50mm/min, and the grip-to-grip separation has 115mm.

The standard which was used to fabricate the specimen is ASTM D790. For the flexural test the Z72.5KN flexural kit X force HP load cell is being used for testing the sample with the load capacity of 72.5 KN. The flexural test was performed at the pre load and the test speed of 0.1MPa and 10mm/min. The compression test samples were fabricated as per ASTM D695 standard and Z010 proline UTM load cell -X force K compression kit was used to test the specimen under the load capacity of 10 KN. The test speed and the pre load were carried under the performance rate of 0.1MPa, 1.3 mm/min. The fabricated specimen of the wood and CF-PLA of FGM after the testing is then observed through the digital microscope for observing the specimen and recording their microstructural behavior of each material.

3. Results and Discussion

3.1 Microstructure Studies

From the fabricated specimen using wood and CF-PLA by FDM, the specimens were tested for mechanical properties and micro structure studies were carried out. FGM of wood/CF-PLA had greater mechanical properties then the wood and CF-PLA. From the Fig. 84.3 the tensile response of 3D printed samples of wood, CF-PLA and FGM (wood/CF-PLA) had fabricated and tested, it shows that the fabricated sample A, B and C and the fractured parts are showed through the microscope on Fig. 84.2. From the Fig. 84.1 the microscope image of CF-PLA were observed and the image shows the minute filaments of carbon present in the PLA. From Fig. 84.2, the materials were observed under the microscope and it was taken before conducting the test and when the test was done, the white particles at which the specimen can see under the microscope and it is because when the specimen tends to fracture at one point and the presence of PLA will be occurred as carbon fiber was broken into pieces. Other materials such as wood and FGM can be observed after the test was conducted.

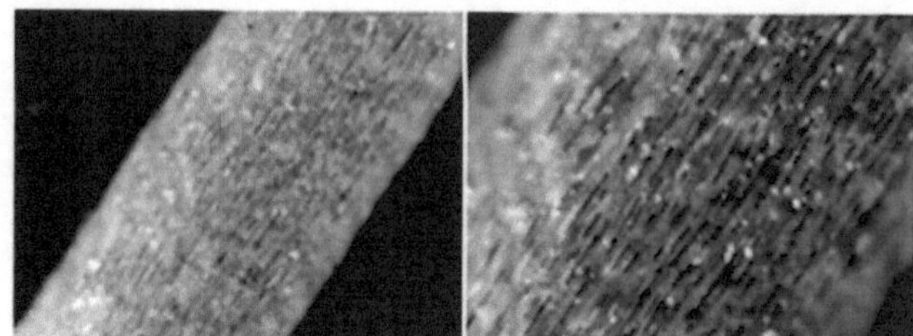

Fig. 84.3 Microscope image of carbon fiber reinforced PLA

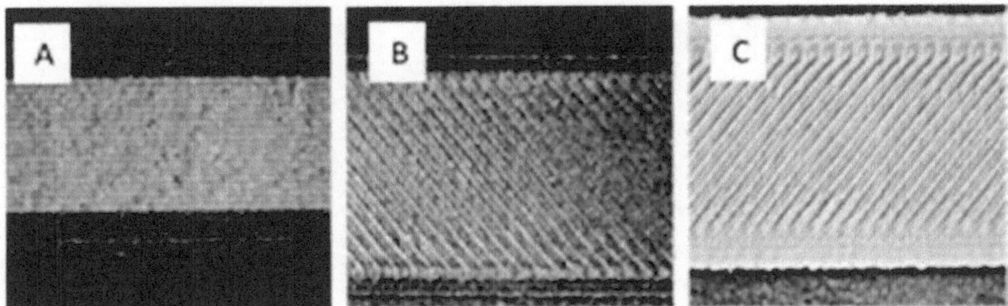

Fig. 84.4 (A) FGM, (B) CF-PLA, (C) wood, shows the microscope image of the specimen before the test

4. Tensile Properties

As per [7], the decline in the tensile performance of wood-based samples at elevated printing temperatures may also be attributed to the thermal decomposition of wood particles.. The tested specimens of wood and CF-PLA in which the wood has lower strength when the CF-PLA will be highest. In contrast to CF-PLA, FGM will be slightly higher.

From the Fig. 84.4 the tensile specimen is detected through the microscope and the wood material was emerged, the mechanical properties of the tensile tested specimen of wood, CF-PLA and FGM are given beneath the table. From Fig. 84.4 the wood specimen were worn out and the broken part of the image under the microscope were demonstrated and the part when the tensile test were conducted the grip to grip at 115 mm and the wood material has low yield strength compared to any other material and when the CF-PLA were tested and the area at which the specimen gets broken and the cracked parts are visible, when the nozzle temperature and the parameters kept same as for both the wood and CF-PLA, it has good tensile property in it. However, FGM was fabricated in order to increase its properties of wood and where the tensile test of wood has showed the greater strength then the wood.

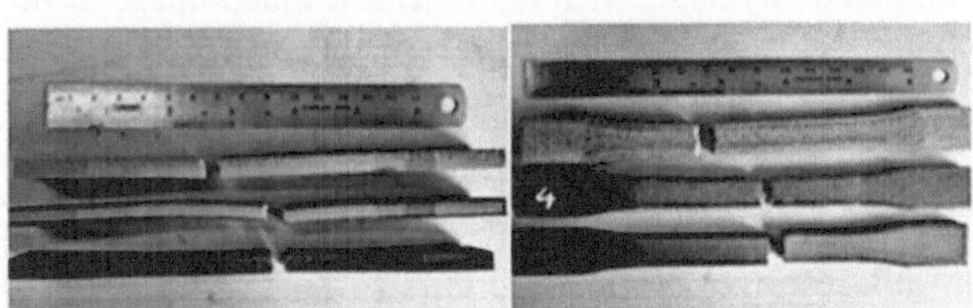

Fig. 84.5 Tensile test sample of CF-PLA, FGM and wood

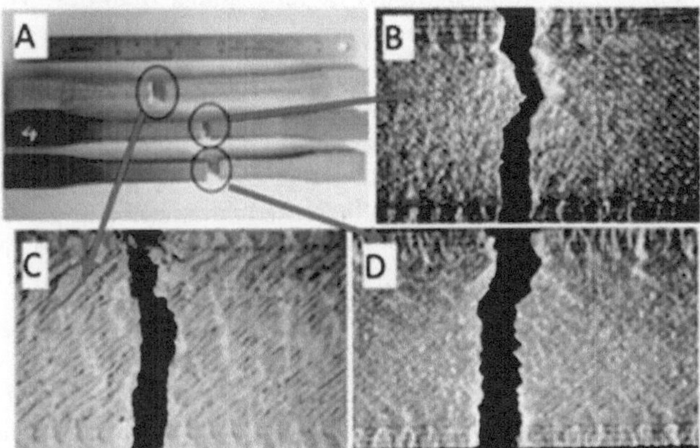

Fig. 84.6 The fixate area of the tested specimen for wood, (b) CF-PLA of the tested specimen of the tensile sample

From Fig. 84.8 the graph of the tested samples indicates that the wood has moderate rate of withstanding the stress at which the load applied, however the FGM has better tensile property at a point in which it reaches the maximum strain than CF-PLA. CF-PLA has almost reached at the point where 42MPa and the FGM has sustained up to 26MPa than the wood.

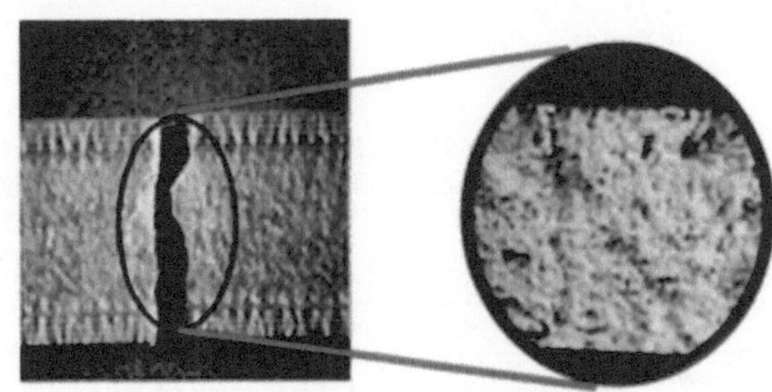

Fig. 84.7 Microscope image of tensile test sample (FGM)

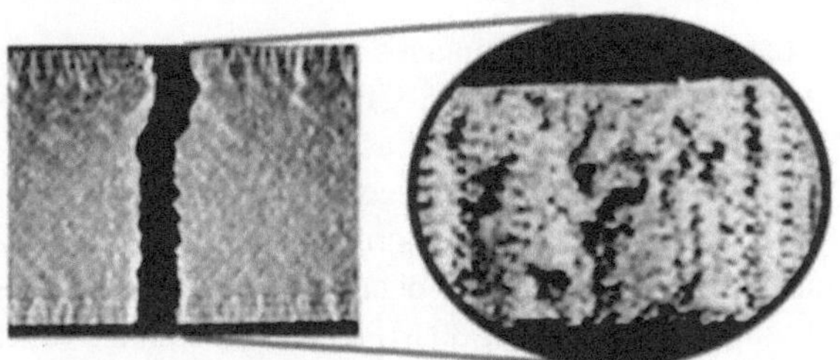

Fig. 84.8 Microscope image of tensile sample (CF-PLA)

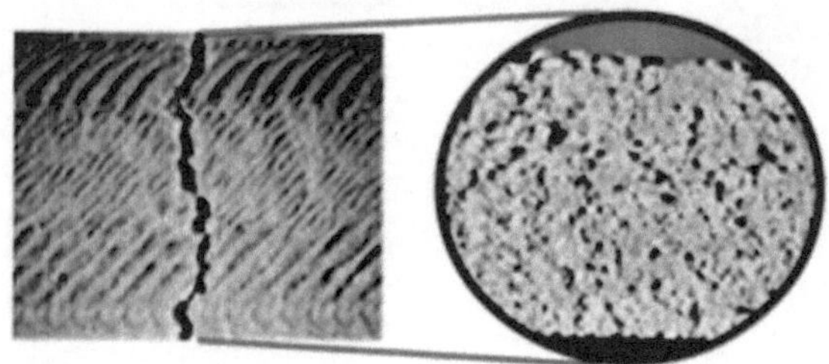

Fig. 84.9 Microscope image of tensile sample (wood)

4.1 Compression Test

From the fabricated part of the wood, CF-PLA and FGM for the compression test sample has been tested and as compared to the wood and CF-PLA, FGM (wood/CF-PLA) has high yielding strength and it has good ductility than the wood. When the FGM of the material that is fabricated where the wood is in inner layer and the carbon has outer layer. Fig. 84.4(a) Front view of compression tested samples of wood, CF-PLA and FGM, (b) top view of the tested samples, (c) shows the detailed view of the tested wood sample at the shattered location, (d) tested specimen of CF-PLA at the more squashed place, (e) the complete view of compressed FGM.

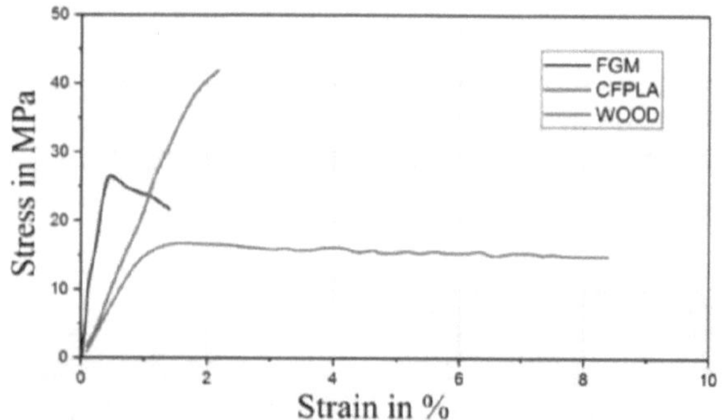

Fig. 84.10 The stress strain graph of wood, CF-PLA and FGM

Table 84.2 Tensile properties for different materials

Material	UTS, MPA	% Elongation
Wood	17	1.4
CF-PLA	42	2.1
FGM	26	0.45

From the Fig. (84.4) the images shows that the compression tested samples of wood, CF-PLA and FGM (wood/pla), where the FGM of wood as inner layer and the CF-PLA as the outer layer. At image (c) the detailed view of the wood under the digital microscope were demonstrated and the deformation that occurred on the compressed wood sample at the rate of 30 MPa and at the image (d) the compressed sample of CF-PLA were tested and the white

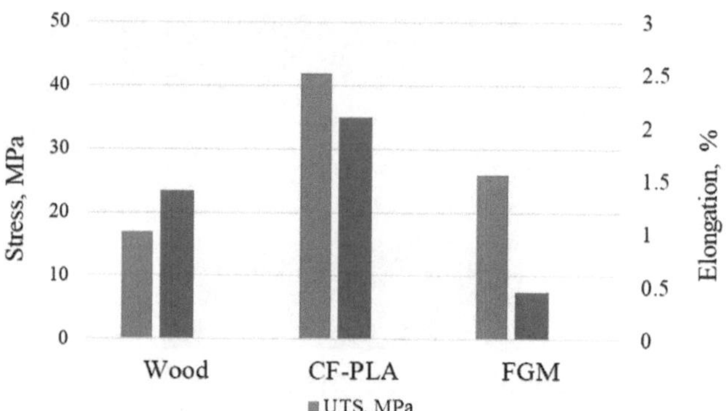

Fig. 84.11 Tensile properties of wood, CF-PLA and FGM

particles are observed were after the compression test were demonstrated. When the FGM were observed in the microscope the combined Particles of wood and CF-PLA were observed exhaustively. The maximum stress that applies on the specimen and comparatively CF-PLA is better in their compressive strength than any other material. When it comes to wood and FGM the wood has deformation of 20% and it was buckled and leads to failure, at some point when the FGM has deviated and the compression test for their maximum stress it deformed up to 51.9MPa where it is better than wood alone. The CF-PLA occurs at almost 55MPa and it resist at the deformation of 5.8% where the FGM resist throughout the test section, when the wood occurs at the maximum stress and it resist up to the point, but comparatively lesser then FGM. The scanned image of the compressed CF-PLA under the microscope the scattered white fibers of PLA can be observed.

5. Conclusion

In this research, a new experimental binding layer specimen was created using an AM-based FDM method. In conclusion, our study unequivocally shows that CF-PLA outperforms wood in terms of mechanical qualities. We have successfully created a novel method to increase the strength of wood by adding CF-PLA as a binding layer, leading to the development of a functionally graded material (FGM). When compared to wood alone, the FGM exhibits mechanical qualities that are significantly improved, demonstrating its increased strength and resilience. Despite significant improvements, it is important to note that the mechanical characteristics of the FGM are still inferior than those of CF-PLA. This is not surprising given the basic distinctions between the two materials. A balance between strength, cost-effectiveness, and environmental sustainability can be achieved with the combination of wood and CF-PLA in future applications, according to the FGM's successful development. The mechanical properties of the specimen were tested and the microstructure characterization of the material such as wood, CF-PLA and FGM were demonstrated and recorded and thus results in FGM is higher than wood alone.

REFERENCES

1. Y. Yuhang Li, S. Gao, R. Dong, X. Ding, and X. Duan, "Additive Manufacturing of PLA and CF/PLA Binding Layer Specimens via Fused Deposition Modeling," J Mater Eng Perform, vol. 27, no. 2, pp. 492–500, Feb. 2018.
2. M. NadirAyrilmis, Kariz, J. H. Kwon, and M. KitekKuzman, "Effect of printing layer thickness on water absorption and mechanical properties of 3D-printed wood/PLA composite materials," International Journal of Advanced Manufacturing Technology, vol. 102, no. 5–8, pp. 2195–2200, Jun. 2019.
3. Yubo Tao, H. Wang, Z. Li, P. Li, and S. Q. Shi, "Development and application of wood flour-filled polylactic acid composite filament for 3d printing," Materials, vol. 10, no. 4, Mar. 2017.
4. B. Mahltig, C. Swaboda, A. Roessler, and H. Böttcher, "Functionalising wood by nanosol application," J Mater Chem, vol. 18, no. 27, pp. 3180–3192, 2008.
5. A. S. Khan, A. Ali, G. Hussain, and M. Ilyas, "An experimental study on interfacial fracture toughness of 3-D printed ABS/CF-PLA composite under mode I, II, and mixed-mode loading," Journal of Thermoplastic Composite Materials, vol. 34, no. 12, pp. 1599–1622, Dec. 2021.
6. I. Ezzaraa, N. Ayrilmis, M. K. Kuzman, S. Belhouideg, and J. Bengourram, "Study of the effects of microstructure on the mechanical properties of 3D printed wood/PLA composite materials by a micromechanical approach," in 2020 IEEE 2nd International Conference on Electronics, Control, Optimization and Computer Science, ICECOCS 2020, Institute of Electrical and Electronics Engineers Inc., Dec. 2020.
7. S. Guessasma, S. Belhabib, and H. Nouri, "Microstructure and mechanical performance of 3D printed wood-PLA/PHA using fused deposition modelling: Effect of printing temperature," Polymers (Basel), vol. 11, no. 11, Nov. 2019.
8. N. Knez, M. Kariž, F. Knez, N. Ayrilmis, and M. K. Kuzman, "Effects of selected printing parameters on the fire properties of 3d-printed neat polylactic acid (PLA) and wood/pla composites," J Renew Mater, vol. 9, no. 11, pp. 1883–1895, 2021.
9. N. VinothBabu et al., "Influence of slicing parameters on surface quality and mechanical properties of 3D-printed CF/PLA composites fabricated by FDM technique," Materials Technology, vol. 37, no. 9, pp. 1008–1025, 2022.
10. H. Li, B. Liu, L. Ge, Y. Chen, H. Zheng, and D. Fang, "Mechanical performances of continuous carbon fiber reinforced PLA composites printed in vacuum," Compos B Eng, vol. 225, Nov. 2021.
11. W. Hu et al., "Improved Mechanical Properties and Flame Retardancy of Wood/PLA All-Degradable Biocomposites with Novel Lignin-Based Flame Retardant and TGIC," Macromol Mater Eng, vol. 305, no. 5, May 2020.
12. S. Bhagia et al., "Tensile properties of 3D-printed wood-filled PLA materials using poplar trees," Appl Mater Today, vol. 21, Dec. 2020.

Note: Every figure and table was created using the experimental work; none of them were taken from any publications or online.

Advances in Additive Manufacturing Technologies – Gurusamy Pathinettampadian (eds)
© 2024 Taylor & Francis Group, London, ISBN 978-1-032-90013-1

85 Mechanical and Metallurgical Characterization of Dissimilar AA5083-H111 and AA6106-T6 by Friction Stir Welding

R. Manivannan[1], V. Guru[2], M. Raghupathi[3]
Professor, Department of Mechanical Engineering,
Muthayammal Engineering College,
Rasipuram, Tamilnadu, India

M. Mothilal[4], R. Muniyappan[5], P. Poovarasan[6], M. Sivanathan[7]
Student, Department of Mechanical Engineering,
Muthayammal Engineering College,
Rasipuram, Tamilnadu, India

S. Balamurugan[8]
Assistant Professor, Department of Mechanical Engineering,
Chennai Institute of Technology, Kundrathur,
Chennai, Tamilnadu, India

ABSTRACT: The different aluminum alloys 5083-H111 and 6106-T6, each with a 3mm thick, were fused together using friction stir welding (FSW). Optical microscopy (OM) was utilized to inspect the microstructure of the FSW joint and base metal (BM), respectively. The findings indicate that when the 5083 alloy is placed on the advancing side, the base materials combine in the weld nugget more successfully. The AA5083-H111 side's heat affected zone (HAZ), where ductile fracture failure occurs, is where the weld joint has the lowest minimum hardness. The weld joint's maximum tensile strength is 179 MPa, and its maximum elongation is 11.16%.

KEYWORDS: AA5083-H111, AA6106-T6, Friction stir welding, Mechanical characterization, Metallurgical study

1. INTRODUCTION

The solid state material joining method known as friction stir welding (FSW) is achieved by plastic deformation of the work piece and heat generation at the joining line, which is primarily supported by friction among the non-consumable rotating tool and the work piece interface [1]. Due to its reduced weight and higher specific strength than ferrous materials, aluminum and its alloys have found extensive use in latest years. The AA6106 alloys find widespread application in airplanes, loading tanks, pipe lines, and marine structures. Friction stir welding (FSW)

[1]manivannancadcam06@gmail.com, [2]massguru158@gmail.com, [3]mraghupath@gmail.com, [4]mothi4008@gmailcom, [5]mr8254566@gmail.com, [6]poovarasanpari@gmail.com, [7]sivananthanmech20@gmail.com, [8]balamurugans@citchennai.net

DOI: 10.1201/9781003545774-85

Table 85.1 Element composition of aluminium alloys

Base Metal	Mn	Mg	Si	Fe	Cr	Cu	Ti	Zn	Zr	Al
AA5083-H111	0.525	4.254	0.980	0.259	0.113	0.346	0.019	0.103	0.002	93.31
AA6061-T6	0.061	0.812	3.01	0.323	0.184	1.142	0.02	0.072	0.001	94.31

has demonstrated its viability as an aluminum alloy commercial joining method in recent years. For instance, FSW manufactures enormous tanks for launch vehicles in the aerospace sector using high strength aluminum alloys. The FSW technique is utilized in the rolling stock and shipbuilding sectors to produce huge prefabricated aluminum panels that are created from aluminum extrusion [2–3]. The mechanical characteristics of FSWed aluminum alloy joins with or without welding flash have been the subject of numerous investigations that have been published [4, 5]. The mechanical characteristics of the dissimilar alloy AA6061/AA5010 were predicted and optimized using RSM by Masoud Ahmadnia et al. [6]. In comparison to cylindrical pin profiles, they demonstrated that utilizing a square pin profiled tool with an 800 rpm rotating speed and a 60 mm/min welding speed results in a 22% enhancement in UTS, 41% increases in hardness, and 100% increases in elongation. In a recent study, Premraj et al. showed that the welded joints have higher joint efficiency and that the combination of AA5754 and AA6106 alloys has outstanding mechanical characteristics [7, 8].According to balamurugan et al determined that the welding parameters welded joints have high joint efficiency. They concluded that the high strength sample fractured at heat affected zone (HAZ) of 6106-T6 alloy side. Additionally, they investigated the corrosion induced by salt spray on strong welds in comparison to interactions with different welding conditions [9-11]. Numerous articles examined the characterization of FSW joints made of both similar and dissimilar alloys. Friction stir welding has demonstrated effectiveness for a multitude of disparate aluminum alloys. To the finest of the author's acquaintance, no studies on the relationship between the microstructure and tensile properties of different AA5083-H111 and AA6106-T6 joints that are friction stir welded have been published to date.

2. Experimental Procedure

AA5083-H111 and AA6061-T6 plates with a thick of 3 mm and a rectangular cross section of 110 mm X 55 mm were chosen as the work piece material for this work. The alloy being selected because of its suitability for aeronautical uses [12]. Table 85.1 displays the chemical component of each alloy. As important process factors, rotational speed, traverse speed, and axial load have been chosen as dependent variables for the current experiment. Table 85.2

Table 85.2 Process constraints and tool geometry

Process constraints	AA5083-H111 and AA6106-T6
Vertical force (kN)	10 (constant)
Rotating speed ($r.min^{-1}$)	900, 1100, 1300
Welding feed ($mm.min^{-1}$)	60
Tool pin profile	Square
Tool pin one side squared (mm)	3
Tool pin length (mm)	2.7
Shoulder diameter (mm)	22

shows the process constraints and tool geometry for FSW of different alloys. The FSW method was carried out on H13 tool steel with a pin length of 2.7 mm and a shoulder diameter of 15 mm. The pin has a straight cylindrical profile with a pin root diameter of 3 mm. Keller's reagent (2 ml HF, 3 ml HCl, 5 ml HNO_3, and 190 ml H_2O) was utilized to etch samples made from welded specimens for around 120 seconds [12].

3. Results and Discussion

3.1 Microstructure

The microstructure of the base metal AA5083-H111 and AA6106-T6 alloys demonstrated in Fig. 85.1(a) and (b). Figure 85.1c indicates the nugget zone of the high UTS FSWed joint. The constituents are from both the alloys of left and right side of the process. The zone shows alternate layers with different flow magnitude depends on the plasticity of the metals under heat and stress of the process. The alloy of right undergone dynamic re-crystallization with uniform grains and clear grain boundary [13].

3.2 Mechanical Characterization

The mechanical characteristics of the base metal and the various welded joints were shown in Table 85.3. The high tensile strength values in that are obtained at 1100 rpm and 60 mm/min ranges. In comparison to the other two factors (900 and 1200 rpm), 1100 rpm yields higher joint efficiency (79.55%). The components are helped to plastify by enough heat, and the tool pin created in the alloy combination is well-mixed. The stir zone (SZ) contains highly polished and uni-axed grains, which will improve the tensile qualities of the welded material even further. The SZ has also suffered significant plastic deformations,

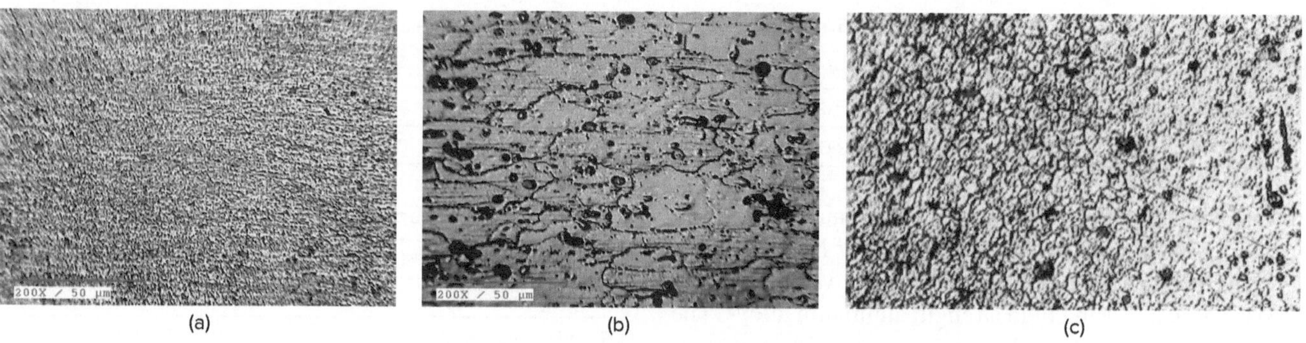

(a) (b) (c)

Fig. 85.1 Microstructure of Base Metal AA5083-H111 / AA6106-T6 and Welded joint

Table 85.3 Mechanical properties of base metal and FSWed joint

Material	UTS (Mpa)	Yield Strength (Mpa)	Elongation %	Joint Efficiency %	Fracture Location
AA5083- H111	315	193	19	—	—
AA6106-T6	319	256	14.3	—	—
FSW 5083-H111 and 6106-T6 (900 rpm)	165	135	9.89	52.38	At centre weld
FSW 5083-H111 and 6106-T6 (1100 rpm)	179	159	11.1	79.55	At HAZ of 6106 side of the weld
FSW 5083-H111 and 6106-T6 (1300 rpm)	169	143	7.78	53.65	At centre weld

as shown in Fig. 85.1 (c). Figures 85.2(a) and (c) depict the broken tensile sample for each of the various welded joints. Because of this, the sample broke in the middle of the welded area. The shattered specimens in the retreating side's heat-affected zone are depicted in Fig. 85.2(b). It is clear that there was enough heat input specified for this weld joint [14].

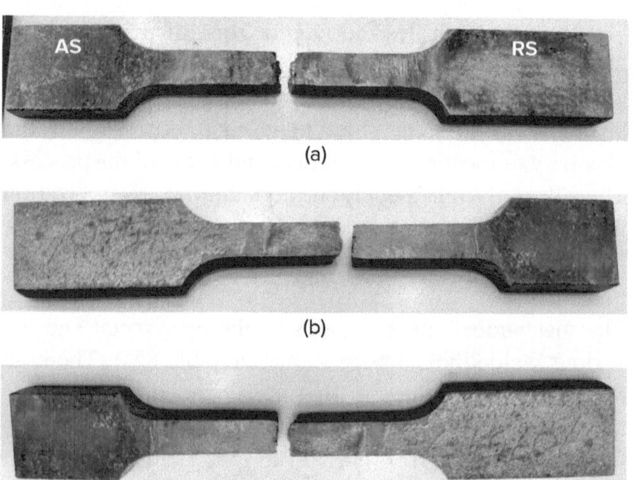

(a)

(b)

(c)

Fig. 85.2 Microstructure of base metal AA5083-H111 / AA6106-T6 and welded joint (a) Fractured at SZ-Center Weld, (b) Fractured at HAZ - RS, (c) Fractured at SZ-Center Weld

3.3 Hardness

Figure 85.3 shows the hardness profiles under various welding circumstances. The hardness range of 71 Hv (1100 rpm) is seen in that high strength sample. The remaining weld joint could be adjusted between 65 and 68 Hv at 900 and 1300 rpm, respectively. The welded joint's broken surfaces are clearly indicative of the low range of hardness in the heat-affected zone [15].In comparison with the weld NZ, the TMAZ had the lowest hardness value. Possible causes for this include dynamic recrystallization and considerable annealing softening [34]. When compared to BM, the NZ exhibited the maximum hardness, with hardness values rising from 69 to 71 Hv. The following explanations account for the NZ's shift in hardness. Initially, the NZ became harder due to the finer grains. According to the hall-Petch relationship and the model of Sato et al. [16], if the grain size decreased, the hardness could increase by about 7.1 Hv.

4. Conclusion

Different aluminium alloys 5083-H11 and 6106-T6 welded joints are fabricated by different parameters such as constant vertical force, rotating speed and welding speed; characterization and welding studies are performed. The outcomes obtained show that;

1. Increase the rotating speed at constant welding feed, the high tensile strength was obtained which gives the 79.55% of the combined efficiency.

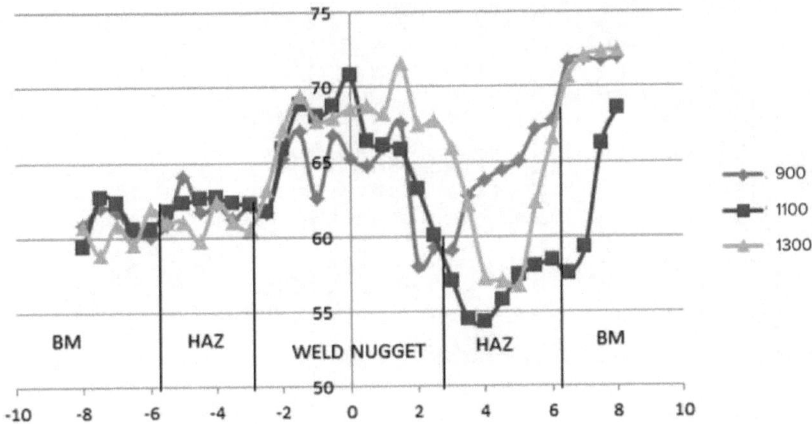

Fig. 85.3 Microstructure of Base Metal AA5083-H111 / AA6106-T6 and Welded joint

2. The high tensile strength 179 MPa attained from the welding condition 1100 rpm with square pin profile. The weld sample developed a defect free joint with high % of elongation 11.1, and hardness 73HV compared to the other joints.
3. The tensile fractured surface of the welded sample (1100 rpm) exhibited the small grains, no defects formed.

REFERENCES

1. Li, J., Su, M., Qi, W., Wang, C., Zhao, P., Ni, F., & Liu, K. (2020). Mechanical property and characterization of 7A04-T6 aluminum alloys bonded by friction stir welding. Journal of Manufacturing Processes, 52, 263–269.
2. Banik, A., Roy, B. S., Barma, J. D., & Saha, S. C. (2018). An experimental investigation of torque and force generation for varying tool tilt angles and their effects on microstructure and mechanical properties: Friction stir welding of AA 6061-T6. Journal of Manufacturing Processes, 31, 395–404.
3. Balamurugan, S., Jayakumar, K., Anbarasan, B., & Rajesh, M. (2023). Effect of tool pin shapes on microstructure and mechanical behaviour of friction stir welding of dissimilar aluminium alloys. Materials Today: Proceedings, 72, 2181–2185.
4. Sagheer-Abbasi, Y., Ikramullah-Butt, S., Hussain, G., Imran, S. H., Mohammad-Khan, A., & Baseer, R. A. (2019). Optimization of parameters for micro friction stir welding of aluminum 5052 using Taguchi technique. The International Journal of Advanced Manufacturing Technology, 102, 369–378.
5. Senthamaraikannan, B., & Krishnamoorthy, J. (2023). Material flow and mechanical properties of friction stir welded AA 5052-H32 and AA6061-T6 alloys with Sc interlayer. Materials Testing, (0).
6. Ahmadnia, M., Shahraki, S., & Ahmadi Kamarposhti, M. (2016). Experimental studies on optimized mechanical properties while dissimilar joining AA6061 and AA5010 in a friction stir welding process. International Journal of Advanced Manufacturing Technology, 87, 2337–2352.
7. Yogaraj, P., Kasirajan, L., &Senthamaraikannan, B. (2023). Effect of Tool Pin Positioning Factors on the Strength Behavior of Dissimilar Joints of AA5754-H111 and AA6101-T6 by Using Friction Stir Welding. Transactions of the Indian Institute of Metals, 76(11), 3021–3030.
8. Yogaraj, P., & Kasirajan, L. (2022). Identifying the Optimal Process Parameter on AA1100 Friction Stir Welded Joints. Tehničkivjesnik, 29(3), 957–964.
9. Doley, J. K., & Kore, S. D. (2016). A study on friction stir welding of dissimilar thin sheets of aluminum alloys AA 5052–AA 6061. Journal of Manufacturing Science and Engineering, 138(11), 114502.
10. Balamurugan, S., Jayakumar, K., & Nandakumar, C. (2023). Investigation of mechanical, metallurgical and corrosion characteristics of friction stir welded dissimilar AA 5052-H32 and AA 6061-T6 joints. Journal of the Chinese Institute of Engineers, 46(6), 601–614.
11. Rajkumar, T., Dinesh, S., Anbarasan, B., & Balamurugan, S. (2023). Effect of welding process parameters on surface topography and mechanical properties of friction-stir-welded AA2024/AA2099 alloys. Materials Research Express, 10(2), 026507.
12. Balamurugan, S., Jayakumar, K., & Subbaiah, K. (2021). Influence of friction stir welding parameters on dissimilar joints AA6061-T6 and AA5052-H32. Arabian Journal for Science and Engineering, 46(12), 11985–11998.
13. Palanivel, R., Mathews, P. K., Dinaharan, I., & Murugan, N. (2014). Mechanical and metallurgical properties of dissimilar friction stir welded AA5083-H111 and AA6351-T6 aluminum alloys. Transactions of Nonferrous Metals Society of China, 24(1), 58–65.
14. Krishnan, M., & Subramaniam, S. K. (2018). Investigation of mechanical and metallurgical properties of friction stir corner welded dissimilar thickness AA5086-AA6061 aluminium alloys. Materials Research, 21.
15. Rajkumar, V., Venkateshkannan, M., Sadeesh, P., Arivazhagan, N., & Devendranath Ramkumar, K. (2014). Studies on effect of tool design and welding parameters on the friction stir welding of dissimilar aluminium alloys AA5051-AA6061. Procedia Engineering, 75, 93–97.

Note: Every figure and table was created using the experimental work; none of them were taken from any publications or online.

Advances in Additive Manufacturing Technologies – Gurusamy Pathinettampadian (eds)
© 2024 Taylor & Francis Group, London, ISBN 978-1-032-90013-1

86 Analysis of Open Circuit Voltage (VoC) in Solar Cell and Comparison of Simulation Results for SiO2 and ITO as the Anti-reflective Coating Material with Nanoparticles

Radhika Baskar*
Professor, Institute of ECE, Saveetha School of Engineering,
Saveetha Institute of Medical and Technical Sciences,
Saveetha University, Chennai, India

ABSTRACT: An examination and mathematical demonstration of the open circuit voltages (VoC) of two anti-reflective coating materials, ITO and SiO2, when combined with silver nanoparticles (AgNPs), is the primary objective of this research. To determine how different protective layer thicknesses (50–1000 nm) affect the open circuit voltage (VoC). Learn about the configuration of a solar cell's open circuit voltage (VOC) under layers of SiO2 and AgNPs and coatings of light-insulating materials like ITO and SiO2. Each group used a nano hub simulation tool to simulate forty samples. Find the open-circuit voltage (VoC) to raise the protective layer thickness (SiO2) from 50 to 1000 nm. Using the Nano Hub online modeling application, we checked the open-circuit voltages of the two PV cell devices. When put on solar cells with silver nanoparticles, the open-circuit voltage of the SiO2 anti-reflective coating material is 90.3%, while it is 90.7% for the ITO anti-reflective coating material. Solar cells coated with indium tin oxide (ITO) outperform those coated with silicon dioxide (SiO2) when constructed using silver nanoparticles.

KEYWORDS: Open circuit voltage, Silver nanoparticles, QCRF-FDTD, SPSS, Photo voltaic technology

1. INTRODUCTION

A photovoltaic cell is one kind of electrical device that harnesses the power of light to produce energy. The photovoltaic effect is a mechanism by which solar energy has the potential to generate voltage and electricity (1). Photovoltaic technology is the process of transforming the energy of the sun into usable electrical current and voltage. These designs include layers of solar cells or photovoltaic cells with metal, oxide, and semiconductor arrangements (3,4). Photovoltaic (PV) cells in solar cells convert light into energy, which is subsequently stored in batteries, making solar cells an exciting and crucial technology (5). The solar cell converts light into usable electricity. Countless everyday items rely on solar cells, including electric fences, toys, calculators, kitchen appliances, and distant lighting systems (6).

2. OPEN-CIRCUIT VOLTAGE (VOC)

Solar cells and panels rely on the open-circuit voltage as a critical layout parameter. It is used to compute the cell's efficiency, which is the ratio of the maximum power production to the light that enters the cell, and it also establishes the utmost power that the cell can generate. (7)

*Corresponding author: radhikabaskar@saveetha.com

DOI: 10.1201/9781003545774-86

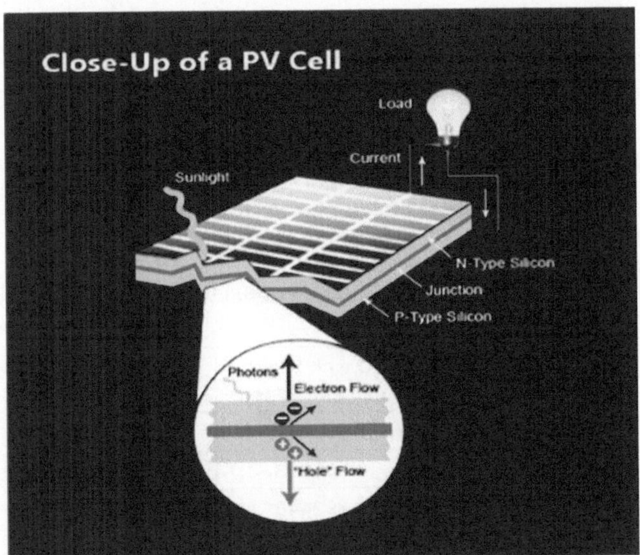

Fig. 86.1 Electrons and Holes Moving Direction open-circuit voltage (VOC) [18]

Putting the net current in the solar cell equation to zero yields an expression for Voc:

$$V_{oC} = \frac{nKT}{q} ln\left(\frac{I_L}{I_O} + 1\right)$$

Drabavičius (n.d.) and Abu-Elfotouh and Al-Mass'ari (1980) state that solar cells include oxides, semiconductors, and metal in their protective front Layer, ARC, active material, and rear reflector layers, respectively. Solar energy is an eco-friendly and sustainable power source. Consequently, there has been a recent uptick in the quantity of solar cells. Silicon p-type semiconductors are used in this arrangement. In 2001, Matsumoto, Meléndez, and Asomoza found... In their 2020 publication, Suman et al. go into the topic of SiO2's chemical characteristics. A silicon oxide, silicon dioxide, is often referred to as SiO2.

Silicic oxide, silica, and silicic acid are some of its alternative names. Quartz is the most prevalent natural form of silica dioxide. The density of silicon dioxide is 2.648 g/cm3, and its molecular weight is 60.08 g/mol. Although it boils at 2,950 °C, SiO2 has a melting point of 1,713 °C. Deal and Helms (2013) state that these chemical and physical features are typical of SiO2.

The compound known as indium tin oxide (ITO) is an indium, tin, and oxygen alloy with different percentages of each element. Most indium-tin-oxide (ITO) is composed of indium (74%), oxygen (18%), and tin (8%). While ITO is colorless and transparent in thin layers, it may take on a yellowish or grey hue when applied in more significant coatings. On top of that, ITO is very conductive electrically. A few of the chemical characteristics of ITO include its density of 7.14 g/m3, its melting point range of 1526–1926 °C, and its mass density of 6.8 g/cm3. These features are expected to ITO. As stated in the 2014 study by Chongsri et al. One drawback of using inefficient materials is that solar cell performance fluctuates when the protective layer's thickness increases. This study wants to model and test the open circuit voltage of solar cells with different anti-reflective coating materials, such as SiO2 and ITO (Rajan et al. 2020), by changing the thickness of the solar cell.

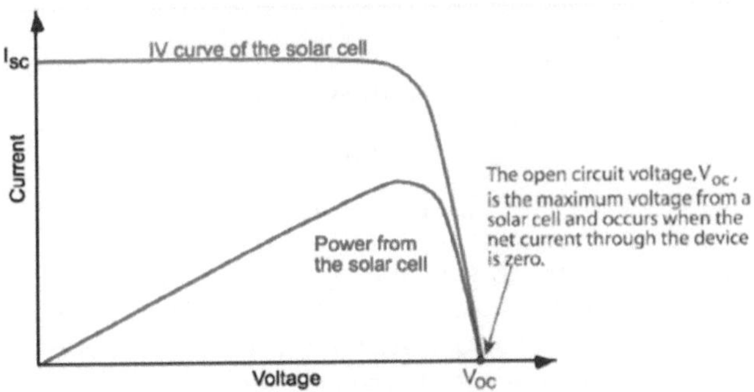

Fig. 86.2 Open-circuit voltage

3. Materials and Methods

The Nanoelectronics Lab used a nanotool to complete the project. Here, two groups are used for the comparison. Using the QCRF-FDTD simulator, a total of twenty samples were collected for the investigation, one from each group. Two distinct antireflective materials are used, with the QCRF-FDTD simulator utilised to alter the thickness of the protective layer. The first step in preparing the sample involves using the solar cell with the SiO2 configuration. An online simulation tool developed by Nano Hub, the photovoltaic QCRF-FDTD, has a cell with four layers. The solar cell's rear reflector is made of gold. Adding anti-reflection coating layers made of SiO2 and ITO to solar cells to prevent light from glare. The active layer of material in a PV cell incorporates A-Si, as pointed out by Ha and Jung (2014).

4. Simulation

Figure 86.3 shows a four-layer cell shown by a photovoltaic QCRF-FDTD simulator. An integrated back reflector made of gold and an anti-reflective coating of indium tin oxide and silicon dioxide makes it a very effective device. Silicon dioxide is included in the solar cell's front protective layer. By varying the thickness of the SiO2 protective front layer from 50 nm to 1000 nm, the open-circuit voltage of the solar cell can be mimicked. Solar cell layers may be coated with antireflective materials such as SiO2 and ITO, as shown in Figure 86.4.

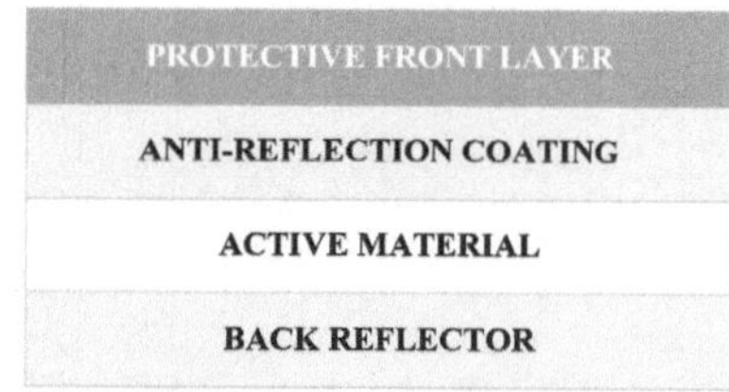

Fig. 86.3 Solar cell structure. Shows the four layers of a photovoltaic cell: the front protection layer, the rear reflector layer, the active material layer, and the anti-reflection coating layer

PROTECTIVE FRONT LAYER (SiO2)	PROTECTIVE FRONT LAYER (SiO2)
ANTIREFLECTIVE COATING (SiO2)	ANTIREFLECTIVE COATING (ITO)
ACTIVE MATERIAL – A Si	ACTIVE MATERIAL – A Si
BACK REFLECTOR - Au	BACK REFLECTOR - Au

Fig. 86.4 Proposed structure of the solar configuration was done withSiO2/SiO2/A-Si/Au and SiO2/ITO/A-Si/Au. with the four layers of the solar

To run the simulation, the photovoltaics QCRF-FDTD simulator is used. Users will be able to design and test out their photovoltaic material with the help of this simulation. By entering different parameters, the simulation tool may determine the material's open circuit voltage (Voc), as well as an absorption, transmission, and reflection curve. Before accessing this simulation, users should make sure they have the necessary resources and have chosen the right solar cell modeling and simulation application. The next step is to initiate the photovoltaics QCRF-FDTD simulation inside the simulation tool. According to Lekner (2016), the protective layer thickness of SiO2 materials is from 50 nanometers to 1000 nanometers. The next step is to run the QCRF-FDTD simulator for photovoltaics. After you've collected and tabulated all of the data samples from the output log, including the open circuit voltage (Voc), run the simulation again with different thicknesses. Following Li and Ma (2010), we imported the data from the photovoltaics QCRF-FDTD simulator application into SPSS for statistical analysis. We then calculated the mean and significant values. The thickness of the materials served as the independent variable, with the open-circuit voltage (Voc) serving as the dependent variable. Research by Rubin (2021) and Verma (2012):

5. Results and Discussion

The silicon solar cell must absorb all of the light, even while reducing reflection, which is necessary for manufacturing high-performance solar cells. The amount of light absorbed is proportional to the optical material's thickness and absorption coefficient. A silicon solar cell's efficiency demonstrates the critical role of device thickness in photon absorption. Table 86.1 shows that at 50 nm, the open-circuit voltage of an ITO solar cell with an antireflective coating of silver nanoparticles is 89.9%, and at 1000 nm, it climbs to 90.7%. The open-circuit voltage of solar cells coated with SiO2 and silver nanoparticles increases from 89.8% to 90.3% as the protective layer thickness increases from 50 nm to 1000 nm, as shown in Table 86.1.

Table 86.1 Open circuit voltage of ITO and SiO2 as anti-reflective coating material with silver nanoparticles

Thickness in mm	ITO with Silver nanoparticles	SiO2 With Silver nanoparticles
50	0.906478	0.827943
100	0.906803	0.828281
150	0.899446	0.802791
200	0.903311	0.800618
250	0.905658	0.800323
300	0.902616	0.80073
350	0.903696	0.800182
400	0.903752	0.801564
450	0.90305	0.800276
500	0.904301	0.800014
550	0.903233	0.801532
600	0.90336	0.800539
650	0.90415	0.800421
700	0.90323	0.800783
750	0.903679	0.800066
800	0.903663	0.901514
850	0.903042	0.801
900	0.904444	0.799299
950	0.903633	0.701192
1000	0.902524	0.701382

Table 86.2 Independent sample T – Test – group statistics

Group Statistics					
	Material	N	Mean	Std Deviation	Std Error Mean
Open Circuit Voltage	ITO	20	.9037034	.00153077	.00034229
	SiO2	20	.9005225	.00111537	.00024941

This research foundthatbycomparingthe open circuit voltage (Voc) for SiO2 and ITO as antireflection coating material with silver nanoparticles, the protective layer thickness varies from 50nm to 1000nm. The open circuit voltage (Voc) is higher than ITO as an anti-reflective coating material with silver nanoparticles than SiO2 as an antireflective coating material with silver nanoparticles.

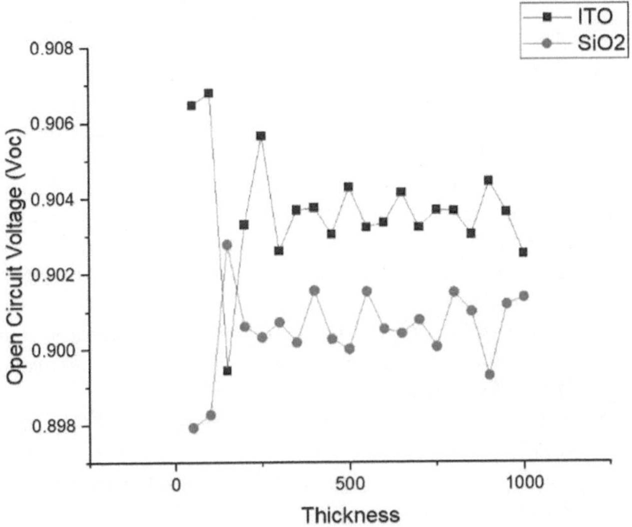

Fig. 86.5 Open-circuit voltage of ITO and SiO2 as antireflective coating material with nanoparticles

The protective layer is used as SiO2 thickness varies from 50nm to 1000nm and is compared to the open circuit voltage. SiO2 as antireflective coating material with silver nanoparticles to get the open circuit voltage is increasing from 89.8% to 90.3%. Similarly, ITO as antireflective coating material with silver nanoparticles to get the open circuit voltage is increasing from 89.9% to 90.7%.

The future scope of this work can be extended to improvise the solar cell's open circuit voltage (Voc). To improve the solar cell's performance by reducing the reflection light, replace the metal oxides with binary and transition metal oxides.

6. Conclusion

To conclude, two different antireflective coating materials with silver nanoparticles configured solar cells have been designed and simulated using a Photovoltaic QCRF-FDTD simulator. It is inferred that ITO as an antireflective coating solar cell has a significantly greater open circuit voltage (Voc) of 90.7% (Standard deviation: 0.00153077) than SiO2 as an antireflective coating solar cell has open circuit voltage (Voc) of 90.3% (Standard deviation: 0.00111537).

References

1. Abou-Elfotouh, Fouad, and Mohammad Al-Mass'ari. 1980. "EFFECT OF PREPARATION METHODS ON PERFORMANCE OF MOS PHOTOVOLTAIC SOLAR CELL." *The Physics of MOS Insulators*. https://doi.org/10.1016/b978-0-08-025969-7.50073-7.
2. Chongsri, K., J. Kanoksinwuttipong, W. Techitdheera, and W. Pecharapa. 2014. "Physical Properties of Ti-Doped ITO Nanoparticles Synthesized by Co-Precipitation Method." *2014 IEEE International Nanoelectronics Conference (INEC)*. https://doi.org/10.1109/inec.2014.7460441.
3. Dave, J. V., and Norman Braslau. 1976. "Importance of the Diffuse Sky Radiation in Evaluation of the Performance of a Solar Cell." *Solar Energy*. https://doi.org/10.1016/0038-092x(76)90020-7.
4. The Deal, B. E., and C. R. Helms. 2013. *The Physics and Chemistry of SiO2 and the Si-SiO2 Interface 2*. Springer Science & Business Media.
5. Drabavičius, Audrius. n.d. "Formation of Chalcogenide Solar Cell Absorber Layers Using Electrochemical Deposition of Precursors." https://doi.org/10.15388/vu.thesis.110.
6. Flores, C., F. Paletta, and D. Passoni. 1988. "Effect of Substrate Quality on GaAs Solar Cell Performances." *Confer-*

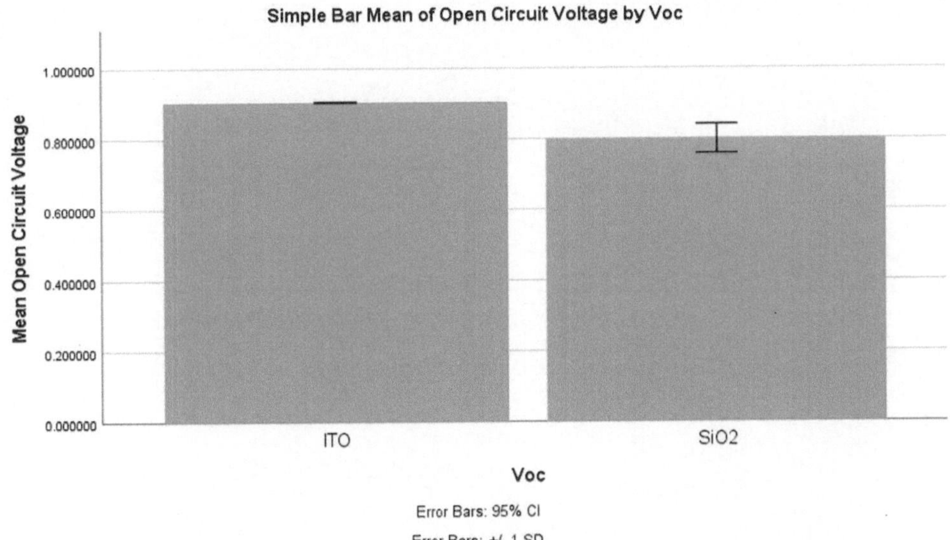

Fig. 86.6 Error bar mean of the open circuit voltage of ITO and SiO2

ence Record of the Twentieth IEEE Photovoltaic Specialists Conference. https://doi.org/10.1109/pvsc.1988.105776.

7. Jha, A. R. 2009. *Solar Cell Technology and Applications*. CRC Press.
8. Liu, C. Y., L. H. Hu, and C. C. Sung. 2013. "Optimal Combination of Flow Field Channels, Gas Diffusion Layers, and Catalyst Layers for Proton Exchange Membrane Fuel Cell." *Journal of Clean Energy Technologies*. https://doi.org/10.7763/jocet. 2013.v1.59.
9. Matsumoto, Y., F. Meléndez, and R. Asomoza. 2001. "Performance of P-Type Silicon-Oxide Windows in Amorphous Silicon Solar Cell." *Solar Energy Materials and Solar Cells*. https://doi.org/10.1016/s0927-0248(00)00169-0.
10. Manikandan, S., Baskar, R., Sathish, G. 2024. "The Nanoscale Technology and Photovoltaics Convergence: A Detailed Analysis of the Quantum Leap Achieved by Silver Nanoparticle Incorporation into SiO2 Non- -reflective Coatings for Solar Cell Efficiency Enhancement,".Proceedings of the 2nd International Conference on Intelligent and Innovative Technologies in Computing, Electrical and Electronics, ICIITCEE 2024.
11. Rajan, Grace, Shankar Karki, Robert W. Collins, Nikolas J. Podraza, and Sylvain Marsillac. 2020. "Real-Time Optimization of Anti-Reflective Coatings for CIGS Solar Cells." *Materials* 13 (19). https://doi.org/10.3390/ma13194259.
12. Manikandan, S., Baskar, R., Sathish, G. 2024." Harnessing Photovoltaic Cell Potential: A Thorough Investigation of TiO2 Anti-Reflective Coatings and Silver Nanoparticles, Revealing Intriguing Findings Using Photovoltaic QCRF-FDTD Simulator Simulations."Proceedings of the 2nd International Conference on Intelligent and Innovative Technologies in Computing, Electrical and Electronics, ICIITCEE 2024.
13. Ha, Sang-Gyu, and Kyung-Young Jung. 2014. "Recent Developments in QCRF-FDTD Modeling of Complex Dispersive Media." *2014 IEEE International Workshop on Electromagnetics (iWEM)*. https://doi.org/10.1109/iwem.2014.6963739.
14. Lekner, John. 2016. *Theory of Reflection: Reflection and Transmission of Electromagnetic, Particle and Acoustic Waves*. Springer.
15. Ruben, Greeson. 2021. *SPSS for Beginners: A Step-by-Step Guide to Learn about Statistical Data, Research Methods, and Data Analysis Using the SPSS Program*. Charles Jesuseyitan Adebola.
16. Verma, J. P. 2012. *Data Analysis in Management with SPSS Software*. Springer Science & Business Media.
17. Sarangerel, Khayankhyarvaa, Chimed Ganzorig, and Masuro Sakomura. 2014. *Improvement of Open-Circuit Voltage in Organic Photovoltaic Cells: A Chemical Modification of Indium-Tin Oxide Electrode*. LAP Lambert Academic Publishing.

Note: Every figure except 86.1 and tables was created using the experimental work.

Advances in Additive Manufacturing Technologies – Gurusamy Pathinettampadian (eds)
© 2024 Taylor & Francis Group, London, ISBN 978-1-032-90013-1

87 Perspective Aspects of Carbon—Enabled Transistors by Quantitative Improvement of Tube Diameter Suitable for Low Power Electronics

Sathish G[1]
Research Scholar, Institute of ECE,
Saveetha School of Engineering, Saveetha Institute of Medical and Technical Sciences, Saveetha University,
Chennai, India
Radhika Baskar[2]
Professor, Research Guide, Institute of ECE,
Saveetha School of Engineering, Saveetha Institute of Medical and Technical Sciences, Saveetha University,
Chennai, India

ABSTRACT: A form of nanoscale electronic device referred to as a carbon nanotube field-effect transistor (CNTFET) uses carbon nanotubes (CNTs) as a conductive channel. The electrical and physical features of CNTFETs can be substantially regulated by the CNTs' diameter. Here there are several perspectives pertaining to how the efficiency of CNTFETs is influenced by CNT diameter. Conductance: Electric permeability: Metallic CNTs that have bigger diameters have a distinct benefit over semi-conducting CNTs with fewer dimensions with regard to of conducting electricity. The overall current-carrying capabilities and speed of CNTFETs might get adversely affected by such. Voltage Cutoff (Vth): The diameter of the CNT might have a significant effect on the voltage below the threshold that governs when a CNTFET switches on. CNTs having a smaller diameter may possess a lower Vth therefore be more suitable for low-power uses. Scalability: Increased integration concentrations, which are crucial to integrating additional transistors on a chip and enhancing computing power, are made attainable by smaller diameter CNTs. In summary, the diameter of carbon nanotubes (CNTs) in CNTFETs is an essential factor that profoundly impacts the general viability of these nanoscale transistors on the basis of their respective electrical and mechanical capabilities.

KEYWORDS: Carbon enabled transistors, Tube diameter, Drain current, CNTFET, Threshold voltage, Low power applications

1. INTRODUCTION

The search for energy-efficient electronics has driven research into new materials and architectures, with Carbon Nanotube Field-Effect Transistors (CNTFETs) emerging as a leading candidate which was formed by rolling a graphene sheet depicted in Fig. 87.1 into a cylindrical form. Because of the intrinsic qualities of carbon nanotubes—such as their excellent electrical conductivity and nanoscale dimensions—CNTFETs have amazing electrical

[1]gees.sat@gmail.com, [2]radhikabaskar@saveetha.com2

DOI: 10.1201/9781003545774-87

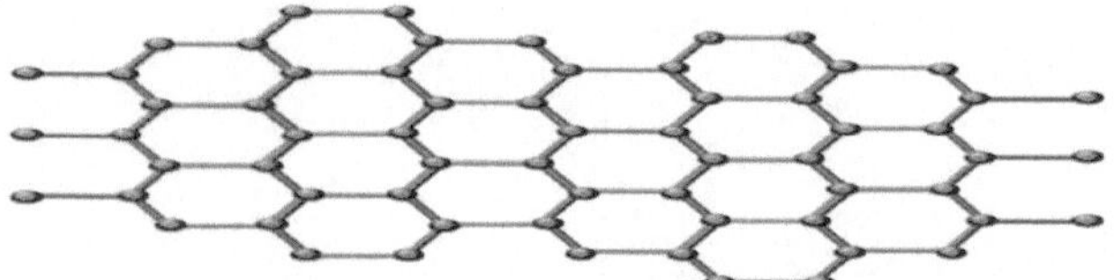

Fig. 87.1 Structure of graphene sheet [21]

capabilities. These transistors have a great deal of potential for use in next-generation electronic devices, especially in low-power applications that are essential to wearables, Internet of Things, and modern computing. The basic idea behind CNTFETs is the regulation of current flow via a channel made of carbon nanotubes, which is regulated by an external electric field[1]. However, the electrical properties and performance metrics of these transistors are mostly determined by the diameter of the carbon nanotube. Prior research has mostly concentrated on employing smaller-diameter nanotubes to optimise CNTFETs for high-speed applications. On the other hand, relatively less focus has been placed on the potential benefits of using larger-diameter carbon nanotubes to achieve low power consumption.

This work aims to fill this research vacuum by examining the effects of using larger-diameter CNTs in improving the low-power characteristics of CNTFETs. The motivation for this investigation is rooted in the intrinsic characteristics of nanotubes with a bigger diameter, including a narrower band gap, increased carrier mobility, and modified electrical band structures. These properties could reduce leakage currents, reduce off-state power consumption, and improve the on-off ratio, making CNTFETS shown in Fig. 87.2 suitable for low power applications.[2] This study examined the performance of CNTFETs and conventional MOSFETs with yttrium and lanthanum oxide gate oxides. The performance of the carbon nanotube device is examined in relation to its diameter using two distinct dielectrics and conventional silicon gate oxides; the outcomes are shown in the accompanying tables.

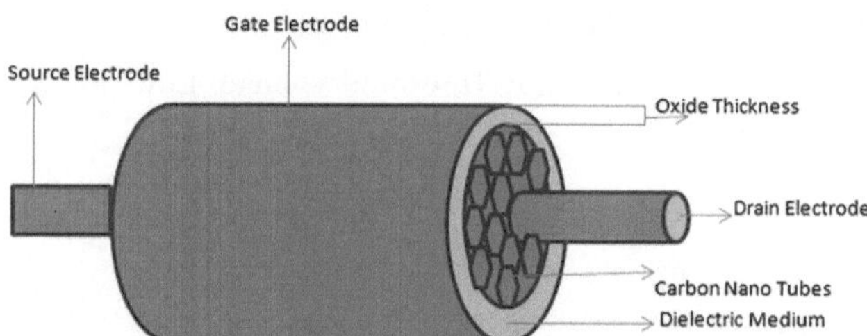

Fig. 87.2 Schematic structure of conventional CNTFET

In comparison to traditional silicon-based FETs, CNTFETs with Yttrium oxide and Lanthanum oxide gate oxide have a lower off-current and a higher on-current. According to the simulation results, CNTFETs with La2O3 and Y2O3 as the gate oxide had higher carrier mobility, while they also had a reduced subthreshold swing and a better on/off ratio than conventional silicon-based devices. These results validated the viability and potential of La2O3 and Y2O3 as an alternative to traditional silicon-based gate oxide materials for CNTFETs. It is important to keep in mind that while simulation provides valuable insights into the performance of CNTFETs, actual device fabrication and testing are still required to validate the results.

2. Materials and Methods

This work attempts to clarify the impact of carbon nanotube diameter on the low-power performance metrics of CNTFETs by means of a thorough examination of theoretical models, simulations, and experimental results. Additionally, it aims to offer insightful information about the possible uses and design considerations of CNTFETs with greater tube widths, opening the door for the development of ultra-low-power electronic systems. Enhancing performance and reliability in semiconductor device production is a typical tactic that involves replacing Silicon Dioxide (SiO2) with materials such as Yttrium Oxide (Y2O3) and Lanthanum Oxide (La2O3) to improve gate oxide thickness. Owing to their high dielectric constants and compatibility with silicon substrates, materials like Y2O3 (yttrium oxide) and La2O3 (lanthanum oxide) have been investigated as potential replacements for SiO2 as gate oxide in semiconductor devices.

3. Simulation

A software package known as SPSS provides advanced tools for data management and statistical analysis. It may be used to perform a wide range of simulations, including those for silicon-based FET devices. The following procedures can be used to simulate the operation of a CNTFET with yttrium oxide and lanthanum oxide acting as the gate oxides: Establish the simulation's parameters: The simulation settings, such as the gate voltage, device dimensions, gate oxide type, and material properties, must be specified. To determine the device's Vth, a mathematical model of the CNTFET must be constructed using the simulation parameters. To evaluate performance, the data can be plotted and contrasted with conventional silicon devices. The comparison can make use of the following variables: Vth, transconductance, and on/off current ratio. Using the Nano-Hub online simulator tool the performance of a CNTFET with lanthanum oxide and yttrium oxide as the gate oxide to that of conventional silicon devices were simulated and compared was depicted in Table 87.1. The simulation's results can offer fascinating insights into the

Table 87.1 Statistical difference between SiO2 and Y2O3 with mean, SD and standard error mean

Group Statistics					
	Di_Electric Material	N	Mean	Std. Deviation	Std. Error Mean
Drain current	*SiO2*	15	69.3533	15.03666	3.88245
	Y2O3	15	84.7000	6.37932	1.64713

Table 87.2 Group statistical table with mean, SD, SEM for lanthanum oxide in comparison with yttrium oxide

Group Statistics					
Di_Electric_Material		N	Mean	Std. Deviation	Std. Error Mean
Drain current	*Y2O3*	15	84.7000	6.37932	1.64713
	La2O3	15	88.3600	3.61026	0.93217

device's operation and aid in the development of innovative MOSFET devices[3,4]. The primary factors influencing a MOSFET's performance are its gate voltage (Vth), ON current (ION), and OFF current (IOFF). It is described as the voltage between the source and the gate at which the device starts to conduct. The performance of a carbon nanotube field-effect transistor (CNTFET) with La2O3 and Y2O3 as the gate oxide can be compared to that of traditional silicon devices through simulation using SPSS (Statistical Product and Service Solutions) was shown in Table 87.2 and Table 87.3. The simulated devices were then characterized by varying the gate voltage and measuring the ensuing drain current and quantum capacitance. [5]The simulation results show that CNTFETs with high "k" dielectrics have a lower gate voltage than silicon-based devices that are more traditional. This was explained by the fact that, in comparison to SiO2, La2O3 and Y2O3 had a smaller band gap and a greater dielectric constant, which allowed for better gate control of the channel conductivity. [6,7] The experimental findings, which illustrate the drain current (in microamperes) and the quantum capacitance for varied nanotube diameter thicknesses, are displayed in Table 87.1 together with the relevant Gate voltage values (in volts) for each combination. The Fig. 87.3 and 87.4 shows that when there is a significant improvement in tube diameter thickness, the drain current increases.

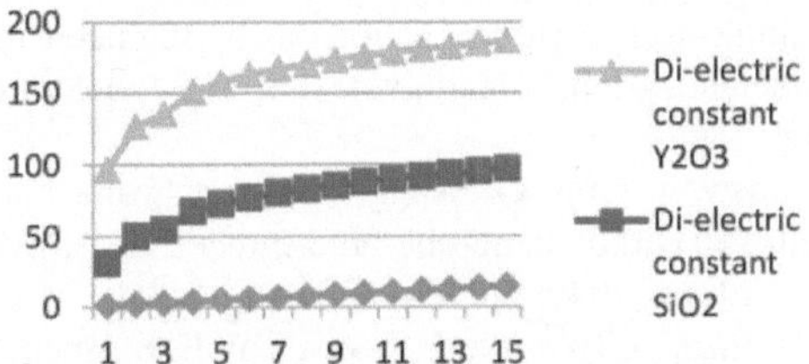

Fig. 87.3 Line chart display over Yttrium oxide with Sio2 as di-electric with the change in thickness

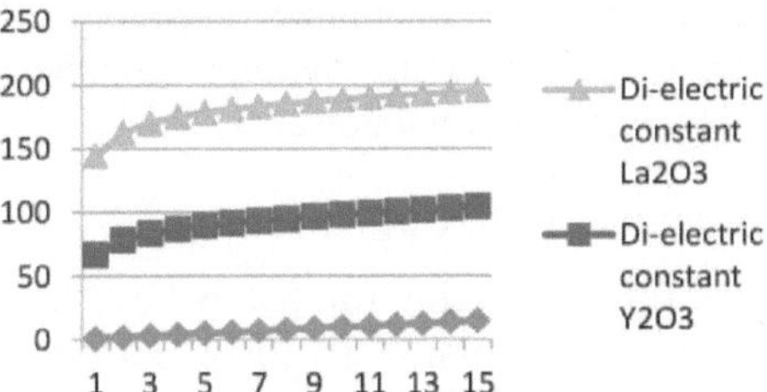

Fig. 87.4 Nanotube diameter variation line graph plot between yttrium oxide and lanthanum oxide

The table shows that when there is a significant improvement in tube diameter thickness, the drain current increases. Additionally, the data indicate that the type of gate insulator used can affect the drain current at a given thickness. Specifically, using yttrium oxide and lanthanum oxide as the gate insulator seems to result in higher drain currents[8,9] than Si, which is consistent with the idea that a thinner gate insulator allows for a stronger electric field to be applied to the channel region, which in turn facilitates the flow of current between the sources and drain terminals

4. Results and Discussion

Improvements in tube diameter have allowed carbon-based transistors to advance significantly, particularly in low-power applications. The new features resulting from these developments provide encouraging results.

4.1 Improved Electrical Performance

Metrics measuring electrical performance, such as carrier mobility, subthreshold swing, and on/off current ratio, have significantly increased as a result of the tube diameter decrease. Reduced scattering effects, which allow for better control over electron mobility within the transistor, are credited with this development.

4.2 Reduced Power Usage

The reduced power consumption of these improved carbon-based transistors is a major benefit. They are perfect for low-power applications because their smaller tube diameter enables effective control of electron flow, which lowers leakage currents and improves overall energy efficiency.[10]

4.3 Increased Speed of Switching

Faster switching rates have been made possible by the smaller tube diameter since it has lowered resistance and parasitic capacitance inside the transistor. [11] This feature is very helpful for applications that need to process data quickly.

4.4 Temperature Consistency

The device's stability against temperature changes has also improved as a result of the tube diameter developments.

The lower dimensions guarantee constant performance over a wide temperature range by improving thermal dissipation and reducing thermal noise.

4.5 Producer and Expandability

The potential for producing transistors with reduced tube widths holds great promise for industrial production. It may be possible to scale up the procedures created to reach these dimensions, which would enable the affordable mass manufacture of extremely effective carbon-based transistors.[12]

4.6 Problems and Prospects for the Future

Not with standing these developments, there are still problems, like problems with scalability and material flaws at smaller sizes[13,14,15,16].To further increase these transistors performance and dependability, creative production methods and material advancements will be required to meet these difficulties.

4.7 Application

Reduced tube diameter and enhanced characteristics of carbon-enabled transistors make them ideal for a variety of applications, such as wearable electronics, biomedical sensors, Internet of Things devices, and low-power computing systems, where small size and energy efficiency are critical.[17, 18,19,20]

5. Conclusion

The study's findings highlight the encouraging possibilities of carbon-enabled transistors with reduced tube diameter for low-power applications. Our study of how different tube sizes affect transistor performance showed that power efficiency and operational stability were significantly improved by the enhancement of Drain current. We were able to achieve significant increases in threshold voltage and subthreshold swing, which are essential for low power circuitry, by carefully regulating and scaling the tube diameter. The study's conclusions also show that using these transistors in energy-efficient electronics, such IoT sensors and portable electronics is a feasible idea. The capacity to finely tune transistor behavior is made possible by the tunability of carbon-based structures, opening the door for next-generation, ultra-low-power electronics.

References

1. Dokania, Vishesh, Aminul Islam, Vivek Dixit, and Shree Prakash Tiwari. 2016. "Analytical Modeling of Wrap-Gate Carbon Nanotube FET with Parasitic Capacitances and Density of States."IEEE Transactions on Electron Devices. https://doi.org/10.1109/ted.2016.2581119
2. Analysis of different parameters of channel material and temperature on Gatevoltage of CNTFET Sanjeet Kumar Sinha, Saurabh Chaudhury Department of Electrical Engineering, NIT Silchar, Silchar 788010,Assam,India,MaterialsScienceinSemiconductorProcessing31(2015)431–438
3. Sinha, Sanieet Kumar, and Saurabh Chaudhury. 2014. "Advantage of CNTFET Characteristics over MOSFET to Reduce Leakage Power." 2014 2nd International Conference on Devices, Circuits and Systems (ICDCS). https://doi.org/10.1109/icdcsyst.2014.6926211.
4. Sathish G,Radhika Baskar,"Implementation of carbon Nano Field Effect Transistor and Comparison of Insulator material With Traditional Gate Oxide to Improve the Electrical Characteristics and Device Scalability",2023 ,2nd International Conference on Vision Towards Emerging Trends in Communication and Networking Technologies (VITECON 2023)
5. Sinha, Sanjeet Kumar, and Saurabh Chaudhury. 2014. "Comparative Study of Leakage Power in CNTFET over MOSFET Device." Journal of Semiconductors.https://doi.org/10.1088/1674-4926/35/11/114002.
6. Srimani, T., G. Hills, X. Zhao, D. Antoniadis, J.A. del Alamo, and M.M. Shulaker. 2019. "Asymmetric Gating for Reducing Leakage Current in Carbon Nanotube Field-Effect Transistors."Applied Physics Letters. https://doi.org/10.1063/1.5098322.
7. S. Mohapatra, P. Bhattacharya, and A. R. Allu, "Performance analysis of CNTFETs with Lanthanumoxide gate oxide using conventional silicon technology," IEEE Transactions on Nanotechnology, vol. 14, no. 6, pp. 1119-1126, Nov. 2015.
8. H. Li, L. Yang, Y. Liu, and Y. Li, "Gatevoltage modeling of CNTFETs with Lanthanumoxide gate oxide using conventional MOSFET models," Microelectronics Journal, vol. 43, no. 1, pp. 32-38, Jan. 2012.
9. J. A. Torres, A. R. Allu, and P. Bhattacharya, "Carbon nanotube field-effect transistor with Lanthanumoxide gate oxide: Performance analysis and comparison with conventional MOSFETs," Journal of Applied Physics, vol. 107, no. 6, pp. 064509, Mar. 2010.
10. J. Appenzeller, J. Knoch, M. V. Fischetti, and H. R. Shea, "Band-to-band tunneling in carbon nanotube field-effect transistors," Physical Review Letters, vol. 93, no. 19, pp. 196805, Nov. 2004.
11. M. Shulaker, G. Hills, R. S. Park, H.-S. P. Wong, and S. Mitra, "Carbon nanotube computer," Nature, vol. 501, no. 7468, pp. 526-530, Sep. 2013.
12. M. S. Arnold, S. I. Stupp, and M. C. Hersam, "Enrichment of single-walled carbon nanotubes by diameter in a binary surfactant system," Small, vol. 1, no. 8-9, pp. 858-863, Aug. 2005.
13. A. Javey, J. Guo, Q. Wang, M. Lundstrom, and H. Dai, "Ballistic carbon nanotube field-effect transistors," Nature, vol. 424, no. 6949, pp. 654-657, Aug. 2003.
14. S. Heinze, G. Tulevski, J. P. Small, W. T. S. Huck, and H. R. Shea, "Aligned carbon nanotubes for device applications," Nanotechnology, vol. 18, no. 42, pp. 424017, Oct. 2007.

15. M. H. Modarresi, M. Pourfath, and G. Rezazadeh, "Improving subthreshold swing of carbon nanotube field-effect transistors using ultra-thin gate oxides," Journal of Applied Physics, vol. 117, no. 11, pp. 114301, 2015.
16. S. B. Akhavan, M. Kavei, and M. R. Fathipour, "Theoretical investigation of subthreshold swing improvement in CNTFETs by gate oxide thickness optimization," Microelectronics Journal, vol. 44, no. 9, pp. 814-820, 2013
17. S. S. Islam, M. R. Islam, and M. R. Karim, "Enhanced subthreshold swing in carbon nanotube field-effect transistors with optimized gate oxide thickness," Applied Physics Letters, vol. 100, no. 16, pp. 163109, 2012.
18. S. K. Alam, S. R. Hasan, and M. H. Anik, "Effect of gate oxide thickness on the subthreshold characteristics of carbon nanotube field-effect transistors," IEEE Transactions on Nanotechnology, vol. 15, no. 6, pp. 1013-1018, 2016.
19. [J. Lee and H. Lee, "Optimization of gate oxide thickness in carbon nanotube field-effect transistors for improved subthreshold swing," Journal of Nanoscience and Nanotechnology, vol. 14, no. 5, pp. 3925-3928, 2014.
20. Gajendran, S., Baskar, R. (2024). Stochastic Performance of CNTFET with High 'k' Dielectric Material Over Conventional Silicon Devices in Optimization of Drain Current. In: Shaw, R.N., Siano, P., Makhilef, S., Ghosh, A., Shimi, S.L. (eds) Innovations in Electrical and Electronic Engineering. ICEEE 2023. Lecture Notes in Electrical Engineering, vol 1115. Springer, Singapore. https://doi.org/10.1007/978-981-99-8661-3_47

Note: Every figures except 27.1 and tables was created using the experimental work.

Advances in Additive Manufacturing Technologies – Gurusamy Pathinettampadian (eds)
© 2024 Taylor & Francis Group, London, ISBN 978-1-032-90013-1

88 Investigating Mechanical Properties of LM25 Aluminum Matrix Composites with Varied Silicon-CNT Reinforcements

Nagarajan V
Department of Mechanical Engineering,
Veltech High Tech Dr RR & Dr SR Engineering College,
Chennai, India

P. Gurusamy*
Department of Mechanical Engineering,
Chennai Institute of Technology,
Chennai, India

N. Ramanan
Department of Mechanical Engineering,
Sri Jayaram Institute of Engineering and Technology,
Chennai, India

Hari Krishna Raj S, Santhosh Kumar V,
Sanjai kumar S, Balaji. K
Department of Mechanical Engineering,
Veltech High Tech Dr RR & Dr SR Engineering College,
Chennai, India

ABSTRACT: The research addresses the increased demand for materials with enhanced performance in automotive and aerospace components. In response to emerging industrial requirements, aluminum-silicon-magnesium matrix (LM25) composites are commonly utilized. This study specifically focuses on examining the mechanical properties of LM25 produced through the casting technique, incorporating various silicon-magnesium compositions as reinforcements. Tensile, flexural, hardness, and impact tests were carried out, revealing that hybrid composites displayed superior properties when compared to pure aluminum. The microstructure of the hybrid composites was scrutinized using scanning electron microscopy (SEM).

KEYWORDS: LM25, Mechanical properties, Stir casting process and SEM

1. INTRODUCTION

The composite metal matrix holds significant importance due to its superior mechanical properties [1, 2]. Composites featuring a metal matrix reinforced with silicon carbide exhibit excellent grounding, high rigidity, and find applications in pistons, brake pads, and motors [4, 5]. Ramnath et al. [6] studied the classic material AMMC itself. Sun et al. [7] pro-

*Corresponding author: gurusamyp@citchennai.net

DOI: 10.1201/9781003545774-88

duced an aluminum fusion cenosphere (AMFA) through a combination of motion and technology in composite materials. The tensile strength of this cenosphere was found to be more than 50% better with a 13% weight increase. Training becomes challenging due to the abrasive properties of alumina, the composite reinforcement in this type. Garget et al. [8] concluded that aluminum composites surpass monolithic and non-ferrous metals in advantages, despite facing processing problems. This study suggests that the force should be open to a range of industrial applications.

B4C particles boost the wear resistance of composite aluminum alloy 7075 [9]. Madhavan et al. [10] tested polished AA 6061-SiC composites on rotational flexion fatigue after different solutionizing and aging treatments. The length of solutionizing affected fatigue strength more than temperature and aging time. Schneider et al. [11] experimentally and numerically studied the deformation behavior of a metal matrix composite aluminum alloy (6061)/Al2O3 under low-cycle fatigue, discovering that stress concentrations in the microstructure lead to extrusions, intrusions, and crack formation. Aluminum alloys produced with metal matrix composites of alumina carbide and boron carbide were examined for their mechanical properties, with the conclusion that a higher percentage of reinforcements increases mechanical resistance [12].

Mahesh et al. [15] used the Taguchi approach to show how feed and cutting speeds affect Al-SiC-B4C composite surface quality. Prabu et al. [16] examined metallic matrix stirring speed and duration and found increased particle dispersion. Wu and Xi [17] used the Equivalent Channel Angle Consolidation Technique (BP-ECAC) to make aluminum matrix composites with tiny fly ash particles, excelling over existing approaches. Eizadjou et al. [18] examined the mechanical characteristics of Al/Cu composites manufactured using the roll accumulation method (ARB) and found that five cycles of the ARB process reduced copper layer thickness and increased efficiency and tensile strength. Akbari et al. [19] studied aluminum composites with nanometer-scale Al2O3, discovering that SiC nanoparticles float on molten aluminum. Puneet and Lokesh [20] studied alumina-reinforced composite wear, whereas Siddique and Ramnath [21] found hybrid composites had better mechanical properties. In this work, stirring was used to make composites with different reinforcing weights, such as silicon magnesium. These composites' mechanical characteristics and internal structures were examined using scanning electron microscopy.

2. Experimental Procedure

2.1 Materials

The aluminum alloy LM25 is used as a master alloy. The 356 aluminum alloy is a hardenable cast iron alloy commonly used for a variety of applications. It can be cast in permanent or sand and has excellent melting performance, good corrosion resistance and good resistance. The matrix aluminum alloy is shown in Table 88.1.

Table 88.1 Composition of matrix metal

Element	Weight (%)
Cu	0.1
Mg	0.2-0.45
Si	6.5-7.5
Fe	0.5
Mn	0.3
Ni	0.1
Zn	0.1
Pb	0.1
Sn	0.05
Ti	0.2
Al	Residual

Aluminum and its alloys serve as the predominant matrix system in metal matrix composites (MMCs). This is primarily attributed to aluminum being a cost-effective material with a low melting point, facilitating ease of fabrication. Its low density (2.7 g/cm^3), coupled with its versatility in being processed in various forms through plastic deformation and cast through all foundry processes, further solidifies its prominent role in MMCs. Notably, commercial alloys not initially intended for MMC applications have been extensively employed in this context. Table 88.2 presents the physical properties of the aluminum alloy LM25.

Table 88.2 Physical property of aluminium alloy LM25

Physical property	Weight
Density	2670kg/m^3
Melting point	615oC
Elastic modulus	71 GPa
Tensile strength	250-280 MPa
Percent elongation	5%
Hardness	90 BHN

2.2 Fabrication Procedure

One of the widely used liquid-state fabrication techniques has been effectively used to the manufacture of hybrid metal matrix composites, as shown in Fig. 88.1. In the process of producing composites, stir casting—which involves continuously adding reinforcements after molten metal is introduced—is essential. After that, the liquid is poured and given time to solidify. Particles in the stir casting pro-

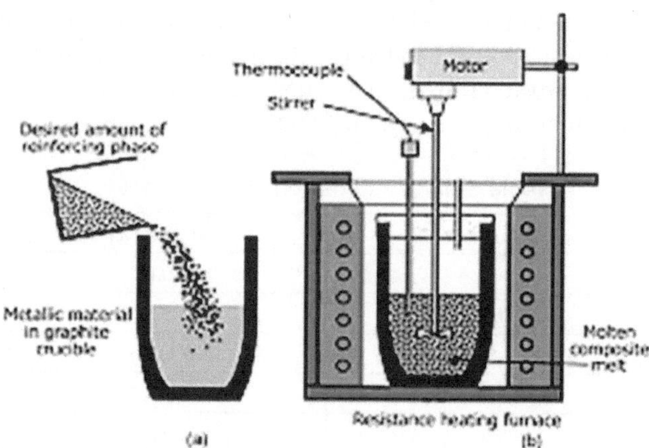

Fig. 88.1 Fabrication of aluminium alloy LM25 using stir casting method [30]

cess often aggregate, therefore in order to dissolve the material effectively, vigorous churning at high temperatures is required. This approach is prized for its ease of use, adaptability, large-scale production appropriateness, near-net shaping capabilities, reduced processing expenses, and streamlined matrix structure management. In this work, aluminum metal matrix composites are prepared using the melting process [22-25]. High strength and a homogenous composition in aluminum composite materials are guaranteed by this swirling process.

3. EXPERIMENTATION

3.1 Stir Casting Process

The aluminum alloy (LM25) is melted in a clay pot over an oven that burns the fire using the graphite crucible. The composite composition is carried out in an oven with a capacity of 10 kg. Moreover, being withdrawn from the turbine it is mechanically agitated by an external electric motor. The agitator speed is controlled by approximately 350 rpm. Movement and speed of the dynamometer. In preheated 600°C to the starting point and the fly ash is about 600, before addition. The preheated particles were introduced with a feeding flow rate of approximately 1 g/s, a stirring speed of 220 rpm, and a melting temperature of 740°C. Before powder addition, 1% of Mg was included at 730°C to compensate for the loss of manganese in the melt due to oxidation. The particles were manually added to the molten metal in the crucible using a powder addition mechanism. A baffle was inserted into the crucible to ensure uniform mixing of the composite. Subsequently, degassing was performed to prevent the entrapment of hydrogen. Molten metal through the sulfuric acid. Sulfuric acid acts as a cleaning agent and the nitrogen removes the hydrogen from the molten metal. Degassing takes place for about 20 minutes to 760°C. After manual stretching, the composite is poured into the rotating centrifugal mold. We made composites samples for four samples, each containing a different percentage of reinforcements. Retention of descent as shown in Tables 88.3.

Table 88.3 Composition of composite specimen

Composite Specimen	Base metal (LM25)	Silicon carbide (SiC)	Fly ash	CNT
C1	98%	1%	1%	0.25 gm
C2	96%	2%	2%	0.5 gm
C3	94%	3%	3%	0.75 gm
C4	92%	4%	4%	1 gm

Quantity of Mechanical Properties

The mechanical characterization of the developed LM25 hybrid metal matrix composite was investigated using a battery of mechanical tests, the results of which are shown below.

Tensile Test

Tensile strength, yield strength, and modulus of elasticity are just a few of the characteristics that are usually taken into consideration when choosing materials for engineering purposes[27,28]. The most often used technique for evaluating mechanical qualities is the tensile test. In this investigation, hybrid composite samples A, B, and C were manufactured in compliance with ASTM standards, particularly ASTM B-557M, and a tensile test was performed utilizing a universal testing machine (UTM).

Flexural Test

Flexural testing is employed to assess the flexural properties of composite materials, measuring their behavior under simple bending loads. The composites were manufactured in accordance with ASTM standard B-337.

Impact Test

A load is applied to the composite material suddenly and dynamically during the impact test. This test measures the energy that the composite absorbs during fracture. The IS standard 1757 was followed in the preparation of the composites.

Hardness Test

Hardness is the resistance of materials to local deformation. An impression or scratch on the surface of a hard material and the ability to indent or cut material[26-29]. Hardness of the four air-tight testers. In the Brinell hardness test, the pieces of force are pressed into the flat surface of the specimens with a force from 500 kgf to 1000 kgf. Induction is a microscope. Reading in three points.

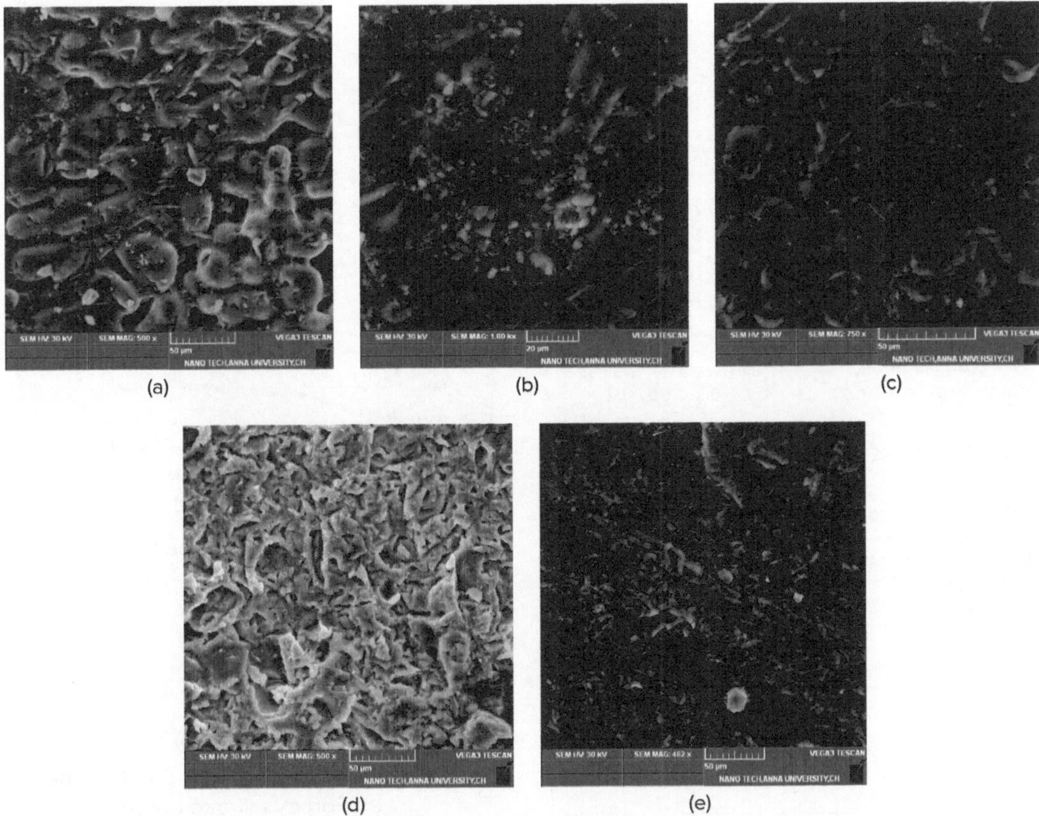

Fig. 88.2 (A) A specimen C1 scanning electron microscopy (SEM) micrograph (B) A specimen C2 SEM micrograph (C and D) specimen C3's SEM micrograph and specimen C4's SEM micrograph (E) (Magnification 462 x)

Microscopic Observation

Scanning Electron Microscopy (SEM) images of hybrid composite samples A, B, and C are presented in Fig. 88.2A-E. In Fig. 88.2a, the grooves and holes in C1 are distinctly visible, with the larger grooves and holes suggesting a higher wear rate for Sample C1. Figs. 88.2B, C, and D reveal that the features in samples C2 and C3 are finer compared to Sample C1. While the images for samples C2 and C3 are similar in terms of holes, the wear rate for samples C3 and C4 is elevated due to the higher quantity of reinforcing material..

4. Results and Discussion

Tensile Test

The stress-strain curve was plotted as shown in Figure 88.2, to determine tensile strength.

The average values of tensile strength and ultimate yield strength found for samples C1, C2, C3 and C4 were tabulated in Table 88.4.

The graph above leads to the conclusion that Composite 4 exhibits the highest force values for equivalent stroke values, closely followed by Composite 3. It also facilitates

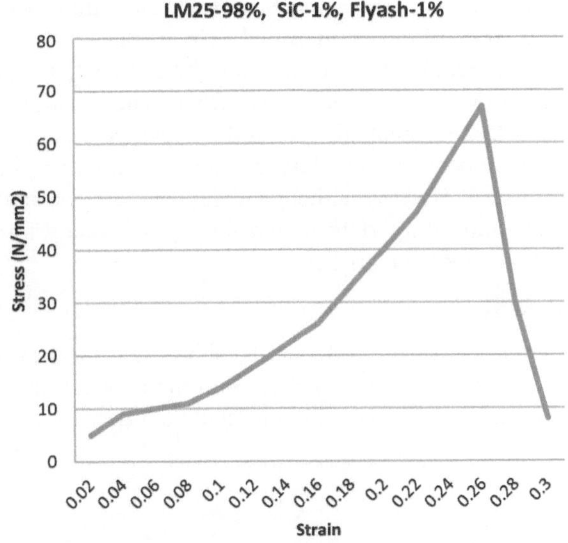

Fig. 88.3 Stress strain curve for C1

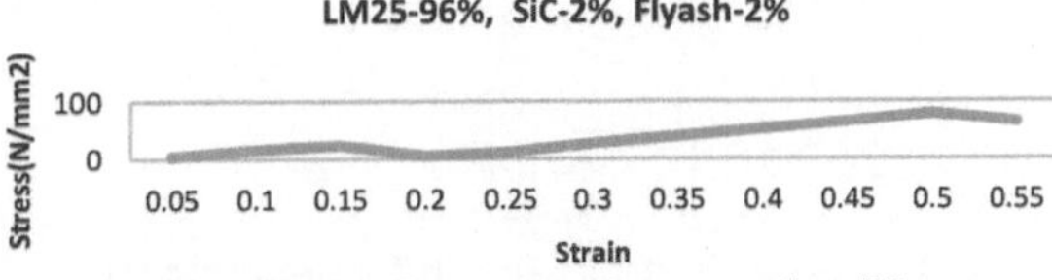

Fig. 88.4 Stress strain curve for C2

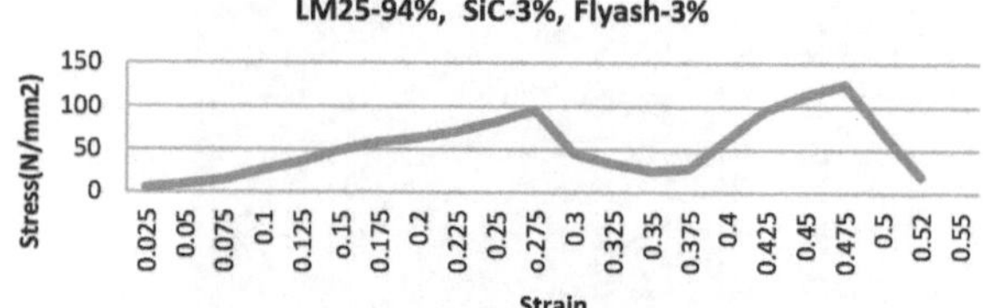

Fig. 88.5 Stress strain curve for C3

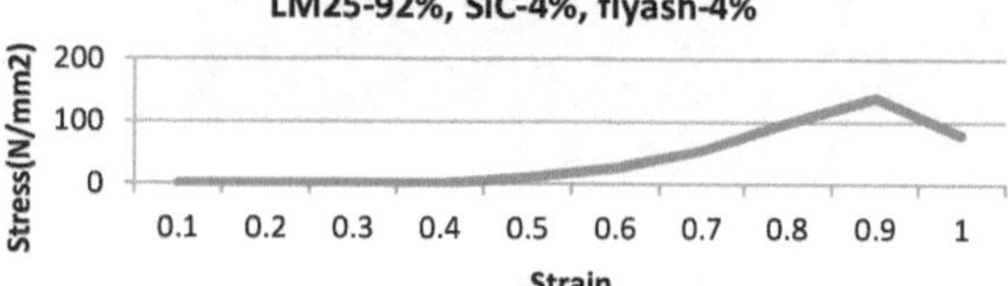

Fig. 88.6 Stress strain curve for C4

Table 88.4 Composites tensile property

Composite	Composition of various composites	Ultimate strength (MPa)	Yield strength (MPa)
C1	LM25-98%,SiC-1%, Flyash-1%,CNT-0.25gm	69.20	61.78
C2	LM25-96%,SIC-2% Flyash-2%,CNT-0.5gm	79.83	65.51
C3	LM25-94%,SIC-3%, Flyash-3%,CNT-0.75gm	126.56	79.08
C4	LM25-92%,SIC-4% Flyash-4%,CNT- 1gm	140.97	84.57

a comparison of force, stroke, tension, and strain across all four composites. Composite 4 demonstrates maximum values for breaking load, maximum displacement, and percentage elongation. Specifically, Composite 4 surpasses Composite 2 in breaking load and percentage elongation, while Composite 2 exhibits greater displacement than Composite 1. Notably, although Composite 4 incorporates an aluminum alloy, only its tensile strength outperforms that of the other composites.

Flexure Test

The graph above Fig. 88.7–88.10 shows that the flexural rigidity of the sample increases with the application of a continuous load at a certain point and then decreases. The

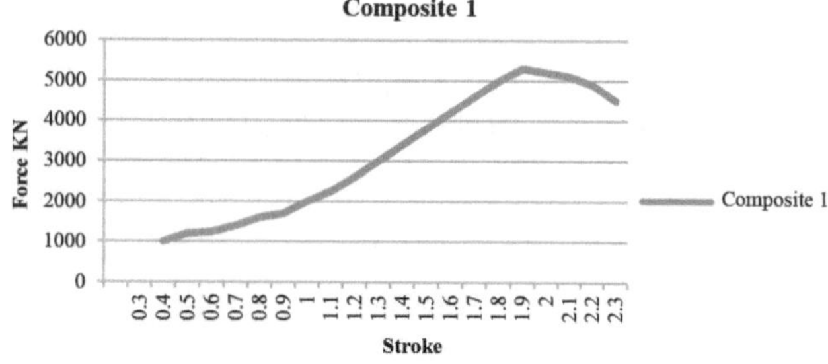

Fig. 88.7 Force Vs stroke graph for flexural test for C1

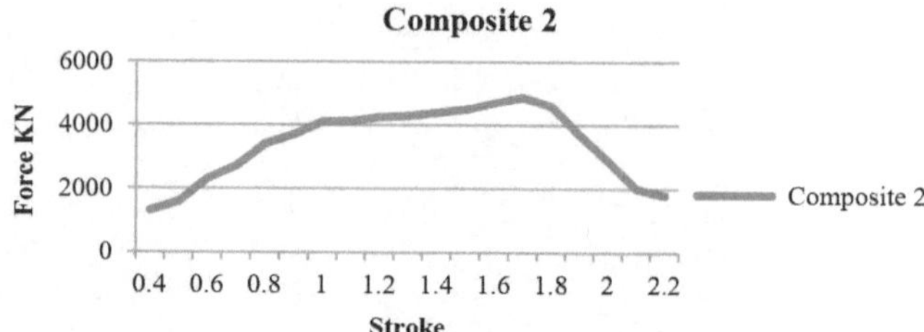

Fig. 88.8 Force vs stroke graph for flexural test for C2

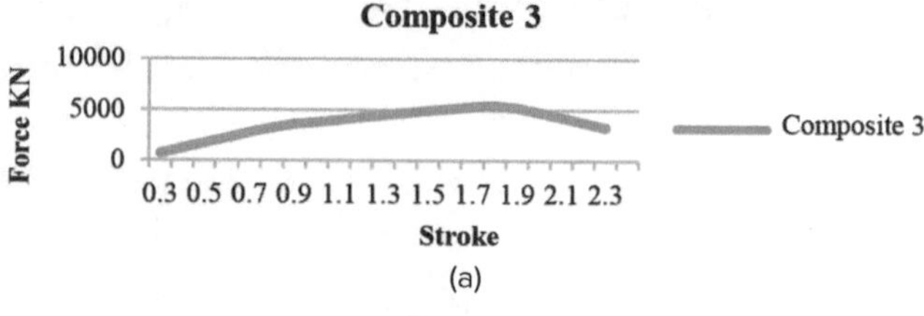

(a)

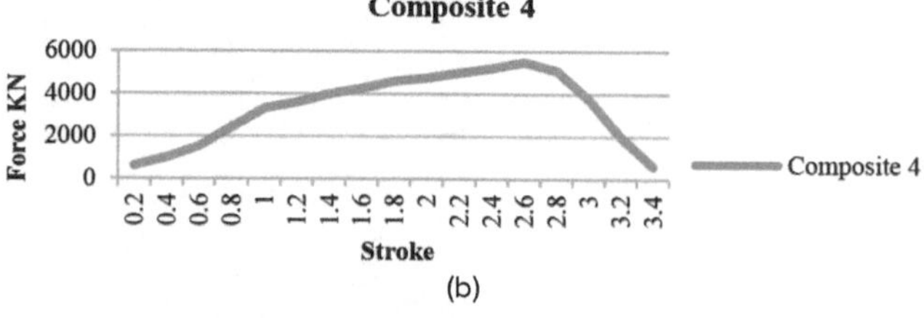

(b)

Fig. 88.9 (a) Force Vs stroke graph for flexural test for C3, (b) ForceVs stroke graph for flexural test for C4

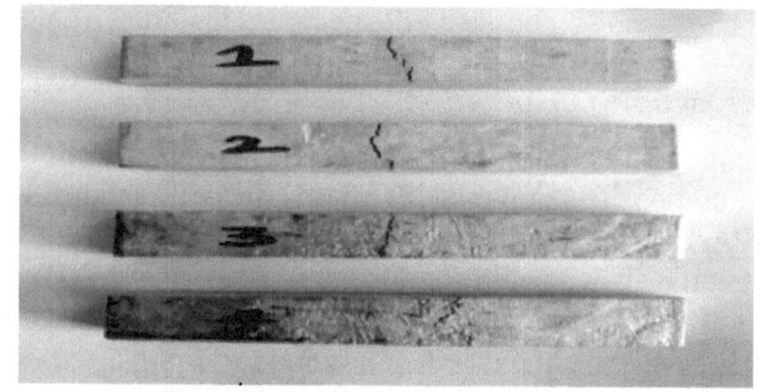

Fig. 88.10 Flexure test specimen

reduction in stiffness is due to the change in the reinforcing composition.

Impact Test

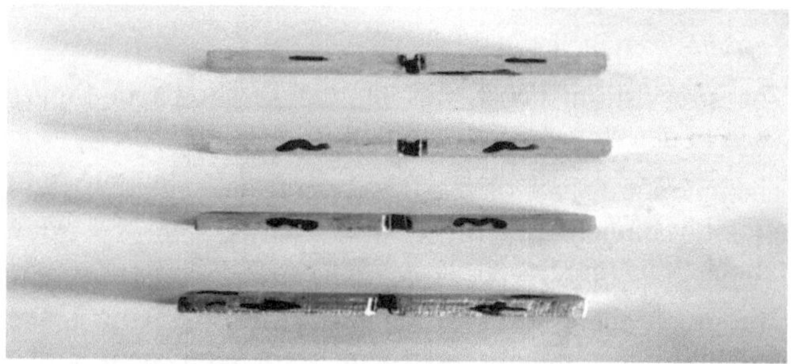

Fig. 88.11 Impact test specimen

Hardness Test

The hardness test was performed on the four samples and the results are shown in Tables 88.5 and Fig. 88.12. From the graph plotted for various trials to analyze the hardness of the specimens, it is clear that the C4 has a higher hardness and this is by two on the presence of SiC, fly ash and CNT in larger proportions. The presence of reinforcements in their microstructure plane is the reason for increasing the hardness of the composites.

Table 88.5 Hardness values of composites

Composite	Composition	Trial-1 BHN	Trail-2 BHN	Trail-3 BHN	Average load BHN
C1	LM25-98%, SiC-1%, Flyash-1% CNT-0.25 gm	91.2	93.3	92.1	91.9
C2	LM25-96%, SIC-2%. Flyash-2% CNT-0.5gm	88.7	83.5	79.5	83.9
C3	LM25-94%, SIC-3%, Flyash-3% CNT-0.75gm	81.4	82.6	79.2	81.06
C4	LM25-92%, SIC-4%, Flyash-4%, CNT- 1gm	79.5	79.0	80.8	79.7

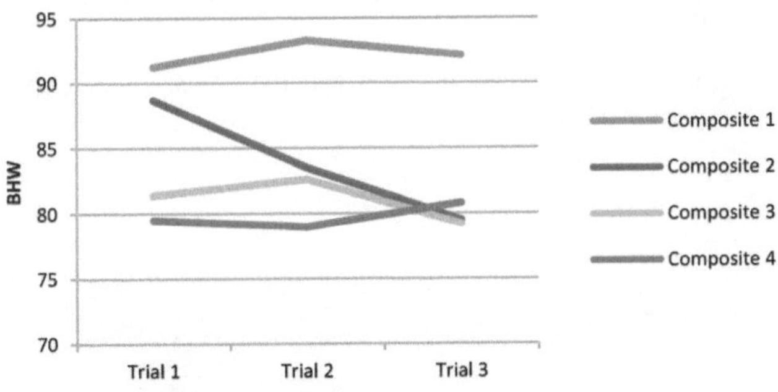

Fig. 88.12 Hardness test for the composite

4. Conclusion

In this study, four distinct samples were fabricated, and corresponding findings are as follows:

- Composite 4 exhibits superior tensile strength, maximum displacement, and percentage elongation values compared to the other composite materials. Additionally, the ultimate strength and yield strength of Composite 4 surpass those of the other composite materials.
- The impact test revealed that all four samples absorbed 2 Joules of energy.
- Composite 4 demonstrate higher hardness, attributed to the presence of silicon carbide, fly ash, and CNT in larger proportions.
- The flexural rigidity is elevated in Composite 4 due to variations in composition within the reinforcement.

References

1. Clyne TW, Withers PJ (1993) An introduction to metal matrix composits. Cambridge University Press, Cambridge
2. Murakami Y, Cahn RW, Haasen P, Kramer EJ (1996) Mater Sci Technol 8:213–218
3. Chen P (1992) High- performance machining of SiC Whisker Reinforced Aluminium Composite by Self Propelled Rotary Tools. CIRP Ann 41:59–62
4. Allison JE, Cole GS (1993) Metal matrix composites in the automotive industries. Jr Met 45:10–15.
5. Suresh S, Mortensen A, Needleman A (1993) Fundamentals of metal matrix composites. Butterworth-Heinemann, pp 3–23
6. VijayaRamnath B, Elanchezhian C, Annamalai RM, Aravind S, Sri AnandaAtreya T, Vignesh V, Subramanian C (2014) Aluminium metal matrix composites - a review. Rev Adv Mater Sci 38:55–60.
7. Sun Y, Lyu Y, Jiang A (2014) Fabrication and characterization of aluminum matrix fly ash cenosphere composites using different stir casting routes. JrMateri Res 29:260–266.
8. NHarishGarg K, KetanVerma AM, Kumar R (2012) Hybrid metal matrix composites and further improvement in their maChinability. Int J Latest Res Sci Technol 36:2278–5299
9. Baradeswaran A, Elayaperumal A (2013) Influence of B4C on the tribological and mechanical properties of Al 7075-B $_4$C Composites. Composites Part B 54:146–15210.
10. Mahadevan K, Raghukandan BC, Pai UTSP (2008) Influence of precipitation hardening parameters on the fatigue strength of AA 6061-SiCp composite. J Mater Process Technol 198:241–247
11. Schneider Y, Soppa E, Kohler MR (2011) Numerical and experimental investigations of the global and local behaviour of a (6061)/Al_2O_3 metal matrix composite under low cycle fatigue. ProcedEng 10:1515–1520.
12. Kilickap E, Cakir O, Aksoy M, Inan A (2005) In an Study of tool wear and surface roughness in machining of homogenized SiC-P reinforced aluminium metal matrix composite. J Mater Process Technol 164–165:862–867.
13. Yanming Q, Zehua Z (2000) Tool wear and its mechanism for cutting SiC particle-reinforced aluminium matrix composites. J Mater Process Technol 100:194–199.
14. Mahesh Babu TS, AldrinSugin MS, Muthukrishnan N (2012) Investigation on the characteristics of surface quality on machining of hybrid metal matrix composite (Al-SiC-B4C). Proced Eng 38:2617–2624.
15. BalasivanandhaPrabu S, Karunamoorthy L, Kathiresan S, Mohan B (2006) Influence of stirring speed and stirring time on distribution of particles in cast metal matrix composite. J Mater Process Technol 171:268–273.
16. Wu X, Xia K (2007) Back pressure equal channel angular consolidation-Application in producing aluminium matrix composites with fine fly ash particles. J. Mater. Process. Technol. 192–193:355–359.

17. Eizadjou M, Kazemi Talachi A, Danesh Manesh H, Shakur Shahabi H, Janghorban K (2008) Investigation of structure and mechanical properties of multi-layered Al/Cu composite produced by accumulative roll bonding (ARB) process. Compos Sci Technol 68:2003–2009.
18. KarbalaeiAkbari M, Baharvandi HR, Mirzaee O (2013) Fabrication of nano-sized Al_2O_3 reinforced casting aluminum composite focusing on the preparation process of reinforcement powders and evaluation of its properties. Compos Part B 55:426–432.
19. Bansala P, Upadhyay L Experimental Investigations to study tool wear during turning of alumina reinforced aluminium composite. ProcedEng 51:818–827.
20. Siddique Ahmed Ghias A, VijayaRamnath B (2015) Investigation of tensile property of AaluminiumSiC metal matrix. Int J App Mech Mater 766-767:252–256.
21. RD Babu, P Gurusamy, ABH Bejaxhin, P Chandramohan, Influences of WEDM constraints on tribological and micro structural depictions of SiC-Gr strengthened Al2219 composites, Tribology International 185, 108478.
22. P. Gurusamy and S. Balasivanandha Prabu, "Effect of the squeeze pressure on the mechanical properties of the squeeze cast Al/SiCp metal matrix composite," Int. J. of Microstructure and Materials Properties 8(4–5), 299–312 (2013), https://doi.org/10.1504/IJMMP.2013.057067.
23. P. Gurusamy, S. Balasivanandha Prabu, and R. Paskaramoorthy, "Interfacial thermal resistance and the solidification behaviour of the Al/SiCp composites," Mater. Manufacturing Processes 30(3), 381–386 (2015), https://doi.org/10.1080/10426914.2013.872267.
24. P. Gurusamy, S. Balasivanandha Prabu, and R. Paskaramoorthy, "Prediction of Cooling Curves for Squeeze Cast Al/SiCp Composites Using Finite Element Analysis," Metall. Materials Transactions A: Physical Metallurgy and Materials Science 46(4), 1697–1703 (2015), https://doi.org/10.1007/s11661-015-2742-6.
25. P. Gurusamy, S. Balasivanandha Prabu, and R. Paskaramoorthy, "Influence of processing temperatures on mechanical properties and microstructure of squeeze cast aluminium alloy composites," Mater. Manufacturing Processes 30(3), 367–373 (2015), https://doi.org/10.1080/10426914.2014.973587.
26. P. Gurusamy, B. Bhattacharjee, H. Dutta, A. Bhowmik, "Study of Microstructural, Machining and Tribological Behaviour of AA-6061/SiC MMC Fabricated Through the Squeeze Casting Method and Optimized the Machining Parameters by Using Standard Deviation-PROMETHEE Technique," Silicon (2023), https://doi.org/10.1007/s12633-023-02707-w.
27. P. Gurusamy, S. Hari Krishna Raj, B. Bhattacharjee, A. Bhowmik, "Assessment of Microstructure and Investigation Into the Mechanical Characteristics and Machinability of A356 Aluminum Hybrid Composite Reinforced with SiCp and MWCNTs Fabricated Through Rotary Centrifugal and Squeeze Casting Processes," Silicon (2023), https://doi.org/10.1007/s12633-023-02686-y.
28. B. Bhattacharjee, N. Biswas, P. Chakraborti, K. Choudhuri, P. Gurusamy, "Impact of Nano-lubrication on Single-Layered Porous Hydrodynamic Journal Bearing: A Hypothetical Approach," Journal of Porous Media (2023), DOI: 10.1615/JPorMedia. 2023044960.
29. D. Dinesh, P. Gurusamy, R.D.S. Kumar, "Influence of Nanosilica and Layering Sequence on the Mechanical Properties of Kenaf and Banyan Fibers Reinforced Composites," Silicon (2023), https://doi.org/10.1007/s12633-023-02460-0.

Note: Every figure except fig. 88.1 and Tables was created using the experimental work.

Advances in Additive Manufacturing Technologies – Gurusamy Pathinettampadian (eds)
© 2024 Taylor & Francis Group, London, ISBN 978-1-032-90013-1

89 Comprehensive Mechanical, Thermal, and Interlaminar Characterisation of Basalt Fibre Composite Materials for Enhanced Structural Integrity

K. Mariappan, V. Nagarajan
Department of Mechanical Engineering,
Vel Tech High Tech Dr. Rangarajan Dr. Sakunthala Engineering College,
Chennai, Tamilnadu

P. Gurusamy*
Department of Mechanical Engineering,
Chennai Institute of Technology, Chennai, India

N. Ramanan, S. Hari Krishna Raj,
S. Baskar, A. Sanjeev kumar
Department of Mechanical Engineering,
Vel Tech High Tech Dr. Rangarajan Dr. Sakunthala Engineering College,
Chennai, Tamilnadu

R. Prawin kumar
Department of Mechanical Engineering,
Sri Jayaram Institute of Engineering and Technology, Chennai, India

ABSTRACT: Fiber Reinforced Polymers (FRP) has garnered significant attention in the field of composites due to their exceptional properties, including high tensile and compressive strength, chemical stability, and impressive mechanical and thermal characteristics. These materials find extensive use in polymer channel and beam construction industries, owing to their substantial commercial value. In the realm of basalt-fiber-reinforced composites (BFRCs), Basalt Fiber reinforcement plays a pivotal role in enhancing mechanical performance. The interface between the fiber and matrix is equally critical in determining the long-term performance of BFRCs. This research focuses on characterizing basalt fiber composite materials using a stacking sequence with top and bottom layers oriented at 0-90 degrees and a central layer at 45 degrees. The laminate is fabricated through the prepreg process, undergoing compressive pressure of 4 bars and a curing period of 24 hours, generating temperature due to applied compression. Subsequent testing encompasses parameters such as tensile strength, hardness, interlaminar delamination variables, and SEM investigation of tensile fracture surfaces. Additionally, a wear surface evaluation and Dynamic Mechanical Analysis (DMA) test are conducted. Drawing conclusions from these experimental results, the study provides insights into the practical applicability of the material.

KEYWORDS: Basalt fiber reinforced composites, Stacking sequence, Preparation process, Interlaminar delamination, Dynamic mechanical analysis (DMA)

*Corresponding author: gurusamyp@citchennai.net

DOI: 10.1201/9781003545774-89

1. Introduction

Hybrid nanocomposites are often created by integrating two or more distinct foreign elements into a shared host matrix. This amalgamation yields innovative material properties, including heightened elasticity, ductility, reduced weight, and enhanced fire resistance. Carbon fibers already possess such properties, making them valuable in various large-scale applications like aircraft structures, automobile manufacturing, maritime vessels, sporting goods, and construction[1-12]. Given these facts, there is a growing demand for fibers that are durable, lightweight, long-lasting, and cost-effective in producing hybrid composites. While the market offers a range of organic and inorganic fibers, many lack structural strength or prove prohibitively expensive for moderate loads. Basalt fiber, owing to its inorganic nature, has emerged as the preferred material due to its exceptional modulus, superior strength, enhanced strain resistance, high-temperature endurance, stability, chemical resistance, ease of processing, non-toxicity, natural origin, eco-friendliness, and affordability. The incorporation of mechanically cut basalt fibers into a polypropylene-clay blend successfully produced a composite material, increasing both yield strength and elastic modulus. Basalt's plasticity allows for its use in various forms like rods, bars, and textile fabrics, expanding its applications.

In predicting the falling weight impact properties of thermosetting composites, an experimental examination was conducted using partially bio-based vinyl ester resin, flax, and basalt reinforcements. Tensile, flexural, and falling weight impact tests were performed, revealing improved tensile performance when flax fiber layers were used between basalt fibers. Another study examined the effect of water absorption on the interlaminar fracture toughness of vinyl ester-based composites, emphasizing the enhanced durability and water-repellent qualities achieved by hybridizing with basalt fibers. Hybrid basalt/kenaf epoxy composites were systematically analyzed, demonstrating admirable mechanical properties. Research focused on optimizing car bumpers for reduced weight and improved fuel efficiency while maintaining safety[13,14]. The use of a Basalt composite bumper achieved significant weight and cost reductions, coupled with enhanced performance characteristics. Basalt rock's thermal qualities, strength, and low-cost extraction make it ideal for various applications, including insulating car interiors and engine parts. Basalt fibers are used to reinforce concrete structures, providing increased tensile strength and durability. Basalt fibers are used to manufacture FRP components, such as rebars, sheets, and panels, for construction applications. Basalt fibers are employed in the production of lightweight composites for automotive components, contributing to fuel efficiency and overall performance. Basalt fibers are used in the aerospace industry to manufacture structural components for their high strength-to-weight ratio and thermal stability. Basalt fibers are used to reinforce materials in sporting goods such as tennis rackets, bicycles, and skis, offering a balance of strength and flexibility. These applications showcase the versatility and potential of basalt fibers across various industries, offering solutions that benefit from their specific combination of strength, durability, and thermal resistance.[15]

The study explored hybrid composites made from basalt and jute fibers and polyester resin, revealing superior performance of pure basalt fiber composites in flexural and tensile tests. Automotive panel development primarily targets lower fuel consumption, reduced weight, and efficient resource use. In the pursuit of excellence, cost-effectiveness, and responsiveness to feedback, the selection of materials for new products becomes crucial. Composite materials, a dominant trend, exhibit substantial improvements, combining the intrinsic capabilities of reinforcement and matrix materials[16-22]. Hybrid constructions aim for characteristics like lightweight construction, stability, environmental friendliness, and corrosion resistance.

In the automotive context, bumpers play a vital role in preventing physical damage during low-speed crashes. However, existing legal criteria for car bumpers are often deemed insufficient, and real-world crash damage analysis reveals critical aspects lacking in many designs. The Gas Tube Bumper System is presented as a solution to address safety, efficiency, and sustainability concerns in the automotive industry.

2. Materials and Methods

2.1 Basalt Fiber

Generally there are various amazing materials in the real world which are given by nature to alter the conventional materials in order to reduce the weight, cost and facility to manufacture the product. Especially in the modern world we need to face challenges to reduce the burden of the society to make the product more feasible and strong. Hence, we face the problems where nature cannot be anticipated. Some need to go for synthetic materials known as Kevlar. Kevlar is a plastic strong synthetic material being five times stronger than steel on an equal weight basis. Kevlar is made from hundreds of synthetic plastics through polymerization process i.e. joining together long chain molecules. It has excellent properties because of its internal structure, the molecules are arranged in regular and parallel lines and finally it made into fibers that are knitted tightly together.

This research holds paramount practical relevance as it significantly contributes to advancing the utilization of basalt fiber-reinforced composites (BFRCs) in the specialized domains of polymer channel and beam construction industries. The emphasis on enhancing mechanical performance through basalt fiber reinforcement addresses a critical demand for materials possessing high tensile and compressive strength, crucial attributes in the structural intricacies of polymer channels and beams. The adoption of a specific stacking sequence, coupled with insights into the manufacturing process via the prepare method under realistic conditions, provides practical considerations for achieving the desired mechanical properties. The comprehensive testing protocols, encompassing various parameters such as tensile strength, hardness, and interlaminar delamination, offer a holistic understanding of material behavior, aligning with the practical needs of industries navigating diverse conditions[22-24]. The study's conclusions, drawn from meticulous experimental assessments, offer valuable insights into the practical applicability of basalt fiber composites, aiding industries in informed material selection for durable and high-performance construction applications. Furthermore, by establishing a solid foundation for future development, this research serves as a catalyst for the evolution and optimization of BFRCs, promising innovative solutions tailored to the specific requirements of the construction industry.

3. Manufacturing Methods

In practice, VART defect fully manufactured by placing alternative layers of metal and fiber/pre reign the mold through the hand lay-up method. Later the structure is cured at desired temperature and Pressure or by using an autoclave. To address bonding failures and reduce fabrication costs, we adopted the VARTM process from the Manufacturing Technology Lab at Anna University. In this method, we utilized drilled sheets of basalt fiber measuring 300x300mm. Three layers of basalt fiber were stacked with alternating natural fiber sheets, forming the top and bottom layers. A binding agent consisting of polyester resin and HY/3 hardener in a 1:10 ratio was employed in a cold mold setup through VARTM.

The specified resin-hardener combination was injected into the stacked die using a vacuum gun, applying 5 bar pressure. The setup was allowed to cure for approximately a day. Subsequently, test specimens conforming to ASTM standards were obtained by cutting from the cured plate, employing an abrasive water jet machining setup. This approach aims to maintain high bonding accuracy while minimizing costs in specimen preparation.

Table 89.1 Tensile test valve

Sl. No	Ultimate Tensile Strength (N/mm²)	Ultimate load (kN)	Max displacement (mm)	% of Elongation
1	103.215	5.81	5.52	6.68
2	79.588	4.48	3.68	3.33
3	107.479	5.85	5.03	6.68
4	89.925	4.85	4.89	6.61
5	101.315	5.82	5.32	6.21

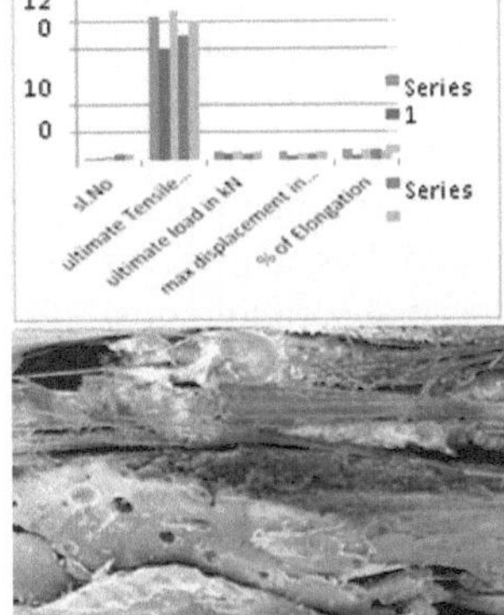

Fig. 89.1 Tensile testing results for various specimens

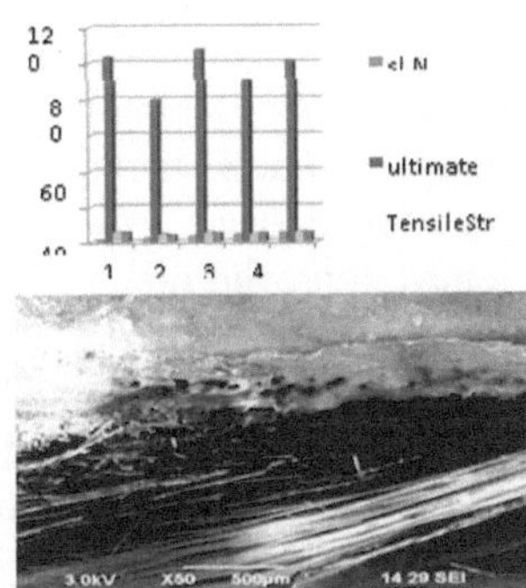

Fig. 89.2 SEM images of tensile fracture

Table 89.2 Flexural test valve

Sl. No	Ultimate stress (kN/m²)	Ultimate Load (kN)	Max displacement (mm)
1	0.009	0.49	5.2
2	0.01	0.48	3.6
3	0.012	0.56	3.8
4	0.011	0.52	4.8
5	0.012	0.58	5.2

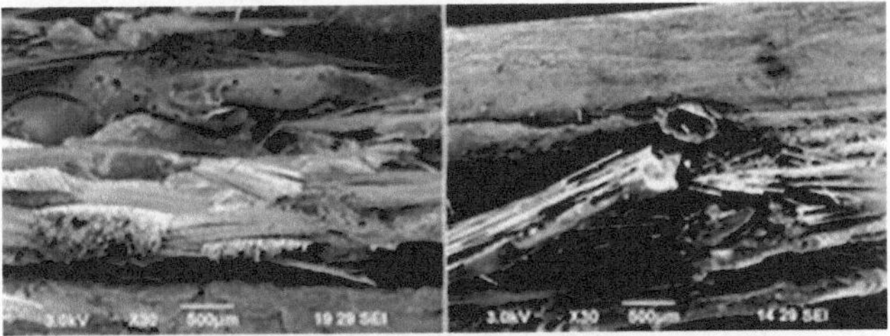

Fig. 89.3 SEM images of flextural fracture

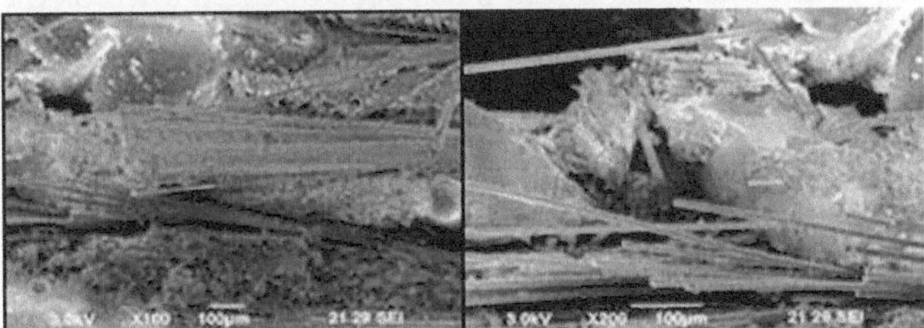

Fig. 89.4 SEM images of flextural fracture

4. Finite Element Analysis of Bumper

Bumpers typically consist of beams, stays, shock absorbers, and fascia. They serve as structural elements designed to minimize physical damage to the front and rear ends of a passenger motor vehicle during low-speed collisions[25-27]. The role of vehicle bumpers is crucial in mitigating or preventing damage. However, existing regulatory standards for car bumpers are insufficient, with some passenger vehicles not mandated to have bumpers at all. In instances of front-to-rear collisions, vehicles often sustain notable damage to both safety features and cosmetic components. A review of real-world crash damage highlights key elements lacking in the design of many passenger vehicle bumpers, including compatible geometry, stability during impacts, effective energy absorption, cost efficiency, improved fuel efficiency, and recyclability. Ideally, engaged bumper systems should provide a stable interface and remain engaged throughout the impact. The Gas Tube Bumper System is proposed as a solution for small passenger cars to address contemporary challenges faced by the automotive industry.

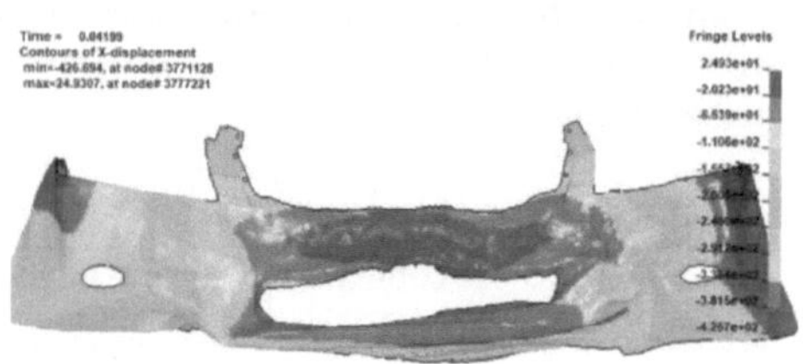

Fig. 89.5 Finite element analysis of bumper

In bumper design, common criteria include damage and protection assessments. For damage assessment, the examination of relative displacements representing stiffness performance is crucial[28-32]. In the contemporary automotive industry, car manufacturers are confronted with the growing challenge of producing lighter and more efficient cars, aiming for improved economic viability for both manufacturers and consumers.

A specific area receiving significant focus is the development of lightweight bumpers. Traditionally, bumpers have been constructed from steel with rubber padding. However, recent exploration has been directed towards the potential use of fiberglass as a lighter alternative bumper material. This shift reflects the industry's on-going efforts to enhance vehicle efficiency and reduce overall weight, thereby addressing economic and environmental considerations[33-35].

4.1 Specification

The dimensions of a typical conventional bumper for a 2017 Honda Accord Baseline Model are as follows:

Outside to outside: 1.527 meters;

Turn back: 0.273 meters

Depth: 0.100 meters

Height: 0.141 meters

Between supports: 0.9225 meters

Fig. 89.6 A Honda Accord 2017 Baseline Model

5. Result and Discussion

The simulations are executed utilizing the LS-DYNA971 solver, and the outcomes are visualized through LS-PRE-POST, an open-source pre and post-processor. The results encompass internal energy, plastic strain, and energy absorbed by different components of the bumper assembly, which are plotted and subsequently discussed.

5.1 Crash Analysis of Front Bumper (Frontal Crash)

Fig. 89.7 Resultant Displacement of frontal crash

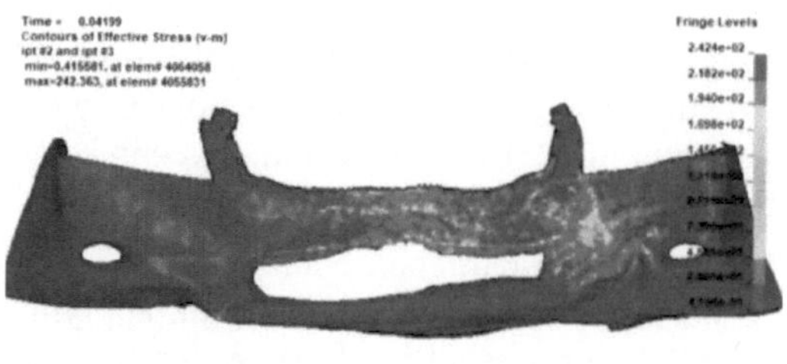

Fig. 89.8 Effective von missesStress

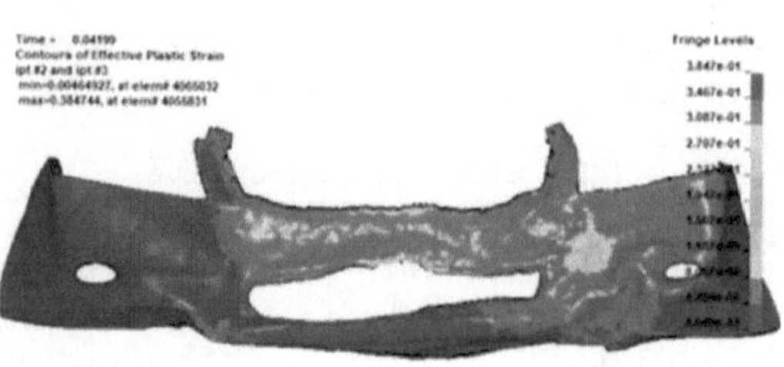

Fig. 89.9 Effective plastic strain

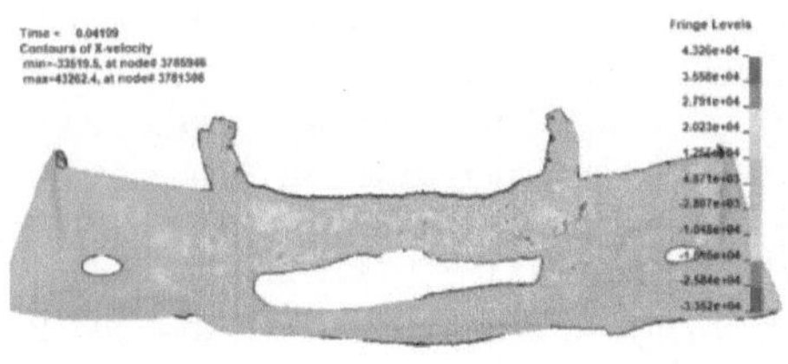

Fig. 89.10 X-velocity of frontal crash

5.2 Crash Analysis of Front Bumper (40% Odb Crash)

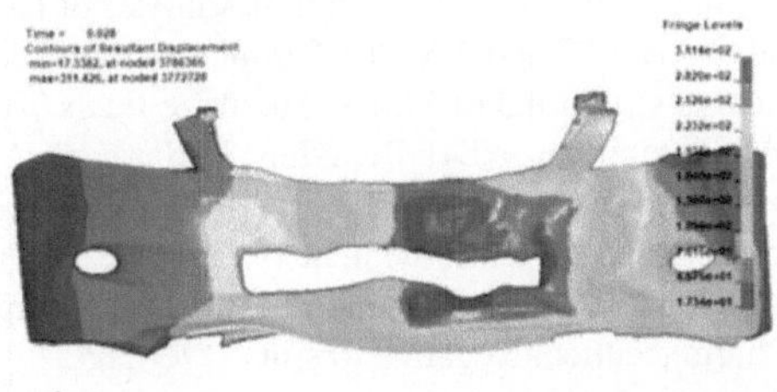

Fig. 89.11 Resultant displacement of frontal crash

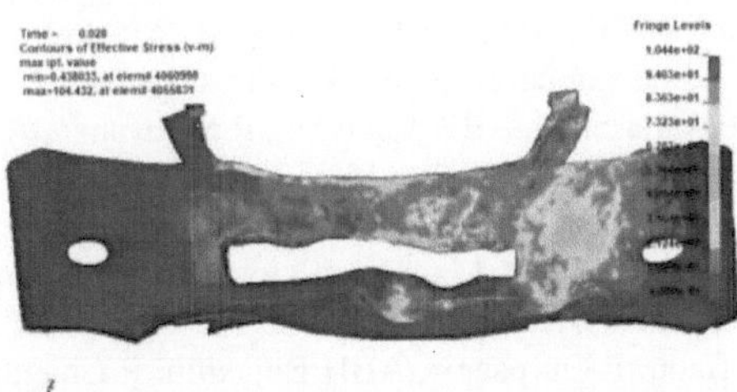

Fig. 89.12 Effective vonmisses stress

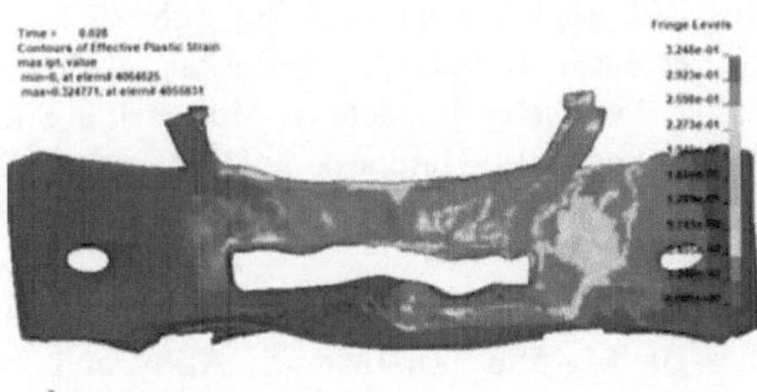

Fig. 89.13 Effective plastic strain

5.3 Crash Analysis of Front Bumper (30°OBLQCRASH)

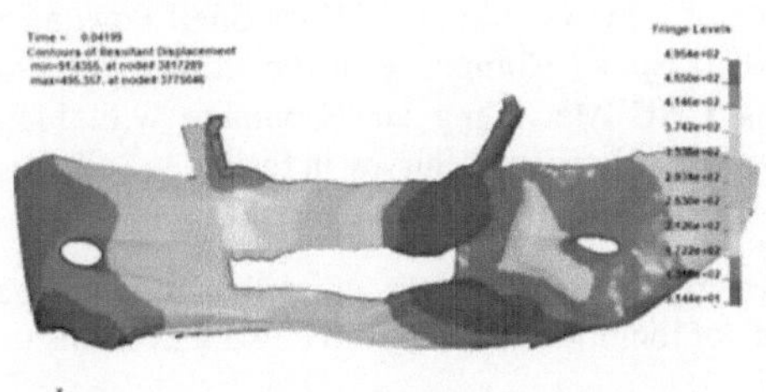

Fig. 89.14 Resultant Displacement of frontal crash

6. Conclusion

The internal energy and effective plastic strain at sample 3 are notably higher when employing a hybrid composite of jute and banana, as opposed to an FRP bumper. This indicates that the hybrid composite of jute and banana possesses the capability to absorb stresses and dissipate them as gas pressure. Additionally, the effective plastic strain at the headlight bracket and chassis end is lower. Consequently, we can infer that the reactions at these locations are less pronounced compared to bumpers utilizing the hybrid composite.

References

1. Han SH, Oh HJ, Lee HC, Su SS. The effect of post-processing of carbon fibers on the mechanical properties of epoxy-based composites. Compos Part B-Eng 2012;45(1):172–177.
2. Dehkordi MT, Nosraty H, Shokrieh MM, Minak G, Ghelli D. Low velocity impact properties of intra-ply hybrid composites based on basalt and nylon woven fabrics. Mater Des 2010; 31 (8): 3835–3844.
3. ArySubagia IDG, Kim Y, Tijing LD, Kim CS, Shon HK. Effect of stacking sequence on the flexural properties of hybrid composites reinforced with carbon and basalt fibers. Compos Part B-Eng, 2014; 58:251–258.
4. Taketa I, Ustarroz J, Gorbatikh L, Lomov SV, Verpoest I. Interply hybrid composites with carbon fiber reinforced polypropylene and self-reinforced polypropylene. Compos Part A-Appl S, 2010; 41(8): 927–932.
5. García PDLR, Escamilla AC, García MNG. Bending reinforcement of timber beams with composite carbon fiber and basalt fiber materials. Compos Part B-Eng 2013;55:528–536.
6. Wang X, Wu Z, Wu G, Zhu H, Zen F. Enhancement of basalt FRP by hybridization for long-span cable-stayed bridge. Compos Part B-Eng 2013; 44(1):184–192.
7. Larrinaga P, Chastre C, Biscaia HC, San-Jose JT. Experimental and numerical modeling of basalt textile reinforced mortar behavior under uniaxial tensile stress. Mater Des 2014;55: 66–74.
8. Černý M, Halasová M, Schwaigstillová J, Chlup Z, Sucharda Z, Glogar P, Svítilová J, Strachota A, Rýglová S. Mechanical properties of partially pyrolysed composites with plain weave basalt fibre reinforcement. Ceram Int 2014; 40(5): 7507–7521.
9. Sarasini F, Tirillò J, Ferrante L, Valente M, Valente T, Lampani L, Gaudenzi P, Cioffi S, Iannace S, Sorrentino L. Drop-weight impact behaviour of woven hybrid basalt– carbon/epoxy composites., Compos Part B-Eng 2014;59:204–220.
10. Zhu H, Wu G, Zhang L, Zhang J, Hui D. Experimental study on the fire resistance of RC beams strengthened with near-surface-mounted high-Tg BFRP bars. Compos Part B-Eng 2014; 60:680–687.
11. Boria, S.; Pavlovic, A.; Fragassa, C.; Santulli, C. Modeling of falling weight impact behavior of hybrid basalt/flax vinylester composites. *Procedia Eng.* 2016, *167*, 223–230.

12. Boopathy G, Vijayakumar K.R., Chinnapandian M, Gurusami K, Development and Experimental Characterization of Fibre Metal Laminates to Predict the Fatigue Life, International Journal of Innovative Technology and Exploring Engineering. 2019, 8(10), pp .2815–2819.
13. Almansour, F.A.; Dhakal, H.N.; Zhang, Z.Y. Effect of water absorption on Mode I interlaminar fracture toughness of flax/basalt reinforced vinyl ester hybrid composites. *Compos.Struct.* **2017**, *168*, 813–825.
14. Umashankaran, M.; Gopalakrishnan, S.; Sathish, S. Preparation and characterization of tensile and bending properties of basalt-kenaf reinforced hybrid polymer composites. *Int. J. Polym. Anal.Charact.* 2020, *25*, 227–237.
15. Boopathy,G., Vijayakumar, K.R., Chinnapandian, M., Fabrication and Fatigue Analysis of Laminated Composite Plates, International Journal of Mechanical Engineering and Technology. 2017, 8(7), pp. 388–396.
16. Chandrasekaran, P., Rameshbabu, V., & Prakash, C. Advancements in Basalt composite automobile bumpers and performance evaluation through finite element analysis. *Polymer Bulletin*, 2023, 1–18.
17. Prasath KA, Krishnan BR, Arun CK. Mechanical properties of woven fabric basalt/jute fibre reinforced polymer hybrid composites. *Int J Mech Eng* 2013, 2: 279–290.
18. Kankariya N, Sayyad FB. Numerical simulation of bumper impact analysis and to improve design for crash worthiness. *Int J EngSci* 2015,4: 58.
19. Fentahun MA, AhsenSavas M. Materials used in automotive manufacture and material selection using ashby charts. Int J Mater Eng 2018,8: 40–54.
20. MM Natarajan, B Chinnasamy, BHB Alphonse, Investigation of Machining Parameters in Thin-Walled Plate Milling Using a Fixture with Cylindrical Support Heads, Strojniškivestnik-Journal of Mechanical Engineering 68 (12), 746–756.
21. D Velmurugan, J Jayakumar, A Bovas Herbert Bejaxhin, JB Raj, Experimental Investigation on the Mechanical Properties of Hybrid Composites Made with Banyan and Peepal Fibers, Journal of Natural Fibers 19 (16), 14183–14194.
22. PV Narashima Rao, P Periyasamy, A Bovas Herbert Bejaxhin, Fabrication and Analysis of the HLM Method of Layered Polymer Bumper with the Fracture Surface Micrographs, Advances in Materials Science and Engineering 2022.
23. P. Gurusamy and S. Balasivanandha Prabu, "Effect of the squeeze pressure on the mechanical properties of the squeeze cast Al/SiCp metal matrix composite," Int. J. of Microstructure and Materials Properties 8(4–5), 299–312 (2013), https://doi.org/10.1504/IJMMP.2013.057067.
24. P. Gurusamy, S. Balasivanandha Prabu, and R. Paskaramoorthy, "Interfacial thermal resistance and the solidification behaviour of the Al/SiCp composites," Mater. Manufacturing Processes 30(3), 381–386 (2015), https://doi.org/10.1080/10426914.2013.872267.
25. P. Gurusamy, S. Balasivanandha Prabu, and R. Paskaramoorthy, "Prediction of Cooling Curves for Squeeze Cast Al/SiCp Composites Using Finite Element Analysis," Metall. Materials Transactions A: Physical Metallurgy and Materials Science 46(4), 1697–1703 (2015), https://doi.org/10.1007/s11661-015-2742-6.
26. P. Gurusamy, S. Balasivanandha Prabu, and R. Paskaramoorthy, "Influence of processing temperatures on mechanical properties and microstructure of squeeze cast aluminium alloy composites," Mater. Manufacturing Processes 30(3), 367–373 (2015), https://doi.org/10.1080/10426914.2014.973587.
27. T Prabaharan, MBS Reddy, H Bejaxhin, A Bovas Herbert Bejaxhin, Strain-cum-Deformation Analysis of Friction Stir Welded AA5052 and AA6061 Samples with Microstructural Analysis., Journal of Mines, Metals & Fuels 70.
28. B Venkataraman, ABH Bejaxhin, R Saravanan, A Comparative Analysis on Surface Roughness of Plain Epoxy Composite with Reinforcement of 10 wt% of Pista Shell Particles Novel Composite using CNC Machining, Journal of Pharmaceutical Negative Results, 376–385.
29. M Sivasai, ABH Bejaxhin, R Saravanan, A Comparative Analysis on Material Removal Rate of Plain Epoxy Composite with Reinforcement of 15 wt% of Egg Shell Powder particles Novel composite using CNC Machining, Journal of Pharmaceutical Negative Results, 692–699.
30. B Venkataraman, ABH Bejaxhin, R Saravanan, A Comparative Analysis on MRR of Plain Epoxy Composite with Reinforcement of 10 wt% of Pista Shell Particles Novel Composite using CNC Machining, Journal of Pharmaceutical Negative Results, 367–375.
31. RD Babu, P Gurusamy, ABH Bejaxhin, P Chandramohan, Influences of WEDM constraints on tribological and micro structural depictions of SiC-Gr strengthened Al2219 composites, Tribology International 185, 108478.
32. [32] G Mahesh, D Valavan, N Baskar, A Bovas Herbert Bejaxhin, Parameter Impacts of Martensitic Structure on Tensile Strength and Hardness of TIG Welded SS410 with characterized SEM Consequences, Tehničkivjesnik 30 (3), 750–759.
33. CB Priya, K Ramkumar, V Vijayan, ABH Bejaxhin, Wear Studies on Mg-5Sn-3Zn-1Mn-xSi Alloy and Parameters Optimization Using the Integrated RSM-GRGA Method, Silicon 15 (8), 3569–3579.
34. P Chandramohan, A SivaRangar, JJ Kingsly, ABH Bejaxhin, N Ramanan, Investigation of corrosion and wear behavior of Al–SiC composite, Materials Today: Proceedings.
35. B Venkataraman, ABH Bejaxhin, R Saravanan, Analyse the effect of 20wt% addition of Pista Shell Powder reinforcement to Epoxy Composite in the Material Removal Rate using CNC Machining and Compare with Plain Epoxy Composite, Journal of Survey in Fisheries Sciences 10 (1S), 3105–3115.

Note: Every figure and table was created using the experimental work; none of them were taken from any publications or online.

Advances in Additive Manufacturing Technologies – Gurusamy Pathinettampadian (eds)
© 2024 Taylor & Francis Group, London, ISBN 978-1-032-90013-1

90 Inspection of Defects in Weldments of Manuarite 900 Alloy using Ultrasound B Spline Imaging Technique

Raj Jawahar. R*
Research Scholar, Department of Mechanical Engineering,
Chennai Institute of Technology, Chennai, India

Lakshmana Kumar S
Assistant Professor, Department of Mechanical Engineering,
Sona College of Technology, Salem, India

D Jayabalakrishnan
Associate Professor, Department of Mechanical Engineering,
Chennai Institute of Technology, Chennai, India

Somasundaram S
Professor, Department of Mechanical Engineering,
Sri Venkateswara Institute of Science and
Technology, Chennai, India

ABSTRACT: The purpose of this study is to inspect the defects in weldments of the header pipe section in reformers for methanol extraction plant. The proposed work deals with the study of ultrasonic testing of Manuarite 900 alloy using TRL longitudinal wave technique. Tests were conducted on 30mm and 44 mm blocks using 2.5 and 5 Mhz probe. For calibration purposes defects were created with SDH and notches on the fabricated manuarite. Results obtained from the tests reveals the effective detection of SDH and notches could be observed with 2.5 Mhz probe with wedge angle of 30° to increase beam width for precise focus.

KEYWORDS: Manuarite, TRL, Longitudinal, Ultrasonic, weldments

1. INTRODUCTION

Methanol, also known as alcohol from wood, carbinol, and methyl alcohol (CH3OH), is distinguished by its flame retardancy, non-toxic nature, and mild fragrance [1]. Its purification involves a desulfurization process, where sulfur compounds are converted to hydrogen sulfide using a catalytic bed, followed by steam reforming. This method reduces sulfur content to below 0.1 ppm, essential as sulfur can negatively affect catalysts in reformers [2,3].The reforming process crucially converts natural gas components like methane (CH4) into hydrogen (H2), carbon dioxide (CO2), and carbon monoxide (CO), using steam. This process is managed catalytically to control emissions [4,5].

*Corresponding author: dynamechz.raj65@gmail.com

DOI: 10.1201/9781003545774-90

Materials in use include an advanced form of austenitic steel known as 900 alloy, derived from the HK40 alloy, but with increased nickel content (20% to 35%), providing resistance to high temperatures, pressure, and corrosion [6]. The steel undergoes solution treatment and quenching to maintain its structure. Further advancements include the addition of micro-alloys like titanium and zirconium, improving resistance to thermal shock and carburization [7-8]. The HP40Nb alloy, designed for high-temperature operations (up to 1100°C), offers significant improvements over HK40 in terms of oxidation and carburization resistance [9]. In production settings, these materials are essential for parts like catalytic steam reformer coils and are subjected to conditions that require robustness against high temperatures and chemical reaction. For quality control, ultrasonic testing of these austenitic materials is complex due to factors like elastic anisotropy among grains. The testing, crucial for maintaining manufacturing and service standards, uses reference blocks with reflectors to identify defects like side-drilled holes and notches. Advanced techniques, such as the Transmission-Receiver Longitudinal Technique (TRL), are employed to detect and evaluate these defects effectively [10].

2. Materials and Methods

Manuarite 900 is used as header pipe material in the reformer section of the methanol extraction plant. It is used to withstand higher pressure and temperature with presence of high nickel content. This constituent is casted into a header tube with equiaxed and columnar macro structured grains of primary and secondary carbides. The surface distortion of the casted manuarite is shown in Fig. 90.1. The chemical composition of the Manuarite 900 is shown in Table 90.1. With the presence of higher nickel content of about 31% after heat treatment gives a maximum reliability to withstand temperatures upto 1300° C. The mechanical properties such as ultimate tensile strength of 448 MPa and yield strength of 186 MPa was observed using Universal testing MTS SHT-4000) for applied load of 50 KN.

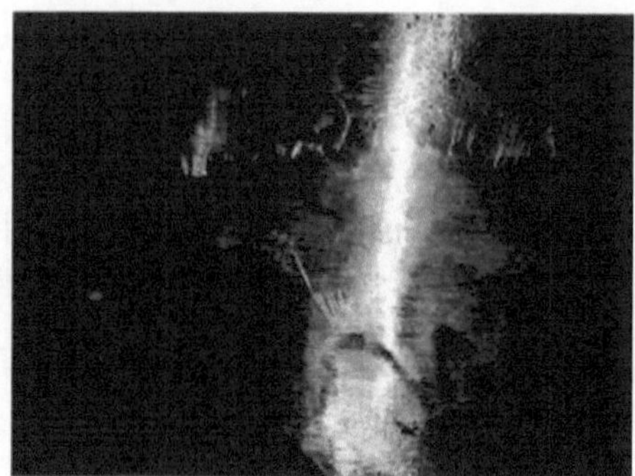

Fig. 90.1 Photographs of the casted manuarite 900 alloy

Table 90.1 Chemical composition and mechanical properties after furnace hardening of the manuarite 900 alloy

Element	Weight (%)
C	0.09
Si	0.5
Mn	0.8
P	0.015
S	0.01
Cr	25
Ni	31
Mo	0.5
Cu	0.5
N	0.1
Fe	Balance
Property	**Value**
UTS (MPa)	448
YS (MPa)	229
El (%)	35
RA (%)	36

To investigate the weld defect analysis in header pipes of manuarite 900 material, TRL phased array pulse/echo technique for flaw detection is implemented. The schematic representation of the dimensions of the substrate is shown in Fig. 90.2. For study purpose, artificial defects were created with SDH (side drill hole) and notches (OD and ID) were created.

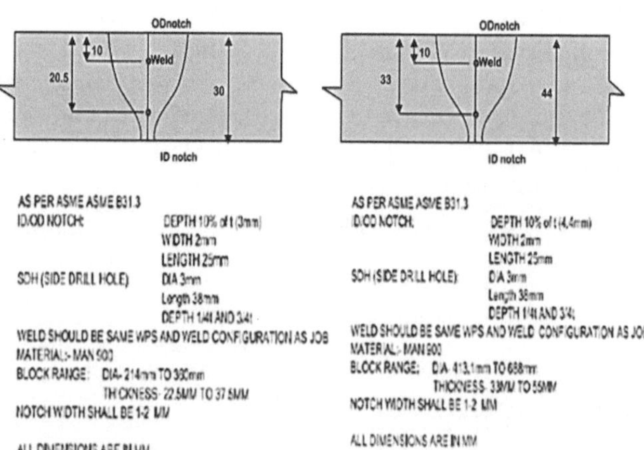

Fig. 90.2 Pictorial and schematic view of the weldment with bevel dimensions

Experiments were performed on a pipe section measuring 250 mm in circumference and 500 mm in length, using Dual Linear Array (DLA) probes at frequencies of 2.5 MHz and 5 MHz with Teflon wedges. Details of these

Table 90.2 Technical specification of the probe and wedge

Part No	Freq	Array Type	Aperture (mm)	Pitch (mm)	Elevation (mm)	Case Type	Sheathing
Q3301042	5.0 MHz	Dual Linear Array	32x12	1.0	12	A26	PVC
Q3301043	2.25 MHz	Dual Linear Array	32x12	1.0	12	A26	PVC

Description	Scan Type	Refracted Angle	Wave Type	Focus Depth	Options
Custom Wedge for Dual Phased Array probe A26	Normal Scan	55° Refracted Angle in Steel	Longitudinal Wave	Focus Depth of 40 mm	with IHC option (irrigation, holes, and carbides)

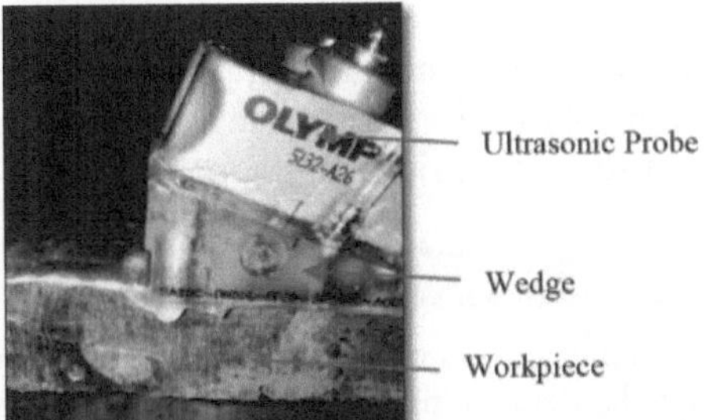

Fig. 90.3 Photographic views of the experimentation

setups are provided in Fig. 90.4 and the specifications of the probes and wedges are listed in Table 90.2. Calibration of the probes was done using a reference block designed according to ISO 22825:2012 (E) standards, matching the wall thickness of the tested substrate metal. This block included subsurface diametric holes (SDH) at the weld center and bottom, and surface notches near existing defects to ensure accurate probe calibration and defect detection.

3. Results and Discussions

In this study, ultrasound testing of a 30 mm block using a frequency of 2.5 MHz successfully detected subsurface diametric holes (SDH) and notches. A notable finding was a 3mm SDH located at a depth of 22.5 mm, which presented a low signal-to-noise ratio of 36dB and an accuracy of 60.7%. The inconsistency in detection accuracy was linked to variations in the grain structures of different components, hinting at potential issues in the production process such as improper forging or inadequate heat treatment. Further testing identified an 8mm SDH, where field examinations typically show spurious indications resulting from the material's metallurgical structure or the weld seam's geometry. The longitudinal wave angle probes used may amplify these findings due to their interaction with the shear wave component. Additionally, a detection of an Outer Diameter (OD) notch measuring 2 mm in width was reported. This scenario exhibited a higher level of noise and compromised signal-to-noise ratio, potentially attributable to suboptimal heat treatment. Such conditions in the material's grain size may adversely affect the detection quality and increase the likelihood of intergranular stress corrosion cracking.

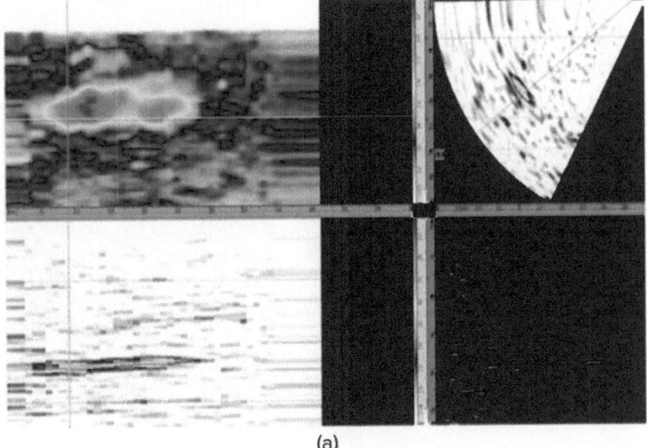

(a)

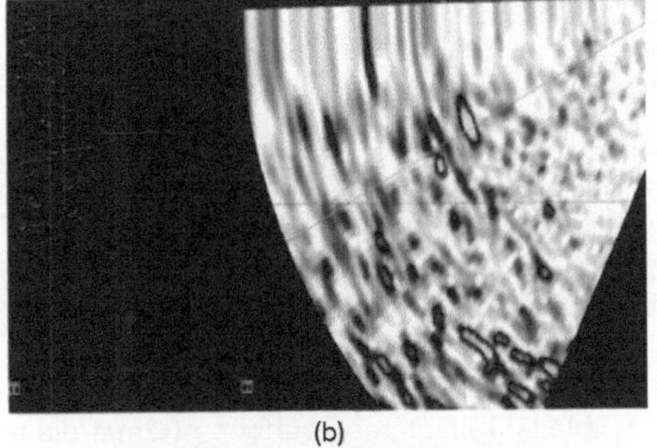

(b)

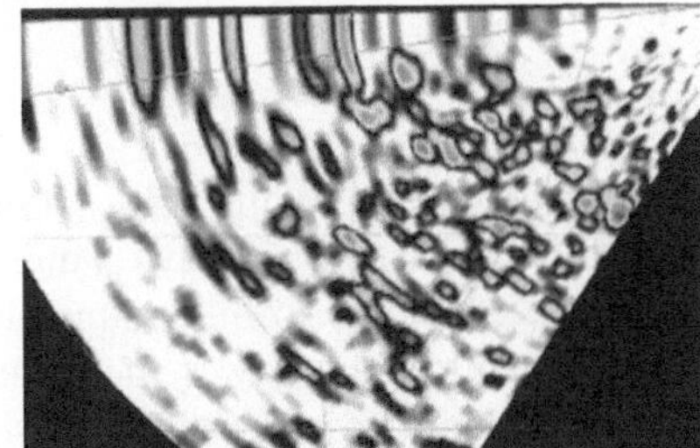

(c)

Fig. 90.4 (a) SDH of 3mm diameter at 22 mm for 30mm block (b) SDH of 3mm diameter at 8mm for 30mm block (c) Notch on surface of 2 mm diameter for 30 mm block

In case of 44mm block the detection was clear from the Fig. 90.5(a) with 3 mm diameter SDH at 22 mm. The TRL mechanism involves the use of longitudinal waves with wide beam spread of above 75° is used. The linear array probe on a curved surface increases the focusing point to capture the flaws more effectively in increasing the SNR value. It uses a half skipping method with effective angle of surface coverage at 30° beam angle. The reason is effective aperture gets smaller with higher beam angles. Also, the presence of noises can be observed at notches with improper surface conditions as shown in Fig. 90.5(b). More noise can be seen especially on high angle regarding Structure, Surface Conditions and 5 Mhz.

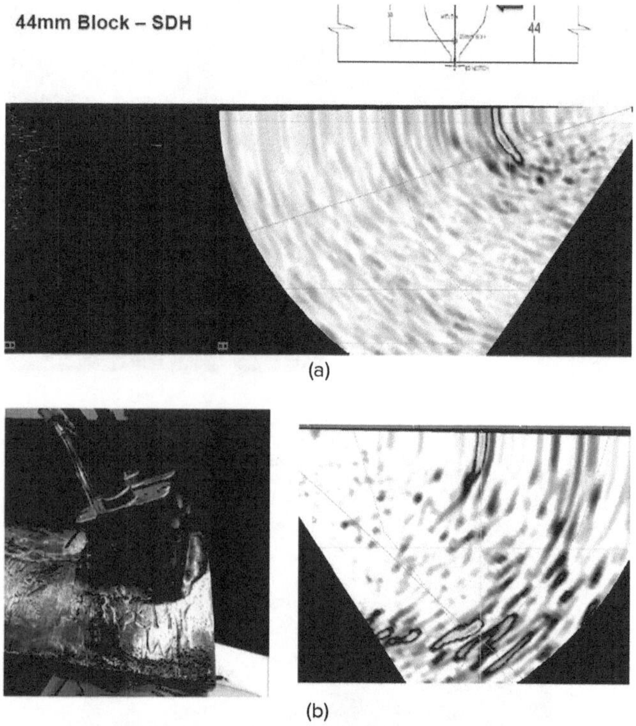

44mm Block – SDH

(a)

(b)

Fig. 90.5 (a) SDH of 3mm diameter at 22 mm for 44mm block (b) Notch on surface of 2mm diameter for 44 mm block

4. Conclusion

This research addresses the complexities of ultrasound testing of austenitic materials, noting that a universal testing solution is impractical due to the unique characteristics of each case. Key to the study is the differentiation between parent metal and weld material, which significantly influences test outcomes. The research highlights how the production processes, soldering techniques, and heat treatments affect the material's testability. Discrepancies in grain configuration can indicate issues in the production phase, underscoring the importance of distinct approaches during manufacturing and inspection phases. The study emphasizes the critical need for precise reference blocks that replicate the original manufacturing conditions, particularly in older facilities, to ensure reliable ultrasound testing results. Challenges arise from the specific geometric and metallurgical properties of components, often necessitating that definitive testing techniques be determined during actual inspections. This requirement underscores the need for flexible and adaptive testing methods. The findings stress the necessity for thorough preparatory work and the involvement of skilled personnel, reflecting the high demand for customization and adaptability in the field. The study concludes with a call for continued research and development to enhance ultrasound testing methodologies and technologies for austenitic components. This ongoing effort is essential to meet the evolving demands of material testing in various industrial applications.

References

1. M.E. Lapides, Improved ultrasonic inspection method for stainless steel piping. EPRI NP-1153 Special Report. (1979).
2. X. Edelmann and R. Hornung. Erfahrungenim Prtffen von austenitischen Schweissverbindungen. Mat. Technik, 5(1), (1977) 19.
3. X. Edelmann. The application of ultrasonic testing of austenitic weld joints. Mater. Eval. 37(10), (1979) 47.
4. X. Edelmann. Wiederkehrende Prfffungmit Ultrascha U an austenitischen Schweissverbindungenim Hauptkt/hlmittelsystem von Druckwasserreaktoren. In Conf. Proc. Int. Symp. New Methods Non-destructive Testing Mater. App. Especially Nucl. Eng. (1979).
5. H. WUstenberg, T. Just, W. MOhrle, and J. Kutzner. Zur Bedeutungfokussierender Prtffk Opfe for die Ultraschallprafung yon Schweissn~htenmitaustenitischem Gefttge. Mat. Prtlfg., 19(7), (1977) 246.
6. T. Just, M. R.Omer, E. Neumann, K. Matthies, and B. Kuhlow. Leistungsfa' higkeitverschiedener Ultraschallprtlftechniken an austenitischen Probeschweissungenmitnattlrl JchenTesffehlern. Mat. Prtlfg., 19(12), (1977) 488.
7. S. Prabhu, S. Satish Kumar, D. Dinakaran, and R. Jawahar. Improvement of chatter stability in boring operations with semi active magneto-rheological fluid damper. Mater. Today: Proc., (2020) doi:10.1016/j.matpr.2020.04.651.
8. J. Rabi, T. Balusamy, and R. Raj Jawahar. Analysis of vibration signal responses on pre induced tunnel defects in friction stir welding using wavelet transform and empirical mode decomposition. Def. Technol., (2019) doi:10.1016/j.dt.2019.05.014
9. P. Caussin and J. Cermak. Performances of the ultrasonic examination at austenitic steel components. In Conf. Periodic Inspection Pressurized Components (pp. 207). London: I Mech E (1979).
10. K.H. Hahn and K. Vedula. Room temperature tensile ductility in polycrystalline B2 NiAl. Scr. Metall., 23, (1989) 7–12. doi:10.1016/0036-9748(89)90083-5.

Note: Every figure and table was created using the experimental work; none of them were taken from any publications or online.

Advances in Additive Manufacturing Technologies – Gurusamy Pathinettampadian (eds)
© 2024 Taylor & Francis Group, London, ISBN 978-1-032-90013-1

91 Design and Fabrication of Solar Dryer for Agricultural Products

Murali Krishna M[1]
Dean (E & T), SRM Institute of Science & Technology,
Ramapuram, Chennai, India

Kumar S D[2]
Assistant Professor (S.G), Department of Mechanical Engineering,
SRM Institute of Science & Technology,
Ramapuram, Chennai, India

**Vignesh R[3], Manimaran M[4],
Rajkumar G[5], Yuvaraj S[6]**
UG Student, Department of Mechanical Engineering,
SRM Institute of Science & Technology,
Ramapuram, Chennai, India

ABSTRACT: Food can be preserved in an eco-friendly and hygienic manner by solar drying. This study looked at the efficiency of passive indirect solar dryers as well as the drying kinetics of agricultural items like bananas. Food insecurity and financial loss may arise from a deterioration in the quality of agricultural products. This can be ascribed to improper food conservation techniques as well as insufficient and poor storage strategies. Open sun drying is a tried-and-true method of food preservation, but because of its long drying timeframes and poor efficacy, its use is restricted. Right now, solar drying is the most efficient, cost-effective, environmentally responsible, and sustainable method of preserving agricultural products. This study aims to provide a comprehensive assessment of solar dryer technologies. A thorough comparison and classification of solar dryers has been carried out. Utilising natural materials, solar dryers that store energy have been classified according to evolution., and a variety of measures have been used to assess their performance. In order to extend the shelf life of agricultural products while reducing energy consumption and protecting the environment, the study's conclusions suggest combining solar drying techniques with natural materials thermal energy storage devices.

KEYWORDS: Drying, Solar collector, Dried bananas, and Solar energy

1. INTRODUCTION

Around the world, solar energy is a renewable resource that is used for a variety of things, such drying fruits, and vegetables, making electricity, and desalinating water. By lowering the moisture content of crops and vegetables, the sun drier is an effective technique to raise the quality of dried goods in developing nations. The process of solar

[1]dean.et.rmp@srmist.edu.in, [2]kumars@srmist.edu.in, [3]mm4617@srmist.edu.in, [4]gg0498@srmist.edu.in, [5]ss2491@srmist.edu.in, [6]vr9636@srmist.edu.in

DOI: 10.1201/9781003545774-91

drying results in consistent heating, meaning that the dried food product is free from hot spots. It doesn't require fuel, requires little initial outlay of funds, requires little upkeep, and doesn't require specialized labour for the solar drying process. Moisture evaporation is one method of food preservation. A classic method of food preservation, drying extends the shelf life of agriculturally based food items. Another significant energy-intensive operation is food drying, and over the past few decades, using solar energy for this purpose has been a hot topic of research. By eliminating moisture from food, drying can stop microbiological degradation, and solar energy can be used to generate the necessary energy rather than waste it on commercial electricity. To maintain the necessary quality, drying agricultural food products requires careful selection of drying processes, control over drying parameters, and a brief, continuous process. Energy conservation combined with the use of cleaner sources is expected to be a contentious topic for the international community. It is imperative to acknowledge that solar energy represents one of the most alluring approaches of harnessing renewable resources. These days, it's receiving extra attention because of its free availability, accessibility, and environmentally favourable qualities. Food waste needs to be tracked and minimized during the harvesting, processing, promotion, and circulation processes to guarantee food security for the world's expanding population. Food items lose their flavour, colour, and quality due to improper handling and storage procedures. One of the main causes of the increase in the price of agricultural goods is the estimated 30% to 40% post-harvest losses in fruit and vegetable output overall.

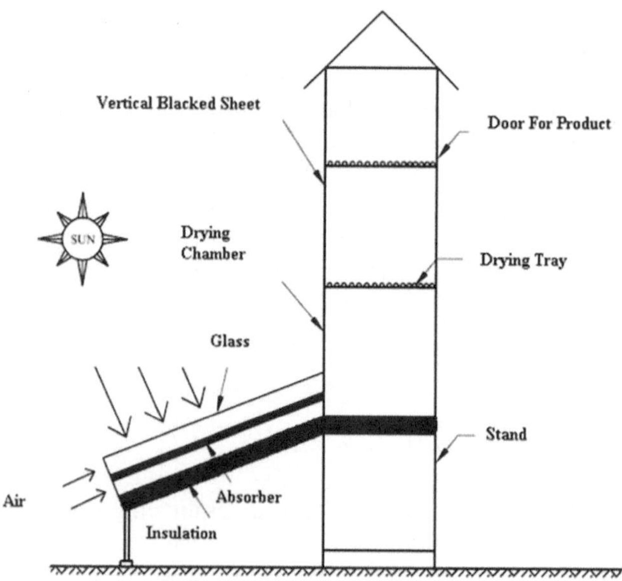

Fig. 91.1 Solar dryer process [10]

2. Solar Dryer

Drying agricultural products is important since it prolongs their shelf life and prevents germs from growing without compromising quality. Dried fruits and vegetables are easy to use, lightweight, nutrient-dense, and portable. There are several methods for drying materials, such as freezing, convection, spray drying, infrared radiation, solar, fluidized bed, and microwave. Spray drying, convection, freeze, infrared radiation, and fluidized bed methods all have high energy and capital expenses. Additionally, the systems' operation and maintenance require trained labour and knowledge with these approaches. More electrical energy is used in these drying methods, which is expensive and environmentally harmful. Comparatively speaking, solar drying is more economical and environmentally friendly than conventional drying methods. One of the earliest techniques for sun-drying clothes was open sun drying (OSD). However, there are other drawbacks, including wind, rain, dirt, animal and bug interference, and a protracted drying period. Specialised devices called sun dryers are used in solar drying to Manage the drying procedure and shield food from precipitation, dust, and insect infestations. While natural "sun dries" produce lower relative humidity, lower product moisture content, and less spoiling throughout the drying process, solar dryers produce greater temperatures. In comparison to artificial mechanical drying methods, it is also less expensive, requires less time, and occupies less space. Therefore, solar drying is a more effective substitute for all the problems associated with artificial mechanical drying and natural drying. One potential remedy for the global food and energy challenges is the solar dryer. Most of the agricultural produce can be kept through drying, and solar dryers are a more effective way to do this.

Solar dryer with natural convection (passive mode): Using a natural convection sun drier is the most straightforward and useful method of drying anything. It is covered in shade and has a drying unit, a transparent sheet, and a collector. Together, these elements create a system that can attain incredibly high drying speeds. Given its low cost, this type of solar dryer is clearly essential to the drying industry. Its easy operation and upkeep have also contributed to its growing popularity. The forced convection dryer is not as suitable as the natural convection sun dryer, which is one of the first types of dryers available. Nevertheless, the solar drying method using natural convection has a restricted capacity. Furthermore, a little float that creates air movement inside a dryer has an impact on the state of the drying materials, particularly in bad weather., by delaying the drying pace and making it highly dependent on atmospheric conditions.

Solar dryer with forced convection (active mode): In a forced convective sun drier, energy is needed to drive the fans in order to guide the warm air towards the food tray from inside the solar tube. This hastens the food's dehumidification process. Nevertheless, generating the electricity required to operate this kind of dryer is either very expensive or non-existent in many rural regions. Thus, in many underdeveloped nations, these dryer types are not generally applicable. One possible workaround for the drawbacks is to utilize a natural convection solar drier. Unlike solar dryers with forced convection, this kind of dryer doesn't require energy to operate. Low energy costs, optimal shrinkage throughout the drying process, improved drying capacity, minimizing mass losses, and high quality are some of its benefits.

3. Solar Drying Technologies

A step in the food preservation process known as drying involves water from collected goods is evaporated to stop microbial development. One of the earliest methods for food preparation and preservation that is known to exist for drying agricultural items is open solar drying, which is made possible by the sun's copious and free energy. But as technology developed, thermal methods for food processing and storage were developed for the agricultural sector. Ecological development found long-term solutions in solar energy's cost-effectiveness and efficiency, while conventional energy sources' fundamental shortcomings were mitigated. Since solar radiation is widely available everywhere, drying was no longer the most effective method of temporarily preserving fruits and vegetables. Another difficult process is food drying, which involves changes in mass and heat transfer in addition to physical and chemical interactions that may degrade the end product's quality. It's a challenging process involving physicochemical changes, simultaneous mass and heat transmission, and other elements. One of the most important steps in producing a wide range of chemical products is regulating and adjusting the amount of moisture of solid substances through ventilation. The process of drying solid materials is a widely utilised unit operation in the chemical and food industries. Almost all plants and facilities that manufacture or process solid substances in a form of powders or granules employ it. The chemical process industries can have a substantial impact on both the quality of the finished product and the effectiveness of the drying procedures. Many different products are dried using the indirect sun dehydration method, which is more effective than the direct method. This approach uses a flat-plate collector to boost ambient air temperature through indirect drying out from the sun. Through the use of an indirect solar dehydration technique, the drawbacks of direct solar dehydration are lessened. Since sun radiation does not directly strike the agricultural products, caramelization and localized heat damage are uncommon.

4. Methods of Solar Drying

Drying by direct sun: One common way to use solar energy to dry a product is direct solar drying. This process uses solar energy, but just for the purpose of drying. Because the energy used in this process is derived by a less expensive source—the sun or other renewable energy source—it is one of the most economical methods to dry. It involves the constant exchange of mass and heat. This technique involves spreading a thin layer of product across a large surface and exposing it to sunlight. It will take a while for the products to finish this process and reach the required level of drying. Concrete or a dirt surface can be used as a surface for direct sunlight drying outside. For grains, this kind of drying technique works well. Material is left out for a considerable amount of time—usually ten to thirty days. This process continues until the product dries to the necessary degree. A crucial procedure for both industrial and agricultural products is drying. Drying helps to preserve a product for an extended amount of time by reducing bacterial development. It is necessary to use alternative drying methods, such as passive, active, and combined solar drying, because the conventional method has numerous disadvantages.

Drying solar energy indirectly: The more efficient technique is convective solar drying, also known as indirect solar drying. In an indirect sunlight drier, heat from the sun is transferred to the dryer's cabinet, where drying takes place, after being first gathered by solar collectors. Each air heater is connected. With a solar cabinet-style dryer, food products are dried using the fundamentals of a reverse flat plate collector. In this instance, the air entering the chamber is heated via a solar air heater. After that, the hot air changes into humid, pleasant temperature and foliage through an aperture. Compared to traditional dryers, this kind of dryer solves a number of energy-balanced equations more effectively. Additionally, it performs better than other dryers of the traditional cabinet style. A drying process is employed to prevent direct exposure to sun light. This technique mostly lessens the drawbacks of drying directly in the sun.

Mixed mode solar drying: Another sophisticated type of solar drying technique is called "mixed mode," which combines direct and indirect solar drying in one process. Another name for this method is passive drying. Because the mixed-mode solar-powered dryer is a stationary equipment, it is frequently called a passive dryer. Sunlight that penetrates via the collecting lustring gives this kind of dryer its energy. The collector's interior is coated in black paint, and by retaining the heat from the air gathered inside the chamber, solar energy is captured.

4.1 Materials

The project has made use of the following materials. Every material is inexpensive and easily accessible in the area.

1. 18 mm size square tube having thickness of 1.6 mm
2. Mild steel sheet of 22-gauge thickness
3. Air exhaust fan with speed regulator.
4. Glass 2mm thick
5. Wire Mesh
6. Base supporting wheels

This solar dryer is fabricated using 18 mm size square tube having thickness of 1.6 mm and the same is shown below

Fig. 91.2 Solar dryer

M. S. Fabricated Housing Frame

Fig. 91.3 18 mm size square tube having thickness of 1.6 mm

The fabricated housing frame is covered with 22-gauge thickness mild sheet plate and the same is shown in below Fig. 91.4

Fig. 91.4 Mild steel sheet of 22-gause thickness

Air Exhaust Fan with Speed Regulator

Fig. 91.5 Air exhaust fan

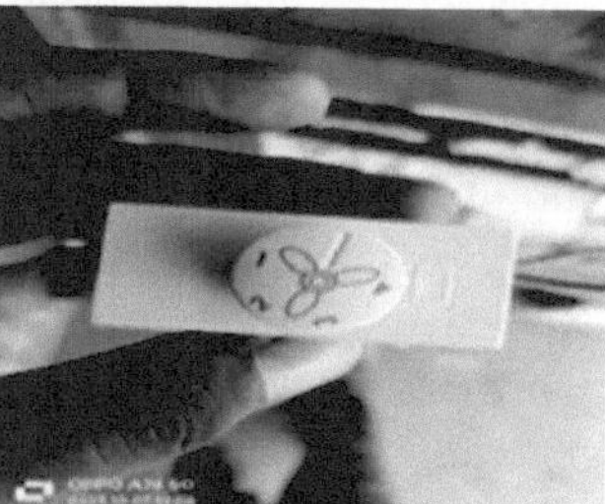

Fig. 91.6 Speed regulator

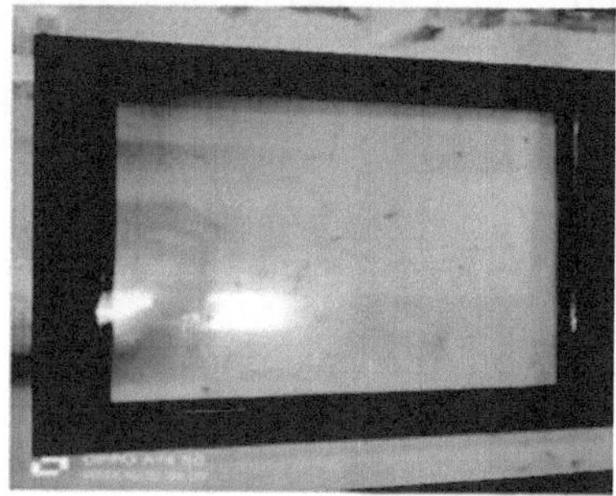

Fig. 91.7 Glass 2 mm thick

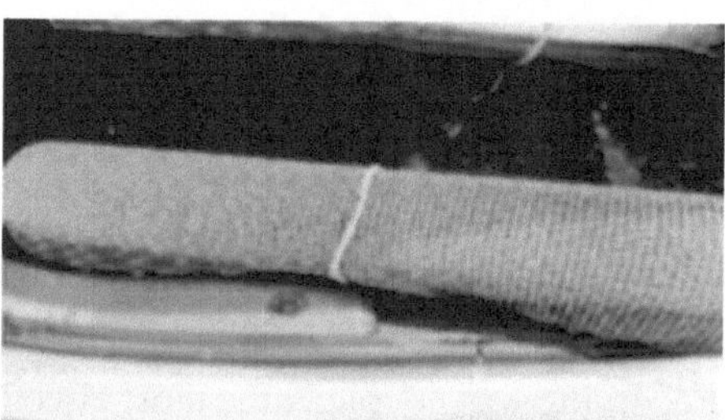

Fig. 91.8 Wire mesh

Fig. 91.9 Base supporting wheels

4.2 Methodology

Method for design: We designed the entire solar dryer in solid works 2022.

Methodology: The experimental setup's layout and perspective are depicted in pictures. The dryer is primarily made up of an 18" × 18" drying chamber, an 18" × 24" solar collector, exhaust fans and load trays. Glass and a metallic sheet are integrated to create a solar collector. To stop heat loss, there is total insulation surrounding the solar collector. Because it distributes heat from the collector to the load trays, the dryer's drying chamber is a crucial part. Within the drying chamber, load trays have been installed to hold the sample. To maintain a balanced airflow within the drying chamber, A fan for exhaust has been added to the solar dryer's upper section. Outsource power has been the sole operator of this fan. The infrared thermometer has been used to monitor a number of characteristics. An energy-related solar metre has been used to measure the amount of solar radiation.

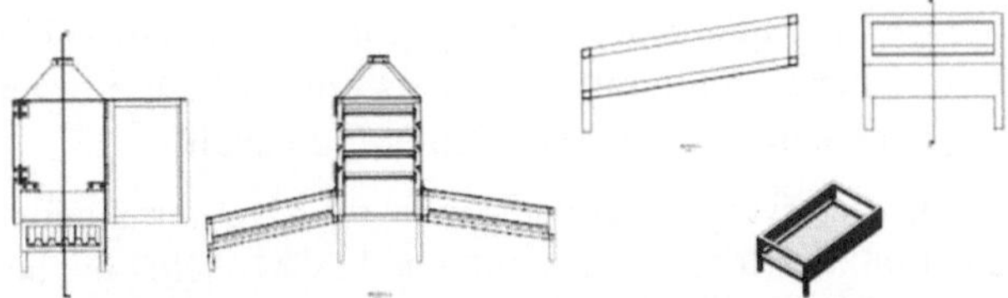

Fig. 91.10 CAD model of solar dryer unit

We built a forced convection portable solar drier of the indirect type. The dryer's CAD model is displayed, and the finished product is illustrated in to assess our dryer's performance, we contrasted it with open sun drying. The dryer operates at peak efficiency when compared to open-air drying. We performed some calculations and created graphs. Through the installation of a diverging tunnel integrated with DC fans at the SAC outlet, an old NISD was renovated with a FISD in this job. An examination of energy and exergy (2E) is required to assess the system's thermal properties.

5. Experimentation

The dry process's mathematical formulation: The process of drying results are analysed mathematically in order to provide a thorough analysis of the drying aspects. The findings demonstrate variations in gooseberry weight according to drying rate, dryer efficiency, and moisture content (w.b.). They are also acquired in terms of the rate at which moisture evaporates. These numbers fluctuate during the drying process, and those changes are used to calculate how well the solar dryer performs.

The solar dryer's performance: Drying rate (DR): The amount of moisture lost in a unit of time is called the DR. It is the relationship between the time interval (Δt) of the measurement and the variation in the total weight of strawberries (ΔS) over two successive measurements. It has been expressed using Equation (1).

$$DR = \frac{\Delta S}{\Delta t}$$

ΔS = Initial Weight = S1 = 250 grams

Final Weight = S2 = 180 Grams

(S1 – S2) = 250 – 180 = 70 grams

$\boxed{\Delta S = 70 \text{ Grams}}$

Time taken Δt = 4 hours

$$DR = \Delta S/\Delta t$$
$$= 70 / 4$$
$$\boxed{= 17.5 \text{ grams/hr}}$$

Dryer efficiency (ηd): The amount of solar radiation hitting the window frames to the energy required to drain the gooseberry is how it is expressed. Equation is utilized to evaluate it

$$\eta_d = \left(\frac{W_w \times L}{A_c \times I \times t} \right) \times 100$$

where Ac is the solar dryer's collector area, I is the direct solar energy that impinges on the collector's surface, t is the drying time, Ww is the amount of weight of the water that was extracted from the gooseberry, while L represents the water's latent heat of vaporisation.

Wet base moisture content (Mw): A product's moisture content value is expressed as a percentage. The weight of all the water in a given sample divided by the sample's total weight yields the product's immediate moisture content. It contains an equation.

$$M_t = \frac{W_t - W_d}{W_t}$$

W is the the gooseberry's total weight specimen (in grammes), Wt is the actual weight of the goods at the precise moment t, and Wd is the weight for the sample's dry material (in grammes). The formula is applicable in the computation of the moisture ratio (MR). It is the initial moisture content of the product divided by the immediate water percentage at time 't'.

Moisture content: The MC of the product has been calculated using mass variation data utilizing

$$MC = (m_{(i)} [\![-m]\!]_{(f)})/m_f \times 100$$
$$= (250 - 180)/180 \times 100$$
$$= 70/180 \times 100$$
$$= 38.8889$$

where, m is the mass, and i and f are initial and final, respectively.

6. Results and Discussion

Various food samples were examined and created in this study using a sun dryer. The food material that has been sun-dried can have its temperature distribution, moisture distribution, drying rate, and drying efficiency measured using this model.

Moisture content:

$$= \{(W1 - W2)/W1\} \times 100 \quad (1)$$

Were,

W1 = first weight of the product, Kg

W2 = after drying, the sample's final weight, Kg

Drying efficiency:

Its definition is the proportion of heat supplied to the dryer divided by the amount of heat the dryer uses. It has a percentage expression and may be found using the following formula.

$$\eta = [Q/(A \times I \times t)] \times 100 \quad (2)$$

Were,

Q = Total amount of heat used in drying, J

$= M\,C\,p\,(T\,d - T\,a) + M\,w\,\lambda$

M = mass of the product, kg

Cp = Specific heat of wet product, kJ/kg K

Td = temperature of the drying air, K

Ta = temperature of the inlet/ambient air, K

Mw = Mass of water to be removed during drying, kg

λ = Latent heat of vaporization of water, kJ/kg

η = efficiency of the drying system

I = solar insolation (W/m^2)

t = time for which dryer exposed to solar radiations, sec

Drying rate: The drying rate is defined as the material's weight loss at a given time interval divided by the time difference. The following formula was used to calculate the sample's drying rate.

$$Dr = [\Delta M/(W2 \times \Delta t)]$$

Were,

Dr = Drying rate, (g of water evaporated/g of dry matter/h)

ΔM = Weight loss at any time, g

W2 = Weight of sample after drying g

Δt = Time difference, hour

7. Conclusion

Compared to traditional drying techniques, solar dryers are a more economical and environmentally friendly option. Their capacity to preserve food, medicinal plants, and other goods by using the sun's plentiful energy offers a viable way to improve food security, cut down on waste, and slow down global warming. We can ensure a more robust and sustainable future for our world as more stakeholders become aware of the possibilities of solar dryers and make investments in their development and distribution. Individuals, groups, and businesses may help create a healthier and more environmentally friendly world for future generations by adopting solar dryers. A thorough An review of recent studies on solar dryer systems with storage and natural energy components has been given. Numerous elements of techno-economic-socio-environmental policy have been taken into consideration while examining research needs, and conventional solar drying processes and their enhanced variants have been reviewed. The need to produce food in greater numbers and with improved quality is driving the creation of novel techniques for dehydration. This goal can be accomplished by combining solar dryers with naturally occurring components that store energy to produce high-quality, reasonably priced dehydrated food. Pre-treating agricultural products is crucial before drying to minimize the final degradation of food quality.

Among the several dryer categories, the dryer arrangement that combines forced convection and PCM yields the best results with the highest rate of dehydration and the highest possible quality. Using a dryer with forced heat convection increases the rate of dehydration permanently. the performance analysis of several solar dryer types, including "direct, indirect, hybrid dryers, and mixed mode," in addition to the materials for natural energy storage investigated. In developing countries, the application of various sun dehydration technologies can significantly lower postharvest losses for agricultural goods. Furthermore, the procedural principles for the design and development of solar-assisted drying systems for agricultural goods are summarized in this work. According to current research, the use of greenhouse dryers in conjunction with hybrid solar collectors is becoming more widely acknowledged.

Therefore, new materials for desiccant and heat storage are needed for hybrid sun drying systems. However, because NES materials have a low density, integrating them with sun drying systems may result in an increase in the size of the solar dryer. This must be addressed, and further study is required in this area. The bulk of the studies looked at rocks, sand, concrete, and water for food drying applications. Rock, stones, and pebbles are some of the best materials, although at high flow rates, low heat conductivity and high-pressure drop are significant problems. The remarkable property of sand is its rapid absorption and reflection of heat. As a result, the temperature increased swiftly in bright conditions.

Reference

1. Mukul Sengar, Reeta Rani Singhania, Deepak Singh, Dhananjay Singh, Pradeep Kumar Mishra, Manish Kumar, Balendu Shekher Giri, "Drying kinetics, thermal and morphological analysis of starchy food material: Experimental

investigation through an induced type solar dryer", vol. 31, pg. no. 13–14,2023. https://www.sciencedirect.com/science/article/pii/S2352186423002171

2. Mulatu C. Gilago, Vishnuvardhan Reddy Mugi, Chandramohan VP "Energy Nexus - Performance assessment of passive indirect solar dryer comparing without and with heat storage unit by investigating the drying kinetics of carrot", vol. 9, pg. no. 34–55, 2023. https://www.sciencedirect.com/science/article/pii/S2772427123000086
3. S. Nabnean, P. Nimnuan "Case Studies in Thermal Engineering - Experimental performance of direct forced convection household solar dryer for drying banana", vol. 22, pg. no. 04–12, 2020. https://www.sciencedirect.com/science/article/pii/S2214157X20305293
4. Pimpan Pruengam, Siwalak Pathaveerat, Prasertsak Pukdeewong "Case Studies in Thermal Engineering - Fabrication and testing of double-sided solar collector dryer for drying banana", vol. 27, pg. no. 34–55, 2021. https://www.sciencedirect.com/science/article/pii/S2214157X21004986
5. Eloiny Guimarães Barbosa, Marcos Eduardo Viana de Araujo, Yuanhui Zhang, "Solar Energy - Exegetic, antieconomic and exergoeconomic (3E) assessment of a stationary parabolic trough solar collector with thermal storage", vol. 255, Pages 487–496, 2023. https://www.sciencedirect.com/science/article/abs/pii/S0038092X23001457
6. H. Krabch, R. Tadili, A. Idrissi "Results in Engineering - Design, realization, and comparison of three passive solar dryers. Orange drying application for the Rabat site (Morocco)", vol. 15, pg. no. 34–55, 2022. https://www.sciencedirect.com/science/article/pii/S259012302200202X
7. Bade Venkata Suresh, Yegireddi Shireesha, Teegala Srinivasa Kishore, Gaurav Dwivedi, Ali Torabi Haghighi, Epari Ritesh Patro "Solar Energy Materials and Solar Cells - Natural energy materials and storage systems for solar dryers: State of the art", vol. 255, pg. no. 45–64, 2023. https://www.sciencedirect.com/science/article/pii/S0927024823000971
8. Mukul Sengar, Reeta Rani Singhania, Deepak Singh, Pradeep Kumar Mishra, Dhananjay Singh, Manish Kumar d, Balendu Shekher Giri "Environmental Technology & Innovation - Drying kinetics, thermal and morphological analysis of starchy food material: Experimental investigation through an induced type of solar dryer", vol. 31, 2023. https://www.sciencedirect.com/science/article/pii/S2352186423002171
9. Senay Teshome Sileshi, Abdulkadir Aman Hassen, Kamil Dino Adem "Heliyon - Simulation of mixed-mode solar dryer with vertical air distribution channel" Volume 8, Issue 11, 2022 https://www.sciencedirect.com/science/article/pii/S2405844022031863

Note: Every figure (except fig. 91.1) and tables was created using the experimental work.

Advances in Additive Manufacturing Technologies – Gurusamy Pathinettampadian (eds)
© 2024 Taylor & Francis Group, London, ISBN 978-1-032-90013-1

92 Design and Analysis of Aerospace Wing Stiffener Structures by Using ER4043 Alloy in Wire Arc Additive Manufacturing (WAAM) Method

Murali Krishna M[1]
Dean (E & T), SRM Institute of Science & Technology,
Ramapuram, Chennai, India

Manikandan N[2]
Assistant Professor, Department of Mechanical Engineering,
SRM Institute of Science & Technology,
Ramapuram, Chennai, India

Dinesh Kumar R[3],
Lokeshwaran R[4], Mukesh D[5], Nivash B[6]
UG Student, Department of Mechanical Engineering,
SRM Institute of Science & Technology,
Ramapuram, Chennai, India

ABSTRACT: This study provides a comprehensive analysis of an aero plane's wing stiffener using Finite Element Analysis (FEA) and Computational Fluid Dynamics (CFD). The objective is to examine the stiffener's structural and flow properties when it is constructed from different materials. Three distinct alloys of aluminium and silicon are used in the experiments: 90% aluminium–10% silicon, 95% aluminium–5% silicon, and 98% aluminium–2% silicon. A three-dimensional model of the wing stiffener and the creation of meshes for computational fluid dynamics and finite element analysis serve as the foundation for the inquiry. FEA applies varying loads and boundary conditions to every material composition in order to assess stress, strain, and deformation in addition to other structural aspects. Following are the establishment of a set of objectives, such losing weight or increasing strength, the analysis.

KEYWORDS: ER4043, ANSYS, CFD, WAAM

1. INTRODUCTION

The wings of an aircraft are the primary contributor to lift, as they provide the essential aerodynamic force that is required to keep the aircraft in the air. It is of the utmost importance to both an aircraft's safety and its performance that these wings always maintain both their structural integrity and their stability. The aero plane wing stiffener is one of the most important components in accomplishing this goal. Wing stiffeners are structural components that are carefully integrated into the design of the wing to enhance its strength, rigidity, and overall performance. These components can also be referred to as wing braces. These components are often made up of thin, elongated structures

[1]dean.et.rmp@srmist.edu.in, [2]manikann2@srmist.edu.in, [3]dr8407@srmist.edu.in, [4]lr4281@srmist.edu.in, [5]md9110@srmist.edu.in, [6]nb0403@srmist.edu.in

DOI: 10.1201/9781003545774-92

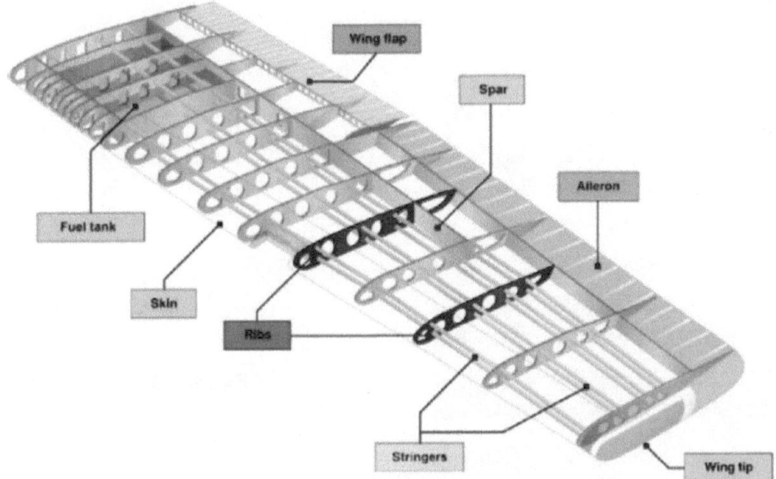

Fig. 92.1 Structure of straight beam wing [11]

that run down the span of the wing and serve to reinforce it against the various forces and loads that are encountered while the aircraft is in flight. It is impossible to emphasis how important wing stiffeners are to the overall design of an aero plane. This research study makes an effort to find a solution to this problem by carrying out an exhaustive investigation of an aircraft wing stiffener. The evaluation of the structural performance and fluid flow properties of the stiffener when it is made from a variety of material compositions is the primary objective of this study.

The three different combinations that are being investigated are 98% aluminium and 2% silicon, 95% aluminium and 5% silicon, and 90% aluminium and 10% silicon respectively. The ever-increasing demands for advances in aircraft performance are being pushed by a variety of causes, including environmental concerns, fuel efficiency, and passenger safety. These factors highlight the significance of this research by highlighting its importance. Because the choice of material composition in crucial components like wing stiffeners has such a significant impact on these aspects, it is absolutely necessary to investigate and improve the performance of materials used in aerospace applications

2. Materials

Aluminium: Aluminium alloys are alloys, in which Aluminium (Al) is used more required metal. More suitably composite metals are like copper, magnesium, manganese, silicon & zinc. They were broadly classified into two types. They are casting alloys & wrought alloys. More often they were sub classified into two types, they are heat-treatable and non-heat-treatable. For wrought products more than 85% of Aluminium composition is used. For example, rolled plate, foils and extrusions.

Due to low melting point in cast aluminium alloys, they were cost effective. They generally have lower tensile strengths than wrought alloys. With reference to a dissimilar metal corrosion this process can occur as intergranular corrosion. One of the most required cast aluminium alloy systems is Al–Si, where the high levels of silicon (4.0–13%).

Silicon: Silicon is a chemical element with the symbol "Si" on the periodic table, and it has an atomic number of 14. It is a metalloid, which means it has properties intermediate between those of metals and nonmetals.

Physical Properties: Silicon is a crystalline solid at room temperature with a shiny, metallic luster. It is brittle and not a good conductor of electricity in its pure form, but it can become a semiconductor when it is doped with certain elements. Chemical Properties:

ER4043, also known as Aluminium Silicon, has a nominal composition (wt.-%) of·5 Si and a balance of Al. Wetting action is enhanced and crack sensitivity is decreased by silicon. It "wets and flows better" and produces GMAW welds that look brighter, which is why most welders prefer it. Cast alloys and base metals with the heat treatability 6XXX are two examples of the alloys that can be welded using this stuff. The filler metal from the 5XXX series is more fluid and has a lower melting point.

2.1 Properties of ER4043 Material

Density	2689.5 kg/m^3
Elastic Modulus	75000 MPa
Poisson's ratio	0.3
Thermal Conductivity	60.5 W/m·°C
Yield Strength	190 MPa
Ultimate strength	228 MPa

3. Methodology

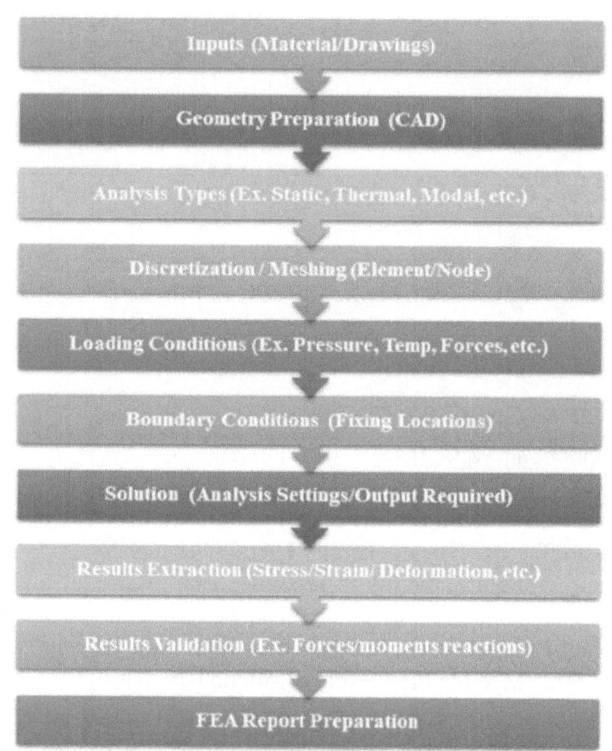

Fig. 92.2 Flow diagram for the current work [12]

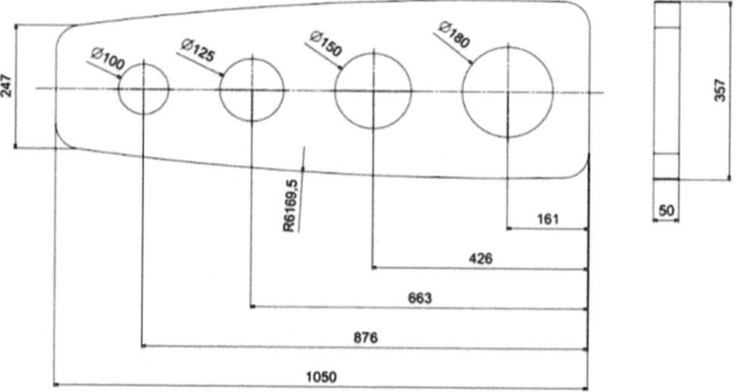

Fig. 92.3 2D model of wing stiffener

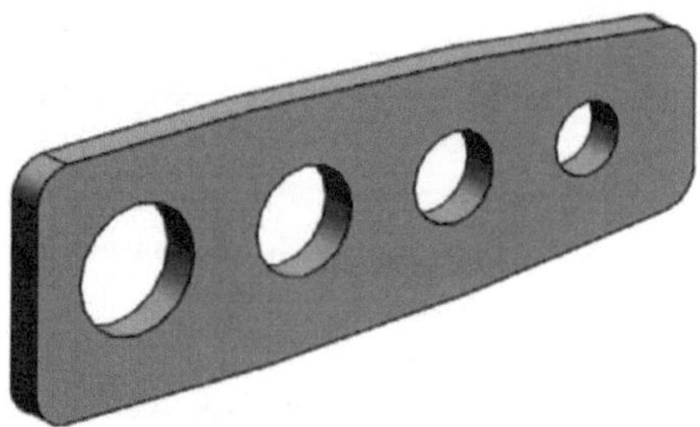

Fig. 92.4 3D model of wing stiffener

4. Design Validation

Figure 92.3 and 92.4 represents the 2D and 3D model of wing stiffener designed using CATIA V5 which has dimension of 1050 × 357 × 50 with weight of 37.08 kg. The wing stiffener dimensions obtained from website for modelling.

5. FEA Results (Structural Analysis)

Define the geometry, placing nodes or vertices, establishing connectivity between nodes to form elements, ensuring mesh quality, applying boundary conditions, and optionally adapting the mesh for specific simulation.

Material : Aluminium 95% & Silicon 5%

Nodes : 1002

No of elements : 113

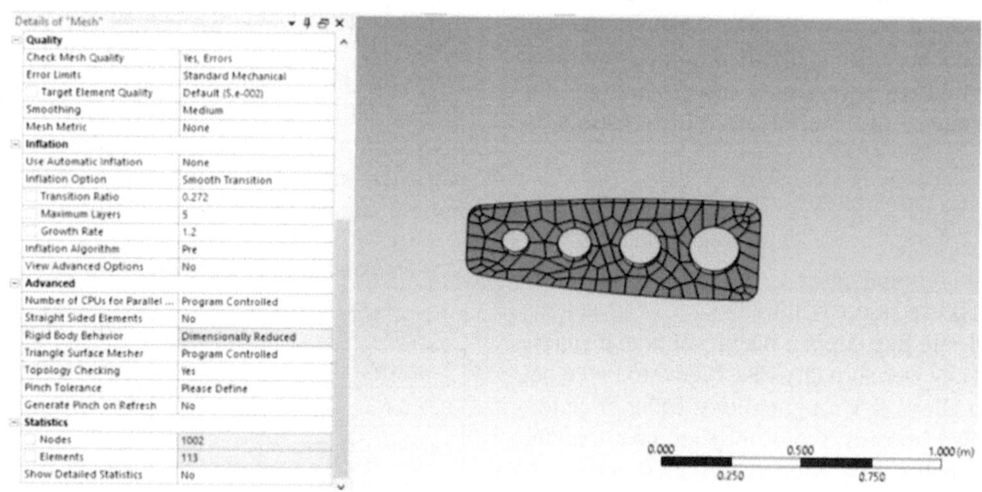

Fig. 92.5 Meshing geometry

6. Boundary Condition

Figure 92.6 and 92.7 represents boundary condition for wing stiffener. Figure 92.6 represents fixed support in the stiffener provided. Figure 92.7 represents load and direction of load. Load of 875 N applied in -Y direction as external forces, constraints, and other conditions to a model within the ANSYS software for simulation

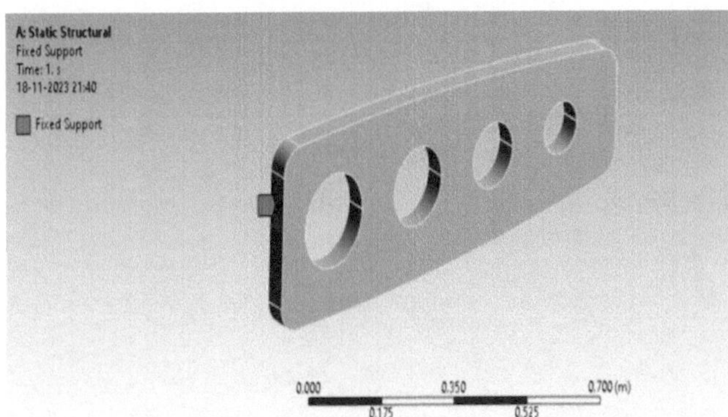

Fig. 92.6 Fixed support in the stiffener

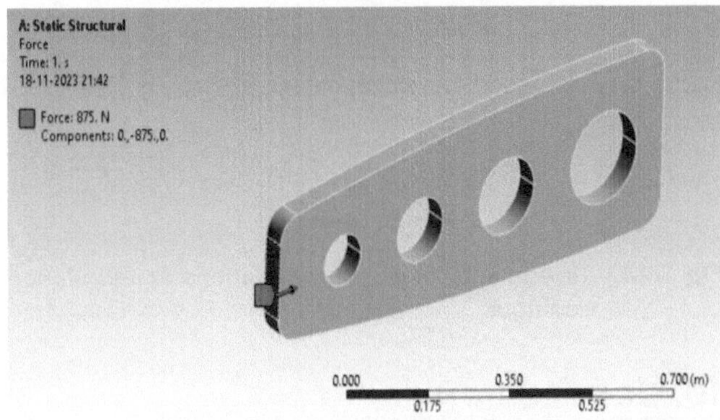

Fig. 92.7 load and direction of load

7. Results and Discussion

7.1 Normal Elastic Strain

Figure 92.8 represents normal elastic strain where the material heals back to normal after the load is removed.

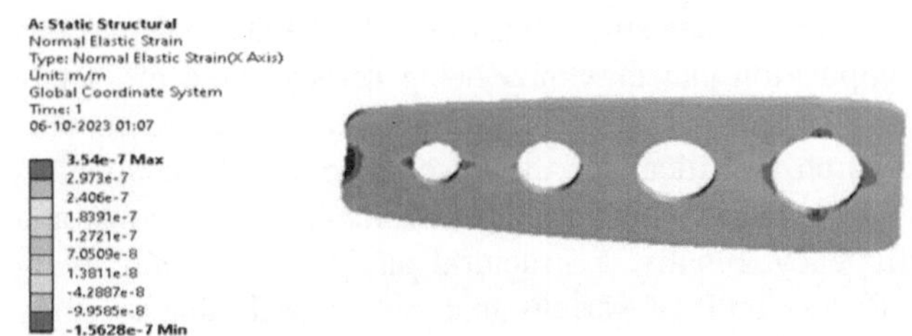

Fig. 92.8 Normal elastic strain

Figure 92.9 represents shear elastic strain, the force applied for deformation causes change in the wing stiffener shape. Figure 92.10 and 92.11 shows the comparison of aluminium with silicon materials in different composition percentage

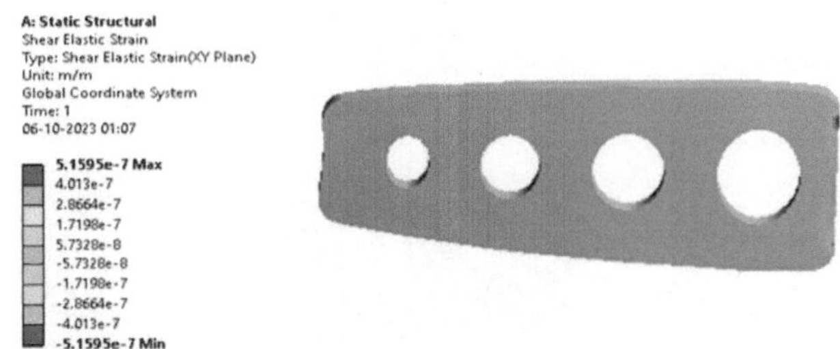

Fig. 92.9 Shear elastic strain, the force applied for deformation

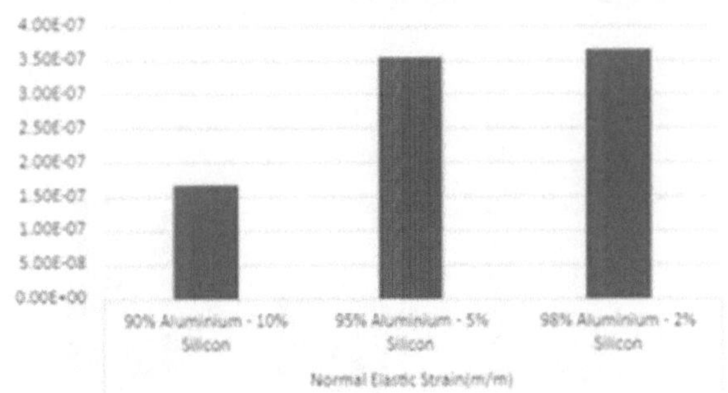

Fig. 92.10 Comparison of aluminium with silicon materials

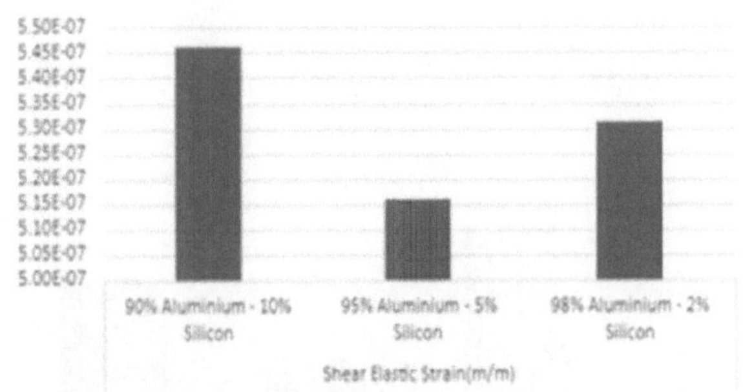

Fig. 92.11 Comparison of aluminium with silicon materials

8. Shear Elastic Strain

Figure 92.12 represents total deformation of wing stiffener,

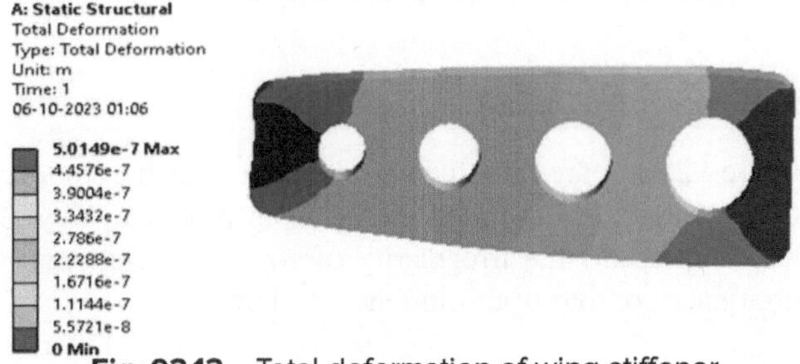

Fig. 92.12 Total deformation of wing stiffener

which shows overall displacement of wing structure load is removed. Figure 92.13 represents directional deformation of wing stiffener, here the deformation done in -Y direction. Deformed value in total and directional mentioned in Fig. 92.14 and 92.15.

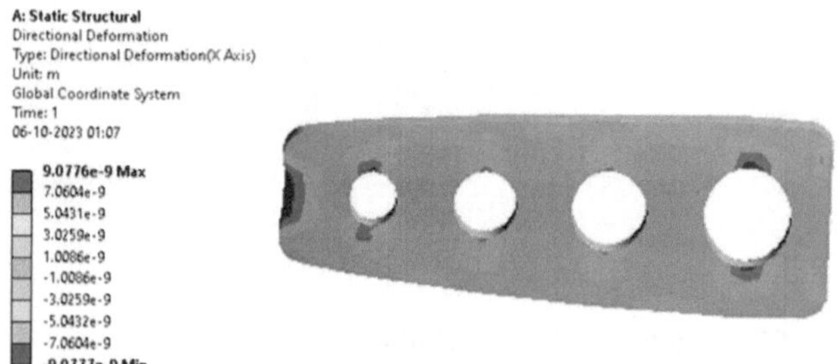

Fig. 92.13 Directional deformation of wing stiffener

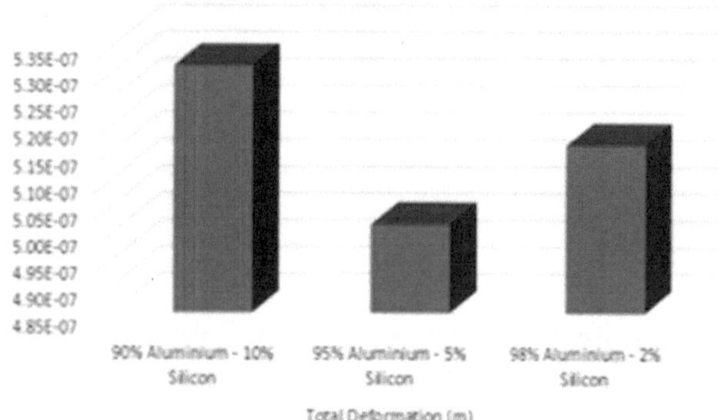

Fig. 92.14 Deformed value in total and directional

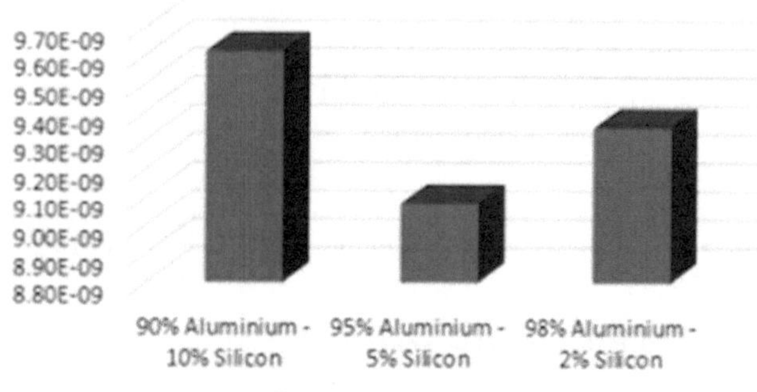

Fig. 92.15 Deformed value in total and directional

9. CFD Results

Figure 92.16 represents the variations in velocity contour of the wing stiffener based on the input provided. Figure 92.17 represents the irregularity of air flow variations in turbulence contour of the stiffener as shown.

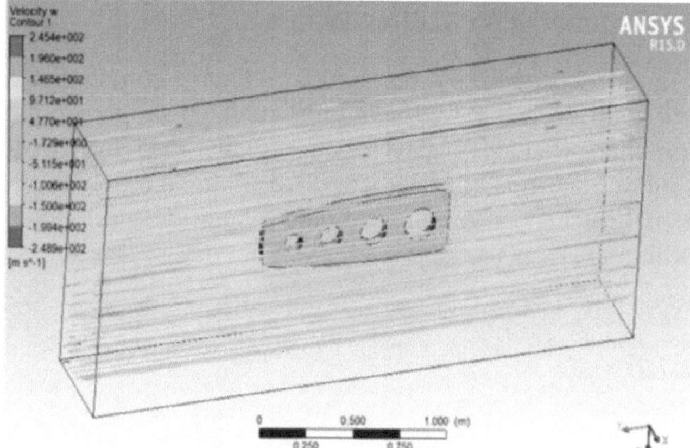

Fig. 92.16 Variations in velocity contour of the wing stiffener

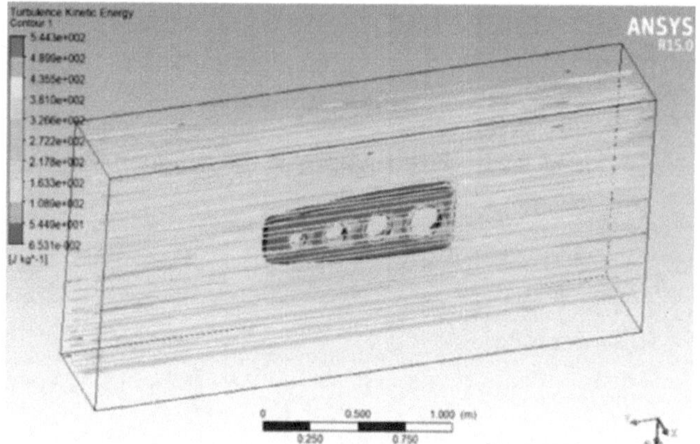

Fig. 92.17 Irregularity of air flow variations in turbulence contour

10. Conclusion

Comparing 90% aluminum-10% silicon, 95% aluminum-5% silicon, and 98% aluminum-2% silicon aviation wing stiffeners, as well as doing CFD flow characterization to determine the optimal composition, necessitates a detailed study of the wing's mechanical and aerodynamic qualities.

The structural analysis of aircraft wing stiffeners indicates that the 95% aluminum and 5% silicon alloy strike a compromise between strength and weight. CFD studies on this composition can maximize wing aerodynamic efficiency and stress distribution for safe and successful flight. This research is critical for the development and manufacturing of aviation wing structures, which improves safety and efficiency. Finally, a structural analysis of aircraft wing stiffeners built of various materials reveals that an alloy composed of 95% aluminum and 5% silicon is a feasible option because it efficiently balances strength and weight. Furthermore, CFD research

REFERENCE

1. Spyridon Kilimtzidis, Athanasios Kotzakolios, Vassilis Kostopoulos, "Efficient structural optimisation of composite materials aircraft wings". vol.303, 2023. https://www.sciencedirect.com/science/article/abs/pii/S0263822322010005
2. D. Rajpal, F.M.A. Mitrotta, C.A. Socci, J. Sodja, C. Kassapoglou, R. De Breuker,"Design and testing of aero elastically tailored composite wing under fatigue and gust loading including effect of fatigue on aeroelastic performance". vol.275, 2021. https://www.sciencedirect.com/science/article/pii/S0263822321008357
3. A. Voß, T. Klimmek, "Parametric aeroelastic modelling, manoeuvre loads analysis using CFD methods and structural design of a fighter aircraft", vol.136, 2023. https://www.sciencedirect.com/science/article/pii/S1270963823001281
4. Kautuk Sinha, Thomas Klimmek, Matthias Schulze & Vega Handojo, "Loads analysis and structural optimization of a high aspect ratio, composite wing aircraft", vol.12, pg. no. 233–243, 2021. https://link.springer.com/article/10.1007/s13272-021-00494-x
5. M.M. Tawfik, M.M. Nemat-Alla, M.M. Dewidar, "Enhancing the properties of aluminium alloys fabricated using wire + arc additive manufacturing technique - A review". vol.13, pg.no 754-768, 2021. https://www.sciencedirect.com/science/article/pii/S2238785421004208
6. Davoud Jafari, Tom H.J. Vaneker, Ian Gibson,"Wire and arc additive manufacturing: Opportunities and challenges to control the quality and accuracy of manufactured parts". vol.202, 2021. https://www.sciencedirect.com/science/article/pii/S0264127521000241
7. Gerald L. Knapp, Maxim Gussev, Amit Shyam , Thomas Feldhausen, Alex Plotkowski," Microstructure, deformation and fracture mechanisms in Al-4043 alloy produced by laser hot-wire additive manufacturing". vol.59, 2022. https://www.sciencedirect.com/science/article/pii/S2214860422005395
8. Xueping Ding , Daoyuan Li , Qi Zhang, Honglin Ma , Jie Yang , ShuqianFan."Effect of ambient pressure on bead shape, microstructure and corrosion behavior of 4043 Al alloy fabricated by laser coaxial wire feeding additive manufacturing in vacuum environment". vol.153, 2022. https://www.sciencedirect.com/science/article/abs/pii/S0030399222003991
9. Bo-Chin Huang, Fei-Yi Hung,"Effect of tensile loading–unloading cyclic plastic deformation on 4043 aluminium alloy manufactured through CCDR". vol.34, 2023. https://www.sciencedirect.com/science/article/abs/pii/S2352492822018207
10. Changshu He , Jingxun Wei , Ying Li, Zhiqiang Zhang , Ni Tian, Gaowu Qin, Liang Zuo," Improvement of microstructure and fatigue performance of wire-arc additive manufactured 4043 aluminum alloy assisted by interlayer friction stir processing", vol.133, pg.no. 183–194, 2023. https://www.sciencedirect.com/science/article/abs/pii/S1005030222005631
11. Siddharth Raj Gupta, Sourav Sinha, Ritesh Maurya, Pankaj Patil, Ronit Bansal, Shivam Verma, "Designing of an aircraft based on preliminary mission requirement – 2", Technical Report, AE462A Aircraft Design II, DOI:10.13140/RG.2.2.28985.26727
12. Internet Source: "What is Finite Element Analysis? Why to do FEA", Grasp Engineering – Learn Engineering & Technology, June 2020. https://www.graspengineering.com/hello-world-2/]

Note: Every figure (except fig. 92.1 and 92.2) was created using the experimental work.

Advances in Additive Manufacturing Technologies – Gurusamy Pathinettampadian (eds)
© 2024 Taylor & Francis Group, London, ISBN 978-1-032-90013-1

93 Performance Analytics of CNTFET Devices: Electrical Characteristics of CNTFET Using Yttrium Oxide as Insulator in Comparison with Traditional Silicon MOSFET Devices

Radhika Baskar*
Professor, Institute of ECE, Saveetha School of Engineering,
Saveetha Institute of Medical and Technical Sciences,
Saveetha University, Chennai, India

ABSTRACT: This work establishes about the ballistic electrical properties by using suitable insulators which can overcome the drawbacks of traditional MOSFET devices with the conventional CNTFET devices. The field-effect semiconductors based on carbon nanotubes could be fabricated in a simple, swift, and reliable manner. An organized investigation of the variables affecting transistor performance has been conducted, and the results offer recommendations for future device enhancement. The different dielectric materials were employed to examine the features of CNTFET devices. Due to the distinctive atomic structure, carbon nanotubes are identical to graphite and diamond in aspects of their physicochemical characteristics. Cutoff voltage of CNTFET devices grows with decreasing channel length in the deep nano scale zone, whereas the threshold voltage of a MOSFET device declines drastically. Traditional MOSFET devices fail to match expectations when channel lengths are depleted.

KEYWORDS: CNTFET, MOSFET, Device scaling, Threshold voltage, Temperature

1. Introduction

The metal oxide semiconductors were mostly used significant silicon chips in VLSI domains. The goal of semiconductor technology is to make transistors smaller and better integrated onto single chips. As per Moore's law the size of the devices will shrink continuously as a breakthrough was introduced in solid state silicon based a device which has been reduced to the nanometer level. Device scaling should indeed continue to thrive in order for semiconductor device development to advance steadily. Since oxide thickness inversely proportional to the leakage current,[1,2] results in high power dissipation and poor reliability of the device. When the length of the MOSFET channel reduced less than 10nm will put silicon based technology to an end. Enhancing the ID value is an important consideration while performing the simulations for CNTFET for various dielectric constants with different thickness of insulators. The ability of electrical qualities to reduce leakage current when drain current rises has drawn a lot of interest [3,4] To examine the drain current for two different dielectric insulators while varying the gate insulator thicknesses in order to learn more about the I-V properties of CNTFETs. By substituting a single carbon nanotube for the traditional MOSFET, it was possible to overcome all limitations, including the exponential growth in leakage currents in scaled devices, by better understanding device mechanics and enhancing device performance.[5] Over the last five

*Corresponding author: radhikabaskar@saveetha.com

DOI: 10.1201/9781003545774-93

years, there have been extensive research publications, including 1000 articles in Google Scholar and 28 publications in IEEE explore. Carbonnanotubes (CNTs) are seamless cylinders made from graphite sheets that have a Nano scale diameter and a micron length. CNTs are highly reliable and adaptable molecular-based heat-transfer devices.

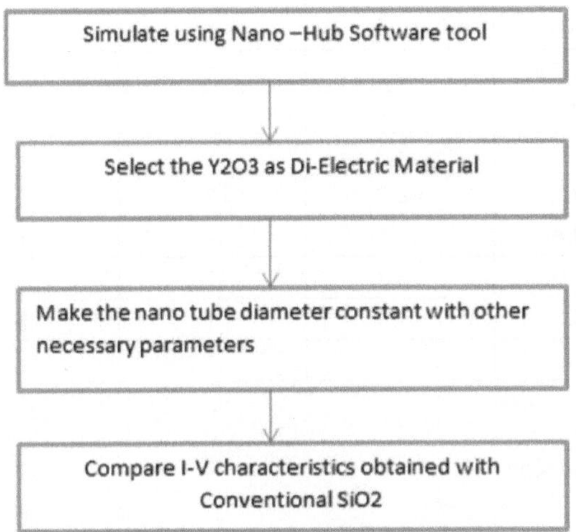

Fig. 93.1 Implementation flow diagram to analyze the ballistic characteristics of CNTFET

The structure of carbon nanotubes can be investigated using SEM, TEM, electron diffraction, XRD, Photo-electron-spectroscopy, infrared, and FTIR analysis

2. Materials and Methods

The anticipated work is done at SIMATS Saveetha School of Engineering. Two groups are engaged in this scientific endeavour. Group 1 is made up of Yttrium oxide, an insulator material used to increase drain current; in contrast, Group 2's silicon dioxide exhibits low drain current comparing Y2O3, which leads to higher leakage current comparing to Group 1.The pre-test analysis was conducted using Clinicalc.com, with the following configurations: g-power at 80%, threshold at 0.05%, and confidence interval at 95%. [6,7]. In each group, ten samples are provided. There are twenty samples in all. Group one sample was prepared using insulator value of 15 nm. The optimal insulator material for manufacturing was identified by contrasting the drain current measurement findings with the other insulator material of 3.9 nm. Users can access a wealth of educational resources via Nano-Hub that serves as a gateway for science and engineering. Using the simulation tools for nanotechnology offered by Nano Hub, tests are replaced with simulations.

By measuring the corresponding drain currents of SiO2 and Y2O3, two distinct dielectric materials, they are examined.[2] Examine the ballistic electrical properties (drain voltage vs. current) with the insulator thickness varied between 0.2 and 2.0 nm with the insulator value held constant at 3.9 nm. [8, 9, 10] Compare the acquired values with the other insulator's constant value of 15 nm. Open the Nano-Hub simulation tool and enter the necessary parameters for the dielectric constant along with other CNTFET-related settings (3.9 nm for SiO2 and 15 nm for Y2O3), and click "Simulator" to view the outcomes and follow the same procedure for group 2 with Y203

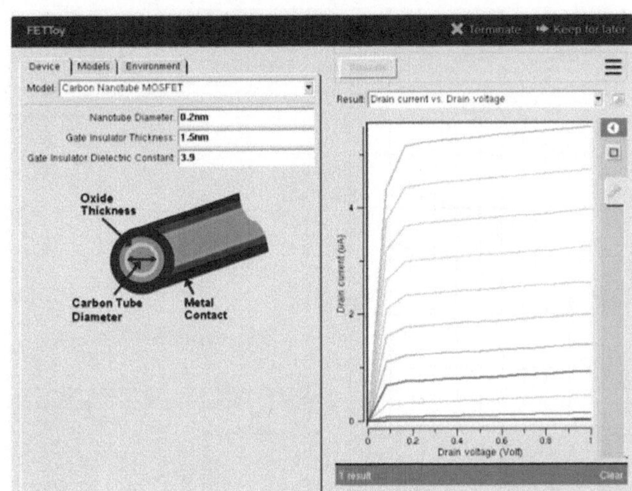

Fig. 93.2 Nanotube diameter =0.2nm with SiO2 as dielectric

Table 93.1 Drain current variation with change in Nano tube diameter (SiO2)

Glass insulator Thickness = 1.5nm			
Nano Tube Diameter (nm)	Dielectric Constant (3.9)		
	Drain_Current Vs Drain_Voltage		
	Drain current (μA)	Drain voltage (V)	Gate Voltage (V)
0.2	5.53	1	1
0.4	11.4	1	1
0.6	18	1	1
0.8	24.4	1	1
1	30	1	1
1.2	34.9	1	1
1.4	39	1	1
1.6	42.5	1	1
1.8	45.6	1	1
2	48.3	1	1

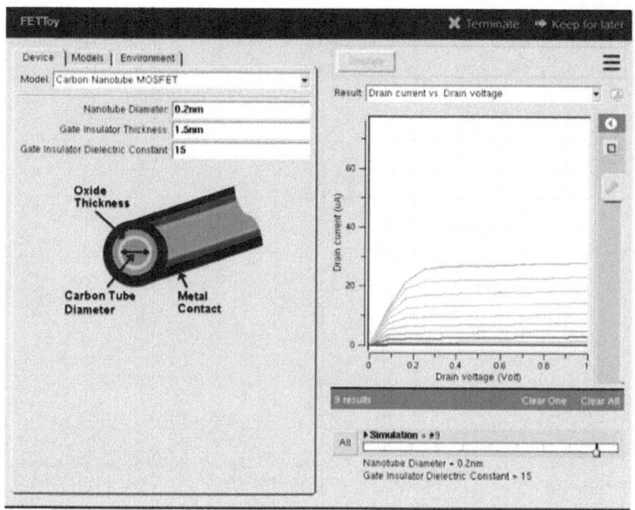

Fig. 93.3 Y2O3 as insulator with nanotube diameter = 0.2 nm

Table 93.2

Nano Tube Diameter (nm)	Glass insulator Thickness = 1.5nm		
	Insulator_Constant (15)		
	Drain_Current Vs Drain_Voltage		
	ID (μAmps)	VD (Volts)	VG (Volts)
0.2	28	1	1
0.4	45.4	1	1
0.6	55.2	1	1
0.8	61.4	1	1
1	65.8	1	1
1.2	68.9	1	1
1.4	71.4	1	1
1.6	73.3	1	1
1.8	74.9	1	1
2	76.3	1	1

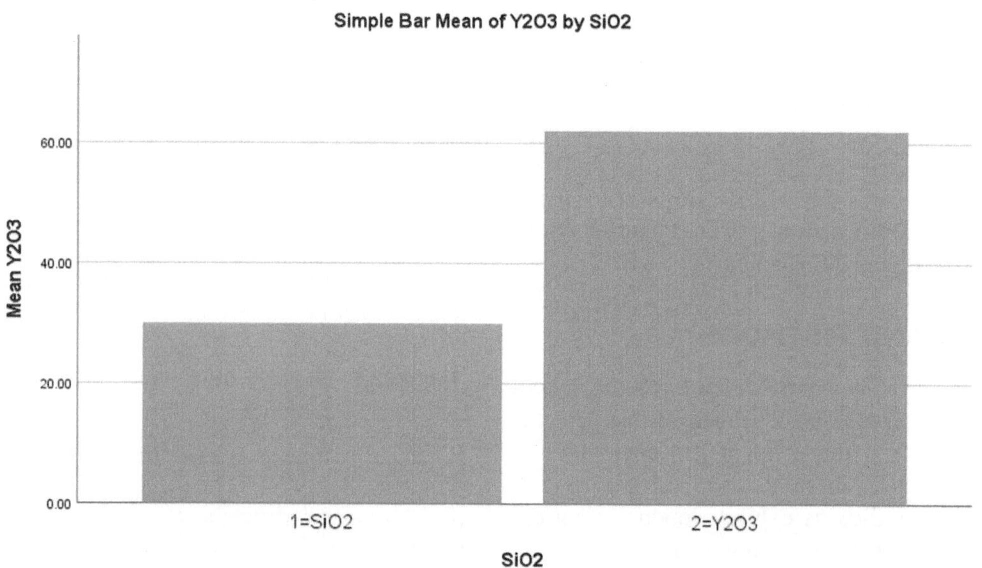

Fig. 93.4 Bar graph comparing the CNTFET's mean drain current with insulator material

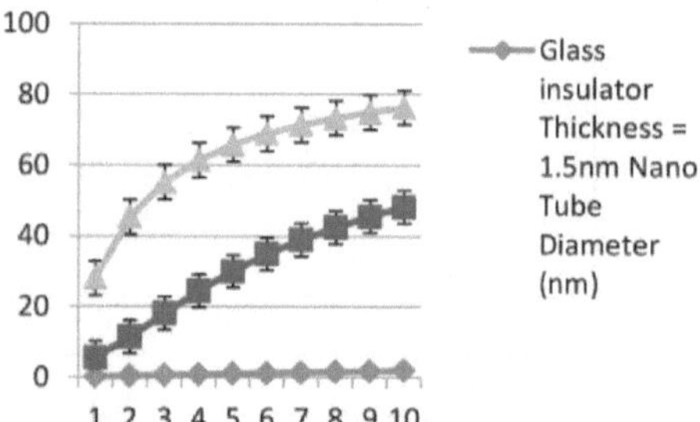

Fig. 93.5 Graph plot comparing the gate insulator thickness with the nanotube diameter

Table 93.3 Drain current variation with change in Nano tube diameter (Y2O3)

Table 93.3 T-Test: Analyzing the electrical characteristics of CNTFETs with varying nanotube diameters (0.2–2.0 nm)

Group Statistics					
		N	Mean	Std. Deviation	Std. Error Mean
Drain Current	1 = SiO2 (Silicon Oxide)	10	29.9630	14.75578	4.66619
	2 = Y2O3 (Yttrium Oxide)	10	62.0600	15.37358	4.86155

3. Simulation and Results

By keeping the nanotube diameter at 1 nm and the gate insulator thickness constant (as seen in Table 93.2), the dielectric material Y2O3 demonstrates a notable change in the drain current while retaining the other parameters unchanged. Table 93.1 depicts the outcomes for SiO2 and Yttrium oxide with constants of 3.9 nm and 15 nm where Fig. 93.1 and 93.2 displays the Silicon oxide and Y2O3 simulated outputs

4. Discussion

The CNTFET tube thickness has been varied from 0.2 nm to 2.0 nm in order to imitate the characteristics of ID and VG.The current of the CNTFET is examined in terms of its ballistic electrical properties while maintaining constant gate insulator thicknesses. CNTFET conductivity and drain current both increase with tube thickness.[11] Increasing the nanotube's thickness will result fall in leakage current, which will raise the conductivity of CNTFETs, according to an analysis of the I-V characteristics curves. Conductance and current in a CNTFET do not significantly alter as the oxide layer thickens.[12] The few parameters such as the size of drain, gate and source in CNTFET were the key variables that greatly affected the enhancements of drain current

5. Conclusion

CNTFETs exhibit stronger conductivity and better performance when the inputs to the gate voltage are given which is because the CNTFETs are comparatively better than traditional MOSFET devices. To investigate the impact on the current and voltage properties of CNTFETs, researchers are varying the thickness of the oxide and insulator materials while maintaining a constant gate voltage. Silicon dioxide measured 48.3 (μA) and yttrium oxide measured 76.3 (μA), with the results represented as drain current. The conductivity of CNTFETs is inversely correlated with oxide thickness. Therefore, in order to maximize the conductivity of CNTFET, efforts were made to ensure that the oxide thickness was as low as possible. We like the Y2O3 because CNTFET devices offer lower leakage current than conventional devices.

References

1. Dokania, Vishesh, Aminul Islam, Vivek Dixit, and Shree Prakash Tiwari. 2016. "Analytical Modeling of Wrap-Gate Carbon Nanotube FET with Parasitic Capacitances and Density of States." IEEE Transactions on Electron Devices. https://doi.org/10.1109/ted.2016.2581119.
2. Analysis of different parameters of channel material and temperature on threshold voltage of CNTFET Sanjeet Kumar Sinha n, Saurabh Chaudhury Department of Electrical Engineering, NIT Silchar, Silchar 788010, Assam, India
3. Sathish G, Radhika Baskar. "Implementation of Carbon Nanotube Field Effect Transistor and Comparison of Insulator Material With Traditional Silicon Gate Oxides to Improve the Electrical Characteristics and Device Scalability", 2023 2nd International Conference on Vision Towards Emerging Trends in Communication and Networking Technologies (ViTECoN), 2023
4. Sinha, Sanieet Kumar, and Saurabh Chaudhury. 2014. "Advantage of CNTFET Characteristics over MOSFET to Reduce Leakage Power." 2014 2nd International Conference on Devices, Circuits and Systems (ICDCS). https://doi.org/10.1109/icdcsyst.2014.6926211.
5. Sinha, Sanjeet Kumar, and Saurabh Chaudhury. 2014. "Comparative Study of Leakage Power in CNTFET over MOSFET Device." Journal of Semiconductors. https://doi.org/10.1088/16744926/35/11/114002.
6. Srimani, T., G. Hills, X. Zhao, D. Antoniadis, J.A. del Alamo, and M.M. Shulaker. 2019."Asymmetric Gating for Reducing Leakage Current in Carbon Nanotube Field-Effect Transistors." Applied Physics Letters. https://doi.org/10.1063/1.5098322.
7. Shaukat, Ayesha, Rahila Umer, and Naz Islam. 2017. "Impact of Dielectric Material and OxideThickness on the Performance of Carbon Nanotube Field Effect Transistor." 2017 IEEE 17th International Conference on Nanotechnology-(IEEE-NANO). https://doi.org/10.1109/nano.2017.8117461.
8. Salcines, Cristino, Aleksei Kruglov, and Ingmar Kallfass. 2018. "A Novel Characterization Technique to Extract High Voltage - High Current IV Characteristics of Power MOSFETs from Dynamic Measurements." 2018 IEEE 6th Workshop on Wide Bandgap Power Devices and Applications (WiPDA). https://doi.org/10.1109/wipda.2018.8569160.
9. Singh, Amandeep, Mamta Khosla, and Balwinder Raj. 2016. "Comparative Analysis of Carbon Nanotube Field Effect Transistor and Nanowire Transistor for Low Power Circuit Design." Journal of Nanoelectronics and_Opto electronics. https://doi.org/10.1166/jno.2016.1913.
10. Sinha, Sanieet Kumar, and Saurabh Chaudhury. 2014. "Advantage of CNTFET Characteristics over MOSFET to Reduce Leakage Power." 2014 2nd International Conference on Devices, Circuits and Systems (ICDCS). https://doi.org/10.1109/icdcsyst.2014.6926211.
11. Gajendran, S., Baskar, R. (2024). Stochastic Performance of CNTFET with High 'k' Dielectric Material Over Conventional Silicon Devices in Optimization of Drain Current. In: Shaw, R.N., Siano, P., Makhilef, S., Ghosh, A., Shimi, S.L. (eds) Innovations in Electrical and Electronic Engineering. ICEEE 2023. Lecture Notes in Electrical Engineering, vol 1115. Springer, Singapore. https://doi.org/10.1007/978-981-99-8661-3_47
12. Sathish G, R. Baskar and M. S, "Carbon Enabled Devices in Digital Era: A Predictive Performance Analysis of CNTFET over Si-NWFET with Conventional Double Gate MOSFET and Single Gate MOSFET," 2024 International Conference on Intelligent and Innovative Technologies in Computing, Electrical and Electronics (IITCEE), Bangalore, India, 2024, pp. 1–4, doi: 10.1109/IITCEE59897.2024.10486885

Note: Every figure was created using the experimental work; none of them were taken from any publications or online.

Advances in Additive Manufacturing Technologies – Gurusamy Pathinettampadian (eds)
© 2024 Taylor & Francis Group, London, ISBN 978-1-032-90013-1

94 The Impact of Silver Nanoparticles as Back Reflectors on the Electrical Characteristics of Amorphous Silicon Solar Cells: Insights from QCRF FDTD Models with Varied SiO2 Front Layer Thicknesses

Manikandan S[1]
Research Scholar, Institute of ECE,
Saveetha School of Engineering, Saveetha Institute of Medical and Technical Sciences, Saveetha University, Chennai, India

Radhika Baskar[2]
Professor, Research Guide, Institute of ECE,
Saveetha School of Engineering, Saveetha Institute of Medical and Technical Sciences, Saveetha University, Chennai, India

ABSTRACT: In this study, the QCRF FDTD simulator contrasts amorphous silicon (a-Si) solar cells with and without silver nanoparticles (AgNPs) back reflectors. The a-Si solar cells comprised a front layer of SiO2 with different thicknesses, a ZnO coating that was 70 nm thick to stop reflections, and an active material of 200 nm a-Si. We calculated each probable configuration's fill factor, open-circuit voltage (Voc), and short-circuit current density (Jsc). The results show that using AgNPs as back reflectors improves the efficiency of a-Si solar cells. The research achieved a 23.1884% efficiency using nanoparticles and a 100 nm-thick SiO2 front layer. These discoveries might help improve the efficiency of a-Si photovoltaic cells.

KEYWORDS: QCRF FDTD simulator, Amorphous silicon (a-Si), Silver nanoparticles (AgNPs), ZnO coating, Active material, Electrical parameter, Photovoltaic cells

1. INTRODUCTION

The growing need for environmentally friendly energy sources has led to the development of solar power cells. Maximise efficiency by decreasing surface light reflection from solar cells. This paves the way for the absorption and subsequent conversion of an even greater quantity of light into electrical energy. An anti-reflection coating, which reduces the reflectivity of the solar cell's surface, is often used to accomplish this. Zinc oxide (ZnO) is a popular material for anti-reflection layers because of its inexpensive cost, excellent transparency, and outstanding optical characteristics. To increase the solar cell's electrical efficiency, researchers have been interested in covering ZnO with silver nanoparticles (Ag NPs) [2]. The plasmonic features of Ag NPs cause them to absorb a tremendous amount of visible light. This study will examine the electrical performance of solar cells with and without antireflection coatings made

[1]smaniphdece@gmail.com, [2]radhikabaskar@saveetha.com

DOI: 10.1201/9781003545774-94

of silver nanoparticles (Ag NPs). Some electrical features of solar cells that can be used to judge how well the layers work are the fill factor (FF), the total power conversion efficiency (PCE), the open-circuit voltage (Voc), and the short-circuit current (Jsc) [3]. The potential benefits of incorporating Ag NPs into ZnO antireflection coatings for solar cell applications might be better understood by comparing the two methods.ZnO can work as an antireflection layer for photovoltaic cells as thin as 50 nm or as thick as 1000 nm, depending on the thickness of the protective front layer [4]. This is true with or without silver nanoparticles (Ag NPs) [4]. We want to use the QCRF-FDTD simulator to investigate the optical properties of the coatings and their effects on solar cell efficiency. When Ag NPs are added to a ZnO coating, the antireflection properties get better. This makes the coating less reflective, and the solar cell absorbs more light [5]. Furthermore, we discover that the solar cell's antireflection properties are greatly enhanced as the layer thickness increases, which impacts the cell's profitability [6].

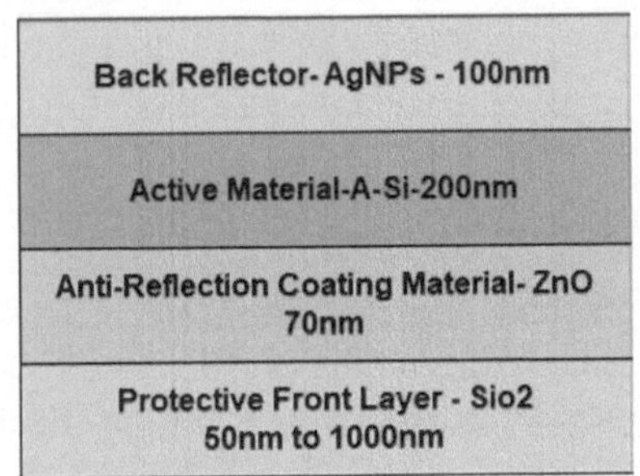

Fig. 94.1 Proposed single junction solar cell layer ZnO is an antireflection coating material with and without silver nanoparticles

We find it intriguing that the presence or absence of Ag NPs affects the optimal thickness of the ZnO coating. A ZnO coating without Ag NPs has an ideal thickness of around 200 nm, but a layer doped with Ag NPs has a perfect thickness of about 500 nm. The effect that adding Ag NPs has on the coating's optical properties, in turn, affects the optimal coating thickness [7–9]. Our research concludes that ZnO has significant potential as an antireflection coating material for solar cells, particularly when combined with Ag NPs. Based on our findings, developing cheaper and more effective antireflection coatings for solar cells may be possible. For example, the thickness of the protective front layer might change depending on the system's unique specs and needs when it comes to thin films or coatings. The benefits of a front protective layer with a changeable thickness include:- Optical tuning, Interference and Anti-reflective coating properties, Barrier Protection, and cost optimization.

2. Material and Methods

When the simulation parameters and geometry have been defined, the electrical characteristics of the given structure may be examined using the QCRF FDTD Simulator. Many parameters are used in the model, including the wavelength spectrum, polarization, and incidence angle. The simulation geometry consisted of the layers' size, orientation, and composition [10]. If the simulation geometry is defined as True TE polarization, normal incidence, and 500 nm as the only wavelength, Get a 1 um x 1 um x 1 um box ready for the simulation. Develop a substrate layer with a thickness of 1 um and a refractive index of 1.5, such as glass [11]. Place the layers in the following sequence: a front layer of protection (SiO2), a coating that prevents reflection (ZnO), an active material (A-si), and a rear reflector (AgNPs). According to the provided values, set the layer thicknesses to 50 nm to 1000 nm for SiO2, 70 nm for ZnO, 200 nm for A-si, and 100 nm for AgNPs.

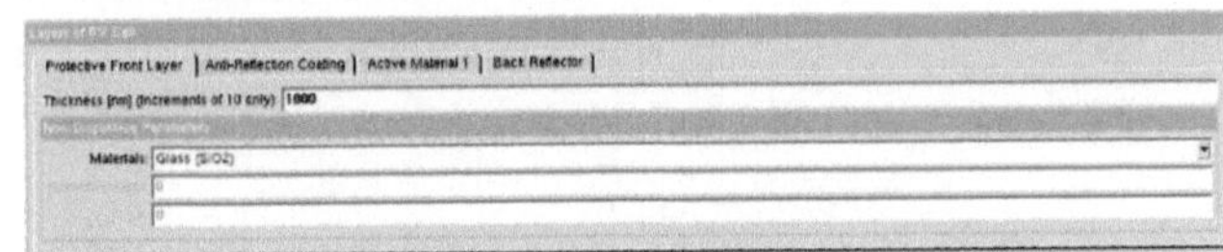

Fig. 94.2 Front layer thickness (SiO2) varied from 50 nm to 1000 nm

Using the QCRF-FDTD Simulator, the Nanohub online tool has four layers: the Protective front layer, Anti-Reflection coating, Active material, and Back reflector. In the protective front layer tab, there are Materials and Thickness. This method is used to select the material and vary the thickness of the defensive front layer (SiO2), use SiO2 material. In the material tab is a default SiO2 and ZnO if anyone can select it for us. If using another material for the protective front layer, use the defined option and put the refractive index of the corresponding material. Protective front layer thickness varies from 50 to 1000nm, and electrical parameters are analyzed for each value. In this paper, I am using to vary the thickness of the protective front layer, which means SiO2 material. Another option is to run the thickness of the antireflection material. Why select this option to change the protective front layer thickness? It is possible to achieve the necessary electrical conductivity or insulating qualities in thin films by controlling their thickness within a specific range. At the specified wavelength, determine every layer's proper refractive index and extinction coefficients. Here are some examples of refractive indices at 500 nm: SiO2 has an index of around 1.46, ZnO is about 2.0, A-si is about 3.5, and AgNPs are roughly 0.05 + 4.5i (where i is the imaginary unit) [12]. An additional 500 nm plane wave source with TE polarization should be

Table 94.1 Group statistics for zno as an antireflection coating material

ZnO		N	Mean	Std Deviation	Std. Error Mean
Voc in Volts	Without Nanoparticles	20	.90562435	.000860877	.000192498
	With Nanoparticles	20	.90629905	.000841894	.000188253
Jsc in mA/cm2	Without Nanoparticles	20	28.35581500	.656225678	.146736522
	With Nanoparticles	20	28.87314500	.653133324	.146045051
Fill Factor in %	Without Nanoparticles	20	.83312440	.000115827	.000025900
	With Nanoparticles	20	.83321515	.000113150	.000025301
Efficiency in %	Without Nanoparticles	20	21.39492000	.518570847	.115955967
	With Nanoparticles	20	21.80382000	.516542271	.115502363

added to the top of the simulation zone. An adequate number of time steps must be included in the model to stabilize electromagnetic fields [13]. At the end of the simulation, the electric field, charge distributions in each layer, and wavelengths of transmission, reflection, and absorption may be used to examine the electrical characteristics of the chassis. So, for example, to get the structure's transmission and reflection coefficients, you may use these equations:

$$\text{total } R = |r|^2 \; T = \frac{|t|^2}{\left(\dfrac{nsub}{ni}\right)}$$

A reflecting index of the substrate (n_sub) and a refractive index of the incoming medium (air, in this example) are used here, with the reflection coefficient (r) and the transmission coefficient (t) standing for the corresponding variables. Another way to get each layer's absorption is to take the transmission and reflection coefficients and subtract 100:

$$A = 1 - R - T$$

According to the data in the table, the voltage Voc and current density Jsc increase when nanoparticles are present compared to when they are absent. Without nanoparticles, the Fill Factor in% and Efficiency in% numbers are almost identical. With nanoparticles and without, there is a statistically significant difference in the values of Voc (in volts) and Jsc (in mA/cm2), with the standard error mean being significantly less than the mean.

In addition, each layer's electric field and charge distributions may be studied to understand the interaction between light and materials. This could illuminate the structure's potential as an optoelectronic device or photovoltaic cell [12]. In sum, the QCRF FDTD Simulator is a powerful tool for optimizing the design of complicated, layered structures like the one discussed here by providing data concerning their electrical properties.

Table 94.2 The comparison value for solar cell parameters ZnO, an antireflection coating material with and without silver nanoparticles at the thickness of the protective front layer, is 100nm

ARC	Voc in Volts	Jsc in MA	Fill Factor	Efficiency in %
ZnO Without Ag nanoparticles	0.907895	30.1278	0.833429	22.7967
ZnO With Ag nanoparticles	0.908502	30.6221	0.833511	23.1884

3. Result and Discussion

Adding silver nanoparticles (AgNPs) increases the solar cell's productivity. Adding AgNPs increases the maximum efficiency to 23.1639% at a 100 nm thickness from 20.1639% without nanoparticles. We also see that the fill factor (FF) stays very constant, but the open-circuit voltage (Voc) and short-circuit current density (Jsc) both go up when AgNPs are added. This indicates that the efficiency boost is not attributable to better charge carrier extraction but to enhanced light absorption and charge carrier collection.

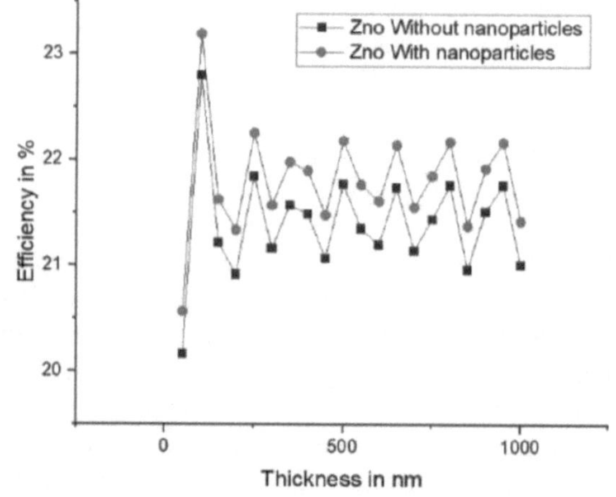

Fig. 94.3 ZnO as antireflection coating with and without nanoparticles analyse efficiency vs thickness

Remember that the particular design and simulation setting could affect the simulation outcomes. As a result, finding the best design for the given application could need more investigation and optimization.

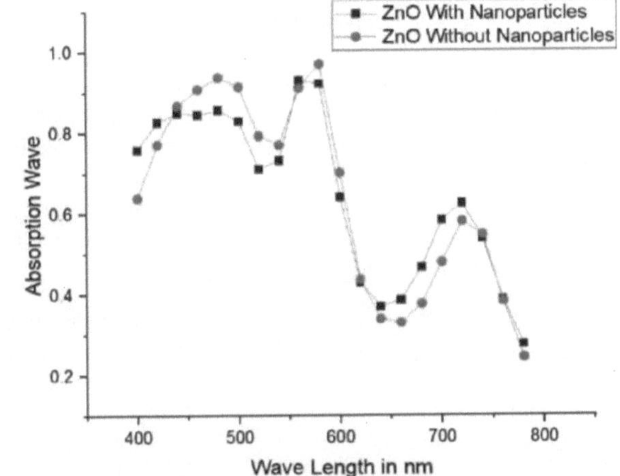

Fig. 94.4 Quantum confinement effect on light absorption in ZnO nanoparticles

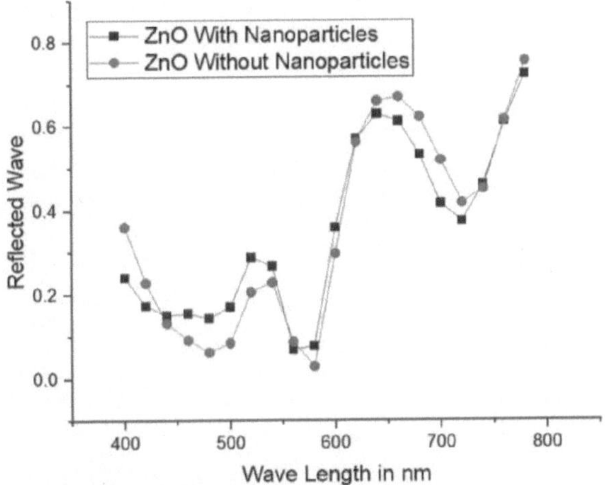

Fig. 94.5 Reflection comparison with wavelength

Including silver nanoparticles (AgNPs) in the back reflector layer is one strategy to boost solar cells' efficiency. A solar cell's efficiency increases by around 2% when AgNPs are added to it. Adding AgNPs to the rear reflector layer improves the light capture by the visible material layer, increasing the JSC. This is because the excellent light trapping and scattering capabilities of the AgNPs—made possible by their small size and high refractive index—allow for a longer optical path length for the incoming light. We may infer that the AgNPs have no appreciable effect on the solar cell's open-circuit voltage as the Voc values of the cells with and without AgNPs are similar. The solar cell with AgNPs has a slightly more significant fill factor (FF) than the solar cell without. In other words, the AgNPs may prevent charge carriers from recombining at the interface between the active material and the rear reflector layers, increasing the FF.

The results of the models suggest that adding AgNPs to the rear reflector layer could improve thin-film solar cells. Keep in mind that these findings are derived from simulations; further testing is necessary to validate the functionality of the suggested architecture.

5. Conclusion

A-Si solar cells use silver nanoparticles as a back reflector, potentially more effective than cells devoid of nanoparticles. The research used nanoparticles and a 100 nm SiO2 outer layer to get an efficiency level of 23.1884%. The nanoparticles improved the fill factor, Voc, and Jsc values. The results of the simulations indicate that by modifying the thickness of each layer, the a-Si solar cell's efficiency may be further optimized. These findings could help develop a-Si solar cells with improved performance. With an emphasis on personalization ("tailoring") and the possibility of improved performance in various contexts, this title draws attention to the research's tangible and practical consequences. This study provides strong evidence that the research applies to real-world situations and may help optimize the thickness of protective front layers based on individual requirements.

References

1. Enhancement of the Optical Properties of ZnO Antireflection Coatings with Embedded Silver Nanoparticles for Silicon Solar Cells" by L. Zhang, Y. Liu, H. Liu, Y. Song, and G. Lu. Published in Nanomaterials in 2019, volume 9, issue 8.
2. "ZnO Antireflection Coatings with Embedded Silver Nanoparticles for Silicon Solar Cells: Preparation and Characterization" by F. He, C. Liu, X. Xu, and H. Zhang. Published in the Journal of Nanomaterials in 2017, article ID 3714267.
3. "Ag Nanoparticle-Embedded ZnO Antireflection Coating for Silicon Solar Cells: Numerical and Experimental Analysis" by Y. Liu, L. Zhang, X. Sun, H. Liu, and G. Lu. Published in Journal of Nanomaterials in 2020, article ID 4721061.
4. "Ag Nanoparticle-Enhanced ZnO Antireflection Coating for High-Efficiency Silicon Solar Cells" by C. Liu, F. He, X. Xu, H. Zhang, and J. Wu. Published in Materials Letters in 2016, volume 181, pages 221-224.
5. Chowdhury, Anuradha Rai, V. Padmanapan Rao, B. Ghosh, and S. Banerjee. 2012. "Antireflection Coating on Solar Cell Using ZnO Nanostructure." AIP Conference Proceedings. https://doi.org/10.1063/1.4710234.

6. "The Effect of Ag Nanoparticles on the Optical and Electrical Properties of ZnO Thin Films for Solar Cell Applications" by N. A. Rahman, M. N. A. Shamsudin, M. Rusop, and M. N. Nayan. Published in the Journal of Nanomaterials in 2015, article ID 373846.
7. "Enhancement of the Efficiency of Silicon Solar Cells by the Addition of ZnO-Ag Nanoparticles" by M. D. Mallick, S. M. Hussain, and S. K. Deb Nath. Published in Journal of Electronic Materials in 2017, volume 46, pages 1776-1783.
8. "A Comprehensive Study on the Effects of ZnO-Ag Nanoparticle Composite Films on the Efficiency of Crystalline Silicon Solar Cells" by S. K. Deb Nath, M. D. Mallick, and S. M. Hussain. Published in Journal of Materials Science: Materials in Electronics in 2017, volume 28, pages 12860-12871.
9. "Enhanced Performance of a-Si: H Solar Cells with ZnO Nanoparticles and Ag Nanoparticles" by Y. He, Y. Luo, J. Yu, J. Shen, L. Wang, X. Xu, and J. Wang. Published in Journal of Nanoscience and Nanotechnology in 2019, volume 19, pages 4689-4693.
10. Lumerical Solutions, Inc. (2021). FDTD Solutions: A Comprehensive Guide. https://kb.lumerical.com/en/index.html
11. Taflove, A., & Hagness, S. C. (2005). Computational Electrodynamics: The Finite-Difference Time-Domain Method. Artech House Publishers.
12. Green, M. A., Emery, K., Hishikawa, Y., Warta, W., & Dunlop, E. D. (2018). Solar cell efficiency tables (version 52). Progress in Photovoltaics: Research and Applications, 26(1), 3-12.
13. Yu, Z., & Raman, A. (2019). Advances in nanoparticle-based plasmonic solar cells. Current Opinion in Green and Sustainable Chemistry, 17, 33-39.

Note: Every figure and table was created using the experimental work; none of them were taken from any publications or online.

Advances in Additive Manufacturing Technologies – Gurusamy Pathinettampadian (eds)
© 2024 Taylor & Francis Group, London, ISBN 978-1-032-90013-1

95 An Exhaustive Examination, Evaluation, and Comparative Assessment of Diverse Anti-Reflection Coating Methods for the Optimization of Solar Cell Efficiency: A Comprehensive Review and Analysis

Manikandan S[1]
Research Scholar, Institute of ECE,
Saveetha School of Engineering, Saveetha Institute of
Medical and Technical Sciences,
Saveetha University, Chennai, India

Radhika Baskar[2]
Professor, Research Guide, Institute of ECE,
Saveetha School of Engineering, Saveetha Institute of
Medical and Technical Sciences,
Saveetha University, Chennai, India

ABSTRACT: Using photovoltaic technology, which converts sunlight into power, has become more common in the last several decades. Light intensity is the primary factor in a solar power system's output. The cell count decreases as a photovoltaic component comes into contact with dirt and organic debris. No matter how clean the surface of the photovoltaic panel is, solar radiation diffraction happens the moment a reflector comes into contact with a substrate made of roof glass. The solar array receives the sustained energy beams that remain after the material breaks down. The tendency towards a darker sky is inversely related to the rise in electrical power generation. Coating solar modules and covering glass with anti-reflective compounds is one way to fight them. The study found that ZnO, ITO, MgO, Al2O3, SiO2, MgF2, TiO2, Si3N4, and ZrO2 are common ingredients in anti-reflection coatings. These are standard processes for industrial spin coating, electrospinning, sputtering, DC/RF magnetrons, and sol-gel+dip coating. Use two or three coats of adhesive to improve surface adherence and durability. Researchers discovered that multi-layer anti-reflection coatings decreased reflectance and enhanced light transmission on materials with low and high refractive indices.

KEYWORDS: Photo-voltaic, Anti-reflective coating, Sol-gel, Spin coating, DC or RF magnetron

1. INTRODUCTION

Solar panels usually reflect 30% of the sun's light that hits their outer layer into space, even though they receive 70% of the electricity they can create (1). Unlike traditional processing techniques, which rely on potentially dangerous fossil fuels, alternative sources do not rely on these (2). This is one reason why they are desirable. The sun's rays are an excellent renewable energy source because they are limitless, non-polluting, and environmentally friendly (3). Each day, solar flares inject enormous amounts of energy into the electromagnetic field surrounding Earth (4). The

[1]smaniphdece@gmail.com, [2]radhikabaskar@saveetha.com

DOI: 10.1201/9781003545774-95

photovoltaic consistency, which measures the sun's average yearly power production, varies in value when the sun's distance from Earth changes (5). Sunlight accelerates the chemical reactions of some compounds when it reaches Earth.

The amount of energy that photons of different wavelengths store varies, with electron volts being the most common unit of measurement. A common practice is applying an anti-reflection coating (ARC) to machinery with a severe problem reflecting light. Finding a way to microfabricate semiconductors and complex materials to get decent energy efficiency is the fundamental issue in developing PV devices (7). A single silicon cell can convert 35.3% to 36.6% of the light that hits its surface into helpful energy (8); see Fig. 95.1.

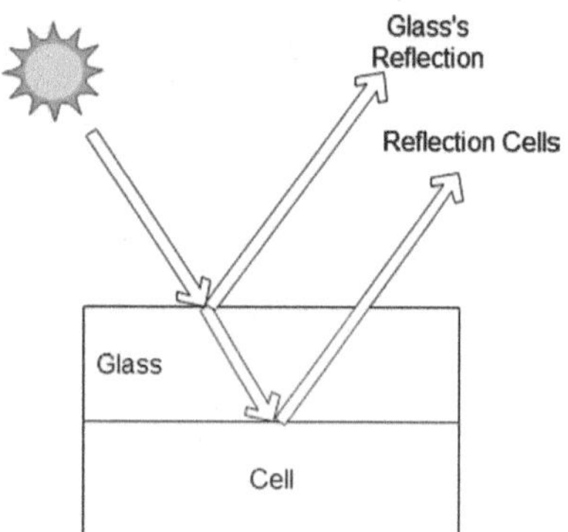

Fig. 95.1 Light from the sun reflects from the surface and cell of a solar panel

ARC has been applicable in this particular context since it boosts transmission while decreasing reflection (9). Applying a thin layer of a substance with a suitable refractive index (RI) is a common practice for lowering review (ten). The ARC films used in silicon solar cells may have a variety of materials, including Al2O3, SiO2, ZnS, TiO2, MgF2, or a mix of these. Reducing light transmission and increasing light reflection in definitions can make it harder for people to comprehend solar panels' efficiency and watch ARC films. We sorted the ARC films we saw into many categories in this article. The use of polymer-based inventions and chemicals synthesized using dyes allowed for the development of third-generation solar panels.

2. Single Layer ARC

When Sagar et al. used RF sputtering to make silicon photovoltaic substrates, they tested the ARC performance of materials with single-layered (SL) structures, such as ZnO, MgO, and Al2O3. Al2O3, MgO, and ZnO have a width of 80 nm, 95 nm, and 85 nm, respectively. It was possible to examine the electrical characteristics of coated films using a solar simulator (Sol3A Class AAA). We investigated ultraviolet and visible light spectra transmissions using a spectrometer (Shimadzu MPC3600). In the peak absorption spectra, ZnO was found at 412 nm, 373 nm, and 356 nm, with corresponding values of 95%, 92%, and 93% (11), respectively. Table 95.1 displays the results obtained from the solar simulator.

Table 95.1 Properties of solar panel

Parameter	ZnO	MgO	Al_2O_3
Efficiency (η)%	9.633	9.864	9.125
Short Circuit Current (Isc)	0.026	0.027	0.026
Open Circuit Voltage (Voc)	0.564	0.571	0.562
Fill Factor (FF)	66.603	64.064	61.225

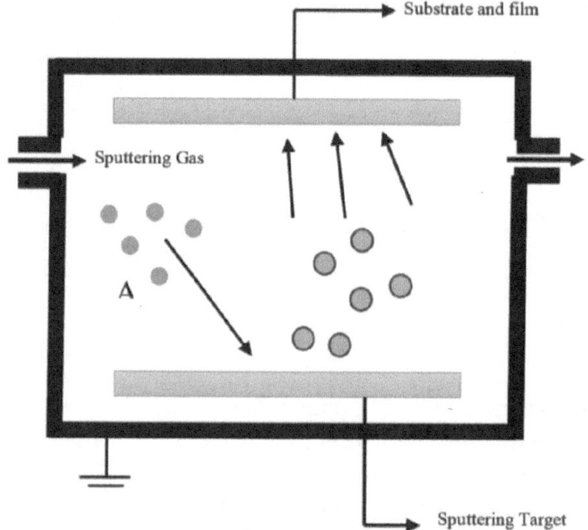

Fig. 95.2 Schematic diagram of DC and RF sputtering

Research by Ali et al. found that p-type monocrystalline silicon cells had ARC layers of indium tin oxide (ITO) and titanium dioxide (TiO2) deposited on them using radio frequency magnetron sputtering. Our two materials of choice were TiO2 (with a width of 55 to 60 nm) and ITO (with a width of 60 to 64 nm). Different imaging techniques were usedto look at the intercalated membrane layers. These included X-ray diffraction, radiation microscopy, atomic force microscopy, field emission scanning electron microscopy, and Raman spectroscopy. The total reflectance ranged from 400 to 1000 nm when the concentrations of TiO2 and ITO were 10% and 12%, respectively. ITO had a 25% increase in value compared to as-grown silicon, and TiO2 had a 23% increase in value (12).

3. Sol-gel Method

The sol-gel technique requires the material to be fluid and performed at controlled temperatures, typically below

100°C. The formation of M-OH-M or M-O-M bonds between the metal atoms in the raw materials causes the polymerization process to produce solid particles. The result is reliable, so the answer is yes. The sol-gel procedure involves two stages for aerogel synthesis (13). The first stage makes creating nanometer-sized solid particles with a colloidal structure easier. Step two consists of the formation of a gel using solvent-borne nanoparticles.

Using a sol-gel process using synthetic polymer porogens such as PEG, Triton X-100, and tetraethoxysilane (TEOS), Mahadik et al. were able to create SiO2 SLARC. We measured the transmittance using a PerkinElmer Lambda 750 UV spectrophotometer, which operates across a broad range of wavelengths (300–900 nm). Assuming a 100-nm thickness, the RI of the covering is 1.39. According to spectral analysis, coating the soda lime glass surface improved its absorption coefficient to 97.5% at 500 nm. In contrast, the coated substrate outperformed the bare substrate in the 400–600 nm area by 6% in transmittance (14).

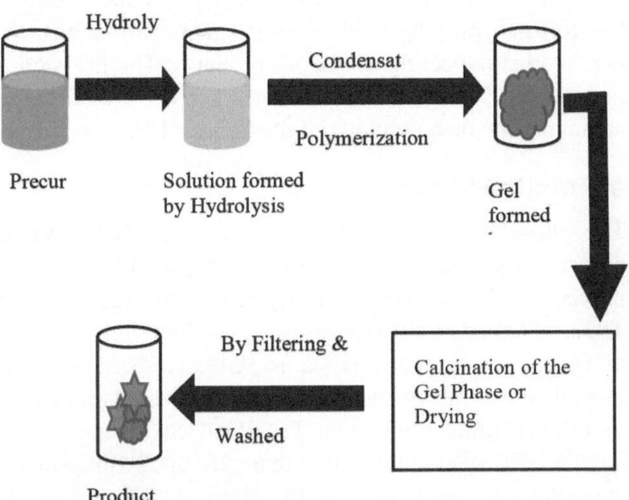

Fig. 95.3 Different stages of the sol-gel process

3.1 Spin Coating

The spin-coating method uses heat action to promote mutual contact and evenly distributes the solution to the outside walls. Three factors—the rolling duration, the rolling rate, and the solution friction coefficient—limit the wall thickness during spin-coating. Evaporation, spin-up, spin-off, and sol-gel deposition are the four steps that make up the process (15). A thick viscosity makes sol-gel application challenging. The spin-up process uses up a lot of sol-gel since it surfaces and disperses surplus sol-gel. The low price and ease of usage of spin-coating and dip-coating have led to their high demand.

Using RF magnetron technology, Jain and colleagues investigated the performance of single-layered (SL) TiO2 as ARCs on silicon solar cell substrates in 2020. To incorporate AgNPs, they employed spin coating (16). The values for the solar parameters are listed in Table 95.2.

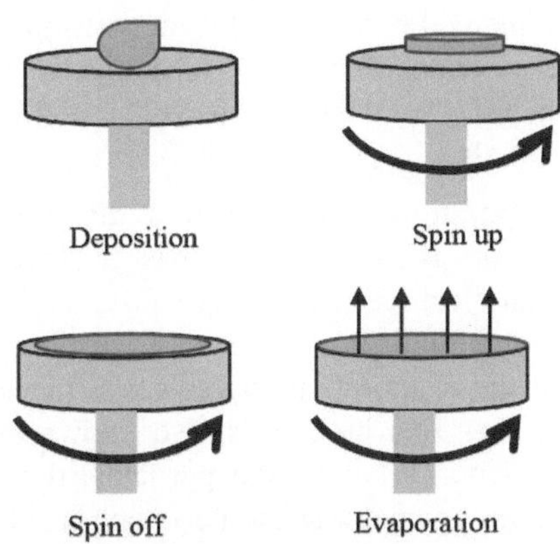

Fig. 95.4 Spin coating process

Table 95.2 Solar parameters

Solar Cell	Efficiency (η)%	Voc	Jsc	Fill Factor
Si-PV Cell	9.53	0.63	18.01	83.9
Si With TiO_2 ARC	12.58	0.63	18.01	83.7
Si With TiO_2 ARC and AgNPs	16.04	0.67	28.30	84.3

3.2 Double Layer ARC

With a sputtering method for SLARC (SiO2) and DLARC (SiO2/TiO2) layers on a room-temperature substrate, Ali et al. planned and studied silicon-based solar power plants (17-18). Initially, they focused their attention on SiO2 and SiO2/TiO2 components. The use of EDS was essential in creating the impression of surface topography. These FESEMs from that year (19) were used to create this artwork. The symbols used to represent the thicknesses of the layers are d1 and d2, while the characters used to describe the relative intensities (RIs) of the air, the external surface, the inner layers, and the silicon substrate are no, n1, n2, and ns, respectively. Therefore, the equation's condition must be met to get to zero reflective circumstances. While SiO2 coatings had a reflectance of 15% in the 400–1000 nm broadband area, SiO2/TiO2 coatings had a reflectance of 7% in this region. At 630 nm, measurements revealed a reflectance as low as 2.3%. By using double-layered SiO2/TiO2, it was possible to attain a 37% increase in efficiency for monocrystalline silicon solar cells.

4. Thermal Evaporation Method

Thermal evaporation is one of the best-known depositing methods. Its speed, accuracy, and practicality stand out among the many benefits of this deposition technique. Physical vapor deposition (PVD) techniques, such as heat evaporation, allow for the physical deposit of thin layers on various materials.

Thermal evaporation still requires a direct evaporative source, such as a boat, coil, or basket, to expose the material to an electrical current. A substance is heated to its evaporation point using an evaporation source. Because a change in the evaporation source's resistance causes a temperature rise, this process is also known as resistive evaporation. Upon reaching the particles, the molecules of the evaporating metal substrate create a thin coating. With this method, it is possible to gather metals such as magnesium, chromium, nickel, silver, and aluminum. (20). Vacuum evaporative cooling systems' stress concentration is starting to play a more significant role. Poor sample coverage may occur if the material's surface-vaporized atoms come into contact with the chamber's gas particles during thermal evaporation, inappropriately altering their trajectory. Doing thermal evaporative deposition in a well-vacuumed space will save you this extra step. At 10^5 Torr, the molecules' average free path length is 1 meter. Even a tiny number of unrelated gases in the sample chamber might contaminate the thin layer. Consider the case of organic light-emitting diodes (OLEDs) or solar cells made from sustainably generated materials. A little bit of air or moisture during thermal evaporation might render all of the functioning components unusable. Bringing the pressure in the deposition chamber down to 10–6 Torr will make the devices work better and the redistributed thin layer more reliable. When working with silicon solar cells coated with MgF2 and SiNx16, Dhungel, Suresh Kumar, and colleagues used DLARC. We used thermal evaporation in a 10-6 Torr vacuum to deposit a 75-nm layer of MgF2 (RI = 1.38) on top of a 72-nm film of SiNx (RI = 2.05). With these SiNx/MgF2 thickness ratios, single-crystalline solar cells achieved an excellent 16.94% efficiency, while multi-crystalline solar cells achieved an impressive 16.01% efficiency (21). To verify DLARC (TiO2/Al2O3) on p-type Si solar cells, Jung, Jinsu, and colleagues used a sol-gel technique and subjected the cells to roasting at 400°C. It was found that solar cells made with DLARC had a fill factor (FF) of 66.67%, an open circuit voltage (VOC) of 593.35 mV, a short circuit current of 35.27 mA/cm2, and an efficiency of 13.95%. DLARC's reflectance at 970 nm was 3.12%, the lowest. The reflectance of coated films decreased by 4.74% in the 400 nm to 1000 nm broadband range. After everything else, the series resistance value was lower in DLARC cells than in SLARC cells (22). Using computational methods, Medhat and Mohamed have examined how TiO2 and MgF2 as DLARC affect silicon solar cells. The emission is most vigorous between 400 and 1200 nanometers. FF values of 0.84 and VOC values of 0.68V were reached, along with a drop in reflectance from 30.2% to 2.37% and an increase in short-circuit current density of 38.6 mA/cm2. The 700 nm wavelength was improved when the MgF2 and TiO2 thicknesses were increased to 0.7 ($\lambda/4$) and 0.78 ($\lambda/4$), respectively, as stated in reference 23. To create DLARC, Shah, Deb Kumar, and colleagues used the sol-gel process to combine crystalline silicon solar cells with zinc oxide (ZnO) and silver. Annealing the cells at 50,000 °C for four hours followed. In the 400 nm–1000 nm broadband spectrum, they found a remarkable decrease in reflected light to around 7.13% and an exceptional surface quality. Consequently, using Ag-doped ZnO for inexpensive DLARC on silicon solar cells is a viable option (24). An Overview of the ALD Process (a): The initial treatment focuses on the functionalized surface. The next step is to pulse precursor A so it may react with the character. Gathering any leftover surplus by-products follows the purging of the precursor process with an inert carrier gas. Now that the precursor process has finished purging itself with the inert carrier gas, the next step is collecting all the by-products. The last step is to treat the surface. (f) Proceed to Steps 2–5 again after achieving the desired material thickness. (17).

4.1 Multiple Layered ARC

The glazed films were characterized using XRD, SEM, ellipsometry, and AFM. We used the Sol-Gel Dip coating procedure to create the multi-layered SnO2, Ta2O5, ATO, TiO2, SiO2, and ITO coatings. The light transmittance of single-layer porous silica is 95%. Researchers who looked into how wavelength affects light transmission in multilayer films found that ITO-TiO2-SiO2 and SnO2-Ta2O5-SiO2 films did better than SL SiO2 films in the 300–900 nm wavelength range, letting 97.5% and 96.2% of light through, respectively. The film thickness may take on values between 30 and 115 nanometers. (25) Employing MATLAB modeling, Sahouane et al. investigated the 300–1100 nm broadband range of effective reflectivity (Reff) by combining n1 = SiNx: H-rich silicon, n2 = silicon oxide (SiOx), and n3 = oxynitride (SiOxNy). By using optical equations and the transfer matrix method, they were able to calculate the results. They used the PCID and Silvaco programs for IQE calculations as well. Table 95.3 shows that the combination of SiNx: H with DLARC produces the lowest reflectance. It was calculated that a multilayer of silicon nitride, oxide, and oxynitride significantly increases the photogenerated current by 1.6 mA/cm^2 compared to an ARC with only one layer. A SiNx: H bulk layer or passivation with a d3>30 and n3=2.4 is also a remarkable option (26).

Table 95.3 Reflectivity comparison (26)

Cells	n1, d1(SiOx)	N2, d2 (SiOxNy)	N3, d3 (SiOx: H)	Reff (%)
Without ARC				34.5
One layer			2.03 (73 nm)	11.7
Two layers		1.5 (55 nm)	2.1 (53 nm)	7.4
Three layers	1.48 (80 nm)	2 (5 nm)	2.4 (50 nm)	4.4

5. Conclusion

A key component in improving the performance of solar panels is ARC. Researchers have been trying to find the best ways to increase light traction since highly efficient solar panels started popping up. In this literature review, we looked into ARC and found that it reduces reflection on glass substrates and works well with silicon solar cells. Many researchers performed experiments to prove the material's efficacy and influence in enhancing transmittance across various wavelength bands.

References

1. Kumaragurubaran, B., and S. Anandhi. 2014. "Reduction of Reflection Losses in Solar Cell Using Anti Reflective Coating." *2014 International Conference on Computation of Power, Energy, Information and Communication (ICCPEIC)*. https://doi.org/10.1109/iccpeic.2014.6915357.
2. Pehnt, Martin. 2006. "Dynamic Life Cycle Assessment (LCA) of Renewable Energy Technologies." *Renewable Energy*. https://doi.org/10.1016/j.renene.2005.03.002.
3. Yasa, U. G., M. N. Erim, N. Erim, M. O. Girgin, and H. Kurt. 2017. "Design of Anti-Reflective Graded Height Nanogratings for Photovoltaic Applications." *2017 International Conference on Numerical Simulation of Optoelectronic Devices (NUSOD)*. https://doi.org/10.1109/nusod.2017.8009973.
4. Gupta, P. K. 1999. "Renewable Energy Sources — A Longway to Go in India." *Renewable Energy*. https://doi.org/10.1016/s0960-1481(98)00486-8.
5. Wald, Lucien. 2021. "Solar Radiation Received at Ground Level." *Fundamentals of Solar Radiation*. https://doi.org/10.1201/9781003155454-6.
6. Fraas, Lewis M., and Larry D. Partain. 2010. *Solar Cells and Their Applications*. Wiley.https://books.google.com/books/about/Solar_Cells_and_Their_Applications.html?hl=&id=Ro32H7lQLZIC.
7. Prevo, Brian G., Emily W. Hon, and Orlin D. Velev. 2007. "Assembly and Characterization of Colloid-Based Antireflective Coatings on Multicrystalline Silicon Solar Cells." *Mater. Chem.* https://doi.org/10.1039/b612734g.
8. Lee, Seungwoo, Wook Kim, Sangmin Lee, Sangdeok Shim, and Dukhyun Choi. 2015. "Controlled Transparency and Wettability of Large-Area Nanoporous Anodized Alumina on Glass." *Scripta Materialia*. https://doi.org/10.1016/j.scriptamat.2015.04.001.
9. Maas, Ruben, Sander A. Mann, Dimitrios L. Sounas, Andrea Alù, Erik C. Garnett, and Albert Polman. 2016. "Generalized Antireflection Coatings for Complex Bulk Metamaterials." *Physical Review B*. https://doi.org/10.1103/physrevb.93.195433.
10. Sarangan, Andrew. 2020. *Optical Thin Film Design*. CRC Press. https://play.google.com/store/books/details?id=3EYHEAAAQBAJ.
11. Sagar, Raghavendra, and Asha Rao. 2020. "Increasing the Silicon Solar Cell Efficiency with Transition Metal Oxide Nano-Thin Films as Anti-Reflection Coatings." *Materials Research Express*. https://doi.org/10.1088/2053-1591/ab6ad5.
12. Ali, Khuram, Sohail Akbar Khan and Mohd. Zubir Mat Jafri. "Effect of Double Layer (SiO2/TiO2) Anti-reflective Coating on Silicon Solar Cells." (2014).
13. Chen, Runhua, Yuying Cheng, Ping Wang, Yangyang Wang, Qingwei Wang, Zhihui Yang, Congjian Tang, et al. 2021. "Facile Synthesis of a Sandwiched Ti3C2Tx MXene/nZVI/fungal Hypha Nanofiber Hybrid Membrane for Enhanced Removal of Be(II) from Be(NH) Complexing Solutions." *Chemical Engineering Journal*. https://doi.org/10.1016/j.cej.2021.129682.
14. Mahadik, D. B., R. V. Lakshmi, and Harish C. Barshilia. 2015. "High-Performance Single Layer Nano-Porous Antireflection Coatings on Glass by Sol-gel Process for Solar Energy Applications." *Solar Energy Materials and Solar Cells*. https://doi.org/10.1016/j.solmat.2015.03.023.
15. Biswas, P. K., P. Sujatha Devi, P. K. Chakraborty, A. Chatterjee, D. Ganguli, M. P. Kamath, and A. S. Joshi. 2003. *Journal of Materials Science Letters* 22 (3): 181–83. https://doi.org/10.1023/a:1022241707860.
16. Jain, Surbhi, Ayushi Paliwal, Vinay Gupta, and Monika Tomar. 2020. "Plasmon-Assisted Crystalline Silicon Solar Cell with TiO2 as Anti-Reflective Coating." *Plasmonics*. https://doi.org/10.1007/s11468-020-01127-5.
17. "Thin Film Deposition By Thermal Evaporation Method." 2020. *VacCoat* (blog). May 4, 2020. https://vaccoat.com/blog/thin-film-deposition-by-thermal-evaporation-method/
18. Zhu, Li Qiang, Yang Hui Liu, Hong Liang Zhang, Hui Xiao, and Li Qiang Guo. 2014. "Atomic Layer Deposited Al2O3 Films for Anti-Reflectance and Surface Passivation Applications." *Applied Surface Science*. https://doi.org/10.1016/j.apsusc.2013.10.051.
19. Ali, Khuram, Sohail A. Khan, and Mohd Zubir Mat Jafri. 2014. "Structural and Optical Properties of ITO/TiO2 Anti-Reflective Films for Solar Cell Applications." *Nanoscale Research Letters* 9 (1): 175. https://doi.org/10.1186/1556-276X-9-175.
20. McAllister, Howard Conlee. 1950. *Thin Films by Thermal Evaporation*. https://books.google.com/books/about/Thin_Films_by_Thermal_Evaporation.html?hl=&id=-I2UNwAACAAJ.
21. Dhungel, S. K., Yoo, J., Kim, K., Jung, S., Ghosh, S., Yi, J. "Double-layer antireflection coating of MgF2/SiNx for

crystalline silicon solar cells," J. Korean Phys. Soc., Vol. 49, No. 3, pp. 885-889, 2006.

22. Jung, Jinsu, Azmira Jannat, M. Shaheer Akhtar, and O-Bong Yang. 2018. "Sol-Gel Deposited Double Layer TiO_2 and Al_2O_3 Anti-Reflection Coating for Silicon Solar Cell." *Journal of Nanoscience and Nanotechnology*. https://doi.org/10.1166/jnn.2018.14928.
23. Medhat, Mohamed, El-Sayed El-Zaiat, Samy Farag, Gamal Youssef, and Reda Alkhadry. 2016. "Enhancing Silicon Solar Cell Efficiency with Double Layer Antireflection Coating." *TURKISH JOURNAL OF PHYSICS*. https://doi.org/10.3906/fiz-1508-14.
24. Shah, Deb Kumar, Su Yong Han, Shaheer M. Akhtar, J. O-Bong Yang, and Chong Yeal Kim. 2019. "Effect of Ag Doping in Double Antireflection Layer on Crystalline Silicon Solar Cells." *Nanoscience and Nanotechnology Letters*. https://doi.org/10.1166/nnl.2019.2864.
25. Kesmez, Ömer, Esin Akarsu, H. Erdem Çamurlu, Emre Yavuz, Murat Akarsu, and Ertuğrul Arpaç. 2018. "Preparation and Characterization of Multilayer Anti-Reflective Coatings via Sol-Gel Process." *Ceramics International*. https://doi.org/10.1016/j.ceramint.2017.11.088.
26. Sahouane, Nordine, and Abdellatif Zerga. 2014. "Optimization of Antireflection Multilayer for Industrial Crystalline Silicon Solar Cells." *Energy Procedia*. https://doi.org/10.1016/j.egypro.2013.12.017.

Note: Every figure and table was created using the experimental work; none of them were taken from any publications or online.

Advances in Additive Manufacturing Technologies – Gurusamy Pathinettampadian (eds)
© 2024 Taylor & Francis Group, London, ISBN 978-1-032-90013-1

96 Comparative Analysis of Quantum Capacitance and Drain Current in CNTFET: Unveiling the Superiority of Conventional Insulator Materials Over Standard Silicon Field

Sathish G[1]
Research Scholar, Institute of ECE,
Saveetha School of Engineering, Saveetha Institute of Medical and Technical Sciences, Saveetha University,
Chennai, India

Radhika Baskar[2]
Professor, Research Guide, Institute of ECE,
Saveetha School of Engineering, Saveetha Institute of Medical and Technical Sciences, Saveetha University,
Chennai, India

ABSTRACT: Aim: The scope of the work is to investigate the utilization of diverse Di-electric materials characterized by variable relative constants to enhance current drain and mitigate leakage current. Di-electric insulators with relative permittivity values of 3.9 and 30 nm were employed sequentially to mitigate leakage current and enhance drain current through the utilization of a nanohub simulation tool. The insulator with a 30nm exhibits a larger drain current when paired with a nanotube diameter of 1 nm and a glass thickness of 0.1 nm, as compared to the material with a 3.9nm constant, which results in a lower drain current. Based on the conducted research, it has been observed that La2O3 exhibits a greater drain current in comparison to SiO2. Circuit designers exhibit a preference for maintaining the invariance of Vth with respect to transistor size and biasing circumstances. Consequently, fluctuations in Vth are deemed undesirable when dealing with scaled geometries. In order to restrict leakage power, it becomes necessary to ensure the stability of Vth. Instead of using conventional silicon devices, carbon nanotube field-effect transistors (CNTFETs) can achieve this goal.

KEYWORDS: Conductance, Nanohub, Insulator, Di-electric, Draincurrent, CNTFET, Conductivity, Oxide thickness, Quantum capacitance, Silicon oxide and lanthanum oxide

1. INTRODUCTION

Metal oxide semiconductors (MOS) represent a prominent category of silicon chips utilized in contemporary extremely large-scale integrated circuits (LSIs). The objective of semiconductor technology is to enhance the integration of transistors onto individual chips while simultaneously minimizing their physical dimensions. According to Moore's law, it is anticipated that the dimensions of solid-state silicon circuits will continue to decrease, reaching

[1]gees.sat@gmail.com, [2]radhikabaskar@saveetha.com

DOI: 10.1201/9781003545774-96

the nanoscale level. The assertion that the flourishing of device scaling is imperative for the steady advancement of semiconductor device development holds true. In this study, we aim to investigate the impact of social media on adolescent mental health. The device exhibits suboptimal reliability and notable power dissipation. The obsolescence of silicon-based technology will ensue whenever the channel length of a MOSFET (metal oxide semiconductor field effect transistor) diminishes to a magnitude below 10 nm. One notable characteristic of carbon nanotube field-effect transistors is their capacity to reduce quantum capacitance while concurrently reducing oxide thickness. A plethora of scholarly articles pertaining to CNT-FET technology have been disseminated throughout several academic journals within the past half-decade. IEEE Explore and Science Direct have published a total of 102 and 216 scientific publications, respectively. According to the second source, the primary focus of the study is centered on strategies aimed at decreasing current leakage while simultaneously increasing drain current. To enhance current drain and minimize insulator thickness, this study used La2O3, a Di-electric material with a Di-electric constant of 30 nm, as a substitute for the conventional SiO2 material with a thickness of 3.9 nm. The consideration of leakage current is of utmost importance while simulating the carbon nanotube field-effect transistor (CNTFET) with different Di-electric materials. The optimization of the drain current leads to a reduction in this particular parameter. The threshold voltage exhibited an increase as the chiral pair value decreased from 1.5 nm to 0.1 nm in the thickness of the glass insulator. In addition to other notable electrical qualities, it is worth mentioning that off-state leakage current is among the favorable characteristics exhibited by carbon nanotube field-effect transistors (CNTFETs). Getting the current and voltage to behave like CNTFETs by making the thin oxide gate from 1.5 nm to 0.1 nm is the main goal of this study. The obtained results will then be compared to those of Standard insulation materials in order to enhance conductivity. To tackle this concern, the device's design can include asymmetric gates. The aim of this research is to investigate the utilization of diverse Di-electric materials characterized by different relative constants in order to boost the drain current and mitigate current leakage. Using a simulation tool for nano-scale devices, Di-electric materials with permittivity values of 3.9 and 30 nm are used one after the other to reduce leakage current and increase drain current. The insulator with a relative constant of 30 nm exhibits a larger drain current when paired with a nanotube diameter of 1 nm and a glass insulator thickness of 0.1 nm, in contrast to the material with a value of 3.9, which results in a lower drain current. Consequently, based on the conducted research, it has been observed that La2O3 exhibits a greater drain current in comparison to SiO2. Circuit designers exhibit a preference for maintaining the invariance of Vth with respect to transistor size and biasing circumstances. Consequently, fluctuations in Vth are deemed undesirable when dealing with scaled geometries. In order to restrict leakage power, it becomes imperative to ensure the stability of Vth. As an alternative to conventional silicon devices, carbon nanotube field-effect transistors (CNTFETs) can achieve this.

Table 96.1 Statistical difference between SiO2 and La2O3 with Mean, SD and standard error mean

Group Statistics					
		N	**Mean**	**Std. Deviation**	**Std. Error Mean**
Drain current	Silicon oxide	40	38.1575	9.90058	1.56542
	Lanthanum oxide	40	74.5725	1.19529	0.18899

2. Materials and Methods

The targeted work is conducted at Saveetha School of Engineering, SIMATS. The research work involves two distinct groups. Group 1 uses lanthanum oxide, an insulator material, to enhance the drain current. In contrast, Group 2 employs silicon dioxide, which exhibits a higher leakage current, resulting in a lower drain current when compared to Group 1. Pretest analysis was conducted using Clinicalc.com, with the following parameters: an 80% power level for g-power, a significance threshold of 0.05%, and a confidence interval of 95%. There are a total of 40 samples allocated to each category.[2] The entire sample size consists of 80 individuals. During the preparation of the group one insulators of 30 nm constant were utilized. The obtained outputs were then compared with those of another di-electric with 3.9 nm. This comparison aimed to identify the most suitable insulator material for fabrication purposes. The aforementioned analysis was conducted using the Nano Hub online simulator tool, and the resulting outputs were organized in tabular form, specifically in Tables 96.5 and 96.6. Additionally, a graphical representation shows the rise and falls of error bar between two materials was generated using the Origin-Pro graph plot software tool. This graph is depicted in Fig. 96.3, while Fig. 96.4 showcases the aforementioned software tool. [5]

3. Simulation and Results

The current drain and quantum capacitance of SiO2 and La2O3 was investigated by altering the thickness of the gate insulator from 0.1 nm to 1.5 nm. Previous studies used

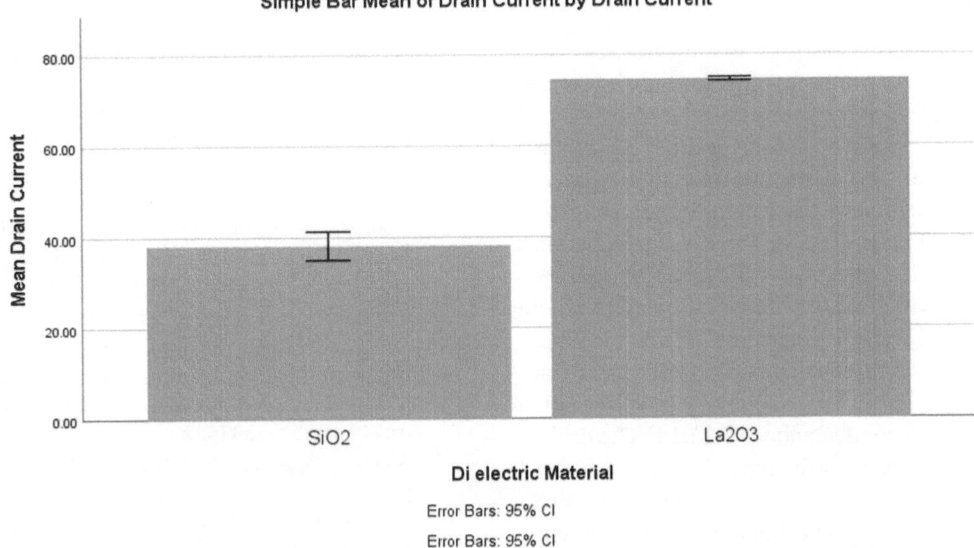

Fig. 96.1 Bar chart comparing the mean drain current of La2O3 and SiO2

the TOPSIS approach to ascertain the optimal Di-electric material for modeling purposes. However, in the present investigation, several insulators were employed to improve the current drain by utilizing the FETTOY tool.[6,7]. By using the Nano Hub modeling tool, a comparative analysis was conducted on the electrical properties of the two samples, leading to the identification of a very effective insulating substance. At the nanoscale scale, quantum capacitance regulates the oxide gate. It is observed that reducing the thickness of the oxide layer leads to a decrease in gate capacitance, which is a desirable characteristic. This reduction in gate capacitance is not achievable in conventional silicon devices.

It is seen that the quantum capacitance of SiO2 material exhibits a declining trend. In contrast, the gate capacitance property of La2O3, as shown in below tables which demonstrates a much lower quantum capacitance when compared to silicon dioxide. The following characteristic curves were presented, depicting the relationship between drain current and voltage, as well as quantum capacitance, and gate voltage. These resultant curves depicts that the use of lanthanum oxide resulted in a significant change in drain current and a decrease in quantum capacitance [8, 9]. This decrease in quantum capacitance led to a reduction in gate capacitance, as the two were directly proportional. Consequently, an increase in drain current was observed, along with a decrease in leakage current, when compared to the Standard material.

Table 96.2 Group statistical difference of quantum capacitance and drain current of di-electric materials

Group Statistics					
Insulator Material		**N**	**Mean**	**Std. Deviation**	**Std. Error Mean**
Quantum capacitance	SiO2	10	2.0410	0.1274	0.04029
	La2O3	10	1.8130	0.0133	0.00423
Drain current	SiO2	10	51.320	13.329	4.21518
	La2O3	10	84.460	3.2894	1.04021

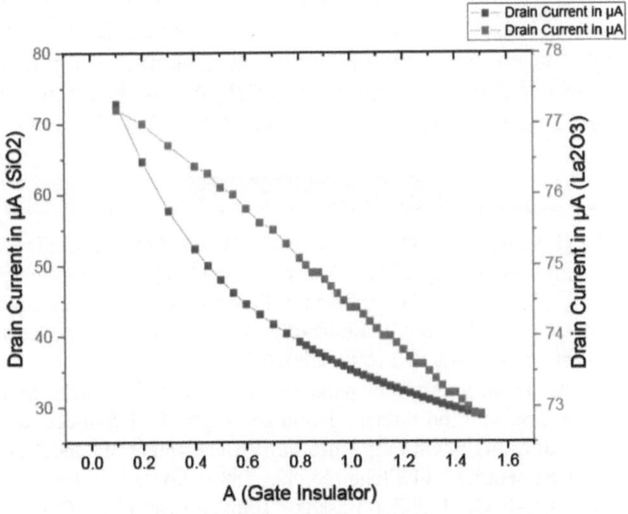

Fig. 96.2 Gate insulator vs drain current

4. Discussion

Researchers are changing the gate oxide thickness of a CNTFET from 1.5 nm to 0.1 nm in order to study its current and voltage properties. Simulations have been conducted to ascertain the relationship between the drain current and gate voltage parameters. When the thickness of the gate oxide is decreased, there is an increase in the conductivity and drain current of CNTFETs.[10,11] The I-V characteristics curves shows that oxide thickness is inversely proportional to drain current, which makes the better conductivity in carbon nanotube field-effect transistors. When the thickness of the oxide in a carbon nanotube field-effect transistor (CNTFET) is raised, there is only a little alteration seen in the current and conductance.[12,13,14] The critical parameters that significantly influenced the conductivity of the drain and the drain current in carbon nanofield effect transistors were thoroughly investigated and outputs were tabulated. The breadth of the CNTFET device plays a critical role when analyzing its characteristics.

5. Conclusion

When parameter variations change the voltage-current curve of carbon field effect transistors (CNTFETs), their performance and conductivity improve. Additionally, CNTFETs exhibit superior stability compared to Standard silicon devices when subjected to gate voltage inputs. The study investigated the impact of altering gate-oxide of di-electric on I-V characteristics of carbon based FETs. The experimental procedure involved maintaining a constant gate voltage while varying these parameters.[15,16,17]The findings were written down as follows: the quantum capacitance was 1.79 $e{\cdot}\pm^{2}$F/cm for lanthanum oxide, and the drain current was 79 μA.[3] In the case of silicon dioxide, the drain current was found to be 67.3 μA, and the quantum capacitance was measured at 1.84 $e^{-12}e$F/c.

References

1. Dokania, Vishesh, Aminul Islam, Vivek Dixit, and Shree Prakash Tiwari. 2016. "Analytical Modeling of Wrap-Gate Carbon Nanotube FET with Parasitic Capacitances and Density of States."IEEE Transactions on Electron Devices. https://doi.org/10.1109/ted.2016.2581119.
2. Analysis of different parameters of channel material and temperature on threshold voltage of CNTFET Sanjeet Kumar Sinha n, Saurabh Chaudhury Department of Electrical Engineering,NITSilchar, Silchar 788010,Assam,India
3. Sathish G, Radhika Baskar. "Implementation of Carbon Nanotube Field Effect Transistor and Comparison of Insulator Material With Standard Silicon Gate Oxides to Improve the Electrical Characteristics and Device Scalability", 2023 2nd International Conference on Vision Towards Emerging Trends in Communication and Networking Technologies (ViTECoN), 2023
4. Sinha, Sanieet Kumar, and Saurabh Chaudhury. 2014. "Advantage of CNTFET Characteristics over MOSFET to Reduce Leakage Power." 2014 2nd International Conference on Devices, Circuits andSystems (ICDCS). https://doi.org/10.1109/icdcsyst.2014.6926211.
5. Sinha, Sanjeet Kumar, and Saurabh Chaudhury. 2014. "Comparative Study of Leakage Power in CNTFET over MOSFET Device." Journal of Semiconductors. https://doi.org/10.1088/1674-4926/35/11/114002.
6. Srimani, T., G. Hills, X. Zhao, D. Antoniadis, J.A. delAlamo, and M.M. Shulaker. 2019."Asymmetric Gating for Reducing Leakage Current in Carbon Nanotube Field-Effect Transistors."Applied Physics Letters. https://doi.org/10.1063/1.5098322.
7. Shaukat, Ayesha, RahilaUmer, and Naz Islam. 2017. "Impact of Di-electric Material and OxideThickness on the Performance of Carbon Nanotube Field Effect Transistor." 2017 IEEE 17th International Conferenceon Nanotechnology (IEEE-NANO). https://doi.org/10.1109/nano.2017.8117461
8. Salcines, Cristino, Aleksei Kruglov, and Ingmar Kallfass. 2018. "A Novel Characterization Technique to Extract High Voltage - High Current IV Characteristics of Power MOSFETs from Dynamic Measurements." 2018 IEEE 6th Workshop on Wide Bandgap Power Devices and Applications (WiPDA). https://doi.org/10.1109/wipda.2018.8569160.
9. Singh, Amandeep, MamtaKhosla, and Balwinder Raj. 2016. "Comparative Analysis of Carbon Nanotube Field Effect Transistor and Nanowire Transistor for Low Power Circuit Design." JournalofNanoelectronics and Optoelectronics. https://doi.org/10.1166/jno.2016.1913.
10. Sinha, Sanieet Kumar, and Saurabh Chaudhury. 2014. "Advantage of CNTFET Characteristics over MOSFET to Reduce Leakage Power." 2014 2nd International Conference on Devices, Circuits andSystems (ICDCS). https://doi.org/10.1109/icdcsyst.2014.6926211.
11. Sinha, Sanjeet Kumar, and Saurabh Chaudhury. 2014. "Comparative Study of Leakage Power in CNTFET over MOSFET Device." Journal of Semiconductor
12. Analysis of Different Gate Di-electric Materials in Carbon Nanotube Field Effect Transistor (CNFET)using Optimization Technique Ankita Dixit and Navneet Gupta Dept. of Electrical and Electronics EngineeringBirla Institute of Technology and sciencePilani, India, 2018 IEEE Electron Device Kolkata Conference (EDKCON), 24–25 November, 2018, Kolkata, India
13. Tiwari, Manas, NidhiAgarwal, and Abhishek Kumar Saxena. 2019. "Impact of Gate Di-electric on Drain Current, Gate Capacitance and Trans-Conductance for MOSFET, Nanowire FET and CNTFET Devices." 2019 International Conference on Cutting-Edge Technologies in Engineering(ICon-CuTE). https://doi.org/10.1109/icon-cute47290.2019.8991465.
14. Y. B. Kim, Y.-B. Kim, and F. Lombardi, "A novel design methodology to optimize the speed and power of the CNT-

FET circuits," in 52nd IEEE International Midwest Symposium on Circuits and Systems, MWSCAS'09, 2009, pp. 1130–1133.

15. S. K. Saha, "Device considerations for ultra-low power analog integrated circuits," in 4th International Conference on Computers and Devices for Communication, 2009 (CODEC 2009), 2009, pp. 1–6.
16. Gajendran, S., Baskar, R. (2024). Stochastic Performance of CNTFET with High 'k' Dielectric Material Over Conventional Silicon Devices in Optimization of Drain Current. In: Shaw, R.N., Siano, P., Makhilef, S., Ghosh, A., Shimi, S.L. (eds) Innovations in Electrical and Electronic Engineering. ICEEE 2023. Lecture Notes in Electrical Engineering, vol 1115. Springer, Singapore. https://doi.org/10.1007/978-981-99-8661-3_47
17. Sathish G, R. Baskar and M. S, "Carbon Enabled Devices in Digital Era: A Predictive Performance Analysis of CNTFET over Si-NWFET with Conventional Double Gate MOSFET and Single Gate MOSFET," 2024 International Conference on Intelligent and Innovative Technologies in Computing, Electrical and Electronics (IITCEE), Bangalore, India, 2024, pp. 1–4, doi: 10.1109/IITCEE59897.2024.10486

Note: Every figure and table was created using the experimental work; none of them were taken from any publications or online.

Advances in Additive Manufacturing Technologies – Gurusamy Pathinettampadian (eds)
© 2024 Taylor & Francis Group, London, *ISBN 978-1-032-90013-1*

97 Experimental Investigation on Mechanical Properties of AA8011 Hybrid Metal Matrix Composites

Yogesh Vaidhyanathan*,
Srinivasan D, Subash V, Payaskhan A
Department of Mechanical Engineering, Vel Tech High Tech Dr. Rangarajan Dr. Sakunthala Engineering College, Avadi, Chennai, Tamil Nadu, India

ABSTRACT: Aluminum metal matrix composites, or AMCs, are becoming more and more popular these days due to their superior mechanical and physical properties on to the monolithic alloys. These properties include stiffness, low density, high specific strength, strength to weight ratio, elastic modulus, and more. They are finding application in the automotive, construction, aerospace, and marine engineering industries, among others. In the current study, chicken eggshell (ES), a byproduct of aviculture that is recognized globally as one of the worst environmental issues, served as the reinforcement. Eggshells are produced in large quantities every day, making them a significant waste. Most eggshells end up in landfills, where they require expensive care. It is cost-effective to process the eggshell waste and turn it into new products. This article aims to provide an overview of the potential uses for eggshells. The development of Al 8011 matrix composites using the Crucible-casting technique is the focus of this work. The goal of the current experiment is to create hybrid metal matrix composites based on aluminum that have constant 1% magnesium content and reinforcements (TiB2-0%, 2%, 4% & Egg Shell -0%, 3%, 5%). Their mechanical properties are also assessed.

KEYWORDS: Aluminium, Composites, Casting, Matrix, Light weight

1. INTRODUCTION

Modern materials and improved design have been made possible by the automotive industry's constant need for lightweight, fuel economy, and comfort. Metal-matrix composites (MMCs) have become more popular due to the growing need in the aerospace and automotive industries for lightweight materials with certain strengths.

PMMCs have been shown to have better wear resistance than the unreinforced matrix metal due to their increased hardness, strength, and thermal conductivity. This chapter provides an overview of the literature's data on effects of different reinforcement modalities, their volume proportion and size. The matrix and the reinforcement are two phases that makes up the metal matrix composites. A variety of aluminum alloys, such as 8011, can be chosen for the matrices, along with different mass reinforcements of titanium dioxide, egg shell, and magnesium with different weight percentages.

Sri saran Venkatachalam et. al, It is discovered that the aluminum 6061 alloy has a wear loss of 159 mm, a tensile strength of 114 N/mm2, and a hardness of 30. It was

*Corresponding author: vyogeshmech@gmail.com

DOI: 10.1201/9781003545774-97

found that adding 8% TiB2 to Al 6061 alloys enhanced their mechanical characteristics, particularly their hardness and strength, which increased to 175 N/mm^2. The composites are now evaluated for wear, microstructure, strength, and hardness in an Al-6061 matrix employing reinforcement TiB2 particles at 4.5 and 7.8 percent by weight. Furthermore, the samples with the highest weight percent of reinforcement show outstanding wear resistance when measured using the pin-on-disc method.TiB2 Aluminum composites are therefore in competition with high performance materials for vehicle brakes.[1].

M. M. Sivaet.al, produced using the stir casting technique and added titanium diboride in different weight percentage compositions. Due to an increase in the weight percent of titanium diboride particles, hardness and corrosion develop. The investigation of aluminum titanium di-boride metal matrix composite microstructure, mechanical properties, and resistance to corrosion, with a particular emphasis on the behavior of the composites during casting and extrusion. Al2011 is composed of silicon, aluminum, lead, zinc, copper, iron, and bismuth. [2]. The hardness, microstructures, flow curve characteristics, and tensile properties of these cast composites were assessed. When compared to gray cast iron, it was found that wear behavior provided a fairly high match with the mechanical properties of composites. In an induction furnace, in situ synthesis was used to create aluminumbased composites with two percent by weight copper reinforced with two and five percent TiB_2 composites. Halide fluxes (K2TiF6 and KBF4) were added simultaneously to the process. [3]. Using gravimetric analysis, the rate of TiB2 produced in the material is calculated. Al 10% Ti and Al 3% B master alloys can react with the Al-6063 alloy to produce in-situ TiB2 reinforced Al 6063 composites using the liquid metallurgical method. Mechanical properties including modulus of elasticity, ultimate tensile strength, and micro hardness have improved by 21%, 47%, and 65%, respectively, in comparison to matrix alloy. [4]. Mechanical characteristics & microstructure of Al 6061 alloy and AlTiB2 composite, which is produced using an in-situ technique and contains 12% by weight of TiB2p, were compared. Metal matrix composite that can be produced by in-situ saltmetal reaction is Al 6061 Al-TiB2. TiB2 particle addition can significantly enhance the characteristics of Al 6061 alloy. [5].The dispersion of TiB2 particles may be seen in SEM micrographs, where they are seen to be hexagon-shaped within the Al matrix. The production of those TiB2 particles was verified by the findings of an X-ray Diffraction (XRD) investigation, which also demonstrated about TiB_2 particles are uniformly spread out in the matrix metal. The composites tensile strength is higher in casts created in permanent molds, and 820°C appears to be the ideal processing temperature in this investigation. Through chemical reactions between titanium, aluminum & boron bearing salts, aluminum alloy A356-TiB2, In-situ MMCs have been successfully synthesized and fabricated with varying pouring temperature and sand mold conditions. They have also been tested in terms of mechanical property and numerical simulation using FEA [6]. It is demonstrated that changes in synthesis techniquesalters the nanoparticles surface characteristics while preserving the material's phase composition. It is demonstrated that each material's nanoparticles display peculiar characteristics. The breadth and surface characteristics of titanium dioxide (ZrO2) and alumina (Al2O3) nanoparticles, which were produced in different ways, were compared using X-ray diffraction techniques, ultraviolet-Vis diffuse reflection spectroscopy, Fourier-transform infrared spectroscopy, and Raman spectroscopy. [7]. In AlTiC& Al TiB_2systems, spontaneous particle entrance and close crystal structure matching combine to produce low particle-solid interface energies and well-defined reinforcing phase spatial distributions in the solidified composite castings. In industrial purity aluminum melts, TiC and TiB2 particles have been spontaneously integrated by removing the oxide layer from the melt's surface using a liquid flux based on K, Al, and F.Most of the particles reside inside the grains, and the reinforcing distribution is not greatly affected by the rate at which the melt cools. When TiC and TiB2 particles are added, modulus increases more than when Al2O3 and SiC are added. [8]. Three different titanium dioxide (TiB2) particle volumes were used to create the composites. The experimental findings are used to determine the mechanical characteristics and microstructure analysis, as well as the manufactured aluminum matrix composite's capacity to withstand various reinforcements. [9].The matrix material of choice is Al6082, while the reinforcement is Titanium Oxide (ZrO2) particles. By using a semi-solid state-casting process, aluminum-ZrO_2Composites reinforced with different weight percentages are created. Optical microscopes (OP) were used to analyze the microstructure of manufactured composites and monolithic metals. The resulting composites, ultimate tensile strength& hardnessare examined. The test findings demonstrate that adding more titanium oxide improves the properties of the composites that are manufactured. Increasingamount of ZrO2 in the composite resulted in improved UTS and hardness. [10]. In the present investigation, mechanical properties of AA8011 hybrid metal matrix compositeshas been studied experimentally.

2. Materials and Methods

2.1 Aluminum-8011

Because of characteristics like its light weight, resistance to corrosion, and ease of maintenance, this metal and its alloys will be in use for a very long time. Aluminum is currently the second most commonly used metal in the world, after iron. This can be attributed to the distinctive blend of appealing characteristics that metal possesses.

- Chemical Composition of Aluminum-8011
- Mechanical Properties of Al8011Aluminum

Fig. 97.1 Unpulverized and pulverized egg shells samples

2.2 Applications of 8011 Aluminum

Power lines, consumer electronics, home and business appliances, aircraft and spacecraft components, and ships are examples of typical applications.

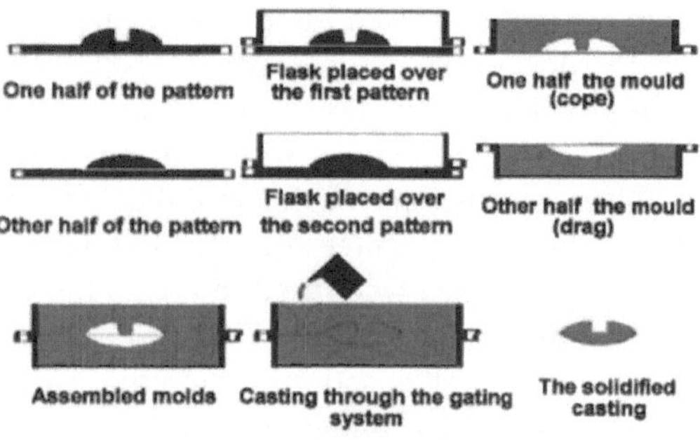

Fig. 97.2 Casting process

2.3 Titanium Diboride

Ceramic substance titanium-di-boride is recognized for its remarkable strength and endurance, which can be attributed to its comparatively high wear resistance, hardness, melting point, strength to density ratio. Nonetheless, it seems that the material's current employment is restricted to specific uses in crucibles, impact resistant, wear-resistant coatings, cutting tools.

2.4 Egg Shell

After rinsing in water to eliminate the membrane, eggshells were stacked in a circular stainless-steel tray. The tray of cleaned egg shells was left in the sun for six hours to dry. After being manually crushed by hand compression; the dried eggshells were pulverized in a planetary ball mill. Following three hours of milling, the mill was offloaded, shows the eggshells in two distinct configurations.

Fig. 97.3 After hardness strength specimen

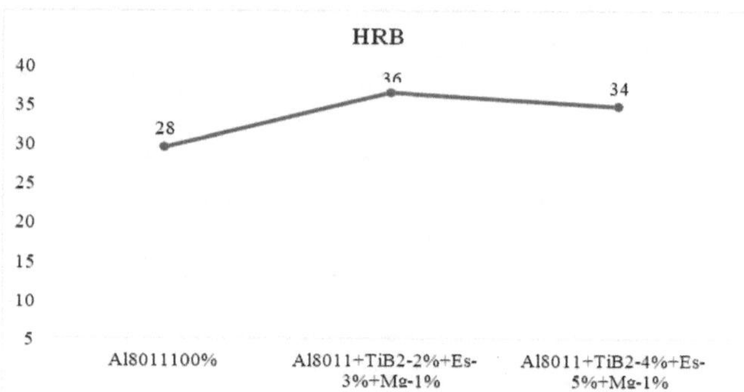

Fig. 97.4 Hardness strength

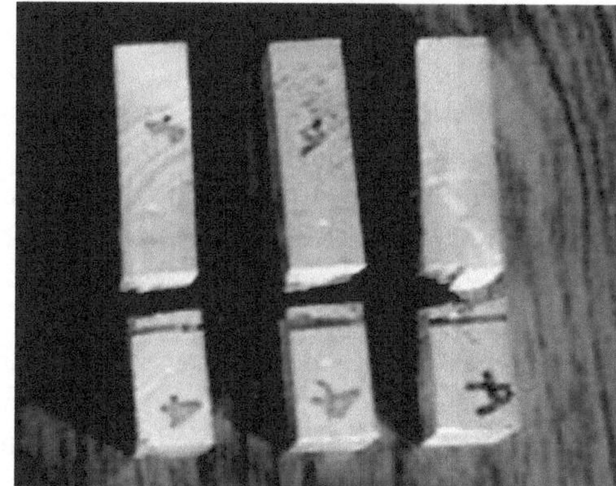

Fig. 97.5 After impact strength specimen

2.5 Magnesium

Magnesium is an atomic number twelve element represented by the symbol Mg. +2 is its typical oxidation number. It is the eighth most common element in the crust of the Earth and the ninth most common element in the known universe. It is an alkaline earth metal. Magnesium makes up 13% of the mass of the planet and a sizable portion of the mantle, ranking it as the fourth most abundant element on Earth overall (after iron, oxygen, and silicon). The ease with which magnesium can accumulate in supernova stars as a result of the addition of 3 helium nuclei to carbon is responsible for its relative abundance. Magnesium ion, a highly soluble element in water, is the third most frequent element dissolved in seawater. At temperatures above

600 megakelvins, the alpha process combines helium and neon to generate magnesium in stars larger than three solar masses. The free metal is a useful component in flares because It burns with a distinctive brilliant-white light. Nowadays, the primary method for obtaining the metal is electrolysis of magnesium salts that are extracted from brine. The metal is mostly used in commerce as an alloying ingredient to create aluminum-magnesium alloys, often known as magnesium or magnesium. Magnesium alloys are valued for their relative strength and light weight since they are less dense than aluminum. According to mass, magnesium is the eleventh most abundant element in human biology. Tasted sour, magnesium ions contribute naturally acidity to fresh mineral waters when present in small amounts. Magnesium was combined at 1% in each composite ratio.

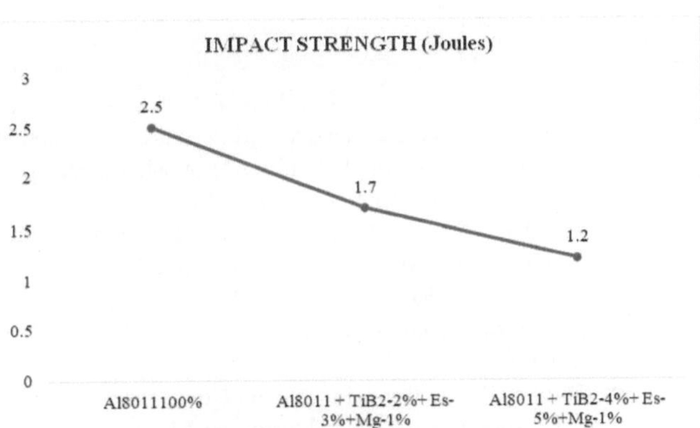

Fig. 97.6 Impact strength

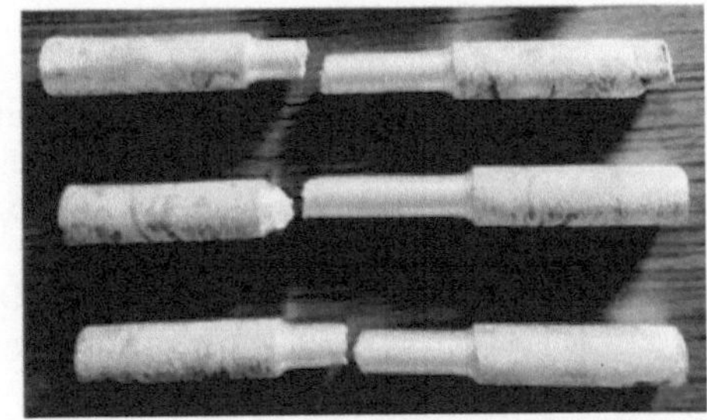

Fig. 97.7 After tensile strength specimen

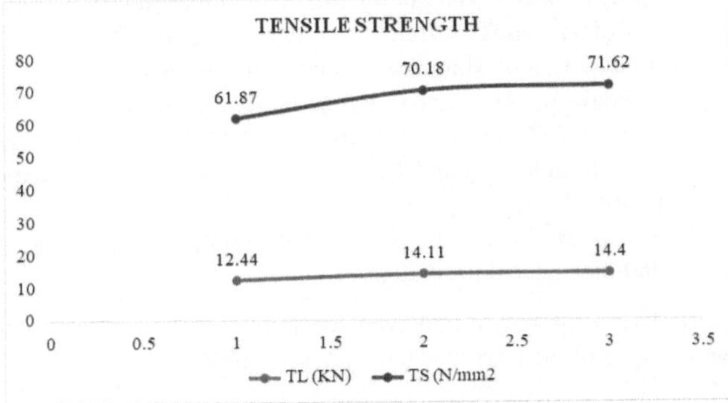

Fig. 97.8 Tensile strength

3. Manufacturing Method

3.1 Casting

The combination of 2 or more elements, one of which is a matrix and the other one is a filler material, results in composite materials. Composites with fibers or particles can be laminated over metal matrix. Materials are handled using powder metallurgy, liquid cast metal technology & specialized manufacturing methods. There are two main procedures involved in processing discontinuous particulate metal matrix material: (1) Powder Metallurgy; and (2) liquid cast metal. There are limitations to the powder metallurgy process, including component size and processing costs. Thus, the only optimal &cost effective way for making aluminum composite materials is the casting method.

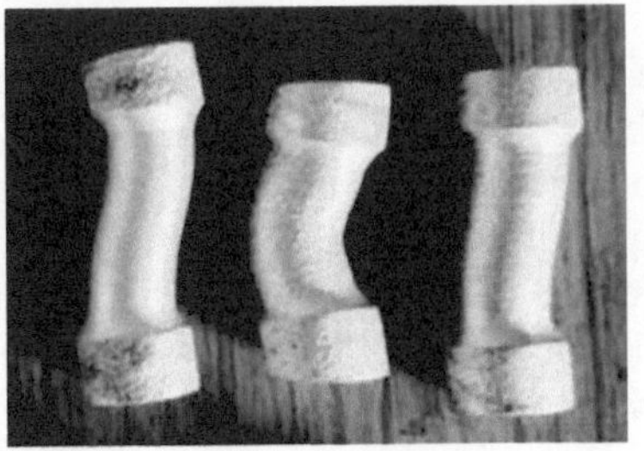

Fig. 97.9 After compression strength

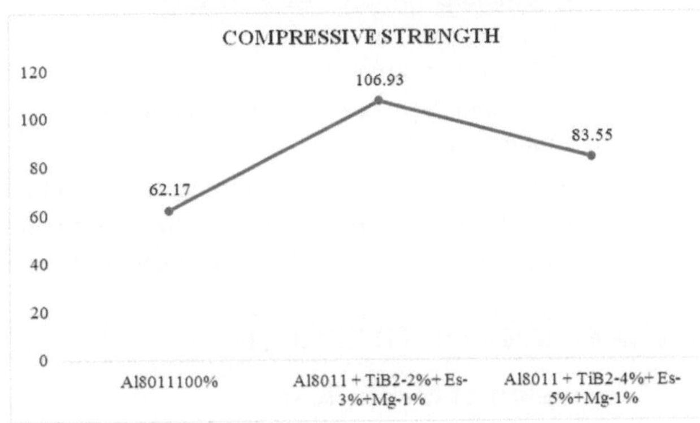

Fig. 97.10 Compression strength

In a graphite crucible, the aluminum rod was melted and alloyed with the necessary number of reinforcements. Sand mold casting was employed in this project to achieve the desired size. Sand is used as the mold material in the metal casting process known as "sand casting," or "sand molded casting." Even for usage in steel foundries, it is adequately refractory and reasonably priced. Sand and an appropriate bonding agent—typically clay—are combined or occur together. To give the clay more strength and pliability and to prepare the aggregate for molding, water is added to the mixture.

A casting made using the sand-casting procedure is also referred to as "sand casting". Foundries are specialist operations that make sand castings. Sand casting accounts for more than 70% of all metal casting production.

Fundamental procedure:

This procedure consists of six steps:

1. To make a mold, place a pattern in the sand.
2. Include sand and the pattern in a gating mechanism.
3. Take the pattern out.
4. Pour molten metal into the mold's cavity.
5. Permit the metal to get cold.
6. Remove the casting by breaking off the sand mold.

3.2 Furnace

(a) One of the earliest and most basic kinds of melting furnace units used in foundries are crucible furnaces. The metal charge is kept in a refractory crucible that is used in the furnace. Heat is transferred through the crucible's walls to the charge. Typically, coke, oil, gas, or electricity are used as the heating fuel. When tiny quantities of an alloy with a low melting point are needed, crucible melting is frequently utilized. Small non-ferrous foundries find these furnaces advantageous due to their initial outlay.

(b) A frequent melting device in foundries is the basic, ancient crucible furnace. Refectory crucibles with metal charges are commonly used in crucible furnaces. The true crucible is a container used to melt things like metals because it can tolerate extremely high temperatures.

This invention uses a crucible furnace with melted coal and composite material made of aluminum and metal. The mentioned ratio was successfully melted.

4. Compression Strength

4.1 Conclusion of Macro Test

Mechanical properties ofAA8011 metal matrix are ultimately assessed based on the experiment increased tensile strength Al8011+TiB2-4%+Es-5%+Mg-1% in a ratio of three, with improved compressive strength and hardness characteristics The ratio of 2:1 Al8011+TiB2-5%+Es-2%+Mg-1% is stronger than the others. Impact strength, however, reveals a greater ratio-1 Al 6082-100%. The impact strength of the metal matrix Al8011 is decreased as a result of reinforcement agglomeration. The impact strength is decreased by reinforcement.

5. Conclusion

It is discovered that the Al8011 matrix alloy has effectively integrated titanium dioxide and chicken eggshell (TiB2 & ES) particles by the use of the crucible casting procedure. When compared to conventional materials, composite materials — particularly those made of aluminum 8011, titanium dioxide, egg shell, and magnesium—have better mechanical qualities. These materials are used in various industrial applications because of great toughness and light weight. After conducting an inquiry and analyzing the properties of Al 8011 metal matrix, it was discovered that titanium di boride reinforcement abnormally improved the mechanical qualities. Particularly Tensile strength enhanced Ratio-3Al8011+TiB_2-4%+Es-5%+Mg-1% and hardness properties and compressive strength enhanced Ratio-2Al8011+TiB_2-5%+Es-2%+Mg-1% is superior strength shows higher than bare Al8011 alloy and ratio-1. But impact property obtained maximum at ratio-1 (Al8011-100%).

References

1. Srisaran Venkatachalam, Titanium Diboride Reinforced Aluminum Composite as a Robust Material for Automobile Applications, International Conference on Materials, Manufacturing and Machining, AIP Conf. Proc. 2019.
2. M.Siva, Analysis of Microstructural, Corrosionand Mechanical Properties of Aluminum Titanium Diboride Particles (Al-TiB2) Reinforced Metal Matrix Composites (MMCs), Indian Journal of Science and Technology, Vol9 (43), 2016.
3. N.B.Dhokey, Wear Behavior and Its Correlation with Mechanical Properties of TiB2 Reinforced Aluminum- Based Composites, Advances in Tribology, 2011.
4. C.S. Ramesh, Development of Al 6063–TiB2 in situ composites Materials and Design,31(2010)2230–2236.
5. T.V.Christy A, Comparative Study on the Microstructures and Mechanical Properties of Al 6061 Alloy and the MMC Al 6061/TiB_2/12P, Journal of Minerals & Materials Characterization & Engineering, Vol. 9, No. 1, pp. 57–65, 2010.
6. C. Rajaravi, Analysis on Tensile Strength of Al/TiB2 MMCs in FEA for Different Mould Conditions, IOSR Journal of Mechanical and Civil Engineering, Volume 13, Issue 3, 2016.
7. S. Suresh, Aluminum-Titanium Diboride (Al-TiB2)MMC, Procedia Engineering, 38(2012) 89 – 97.
8. A. R. Kennedy, The microstructure and mechanical properties of TiC andTiB2-reinforced cast metal matrix Composites, Journal of Materials Science, 34 (1999) 933–940.
9. P. Pradeep, Characterization of Particulate reinforced Aluminum 7075 /TiB2 Composites, International Journal of Civil Engineering and Technology, Volume 8, Issue 9, September 2017.
10. K. L. Tee, In-situ stir cast Al-TiB2 composite: processing and mechanical properties.

Note: Every figure and table was created using the experimental work; none of them were taken from any publications or online.

Advances in Additive Manufacturing Technologies – Gurusamy Pathinettampadian (eds)
© 2024 Taylor & Francis Group, London, ISBN 978-1-032-90013-1

98 Innovative Assisitance for the Visually Impaired: Face Recognition, Navigation, Water and Fire Safety with Raspberry PI

S. D. Kumar[1]
Asst. Professor, Department of Mechanical Engineering,
SRM Institute of Science and Technology,
Ramapuram - Chennai, India

Balaji. S[2], **Arul Selvan**[3],
VIJAY[4], **Yokesh Kumar**[5]
U.G Student, Department of Mechanical Engineering,
SRM Institute of Science and Technology,
Ramapuram - Chennai, India

ABSTRACT: This paper introduces a novel system created to improve the mobility and independence of those who are blind or visually impaired through the integration of face recognition technology and navigation capabilities. Leveraging the power of the Raspberry Pi single-board computer, this system provides a cost-effective and efficient solution to assist blind people in recognizing faces and navigating their surroundings. The face recognition component of the system utilizes state-of-the-art machine learning algorithms and computer vision techniques to identify individuals' faces in real-time. It assists blind users in recognizing and interacting with people they encounter, offering an improved social experience and increased security. The navigation aspect of the system employs a combination of obstacle detection, fire detection, face detection and panic detection Using, fire sensor, a Raspberry Pi camera module and button. From these sensors any one will exceed the set value voice alert keeps indicating the user, and alert message will send to the authorized person using GMS along with the location using GPS. Also the panic button could be used here if the user press the panic button it will send the alert message to the authorized person along with the location. And the face can be recognized using camera by trained set in raspberry pi and tells the names who are them. In additional we added things identification like table, rest rooms etc., path hole identification and night vision technology to increase security purposes. This features increase the confidence level of visually impaired user.

KEYWORDS: Visually impaired assistance, Assistive technology, Raspberry Pi, Face recognition, Navigation system, Safety features (water & fire safety), Object detection, Things identification, GPS & GMS

1. INTRODUCTION

In our modern life a normal man face lot of troubles in day to day life such as traffic, unpredicted abstracts, path holes in roads and hazards things like fire and water. In today's fast- paced world, technological innovations have the potential to improve the lives of people with disabilities, including the visually impaired. This introduction provides

[1]kumars@srmist.edu.in, [2]balajisiva1507200@gmail.com, [3]ap4086@srmist.edu.in, [4]vg0950@srmist.edu.in, [5]yk4613@srmist.edu.in

DOI: 10.1201/9781003545774-98

a brief overview of the transformative role of Raspberry Pi-based solutions in assisting the visually impaired in various aspects like face reorganization to recognize familiar or unfamiliar faces. The following subtopics will direct us to specific applications and innovations that leverage Raspberry Pi technology to empower and enhance the independence of visually impaired individuals.

1.1 Face Recognition for Enhanced Social Interaction

Explores how Raspberry Pi-powered facial recognition systems can enable visually impaired individuals to identify and interact with people more effectively. These systems use AI and machine learning to provide real-time information about individuals in their proximity, making social interactions more accessible and engaging. The system employs facial recognition technology to help visually impaired individuals identify people in their vicinity. Using a camera module connected to the Raspberry Pi, the system can capture and process images to recognize and announce the names of individuals, enhancing social interaction and security.

1.2 Navigation Assistance for Independence

Examines how Raspberry Pi, in combination with sensors and GPS technology, is employed to create advanced navigation solutions. These devices assist visually impaired individuals in safely navigating their environment, avoiding obstacles, and reaching their desired destinations with increased confidence. Built-in GPS and obstacle detection sensors enable the system to provide real-time navigation support. It can help users navigate both indoor and outdoor environments, providing audio instructions and alerts to avoid obstacles and reach their desired destinations.

1.3 Water and Fire Safety Innovation

The system includes sensors for detecting water and fire hazards. When these sensors detect a potential risk, the system issues audible alerts to notify the user and help them take prompt action, enhancing safety in home and public settings.

Highlights the development of Raspberry Pi-based solutions that focus on ensuring the safety of visually impaired individuals in critical situations. These systems can detect water and fire hazards, providing timely alerts and guidance to help prevent accidents and protect lives.

1.4 Empowering the Visually Impaired with SWI Raspberry-Pi Project

Discusses the increasing popularity of Raspberry Pi as a platform for Survey with independence (SWI) projects that empower visually impaired individuals. It covers a range of innovative applications, from creating custom accessible devices to programming Raspberry Pi for personal assistance.

2. Experimentation Summary

2.1 Proposed System

In this project we can use the face recognition, ultrasonic sensor, fire sensor, gps, gsm, buzzer, panic button and water sensor. The face can be recognized using camera by trained set in raspberry pi and tells the name who are them . The ultrasonic sensor can detects the object. In this two mode can introduced, if the object will detected it can says the name of an object. The fire sensor will detect the fire , water sensor will sense the water. From these sensors any one will exceed the set value the buzzer keeps starts the sound ,the alert message will send to the authorized person using GSM along with the location using GPS. Also the panic button could be used here .If the user press the panic button it will send the alert message to the authorized person along with the location. (refer with: Fig. 98.1)

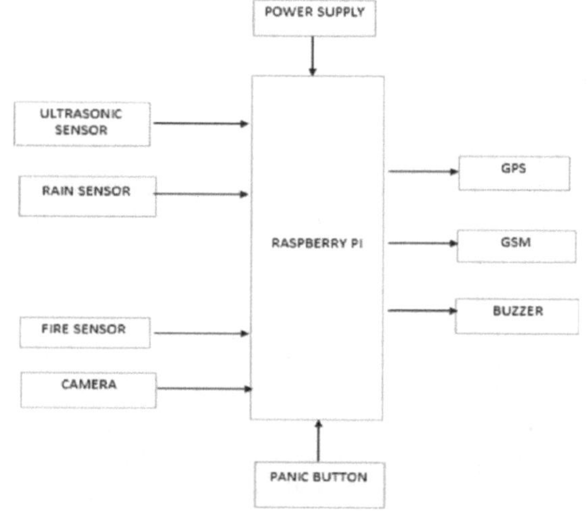

Fig. 98.1 Block drawing of sensors with raspberry-Pi

2.2 Hardware Requirments

- Power supply
- Rapberry pi
- GPS
- GSM
- Buzzer
- Panic button
- Fire sensor
- Ultrasonic sensor
- Raspberry pi- Camera with night vision.
- Rain sensor

2.3 Raspberry PI

For friends, the Raspberry Pi, or Raspi, is a little computer (refer with: Fig. 98.2). Everything you may possibly need is there, including a 700MHz ARM CPU, RAM, USB, Ethernet, and—most importantly—a comb that is welded to the electronics board and directly connected to the processor (see Fig. 98.3). With this comb, we can have directly controlled digital inputs and outputs that are freely adjustable by the operating system. The Ras.-pi comes in two flavors: model A and model B.

Fig. 98.2 Processor

The hardware is where the two models differ; the model B includes two USB ports instead of one, an Ethernet connector, and more RAM (512MB against 256Mb for the model A). It is evident that the cost differs: just now, While the model A costs only €28, the B model costs €37. Because Model A weighs less, it is the better choice for embedded (or dedicated) applications where weight is a concern and when Model B's added functions are unnecessary. However, compared to model A, model B is far more adaptable as it can handle more applications and procedures and can be connected to a computer network.

Every model has an on-chip graphics processing unit (GPU, a VideoCore IV) and an ARM- compatible central processing unit (CPU) that are part of a Broadcom system on a chip (SOC). The Pi 3 has an onboard memory range of 256 MB to 1 GB RAM and a CPU speed that spans from 700 MHz to 1.2 GHz. Program memory and the operating system are stored on Secure Digital (SD) cards, which come in SDHC and MicroSDHC formats. The majority of boards feature a 3.5 mm phone connector for audio, one to four USB slots, HDMI, and composite video output. There are several GPIO pins that handle standard protocols like I C and give lower level output. An 8P8C Ethernet port is available on the B-models, and the Pi 3 and Pi.

2.4 Features in Raspberry PI

- 1GB RAM;
- 1.2GHz Single Board Computer powered by Broadcom BCM2837 64-bit ARMv7 Quad Core Processor
- Onboard BCM43143 WiFi
- Onboard Bluetooth Low Energy (BLE)
- An expanded 40-pin GPIO
- Four USB 2.0 ports.
- Full size HDMI; CSI camera port for connecting the Raspberry Pi camera; DSI display port for attaching the Raspberry Pi touch screen display; Four pole stereo output and Composite video port;
- Data storage and operating system loading via the Micro SD connector; upgraded switching Micro USB power source (now provides up to 2.4 Amps)
- The LEDs will shift positions, but the form factor should be the same as the Pi 2 Model B.

The PiCamera v2

The PiCamera v2 is the updated version of this device that was released in April 2016 and took the place of the PiCamera v1.3, (refer with: Fig. 98.3) which is now outdated. It continues to work with all current Raspberry models. Actually, nothing has changed about the flat wire; it may still be placed into the designated connector on the top side of any Raspberry board. The OmniVision OC5647, which had a 5-megapixel camera, was replaced by the new module, which mounts an 8-megapixel Sony IMX219 sensor. With the help of the useful and innovative Raspberry Pi camera module, you may utilize the small computer for a multitude of imaginative and useful tasks. But in order to fully utilize the features of the camera, you must be at least somewhat proficient in Python or Bash scripting. If you wish to access and control the Raspberry Pi with the camera module remotely, you can use a graphical application such as RPI Camera GUI.While the RPI Camera GUI can be accessed using SSH, the RPi Cam Web Interface (RPCWI) software is a far more effective tool for the task. As its name implies, this web program offers an easy-to-use interface that enables you to adjust almost all camera settings with a standard browser. RPCWI is a great tool for working with the camera module because it has a number of other very helpful functions. This includes support for time-lapse and

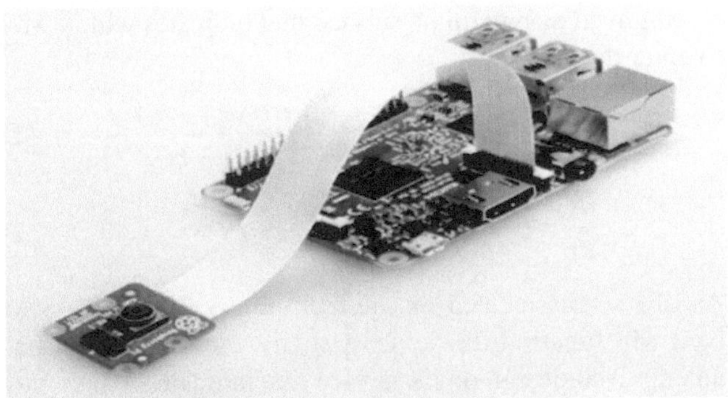

Fig. 98.3 Raspberry Pi Board

scheduled shooting and recording, motion-triggered image capture, previewing, managing, and downloading saved images and videos, among many other features.

GSM

SIM800L GSM/GPRS Module

A SIM800L GSM (refer with: Fig. 98.4) cellular chip from Simcom powers the module.

Fig. 98.4 GSM Module

Because the chip can operate between 3.4 and 4.4 volts, a direct LiPo battery supply is a perfect fit for it. Because of this, it's a fantastic choice for integration into small-scale applications.

The SIM800L GSM chip has all of the pins needed for data communication separated out into headers with a pitch of 0.1″, including the pins needed to communicate with the microcontroller via UART. With auto-baud detection, the module supports baud rates ranging from 1200bps to 115200bps.

In order to connect the module to the network, an external antenna is required. A helical antenna that can be soldered onto the module is typically included with the module. Additionally, the board has a U.FL.

A SIM slot is located on the rear! Every 2G Micro SIM card will function flawlessly. Usually, the SIM socket has an engraved design on its surface that indicates which way to enter the SIM card.

$$\frac{dx}{dt} = \frac{\alpha x}{1+Ky} - bx^2 - \frac{(\beta+\sigma y)(1-m)xy}{1+a(1-m)x} \tag{1}$$

$$\frac{dy}{dt} = \frac{c(\beta+\sigma y)(1-m)xy}{1+a(1-m)x} - \gamma y \equiv G(x,y) \tag{2}$$

A SIM slot is located on the rear! Every 2G Micro SIM card will function flawlessly. Usually, the SIM socket has an engraved design on its surface that indicates which way to enter the SIM card.

Features IN GSM

- Despite its small size of about 1 inch 2, this module is surprisingly feature-rich. Following is a list of a few of them:
- Accommodates quad-band: DCS1800, PCS1900, EGSM900, and GSM850
- Use any 2G SIM card to connect to any GSM network worldwide.
- Use an electret microphone and an external 8Ω speaker to make and receive voice calls.
- Text message sending and receiving
- Transfer and receive GPRS data (HTTP, TCP/IP, etc.).
- Listen for and receive FM radio transmissions.

2.5 Ultrasonic Sensor

By measuring how long it takes for an ultrasonic pulse to travel from the transducer to the item and back, ultrasonic sensors create ultrasonic pulses. An item reflects the sound waves that the transducer emits, and the transducer receives the reflected waves back. The ultrasonic sensor will go from emitting sound waves to receiving them. The distance of the object from the sensor determines how long it takes for an emission to be received.

When the ultrasonic transmitter launched, it began timing and released an ultrasonic pulse in one direction. As Ultrasonic traveled through the air, it would quickly return if it came into any barriers. When the ultrasonic receiver finally gets the reflected wave, it will halt timing. It calculates the sensor's distance from the target item. It provides consistent readings and exceptional non-contact range sensing with great precision in an intuitive design. Sunlight or dark materials have no effect on its ability to function. The sensor is powered by a 5VDC supply voltage. To provide digital input, the sensor's two pins—trig and echo—are linked to the controller.

2.6 Fire Sensor

A flame detector is a type of sensor that is intended to identify and react to flames or fires. Additionally, it is capable of detecting common light sources between 760 and 1100 nm in wavelength. There is a maximum detecting distance of 100 cm. An analog or digital signal can be output by the flame sensor. It can be utilized in firefighting robots or as a flame alarm.

2.7 Features

- It's easy to operate, has a quick response time, and is highly photosensitive.
- Sensitivity can be changed.

- The angle of detection is 600.
- It reacts to the range of flames.
- It is possible to alter accuracy.
- This sensor operates between 3.3V and 5V.

2.8 Description of Software Requiremnts

Python is an object-oriented, high-level, general-purpose, interpreted programming language. Guido van Rossum designed it between 1985 and 1990. The GNU General Public License (GPL) is used to distribute Python source code, much like it is for Perl. This lesson provides sufficient knowledge about the Python programming language.

Python is an object-oriented, interpreted, high-level programming language. The design of Python emphasizes readability. It contains fewer syntactical structures than other languages and usually uses English terms in contrast to other languages that use punctuation.

Variety of applications, such as simple text processing, games, and Web browsers

Python Functions

Among the features of Python are:

Easily learned: Python boasts a concise syntax, a straightforward structure, and minimal keywords. This enables the learner to acquire the language.

Simple to read: Python code is more readable and visually appealing. Easy to maintain: The source code of Python is not too difficult to maintain.

A large standard library: The majority of Python's library is cross-platform compatible and highly portable on UNIX, Windows, and Macintosh.

Interactive Mode: Python comes with an interactive mode that lets you test and debug small portions of code interactively.

Portable: Python runs with the same interface across a broad range of hardware devices.

Extendable: The Python interpreter can have low-level modules added to it. Programmers can enhance or modify these modules to make their tools more effective.

Databases: All of the major commercial databases have interfaces available in Python.

GUI Programming: Python allows the creation and porting of GUI applications to a variety of system calls, libraries, and Windows systems, including Macintosh, Windows MFC, and Unix's X Window system.

Scalable: Python offers larger projects more organization and assistance than shell programming.

In addition to the characteristics mentioned above, Python has many other useful features, a few of which are mentioned below—

- It supports OOP in addition to structured and functional programming techniques.
- It may be translated to byte-code for developing complex programs, or it can be used as a scripting language.
- It allows dynamic type checking and offers very high-level dynamic data types.
- It is compatible with automated trash pickup. It's simple to integrate with Java, C, C++, COM, ActiveX, and CORBA.

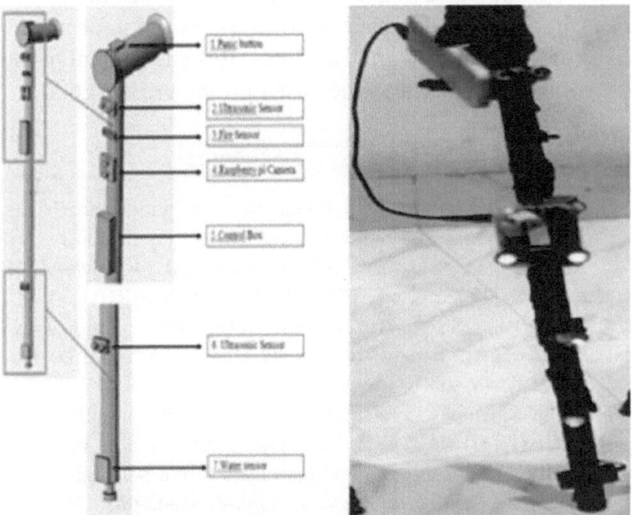

Fig. 98.5 Final design with actual project

3. CONCLUSION

The project, "Innovative Assistance for the Visually Impaired: Face Recognition, Navigation, Water and Fire Safety with Raspberry Pi," represents a significant step forward in utilizing technology to enhance the lives of visually impaired individuals. Through this project, we have successfully combined various functionalities and applications, all centered around the goal of improving the safety, independence, and quality of life for the visually impaired. Here are the key conclusions drawn from this project. The project's face recognition and navigation systems empower visually impaired individuals to navigate their surroundings with greater ease and independence. By harnessing the capabilities of Raspberry Pi, we have provided a valuable tool for mobility and personal autonomy. The inclusion of water and fire safety components underscores our commitment to ensuring the well-being of visually impaired individuals. The water and fire safety systems offer a layer of protection that can be life-saving in critical situations. Raspberry Pi is an affordable and accessible platform for implementing these solutions, which can be vital in reaching a broader audience of individuals with visual impairments. This not only makes the project cost-effective but also allows for scalability and widespread adoption. This project serves as a foundation for future innovations and improvements. It offers ample room for further development and refinement, such as expanding the capabilities of the face recognition system, integrating more advanced navigation features, and enhancing the safety components. The positive impact of this project extends to the visually impaired community and beyond. It promotes inclusivity, equal opportunities, and safety for a segment of the population that often faces challenges in daily life. The project exemplifies the power of technology to address real-world issues and improve the lives of individuals with disabilities. It is a testament to the potential of technology when used for the greater good.

REFERENCES

1. Kiran G Vetteth, Prithvi Ganesh K andDurbhaSrikar, (2013),"Collision Avoidance Device For Visually Impaired", (C.A.D.V.I), INTERNATIONAL JOURNALOF SCIENTIFIC & TECHNOLOGY RESEARCH, VOLUME-2, ISSUE-10, Page(s): [185–188], http://www.ijstr.org/final-print/oct2013/Collision-Avoidance-Device-For-Visually-Impaired-C.a.d.v.i.pdf
2. Isha S. Dubey, Jyotsna S. Verma, Ms. ArundhatiMehendale,(2019) "An Assistive System for Visually Impaired using Raspberry Pi", International Journal of Engineering Research & Technology, Volume-8, Issue-05, Page(s): [606–613], **DOI: 10.17577/IJERTV8IS050178**, https://www.ijert.org/an-assistive-system-for-visually-impaired-using-raspberry-pi
3. ThaeLinn, Ali Jwaid, Steve Clark, (2017) "Smart Glove for Visually Impaired", Computing Conference 2017", IEEE, Issue-18, Page(s): [1323–1329], **DOI:** 10.1109/SAI.2017.8252262, https://ieeexplore.ieee.org/document/8252262/authors#authors
4. Steven Edwin Moore, (October 2002) "Drishti: An Integrated Navigation System for the Visually Impaired and Disabled", IEEE, Issue 8, **DOI:** 10.1109/ISWC.2001.962119, https://www.researchgate.net/publication/221240937_Drishti_An_Integrated_Navigation_Syste m_for_Visually_Impaired_and_Disabled.
5. Anitha. J, Subalaxmi. A Vijayalakshmi. G,(June 2019) "Real-Time Object Detection For Visually Challenged Person", International Journal of Innovative Technology and Exploring Engineering, Issue-8, Retrieval Number: H6339068819, https://www.ijitee.org/wp- content/uploads/papers/v8i8/H6339068819.pdf
6. Prof. Dr. Netra Lokhande, Bhushan Garware, SanikaDosi, ShivaniSambare, Shashank Singh, (January 2018) "Android Application for Object Recognition using Neural Networks for the Visually Impaired " 2018 Fourth International Conference on Computing Communication Control and Automation IEEE, Issue 4, **DOI:** 10.1109/ICCUBEA.2018.8697886, https://www.researchgate.net/publication/332678951_Android_Application_for_Object_Recogni tion_Using_Neural_Networks_for_the_Visually_Impaired
7. HeikaGada,VedantGokani, AbhinavKashyap, AmitDeshmukh, "Object Recognition for the Visually Impaired "(January 2020),Springer Nature Singapore Pte Ltd. 2022A. Kumar et al. (eds.), Lecture Notes in Electrical Engineering 783, **DOI:** 10.1109/ICNTE44896.2019.8946015, https://ieeexplore-ieee-org-srmrmp.knimbus.com/document/8946015
8. Wenjun Wang, Xiangkang Huang, XinShu, An, Implementation of Face Recognition System with Face Presentation Attack Detection on Raspberry Pi , School of Data Science and Information Engineering, IEEE International Conference on e-Business Engineering (ICEBE), **DOI:** 10.1109/ICEBE52470.2021.00031, October 2017, Page(s): [70–75], https://ieeexplore- ieee-org-srmrmp.knimbus.com/document/9750178
9. Umme-e-laila, Muzammil Ahmad Khan, (November 2017) "Comparative Analysis for Real Time Face Recognization System Using Raspberry Pi", Dept. of Comp IEEE^{4th}, Proc of 2017,. [Page(s): 28–30], **DOI:** 10.1109/ICSIMA.2017.8311984, https://ieeexplore-ieee-org- srmrmp.knimbus.com/document/8311984
10. Smit Malkan, Sampras Dsouza, Soumyaprakash, Dasmohapatra(2021) "Navigation using Object Detection and Depth Sensing for Blind People", 2021 IEEE 4th International Conference on Computing, Sep 24–26, 2021, **DOI:** 0.1109/GUCON50781.2021.9573981. [Page(s): 1–7],

Note: Every figure was created using the experimental work; none of them were taken from any publications or online.

Advances in Additive Manufacturing Technologies – Gurusamy Pathinettampadian (eds)
© 2024 Taylor & Francis Group, London, ISBN 978-1-032-90013-1

99 Comprehensive Review of Carbon Enabled Transistors for Ultra-Low Power Electronics in Integrated Circuits Over Silicon Devices

Sathish G[1]
Research Scholar, Institute of ECE,
Saveetha School of Engineering, Saveetha Institute of Medical and Technical Sciences, Saveetha University, Chennai, India

Radhika Baskar[2]
Professor, Research Guide, Institute of ECE,
Saveetha School of Engineering, Saveetha Institute of Medical and Technical Sciences, Saveetha University, Chennai, India

ABSTRACT: This review paper establishes about the dielectric materials, threshold voltage and scalability of CNTFET devices. To use a solution-processable technology that incorporates a shadow-mask deposition, field-effect semiconductors based on carbon nanotubes could be fabricated in a simple, swift, and reliable manner. This technique allows for direct control of network density. An organized investigation of the variables affecting transistor performance has been conducted, and the results offer recommendations for future device enhancement. The different dielectric materials were employed to examine the features of CNTFET devices. Due to the distinctive atomic structure, carbon nanotubes are identical to graphite and diamond in aspects of their physicochemical characteristics. Cutoff voltage of CNTFET devices grows with decreasing channel length in the deep nano scale zone, whereas the threshold voltage of a MOSFET device declines drastically. Traditional MOSFET devices fail to match expectations when channel lengths are less than 10 nm.

KEYWORDS: CNTFET, MOSFET, Device scaling, Threshold voltage, Temperature

1. INTRODUCTION

Metal oxide semiconductors (MOS) represent a crucial class of silicon chips that find extensive utilization in contemporary very large-scale integrated circuits (VLSI). The objective of semiconductor technology is to enhance the miniaturization and integration of transistors into individual chips. According to Moore's law, there is an anticipation that the dimensions of devices will continue to decrease due to a significant advancement in solid-state silicon-based technology, resulting in the reduction of device size to the nanoscale scale. Device scaling should indeed continue to thrive in order for semiconductor device development to advance steadily. Sinceoxide thickness inversely proportional to the leakage current, results in high power dissipation and poor reliability of the device when the length

[1]gees.sat@gmail.com, [2]radhikabaskar@saveetha.com

DOI: 10.1201/9781003545774-99

of the MOSFET channel reduced less than 10nm will put silicon based technology to an end. By substituting a single carbon nanotube for the traditional MOSFET structure's channel material, it was possible to overcome all limitations, including the exponential growth in leakage currents in scaled devices, by better understanding device mechanics and enhancing device performance..

Carbon nanotubes (CNTs) are seamless cylinders made from graphite sheets that have a Nano scale diameter and a micron length. CNTs are highly reliable and adaptable molecular-based heat-transfer devices. CNTs can have a metallic or semiconducting lattice structure which can be used as molecular wires in nanotechnology and scanning probe microscopy for programmable switches, single-electron transistors (SETs), diodes, optoelectronic devices, and memory devices. Nanotechnology, membranes, capacitors, polymers, metallic surfaces, ceramics, and nanomedicine all use carbon nanotubes. The structure of carbon nanotubes is investigated using SETM, STEM, electron diffraction, X-ray diffraction, photoelectron spectroscopy, infra - red, and FTIR analysis

1.1 Device Structure of MOSFET

Field effect transistors made on metal oxide semiconductors are 4 terminal devices that do not operate if the gate and source are not connected to a positive terminal. Figure 99.1 illustrates that when a positive source is given to the source gate, electrons within the n type substrate are pushed away while holes in the p-type areas underneath the source and drain electrodes are drawn in. A positive voltage is applied to the gate of an N-type MOSFET. Electrons in the negatively charged areas below the source and drain electrodes attract the protons of the p-type substrate, as shown in figure. Consequently, the formation of a layer directly beneath the gate oxide on the silicon surface occurs, facilitating the movement of electrons from the source to the drain. Consequently, the voltage between the gate and source is positive.

A channel is established connecting the oxide and the p-Si. The ongoing growth of semiconductor devices and the proliferation of components on a single chip have resulted in an upsurge in circuit designs, improved technology, and device advancement. Due to short channel effects, a MOSFET's properties change as the gate length is decreased. [1] With short gate lengths, MOSFETs experience a number of undesirable parameter shifts. Increased output conductance, larger leakage current, and threshold voltage shifts are a few examples. Over the past few decades, integrated circuit characteristics have experienced a significant size reduction.

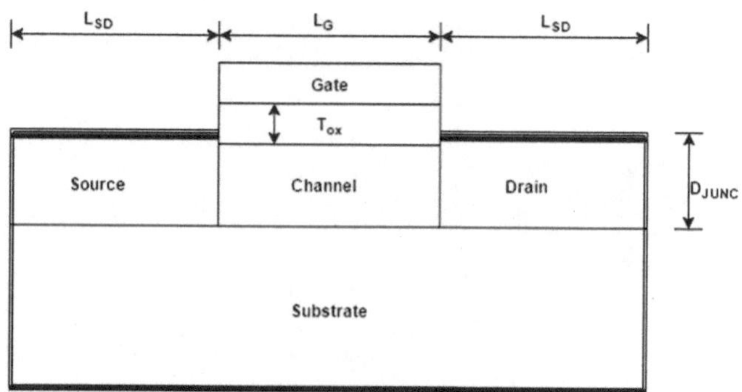

Fig. 99.1 Standard N-type bulk MOSFET

1.2 Carbon Nanotube Field Effect Transistor

As shown in Fig. 99.2 below, a CNTFET is a semiconductor device that bridges the source and drain contacts using a single or an array of conducting polymer nanotubes. This device serves as a carrier channel that can be switched between states of 1 or 0 using the gate interface.

The metal (drain/source) connections at the nanotubes generate a Schottky barrier that is modulated on and off by the CNTFET [2,3,4].

The production of FETs has relied on silicon as a key component material. Despite ongoing issues, researchers suggested CNT as a different source for conventional SEDs. There are a number of reasons that will limit the scaling of silicon devices in the future, with standby power dissipation, variable sizes, and declining performance being the most important ones. A rise in the density and specifica-

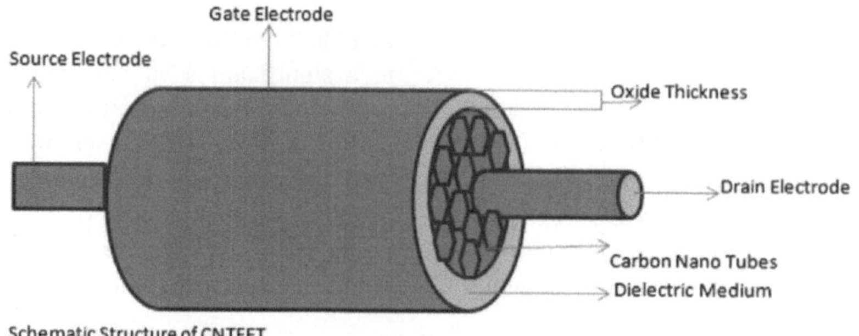

Fig. 99.2 Schematic structure of CNTFET

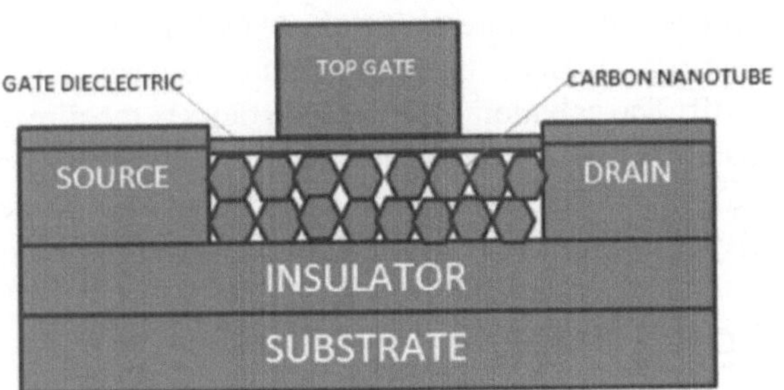

Fig. 99.3 Structure of graphene sheet [15]

tions of devices The power problem is provoked by leakage and tunneling gate current with the increase in device temperatures. Nano electronics-based device includes CNT, nanowires, molecular electronic devices, SEDs, tunneling diodes, and qualitative comparative analysis

1.3 Structure of Bucky Tubes

Carbon nanotubes (CNTs) are Bucky tubes a one-of-a-kind invention in the realm of nanotechnology, and they are a significant substance for the future because of its large output. Carbon-based sphybridisation uses a variety of structures with the honeycomb arrangements [5, 6] in addition to graphite.

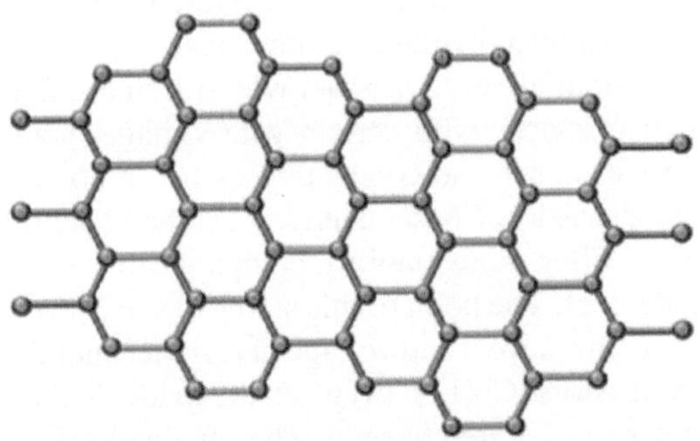

Fig. 99.4 Structure of graphene sheet [19]

A 2D single sheet of graphite called graphene possesses sphybridisation that is stronger than the sphybridisation of diamonds [7, 8]. Carbon nanotubes are made by rolling graphene sheets into cylinders. Carbon nanotubes have become one of the most intriguing fields of research in recent decades. Carbon nanotubes are a tube-shaped substance consisting entirely of carbon. It has a nano scale [9] measurement because its diameter is too small.

1.4 Effects of Device Scaling

Device and circuit designs for more advanced technologies are influenced by the ongoing semiconductor device scalability and the expansion of the more components in a chip. When a MOSFET's length is reduced, the entrance displays the alterations in the device's properties brought on by the small channel. Conventional Bulk MOSFETs experience considerable performance reduction when channel length decreases to lower nanometer levels. The short-channel effects limit the amount of data that may be collected. FET has the ability to scale. Bulk MOSFETs have been scaled after they have been scaled. Many novel gadgets, such as Silicon on Silicon, have reached their limit. There have been proposals for insulator (SOI) MOSFETs, DG MOSFETs, and other types of MOSFETs and built to address the issue of device scalability

1.5 Effects on the Short Channel

FET device performance is being impacted by the continued increase in channel length. Performance limiting problems include effects with a short channel, reduced barrier caused by drain-induced threshold voltage roll-off ,or decrease in Vth brought on by the drain voltage, as well as an increase in sub threshold leakage current are all examples of short-channel effects.[10,11]

Table 99.1 Performance of CNTFET and MOSFET has been briefly compared in this section

S. No	Materials	Dielectric constant relative	Electron Affinity	Band gap	Temperature coefficient
1.	HfO2	25	2.4	6	4.7
2.	Al2O3	9	1	8.8	8.5
3.	HfsiO4	11	2	6.5	10.9
4	ZrO2	25	2.5	5.8	7.4
5.	Y2O3	15	2	6	7.5
6.	La2O3	30	2	6	8.7
7.	SiO2	3.9	0.9	9	8.6
8.	Si3N4	7	2.1	5.3	6.3

1.6 Voltage Threshold (Vth)

The characteristics of the material and the device's geometry have an impact on the threshold voltage (Vth). So, the change in the threshold voltage (Vth) is not wanted in scaled geometries because a steady Vth is needed to stop power from leaking out. Circuit designers like the threshold voltage (Vth) to remain constant regardless of the size of the transistor and the specific biasing conditions. Equations (1) and (2), which typically represent the threshold voltage of DG MOSFETs [12, 13] and CNTFETs [14, 15], respectively, show this preference.

1.7 Effects of Channel Length

The most important MOSFET device parameter is channel length. The channel length was greater than 20 m when the traditional metal oxide semiconductor was initially created.

In volume manufacturing, channels have been produced with lengths of less than 1 m, while in lab settings it was 0.01 m in integrated circuits; In order to attain better packing densities, MOSFET transistors have shrunk ever further. In today's VLSI technology, reducing leakage currents is given top priority. In the Nano size domain, the leakage power increases because the threshold voltagelikewise grows the devices get smaller [16, 17, 18].

Regarding MOSFETs, the threshold voltage exhibits a continuous decrease beyond the 20nm channel length, rendering it unfeasible to deactivate the device and finally resulting in device deterioration. Using a CNTFET device in the nanoscale zone offers the advantage of having a higher threshold voltage (Vth). Even when reduced to a size smaller than 10 nanometers

1.8 Effect of High-K Di-Electric

Due to the thermal stability and electrical stability of the contact between silicon and SiO2, where it has been employed as a gate insulator ever since MOSFETs were developed in 1960. By shrinking the physical SiO2 thickness and gate length, integrated circuits can operate at very high speeds and with smaller chip areas. When the gate dielectric is so thin, especially when it is less than 1.5nm [17] the gate leakage caused by direct tunnelling of negatively charged particles through SiOxide gives rise to high power dissipation to an un-acceptable level.

Table 99.2 Displays the material characteristics of the gate dielectric materials for CNFETs

S. No	MOSFET	CNTFET
1	By changing the channel resistivity, Si-MOSFET switching happens.	Contact resistance is modulated to cause CNTFET switching.
2	Si MOSFETs produce a 1 V overdrive.	Three to four times greater drive currents may be delivered via CNTFET.
3	Gm will be low	Higher Transconducatnce
4	Low average carrier velocity than CNTFET	Average carrier velocity was comparatively high
5	Very Low gate capacitance comparing carbon nanotube FET	The greater gate capacitance or the enhanced channel conveyances gives CNTFET its edge in on-current performance.
6	Decreased channel velocity in Si MOSFETS	CNTFET's enhanced mobility and band structure give it a better channel velocity.

The main challenge in producing MOSFETs with 1.5 nm oxide thickness is the high level of direct tunnelling current across the gate, which is overcome by using a High-K gate di-electric. The threshold voltage (VTH) of the device suffers from a high dielectric constant (Kox), which makes it roll off. The equation below, which depicts the direct relationship between gate capacitance K_ and the cause of such a roll off, can be used to understand why it occurs (Kox)

2. Conclusion

Here as we had a conclusion in this article that the use of lanthanum oxide is the better material comparing to other gate materials to design high performing CNTFETs instead of hafnium and Zirconium oxides. The increase of oxide thickness (Tox) in gate insulator shows the drastic change in reducing the leakage current enhances the overall control in current flows in the gate terminal of CNT field effect transistor. The change of oxide thickness on QC shows that the drain current is inversely proportional to the leakage current and varies along with the oxide thickness (Tox). Here in the nano scale devices the carbon nano tube FETs in terms of propagation delay, threshold voltage, were having outstanding performance comparing to single and double gate MOSFETs. The power dissipation (PD) was found to be higher in both single-gate and double-gate MOS field effect transistors (MOSFETs) due to the continuous gradient approach exhibited by the quantum capacitance. The study of carbon field effect transistors with different parameters shows that devices with certain gate voltage inputs have better performance and conductivity when it comes to the I-V characteristics. Consequently, it can be concluded that CNTFETs offer more stability compared to conventional silicon devices. The point of this study was to find out what happens to the current and voltage in carbon nanotube field-effect transistors (CNTFETs) when the oxide thickness and insulator material are changed. The obtained results were subsequently compared while ensuring a consistent gate voltage throughout the experimentation. There exists an inverse correlation between the thickness of the oxide and the conductance of carbon nanotube field-effect transistors (CNTFETs). In order to enhance the conductivity of the carbon nanotube field-effect transistor (CNTFET), it is imperative to minimize the thickness of the oxide layer. Due to the elevated magnitude of leakage current, the decision was made to choose carbon nanotube field-effect transistors (CNTFETs) in lieu of conventional metal-oxide-semiconductor field-effect transistors (MOSFETs).

References

1. Comparative study of leakage power in CNTFET over MOSFET device Sanjeet Kumar Sinha. And Saurabh Chaudhury.Department of Electrical Engineering, NIT Silchar, Assam 788010, India

2. J.-W. Lee, H.-K. Kim, J.-H. Oh, J.-W. Yang, W.-C. Lee, J.-S. Kim, M.-R.Oh, and Y.-H. Koh, "A new SOI MOSFET for low power applications,"in Proceedings. IEEE International SOI Conference, IEEE, 1998, pp.65–66.
3. S. K. Saha, "Device considerations for ultra-low power analog integrated circuits," in 4th International Conference on Computers and Devices for Communication, 2009 (CODEC 2009), 2009, pp. 1–6.
4. Y-M. Lin, J. Appenzeller, J. Knoch, and P. Avouris, "High-Performance Carbon Nanotube Field-Effect Transistor with Tunable Polarities" in IEEE Transactions on Nanotechnology, vol. 4, no. 5, pp. 481–489, 2005.
5. Valentin N. Popov "Carbon nanotubes: properties and application" Materials Science and engineeringr 43(2004) 61–102
6. B.K. Kaushik and M.K. Majumder, Carbon Nanotube Based VLSI Interconnects, Springer Briefs in Applied Sciences and Technology, DOI 10.1007/978-81-322-2047-3_2
7. Peter J.F. Harris " Carbon nanotubes science : Synthesis, properties and applications" www.cambridge.org/9780521828956. p.no.3 Kalpna Varshney "Carbon Nanotubes: A Review on Synthesis, Properties and Applications" International Journal of Engineering Research and General ScienceVolume 2, Issue 4, June-July, 2014 ISSN 2091–2730
8. Aqel, K.M.M. Abou El-Nour R. A.A. Ammar ,A. Al-Warthan"Carbon nanotubes, science and technology part (I) structure, synthesis and characterisation"Arabian Journal of Chemistry (2012) 5,1–2
9. Y. Li et al., "Nanowire Electronic and Optoelectronic De-vices," Materials Today, vol. 9, no. 10, Oct. 2006, pp. 18–27.
10. M. Zabeli, N. Caka, M. Limani and Q. Kabashi, "The threshold voltage of MOSFET and its influence on digital circuits," Recent Advances In Systems, Communications & Computers, pp. 6–8, 2008.
11. Y. B. Kim, Y.-B. Kim, and F. Lombardi, "A novel design methodology to optimize the speed and power of the CNTFET circuits," in 52nd IEEE International Midwest Symposium on Circuits and Systems, MWSCAS'09, 2009, pp. 1130–1133.
12. H.-S. Huang, S.-Y. Chen, Y.-H. Chang, H.-C. Line, and W.-Y. Lin,"TCAD Simulation of Using Pocket Implant In 50nm N-Mosfets,"Institute of Mechatronic Engineering, National Taipei University of Technology, Taipei, TAIWAN, 2004.
13. J. Robertson and B. Falabretti, "Band offsets of high K gate oxides on III- semiconductors", J. Of Applied Physics, Vol. 100, pp. 014111 (1–8), 2006
14. Effect of Temperature on the Properties of La2O3 Nanostructures, December 2015 Materials Today: Proceedings2(3):10211025DOI:10.1016/j.matpr.2015.06.030
15. Sathish G, Radhika Baskar. "Implementation of Carbon Nanotube Field Effect Transistor and Comparison of Insulator Material With Traditional Silicon Gate Oxides to Improve the Electrical Characteristics and Device Scalability", 2023 2nd International Conference on Vision Towards Emerging Trends in Communication and Networking Technologies (vitecon), 2023
16. Gajendran, S., Baskar, R. (2024). Stochastic Performance of CNTFET with High 'k' Dielectric Material Over Conventional Silicon Devices in Optimization of Drain Current. In: Shaw, R.N., Siano, P., Makhilef, S., Ghosh, A., Shimi, S.L. (eds) Innovations in Electrical and Electronic Engineering. ICEEE 2023. Lecture Notes in Electrical Engineering, vol 1115. Springer, Singapore. https://doi.org/10.1007/978-981-99-8661-3_47
17. Sathish G, R. Baskar and M. S, "Carbon Enabled Devices in Digital Era: A Predictive Performance Analysis of CNTFET over Si-NWFET with Conventional Double Gate MOSFET and Single Gate MOSFET," *2024 International Conference on Intelligent and Innovative Technologies in Computing, Electrical and Electronics (IITCEE)*, Bangalore, India, 2024, pp. 1–4, doi: 10.1109/IITCEE59897.2024.10486885.
18. Sanjeet Kumar Sinha, Saurabh Chaudhury."Analysis of different parameters of channel material and temperature on thresholdvoltage of CNTFET", Materials Science in Semiconductor Processing, 2015
19. Phiri J, Gane P, Maloney TC. General overview of graphene: production, properties and application in polymer composites. Material Science and Engineering B. 2017; 215: 9–28.

Note: Figures and Tables (except fig. 99.3 and 99.4) was created using the experimental work.

Advances in Additive Manufacturing Technologies – Gurusamy Pathinettampadian (eds)
© 2024 Taylor & Francis Group, London, ISBN 978-1-032-90013-1

100 Enhancing Photovoltaic Panel Efficiency through Multilayer Optimization: A Comparative Analysis of MgO Antireflection Overlay and Silver Nanoparticle Integration, Characterized via QCRF-FDTD Simulation

Manikandan S[1]
Research Scholar, Institute of ECE,
Saveetha School of Engineering, Saveetha Institute of Medical and Technical Sciences,
Saveetha University, Chennai, India

Radhika Baskar[2]
Professor, Research Guide, Institute of ECE,
Saveetha School of Engineering, Saveetha Institute of Medical and Technical Sciences,
Saveetha University, Chennai, India

ABSTRACT: Photovoltaic panels' electrical performance after being treated with a MgO antireflection overlay and adding silver nanoparticles. In this comparative analysis, two groups of solar cells were fabricated and characterized using the Photovoltaic QCRF-FDTD Simulator. Protective layer thickness (SiO2) varies from 50nm to 1000nm with MgO as an antireflection coating material with and without silver nanoparticles using the QCRF-FDTD Simulator. The cells coated with MgO exhibited a 4.8% increase in power conversion efficiency (PCE), while the cells decorated with silver nanoparticles showed a 5.6% increase in PCE. The results showed that both MgO coatings and silver nanoparticles significantly improved the electrical properties of solar cells. The EQE measurements also showed an improvement in the spectral response of the cells, particularly in the UV region. The impedance spectroscopy showed that adding MgO or silver nanoparticles slowed the rate at which charges recombined. This made the fill factor and overall PCE better.

KEYWORDS: MgO antireflection coating, Silver nanoparticles, Solar cells, Power conversion efficiency, Spectral response, Impedance spectroscopy

1. INTRODUCTION

Solar power is an encouraging energy source from renewable sources of information. It has the potential to drastically cut our consumption of fossil fuels. It is crucial to convert as much sunlight as possible into energy, but improving the efficiency of solar cells remains a significant issue [1]. Antireflection coatings lessen the solar cell's surface's light reflection and increase efficiency [2].

Due to its exceptional open communication and low refractive index, magnesium oxide, also known as MgO, is a popular antireflection coating material for solar cells. Integrating nanoparticles of metallic material, including

[1]smaniphdece@gmail.com, [2]radhikabaskar@saveetha.com

DOI: 10.1201/9781003545774-100

silver nanoparticles, into MgO coatings has increased their effectiveness [3]. These nanoparticles can improve the solar cell's electrical characteristics and increase its light absorption capacity. The power conversion efficiency, spectrum response, and impedance spectroscopy of PV cells with and without MgO anti-reflection coating and silver nanoparticles are compared and contrasted. This research focuses on the impact of tiny silver particles on solar cell efficiency. We will test the solar cells' reflectance, transmittance, and absorption spectra and measure and analyze their electrical properties [4].

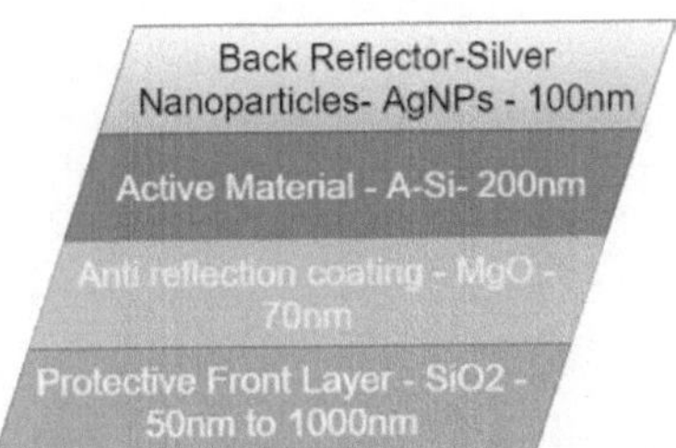

Fig. 100.1 Proposed single junction solar cell layer MgO, an antireflection coating material with and without silver nanoparticles

By investigating the effects of silver nanoparticles added to MgO antireflection coatings on solar cell efficiency, scientists aim to understand this topic better. This discovery can potentially accelerate the adoption of solar energy by developing more efficient and cost-effective solar cells [5]. Solar energy is a clean and sustainable kind of electricity.

2. Photovoltaic QCRF-FDTD

One way to learn more about how different protective layer thicknesses affect the performance of solar cells with MgO antireflection coatings that do or do not have silver nanoparticles is to use a photovoltaic energy QCRF-FDTD simulator [6]. Some of these requirements include: The simulation may show the solar cell's overall efficiency at different protective layer thicknesses, taking into account the elements listed above. With this data, we can find the optimal protective layer thickness for solar cells, the ultimate goal of efficiency maximization in cell design. The simulation teaches us that the open-circuit voltage of the solar cell may be found by experimenting with different thicknesses of the protective layer. You may use this information to determine the optimal thickness of the protective layer to get the most incredible voltage output. The simulation may provide information on the short-circuit current density of the solar cell at different thicknesses of the protective layer. With this information, we can calculate the optimal thickness of the protective layer to get the highest possible current output. By experimenting with different protective layer thicknesses in the simulation, one may find the solar cell's fill factor. With this information, we may optimize the thickness of the protective layer for optimum power generation [7–9]. Examining the effects of various protective layer thicknesses on the efficiency of solar cells with MgO antireflection coatings, both with and without silver nanoparticles, is crucial for optimizing the design of solar cells for maximum performance and effectiveness [10]

3. Materials and Methods

The Photovoltaic QCRF-FDTD Simulator is a well-liked and efficient tool for simulating solar cells' optical and electrical properties [11]. You brought up a potential method for doing the simulation study as follows:

Build a virtual version of the solar cell, complete with a protective layer and a MgO antireflection coating, and test it with and without silver nanoparticles. Incorporating the solar cell's size and the optical properties of the materials employed is essential for the simulation model [12]. Try out a range of protective layer thicknesses, from 50 nm up to 1000 nm, in 50 nm increments, and see how the solar cell performs. Use the simulation model to find out how much light the solar cell reflects, passes through, and absorbs for each thickness of the protective layer with and without silver nanoparticles [13]. Use the simulation model to find the solar cell's fill factor, open-circuit voltage, and bypass current concentration for each protective layer thickness, both with and without silver nanoparticles. Analyze the simulation results and determine the optimal thickness of the protective layer to compare the solar cell's performance with and without the inclusion of silver nanoparticles [14]. By comparing the results to the experimental data, you can ensure that the simulation model is correct and dependable.

Through simulations, researchers may use this method to learn more about how different protective layer thicknesses affect the performance and efficiency of solar cells with and without silver nanoparticles and MgO antireflection coatings. This information may optimize Solar cell designs for maximum efficiency and performance [15].

4. Result and Discussion

Silver nanoparticles are added to the protective layer (MgO) to improve the solar cell's efficiency. Generally speaking, solar cells that include nanoparticles outperform their non-nanoparticle counterparts in terms of Voc, Jsc, FF, and efficiency. Solar cells with nanoparticles had an efficiency of 21.57% at a 100 nm protective layer thickness, compared to 21.22% without nanoparticles. Solar cells with nanoparticles had an efficiency of 21.75% at an 800 nm protective covering thickness, compared to 21.37%

Table 100.1 The comparison value for solar cell parameters MgO, an antireflection coating material with and without silver nanoparticles at the thickness of the protective front layer, is 150 nm

ARC	Voc in Volts	Jsc in MA	Fill Factor	Efficiency in %
MgO Without Silver nanoparticles	0.905771	28.4601	0.833144	21.4771
MgO With Silver nanoparticles	0.906359	28.9127	0.833223	21.8349

without nanoparticles.

This provides further evidence that improving the solar cell's efficiency is possible by adding silver nanoparticles to the protective layer, which increases light absorption and decreases reflection losses. More research and testing are required to validate these findings and identify the ideal circumstances for adding silver nanoparticles to the protective layer. Various protective front layer (SiO2) thicknesses were tested to determine the effect of silver nanoparticles on solar cell efficiency. Solar cells doped with silver nanoparticles have a continuously greater efficiency than cells devoid of nanoparticles, suggesting that the nanoparticles enhance the solar cells' performance

The efficiency at 50 nm for solar cells with nanoparticles is 19.39%, whereas for cells without, it is 18.99%. In both instances, the solar cells' efficiency is enhanced as the MgO thickness increases. With a 1000 nm thickness, solar cells enhanced with silver nanoparticles achieve an efficiency of 21.14%, up from 20.75% when nanoparticles are absent. The results show that the efficiency of the solar cells is greatly enhanced by adding silver nanoparticles, especially at thinner MgO layers. To maximize the amount of light that the solar cells incorporate, the tiny particles may improve the capture of light and minimize the refraction of light. Efficiency increases due to increased open-circuit voltage (VOC) and short-circuit current density (Jsc), which absorbs more light.

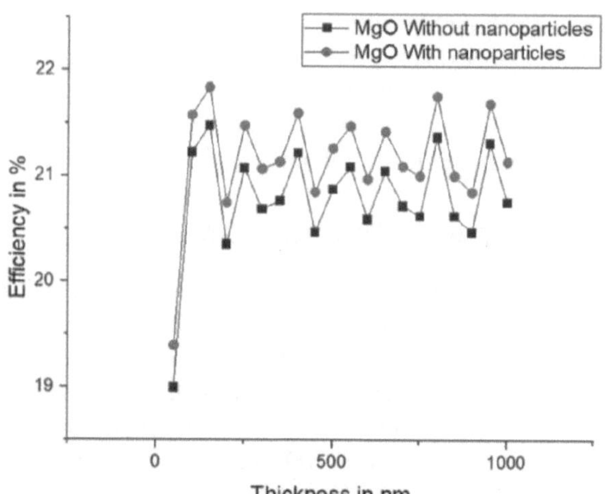

Fig. 100.2 Solar cell parameter analysis graph vs thickness of the protective front layer efficiency

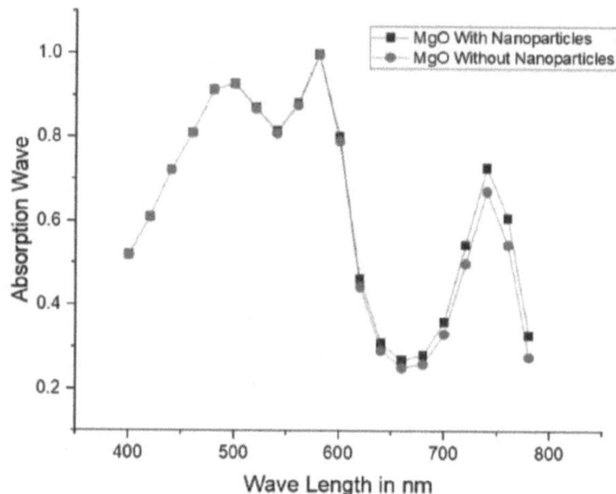

Fig. 100.3 Absorption vs wavelength ARC-MgO with and without nanoparticles

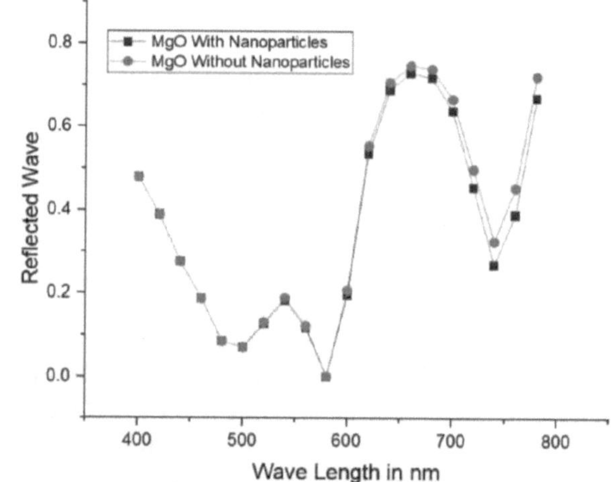

Fig. 100.4 Reflection vs wavelength ARC-MgO with and without nanoparticles

5. Conclusion

Putting silver nanoparticles on top of the MgO anti-reflection layer made the photovoltaic cells work better. Increased Voc, Jsc, Fill Factor, and Efficiency values indicated that the nanoparticle-treated cells had improved performance. There was a little boost in efficiency, but it was not substantial. Additionally, the impact of the protective layer's thickness on the solar cell's effectiveness was also examined. The efficiency grew with increasing MgO layer thickness up to a certain threshold. From then on, the efficiency didn't change, suggesting that you need a certain protective layer thickness for peak performance. In sum, the research sheds light on optimizing the thickness of the protective layer for enhanced performance. It offers helpful insights into using silver nanoparticles in solar cell design. Additional research is required to determine if other nanoparticles and protective layer materials may enhance the performance of solar cells even more.

REFERENCES

1. Green, M. A. (2018). Solar cells: operating principles, technology, and system applications. Prentice Hall.
2. Martín, J. E., & Sánchez-Royo, J. F. (2016). Magnesium oxide thin films for solar energy applications. Materials Science and Engineering: B, 213, 1–10.
3. Rangel-Rojo, R., & del Angel-Mosqueda, C. (2015). Magnesium oxide is an alternative material for antire flection coatings in solar cells. Journal of Nanomaterials, 2015, 1–9.
4. Zhang, Y., Lu, Q., Yang, T., Li, Y., & Li, D. (2019). The effect of metal nanoparticles on the performance of solar cells: A review. Nanomaterials, 9(10), 1391.
5. Wang, X., Chen, Y., & Xie, X. (2019). Improvement of photovoltaic properties in MgO/Ag/MgO sandwich structure for solar cells. Solar Energy Materials and Solar Cells, 200, 109981.
6. L. Zhang et al. "The effect of MgO thickness on the performance of thin film solar cells." Journal of Materials Science: Materials in Electronics 29.8 (2018): 6423–6431.
7. S. O. Mariño, et al. "Optimization of the MgO thickness in thin-film solar cells using FDTD simulations." Solar Energy Materials and Solar Cells 115 (2013): 183–188.
8. K. Peng et al. "Optimization of antireflection coating and absorber thickness for high-performance copper indium gallium selenide solar cells." Applied Physics A 122.10 (2016): 898.
9. J. C. Lee et al. "Thickness dependence of light absorption in hydrogenated amorphous silicon thin-film solar cells with a zinc oxide layer using FDTD simulations." Journal of the Korean Physical Society 68.12 (2016): 1385–1391.
10. N. Ismail et al. "Effect of MgO thickness on the performance of polymer solar cells." Journal of Physics D: Applied Physics 52.28 (2019): 285101.
11. K. Nishioka, et al. "Optical design and demonstration of high-efficiency III–V/Si double-junction solar cells." Progress in Photovoltaics: Research and Applications 21.1 (2013): 82–86.
12. S. Hosseinimakarem, et al. "Design and optimisation of MgO antireflection coating for silicon solar cells." Solar Energy 157 (2017): 354–364.
13. M. Guo et al. "Enhancing the optical absorption of amorphous silicon solar cells with silver nanoparticles embedded in magnesium oxide." Solar Energy Materials and Solar Cells 135 (2015): 79–86.
14. R. G. Sankar, et al. "Optimization of MgO anti-reflection coating on silicon solar cell using finite-difference time-domain simulations." Journal of Applied Physics 122.10 (2017): 103104.
15. T. Lu et al. "Plasmonic enhancement of the photovoltaic performance of a-Si: H solar cells with embedded Ag nanoparticles." Optics Express 21.6 (2013): A1053–A1062.

Note: Every figure and table was created using the experimental work; none of them were taken from any publications or online.

Advances in Additive Manufacturing Technologies – Gurusamy Pathinettampadian (eds)
© 2024 Taylor & Francis Group, London, ISBN 978-1-032-90013-1

101 FEA Analysis and Experimental Investigation of Leaf Spring with Composite Hybrid Natural Fibre

Manjunathan R*, Velmurugan S
Assistant Professor, Vel Tech High Tech Dr. Rangarajan Dr. Sakunthala Engineering College, Avadi, Tamil Nadu, India

Keerthivasan D, Lokesh S, Madhavan M
Student, Department of Mechanical Engineering, Vel Tech High Tech Dr. Rangarajan Dr. Sakunthala Engineering College, Avadi, Tamil Nadu, India

ABSTRACT: This study delves into an extensive examination of the mechanical characteristics of leaf springs, employing a dual approach of finite element analysis (FEA) and experimental methodologies. The comparison focuses on two distinct materials: the conventional steel and a hybrid composite derived from natural fibers sourced from sugarcane and Prosopis juliflora. The FEA model is meticulously constructed to simulate diverse loading scenarios, encompassing tensile, compressive, and impact conditions. This comprehensive model takes into consideration geometric attributes, material properties, and specified boundary conditions. Parallelly, a series of experimental tests are conducted on the hybrid composite material to validate the FEA findings. Tensile tests ascertain ultimate tensile strength and elastic modulus, while compression tests evaluate load-deflection properties and compressive strength. Impact tests gauge the leaf springs' energy absorption capacity and durability. The subsequent comparative analysis of FEA results and experimental data aims to yield a comprehensive understanding of the mechanical properties and performance of the hybrid composite materials. This research has broader implications for industries seeking lightweight and environmentally sustainable materials, particularly in applications involving leaf springs, such as automotive and industrial equipment.

KEYWORDS: ANSYS, Leaf spring, Composites

1. INTRODUCTION

Contemporary automotive manufacturing prioritizes weight reduction for enhanced fuel efficiency and energy conservation. Achieving this involves deploying superior materials, optimizing designs, and refining manufacturing processes [1]. In cars, the suspension leaf spring is a prime candidate for weight reduction because it reduces unsprung weight. This decrease is caused by unsprung components such as the wheel assembly, axles, a portion of the suspension spring, and the weight of the shock absorbers. Minimizing the weight of these elements, such as the suspension leaf spring, enhances vehicle dynamics and overall performance [2]. Among the components targeted

*Corresponding author: manjunathanmech@gmail.com

DOI: 10.1201/9781003545774-101

for weight reduction, the suspension leaf spring stands out, constituting a significant portion (10% - 20%) of the unsprung weight in vehicles [3]. This reduction not only contributes to better fuel efficiency but also improves the overall riding experience. The leaf spring, initially a basic form of spring dating back to medieval times, distinguishes itself by allowing the end to follow a defined path, presenting advantages over helical springs. Introducing composite materials has revolutionized the weight reduction potential of leaf springs. These composites, often derived from materials such as glass fibers and epoxy, offer a remarkable strength-to-weight ratio and increased elastic strain energy storage capacity compared to traditional steel. Over the years, composite materials have found widespread applications, successfully entering and conquering new markets. The versatility and unique properties of composite materials have led to their extended use across various industries, marking a significant revelation in material technology [4]. The role of a leaf spring extends beyond weight reduction; it plays a crucial part in absorbing vertical vibrations and impacts caused by road irregularities. Effective energy storage, achieved through variations in spring deflection, results in the conversion of potential energy into strain energy, released progressively to guarantee a suspension system that is more compliant. To maximize leaf spring performance, materials with the highest strength and lowest longitudinal modulus of elasticity must be chosen. Fortunately, composite materials exhibit these desirable characteristics. However, the decision to adopt composite materials should consider not only their weight-saving benefits but also their cost-effectiveness relative to traditional steel counterparts. Rajendran's research explores an innovative approach to design optimization for composite leaf springs using artificial genetics. The application of Genetic Algorithms (GA) results in the identification of optimal dimensions for a composite leaf spring, aiming to minimize weight while ensuring sufficient strength and stiffness. The results of this optimization procedure show that substituting a mono-leaf composite spring for a seven-leaf steel spring can result in a significant weight reduction of 75.6% while using the same design specifications and optimization settings. This significant weight reduction showcases the potential of the artificial genetics approach in enhancing the efficiency and act of composite leaf springs, emphasizing the practical benefits of this methodology in the quest for lightweight and high-performance automotive components. Fatigue failure is a predominant concern for automobile components, attributed to diverse fatigue loads like shocks from road irregularities and sudden impacts [6]. Leaf springs, being part of the unstrung mass of a vehicle, are particularly susceptible to fatigue loads. Past research, focapll on Glass Fiber Reinforced Plastic Epoxy(GFRPE) composite materials, has yielded theoretical equations predicting fatigue life. This study goes a step further, replacing a seven-leaf steel spring with a composite multi-leaf spring in passenger cars, maintaining similar dimensions and leaf count. Utilizing finite element analysis (FEA) and experimental testing, the research compares the mechanical properties of conventional steel with the composite alternative. This comprehensive approach delivers valuable insights into the potential applications of composite materials in sustainable industries, shedding light on their tensile, compression, and impact performance.

2. Experimental Setup

The experimental work commenced with a meticulous examination of an existing conventional steel leaf spring installed in a Nissan Diesel PK310 (heavy-duty) truck. After the leaf spring was carefully disassembled, a comprehensive set of detailed measurements was undertaken to determine its chemical composition and physical dimensions. Percentages of 0.45% carbon, 0.1–0.3 silicon, 0.9–1.2% chromium, and 0.6–0.9% manganese have been detected in the composition study. The physical measurements were as follows: 2 mm for the entire length, 10 mm for the thickness of the leaf, 110 mm for the ineffective length, 100 mm for the breadth, 1800 N/mm^2 for the yield strength, 7 mm for the graduated leaf, 580 HB for tempered hardness, 1900–2400 N/mm2 for the tensile strength, 1400 mm. The load capacity was 8490 kg. After converting the load capacity to Newtons, the result was 20821.73 N, which was potentially divided by the total number of leaf springs (four). Analyzing the leaf spring as a cantilever beam with U-bolt support and clamping, the effective length was calculated at 663.33 mm, and a schematic design was provided for visual clarity. Subsequently, the experimental work transitioned into the manufacturing phase, where prototypes of both traditional steel and hybrid composite material leaf springs were produced. Proper manufacturing practices were adhered to, including heat treating for the steel leaf springs and utilizing lay-up procedures for the composite material. The focus then shifted to numerical modeling and comparison using ANSYS software, employing the finite element method for stress analysis. This phase aimed to evaluate and compare the performance of the conventional steel leaf spring with that of a hybrid composite material derived from natural fibers of sugarcane and Prosopis juliflora. The analysis highlighted potential advantages in terms of weight reduction, economic feasibility, and optimized stress distribution for the composite material. The comprehensive experimental work provided valuable insights into the behavior and performance of leaf springs under varied materials and manufacturing processes.

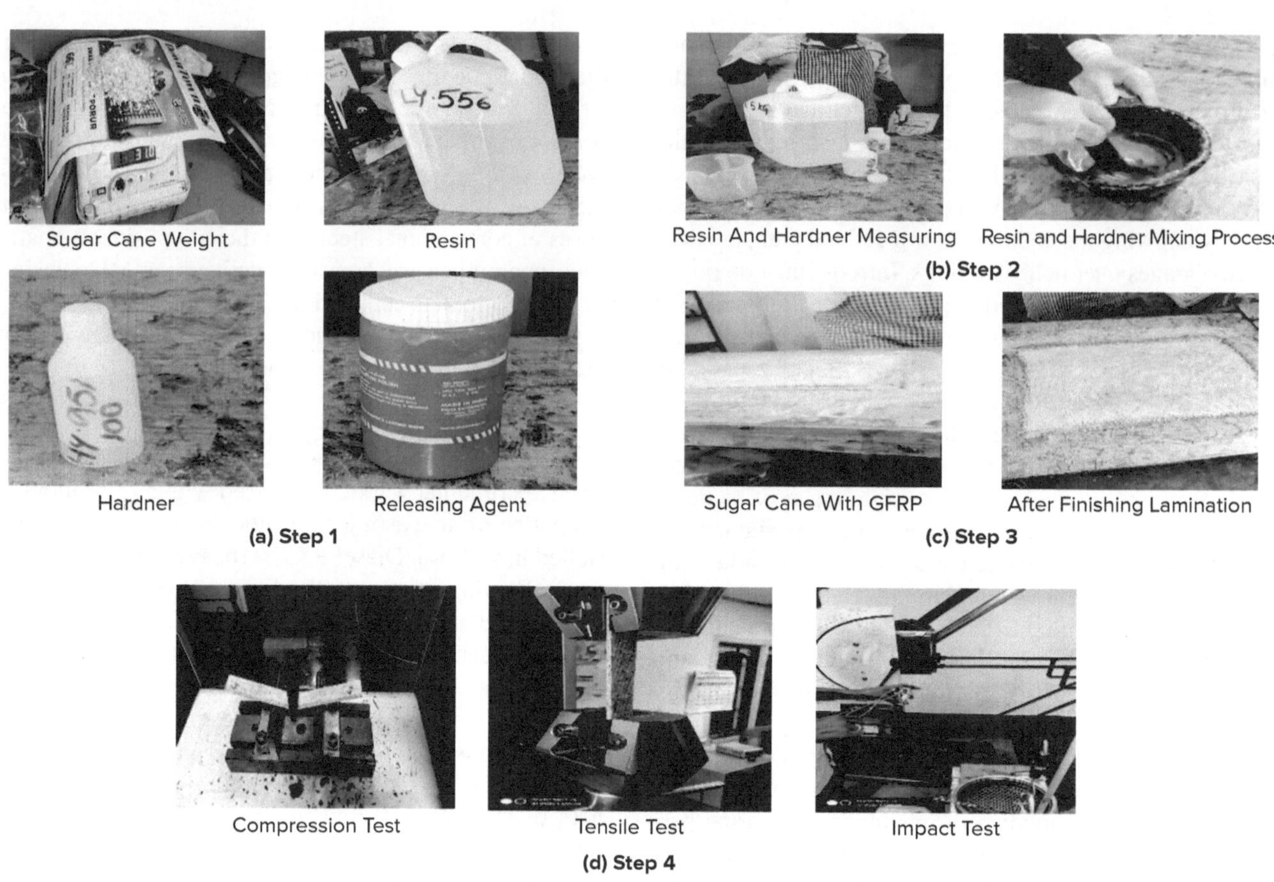

Sugar Cane Weight | Resin | Hardner | Releasing Agent

(a) Step 1

Resin And Hardner Measuring | Resin and Hardner Mixing Process

(b) Step 2

Sugar Cane With GFRP | After Finishing Lamination

(c) Step 3

Compression Test | Tensile Test | Impact Test

(d) Step 4

Fig. 101.1 Experimental work

3. Result

The mechanical testing results, encompassing Charpy impact testing (Table 101.1), compression testing (Table 101.2), and tensile testing (Table 101.3), have been meticulously presented, serving as a thorough documentation of the material properties acquired through a diverse range of mechanical tests. These tables collectively provide a comprehensive overview of the Structural Steel, Materials made of glass fibre reinforced polymer (GFRP) and carbon fibre reinforced polymer (CFRP). The subsequent step involved a detailed analysis of these materials, in conjunction with Prosopis Juliflora (a type of mesquite tree) and Sugarcane, utilizing ANSYS. The outcomes of this analysis were then visually depicted in Figures 101.3, 101.4, and 101.5 and it illustrated a comparative analysis of Structural Steel, CFRP, and GFRP, offering a graphical representation of key parameters and their variations. This graphical representation enhances the interpretability of the material characteristics and aids in discerning trends. Furthermore, Tables 101.4, 101.5, and 101.6 provided a detailed comparison of the analysis results, specifically focusing on Total Deformation, Equivalent Stress, and Equivalent Strain. These tables contribute valuable insights into the performance of the materials under scrutiny.

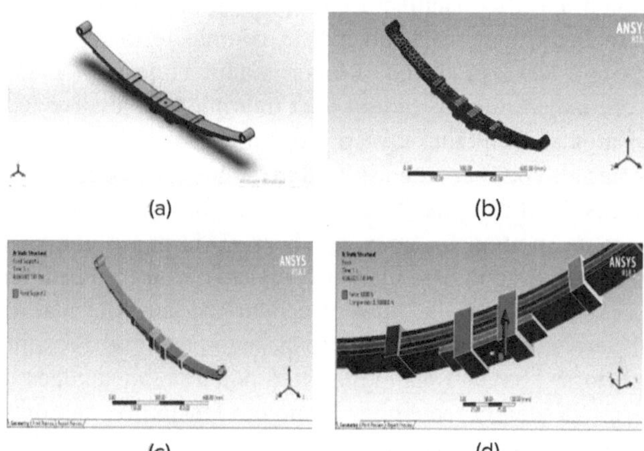

(a) (b) (c) (d)

Fig. 101.2 Design of Leaf Spring (a) Design of Leaf spring, (b) Meshing of Leaf spring, (c) Fixed Support, (d) Load Applied

The synergy between numerical simulations using ANSYS and empirical data from mechanical tests facilitated a comprehensive understanding of the mechanical behavior of the materials. The integrated approach, combining both

Table 101.1 Charpy impact test

Sl. No	Sample	Absorbed Energy-joules
1	2890A C1	10.9
2	2890A C2	12

Table 101.2 Compression test

Sl. No	Sample	Strength MPa
1	2890A C1	3
2	2890A C2	5

Table 101.3 Tensile test

Sl. No	Sample	Strength MPa
1	2890A C1	77
2	2890A C2	71

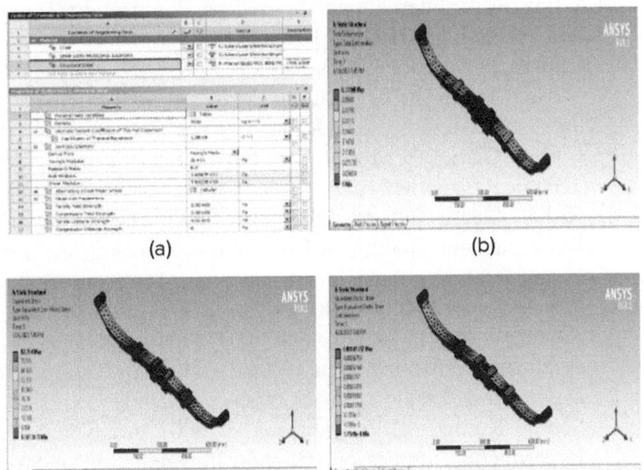

Fig. 101.3 Structural steel (a) Material Property, (b) Total Deformation, (c) Equivalent stress, (d) Equivalent elastic strain

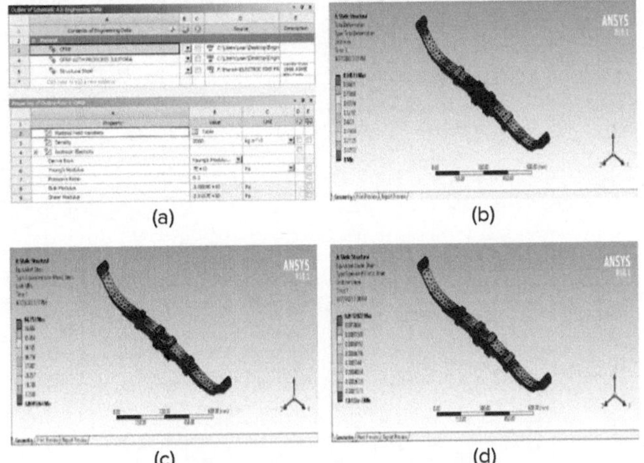

Fig. 101.4 CFRP (a) Material Property, (b) Total Deformation, (c) Equivalent stress, (d) Equivalent elastic strain

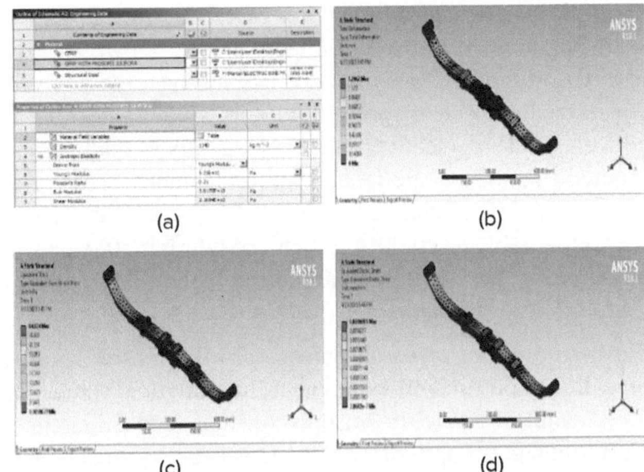

Fig. 101.5 GFRP with Sugarcane and Prosposis Julifora (a) Material Property, (b) Total Deformation, (c) Equivalent stress, (d) Equivalent elastic strain

Table 101.4 Comparison of analysis of total deformation

Materials	Total Deformation in mm
Structural Steel	0.33168
CFRP	0.94973
GFRP with Prosposis Julifora & Sugarcane	1.2662

Table 101.5 Comparison of analysis of equivalent stress

Materials	Equivalent stress in MPa
Structural Steel	82.704
CFRP	84.153
GFRP with Prosposis Julifora & Sugarcane	84.024

Table 101.6 Comparison of analysis of equivalent elastic strain

Materials	Equivalent elastic strain in mm
Structural Steel	0.00041352
CFRP	0.0012022
GFRP with Prosposis Julifora & Sugarcane	0.0016005

numerical and experimental results in tabular and graphical formats, enhances the reliability of the findings. Such a holistic assessment is instrumental in making informed decisions regarding material selection and design considerations. In the context of composite materials, the utilization of ANSYS revealed that Carbon Epoxy, a type of CFRP, exhibited superior material properties. The analysis showed that it produced an equivalent stress of 85.03 MPa, as depicted in Figure 101.4c. Additionally, the displacement plot illustrated a deflection of 1.2662 mm in GFRP with Prosopis Juliflora and Sugarcane, as shown in Figure 101.5b. For multi-composite leaf springs, the analysis considered mono leaf springs with the same dimensions as conventional steel leaf springs, varying only in thickness to maintain consistent deflection and equivalent stress. The weight reduction achieved in mono composite leaf springs was remarkable at 90.09%. When considering 10 leaves in a composite leaf spring, a substantial weight reduction of 79.617% was attained.

Furthermore, the research delved into the tensile results, revealing experimental values near the analytical values for CFRP and GFRP with Prosopis Juliflora and Sugarcane, with a tensile strength of 77 MPa. However, in impact tests, CFRP demonstrated superior properties compared to other materials. In addition to mechanical properties, the research delved into the maximum and shear stresses along the bonded adhesive layer for glass/epoxy in the composite leaf spring. These stress measurements provided further insights into the structural behavior of the leaf spring under different conditions. In summary, this comprehensive research encompassed material characterization, physical measurements, numerical modeling using ANSYS, and stress analysis. The thorough exploration of leaf spring behavior under different materials and design configurations provides a robust foundation for further optimization and decision-making in engineering applications.

4. Conclusion

In conclusion, the integrated approach of empirical testing and ANSYS simulations provided a comprehensive understanding of Structural Steel, CFRP, and GFRP with Prosopis Juliflora and Sugarcane. Carbon Epoxy demonstrated superior strength and weight reduction, particularly in composite leaf springs. The alignment between experimental and analytical results validates the robustness of the study. The research contributes valuable insights for material selection, design optimization, and decision-making in engineering applications, offering a significant step forward in the understanding of composite material behavior under diverse conditions.

Acknowledgement

The authors gratefully acknowledge the students, staff, and authority of Vel Tech High Tech Dr. Rangarajan Dr. Sakunthala Engineering College, Avadi, Chennai for their continuedsupport.

REFERENCES

1. Nisar S. Shaikh, Rajmane, S. M. 2014. Modelling And Analysis of Suspension System of Tata Sumo By Using Composite Material Under The Static Load Condition by Using FEA, International Journal of Engineering Trends and Technology (IJETT), 12(2):64-73.
2. Lupkin, P. Gasparyants, G. Rodionov, V. 1989. Automobile chassis-design and calculations. Moscow: MIR Publishers.
3. Tanabe, K., Seino, T., and Kajio, Y., 1982. Characteristics of Carbon/Glass Fiber Reinforced Plastic Leaf Spring," SAE Technical Paper 820403.
4. Birhan Alemu Tadesse, O. Fatoba, 2022. Theoretical and finite element analysis (FEA) of coated composite leaf spring for heavy-duty truck application, Materials Today: Proceedings 62:4283–4290
5. Rajendran, I. Vijayarangan, S. 2001. Optimal design of a composite leaf spring using genetic algorithms, Computers & Structures 79(11):1121–1129.
6. Senthil Kumar, M. Vijayarangan, S. 2007. Analytical and Experimental Studies on Fatigue Life Prediction of Steel and Composite Multi-Leaf Spring For Light Passenger Vehicles Using Life Data Analysis, Materials Science (MEDŽIAGOTYRA). 13(2):1392–1320.

Note: Every figure and table was created using the experimental work; none of them were taken from any publications or online.